Epstein-Barr Virus

Edited by

Erle S. Robertson

Department of Microbiology and the Abramson Comprehensive Cancer Center, University of Pennsylvania Medical School, Philadelphia, Pennsylvania 19104, USA

Copyright © 2005

Caister Academic Press
32 Hewitts Lane
Wymondham
Norfolk NR18 0JA
England

www.caister.com

British Library Cataloguing-in-Publication Data

A catalogue record for this book is available from the British Library

ISBN: 1-904455-03-4

Printed and bound in Great Britain

Contents

Books of Related Interest

Antimicrobial Peptides in Human Health and Disease	2005
Biodefense: Principles and Pathogens	2005
Campylobacter: Molecular and Cellular Biology	2005
Cancer Therapy: Molecular Targets in Tumour-Host Interactions	2005
Cytomegaloviruses: Molecular Biology and Immunology	2005
Dictyostelium Genomics	2005
Evolutionary Genetics of Fungi	2005
Foodborne Pathogens: Microbiology and Molecular Biology	2005
HIV Chemotherapy: A Critical Review	2005
Papillomavirus Research: From Natural History To Vaccines	2005
Microbe-Host Interface in Respiratory Tract Infections	2005
Mycobacterium: Molecular Biology	2005
Mycoplasmas: Molecular Biology, Pathogenicity and Strategies for Control	2005
Probiotics and Prebiotics: Scientific Aspects	2005
SAGE: Current Technologies and Applications	2005
Microbial Toxins: Molecular and Cellular Biology	2005
Vaccines: Frontiers in Design and Development	2005
Bacterial Spore Formers: Probiotics and Emerging Applications	2004
Brucella: Molecular and Cellular Biology	2004
Computational Genomics: Theory and Application	2004
DNA Amplification: Current Technologies and Applications	2004
Ebola and Marburg Viruses: Molecular and Cellular Biology	2004
Foot and Mouth Disease: Current Perspectives	2004
Internet for Cell and Molecular Biologists (2nd Edition)	2004
Malaria Parasites: Genomes and Molecular Biology	2004
Metabolic Engineering in the Post Genomic Era	2004
Pathogenic Fungi: Structural Biology and Taxonomy	2004
Pathogenic Fungi: Host Interactions and Emerging Strategies for Control	2004
Peptide Nucleic Acids: Protocols and Applications (2nd Edition)	2004
Prions and Prion Diseases: Current Perspectives	2004
Protein Expression Technologies: Current Status and Future Trends	2004
Strict and Facultative Anaerobes: Medical and Environmental Aspects	2004
Real-Time PCR: An Essential Guide	2004
Sumoylation: Molecular Biology and Biochemistry	2004
Tuberculosis: The Microbe Host Interface	2004
Yersinia: Molecular and Cellular Biology	2004
Bioinformatics and Genomes: Current Perspectives	2003
Bioremediation: A Critical Review	2003
Frontiers in Computational Genomics	2003
Genome Mapping and Sequencing	2003
MRSA: Current Perspectives	2003
Multiple Drug Resistant Bacteria: Emerging Strategies	2003
Transgenic Plants: Current Innovations and Future Trends	2003
Regulatory Networks in Prokaryotes	2003
Vaccine Delivery Strategies	2003

Full details of all these books at: www.horizonbioscience.com

Contributors

Hratch Arbach
INSERM U716
IFR Saint-Louis
27 rue Juliette Dodu
75010 Paris
France

Georg W. Bornkamm
GSF-National Research Center for Environment
and Health
Institute of Clinical Molecular Biology
Marchioninistr. 25
D-81377 München
Germany
Email: bornkamm@gsf.de

Scott R. Burrows
Infectious Disease and Immunology Division
Queensland Institute of Medical Research
Herston
Queensland
Australia

Ellen Cahir-McFarland
Department of Medicine
Brigham and Women's Hospital
Harvard Medical School and University
181 Longwood Avenue
Boston
MA 02115, USA

Jeffrey I. Cohen
Laboratory of Clinical Infectious Diseases
Bldg. 10, Room 11N228
National Institutes of Health
10 Center Drive
Bethesda
MD 20892, USA
Email: jcohen@niaid.nih.gov

Guy de Thé
Pasteur Institute
EPVO Unit
28 rue Dr Roux
Paris 75015
France
Email: dethe@pasteur.fr

Gianpietro Dotti
Center for Cell and Gene Therapy
Baylor College of Medicine
6621 Fannin Street
MC 3-3320
Houston
TX 77030, USA
Email: gdotti@bcm.tmc.edu

M. Anthony Epstein
Nuffield Department of Clinical Medicine
University of Oxford
John Radcliffe Hospital
Oxford
OX3 9DU, U.K.

Paul J. Farrell
Department of Virology
Faculty of Medicine
Imperial College
St Mary's Campus
Norfolk Place
London W2 1PG, UK
Email: p.farrell@imperial.ac.uk

Makoto Fukuda
Department of Microbiology and Immunology
Feinberg School of Medicine
Northwestern University
Ward 6-231
303 East Chicago Ave.
Chicago
IL 60611, USA

Beverly E. Griffin
Viral Oncology Unit
Division of Medicine
Imperial College at St. Mary's
Norfolk Place
London W2 1PG, U.K
Email: b.griffin@imperial.ac.uk

Helen Heslop
Center for Cell and Gene Therapy
Baylor College of Medicine
6621 Fannin Street
MC 3-3320
Houston
TX 77030, USA

Brigitte T. Huber
Tufts University School of Medicine
Jaharis 906
150 Harrison Avenue
Boston
MA 02111, USA
Email: brigitte.huber@tufts.edu

Lindsey Hutt-Fletcher
Department of Microbiology and Immunology
Feist-Weiller Cancer Center
Louisiana State University Health Science Center
1501 Kings Highway
Shreveport
LA 71130, USA
Email: lhuttf@lsuhsc.edu

Masato Ikeda
Department of Microbiology and Immunology
Feinberg School of Medicine
Northwestern University
Ward 6-231
303 East Chicago Ave.
Chicago
IL 60611, USA

Bruce F. Israel
Department of Medicine
The University of North Carolina
Lineberger Comprehensive Cancer Center 7295
Chapel Hill
NC 27599-7295, USA

Dai Iwakiri
Department of Tumor Virology
Institute for Genetic Medicine
Hokkaido University
N15 W7 Kita-Ku
Sapporo 060-0815
Japan

Kenneth M. Izumi
Department of Microbiology and Immunology
University of Texas Health Science Center at San
Antonio
7703 Floyd Curl Drive
San Antonio
Texas 78229-3900, USA
Email: izumi@uthscsa.edu

Irène Joab
INSERM U716
IFR Saint-Louis
27 rue Juliette Dodu
75010 Paris
France
Email: Irene.Joab@stlouis.inserm.fr

Bettina Kempkes,
GSF-National Research Center for Environment
and Health
Institute of Clinical Molecular Biology
Marchioninistr. 25
D-81377 München
Germany
Email: kempkes@gsf.de

Shannon C. Kenney
Departments of Medicine, Microbiology &
Immunology
The University of North Carolina
Lineberger Comprehensive Cancer Center 7295
Chapel Hill
NC 27599-7295, USA
Email: shann@med.unc.edu

Rajiv Khanna
Infectious Disease and Immunology Division
Queensland Institute of Medical Research
Herston
Queensland
Australia

Elliott Kieff
Deptartment of Medicine and Microbiology and
Molecular Genetics
Channing Laboratory
Harvard MUniversity
181 Longwood Avenue
Boston
MA 02115, USA
Email: ekieff@rics.bwh.harvard.edu

George Klein
Microbiology and Tumor Biology Center
Karolinska Institute
Stockholm
Sweden

Jason S. Knight
Department of Microbiology and the Abramson
Comprehensive Cancer Center
University of Pennsylvania Medical School
201E Johnson Pavilion
3610 Hamilton Walk
Philadelphia
PA 19104, USA

Richard Longnecker
Department of Microbiology and Immunology
Feinberg School of Medicine
Northwestern University
Ward 6-231
303 East Chicago Ave.
Chicago
IL 60611, USA
Email: r-longnecker@northwestern.edu

Amanda A. Mack
McArdle Laboratory for Cancer Research
Department of Oncology
University of Wisconsin Medical School
1400 University Ave.
Madison, WI 53706
USA

Cary A. Moody
Department of Microbiology and Immunology
Feist-Weiller Cancer Center
Louisiana State University Health Sciences
Center
1501 Kings Highway
Shreveport
LA 71130, USA

Denis J. Moss
Epstein-Barr Virus Unit
Queensland Institute of Medical Research
300 Herston Rd
Brisbane
Australia 4029.
Email: denisM@qimr.edu.au

Christian Münz
Laboratory of Viral Immunobiology
The Rockefeller University
1230 York Avenue
New York
NY 10021-6399, USA
Email: munzc@rockefeller.edu

P.G. Murray
The Cancer Research UK Institute for Cancer
Studies, Medical School
University of Birmingham
Birmingham
B15 2TT, UK
Email: p.g.murray@bham.ac.uk

Tadamasa Ooka
Laboratoire de Virologie Moléculaire
UMR5537, CNRS
Faculté de Médecine Laennec,
Université de Lyon-1
69372 Lyon Cedex 08
France
Email: ooka@sante.univ-lyon1.fr

Joseph S. Pagano
Lineberger Cancer Center and Department of
Microbiology
University of North Carolina at Chapel Hill
CB #7295
Chapel Hill
NC 27599-7295, USA
Email: joseph_pagano@med.unc.edu

Nancy Raab-Traub
Lineberger Comprehensive Cancer Center
CB# 7295
University of North Carolina
Chapel Hill
North Carolina
NC 27599-7295, USA
Email: nrt@med.unc.edu

Alan Rickinson
Inst for Cancer Studies
University of Birmingham
Vincent Drive
Edgbaston
Birmingham B15 2TT
UK
Email: A.B.RICKINSON@bham.ac.uk

Erle S. Robertson
Department of Microbiology and the Abramson
Comprehensive Cancer Center
University of Pennsylvania Medical School
201E Johnson Pavilion
3610 Hamilton Walk
Philadelphia
PA 19104, USA
Email: erle@mail.med.upenn.edu

Rosemary Rochford
Department of Microbiology and Immunology
SUNY Upstate Medical University
750 East Adams St.
Syracuse
NY 13210, USA

Cliona Rooney
Center for Cell and Gene Therapy
Baylor College of Medicine
6621 Fannin Street
MC 3-3320
Houston
TX 77030, USA
Email: cmrooney@txccc.org

Lars Rymo
Göteborg University
The Sahlgrenska Academy
Department of Clinical Chemistry and
Transfusion Medicine
Sahlgren's University Hospital
S-413 45 Göteborg
SWEDEN
Email: lars.rymo@clinchem.gu.se

Rona S. Scott
Department of Microbiology and Immunology
Feist-Weiller Cancer Center
Louisiana State University Health Sciences
Center
1501 Kings Highway
Shreveport
LA 71130, USA

John W. Sixbey
Department of Microbiology and Immunology
Feist-Weiller Cancer Center
Louisiana State University Health Sciences
Center
1501 Kings Highway
Shreveport
LA 71130, USA
Email: jsixbe@lsuhsc.edu

Samuel H. Speck
Center for Emerging Infectious Diseases
Yerkes National Primate Research Center
Emory University School of Medicine
954 Gatewood Rd NE
Atlanta
GA 30329, USA
Email: sspeck@rmy.emory.edu

Lothar J. Strobl
GSF-National Research Center for Environment
and Health
Institute of Clinical Molecular Biology
Marchioninistr. 25
D-81377 München
Germany
Email: strobl@gsf.de

Bill Sugden
McArdle Laboratory for Cancer Research
Department of Oncology
University of Wisconsin Medical School
1400 University Ave.
Madison
WI 53706, USA
Email: sugden@oncology.wisc.edu

Natalie Sutkowski
Microbiology and Immunology
Medical University of South Carolina
PO Box 250504
173 Ashley Avenue
Charleston, SC 29425, USA
Email: sutkows@musc.edu

Sankar Swaminathan
UF Shands Cancer Center and
Division of Infectious Diseases
1600 SW Archer Road
Gainesville
FL 32610-0232, USA
Email: sswamina@ufl.edu

Kenzo Takada
Department of Tumor Virology
Institute for Genetic Medicine
Hokkaido University
N15 W7 Kita-Ku
Sapporo 060-0815
Japan
Email: Kentaka@igm.hokudai.ac.jp

David A. Thorley-Lawson
Department of Pathology
Tufts University School of Medicine
Jaharis Builiding
150 Harison Ave
Boston
MA 02111, USA
Email: david.thorley-lawson@tufts.edu

Naohiro Wakisaka
Lineberger Cancer Center and Department of
Microbiology
University of North Carolina at Chapel Hill
Chapel Hill
NC 27599-7295, USA

Fred Wang
Harvard Medical School
Brigham & Women's Hospital
Channing Laboratory
181 Longwood Avenue
Boston
MA 02115, USA
Email: fwang@rics.bwh.harvard.edu

Tomokazu Yoshizaki
Lineberger Cancer Center and Department of
Microbiology
University of North Carolina at Chapel Hill
Chapel Hill
NC 27599-7295, USA

L.S. Young
The Cancer Research UK Institute for Cancer
Studies
Medical School
University of Birmingham
Birmingham
B15 2TT, UK
Email: l.s.young@bham.ac.uk

Henrik Zetterberg
Department of Clinical Chemistry and
Transfusion Medicine
Sahlgrenska University Hospital
S-413 45 Göteborg
SWEDEN

Ursula Zimber-Strobl
GSF-National Research Center for Environment
and Health
Institute of Clinical Molecular Biology
Marchioninistr. 25
D-81377 München
Germany

Harald zur Hausen
Deutsches Krebsforschungszentrum
Ím Neuenheimer Feld 280
69120 Heidelberg
Germany
Email: ZurHausen@DKFZ.de

Preface

My initial forays into the study of Epstein Barr Virus (EBV) began in 1992 as a postdoctoral fellow in the laboratory of Elliott Kieff at Harvard University School of Medicine, Brigham and Women's Hospital. This was professionally, a difficult transition for me as I had no previous experience in laboratory investigations of human viruses. I came from a predominantly prokaryotic background studying bacteriophage T7 genetics and biochemistry. Reading the two chapters in Fields Virology, edited by Fields, Knipe and Howley, on the biology and replication of EBV was informative but quite a daunting task as it did not provide a clear view from discovery to current trends in the field. There was no comprehensive text to provide novices to the field a cohesive blueprint which represented the body of work in EBV. I had promised myself then that one of my future goals would be to attempt at putting together a comprehensive text on the subject that would be a must read for students, postdocs, new investigators and established investigators, current as well as newcomers to the field. This would serve as a core reference guide for the field of EBV. This text represents the first attempt at such a contribution which is intended to bring a sense of discovery, discussion of initial hypotheses, current studies and vaccine development.

This textbook on EBV contains ten individual sections which are discovery, history and seroepidemiology; EBV and human diseases; EBV and immune responses; EBV genetics; EBV infection and persistence; EBV latency; EBV lytic proteins and reactivation control; EBV vaccine approaches; animal models and a final section on the future of EBV studies. These thirty three individual chapters which make up the ten sections are all written by investigators with substantial expertise in these specific areas of study and who are well recognized for their contributions. Each chapter is aimed to be as comprehensive as possible to provide the reader with the necessary background to quickly grasp the new and up to date information on the major concepts of infection, pathogenesis and control.

It should be acknowledged that every attempt was made to be as comprehensive as possible. However, in the area of lytic replication this text can definitely be expanded. The chapters 27-29 by Drs. Kenney, Ooka and Swaminathan which covers reactivation, BARF1 function and post-transcription gene regulation by the EBV SM protein are substantial, but there is a lack of chapters dealing with additional lytic proteins and their functions in lytic replication and packaging. My hope is that these areas will be addressed in the future as we learn more about specific EBV antigens required for lytic replication and packaging and the cellular pathways and molecules that are involved.

The authors have all made sacrifices in time and effort in getting each chapter to me in a timely fashion and I am happy to say that even with some delays we were able to get all the chapters reviewed and modified in a reasonable time frame. I would like to thank all the contributors as well as others in the field who has given their time to making this effort possible and I hope that everyone is pleased with the final outcome. I would specifically like to thank Sir Anthony Epstein for providing the introductory chapter which provides a first hand account of the beginning, and

Dr. Alan Rickinson who has been calm during my many phone calls nudging him to get his summary chapter which is quite impressive. I hope that this text is helpful and of interest to the many established and new investigators studying EBV and that the contributors feel a sense of accomplishment in seeing the final product. Thanks to everyone who have made this possible.

Erle S. Robertson Ph.D.

From: Epstein-Barr Virus. Edited by: Erle S. Robertson

Chapter 1

The Origins of EBV Research: Discovery and Characterization of the Virus

M. Anthony Epstein

INTRODUCTION

In the years following the first report of the finding of what became known as Epstein-Barr virus (EBV) (Epstein et al., 1964), various limited aspects of the discovery were described in book chapters and published lectures (Epstein, 1970; Epstein and Achong, 1979a; Epstein, 1984; 1985) with rather moderate circulations. Nevertheless, the outlines of the story could be pieced together by those interested in knowing the events which led to a whole new area in virology with important implications for viral carcinogenesis, immunology, and later even the treatment and prevention of certain human malignancies. However, despite this, serious articles did not hesitate to present sketchy and inadequate accounts of what took place.

Because of the persistence of inaccuracies it was considered worthwhile in the thirty fifth anniversary year of the discovery to provide a definitive history of the circumstances in which EBV was first recognized (Epstein, 1999), yet even after this an irresponsibly concocted version appeared in a supposedly authoritative review. It is therefore timely to repeat the story, while this is still possible as a personal memoir, at the outset of this present comprehensive textbook which it is anticipated will be widely read and quoted.

THE START OF THE SEARCH FOR EBV

In 1941, in the darkest days of World War II, a young British surgeon, Denis Burkitt, was commissioned into the Royal Army Medical Corps and served in Southern England with 219 Field Ambulance before being posted in 1943 to British Colonial African troops in East Africa. Here he was stationed in Kenya and Somaliland where there had been a campaign against Italian forces; more importantly for his future he also saw something of Uganda while on local leave. Uganda, then often described as "the pearl of Africa", was at that time both a prosperous and peaceful country with much Christian activity which appealed to Burkitt's deeply religious convictions. He decided therefore, that after the war he would devote his life to helping the people of Uganda and on demobilization in 1946 he applied to the British Colonial Office to join its Medical Service in Uganda. He was duly accepted and went out as a government servant, not as a missionary as is so often wrongly stated. His first appointment was in Lira, Lango District, where he was in sole medical charge of a population of about 300,000 with nothing more than a 100 bed bush hospital to cover all branches of medicine and surgery. While in Lira he collected data on 200 cases of hydrocele and made his first contribution to the geography of disease by showing that almost all came from East Lango District and almost none from West Lango (Burkitt, 1951); this finding was later explained by others as being due to the distribution of filariasis.

In 1957 Denis Burkitt was working in Mulago Hospital in Kampala, the capital of Uganda, when he was consulted by a colleague about a child with puzzling swellings in all four angles of the jaws which seemed to be neither infection nor cancer, but despite the enigma Burkitt kept typically meticulous notes and photographs of the case. Then shortly afterwards, quite by chance while visiting another hospital, he noticed another child with swellings in all four jaw angles and immediately realized that this was something special; the boy also had abdominal tumours. As he described it much later "a curiosity can occur once, but two cases indicated something more than a curiosity" (Burkitt, 1983). Having an intensely enquiring mind, Burkitt took the details of these two cases to the records department of Mulago Hospital and his search there for children's cancers showed that jaw tumours were common, were often associated with other tumours at unusual sites, and that the latter sometimes presented without involvement of the jaws: he concluded that these apparently different childhood cancers were all manifestations of a single, hitherto unrecognized tumour complex which was exceedingly common in the first decade of life in Uganda.

The first publication on the tumour (Burkitt, 1958) failed to attract any attention at all, and it had to wait until 1983 to become a Citation Classic. It was only when Denis Burkitt enlisted collaboration from a pathological colleague who diagnosed the tumour as a highly unusual lymphoma that their linked papers (Burkitt and O'Conor, 1961; O'Conor, 1961) led to the beginnings of interest in what ultimately became known as Burkitt's lymphoma (BL). In the meantime, perhaps motivated by his geographical work on hydrocele in Lira, Burkitt began to map the distribution of the tumour and found that this was apparently dependent on conditions of temperature and rainfall in Uganda, and maybe elsewhere in Africa.

While Burkitt was engaged on these studies in East Africa completely unknown to the world scientific community, I had been working for some years on the Rous sarcoma virus (Epstein, 1956; 1958a; 1958b) at the Middlesex Hospital Medical School in London. Although this was a topic of interest then to perhaps only about a dozen individuals world-wide, it was one which guaranteed a keen awareness of viruses which might cause cancer, a concept considered rather unfashionable at that time. By an extraordinarily fortunate coincidence, Denis Burkitt had a connection with the Academic Surgical Unit at the Middlesex Hospital and when on home leave in UK was customarily invited to lecture to students and staff on his experiences in Uganda, presenting the exotic and bizarrely extreme cases encountered in a developing country.

Early in 1961, Burkitt came again to London on leave but on this occasion at the Surgical Unit he gave the very first talk outside Africa on the new lymphoma

Figure 1. Denis Burkitt in East Africa with two lymphoma patients; a simple "bush surgeon" at work (from Epstein, M.A., 1985, courtesy of the publisher).

on 22 March 1961; the talk was entitled "The commonest children's cancer in Tropical Africa. A hitherto unrecognized syndrome". Quite randomly I saw the title of the talk on a notice board at the Hospital and for reasons I am to this day unsure about, but probably curiosity, I attended and was electrified from the start by the strange nature and rapid progression of the tumour, and even more so by the unprecedented preliminary findings on epidemiology which seemed to show that its distribution depended on such climatic factors as temperature and rainfall. For, dependence on climate indicated that some biological agent must play a part in the aetiology and with my knowledge of tumour viruses I postulated in my mind that a climate-dependent arthropod vector might be involved in spreading an oncogenic virus; even as Burkitt was talking, I resolved to investigate the tumour for unusual viruses. The lack of interest at that time in Burkitt's work is exemplified by the fact that when he repeated the talk a few days later in a large lecture theatre at the Royal College of Surgeons, also in London, it was attended by about 12 people. But then as he subsequently said, he was "just an unknown bush surgeon" (Figure 1). So impressed was I by Burkitt's presentation that I took a copy of the announcement off a notice board and it is still in existence as the first item in my original BL/ EBV file. A full account of Denis Burkitt's life and work has been given elsewhere (Epstein and Eastwood, 1995).

Although it later became clear that it was co-factors which determine the climate dependence of BL (Burkitt, 1969), my original hypothesis was responsible for getting my research programme going and focused it correctly on the search for a virus cause. The programme actually began in discussions with Denis Burkitt a few days after the talk when it was agreed that he would make biopsy samples from his lymphoma patients available for me to work on in UK. Despite unseemly wrangling and actual obstruction arising from politics at the very top of the British establishment concerned with cancer research, the then British Empire Cancer Campaign (now Cancer Research UK) generously funded a fact-finding trip for me to Uganda which resulted in arrangements for a steady supply of BL biopsies to be flown overnight from Kampala to my laboratory in London. These events have pertinently been commented upon by L S Young (1999) who questioned the ease with which such research would be funded today: "it's hard to imagine any funding agency supporting a project based on the 'gut feeling' of a young research worker without any 'preliminary data'… Thank goodness that was not the case 35 years ago".

EARLY INVESTIGATIONS

For almost two years the standard techniques for virus detection current at the time were applied to the lymphoma samples with uniformly negative results. The tumour material was inoculated into various kinds of cell cultures, embryonated hen eggs, and newborn mice without showing any effects, and in addition, examination of the biopsies directly in the electron microscope likewise proved fruitless. I had had the good fortune to work with George Palade (Nobel Laureate, 1974) in 1956 at the then Rockefeller Institute (now the Rockefeller University) in New York, thanks to greatly appreciated support from the Anna Fuller Fund of New Haven, at a seminal time as regards his contributions to the earliest phases of biological electron microscopy and I therefore knew the possibilities of the technique; failure to gain anything from this new tool in relation to the lymphoma was therefore especially disappointing. Furthermore, the long period without results was both depressing and alarming at a particularly insecure stage of my career.

Two things changed the outlook at this point, one obvious at the time and the other only evident when it led to success. First a grant of US$45,000 (relatively small even then) from the US National Cancer Institute allowed me to recruit Yvonne Barr and Bert Achong* to join the project at the end of 1963; and secondly, it occurred to me that if we could grow BL cells *in vitro* away from host defences a latent oncogenic virus might be activated and become apparent (Epstein et al., 1964a), as I knew could happen with cultured cells from certain chicken tumours (Bonar et al., 1960) in which the avian tumour virus was not always detectable *in vivo*.

However, the likelihood of being able to culture the cells from a human lymphoid tumour seemed dauntingly low since no single member of the human lymphocytic series had ever been grown *in vitro* for more than a very brief time despite innumerable attempts ever since tissue culture techniques became available at the beginning of the twentieth century (Woodliff, 1964). As might have been foreseen, repeated efforts using floating lymphoma fragments, fragments in plasma clots, fragments on rafts, and so on, failed with disheartening regularity.

---

* *"The author is sad to record that Dr B. G. Achong died in November 1996"*

THE SUCCESSFUL ESTABLISHMENT OF BL CELL LINES

On Friday 3 December 1963, a flight from Kampala was diverted to Manchester by fog and it was only late in the afternoon that we were able to retrieve the biopsy sample on board when the plane finally reached London Airport. The tissue was, as usual, floating in guinea pig serum/saline transit fluid but unusually the latter was cloudy. At this point there was a general feeling that we could leave the lab for the weekend since it was assumed that the specimen was useless because contaminating bacteria had proliferated and caused the turbidity on the extra long journey. But instead of discarding the specimen and going home I put a drop of transit fluid on a slide and examined it directly as a wet preparation under the light microscope. Rather than seeing bacteria I was astonished to find that the cloudiness was due to large numbers of round, viable looking, free-floating tumour cells which must have been shaken free during transit from the cut edges of the lymphoma sample. I was immediately reminded of something I had learnt earlier in the year on a visit to Yale University School of Medicine where there was an ongoing programme concerned with murine lymphoid tumours. There G A Fischer had only succeeded in growing malignant mouse lymphoblasts in continuous culture by using as starting material suspensions of dispersed single cells (Fischer, 1957; 1958) obtained in his case from the ascites form of the tumour. Because of this, the free-floating cells from our delayed biopsy sample, described at once in my laboratory note book as "like ascites" (ref. TC149/150 of December 1963 – illustrated in Epstein, 1999), were set up in suspension culture for the first time; shortly afterwards the first BL-derived continuous cell line designated EB1 grew out and was reported briefly (Epstein and Barr, 1964).

This method of suspension culture rapidly gave us further BL-derived cell lines (EB2, EB3) (Epstein et al., 1964b; 1965a) and using it others were shortly afterwards able to do the same (Stewart et al., 1965; Rabson et al., 1966). The suspension culture method has remained the standard routine technique for human lymphoid cells for a huge range of studies world-wide to this day. However, when an account of the unsuccessful preliminary attempts together with full details of the successful procedure was submitted to a leading journal the paper encountered serious difficulties before final publication (Epstein and Barr, 1965) because expert reviewers were unwilling to accept that human lymphoid cells could be grown long term *in vitro* at all.

At the same time as we succeeded in growing BL cells, and quite unbeknown to us, BL cells were being placed in short term cultures in Nigeria to study their cytological features by phase contrast microscopy, and data on this was reported in the same journal as our first paper (Epstein and Barr, 1964) and in the same issue (Pulvertaft, 1964). The description of the cytology included in passing a single sentence stating that a strain of the cells had grown out in culture for ten weeks, but no details were given either of the cells or the methods used to grow them. It was not until over a year later, after a sample of these cells had been sent to us in London under the designation "Raji" and studied by us in collaboration with colleagues in the US, that any account of the cell line was provided (Epstein et al., 1966) or that we were able to make Raji cells available to the scientific community. At the time I received the sample the Raji cells were stated "to differ from their first form", perhaps as a result of freezing and thawing in Nigeria.

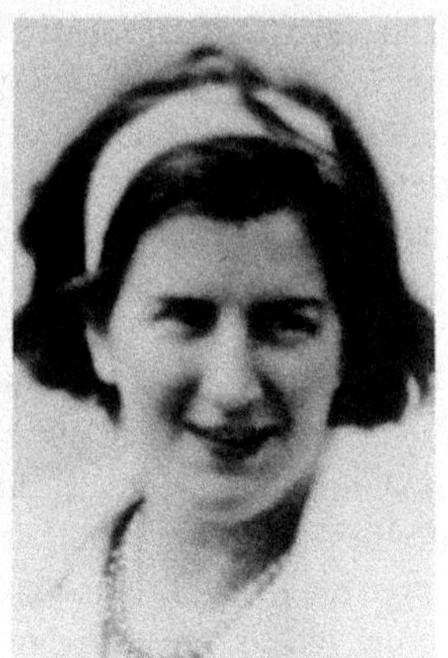

Figure 2. M. A. Epstein, B. G. Achong, and Y. M. Barr in 1964 (from Galloway, J., 1989, courtesy of the publisher).

THE DISCOVERY OF EBV

As soon as material could be spared from the EB1 BL-derived cell line, it was pelleted, fixed, embedded, and sectioned for electron microscopy. This was certainly not accepted then as a method for demonstrating viruses; dogma required that the presence of viruses must be shown by their effects on suitable laboratory animal or tissue culture systems, or by finding the antibodies they induced. It was not credited that they could be distinguished by their morphology even though this had been acceptable for many families of bacteria using the light microscope for nearly 100 years. Indeed, for many at the time, the images obtained of biological material with the electron microscope were considered as artefacts of fixation and processing. However, my experiences with George Palade had convinced me that the structural features of viruses could validly be recognized and used for classification (Epstein, 1961; 1962a).

I examined the first EB1 preparation with the electron microscope on 24 February 1964 and was exhilarated to observe unequivocal virus particles in a cultured BL cell in the very first grid square to be searched. Understandably, I was extremely agitated in case the specimen might burn up in the electron beam; I therefore switched off, walked round the block in the snow wearing nothing more sensible than a light lab coat, and returned only when somewhat calmer to take electron micrographs of what I had seen. I recognized the virus at once as having the typical morphology of the herpes group with which I was familiar (Epstein, 1962b; 1962c) and recorded it as such in my laboratory notebook ("virus, like herpes" ref. EM584 of February 1964 – illustrated in Epstein, 1999), but there was naturally no means of knowing which herpesvirus it was although it did seem extraordinary that a member of the herpes family was producing virions in a cell line yet was so biologically inert as not to be causing destruction of the cultures. Nevertheless, I rapidly set about reporting the discovery with Dr. Achong and Miss Barr (as she then was) (Figure 2) (Epstein et al., 1964c – became a Citation Classic in 1979).

The peculiar inertness of the virus was reinforced further when all our efforts to show biological activity in cell cultures, hen eggs, and new born mice again proved negative. At this stage I became concerned in case we were doing something unnoticed in our procedures which was inactivating the virus. Accordingly it became imperative to have our tests repeated in some other laboratory; two separate leading British herpes virologists were approached but neither was interested in our unorthodox findings and I therefore contacted Werner and Gertrude Henle in Philadelphia.

CONFIRMATION OF THE UNIQUENESS OF EBV AND ORIGIN OF ITS NAME

Werner Henle had qualified in medicine at the University of Heidelberg, Germany, in 1934 at the beginning of the Nazi era and because he had a Jewish grandfather, the famous 19th Century anatomist Jacob Henle, his career was blocked and he emigrated to the US in 1936. The following year, Gertrude Szpingier also graduated in medicine and having known Werner for some time followed him in April 1937. They were married in New York the day after her arrival and so Gertrude Szpingier became Gertrude Henle; they worked together until Werner's sad death in 1987 and were soon recognized as one of those rare, famous, husband and wife scientific research partnerships. The turning point in their careers came when they were invited to the Children's Hospital of Philadelphia to study influenza virus and thereafter set up one of the foremost US virus diagnostic and research laboratories. Thus it was that on my first visit to America in 1956 when making a tour of leading virology centres the Henles were amongst those not to be missed along with Wendell Stanley, Renato Dulbecco, John Bittner, Thomas Francis, Jonas Salk, Albert Sabin, John Enders, and Hilary Koprowski. It was because I had remained in touch with the Henle's through many subsequent reciprocal laboratory visits and they had become friends that I turned to them when approaches to British virologists failed.

EBV-carrying cells were flown from my laboratory in London to Philadelphia, the biological inertness of the virus was soon confirmed there, and we were able to report jointly that this was indeed a new human herpesvirus (Epstein et al., 1965b). It was in this way that the Henle laboratory started on its long and important involvement with EBV (see Zur Hausen Chapter 2). Shortly after our joint paper, the immunological uniqueness of the virus was demonstrated both by the Henle's (Henle and Henle, 1966a; 1966b) and by different methods in our laboratory (Epstein and Achong, 1967; 1968a; 1968b) and a year or two later its biochemical singularity was also established (zur Hausen et al., 1970).

In the light of later knowledge on the extremely limited range of cells with receptors for the virus our failure to show biological activity was readily understandable, but it was certainly very puzzling at the time. It was indeed fortunate that the work on BL cells and the search for a virus was going forward in a laboratory where the electron microscope, a rather rare tool at the time, was in daily use as otherwise the biological inertness could well have left the agent undiscovered. In this connection it is worth noting that EBV appears to be the first virus to have been found solely by electron microscopy.

For several years after the discovery of the virus we and others always referred to it as the "herpes-type" or "herpes-like" agent from BL and it was only later when its uniqueness was becoming evident that the Henle's applied the designation "EBV" (Henle et al., 1968) after the "EB" BL-derived cell lines in which it had been found and in which I had sent it to them in the first place for our joint investigations (Epstein et al., 1965b).

RETROSPECTIONS ON THE FIRST PHASES OF THE EBV VACCINE PROGRAMME

Another aspect of EBV research whose early stages are often forgotten concerns the development of a vaccine against the virus. This is not just a matter of historical interest but has relevance at the present time when the whole programme, spanning several decades, has recently come to fruition and there are current expectations that vaccine intervention can be used to protect against EBV-induced disease, for example in the case of seronegative children receiving organ grafts from seropositive donors.

Already less than 10 years after EBV was discovered, it was clear that the known and suspected involvements of the virus in human diseases (Epstein and Achong, 1973; Klein, 1973) made consideration of a vaccine against it a perfectly reasonable proposition, particularly in the light of its possible carcinogenic role. The first suggestion for this (Epstein, 1975) was soon followed by a detailed description of a possible approach (Epstein, 1976); this identified an appropriate immunogen, the EBV-determined membrane antigen (MA), proposed purification methods for it, recognized the necessary susceptible animals for vaccine testing, and foresaw the possibility of using such a vaccine in the first place to protect against infectious mononucleosis, well known by then to be caused by EBV (Henle et al., 1968). EBV MA was selected because antibodies directed against it were virus-neutralizing (Pearson et al., 1970; de Schryver et al., 1974) and might be expected to protect against or modify infection; it had also been shown that antigen containing membranes from cells infected with Marek's disease herpesvirus (MDHV) had markedly reduced the incidence of MDHV-induced lymphomas when used as an experimental vaccine in chickens (Kaaden and Dietzschold, 1974). Indeed, success in protecting against Marek's lymphomas had even been achieved experimentally with a vaccine consisting of soluble antigens extracted from such cells (Lesnick and Ross, 1975). The need to ensure the supply of experimental animals susceptible to EBV was also a crucial point since the only animal known to respond regularly with lesions after injection of EBV was the cottontop tamarin (Shope et al., 1973) a small South American non-human primate (*Saguinus oedipus oedipus*) which, in 1976, was about to be placed on the endangered species list and thus likely to become unavailable. Once again, the then Cancer Research Campaign (the former British Empire Cancer Campaign, now Cancer Research UK) responded positively to my proposals, this time to develop an EBV vaccine, and joined with the British Medical Research Council to allocate generous funds. I was thus able to recruit appropriate collaborators and set about providing the essential requirements for a vaccine programme.

Work in several laboratories demonstrated that MA consists of two, antigenically related, large glycoprotein components of 340kD and 270kD (MA gp340 and gp270) the latter being a precursor of the former (reviewed in Epstein, 1984); the larger of these was therefore selected for study. As regards other requirements, supplies of cottontop tamarins had to be ensured, and a breeding colony was therefore set up and the diet, bioenergetics, husbandry, and handling procedures for reproduction of this little known species had to be worked out (Kirkwood et al., 1983; 1985); it was also necessary to establish a challenge dose of EBV which would ensure the induction of tumours in 100% of animals of this outbred population (Epstein, 1984). During the course of this work it was ascertained en passant that the EBV-induced tamarin tumours were not just mere polyclonal lymphoproliferations but that each mass of the multifocal lesions was a true malignant lymphoma arising from a separate and different B cell clone (Cleary et al., 1985). A sensitive test for the gp340 antigen was likewise essential to optimize yields and a radio-immunoassay was therefore elaborated (North et al., 1982a); with it a workable MW-based preparative procedure was introduced (Morgan et al., 1983). The purified gp340 was incorporated into artificial liposomes which greatly increased immunogenicity (North et al., 1982b) and, finally, the antibodies induced were monitored by a specifically designed ELISA (Randle and Epstein, 1984) together with standard virus-neutralization tests (Moss and Pope, 1972; de Schryver et al., 1974). As soon as these advances could be exploited, a prototype vaccine was validated in cottontop tamarins by showing its ability to protect vaccinated animals against the 100% carcinogenic dose of challenge EBV (Epstein et al., 1985; Epstein, 1986).

Developments from this base-line led to the introduction of a more efficient gp340 preparation method using FPLC which readily yielded the antigen in tractable amounts (David and Morgan, 1988), and to the demonstration of the striking protective effect of such material when employed to immunize cottontop tamarins along with powerful new adjuvants (Morgan et al., 1988; 1989).

Subsequent progress was held up for some years by legal considerations regarding patents, but once these were resolved a major international pharmaceutical company pressed the programme forward by extending it to human testing. By this stage it was clear that a successful vaccine would not only be of value against infectious mononucleosis in young people in developed countries but, more importantly, against at least some of the increasing number of human tumours with which EBV is currently thought to make a contribution to causation. For the human trials, gp340 has been purified from cultures of mammalian cells into which the gp340 gene had been engineered (Jackman et al., 1999) and the product has been used with human volunteers in a Phase I trial, a successful Phase II trial (Bollen, 2002), and a Double Blind Placebo trial which has just been completed as this Chapter was finished.

At the time the prototype gp340 EBV vaccine was shown to be effective (Epstein, 1984; Epstein et al., 1985) other approaches to the use of gp340 began to be pursued in several laboratories; the gene for gp340 was engineered into a variety of cultured cell systems with a view to production *in vitro*, and the use of viral vectors has also been explored (see Moss and Khanna, Chapter 30). In addition, attempts to stimulate specific cytotoxic T cell activity against EBV-induced antigens using artificial polyepitope molecules in order to control existing EBV disease have recently been reported (see Chapter 30).

CONCLUDING REMARKS

Besides prejudice against the finding that human lymphoid cells could be grown *in vitro* and dogma against viruses being recognized by their morphology with the electron microscope, the early work on EBV was received with a great deal of additional scepticism. If the images reported were not artefacts they might at best be considered "virus-like particles" – whatever that was thought to mean; as for a hitherto unknown virus without biological activity, that was asking too much. Even after EBV was reluctantly accepted as a virus, it was usually discounted as a banal contaminant despite having been shown to be biologically, immunologically, and biochemically unique. Furthermore, the climate of opinion at the time of the discovery was generally unsympathetic to the idea that viruses might play a role in human cancer causation. Thus, recognition of the relevance of EBV as a possible human tumour virus, even if only in the case of BL, took very much longer and only began very slowly once a great body of information had built up on it and its behaviour (reviewed in Epstein and Achong, 1979b) and after the outstanding seven-year prospective study of 42,000 children in the West Nile District of Uganda by the World Health Organization's International Agency for Research on Cancer (de Thé et al., 1978). Additional impetus for change came from the discovery of the unequivocally oncogenic herpesviruses of Marek's disease (reviewed by Biggs, 2001) and of New World non-human primates (reviewed by Fickenscher and Fleckenstein, 2001).

The change in attitudes to human tumour viruses over the past few decades is exemplified by the reception given to the finding of EBV in 1964 by unconventional methods as compared to that given thirty years later to the most recently discovered oncogenic γ-herpesvirus of man, Kaposi's sarcoma herpesvirus (KSHV) (Chang et al., 1994). KSHV was also found by an unusual new technique, representational difference analysis (Lisitsyn et al., 1993) which had only been introduced barely a year before, yet the evidence was accepted immediately and KSHV was viewed as a human tumour virus shortly afterwards.

Since the early days of EBV when it was aptly remarked in conversation by de Thé "ce virus est mal aimé", interest in EBV has undergone great changes as attested by the regular cycle of international meetings dealing with many aspects of the virus, by the great and increasing number of publications devoted to it each year, and indeed by the clear need for a new, up-to-date, and authoritative monograph on it such as the present volume provides.

References

Biggs, P.M. (2001). The history and biology of Marek's disease. In Current Topics in Microbiology and Immunology, K.Hirai, ed. (Berlin, Heidelberg and New York: Springer), pp. 1-24.

Bollen, A. (2002). Encouraging results from Epstein-Barr virus vaccine Phase II trial. Henogen Press Release, Charlerois, Belgium, 3 October 2002.

Bonar, R.A., Weinstein, D. Sommer, J.R., Beard, D. and Beard, J.W. (1960). Virus of avian myeloblastosis XVII. Morphology of progressive virus-myeloblast interactions *in vitro*. U.S. Nat. Cancer Inst. Monogr. *4*, 251-290.

Burkitt, D.P. (1951). Primary hydrocele and its treatment: review of 200 cases. Lancet *i*, 1341-1343.

Burkitt, D. (1958). A sarcoma involving the jaws of African children. Brit. J. Surg. *46*, 218-223.

Burkitt, D.P. (1969). Etiology of Burkitt's lymphoma – an alternative hypothesis to a vectored virus. J. U.S. Nat. Cancer Inst. *42*, 19-28.

Burkitt, D.P. (1983). The discovery of Burkitt's lymphoma. Cancer, *51*, 1777-1786.

Burkitt, D.P. and O'Conor, G.T. (1961). Malignant lymphoma in African children. I. A clinical syndrome. Cancer *14*, 258-269.

Chang, Y., Cesarman, E., Pessin, M.S., Lee, F., Culpepper, J., Knowles, D.M. and Moore, P.S. (1994). Identification of herpesvirus-like DNA sequences in AIDS associated Kaposi's sarcoma. Science *266*, 1865-1869.

Cleary, M.L., Epstein, M.A., Finerty, S., Dorfman, R.F., Bornkamm, G.W., Kirkwood, J.K., Morgan, A.J. and Sklar, J. (1985). Individual tumors of multifocal EB virus-induced malignant lymphomas in tamarins arise from different B cell clones. Science *228*, 722-724.

David, E.M. and Morgan, A.J. (1988). Efficient purification of Epstein-Barr virus membrane antigen gp340 by fast protein liquid chromatography. J. Immunol. Methods *108*, 231-236.

de Schryver, A., Klein, G., Hewetson, J., Rocchi, G., Henle, W., Henle, G. Moss, D.J. and Pope, J.H. (1974). Comparison of EBV neutralization tests based on abortive infection or transformation of lymphoid cells and their relation to membrane reactive antibodies (anti MA). Int. J. Cancer *13*, 353-362.

de Thé, G., Geser, A., Day, N.E., Tukei, P.M., Williams, E.H., Beri, D.P., Smith, P.G., Dean, A.G., Bornkamm, G.W., Feorino, P., Henle, W. (1978). Epidemiological evidence for causal relationship between Epstein-Barr virus and Burkitt's lymphoma: results of the Ugandan prospective study. Nature *274*, 756-761.

Epstein, M.A. (1956). The intra-cellular identification of the Rous virus. Nature, *178*, 45-46.

Epstein, M.A. (1958a). Observations on the Rous virus; purification and identification of particles from solid tumours. Brit. J. Cancer *12*, 248-255.

Epstein, M.A. (1958b). Composition of the Rous virus nucleoid. Nature *181*, 1808.

Epstein, M.A. (1961). The electron microscopy of viruses. Proc. Roy. Soc. Med., *54*, 1071-1078.

Epstein, M.A. (1962a). Functional aspects of the structure of some animal viruses. Brit. Med. Bull. *18*, 183-186.

Epstein, M.A. (1962b). Observations on the fine structure of mature herpes simplex virus and on the composition of its nucleoid. J. Exp. Med. *115*, 1-12.

Epstein, M.A. (1962c). Observations on the mode of release of herpes virus from infected HeLa cells. J. Cell. Biol. *12*, 589-597.

Epstein, M.A. (1970). Long term tissue culture of Burkitt's lymphoma cells. In Burkitt's Lymphoma, D.P. Burkitt and D.H. Wright eds. (Edinburgh and London; E. and S. Livingstone Ltd.) pp 148-157.

Epstein, M.A. (1975). Towards an anti-viral vaccine for a human cancer. Nature *253*, 6.

Epstein, M.A. (1976). Epstein-Barr virus – is it time to develop a vaccine program? J. U.S. Nat. Cancer Inst. *56*, 697-700.

Epstein, M.A. (1984). A prototype vaccine to prevent Epstein-Barr (EB) virus- associated tumours. Proc. Roy. Soc. B *221*, 1-20.

Epstein, M.A. (1985). Historical background: Burkitt's lymphoma and Epstein-Barr virus. In Burkitt's lymphoma: a human cancer model, G. Lenoir, G. O'Conor and C.L. Olweny eds. (Lyon: Internat. Agency for Research on Cancer), pp 17-27.

Epstein, M.A. (1986). Vaccination against Epstein-Barr virus: current progress and future strategies. Lancet *ii*, 1425-1427.

Epstein, A. (1999). On the discovery of Epstein-Barr virus: a memoir. Epstein-Barr Virus Report *6*, 58-63.

Epstein, M.A., and Achong, B.G. (1967). Immunologic relationship of the herpes-like EB virus of cultured Burkitt lymphoblasts. Cancer Res. *27*, 2489-2493.

Epstein, M.A., and Achong, B.G. (1968a). Specific immunofluorescence test for the herpes-type EB virus of Burkitt lymphoblasts, authenticated by electron microscopy. J. U.S. Nat. Cancer Inst. *40*, 593-607.

Epstein, M.A., and Achong, B.G. (1968b). Observations on the nature of the herpes-type EB virus in cultured Burkitt lymphoblasts, using a specific immunofluorescence test. J. U.S. Nat. Cancer Inst. *40*, 609-621.

Epstein, M.A., and Achong, B.G. (1973). The EB virus. Ann. Rev. Microbiol. *27*,413-436.

Epstein, M.A., and Achong, B.G. (1979a). Introduction: discovery and general biology of the virus. In The Epstein-Barr Virus, M.A. Epstein and B.G. Achong, eds. (Berlin, Heidelberg and New York: Springer) pp 1-22.

Epstein, M.A., and Achong, B.G. (1979b). The Epstein Barr Virus (Berlin, Heidelberg and New York: Springer) pp 1-459.

Epstein, M.A., and Barr, Y.M. (1964). Cultivation *in vitro* of human lymphoblasts from Burkitt's malignant lymphoma. Lancet *i,* 252-253.

Epstein, M.A., and Barr, Y.M. (1965). Characteristics and mode of growth of a tissue culture strain (EB1) of human lymphoblasts from Burkitt's lymphoma. J. U.S. Nat. Cancer Inst. *34*, 231-240.

Epstein, M.A., and Eastwood, M.A. (1995). Denis Parsons Burkitt: 28 February 1911 – 23 March 1993. Biograph. Memoirs of Fellows of the Roy. Soc. *41*, 87-102.

Epstein, M.A., Achong, B.G., and Barr, Y.M. (1964c) Virus particles in cultured lymphoblasts from Burkitt's lymphoma. Lancet *i*, 702-703.

Epstein, M.A., Achong, B.G., Barr, Y.M., Zajac, B., Henle, G., and Henle, W. (1966). Morphological and virological investigations on cultured Burkitt tumor lymphoblasts (strain Raji). J. U.S. Nat. Cancer Inst. *37*, 547-559.

Epstein, M.A., Barr, Y.M., and Achong, B.G. (1964a). Avian tumor virus behavior as a guide in the investigation of a human neoplasm. U.S. Nat. Cancer Inst. Monogr. *17*, 637-650.

Epstein, M.A., Barr, Y.M., and Achong, B.G. (1964b). A second virus-carrying tissue culture strain (EB2) of lymphoblasts from Burkitt's lymphoma. Path. Biol. (Paris) *12*, 1233-1234.

Epstein, M.A., Barr, Y.M., and Achong, B.G. (1965a). Studies with Burkitt's lymphoma. Wistar Inst. Symp. Monogr. *4*, 69-82.

Epstein, M.A., Henle, G, Achong, B.G., and Barr, Y.M. (1965b). Morphological and biological studies on a virus in cultured lymphoblasts from Burkitt's lymphoma. J. Exp. Med. *121*, 761-770.

Epstein, M.A., Morgan, A.J., Finerty, S., Randle, B.J. and Kirkwood, J.K. (1985). Protection of cottontop tamarins against Epstein-Barr virus-induced malignant lymphoma by a prototype subunit vaccine. Nature *318*, 287-289.

Fickenscher, H. and Fleckenstein, B. (2001). Herpesvirus saimiri. In Oncogenic γ-herpesviruses: an expending family. M.A. Epstein, A.B. Rickinson and R.A. Weiss eds. (London: Theme Issue Phil. Trans. Royal Society B 356) pp 545-567.

Fischer, G.A. (1957). Tissue culture of mouse leukemic cells. Proc. Amer. Assn. Cancer Res. *2*, 201.

Fischer, G.A. (1958). Studies of the culture of leukemic cells *in vitro*. Ann. N.Y. Acad. Sci., *76*, 673-680.

Galloway, J. (1989). Chance and Felicity. Nature *338*, 463-464.

Henle, G. and Henle, W. (1966a). Studies on cell lines derived from Burkitt's lymphoma. Trans. N.Y. Acad. Sci. *29*, 71-79.

Henle, G. and Henle, W. (1966b). Immunofluorescence in cells derived from Burkitt's lymphoma. J. Bact. *91*, 1248-1256.

Henle, G., Henle, W. and Diehl, V. (1968). Relation of Burkitt's tumor-associated herpes-type virus to infectious mononucleosis. Proc. Nat. Acad. Sci. U.S. *59*, 94-101.

Jackman, W.T., Mann, K.A., Hoffmann, H.J. and Spaeter, R. (1999). Expression of Epstein-Barr virus gp350 as a single chain glycoprotein for an EBV subunit vaccine. Vaccine *17*, 660-668.

Kaaden, O.R. and Dietzschold, B. (1974). Alterations of the immunological specificity of plasma membranes of cells infected with Marek's disease and turkey herpes viruses. J. Gen. Virol. *25*, 1-10.

Kirkwood, J.K., Epstein, M.A. and Terlecki, A.J. (1983). Factors influencing population growth of a colony of cotton-top tamarins. Lab. Animals *17*, 35-41.

Kirkwood, J.K., Epstein, M.A., Terlecki, A.J. and Underwood, S.J. (1985). Rearing a second generation of cotton-top tamarins (*Saguinus Oedipus Oedipus*) in captivity. Lab. Animals *19*, 269-272.

Klein, G. (1973). The Epstein-Barr virus. In The Herpesviruses, A. Kaplan ed. (London and New York: Academic Press) pp 521-555.

Lesnick, F. and Ross, L.J.N. (1975). Immunization against Marek's disease using Marek's disease virus-specific antigens free from infectious virus. In. J. Cancer *16*, 153-163.

Lisitsyn, N., Lisitsyn, N. and Wigler, M. (1993). Cloning differences between two complex genomes. Science *259*, 946-951.

Morgan, A.J., Allison, A.C., Finerty, S., Scullion, F.T., Byars, N.E. and Epstein, M.A. (1989). Validation of a first generation Epstein-Barr virus vaccine preparation suitable for human use. J. Med. Virol. *29*, 74-78.

Morgan, A.J., Finerty, S., Lovgren, K., Scullion, F.T. and Morein, B. (1988). Prevention of Epstein-Barr (EB) virus-induced lymphoma in cottontop tamarins by vaccination with the EB virus envelope glycoprotein gp340 incorporated into immune-stimulating complexes. J. Gen. Virol. *69*, 2093-2096.

Morgan, A.J., North, J.R. and Epstein, M.A. (1983). Purification and properties of the gp340 component of Epstein-Barr (EB) virus membrane antigen (MA) in an immunogenic form. J. Gen. Virol. *64*, 455-460.

Moss, D.J. and Pope, J.H. (1972). Assay of the infectivity of Epstein-Barr virus by transformation of human leucocytes *in vitro*. J. Gen Virol. *17*, 233-236.

North, J.R., Morgan, A.J., Thompson, J.L. and Epstein, M.A. (1982a). Quantification of an EB virus-associated membrane antigen (MA) component. J. Virol. Methods *5*, 55-65.

North, J.R., Morgan, A.J., Thompson, J.L. and Epstein, M.A. (1982b). Purified EB virus gp340 induces potent virus-neutralising antibodies when incorporated in liposomes. Proc. Nat. Acad. Sci. U.S. *79*, 7504-7508.

O'Conor, G.T. (1961). Malignant lymphoma in African children. II. A pathological entity. Cancer *14*, 270-283.

Pearson, G., Dewey, F., Klein, G., Henle, G. and Henle, W. (1970). Relation between neutralization of Epstein-Barr virus and antibodies to cell-membrane antigens induced by the virus. J. U.S. Nat. Cancer Inst. *45*, 989-995.

Pulvertaft, R.J.V. (1964). Cytology of Burkitt's tumour (African lymphoma). Lancet *i*, 238-240.

Rabson, A.S., O'Conor, G.T., Baron, S., Whang, J.J. and Legallais, F.Y. (1966). Morphologic, cytogenetic and virologic studies *in vitro* of a malignant lymphoma from an African child. Int. J. Cancer *1*, 89-106.

Randle, B.J. and Epstein, M.A. (1984). A highly sensitive enzyme-linked immunosorbent assay to quantitate antibodies to Epstein-Barr virus membrane antigen gp340. J. Virol. Methods *9*, 201-208.

Shope, T., Dechairo, D. and Miller, G. (1973). Malignant lymphoma in cottontop marmosets after inoculation with Epstein-Barr virus. Proc. Nat. Acad. Sci. U.S. *70*, 2487-2491.

Stewart, S.E., Lovelace, E., Whang, J.J. and Ngu, V.A. (1965). Burkitt tumor: tissue culture, cytogenetic and virus studies. J. U.S. Nat. Cancer Inst. *34*, 319-327.

Woodliff, H.J. (1964). Blood and bone marrow cell culture. (London: Eyre and Spottiswoode) pp 1-141.

Young, L.S. (1999). Editorial Comment. Epstein-Barr Virus Report *6*, 57.

zur Hausen, H., Schulte-Holthausen, H., Klein, G., Henle,W., Henle, G., Clifford, P. and Santesson, L. (1970). EBV DNA in biopsies of Burkit tumours and anaplastic carcinomas of the nasopharynx. Nature *228*, 1056-1058.

Chapter 2

The Early Days of Epstein-Barr Virus Research: The Henle Years

*Harald zur Hausen**

ABSTRACT

Werner and Gertrude Henle were early pioneers in Epstein-Barr virus (EBV) research. Their seroepidemiological studies resulted in the discovery of EBV as causative agent of infectious mononucleosis. Between the years 1966 and 1987 they characterized the seroresponse to viral capsid and early antigens, demonstrated the ubiquity and global distribution of EBV infections, and identified specific antibody patterns in patients with EBV-linked tumors. In addition, they were the first to demonstrate the transforming properties of EBV. Their ground-breaking work formed the basis for the large number of subsequent studies on the role of EBV in human pathogenesis and human tumors.

INTRODUCTION: AN INITIAL STIMULUS

In 1939, Werner Henle started his career as virologist in the University's department of pediatrics at the Philadelphia Children's Hospital, joined from the beginning by his wife Gertrude (see Figure 1). For quite some years they both worked on influenza viruses and influenza vaccines development. Studies on paramyxoviruses followed with a specific interest in vaccination against mumps and in persistent infections by these agents. They also attempted to transmit Hepatitis viruses to cell cultures and embryonated eggs. Not all of these studies have been equally successful.

Particularly, the Henles' worked with cells persistently infected with mumps and Newcastle disease virus, and this evoked their interest in persistent virus infections. In 1963 they were approached by the surgeon-in-chief of the Children's Hospital in Philadelphia, C. Everett Koop, the later US Surgeon-General, who had returned from a meeting in Africa. He told them (quoted by Werner Henle) *"Henles, if you really want to work on a tumor which most likely is caused by a virus, you should work on Burkitt's lymphoma"*. Since Tony Epstein in 1964, after observing herpes-like particles in suspension cultures derived from Burkitt's tumors, could not enlist the help of British virologists, he sent two cultures (EB-1 and EB-2) to a colleague of the Henles', Klaus Hummeler, who was in charge of the diagnostic service at the Children's Hospital in Philadelphia. Hummeler asked the Henles for

*For correspondence email ZurHausen@DKFZ.de

Figure 1. Shows a photograph of the Henles, Werner and Gertrude (in the middle), Alan Granoff on the right and Harald zur Hausen partially on the left which was taken in 1971 (*courtesy of Harald zur Hausen*).

support, and they immediately became excited by the possibility of working on this tumor and established collaborative contacts with Tony Epstein.

PERSONAL RECOLLECTION

My personal first recollection of their interest dates back to 1965. I had applied for a postdoctoral position in the Henle lab and in the late summer of 1965 they visited Germany and asked me to come to Heidelberg for an interview. I met them in the most prestigious hotel there, the "Europäischer Hof", where, particularly Gertrude Henle, talked intensively about the "*EB virus*" and flooded me with information about an agent I had never heard of before. They accepted me as a postdoc who was supposed to work on EBV. Thus, in the end of December 1965 I flew to Philadelphia and started to work in the Henle laboratory on January 2, 1966.

In the meantime the Henles had developed an immunofluorescent test to detect viral antigens in virus-producing cells which greatly facilitated further studies (Henle and Henle, 1966). The availability of this test system resulted in the detection of highly elevated antibody titers against viral antigens in patients with Burkitt's lymphoma (Henle and Henle, 1966; Henle et al., 1969) and subsequently also in a second human cancer, in nasopharyngeal carcinomas (Old et al. 1966; de Schryver et al., 1969). My own early contribution to the system was the identification EBV particles in individual fluorescence-positive cells, revealing the specificity of the Henle test system (zur Hausen et al., 1967). The seroepidemiological studies or Werner and Gertrude Henle made it clear in these years that EBV-infections are ubiquitous and that the virus must be widely spread in all human populations (Henle et al., 1969). Close to 90% of the population became infected during childhood or adolescence. The subsequent development of an immunofluorescent test to detect early EBV-induced antigens (Henle et al., 1970; 1971) showed again the specificity of a highly elevated immune response against EBV in patients with Burkitt's lymphoma or nasopharyngeal carcinoma.

EBV PERSISTENCE IN BURKITT'S LYMPHOMA CELLS

For quite some time Werner Henle and I discussed the question of whether the presence of EBV particles in a very limited number of Burkitt tumor cells kept in tissue culture was the consequence of a persistent infection or of an activated state of otherwise latently infected cells. At this time Werner was very much influenced by data which he and Gertrude had obtained earlier in studying persistent mumps virus infections. Thus, he reasoned that the situation in Burkitt's lymphoma cells might be analogous to those persistent infections. My personal view was different. I was fascinated by bacterial systems carrying lysogenic phages and suspected that the fluorescent cells may indeed result from a spontaneous activation from a state of latency. During the years 1967-1969 I tried to acquire the necessary molecular skills to prove this directly and worked during this period on the integrational pattern of adenovirus type 12 in abortive infection of Syrian hamster cells. I returned in March 1969 to Germany and felt sufficiently confident that I could purify EBV DNA from the EBV producer line, P3HR-1. Fortunately, the first attempts yielded already positive results (Schulte-Holthausen and zur Hausen, 1969). Subsequently it was indeed possible to detect EBV DNA in the non-virus-produced line Raji by DNA-DNA hybridization (zur Hausen and Schulte-Holthausen, 1970) and to demonstrate EBV DNA in primary biopsies from Burkitt's lymphoma and from nasopharyngeal carcinomas (zur Hausen et al., 1970). Three years later we also showed that EBV is not only persisting in cells of lymphatic origin, but also in epithelial nasopharyngeal carcinoma cells (Wolf et al., 1973). Viral DNA persistence in Burkitt's lymphomas was soon confirmed by Nonoyama and Pagano (1973), in epithelial cells of nasopharyngeal carcinomas by Klein et al. (1974).

The finding of EBV DNA in "virus-negative" Burkitt tumor cells settled the question of the mode of viral DNA persistence. It was at that time not totally unexpected, since Pope and colleagues reported in 1969 a specific complement-fixing antigen in the non-virus producing Raji cells. This antigen was later visualized by Reedman and Klein (1973) and designated as EBNA (Epstein-Barr virus nuclear antigen).

The Henles and colleagues (1967) were the first to demonstrate transformation of human lymphocytes by co-cultivation with X-irradiated EBV-producing Burkitt's lymphoma cells. They irradiated an EBV-producer Burkitt lymphoma line (Jijoye) and in parallel a non-produced culture (Raji) each with 6000 r which prevented further cell divisions and resulted in cell death 3-5 weeks later. Both of these lines were derived from male patients. They were then co-cultivated with peripheral leukocytes from healthy female donors. Continuously proliferating cultures were only established after co-cultivation with the irradiated Jijoye, but not with Raji cells. The outgrowing cells revealed a female karyotype and a small percentage of cells were EBV-VCA positive. This was the earliest demonstration of a biological function exerted by EBV infection. Shortly thereafter similar results were obtained with cell-free filtrates of EBV-positive cells (Pope et al., 1968). These data opened the path for the large number of subsequent papers by other groups to analyze the transforming properties of EBV. Obviously, the virus is extremely effective

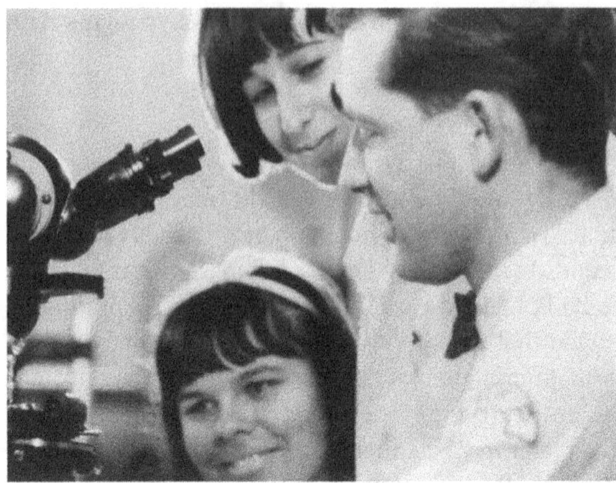

Figure 2. Shows a photograph which was taken in the Henle's laboratory in Philadelphia, Pennsylvania USA in 1967 with two of the technicians in the laboratory one of which was Elaine whose infectious mononucleosis provided the clue for the role of EBV in IM (*courtesy of Harald zur Hausen*).

in immortalizing B lymphocytes. Since most human adults carry latently infected B cells in their peripheral blood, outgrowth of these cells can be readily achieved, particularly by depleting or inactivating accompanying T cells (von Knebel Doeberitz et al., 1983).

EBV THE CAUSATIVE AGENT OF INFECTIOUS MONONUCLEOSIS

In 1968 the Henles identified EBV as the causative agent of infectious mononucleosis (Henle et al., 1968). As Werner Henle wrote himself in his "Annotated Bibliography" *"The answer to our search* (for an illness caused by EBV) *came quite unexpectedly right in our laboratory when one of our technicians reported ill in August 1967. She was seronegative* (for EBV) *and had been a frequent, more or less willing donor for leukocytes for EBV transmission/transformation studies carried out by Volker Diehl, a German postdoctoral fellow. Our technician returned to work 5 days later but developed under our eyes a rubella-like rash. Her physician had considered either rubella or IM as probable diagnoses. We obtained blood from her for serologic tests as well as yet another leukocyte culture thinking more of rubella than IM. The heterophil antibody test, however, turned out to be positive. The rash was thus not due to rubella but ampicillin which she had received and which for unknown reasons can be the cause of a rash in IM patients. We noted next with astonishment that our technician (see Figure 2) had developed antibodies to EBV since the last time we tested her and observed in due course her leukocytes which formerly never survived in culture grew this time into a permanent lymphoblastoid cell line with a small percentage of EBV-producing cells. These were the first exciting clues that EBV was the long sought cause of IM. It was fortunate that we never took our summer vacation in August since in our absence this event would probably have passed unnoticed".*

In a subsequent collaboration with Jim Niederman and Bob McCollum (1968) the testing of "pre-sera" from college students who later developed infectious mononucleosis and of sera obtained thereafter confirmed the observed seroconversion. This left no doubt that EBV is indeed the causative agent of IM. It represented a major breakthrough in EBV research and permitted a new understanding of EBV-linked pathogenesis.

In the following years the Henles were occupied with further analyzing the pathogenesis of infectious mononucleosis, but even more so by studying the seroepidemiology of human tumors in relation to EBV. Their main emphasis was Burkitt's lymphoma. Extensive studies carried out with George Klein and numerous other colleagues revealed that Burkitt's lymphoma patients had generally high antibody titers against EBV in comparison to various control sera (Henle et al., 1969a). The discovery of early antigens was quickly applied to Burkitt's lymphoma patients. The differentiation between "*diffuse*" (D) and "*restricted*" (R) components showed that most BL patients gave a different immunofluorescent pattern (R) from IM or nasopharyngeal carcinoma patients (Henle et al., 1971).

EBV ANTIBODY DETECTION AND HUMAN MALIGNANCIES

The seroepidemiology of EBV in nasopharyngeal cancer was studied jointly with the Stockholm group of George Klein (Andersson-Anvret et al., 1977) and with Anne Lanier of the Alaskan Branch of the Center of Disease Control, the latter study on NPC in Eskimos, Aleuts and Indians (Lanier et al., 1981). They were also extended to Hong Kong with a high frequency of nasopharyngeal cancer cases (Henle et al., 1973a). Basically, tumor-bearing patients revealed higher antibody titers as the total tumor burden increased from stage I to later stages of the disease. In addition, the prevalence of anti-D early antibodies, as well as of EBV-specific IgA antibodies, was characteristic for this condition (Henle and Henle, 1976).

A number of studies followed on EBV antibody patterns in disease-related immunodeficiencies like sarcoidosis (Wahren et al., 1971), rheumatoid arthritis (Alspaugh et al., 1981) and Hodgkin's disease (Johansson et al., 1970; Henle and Henle, 1973). The first two diseases provided no conclusive results. In Hodgkin's disease, however, the patients revealed increased antibody titers in comparison to controls. The Henles initially interpreted this as a reactivation of latent EBV infection. In addition to Hodgkin's disease; also 40% of patients with chronic lymphocytic leukemia showed relatively high EBV reactivities (Johansson et al., 1971).

In the early 1980's the Henles turned their interest also to the acquired immunodeficiency syndrome (AIDS) and Kaposi sarcomas (Henle and Henle, 1981; Kaminsky et al., 1985) without results pointing to an involvement of EBV in these conditions. Werner Henle contributed, however to a study describing for the first time familial occurrence of fatal infectious mononucleosis (Bar et al., 1974), later named by Purtillo the X-linked lymphoproliferative (XLP) syndrome. The Henles also studied the EBV seroepidemiology patients with renal, bone marrow or cardiac allografts (Lange et al., 1980, Schooley et al., 1983, Bieber et al., 1984). A review covering their data on the seroepidemiology of EBV was published in 1980 (Henle and Henle).

FINAL COMMENTS

In retrospect it remains amazing how substantially and at the same time how elegantly the Henles contributed the Epstein-Barr virus field (see Figure 3). Their ground-breaking work formed the basis for the large number of subsequent studies on the role of EBV in human pathogenesis and human tumors. Within a period of slightly more than 20 years this dedicated couple collaborated with numerous

Figure 3. Shows the Henles in Sapporo in Japan in 1977. From left to right: Tony Epstein, Harald zur Hausen, Werner Henle, Gary Pearson and Gertrude Henle at a conference of Epstein Barr virus and related tumors (*courtesy of Harald zur Hausen*).

laboratories across the world and published approximately 300 manuscripts, most of them based on straightforward application of immunofluorescence tests. When Werner Henle died on July 6 in 1987 at the age of seventy-six, his wife Gertrude retired from her most active career as a virologist. Both of them are clearly missed today.

References

Alspaugh, M.A., Henle, G., Lennette, E.T., and Henle, W. (1981). Elevated levels of antibodies to Epstein-Barr virus antigens in sera and synovial fluids of patients with rheumatoid arthritis. J. Clin. Invest. 67, 1134-1140.

Andersson-Anvret, M., Forsby, N., Klein, G. and Henle, W. (1977). Relationship between the Epstein-Barr virus and undifferentiated nasopharyngeal carcinomas, Correlated nucleic acid hybridization and histopathological examination. Int. J. Cancer *20*, 486-494.

Bar, R.S., DeLor, C.J., Clausen, K.P., Hurtubise, P., Henle, W., and Hewetson, J.F. (1974). Fatal infectious mononucleosis in a family. New Engl. J. Med. *290*, 363-367.

Bieber, C.P., Heberling, R.I., Jamieson, S.W., Oyer, P.E., Cleary, M., Warnke, R., Saemundsen, A., Klein, G., Henle, W., and Stinson, E.B. (1984). Lymphoma in cardiac transplant recipients associated with cyclosporine A, prednisone, and antithymocytic globulin (ATG). In Immune Deficiency and Cancer. D. Purtillo, ed. (Plenum Press, New York), pp. 309-320.

de Schryver, A., Friberg, S. Jr., Klein, G., Henle, W., Henle, G., de Thé, G., Clifford, P., and Ho, H.C. (1969). Epstein-Barr virus-associated antibody patterns in carcinoma of the post-nasal space. Clin. Exp. Immunol. *5*, 443-459.

Epstein, M.A., Achong, B.G., and Barr, Y.M. (1964). Virus particles in cultured lymphoblasts from Burkitt's lymphoma. Lancet *1*, 702-703.

Henle, G. and Henle, W. (1966). Immunofluorescence in cells derived from Burkitt's lymphoma. J. Bacteriol. *91*, 1248-1256.

Henle, W., Diehl, V., Kohn, G., zur Hausen, H., and Henle, G. (1967). Herpes-type virus and chromosome marker in normal leukocytes after growth with irradiated Burkitt cells. Science 157, 1064-1065.

Henle, G., Henle, W., and Diehl, V. (1968). Relation of Burkitt's tumour associated herpes-type virus to infectious mononucleosis. Proc. Natl. Acad. Sci. 59, 94-101.

Henle, G., Henle, W., Clifford, P. Clifford, P., Diehl, V., Kafuko, G.W., Kirya, B.G., Klein, G., Morrow, R.H., Munube, G.M.R., Pike, P., Tukei, P.M., and Ziegler, J.L. (1969a). Antibodies to EB virus in Burkitt's lymphoma and control groups. J. Natl. Cancer Inst. 43, 1147-1157.

Henle, W. and Henle, G. (1969). The relation between Epstein-Barr virus and infectious mononucleosis, Burkitt's lymphoma and cancer of the postnasal space. East Afr. Med. J. 46, 402-406

Henle, W., Henle, G., Zajac, B.A., Pearson, G., Waubke, R., and Scriba, M. (1970). Differential reactivity of human serums with early antigens induced by Epstein-Barr virus. Science 169, 188-190.

Henle, G., Henle, W., and Klein, G. (1971). Demonstration of two distinct components in the early antigen complex of Epstein-Barr virus infected cells. Int. J. Cancer 8, 272-282.

Henle, W., Ho, H.C., Henle, G., and Kwan, H.C. (1973a). Antibodies to Epstein-Barr virus-related antigens in nasopharyngeal carcinoma. Comparison of active cases and long-term survivors. J. Nat. Cancer Inst. 361-369.

Henle, W. and Henle, G. (1973). Epstein-Barr virus-related serology in Hodgkin's disease. Nat. Cancer Inst. Monograph. 36, 79-84.

Henle, G. and Henle, W. (1976). Epstein-Barr virus-specific IgA antibodies as an outstanding feature of nasopharyngeal carcinoma. Int. J. Cancer 17, 1-7.

Henle W, Henle G. (1980). Epidemiologic aspects of Epstein-Barr virus (EBV)-associated diseases Ann. N. Y. Acad. Sci. 354, 326-331.

Henle W, Henle G. (1981). Epstein-Barr virus-specific serology in immunologically compromised individuals. Cancer Res. 41, 4222-4225.

Johansson, B., Klein, G., Henle, W., and Henle, G. (1970). Epstein-Barr virus (EBV) associated antibody patterns in malignant lymphoma and leukemia. I. Hodgkin's disease. Int. J. Cancer 6, 450-462.

Johansson, B., Klein, G., Henle, W., and Henle, G. (1971). Epstein-Barr virus (EBV)-associated antibody patterns in malignant lymphoma and leukemia. II. Chronic lymphatic leukaemia and lymphocytic lymphoma. Int. J. Cancer 8, 475-486.

Kaminsky LS, McHugh T, Stites D, Volberding P, Henle G, Henle W, Levy JA. (1985). High prevalence of antibodies to acquired immune deficiency syndrome (AIDS)-associated retrovirus (ARV) in AIDS and related conditions but not in other disease states. Proc Natl Acad Sci U S A. 82, 5535-5539.

Klein, G., Giovanella, B.C., Lindahl, T., Fialkow, P.J., Singh, S., and Stehlin, J.S. (1974). Direct evidence for the presence of Epstein-Barr virus DNA and nuclear antigen in the malignant epithelial cells from patients with poorly differentiated carcin oma of the nasopharynx. Proc. Natl. Acad. Sci. 71, 4737-4741.

Lange, B., Henle, W., Meyers, J.D., Yang, L.C., August, C., Koch, P., Arbeter, A., and Henle, G. (1980). Epstein-Barr virus-related serology in marrow transplant recipients. Int. J. Cancer 26, 151-158.

Lanier, , A.P., Bornkamm, G.W., Henle, W., Henle, G., Bender, T.R., Talbot, M.L., and Dohan, P.H. (1981). Association of Epstein-Barr virus with nasopharyngeal carcinoma in Alaskan native patients, Serum antibodies and tissue EBNA and DNA. Int. J. Cancer 28, 301-305.

Niederman, J.C., McCollum, R.W., Henle, G., and Henle, W. (1968). Infectious mononucleosis. Clinical manifestations in relation to Epstein-Barr virus. J. Am. Med. Assoc. 203, 139-143.

Nonoyama, M., Huang, C.H., Pagano, J.S., Klein, G., and Singh, S. (1973). DNA of Epstein-Barr virus detected in tissue of Burkitt's lymphoma and nasopharyngeal carcinoma. Proc. Natl. Acad. Sci. *70*, 3265-3268.

Old, L.J., Boyse, E.A., Oettgen, H.F., De Harven, E., Geering, G., Williamson, B., and Clifford, P. (1966). Precipitating antibody in human serum to an antigen present in cultured Burkitt's lymphoma cells. Proc. Natl. Acad. Sci. *56*, 1699-1704.

Pope, J.H., Horne, M.K., and Scott, W. (1968). Transformation of foetal human leukocytes *in vitro* by filtrates of a human leukaemic cell line containing herpes-like virus. Int. J. Cancer *3*, 857-866.

Pope, J.H., Horne, M.K., and Wetters, E.J. (1969). Significance of a complement-fixing antigen associated with herpes-like virus and detected in the Raji cell line. Nature *222*, 186-187.

Reedman, B.M. and Klein, G. (1973). Cellular localisation of an Epstein-Barr virus (EBV)-associated complement-fixing antigen in producer and non-producer lymphoblastoid cell lines. Int. J. Cancer *11*, 599-620.

Schooley, R.T., Hirsch, M.S., Colvin, R.B., Cosimi, A.B., Tolkoff-Rubin, N.E., McCluskey, R.T., Burton, R.C., Russell, P.S., Herrin, J.T., Delmonico, F.A., Giorgi, J.V., Henle, W., and Rubin, R.H. (1983). Association of herpesvirus infections with T-lymphocytes subset alterations, glomerulopathy and opportunistic infections following renal transplantation. New Engl. J. Med. 308, 307-313.

Schulte-Holthausen, H. and zur Hausen, H. (1970). Partial purification of the Epstein-Barr virus and some properties of its DNA. Virology 40, 776-779.

von Knebel Doeberitz. M., Bornkamm, G.W., and zur Hausen, H. (1983). Establishment of spontaneously outgrowing lymphoblastoid cell lines with Cyclosporin A. Med. Microbiol. Immunol. (Berl.) *172*, 87-99.

Wahren, B., Carlens, E., Espmark, A., Lundbeck, H., Lofgren, S., Madar, E., Henle, G., and Henle, W. (1971). Antibodies to various herpesviruses in sera from patients with sarcoidosis. J. Nat. Cancer Inst. *47*, 747-756.

Wolf, H., zur Hausen, H., and Becker, V. (1973). EB viral genomes in epithelial nasopharyngeal carcinoma cells. Nature 244, 245-247.

zur Hausen, H., Henle, W., Hummler, K., Diehl, V., and Henle, G. (1967). Comparative study of cultured Burkitt tumor cells by immunofluorescence, autoradiography and electronmicroscopy. J. Virol. *1*, 830-837.

zur Hausen, H. and Schulte-Holthausen, H. (1970). Presence of EB virus nucleic acid homology in a "virus-free" line of Burkitt tumour cells. Nature *227*, 245-247.

zur Hausen, H., Schulte-Holthausen H., Klein, G., Henle, W., Henle, G., Clifford, P., and Santesson, L. (1970). EB-virus DNA in biopsies of Burkitt tumours and anaplastic carcinomas of the nasopharynx. Nature *228*, 1056-1058,

From: Epstein-Barr Virus. Edited by: Erle S. Robertson

Chapter 3

EBV and the Tumor Virus Context

*George Klein**

ABSTRACT

The initial studies of Epstein-Barr virus (EBV) were a difficult foray into the study of human cancer and the contributions of viruses to the development of these cancers. Accepting the fact that viruses were a major cofactor in contributing to the development of cancers had little support from some factions. However, the discovery of herpesvirus particles in cells isolated from a human Burkitt's lymphoma was nonetheless thrilling. This chapter is meant to consider the EBV fields in the context of existing knowledge on small DNA tumor viruses.

HISTORY AND INTRODUCTION

Altered genes or environmental factors are usually considered as major risk factors for tumor development. However, the strategy of certain viruses may constitute a risk factor in itself. Tumor associated viruses in man and other species have a survival strategy like other viruses, aiming to maintain, replicate and propagate their genomes, but some features of this strategy entail a risk to initiate or favour tumor development under certain circumstances. This implies that only a small minority of the infected cells enter the pathway towards a malignant tumor and even fewer succeed.

Views on the role of viruses in the etiology of cancer have been polarized between two extreme positions during the major part of the last century. The belief that viruses have nothing to do with cancer was as widespread at certain times, as the suspicion that most and perhaps all tumors are caused by virus at other times. The field started with the discovery of Peyton Rous in 1911 that chicken sarcomas could be transmitted with cell free filtrates (Rous, 1911). The tumors arose at the site of inoculation and were of the same histological type as the original sarcoma. This created great excitement: the cancer problem was solved! The enthusiasm subsided rapidly, however, when mouse and rat tumor filtrates failed to induce tumors. In retrospect we may see this as the consequence of exaggerated expectations, hasty experiments and increasing lack of confidence. It became the prevalent view that viruses may play a role for tumors in birds, but not in mammals.

*For correspondence email Elsbeth.Bengtsson@mtc.ki.se

Two decades later, Richard Shope found that benign warts could be transmitted from the wild cottontail to the domestic rabbit by cell free filtrates (Shope, 1933). This did not change the climate of opinion. The rabbit was mammalian but warts were benign tumors, not cancers. However, several important points were overlooked by outside commentators. The initially benign rabbit papillomas turned occasionally into carcinomas. This could be accelerated by the topical application of chemical carcinogens. The term **tumor progression** was originally coined by Rous to designate this transition, or, in its generalized form, the process whereby "tumors go from bad to worse". Later, Leslie Foulds defined and extended the term (Foulds, 1958). It refers to the development of tumors by multiple, stepwise changes in several "unit characteristics". Today we see them as distinct phenotypic traits. They are individually variable and reassort independently of each other. Tumor progression can therefore proceed along several alternative pathways and each tumor becomes individually unique from the biological point of view.

The **DNA tumor viruses** belong to several unrelated virus families. In contrast to the RNA tumor viruses that belong to the single retrovirus family and can replicate in growing cells without killing them, the **DNA tumor viruses kill the cells in which they replicate**. Their tumorigenic activity depends therefore on the blocking of the lytic viral cycle. This may occur in cells that are non-permissive for the lytic cycle due to their species and/or tissue derivation.

The transforming genes of all DNA tumor viruses are genuine constituents of the viral genome. The number of virally encoded transforming genes varies between one (SV40), two (adeno- and papillomaviruses) and six (EBV). The virally encoded transforming proteins are immunogenic, as a rule. The challenge of viral transformation is met by the immune surveillance of the host. Cells transformed by these viruses grow usually only in immunosuppressed hosts. They represent the major part of the "opportunistic tumors" that arise exclusively or predominantly in congenitally, iatrogenically (as after organ transplantation) or virally (*e.g.* by HIV) immunosuppressed persons.

CONTRASTS BETWEEN EBV AND THE SMALL DNA TUMOR VIRUSES

There is a fundamental difference between the small DNA tumor viruses and the herpesviruses, with regard to potentially tumorigenic interactions with their host cell. Permissiveness for the early but not for the late (lytic) steps of the viral cycle constitutes the main risk for the former group. Convergent evolution has provided SV40, the transforming adenoviruses and the tumor associated papillomaviruses with the ability to inactivate two of the main tumor suppressor pathways that involve p53 and Rb, respectively. Interestingly, inactivation of the same two pathways appears to be mandatory for non virally related tumor development as well. This impairs two of the main controlling functions that prevent the replication of cells driven by illegitimately activated oncogenes.

In the case of the tumor associated herpesviruses and particularly of EBV, the possible tumorigenic contributions of the virus need to be considered in relation to the different forms of non-productive virus-cell interactions. EBV has evolved mechanisms to activate and expand its primary host cell population, the human B-lymphocyte. It is also capable of switching off its B-cell activating program and remains latent in long-lived resting memory B cells. The virus can thus use several non-lytic interaction programs, tailored to different B cell subclasses. They lead to different programs of viral expression that also differ in their ability to induce a CD8+ T-cell mediated immune response in the host that prevents the excessive proliferation of the virally transformed immunoblasts.

Unlike the small DNA tumor viruses where tumorigenicity is favored by the structural or regulatory impairment of the lytic genes, the latent EBV-B cell interaction that occurs without any genetic defect in the virus is a potential tumor risk in the immunodeficient host. This is due to the fact that apart from the occasional activation of the viral cycle in EBV-carrying B cells, the interaction is largely non-lytic. In EBV carrying immunoblastomas that arise in transplant recipients, or in certain congenitally T-cell defective patients (XLP in particular) as well as in part of the AIDS associated lymphomas (with immunoblastic morphology), the interaction of the virus with proliferating immunoblasts is comparable to or identical with the usual interaction of the virus with normal B cells. The "tumorigenic accident" occurs at the level of the host and not at the level of the virus or the cell.

None of the other EBV associated B-cell derived malignant lymphomas, such as BL, HL, or PEL or the unusual; EBV carrying T cell lymphomas express the proliferation driving, blastogenic program of the virus. The virus expresses only minimalistic programs designated as latency I or II, in contrast to the full immunoblastic program (III). The tumor promoting contribution of the virus must therefore be sought in other, less direct effects.

Normal B cell physiology is tightly regulated. The generation of B-cell receptor diversity by immunoglobulin gene rearrangements, activation of resting B cells by cognate antigen and associated molecules, expansion of the activated population, migration through germinal centers and concomitant hypermutation, generation of long-lived memory cells and differentiation into secretory subtypes, plasma cells in particular, is regulated both by internal programs and external signals, including antigens, ligands and cytokines that influence homing, proliferation, differentiation and death of the cells. EBV has a whole gamut of highly refined mechanisms that exploit normal B cell physiology. Unlike many other viruses, its strategy is not limited to the turning of its host cell into a viral protein factory with the single purpose of virus production and release to the environment, although it can switch on that mechanism when it enters the lytic cycle. Rather, the virus follows a "live and let live" principle. Its vast success in infecting all human populations and persisting in latent form over the lifetime of the host without causing disease, except by accident, testifies to the validity of this strategy, both for the virus and for the cells.

In BL, one needs to depart from the fact that the lymphoma originates in the post -GC, centroblastic or centrocytic cell, that is either resting or is on its way to a long-lived resting memory cell, but cannot leave the cycling compartment because it is driven by an Ig/myc translocation. The translocation results from a faulty recombination that occurs in the course of physiological Ig gene rearrangement. Conceivably, EBV may contribute to the emergence of the virus carrying BL clone by expanding the original population or, alternatively, or in addition, by protecting the myc-driven, apoptosis prone cell from apoptosis.

Table 1. EBV-associated tumors in man. The percentage figures indicate the frequency of EBV carrying tumors

Lymphoid tissues		Epithelial tissues	
Burkitts lymphoma, endemic	98%		
Burkitts lymphoma, sporadic	25%	Gastric adenocarcinoma	5-10%
AIDS-immunoblastic lymphoma	60%	Nasopharyngeal carcinoma	
-in CNS	100%	undifferentiated	100%
Post-transplant lymphoma	100%	Salivary gland carcinomas	100%
Hodgkin's lymphoma	50%	Leiomyosarcoma in immunosuppressed	100%
T-cell lymphomas	10-30%		
Lethal midline granuloma	>90%		

EBV is the causative agent of a self-limiting lymphoproliferative disease, infectious mononucleosis. In immunodefectives, the proliferation may proceed to progressively growing immunoblastomas, as already mentioned. Multiple viral genomes, derived from a single infectious event, are regularly found in high endemic Burkitt lymphomas (BLs) and in low differentiated or anaplastic nasopharyngeal carcinomas (NPCs). EBV is also found, although less regularly, in Hodgkin's lymphomas, nasal T-cell lymphomas, gastric carcinomas, salivary gland tumors and leiomyosarcomas (Table 1).

VIRAL INACTIVATION OF MAJOR TUMOR SUPPRESSOR PATHWAYS

From its earliest beginnings, cancer research has been looking for "the" fundamental change in cancer cells, the ultimate common denominator. The idea that such a change must exist, was, if not abandoned, substantially mollified by the increasing realization, from the late 1950s, that cancer development is a multistep process, based on the individual reassortment of several unit characteristics (Foulds, 1958) or, as we now say, phenotypic traits.

In spite of this diversity, the molecular era has brought forward a new unifying concept. The two major tumor suppressor pathways, Rb and p53 are inactivated, at one point or another, in most and perhaps all cancer cells. Rb and p53 are individual components of extensive regulatory networks with many participating molecules. It is therefore appropriate to ask whether their early discovery may have biased the subsequent development of the field. Both genes have been found through indirect and quite tortuous routes, led by accidental findings and experimental convenience. Has there been undue focus on these two genes? Is it possible that many others could have turned out to be equally, if not more important?

How can you approach such a question? The DNA tumor viruses have given us some clues, as mentioned above. None of the oncogenic papova-, adeno- or herpesviruses has evolved to cause tumors. Lethal viruses are at selective disadvantage, compared to their low or non-pathogenetic relatives. They are only tumorigenic under experimental conditions: *in vitro*, in non-natural hosts or, sometimes, in special categories of natural hosts. They have, nevertheless, evolved the ability to encode specific proteins that can break down or otherwise inactivate Rb and p53. Since several of the viruses are unrelated and have different strategies, this is a true example of convergent evolution. The viral proteins responsible for these effects are also responsible for the transforming action of each virus. In most cases, the natural host of the virus responds with effective immunological responses that can keep the transformed cells under effective control. Tumor development occurs then only as an accident of immunosuppression or cytogenetic changes.

One common denominator is the need to expand the virus carrying cell population. In addition, the integrating viruses, such as SV40 and adeno, must induce an S-phase in their target cells, in order to open the chromatin before they can integrate their proviral DNA within the cellular genome. Episomal viruses like HPV may also have to put or to keep their target cells in the cycle, in order to establish the correct episomal-chromosomal balance. However this may be, the fact remains that unrelated viruses have evolved the capacity to inactivate the Rb and p53 proteins, rather than other members of the corresponding pathways. Is the dominating position of these two key proteins the reason for their multifunctionality?

Coordinated or at least parallel dysregulation of the Rb and p53 pathways is not only a convergent feature of DNA tumor virus strategies but is also an important characteristic of spontaneous tumor development and progression. The impairment of both pathways by genetic and/or epigenetic mechanisms has emerged as a common marker of most and perhaps all cancer cells. It is now a central feature of cancer cell biology, and one of the few generalizations that can be made. It makes sense. Dysregulation of the Rb dependent normal control of the cell cycle and the crippling of one of the most important apoptotic mechanisms by the impairment of two independent but intercommunicating pathways is an efficient way to favor the evolution of the cancer cell phenotype.

In view of its virtually universal inactivation in tumors, the p53 pathway is probably at the bottom of this "sculpting", perhaps because it is eminently efficient in nipping clones, driven to expansion by illegitimately activated oncogenes, in the bud. It may be added that the **early** impairment of p53 effector functions during tumorigenesis are context dependent, since they vary with cell type and oncogenic insult. They can be phenocopied by the loss of specific modifiers of p53 function. The p53 effector functions lost as **a consequence** of tumor evolution or antitumor therapy are context independent and involve primary mutation or deletion of p53.

The context dependence of p53 impairment is also reflected by the tissue predilections of the oncogenic effect. Oncogene induced, p53 dependent apoptosis is apparently critical in preventing the development of B-cell derived tumors, whereas the roles of the p53 pathway in arresting the cell cycle and the maintenance of genomic stability are important for suppressing malignancies originating in other tissues. Moreover, the efficiency of oncogene induced apoptosis varies between different illegitimately activated oncogenes.

In BL, the Rb pathway is inactivated by p16 methylation, as a rule, whereas the p53 pathway is inactivated by either p53 mutation, ARF deletion of MDM2 amplification (Lindström et al. 2001). None of these events occur in EBV-transformed immunoblasts, where the full viral program upregulates cellular bcl-2 and other antiapoptotic genes. The question whether and to what extent the full viral growth transformation program influences the Rb and p53 pathway in other ways in the immunoblastomas has not been fully clarified.

References

Foulds, L. (1958) The natural history of cancer. J. Chronic Dis. *8*, 2-37.

Lindström, M., Klangby, U., Wiman, KG. (2001). p14ARF homozygous deletion or MDM2 overexpression in Burkitt lymphoma lines carrying wild type p53. Oncogene *19*, 2171-2177.

Rous, P. (1911) A sarcoma of fowl transmissible by an agent from the tumor cells. J. Exp. Med. *13*, 397-411.

Shope, R.E. (1933) Infectious papillomatosis of rabbits; with a note on the histopathology. J. Exp. Med. *68*, 607-624.

Chapter 4

Sero Epidemiology of EBV and Associated Malignancies

*Guy de Thé**

ABSTRACT

EBV is an ubiquitous virus, which under different environmental conditions, initiates or promotes various malignant proliferations. Sero epidemiology has played a key role in providing evidence for an etiological role of EBV in the development of three preeminent diseases, namely teenagers' infections mononucleosis in western countries, B cell lymphomas as described by Burkitt (BL) in sub-Sahara African children, and nasopharyngeal carcinoma (NPC) in South East Asia among Cantonese Chinese.

INFECTIOUS MONONUCLEOSIS

In 1968, Henle et al. observed that IM was regularly associated with EBV sero-conversion. Proof that it was the causative agent was provided by a prospective sero epidemiological survey carried out by Evans et al. (1968) in Yale students showing that only seronegative students were candidate for subsequent development infectious mononucleosis (IM). Patients would first develop IgM then IgG EBV antibodies against structural antigens (VCA*), early non structural (EA*), and membrane antigens (MA*) (Henle et al. 1968).

The factors influencing the viral infection by EBV are related to hygiene habits, and favor exposure to saliva carrying infectious EBV. The factors influencing the clinical development of IM, however, are related to the age at infection, the host immune status and genetic predisposition or resistance to the disease. In fact, all children in developing countries and most adolescents in industrialized countries carry a latent EBV infection as shown by the presence of IgG antibodies to VCA, without antibodies to EA. Transmission studies in monkeys and in man have established the causal implication of EBV in heterophile positive infectious mononucleosis syndromes (Evans et al., 1983). The age of acquisition of EBV

*EA = Early Antigens
*VCA = Viral Capsid Antigens
*MA = Membrane Antigens

*For correspondence email dethe@pasteur.fr

antibdies varies greatly in different geographic areas. In Barbados, by the time they are two years old, 75% of the babies are already infected by EBV, while less than 10% in Connecticut are in this category (Evans, 1974). The route of transmission of the viral infection involves primarily saliva, but also, milk, blood and vaginal secretions containing EBV infected lymphocytes.

EBV RELATED MALIGNANT LYMPHOMAS

Denis Burkitt, a missionary oriented pediatric surgeon at Makarere University Hospital in Kampala, Uganda, made an historical observation in 1958. He described a childhood lymphoma occurring in children aged 5 to 9 years. He also found this tumor to be geographically limited to intertropical Africa in holoendemic malaria regions. Clinically and pathologically Burkitt lymphoma is a rapidly growing B cell lymphoma which develops, characteristically, in the jaw, and is very sensitive to chemotherapy.

The International Agency for Research on Cancer (IARC), with support from the Virus Cancer Program of the NCI/NIH, carried out a major sero epidemiological prospective study involving 42,000 children, aged 6 months to eight years who, after the agreement by the parents, were bled and regularly followed up for nine years to determine the relationship of infection by EBV to BL. Sixteen of these children developed BL. They were then bled again to determine their serological profile at time of diagnosis. The parallel EBV serological testing of pre and post BL of these emerging cases together with a number of control sera was carried out comparatively in three laboratories (G. Klein in Stockholm, W. Henle in Philadelphia and G. de Thé in Lyon). The results showed that pre-BL sera, taken 7 to 54 months prior to diagnosis, had significantly higher VCA titers than any control. The increased risk of developing BL was approximately 30 times for children with VCA titers two dilutions or more above the geometric mean titer of the corresponding normal population standardized for age, sex and locality (de-Thé et al, 1978). Taken together, these data provided evidence for a causal relationship between an early infection by EBV (most probably due to mother to child transmission), with the development of a Burkitt's lymphoma 5 to 9 years later.

There appeared to be three steps involved in BL pathogenesis: a massive EBV primary infection in the first months of life, together with severe malaria infection, depressing specifically the cellular immunity directed to EBV infected lymphocytes, resulting in an expansion of EBV infected B cells clones. Later on, and under as yet undetermined causes, an oncogenic event linked to an 8-14 chromosomal translocation, takes place within this subpopulation of EBV infected B cells. Such a translocation exposes the C-myc proto oncogene inappropriately expressed in BL cells. The mutation in the *p53* gene observed in 40% of BL may be a co-factor or a result of genetic disorder of the tumor clones.

An sero-epidemiological survey was also instrumental in evaluating the impact of an antimalaria chemoprophylaxis project carried out in a cohort of children in the 0-10 year's age group in the North Mara region of Tanzania. This intervention lowered incidence of BL, but without effecting EBV antibodies GMT titers in the cohort under prophylaxis. Sero epidemiology of EBV will also be required to assess the effectiveness of any EBV vaccine, but the very high cost of vaccine trials makes such a project unrealistic at present, when one considers the relative low cost of BL chemotherapy and its high cure rate.

Hodgkin's disease (HD) is a lymphoma characterized by the specific giant Reed-Sternberg cells (RSC) tumor cell exhibiting a low mitotic index, and surrounded by activated T-helper cells. Sero-epidemiology comparing HD patients versus controls showed elevated antibody titers against EBV/VCA, EA-D and EA-R antigens in HD patients. But do the elevated EBV antibodies titers represent markers of the immune dysfunction characteristic of HD? or do they reflect a pre HD status? as it is the case in BL. In order to get pre-HD sera, large population based serum banks identified blood specimens drawn at different time prior to HD onset. The resulting data indicated that an enhanced level of EBV replication was detectable serologically at least 4 years before the diagnosis of HD (Mueller et al., 1989). At time of diagnosis monoclonal EBV/DNA was present in RSC cells (Weiss et al., 1989) of about 30 to 50% of HD cases.

Three groups of patients with severe immuno-deficiency, namely the X-linked lymphoproliferative syndrome of Purtillo, the transplant patients given immunosuppressive drugs to prevent graft rejection and the HIV-AIDS patients , are known to develop polyclonal expansions of EBV infected B cells leading to severe and often deadly lymphoblastic proliferations. In the latter, high grade diffuse B cell lymphomas, localized in the brain predominate, but true BL are also observed with typical chromosomal translocation (Steel et al, 1985). In AIDs patients, nearly all CNS lymphomas, about 40-50% of BL type lymphomas and 50% of non-Hodgkin's lymphomas are EBV genome positive. (Luka et al., 1978, Hamilton-Dutoit et al., 1993).

NASOPHARYNGEAL CARCINOMA

Sero epidemiology has been critical for establishing the association between EBV and NPC, a major cause of mortality for nearly 200 millions Cantonese Chinese in South China and South East Asia. Furthermore, a serological test for early detection of the tumor allowed the development of a critical step in the control of NPC.

It was a serendipitous serological observation by Old et al. that opened the EBV-NPC field in 1966. They found that sera from Kenyan NPC patients, selected to serve as controls for a study on BL, had much higher precipitating antibodies than BL sera. Then, two serological tests, namely the EBV immuno fluorescence tests developed by the Henles (1966) allowed the titering of antibodies to structural VCA and early antigens-EA-D or EA-R, and the EBNA test of Reedman and Klein (1973), were most instrumental in NPC research.

Undifferentiated carcinoma of the nasopharynx (NPC) develops in the posterior and lateral walls of the nasopharynx where the ciliary respiratory epithelium lays over a bed of lymphoplasmocytic tissue, suggesting that both cell types interact in both the normal and in the oncogenic process. In contrast to Burkitt's lymphoma, NPC exhibits a high EBV serological profile wherever it occurs around the world, in high risk groups (Cantonese Chinese), intermediate (North Africa) or low risk areas (Western countries). High antibodies titers directed to a large spectrum of EBV antigens: VCA, EA, (mostly EA-D), EBNA, ZEBRA, of both IgG and IgA classes (Desgranges et al, 1977) are characteristics of the disease (see review by de Thé, 1997). Of special clinical and public health value are the IgA antibodies directed to

VCA and EA. A series of sero epidemiological investigations carried out in South China by Zeng et al. demonstrated that simple serological IgA / VCA and IgA/ EA tests were most instrumental for diagnosis of early N PC clinical stages with high sensitivity and high specificity (Zeng et al., 1983; 1993). We observed that these IgA/VCA antibodies originated in the lympho-plasmocytic component of the tumor, while the epithelial tumor cells exhibited the secretory piece of IgA at their surface (Desgranges et al., 1977). The tight EBV-NPC association was confirmed by regular presence of EBV-DNA fingerprints in NPC tumor cells observed either in high intermediate or low risk areas (Wolf et al., 1973; Desgranges et al., 1975). Viral genes, such as EBER, EBNA-I, LMP-I are usually transcribed in NPC tumor cells.

The consistency of both the serological and molecular associations between EBV and undifferentiated carcinoma of the nasopharynx (NPC), together with the lack of similar reactivities in patients with other malignancies at the same site, as well as in patients with cancers in the oro and hypopharynx, indicates that a peculiar relationship, most likely of causal nature, does exist between EBV and NPC.

Most viral associated cancers are multifactorial in their etiology, with a multistep oncogenic process. NPC is no exception, and we tried to decipher the interactions between EBV, environmental chemical carcinogens, and disease susceptibility genes. Traditional foods were suspected to be etiologically involved through a number of epidemiological studies. Extracts of traditional Cantonese Chinese salted fish preparations, and traditional Tunisian foods were investigated for their *in vitro* activity in reactivating latent EBV infected B lymphocytes. Volatile nitrosamines, genotoxins, and inducers *in vitro* reactivation of EBV were found in traditional foods of both Cantonese and Tunisian, the most active being a lignin component extracted from traditional Harissa.

In order to search for genetic factors in NPC etiology, we studied families with multiple NPC cases among sib ships. Members of these families sharing the HLA profile with the sib with NPC were shown to have a 21 fold increased risk of the disease (Lu et al., 1990), but a different HLA profile was involved for each family. The genes involved remain unknown, but a number of studies explore different clues.

The role of EBV serology in the management and control of NPC has been and remains critical. The follow up of IgA positive individuals made possible to detect asymptomatic preclinical lesions and to treat efficiently very early tumors. In their 10 year prospective seroepidemiological study involving 1,138 IgaA positive individuals, in Zang Wu county in South China, Zeng et al. clearly showed that the risk of developing NPC increased in parallel with the raising IgA antibodies titers among asymtomatic individuals, while decrease or loss of these IgA IgA/VCA annihilated the risk (Zeng et al., 1983; 1993) The same IgA test was shown to be a prognostic marker in the management of NPC in intermediate and low incidence regions of the world (de Vathaire et al., 1988)

References

Burkitt, D. P. (1958). A sarcoma involving the jaws in African children. Br. J. Surg. 46, 218-223.

De-Thé, G. Nasopharyngeal Carcinoma (1997). In Viral Infections of Humans- Epidemiology and Control. 4th Edition. A.S. Evans and A. Kaslow. Plenum Medical Publ. pp 935-967.

De-Thé, G., Geser, A., Day, N.. E., Dean, A., Bornkamm, G., Feorino, P., and Henle, W. (1978). Epidemiological evidence for causal relationship between Epstein-Barr virus and Burkitt's lymphoma from Ugandan prospective study. Nature *274*, 756-761.

De Vathaire, F., Sancho-Garnier, H., De-Thé, H., Puo-Deloup, C., Schwaab, G., Ho, J. H. C., Ellouz, R., Michrau, C., Cammoun, M., Cachin, Y., and De-Thé, G. (1988). Prognostic value of EBV markers in the clinical management of nasopharingeal carcinoma, A multicenter follow-up study. Int. J. Cancer *42*, 176-181.

Desgranges, C., De-Thé, G., and Ho, J. H. C. (1977). Epstein-barr associated antibody pattrens in carcinoma (NPC) patients. Int. J. Cancer *19*, 627-633.

Desgranges, C., Li, J. Y., and De-Thé, G. (1977). EBV specific secretory IgA in saliva of NPC patients , presence of secretory piece in epithelial malignant cells. Int. J. Cancer *20*, 881-886.

Desgranges, C., Wolf, H., De-Thé, G., Shanmugaratnam, K., Ellouz, R., Cammoun, N., klein, G., and zur Hausen, H. (1975). Nasopharingeal carcinoma X-presence of Epstein-barr genomes in epithelial cells of tumours from high and medium risk areas. Int. J. Cancer *16*, 7-15.

Evans, A. S., J. C., and McCollum, R. W. 7 (1968). Sero-epidemiological studies of infectious mononucleosis with EB virus. N. Engl. J. Med. *279*, 1121-1122.

Evans, A. S. 7 (1968). New discoveries in infectious mononucleosis. Mod. Med. *1*, 18-24x.

Evans A. S., Wanat, J; and Niederman, J.C. (1983). Failure to demonstrate concomitant antibody changes to viral antigens other than EBV during or after infectious mononucleosis, Yale J. Biol. Med. *56*, 203-209.

Hamilton-Dutoit, S. J., Raphael, M., Audouin, J., Diebold, J., Lisse, I., Pedersen, C., Oksenhendler, E., Marelle, L., and Pallesen, G. (1993). *In situ* demonstration of Epstein-Barr virus small RNAs 'EBER 1) in acquired immunodeficiency syndrome-related lymphomas , correlation with tumor morphology and primary site. Blood *82*, 619-624.

Henle, G., Henle, W., and Diehl, V. (1968). Relation of Burkitt's tumor-associated herpes-type virus to infectious mononucleosis. Proc. Natl. Acad. Sci. USA *59*, 94-101.

Henle, G., and Henle, W. (1966). Immunofluorescence in cells derved from Burkitt's lymphoma. J. Bacteriol. *91*, 1248-1256.

Klein, E., Klein, G., Nadkarm, J. S., Wigzill, H., and Clifford, P. (1967). Surface IgM-kappa specificity on cells derived from a Burkitt's lymphoma. Lancet *2*,1068-1070.

Luka, J., Lindahl, T., and klein, G. (1978) Purification of the Epstein-Barr virus nuclear antigen from transformed human lymphoid cell lines. J. Virol. *27*, 604-611.

Lu, S. J., Day, N. E., Degos, I., Lepage, V., Wang, P. H., Chan, S., Simons, H., Macknight, B., Easton, D., Zeng, Y., and De-Thé, G. (1990). The genetic basis for carcinoma of the nasopharynx (NPC), Evidence to linkage to the HLA region. Nature *346*,470-471.

Mueller, N., Evans, A., Harris, N.L., Comstock, G.W., Jellum, E., Magnus, K., Orientreich, N., Polk, B.F., and Vogelman, K. (1989). Hodgkin's disease and the EBV, Altered antibody patterns before diagnosis. N. Engl. J. Med. *320*, 696-701.

Old, L. J., Boyse, E. A., Oettigen, H. F., De Harven, E., Geering, C., Williamson, B., and Clifford, L. (1966). Precipitating antibody in human serum to an antigen present in cultured Burkitt's lymphoma cells. Proc. Natl. Acad. Sci. USA *56*, 1699-1704.

Reedman, B. M., and Klein, G. (1973). Cellular localization of anEpstein-Barr virus (EBV) associated complement-fixing antigen in producer and nonproducer lymphoblastoid cell lines. Int. J. Cancer *11*, 499-520.

Steel, C. M., Morten, J. E. N., and Foster, E., (1985). The cytogenetics of human lymphoid malignancy. Studies in Burkitt's lymphoma and Epstein-Barr virus transformed lymphoblastoid cell lines. In Burkitt's Lymphoma, A Human Cancer Model (G. M. Lenoir, G. T. O'Conor, and C. L. M. Olweny, Eds.). IARC, Lyon, France. pp. 265-292.

Weiss, L. M., Movahed, L. A., Warnke, R. A., and Sklar, J. (1989). Detection of Epstein-Barr viral genomes in Reed-Sternberg cells of Hodgkin'S disease, *N. Engl. J. Med. 320*, 5026506.

Wolf, H., zur Hausen, H., and Becker, V. (1973). EB viral genomes in epithelial nasopharingeal carcinoma cells, Nature (New Biol.) *244*, 245-257.

Zeng, Y., Hong, D., Jianming, Z., Naiqin, H., Pingjun, L., Wenjun, P., Yuying, H., Yue, L., Peizhong, W., and De-Thé, G. (1993). A 10 year prospective study on nasopharingeal carcinoma in Wuzhou city and Zanfwu County, Guanxi, china. In The Epstein-barr Virus and Aassociated Diseases , Proceedings of the Vth International Symposium (T. Tursz, J. S. Pagano, D. V. Ablashi, G. De-Thé, G. Lenoir, and G. R. Pearson, EDS.). Colloques INSERM ,John Libbey Eurotext Publishing, Paris. pp. 735-742.

Zeng, Y., Zhong, J. M., Li, L. Y., Wang, P.Z., Tang, H., Ma, Y. R., Zhu, J. S., Pan, W. J., Lio, Y. X., Wei, Z. N., Chen, J. Y., Mo, Y. K., Li, E. J., and Tan, B. F. (1983). Follow-up studies on Epstein-Barr virus IgA/VCA antibody positive persons in Zangwu County, China. Intervirology *20*, 190-194.

From: Epstein-Barr Virus. Edited by: Erle S. Robertson

Chapter 5

Clinical Aspects of Epstein-Barr Virus Infection

*Jeffrey I. Cohen**

ABSTRACT

Epstein-Barr virus (EBV) is the cause of heterophile-positive infectious mononucleosis. Symptoms are predominantly due to proliferation of T cells reacting to virus-infected B cells. Most cases are self limited and do not require therapy. Patients with chronic active EBV have persistent disease due to inability to control the virus and often die of B or T cell lymphoproliferative disease. Patients with the X-linked lymphoproliferative syndrome infected with EBV usually succumb to fulminant infectious mononucleosis. Organ or stem cell transplant recipients may develop EBV lymphoproliferative disease and reduction of immunosuppression, interferon-alpha, anti-CD20 antibody, or infusions of EBV-specific T cells have been effective in many patients. A number of malignancies have been associated with EBV including Burkitt's lymphoma, nasopharyngeal carcinoma, Hodgkin's disease, non-Hodgkin's lymphoma in patients with AIDS, T/NK cell lymphoma, gastric carcinoma, smooth muscle tumors, and lymphomatoid granulomatosis. Many of these tumors show different patterns of latent EBV gene expression.

INTRODUCTION

Epstein-Barr virus (EBV) was initially isolated from Burkitt's lymphoma tissue (Epstein et al., 1964). Subsequent studies showed that it was the etiologic agent of heterophile-positive infectious mononucleosis (Henle et al. 1968). Rare persons are unable to control their primary infection and develop chronic active EBV infection, or fulminant infectious mononucleosis associated with the X-linked lymphoproliferative syndrome or with no known underlying cause. The virus has been linked to a number of malignancies including nasopharyngeal carcinoma (zur Hausen et al., 1970), Burkitt's lymphoma (Epstein et al., 1964), Hodgkin's disease (Weiss et al., 1989), lympholiferative disease in immunosuppressed persons, T/NK cell lymphomas, gastric carcinoma, non-Hodgkin's lymphomas in patients with AIDS, leiomyosarcomas and smooth muscle tumors in immunocompromised persons, and lymphomatoid granulomatosis. The role of EBV in nasopharyngeal carcinoma, Burkitt's lymphoma, and Hodgkin's disease are discussed elsewhere in

*For correspondence email jcohen@niaid.nih.gov

this volume. This chapter will focus on nonmalignant conditions associated with EBV as well as malignant conditions not covered elsewhere in this book.

INFECTIOUS MONONUCLEOSIS
Signs and symptoms

Most infections with EBV occur during infancy and childhood and are either asymptomatic or result in non-specific symptoms. Adolescents and young adults infected with EBV often develop infectious mononucleosis. Most of these patients present with the triad of fever, sore throat, and lymphadenopathy (Carter and Penman, 1969; Schlossberg 1989; Straus et al. 1993). Fever usually persists for two weeks, but may last over a month. Lymphadenopathy usual involves the posterior cervical lymph nodes, but generalized lymphadenopathy may occur. Other common signs and symptoms include headache, fatigue, splenomegaly, and hepatomegaly. Less frequently, patients present with abdominal pain, nausea, vomiting, or chills. Other less common findings on physical examination may include periorbital edema, rash, enathem of the palate, or jaundice. Most patients with infectious mononucleosis who are treated with ampicillin develop a rash. The symptoms of infectious mononucleosis are due to lymphocytic proliferation associated with the virus, rather than to virus lysis of infected B lymphocytes. While most patients with infectious mononucleosis make a full recovery in two to four weeks, some patients have persistent malaise and fatigue that can last several months. In one study fatigue was the most common persistent symptom, with 28% of patients noting fatigue at 1 month and 13% at 6 months after presentation (Rea et al., 2001).

Infectious mononucleosis is occasionally associated with complications (Jensen 2000). Massive enlargement of tonsils and adenoids, as well as inflammation and edema of the throat can result in life-threatening upper airway obstruction. While most patients with a sore throat have an exudative pharyngitis that is negative for beta-hemolytic streptococci, about 10% develop streptococcal pharyngitis. Splenic rupture can occur associated with enlargement of the spleen in about 0.5% of patients. Fatal bacterial superinfections have also been reported.

Central nervous system complications are a frequent cause of hospitalization associated with acute EBV (Hoover et al., 2004). EBV can cause aseptic meningitis or encephalitis. The former patients present with fever, headache, and stiff neck, while the latter have altered mental status, personality changes, cerebellar ataxia, or seizures. Bells' palsy, due to inflammation of cranial nerve VII, has been reported during infectious mononucleosis. Many other cranial neuropathies have also been reported. Other complications may include transverse myelitis, Guillain-Barre syndrome, or peripheral neuropathy. Most patients with central nervous system complications of EBV infection recover completely. Many patients with central nervous system complications of acute EBV infection do not present with infectious mononucleosis.

EBV infection has been associated with a number of hematologic complications (Cohen, 2003). Coombs positive hemolytic anemia is present in about 2% of patients. Aplastic anemia is a rare complication of acute EBV infection. Neutropenia has also been reported and in rare patients agranulocytosis occurs. A mild thrombocytopenia is common; however, rare patients develop life threatening thrombocytopenia. Other less common hematologic complications include

Table1. Pattern of EBV gene expression with different forms of latency

Latency	EBNA-1	EBNA-2	LMP-1	LMP-2	EBER	BARTs
I	+	-	-	-	+	+
II	+	-	+	+	+	+
III	+	+	+	+	+	+
0	-	-	-	-	+	ND

ND=not determined

pancytopenia, thrombocytopenic purpura, disseminated intravascular coagulation, and hemophagocytic syndrome. The latter syndrome is due to phagocytosis of red blood cells, platelets, or mononuclear cells by histiocytes in the bone marrow or spleen.

Most patients with acute EBV infection have elevation of the alkaline phosphatase and many have mild elevations of serum transaminases or bilirubin. Older patients may present with hepatitis or cholestasis. Rare patients have developed hepatic necrosis. Other, less frequent complications of infectious mononucleosis include myocarditis, pericarditis, interstitial nephritis, glomerulonephritis, interstitial pneumonitis, vasculitis, and genital ulcers.

Table 2. Patterns of EBV latency in different diseases

Disease	Predominant EBV Latency Pattern
Burkitt's lymphoma	I
AIDS-Burkitt's lymphoma	I/II
Hodgkin's disease	II
Nasopharyngeal carcinoma	II
T/NK lymphoma	II
Peripheral T cell lymphoma	II
Gastric carcinoma	II
Primary effusion lymphoma	II
AIDS-immunoblastic or large cell lymphoma	II/III
Infectious mononucleosis	III
Chronic active EBV	III
X-linked lymphoproliferative disease	III
Post-transplant lymphoproliferative disease	III
AIDS-CNS lymphoma	III
Lymphomatoid granulomatosis	III
Smooth muscle tumor	various

Laboratory findings

Most patients with infectious mononucleosis have a leukocytosis during the second week of the disease, along with a mild thrombocytopenia and neutropenia (Enberg et al. 1974). The number of T cells increases, while the number of B cells increases slightly or remains unchanged. The CD8 to CD4 T cell ratio and the number of activated (HLA-DR positive) T cells are increased (Tompkinson et al., 1987). During the early stages of infectious mononucleosis EBV-infected peripheral blood B cells express each of the latency associated genes (EBV nuclear antigens [EBNAs], latent membrane proteins [LMPs], EBV encoded RNAs (EBERs), BamHI A rightward transcripts [BARTs]) termed the type III pattern of latency (Tables 1 and 2). After the acute disease has resolved and the patient is a carrier of EBV, most of these latency transcripts are no longer detected in resting EBV-infected B cells and this stage is referred to as the type 0 pattern of latency (Hochberg et al., 2004; Thorley-Lawson and Gross, 2004).

Atypical lymphocytes are a common feature of infectious mononucleosis. These cells are predominantly T cells that proliferate in response to virus-infected B cells. Atypical lymphocytes have a large cytoplasm frequently with vacuoles, indented edges, and basophilic cytoplasm (Figure 1).

Patients with infectious mononucleosis have a general reduction in cellular immune responses. In vitro stimulation of T cells from patients with infectious mononucleosis results in impaired responses to antigens and mitogens, compared with healthy controls. Patients show reduced delayed type hypersensitivity when challenged with recall antigens. Elevated levels of IL-10 are present in the serum of patients with infectious mononucleosis (Taga et al., 1995). Since IL-10 suppresses Th1 cytokine synthesis, this cytokine might contribute to the impairment in cellular immunity observed with infectious mononucleosis. Polymorphisms in IL-10 correlate with symptoms of infectious mononucleosis and susceptibility to infection (Helminen et al. 2001).

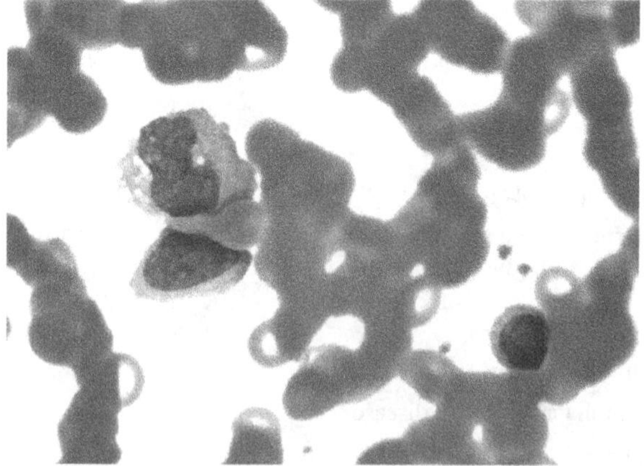

Figure 1. Atypical lymphocytes in a patient with infectious mononucleosis.

Polyclonal B cell activation that occurs during infectious mononucleosis often results in production of autoantibodies. These include cryoglobulins, cold agglutinins, rheumatoid factor, anti-nuclear antibodies, as well as heterophile antibodies (see below). EBV-infected B cells typically express viral lytic proteins and each of the latency associated proteins (Tables 1, 2).

Diagnosis

The heterophile test is used in most patients for the diagnosis of infectious mononucleosis. The classic heterophile test measures IgM antibodies to sheep, horse, or beef red blood cells after the serum has been adsorbed with guinea pig kidney. Thus, this test does not measure antibodies that are specific for EBV. Simpler tests to measure heterophile antibodies, such as the monospot test, are now usually used. These tests are more sensitive and more rapid to perform than the classic heterophile test. False negative tests are more common in young children and in older adults. False positive tests are more frequent in patients who do not have the classic symptoms of infectious mononucleosis. The heterophile test usually remains positive for 3 months after acute infection. Patients with symptoms typical for infectious mononucleosis, but who have a negative heterophile test at the onset of disease, should have a repeat test performed in three weeks.

Virus specific antibody tests are recommended for patients suspected of having acute EBV infection, but who have a negative heterophile test. Antibody to the viral capsid antigen (VCA) IgM is usually present at the onset of disease, and remains positive for two months after the onset of infection. Antibody to EBV VCA IgG is usually present at the onset of symptoms and remains detectable for life. Antibody to EBNA becomes detectable three to six weeks after the onset of disease and persists for life. Antibodies to early antigens are present in about three-fourths of patients with infectious mononucleosis. Antibodies to early antigens restricted to the cytoplasm (EA-R) are present in many patients with Burkitt's lymphoma, while antibody to early antigen diffusely present in the cell (EA-D) is detected in many patients with EBV-positive nasopharyngeal carcinoma. EBV-specific antibody tests are especially useful in young children and older adults who frequently lack heterophile antibodies.

Patients who present with symptoms of mononucleosis with a positive heterophile test and atypical lymphocytes are considered to have a diagnosis of infectious mononucleosis. In heterophile-negative patients, the diagnosis of acute EBV is usually made by a positive EBV VCA IgM antibody, but can also be made by seroconversion from negative to positive EBNA or VCA IgG antibody.

The differential diagnosis of infectious mononucleosis includes acute infection with cytomegalovirus, human herpesvirus 6, toxoplasmosis, and human immunodeficiency virus. Patients with cytomegalovirus mononucleosis usually present at an older age and have less pharyngitis and lymphadenopathy than those with disease due to EBV. Patients with human herpesvirus 6 mononucleosis often have fewer atypical lymphocytes than those with EBV-positive infectious mononucleosis. Acute infection with toxoplasmosis may also resemble EBV mononucleosis, although splenomegaly and pharyngitis are less common with toxoplasmosis. The acute retroviral syndrome, due to infection with HIV may also present with many of the symptoms of infectious mononucleosis; however aseptic meningitis and diarrhea are more common with HIV. Other, less frequent diseases that may be confused with EBV mononucleosis include lymphoma and drug reactions.

Therapy

Most patients with infectious mononucleosis require only supportive therapy and antipyretics for fever. Erythromycin or azithromycin are used for patients with concomittent streptococcal pharyngitis; ampicillin should not be given, since it usually results in a rash in patients with infectious mononucleosis. While acyclovir inhibits the replication of EBV in cell culture, it does not reduce the duration of symptoms of infectious mononucleosis. Studies of intravenous and oral acyclovir show that while the drug reduces virus shedding in saliva, acyclovir does not result in clinical benefits (Andersson et al., 1986; van der Horst et al., 1991). A meta-analysis of acyclovir based on five double-blind, placebo controlled trials showed no significant difference in clinical parameters for patients treated with the drug versus those receiving placebo (Torre and Tambini, 1999). Since the symptoms of infectious mononucleosis are due to the proliferation of lymphocytes in response to the infection, rather than to lysis of cells by replicating virus, it is not surprising that acyclovir does not benefit patients with infectious mononucleosis.

Corticosteroids are not recommended in patients with uncomplicated infectious mononucleosis. Placebo controlled trials of corticosteroids show that they have no effect on lymphadenopathy or splenomegaly; however, corticosteroids do shorten the duration of pharyngitis and fever (reviewed in Straus et al., 1993). Corticosteroid therapy for infectious mononucleosis has occasionally been associated with complications, including bacterial infections. Corticosteroids may alter the initial immune response to the virus, and since the virus remains in the body for life, corticosteroids are not indicated for the vast majority of patients with EBV infectious mononucleosis.

Corticosteroids are used for patients with severe complications of infectious mononucleosis. These include upper airway obstruction due to enlarged lymphoid tissue, severe autoimmune hemolytic anemia, and severe thrombocytopenia (Andersson and Ernberg, 1988). Rare patients with EBV-associated myocarditis, pericarditis, or encephalitis are sometimes treated with corticosteroids (McGowan et al., 1992). In these cases prednisone is often given at 1 mg per kg for three to four days followed by a rapid taper over 7 to 10 days. Some authorities recommend adding acyclovir to corticosteroids if the latter are used, in an attempt to limit virus replication that might be enhanced with the addition of cortiocosteroids.

A double-blind, placebo controlled study combining oral acyclovir and prednisolone for 10 days resulted in reduction virus shedding from the throat (during therapy only), but no significant improvement in clinical parameters (Tynell et al. 1996).

CHRONIC ACTIVE EBV

Chronic active EBV is a very rare disorder in which primary infection with EBV is not controlled and patients develop persistent mononucleosis-like symptoms with organ disease. The disease is more common in Asians than in Europeans and Americans.

Signs and symptom

Most patients present with persistent fever, lymphadenopathy, splenomegaly, and liver dysfunction (Kimura et al., 2001, Straus 1992). Other frequent findings include thrombocytopenia, anemia, and rash. Organs are infiltrated with EBV-infected B, T, or NK cells, resulting in hepatitis, interstitial pneumonitis, central nervous system disease, or uveitis due to EBV-infected cells in the organs. Hemophagocytosis, which presents with fever and pancytopenia, is due to ingestion of red and white blood cells by mononuclear cells. Asian patients with chronic active EBV often have hypersensitivity reactions to mosquito bites, unlike those from the United States. Less common findings include coronary artery aneurysms and oral ulcers.

Patients with chronic active EBV usually have a progressive decline in immune function with loss of B cells, NK cells, and hypogammaglobulinemia. They ultimately develop fatal opportunistic infections or B or T cell lymphoproliferative disease. Chronic active EBV is a distinct entity unrelated to chronic fatigue syndrome.

Laboratory findings

EBV is often present in T or NK cells of patients with chronic active EBV, rather than B cells as in persons with infectious mononucleosis. Activated (HLA-DR positive) T cells are often present in the peripheral blood. Patients with chronic active EBV and hemophagocytosis have elevated levels of TNF-alpha and interferon gamma in the blood (Lay et al. 1997). The number of B cells and NK cells is often diminished and over time B cells may be undetectable in the peripheral blood. EBV serologies are positive, often at very elevated levels (see below). Some patients fail to make antibody to EBNA.

The etiology of chronic active EBV is not certain. While earlier studies suggested that the disease might be due to an abnormal strain of virus that was impaired for latency, more recent studies indicate that patients with chronic active EBV have dysregulation of the cytokine response, often with elevations of both TH1 and TH2 cytokines simultaneously (Ohga et al. 2001). One patient was recently reported to have mutations in both alleles of his perforin gene resulting in an immature form of the protein with impaired cellular cytotoxicity (Katano et al. 2004).

Diagnosis

Criteria for the diagnosis of chronic active EBV have included (a) disease lasting at least six months that occurred after acute infection with EBV or is associated with markedly elevated EBV antibody titers, (b) evidence of organ disease with infiltration of tissues by cells containing EBV DNA, RNA, or proteins, and (c) no other known immunocompromising condition (Straus 1992). Recently, a marked elevated in EBV viral load in peripheral blood mononuclear cells was proposed as an alternative criterion to detection of EBV nucleic acid or protein in the tissues (Kimura et al. 2001).

Therapy

Therapy for chronic active EBV is unsatisfactory. Many of the symptoms of disease can be temporarily palliated with immunosuppressive therapy; however, this does not resolve the underlying problem and the disease progresses to a fatal conclusion. While initial anecdotes suggested that acyclovir may have a role in some patients, particularly those with interstitial pneumonitis (Schooley et al., 1986), acyclovir has not been effective in most recent studies. Patients with hemophagocytic syndrome often have a rapidly fatal course; in some cases etoposide with or without cyclosporine has been reported to be effective. While interferon-alpha or interferon-gamma has been reported to be successful anecdotally, most reports fail to show an effect with these cytokines.

Infusion of autologous cytotoxic T cells has been reported to be successful in 5 cases of chronic active EBV; one patient had recurrence of symptoms one year later. In these cases, patient's T cells were stimulated with their EBV-transformed B cells and expanded *in vitro*; the cells were then infused back into the patients (Salvodo et al., 2002). Allogeneic stem cell transplantation has been reported to be successful in several cases of chronic active EBV (Kimura et al., 2001; Okamura et al., 2000). In most cases the stem cells were obtained from unaffected siblings.

X-LINKED LYMPHOPROLIFERATIVE DISEASE

X-linked lymphoproliferative disease (XLPD) is a rare genetic disorder in which otherwise healthy boys become infected with EBV and usually develop fatal infectious mononucleosis.

Signs and symptoms

Most patients with XLDP develop fulminant infection with EBV (Seemayer et al., 1995). Infection results in infiltration of various organs with B and T cells and marked macrophage activation that often results in severe hepatitis, bone marrow failure, or hemophagocytic syndrome. Death is often due to liver failure, bleeding, or uncontrolled infection. Rare patients that survive usually have abnormalities in their immunoglobulins with reduced levels of IgG, and elevated levels of IgM and IgA. Over time these patients may develop B (or rarely T) cell lymphomas, aplastic anemia, or vasculitis. Few patients survive beyond age 40.

Recent studies indicate that some boys with XLPD develop complications even in the absence of EBV infection. This is not surprising in view of the mutation that causes XLPD (see below), which affects B, T, and NK cell activities. These complications include B cell lymphomas, vasculitis, and abnormalities in immunoglobulins.

Laboratory findings and diagnosis

Patients with a familial history of XLPD nearly always have mutations in the SAP (SLAM-associated protein) gene (Sumegi et al. 2000). In contrast, patients with fulminant mononucleosis, but who lack a family history of the fatal disease, generally do not have mutations in SAP. SAP, which is expressed in both T and NK cells, interacts with proteins on the surface of T, B, NK, and dendritic cells. The interaction of SAP with SLAM (signaling lymphocyte activation molecule) on T cells inhibits interferon-gamma production, resulting in proper regulation of the cytokine. The interaction of SAP with the 2B4 and NTB-A proteins on NK cells is important for NK cell killing. Thus mutations in SAP result in impaired regulation of cytokines as well as reduced NK cell activity.

Therapy

Treatment for patients with XLPD and fulminant infectious mononucleosis is usually unsuccessful (Gaspar et al., 2002). Antiviral therapy and cytotoxic chemotherapy are frequently used, but most patients succumb to their disease. Patients who are identified prior to infection with EBV are usually given immunoglobulin infusions which contain neutralizing antibodies to EBV; however, some patients receiving these infusions have become infected with EBV and died. Stem cell transplantation has been used in a few cases of XLPD and is curative.

SELECTED MALIGNANCIES ASSOCIATED WITH EBV
Post-transplant EBV lymphoproliferative disease

EBV lymphoproliferative disease has been reported in patients after organ or stem cell transplant, as well as patients with congenital or other acquired immunodeficiencies (Cohen 1991). These include patients with severe combined immunodeficiency, Wiskott-Aldrich syndrome, ataxia telangiectasia, acute lymphoblastic leukemia or AIDS (see "Non-Hodgkin's lymphomas in patients with AIDS" below). Patients with rheumatoid arthritis or polymyositis who were treated with methotrexate have developed EBV lympholiferative disease; in some of these patients the disease resolved when methotrexate was discontinued.

Risk factors for EBV lymphoproliferative disease in transplant recipients include treatments that further impair T cell immunity (*e.g.* anti-T cell antibodies), graft-versus-host disease, concurrent cytomegalovirus infection, receipt of T cell depleted stem cells, and EBV seronegativity prior to transplantation.

Signs and symptoms

Patients with post-transplant EBV lymphoproliferative disease can present with symptoms of infectious mononucleosis including fever and lymphadenopathy, or with fever and mass lesions due to EBV in various organs. Frequently the transplanted organ is the site of these lesions. Other frequent sites include the lymph nodes, small intestine, liver, lung, kidney, or central nervous system. Death is usually due to tumor progression, organ rejection (in the case of transplant recipients), or fatal opportunistic infections.

Laboratory findings and diagnosis

While the vast majority of cases of EBV lymphoproliferative disease are due to B cells tumors, rare cases have occurred with T cell lymphomas (Stadlmann et al., 2001). Many patients with B cell EBV lymphoproliferative disease have elevated levels of EBV DNA in peripheral blood mononuclear cells prior to and during their disease (Riddler et al., 1994). Measurement of the viral load has been used to predict which patients might develop disease, as well as to follow-up patients after treatment. The tumor cells generally express each of the EBV latency genes.

Tissue biopsies show EBV DNA, RNA (Figure 2), or proteins in B cells infiltrating the tissues. Histology can show B cell or plasma cell hyperplasia, or B cell lymphoma. The tumors can be monomorphic or polymorphic, as well as polyclonal or monoclonal. Distinct EBV-positive B cell clones can arise at multiple sites. Patients with polymorphic lesions generally have a better prognosis. Most of the tumors in transplant recipients are immunoblastic lymphomas.

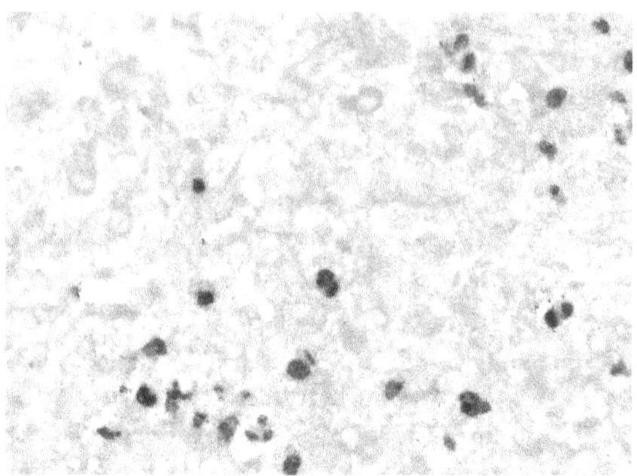

Figure 2. EBV encoded RNA (EBER) staining in a biopsy from a patient with lymphoproliferative disease.

Therapy

Lymphoproliferative disease in organ transplant recipients often responds to reduction in immunosuppressive therapy (Paya et al., 1999). A variety of therapies have been used for patients with EBV lymphoproliferative disease who do not respond to reduction in immunosuppression. Interferon-alpha therapy has been successful in some cases (Davis et al., 1998); however, treatment must be balanced against the possibility of organ rejection. Rituximab, an anti-CD20 (B cell) antibody, results in complete remissions in about 50% of patients (Faye et al., 2001). Other treatments include radiation therapy (especially for central nervous system lesions), excision of lesions (particularly for localized lesions in the small intestine), and cytotoxic chemotherapy. Antiviral therapy has been ineffective for the treatment of EBV lymphoproliferative disease, since most lesions are due to B cells latently infected with virus, rather than to EBV replication in the cells.

Infusions of EBV-specific T cells have been effective in several cases of lymphoproliferative disease after organ transplantation (Haque et al., 2001; Khanna et al., 1999). Since EBV lymphomas in organ transplant patients are usually of recipient origin, the T cells must be HLA-matched to those of the recipient. The source of these cells may be HLA-matched relatives of the organ recipient, HLA-matched cells from unrelated donors, or autologous EBV-specific T cells that are amplified ex vivo from the transplant recipient.

In contrast to organ transplant recipients, EBV lymphomas in stem cell transplant recipients are usually of donor origin. In this setting, infusion of non-irradiated HLA-identical donor lymphocytes, which contain EBV-specific T cells, has been curative in many patients (Gottschalk et al. 2002, O'Reilly et al., 1997). Repeat infusions of cells may be required and graft-versus host disease may occur, especially after infusion of large numbers of cells.

Prevention

Since elevated EBV viral loads in the peripheral blood in organ and stem cell transplant recipients are predictive for the development of EBV lymphoproliferative disease, a number of regimens have been proposed to prevent the development of this disorder. Antiviral therapy, using acyclovir or ganciclovir, has been shown in some (McDiarmid et al., 1998), but not other studies (Green et al., 1997) to reduce the development of EBV lymphoproliferative disease. Infusions of EBV-cytotoxic T cells have been used in some studies to prevent disease in persons with rapidly rising viral loads (Rooney et al., 1998). Pre-emptive rituximab has also been used in some patients with elevated viral loads (van Esser et al., 2002); however, the numbers of patients treated has been limited, and it is unclear if rituximab is effective in preventing lymphoproliferative disease.

Non-Hodgkin's lymphomas in patients with AIDS

About half of non-Hodgkin's lymphomas in patients with AIDS are EBV positive (Cohen 1991). Most of these tumors are immunoblastic lymphomas or Burkitt's lymphomas; less frequently EBV-positive large cell lymphomas are present. Burkitt's lymphomas and large cell lymphomas tend to occur when the immune system is less impaired, and the disease often presents in the lymph nodes. In contrast, immunoblastic lymphomas occur later during the course of AIDS, are extranodal, and are more frequently EBV-positive. Most of these tumors are monoclonal and do not contain HIV DNA or RNA. Different patterns of latent EBV gene expression have been described (Rea et al., 1994; Hamilton-Dutoit, et al., 1993). Immunoblastic lymphomas and large cell lymphomas usually show a type II (EBNA-1, LMP-1, LMP-2) or III (all latency genes) pattern of latency, while Burkitt's lymphomas show a type 1 (EBNA-1 alone) or less commonly type II pattern of latency (Tables 1, 2).

Virtually all central nervous system lymphomas in patients with AIDS are EBV-positive (MacMahon et al., 1991). These tumors occur late in the course of HIV infection and express the full complement of EBV latent proteins. Detection of EBV DNA in the cerebrospinal fluid of AIDS patients with central nervous system mass lesions has been predictive of EBV lymphoma in some studies (Cinque et al., 1996). A recent small study suggests that the positive predictive value of the CSF PCR for EBV lymphoma is only 29% (Ivers et al. 2004). A positive PET or SPECT scan in combination with EBV DNA in the cerebrospinal fluid enhances the likelihood of central nervous system lymphoma.

Primary effusion lymphomas also occur in patients with AIDS. Tumor cells are located in the pleural, pericardial, or peritoneal cavities and patients present with dyspnea, impaired cardiac function, or ascites, respectively. These tumors contain Kaposi's sarcoma associated herpesvirus DNA and often EBV DNA (Cesarman et al., 1995). The tumors express EBNA-1 and LMP-1 (Callahan et al., 1999).

EBV T/NK cell and T cell lymphomas

T/NK cell lymphomas are strongly associated with EBV (Jaffe et al, 1996; Jaffe et al., 2003; Yachie et al., 2003). The tumors are more common in Asia and South America than in Europe or the United States, and males are more frequently affected than females. They were previously referred to as lethal midline granuloma. Most lesions are present in the midline of the face at the nose; less frequently lesions are present in the upper airway, skin, gastrointestinal tract, or testes. Patients frequently present with destructive lesions of the palate, mass lesions in the nose, or swelling of the ocular orbit. Some patients develop hemophagocytic syndrome.

Rare patients develop EBV-positive NK cell leukemia. Pathology frequently shows angioinvasion and necrosis. The tumors usually express EBNA-1, LMP-1, and LMP-2, but not EBNA-2 (Chiang et al., 1996; Xu et al., 2001). The lesions often show expression of TNF-alpha with activation of NF-κB. While localized lesions often respond to radiation, disseminated lesions are treated with cytotoxic chemotherapy and the prognosis is poor.

T cell lymphomas occur in patients with chronic active EBV or very rarely after initial infection with the virus (Jones et al., 1988; Quintanilla-Martinez et al., 2000). These tumors are EBV-positive, monoclonal, and are often associated with hemophagocytic syndrome, pancytopenia, fever, and hepatic failure. Patients are more likely to be of Asian descent and often have hepatosplenomegaly, but little lymphadenopathy. High levels of TNF-alpha are present in the serum of many of these patients and they correlate with a poorer prognosis (Mori et al., 2003). EBV T cell lymphomas occur more commonly in Asians and Mexicans. Peripheral T cell lymphomas express EBNA-1, LMP-1, and LMP-2, but not EBNA-2 (Chen et al., 1993). The prognosis is poor; the mean survival once a diagnosis of T cell lymphoma is made is 1 month, with a range of several days to one year (Quintanilla-Martinez et al. 2000).

Gastric carcinoma

EBV was first reported in lymphoepithelial carcinoma of the stomach in 1990 (Burke et al., 1990). Subsequently, EBV was reported in 16% (22/138) of gastric adenocarcinomas (Shibata and Weiss, 1992). In a study of 1,000 patients with gastric cancer, 89% of lymphoepithelioma-like carcinomas, 7% of moderately to well-differentiated adenocarcinomas, and 6% of poorly differentiated adenocarcinomas of the stomach were EBV positive (Imai et al., 1994). Gastric lymphoepitheliomas share some of the features seen in nasopharyngeal carcinoma; both tumors are highly undifferentiated and have prominent lymphocytic infiltration. EBV DNA is present both in the epithelial cells of the tumor tissue and in the nearby dysplastic cells as well as metastases, but not in the adjacent lymphocytes or normal stromal or gastric epithelial cells (Shibata et al., 1991). A prominent CD8 T cell infiltrate is frequently present in EBV-positive gastric carcinomas (Saiki et al., 1996). EBV DNA has generally not been detected in gastric ulcers or esophageal carcinomas. EBV DNA is present in all of the tumor cells in an episomal form, and the tumors are monoclonal or oligoclonal. Only latent viral transcripts are expressed. EBNA-1, LMP-2, but not EBNA-2 transcripts are usually detected in the tumor cells (Sugiura et al., 1996). Since gastric epithelial cells lack the EBV receptor, CD21, they may express another receptor for the virus.

EBV VCA and EA IgG antibody titers are higher in patients with EBV-positive gastric carcinoma, compared with patients having EBV-negative gastric carcinoma or healthy controls (Imai et al., 1994). These elevations in antibody titers resemble those seem with other EBV-associated cancers such as nasopharyngeal carcinoma or Burkitt's lymphoma. Serum levels of EBV VCA IgA antibodies are also higher in persons who subsequently develop EBV-positive gastric carcinoma, than in those with EBV-negative gastric carcinoma or healthy controls (Levine et al., 1995).

EBV-positive gastric adenocarcinoma is more common in men and younger persons compared with EBV-negative adenocarcinomas. EBV-positive tumors are more often in the gastric antrum or corpus, and less likely to metastasize to lymph nodes when compared with EBV-negative gastric carcinoma (van Beek et al., 2004). Patients with EBV-positive gastric adenocarcinomas have a better prognosis with a longer disease-free period and improved cancer-related survival than those with EBV-negative gastric carcinoma in some (van Beek et al., 2004), but not all (Kijima et al., 2003) studies. Multiple tumors, derived from separate clones, may arise within the stomach. Patients with EBV-positive gastric carcinomas often have atrophic gastritis.

EBV-positive lymphoepithelial-like thymic (Dimery et al., 1988), salivary gland (Krishnamurthy et al., 1987), and lung (Wong et al., 1995) tumors have also been reported, particularly in areas of the world where nasopharyngeal carcinoma is common. These tumors share many of the histopathologic features of nasopharyngeal carcinoma.

Smooth muscle tumors in immunocompromised persons

Patients with AIDS and organ transplant recipients have been reported to develop EBV-positive smooth muscle tumors. McClain et al. (1995) reported EBV-positive leiomyosarcomas and leiomyomas in six HIV-infected persons, but EBV was not detected in smooth muscle tumors from patients without HIV. Monoclonal tumors were present in the two EBV-positive lesions that were tested. At least 17 other cases of EBV-associated smooth muscle tumors have been reported in patients with AIDS with a mean CD4 count of 63 (range 0 to 330) (Cheuk et al. 2002). EBV-positive smooth muscle tumors in the lung, stomach, intestine, colon, central nervous system, kidney, lymph nodes, and pericardium have been reported.

Lee et al. (1995) described smooth muscle tumors in three children after liver transplantation. The tumors contained EBV DNA and were clonal. Tumors were present in the liver, lung, stomach, and colon. Smooth muscle tumors have been described in other organ transplant recipients at a mean of 40 (range 1 to 66) months after transplant (Cheuk et al. 2002). EBV-positive tumors were reported in the liver, lung, stomach, colon, spleen, heart, and bone. EBV-positive smooth muscle tumors can be of donor or recipient in origin. Some of these tumors have regressed with reduction in immunosuppression. EBV-positive smooth muscle tumors have also been reported in three patients with congenital immunodeficiencies.

Different studies have shown varying patterns of EBV latent gene expression in patients with smooth muscle tumors. Some cases have shown expression of EBNA-2 without LMP-1 (Lee et al., 1995), while another showed both EBNA-2 and LMP-1 (Rogatsch et al., 2000).

Lymphomatoid granulomatosis

Lymphomatoid granulomatosis is an EBV-positive B cell disorder in which angiocentric and angiodestructive lesions occur at extranodal sites such as the lungs, kidneys, skin, and central nervous system (Jaffe and Wilson, 1997). Less commonly lesions are present in other organs such as the liver, upper respiratory tract, and gastrointestinal tract. The lesions have few EBV-positive B cells, but a markedly reactive T cell infiltrate. Most patients have impairment in cellular immunity. Distinct EBV-positive B cell clones can arise at multiple sites. EBV LMP-1 and EBNA-2 protein expression has been detected in the lesions (Taniere et al.1988). The disease may be indolent or it can evolve to an aggressive malignancy. Interferon-alpha has been effective in some cases for low grade lesions (few EBV positive B cells, no atypia, minimal necrosis), while high grade lesions (numerous EBV-positive B cells, large atypical lymphoid cells, large areas of necrosis) require cytotoxic chemotherapy (Wilson et al., 1996).

Oral hairy leukoplakia

Oral hairy leukoplakia (OHL) occurs in severely immunocompromised patients, especially those with AIDS (Greenspan et al., 1985; Triantos et al., 1997). Patients present with painless white corrugated, "hairy," lesions on the side of the tongue or occasionally on other portions of the tongue or the buccal mucosa. Biopsies of the lesions show epithelial hyperplasia with acanthosis and viral intranuclear inclusions. EBV replicating antigens (*e.g.* VCA, EA, BZLF1), latent antigens (EBNA-1, EBNA-2, LMP-1), linear viral DNA, and virions are present in the epithelial cells in the outer layer of the lesions (Webster-Cyriaque et al., 2000; Walling et al. 2004b). Unlike nearly all of the other EBV-associated diseases discussed in this chapter, OHL is due to actively replicating virus. EBV is thought to enter the tongue from EBV-infected B cells circulating in the blood (Walling et al., 2004a).

The diagnosis is often made by clinical examination in patients with AIDS. Brushings, scrapings, or biopsies that show EBV DNA or antigens are useful when the diagnosis is uncertain. Multiple EBV strains have been detected in the same lesion from a given patient (Walling et al., 1992).

While many cases do not require therapy, antiviral therapy (*e.g.* acyclovir, valacyclovir) results in regression of the lesions (Walling et al., 2003). However, the virus persists in the lesions in a nonproductive state, and recurrences are common when antiviral therapy is stopped. Highly active antiretroviral therapy in patients with AIDS has been reported to cause resolution of lesions.

Breast cancer

While EBV DNA has been detected in nearly 50% of breast carcinomas in some reports (Labrecque et al., 1995; Luqmani and Shousha, 1995; Bonnet et al., 1999), other studies have failed to show an association of EBV with breast cancer (Chu et al., 2001, Deshpande et al., 2002; Murray et al. 2003). In some studies in which EBV was associated with breast cancer, PCR or Southern blotting of whole tissues were positive, which could have been due to small numbers of EBV-infected malignant or nonmalignant cells that are present in the tissue, rather than EBV infection in the majority of the breast cancer cells. When immunohistochemistry for EBV protein was performed, only the minority of the tumor cells were positive (Bonnet et al., 1999; Luqmani and Shousha, 1995), and when in situ hybridization for EBV encoded RNA was performed none (Deshpande et al., 2002; Murray et al., 2003) or a minority (Labrecque et al., 1995) of the tumor cells were positive. This contrasts with other EBV tumors such as Burkitt's lymphoma, nasopharyngeal carcinoma, and the Reed-Sternberg cells of Hodgkin's disease in which all of the tumor cells are EBV-positive. While EBV EBNA-1 antibody staining was reported to be positive in some studies (Bonnet et al., 1999), the antibody used has been shown to cross-react with proteins in EBV-negative cells (Murray et al., 2003).

More recent quantitative studies suggest that <1% of cells in breast carcinoma biopsies may be infected with EBV, that the malignant cells do not contain EBV DNA (Murray et al., 2003), and that the infection in these tissues is a viral lytic process (Huang, et al. 2003), rather than a latent infection which is characteristic of EBV-associated malignancies. Thus, at present, EBV has not been definitively linked with breast carcinoma.

References

Andersson, J. Britton, S., Ernberg, I., Andersson, U., Henle, W., Skoldenberg, B., and Tisell, A. (1986). Effect of acyclovir on infectious mononucleosis, a double-blind, placebo-controlled study. J. Infect. Dis. *153*, 283-290.

Andersson, J. and Ernberg, I. (1988). Management of Epstein-Barr virus infections. Am. J. Med. *85*, 107-115.

Bonnet, M., Guinebretiere, J.-M., Kremmer, E., Grunewald , V., Benhamou, E., Contesso, G., and Joab I. (1999). Detection of Epstein-Barr virus in invasive breast cancers. J. Natl. Cancer Instit. *91*, 1376-1381.

Burke, A.P., Yen, T.S., Shekitka, K.M., and Sobin, L.H. (1990). Lymphoepithelial carcinoma of the stomach with Epstein-Barr virus demonstrated by polymerase chain reaction. Mod. Pathol. *3*, 377-380.

Callahan, J., Pai, S., Cotter, M., Robertson, E.S. (1999). Distinct patterns of viral antigen expression in Epstein-Barr virus and Kaposi's sarcoma-associated herpesvirus coinfected body-cavity-based lymphoma cell lines, potential switches in latent gene expression due to coinfection. Virology *262*, 18-30.

Carter, R.L. and Penman, H.G. (1969) Infectious mononucleosis. (Oxford, U.K., Blackwell Scientific).

Cesarman, E., Moore, P.S., Rao, P.H., Inghirami, G., Knowles, D.M., and Chang, Y. (1995). In vitro establishment and characterization of two acquired immunodeficiency syndrome-related lymphoma cell lines (BC-1 and BC-2) containing Kaposi's sarcoma-associated herpevirus-like (KSHV) DNA sequences. Blood *86*, 2708-2714.

Chiang, A.K., Tao, Q., Srivastava, G., Ho, F.C. (1996). Nasal NK- and T-cell lymphomas share the same type of Epstein-Barr virus latency as nasopharyngeal carcinoma and Hodgkin's disease. Int. J. Cancer *68*, 285-290.

Chen, C.L., Sadler, R.H., Walling, D.M., Su. I.J., Hsieh, H.C., and Raab-Traub, N. (1993). Epstein-Barr virus gene expression in EBV-positive peripheral lymphomas. J. Virol. *67*, 6303–6308.

Cheuk, W., Li, P.C.K., and Chang, J.K.C. (2002). Epstein-Barr virus-associated smooth muscle tumour, a distinctive mesenchymal tumour of immunocompromised individuals. Pathology *34*, 245-249.

Chu, P.G., Chang, K.L., Chen, Y-Y., Chen, W.G., and Weiss, L M. (2001). No significant association of Epstein-Barr virus infection with invasive breast carcinoma. Am. J. Pathol. *159*, 571-578.

Cinque, P., Vago, L., Dahl, H., Brytting, M., Terreni, M.R., Fornara, C., Racca, S., Castagna, A., Monforte, A.D., Wahren, B., Lazzarin, A., and Linde, A. (1996). Polymerase chain reaction on cerebrospinal fluid for diagnosis of virus-associated opportunistic diseases of the central nervous system in HIV-infected patients. AIDS *10*, 951-958.

Cohen, J.I. (1991). Epstein-Barr virus lymphoproliferative disease associated with acquired immunodeficiency. Medicine *70*, 137-160, 1991.

Cohen, J.I. (2003) Benign and malignant Epstein-Barr virus associated B-cell lymphoproliferative diseases. Sem. Hematol. *40*, 116-123.

Davis, C.L., Wood, B.L., Sabath, D.E., Joseph, J.S., Stehman-Breen, C., and Broudy, V.C. (1998). Interferon-alpha treatment of posttransplant lymphoproliferative disorder in recipients of solid organ transplants. Transplantation *66*, 1770-1779.

Deshpande, C.G., Badve, S., Kidwai, N., and Longnecker, R. (2002). Lack of expression of the Epstein-Barr virus (EBV) gene products, EBERs, EBNA1, LMP1, and LMP2A, in breast cancer cells. Lab. Invest. *82*, 1193-1199.

Dimery, I.W., Lee, J.S., Blick, M., Pearson, G., Spitzer, G., and Hong, W.K. (1988). Association of the Epstein-Barr virus with lymphoepithelioma of the thymus. Cancer *61*, 2475-2480.

Enberg, R.N., Eberle, B.J., and Williams, R.C. (1974). T and B cells in the peripheral blood during infectious mononucelosis. J. Infect. Dis. *130*, 104-111.

Epstein, M.A., Achong, B.G., and Barr, Y.M. (1964). Virus particles in cultured lymphoblasts from Burkitt's lymphoma. Lancet *1*, 702-703.

Faye, A., Quartier, P., Reguerre, Y., Lutz, P., Carret, A.S., Dehee, A., Rohrlich, P., Peuchmaur, M., Matthieu-Boue, A., Fischer, A., and Vilmer, E. (2001). Chimaeric anti-CD20 monoclonal antibody (rituximab) in posttransplant B-lymphoproliferative disorder following stem cell transplantation in children. Br. J. Haematol. *115*, 112-118.

Gaspar, H.B., Sharifi, R., Gilmour, K.C., and Thrasher, A.J. (2002). X-linked lymphoproliferative disease, clinical, diagnostic and molecular perspective. Br. J. Haematol. *119*, 585-95.

Gottschalk, S., Heslop, H.E., and Rooney, C.M. (2002). Treatment of Epstein-Barr virus-associated malignancies with specific T cells. Adv. Cancer Res. *84*, 175-201.

Green, M., Kaufmann, M., Wilson, J., and Reyes, J. (1997). Comparison of intravenous ganciclovir followed by oral acyclovir with intravenous ganciclovir alone for prevention of cytomegalovirus and Epstein-Barr virus disease after liver transplantation. Clin. Infect. Dis. *25*, 1344-1349.

Greenspan, J.S., Greenspan, D., Lennette, E.T., et al. (1985). Replication of Epstein-Barr virus within the epithelial cells of oral "hairy" leukoplakia, an AIDS-associated lesion. N. Engl. J. Med. *313*, 1564-71.

Hamilton-Dutoit, S.J., Rea, D., Raphael M., Sandvej, K., Delecluse H.J., Gisselbrecht, C., Marelle, L., van Krieken H.J.M., and Pallesen, G. (1993). Epstein-Barr virus-latent gene expression and tumor cell phenotype in acquired immunodeficiency syndrome-related non-Hodgkin's lymphoma. Am. J. Pathol. *143*, 1072-1085.

Haque, T., Taylor, C., Wilkie, G.M., Murad, P., Amlot, P.L., Beath, S., McKiernan, P.J., and Crawford, D.H. (2001) Complete regression of posttransplant lymphoproliferative disease using partially HLA-matched Epstein-Barr virus-specific cyototoxic T cells. Transplantation *72*, 1399-1402.

Helminen, M.E., Kilpinen, S., Virta, M., and Hurme, M. (2001). Susceptibility to primary Epstein-Barr virus infection is associated with interleukin-10 gene promoter polymorphism. J. Infect. Dis. *184*, 777-780.

Henle, G., Henle, W., and Diehl, V. (1989). Relationship of Burkitt's tumor associated herpes-type virus to infectious mononucleosis. Proc. Natl. Acad. Sci. USA *59*, 94-101.

Hochberg D, Middeldorp JM, Catalina M, Sullivan JL, Luzuriaga K, Thorley-Lawson DA. (2004). Demonstration of the Burkitt's lymphoma Epstein-Barr virus phenotype in dividing latently infected memory cells *in vivo*. Proc Natl Acad Sci USA. 101, 239-44.

Hoover, S.E., Ross, J.R., and Cohen, J.I. (2004). Epstein-Barr virus. In Infections of the Central Nervous System, 3rd Edition. W.M. Scheld, R.J. Whitley R, C.M. Marra (eds) Philadelphia, Lippincott, Williams, and Wilkins. pp 175-183.

Huang, J., Chen, H., Hutt-Fletcher, L., Ambinder, R.F., and Hayward, S.D. (2003). Lytic viral replication as a contributor to the detection of Epstein-Barr virus in breast cancer. J. Virol. *77*, 13267-13274.

Imai, S., Koizumi, S., Sugiura, M., Tokunaga, M., Uemura, Y., Yamamoto, N., Tanaka, S., Sato, E., and Osato, T. (1994). Gastric carcinoma, monoclonal epithelial malignant cells expressing Epstein-Barr virus latent infection protein. Proc. Natl. Acad. Sci. USA *91*, 9131-9135.

Ivers, L.C., Kim, A.Y., and Sax, P.E. (2004). Predictive value of polymerase chain reaction of cerebrospinal fluid for detective of Epstein-Barr virus to establish the diagnosis of HIV-related primary central nervous system lymphoma. Clin. Infect. Dis. *38*, 1629-1632.

Jaffe, E.S., Chan, J.K.C., Su, I-J., Frizzera, G., Mori, S., Feller, A.C., and Ho, F.C.S. (1996). Report of a workshop on nasal and related extranodal angiocentric T/natural killer cell lymphomas. Am. J. Surg. Pathol. *20*, 103-111.

Jaffe, E.S., Krenacs, L., Raffeld, M. (2003). Classification of cytotoxic T-cell and natural killer cell lymphomas. Sem. Hematol. *40*, 175-184.

Jaffe, E.S., and Wilson, W.H. (1997). Lymphomatoid granulomatosis, pathogenesis, pathology and clinical implications. Cancer Surv. *30*, 233-48.

Jenson, H.B. (2000). Acute complications of Epstein-Barr virus infectious mononucleosis. Curr. Opin. Pediatr. *12*, 263-268.

Jones, J.F., Shurin, S., Abramowsky, C., Tubbs, R.R., Sciotto, C.G., Wahl, R., Sands, J., Gottman, D., Katz, B.Z., and Sklar, J. (1988). T-cell lymphomas containing Epstein-Barr viral DNA in patients with chronic Epstein-Barr virus infections. N. Engl. J. Med. *318*, 733-41.

Katano, H., Ali, M.A., Patera, A.C., Catalfamo, M., Jaffe, E.S., Kimura, H., Dale, J.K., Straus, S.E., and Cohen, J.I. (2004). Chronic active Epstein-Barr virus infection associated with mutations in perforin that impair its maturation. Blood *103*, 144-1252.

Khanna, R., Bell, S., Sherritt, M., Galbraith, A., Burrows, S.R., Rafter, L., Clarke, B., Slaughter, R., Falk, M.C., Douglass, J., Williams. T., Elliott, S.L., and Moss, D.J. (1999). Activation and adoptive transfer of Epstein-Barr virus-specific cytotoxic T cells in solid organ transplant patients with posttransplant lymphoproliferative disease. Proc. Natl. Acad. Sci. USA *96*, 10391-10396.

Kijima, Y., Ischigmai, S., Hokita, S., Koriyama, C., Akiba, S., Eizuru, Y., and Aikou, T. (2003). The comparison of the prognosis between Epstein-Barr virus (EBV)-positive gastric carcinomas and EBV-negative ones. Cancer Letters *200*, 20-33.

Kimura, H., Hoshino, Y., Kanegane, H., Tsuge, I., Okamura, T., Kawa, K., and Morishima, T. (2001) Clinical and virologic characteristics of chronic active Epstein-Barr virus infection. Blood *98*, 280-286.

Krishnamurthy, S., Lanier, A.P., Dohan, P., Lainer, J.F., and Henle, W. (1987). Salivary gland cancer in Alaskan natives, 1966-1980. Hum. Pathol. *18*, 986-996.

Labrecque, L.G., Barnes, D.M., Fentiman, I.S., and Griffin BE. (1995). Epstein-Barr virus in epithelial cell tumors, a breast cancer study. Cancer Res. *55*, 39-45.

Lay, J.D., Tsao, C.J., Chen, J.Y., Kadin, M.E., and Su, I.J. (1997). Upregulation of tumor necrosis factor-alpha gene by Epstein-Barr virus and activation of macrophages in Epstein-Barr virus-infected T cells in the pathogenesis of hemophagocytic syndrome. J. Clin. Invest. *100*, 1969-1979.

Lee, E.S., Locker, J., Nalesnick, M., Ryes, J., Jaffe, R., Alashari, M., Nour, B., Tzakis, A., and Dickman, P.S. (1995). The association of Epstein-Barr virus with smooth-muscle tumors occurring after organ transplantation. N. Eng. J. Med. *332*, 19-25.

Levine, P.H., Stemmermann, G., Lennette, E.T., Hildesheim, A., Shibata, D., and Nomura, A. (1995). Elevated antibody titers to Epstein-Barr virus prior to the diagnosis of Epstein-Barr-virus-associated gastric adenocarcinoma. Int. J. Cancer *60*, 642-4.

Luqmani, Y., and Shousha, S. (1995). Presence of Epstein-Barr virus in breast carcinoma. Int. J. Oncol. *6*, 899-903.

MacMahon, E.M., Glass, J.D., Hayward, S.D., Mann, R.B., Becker, P.S., Charache, P., McArthur, J.C., and Ambinder, R.F. (1991). Epstein-Barr virus in AIDS-related primary central nervous system lymphoma. Lancet *338*, 969-973.

McClain, K.L., Leach, C.T., Jenson, H.B., Joshi, V.V., Pollock, B.H., Parmley, R.T., DiCarlo, F.J., Chadwick, E.G., and Murphy S.B. (1995). Association of Epstein-Barr virus with leiomyosarcomas in young people with AIDS. N. Engl. J. Med. *332*, 12-18.

McDiarmid, S.V., Jordan, S., Kim, G.S., Toyoda, M., Goss, J.A., Vargas, J.H., Martin, M.G., Bahar, R., Maxfield, A.L., Ament, M.E., Busuttil, R.W., and Lee, G.S. (1998). Prevention and preemptive therapy of posttransplant lymphoproliferative disease in pediatric liver recipients. Transplantation *66*, 1604-1611.

McGowan, J.E. Jr., Chesney, P.J., Crossley, K.B., and LaForce, F.M. (1992). Guidelines for the use of systemic glucocorticoids in the management of selected infections. J. Infect. Dis. *165*, 1-13.

Mori, A., Takao, S., Pradutkanchana, J., Kietthubthew, S., Mitarnun, W., and Ishida, T. (2003). High tumor necrosis factor-alpha levels in the patients with Epstein-Barr virus-associated peripheral T-cell proliferative disease/lymphoma. Leuk. Res. 27, 493-498.

Murray, P.G., Lissauer, D., Junying, J., Davies, G.., Moore, S., Bell, A., Timms, J., Rowlands, D., McConkey, C., Reynolds, G. M., Ghataura, S., England, D., Caroll, R., and Young, L.S. (2003). Reactivity with a monoclonal antibody to Epstein-Barr virus (EBV) nuclear antigen 1 defines a subset of aggressive breast cancers in the absence of the EBV genome. Cancer Res. *63*, 2338-2343.

Ohga, S., Nomura, A., Takada, H., Ihara, K., Kawakami, K., Yanai, F., Takaha, Y., Tanaka, T., Kasuga, N, and Hara, T. (2001). Epstein-Barr virus (EBV) load and cytokine gene expression in activated T cells of chronic active EBV infection. J. Infect. Dis. *183*, 1-7.

Okamura, T., Hatsukawa, Y., Arai, H., Inoue, M., and Kawa, K. (2000). Blood stem-cell transplantation for chronic active Epstein-Barr virus with lymphoproliferation. Lancet *356*, 223-224.

O'Reilly, R.J., Small, T.N., Papadopoulos, E., Lucas, K., Lacerda, J., and Koulova, L. (1997). Biology and adoptive cell therapy of Epstein-Barr virus-associated lymphoproliferative disorders in recipients of marrow allografts. Immunol Rev. *157*, 195-216.

Paya, C.V., Fung, J.J., Nalesnik, M.A., Kieff, E., Green, M., Gores, G., Habermann, T.M., Wiesner, P.H., Swinnen, J.L., Woodle, E.S., and Bromberg, J.S. (1999). Epstein-Barr virus-induced posttransplant lymphoproliferative disorders. Transplantation *68*, 1517-1525.

Quintanilla-Martinez, L., Kumar, S., Fend, F., Reyes, E., Teruya-Feldstein, J., Kingma, D., Sorbara, L., Raffeld, M., Straus, S.E., and Jaffee, E.S. (2000). Fulminant EBV(+) T-cell lymphoproliferative disorder following acute/chronic EBV infection, a distinct clinicopathologic syndrome. Blood *96*, 443-451.

Rea, D., Delecluse, H.-J., Hamilton-Dutoit, S.J., Marelle, L., Joab, I., Edelman, L., Finet, J.-F., and Raphael, M. (1994). Epstein-Barr virus latent and replicative gene expression in post-transplant lymphoproliferative disorders and AIDS-related non-Hodgkin's lymphomas. Ann. Oncology *5 (Suppl 1)*, S113-116.

Rea, T.D., Russo, J.E., Katon, W., Ashley, R.L. and Buchwald, D.S. (2001). Prospective study of the natural history of infectious mononucleosis caused by Epstein-Barr virus. J. Am. Board Fam. Pract. *14*, 234-242.

Riddler, S.A., Breinig, M.C., and McKnight, J.L.C. (1994). Increased levels of circulating Epstein-Barr virus (EBV)-infected lymphocytes and decreased EBV nuclear antigen antibody responses are associated with the development of posttransplant lymphoproliferative disease in solid-organ transplant patients. Blood *84*, 972-984.

Rogatsch, H., Bonatti, H., Menet, A., Larcher, C., Feichtinger, H., and Dirnhofer, S. (2000). Epstein-Barr virus-associated multicentric leiomyosarcoma in an adult patient after heart transplantation, case report and review of the literature. Am. J. Surg. Pathol. *24*, 614-621.

Rooney, C.M., Smith, C.A., Ng, C.Y., Loftin, S.K., Sixbey, J.W., Gan, Y., Srivastava, D.K., Bowman, L.C., Krance, R.A., Brenner, M.K., and Heslop, H.E. (1998). Infusion of cytotoxic T cells for the prevention and treatment of Epstein-Barr virus-induced lymphomas in allogeneic transplant recipients. Blood *92*, 1549-1555.

Saiki, Y., Ohtani, H., Naito, Y., Miyazawa, M., and Nagura, H. (1996). Immunophenotypic characterization of Epstein-Barr virus-associated gastric carcinoma, massive infiltration by proliferating CD8+ T-lymphocytes. Lab. Invest. *75*, 67-76.

Savoldo, B., Huls, M.H., Liu, Z., Okamura, T., Volk, H.D., Reinke, P., Sabat, R., Babel, N., Jones, J.F., Webster-Cyriaque, J., Gee, A.P., Brenner, M.K., Heslop, H.E., and Rooney, C.M. (2002). Autologous Epstein-Barr virus (EBV)-specific cytotoxic T cells for the treatment of persistent active EBV infection. Blood *100*, 4059-66.

Schlossberg, D. (1989). Infectious mononucleosis, 2nd ed. (New York, Springer-Verlag).

Schooley, R.T., Carey, R.W., Miller, G., Henle, W., Eastman, R., Mark, E.J., Kenyon, K., Wheeler, E.O., and Rubin, R.H. (1986). Chronic Epstein-Barr virus infections associated with fever and interstitial pneumonitis. Ann. Intern. Med. *104*, 636-643.

Seemayer, T.A., Gross, T.G., Egeler, R.M., Pirruccello, S.J., Davis, J.R., Kelly, C.M., Okano, M., Lanyi, A., and Sumegi J. (1995). X-linked lymphoproliferative disease, twenty-five years after the discovery. Ped. Res. *38*, 471-478, 1995.

Shibata, D., Tokunaga, M., Ueura, Y., Sato, E., Tanaka, S., and Weiss, L.M. (1991). Association of Epstein-Barr virus with undifferentiated gastric carcinomas with intense lymphoid infiltration. Am. J. Pathol. *139*, 469-474

Shibata, D. and Weiss, L.M. (1992). Epstein-Barr virus-associated gastric adenocarcinoma. Am. J. Pathol. *140*, 769-774.

Stadlmann, S., Fend, F., Moser, P., Obrist, P., Greil, R., and Dirnhofer, S. (2001). Epstein-Barr virus-associated extranodal NK/T cell lymphoma, nasal type of the hypopharynx, in a renal allograft recipient, case report and review of the literature. Mod. Pathol. *32*, 1264-1268.

Straus SE. (1992). Acute progressive Epstein-Barr virus infections. Ann. Rev. Med. *43*, 437-449.

Straus, S.E., Cohen, J.I., Tosato, G., and Meier, J. (1993). Epstein-Barr virus infections, biology, pathogenesis, and management. Ann. Intern. Med. *118*, 45-58.

Sugiura, M., Imai, S., Tokunaga, M. Koizumi, S., Uchizawa, M., Okamoto, K., and Osato, T. (1996). Transcriptional analysis of Epstein-Barr virus gene expression in EBV-positive gastric carcinoma, unique viral latency in the tumour cells. Br. J. Cancer *74*, 625-631.

Sumegi, J., Huang, D., Lanyi, A., Davis, J.D., Seemayer, T.A., Maeda, A., Klein, G., Seri, M., Wakiguchi, H., Purtilo, D.T., and Gross, T.G. (2000). Correlation of mutations of the SH2D1A gene and Epstein-Barr virus infection with clinical phenotype and outcome in X-linked lymphoproliferative disease. Blood *96*, 3118-3125.

Taga, H., Taga, K., Wang, F., Chretien, J., and Tosato, G. (1995). Human and viral interleukin-10 in acute Epstein-Barr virus-induced infectious mononucleosis. J. Infect. Dis. *171*, 1347-1350.

Taniere, P., Thivolet-Bejui, F., Vitrey, D., Isaac, S., Loire, R., Cordier, J.F., and Berger, F. (1998). Lymphomatoid granulomatosis--a report on four cases, evidence for B phenotype of the tumoral cells. Eur. Respir. J.*12*, 102-106.

Thorley-Lawson, DA, and Grose, A. (2004). Persistence of the Epstein-Barr virus and the origins of associated lymphomas. N. Eng. J. Med. *350*, 1328-1337.

Tomkinson, B.E., Wagner, D.K., Nelson, D.L., and Sullivan, J.L. (1987). Activated lymphocytes during acute Epstein-Barr virus infection. J. Immunol. *139*, 3802-3807.

Torre, D., and Tambini, R. (1999). Acyclovir for treatment of infectious mononucleosis, a meta-analysis. Scand. J. Infect. Dis. *31*, 543-547.

Triantos, D., Porter, S.R., and Scully, C. (1997). Oral hairy leukoplakia, clinicopathologic features, pathogenesis, diagnosis, and clinical significance. Clin Infect Dis *25*, 1392-1396.

Tynell, E., Aurelius, E., Brandell, A., Julander, I., Wood, M., Yao, Q.Y., Rickinson, A., Akerlund, B., and Andersson, J. (1996). Acyclovir and prednisolone treatment of acute infectious mononucleosis, a multicenter, double-blind, placebo-controlled study. J. Infect. Dis. *174*, 324-331.

Van Beerk, J., zur Hausen, A., Kranenbarg, E.K., van de Velde, C.J.H., Middeldorp, J.M., van den Brule, A.J.C., Meijer, C.J.L.M., and Bloemena, E. (2004). EBV-positive gastric adenocarcinomas, a distinct clinicopathologic entitity with a low frequency of lymph node involvement. J. Clin. Oncol. *22,* 664-670.

van der Horst, C., Joncas, J., Ahronheim, G., Gustafson, N., Stein, G., Gurwith, M., Fleisher, G., Sullivan, J., Sixbey, J., and Roland, S., et al. (1991). Lack of effect of peroral acyclovir for the treatment of acute infectious mononucleosis. J. Infect. Dis. 1991, *164,* 788-792.

vanEsser, J.W.J., Niesters, H.G., van derHolt, B., Meijer, E., Osterhaus, M.E., Gratama, J.W., Verdonck, L.F., Lowenberg, B., and Cornelissen, J.J. (2002). Prevention of Epstein-Barr virus-lymphoprolfierative disease by molecular monitoring and preemptive rituximab in high-risk patients after allogeneic stem cell transplantation. Blood *99*, 4364-4369.

Walling, D.M., Edmiston, S.N., Sixbey, J.W., Adel-Hamid, M., Resnick, L., and Raab-Traub-N. (1992). Coinfection with multiple strains of Epstein-Barr virus in human immunodeficiency virus-associated hairy leukoplakia. Proc. Natl. Acad. Sci. USA *89*, 6560-6564.

Walling, D.M., Etienne, W., Ray, A.J., Flaitz, C.M., and Nichols, C.M. (2004a). Persistence and transition of Epstein-Barr virus genotypes in the pathogenesis of oral hairy leukoplakia. J. Infect. Dis. *190*, 387-395.

Walling, D.M., Flaitz, C.M., and Nichols, C.M. (2003). Epstein-Barr virus replication in oral hairy leukoplakia, response, persistence, and resistance to treatment with valacyclovir. J. Infect. Dis. *188*, 883-890.

Walling, D.M., Ling, P.D., Gordafze, A.V., Montes-Walters, M., Flaitz, C.M., and Nichols, C.M. (2004b). Expression of Epstein-Barr virus latent genes in oral epithelium determines the pathogenesis of oral hairy leukoplakia. J. Infect. Dis. *190*, 396-399.

Webster-Cyriaque, Middeldorp, J., and Raab-Traub, N. (2000). Hairy leukoplakia, an unusual combination of transforming and permissive Epstein-Barr virus infections. J. Virol. *74*, 7610-7618.

Weiss, L.M., Movahed, L.A., Warnke, R.A., and Sklar, J. (1989). Detection of Epstein-Barr viral genomes in Reed-Sternberg cells of Hodgkin's disease. N Engl. J. Med. *320*, 502-506.

Wong, M.P., Chung, L.P., Yuen, S.T., Leung, S.Y., Chan, S.Y., Wang, E., and Fu, K.H. (1995). In situ detection of Epstein-Barr virus in non-small cell lung carcinomas. J. Pathol. *177*, 233-240.

Wilson, W.H., Kingma, D.W., Raffeld, M., Wittes, R.E., and Jaffe, E.S. (1996). Association of lymphomatoid granulomatosis with Epstein-Barr viral infection of B lymphocytes and response to interferon-alpha 2b. Blood *87*, 4531-4537.

Xu, Z.G., Iwatsuki, K., Oyama, N., Ohtsuka, M. Satoh, M., Kikuchi, S., Akiba, H., and Kaneko, F. (2001). Br. J. Ca. *84*, 920-925.

Yachie, A., Kanegame, H., and Kasahara, Y. (2003). Epstein-Barr virus-associated T-/natural killer cell lymphoproliferative diseases. Sem. Hematol. *40*, 124-132.

zur Hausen, H., Schulte-Holthausen, H., Klein, G., Henle, W., Henle, G., Clifford, P., and Santesson, L. (1970). EBV DNA in biopsies of Burkitt's tumors and anaplastic carcinoma of the nasopharynx. Nature *228*, 1056-1058.

Chapter 6

Epstein-Barr Virus and Oral Malignancies

*Rona S. Scott, Cary A. Moody and John W. Sixbey**

ABSTRACT

Epstein-Barr virus (EBV) is transmitted through saliva and is periodically shed throughout the life of virus carriers in oral secretions. The well-known association of EBV with lymphoid and epithelial cell malignancies has prompted numerous investigations into whether malignancies of the oral cavity, occurring in proximity of sites of natural EBV replication, preferentially harbor EBV. Attempts to define pathologic relationships in the context of an anatomic site with intrinsic viral activity highlight difficulties inherent in establishing valid clinical correlations between this ubiquitous herpesvirus and disease. Recent literature on EBV in oral cavity malignancies underscores, on the one hand, the technical challenges that confound the conclusive delineation of such links and, on the other, has suggested alternative roles for EBV in human malignancy as a guide to future clinical investigation.

INTRODUCTION

The oral cavity is the site for entry and egress of infectious Epstein-Barr virus (EBV). Participation of both lymphoid as well as epithelial components of the oropharynx at some point in the natural life cycle of EBV infection has become increasingly apparent (Borza et al., 2002; Tugizov et al., 2003). While the exact nature and timing of viral exchange between tissues remains at issue (Lemon et al., 1977; Sixbey et al., 1984; Anagnostopoulos et al., 1995; Thorley-Lawson et al., 1996; Karajannis et al., 1997; Niedobitek et al., 2000; Herrmann et al., 2002; Pegtel et al., 2004), the capability of epithelial cells *in vivo* to support latent and lytic infection and the status of memory B lymphocytes as a reservoir for persistent infection go unquestioned (Greenspan et al., 1985; Babcock et al., 1998; Webster-Cyriaque et al., 2000; Walling et al., 2001; Raab-Traub, 2002; Huang et al., 2003). As a site of life-long periodic EBV shedding (Gerber et al., 1972; Miller et al., 1973; Chang et al., 1973), the oral cavity would seem to offer unique opportunities to address questions provoked by the growing number and diversity of neoplastic lesions linked to EBV infection.

Aside from its involvement with tumors such as Burkitt's lymphoma, Hodgkin's lymphoma, gastric carcinoma and nasopharyngeal carcinoma, EBV has also been

*For correspondence email jsixbe@lsuhsc.edu

implicated in more controversial settings. Unlike the well-established association of EBV with undifferentiated nasopharyngeal carcinoma of the posterior pharynx (Raab-Traub, 2002), for example, the literature on oral squamous cell carcinomas is contradictory and often inconclusive, confounded by the use of assays with markedly different sensitivities and an apparent lack of a standard definition of what constitutes EBV positivity within tumor tissue. Is infection of oral cavity malignancies coincidental? Is the incidence of EBV in epithelial and lymphoid malignancies altered in microenvironments with enhanced viral activity? What is the likelihood of secondary infection of pre-malignant lesions or established tumors, with EBV acting either as passenger or agent of tumor progression? With these issues in mind, we focus on recent reports of EBV DNA in malignancies presenting in the oral cavity.

EBV INFECTION IN NON-HODGKIN'S LYMPHOMA OF THE ORAL CAVITY

Although non-Hodgkin's lymphomas presenting in the oral soft tissues are rare, when they do occur they are typically high grade, aggressive histological subtypes such as diffuse large B cell, immunoblastic or Burkitt's lymphomas. At other anatomic locations, such high grade tumor types have well-established, albeit erratic, associations with EBV. In comparison to low grade counterparts such as mantle cell, follicular cell and MALT lymphomas, high grade tumors are three times more likely to contain EBV (Hummel et al., 1995). Despite the presence of EBV in oral secretions, however, virus appears to be no more common in primary oral cavity lymphomas than in comparable tumors arising at other sites, regardless of host immune status (Gulley et al., 1995; Hummel et al., 1995; Leong et al., 2001; Solomides et al., 2002).

Viral association with oral non-Hodgkin's lymphomas of the immunocompetent host

In three studies that utilized the revised European-American classification scheme for diagnosis of oral lymphoma in combination with *in situ* hybridization to detect the abundant small non-polyadenylated RNAs (EBERs 1 and 2) encoded by EBV, virus was associated with 19% (4/21), 9% (4/46), and 14% (9/65) of cases of oral lymphoma in immunocompetent patients (Gulley et al., 1995; Leong et al., 2001; Solomides et al., 2002). These levels of EBV detection are not markedly different from the 13% (27/208) association described for non-Hodgkin's lymphomas occurring at systemic sites (Hummel et al., 1995). In all three patient series, oral lymphomas were largely of B cell origin and histologically of the diffuse large B cell type.

EBV and oral non-Hodgkin's lymphoma in HIV-induced immunodeficiency

Lymphomas are up to 60- to 100-fold more common in patients with AIDS compared to the general population and approximately 60% are EBV-associated (Shibata et al., 1993; Carbone, 2003). Up to 5% of lymphomas in this patient group present as oral lesions (Ziegler et al., 1984), and the majority are linked to EBV (Green and Eversole, 1989; Gulley et al., 1995; Leong et al., 2001; Iamaroon et al., 2003). As in immune competent patients, the frequency of EBV involvement in oral lymphomas of AIDS appears to be no greater than at anatomic sites where replication is considered infrequent (Gulley et al., 1995). Oral T cell lymphomas occur more frequently in the immunosuppressed, however, comprising up to two-thirds of oral lymphomas in some series, all of which contained EBV (Thomas et al., 1993; Leong *et al*, 2001). A relatively new entity, plasmablastic lymphoma of the oral cavity, is closely associated with patients infected by HIV. It appears to be a variant of diffuse large B cell lymphomas, and is characterized by immunoblastic morphology and a plasma cell immunophenotype (Delecluse et al., 1997; Carbone et al., 1999; Giadano et al., 2002; Flaitz et al., 2002). Tumor cells have weak or absent expression of the most common B cell-associated surface antigens, CD20 and CD45, and strong immunostaining with plasma cell-reactive antibodies VS38c and anti-CD79a. Although the reported series are small, an association with EBV has been detected in from 60-84% of cases (Delecluse et al., 1997; Giadano et al., 2002; Colomo et al., 2004).

Patterns of infection within individual oral lymphomas

A remarkable finding not unique to oral lymphomas is the apparent presence of virus in only a portion of tumor cells as determined by EBER *in situ* hybridization (Leong, et al., 2001; Solomides et al., 2002). In some lymphomas, virtually all identifiable tumor cells express EBERs. In others, varying fractions of tumor cells are EBER-positive. These patterns of infection concur with previous descriptions of non-Hodgkin's lymphomas occurring outside the oral cavity, where EBV-positive tumors were grouped into those in which EBV may have contributed to the malignant phenotype (80-100% EBER-positive tumor cells) versus those in which infection was assumed to be of little significance to tumorigenesis (under 80% EBER-positive tumor cells) (Hummel et al., 1995). Such thresholds are arbitrary, have varied with investigator, and overlook potential pathogenic implications of the finding. When analyzed at a single cell level by micromanipulation, the partial EBER staining in tumors often attributed to technical artifact or downregulation of EBER gene expression in a portion of malignant cells indeed reflects a mixed population of tumor cells, some bearing EBV and others lacking virus. By single cell PCR analysis, the EBER-negative subset of tumor cells was uniformly EBV DNA negative (Hänel et al., 2001). Infection of only a proportion of a neoplastic clone suggests secondary infection of an established tumor or, alternatively, the loss of virus in the course of tumor expansion, both of which have profound implications regarding our understanding of the fundamental contribution of EBV to oncogenesis.

EBV INFECTION OF ORAL EPITHELIAL LESIONS

Among head and neck carcinomas, nasopharyngeal carcinoma stands out as having a strong association with EBV that is deemed critical for transforming nasopharyngeal epithelial cells into invasive cancer (see Chapter 7 by Raab-Traub). Recent findings have highlighted possible genetic and epigenetic alterations in epithelial cells that may precede and possibly facilitate infection. For example, molecular alterations have been reported in histologically normal epithelium of southern Chinese at high risk for development of EBV-associated nasopharyngeal carcinoma (Chan et al., 2000; 2002). Frequent genetic losses on chromosome 3p and 9p were detected in normal epithelia, dysplastic epithelia, and nasopharyngeal carcinoma in high risk areas but not in healthy tissues from individuals inhabiting low risk regions. EBV infection was detected in all high grade dysplastic lesions as well as nasopharyngeal carcinoma but not in any nasopharyngeal tissue of normal adults, suggesting that early genetic alterations may take place prior to EBV infection. These may increase epithelial cell susceptibility to infection, as well as enhance their ability to maintain the virus and to undergo subsequent clonal expansion (Lo et al., 2004).

One compelling hypothesis is that aberrant STAT (signal transducers and activators of transcription) activation may be a predisposing event for EBV-driven carcinogenesis, the promoters (Qp and L1-TR) of two viral genes expressed in nasopharyngeal carcinoma, Epstein-Barr nuclear antigen 1 (EBNA1) and latent membrane protein 1 (LMP1), respectively, being regulated by STAT3 (Chen et al., 2001; 2003). In particular, Qp-driven EBNA1 expression is required for episomal propagation and maintenance within the latently infected cell. LMP1 is an integral membrane protein with transforming potential explicable in large part by its functioning as a constitutively activated tumor necrosis factor receptor. Since activated STATs are usually not found in normal epithelium, entry of EBV into a cell with dysregulated STAT signaling may facilitate both viral retention (EBNA1 expression) and subsequent tumor progression (LMP1 expression). STAT3 activation is an early event in head and neck squamous cell carcinogenesis (Song and Rubin Grandis, 2000), and is one component of molecular changes termed "field cancerization" that become widespread in histologically normal oral mucosa during carcinogenesis (Tabor et al., 2001; Braakhuis et al., 2003; 2004).

Viral clonality in tumors remains the most persuasive argument against infection being a late event, the structure of the termini of EBV serving as marker of clonal cellular proliferation. When EBV enters cells, its linear genome circularizes by terminal repeat sequences to generate a fused repeat fragment of variable length in each infected cell. If virus was present at the initiation of clonal expansion, repeat number should be uniform in every tumor cell. If cell transformation were to precede infection in multistep carcinogenesis, the prediction is that tumors would be polyclonal by EBV terminal repeat analysis and, in many malignant cells, there would be no EBV altogether. In fact, EBV-associated tumors as well as some pre-invasive lesions do contain monoclonal episomes, supporting an early (if not initiating) role for EBV in tumorigenesis ((Raab-Traub and Flynn, 1986; Pathmanathan et al., 1995).

In vitro data raise an alternate interpretation of viral clonality which would not require presence of EBV at the genesis of a tumor. In newly infected cancer cell lines, there is a rapid evolution from polyclonal infection towards a predominant episomal form, reflecting selection for a terminal repeat number that optimizes EBV latency gene-enhanced cell growth (Moody et al., 2003). Consistent with that notion are clinical studies that have shown the EBV genome in only a portion of tumor cells in early NPC (without microinvasion) and progressively increased numbers of infected cells at more advanced stages of disease (with microinvasion)

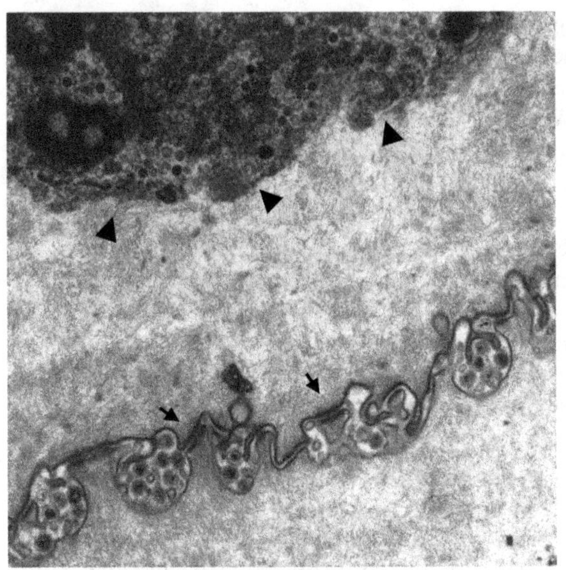

Figure 1. Electron micrograph of oral hairy leukoplakia. Abundant EBV nucleocapsids in the epithelial cell nucleus and presence of completed virions in intracellular spaces are consistent with exuberant viral replication characteristic of this benign epithelial cell lesion. Arrows indicate desmosomes; arrowheads indicate nuclear membrane.

(Yeung et al., 1993). Similarly, absence of EBV in premalignant gastric lesions lends credence to the view that virus infection may be a relatively late event, at least in EBV-associated gastric carcinogenesis (zur Hausen et al., 2004).

Oral hairy leukoplakia

Just as EBV-induced infectious mononucleosis has been informative with respect to the capacity of this gamma-herpesvirus to drive lymphoproliferation, the benign oral epithelial lesion, oral hairy leukoplakia (HLP) may likewise be instructive in gauging aspects of EBV biology that could be instrumental in oral carcinogenesis. HLP often occurs in patients infected with human immunodeficiency virus or those with severe immunodeficiency. HLP lesions appear as elevated, white patches on the lateral borders and dorsum of the tongue (see Figure 1). Histologically, acanthosis (hypertrophy of the stratum spinosum) and koilocytic changes are observed in the absence of inflammatory infiltrates.

HLP is unique from a variety of perspectives. First, HLP is the only pathologic manifestation of a permissive EBV infection, with exuberant viral replication occurring within squamous epithelial cells of the tongue. HLP is a direct result of EBV permissive replication since treatment with inhibitors of viral replication ameliorates the lesion (Greenspan et al., 1985; Greenspan and Greenspan, 1989; Walling et al., 2003). Contrary to permissive infections with other herpesviruses, the profuse replication of EBV within epithelial cells of HLP is non-cytolytic. This may be a feature of the concurrent expression of replicative and transforming proteins within the infected cell, enabling epithelial cell survival in face of abundant replication (Webster-Cyriaque et al., 2000; Walling et al., 2004a). For instance, latency gene LMP-1 is known to induce expression of the antiapoptotic cellular gene A20, which may extend cell survival in HLP (Fries et al., 1996). Likewise, the EBV BHRF1 gene product, a homologue of Bcl-2, is expressed at high levels in HLP and may inhibit apoptosis as well as delay epithelial cell differentiation.

Characteristic koilocytic changes of HLP are thought to result from the high-level of virus replication, while acanthosis may result from the expression of EBV transforming genes that extend cell survival (Becker et al., 1991; Dawson et al., 1995; Triantos et al., 1997; Webster-Cyriaque and Raab-Traub, 1998). Although BHRF1 expression in epithelium has been postulated to contribute to early stages in cancer development (Horner et al., 1995), HLP has not shown the predisposition to undergo malignant transformation that is characteristic of oral leukoplakia.

Another feature of HLP is its frequent co-infection with different EBV types and strain variants, as well as ongoing intra- and interstrain recombination (Walling et al., 1992; Walling and Raab-Traub, 1994). A rearranged, defective EBV genome known to disrupt latency, induce permissive EBV replication, and accelerate loss of parental viral episomes has also been detected in HLP lesions (Patton et al., 1990; Gan et al., 1993). Viral genotyping based on sequence polymorphisms within LMP1 have corroborated the co-infection with multiple EBV strains in HLP lesions (Sitki-Green et al., 2002; Walling et al., 2004b). Given the unusual combination of transforming and permissive EBV infection manifest as HLP, it is likely that expression of latency gene EBNA2 may upregulate the virus receptor CD21 on infected epithelial cells, permitting superinfection by multiple virus strains in this replicative environment and subsequent generation of intertypic viral recombinants (Webster-Cyriaque and Raab-Traub, 1998). That HLP may embody mechanisms generally operative in EBV pathobiology is suggested by detection of similar recombinants in a variety of immunocompetent hosts (Burrows et al., 1996; Midgley et al., 2000; Gan et al., 2002; Kelly et al., 2002).

Salivary gland lymphoepitheliomas

Salivary gland tumors are a relatively rare, morphologically diverse group of malignancies comprising less than 5% of all head and neck cancers (Speight and Barrett, 2002). Within this group of tumors, lymphoepithelioma of the parotid gland has been linked to EBV infection (Saemundsen et al., 1982; Iezzoni et al., 1995) . These tumors are of the same histogenetic type as nasopharyngeal carcinoma, being defined as undifferentiated squamous cell carcinomas with lymphoid infiltration. They have the highest incidence among the Greenland and North American Inuit/Eskimo and southern Chinese populations (Albeck et al., 1993; Leung et al., 1995; Kuo and Hsueh, 1997), where virtually all tumors are associated with EBV (Iezzoni et al., 1995; Jen et al., 2003). In other ethnic groups, the association is infrequent (Gallo et al., 1994; Kotsianti et al., 1996). *In situ* hybridization has demonstrated EBV DNA localized specifically to the malignant epithelial cells, whereas virus is not detected in the lymphocytes or in the adjacent benign tissue (Huang et al., 1988; Lanier et al., 1991; Hamilton-Dutoit et al., 1991; Iezzoni et al., 1995; Jen et al., 2003). Similar to nasopharyngeal carcinoma, viral DNA is present in salivary gland lymphoepitheliomas as a clonal episomal genome and infected cells express EBV latency genes (Raab-Traub et al., 1991; Sheen et al., 1997; Jen et al., 2003). A search for EBV in other salivary gland tumors, including mucoepidermal carcinomas, adenoid cystic carcinomas, acinic cell carcinomas and adenocarcinomas has been without success (Tsai et al., 1996; Wen et al., 1997; Atula et al., 1998; Venkateswaran et al., 2000).

Oral squamous cell carcinomas (OSCC)

In light of the ability of EBV to infect oral epithelial cells and contribute to the genesis of nasopharyngeal carcinoma, multiple attempts have been made to establish a role for EBV in conventional head and neck squamous cell carcinoma, with variable and conflicting results (summarized in Table 1). PCR targeted to various regions of the EBV genome has detected virus at a higher frequency in OSCC as compared to normal control tissue (Mao and Smith, 1993; Cruz et al., 1997; D'Costa et al., 1998; Sand et al., 2002; Szkaradkiewicz et al., 2002). However, EBV cannot always be localized to tumor cells by less sensitive *in situ* hybridization techniques. Moreover, when positive hybridization is obtained, signals often identify rare infiltrating lymphocytes rather than tumor cells (Cruz et al., 1997; Goldenberg et al., 2004). In individual cases, convincing focal staining for EBERs has been demonstrated (Goldenberg et al., 2000). Others report an absence of EBER transcripts, with concomitant detection of the LMP1 protein by immunohistochemistry (Kobayashi et al., 1999; Gonzalez-Moles et al., 2002). Although detection of EBERs by *in situ* hybridization has become the standard approach to detect EBV in tumor tissue, the possibility that EBER-negative forms of latency exist in as-yet unsuspected EBV-associated malignancies has been raised (Sugawara et al., 1999; Bonnet et al., 1999). In a recent study that used real-time quantitative PCR, 24/113 OSCCs were positive for EBV (Goldenberg et al., 2004), a percentage much like that reported in investigations employing standard PCR (Table 1). However, most positive tumors (22/24) had just trace levels of the virus (<0.01 EBV copies/cell). Only two OSCC were overtly positive, containing >0.1 EBV copies/cell genome. The lack of concordance of data between studies, together with the heavy reliance on PCR for determinations of a viral association, makes a putative role for EBV in oral squamous cell carcinoma at best inconclusive, with findings in some cases consistent with baseline virus activity expected within the oral cavity.

CONCLUSION

Designation of a tumor as "EBV-associated" currently requires the unambiguous demonstration of EBV DNA and/or viral gene products in a majority of cells within the tumor. Ubiquity of EBV infection in human populations, with the accompanying probability of contamination of tumor tissue by infected bystander cells, must always be considered when interpreting results, particularly data derived using sensitive PCR approaches. Because tumors are often more cellular than normal tissue, side by side comparisons of viral burden should include normalization to number of cells assayed, as determined by PCR of a control cellular gene such as actin.

As implied by several recent findings described above, these caveats do not negate, in every case, the possible biologic relevance of a low level of infection in the course of tumor evolution. At present, however, such results must be viewed with considerable caution and best serve to provoke studies that query two possibilities. One is the contention that secondary infection of a primarily EBV-negative malignancy may contribute to malignant progression. Such a premise assumes that there is a source for infectious EBV in the vicinity of the tumor, that

Table 1. Reported prevalence of EBV in oral squamous cell carcinoma (OSCC)

EBV-positive OSCC/total	EBV-positive samples/total normal controls	Method of EBV detection	Reference
0/20	0/15	DNA ISH	Talacko et al., 1991
10/20 (50%)	15/60 (25%)	PCR to BamHI W	Mao and Smith, 1993
19/36 (53%) 10/36 (27%) 10/36 (27%)		PCR to BamHI W DNA ISH EBER ISH	Horiuchi et al., 1995
22/90 (24%)	11/30 (37%) non-malignant lesions	PCR to BamHI W	van Heerden et al., 1995
36/36 (100%) 18/36 (50%) 0/11	1/12 (8%)	PCR to BamHI W PCR to BNLF1 EBER ISH	Cruz et al., 1997
25/103 (24%)	3/76 (4%)	PCR to BamHI L	D'Costa et al., 1998
29/45 (65%)		PCR	Maeda et al., 1998
0/36		PCR	Mizugaki et al., 1998
19/108 (17%)		PCR	Gonzalez-Moles et al., 1998
7/46 (15%) 6/46 (13%) 0/46		PCR LMP1 IHC EBER ISH	Kobayashi et al., 1999

11/39 (38%)	5/67 (7%)	PCR to BamHI W	Sand et al., 2002
15/78 (19%)		PCR to BALF5	Gonzalez-Moles et al., 2002
12/78 (15%)		LMP1 IHC	
0/78		EBER ISH	
24/28 (85%)	2/12 (16%)	PCR	Szkaradkiewicz et al., 2002
26/26 (100%)	4/4 (100%)	PCR to BamHI W	Shimakage et al., 2002
16/24 (66%)	0/2	EBER ISH	
9/12 (75%)	0/1	EBNA2 ISH	
15/29 (51%)	0/2	LMP1 IFA	
14/28 (50%)	0/2	BZLF1 IFA	
0/26		PCR to BNLF1	Tsang et al., 2003
0/24		EBER ISH	Iamaroon et al., 2004
24/113 (21%)		Q-PCR to BamHI W and EBNA1 regions	Goldenberg et al., 2004

ISH=*in situ* hybridization, IHC=immunohistochemistry, IFA=immunofluorescence, Q-PCR=quantitative PCR neoplastic cells are permissive for EBV infection, and that virus infection may confer a selective advantage to newly infected tumor cells. An alternative view is that viral episomes may be inefficiently retained during rapid clonal expansion. This notion presupposes a viral contribution at a very early stage of malignant transformation, with virus becoming redundant upon subsequent genetic and epigenetic changes. Documentation of either scenario in a relevant clinical setting will require imaginative use of clinical materials in future studies, coupled with the creative application of investigative techniques. Such efforts could reveal fundamentals of pathogenesis involving this persistent opportunist that would reshape current views on virus-associated disease.

References

Albeck, H., Bentzen, J., Ockelmann, H.H., Nielsen, N.H., Bretlau, P., and Hansen, H.S. (1993). Familial clusters of nasopharyngeal carcinoma and salivary gland carcinomas in Greenland natives. Cancer *72*, 196-200.

Anagnostopoulos, I., Hummel, M., Kreschel, C., and Stein, H. (1995). Morphology, immunophenotype, and distribution of latently and/or productively Epstein-Barr virus infected cells in acute infectious mononucleosis: implications for the interindividual infection route of Epstein-Barr virus. Blood *85*, 744-750.

Atula, T., Grenman, R., Klemi, P., and Syrjanen, S. (1998). Human papillomavirus, Epstein-Barr virus, human herpesvirus 8 and human cytomegalovirus involvement in salivary gland tumours. Oral Oncol. *34*, 391-395.

Babcock, G.J., Decker, L.L., Volk, M., and Thorley-Lawson, D.A. (1998). EBV persistence in memory B cells *in vivo*. Immunity *9*, 395-404.

Becker, J., Leser, U., Marschall, M., Langford, A., Jilg, W., Gelderblom, H., Reichart, P., and Wolf, H. (1991). Expression of proteins encoded by Epstein-Barr virus trans-activator genes depends on the differentiation of epithelial cells in oral hairy leukoplakia. Proc. Natl. Acad. Sci. USA *88*, 8332-8336.

Bonnet, M., Guinebretiere, J.M., Kremmer, E., Grunewald, V., Benhamou, E., Contesso, G., Joab, I. (1999). Detection of Epstein-Barr virus in invasive breast cancers. J. Natl. Cancer Inst. *91*, 1376-1381.

Borza, C.M., and Hutt-Fletcher, L.M. (2002). Alternate replication in B cells and epithelial cells switches tropism of Epstein-Barr virus. Nat. Med. *8*, 594-599.

Braakhuis, B.J., Leemans, C.R., and Brakenhoff, R.H. (2004). A genetic progression model of oral cancer: current evidence and clinical implications. J. Oral. Pathol. Med. *33*, 317-322.

Braakhuis, B.J.M., Tabor, M.P., Kummer, J.A., Leemans, C.R., and Brakenhoff, H. (2003). A genetic explanation of Slaughter's concept of field cancerization: evidence and clinical implications. Cancer Res. *63*, 1727-1730.

Burrows, J.M., Khanna, R., Sculley, T.B., Alpers, M.P., and Moss, D.J. (1996). Identification of a naturally occurring recombinant Epstein-Barr virus isolate from New Guinea that encodes both type 1 and type 2 nuclear antigen sequences. J. Virol. *70*, 4829-4833.

Carbone, A. (2003). Emerging pathways in the development of AIDS-related lymphomas. The Lancet Oncology *4*, 22-29.

Carbone, A., Gaidano, G., Gloghini, A., Ferlito, A., Rinaldo, A., and Stein, H. (1999) AIDS-related plasmablastic lymphomas of the oral cavity and jaws: a diagnostic dilemma. Ann. Otol. Rhinol. Laryngol. *108*, 95-99.

Chan, A.S.C., To, K.F., Lo, K.W., Ding, X., Li, P., Johnson, P., and Huang, D.P. (2002). Frequent chromosome 9p losses in histologically normal nasopharyngeal epithelia from southern Chinese. Int. J. Cancer *102*, 300-303.

Chan, A.S.C., To, K.F., Lo, K.W., Mak, K.F., Pak, W., Chiu, B., Tse, G.M.K., Ding, M., Li, X., Lee, J.C.K., and Huang, D.P. (2000). High frequency of chromosome 3p deletion in histologically normal nasopharyngeal epithelia from southern Chinese. Cancer Res. *60*, 5365-5370.

Chang, R.S., Lewis, J.P., and Abildgaard, C.F. (1973). Prevalence of oropharyngeal excretors of leukocyte-transforming agents among a human population. N. Engl. J. Med. *289*, 1325-1329.

Chen, H., Hutt-Fletcher, L.M., Cao, L. and Hayward, S.D. (2003). A positive regulatory loop of LMP1 expression and STAT activation in epithelial cells latently infected with Epstein-Barr virus. J. Virol. *77*, 4139-4148.

Chen, H., Lee, J.M., Zong, Y., Borowitz, M., Ng, M.H., Ambinder, R.F., and Hayward, S.D. (2001). Linkage between STAT regulation and Epstein-Barr virus gene expression in tumors. J. Virol. *75*, 2929-2937.

Colomo, L., Loong, F., Rives, S., Pittaluga, S., Martinez, A., Lopez-Guillermo, A., Ojanguren, J., Romagosa, V., Jaffe, E.S., and Campo, E. (2004). Diffuse large B-cell lymphomas with plasmablastic differentiation represent a heterogeneous group of disease entities. Am. J. Surg. Pathol. *28*, 736-47.

Cruz, I., Van den Brule, A.J., Steenbergen, R.D., Snijders, P.J., Meijer, C.J., Walboomers, J.M., Snow, G.B., and Van der Waal, I. (1997). Prevalence of Epstein-Barr virus in oral squamous cell carcinomas, premalignant lesions and normal mucosa - a study using the polymerase chain reaction. Oral Oncol. *33*, 182-188.

Dawson, C.W., Eliopoulos, A.G., Dawson, J., and Young, L.S. (1995). BHRF1, a viral homologue of the Bcl-2 oncogene, disturbs epithelial cell differentiation. Oncogene *10*, 69-77.

D'Costa, J., Saranath, D., Sanghvi, V., and Mehta, A.R. (1998). Epstein-Barr virus in tobacco-induced oral cancers and oral lesions in patients from India. J. Oral. Pathol. Med. *27*, 78-82.

Delecluse, H.J., Anagnostopoulos, I., Dallenbach, F., Hummel, M., Marafioti, T., Schneider, U., Huhn, D., Schmidt-Westhausen, A., Rreichart, P.A., Gross, U., and Stein, H. (1997). Plasmablastic lymphomas of the oral cavity: a new entity associated with human immunodeficiency virus infection. Blood *89*, 1413-1420.

Flaitz, C.M., Nichols, C.M., Walling, D.M., and Hicks, M.J. (2002). Plasmablastic lymphoma: an HIV-associated entity with primary oral manifestations. Oral Oncol. *38*, 96-102.

Fries, K.L., Miller, W.E., and Raab-Traub, N. (1996). Epstein-Barr virus latent membrane protein 1 blocks p53-mediated apoptosis through the induction of the A20 gene. J. Virol. *70*, 8653-8659.

Gaidano, G., Cerri, M., Capello, D., Berra, E., Deambrogi, C., Rossi, D., Larocca, L.M., Campo, E., Gloghini, A., Tirelli, U., and Carbone, A., (2002). Molecular histogenesis of plasmablastic lymphoma of the oral cavity. Brit. J. Haem. *119*, 622-628.

Gallo, O., Santucci, M., Calzolari, A., and Storchi, O.F. (1994). Epstein-Barr virus (EBV) infection and undifferentiated carcinoma of the parotid gland in Caucasian patients. Acta Otolaryngol. *114*, 572-575.

Gan, Y.J., Razzouk, B.I., Su, T., and Sixbey, J.W. (2002) A defective, rearranged Epstein-Barr virus genome in EBER-negative and EBER-positive Hodgkin's disease. Am. J. Pathol. *160*, 781-786.

Gan, Y.J., Shirley, P., Zeng, Y., and Sixbey, J.W. (1993). Human oropharyngeal lesions with a defective Epstein-Barr virus that disrupts viral latency. J. Infect. Dis. *168*, 1349-1355.

Gerber, P., Nonoyama, M., Lucas, S., Perlin, E., and Goldstein, L.I. (1972). Oral excretion of Epstein-Barr virus by healthy subjects and patients with infectious mononucleosis. Lancet *2*, 988-989.

Goldenberg, D., Benoit, N.E., Begum, S., Westra, W.H., Cohen, Y., Koch, W.M., Sidransky, D., and Califano, J.A. (2004). Epstein-Barr virus in head and neck cancer assessed by quantitative polymerase chain reaction. Laryngoscope *114*, 1027-1031.

Goldenberg, D., Rachmiel, A., Yitzhak, O.B., Peled, M., Golz, A., Laufer, D., Joachims, H.Z. (2000). Triple primary malignancies of the head and neck. J. Otolaryngology *29*, 51-54.

Gonzalez-Moles, M.A., Gutierrez, J., Rodriguez, M.J., Ruiz-Avila, I., and Rodriguez-Archilla, A. (2002). Epstein-Barr virus latent membrane protein-1 (LMP-1) expression in oral squamous cell carcinoma. Laryngoscope *112*, 482-487.

Gonzalez-Moles, M., Gutierrez, J., Ruiz, I., Fernandez, J.A., Rodriguez, M., and Aneiros, J. (1998). Epstein-Barr virus and oral squamous cell carcinoma in patients without HIV infection: viral detection by polymerase chain reaction. Microbios. *96*, 23-31.

Green, T.L. and Eversole, L.R. (1989) Oral lymphomas in HIV-infected patients: association with Epstein-Barr virus DNA. Oral Surg. Oral Med. Oral Pathol. *67*, 437-442.

Greenspan, J.S. and Greenspan, D. (1989). Oral hairy leukoplakia: diagnosis and management. Oral Surg. Oral Med. Oral Pathol. *67*, 396-403.

Greenspan, J.S., Greenspan, D., Lennette, E.T., Abrams, D.I., Conant, M.A., Petersen, V., and Freese, U.K. (1985). Replication of Epstein-Barr virus within the epithelial cells of oral "hairy" leukoplakia, an AIDS-associated lesion. N. Engl. J. Med. *313*, 1564-1571.

Gulley, M.L., Sargeant, K.P., Grider, D.J., Eagan, P.A., Davey, D.D., Damm, D.D., Robinson, R.A., Vandersteen, D.P., McGuff, H.S., and Banks, P.M. (1995). Lymphomas of the oral soft tissues are not preferentially associated with latent or replicative Epstein-Barr virus. Oral Surg. Oral Med. Oral Pathol. Oral Radiol. Endod. *80*, 425-431.

Hamilton-Dutoit, S.J., Therkildsen, M.H., Neilsen, N.H., Jensen, H., Hansen, J.P., and Pallesen, G. (1991). Undifferentiated carcinoma of the salivary gland in Greenlandic Eskimos: demonstration of Epstein-Barr virus DNA by in situ nucleic acid hybridization. Hum. Pathol. *22*, 811-815.

Hänel, P., Hummel, M., Anagnostopoulos, I., and Stein, H. (2001). Analysis of single EBER-positive and negative tumour cells in EBV-harbouring B-cell non-Hodgkin lymphomas. J. Pathol. *195*, 355-360.

Herrmann, K., Frangou, P., Middeldorp, J., and Niedobitek, G. (2002). Epstein-Barr virus replication in tongue epithelial cells. J. Gen. Virol. *83*, 2995-2998.

Horiuchi, K., Mishima, K., Ichijima, K., Sugimura, M., Ishida, T., and Kirita, T. (1995). Epstein-Barr virus in the proliferative diseases of squamous epithelium in the oral cavity. Oral Surg. Oral Med. Oral Pathol. Oral Radiol. Endod. *79*, 57-63.

Horner, D., Lewis, M., and Farrell, P.J. (1995). Novel hypotheses for the roles of EBNA-1 and BHRF1 in EBV-related cancers. Intervirology *38*, 195-205.

Huang, D.P., Ng, H.K., Ho, Y.H., and Chan, K.M. (1988). Epstein-Barr virus (EBV)-associated undifferentiated carcinoma of the parotid gland. Histopathology *13*, 509-517.

Huang, J., Chen, H., Hutt-Fletcher, L., Ambinder, R.F., and Hayward, S.D. (2003). Lytic viral replication as a contributor to the detection of Epstein-Barr virus in breast cancer. J. Virol. *77*, 13267-13274.

Hummel, M., Anagnostopoulos, I., Korbjuhn. P., and Stein, H. (1995). Epstein-Barr virus in B-cell non-Hodgkin's lymphomas: unexpected infection patterns and different infection incidence in low- and high-grade types. J. Pathol. *175*, 263-271.

Iamaroon, A., Khemaleelakul, U., Pongsiriwet, S., and Pintong, J. (2004). Co-expression of p53 and Ki67 and lack of EBV expression in oral squamous cell carcinoma. J. Oral Pathol. Med. *33*, 30-36.

Iamaroon, A., Pongsiriwet, S., Mahanupab, P., Kitikamthon, R., and Pintong, J. (2003). Oral non-Hodgkin lymphomas: studies of EBV and p53 expression. Oral Dis. *9*, 14-18.

Iezzoni, J.C., Gaffey, M.J., and Weiss, L.M. (1995). The role of Epstein-Barr virus in lymphoepithelioma-like carcinomas. Am. J. Clin. Pathol. *103*, 308-315.

Jen, K.Y., Cheng, J., Li, J., Wu, L., Li, Y., Yu, S., Lin, H., Chen, Z., Gurtsevitch, V., and Saku, T. (2003). Mutational events in LMP1 gene of Epstein-Barr virus in salivary gland lymphoepithelial carcinomas. Int. J. Cancer *105*, 654-660.

Karajannis, M.A., Hummel, M, Anagnostopoulos, I., and Stein, H. (1997). Strict lymphotropism of Epstein-Barr virus during acute infectious mononucleosis in nonimmunocompromised individuals. Blood *89*, 2856-2862.

Kelly, B.G., Lok, S.S., Hasleton, P.S., Egan, J.J., and Stewart, J.P. (2002). A rearranged form of Epstein-Barr virus DNA is associated with idiopathic pulmonary fibrosis. Am. J. Respir. Crit. Care Med. *166,* 510-513.

Kobayashi, I., Shima, K., Saito, I., Kiyoshima, T., Matsuo, K., Ozeki, S., Ohishi, M., and Sakai, H. (1999). Prevalence of Epstein-Barr virus in oral squamous cell carcinoma. J. Pathol. *189*, 34-39.

Kotsianti, A., Costopoulos, J., Morgello, S., and Papadimitriou, C. (1996). Undifferentiated carcinoma of the parotid gland in a white patient: detection of Epstein-Barr virus by in situ hybridization. Hum. Pathol. *27*, 87-90.

Kuo, T. and Hsueh, C. (1997). Lymphoepithelioma-like salivary gland carcinoma in Taiwan: a clinicopathological study of nine cases demonstrating a strong association with Epstein-Barr virus. Histopathology *31*, 75-82.

Lanier, A.P., Clift, S.R., Bornkamm, G., Henle, W., Goepfert, H., and Raab-Traub, N. (1991). Epstein-Barr virus and malignant lymphoepithelial lesions of the salivary gland. Arctic Med. Res. *50*, 55-61.

Lemon, S.M., Hutt, L.M., Shaw, J.E., Li, J.-L.H., and Pagano, J.S. (1997) Replication of EBV in epithelial cells during infectious mononucleosis. Nature *268,* 268-270.

Leong, I.T., Fernandes, B.J., and Mock, D. (2001). Epstein-Barr virus detection in non-Hodgkin's lymphoma of the oral cavity: an immunocytochemical and in situ hybridization study. Oral Surg. Oral Med. Oral Pathol. Oral Radiol. Endod. *92*, 184-193.

Leung, S.Y., Chung, L.P., Yuen, S.T., Ho, C.M., Wong, M.P., and Chan, S.Y. (1995). Lymphoepithelial carcinoma of the salivary gland: in situ detection of Epstein-Barr virus. J. Clin. Pathol. *48*, 1022-1027.

Lo, K.W., To, K.F., and Huang, D.P. (2004). Focus on nasopharyngeal carcinoma. Cancer Cell *5*, 423-428.

Maeda, T., Hiranuma, H., Matsumura, S., Furukawa, S., and Fuchihata, H. (1998). Epstein-Barr virus infection and response to radiotherapy in squamous cell carcinoma of the oral cavity. Cancer Lett. *125*, 25-30.

Mao, E.J., and Smith, C.J. (1993). Detection of Epstein-Barr virus (EBV) DNA by the polymerase chain reaction (PCR) in oral smears from healthy individuals and patients with squamous cell carcinoma. J. Oral Pathol. Med. *22*, 12-17.

Midgley, R.S., Blake, N.W., Yao, Q.Y., Croom-Carter, D., Cheung, S.T., Leung, S.F., Chan, A.T.C., Johnson, P.J., Huang, D., Rickinson, A.B., and Lee, S.P. (2000). Novel intertypic recombinants of Epstein-Barr virus in the Chinese population. J. Virol. *74,* 1544-1548.

Miller, G., Niederman, J.C., and Andrews, L.L. (1973). Prolonged oropharyngeal excretion of Epstein-Barr virus after infectious mononucleosis. N. Engl. J. Med. *288,* 229-232.

Mizugaki, Y., Sugawara, Y., Shinozaki, F., and Takada, K. (1998). Detection of Epstein-Barr virus in oral papilloma. Jpn. J. Cancer Res. *89*, 604-607.

Moody, C.A., Scott, R.S., Su, T., Sixbey, J.W. (2003). Length of Epstein-Barr virus termini as a determinant of epithelial cell clonal emergence. J.Virol. *77*, 8555-8561.

Niedobitek, G., Agathanggelou, A., Steven, N., and Young, L.S. (2000). Epstein-Barr virus (EBV) in infectious mononucleosis: detection of the virus in tonsillar B lymphocytes but not in desquamated oropharyngeal epithelial cells. Mol. Pathol. *53*, 37-42.

Pathmanathan, R., Prasad, U., Sadler, R., Flynn, K., and Raab-Traub, N. (1995). Clonal proliferations of cells infected with Epstein-Barr virus in pre-invasive lesions related to nasopharyngeal carcinoma. N. Engl. J. Med. *333,* 693-698.

Patton, D.F., Shirley, P., Raab-Traub, N., Resnick, L., and Sixbey, J.W. (1990). Defective viral DNA in Epstein-Barr virus-associated oral hairy leukoplakia. J. Virol. *64*, 397-400.

Pegtel, D.M., Middeldorp, J., and Thorley-Lawson, D.A. (2004). Epstein-Barr virus in *ex-vivo* tonsil epithelial cell cultures of asymptomatic carriers. J. Virol. 78, 12613-12624.

Raab-Traub, N. (2002). Epstein-Barr virus in the pathogenesis of NPC. Semin. Cancer Biol. *12*, 431-441.

Raab-Traub, N. and Flynn, K. (1986). The structure of the termini of the Epstein-Barr virus as a marker of clonal cellular proliferation. Cell *47*, 883-889.

Raab-Traub, N., Rajadurai, P., Flynn, K., and Lanier, A.P. (1991). Epstein-Barr virus infection in carcinoma of the salivary gland. J. Virol. *65*, 7032-7036.

Saemundsen, A.K., Albeck, H., Hansen, J.P., Nielsen, N.H., Anvret, M., Henle, W., Henle, G., Thomsen, K.A., Kristensen, H.K., and Klein, G. (1982). Epstein-Barr virus in nasopharyngeal and salivary gland carcinomas of Greenland Eskimoes. Br. J. Cancer *46*, 721-728.

Sand, L.P., Jalouli, J., Larsson, P.A., and Hirsch, J.M. (2002). Prevalence of Epstein-Barr virus in oral squamous cell carcinoma, oral lichen planus, and normal oral mucosa. Oral Surg. Oral Med. Oral Pathol. Oral Radiol. Endod. *93*, 586-592.

Sheen, T.S., Tsai, C.C., Ko, J.Y., Chang, Y.L., and Hsu, M.M. (1997). Undifferentiated carcinoma of the major salivary glands. Cancer *80*, 357-363.

Shibata, D., Weiss, L.M., Hernandez, A.M., Nathwani, B.N., Bernstein, L., and Levine, A.M. (1993). Epstein-Barr virus-associated non-Hodgkin's lymphoma in patients infected with human immunodeficiency virus. Blood *83*, 2102-2109.

Shimakage, M., Horii, K., Tempaku, A., Kakudo, K., Shirasaka, T., and Sasagawa, T. (2002). Association of Epstein-Barr virus with oral cancers. Hum. Pathol. *33*, 608-614.

Sitki-Green, D., Edwards, R.H., Webster-Cyriaque, J., and Raab-Traub, N. (2002). Identification of Epstein-Barr virus strain variants in hairy leukoplakia and peripheral blood by use of a heteroduplex tracking assay. J. Virol. *76*, 9645-9656.

Sixbey, J.W., Nedrud, J.G., Raab-Traub, N., Hanes R.A., and Pagano, J.S. (1984). Epstein-Barr virus replication in oropharyngeal epithelial cells. N. Engl. J. Med. *310*, 1225-1230.

Solomides, C.C., Miller, A.S., Christman, R.A., Talwar, J., and Simpkins, H. (2002). Lymphomas of the oral cavity: histology, immunologic type, and incidence of Epstein-Barr virus infection. Human Pathol. *33*, 153-157.

Song, J.I. and Rubin Grandis, J. (2000). STAT signaling in head and neck cancer. Oncogene *19*, 2489-2495.

Speight, P. M. and Barrett, A. W. (2002). Salivary gland tumours. Oral Dis. *8*, 229-240.

Sugawara, Y., Mizugaki, Y., Uchida, T., Torii, T., Imai, S., Makuuchi, M., and Takada, K. (1999). Detection of Epstein-Barr virus (EBV) in hepatocellular carcinoma tissue: a novel EBV latency characterized by the absence of EBV-encoded small RNA expression. Virology *256*, 196-202.

Szkaradkiewicz, A., Kruk-Zagajewska, A., Wal, M., Jopek, A., Wierzbicka, M., and Kuch, A. (2002). Epstein-Barr virus and human papillomavirus infections and oropharyngeal squamous cell carcinomas. Clin. Exp. Med. *2*, 137-141.

Tabor, M.P., Brakenhoff, R.H., van Houten, V.M.M., Kummer, J.A., Snel, M.H.J., Snijders, P.J.F., Snow, G.B., Leemans, C.R., and Braakhuis, B.J.M. (2001). Persistence of genetically altered fields in head and neck cancer patients: biological and clinical implications. Clin. Cancer Res. *7*, 1523-1532.

Talacko, A.A., Teo, C.G., Griffin, B.E., and Johnson, N.W. (1991). Epstein-Barr virus receptors but not viral DNA are present in normal and malignant oral epithelium. J. Oral Pathol. Med. *20*, 20-25.

Thomas, J.A., Cotter, F., Hanby, A.M., Long, L.Q., Morgan, P.R., Bramble, B., and Bailey, B.M.W. (1993). Epstein-Barr virus-related oral T-cell lymphoma associated with human immunodeficiency virus immunosuppression. Blood *81*, 3350-3356.

Thorley-Lawson, D.A., Miyashita, E.M., and Khan, G. (1996). Epstein-Barr virus and the B cell: that's all it takes. Trends Microbiol. *4*, 204- 208.

Triantos, D., Porter, S.R., Scully, C., and Teo, C.G. (1997). Oral hairy leukoplakia: clinicopathologic features, pathogenesis, diagnosis, and clinical significance. Clin. Infect. Dis. *25*, 1392-1396.

Tsai, C.C., Chen, C.L., and Hsu, H.C. (1996). Expression of Epstein-Barr virus in carcinomas of major salivary glands: a strong association with lymphoepithelioma-like carcinoma. Hum. Pathol. *27*, 258-262.

Tsang, N.M., Chang, K.P., Lin, S.Y., Hao, S.P., Tseng, C.K., Kuo, T T., Tsai, M.H., and Chung, T.C. (2003). Detection of Epstein-Barr virus-derived latent membrane protein-1 gene in various head and neck cancers: is it specific for nasopharyngeal carcinoma? Laryngoscope *113*, 1050-1054.

Tugizov, S.M., Berline, J.W., Palefsky, J.M. (2003). Epstein-Barr virus infection of polarized tongue and nasopharyngeal epithelial cells. Nat. Med. *9*, 307-314.

van Heerden, W.E., van Rensburg, E.J., Engelbrecht, S., and Raubenheimer, E.J. (1995). Prevalence of EBV in oral squamous cell carcinomas in young patients. Anticancer Res. *15*, 2335-2339.

Venkateswaran, L., Gan, Y.J., Sixbey, J.W., and Santana, V.M. (2000). Epstein-Barr virus infection in salivary gland tumors in children and young adults. Cancer *89*, 463-466.

Walling, D.M., Edmiston, S.N., Sixbey, J.W., Abdel-Hamid, M., Resnick, L., and Raab-Traub, N. (1992). Coinfection with multiple strains of the Epstein-Barr virus in human immunodeficiency virus-associated hairy leukoplakia. Proc. Natl. Acad. Sci. USA *89*, 6560-6564.

Walling, D.M., Etienne, W., Ray, A.J., Flaitz, C.M., and Nichols, C.M. (2004b). Persistence and transition of Epstein-Barr virus genotypes in the pathogenesis of oral hairy leukoplakia. J. Infect. Dis. *190*, 387-395.

Walling, D.M., Flaitz, C.M., and Nichols, C.M. (2003). Epstein-Barr virus replication in oral hairy leukoplakia: response, persistence, and resistance to treatment with valacyclovir. J. Infect. Dis. *188*, 883-890.

Walling, D.M., Flaitz, C.M., Nichols, C.M., Hudnall, S.D., Adler-Storthz, K. (2001). Persistent productive Epstein-Barr virus replication in normal epithelial cells *in vivo*. J Infect. Dis. *184*, 1499-1507.

Walling, D.M., Ling, P.D., Gordadze, A.V., Montes-Walters, M., Flaitz, C.M., and Nichols, C.M. (2004a). Expression of Epstein-Barr virus latent genes in oral epithelium: determinants of the pathogenesis of oral hairy leukoplakia. J. Infect. Dis. *190*, 396-399.

Walling, D.M., and Raab-Traub, N. (1994). Epstein-Barr virus intrastrain recombination in oral hairy leukoplakia. J. Virol. *68*, 7909-7917.

Webster-Cyriaque, J., Middeldorp, J., and Raab-Traub, N. (2000). Hairy leukoplakia: an unusual combination of transforming and permissive Epstein-Barr virus infections. J. Virol. *74*, 7610-7618.

Webster-Cyriaque, J. and Raab-Traub, N. (1998). Transcription of Epstein-Barr virus latent cycle genes in oral hairy leukoplakia. Virology *248*, 53-65.

Wen, S., Mizugaki, Y., Shinozaki, F., and Takada, K. (1997). Epstein-Barr virus (EBV) infection in salivary gland tumors: lytic EBV infection in nonmalignant epithelial cells surrounded by EBV-positive T-lymphoma cells. Virology *227*, 484-487.

Yeung, W.M., Zong, Y.S., Chiu, C.T., Chan, K.H., Sham, J.S., Choy, D.T., Ng, M.H. (1993). Epstein-Barr virus carriage by nasopharyngeal carcinoma in situ. Int. J. Cancer *53*, 746-750.

Ziegler, J.L., Beckstead, J.A., Volberding, P.A., Abrams, D.I., Levine, A.M., Lukes, R.J., Gill, P.S., Burkes, R.L., Meyer, P.R., Metroka, C.E., Mouradian, J., Moore, A., Riggs, S.A., Butler, J.J., Cabanillas, F.C., Hersh, E., Newell, G.R., Laubenstein, L.J., Knowles, D., Odajnyk, C., Raphael, B., Koziner, B., Urmacher, C., Clarkson, B.D. (1984). Non-Hodgkin's lymphoma in 90 homosexual men. N. Engl. J. Med. *311,* 565-570.

zur Hausen, A., van Rees, B.P., van Beck, J., Craanen, M.E., Bloemena, E., Offerhaus, G.J.A., Meijer, C.J.L.M., and van den Brule, A.J.C. (2004). Epstein-Barr virus in gastric carcinoma and gastric stump carcinomas: a late event in gastric carcinogenesis. J. Clin. Pathol. *57,* 487-491.

Chapter 7

Epstein Barr Virus in the Pathogenesis of NPC

*Nancy Raab-Traub**

ABSTRACT

Nasopharyngeal carcinoma (NPC) is an epithelial tumor that is characterized by marked geographic and population differences in incidence and is consistently associated with the Epstein-Barr virus. The EBV genome is clonal within the tumor suggesting that the tumor is a clonal proliferation of an EBV infected cell. Within the tumor, the viral infection is latent with expression of the EBV proteins EBNA1, LMP1, LMP2 and the EBER and BamHI A RNAs. LMP1 and LMP2 have profound effects on cell growth regulation. LMP1 activates NFkB and a specific form of NFkB is detected in NPC tumors. LMP2 activates the Akt kinase which inactivates the GSK3β kinase and induces β-catenin regulated expression in epithelial cells. In NPC tumors activated GSK3β and nuclear β-catenin are detected. the malignant cells continue to express multiple viral proteins. The continued expression of these viral proteins provides multiple opportunities to target the viral proteins and their properties using immuno-therapy, inhibitors of critical activated pathways, or specific molecular therapy directed toward the viral functions.

INTRODUCTION

Epstein-Barr virus (EBV) is a ubiquitous infectious agent, infecting greater than 90% of the world's population (Rickinson, 2001). However, EBV is also a major human tumor virus that is linked to the development of many specific cancers. Many of the malignancies associated with EBV develop in the setting of immunosuppression or have endemic patterns of incidence (de The et al., 1975; Raab-Traub, 1996). Malignancies associated with immunosuppression include post-transplant and AIDS-associated lymphomas. Endemic patterns of incidences characterize the unusual childhood tumor, African Burkitt's lymphoma (BL) from which EBV was originally isolated, the epithelial cancer nasopharyngeal carcinoma (NPC), carcinoma of the parotid gland, and T cell lymphomas, including nasal midline granuloma and peripheral T cell lymphoma. Cancers linked to EBV also occur sporadically throughout the world and approximately 50% of Hodgkin's lymphomas and 10% of gastric carcinomas contain EBV. As EBV infection is

*For correspondence email nrt@med.unc.edu

almost universal, the development of cancer in a subset of infected individuals suggests that there are contributing factors and perhaps unique aspects of infection in those who develop cancers. The study of EBV and its associated tumors points to specific interactions between environmental, genetic, and viral factors.

Much of our understanding of the pathogenesis of NPC has been revealed by the study of biopsy samples of NPC (Raab-Traub, 2002). Such studies have identified key genetic components in the development of NPC and have provided insight to the contribution of EBV to the development of this cancer (Huang D, 1994; Huang et al., 1991; Raab-Traub and Flynn, 1986; Raab-Traub et al., 1983). Initial studies revealed some surprising differences in viral expression from B-lymphocytes transformed *in vitro* and ongoing studies continue to find differences between effects of EBV proteins on epithelial and lymphoid cell growth. This chapter will review the characteristics of EBV infection in NPC; the effects of the individual viral proteins expressed in NPC on epithelial cell growth regulation, and summarize our current understanding of virologic and genetic aspects that contribute to the genesis of NPC.

NASOPHARYNGEAL CARCINOMA

Nasopharyngeal carcinoma (NPC) is an epithelial tumor that is characterized by marked geographic and population differences in incidence (Nielsen et al., 1977; Shanmugaratnam, 1970). It develops with exceptionally high incidence in southern China and Southeast Asia where it may represent 20% of all cancer cases, occurring at a rate of $100/10^5$/year in some regions . The tumor also frequently develops in Eskimo populations and occurs with elevated incidence in Mediterranean Africa (Nielsen et al., 1977). The incidence of NPC is low in Western populations where it is only 0.25% of all cancers occurring with a rate of $0.1/10^5$/yr (Buell, 1974). However in contrast to the relationship of EBV to Burkitt's lymphoma, EBV is consistently detected in NPC regardless of geographic distribution or racial background (Raab-Traub et al., 1987). The consistent association with a ubiquitous herpesvirus and the remarkable patterns of incidence suggests that other co-factors contribute to the development of this cancer.

Early seroepidemiologic studies indicated that, similarly to patients with BL, patients with NPC had elevated antibody titers to the EBV capsid antigen VCA and a group of antigens synthesized early in the viral replication process (EA) (Henle et al., 1973). However, NPC patients uniquely had elevated IgA antibodies to these antigens (Henle and Henle, 1975). Detection of IgA antibodies has been of great diagnostic and prognostic value as the occurrence may predict the development of NPC by several years and also prognostically correlates with tumor burden and recurrence. Subsequent molecular analyses revealed that the viral DNA and the EBV nuclear antigen, EBNA, were present within the malignant epithelial cells rather than in the abundant infiltrating lymphoid cells (Wolf et al., 1975). This was the first detection of EBV within epithelial cells.

The tumor presents with varying degrees of differentiation and has been classified by the World Health Organization into three categories (Shanmugaratnam, 1978). Squamous cell carcinomas, WHO1 tumors, are highly differentiated with characteristic epithelial growth patterns and keratin filaments. Non-keratinizing WHO2 carcinomas retain epithelial cell shape and growth patterns. Undifferentiated carcinomas, WHO3, do not produce keratin and lack a distinctive growth pattern. In addition many tumors have mixed degrees of differentiation or may present as WHO3 at the primary site with increased differentiation in metastatic lymph nodes. Initial studies indicated that patients with WHO2 and WHO3 tumors had elevated IgG and IgA titers to VCA and EA whereas the patients with WHO1 tumors had EBV serologic profiles similar to control populations (Krueger et al., 1981; Pearson et al., 1983). Other studies suggested that patients with all three forms of NPC had elevated IgA titers to VCA (Sam et al., 1989).

Unlike BL, all samples of undifferentiated NPC from endemic areas of high incidence to samples from areas of intermediate or low incidence had EBV DNA (Desgranges et al., 1975; Pagano et al., 1975; Raab-Traub et al., 1987). The association of EBV with differentiated Type 1 NPC is less consistent. Differentiated NPC tumors represent less than 1% of NPC in endemic areas of high incidence; therefore fewer of these tumors were analyzed in early studies. With the availability of cloned probes, several studies have consistently detected EBV genomes in the squamous cell Type 1 NPC (Raab-Traub et al., 1987; Wu et al., 1991). while other studies using *in situ* hybridization did not detect the viral genome in squamous NPC (Niedobitek et al., 1991). In a study of NPC in Malaysia, EBV DNA was detected on Southern blots and EBER RNAs and LMP1 were detected by *in situ* hybridization and immunohistochemistry (Pathmanathan et al., 1995). In areas of differentiation, LMP1 and EBER expression were decreased. The decreased expression of these EBV markers may account for the reported differences in EBV association in differentiated NPC. The difference in association may also reflect differences in the sample populations as most studies in Oriental populations, where undifferentiated NPC is the prevalent type, have detected EBV in all forms of the tumor while studies of European or American NPC samples have had less consistent detection of EBV DNA in differentiated NPC (Niedobitek et al., 1991).

Latent, clonal EBV infection in NPC

In latent infection, EBV DNA persists as an extrachromosomal episome while in permissive infection; the episome is converted into replicative intermediates that are cleaved into linear DNA that is packaged into virions (Pritchett et al., 1975). In NPC samples, EBV episomes were initially detected by electron microscopy, indicating that NPC represented a latent infection (Kaschka-Dierich et al., 1976). The presence of episomal DNA was confirmed by using restriction enzyme analysis to distinguish linear from circular forms of DNA. A homologous direct repeat 500 bp sequence is present at the termini of the linear DNA (Given et al., 1979). The number of terminal repeats (TR) differs between individual DNA molecules such that identification of the restriction enzyme fragments that contain the TR are heterogeneous in size. On Southern blots, the terminal restriction fragments form a ladder array representing molecules that have differing numbers of the TR element. Upon entry into the cell, the viral DNA circularizes through the TR to form the intracellular, extrachromosomal episome. This circularization produces a fused restriction enzyme fragment that can be identified with DNA probes representing unique DNA adjacent to the TR from either ends of the linear genome. A single fused restriction enzyme fragment representing the EBV episomal DNA was detected in all samples of NPC and in BL (Raab-Traub and Flynn, 1986). This finding indicated that within each cell, every copy of the viral genome was identical with regard to the number of TR molecules and within the tumor, the EBV episomes in every cell were identical to those in other cells. This clonality of the viral genome indicated that the NPC tumors represent the outgrowth of a single cell which was infected with EBV prior to clonal expansion.

In addition, in some tumor samples, particularly from the endemic area, ladder arrays of fragments representing linear genomes were also detected (Raab-Traub and Flynn, 1986). This suggested that in an occasional cell within the tumor, the virus may reactivate from latent infection and produce virus. It is likely that this sporadic replication is the source of the antigenic stimulus that provokes the elevated antibody response to viral replicative antigens that are indicative of tumor development or relapse.

The presence of multiple copies of episomal EBV DNA has obscured the clear detection of integrated forms of EBV DNA. Pulse field electrophoresis studies have detected evidence of EBV integration in 4 of 17 NPC samples (Kripalani-Joshi and Law, 1994). It is possible that integration could affect cellular expression and contribute to tumor development; however, it is probably not a necessary event for development of NPC.

Premalignant lesions

In contrast to many cancers, early manifestations of malignancy in the development of NPC such as dysplasia, or carcinoma *in situ* (CIS) are rare. The very low incidence of coexistent nasopharyngeal intraepithelial neoplasia with invasive cancer, approximately 3%, the extreme rarity of lesions without concomitant carcinoma, and the follow-up data all indicated a rapid progression of the initiated cell through the dysplasia, carcinoma *in situ*, and invasive cancer sequence (Pathmanathan et al., 1995). Thus the biologic behavior of NPC is quite distinct from the malignant transformation and progression of mammary carcinoma or cancer of the uterine cervix where intraductal cancer of the breast or carcinoma *in situ* of the cervix may persist for years. EBV has been detected in all of the samples of preinvasive carcinoma *in situ* with all cells staining positive for LMP1 and with detection of the EBERs in the majority of cells. Expression of EBNA1, LMP1, LMP2A, and transcription from BamHI A was detected in all of the early lesions (Pathmanathan et al., 1995). In all of the samples tested, a single restriction enzyme fragment representing the fused termini of the EBV episome was detected on Southern blots. These data indicated that preinvasive lesions contain clonal EBV and represent a focal cellular growth that developed from an EBV infected cell. The rarity of preinvasive lesions and the more frequent detection of CIS concomitant with invasive cancer suggests that the clonal infection by EBV is an early event in the development of malignancy.

In other similar studies, examples of milder dysplasia were identified that contained EBV in only a portion of cells (Chan ACS, 2000). This finding suggested that EBV infection occurred after some stimulus affected cell growth. It is thought that EBV infection in epithelial cells usually results in viral replication. It is also possible that some preceding genetic change affects viral infection such that a latent EBV infection is established with expression of potent genes such as LMP1 and LMP2 that significantly affect cell growth. Recent findings have identified a loss of the p16 tumor suppressor in early, EBV-negative lesions (Lo, 2002). The expression of EBV transforming proteins, in combination with p16 loss, could result in rapid progression to invasive malignancy.

EBV expression in NPC

Initial studies of EBV transcription in NPC identified differences between EBV expression in transformed lymphocytes and in NPC in that sequences later shown to encode EBNA2 and EBNA3 were not transcribed (Raab-Traub et al., 1983). Multiple studies have identified transcription of the EBV-encoded small nuclear RNAs (EBERs), the genes encoding EBNA1, LMP1, and LMP2, and previously unidentified rightward transcripts through the *Bam*HI A restriction fragment (Busson et al., 1992; Gilligan et al., 1990; 1991; Pathmanathan et al., 1995; Tsai et al., 1996). This state of expression is characteristic of many of the EBV associated tumors and is termed Latency 2. At the protein level EBNA1 has been identified and LMP1 and LMP2 have been detected in approximately 50% of tumors (Heussinger, 2004; Young et al., 1988). It is possible that LMP1 and LMP2 are expressed below the level of detection or that as the cancer evolves, expression of the viral proteins is no longer required.

The underlying molecular reasons for some of the differences in expression have been identified. One factor is different promoter usage for EBNA expression. In latency 1 and 2, a promoter in BamHI F/Q is used for EBNA1 transcription, while in latency 3, expression of EBNA2 apparently activates a promoter in BamHI C for expression of all the EBNAs (Henkel et al., 1994; Smith and Griffin, 1992). In lymphoid cells, LMP1 expression is dependent upon transactivation of its promoter by EBNA2 and EBNA-LP (Wang et al., 1990). The factors that regulate the usual LMP1 promoter in epithelial cells are not known, however in NPC and in an unusual epithelial lesion that develops in AIDs patients, hairy leukoplakia, LMP1 is expressed from a larger mRNA that initiates within the first terminal repeat from a promoter that is regulated by SP1 and STAT3 (Gilligan et al., 1990; Sadler and Raab-Traub, 1995). This promoter is considerably more active in epithelial cells and STAT3 has been shown to be constitutively activated in NPC (Chen, 2003). Thus factors that lead to the activation of SP1 or STAT3 may contribute to the development of NPC by inducing expression of LMP1.

Latent membrane protein 1

LMP1 is considered the EBV oncogene as it has transforming ability in rodent fibroblasts, transforming Rat-1 cells *in vitro* to anchorage-independent growth and tumorigenicity in nude mice (Wang et al., 1985). Expression of LMP1 also has significant effects on epithelial cell growth and inhibits differentiation and induces morphologic transformation of some cell lines (Dawson et al., 1990; Fahraeus et al., 1992). LMP1 induces expression of the epidermal growth factor receptor (EGFR) in epithelial cells and EGFR is expressed at high levels in NPC (Miller et al., 1995). The induction of EGFR expression in NPC may be an important contributing factor to transformation in this epithelial tumor. LMP1 also induces expression of CD40 and secretion of IL-6 in epithelial cells and decreases expression of cytokeratins

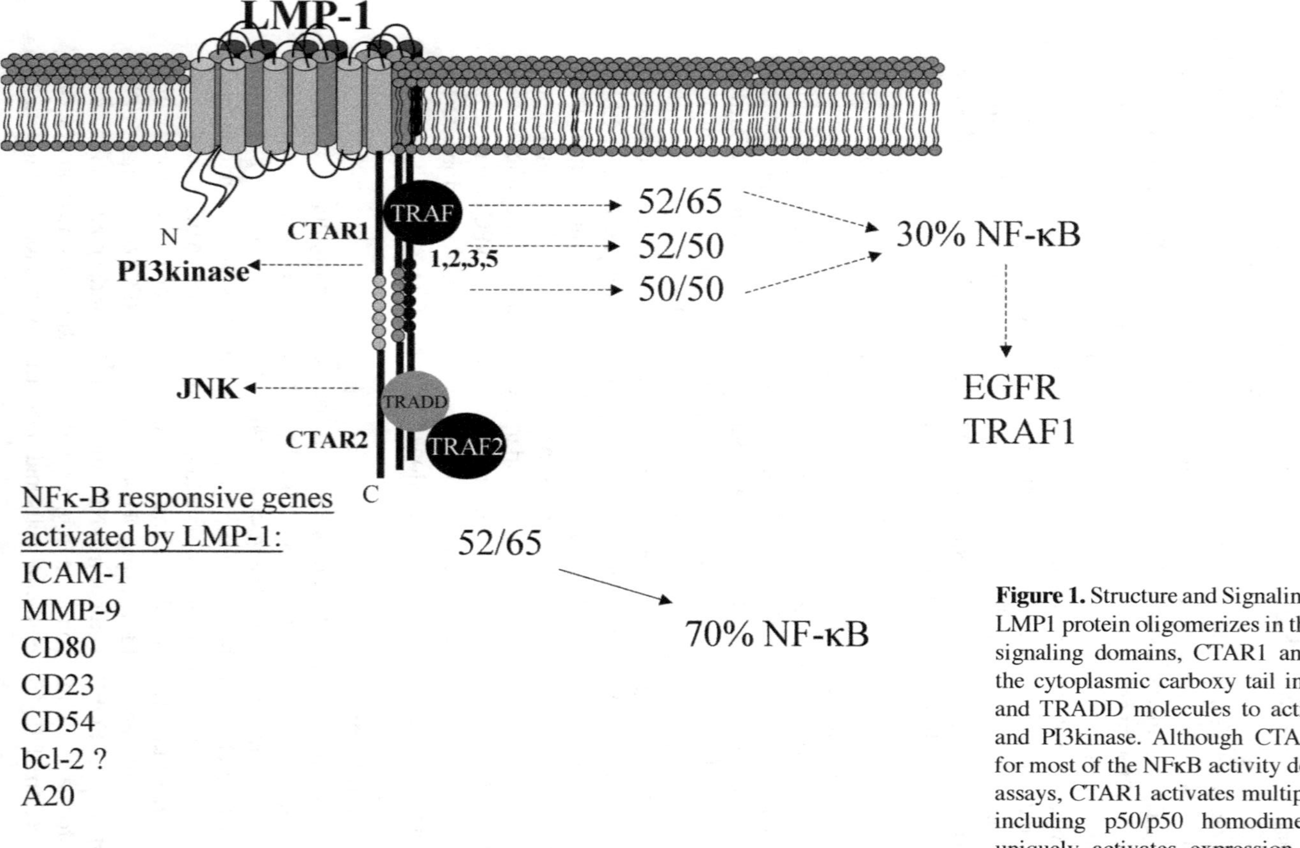

Figure 1. Structure and Signaling from LMP1. The LMP1 protein oligomerizes in the membrane. Two signaling domains, CTAR1 and CTAR2, within the cytoplasmic carboxy tail interact with TRAF and TRADD molecules to activate NFκB, JNK, and PI3kinase. Although CTAR2 is responsible for most of the NFκB activity detected by reporter assays, CTAR1 activates multiple forms of NFκB including p50/p50 homodimers. CTAR1 also uniquely activates expression of the epidermal growth factor receptor (EGFR) and TRAF1.

and E-cadherin (Fahraeus et al., 1990). LMP1 has been shown to inhibit apoptosis in B lymphocytes which may be due to from its induction of expression of the bcl-2 oncogene (Henderson et al., 1991). In epithelial cells, LMP1 specifically inhibits p53-mediated apoptosis but not p53-induced cell cycle arrest through induction of expression of the A20 ring finger protein (Fries et al., 1996). This protection from p53-mediated apoptosis may be responsible for the lack of p53 mutations in EBV associated cancers that express LMP1 such as nasopharyngeal carcinoma.

LMP1 is an integral membrane protein with a complex molecular structure containing a cytoplasmic amino terminus, six transmembrane domains, and a long cytoplasmic carboxy terminal portion (Figure 1) (Fennewald et al., 1984). LMP1 self aggregates through its transmembrane domains and apparently acts as a constitutively activated member of the TNF receptor family with activation of NFκB, JNK, and PI3 kinase (Mosialos et al., 1995). Two domains, CTAR1 and CTAR2, have been identified within the cytoplasmic carboxy terminus of LMP1, that both can activate the NFκB transcription factor (Figure 1) (Huen et al., 1995). The membrane proximal domain, CTAR1, interacts with the sane cellular molecules that mediate signals from the tumor necrosis factor family of receptors including CD40 (Eliopoulos et al., 1997). These molecules, entitled TRAFs, form heteromeric complexes that transduce signals that depending on the receptor may activate NFκB, induce cellular growth, or induce apoptosis. TRAF1, 2, and 3 assemble on the TRAF interacting domain in CTAR1, while the TRAF adaptor proteins, TRADD and RIP, bind to CTAR2 and interact with TRAF2 (Izumi and Kieff, 1997; Miller et al., 1998; Mosialos et al., 1995).

Although CTAR2 has the most pronounced activation of NFκB in reporter assays, CTAR1 activates multiple distinct forms of NFκB. In epithelial cells CTAR1 activates p50/p50 homodimers and p50/p52 and p52/p65 heterodimers while CTAR2 only activates p52/p65 heterodimers (Miller et al., 1998). CTAR1 is necessary for B-lymphocyte transformation while CTAR2 is dispensable . CTAR1 also retains the unique ability to upregulate the epidermal growth factor receptor (EGFR) through its interactions with TRAFs (Miller et al., 1998; Miller et al., 1997). Because only LMP1 CTAR1 activates p50/p50 homodimers and p50/p52 heterodimers, these dimeric forms may contribute to the CTAR1-specific effects on the EGFR. Although neither p50 or p52 have transactivation domains, both can bind the proto-oncogene, Bcl-3 (Bours, 1993; Nolan, 1993). Bcl-3 is an unusual member of the IκB family in that it is most commonly expressed in the nucleus and it has a transactivation domain that can provide transactivating function to bound p50 or p52. In NPC, analysis of NFκB activation using electrophoretic mobility shift assays (EMSA), identified specific activation of p50/p50 homodimers that were in a complex with Bcl-3 (Thornburg et al., 2003). Using chromatin immunoprecipitation (Chip), the p50/p50/Bcl-3 complexes were shown to be bound to the NF-κB sites within the *egfr* promoter. These findings suggest that the unique ability of CTAR1 to upregulate the EGFR is linked to its ability to activate p50/p50/bcl3 complexes. Interestingly, in the LMP1 positive NPC xenograft, C15, NF-κB was activated in the presence of LMP1 and IκBα was expressed at wild type levels while in the LMP1 negative xenograft, C17, NF-κB was activated in the absence of LMP1 but IκBα was not expressed (Thornburg et al., 2003). This is similar to the status of NF-κB in Hodgkin's lymphoma where NF-κB is activated in both EBV positive and negative samples via different molecular mechanisms (Emmerich et al., 1999). This activation can occur by expression of LMP1 or by expression of non-functional IκBα

LMP1-CTAR1 has also been shown to uniquely induce expression of two members of the inhibitor of DNA binding or inhibitor of differentiation (Id) family of proteins, Id1 and Id3 (Everly DN Jr, 2004). These proteins, Id1, Id2, Id3, and Id4, all contain helix-loop-helix domains that allow them to bind to helix-loop-helix transcription factors, called E-box proteins. However, the Id proteins lack DNA binding domains and act as dominant-negative transcription factors. Id proteins are highly expressed in proliferating cells and some cancers, and downregulated in differentiating cells and during senescence. Increased Id1 protein was detected by immunofluorescence in Rat-1 foci induced by transformation with LMP1 and Id1 and Id3 protein levels were increased in Rat-1 LMP1 stable cell lines. The Id proteins are potent regulators of cellular differentiation and cell cycle progression, and in the Rat-1 stable cell lines cdkI p27 protein levels were reduced while levels of cyclin dependent kinase 2 (Cdk2) and phosphorylated retinoblastoma (Rb) protein were increased. The LMP1 mediated effects upon Id proteins and cell cycle proteins likely contribute to transformation and oncogenesis. It is likely that the effects of LMP1 on Id proteins, downregulation of p27, increased Cdk2, and phosphorylation of Rb are important for the transformation of rodent fibroblasts and in the development of NPC.

Latent membrane protein 2

The LMP2 proteins are encoded by highly spliced mRNAs that contain exons located at both ends of the linear EBV genome (Sample et al., 1989). The two forms of LMP2 (LMP2A and 2B) differ in that only LMP2A has a 119 amino acid N-terminal cytoplasmic domain. This domain contains nine tyrosine residues with two of the tyrosines forming an immunoreceptor-tyrosine-based activation motif (ITAM) . The B and T cell receptors contain ITAMs that are phosphorylated by src family kinases and bind syk and ZAP70 leading to activation of multiple signal transduction cascades. LMP2A sequesters lyn and syk kinases and leads to their degradation. This block of lyn and syk inhibits lymphocyte activation which would activate the EBV replicative cascade. Thus, expression of LMP2A helps maintain a latent infection in B lymphocytes (Miller et al., 1995).

In epithelial cells, LMP2A becomes phosphorylated on tyrosines upon adhesion to extracellular matrix, ECM, proteins that are ligands for integrin receptors (Scholle et al., 1999). Integrins that are expressed in the proliferating, basal layer of the epithelium are in contact with the basement membrane and integrin ligands. As cells move from the basal layer to form the upper epithelium, integrin signaling is affected and the cells begin to differentiate. LMP2 inhibits differentiation of the HaCat epithelial cell line in organotypic raft cultures. The LMP2A expressing HaCat cells also formed colonies in soft agar indicating that they continued to proliferate without attachment to matrix and these cells formed aggressive tumors in nude mice (Scholle et al., 2000). The cells had activated Akt kinase suggesting activation of PI3 kinase and inhibition of PI3 kinase blocked their ability to grow

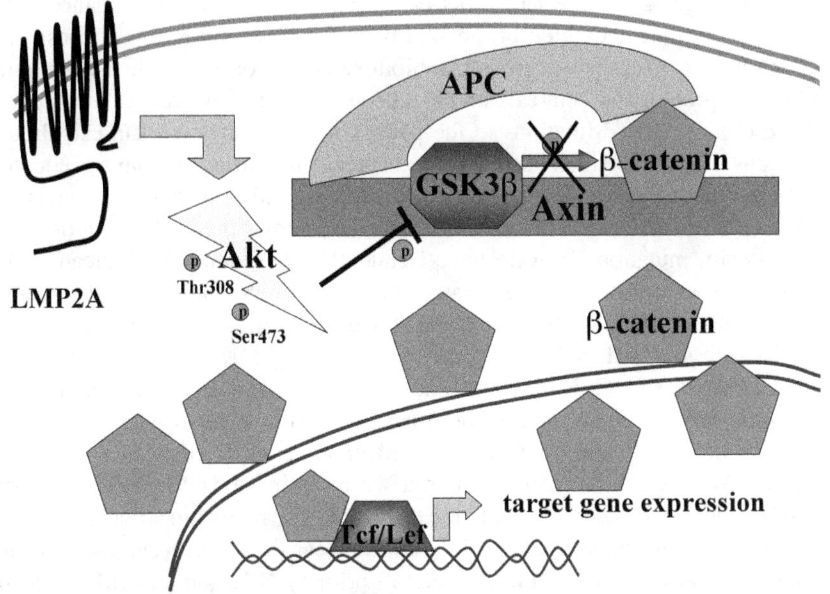

Figure 2. Activation of β-catenin signaling by LMP2. LMP2 activates the Akt kinase which phosphorylates and inactivates its target, GSK3β. The regulatory Axin complex binds APC, GSK3β, and β-catenin. Within this complex, GSK3β phosphorylates β-catenin which results in its degradation by the proteasome complex. Activation of Akt by LMP2 inactivates GSK3β and induces accumulation of β-catenin in the cytoplasm and its transport to the nucleus. In the nucleus, β-catenin activates the Tcf/Lef transcription factors to regulate cell gene expression.

in soft agar. Thus, in epithelial cell lines LMP2A may be a transforming protein possibly due to activation of Akt.

Targets of activated Akt include several pro-apoptotic proteins such as the Forkhead family of transcription factors, Bad, and pro-caspase 9. These and other pro-apoptotic proteins are inactivated upon phosphorylation by Akt, designating a prominent role for Akt in cell survival. Akt also plays a role in glucose metabolism. Activated Akt can phosphorylate and inactivate glycogen synthase kinase 3β (GSK3β), a serine/threonine kinase functions in the Wnt signaling pathway which is dysregulated in over 90% of human colon cancers. GSK3β exists in a complex in the cytoplasm with the tumor suppressors adenomatous polyposis coli (APC), axin and the proto-oncogene β-catenin (Figure 2). In the absence of Wnt signal, GSK3β constitutively phosphorylates β-catenin, leading to its ubiquitination and proteasome-mediated degradation. In the presence of a Wnt signal, signaling of the Disheveled protein is activated and leads to a block in the ability of GSK3β to associate with and phosphorylate β-catenin. As a result, β-catenin is stabilized, accumulates in the cytoplasm, and translocates into the nucleus. In the nucleus, β-catenin interacts with members of the T cell factor (TCF)/lymphoid enhancer factor (LEF) family of transcription factors to activate expression of target genes such as cyclin D1 and *c-myc*. Constitutive nuclear β-catenin and expression of

TCF/LEF target genes are hallmarks of the vast majority of colon cancers that have inappropriate activation of the Wnt pathway. APC and Axin are considered tumor suppressor genes due to their inhibitory effects on this pathway while the wnt ligand protein and beta-catenin have been shown to have oncogenic capacity (Morin et al., 1997). Mutations in the phosphorylation sites on beta-catenin and mutations in APC, which also affects beta catenin turnover, have frequently been identified in colon, prostate, and skin cancers and in medulloblastomas and hepatocellular carcinomas (Morin, 1999). In carcinomas that lack mutations APC or β-catenin, mutations in Axin have frequently been identified, indicating that constitutive activation of this pathway can occur through multiple genetic events.

In normal epithelial cells immortalized with the catalytic subunit of telomerase, LMP2 expression induced the activation of Akt via PI3K signaling and the Akt targets, FKHR and GSK3β, were phosphorylated. LMP2 expression resulted in dramatically increased levels of β-catenin in the cytoplasm and increased translocation of β-catenin to the nucleus (Morrison et al., 2003). The translocated β-catenin was functional and in reporter assays using a TCF-sensitive reporter, LMP2A increased activation of the reporter by β-catenin. Phosphorylated, activated Akt, phosphorylated GSK3β, and nuclear β-catenin were also detected in the xenografted NPC tumor, C15, and in primary NPC samples (Morrison JA, 2004). These findings indicate that the β-catenin/Wnt pathway is activated in NPC. Expression of LMP2 and activation of β-catenin regulated expression are likely important components of EBV-effects on epithelial cell growth.

A recent study also revealed potentially powerful effects of LMP2 on epithelial cell growth (Moody CA, 2003). In primary infected epithelial cultures, a few EBV infected clones rapidly overtook the culture. The clones that had this growth advantage had fewer numbers of TR and it was shown that this property that resulted in increased expression of LMP2. Thus cells that expressed more LMP2 as a result of fewer terminal repeats grew significantly faster in culture.

BamHI A rightward transcripts

A family of rightward transcripts from the *Bam*HI A region was initially identified in cDNA libraries from NPC (Gilligan et al., 1990; 1991; Hitt et al., 1989). These RNAs are transcribed rightward through the BamHI A fragments (BARTs) and are antisense to multiple replicative functions. At least three mRNAs are abundantly expressed in NPC and consistently detected on Northern blots, while in lymphoid cell lines; these RNAs are only detectable by PCR amplification of cDNAs (RT-PCR) (Gilligan et al., 1990; 1991; Smith et al., 1993). The transcripts are 3′-end co-terminal and all contain an open reading frame (ORF) *Bam*HI A rightward frame 0 (BARF0) at the 3′-end (Sadler and Raab-Traub, 1995). Cloning and sequencing cDNAs from the C15 tumor revealed patterns of alternate splicing of at least 7 exons that form several ORFs (Sadler and Raab-Traub, 1995; Smith et al., 2000).

It is unclear if the BamHI A mRNAs actually encode protein as none of the proteins have been detected. However, several interesting properties have been identified for the putative proteins. The BARF0 ORF is contained within all of the transcripts; however it is unusual in that the stop codon is embedded in the polyadenylation codon which might affect efficient protein synthesis from this mRNA. An additional cDNA, RK-BARF0, was identified that contained

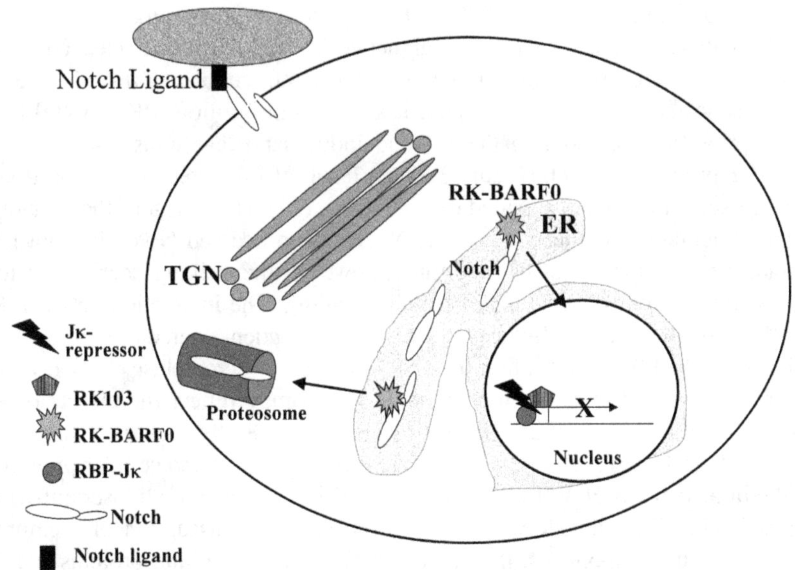

Figure 3. Inhibition of Notch Signaling by the BamHI A proteins, RK-BARF0 and RK103. The Notch receptor is synthesized and then processed in the trans Golgi network (TGN) where it is cleaved and transported to the plasma membrane. Upon interacting with Notch ligands, the intracellular portion of Notch is cleaved where it is transported to the nucleus to interact with the DNA binding protein, RBPJκ. The interaction of RK-BARF0 with full length Notch in the TGN induces its degradation by the proteasome complex. The RK103 protein (RPMS1) also interacts with RBPJκ and stabilizes its interaction with its repressor complex. Thus RK-BARF0 and RK103 inhibit Notch signaling by two distinct mechanisms.

sequences spliced into the 5′ end of the BARF0 ORF. These sequences potentially contain an endoplasmic reticulum (ER) targeting signal peptide sequence. Another ORF, originally called RPMS1 or RK103 contains a WW motif, similar to the sequences within the EBNA2 protein that interact with the cellular protein RBPJκ. Two cDNAs have been obtained, RB2 and A73 that potentially represent the most abundant RNA. These cDNAs contain a large ORF of 126 codons that is formed by four exons(Sadler and Raab-Traub, 1995).

To determine the possible functions of these ORFs, the yeast two-hybrid system has been used to identify by identifying interacting cellular proteins. The fragment of the RPMS1 or RK103 ORF that contains the WW motif does interact with RBPJκ as predicted (Zhang et al., 2001). The interaction of RPMS1 with RBPJκ may stabilize the RBPJκ interaction with the co-repressor complex and block EBNA2 or Notch effects on RBPJκ-regulated genes (Figure 3). The A73 or RB2 ORF interacts with the receptor for activated cell kinase (RACK1) with as yet unknown consequences on RACK1 or src function (Smith et al., 2000).

Yeast two hybrid analysis of the RK-BARF0 peptide identified several cellular proteins including the Notch family proteins, Notch3 and Notch4, human I-mfa domain containing protein (HIC), epithelin, and scramblase (Kusano and Raab-

Traub, 2001; Thornburg NJ, 2004). During Notch processing and signaling, Notch is processed in the transGolgi network, embedded and cleaved at the plasma membrane upon ligand binding, and the intracellular portion is released and translocated to the nucleus where it affects transcription. RK-BARF0 binds Notch prior to processing in the ER and induces nuclear translocation of full length, unprocessed Notch (Figure 3). As a result of this interaction, a portion of unprocessed Notch is translocated to the nucleus (17) while most of the remaining Notch is degraded by the proteasome. Nuclear unprocessed Notch has not been characterized and may not be functional; however, the striking decrease of total Notch protein would likely inhibit Notch signaling. The interaction between RK-BARF0 and epithelin also induced proteasome-dependent degradation of epithelin while RK-BARF0 did not affect the levels of HIC or scramblase. Low levels of Notch are detected in EBV infected cells which may reflect the effects of RK-BARF0 on Notch (Thornburg NJ, 2004). HIC was initially identified through its ability to bind HIV Tat and regulates expression from viral and cellular promoters. Epithelin also binds HIV Tat, and is involved in cellular growth. Scramblase is a non-specific lipid translocase and is responsible for disruption of membrane asymmetry during apoptosis. Interactions between these cellular proteins and RK-BARF0 could affect a wide variety of cellular functions including transcription, cell growth, and apoptosis.

The EBV encoded noncoding RNAs - EBERs

The most abundant RNAs in EBV infected cells are small nuclear EBER RNAs that are present at approximately 10^5 copies per cells but are not necessary for lymphocyte transformation (Arrand and Rymo, 1982; Swaminathan et al., 1992). The EBERs are expressed in many of the malignancies linked to EBV and presumably contribute in some way to the maintenance of latency *in vivo*. The EBERs are expressed in most examples of NPC but are not necessarily detected in every cell. Their expression is also apparently downregulated during differentiation such that NPC examples that have differing degrees of differentiation lack EBER expression in differentiated area (Pathmanathan et al., 1995). The EBER RNAs are also not detected in the permissive EBV infection, hairy leukoplakia, and are downregulated during viral replication (Gilligan et al., 1990).

Derivatives of the EBV positive Burkitt lymphoma cell line, Akata, can be isolated that have lost the EBV genome. The EBV negative Akata cells do not grow in soft agar and cannot form tumors in nude mice. Reinfection with EBV restores these properties (Komano et al., 1998). The viral protein(s) that are responsible for the tumorigenic phenotype have not been identified, however, it has been demonstrated that expression of the EBER RNAs partially restores growth transformation through effects on bcl2 (Komano et al., 1999). Expression of the EBERs also affected the growth properties of a gastric epithelial cell line through effects on the insulin growth factor complex 1 (Iwakiri et al., 2003). It is unknown if the EBERs similarly affect IGF1 in NPC.

EBV strain variation

The endemic patterns of incidence of NPC and other EBV-associated malignancies have prompted studies to discern if there are distinct strains of EBV with distinct biologic properties that might contribute to these differences in disease incidence. The EBV strains in epithelial and lymphoid malignancies from different geographic regions were initially distinguished by restriction enzyme polymorphisms (Abdel-Hamid et al., 1992). A strain that was prevalent in Chinese populations had lost the *Bam*HI restriction site between the W1' and I1' fragments and had an additional cut site within BamHI F (Lung et al., 1990). The Chinese strain was also marked by the loss of an Xho1 restriction enzyme polymorphism within the LMP1 gene (Hu et al., 1991).

Multiple studies have analyzed variants of LMP1 and the sequence variation in LMP1 was shown to vary independently of the EBV type determined by EBNA2 and EBNA3 sequence variation (Miller et al., 1994). Variation in LMP1 includes variable numbers of an 11-amino acid repeat element that is not strain specific, a possible 30 base pair deletion, and distinct signature sequence changes (Miller et al., 1994). Strains with this variation have been suggested to result in a more aggressive phenotype and LMP1 with this deletion has been shown to have greater transforming ability in some studies (Knecht et al., 1993). Sequence analysis identified seven variants that consistently associated in all the trees drawn from the phylogenetic analysis (Edwards et al., 1999). The two critical motifs that bind the tumor necrosis receptor associated factors; TRAF and TRADD were unchanged in all 84 isolates. NPC samples from the endemic region in southern China usually contained the strain termed China1 (Ch1) which has the 30 bp deletion. Another variant was also identified in some samples that was undeleted and was termed China2 (Ch2). The China2 variant was found in samples that contained either EBNA Type 1 or 2 (Sung et al., 1998). NPC samples from North Africa and Southern Europe frequently contained the Mediterranean strain (Med)

A heteroduplex tracking assay (HTA) was developed based on strain defining changes in the carboxy terminus of LMP1 in which each of seven strains had specific migration patterns. This approach revealed that most individuals harbored multiple EBV variants and that the relative abundance or presence of the strains differed over time (Sitki-Green et al., 2003). The prevalent EBV strain was usually distinct in the oral cavity from that in the peripheral blood; however, all strains are eventually detected in saliva and blood, suggesting transmission between the compartments. During infectious mononucleosis, individuals were also usually infected with multiple strains (Sitki-Green DL, 2004).

HTA analysis was used to identify and compare the LMP1 variant in NPC samples and in matching blood samples from the endemic southern region and nonendemic northern region of China (Edwards RH, 2004). The majority of the tumor samples had Ch1 while the blood samples from the endemic region largely contained the B95 strain. It was highly statistically significant that the B958 strain that was consistently detected in the blood was never detected in the tumors of the endemic NPCs (0/26) (p < 0.0001), and the Med- strain present in blood (4/11) was not detected in the endemic tumors (0/26) (p = 0.005). These compartmental differences suggested a possible selection against these strains in NPC. It is possible that immune selection of LMP1 variants would affect the ability of the

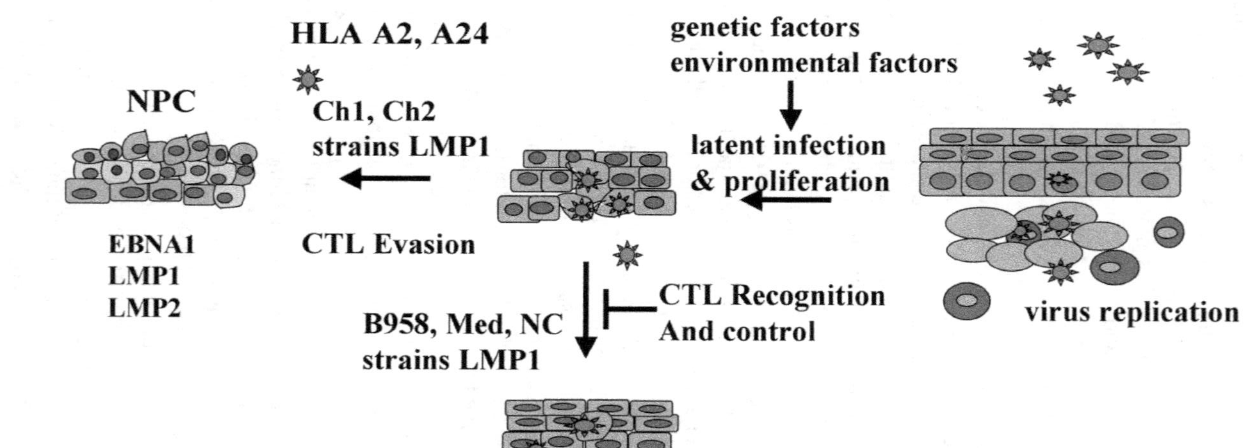

Figure 4. Potential Immune Selection of LMP1 Variants in NPC. The development of NPC is marked by elevated antibody titers to EBV replicative genes prior to the development of the cancer. This suggests that EBV reactivation and replication precedes NPC and may contribute to increased epithelial cell infection. Genetic and/or environmental factors possibly affect the outcome of infection and influence the establishment of a latent, transforming infection. Strains that express the China1 (Ch1) and China 2 (Ch2) LMP1 variants would not be presented by MHC HLA A2 or A24. Strains that express the B95, Mediterranean (Med) and North Carolina (NC) would be presented and the infected cells would be eliminated by cytotoxic lymphocytes (CTL).

B95 strain to express LMP1 and induce tumor formation. The known A2 restricted epitope, YLLEMLWRL (YLL) scored the highest of all computer-predicted LMP1 epitopes and was predicted not to be recognized in the Ch1, Ch2, AL, and NC strains. When the scores of the computer predicted epitopes from the full length LMP1 sequences of the tumors were compared to B958 sequence, which was not present in the tumors, marked differences could be found. The majority of NPC samples were HLA A24 or A2 restricted and many of the predicted A2 or A24 epitopes were changed in the Ch1 and Ch2 strains. Most interestingly, a single NPC sample from an HLA A2 person had the Mediterranean (Med-) variant. In this sample, the Med- LMP1 sequence had an additional non-consensus change at position 130 that reduced the predicted presentation score of this epitope. This finding suggested that the Med- strain in this NPC was possibly an LMP1 immune escape mutant (Edwards RH, 2004).

LMP1 is not expressed in latently infected peripheral blood lymphocytes, but is detected in a considerable proportion of NPC and in early, preinvasive examples of NPC (Pathmanathan et al., 1995). It is possible that infected epithelial cells that express LMP1 but are not recognized due to epitope changes in LMP1 would persist while those cells that become infected with variants whose LMP1 epitopes are presented would be eliminated (Figure 4). This negative selection of the immune system on strains detected in the blood would be reflected in the striking predominance of the Ch1 strain in the tumors. The additional mutation in the Med-strain in an A2 restricted sample suggests that sporadic mutations may also occur in LMP1 that enable immune escape.

These findings suggest that there the prevalence of specific variants of LMP1 in various malignancies may reflect an immune selection for strains of EBV that express an LMP1 that is not presented by certain HLA types. This type of selection could also contribute to the development of other cancers that are more sporadically associated with EBV where sporadic changes in combination with a particular HLA type would enable transformed cells expressing LMP1 to escape immune recognition.

CONTRIBUTING FACTORS

The association of a ubiquitous herpesvirus with a tumor that develops with striking patterns of incidence suggests that other genetic or environmental factors contribute to tumor development. One dietary component that has been suggested is exposure to salted fish at an early age (Zou XN, 1994). Tumor promoting compounds have also been identified in food products in other populations with elevated incidence.

Extensive surveys for activated oncogenes or tumor suppressor loss have not identified characteristic translocations, mutations in p53 or Rb alterations, or activating ras mutations (Raab-Traub, 2002). This suggests that viral genes affect these pathways, directly or indirectly, such that there is no selection for mutation in these genes. However, genetic changes could influence the establishment of latent infection, affect viral expression, or contribute to tumor progression and metastasis.

NPC samples have been analyzed in multiple detailed studies to identify loss of heterozygosity of cellular sequences. This approach has shown that the p16 cyclin dependent kinase inhibitor at 9p21 and the RASSF1a gene at 3p14 are frequently lost in NPC samples (Chow LS, 2004).

Reintroduction of the RASSF1a gene into an NPC cell line inhibited growth. This finding suggests that activation of this ras isoform is an important contributor to NPC development. In addition to loss of the gene, p16 may also not be expressed in NPC due to specific methylation of the promoter (Lo et al., 1995). The loss of p16 may be highly synergistic with EBV expression in the induction of NPC.

SUMMARY

The consistent expression of specific viral genes in every cell in NPC samples and in premalignant lesions suggests that these viral gene products contribute to the abnormal proliferation. It is perhaps the establishment of latent infection and the expression of these critical viral transforming functions in epithelial cells that are the rare events that lead to the development of NPC. After reactivation from latency in lymphoid cells, EBV may enter normal epithelial cells and replicate. In the development of NPC, the cascade of expression to viral replication does not proceed but instead stalls or switches and latent, transforming genes are expressed and an infected cell begins to proliferate. Preexisting genetic changes may affect the establishment of latent or replicative infection. Depending on the HLA type of the individual and the presentation of viral epitopes of proteins such as LMP1 and LMP2, the infected cells may be eliminated by cytotoxic lymphocytes or the proliferation may continue. The rarity of premalignant lesions and the concurrent detection of dysplasia and invasive carcinoma suggest that expression of the viral transforming genes is a critical event such that the abnormal growths are almost immediately malignant and invasive.

The compelling epidemiologic and molecular data indicate that EBV is an essential factor in the development of NPC as the genome is maintained and the malignant cells continue to express multiple viral proteins. This provides multiple opportunities to specifically target the viral proteins and their properties using immuno-therapy, inhibitors of critical activated pathways, or specific molecular therapy directed toward the viral functions. The continued study of determinants of immune control and the biochemical properties of the viral genes will in the long term enable us to control viral infection or eliminate the pathogenic consequences of infection. The identification of the interactions of viral and cellular proteins will also provide a new understanding of critical cellular pathways that are affected in human cancers.

References

Abdel-Hamid, M., Chen, J. J., Constantine, N., Massoud, M., and Raab-Traub, N. (1992). EBV strain variation: geographical distribution and relation to disease state. Virol. *190*, 168-175.

Arrand, J. R. and Rymo, L. (1982). Characterization of the major Epstein-Barr virus-specific RNA in Burkitt lymphoma-derived cells. J. Virol. *41*, 376-389.

Bours, V., Franzoso, G., Azarenko, V., Park, S., Kanno, T., Brown, K., and Siebenlist, U. (1993). The oncoprotein Bcl-3 directly transactivates through kappa B motifs via association with DNA-binding p50B homodimers. Cell *72*, 729-739.

Buell, P. (1974). The effect of migration on the risk of developing nasopharyngeal carcinoma. Cancer Res. *34*, 1189-1191.

Busson, P., McCoy, R., Sadler, R., Gilligan, K., Tursz, T., and Raab-Traub, N. (1992). Consistent transcription of the Epstein-Barr virus LMP2 gene in nasopharyngeal carcinoma. J. Virol. *66*, 3257-62.

Chan, ACS, T. K. and Lo, K.W. (2000). High frequency of chromosome 3p deletion in histologically normal nasopharyngeal epithelia from Southern Chinese. Cancer Res. *60*, 5365-5370.

Chen H, H.-F. L., Cao L., Hayward, S.D. (2003). A positive autoregulatory loop of LMP1 expression and STAT activation in epithelial cells latently infected with Epstein-Barr virus. J. Virol. *77*, 4139-4148.

Chow, L.S., L. K. Kwong, J. To, K.F., Tsang, K.S., Lam, C.W., Dammann, R., Huang, D.P. (2004). RASSF1A is a target tumor suppressor from 3p21.3 in nasopharyngeal carcinoma. Int. J. Cancer *109*, 839-847.

Dawson, C. W., Rickinson, A. B., and Young, L. S. (1990). Epstein-Barr virus latent membrane protein inhibits human epithelial cell differentiation. Nature *344*, 777-780.

de The, G., Day, N., Geser, A., Ho, J. H., Simons, M. J., Sohier, R., Tukei, P., and Vonka, V. (1975). Epidemiology of the Epstein-Barr virus infection and associated tumors in man. Bibl. Haematol. 216-220.

Desgranges, C., de-The, G., Wolf, H., and zur Hausen, H. (1975). Further studies on the detection of the Epstein-Barr virus DNA in nasopharyngeal carcinoma biopsies from different parts of the world. IARC Sci. Publ. *11*, 191-193.

Desgranges, C., Wolf, H., De-The, G., Shanmugaratnam, K., Cammoun, N., Ellouz, R., Klein, G., Lennert, K., Munoz, N., and zur Hausen, H. (1975). Nasopharyngeal carcinoma. X. Presence of epstein-barr genomes in separated epithelial cells of tumours in patients from Singapore, Tunisia and Kenya. Int. J. Cancer *16*, 7-15.

Edwards, R. H., Seillier-Moiseiwitsch, F., and Raab-Traub, N. (1999). Signature amino acid changes in latent membrane protein 1 distinguish Epstein-Barr virus strains. Virol. *261*, 79-95.

Edwards RH, S.-G. D., Moore DT, Raab-Traub N. (2004). Potential selection of LMP1 variants in nasopharyngeal carcinoma. J. Virol. *78*, 868-881.

Eliopoulos, A. G., Stack, M., Dawson, C. W., Kaye, K. M., Hodgkin, L., Sihota, S., Rowe, M., and Young, L. S. (1997). Epstein-Barr virus-encoded LMP1 and CD40 mediate IL-6 production in epithelial cells via an NF-kappaB pathway involving TNF receptor-associated factors. Oncogene *14*, 2899-2916.

Emmerich, F., Meiser, M., Hummel, M., Demel, G., Foss, H. D., Jundt, F., Mathas, S., Krappmann, D., Scheidereit, C., Stein, H., and Dorken, B. (1999). Overexpression of I kappa B alpha without inhibition of NF-kappaB activity and mutations in the I kappa B alpha gene in Reed-Sternberg cells. Blood *94*, 3129-3134.

Everly DN Jr, M. B., and Raab-Traub N. (2004). Induction of Id1 and Id3 by latent membrane protein 1 of Epstein-Barr virus and regulation of p27/Kip and cyclin-dependent kinase 2 in rodent fibroblast transformation. J. Virol. *78*, 13470-8.

Fahraeus, R., Chen, W., Trivedi, P., Klein, G., and Obrink, B. (1992). Decreased expression of E-cadherin and increased invasive capacity in EBV-LMP-transfected human epithelial and murine adenocarcinoma cells. Int. J. Cancer *52*, 834-838.

Fahraeus, R., Rymo, L., Rhim, J. S., and Klein, G. (1990). Morphological transformation of human keratinocytes expressing the LMP gene of Epstein-Barr virus. Nature *345*, 447-449.

Fennewald, S., van Santen, V., and Kieff, E. (1984). Nucleotide sequence of an mRNA transcribed in latent growth-transforming virus infection indicates that it may encode a membrane protein. J. Virol. *51*, 411-419.

Fries, K. L., Miller, W. E., and Raab-Traub, N. (1996). Epstein-Barr virus latent membrane protein 1 blocks p53-mediated apoptosis through the induction of the A20 gene. J. Virol. *70*, 8653-8659.

Gilligan, K., Rajadurai, P., Resnick, L., and Raab-Traub, N. (1990). Epstein-Barr virus small nuclear RNAs are not expressed in permissively infected cells in AIDS-associated leukoplakia. Proc. Natl. Acad. Sci. USA *87*, 8790-8794.

Gilligan, K., Sato, H., Rajadurai, P., Busson, P., Young, L., Rickinson, A., Tursz, T., and Raab-Traub, N. (1990). Novel transcription from the Epstein-Barr virus terminal *Eco*RI fragment, DIJhet, in a nasopharyngeal carcinoma. J. Virol. *64*, 4948-9456.

Gilligan, K. J., Rajadurai, P., Lin, J. C., Busson, P., Abdel-Hamid, M., Prasad, U., Tursz, T., and Raab-Traub, N. (1991). Expression of the Epstein-Barr virus BamHI A fragment in nasopharyngeal carcinoma: evidence for a viral protein expressed *in vivo*. J. Virol. *65*, 6252-6259.

Given, D., Yee, D., Griem, K., and Kieff, E. (1979). DNA of Epstein-Barr virus. V. Direct repeats of the ends of Epstein-Barr virus DNA. J. Virol. *30*, 852-862.

Henderson, S., Rowe, M., Gregory, C., Croom-Carter, D., Wang, F., Longnecker, R., Kieff, E., and Rickinson, A. (1991). Induction of bcl-2 expression by Epstein-Barr virus latent membrane protein 1 protects infected B cells from programmed cell death. Cell *65*, 1107-1115.

Henkel, T., Ling, P. D., Hayward, S. D., and Peterson, M. G. (1994). Mediation of Epstein-Barr virus EBNA2 transactivation by recombination signal-binding protein J kappa. Science *265*, 92-95.

Henle, G., and Henle, W. (1975). Serum IgA antibodies of Epstein-Barr virus (EBV)-related antigens. A new feature of nasopharyngeal carcinoma. Bibl. Haematol. 322-325.

Henle, W., Ho, H. C., Henle, G., and Kwan, H. C. (1973). Antibodies to Epstein-Barr virus-related antigens in nasopharyngeal carcinoma. Comparison of active cases with long-term survivors. J. Natl. Cancer Inst. *51*, 361-369.

Heussinger, N., B. M., Ott, G., Brachtel, E., Pilch, B.Z., Kremmer, E., and Niedobitek, G. (2004). Expression of the Epstein-Barr virus (EBV)-encoded latent membrane protein 2A (LMP2A) in EBV-associated nasopharyngeal carcinoma. J. Pathol. *203*, 696-699.

Hitt, M. M., Allday, M. J., Hara, T., Karran, L., Jones, M. D., Busson, P., Tursz, T., Ernberg, I., and Griffin, B. E. (1989). EBV gene expression in an NPC-related tumour. Embo J. *8*, 2639-2651.

Hu, L. F., Zabarovsky, E. R., Chen, F., Cao, S. L., Ernberg, I., Klein, G., and Winberg, G. (1991). Isolation and sequencing of the Epstein-Barr virus BNLF-1 gene (LMP1) from a Chinese nasopharyngeal carcinoma. J. Gen. Virol. *72*, 2399-2409.

Huang, D., L. K., van Hassel,t A., Woo, J.K.S., Choi, P.H.K., Leung, S.F., Cheung, S.T., Cairns, P., Sidransky, D., and Lee, J.C.K. (1994). A region of homozygous deletion on chromosome 9p21-22 in primary nasopharyngeal carcinoma. Cancer Res. *54*, 4003-4006.

Huang, D. P., Lo, K. W., Choi, P. H., Ng, A. Y., Tsao, S. Y., Yiu, G. K., and Lee, J. C. (1991). Loss of heterozygosity on the short arm of chromosome 3 in nasopharyngeal carcinoma. Cancer Genet. Cytogenet. *54*, 91-99.

Huen, D. S., Henderson, S. A., Croom-Carter, D., and Rowe, M. (1995). The Epstein-Barr virus latent membrane protein-1 (LMP1) mediates activation of NF-kappa B and cell surface phenotype via two effector regions in its carboxy-terminal cytoplasmic domain. Oncogene *10*, 549-560.

Iwakiri, D., Eizuru, Y., Tokunaga, M., and Takada, K. (2003). Autocrine growth of Epstein-Barr virus-positive gastric carcinoma cells mediated by an Epstein-Barr virus-encoded small RNA. Cancer Res. *63*, 7062-7067.

Izumi, K. M., and Kieff, E. D. (1997). The Epstein-Barr virus oncogene product latent membrane protein 1 engages the tumor necrosis factor receptor-associated death domain protein to mediate B lymphocyte growth transformation and activate NF-kappaB. Proc. Natl. Acad. Sci. USA *94*, 12592-12597.

Sadler, R. H., and Raab-Traub, N. (1995). The Epstein-Barr virus 3.5-kilobase latent membrane protein 1 mRNA initiates from a TATA-Less promoter within the first terminal repeat. J. Virol. *69*, 4577-4581.

Sadler, R. H., and Raab-Traub, N. (1995). Structural analyses of the Epstein-Barr virus BamHI A transcripts. J. Virol. *69*, 1132-1141.

Sam, C. K., Prasad, U., and Pathmanathan, R. (1989). Serological markers in the diagnosis of histopathological types of nasopharyngeal carcinoma. Eur J Surg Oncol *15*, 357-60.

Sample, J., Liebowitz, D., and Kieff, E. (1989). Two related Epstein-Barr virus membrane proteins are encoded by separate genes. J. Virol. *63*, 933-937.

Scholle, F., Bendt, K. M., and Raab-Traub, N. (2000). Epstein-Barr virus LMP2A transforms epithelial cells, inhibits cell differentiation, and activates Akt. J. Virol. *74*, 10681-10689.

Scholle, F., Longnecker, R., and Raab-Traub, N. (1999). Epithelial cell adhesion to extracellular matrix proteins induces tyrosine phosphorylation of the Epstein-Barr virus latent membrane protein 2: a role for C-terminal Src kinase. J. Virol. *73*, 4767-4775.

Shanmugaratnam, K. (1978). Histological typing of nasopharyngeal carcinoma. IARC Sci. Publ. *20*, 3-12.

Shanmugaratnam, K. a. T. C. (1970). A study of nasopharyngeal carcinoma among Singapore Chinese with special reference to migrant status and specific community. J. Chronic Dis. *23*, 433-441.

Sitki-Green, D., Covington, M., and Raab-Traub, N. (2003). Compartmentalization and transmission of multiple epstein-barr virus strains in asymptomatic carriers. J. Virol. *77*, 1840-1847.

Sitki-Green DL, E. R., Covington MM, and Raab-Traub N. (2004). Biology of Epstein-Barr virus during infectious mononucleosis. J Infect Dis. *189*, 483-492.

Smith, P. R., de Jesus, O., Turner, D., Hollyoake, M., Karstegl, C. E., Griffin, B. E., Karran, L., Wang, Y., Hayward, S. D., and Farrell, P. J. (2000). Structure and coding content of CST (BART) family RNAs of Epstein-Barr virus. J. Virol. *74*, 3082-3092.

Smith, P. R., Gao, Y., Karran, L., Jones, M. D., Snudden, D., and Griffin, B. E. (1993). Complex nature of the major viral polyadenylated transcripts in Epstein-Barr virus-associated tumors. J. Virol. *67*, 3217-3225.

Smith, P. R. and Griffin, B. E. (1992). Transcription of the Epstein-Barr virus gene EBNA-1 from different promoters in nasopharyngeal carcinoma and B-lymphoblastoid cells. J. Virol. *66*, 706-14.

Sung, N. S., Edwards, R. H., Seillier-Moiseiwitsch, F., Perkins, A. G., Zeng, Y., and Raab-Traub, N. (1998). Epstein-Barr virus strain variation in nasopharyngeal carcinoma from the endemic and non-endemic regions of China. Int. J. Cancer *76*, 207-215.

Swaminathan, S., Huneycutt, B. S., Reiss, C. S., and Kieff, E. (1992). Epstein-Barr virus-encoded small RNAs (EBERs) do not modulate interferon effects in infected lymphocytes. J. Virol. *66*, 5133-5136.

Thornburg, N. J., Pathmanathan, R., and Raab-Traub, N. (2003). Activation of nuclear factor-kappaB p50 homodimer/Bcl-3 complexes in nasopharyngeal carcinoma. Cancer Res. *63*, 8293-8301.

Thornburg, N.J., Kusano, S., and Raab-Traub, N. (2004). Identification of Epstein-Barr virus RK-BARF0-interacting proteand ins and characterization of expression pattern. J. Virol. *78*, 12848-12856.

Tsai, S. T., Jin, Y. T., and Su, I. J. (1996). Expression of EBER1 in primary and metastatic nasopharyngeal carcinoma tissues using *in situ* hybridization. A correlation with WHO histologic subtypes. Cancer *77*, 231-236.

Wang, D., Liebowitz, D., and Kieff, E. (1985). An EBV membrane protein expressed in immortalized lymphocytes transforms established rodent cells. Cell *43*, 831-840.

Wang, F., Tsang, S. F., Kurilla, M. G., Cohen, J. I., and Kieff, E. (1990). Epstein-Barr virus nuclear antigen 2 transactivates latent membrane protein LMP1. J. Virol. *64*, 3407-16.

Wolf, H., Werner, J., and zur Hausen, H. (1975). EBV DNA in nonlymphoid cells of nasopharyngeal carcinomas and in a malignant lymphoma obtained after inoculation of EBV into cottontop marmosets. Cold Spring Harb Symp Quant Biol *39*, 791-76.

Wu, T. C., Mann, R. B., Epstein, J. I., MacMahon, E., Lee, W. A., Charache, P., Hayward, S. D., Kurman, R. J., Hayward, G. S., and Ambinder, R. F. (1991). Abundant expression of EBER1 small nuclear RNA in nasopharyngeal carcinoma. A morphologically distinctive target for detection of Epstein-Barr virus in formalin-fixed paraffin-embedded carcinoma specimens. Am. J. Pathol. *138*, 1461-1469.

Young, L. S., Dawson, C. W., Clark, D., Rupani, H., Busson, P., Tursz, T., Johnson, A., and Rickinson, A. B. (1988). Epstein-Barr virus gene expression in nasopharyngeal carcinoma. J. Gen. Virol. *69*, 1051-1065.

Zhang, J., Chen, H., Weinmaster, G., and Hayward, S. D. (2001). Epstein-Barr virus BamHi-a rightward transcript-encoded RPMS protein interacts with the CBF1-associated corepressor CIR to negatively regulate the activity of EBNA2 and NotchIC. J. Virol. *75*, 2946-2956.

Zou, X.N., L. S., and Liu, B. (1994). Volatile N-nitrosamines and their precursors in Chinese salted fish--a possible etological factor for NPC in china. Int. J. Cancer *59*, 155-158.

From: Epstein-Barr Virus. Edited by: Erle S. Robertson

Chapter 8

Hodgkin's Lymphoma: Molecular Pathogenesis and the Contribution of the Epstein-Barr Virus

*P.G. Murray and L.S. Young**

ABSTRACT

Although the morphology of the pathognomonic Reed-Sternberg cells of Hodgkin's lymphoma (HL) was described over a century ago, it was not until recently that the origin of these cells from germinal centre B cells was recognised. The demonstration that a proportion of HL tumours harbours the Epstein-Barr virus (EBV) and that its genome is monoclonal in these tumours suggests that the virus contributes to the development of HL in some cases. This review summarises current knowledge of the pathogenesis of HL with particular emphasis on the association with EBV.

INTRODUCTION AND HISTOLOGICAL FEATURES OF HODGKIN'S LYMPHOMA

Hodgkin's lymphoma (HL) tissues are characterised by the disruption of normal lymph node architecture and the presence of a minority, usually less than 1 to 2% of the total tumour mass, of malignant Hodgkin/Reed Sternberg (HRS) cells amid a background of non-neoplastic cell populations (Harris et al., 1994). The typical cellular background comprises T- and B-lymphocytes, eosinophils, neutrophils, plasma cells, histiocytes, fibroblasts and stromal cells, which either surround the HRS cells or accumulate in their close vicinity. The HRS cells and their reactive neighbouring cells are able to cross-talk via a complex of cytokine and cell contact dependent interactions and these probably include proliferative and anti-apoptotic signals favouring tumour cell survival and expansion (Pinto et al., 1998).

The recognition of considerable histological diversity within the broad diagnosis of HL has led to the development of numerous histological classifications. The first was proposed by Rosenthal in 1936 (Rosenthal, 1936) and was based on the observation that prognosis for Hodgkin's lymphoma patients is related to the number of lymphocytes in tissue sections of their tumours. Other systems proposed after Rosenthal's initial work included Jackson and Parker's in 1944 (Jackson and Parker, 1944) and Lukes and Butler's in 1966 (Lukes and Butler, 1966), the latter being modified at the Rye conference. The Rye modification divided HL into

*For correspondence email l.s.young@bham.ac.uk

Table 1. Comparison of classical and lymphocyte predominant Hodgkin's lymphoma

REAL/WHO classification	Morphology /Immunophenotype of HRS cells	EBV association	Ig status
Classical Hodgkin's lymphoma Nodular sclerosis Mixed cellularity Lymphocyte depletion Lymphocyte-rich classical	Typical HRS cells which are CD15+, CD20-, CD30+, CD45-	Positive or negative	Lack BCR expression Destructive or non-functional IgH rearrangements or loss of Ig-specific transcription factors
Lymphocyte-predominant Hodgkin's lymphoma	Atypical 'popcorn' cells which are CD15-, CD20+, CD30-, CD45+,	negative	Express BCR Functional rearrangements of Ig genes Evidence of intraclonal diversity indicating ongoing somatic hypermutation

four histological subtypes: lymphocyte predominant (LP), nodular sclerosis (NS), mixed cellularity (MC) and lymphocyte depletion (LD). The more recent Revised European American Lymphoma (REAL)/WHO classification separates HL into classical and LP (LPHL) types; the former comprises the subtypes of NS, MC, LD and a newly defined entity referred to as 'lymphocyte rich classical' Hodgkin's lymphoma. There are numerous differences between the classical and LP forms of HL (Table 1). For example, the HRS cells of LPHL have a different morphology to those of classical HL, being referred to as 'popcorn' cells. Furthermore, these cells rarely express the classical HL markers CD15 or CD30, but regularly express B cell antigens such as CD20 which are usually absent in HRS cells of classical HL.

ORIGIN OF HRS CELLS

Early phenotypic studies suggested that HRS cells might be derived from macrophages, dendritic reticulum cells or granulocytes. However, the identification of clonal rearranged immunoglobulin (Ig) genes by PCR analysis of single HRS cells micromanipulated from HL tissues provided evidence not only of their malignant character but also of their origin from B cells (Kuppers et al., 1994). Further analysis revealed that HRS cells carry high loads of somatic Ig mutations indicating that they originate from germinal centre or post-germinal centre B cells. In the majority of cases it was shown that the V_H gene rearrangements were non-functional, suggesting that HRS cells can bypass the apoptosis that would otherwise eliminate B cells with non-functional Ig genes (Kanzler et al., 1996). In fact, the lack of expression of functional surface immunoglobulin (B cell receptor; BCR) is the hallmark of classical HL. While in some cases this is certainly due to non-functional or destructive rearrangements, other mechanisms can also account for the loss of functional BCR. For example, the loss of immunoglobulin-specific transcription factors, BOB-1, OCT2 and PU.1 in HRS cells has been reported (Re et al., 2001; Jundt et al., 2002; Torlakovic et al., 2001; Marafioti et al., 2000). It has also been shown that in a minority of cases HRS cells carry mutations in the octamer region of the immunoglobulin gene promoter, which can prevent binding of Ig-specific transcription factors (Jox et al., 1999; Theil et al., 2001). Interestingly, gene expression profiling has recently revealed the downregulation of a number of B cell specific genes in classical HL (Schwering et al., 2003). Indeed, many genes important in BCR signalling were affected by silencing of the B-cell specific transcription programme.

The absence of Ig gene expression in classical HL contrasts with results from analysis of LPHL, where the tumour cell population often expresses Ig (Stoler et al.,1995). Furthermore, whereas intraclonal diversity is uncommon in classical HL it is frequent in LPHD, suggesting the presence of ongoing somatic mutations indicative of selection for antigen in the latter but not the former (Marafioti et al., 1997).

Occasionally, HL tumours express T cell antigens, including granzyme B and T cell intracellular antigen (TIA)-1. In some cases this has been shown to represent aberrant expression of T cell antigens by HRS cells that show evidence of IgH gene rearrangement and are thus assumed to be B cell in origin (Muschen et al., 2000). However, in the same study a single HL case showed expression of T cell markers and also TCR gene rearrangements, indicating that at least a minority of HL tumours is genuinely of T cell origin.

EPIDEMIOLOGY OF HODGKIN'S LYMPHOMA

Early studies of the epidemiology of HL identified a bimodal age distribution in the United States, with the first peak of onset occurring between 15 and 34 years, and the second after 50 years of age (MacMahon, 1966). Three age periods were distinguished: 0-14, 15-34 and 50 years and above. It was noted that Hodgkin's lymphoma of childhood was more common in boys under 10 years of age. In young adults it was hypothesised that the disease was probably infectious in nature, with low infectivity. This was supported by data from some families, which had more than one affected member of different ages at the same time (Devor and Doan, 1957; Razis et al., 1959), and the identification of an excess of cases in the winter months in some studies (Cridland, 1961; Fraumeni and Li, 1969). The peak incidence in young adults was between 25 and 30 years of age, the sex ratio was almost equal at this time and the disease was associated with high socioeconomic status. In contrast, HL in the elderly showed increasing incidence with age, a male-to-female ratio of 2:1 and epidemiological features similar to other neoplastic diseases.

Subsequently, the concept of at least three epidemiological patterns of Hodgkin's lymphoma based upon country of residence was introduced (Correa and O'Conor, 1971). A type I pattern which prevails in developing countries is characterised by relatively high incidence rates in male children, low incidence in the third decade and a second peak of high incidence in older age groups. The most common histological subtype seen in these populations is the MC form. Type III is the converse of the Type I pattern, being characterised by low rates in children and a pronounced initial peak in young adults. NSHL is common and this pattern is typical of developed countries. Type II is an intermediate pattern found, for example, in rural areas of developed countries and reflects a transition between Type I and Type III. These patterns were interpreted as the result of the interplay of environmental and host factors influencing the natural history of a single disease similar to that observed for tuberculosis. In underprivileged communities, tuberculosis has high rates in children and the disease presents itself in the more serious pneumonic form. When economic conditions improve, childhood tuberculosis becomes less common and most cases in young adults are of the more benign pulmonary form. This led to the hypothesis that in a given population susceptibility to the agent or agents which cause Hodgkin's lymphoma is related to immunocompetence and host response, the level of which is, in turn, dependent on environmental and socio-economic factors. Hence, there is an alternative to the dual aetiology explanation of bimodality; that of a single aetiological process that is affected by variations in host response over age.

Childhood social environment has been suggested to play an important role in influencing the risk of Hodgkin's lymphoma among young adults (Gutensohn and Cole, 1980). Higher risk being associated with factors that diminish or delay exposure to infectious agents such as higher social class, more education, small family size and early birth-order position. These are consistent with a virus-induced pathogenesis, with greater risk of Hodgkin's lymphoma occurring with increasing age at infection.

The epidemiology of Hodgkin's lymphoma in young adults was likened to that of paralytic poliomyelitis in the pre-vaccine era (Gutensohn and Cole, 1981). In both diseases, age of peak incidence is delayed as living conditions improve. For both, increased risk is associated with higher social class and small family size. It was suggested that Hodgkin's lymphoma is a rare consequence of a common infection, with the probability of oncogenesis increasing as age at the time of infection increases.

These findings led to the premise that the variation of the bimodal age incidence curve of Hodgkin's lymphoma is related to the age at primary infection with a common virus. As a population moved towards a higher standard of living, an initial early peak among young boys disappeared and produced the characteristic young adult peak (Gutensohn and Cole, 1980). The data from the study of factors in childhood environment that influence the age of infection are consistent with this idea. The incidence of disease in the older age group varies little between populations and is not associated with social class factors (Gutensohn, 1982).

EBV AND HODGKIN'S LYMPHOMA

First evidence that EBV might be involved in the pathogenesis of HL was provided by the detection of raised antibody titres to EBV antigens in HL patients when compared with other lymphoma patients (Levine et al., 1971) and further that these raised levels preceded the development of Hodgkin's lymphoma by several years (Mueller et al., 1989). In addition, the relative risk of developing Hodgkin's lymphoma in individuals with a history of infectious mononucleosis (IM), relative to those with no prior history was shown to range between 2.0 and 5.0 (Gutensohn and Cole, 1980). It has recently been shown that the risk of EBV-positive HL is increased four-fold after IM, whereas the risk of EBV-negative HL is not increased (Hjalgrim et al., 2003)

With the advent of cloned viral probes and Southern blot hybridisation methods; EBV DNA was initially detected in 20-25% of Hodgkin's lymphoma tumour specimens (Weiss et al., 1987). However, this approach could not determine the locality of the EBV genome in tissues. *In situ* hybridisation methods to detect EBV DNA provided the first demonstration of its existence in the HRS cells (Anagnostopoulos et al., 1989; Weiss et al., 1989). Subsequently, the demonstration of the abundant EBV encoded RNAs (EBER1 and EBER2) in HRS cells provided a sensitive method for detecting latent infection *in situ*. This technique is now generally accepted as the 'gold standard' for the detection of latent EBV infection in clinical samples (Wu et al., 1990) (Figure 1a).

In EBV-associated Hodgkin's lymphoma the viral genomes are found in monoclonal form indicating that infection of the tumour cells has occurred prior to their clonal expansion (Anagnostopoulos et al., 1989). In the majority of cases EBV appears to persist throughout the course of HL and is also found in multiple sites of HL (Coates et al., 1991). Furthermore, EBV genome copy number within HRS cells varies between patients but appears constant within individual patients with HL (Coates et al., 1991).

The possibility that EBV may contribute to the pathogenesis of HL early in the transformation of the progenitor cells but is subsequently lost ('hit and run') has prompted the search for evidence of defective re-arranged EBV DNA in tumours that by conventional testing (*e.g.* by detection of the EBERs) are seemingly virus

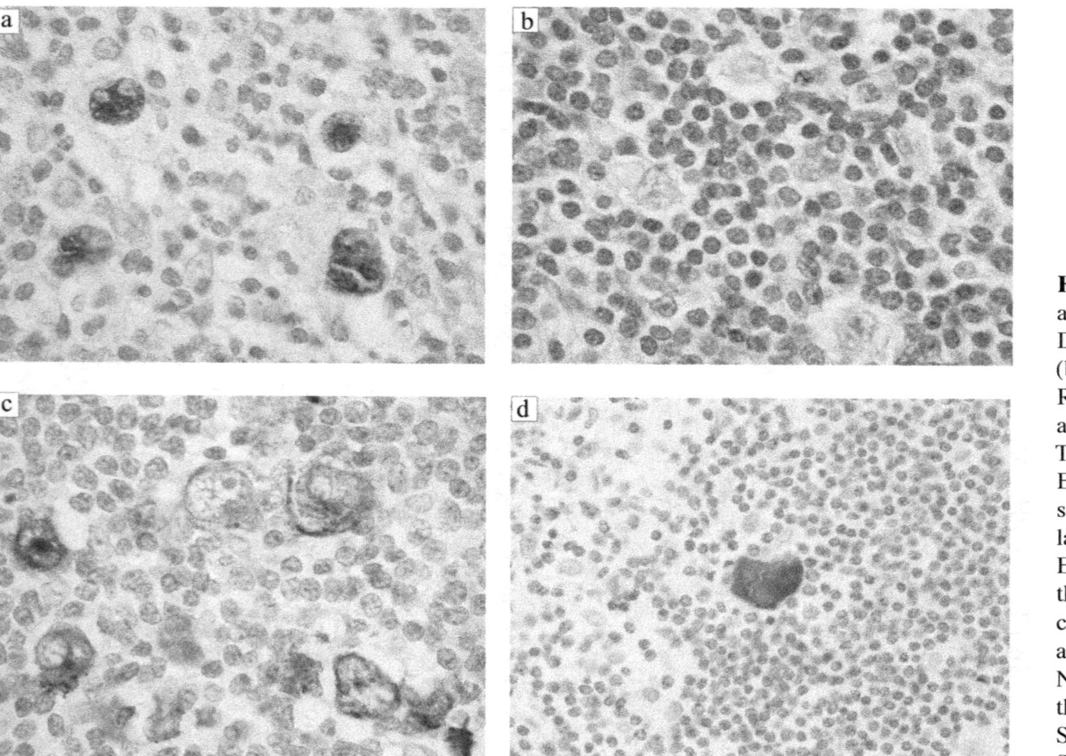

Figure 1. Gene expression in EBV-associated Hodgkin's lymphoma. a) Demonstration of EBER expression (brown staining) in the nuclei of Hodgkin's/Reed-Sternberg (HRS) cells from an EBV-associated Hodgkin's lymphoma biopsy. The use of *in situ* hybridisation to detect EBER expression is the most reliable and sensitive method to detect the presence of latent EBV infection in clinical samples. b) EBV-positive HRS cells lack expression of the transforming EBNA2 protein, whereas c) the latent membrane protein-1 (LMP1) and d) LMP2 are both highly expressed. Not shown is the consistent expression of the EBV maintenance protein, EBNA1. See this figure in colour in the Colour Plate Section at the back of the book

negative. Such defective rearranged viral DNA has been previously detected in some cases of virus-negative sporadic Burkitt's lymphoma (Razzouk et al., 1996) and could result in partial elimination of EBV episomes from infected cells through the expression of BZLF1. In support of this possibility, Gan and colleagues amplified sequences that span abnormally juxtaposed BamH1W and Z fragments (which characterise defective heterogeneous EBV DNA) from 2/24 EBER-negative HL tumours in which the standard viral genome could not be detected (Gan et al., 2002). However, others using fluorescence in situ hybridisation (FISH) analysis found no evidence of integrated EBV genomes in EBV-negative HL (Staratschek-Jox et al., 2000) Furthermore, using quantitative PCR assays that spanned the whole genome, no evidence of deletion of EBV genomes in EBV-positive HL nor retention of EBV genomes in EBV-negative HL tissues was found (Gallagher et al., 2003) Therefore, at this time it seems less probable that EBV also contributes to the development of EBV-negative HL.

The detection of EBV in Hodgkin's lymphoma seems to depend on factors such as country of residence, histological subtype, sex, ethnicity and age. In particular, EBV-positive Hodgkin's lymphoma tumours appear to be less common in developed populations, with percentages of between 20-50% for North American and European cases (Wu et al., 1990; Weiss et al.,1991; Herbst et al., 1992; Hummel et al., 1992), 57% for Hodgkin's lymphoma in China (Zhou et al.,1993) but much higher rates in underdeveloped countries (Chang et al., 1993; Weinreb et al., 1996). The increased incidence of EBV-positive Hodgkin's lymphoma in underdeveloped countries could be due to the existence of an underlying immunosuppression similar to that observed for African Burkitt's lymphoma in a malaria-infected population. This is supported by the higher EBV-positive rates in Hodgkin's lymphoma from HIV-infected patients (Uccini et al., 1990). Alternatively, the timing of EBV infection (which is likely to occur earlier in developing populations) might also be important.

EBV is more commonly associated with the mixed cellularity subtype and less frequently with the other forms of this disease (Pallesen et al., 1991; Murray et al., 1992). Furthermore, it is now generally accepted that the LP form of HL as defined by the REAL classification is an EBV-negative disease (Chan et al., 1999). Hodgkin's lymphoma in the older age groups and in children, especially boys under 10 years, has been shown to be more likely to be EBV-associated than Hodgkin's lymphoma in young adults (Glaser et al., 1997; Jarrett et al., 1991; Armstrong et al., 1998a). This has led to the suggestion that HL consists of three disease entities; HL of childhood (EBV-positive, MC type), HL of young adults (EBV-negative, NS type) and HL of older adults (EBV-positive, MC type) (Armstrong et al., 1998a). The infrequent association of EBV with Hodgkin's lymphoma in young adulthood prompted the suggestion that a second virus might be involved, although there is little direct evidence to support this at present (Armstrong et al., 1998b). EBV-positive rates are generally higher in males than in females. EBV-positive Hodgkin's lymphoma also affects more Asians and Hispanics than whites or blacks (Glaser et al., 1997) and in the UK is more common in South Asian children compared with non-South Asian children (Flavell et al., 2001).

CONTRIBUTION OF VIRUS GENES TO THE PATHOGENESIS OF EBV-POSITIVE HODGKIN'S LYMPHOMA

EBV-positive HRS cells exhibit a type II form of latency, virus gene expression being limited to the EBERs, Epstein-Barr nuclear antigen-1 (EBNA1) (Grasser et al., 1994), latent membrane protein 1 (LMP1) (Pallesen et al., 1991; Murray et al., 1992; Deacon et al., 1993), LMP2 (Deacon et al., 1993; Niedobitek et al., 1997), (Figure 1b-d), and the Bam H1A rightward transcripts (BARTs) (Deacon et al., 1993). The particularly high levels of LMP1 expression in HRS cells suggest it plays an important role in the pathogenesis of EBV-associated cases. Previous studies have established that LMP1 induces many of the phenotypic changes observed in EBV-infected B cells, including expression of the B cell activation markers, CD23 and CD40, IL-10 production, upregulation of cell adhesion molecules such as ICAM1, LFA1 and LFA3, and downregulation of CD99 (Wang et al., 1990; Nakagomi et al., 1994; Kim et al., 2000). LMP1 also protects B cells from cell death by the upregulation of several anti-apoptosis genes including Bcl-2, Mcl-1 and A20 (Henderson et al., 1991; Laherty et al., 1992; Wang et al., 1996; Rowe et al., 1994). LMP1 functions as a constitutively activated member of the tumour necrosis factor receptor (TNFR) superfamily activating a number of signalling pathways in a ligand-independent manner. Two distinct functional domains referred to as C-terminal activation regions 1 and 2 (CTAR1 and CTAR2) have been identified on the basis of their ability to activate the NF-κB transcription factor pathway (Huen et al., 1995). More recent work demonstrates that LMP1 can also regulate the processing of p100 NF-κB2 to p52 and that this is independent of the pathways responsible for controlling the canonical NF-κB pathway (Eliopoulos et al., 2003). LMP1 is also able to engage the MAP kinase cascade resulting in activation of ERK, JNK and p38 and to stimulate the JAK/STAT pathway (Eliopoulos et al., 1999; 1998; Gires et al., 1999; Roberts and Cooper et al., 1998; Keiser et al., 1997).

Constitutive activation of several of the pathways known to be activated by LMP1 is also observed in HRS cells. In particular, constitutive activation of NF-kB is a regular feature of HRS cells and inhibition of this pathway in HL cell lines leads to their increased sensitivity to apoptosis after growth factor withdrawal and impaired tumourigenicity in severe combined immunodeficiency (SCID) mice (Bargou et al., 1996; 1997). Although NF-kB activation is a common finding in HRS cells the molecular routes to this activation may be different between EBV-positive and EBV-negative HL. IkBa mutations have been reported in EBV-negative HRS cells (Emmerich et al., 1999; Cabannes et al., 1999) and in 2/3 cases of EBV-negative primary HL, but not in the two EBV-positive cases examined (Jungnickel et al., 2000). More recently, Jarrett et al., found IkBa mutations in 15/26 primary HL tumours, 11 of which were EBV-negative (Jarrett et al., 2002). In addition, comparative genomic hybridisation (CGH) and FISH analysis has demonstrated frequent amplification of the c-REL locus at 2p13-p12 in HRS cells of classical HL (Martin-Subero et al., 2002; Barth et al., 2003). This is an alternative mechanism that could lead to constitutive NF-kB activation, although it is yet to be shown whether this is mutually exclusive with an EBV-positive status. Kube and colleagues have demonstrated the constitutive activation of STAT3 in HRS cells (Kube et al., 2001). STAT6 and STAT5a constitutive activation has also been reported (Skinnider et al., 2002; Hinz et al., 2002). Amplification of the JAK2 locus has been reported in HRS cells in some cases, providing a mechanistic explanation for the STAT activation (Joos et al., 2003). Furthermore, it has been shown HRS cells have constitutively activated AP-1 with c-Jun and JunB overexpression (Mathas et al., 2001). Interestingly, a similar AP-1 activation was present in anaplastic large cell lymphoma (ALCL), which is closely related to HRS cells, but was absent in other lymphoma types.

The identification that a virus strain carrying a 30bp deletion in the LMP1 gene was more tumourigenic than the prototype B95.8 LMP1 (Hu et al., 1993) led to numerous studies of the prevalence of this virus strain within EBV-associated cancers, including Hodgkin's lymphoma. The general findings from these studies were that healthy virus carriers have a similar frequency of mutations as virus-positive tumour patients from the same geographical region (Khanim et al., 1996). However, the exception to this appears to be Hodgkin's lymphoma (Knecht et al., 1993) where some studies have shown an increased incidence of this deletion variant, in HIV-positive Hodgkin's lymphoma compared to HIV-negative Hodgkin's lymphoma (Bellas et al., 1996) and also in paediatric Hodgkin's lymphoma compared to normal controls (Santon et al., 1998).

The other EBV-encoded latent membrane protein, LMP2A is also highly expressed in EBV-positive HRS cells (Deacon et al., 1993). LMP2A competes for the binding of the Src family protein tyrosine kinases and the Syk protein tyrosine kinases. This blocks signalling through the BCR and prevents transition of the EBV-infected B cell into the lytic cycle and thus maintains EBV latency (Miller et al., 1995). Expression of LMP2A in the B cells of transgenic mice abrogates normal B cell development allowing immunoglobulin-negative cells to colonise peripheral lymphoid organs (Caldwell et al., 1998). This suggests that LMP2A can drive the proliferation and survival of B cells in the absence of signalling through the BCR. More recently, it has been shown that LMP2A expression in B cells results in the downregulation of many of the B cell factors previously shown to be absent or expressed at low levels in HRS cells (*e.g.* early B cell factor, PU.1, CD19, CD20) (Portis et al., 2003). In addition, LMP2A expression induces the upregulation of genes involved in proliferation (*e.g.* Ki67, PCNA), protection from apoptosis (Bcl-xL, survivin) and suppression of cell-mediated immunity (*e.g.* IL-13R, EBI3) (Portis et al., 2003). Many of the transcriptional changes seen in response to LMP2A are also observed in EBV-negative HRS cells, suggesting that alternative mechanisms may contribute similar effects when EBV is absent.

Data from studies of the patterns of EBV gene expression during B cell differentiation *in vivo* are consistent with a role for LMP1 and LMP2A in the early stages of the pathogenesis of HL. Thus, in normal carriers EBV exists in IgD⁻ memory B cells, EBV gene expression being restricted to LMP2A and possibly EBNA1 (Babcock et al., 1998). The detection of both LMP1 and LMP2 in tonsillar memory B cells and germinal centre B cells (Babcock et al., 2000) suggests a model whereby these viral proteins through surrogate T cell help (LMP1) and BCR engagement (LMP2A), provide the necessary signals for EBV-infected memory B cells to undergo antigen-independent proliferation in the germinal centre, in turn leading to replenishment of the pool of EBV-positive memory B cells. However, expression of both of these virus proteins, together with as yet undefined cellular alterations, might also favour the neoplastic transformation of germinal centre B cells ultimately leading to the development of EBV-positive HL. However, the exact contribution of EBV to the pathogenesis of classical HL remains to be established.

FAILURE OF THE IMMUNE RESPONSE TO ELIMINATE EBV-INFECTED HRS CELLS AND ATTEMPTS TO DEVELOP IMMUNOTHERAPEUTIC STRATEGIES

LMP2A, and to a lesser extent LMP1, are targets for cytotoxic T lymphocytes in association with different MHC class I restriction elements *in vitro* (Khanna et al., 1998; Lee et al.,1997). The survival of EBV-infected HRS cells expressing high levels of LMP1 and LMP2A *in vivo* suggests that mechanisms to avoid or prevent immune recognition are manifest in HL patients. These include the existence of specific immunological defects present in Hodgkin's lymphoma patients that permit the growth of the neoplastic cells as well as the development of specific immune evasion strategies by EBV-infected HRS cells. Support for the latter is provided by the finding that IL-10 is more frequently expressed in EBV-infected HRS cells when compared to their EBV-negative counterparts and this has been suggested to account for the failure of these cells to be recognised by EBV-specific cytotoxic T lymphocytes (CTLs) (Herbst et al., 1996)· Furthermore, HRS cells express the thymus and activated regulated chemokine (TARC) which together with other cytokines, such as IL-10, IL-13 and TGFβ, attract Th2 cells and contribute to a shift in the local environment away from a Th1 towards a Th2 predominated response (Poppema et al., 1998; Kapp et al., van den Berg et al., 1999). The contribution of the microenvironment is underlined by the observation that tumour-derived T lymphocytes from EBV-negative HL show EBV-specific cytotoxicity, whereas corresponding lymphocytes from EBV-positive HL lesions do not (Frisan et al., 1995) Surprisingly, EBV-positive cases of HL have been shown to contain more activated CTLs and express relatively higher levels of MHC class I than EBV-negative cases (Oudejans et al., 1996; Murray et al., 1998; Oudejans et al., 1997; Lee et al., 1998). Despite this, there is clearly a failure to elicit an effective anti-EBV specific CTL response.

In vitro, HL cells are able to process and present epitopes from LMP1 and LMP2A in the context of multiple class I alleles including HLA A2 and are sensitive to lysis by EBV-specific CTLs (Lee et al., 1998; Sing et al., 1997). Despite concerns that cytotoxicity might be compromised *in vivo,* the use of donor-derived EBV-specific CTLs has been investigated in the treatment of EBV-positive HL patients (Roskrow et al., 1998). In this study, EBV-specific CTLs could be generated from patients with advanced HL, albeit at lower frequency than normal controls and EBV-specific CTLs survived and had antiviral activity *in vivo*. Despite the resolution of some symptoms and the stabilisation of disease, all patients unfortunately failed to recover from their disease.

Dendritic cells transduced with LMP2A have been used to stimulate and expand LMP2-specific CTLs that had increased cytolytic activity to LMP2-positive targets *in vitro* compared with EBV-specific CTLs (Su et al., 2001). Various other approaches, such as the use of a recombinant poxvirus vaccine that encodes a polypeptide protein comprising six HLA A2 restricted epitopes derived from LMP1 ('LMP1 polyepitope vaccine'), may prove useful in generating stronger anti-LMP1 responses (Duraiswamy et al., 2003). Together with the development of alternative strategies to alleviate the CTL inhibitory effect of the Th2 environment (Wagner et al., 2004; Bollard et al., 2002) it is possible that immunotherapy may have a role alongside standard approaches in the future treatment of EBV-positive HL.

Two large studies investigating the influence of EBV infection in HL have shown improved outcomes for EBV-positive patients compared to their EBV-negative counterparts (Morente et al., 1997; Murray et al., 1999). This is somewhat surprising when one considers that the oncogenic LMP1 protein is highly expressed in EBV-infected HRS cells, but possible explanations could be that the malignant cells of EBV-positive HL may be more sensitive to chemotherapy agents or that the EBV-positive cells might be targets for immune cytolysis particularly after cytoreduction by chemotherapy. However, it appears that the beneficial effects of an EBV-positive status may be restricted to younger adults, since compared with EBV-negative HL, EBV-positive tumours carry a poorer prognosis in older adults (Glavina-Durdov et al., 2001; Clarke et al., 2001; Stark et al., 2002).

CONCLUSIONS

The tumour cells of classical HL are derived from germinal centre B cells that lack functional B cell receptors and would normally be eliminated by apoptosis during the germinal centre reaction. Recent studies have identified the dysregulation of a number of cellular pathways as key events that enable the survival and growth of these malignant cells. A proportion of HL tissues harbours EBV within tumour cells. Superficially at least, there are surprisingly few differences between EBV-positive and EBV-negative HRS cells. However, emerging evidence suggests that while EBV is able to subvert cellular processes in favour of growth and survival, cellular genetic events are required when EBV is absent. Age-related differences in the prognostic significance of EBV infection in HL tumours suggest a complex interplay between EBV infection and underlying genetic abnormalities in HRS precursor cells. The challenge is to unravel this complexity by detailed consideration of the function of latent EBV genes in the appropriate cellular context. It is hoped that this approach will reveal more fundamental aspects of HL pathogenesis and pave the way to more targeted therapies for HL patients.

Acknowledgements

We are grateful to the Leukaemia Research Fund and to Cancer Research UK for funding.

References

Anagnostopoulos, I., Herbst, H., Niedobitek, G., and Stein, H. (1989). Demonstration of monoclonal EBV genomes in Hodgkin's disease and Ki-1 positive anaplastic large cell lymphoma by combined Southern blot and *in situ* hybridization. Blood *74*, 810-816.

Armstrong, A.A., Alexander, F.E., Cartwright, R., Angus, B., Krajewski, A.S., Wright, D.H., Brown, I., Lee, F., Kane, E., and Jarrett, R.F. (1998a). Epstein-Barr virus and Hodgkin's disease: further evidence for the three disease hypothesis. Leukemia *12*, 1272-1276.

Armstrong, A.A., Shield, L., Gallagher, A., and Jarrett, R.F. (1998b). Lack of involvement of known oncogenic DNA viruses in Epstein-Barr virus-negative Hodgkin's disease. Brit J Cancer 77, 1045-1047.

Babcock, G.J., Decker, L.L., Volk, M., and Thorley-Lawson, D.A. (1998). EBV persistence in memory B cells *in vivo*. Immunity 9, 395-404.

Babcock, G.J., and Thorley-Lawson, D.A. (2000). Tonsillar memory B cells, latently infected with Epstein-Barr virus, express the restricted pattern of latent genes previously found only in Epstein-Barr virus-associated tumours. Proc. Natl. Acad. Sci. USA 97, 12250-12255.

Bargou, R.C., Leng, C., Krappmann, D., Emmerich, F., Mapara, M.Y., Bommert, K., Royer, H.D., Scheidereit, C., and Dörken, B. (1996). High-level nuclear NF-kappa B and Oct-2 is a common feature of cultured Hodgkin/Reed-Sternberg cells. Blood 87, 4340-4347.

Bargou, R.C., Emmerich, F., Krappmann, D., Bommert, K., Mapara, M.Y., Arnold, W., Royer, H.D., Grinstein, E., Greiner, A., Scheidereit, C., and Dörken, B. (1997). Constitutive nuclear factor-kappaB-RelA activation is required for proliferation and survival of Hodgkin's disease tumor cells. J. Clin. Invest. 100, 2961-2969.

Barth, T.F., Martin-Subero, J.I., Joos, S., Menz, C.K., Hasel, C., Mechtersheimer, G., Parwaresch, R.M., Lichter, P., Siebert, R., and Mooller, P. (2003). Gains of 2p involving the REL locus correlate with nuclear c-Rel protein accumulation in neoplastic cells of classical Hodgkin lymphoma. Blood 101, 3681-3686.

Bellas, C., Santon, A., Manzanal, A., Campo, E., Martin, C., Acevedo, A., Varona, C., Forteza, J., Morente, M., and Montalban, L. (1996). Pathological, immunological, and molecular features of Hodgkin's disease associated with HIV infection comparison with ordinary Hodgkin's disease. Am. J. Surg. Pathol. 20, 1520 -1524.

Bollard, C.M., Rossig, C., Calonge, M.J., Huls, M.H., Wagner, H.J., Massague, J., Brenner, M.K., Heslop, H.E., and Rooney, C.M. (2002). Adapting a transforming growth factor beta-related tumor protection strategy to enhance antitumor immunity. Blood 99, 3179-3187.

Cabannes, E., Khan, G., Aillet, F., Jarrett, R.F., and Hay, R.T. (1999). Mutations in the IkBa gene in Hodgkin's disease suggest a tumour suppressor role for IkappaBalpha. Oncogene 18, 3063-3070.

Caldwell, R.G., Wilson, J.B., Anderson, S.J., and Longnecker, R. (1998). Epstein-Barr virus LMP2A drives B cell development and survival in the absence of normal B cell receptor signals. Immunity 9, 405-411.

Chan, W.C. (1999). Cellular origin of nodular lymphocyte-predominant Hodgkin's lymphoma: immunophenotypic and molecular studies. Semin. Hematol. 36, 242-252.

Chang, K.L., Albujar, P.F., Chen, Y-Y., Johnson, R.M., and Weiss, L.M. (1993). High prevalence of Epstein-Barr virus in the Reed-Sternberg cells of Hodgkin's disease occurring in Peru. Blood 81, 496-501.

Coates, P.J., Slavin, G., and D'Ardenne, A.J. (1991). Persistence of Epstein-Barr virus in Reed-Sternberg cells throughout the course of Hodgkin's disease. J. Pathol. 164, 291-297.

Correa, P., and O'Conor, G.T. (1971). Epidemiologic patterns of Hodgkin's disease. Int. J. Cancer 8, 192-201.

Clarke, C.A., Glaser, S.L., Dorfman, R.F., Mann, R., DiGiuseppe, J.A., Prehn, A.W., and Ambinder, R.F. (2001). Epstein-Barr virus and survival after Hodgkin disease in a population-based series of women. Cancer 91, 1579-1587.

Cridland, M.D. (1961). Seasonal incidence of clinical onset of Hodgkin's disease. Brit. Med. J. 2, 621-623.

Deacon, E.M., Pallesen, G., Niedobitek, G., Crocker, J., Brooks, L., Rickinson, A.B., and Young, L.S. (1993). Epstein-Barr Virus and Hodgkin's Disease: Transcriptional Analysis of Virus Latency in the Malignant Cells. J. Exp. Med. 177, 339-349.

Devore, J.W., and Doan, C.A. (1957). Studies in Hodgkin's Syndrome- XII. Hereditary and Epidemiologic Aspects. Ann. Intern. Med. *47*, 300-316.

Duraiswamy, J., Sherritt, M., Thomson, S., Tellam, J., Cooper, L., Connolly, G., Bharadwaj, M., and Khanna, R. (2003). Therapeutic LMP1 polyepitope vaccine for EBV-associated Hodgkin disease and nasopharyngeal carcinoma. Blood *101*, 3150-3156.

Eliopoulos, A.G., Caamano, J.H., Flavell, J.R., Reynolds, G., Murray, P.G., Poyet, J-L., and Young, L.S. (2003). Epstein-Barr virus-encoded latent infection membrane protein 1 regulates the processing of p100 NFkB2 to p52 via an IKKg/NEMO-independent signalling pathway. Oncogene *22*, 7557-7569.

Eliopoulos, A.G., Gallagher, N.J., Blake, S.M., Dawson, C.W., and Young, L.S. (1999). Activation of the p38 mitogen-activated protein kinase pathway by Epstein-Barr virus-encoded latent membrane protein 1 coregulates interleukin-6 and interleukin-8 production. J. Biol. Chem. *274*, 16085-16096.

Eliopoulos, A.G., and Young, L.S. (1998). Activation of the cJun N-terminal kinase (JNK) pathway by the Epstein-Barr virus-encoded latent membrane protein 1 (LMP1). Oncogene *16*, 1731-1742.

Emmerich, F., Meiser, M., Hummel, M., Demel, G., Foss, H.D., Jundt, F., Mathas, S., Krappmann, D., Scheidereit, C., Stein, H., and Dorken, B. (1999). Overexpression of I kappa B alpha without inhibition of NF-kappaB activity and mutations in the I kappa B alpha gene in Reed-Sternberg cells. Blood *94*, 3129-3134.

Flavell, K.J., Biddulph, J.P., Powell, J.E., Parkes, S.E., Redfern, D., Weinreb, M., Nelson, P., Mann, J.R., Young, L.S., and Murray, P.G. (2001). South Asian ethnicity and material deprivation increase the risk of Epstein-Barr virus infection in childhood Hodgkin's disease. Br. J. Cancer *85*, 350-356.

Fraumeni, J.F. Jr., and Li, F.P. (1969). Hodgkin's disease in childhood: An epidemiologic study. J. Natl. Cancer Inst. *42*, 681-691.

Frisan, T., Sjoberg, J., Dolcetti, R., Boiocchi, M., De Re, V., Carbone, A., Brautbar, C., Battat, S., Biberfeld, P., and Eckman, M. (1995). Local suppression of Epstein-Barr virus (EBV)-specific cytotoxicity in biopsies of EBV-positive Hodgkin's disease. Blood *86*, 1493-1501.

Gan, Y.J., Razzouk, B.I., Su, T., and Sixbey, J.W. (2002). A defective, rearranged Epstein-Barr virus genome in EBER-negative and EBER-positive Hodgkin's disease. Am. J. Pathol. *160*, 781-786.

Gallagher, A., Perry, J., Freeland, J., Alexander, F.E., Carman, W.F., Shield, L., Cartwright, R., and Jarrett, R.F. (2003). Hodgkin lymphoma and Epstein-Barr virus (EBV): no evidence to support hit-and-run mechanism in cases classified as non-EBV-associated. Int. J. Cancer *104*, 624-630.

Gires, O., Kohlhuber, F., Kilger, E., Baumann, M., Kieser, A., Kaiser, C., Zeidler, R., Scheffer, B., Ueffing, M., and Hammerschmidt, W. (1999). Latent membrane protein 1 of Epstein-Barr virus interacts with JAK3 and activates STAT proteins. EMBO J. *18*, 3064-3073.

Glaser, S.L., Lin, R.J., Stewart, S.L., Ambinder, R.F., Jarrett, R.F., Brousset, P., Pallesen, G., Gulley, M.L., Khan, G., O'Grady, J., Hummel, M., Preciado, M.V., Knecht, H., Chan, J.K., and Claviez, A. (1997). Epstein-Barr virus-associated Hodgkin's disease: epidemiologic characteristics in international data. Int. J. Cancer *70*, 375-382.

Glavina-Durdov, M., Jakic-Razumovic, J., Capkun, V., and Murray, P. (2001). Assessment of the prognostic impact of the Epstein-Barr virus-encoded latent membrane protein-1 expression in Hodgkin's disease. Br. J. Cancer *84*, 1227-1234.

Grässer, F.A., Murray, P.G., Kremmer, E., Klein, K., Remberger, K., Feiden, W., Reynolds, G., Niedobitek, G., Young, L.S., and Mueller-Lantzsch, N. (1994). Monoclonal antibodies directed against the Epstein-Barr virus-encoded nuclear antigen 1 (EBNA 1): Immunohistologic detection of EBNA 1 in the malignant cells of Hodgkin's Disease. Blood *84*, 3792-3798.

Gutensohn, N., and Cole, P. (1980). Epidemiology of Hodgkin's disease. Semin. Oncol. 7, 92-102.

Gutensohn, N., and Cole, P. (1981). Childhood social environment and Hodgkin's disease. New Engl. J. Med. *304*, 135-140.

Gutensohn, N.M. (1982). Social class and age at diagnosis of Hodgkin's disease; New epidemiologic evidence on the "two-disease" hypothesis. Cancer Treat. Rep. *66*, 689-695.

Harris, N.L., Jaffe, E.S., Stein, H., Banks, P.M., Chan, J.K.C., Cleary, M.L., Delsol, G., De Wolf-Peeters, C., Falini, B., Gatter, K.C., Grogan, T.M., Isaacson, P.G., Knowles, D.M., Mason, D.Y., Muller-Hermelink, H., Pileri, S.A., Piris, N.A., Ralfkiaer, E., Warnke, R.A. (1994). A revised European-American classification of lymphoid neoplasms: A proposal from the International Lymphoma Study Group. Blood 84, 1361-1392.

Henderson, S., Rowe, M., Gregory, C., Croom-Carter, D., Wang, F., Longnecker, R., Kieff, E., and Rickinson, A. (1991). Induction of bcl-2 expression by Epstein-Barr virus latent membrane protein 1 protects infected B cells from programmed cell death. Cell *65*, 1107-1115.

Herbst, H., Steinbrecher, E., Niedobitek, G., Young, L.S., Brooks, L., Müller-Lantzsch, N., and Stein, H. (1992). Distribution and phenotype of Epstein-Barr virus-harboring cells in Hodgkin's disease. Blood *80*, 484-491.

Herbst, H., Foss, H.D., Samol, J., Araujo, I., Klotzbach, H., Krause, H., Agathanggelou, A., Niedobitek, G., and Stein, H. (1996). Frequent expression of interleukin-10 by Epstein-Barr virus-harboring tumor cells of Hodgkin's disease. Blood *87*, 2918-2929.

Hinz, M., Lemke, P., Anagnostopoulos, I., Hacker, C., Krappmann, D., Mathas, S., Dorken, B., Zenke, M., Stein, H., and Scheidereit, C. (2002) Nuclear factor kappaB-dependent gene expression profiling of Hodgkin's disease tumor cells, pathogenetic significance, and link to constitutive signal transducer and activator of transcription 5a activity. J. Exp. Med. *196,* 605-617.

Hjalgrim, H., Askling, J., Rostgaard, K., Hamilton-Dutoit, S., Frisch, M., Zhang, J.S., Madsen, M., Rosdahl, N., Konradsen, H.B., Storm, H.H., and Melbye, M. (2003) Characteristics of Hodgkin's lymphoma after infectious mononucleosis. N. Engl. J. Med. *349,* 1324-1332.

Hu, L.F., Chen, F., Zheng, X., Ernberg, I., Cao, S.L., Christensson, B., Klein, G., and Winberg, G. (1993). Clonability and tumorigenicity of human epithelial cells expressing the EBV encoded membrane protein LMP1. Oncogene *8*, 1575-1583.

Huen, D.S., Henderson, S.A., Croom-Carter, D., and Rowe, M. (1995). The Epstein-Barr virus latent membrane protein-1 (LMP1) mediates activation of NF-kappa B and cell surface phenotype via two effector regions in its carboxy-terminal cytoplasmic domain. Oncogene *10*, 549-560.

Hummel, M., Anagnostopoulos, I., Dallenbach, F., Korbjuhn, P., Dimmler, C., and Stein, H. (1992). EBV infection patterns in Hodgkin's disease and normal lymphoid tissue: expression and cellular localization of gene products. Brit. J. Haematol. *82*, 689-694.

Jackson, H., and Parker, F. (1944). Hodgkin's disease: II. Pathology. N. Engl. J. Med. *231*, 35-44.

Jarrett, R.F., Gallagher, A., Jones, D.B., Alexander, F.E., Krajewski, A.S., Kelsey, A., Adams, J., Angus, B., Gledhill, S., Wright, D.H., Cartwright, R.A., and Onions, D.E. (1991). Detection of Epstein-Barr virus genomes in Hodgkin's disease: relation to age. J. Clin. Pathol. *44*, 844-848.

Jarrett, R.F., Lake, A., Andrew, L., et al. (2002). Somatic IkBa mutations are a frequent occurrence in Hodgkin's lymphoma. Blood *100*, 4333 (abstr).

Joos, S., Granzow, M., Holtgreve-Grez, H., Siebert, R., Harder, L., Martin-Subero, J.I., Wolf, J., Adamowicz, M., Barth, T.F., Lichter, P., and Jauch, A. (2003). Hodgkin's lymphoma cell lines are characterized by frequent aberrations on chromosomes 2p and 9p including REL and JAK2. Int. J. Cancer *103*, 489-495.

Jox, A., Zander, T., Kuppers, R., Irsch, J., Kanzler, H., Kornacker, M., Bohlen, H., Diehl, V., and Wolf, J. (1999). Somatic mutations within the untranslated regions of rearranged Ig genes in a case of classical Hodgkin's disease as a potential cause for the absence of Ig in the lymphoma cells. Blood *93*, 3964-3972.

Jundt, F., Kley, K., Anagnostopoulos, I., Schulze Probsting, K., Greiner, A., Mathas, S., Scheidereit, C., Wirth, T., Stein, H., and Dorken, B. (2002). Loss of PU.1 expression is associated with defective immunoglobulin transcription in Hodgkin and Reed-Sternberg cells of classical Hodgkin disease. Blood *99*, 3060-3062.

Jungnickel, B., Staratschek-Jox, A., Bräuninger, A., Spieker, T., Wolf, J., Diehl, V., Hansmann, M-L., Rajewsky, K., and Küppers, R. (2000). Clonal deleterious mutations in the IkBa gene in the malignant cells in Hodgkin's lymphoma. J. Exp. Med. *191*, 395-401.

Kanzler, H., Kuppers, R., Hansmann, M.L., and Rajewsky, K. (1996). Hodgkin and Reed-Sternberg cells in Hodgkin's disease represent the outgrowth of a dominant tumor clone derived from (crippled) germinal center B cells. J. Exp. Med. *184*, 1495-1505.

Kapp, U., Yeh, W-C., Patterson, B., Elia, A.J., Kagi, D., Ho, A., Hessel, A., Tipsword, M., Williams, A., Mirtsos, C., Itie, A., Moyle, M., and Mak, T.W. (1999). Interleukin 13 is secreted by and stimulates the growth of Hodgkin and Reed-Sternberg cells. J. Exp. Med. *189*, 1939-1946.

Khanna, R., Burrows, S.R., Nicholls, J., and Poulsen, L.M. (1998). Identification of cytotoxic T cell epitopes within Epstein-Barr virus (EBV) oncogene latent membrane protein 1 (LMP1): evidence for HLA A2 supertype-restricted immune recognition of EBV-infected cells by LMP1-specific cytotoxic T lymphocytes. Eur. J. Immunol. *28*, 451-458.

Khanim, F., Yao, Q-Y., Niedobitek, G., Sihota, S., Rickinson, A.B., and Young, L.S. (1996). Analysis of Epstein-Barr virus gene polymorphisms in normal donors and in virus-associated tumors from different geographic locations. Blood *88*, 3491-3501.

Kieser, A., Kilger, E., Gires, O., Ueffing, M., Kolch, W., and Hammerschmidt, W. (1997). Epstein-Barr virus latent membrane protein-1 triggers AP-1 activity via the c-Jun N-terminal kinase cascade. EMBO J. 16, 6478-6485.

Kim, S.H., Shin, Y.K., Lee, I-S., Bae, Y.M., Sohn, H.W., Suh, Y.H., Ree, H.J., Rowe, M., and Park, S.H. (2000). Viral latent membrane protein 1 (LMP-1) – induced CD99 down-regulation in B cells leads to the generation of cells with Hodgkin's and Reed-Sternberg phenotype. Blood *95*, 294-300.

Knecht, H., Bachmann, E., Brousset, P., Sandvej, K., Nadal, D., Bachmann, F., Odermatt, B.F., Delsol, G., and Pallesen, G. (1993). Deletions within the LMP1 oncogene of Epstein-Barr virus are clustered in Hodgkin's disease and identical to those observed in nasopharyngeal carcinoma. Blood *82*, 2937-2942.

Kube, D., Holtick, U., Vockerodt, M., Ahmadi, T., Haier, B., Behrmann, I., Heinrich, P.C., Diehl, V., and Tesch, H. (2001). STAT3 is constitutively activated in Hodgkin cell lines. Blood *98*, 762-770.

Küppers, R., Rajewsky, K., Zhao, M., Simons, G., Laumann, R., Fischer, R., Hansmann, M. (1994). Hodgkin disease: Hodgkin and Reed-Sternberg cells picked from histological sections show clonal immunoglobulin gene rearrangements and appear to be derived from B cells at various stages of development. Proc. Natl. Acad. Sci. USA *91*, 10962-10966.

Laherty, C.D., Hu, H.M., Opipari, A.W., Wang, F., and Dixit, V.M. (1992). The Epstein-Barr virus LMP1 gene product induces A20 zinc finger protein expression by activating nuclear factor kappa B. J. Biol. Chem. *267*, 24157-24160.

Lee, S.P., Tierney, R.J., Thomas, W.A., Brooks, J.M., and Rickinson, A.B. (1997). Conserved CTL epitopes within EBV latent membrane protein 2: a potential target for CTL-based tumor therapy. J. Immunol. *158*, 3325-3334.

Lee, S.P., Constandinou, C.M., Thomas, W.A., Croom-Carter, D., Blake, N.W., Murray, P.G., Crocker, J., and Rickinson, A.B. (1998). Antigen presenting phenotype of Hodgkin Reed-Sternberg cells: analysis of the HLA class I processing pathway and the effects of interleukin-10 on Epstein-Barr virus-specific cytotoxic T-cell recognition. Blood 92, 1020-1030.

Levine, P.H., Ablashi, D.V., Berard, C.W., Carbone, P.P., Waggoner, D.E., and Malan, L. (1971). Elevated antibody titers to Epstein-Barr virus in Hodgkin's disease. Cancer 27, 416-421.

Lukes, R.J., and Butler, J.J. (1966). The pathology and nomenclature of Hodgkin's disease. Cancer Res. *26*, 1063-1081.

MacMahon, B. (1966). Epidemiology of Hodgkin's disease. Cancer Res. *26*, 1189-1200.

Marafioti, T., Hummel, M., Anagnostopoulos, I., Foss, H.D., Falini, B., Delsol, G., Isaacson, P.G., Pileri, S., and Stein, H. (1997). Origin of nodular lymphocyte-predominant Hodgkin's disease from a clonal expansion of highly mutated germinal-center B cells. New Engl. J. Med. *337*, 453-458.

Marafioti, T., Hummel, M., Foss, H-D., Laumen, H., Korbjuhn, P., Anagnostopoulos, I., Lammert, H., Demel, G., Theil, J., Wirth, T., and Stein, H. (2000). Hodgkin and Reed-Sternberg cells represent an expansion of a single clone originating from a germinal centre B-cell with functional immunoglobulin gene rearrangements but defective immunoglobulin transcription. Blood *95*, 1443-1450.

Martin-Subero, J.I., Gesk, S., Harder, L., Sonoki, T., Tucker, P.W., Schlegelberger, B., Grote, W., Novo, F.J., Calasanz, M.J., Hansmann, M.L., Dyer, M.J., and Siebert, R. (2002). Recurrent involvement of the REL and BCL11A loci in classical Hodgkin lymphoma. Blood 99, 1474-1477.

Mathas, S., Hinz, M., Anagnostopoulos, I., Krappmann, D., Lietz, A., Jundt, F., Bommert, K., Mechta-Grigoriou, F., Stein, H., Dorken, B., and Scheidereit, C. (2001). Aberrantly expressed c-Jun and JunB are a hallmark of Hodgkin lymphoma cells, stimulate proliferation and synergize with NF-kappa B. EMBO J. *21*, 4104-4113.

Miller, C.L., Burkhardt, A.L., Lee, J.H., Stealey, B., Longnecker, R., Bolen, J.B., and Kieff, E. (1995). Integral membrane protein 2 of Epstein-Barr virus regulates reactivation from latency through dominant negative effects on protein-tyrosine kinases. Immunity *2*, 155-166.

Morente, M.M., Piris, M.A., Abraira, V., Acevedo, A., Aguilera, B., Bellas, C., Fraga, M., Garcia-Del-Moral, R., Gomez-Marcos, F., Menarguez, J., Oliva, H., Sanchez-Beato, M., and Montalban, C. (1997). Adverse clinical outcome in Hodgkin's disease is associated with loss of retinoblastoma protein expression, high Ki67 proliferation index, and absence of Epstein-Barr virus-latent membrane protein 1 expression. Blood *90*, 2429-2436.

Mueller, N., Evans, A., Harris, N.L., Comstock, G.W., Jellum, E., Magnus, K., Orentreich, N., Polk, B.F., and Vogelman, J. (1989). Hodgkin's disease and Epstein-Barr virus: Altered antibody pattern before diagnosis. New Engl. J. Med. *320*, 689-695.

Murray, P.G., Young, L.S., Rowe, M., and Crocker, J. (1992). Immunohistochemical demonstration of the Epstein-Barr virus - encoded latent membrane protein in paraffin sections of Hodgkin's disease. J. Pathol. *166*, 1-5.

Murray, P.G., Constandinou, C.M., Crocker, J., Young, L.S., and Ambinder, R.F. (1998). Analysis of major histocompatibility complex class I, TAP expression, and LMP2 epitope sequence in Epstein-Barr virus-positive Hodgkin's disease. Blood 92, 2477-2483.

Murray, P.G., Billingham, L.J., Hassan, H.T., Flavell, J.R., Nelson, P.N., Scott, K., Reynolds, G., Constandinou, C.M., Kerr, D.J., Devey, E.C., Crocker, J., and Young, L.S. (1999). Effect of Epstein-Barr virus infection on response to chemotherapy and survival in Hodgkin's disease. Blood *94*, 442-447.

Müschen, M., Rajewsky, K., Bräuninger, A., Baur, A.S., Oudejans, J.J., Roers, A., Hansmann, M.L., and Küppers, R. (2000). Rare occurrence of classical Hodgkin's disease as a T cell lymphoma. J. Exp. Med. *191*, 387-394.

Niedobitek, G., Kremmer, E., Herbst, H., Whitehead, L., Dawson, C.W., Niedobitek, E., von Ostau, C., Rooney, N., Grässer, F.A., and Young, L.S. (1997). Immunohistochemical detection of the Epstein-Barr Virus - Encoded Latent Membrane Protein 2A in Hodgkin's Disease and infectious mononucleosis. Blood *90*, 1664-1672.

Nakagomi, H., Dolcetti, R., Bejarano, M.T., Pisa, P., Kiessling, R., and Masucci, M.G. (1994). The Epstein-Barr-virus latent membrane protein-1 (LMP1) induces interleukin-10 production in Burkitt-lymphoma lines. Int. J. Cancer *57*, 240-244.

Oudejans, J.J., Jiwa, N.M., Kummer, J.A., Horstman, A., Vos, W., Baak, J.P., Kluin, P.M., van der Valk, P., Walboomers, J.M., and Meijer, C.J.L.M. (1996). Analysis of major histocompatibility complex class I expression on Reed-Sternberg cells in relation to the cytotoxic T-cell response in Epstein-Barr virus-positive and -negative Hodgkin's disease. Blood *87*, 3844-3851.

Oudejans, J.J., Jiwa, N.M., Kummer, J.A., Ossenkoppele, G.J., van Heerde, P., Baars, J.W., Kluin, P.M., Kluin-Nelemans, J.C., van Diest, P.J., Middeldorp, J.M., and Meijer, C.J. (1997). Activated cytotoxic T cells as prognostic marker in Hodgkin's disease. Blood *89*, 1376-1382.

Pallesen, G., Hamilton-Dutoit, S.J., Rowe, M., and Young, L.S. (1991). Expression of Epstein-Barr virus latent gene products in tumour cells of Hodgkin's disease. Lancet *337*, 320-322.

Pinto, A., Gattei, V., Zagonel, V., Aldinucci, D., Degan, M., De Iuliis, A., Rossi, F.M., Tassan Mazzocco, F., Godeas, C., Rupolo, M., Poletto, D., Gloghini, A., Carbone, A., and Gruss, H.J. (1998). Hodgkin's disease: a disorder of dysregulated cellular cross-talk. Biotherapy *10*, 309-320.

Poppema, S., Potters, M., Visser, L., and van den Berg, A.M. (1998). Immune escape mechanisms in Hodgkin's disease. Ann. Oncol. *9,* Suppl 5: S21-24.

Portis, T., Dyck, P., and Longnecker, R. (2003). Epstein-Barr Virus (EBV) LMP2A induces alterations in gene transcription similar to those observed in Reed-Sternberg cells of Hodgkin lymphoma. Blood *102,* 4166-4178.

Razis, D.V., Diamond, H.D., and Craver, L.F. (1959). Familial Hodgkin's disease: Its significance and implications. Ann. Intern. Med. *51*, 933-971.

Razzouk, B.I., Srinivas, S., Sample, C.E., Singh, V., and Sixbey, J.W. (1996). Epstein-Barr Virus DNA Recombination and Loss in Sporadic Burkitt's Lymphoma. J. Infect. Dis. *173*, 529-535.

Re, D., Muschen, M., Ahmadi, T., Wickenhauser, C., Staratschek-Jox, A., Holtick, U., Diehl, V., and Wolf, J. (2001). Oct-2 and Bob-1 deficiency in Hodgkin and Reed Sternberg cells. Cancer Res. *61*, 2080-2084.

Roberts, M.L., and Cooper, N.R. (1998). Activation of a ras-MAPK-dependent pathway by Epstein-Barr virus latent membrane protein 1 is essential for cellular transformation. Virology *240,* 93-99.

Roskrow, M.A., Suzuki, N., Gan, Y.J., Sixbey, J.W., Ng, C.Y., Kimbrough, S., Hudson, M., Brenner, M.K., Heslop, H.E., and Rooney, C.M. (1998). Epstein-Barr virus (EBV)-specific cytotoxic T lymphocytes for the treatment of patients with EBV-positive relapsed Hodgkin's disease. Blood *91*, 2925-2934.

Rosenthal, S.R. (1936). Significance of tissue lymphocytes in the prognosis of lymphogranulomatosis. Arch. Pathol. *21*, 628-631.

Rowe, M., Peng-Pilon, M., Huen, D.S., Hardy, R., Croom-Carter, D., Lundgren, E., Rickinson, A.B. (1994). Upregulation of bcl-2 by the Epstein-Barr virus latent membrane protein LMP1: a B-cell-specific response that is delayed relative to NF-kappa B activation and to induction of cell surface markers. J. Virol. *68*, 5602-5612.

Santon, A., Martin, C., Manzanal, A.I., Preciado, M.V., and Bellas, C. (1998). Paediatric Hodgkin's disease in Spain: association with Epstein-Barr virus strains carrying latent membrane protein-1 oncogene deletions and high frequency of dual infections. Brit. J. Haematol. *103*, 129-136.

Schwering, I., Brauninger, A., Klein, U., Jungnickel, B., Tinguely, M., Diehl, V., Hansmann, M.L., Dalla-Favera, R., Rajewsky, K., and Kuppers, R. (2003). Loss of the B-lineage-specific gene expression program in Hodgkin and Reed-Sternberg cells of Hodgkin lymphoma. Blood *101*, 1505-1512.

Skinnider, B.F., Elia, A.J., Gascoyne, R.D., Patterson, B., Trumper, L., Kapp, U., and Mak, T.W. (2002). Signal transducer and activator of transcription 6 is frequently activated in Hodgkin and Reed-Sternberg cells of Hodgkin lymphoma. Blood *99*, 618-626.

Sing, A.P., Ambinder, R.F., Hong, D.J., Jensen, M., Batten, W., Petersdorf, E., and Greenberg, P.D. (1997). Isolation of Epstein-Barr virus (EBV)-specific cytotoxic T lymphocytes that lyse Reed-Sternberg cells: implications for immune-mediated therapy of EBV+ Hodgkin's disease. Blood *89*, 1978-1986.

Staratschek-Jox, A., Kotkowski, S., Belge, G., Rudiger, T., Bullerdiek, J., Diehl, V., and Wolf, J. (2000). Detection of Epstein-Barr virus in Hodgkin-Reed-Sternberg cells: No evidence for the persistence of integrated viral fragments in Latent membrane protein-1 (LMP-1)-negative classical Hodgkin's disease. Am. J. Pathol. *156*, 209-216.

Stark, G.L., Wood, K.M., Jack, F., Angus, B., Proctor, S.J., Taylor, P.R.; Northern Region Lymphoma Group (2002). Hodgkin's disease in the elderly: a population-based study. Br. J. Haematol. 119, 432-440.

Stoler, M.H., Nichols, G.E., Symbula, M., and Weiss, L.M. (1995). Lymphocyte predominance Hodgkin's disease - evidence for a kappa light chain-restricted monotypic B-cell neoplasm. Am. J. Pathol. *146*, 812-818.

Su, Z., Peluso, M.V., Raffegerst, S.H., Schendel, D.J., and Roskrow, M.A. (2001). The generation of LMP2a-specific cytotoxic T lymphocytes for the treatment of patients with Epstein-Barr virus-positive Hodgkin disease. Eur. J. Immunol. *31*, 947-958.

Torlakovic, E., Tierens, A., Dang, H.D., and Delabie, J. (2001). The transcription factor PU.1, necessary for B-cell development is expressed in lymphocyte predominance, but not classical Hodgkin's disease. Am. J. Pathol. *159*, 1807-1814.

Theil, J., Laumen, H., Marafioti, T., Hummel, M., Lenz, G., Wirth, T., and Stein, H. (2001). Defective octamer-dependent transcription is responsible for silenced immunoglobulin transcription in Reed-Sternberg cells. Blood *97*, 3191-3196.

Uccini, S., Monardo, F., Stoppacciaro, A., Gradilore, A., Agliano, A.M., Faggioni, A., Mazari, V., Vaso, L., Costanzi, G., Ruco, P., and Baroni, C.D. (1990). High frequency of Epstein-Barr virus-genome detection in Hodgkin's disease of HIV-positive patients. Int. J. Cancer *46*, 581-585.

van den Berg, A., Visser, L., and Poppema, S. (1999). High expression of the CC chemokine TARC in Reed-Sternberg cells. A possible explanation for the characteristic T-cell infiltratein Hodgkin's lymphoma. Am. J. Pathol. *154*, 1685-1691.

Wagner, H.J., Bollard, C.M., Vigouroux, S., Huls, M.H., Anderson, R., Prentice, H.G., Brenner, M.K., Heslop, H.E., and Rooney, C.M. (2004). A strategy for treatment of Epstein-Barr virus-positive Hodgkin's disease by targeting interleukin 12 to the tumor environment using tumor antigen-specific T cells. Cancer Gene Ther. *11*, 81-91.

Wang, F., Gregory, C., Sample, C., Rowe, M., Liebowitz, D., Murray, R., Rickinson, A., and Kieff, E. (1990). Epstein-Barr virus latent membrane protein (LMP1) and nuclear proteins 2 and 3C are effectors of phenotypic changes in B lymphocytes: EBNA-2 and LMP1 cooperatively induce CD23. J. Virol. *64*, 2309-2318.

Wang, S., Rowe, M., and Lundgren, E. (1996). Expression of the Epstein Barr virus transforming protein LMP1 causes a rapid and transient stimulation of the Bcl-2 homologue Mcl-1 levels in B-cell lines. Cancer Res. *56*, 4610-4613.

Weinreb, M., Day, P.J.R., Niggli, F., Powell, J.E., Raafat, F., Hesseling, P.B., Schneider, J.W., Hartely, P.S., Tzortzatous-Stathopoulou, F., Khalek, E.R.A., Mangoud, A., El-Safy, U.R., Madanat, F., Sheyyab, M.A., Mpofu, C., Revesz, T., Rafii, R., Tiedemann, K., Waters, K.D., Barrantes, J.C., Nyongo, A., Riyat, M.S., and Mann, J.R. (1996). The role of Epstein-Barr virus in Hodgkin's disease from different geographical areas. Arch. Dis. Child. *74*, 27-31.

Weiss, L.M., Strickler, J.G., Warnke, R.A., Purtilo, D.T., and Sklar, J. (1987). Epstein-Barr viral DNA in tissues of Hodgkin's disease. Am. J. Pathol. *129*, 86-91.

Weiss, L.M., Movahed, L.A., Warnke, R.A., and Sklar, J. (1989). Detection of Epstein-Barr viral genomes in Reed-Sternberg cells of Hodgkin's disease. New Engl. J. Med. *320*, 502-506.

Weiss, L.M., Chen, Y-Y., Liu, X-F., and Shibata, D. (1991). Epstein-Barr virus and Hodgkin's disease: A correlative *in situ* hybridization and polymerase chain reaction study. Am. J. Pathol. *139*, 1259-1265.

Wu, T.C., Mann, R.B., Charache, P., Hayward, S.D., Staal, S., Lambe, B.C., and Ambinder, R.F. (1990). Detection of EBV gene expression in Reed-Sternberg cells of Hodgkin's disease. Int. J. Cancer *46*, 801-804.

Zhou, X-G., Hamilton-Dutoit, S.J., Yan, Q-H., and Pallesen, G. (1993). The association between Epstein-Barr virus and Chinese Hodgkin's disease. Int. J. Cancer *55*, 359-363.

From: Epstein-Barr Virus. Edited by: Erle S. Robertson

Chapter 9

Endemic Burkitt's Lymphoma

Beverly E. Griffin and Rosemary Rochford*

ABSTRACT

Endemic Burkitt's lymphoma (BL) is sometimes called 'African Burkitt's lymphoma' because it occurs at the highest frequencies in children in Equatorial Africa. It also is a major childhood cancer, albeit occurring in lower frequencies, in locations as diverse as Papua New Guinea and Bahia, northeastern Brazil. This intriguing geographical bias is only partly explainable. Endemic BL, as with sporadic and immunodeficiency-related BL, shares translocations that involve the oncogenic c-*myc* and immunoglobulin genes, but how, and to what extent, these tumors differ in their properties is still a matter for debate and exploration. Wright (1999) asked the question, 'what is Burkitt's Lymphoma and when is it endemic'? We address in more detail this query but, like him, cannot provide a wholly satisfactory answer. For the moment, the numbers of children who still succumb to BL are great enough that other topics, from an humanitarian point of view, may rightly be deemed to take precedence in dealing with this malignancy. However, the control and/or containment of this tumor, in the longer term, may rely on solving some of the problems that deal with endemic aspects of the disease.

INTRODUCTION AND HISTORICAL OBSERVATIONS

Reviewing the history of BL involves reading about this childhood cancer in sub-Saharan Africa, starting with the publication by Denis Burkitt in 1958 on the lymphoma, called by him a lymphosarcoma, that now bears his name (Burkitt, 1958) and following the topic for the next ten or so years. This is well worth the effort since much of what we know today on the subject, in broad outline, initiated from this period. Background reading is made easy if one can obtain access to the book, 'Cancer in Africa', published in 1968 by the East African Medical Journal (Clifford, 1968). There, the role of infectious agents, peak incidences, age and sex distribution, geographical 'hotspots', preferred treatment protocols, *et cetera*, is discussed by pioneers on this topic. They define the Burkitt's 'lymphoma belt' of Africa as extending from about 10° degrees north to 10° degrees south of the equator, with an edge in East Africa that runs south to about 15° degrees (Figure 1). This encompasses the so-called 'Great Lakes' region of Africa. Observations of

*For correspondence email b.griffin@imperial.ac.uk

Figure 1. Map of Africa with "Lymphoma Belt". Shown is a current map of Africa. The region within the bold line highlights the countries that Burkitt originally identified as the "lymphoma belt", *i.e.* countries where BL tumors were commonly seen in hospitals (Burkitt, 1962).

the geographic distribution, defined by altitude, rainfall and temperature, prompted the suggestion by Dalldorf, an American microbiologist, that BL represents a malignancy induced by a mosquito-borne agent (Dalldorf and Barnhart, 1972; Dalldorf et al., 1964). Burkitt, credited with the discovery of the tumor, stated later in an interesting personal history, that during his mapping journeys, whereas he thought he was identifying the location and extent of the tumor itself, what he was really doing was producing a map of malaria transmission in sub-Saharan Africa as the BL cases he identified were found in areas holoendemic for malaria (Burkitt, 1987). Wright (1971), in a review of the existing data on BL, suggested that three infectious agents, the Epstein-Barr virus (EBV), reovirus-3, and *Plasmodium falciparum*, were emerging as having possible etiologic importance. Whereas two of these agents remain relevant to the genesis of BL, as will be discussed below, reovirus-3 has been discarded as having any role in this malignancy.

A remarkably prevalent malignant tumor of children, endemic BL was originally designated as a sarcoma of the jaw, but was later determined to be a lymphatic tumor that could also arise elsewhere in the body (O'Conor and Davies, 1960). The following description of BL appears in the early literature (Editorial,

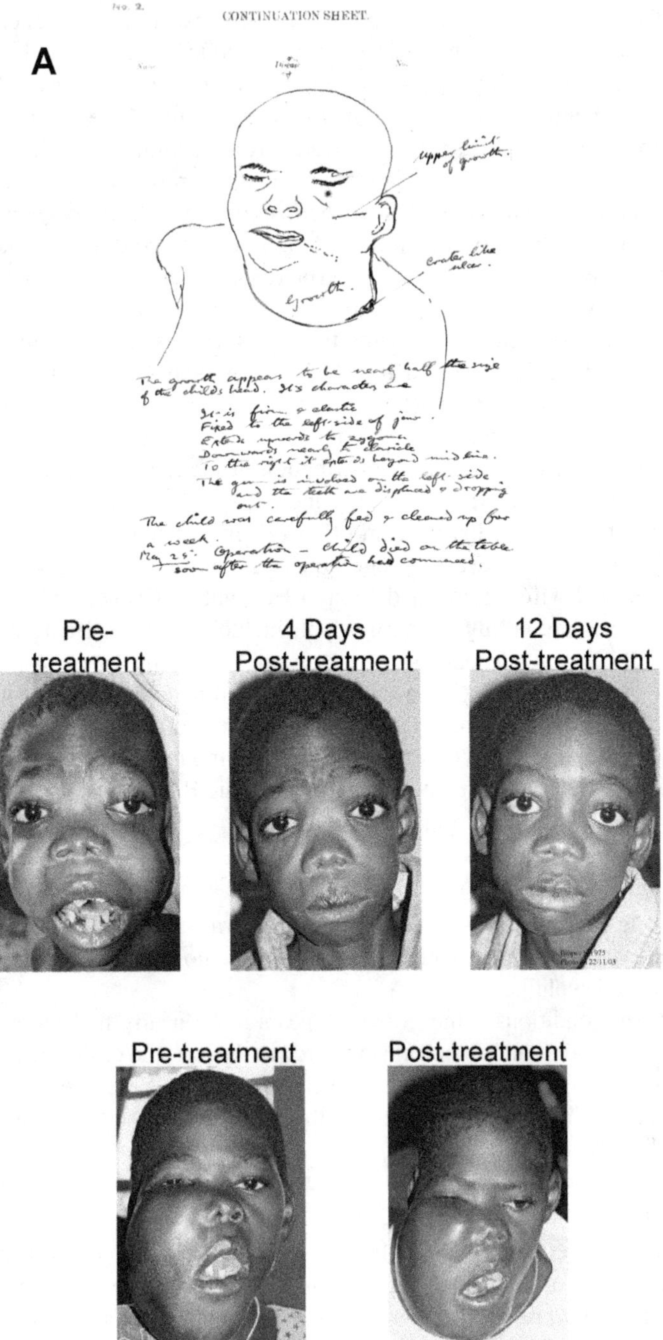

Figure 2. Clinical picture of BL. (A) Copy of original drawing from the records of the Mengo Hospital, 1899 (Davies et al., 1964. Orignal from 1910 in Albert Cook medical library, Kampala, Uganda). This is likely the first recorded drawing of a patient with BL. (B) Child with BL successfully treated with chemotherapy. (C) Child with BL refractory to chemotherapy.

1964), as follows: 'The commonest tumor in the Kampala (Uganda) cancer registry is a lymphosarcoma of young children...now called the African lymphoma or the Burkitt tumor...found almost entirely in children aged 2 to 14 years. It usually presents as a swelling of the jaw, often affecting both mandible and maxilla. The teeth are loosened, and as the orbit is invaded (this) culminates in the loss of the eye. The first indication may be an abdominal mass...probably due to involvement of the para-aortic lymph nodes (which when) spread to the vertebral column leads to paraplegia. Occasionally the kidneys, long bones, thyroid, testes and salivary glands are massively infiltrated...the blood picture is not characteristic of leukaemia. Death usually occurs within four to six months, but dramatic improvement may follow...therapy, and remissions lasing two years have occurred.' This remains a fairly accurate clinical picture of BL in Africa even today (Figure 2).

Burkitt proposed that BL must be an 'old' tumor, and not one that had suddenly arisen in Africa due to ecological, economic or other changes, since on a ten week safari taken to map the geographic distribution of malignancy (with T. Williams and C. Nelson) – using $1500 of grant money – he saw old wooden carvings reminiscent of BL. Long before Burkitt's work in Uganda, in 1897, the British surgeon Albert Cook funded by a missionary society and, accompanied by his medically-trained wife, established Mengo Hospital in Uganda, where he dealt with all diseases presenting to the hospital, including cancer. Being meticulous in approach, Cook made extensive records of interesting cases. Among them was a patient who was carrying a tumor that would undoubtedly later have been designated 'a BL'. Cook's records, re-evaluated in two early publications in the British Medical Journal (Davies et al., 1964), have among them a case report which we think is worthy of note in any chapter on endemic BL. It says:

Name: Jejefu Omukalazi, male child

Date of admission: 18 May, 1910

Date of death: 25 May, 1910

History: Present illness – swelling began underneath the jaw one month ago. The eye began to be affected two weeks ago...no coughing. Previous illness – has had syphilis.

Present conditions – the growth appears to be nearly half the size of the child's head. It is firm and elastic, fixed to the left side of the jaw, extending upwards to zygoma, downward to clavical. To the right, it extends beyond medline. The gum is involved on the left side and the teeth are displaced and dropping out.

May 25, operation: Child died on the table soon after the operation had commenced.

The child, Jejefu Omukalazi, 1910, must be among the earliest medically documented cases of BL. Since syphilis, as well as small pox, sleeping sickness, and infant mortality, were such commonplace events in Cook's experience (Cook, 1945), he may have seen this case as another manifestation of syphilis, but Davies et al. (1964) arguably gave credit to him for first documenting a case of 'BL'. Cook's original records are housed in the Albert Cook Library of Makere University, Uganda, with copies in the Wellcome Trust Library, London, U.K.

The "lymphoma belt" as first described for African BL can be extended to other countries spanning the equator. The first cases of BL in Papua New Guinea were reported as long ago as 1966 (Ten Seldam et al., 1966) and in Brazil in 1968 (Fagundes et al., 1968). Although the incidence of BL is not as high in Papua New Guinea as in sub-Saharan Africa, there too BL remains the commonest tumor of childhood (Tefuarani et al., 1988; Winnett et al., 1997).

The global epidemiology of BL to 1967 was reviewed by Wright (1967). A more recent review can be found in 'Cancer in Africa –epidemiology and prevention' (2003). However, many technical issues remain problematic in any comprehensive epidemiologic description of endemic BL. For example, a lack of cancer registries in many African countries results in underreporting of total numbers of cases. Other factors include the involvement of 'traditional healers', lack of patient access to hospital as many cases arise from rural areas with poor transport, and a reliance on hospital-based, often inadequate, data for estimation of BL incidence rates. Definition of case frequently is based on clinical features without good diagnostic facilities, which can also complicate estimates of BL cases. Nonetheless, the available data clearly show that endemic BL remains the most common childhood cancer in Africa, and is a high frequency tumor from east to west in sub-Saharan African. It remains to be seen how the epidemic of AIDS will impact on this disease. Although the available data suggest that it is not a "risk factor" for endemic BL, it clearly affects mortality.

CLINICAL PATTERNS OF BL

There has always been an element of uncertainty about whether BL constitutes 'a syndrome', that is, whether it is a unique clinical entity that can be separated from other lymphomas on the basis of morphologic and pathologic features. It is designated a 'non-Hodgkin's lymphoma'. Histologically, BL is classically characterized by a 'monotonous' proliferation of B-cells, 10-25 μm in diameter, with round or oval nuclei and prominent basophilic nucleoli. Benign 'starry-sky' macrophages are often, but not invariably, present among the tumor population. The tumors have an extremely high rate of proliferation, both clinically, and as characterized histologically by their high mitotic indices. In a World Health Organization 2001 classification, *endemic* is one of three sub-categories of BL, the other two being sporadic (or non-endemic) and AIDS-associated BL (Bellan et al., 2003). All BLs carry a chromosomal translocation that results in deregulation (essentially, upregulation) of the c-*myc* oncogene. With endemic BL, the association of Epstein-Barr virus (EBV) with primary tumors in some parts of Equatorial Africa approaches 100 percent (Labrecque et al., 1994), which is not the case with the other categories. Sporadic BL rarely is associated with EBV-infection, while AIDS-BL in US or European patients is associated with EBV infection in roughly 30% of the cases. A recent study of BL in Brazil suggests that there is a strong association with EBV at levels nearing those seen in the Equatorial African cases (Araujo et al., 1996; Bacchi et al., 1996). There may be other differences in EBV association. For example, our data suggest that EBV may not be associated with all BLs that occur as relapses following treatment, even in cases where clinically and morphologically the primary and relapse tumors from a single patient appear indistinguishable (Xue et al., 2002). Whether this means that the relapsed tumor is a non-endemic form of BL arising from a separate malignant event, or that the relapse reflects a 'hit and run' event by the virus, has not been investigated.

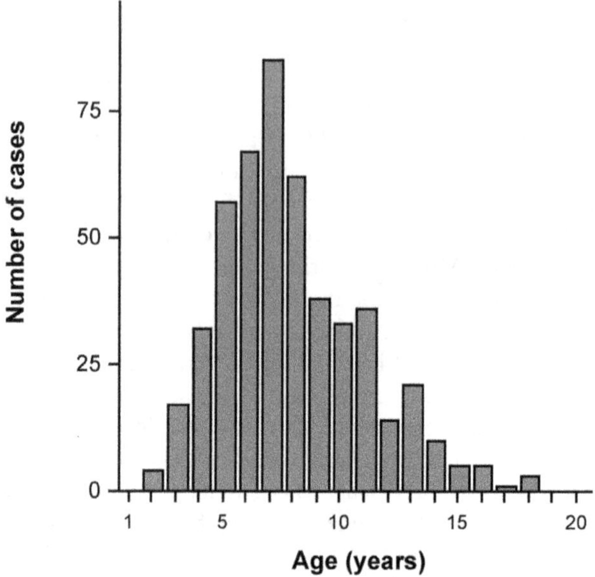

Figure 3. Age distribution of Burkitt's lymphoma patients. Cytologically proven cases of BL were compiled from 01/01/1990 to 08/20/2004 from the Burkitt's Lymphoma Unit, Kamuzu Central Hospital, Lilongwe, Malawi.

The variation in EBV-association with the different forms of BL has led some to speculate that EBV is merely a passenger in this malignancy. However, as will be discussed below, there is strong evidence that EBV-infection in primary cases of endemic BL is linked to the etiology of this malignancy.

The cellular origin of the endemic BL has long been thought to be a germinal center B-cell based on phenotypic expression of the B-cell surface marker, CD10 (Gregory et al., 1987), as well as genetic evidence of somatic hypermutation in the variable region of the immunoglobulin genes (Kuppers et al., 1999). More recently, the argument was made that BL might be a tumor of latently infected memory B-cells since such EBV-infected cells, with a similar EBV expression profile to BL, have been identified in the peripheral blood of 'normal' individuals (Hochberg et al., 2004). However, these studies were carried out on individuals with acute infectious mononucleosis (AIM), so it is not clear that this argument can be extended to endemic BL. The question still remains why a tumor such as BL, if derived from a germinal center B cell (ie., found in lymph nodes) is found extranodally. It may be relevant to keep in mind that a tumor cell may have evolved properties that differ from those of its progenitor, which complicates questions of origin.

Although there are minor variations among 'lymphoma belt' African countries, the general picture of BL is one of a predominance of male to females (roughly 2:1) and an age profile with a major peak (greater than 80% of patients) between 5-9 years of age and a second (minor) peak between about 11-14 years (Labrecque et al., 1999; Morrow et al., 1976; Mwanda et al., 2004; Nkrumah and Olweny, 1985) (see Figure 3). In Malawi, a BL case series (>1000 cases) covering a 14 year period (1990-2003, inclusive) was made, based on diagnosis by clinical and histopathological evaluation. Eight-seven percent of BLs fell within the major age peak - highest incidence age 7 years – with 13% of BLs in the second age peak - highest incidence age 12-13 years (J. Phillips, personal communication). In all published histograms covering age incidence, this double peak pattern is seen, although it has not been studied in detail to determine whether it reflects a significant etiology.

A unique feature of endemic BL, as noted, is the presentation of this cancer in extranodal sites and absence of lymph node involvement. An early clinical study on 110 BL patients in Ghana (Nkrumah and Perkins, 1976) showed a characteristic clinical and anatomical pattern (discussed, Editorial, 1964). In this report, the overwhelming majority of endemic BL patients had facial (jaw) and abdominal involvement. Among the patients, only 33% of age 6 BL patients presented, although not exclusively, with abdominal tumors, a figure that rose to nearly 50% at age 12, and to 60% by age 14. Only rarely (2%) did a patient present with a tumor at a site exclusive of the facial or abdominal region. Data from Ghana (Nkrumah and Olweny, 1985) and elsewhere (Berard, 1985) showed jaw presentations to be more common in younger age groups, whereas abdominal involvement increased in incidence among older patients. One suggestion for the high prevalence of jaw tumors in younger children is the association with developing tooth buds in this age group, as noted by a number of investigators. Why this would favour emergence of a lymphoma at that site is unknown.

In reality, BL patients often have tumors at more than one site, making clear pattern assessments difficult. The problem of diagnosis in endemic regions also creates problems for classical tumor staging. The Ugandan BL staging protocol (Olweny and Nkrumah, 1985), adopted in much early work and still widely used in Africa today, sometimes modified (Magrath, 1991), is based on clinical data of BL at presentation. This staging has the merit of being straight-forward, and easy to follow and teach. Stage A (or I) in this protocol consists of a single facial tumor, stage B (II) includes multiple facial tumors or other sites except intra-abdominal and intrathoracic tumors or central nervous system (CNS) involvement. Stage C (III) has tumors at any intrathoracic, intra-abdominal or bone sites except the CNS, and stage D (IV) includes CNS involvement. In a Ugandan study on 280 patients for which there was sufficient data to assume accurate staging, over 60% of cases were either stage C or D on admission (Olweny et al., 1980). A more comprehensive and complicated staging protocol (the St. Jude protocol) is often used for all non-Hodgkin's lymphomas (which includes BL) in more affluent settings. As discussed by Shad and Magrath (1998), no staging protocol is perfect and all leave room for error. For example, the Ugandan staging does not distinguish patients with bone marrow disease from those with CNS disease. This may, however, be more a problem for AIDS-related than endemic BL. In the future, it may prove possible to add molecular markers to clinical staging, to improve their utility. Moreover, in sub-Saharan Africa, since many - indeed most - patients are initially seen at district hospitals, improved skills at these sites should lead to earlier referral of children to the main urban hospitals where problems associated with disease staging can be better managed.

TREATMENT OF ENDEMIC BL

Peter Clifford, a British surgeon working in Nairobi, Kenya–who provided tumor samples that led to the establishment of many of the currently studied BL lines–published papers on the preferred treatment protocols for BL patients (Clifford, 1966; Clifford et al., 1967) following earlier collaborative studies with Burkitt and others (Burkitt, 1967; Oettgen et al., 1963). By then, the general conclusion was that surgery, although essential for taking confirmatory biopsies and on occasion reducing tumor bulk, played no curative role in this disease. Today, fine needle aspirates, FNAs, have largely replaced surgery for obtaining confirmatory biopsies, although surgery may be advised still for 'de-bulking' very large tumors (see Figure 2C) *prior to* treatment. The necessity of testing alternative treatment protocols that did not rely exclusively on radiotherapy, the paradigm of cancer treatment at the time, led to the first demonstration that a human cancer could be 'cured' by chemotherapy alone. Clifford concluded that 'response was not related to tumor size or site or to dosage' and that observed patient variation in response to treatment might reflect the reactivity of the host to the tumor. Similar observations have been made more recently on patients in Malawi (P. Kazembe, personal communication). But to date, the reasons for this variation remain unidentified. The main problems in treatment of BLs then – diagnosis, scarcity of drugs, patient compliance and variation in host response – still remain major problems in the management of endemic BL.

The seminal papers in the field of treatment of this malignancy, however, are attributable to an American, John Ziegler, and his colleagues, working on BL in Uganda (Olweny et al., 1980; Ziegler, 1972; Ziegler et al., 1970). Their treatment protocol was one based on single doses of the alkylating agent, cyclophosphamide, given at 40 mg/kg patient's body weight with 6 doses administered at 2-3 week intervals. This protocol has many advantages in terms of cost and reduced toxicity to the patient. As a first line treatment, it can result in tumor responsiveness in 50-60% of cases. In the Ugandan studies, about 50% of the responsive patients ultimately presented with relapse tumors with roughly equal numbers before 3 months, or much later. Overall, in the Ugandan experience, about 25% of all treated patients survived free of disease beyond 5 years and were assumed to be 'cured' (Ziegler, 1981).

There remains a fair amount of 'trial and error' in the treatment of BL cases in Equatorial Africa, often as a consequence of drug shortages. Methotrexate, administered alone or usually in combination with other drugs, has some advantages in the treatment of non-responsive (to cyclophosphamide) or relapse BL cases. A recent field study in Malawi shows that the cyclophosphamide monotherapy protocol, with careful patient management, can result in up to 50% 'cures' being effected (Kazembe et al., 2003). In attempts to improve on this, a multi-drug experimental protocol, carried out with international financial support, achieved a nearly 90% survival rate for stage I (or A) patients, but less than 60% where patients were in stages II or III (B, C). Stage IV (CNS involvement) patients were not included in this trial (Hesseling et al., 2003). Problems with toxicity—a serious issue where supportive care is limited— and patient 'drop-out' from the trial were experienced in this study. Outside Africa, such protocols have been successfully adopted, and have achieved nearly 90% success rate. In general, the cure rates for treatment have appeared to vary depending on the stage at presentation with the most successful rates seen in those with Stage I or II BL (Spreafico et al., 2002). However, in Africa, Stage III and IV BL are the most frequent patterns observed and the response rate for Stage III can be as low as 50%, with multidrug programs (Hesseling et al., 2003). Stage III/IV toxicity ranges up to 41% for infection have been reported, 36% for hematologic toxicity, and 17% for gastrointestinal toxicity (Hesseling et al., 2003). Relapse and resistance to first-line chemotherapy drugs are also frequently observed (Olweny and Nkrumah, 1985). Thus, multidrug chemotherapy based on available reagents does not at this stage seem the clinical solution for endemic BL in Equatorial Africa (Rochford et al., 2005).

In the EBV field, new approaches for treating patients refractive to chemotherapy are being explored. Such strategies include prevention of viral oncogene expression, induction of EBV episome loss, or alternatively, induction of EBV lytic cycle-related gene expression to enhance a host immune response to virally-encoded antigens [reviewed (Israel and Kenney, 2003)]. Notably, in endemic BLs, the EBV genome is often present in extraordinarily high copy numbers in virtually 100% of tumors (Stevens et al., 1999), although found only in a small minority of circulating B-lymphocytes in normal individuals (Thorley-Lawson and Gross, 2004). This heavy viral load leads to the possibility that EBV might be harnessed to produce auto-destruction of the tumor, a process called 'oncolysis'. So-called 'oncolysis' approaches are currently in clinical trials aimed at destroying cancer cells by injection of replication competent viruses, such as herpes simplex, into tumors. A pilot study underway in Malawi, at Kamuzu Central Hospital in Lilongwe, involves the use of an adjuvant therapy. In this case, a chemical reagent that is capable of inducing the EBV lytic cycle is given in addition to cyclophosphamide, for the treatment of patients who fail to respond to the cyclophosphamide alone. The rationale of this study is to invoke involvement of the immune system (by activating genes recognized by the host) as well as induce oncolysis in tumors. Early pilot studies show this approach to be essentially non-toxic, and to show promise. A similar conclusion comes from studying SCID mice models carrying EBV-associated tumors, where stimulating viral lytic gene expression was found to enhance therapeutic efficiencies of chemotherapeutic reagents (Feng et al., 2004; Westphal et al., 2000). A pilot anti-viral therapy study on AIDS-associated BL involves treating tumors with the drug AZT to induce NF-κB; using this approach, efficacy has been demonstrated for relapsed BLs (Kurokawa et al., 2005). No therapeutic approach, however, in the long term would seem preferable to a vaccine which interfered with EBV infection of children in endemic regions of the world, and by inference, blocked tumor formation. Recent data from Phase II clinical trials using an anti-gp350 (neutralizing antibody) vaccine have shown some success in prevention of EBV infection in teenagers. Whether such a vaccine could be used to prevent EBV infection in children in endemic BL areas remains to be seen.

ETIOLOGY OF BL
A role for malaria
Based on the restricted geographic distribution of BL and the age of onset, early investigators hypothesized that an infectious agent, possibly a vectored virus such

as an arbovirus, might be etiologically linked with this cancer, (Burkitt, 1962; Haddow, 1963; Dalldorf et al. 1964). It quickly became evident that BL occurred at a high incidence in regions where malaria transmission by *Plasmodium falciparum* was sustained and intense, *i.e.* holoendemic (Wright, 1967), pointing to a role for it in the malignancy. Because Denis Burkitt's early mapping studies of BL cases were relatively broad, Morrow (1985) re-examined BL cases and malaria parasitemia rates (indicative of holoendemic malaria) at a district level in Uganda. In this study, high levels of malaria parasitemia correlated with high BL incidence rates, further supporting a role for *P. falciparum* transmission with increased risk for this malignancy.

To understand the potential role of malaria in the etiology of BL, it is first essential briefly to discuss malaria pathogenesis: Malaria results from infection with a protozoan parasite of the *Plasmodium* family and even today, it remains one of the two or three numerically most relevant diseases in Africa; roughly 300 million cases of clinical malaria are reported each year with over 900,000 cases resulting in death. Severe morbidity and mortality are most commonly associated with *P. falciparum* infection. Malaria transmission dynamics vary among geographic locales, with people living in some areas experiencing sporadic (unstable, less intense) exposure, whereas others are continually repeatedly exposed (*e.g.* holoendemic malaria) (Mbogo et al., 1995; Snow et al., 1997). Differences in malaria transmission intensity result in host response immunity and disease severity variations. Unlike many viral diseases, malaria only elicits protective immunity after several years of continuous exposure, during which time recurring infections and illness occur (Marsh et al., 1989). Because of this, children aged 1-5 years living in holoendemic malaria regions are at the highest risk of morbidity and mortality associated with *P. falciparum* infections (Snow et al., 1997). In these regions, greater than 70% of children are parasitemic at any given time (Hogh, 1996). Under these conditions, *Plasmodium* infection in children is a recurrent, even chronic infection, and during this period the child's immune system is under constant stress from repeated infections.

The geographic variability in BL onset age correlates with malaria exposure. For example, it was found that patients living in holoendemic malaria areas develop BL younger than those living in sporadic or seasonal malaria areas (Morrow, 1985). A ten-year study in Uganda during the 1960s detected a median onset age of 12 years in areas where malaria transmission was weak or inconsistent (hypo-endemic), compared to 8 and 6 years, respectively, in areas with meso- and hyper-endemic transmission (Morrow et al., 1976). Burkitt and Wright detected an equal susceptibility, but older onset age, among persons migrating from low to high-risk BL areas (that is, ecologically consistent with non-malarious and endemic malaria zones, respectively) (Burkitt and Wright, 1966). This BL age-dependence might also be linked to the age-dependence in acquisition of immunity to malaria in holoendemic regions.

Despite the strong epidemiologic link between *P. falciparum* malaria and BL (Morrow, 1985), the mechanism for the interaction between these two diseases remains unknown. *P. falciparum* malaria is causally implicated in defects of cell-mediated immune responses (Ho et al., 1988; Ho et al., 1986; Hviid et al., 1992) and in alterations in B cell homeostasis (Nagaoka et al., 2000), acting as a B cell mitogen and inducing hypergammaglobulinemia (Greenwood, 1974). Based on these known effects on the host immune system, two possible but not mutually exclusive mechanisms have been proposed for the role of *P. falciparum* malaria in the etiology of BL, that is, suppression of T-cell immunity or, alternatively, activation and expansion of B cells. With regard to immunosuppression, Whittle et al. (1984), using a standard regression assay to assess EBV-specific T cell function, demonstrated that peripheral blood lymphocytes isolated from adult patients with acute malaria were unable to control outgrowth of EBV-transformed cells. Moss et al. (1983), with a similar assay, showed that healthy adults living in malaria holoendemic regions of Papua New Guinea had impaired EBV-specific T cell responses. In children experiencing an episode of acute malaria, spontaneous outgrowth of EBV-transformed B cells *ex vivo* occurred at even greater frequency than in children recovering from malaria (Whittle et al., 1990). Using more sensitive assays and peptides specific for both EBV-lytic and latent antigens, suppressed EBV-specific T cell immunity in children living in malaria holoendemic regions is observed (Moormann et al., 2005).

As early as 1970, O'Conor had proposed that chronic antigenic stimulation from malaria infection might played a role in the emergence of pre-malignant lesions (O'Conor, 1970). Others argued that the immunosuppressive effects of malaria are insufficient to explain a correlation between holoendemic malaria and BL (Rickinson and Gregory, 1988). Lam et al. (1991) demonstrated that the number of EBV-infected B cells increases during single episodes of acute *P. falciparum* malaria. In a recent study, healthy children (age 1-4 years) living in a malaria holoendemic region in Kenya were found to have EBV viral loads typical of a chronic viral infection (Moormann et al., 2005). Whether malaria and perinatal infection with EBV are affecting the establishment or maintenance of viral persistence remains an open question, as do topics relating to how recurrent malaria infections affect EBV-specific T cell responses, and the long-term effect of repeated malarial infections on B cell homeostasis in children at risk for developing BL.

A role for Epstein-Barr virus (EBV)

Following the isolation of EBV from a BL-derived cell line (Epstein et al., 1964), many studies demonstrated the presence and expression of this human herpesvirus in tumors from BL endemic areas. Labrecque et al. (1994) studying tumors in Malawian children found that EBV gene expression could be detected in greater than 90 percent of them; cases of apparently EBV-negative BLs were mainly found among relapse tumors from treated patients. The virus was clonal in BLs suggesting that infection preceded expansion of the malignant clone (Neri et al., 1991). Finally, the ability of EBV to transform B lymphocytes in culture reinforced credence in the possibility of EBV being an etiological agent for this malignancy. A causal relationship between EBV and BL received strong support from a large-scale prospective study from Uganda in the 1970s (de-Thé et al., 1978). Here, greater than 200,000 children were pre-bled, serum was stored and, in children that subsequently developed BL, very high antibody titers against the EBV viral capsid antigen (VCA) were subsequently found. The elevated VCA titers, and the stability of the VCA antibodies, led de-Thé et al. (1977) to suggest that perinatal infection with EBV, as is frequent in Africa (Biggar et al., 1978b), could result in an infection that was poorly controlled by the host and increased the risk for BL.

Although most data, obtained over a nearly 50 year period, suggest EBV to be a co-factor in the etiology of endemic BL, it is still debatable how EBV contributes to BL tumorigenesis at the cellular level. Takada and colleagues have demonstrated that a BL-derived cell line lacking expression of the small non-coding viral transcripts, called EBERs, is more sensitive to certain apoptotic stimuli than EBER-positive cells. Thus they argue that EBV contributes to BL tumorigenesis by blocking cellular apoptosis (Takada, 2001). Similarly, the expression of a protein, Tcl-1, derived originally from T-cells, is found to be stimulated by EBV in BLs and to promote cell survival (Kiss et al., 2003). Based on the ability of EBV to immortalize B lymphocytes and drive cellular proliferation, another possibility exists. That is during the expansion of the pool of B-cells that follows on from EBV infection, a stochastic event occurs which results in the emergence of a B-cell with the *c-myc* oncogene translocation, so rendering the expression of further EBV gene functions redundant. Notably, AIDS-associated and sporadic BLs are not invariably EBV-positive, but have *c-myc* translocations similar to those in endemic BL. Inconclusive evidence to date thus suggests that EBV plays a role as an initiator of tumorigenesis, rather than being essential for maintenance of the tumor phenotype. Klein originally described BL as "a diversity of initiation (events) followed by convergent cytogenetic evolution" (Klein, 1979), a statement seemingly as relevant today as when it was made. For understanding endemic BL, it is obvious that the focus needs to be on factors that potentiate tumor development within the environment of children at risk for developing this tumor. In a ubiquitous virus such as EBV, it is necessary to invoke an unusual viral expression pattern in cells in these children, either quantitatively or qualitatively, to explain its role in malignancy. What seems clear from all data is that EBV is a contributor to endemic BL, but only as one of several cofactors. Only its eradication, as through vaccination for example, will solve the question of how important a contributor it is.

A role for other cofactors

The data overwhelmingly support the argument that other environmental risk factors must exist for BL. Other contributing agents that have been explored include plant exposure (see below), infection with arboviruses, or dietary consumption of smoked fish (Griffin, 2000; Imai et al., 1994a; MacNeil et al., 2003; van den Bosch et al., 1993a; van den Bosch and Lloyd, 2000). Moreover, whereas the geographic distribution of high incidence rates of BL occurs in regions where malaria is holoendemic, BL does occur, albeit at a lower incident rate, within other areas where malaria transmission is only epidemic, raising the possibility that some other environmental agent may serve the same function as malaria.

One possibility being considered is that of a succulent plant, *Euphorbia tirucalli*, indigenous to regions in Africa that include those where BL is prevalent. This plant, a member of the Euphorbiaceae family, grows easily in areas with significant sunlight, adequate drainage, slightly acidic soil, and elevation below 1,500 m (Abdalla, 1992). *E. tirucalli* has been proposed as a co-factor in development of BL

(Osato et al., 1987; van den Bosch et al., 1993a) based on its biological properties, its abundance, as well as its potential for exposure to children. *E. tirucalli* is widely found in areas of Kenya and Tanzania with high rates of BL - most notably in the Lake Victoria basin - but is not seen in areas where BL is uncommon (Mizuno et al., 1983; Osato et al., 1987). Children with BL in Malawi were found to have a significantly greater incidence of *E. tirucalli* growing around their homes relative to controls (van den Bosch et al., 1993b). Anecdotal reports suggest that this plant is used as medication in the treatment of a variety of ailments (Osato et al., 1990; 1987) and we (Rochford and colleagues) found, as did Osato et al., that in the Lake Victoria region in Kenya it was used as a herbal medicine. It was also observed that children play with the latex sap of the plant, or use it for glue, and that its wood is often used for smoking foodstuffs, including fish. No studies have yet been carried out to see whether there is a connection between the latter use and the high levels of soluble nitrosamines that were observed in fish from Lake Malawi (Griffin, 2000). The biologic plausibility of *E. tirucalli* as a BL co-factor lies in the fact that methanol extracts of the plant's latex contain known tumor promoting agents. These have the ability to enhance EBV-mediated cell transformation (Mizuno et al., 1983), modulate EBV-specific T cell activity, and induce chromosomal translocations in B cells (Imai et al., 1994b). Moreover, the unpurified plant latex can reactivate the EBV lytic cycle at dilutions up to a million fold (MacNeil et al., 2003).

BL has been cited many times as a classic model of multi-step tumorigenesis. The present data allow for an etiological mechanism where the first step is early EBV infection, followed by subsequent repeated malaria infections, with a final stage in oncogenesis being one of the chromosomal translocations that result in over-expression of the *c-myc* oncogene. However, many questions remain. For example, malaria and EBV are both prevalent in areas where BL is endemic, and yet the occurrence of the tumor is a relatively rare event (<20 in 10^5 children of the appropriate age). Why don't more children develop BL? The answer may lie in the fact that another factor(s), as yet not identified, is an essential part of the mechanism. We offer various suggestions for the identity of this other factor in the lymphoma belt of Africa and possibly elsewhere, but continued epidemiological studies relative to the etiology of this childhood cancer are needed to provide the answers.

MOLECULAR DESCRIPTION OF BL
C-*myc* gene expression
In addition to EBV-positivity, another molecular hallmark of BL is the activation of the c-*myc* oncogene on chromosome-8 through reciprocal translocations that result in its translocation to sites controlling immunoglobulin gene expression on chromosomes-2,-14 or –22 (reviewed, Magrath, 1990). These translocations result in aberrant expression of the c-*myc* gene, one of a family of basic helix-loop-helix transcription factors that regulate cell growth, differentiation and apoptosis. This MYC protein family forms heterodimers with another host protein, MAX. Together, MYC-MAX activate transcription by binding to so-called E-box, CACGTG, promoter motifs (Grandori et al., 2000). In a process required for its oncogenic activity, the N-terminal domain of c-MYC interacts with TRRAP, a

protein which, among its functions, recruits the enzyme histone acetyltransferase, thereby altering expression of genes that come under its control. MYC-MAX heterodimers influence expression not only of host cell functions, but may be involved in chromatin remodeling that could also affect EBV gene expression. Relevant to endemic BL is the fact that the c-*myc* gene is expressed at abnormally high levels. However, this is a property of a variety of other tumors. Thus, although c-*myc* rearrangements are sensitive markers for BL, they are not specific to it. As discussed elsewhere (Hecht and Aster, 2000; Lindstrom and Wiman, 2002), c-*myc* activation may drive BL development through a variety of cellular mechanisms, including stimulating the host cell metabolism and cell cycle progression, as well as down-regulating functions associated with cell cycle inhibition. On the other hand, c-*myc* expression can also induce apoptosis partially through activation of the p53 pathway. *C-myc* activation may in part, if not exclusively, account not only for the exceptional growth capacity of BL but also for the fact that a considerable amount of apparent apoptotic material coexists with tumor in some, though not all, BLs. Model mouse studies have suggested, however, that the main role contributed by c-*myc* to lymphomagenesis is one of a dominant mutator function which induces chromosomal instability (Rockwood et al., 2002). Sample and colleagues propose that c-*myc* expression is driven by an EBV gene or genes (Ruf et al., 2001). It is obvious from this discussion that c-*myc* is a key player in the genesis of BL and that an interaction with EBV may be a component of its role, but much work yet remains to sort out the details of what is undoubtedly a complex series of events.

EBV gene expression

It has been argued that the two small untranslated nuclear RNAs (EBERs) expressed by EBV, may protect against tumor apoptosis (Klein, 2000; Takada, 2001). It is often assumed that EBER expression is another hallmark of EBV-associated malignancies. In some tumors, their levels are indeed so high that they become useful markers for the presence (and expression) of EBV. For African BLs, expression of EBERs does not appear to be a reliable marker for the presence of EBV, however. As described by Labrecque et al. (1999), in some primary Malawian BL biopsies, EBERs were detected (by sensitive *in situ* hybridization methods) in most of the cells in a tumor population, whereas in others, these transcripts were present in sub-populations of cells only. Notably, in a third category, although the viral genome was present, there was no apparent EBER expression. None of these tumors were, however, simultaneously analyzed for their c-*myc* expression levels, an experiment that might have produced interesting results, if the activities of EBERS and c-*myc in vivo* are indeed in opposition. One of the continuing problems in addressing BL at the molecular level is the fact that, even for c-*myc*, so few studies have been carried out on the tumors themselves. As argued elsewhere (Hecht and Aster, 2000), and explicit in the other work (Labrecque et al., 1994; 1999; Xue et al., 2002), it is an open question how relevant observations made in experimental assays - although of undoubted value in identifying properties of individual viral genes - are to the genuine etiology of endemic BL. Further to this argument, studies on BL-derived cultured cells, which have allowed for EBV-associated lymphomas to be categorized on the basis of morphology and gene expression patterns (Rickinson and

Kieff, 2001; Rowe et al., 1987), do not readily extrapolate to primary BL biopsies. Simplistically, this categorization, which is descriptively useful, is as follows: A 'latency I'profile or Group I cells (Rowe et al., 1987), are proposed to represent endemic BLs. Typical BL cell lines microscopically consist of single round cells which express three viral entities, the EBERs, the nuclear antigen, EBNA1, and a family of antisense transcripts, sometimes designated as BamHI A transcripts (although originating from promoters in the EBV BamHI I fragment; Smith et al., 1993). In earlier categorization versions, the BamHI A (or complementary, CST) transcripts were omitted from the latency categories, as they were only discovered in 1989, initially in nasopharyngeal carcinoma, NPC (Hitt et al., 1989). Latency II includes, in addition to the three viral genes that make up latency I, expression of the membrane protein genes, LMP1 and LMP2 (Rickinson and Kieff, 2001). Latency II is characteristic of other tumor types associated with EBV, such as NPC or Hodgkin's disease, which is also found among African children. This category has been sub-divided so that LMP1 (but not apparently LMP2) appears to be invariably expressed in Hodgkin's disease. Latency III profiles, in group III cells, characteristic of lymphoblastoid lines, frequently with 'clumping cells' further express all the other EBV latent genes and proteins, as well as cellular adhesion molecules. Notably, EBV genes that have been included in these categories come almost exclusively from functions designated as "latent", although a few of the genes have been identified as expressed also during the lytic cycle.

These categories present two main problems where endemic African BL is concerned. First, light microscopic examination of cells gently shaken from these tumors can morphologically reflect both latency I or III cells, or mixtures of them, that is, single and clumped populations. Interestingly, a preliminary study of Malawian BLs identified the former mainly among the major age (young children) profile group and the latter mainly in the second (older) peak of patients. Secondly, in terms of viral gene expression, a variety of patterns emerge from studies on primary tumor biopsies, some individually appearing to fall into either latency I or II categories (Xue et al., 2002). These and other data raise questions about the validity of applying the categories, as currently used to descirbe EBV-associated malignancies, to endemic BLs. On the other hand, the variations in viral gene expression could account for, and may ultimately prove useful in explaining, clinical differences that exist among BLs. The data compellingly argue for a closer relationship between scientists and clinicians in order to understand not only which genes may contribute to tumor variability, but also how to take advantage of such variation. Work of Kelly et al. (2002) appears to 'presume' a lymphoblastoid presursor for endemic BL and has in part addressed the question of whether, for example, a latency I profile characteristic of the tumor may have been selected from a Latency III progenitor. Their comparative and *in vitro* studies suggest that the latter can occur, either by an alteration in viral transcriptional promoter usage or the loss, by deletion, of a key viral transactivator gene, EBNA2, a prime oncogene component in the immortalization of B-lymphocytes by EBV in culture. Their data support the hypothesis that the well-known African BL-derived cell line, Daudi, or the culture derived line, P3HR-1 from another endemic BL, are not unique in having 'lost' their EBNA2 gene through a deletion event, and point to yet more

Table 1. Summary of EBV transcriptional expression in BL tumour samples*

Tumor	EBNA1		CSTs	IR4	IR2	BARF1	BHRF1		BZLF1	LMP1	LMP2A	LMP2B	*v*IL-10
	Qp	Cp/Wp					lyt	lat					
SD	+	+	+	+	+	+	-	-	+	-	+	+	-
IB	+	-	+	+	+	+	+	-	+	-	+	-	+
LC	+	+	+	+	+	+	+	-	-	-	+	-	-
TJ	+	-	+	+	+	+	+	-	+	-	+ (w)	-	-
GE	+	-	+	+	+	+	+	-	-	-	+	-	-
BL-1**	+	-	+	+	+	+	+	+	+	+	+ (w)	-	+
BL-2§	+	-	+	+	-	-	-	-	-	-	-	-	-
BL-3	+	-	+	+	+	+	+	-	+	+	-	-	+
BL-4	+	-	+	+	+ (w)	+	-	-	-	+	-	-	-
BL-5	+	-	+	+	+	+	+	-	+	+	+	+	-
BL-6	+	-	+	+	+	+	+	+	+	+	+	-	-
BL-7	+	-	+	+	+	+	+	-	+	+	+	-	-
CC (control)	-	-	-	-	-	-	-	-	-	-	-	-	-

*Taken from fresh frozen materials. **HIV+. §probably not a BL. w = weak. (From Xue et al., 2002) interesting diversities among BLs. If the functions of EBNA2 and *c-myc* are truly in opposition in the case of BL, these data emphasize the key activity of the latter oncogene in promoting tumor formation.

Of perhaps greater interest, at least with regard to planning future therapies for controlling Endemic BL, are the findings that African BL biopsies may express viral genes that fall outside the latency designations, generally being assumed to be associated exclusively with the viral lytic cycle. Here we refer to viral genes such as the immediate early gene, BZLF1 which, like *c-myc*, is a potent transcription factor and is one of the initiators of lytic replication. It may regulate expression of the viral homologs, BCRF1 and BHRF1, of cellular genes interleukin-10 and *bcl*-2, of the early proteins, BHLF1 and RF3, associated with viral lytic replication, and an oncogenic still ill-defined function, BARF1. These genes, and their functions, as identified to date, have been dealt with in other chapters, and their expression as transcripts, and in some cases as proteins in endemic BLs, are also described in Labrecque et al. (1994; 1999), Ong et al. (2001) and Xue et al., (2002). In clinically diagnosed endemic BLs, closely-related as well as divergent patterns of viral gene expression are obseved (Table 1). Some of these patterns have proved diagnostically interesting. For example, EBV gene expression coupled with histopathologic re-evaluation of several initially assumed endemic BLs, showed them rather to be rhabdomyosarcoma, a tumor not previously associated with EBV (Ong et al., 2001; Xue et al., 2002). Another interesting variation came from an HIV-infected child and would not therefore be classified as 'endemic BL'. This tumor differed in its gene expression profile from other BLs, raising the question of HIV-induced gene alterations of EBV expression in B-cells, and whether and how this might occur (Xue et al., 2002). A third anomaly involved expression of the host IL-10 gene and its viral homolog, vIL-10 (BCRF1) - not previously identified as expressed in BL - in two characteristic endemic BLs, and in a third tumor derived from the HIV-positive child. A diversity of viral gene expression patterns may contribute to the apparent differences observed among BLs, both with regard to their growth properties and their response to therapy. Where antibodies were used to examine protein expression as well as viral transcription, among the tumor cells themselves expression heterogeneity was also observed. The question can be raised whether a tumor that grows as rapidly as BL (with measured doubling times ranging from 18-60 hrs) remains absolutely 'clonal' in its later growth stages, whether heterogeneity of gene expression may reflect differential chromosomal changes during tumor development, or whether (see Kelly et al., 2002) the tumor population may have lymphoblastoid elements.

In the early work on endemic BL, a number of reports dealt with the prognostic value of EBV antibodies as found in these children. Already noted is the Ugandan prospective study which showed that high serum titres to EBV antigens, EBNAs, and early functions, so-called EA-R (for nuclear localization restriction) and EA-D (diffuse, found in both the cytoplasm and nucleus) and viral capsid antigens, VCA, preceeded tumor formation. The risk of developing BL was estimated at 30 times higher for children with titers two dilutions (ie., 100-fold) or more above the normal population (de-Thé et al., 1978). Earlier studies (Henle et al., 1973) correlated antibody titres to VCA and the early antigens with patient parameters, such as clinical status (disease stage) at admission, subsequent clinical presentation,

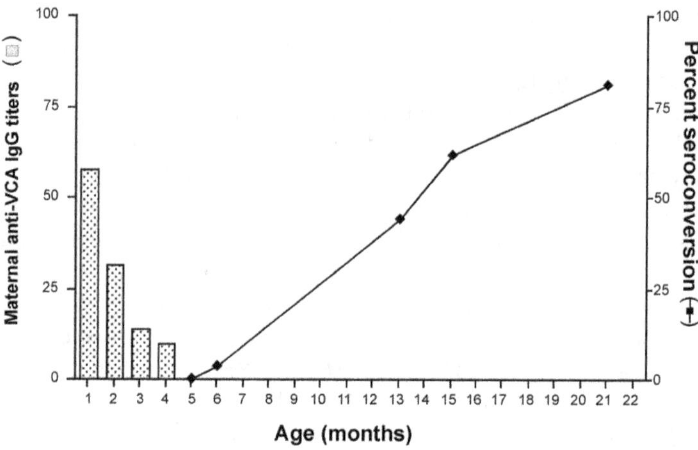

Figure 4. EBV infection in infants. Comparison of maternal antibodies against EBV and time of primary infection with EBV. (Adapted from Henle et al. 1979).

and therapy used. These data were used to argue for a role of EBV in the genesis of BL, rather than as a mere 'passenger'. The work of Labrecque et al. (1994; 1999) showed that with cyclophosphamide therapy, the breaking of latency and expression of lytic cycle-related genes in the tumor correlated with a better prognosis for the patient. Such data suggest that whereas EBV is a risk factor for BL, its genes might be harnessed in a manner that could be of theraupetic benefit in treatment.

SUMMARY

This chapter deals with endemic Burkitt's lymphoma, as it was in the past and as it appears today. Most of the discussion concerns the malignancy seen in children in Equatorial Africa, where the disease is most prevalent and has been studied in greatest detail. For the future, comparisons among endemic BLs seen in East and West African countries, and between West Africa and Brazil – the site of the early slave trade that carried individuals from Africa to its east coast - could be of value in identifying local co-factors that influence tumorigenesis. Where endemic BL is concerned, we may need reminding that the ubiquitous factors, the virus and malaria, cannot alone define its etiology, and other cofactors such as lifestyle, diet, genetics, other infectious agents, *et cetera*, deserve equal consideration. In future programs aimed at erradication of BL – a desirable hurdle - should not research on these topics take its place alongside the EBV and malaria as a major focus for investigation? Once identified, a non-infectious co-factor might prove easier to contain than infectious agents, as experience illustrates.

Some challenging items are missing from our main text, either because they have not been given great attention in the literature, or they did not fall naturally into our major subheadings. One is the question of perinatal (or neonatal) EBV infection as a risk factor for endemic BL as posited by de-Thé (1977). In Africa maternal antibodies to EBV are observed in the children in their first few months of life (Figure 4). These reach a trough at about 7 months; primary EBV infection begins soon after, with greater than 90% of children having evidence of EBV infection by 3 years of age (Biggar et al., 1978a; 1978b; Henle et al., 1969; 1979) (Figure 4). It may well be the case that the age at time of infection constitutes the main risk factor for the development of endemic BL. de-Thé (1977; 1993) discusses 'risk factors' for endemic BL, and draws interesting analogies with other viruses where, in animal models, neonatal infection occurs as an important *tumor risk*.

Then there is the continuing question of a genetic component to at least some endemic BLs. In a recent follow-up study of patients in Malawi (Kazembe et al., 2003), among the notes of nearly 100 children, 6 were identified as having either a sibling pair, or a near relative, with this malignancy (our unpublished data). Similar findings have been reported in the earlier African literature, and familial BL has also been identified in Papua New Guinea (Winnett et al., 1997). This topic is waiting to be further explored by good molecular epidemiological investigations backed up by diagnostic tools that reliably differentiate BL from the other childhood malignancies in high risk tumor regions of the world.

One of the observations that intrigued early BL investigators was the very low incidence of acute lymphocytic leukemia (ALL) seen then - and now as well (Parkin et al., 2003) - in the lymphoma belt of Africa. This led Dalldorf, a microbiologist, to postulate that ALL, seen in high frequency in Western countries, was presenting as BL in the African population, with holoendemic malaria providing the trigger for altering the pathology. His arguments received serious consideration in the 1960s and 1970s with a publication which showed that among Kenyan patients, there was no evidence of malaria in 13 identified with leukemia, whereas in patients of like age with BL, malaria could be identified in 8/9 of them (Dalldorf and Barnhart, 1972). This argument could be seen to be consistent with the postulate of early EBV as an initiating event in BL, with B-cell expansion following malarial infection, and *c-myc* translocations, resulting in frank malignancies, occurring after. Its weakness is that no infectious agent has yet been associated with ALL, although one has frequently been postulated to exist, and sought. It is possible that EBV is present in such low levels as to be virutally undetectable in ALL, or that it has been lost, acting only in a 'hit and run' capacity in the leukemia, or alternatively, that there is yet an unknown infectious agent to be found. But the latter would not be consistent with Dalldorf's argument. This intriguing topic receives renewed interest from studies on a T-cell oncogene, Tcl-1, which has been found expressed not only in ALL but also in BL, and, moreover, found to be upregulated by EBV (Kiss et al., 2003). Perhaps the inverse correlation between the two malignancies is ready for a resurgence in interest?

We conclude with repeating arguments of de-Thé (1993), an epidemiologist "that many dogmas have been overturned by knowledge gained from research on EBV, that for BL (as for a number of other human tumors) this virus has been shown to be a necessary cofactor in pathogenesis but not in itself fully sufficient, and that, whereas etiologically related to BL, EBV does not appear to represent the causative agent (in the Pasteurian sense of 'one agent for one disease')". For the future, our current knowledge suggests that intervention against EBV may prevent virus-associated cancers prevalent in the Developing World. We hope that this discussion will stimulate others to search for routes to accomplish this.

We acknowledge our great debt to the many clinicians and scientists who have come before us, and offer our apologies to others whose contributions space has forced us to neglect. We especially thank our colleagues in Malawi and Kenya for sharing their enthusiasm and knowledge on Endemic BL with us.

References

Abdalla, S. H. (1992). Euphorbia species. Lancet *339*, 434.

Araujo, I., Foss, H. D., Bittencourt, A., Hummel, M., Demel, G., Mendonca, N., Herbst, H., and Stein, H. (1996). Expression of Epstein-Barr virus-gene products in Burkitt's lymphoma in Northeast Brazil. Blood *87*, 5279-5286.

Bacchi, M. M., Bacchi, C. E., Alvarenga, M., Miranda, R., Chen, Y. Y., and Weiss, L. M. (1996). Burkitt's lymphoma in Brazil: strong association with Epstein-Barr virus. Mod Pathol. *9*, 63-67.

Bellan, C., Lazzi, S., De Falco, G., Nyongo, A., Giordano, A., and Leoncini, L. (2003). Burkitt's lymphoma: new insights into molecular pathogenesis. J. Clin. Pathol. *56*, 188-192.

Berard, C. (1985). Morphological defintion of Burkitt's tumor: historical review and present status. In Burkitt's Lymphoma, G. Lenoir, G. O'Connor, and C. Olweny, eds. (Lyon, IARC Scientific Publication), pp. 30-35.

Biggar, R. J., Henle, G., Bocker, J., Lennette, E. T., Fleisher, G., and Henle, W. (1978a). Primary Epstein-Barr virus infections in African infants. II. Clinical and serological observations during seroconversion. Int. J. Cancer *22*, 244-250.

Biggar, R. J., Henle, W., Fleisher, G., Bocker, J., Lennette, E. T., and Henle, G. (1978b). Primary Epstein-Barr virus infections in African infants. I. Decline of maternal antibodies and time of infection. Int. J. Cancer *22*, 239-243.

Burkitt, D. (1958). A sarcoma involving the jaws in African children. Br. J. Surg *46*, 218-223.

Burkitt, D. (1962). Determining the climatic limitations of a children's cancer common in Africa. Br. Med. J. *5311*, 1019-1023.

Burkitt, D. (1967). Long-term remissions following one and two-dose chemotherapy for African lymphoma. Cancer *20*, 756-759.

Burkitt, D. (1987). Discovering Burkitt's lymphoma: a special address to the second internation symposium on the Epstein-Barr virus and associated diseases. In Epstein-Barr virus and human disease, P. Levine, D. Ablashi, M. Nonoyama, G. Pearson, and R. Glaser, eds. (Clifton, NJ, Humana Press), pp. xxi-xxxi.

Burkitt, D., and Wright, D. (1966). Geographical and tribal distribution of the African lymphoma in Uganda. Br. Med. J. *5487*, 569-573.

Clifford, P., Linsell, C.A., and Timms, G.L. (eds.). (1966). Further studies on the treatment of Burkitt's lymphoma. East Afr Med. J. *43*, 179-199.

Clifford, P. (1968). Cancer in Africa (Nairobi, East African Medical Publishing Co).

Clifford, P., Singh, S., Stjernsward, J., and Klein, G. (1967). Long-term survival of patients with Burkitt's lymphoma: an assessment of treatment and other factors which may relate to survival. Cancer Res. *27*, 2578-2615.

Cook, A. (1945). Uganda Memories (1897-1940). The Uganda Society, Kampala.

Dalldorf, G., and Barnhart, F. E. (1972). Childhood leukemia, malaria and Burkitt's lymphoma. N. Engl. J. Med. *286*, 1216.

Dalldorf, G., Linsell, C. A., Barnhart, F. E., and Martyn, R. (1964). An epidemiologic approach to the lymphomas of African children and Burkitt's sarcoma of the jaws. Perspectives in Biology and Medicine *36*, 435-449.

Davies, J. N., Elmes, S., Hutt, M. S., Mtimavalye, L. A., Owor, R., and Shaper, L. (1964). Cancer in an African Community, 1897--1956. An Analysis of the Records of Mengo Hospital, Kampala, Uganda. Br. Med. J. *5378*, 259-264.

de-Thé, G. (1977). Is Burkitt's lymphoma related to perinatal infection by Epstein-Barr virus? Lancet *1*, 335-338.

de-Thé, G. (1993). The etiology of Burkitt's lymphoma and the history of the shaken dogmas. Blood Cells *19*, 667-673.

de-Thé, G., Geser, A., Day, N. E., Tukei, P. M., Williams, E. H., Beri, D. P., Smith, P. G., Dean, A. G., Bronkamm, G. W., Feorino, P., and Henle, W. (1978). Epidemiological evidence for causal relationship between Epstein-Barr virus and Burkitt's lymphoma from Ugandan prospective study. Nature *274*, 756-761.

Editorial (1964). Pattern of cancer in central Africa. Br. Med. J. *1*, 321-322.

Epstein, M. A., Barr, Y. M., and Achong, B. G. (1964). Virus particles in cultured lymphoblasts from Burkitt's lymphoma. Lancet *i*, 702-703.

Fagundes, L. A., de Oliveira, R. M., and Amaral, R. (1968). Childhood lymphosarcoma in the state of Rio Grande do Sul, Brazil. Report of 20 cases histologically similar to Burkitt's tumor. Cancer *22*, 1283-1291.

Feng, W. H., Hong, G., Delecluse, H. J., and Kenney, S. C. (2004). Lytic induction therapy for Epstein-Barr virus-positive B-cell lymphomas. J. Virol. *78*, 1893-1902.

Grandori, C., Cowley, S. M., James, L. P., and Eisenman, R. N. (2000). The Myc/Max/Mad network and the transcriptional control of cell behavior. Annu. Rev. Cell Dev. Biol. *16*, 653-699.

Greenwood, B. M. (1974). Possible role of a B-cell mitogen in hypergammaglobulinaemia in malaria and trypanosomiasis. Lancet *1*, 435-436.

Gregory, C. D., Tursz, T., Edwards, C. F., Tetaud, C., Talbot, M., Caillou, B., Rickinson, A. B., and Lipinski, M. (1987). Identification of a subset of normal B cells with a Burkitt's lymphoma (BL)-like phenotype. J. Immunol. *139*, 313-318.

Griffin, B. E. (2000). Epstein-Barr virus (EBV) and human disease: facts, opinions and problems. Mutat Res. *462*, 395-405.

Haddow, A. J. (1963). An improved map for the study of Burkitt's lymphoma syndrome in Africa. East Afr Med. J. *40*, 429-432.

Hecht, J. L., and Aster, J. C. (2000). Molecular biology of Burkitt's lymphoma. J. Clin. Oncol *18*, 3707-3721.

Henle, G., Henle, W., Clifford, P., Diehl, V., Kafuko, G. W., Kirya, B. G., Klein, G., Morrow, R. H., Munube, G. M., Pike, P., et al. (1969). Antibodies to Epstein-Barr virus in Burkitt's lymphoma and control groups. J. Natl. Cancer Inst *43*, 1147-1157.

Henle, W., Henle, G., Gunven, P., Klein, G., Clifford, P., and Singh, S. (1973). Patterns of antibodies to Epstein-Barr virus-induced early antigens in Burkitt's lymphoma. Comparison of dying patients with long-term survivors. J. Natl. Cancer Inst *50*, 1163-1173.

Henle, W., Henle, G., and Lennette, E. T. (1979). The Epstein-Barr virus. Scientific American, 40-51.

Hesseling, P. B., Broadhead, R., Molyneux, E., Borgstein, E., Schneider, J. W., Louw, M., Mansvelt, E. P., and Wessels, G. (2003). Malawi pilot study of Burkitt lymphoma treatment. Med. Pediatr. Oncol *41*, 532-540.

Hitt, M.M., Allday, M.J., Hara, T., Karran, L., Jones, M.D., Busson, P., Tursz T., Ernberg, I. and Griffin, B.E. (1989). EBV gene expression in an NPC-related tumor. EMBO J. *8*, 2639-2651.

Ho, M., Webster, H. K., Green, B., Looareesuwan, S., Kongchareon, S., and White, N. J. (1988). Defective production of and response to IL-2 in acute human falciparum malaria. J. Immunol. *141*, 2755-2759.

Ho, M., Webster, H. K., Looareesuwan, S., Supanaranond, W., Phillips, R. E., Chanthavanich, P., and Warrell, D. A. (1986). Antigen-specific immunosuppression in human malaria due to *Plasmodium falciparum*. J. Infect. Dis. *153*, 763-771.

Hochberg, D., Middeldorp, J. M., Catalina, M., Sullivan, J. L., Luzuriaga, K., and Thorley-Lawson, D. A. (2004). Demonstration of the Burkitt's lymphoma Epstein-Barr virus phenotype in dividing latently infected memory cells *in vivo*. Proc. Natl. Acad. Sci. USA. *101*, 239-244.

Hogh, B. (1996). Clinical and parasitological studies on immunity to *Plasmodium falciparum* malaria in children. Scand. J. Infect. Dis. Suppl *102*, 1-53.

Hviid, L., Jakobsen, P. H., Abu-Zeid, Y. A., and Theander, T. G. (1992). T-cell responses in malaria. APMIS *100*, 95-106.

Imai, S., Sugiura, M., Mizuno, F., Ohigashi, H., Koshimizu, K., Chiba, S., and Osato, T. (1994a). African Burkitt's lymphoma: a plant, *Euphorbia tirucalli*, reduces Epstein-Barr virus-specific cellular immunity. Cancer Research *14*, 933-936.

Imai, S., Sugiura, M., Mizuno, F., Ohigashi, H., Koshimizu, K., Chiba, S., and Osato, T. (1994b). African Burkitt's lymphoma: a plant, *Euphorbia tirucalli*, reduces Epstein-Barr virus-specific cellular immunity. Anticancer Res. *14*, 933-936.

Israel, B. F., and Kenney, S. C. (2003). Virally targeted therapies for EBV-associated malignancies. Oncogene *22*, 5122-5130.

Kazembe, P., Hesseling, P. B., Griffin, B. E., Lampert, I., and Wessels, G. (2003). Long term survival of children with Burkitt lymphoma in Malawi after cyclophosphamide monotherapy. Med. Pediatr. Oncol *40*, 23-25.

Kelly, G., Bell, A. and Rickinson, A. (2002). Epstein-Barr virus-associated Burkitt lymphomagenesis selects for downregulation of the nuclear antigen EBNA2. Nature Med. *8*, 1098-1104.

Kiss, C., Nishikawa, J., Takada, K., Trivedi, P., Klein, G., and Szekely, L. (2003). T cell leukemia I oncogene expression depends on the presence of Epstein-Barr virus in the virus-carrying Burkitt lymphoma lines. Proc. Natl. Acad. Sci. USA. *100*, 4813-4818.

Klein, G. (1979). Lymphoma development in mice and humans: Diversity of initiation is followed by convergent cytogenetic evolution. Proc. Natl. Acad. Sci. USA. *76*, 2442-2446.

Klein, G. (2000). Dysregulation of lymphocyte proliferation by chromosomal translocations and sequential genetic changes. Bioessays *22*, 414-422.

Kuppers, R., Klein, U., Hansmann, M. L., and Rajewsky, K. (1999). Cellular origin of human B-cell lymphomas. N. Engl. J. Med. *341*, 1520-1529.

Kurokawa, M., Ghosh, S.K., Ramos, J.C., Mian, A.M., Toomey, N.L., Cabral, L., Whitby, D., Barber, G.N., Dittmer, D.P., and Harrington, Jr, W.J. (2005). Azidothymidine inhibits NF-κB and induces Epstein-Barr virus gene expression in Burkitt lymphoma. Blood. Mar 24 (In Press).

Labrecque, L. G., Lampert, I., Kazembe, P., Philips, J., and Griffin, B. E. (1994). Correlation between cytopathological results and in situ hybridisation on needle aspiration biopsies of suspected African Burkitt's lymphomas. Int. J. Cancer *59*, 591-596.

Labrecque, L. G., Xue, S. A., Kazembe, P., Phillips, J., Lampert, I., Wedderburn, N., and Griffin, B. E. (1999). Expression of Epstein-Barr virus lytically related genes in African Burkitt's lymphoma: correlation with patient response to therapy. Int. J. Cancer *81*, 6-11.

Lam, K. M., Syed, N., Whittle, H., and Crawford, D. H. (1991). Circulating Epstein-Barr virus-carrying B cells in acute malaria. Lancet *337*, 876-878.

Lindstrom, M. S., and Wiman, K. G. (2002). Role of genetic and epigenetic changes in Burkitt lymphoma. Semin Cancer Biol. *12*, 381-387.

MacNeil, A., Sumba, O. P., Lutzke, M. L., Moormann, A., and Rochford, R. (2003). Activation of the Epstein-Barr virus lytic cycle by the latex of the plant *Euphorbia tirucalli*. Br. J. Cancer *88*, 1566-1569.

Magrath, I. (1990). The pathogenesis of Burkitt's lymphoma. Adv. Cancer Res. *55*, 133-269.

Magrath, I. T. (1991). African Burkitt's lymphoma. History, biology, clinical features, and treatment. Amer. J. Pediatr. Hematol Oncol *13*, 222-246.

Marsh, K., Otoo, L., Hayes, R. J., Carson, D. C., and Greenwood, B. M. (1989). Antibodies to blood stage antigens of *Plasmodium falciparum* in rural Gambians and their relation to protection against infection. Trans R Soc. Trop Med. Hyg *83*, 293-303.

Marsh, K., and Snow, R. W. (1999). Malaria transmission and morbidity. Parassitologia *41*, 241-246.

Mbogo, C. N., Snow, R. W., Khamala, C. P., Kabiru, E. W., Ouma, J. H., Githure, J. I., Marsh, K., and Beier, J. C. (1995). Relationships between *Plasmodium falciparum* transmission by vector populations and the incidence of severe disease at nine sites on the Kenyan coast. Amer. J. Trop Med. Hyg *52*, 201-206.

Mizuno, F., Koizumi, S., Osato, T., Kokwaro, J. O., and Ito, Y. (1983). Chinese and African Euphorbiaceae plant extracts: markedly enhancing effect on Epstein-Barr virus-induced transformation. Cancer Lett. *19*, 199-205.

Moormann, A.M., Chelimo, K., Sumba, O. P., Lutzke, M. L., Kazura, J., and Rochford, R.. (2005). Exposure to holoendemic malaria results in elevated Epstein-Barr virus loads in children. J. Inf. Diseases *191*, 1233-1238.

Morrow, R. H. (1985). Epidemiological evidence for the role of faciparum malaria in the pathogenesis of Burkitt's lymphoma. In Burkitt's Lymphoma: a human cancer model, G. Lenoir, G. O'Connor, and C. Olweny, eds. (Lyon, IARC Press), pp. 177-185`.

Morrow, R. H., Kisuule, A., Pike, M. C., and Smith, P. G. (1976). Burkitt's lymphoma in the Mengo Districts of Uganda: epidemiologic features and their relationship to malaria. J. Natl. Cancer Inst *56*, 479-483.

Moss, D. J., Burrows, S. R., Castelino, D. J., Kane, R. G., Pope, J. H., Rickinson, A. B., Alpers, M. P., and Heywood, P. F. (1983). A comparison of Epstein-Barr virus-specific T-cell immunity in malaria- endemic and -nonendemic regions of Papua New Guinea. Int. J. Cancer *31*, 727-732.

Mwanda, O.W., Rochford, R., Moormann, A.M., Macnei,l A., Whalen, C., and Wilson, M.L. (2004). Burkitt's lymphoma in Kenya: age, gender and ethnic distribution. East Afr Med J. S111-116.

Nagaoka, H., Gonzalez-Aseguinolaza, G., Tsuji, M., and Nussenzweig, M.C. (2000). Immunization and infection change the number of recombination activating gene (RAG)-expressing B cells in the periphery by altering immature lymphocyte production. J. Exp. Med. *191*, 2113-2120.

Neri, A., Barriga, F., Inghirami, G., Knowles, D. M., Neequaye, J., Magrath, I. T., and Dalla-Favera, R. (1991). Epstein-Barr virus infection precedes clonal expansion in Burkitt's and acquired immunodeficiency syndrome-associated lymphoma. Blood *77*, 1092-1095.

Nkrumah, F. K., and Olweny, C. L. (1985). Clinical features of Burkitt's lymphoma: the African experience. IARC Sci. Publ. (Lyon), 87-95.

Nkrumah, F. K., and Perkins, I. V. (1976). Burkitt's lymphoma: a clinical study of 110 patients. Cancer *37*, 671-676.

O'Conor, G. T. (1970). Persistent immunologic stimulation as a factor in oncogenesis with special reference to Burkitt's tumor. Amer J. Med. *48*, 279-285.

O'Conor, G. T., and Davies, J. N. (1960). Malignant tumors in African children. With special reference to malignant lymphoma. J. Pediatr. *56*, 526-535.

Oettgen, H. F., Clifford, P., and Burkitt, D. (1963). Malignant lymphoma involving the jaw in African children: treatment with alkylating agents and actinomycin D. Cancer Chemother Rep. *28*, 25-34.

Olweny, C. L., Katongole-Mbidde, E., Otim, D., Lwanga, S. K., Magrath, I. T., and Ziegler, J. L. (1980). Long-term experience with Burkitt's lymphoma in Uganda. Int. J. Cancer *26*, 261-266.

Olweny, C. L., and Nkrumah, F. K. (1985). Treatment of Burkitt's lymphoma: the African experience. IARC Sci. Publ. (Lyon), 375-382.

Ong, S. K., Xue, S. A., Molyneux, E., Broadhead, R. L., Borgstein, E., Ng, M. H., and Griffin, B. E. (2001). African Burkitt's lymphoma: a new perspective. Trans R Soc. Trop Med. Hyg *95*, 93-96.

Osato, T., Imai, S., Kinoshita, T., Aya, T., Sugiura, M., Koizumi, S., and Mizuno, F. (1990). Epstein-Barr virus, Burkitt's lymphoma, and an African tumor promoter. Adv. Exp. Med. Biol. *278*, 147-150.

Osato, T., Mizuno, F., Imai, S., Aya, T., Koizumi, S., Kinoshita, T., Tokuda, H., Ito, Y., Hirai, N., Hirota, M., et al. (1987). African Burkitt's lymphoma and an Epstein-Barr virus-enhancing plant *Euphorbia tirucalli*. Lancet *1*, 1257-1258.

Parkin, D.M., Ferlay, J, Hamdi-Cheerif, M., Sitas, F., Thomas, J.C., Wabinga, H. and Whelan, S.L. (eds.). (2003). Cancer in Africa. Epidemiology and Prevention. Vol 153 (Lyon, IARC Press).

Rickinson, A. B., and Gregory, C. D. (1988). Burkitt's lymphoma. Trans R Soc. Trop Med. Hyg *82*, 657-659.

Rickinson, A. B., and Kieff, E. (2001). Epstein-Barr virus. In Fields Virology, D. M. Knipe, and P. M. Howley, eds. (Philadelphia, Lippincott Williams and Wilkins), pp. 2575-2627.

Rochford, R., G. Feuer, J. Orem, C. Banura, E. Katongole-Mbidde, M. W. Otieno, A. Moormann, W. Harrington, S. C. Remick. (2005). Challenges of combination chemotherapy in the resource constrained setting: Clinical and preclinical strategies to overcome myelotoxic therapy for the treatment of Burkitt's and AIDS-related non-Hodgkin's lymphoma. East African Med. J. In Press.

Rowe, M., Rowe, D. T., Gregory, C. D., Young, L. S., Farrell, P. J., Rupani, H., and Rickinson, A. B. (1987). Differences in B cell growth phenotype reflect novel patterns of Epstein-Barr virus latent gene expression in Burkitt's lymphoma cells. EMBO J. *6*, 2743-2751.

Rockwood, L.G., Torrey, T.A., Kim, J.S., Coleman, A.E., Coleman, A.E., Kovalchuk, A.L., Xiang, S., Ried, T., Morse, H.C. , and Janz, S. (2002). Genetic instability in mouse Burkitt lymphoma is dominated by illegitimate genetic recombinaions, not point mutations. Oncogene *21*, 7235-7240.

Ruf, I.K., Rhyne, P.W., Yang, H., Borza, C.M., Hutt-Fletcher, L.M, Cleveland, J.L., and Sample, J.T. (2001). EBV regulates c-MYC, apoptosis, and tumorigenicity in Burkitt's lymphoma. Curr. Top Microbiol. Immunol. *258*, 153-160.

Shad, A., and Magrath, I. (1998). Diagnosis and treatment of non-Hodgkin's lymphoma in children. In Neoplastic Diseases of the Blood. (P. Wiennik, ed), pp. 925-961.

Smith, P.R., Gao, Y., Karran, L., Jones, M.D., Snudden, D., and Griffin, B.E. (1993). Complex nature of the major viral polyadenylated transcripts in Epstein-Barr virus-associated tumors. J. Virol. *67*, 3217-3225.

Snow, R. W., Omumbo, J. A., Lowe, B., Molyneux, C. S., Obiero, J. O., Palmer, A., Weber, M. W., Pinder, M., Nahlen, B., Obonyo, C., et al. (1997). Relation between severe malaria morbidity in children and level of *Plasmodium falciparum* transmission in Africa. Lancet *349*, 1650-1654.

Spreafico, F., Massimino, M., Luksch, R., Casanova, M., Cefalo, G. S., Collini, P., Ferrari, A., Polastri, D., Terenziani, M., Gasparini, M., and Fossati-Bellani, F. (2002). Intensive, very short-term chemotherapy for advanced Burkitt's lymphoma in children. J. Clin. Oncol *20*, 2783-2788.

Stevens, S. J., Vervoort, M. B., van den Brule, A. J., Meenhorst, P. L., Meijer, C. J., and Middeldorp, J. M. (1999). Monitoring of Epstein-Barr virus DNA load in peripheral blood by quantitative competitive PCR. J. Clin. Microbiol. *37*, 2852-2857.

Takada, K. (2001). Role of Epstein-Barr virus in Burkitt's lymphoma. Curr. Top Microbiol. Immunol. *258*, 141-151.

Tefuarani, N., Vince, J. D., Murthy, D. P., Sengupta, S. K., and White, J. C. (1988). Childhood malignant tumors in Papua New Guinea. Ann. Trop Paediatr. *8*, 201-206.

Ten Seldam, R., Cook, R., and Atkinson, L. (1966). Childhood lymphomas in the territories of Papua and New Guinea. Cancer *19*, 437-446.

Thorley-Lawson, D. A., and Gross, A. (2004). Persistence of the Epstein-Barr virus and the origins of associated lymphomas. N. Engl. J. Med. *350*, 1328-1337.

van den Bosch, C., Griffin, B. E., Kazembe, P., Dziweni, C., and Kadzamira, L. (1993a). Are plant factors a missing link in the evolution of endemic Burkitt's lymphoma? Br. J. Cancer *68*, 1232-1235.

van den Bosch, C., Hills, M., Kazembe, P., Dziweni, C., and Kadzamira, L. (1993b). Time-space case clusters of Burkitt's lymphoma in Malawi. Leukemia *7*, 1875-1878.

van den Bosch, C., and Lloyd, G. (2000). Chikungunya fever as a risk factor for endemic Burkitt's lymphoma in Malawi. Trans R Soc. Trop Med. Hyg *94*, 704-705.

Westphal, E. M., Blackstock, W., Feng, W., Israel, B., and Kenney, S. C. (2000). Activation of lytic Epstein-Barr virus (EBV) infection by radiation and sodium butyrate *in vitro* and *in vivo*: a potential method for treating EBV-positive malignancies. Cancer Res. *60*, 5781-5788.

Whittle, H. C., Brown, J., Marsh, K., Blackman, M., Jobe, O., and Shenton, F. (1990). The effects of *Plasmodium falciparum* malaria on immune control of B lymphocytes in Gambian children. Clin. Exp. Immunol. *80*, 213-218.

Whittle, H. C., Brown, J., Marsh, K., Greenwood, B. M., Seidelin, P., Tighe, H., and Wedderburn, L. (1984). T-cell control of Epstein-Barr virus-infected B cells is lost during *P. falciparum* malaria. Nature *312*, 449-450.

Winnett, A., Thomas, S. J., Brabin, B. J., Bain, C., Alpers, M. A., and Moss, D. J. (1997). Familial Burkitt's lymphoma in Papua New Guinea. Br. J. Cancer *75*, 757-761.

Wright, D. H. (1967). The epidemiology of Burkitt's tumor. Cancer Res. *27*, 2424-2438.

Wright, D. H. (1971). Burkitt's lymphoma: a review of the pathology, immunology, and possible etiologic factors. Pathol. Annu. *6*, 337-363.

Wright, D. H. (1999). What is Burkitt's lymphoma and when is it endemic? Blood *93*, 758.

Xue, S. A., Labrecque, L. G., Lu, Q. L., Ong, S. K., Lampert, I. A., Kazembe, P., Molyneux, E., Broadhead, R. L., Borgstein, E., and Griffin, B. E. (2002). Promiscuous expression of Epstein-Barr virus genes in Burkitt's lymphoma from the central African country Malawi. Int. J. Cancer *99*, 635-643.

Ziegler, J. L. (1972). Chemotherapy of Burkitt's lymphoma. Cancer *30*, 1534-1540.

Ziegler, J. L. (1981). Burkitt's lymphoma. N. Engl. J. Med. *305*, 735-745.

Ziegler, J. L., Morrow, R. H., Jr., Fass, L., Kyalwazi, S. K., and Carbone, P. P. (1970). Treatment of Burkitt's lymphoma with cyclophosphamide. Bibl. Haematol. 701-705.

From: Epstein-Barr Virus. Edited by: Erle S. Robertson

Chapter 10

EBV and Breast Cancer: Questions and Implications

*Hratch Arbach and Irène Joab**

ABSTRACT

Epstein-Barr virus (EBV), which is known to be associated with nasopharyngeal carcinoma, has recently become linked with a growing list of carcinomas. The presence of the EBV genome in a subset of breast cancers has been detected but the results remain disparate. When detected, the overall viral load was found to be low and heterogeneous within the same tumor as well as amongst different tumors. The type of cells the virus has infected remains a matter of debate although different data favor the inference that EBV is harbored by epithelial cells: i) The EBV genome was detected in tumor cells isolated by laser capture microdissection ii) the pattern of EBV gene expression observed in breast tumors is different to what was described in peripheral blood lymphocytes suggesting that EBV expression was due to virus from epithelial cells and not from infiltrating lymphocytes. The possible role of EBV in breast cancer development or progression is discussed.

INTRODUCTION

Persistent infection with Epstein-Barr virus (EBV) was initially associated with Burkitt's lymphoma. Subsequently, the virus was shown to be associated with nasopharyngeal carcinoma (NPC) especially the undifferentiated type. EBV has now become linked with a growing list of carcinomas: in NPC-like tumors (lymphoepitheliomas) of the stomach, salivary gland, lung and thymus and also in a subset of gastric adenocarcinomas. Moreover, Grinstein et al. (2002) also described EBV in a subset of colon and prostate adenocarcinomas. Since 1995, different studies report the detection of EBV in breast cancer; however the results remain disparate, and their interpretation is a matter of debate.

Breast cancer is the most common neoplastic disease of women in the Western world. In developed countries, more than 200 cases per 100,000 women are diagnosed per year. The etiology of breast cancer is understood poorly; however genetic background and hormonal effects are believed to play important roles in its development. Late exposure to a common virus such as human cytomegalovirus has been suggested as a risk factor for breast carcinomas (Richardson et al., 1997), and

*For correspondence email Irene.Joab@stlouis.inserm.fr

Table 1. Characteristics and results of PCR studies to detect the EBV genome in breast cancer biopsies

Authors	Sample	Amplified fragment → Positive/total	Cycle number
Gaffey et al., 1993	paraffin	EBNA1→0/35 medullary carcinoma	
Lespagnard et al., 1995	paraffin	BamH1W →0/10 medullary carcinoma	35+southern-blotting
Labreque et al., 1995	Frozen	BamH1W →21/91 BamH1C→ 17 of the 19 EBV+ (BamH1W)	35+southern-blotting
Luqmani et al., 1995	Paraffin	BamH1W →15/28	Nested, 2x40+southern-blotting
	frozen	BamH1W →19/48	Nested 2x40
Bonnet et al., 1999	Frozen	EBER→ 51/100 LMP1→ 51/100 BZLF1 →51/100 51 were positive for the 3 PCR amplifications	40 40 nested 2x40
Brink et al., 2000	Frozen	BamH1W →5/24 LMP1→2/24 (the 2 of the 5 positive for BamH1W)	NS
Chu et al., 2001	Frozen	EBNA3B →5/48 LMP1→5/48 3/48 samples were positive in both PCR amplification	45
Fina et al., 2001	Frozen	BamH1C→162/509	38
Grinstein et al., 2002	paraffin	EBER→14/14 EBNA1+ (by IHC with 2B4)	40
Herrmann et al., 2003	Paraffin	BamH1W →4/59	35
Murray et al., 2003	Paraffin	DNA polymerase →19/92	Q-PCR

NS: not specified

Yasui et al. (2001) showed that data on breast cancer exhibit similarities to those on Hodgkin's disease in young adulthood and suggested that delayed infection with EBV may be a risk for breast cancer.

An association with breast cancer would have potential relevance to its prevention and/or treatment of such cancers. Until now, no definitive consensus opinion is shared in the field, since the results are conflicting giving rise to different interpretations of the data (Glaser et al., 2004).

Different questions have to be considered:
- Is the EBV genome detectable in breast cancer? What is the viral load?
- What is the prevalence of EBV-positive breast tumors?
- Is the virus localized in epithelial tumor cells, and if so, what percentage of cells are infected?
- Is the EBV genome expressed: which are the viral genes expressed?
- What could be the relevance of EBV infection detected in a small fraction of tumor cells?
- Can EBV behave differently in various types of epithelial cells? That is, can EBV infection be an early event in the molecular pathogenesis of NPC, and a late event in breast cancer development?
- Can the EBV genome, which is essential in NPC cells, be lost from breast cancer cells?
- Can EBV be important for tumor initiation and/or development of NPC, whereas it would be important for progression of breast cancer?

DETECTION OF THE EBV GENOME AND ITS EXPRESSION

Results of the different studies performed for EBV detection, are conflicting: the EBV genome was totally absent or found in up to 51% of the breast tumors (Table 1). The discrepancies could arise from technical problems. Therefore it is important to carefully compare the methods used in the different studies in order to understand why the negative as well as positive results are disparate.

In most of the studies, invasive carcinomas of different types were examined, ductal being the most frequent. Table 1 shows amplification by PCR of the EBV genome performed by different groups. Using frozen biopsies, the fraction of EBV positive breast tumors found varied from 10 to 51% (10, 20, 23, 31, 39, and 51% in the study performed respectively by Chu et al. (2001), Brink et al. (2000), Labrecque et al. (1995), Fina et al. (2001), Lucqmani et al. (1995), Bonnet et al. (1999). From paraffin embedded sections, the EBV genome was detected using conventional PCR in 6.7% (Herman et al. (2003)), and using quantitative PCR (Q-PCR) in 10% (Murray et al. (2003)) of the samples. The results obtained using frozen tissue samples (in which the effects of fixative procedures and long term storage are eliminated) gave the highest % of EBV-positive tumors. Exceptions are the studies performed by Luqmani and Shousha (1995) (utilizing nested-PCR followed by southern blotting) and by Grinstein et al. (2002) who could amplify EBER sequence from EBNA1 positive samples (EBNA1 was detected by labeling with the 2B4 monoclonal antibody (Grasser et al., 1994)). In two studies using fixed medullary carcinomas, no EBV DNA was detected using PCR (Gaffey et al., 1993; Lespagnard et al., 1995). Medullary carcinoma is a very rare type of breast

Table 2. Characteristics and results of ISH studies to detect the EBV genome in breast cancer biopsies

Authors	Sample	ISH → Positive/total	Comments
Lespagnard et al., 1995	paraffin	DNA ISH (Biohit) →0/10	Medullary carcinomas
Labreque et al., 1995	Frozen	DNA ISH BamH1W probe (*)	12/19 PCR+ were positive for ISH signal restricted to a small proportion of the cells.
Herrmann et al., 2003	Paraffin	DNA ISH BamH1W 35SdCTP probe (*)	0/4 PCR+

(*)protocol developed in the laboratory

Table 3. Characteristics and results of southern-blot studies to detect the EBV genome in breast cancer biopsies

Authors	DNA load on the gel	Probe	Positive/total
Bonnet et al., 1999	40ug	BamH1W	7/7 **
Chu et al., 2001	16µg	EBV clonality	0/6

•**the samples used were those which gave the highest PCR amplification, estimated by ethidium bromide staining cancer. In order to study enough samples, the authors had to use biopsies with had been stored for a long time and maybe fixed with procedures which evolved with time. Storage and fixation procedures might have altered the threshold of detection. Another study, (Bonnet et al., 1999) involved one case of medullary carcinoma which was found EBV positive. The analysis of fixed tissue may result in lower sensitivity. This possibility is supported by comparison of the viral load found in NPC biopsies. Murray et al. (2003) found 2.4 and 3.2 copies per cell in fixed tissues in contrast to 25 and 27 copies per cell that we detected in frozen NPC samples (Arbach et al., unpublished). Side by side comparison of viral load measured in extracts from the same tissue either frozen or fixed would be informative.

The discrepancies in the results described above could also arise on sensitivity of the PCR methods used and/or the number of amplifications (from 35 to 45 cycles). Alternatively, the DNA preparations might be different or the results may reflect epidemiologic diversity among study samples (Glaser et al. 2004). It is detailed below that the viral load is low and heterogeneous within the same tumor, as well as amongst different tumors; this heterogeneity might also be reflected by the differences among the reports.

Indeed PCR cannot indicate what type of cells the virus has infected and as EBV might infect infiltrating lymphocytes; this method may overestimate EBV association. On the other hand, heterogeneity of the virus in a tumor may lead a false negative result obtained by assaying areas of a biopsy devoid of EBV while the virus might be present somewhere else.

DNA *in situ* hybridization (ISH) and southern blots to detect the EBV genome in breast cancer biopsies

DNA ISH has been used by several authors (Table 2). A small number of paraffin-embedded samples that had been screened previously by PCR, were found to be EBV negative by ISH (Lespagnard et al., 1995; Herrmann et al., 2003). In contrast, Labrecque et al. (1995) detected the EBV genome in a small proportion of tumor cells by ISH assay in 12/19 PCR-positive frozen tissue.

Bonnet et al. (1999) and Chu et al. (2001) performed Southern blots to detect EBV DNA in breast cancer tissues (Table 3). While Chu et al. (2001) did not detect viral DNA by this method, a positive signal was obtained in each of the 7 samples tested by Bonnet et al. (1999). These samples were chosen because they gave the highest PCR amplification estimated by ethidium bromide staining. In comparison with Chu's study, Bonnet loaded a larger amount of total DNA and the blot was probed with repeat region (BamH1W) of the EBV genome instead of a unique sequence. This optimization might explain the discrepancy between Bonnet's results and those of Chu et al. (2001).

EBV viral load is low in breast cancer tissues

The copy number estimated by Luqmani and Shousha (1995) was 20-500 copies per microgram of tissue DNA by comparison with EBV containing plasmid. If one assumes that 1 cell contains 6 pg of DNA, this number would correspond to 0.003-0.08 copies per 1000 cells: this was an approximate figure since quantitative PCR (Q-PCR) was not available when the study was performed. Using Q-PCR, the average viral load was shown to be low in breast cancer biopsies (Murray

Table 4. Characteristics and results of PCR studies to detect the EBV genome in isolated tumor cells

Authors	Sample	Method	PCR amplification	Comment
Fina et al., 2001	frozen	7 independent microdissections obtained from 2 cases	Q-PCR	Case a: 3 positive/7 clusters
				Case b: 3 positive/7 clusters
McCall et al., 2001	paraffin	5′ nuclease PCR EBNA1	50 cycles	1 positive/115 cases
Xue et al., 2003	frozen	PCR LF3	39 cycles	6/15
Murray et al., 2003	paraffin	Pool of multiple clusters of tumor cells	Q-PCR	0/19 PCR positive cases
Arbach et al., (unpublished)	frozen	2 to 4 independent microdissections obtained from 4 different cases	Q-PCR	1 to 4 positive clusters for each case

et al,. 2003, Arbach et al., unpublished). Murray et al. (2003) used sections of paraffin-embedded breast cancer tissues. They measured, by Q-PCR two, twelve and five samples, respectively with less than 0.1, 0.1-0.9 and 1-7 copies of the EBV genome per 1000 cells. In another study of 95 samples, similar results were obtained (Arbach et al., unpublished). Comparing the viral load that Murray et al. (2003) measured in paraffin embedded breast cancer to results obtained with similarly fixed NPC tissue; one can estimate that the level of EBV DNA is 100-10,000 lower overall for breast cancer than for NPC. This estimate implies that detection of the genome and its expression has to be optimized. Huang et al. (2003) suggested that lytic viral replication might contribute to the detection of the EBV genome in breast cancer. However it is unlikely that all the EBV genome copies detected reflect only replicating DNA, since Xue et al. (2003) showed that EBV latent CST/BART transcripts were detected in breast cancer biopsies specimens (see below), which excludes the possibility that the lytic cycle was the only source of EBV genomes in breast cancer tissues. Moreover, it is known that EBV reactivation also occurs in a few cells in NPC (Cochet et al., 1993) in which infection is predominantly latent. Furthermore, it has been shown that lytic gene expression may also contribute to invasion and metastasis of NPC (Yoshizawaki et al., 1999; Mauser et al., 2002). Lytic gene expression in breast cancer cells, might also contribute to tumor progression.

The presence of virus is heterogeneous within the tumor

Murray et al. (2003) failed to detect EBV genomes in microdissected tissue. However, they did not sample different areas of the same tumor, but pooled all the microdissected cells from each specimen. Thus, dilution of EBV-positive cells by non-infected cells would result. Moreover they analyzed fixed tissue which may have resulted in lower sensitivity. Similarly, McCall et al. (2001), detected EBV in 1/115 cases of tumor cells isolated from paraffin embedded sections. However, Q-PCR performed on epithelial tumor cells purified from EBV-positive tumors by microdissection gave positive results, when performed with frozen samples (Table 4): six to seven microdissections were performed in the two biopsied specimens of breast cancer tissue by Fina et al. (2001). Their results also showed heterogeneity of the viral load in samples containing isolated epithelial cancer cells: in one biopsy, 3/7 of the epithelial clusters isolated were EBV positive with a viral load of 128-5816 copies/100 ng GAPDH whereas 4 other clusters were negative. From another biopsy, 3/6 clusters isolated were EBV positive with a viral load of 272-1532 copies/100 ng GAPDH whilst 3 other clusters were negative. Arbach et al. (unpublished) have also performed quantification of EBV DNA in microdissections of different tumors and observed a large heterogeneity in distribution of viral genomes from one region to another within the same tumor, as well as amongst different tumors. These observations may account for the variable distribution of EBNA1 protein detected by immunohistochemistry (IHC) with two different monoclonal antibodies in breast cancer biopsies (Bonnet et al., 1999). It is well known that the distribution of different biologically relevant cellular markers characteristic of breast cancer (estrogen receptor (ER), progesterone receptor (PR), epidermal growth factor receptor (EGFR), metalloprotease 2 (MMP-2) (van Netten

Table 5. Characteristics and results of studies to detect the EBV expression in breast carcinomas, normal tissue and benign tumors

Authors	Sample	Gene	Comments
Lespagnard et al., 1995	paraffin	EBERs →0/10 LMP-1→0/10	Medullary carcinomas
Labreque et al., 1995	paraffin	EBERs *	Signal restricted to a small proportion of epithelial cells
Luqmani et al., 1995	Paraffin	EBER FITC labeled probe → no signal observed LMP-1 CS1-4→positive scattered tumor cells were observed	
Chu et al., 1998	paraffin	ISH EBER→0/60 IHC EBNA-2 (Dako) →0/60 IHC LMP-1 (Dako) →0/60	
Glaser et al., 1998	paraffin	ISH EBER→0/107	
Bonnet et al., 1999	paraffin Frozen paraffin	EBNA-1 HIC with 2B4 positive EBNA-1 HIC with IH4 positive EBER (Dako) negative	Heterogeneous labeling pattern of epithelial tumor cells
Brink et al., 2000	Frozen	EBNA-1 RT-PCR→0/5 BARTs RT-PCR→0/5 IHC EBNA-1 2B4→1/5 PRC positive cases	
Chu et al., 2001	Frozen	EBER ISH (*)→5/48 EBNA-1 2B4 IHC→12/48 LMP-1 CS1-4 (Dako) →0/48 ZEBRA BZ1 (Dako) →0/48	+<0.1% of the total cell population
Fina et al., 2001	Frozen , paraffin	EBERs radiolabeled probes autoradiography (1week-1month) (*)	Heterogeneous labeling pattern of epithelial tumor cells
McCall et al., 2001	Paraffin microdissected samples	EBER ISH (Biogenex) →0/6 PCR+ LMP-1 CS1-4 (Dako) →0/6	1 patient →EBER and LMP1+ (nuclear staining) in normal and cancer tissue
Dadmanesh et al., 2001	Paraffin	EBER ISH (Dako) →0/6	Lymphoepithelioma-like carcinoma

Reference	Sample	Method → Results	Comments
Grinstein et al., 2002	paraffin	EBNA-1 HIC with 2B4 →14/33 EBERs ISH (Dako) →0/14 LMP-1 (CS1-4) (Dako) →0/14 BZLF1 (BZ1) (Dako) →0/14	Total correlation between EBNA-1 and PCR positive results
Deshpande et al., 2002	paraffin	EBERs ISH (Novocasta) →0/43 RT-PCR IHC EBNA-1→ 0/43 2B4 LMP-1(CS1-4) (Dako) →0/14 LMP-2A IHC with rat 15F9→0/14	breast carcinoma
Kleer et al., 2002	Paraffin	LMP-1 (CS1-4) HIC→9/13 EBV+ samples LMP-1 RT-PCR→0/120	fibroadenomas in immunosuppressed hosts LMP-1 staining in the ductal epithelial cells
Herrmann et al., 2003	Paraffin	EBERs ISH (*) 35S probe→0/4 PCR positive samples EBNA-1 1H4→0/4 PCR positive samples EBNA-1 2B4→0/4 PCR positive samples	1H4 is less efficient on paraffin embedded samples than on frozen samples
Murray et al., 2003	Paraffin	EBNA-1 HIC with 2B4 25 cases →18/25 13/19←Q-PCR+ 5/6←Q-PCR- 47/153 2B4 positive	EBV DNA was not found in isolated tumor cells of EBNA-1 + tumors More likely ER negative
Xue et al., 2003	frozen	RT-PCR EBERs→0/6 RT-PCR EBNA-1 (Qp)→3/6 RT-PCR CST/BARTs→6/6 RT-PCR BARF1→3/6 RT-PCR BZLF1→1/6 RT-PCR BHLF1/IR2→0/6 RT-PCR LMP-2A→0/6 RT-PCR LMP-1→0/6	RNA isolated from excised sections devoid of lymphocytes
Huang et al., 2003	Frozen breast carcinoma	RT-PCR BZLF1→ 6/10	
	Frozen normal tissue	RT-PCR BZLF1→ 3/10	
	paraffin	ZEBRA IHC→positive in cancer tissue	
Ribeiro-Silva et al., 2004	paraffin	IHC EBNA-1 detected with 2BA→32/85	Correlation between EBNA-1 and p63 expression

(*) protocol developed in the laboratory

et al., 1987; Talvensaari-Mattila et al., 2003) is heterogeneously distributed within breast cancer tissue. Heterogeneity within other EBV-associated tumors has been described: Yao et al. (2000) reported morphological differences among tumor cells in NPC, with some tumors expressing HLA class-I and/or -II, whilst in others such expression is down-regulated or negative. Individual NPC may or may not express the EBV-encoded latent membrane protein 1 (LMP1 protein), and individual tumor cells may express high levels of EBERs (Epstein-Barr virus encoded small RNAs), yet adjacent tumor cells express very little or none, showing extensive phenotypic variation in NPC. In conclusion, it appears that EBV genomes are also heterogeneously distributed within a subset of breast tumors. Whether there is any correlation between distribution of the cellular and viral markers is unknown.

Expression of the EBV genome in breast cancer biopsies

Expression of the EBV genome at the RNA and protein level has been investigated. The results are summarized in Table 5. *In situ* hybridization (ISH) with EBER probes, a highly sensitive method for the detection of EBV in latently infected cells, is widely used because of the high EBER RNA copy number. Labrecque et al. (1995), Fina et al. (2001) identified EBER RNA in a fraction of malignant cells in different breast tumors. However, in other studies, EBER transcripts could not be detected in breast cancer, however the authors did not rule out the possibility that EBV was in these samples and in spite of the lack of EBER detection (Luqmani, and Shousha, 1995; Chu et al., 1998; Glaser et al., 1998; Bonnet et al., 1999). Technical problems (related to tissue fixation, probe penetration in breast tissue) may contribute to these differences. In addition, the regulation of transcription of EBERs remains poorly understood, and their high level expression in infected cells might not be universal. NPC that have varying degrees of differentiation lack EBER expression in some areas have been described (Pathmanathan et al., 1995). Takeuchi et al. (1997) did not detect any EBER-1 expression in some EBV-positive NPC cases, and Yao et al. (2000) observed heterogeneity in EBER expression in different areas of NPC. It is therefore possible that EBERs are not expressed in breast cancer or, if they are, only at a lower level than in other tissue. Contrary to other EBV-associated diseases, it appears that EBER ISH may not be the best method to identify infected cells in breast cancers.

Immunohistochemical studies on 60 invasive breast cancers were performed by Chu et al. (1998). They did not detect Epstein-Barr nuclear antigen 2 (EBNA2) which is only expressed in lymphoproliferative disorders and lymphomas of immunocompromised patients (1). The same result was obtained with the latent membrane protein 1 (LMP1). However, Luqmani et al. (1995) showed a distinct staining of scattered tumor epithelial cells in several of the sections examined for LMP1. The fact that few cells were stained, the low sensitivity of the immunochemical methods, and the epidemiologic differences of the samples analyzed could explain the divergence between these two results.

EBNA1 protein is essential for the maintenance of the viral episome in infected cells and is constantly expressed in all EBV infections (1). In three studies, EBNA1 protein was detected by IHC, but the staining was thought to be nonspecific because EBNA1 transcripts were not detected and the specificity of the 2B4 antibody has been questioned (Murray et al. 2003). Bonnet et al. (1999) have detected EBNA1

Table 6. Characteristics and results of PCR studies to detect the EBV genome in normal tissue and benign tumors

Authors	Sample	Amplified fragment: Positive/total	Cycle number
Labreque et al., 1995	Frozen Normal (1 case) Benign (no atypia) 20	BamH1W → 0/21	35+southern-blotting
Luqmani et al., 1995	Normal tissue Paraffin	BamH1W →0/12	Nested, 2x40+southern-blotting
Bonnet et al., 1999	Normal tissue taken next to the tumor Frozen	EBER→ 3/30 LMP1→ 3/30 BZLF1 →3/30 3 were positive for the 3 PCR amplifications	40 40 nested 2x40
Kleer et al., 2002	Paraffin fibroadenomas in immunosuppressed hosts	EBER→13/18	40

staining with another EBNA1 monoclonal antibody (1H4) in frozen sections of breast cancer tumors (as well as the antibody used in the other studies) (Bonnet et al. 1999) and we and Xue et al. (2003) also detected EBNA1 transcripts (Arbach et al., unpublished).

Xue et al. (2003) detected EBV transcripts in samples sorted for the absence of lymphocytes. They not only detected EBNA1 (in 3/6 cases) but also CST/BARTs, and BARF-1 RNA in three of six cases. BARF-1 is able to confer transforming properties in cells (Ooka, 2001). BARF-1 is expressed in other carcinomas, namely, in NPC and in gastric cancers. Xue et al. (2003) amplified BZLF1 transcripts in 1/6 cases. Huang et al. also detected BZLF1 transcripts in 6/10 breast carcinomas as well as 3/10 normal tissues although the protein was detected only in a few tumor cell specimens. As discussed earlier, lytic expression in a few cells does not exclude latent infection in others.

In conclusion latent EBV gene expression is detected in a subset of breast cancer tissue and lytic expression in a few breast cancer cells. Moreover, the expression pattern observed in breast cancer biopsies was never seen in circulating lymphocytes since peripheral memory cells express a more restricted pattern where no latent genes are expressed, with the possible exception of LMP2 (Babock et al., 2000). It seems then that EBV in epithelial tumor cells is transcriptionally active.

Detection and expression of EBV in normal tissue or benign tumors

Normal tissue and benign tumors have been investigated by several authors. PCR studies to detect the viral genome are summarized in Table 6. In the studies performed by Labrecque et al.,(1995), Luqmani and Shousha, (1995), with benign tumors or normal breast tissue taken from immunocompetent patients no EBV DNA was detected. Bonnet et al., (1999) amplified EBV DNA from 3/30 normal tissue samples taken adjacent to the tumor, in which the EBV genome could be provided by the a few invading tumor cells in the sample. In contrast, 72% of fibroademonas taken from immunocompromised patients (Kleer et al., 2002) were positive for the EBV genome. Moreover, 45% were concordantly positive for LMP1 staining by HIC. The immunological state of the host plays a critical role in the development of some EBV-related diseases, namely lymphomas. When patients become immunosuppressed the viral load increases in the peripheral blood (Babcock, et al., 1999). Expansion of proliferating lymphoblasts due to suppressed CTL response is believed to account for this increase and is considered to be a major risk factor for post-transplant lymphoproliferative disease (PTLD) and AIDS-associated B cell lymphoma. The study described by Kleer et al. (2002) shows that immunosuppression might also be a risk factor for breast fibroadenomas and might increase the likelihood of permanently infecting epithelial mammary cells.

EBV DETECTION AND BREAST CANCER PROGNOSTIC FACTORS

A statistically significant relation between the presence of EBV and several poor prognostic factors for breast carcinomas could be observed (Bonnet et al. 1999). EBV is detected more frequently in steroid hormone receptor-negative tumors, which are associated with a poor outcome. In addition, the EBV genome is preferentially detected in high histological SBR (Scarff-Bloom and Richardson classification) grade tumors, which are characterized by a high mitotic index and a lesser degree of differentiation. Similarly, the most undifferentiated forms of NPC and gastric carcinomas are more frequently associated with EBV (Tokunaga et al., 1993). EBV-positive breast tumors were frequent (72%) among patients having more than 3 metastatic lymph nodes. The association with axillary node invasion suggests that infection by EBV may be related to high metastatic potential of the tumor. Moreover, EBV genome was also detected in metastatic lymph nodes, as well as in the primary tumor, suggesting that EBV could already be present in the tumor cells prior to their migration (Bonnet et al. 1999). Murray et al. found that 2B4 (monoclonal antibody which recognize EBNA1) positivity, correlated with aggressive breast cancer. Ribeiro-Silva et al. found a correlation between staining with the 2B4 monoclonal antibody and p63 expression. In NPC, immunohistochemistry consistently revealed that all the tumour cells express substantial amounts of the p53-related protein p63 (Crook et al., 2000). The deleted form of p63 may be a suppressor of wild type p53 function. The correlation between staining with the 2B4 monoclonal antibody and p63 expression might reflect a clue of tumorigenesis.

IN VITRO INFECTION OF MAMMARY EPITHELIAL CELLS BY EBV

In vitro infection of mammary epithelial cells has been performed by several groups. Speck and Longnecker (2000) could infect, by cell-cell contact, the mammary carcinoma cell lines T47D and MCF7 *in vitro*; they observed virus entry 24h post infection. Huang et al. infected MDAMB231, MDA-MB435, and MDA-MB468 breast carcinoma cell lines. They characterized the last one which exhibits latent as well as lytic infection. We infected MDA-MB231 cells (Wakisaka et al. 2004) in which EBV is tightly latent *in vitro*. The EBV-infected clones exhibit 11-15 genome copies per cell (Bonnet-Duquennoy et al., 2003) and express hypoxia-inducible factor 1 HIF-1 alpha which in turn induces VEGF, a major factor for angiogenesis (Wakisaka et al., 2004). EBV is then able to infect breast epithelial cells suggesting that these cells or the tumor cells themselves could be a natural host for the virus and confer properties that are important for tumor progression.

CONCLUSION

EBV is associated with the development of different malignancies. In the last few years, EBV genomes and gene expression have been detected in a subset of breast tumors by several groups while others had negative results. The discrepancies between results need to be resolved since an association of EBV with breast cancer has potential relevance to its early detection, treatment and even prevention. The evidence regarding a possible association of EBV with breast carcinomas requires clarification in view of the high incidence of the tumor.

Identification of EBV-infected cells in breast cancer has given conflicting results in the literature since *in situ* markers are not available: EBER expression is low or absent, and results of EBNA1 detection by 2B4 antibody has been questioned (Murray et al., 2003). However, detection of the EBV genome in tumor cells isolated by laser capture microdissection (LCM), and the observed pattern of EBV expression in breast tumor tissue, favors the inference that EBV is harbored by epithelial cancer cells.

Implication of EBV in breast cancer has been questioned since the virus is detected in only a subset of tumor cells (Herrmann and Niedobitek, 2003). However, EBV might not be etiologic but instead have a role at an early step in carcinogenesis by altering the behavior of a subset of tumor cells so that they acquire a more aggressive phenotype. This hypothesis may be in agreement with Grinstein's et al. (2002) observation of EBNA1 expression in precursor lesions. Alternatively, infection with EBV at a late state of tumor development might enhance oncogenic properties such as invasiveness, angiogenesis and metastasis. For example, it has been shown that LMP1 induces matrix metalloproteinase-9 (MMP9), cyclooxygenase-2 (COX2), vascular endothelial growth factor (VEGF), and induces and causes release of FGF-2 in human epithelial cells (reviewed by Wakisaka, and Pagano, 2003). Moreover a correlation between LMP1 expression and presence of COX2 in NPC tissue was observed, and HIF-1 is increased in diverse Epstein-Barr virus (EBV)-infected type II and III cell lines, which express EBV latent membrane protein 1 (LMP1) (Wakisaka et al., 2004). However because EBV is detected in only a subset of breast cancer and in a limited number of breast cancer cells, it seems unlikely to be a primary etiologic agent. In any case, an association of breast cancer cells with EBV would be neither simple nor conventional, so that it should be considered in a different light from other EBV malignancies such as NPC.

The possibility that EBV may be involved in the pathogenesis of breast carcinomas has to be taken into account. However, whether EBV might have a role in breast cancer development or progression now needs to be thoroughly investigated with appropriate cell biologic tools. The question of EBV as an etiologic factor remains to be answered. In any case the virus may be a useful prognostic marker or provide molecular targets for therapy.

Acknowledgments

I thank Darren Powell for his kind suggestions and Joseph Pagano for careful reading of the manuscript. Our work is supported by ARC 4451.

References

Babcock, G.J., Decker, L.L., Freeman, R.B., and Thorley-Lawson, D.A. (1999). Epstein-barr virus-infected resting memory B cells, not proliferating lymphoblasts, accumulate in the peripheral blood of immunosuppressed patients. J. Exp. Med. *190*, 567-576.

Babcock, G.J., Hochberg, D., and Thorley-Lawson, A. D. (2000). The expression pattern of Epstein-Barr virus latent genes *in vivo* is dependent upon the differentiation stage of the infected B cell. Immunity *13*, 497-506.

Bonnet, M., Guinebretiere, J.M., Kremmer, E., Grunewald, V., Benhamou, E., Contesso, G., and Joab, I. (1999). Detection of Epstein-Barr virus in invasive breast cancers. J. Natl. Cancer Inst. *91*, 1376-1381.

Bonnet-Duquennoy, M, Arbach, H, Vandest, B, Lefeu, F and Joab, I (2003). Re : absence of Epstein-Barr virus genome in breast cancer-derived cell lines. J. Nat.l Cancer Inst. *95*, 1254-1255

Brink, A.A., van Den Brule, A.J., van Diest, P., and Meijer, C.J. (2000). Re: detection of Epstein-Barr virus in invasive breast cancers. J. Natl. Cancer Inst. *92*, 655-656.

Chu, J.S., Chen, C.C. and Chang, K.J. (1998). *In situ* detection of Epstein-Barr virus in breast cancer. Cancer Lett. *124*, 53-57.

Chu, P.G., Chang, K.L., Chen, Y.Y., Chen, W.G., and Weiss, L.M. (2001). No significant association of Epstein-Barr virus infection with invasive breast carcinoma. Am. J. Pathol., *159,* 571-578.

Cochet, C., Martel-Renoir, D., Grunewald, V., Bosq, J., Cochet, G., Schwaab, G., Bernaudin, J.F., and Joab, I. (1993). Expression of the Epstein-Barr virus immediate early gene, BZLF1, in nasopharyngeal carcinoma tumor cells. Virology. *97,* 358-365.

Crook, T., Nicholls, J.M., Brooks, L., O'Nions, J., and Allday, M.J. (2000). High level expression of deltaN-p63: a mechanism for the inactivation of p53 in undifferentiated nasopharyngeal carcinoma (NPC)? Oncogene 19, 3439-3444.

Dadmanesh, F., Peterse, J.L., Sapino, A., Fonelli, A. and Eusebi, V. (2001). Lymphoepithelioma-like carcinoma of the breast: lack of evidence of Epstein-Barr virus infection. Histopathology *38,* 54-61.

Deshpande, C.G., Badve, S., Kidwai, N., and Longnecker, R. (2002). Lack of expression of the Epstein-Barr Virus (EBV) gene products, EBERs, EBNA1, LMP1, and LMP2A, in breast cancer cells. Lab. Invest. *82,*1193-1199.

Fina, F., Romain, S., Ouafik, L., Palmari, J., Ben Ayed, F., Benharkat, S., Bonnier, P., Spyratos, F., Foekens, J.A., Rose, C., Buisson, M., Gerard, H., Reymond, M.O., Seigneurin, J.M., and Martin, P.M. (2001). Frequency and genome load of Epstein-Barr virus in 509 breast cancers from different geographical areas. Br. J. Cancer. *84,* 783-790.

Gaffey, M.J., Frierson, H.F., Jr., Mills, S.E., Boyd, J.C., Zarbo, R.J., Simpson, J.F., Gross, L.K., and Weiss, L.M. (1993). Medullary carcinoma of the breast. Identification of lymphocyte subpopulations and their significance. Mod. Pathol. *6,* 721-728.

Glaser, S.L., Ambinder, R.F., Digiuseppe, J.A., Hornross, P.L. and Hsu, J.L. (1998). Absence of Epstein-Barr virus EBER-1 transcripts in an epidemiologically diverse group of breast cancers. Int. J. Cancer. *75,* 555-558.

Glaser, S.L., Hsu, J.L., and Gulley, M.L. (2004). Epstein-Barr virus and breast cancer: state of the evidence for viral carcinogenèses. Cancer Epidemiol. Biomarkers Prev. *13,* 688-697.

Grasser, F.A., Murray, P.G., Kremmer, E., Klein, K., Remberger, K., Feiden, W. et al. (1994). Monoclonal antibodies directed against the Epstein-Barr virus-encoded nuclear antigen 1 (EBNA1): Immunohistologic detection of EBNA1 in the malignant cells of Hodgkin's disease. Blood. *84,* 3792-3798.

Grinstein, S., Preciado, M.V., Gattuso, P., Chabay, P.A., Warren, W.H., De Matteo, E., and Gould, V.E. (2002). Demonstration of Epstein-Barr virus in carcinomas of various sites. Cancer Res. *62,* 4876-4878.

Herrmann, K. and Niedobitek, G. (2003). Lack of evidence for an association of Epstein-Barr virus infection with breast carcinoma. Breast Cancer Res. *5,* R13-7

Huang, J., Chen, H., Hutt-Fletcher, L., Ambinder, R.F., and Hayward, S.D. (2003). Lytic viral replication as a contributor to the detection of Epstein-Barr virus in breast cancer. J. Virol. *77,* 13267-13274.

Joab, I., (2000). Re : Detection of Epstein-Barr virus in invasive breast cancers, J. Natl. Cancer Inst. *92,* 655.

Kijima, Y. Hokita, S., Takao, S., Baba, M., Natsugoe, S., Yoshinaka, H., Aridome, K., Otsuji, T., Itoh, T., Tokunaga, M., Eizuru, Y., and Aikou, T. (2001). Epstein-Barr virus involvement is mainly restricted to lymphoepithelial type of gastric carcinoma among various epithelial nicolaïsmes. J. Med. Virol. *64,* 513-518.

Kleer, C.G., Tseng, M.D., Gutsch, D.E., Rochford, R.A., Wu, Z., Joynt, L.K., Helvie, M.A., Chang, T., Van Golen, K.L., and Merajver, S.D. (2002). Detection of Epstein-Barr virus in rapidly growing fibroadenomas of the breast in immunosuppressed hosts. Mod. pathol. *15,* 759-764

Labrecque, L.G., Barnes, D.M., Fentiman, I.S., and Griffin, B.E. Epstein-Barr virus in epithelial cell tumors: a breast cancer study. (1995). Cancer Res. *55,* 39-45.

Lespagnard, L. Cochaux, P., Larsimont, D., Degeyter, M., Velu, T., and Heimann, R. (1994). Absence of Epstein-barr virus in medullary carcinoma of the breast as demonstrated by immunophenotyping, *in situ* hybridization and polymerase chain reaction. Anatomic Pathol. *103*, 449-452.

Luqmani, A. and Shousha, S. (1995). Presence of Epstein-Barr virus in breast carcinoma. Int. J. Oncol. *6*, 899-903.

Mauser, A., Holley-Guthrie, E., Zanation, A., Yarborough, W., Kaufmann, W., Klingelhutz, A., Seaman, W.T., and Kenney, S. (2002). The Epstein-Barr virus immediate-early protein BZLF1 induces expression of E2F-1 and other proteins involved in cell cycle progression in primary keratinocytes and gastric carcinoma cells. J. Virol. 76, 12543-12552.

McCall, S.A., Lichy, J.H., Bijwaard, K.E., Aguilera, N.S., Chu, W.S., and Taubenberger, J.K. (2001). Epstein-Barr virus detection in ductal carcinoma of the breast. J. Natl. Cancer Inst. *93*, 148-150.

Murono, S., Inoue, H., Tanabe, T., Joab, I., Yoshizaki, T., Furukawa, M., and Pagano, J.S. (2001). Induction of cyclooxygenase-2 by Epstein-Barr virus latent membrane protein 1 is involved in vascular endothelial growth factor production in nasopharyngeal carcinoma cells. Proc. Natl. Acad. Sci. USA, *98*, 6905-6910..

Murray, P.G., Lissauer, D., Junying, J., Davies, G., Moore, S., Bell, A., Timms, J., Rowlands, D., McConkey, C., Reynolds, G.M., Ghataura, S., England, D., Caroll, R., and Young, L.S. (2003). Reactivity with A monoclonal antibody to Epstein-Barr virus (EBV) nuclear antigen 1 defines a subset of aggressive breast cancers in the absence of the EBV genome. Cancer Res. *63*, 2338-2343.

Ooka, T. (2001). BARF-1 Gene as an EBV Encoded Oncogene. Epstein-Barr Virus Report, *8*, 177-182.

Speck, P., and Longnecker, R. (2000). Infection of breast epithelial cells with Epstein-Barr virus via cell-to-cell contact. J. Natl. Cancer Inst. *92*, 1849-1850.

Ribeiro-Silva, A., Ramalho, L.N., Garcia, S.B., and Zucoloto, S. (2004). Does the correlation between EBNA1 and p63 expression in breast carcinomas provide a clue to tumorigenesis in Epstein-Barr virus-related breast malignancies? Braz. J. Med. Biol. Res. *37*, 89-95

Richardson, A. Is breast cancer caused by late exposure to a common virus? (1997). Med. Hypotheses. *48*, 491-497.

Talvensaari-Mattila, A., Paakko, P., and Turpeenniemi-Hujanen, T. (2003). Matrix metalloproteinase-2 (MMP-2) is associated with survival in breast carcinoma. Br. J. Cancer, *89*, 1270-1275.

Tokunaga, M., Uemura, Y., Tokudome, T., Ishidate, T., Masuda, H., Okazaki, E. et al. (1993). Epstein-Barr Virus Related Gastric Cancer in Japan - A Molecular Patho-Epidemiological Study. Acta Pathol. Japon. *43*, 574-581.

Van Netten, J.P., Thornton, I.G., Carlyle, S.J., Brigden, M.L., Coy, P., Goodchild, N.L., Gallagher, S., and George, E.J. (1987). Multiple microsample analysis of intratumor estrogen receptor distribution in breast cancers by a combined biochemical/immunohistochemical method. Eur. J. Cancer Clin. Oncol. *23*, 1337-1342.

Wakisaka, N. and Pagano, J.S. (2003). Epstein-Barr virus induces invasion and metastasis factors. Anticancer Res. *23*, 2133-2138.

Wakisaka, N., Kondo, S., Yoshizaki, T., Murono, S., Furukawa, M., and Pagano, J.S. (2004). Epstein-Barr virus latent membrane protein 1 induces synthesis of hypoxia-inducible factor 1 alpha. Mol. Cell Biol. *24*, 5223-5234.

Xue, S.A., Lampert, I.A., Haldane, J.S., Bridger, J.E., and Griffin, B.E. (2003). Epstein-Barr virus gene expression in human breast cancer: protagonist or passenger? Br. J. Cancer. *89*, 113-119.

Yao, Y., Minter, H.A., Chen, X., Reynolds, G.M., Bromley, M., and Arrand, J.R. (2000). Heterogeneity of HLA and EBER expression in Epstein-Barr virus-associated nasopharyngeal carcinoma. Int. J. Cancer. *88*, 949-55.

Yoshizaki, T., Sato, H., Murono, S., Pagano, J.S., and Furukawa, M. (1999). Matrix metalloproteinase 9 is induced by the Epstein-Barr virus BZLF1 transactivator. Clin. Exp. Metastasis. *17*, 431-436.

Chapter 11

Epstein-Barr Virus and Gastric Cancers

*Dai Iwakiri and Kenzo Takada**

ABSTRACT

Recent studies have revealed the association of the Epstein-Barr virus (EBV) with about 10% of gastric cancer cases worldwide. EBV is found in 100% of cancer cells of the EBV-associated gastric cancer cases. In addition, EBV DNA in cancer biopsies shows monoclonality, indicating that carcinoma arises from a single EBV-infected cell. These findings suggest that EBV plays an important role in the development of EBV-positive gastric cancers. The difficulty of establishment of EBV infection of epithelial cells *in vitro* has hampered the study of the role of EBV in epithelial malignancies. The development of an *in vitro* model of EBV-infection with recombinant EBV carrying a selectable marker has enabled this difficulty to be overcome and has made it possible to establish EBV-infected primary cultured gastric epithelial cells. The cells express EBV genes identical to those typically observed in EBV-positive gastric cancer, and show accelerated malignant properties, suggesting that EBV contributes to the maintenance of the malignant phenotype of EBV-positive gastric cancer. Recently, functional studies of EBV genes in epithelial cells have been carried out to clarify the mechanisms of EBV-associated gastric cancer development.

INTRODUCTION

The Epstein-Barr virus (EBV) is well known for its association with several human malignancies, including epithelial neoplasms (Rickinson and Kieff, 2001). Oral hairy leukoplakia is a lesion caused by EBV replication. Nasopharyngeal carcinoma (NPC) is strongly related to EBV. In addition, the development of polymerase chain reaction (PCR) and in situ hybridization (ISH) techniques revealed the association of EBV with many other malignancies, including gastric cancer. Approximately 10% of gastric cancers throughout the world are EBV-positive; therefore the worldwide occurrence of EBV positive gastric cancer is estimated at more than 50,000 cases/year.

EPIDEMIOLOGY

EBV is associated with undifferentiated nasopharyngeal carcinomas. These nasopharyngeal carcinomas display a number of characteristic morphological

*For correspondence email kentaka@igm.hokudai.ac.jp

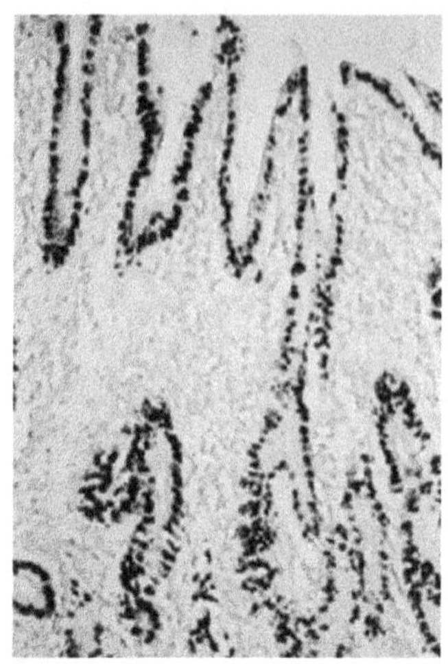

Figure 1. Detection of Epstein-Barr virus (EBV) encoded small RNA 1 (EBER1) in gastric carcinoma of the differentiated type. All carcinoma cells are positive for EBER1.

features, including intense lymphoid infiltration (termed lymphoepithelioma). Carcinomas with a similar histological feature (termed lymphoepithelioma-like carcinomas) from other organs such as the salivary glands, thymus, lungs, etc., have been analyzed for the presence of EBV and most cases are EBV positive (Rickinson and Kieff, 2001). Burke et al. (1990) first reported the association between EBV and gastric cancer presenting this particular histological type. It was demonstrated that EBV DNA was positive in more than 80% of lymphoepithelioma-type gastric cancers by PCR and ISH (Shibata et al., 1991; Oda et al., 1993; Takano et al., 1994; Nakamura et al., 1994). Afterward, EBV infection was demonstrated in gastric adenocarcinomas of the ordinary histological type (Figure 1). Shibata and Weiss (1992) reported that EBV is present in almost all cancer cells in EBV-positive cases. Interest has increased with a report that 69 of 999 (6.9%) cases of gastric cancer from Japan were EBV positive using ISH for the EBV-encoded small RNAs (EBERs) (Tokunaga et al., 1993). Among them, the lymphoepithelioma type constituted only eight cases and the remaining cases were typical adenocarcinomas.

EBV-positive gastric cancer is a non-endemic disease distributed throughout the world, while Burkitt's lymphoma and nasopharyngeal carcinoma, are endemic in equatorial Africa and Southern China, respectively (Rickinson and Kieff, 2001). However, there are some regional differences in the incidence of EBV-positive gastric cancer as a proportion of all cases of gastric cancer, with the highest (16–18%) in the USA and Germany, and the lowest (4.3%) in China (Figure 2) (Shibata and Weiss, 1992; Tokunaga et al., 1993; Leoncini et al., 1993; Shibata et al., 1993; Ott et al., 1994; Yuen et al., 1994; Nourad et al., 1994; Harn et al., 1995; Shin et al., 1996; Tashiro et al., 1998). Tokunaga et al. (1993) studied the incidence of EBV-positive cases in gastric cancers of nine Japanese cities. They observed the highest incidence of EBV-positive gastric cancer in Okinawa, which has the lowest gastric cancer mortality rate in Japan, and the lowest incidence in Niigata, which has the highest gastric cancer mortality rate. These observations suggested an inverse correlation between the incidence of gastric carcinoma and mortality from the disease. They also reported that the incidence of EBV-positive gastric cancer was

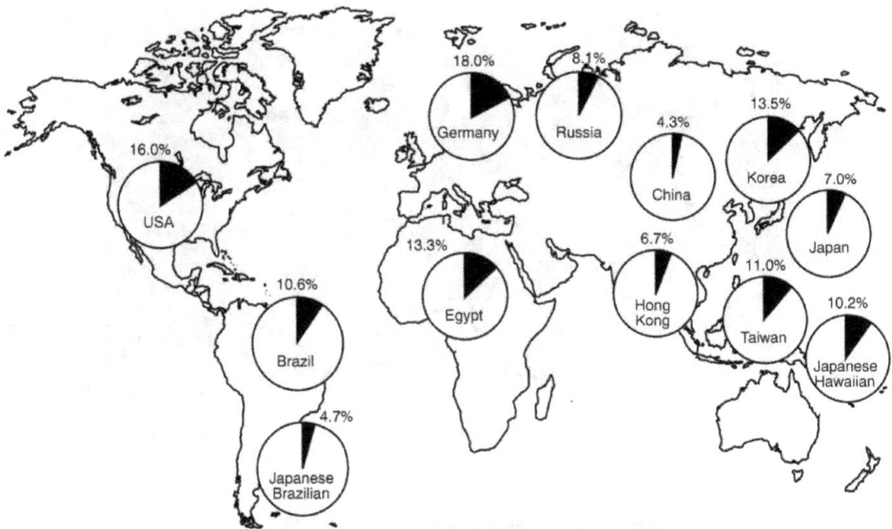

Figure 2. World distribution of Epstein-Barr virus (EBV) positive gastric carcinoma.

three times higher in men than in women in Japan. EBV-positive cases are more frequent in the proximal portion of the stomach, the cardia (8.1%), middle stomach (10.6%) and gastric stump (12.5%) than in the distal portion, the antrum (3.9%). EBV involvement is observed in papillary as well as tubular adenocarcinomas. There is no difference in EBV prevalence between intestinal and diffuse types of cancer.

THE ASSOCIATION OF EBV WITH GASTRIC CANCER

In gastric cancer cells, EBV is not integrated into the cellular DNA but is maintained as a plasmid. Since the numbers of terminal repeats (TRs) in EBV-positive gastric cancers are always uniform, the tumors arise from a single EBV-infected cell (Figure 3) (Shibata and Weiss, 1992; Leoncini et al., 1993; Imai et al., 1994; Fukayama et al., 1994). Furthermore, 100% of cancer cells in EBV-positive gastric cancer are EBV- positive. These findings suggest that EBV plays a significant role in the cancer development. Analysis of EBV gene expression in gastric cancer cells demonstrated that the EBV gene expression pattern was similar to that of Burkitt's lymphoma. Among EBV determined nuclear antigens (EBNAs), only EBNA1 is expressed, and among the three latent membrane proteins (LMPs), LMP1 and LMP2B are not expressed, while LMP2A is expressed in about 43% of gastric cancer cases (Imai et al., 1994). In addition, the BARF0 gene from the BamHI-A region and the EBERs are always expressed (Imai et al., 1994; Sugiura et al., 1996).

PRESENCE OF EBV IN NON-NEOPLASTIC GASTRIC MUCOSA

It has been reported that scattered EBV-positive cells are present in non-neoplastic gastric mucosa. EBER-positive cells can be detected in dysplastic mucosa bordering the tumors by means of ISH. However, EBV-positive cells are absent in surrounding lymphocytes, other normal stromal cells, intestinal metaplasia, and

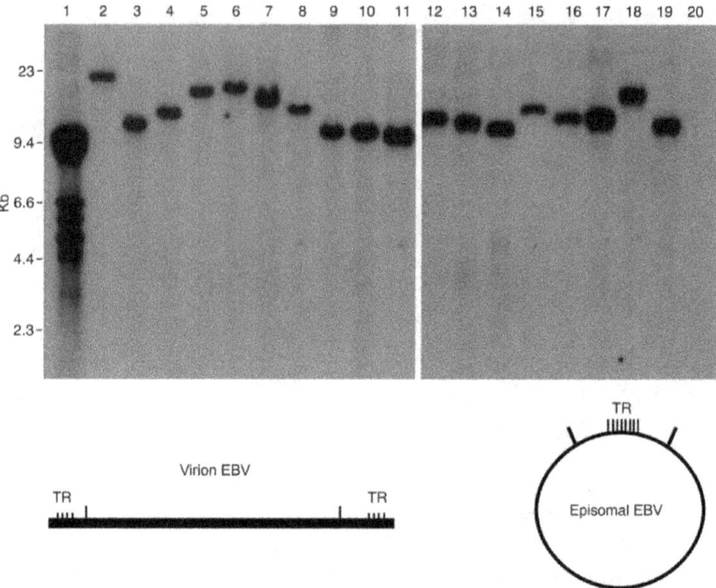

Figure 3. Epstein-Barr virus (EBV) DNA detection in EBV positive gastric carcinoma biopsies by Southern blotting. The EBV DNA terminal repeat (TR) was used as a probe. A single band was detected for each gastric carcinoma tissue, indicating that the gastric carcinoma was formed by proliferation of one EBV infected cell. Lane 1, EBV producing cell line B95-8 (multiple bands because the number of viral TRs is different for each virus particle); lane 2, Burkitt's lymphoma cell line Raji (single band); lanes 3–19, EBV positive gastric carcinoma samples (single bands); lane 20, EBV negative gastric carcinoma. Kindly provided by Dr S Imai, Kochi University, Kochi, Japan

normal gastric mucosa (Shibata and Weiss, 1992; Leoncini et al., 1993; Yuen et al., 1994; Harn et al., 1995; Fukayama et al., 1994; Gulley et al., 1996). These observations suggest that EBV infection occurs in the dysplastic stage and that an obvious growth advantage is conferred by the EBV infection. On the other hand, a sensitive DNA ISH method revealed frequent EBV infection in non-neoplastic gastric epithelium including intestinal metaplasia, whereas EBER expression was not seen (Yanai *et al*, 1997). This highlights the danger of relying on EBER detection as the sole marker of latent EBV infection, and the presence of EBV in non-neoplastic gastric epithelium indicates that EBV may play a role in the early stage of gastric cancer development.

EBV-SPECIFIC IMMUNITY IN GASTRIC CANCER PATIENTS

High serum IgG antibody titers against EBV capsid antigens (VCAs) and early antigens (EAs) were detected in patients with EBV-positive gastric cancer. Although IgA antibodies against VCAs are detected in about 60% of cases (Imai et al., 1994; Fukayama et al., 1994), their diagnostic value is limited since the titers are much lower than those in nasopharyngeal carcinoma. Levine et al. (1995) have reported that high antiviral titers are observed before the diagnosis of EBV-positive gastric cancer. These observations suggest that active EBV infection happens before the development of EBV-positive gastric cancer, although EBV-specific cellular immunity is not greatly reduced (Imai et al., 1994).

EBV-positive gastric cancers are frequently associated with relatively strong infiltration by lymphocytes. Saiki et al. (1996) have reported that lymphocytes infiltrating EBV-positive tumor nests are predominantly CD8-positive cytotoxic T lymphocytes (CTLs). More recently, Kuzushima et al. (1999) demonstrated that the tumor infiltrating lymphocytes (TILs) isolated from an EBV-positive gastric cancer consisted of HLA class I–restricted CD8-positive CTLs, which specifically recognized and killed autologous EBV-immortalized cells. However, the TILs did not kill HLA-A24-positive fibroblasts infected with vaccinia recombinant virus expressing each of the EBV latent proteins. These findings suggest that some cellular proteins might be involved in the strong T cell response to EBV-positive gastric cancer.

EBV AND GASTRIC CANCER-RELATED FACTORS

Helicobacter pylori infection leads to chronic atrophic gastritis and subsequent intestinal metaplasia, and is associated with gastric cancer (Parsonnet et al., 1991). Gastric cancer tissues are surrounded by atrophic mucosa and intestinal metaplasia in both EBV-positive and EBV-negative cases (Yanai et al., 1999). Since severe atrophic gastritis is frequently associated with EBV-positive gastric cancers, it is possible that EBV infection occurs in the atrophic epithelium and leads to cancer development. No significant difference was observed in the frequency of *H. pylori* infection between EBV- positive and EBV-negative gastric cancers (Yanai et al., 1999; Wu *et al*, 2000).

After partial gastrectomy for benign diseases, carcinomas occur at a high frequency in the remaining stomach (Caygill et al., 1986). This is referred to as gastric remnant cancer or stump cancer. The prevalence of EBV involvement in remnant cancers is significantly higher (27%) than in non-remnant cancers (Yamamoto et al., 1994).

No specific linkage to any HLA subtype has been reported in patients with EBV-positive gastric cancer (Qiu et al., 1997). EBNA3B is a major antigenic target that elicits an HLA-A11 CTL response to EBV infection. Mutations in the HLA-11 epitopes of EBNA3B are common in EBV-associated gastric carcinomas, but are not related to the patient's HLA status (Chu et al., 1999).

GENE ABNORMALITIES IN EBV-ASSOCIATED GASTRIC CANCER

Studies of gene abnormalities in EBV-associated gastric cancers have been reported. p53 mutation and overexpression are common in gastric cancer, and are detected not only in cancerous regions but also in areas of precancerous dysplasia and metaplasia (Shiao et al., 1994). Thus, p53 mutation might be an early event in gastric carcinogenesis. It was reported that no significant difference was found in the frequency of p53 overexpression between EBV-positive and EBV-negative gastric cancer (Ojima et al., 1997). Leung et al. (1998) reported that nearly all EBV-positive gastric cancers showed weak to moderate p53 expression in a variable proportion of cancer cells, and they suggested a role for EBV in a non-mutational mechanism of p53 overexpression.

It has been reported that the number of apoptotic tumor cells in EBV-positive gastric cancer is lower than that in EBV-negative cases (Ohfuji et al., 1996; Kume et al., 1999). In addition, Kume et al. (1999) have reported that bcl-2 expression in EBV-positive gastric cancer is higher than in EBV-negative cases, which suggests that high *bcl-2* expression protects tumor cells from apoptosis.

Osawa et al. (2002) have reported that p16 expression is downregulated by methylation of the promoter in EBV-positive gastric cancer compared to EBV-negative cases. This suggests that epigenetic changes affect the development of EBV-associated gastric cancer.

MECHANISM OF EBV INFECTION OF EPITHELIAL CELLS

Many cases of EBV infection of epithelial cells *in vivo* are known, including AIDS-related hairy leukoplakia (Greenspan et al., 1985) and EBV infection of the salivary gland epithelium (Wen et al., 1996), in addition to nasopharyngeal carcinoma and gastric cancer. EBV adheres to B cells through their surface CD21 molecules and this determines the B cell tropism (Fingeroth et al., 1984). Since CD21 expression is very low or negative in epithelial cells, it has been considered that EBV infects epithelial cells through mechanism distinct from B-cells.

Studies on EBV infection of epithelial cells have been limited by the lack of infection systems *in vitro*. Recently, the development of recombinant virus-carrying drug-resistant genes and fluorescence proteins has made it possible to establish *in vitro* EBV infection of various epithelial cell lines, and it has been reported that EBV may infect via more than one mechanism. Yoshiyama et al. (1997) have demonstrated successful EBV infection of CD21-negative gastric epithelial cell line AGS, using recombinant EBV (Akata strain) carrying a selectable marker gene (Yoshiyama et al.; 1995, Shimizu et al., 1996). This infection was not inhibited by anti-CD21 monoclonal antibodies (OKB7) (Yoshiyama et al., 1997) or soluble forms of gp350 (Maruo et al., 2001). Imai et al. (1998) have reported that EBV can infect a number of epithelial cell lines without CD21 expression. These findings indicate that EBV infects epithelial cells through a receptor other than CD21. On the other hand, Fingeroth et al. (1999) have reported that EBV can infect human embryonic kidney epithelial cell line 293 in a CD21-dependent manner.

Sixbey and Yao (1992) have reported that, in the pathway whereby IgA in the blood is processed into the secreted form by epithelial cells, EBV is incorporated into the epithelial cells in a bound state with an IgA antibody against envelope protein gp350/220. This phenomenon could explain the involvement of EBV infection in the development of nasopharyngeal carcinoma and possibly of gastric cancer, which are typically preceded and accompanied by the appearance of virus-specific IgA in serum. Imai et al. (1998) have shown that mixed culture with recombinant EBV-producing Akata cells is an efficient method of infection *in vitro*. They used numerous human cancer cell lines and got more efficient infection than with the method using a high- titer virus supernatant. This efficient 'cell to cell' contact suggests that there is another mechanism of infection *in vivo*, since the virus donors are most likely EBV-infected B cells migrating into the epithelial stroma or intraepithelial space. More recently, Tugizov et al. (2003) have reported a study on EBV infection of human polarized tongue and pharyngeal epithelial cells. In their report, they indicated that EBV infects these cells through three CD21-independent pathways: (i) by direct 'cell to cell' contact of apical cell membranes with EBV-infected lymphocytes; (ii) by entry of cell-free virions through basolateral membranes, mediated in part through an interaction between β- or α5β1-integrin and the EBV BMRF-2 protein; and (iii) after initial infection, by virus spread directly across lateral membranes to adjacent epithelial cells.

Several studies have suggested that EBV glycoprotein complex gp85/gp25 is involved in the EBV infection of epithelial cells. It has been reported that a monoclonal antibody that reacts with the gp85/gp25 complex, but not the gp85/gp25/gp42 complex, neutralizes infection of epithelial cells. (Li et al., 1995 Wang et al., 1998). In contrast, a recombinant EBV lacking gp42 could infect epithelial cells (Wang and Hutt-Fletcher, 1998). Furthermore, it has been shown that attachment of a recombinant EBV lacking gp85 to CD21-negative epithelial cells is impaired (Molesworth et al., 2000; Oda et al., 2000). These findings suggest the critical role of the gp85/gp25 complex in the infection of epithelial cells. However, the cellular surface molecule that interacts with gp85/gp25 remains unknown.

IN VIVO AND *IN VITRO* MODELS OF EBV INFECTION

Several model systems of EBV infection of epithelial cells have been established to study the role of EBV in epithelial malignancies. A recombinant EBV (Akata strain) carrying a selectable marker gene has been used to infect various epithelial cells *in vitro*. This system is useful in that EBV-positive and EBV-negative epithelial cell pairs with identical cell backgrounds are available to make a comparison (Takada, 2000). KT is a transplantable human EBV-associated gastric cancer propagated in severe combined immunodeficient (SCID) mice (Iwasaki et al., 1998). This cell line retains the same clonal EBV genome and EBV gene expression pattern as the original tumor biopsy and can serve as a good *in vivo* model for EBV-associated gastric cancer. Chong et al. (2002) have reported that the IL-1β expression is higher in KT tumor cells than in an EBV-negative gastric carcinoma tumor strain. Two EBV-positive gastric cancer cell lines have been established from noncancerous gastric tissues of two patients with gastric cancer (Tajima et al., 1998; Takasaka et al., 1998). Both cell lines express EBNA2 and LMP1, and spontaneously produce infectious viruses. These viruses are unique in that they have the ability to transform primary B cells and induce EA in latently EBV-infected Raji cells. More recently, Oh et al. (2004) have reported the establishment of a gastric cancer cell line naturally infected with EBV, derived from a primary gastric tumor that was classified as a moderately differentiated lymphoepithelial carcinoma. In this cell line, EBNA1 and LMP2A are expressed, while EBNA2 and LMP1 are negative. The expression pattern of EBV genes is similar to that of most EBV-associated gastric cancers.

ROLE OF EBV IN GASTRIC CANCER DEVELOPMENT

Although EBNA2 and LMP1 appear to play the most important roles in the immortalization of lymphocytes (Kieff, 2001), these molecules are not expressed in EBV-associated gastric cancer with type I EBV latency. This may give rise to doubts about the importance of the presence of EBV. Studies using the Akata Burkitt's lymphoma cell line, which retains type I EBV latency *in vitro*, have demonstrated that the presence of EBV is essential for the maintenance of the

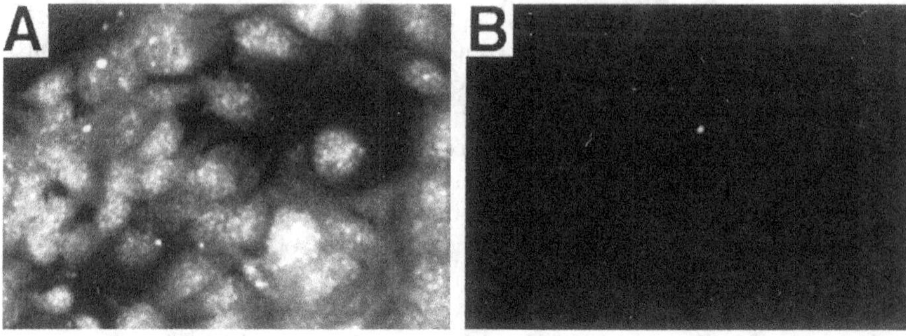

Figure 4. Detection of Epstein-Barr virus (EBV) in neomycin resistant gastric epithelial cell clones. (A) Immunofluorescent staining of EBV determined nuclear antigen (EBNA) in a neomycin-resistant gastric epithelial cell clone with EBV-seropositive human serum. Original magnification, x400. (B) Immunofluorescent staining of the same clone with EBV-seronegative human serum as a control. Original magnification, x400.

malignant phenotype and resistance to apoptosis (Shimizu et al., 1994; Komano et al., 1998; Chodosh et al., 1998; Ruf et al., 1999). These findings suggest that EBV contributes to the development and maintenance of gastric cancers.

Nishikawa et al. (1999) have reported the establishment of gastric primary culture cells infected with EBV. Primary cell cultures from healthy gastric mucosal biopsies were infected with recombinant EBV carrying a neomycin-resistance gene. As a result, they could get cell clones that could be maintained for at least 300 generations. These cell clones were all 100% EBNA positive (Figure 4) and expressed only the EBV genes EBNA1, EBER, BARF0, and LMP2A, similar to EBV-positive gastric cancers. The EBV-infected clones had higher proliferation rates and at least twice the cell saturation density in 10% serum than non-infected clones, and also showed the malignant phenotype in colony formation assay in soft agar. They also reported that the cultured cells from which the EBV-infected clones were obtained were successively subcultured to a maximum of 60 generations without EBV infection, whereas normal gastric mucosal primary culture cells die after subculturing for a few generations. Furthermore, the cultured cells showed p53 overexpression. These results suggest that EBV alone is not sufficient for oncogenesis, but can contribute to oncogenesis together with p53 or other gene abnormalities.

Growth promotion by EBV infection was also observed in some gastric cancer cell lines, with the NU-GC-3 line showing the greatest difference. These results suggest that EBV promotes growth of gastric epithelial cells. Iwakiri et al. (2003) have demonstrated that EBV infection induces insulin-like growth factor (IGF)-1 expression and promotes growth of infected cells since IGF-1 acts as an autocrine growth factor. Furthermore, they have shown that among the EBV latent genes expressed in EBV-infected cells, EBERs are responsible for IGF-1 induction. EBERs have been reported to have an oncogenic role in B lymphocytes (Komano et al., 1998, Kitagwa et al., 2000, Nanbo et al., 2002). These findings suggest that EBERs play a significant role in the development of EBV-associated malignancies, including gastric cancer.

Studies on the functions of other EBV genes expressed in gastric cancer have been reported. Zur Hausen et al. (2000) have reported that the BARF1 gene is expressed in EBV-associated gastric adenocarcinomas. Since the BARF1 gene has been shown to act as an oncogene in several cell lines (Wei and Ooka, 1989), it may play an oncogenic role in EBV-positive gastric cancers. The LMP2A gene, which is known to block B cell antigen-receptor signaling (Miller et al., 1994), is also expressed in some cases of EBV-associated gastric cancer. Recent findings that LMP2A has the ability to activate the phosphatidylinositol 3-kinase (PI3K) and Wnt signaling pathways in epithelial cells (Scholle et al., 2000, Morrison et al., 2003) suggest that LMP2A contributes to gastric cancer development.

CONCLUSIONS

It is clear that EBV is associated with some cases of gastric cancer. Clarification of the mechanisms of EBV infection of gastric epithelial cells *in vivo* and the roles of EBV genes in gastric cancer development are topics for future research.

References

Burke, A.P., Yen, T.S.B., Shekitka, K.M., and Sobin, L.H. (1990). Lymphoidepithelial carcinoma of the stomach which Epstein-Barr virus demonstrated by polumerase chain reaction. Mod. Pathol. *3*, 377-380.

Caygill, C.P., Hill, M.J., Kirkham, J.S., and Northfield T.C. (1986) Mortality from gastric cancer following gastric surgery for peptic ulcer. Lancet *1(8487)*, 929–931.

Chu, P.G., Chang, K.L., Chen, W.G., Chen, Y.Y., Shibata, D., Hayashi, K., Bacchi, C., Bacchi, M., and Weiss, L.M. (1999). Epstein-Barr Virus (EBV) nuclear antigen (EBNA)-4 mutation in EBV-associated malignancies in three different populations. Am. J. Pathol. *155*, 941-947.

Chodosh, J., Holder, V.P., Gan, Y.J., Belgaumi, A., Sample, J., and Sixbey, J.W. (1998). Eradication of latent Epstein-Barr virus by hydroxyurea alters the growth-transformed cell phenotype. J. Infect. Dis. *177*, 1194–1201.

Chong, J.M., Sakuma, K., Sudo, M., Osawa, T., Ohara, E., Uozaki, H., Shibahara, J., Kuroiwa, K., Tominaga, S., Hippo, Y., Aburatani, H., Funata, N., and Fukayama, M. (2002). Interleukin-1beta expression in human gastric carcinoma with Epstein-Barr virus infection. J.Virol. *76*, 6825-6831.

Fingeroth, J.D., Weis, J.J., Tedder, T.F., Strominger, J.L., Biro, P.A., and Fearon, D.T. (1984). Epstein-Barr virus receptor of human B lymphocytes is the C3d receptor CR2. Proc. Natl. Acad. Sci. USA *81*, 4510–4514.

Fingeroth, J.D., Diamond, M.E., Sage, D.R., Hayman, J., and Yates, J.L. (1999). CD21 dependent infection of an epithelial cell line, 293, by Epstein-Barr virus. J. Virol. *73*, 2115–2125.

Fukayama, M., Hayashi, Y., Iwasaki, Y., Chong, J.M., Ooba, T., Takizawa, T., Koike, M., Mizutani, S., Miyaki, M., and Hirai, K. (1994). Epstein-Barr virus-associated gastric carcinoma and Epstein-Barr virus infection of the stomach. Lab. Invest. *73*, 73–81.

Greenspan, J.S., Greenspan, D., Lennette, E.T., Abrams, D.I., Conant, M.A., Petersen, V., and Freese, U.K. (1985). Replication of Epstein-Barr virus within the epithelial cells of oral "hairy" leukoplakia, an AIDS-associated lesion. N. Engl. J. Med. *313*, 1564–1571.

Gulley, M.L., Pulitzer, D.R., Eagan, P.A., and Schneider, B.G. (1996). Epstein-Barr virus infection is an early event in gastric carcinogenesis and is independent of bcl-2 expression and p53 accumulation. Hum. Pathol. 27, 20–27.

Harn, H.J., Chang, J.y., Wang, Mw.., Ho, L.I., Lee, H.S., Chiang, J.H., and Lee, W.H. (1995). Epstein-Barr virus-associated gastric adenocarcinoma in Taiwan. Hum. Pathol. *26*, 267–271.

Imai, S,, Koizumi, S., Sugiura, M.., Tokunaga, M., Umemura, Y., Yamamoto, S., Tanaka, S., Sato, E., and Osato, T. (1994). Gastric carcinoma: monoclonal epithelial malignant cells expressing Epstein-Barr virus latent infection protein. Proc. Natl. Acad. Sci. USA *91*, 9131–9135.

Imai, S., Nishikawa, J., and Takada, K. (1998). Cell-to-cell contact as an efficient mode of Epstein-Barr virus infection of diverse human epithelial cells. J. Virol. *72*, 4371-4378.

Iwakiri, D., Eizuru, Y., Tokunaga, M., and Takada, K. (2003). Autocrine growth of Epstein-Barr virus-positive gastric carcinoma cells mediated by an Epstein-Barr virus-encoded small RNA. Cancer Res. *63*, 7062-7067.

Iwasaki, Y., Chong, J.M., Hayashi, Y., Ikeno, R., Arai, K., Kitamura, M., Koike, M., Hirai, K., and Fukayama, M. (1998). Establishment and characterization of a human Epstein-Barr-virus-associated gastric carcinoma in SCID mice. J.Virol.*72*, 8321–8326.

Kieff E. and Rickinson A.B. (2001). Epstein-Barr virus and its replication. In: B.N. Fields, D.M. Knipe, P.M. Howley, eds. *Virology*, 4th ed. (Philadelphia: Lippincott- Raven), 2511-2573.

Kitagawa, N., Goto, M., Kurozumi, K., Maruo, S., Fukayama, M., Naoe, T., Yasukawa, M., Hino, K., Suzuki, T., Todo, S., and Takada, K. (2000). Epstein-Barr virus-encoded poly(A)- RNA supports Burkitt's lymphoma growth through interleukin-10 induction. EMBO J. *19*, 6742-6750.

Komano, J., Maruo, S., Koizumi, K., Oda, T., and Takada, K. (1999). Oncogenic role of Epstein-Barr virus-encoded RNAs in Burkitt's lymphoma cell line Akata. J. Virol., *73*, 9827-9831.

Komano, J., Sugiura, M., and Takada,K. (1998). Epstein-Barr virus contributes to the malignant phenotype and to apoptosis resistance in Burkitt's lymphoma cell line Akata. J. Virol.*72*, 9150–9156.

Kume, T., Oshima, K., Shinohara, T., Takeo ,H., Yamashita ,Y., Shirakusa, T., and Kikuchi, M. (1999). Low rate of apoptosis and overexpression of bcl-2 in Epstein-Barr virus-associated gastric carcinoma. Histopathology *34*, 502–509.

Kuzushima, K., Nakamura,S., Nakamura, T., Yamamura, Y., Yokoyama, N., Fujita, M. Kiyono, T., and Tsurumi, T. (1999). Increased frequency of antigen-specific CD8(+) cytotoxic T lymphocytes infiltrating an Epstein-Barr virus-associated gastric carcinoma. J. Clin. Invest. *104*,163–171.

Leoncini, L,, Vindigni, C,, Megha, T., Funto, I., Pacenti, L., Musaro, M., Renieri, A., Seri, M.,Anagnostopoulos, J., and Tosi, P. (1993). Epstein-Bar virus and gastric cancer: data and unanswered questions. Int. J. Cancer *53*, 898–901.

Leung, S.Y., Chau, K.Y., Yuen, S.T., Chu, K.M., Branicki, F.J., and Chung, L.P. (1998). p53 overexpression is different in Epstein-Barr virus-associated and Epstein-Barr virus-negative carcinoma. Histopathology *33*, 311–317.

Levine, P.H., Stemmermann, G., Lennette, E.T., Hildesheim, A., Shibata, D., and Nomura, A. (1995). Elevated antibody titers to Epstein-Barr virus prior to the diagnosis of Epstein-Barr-virus-associated gastric adenocarcinoma. Int. J. Cancer *60*, 642–644.

Li, Q., Turk, S. M., and Hutt-Fletcher, L. M. (1995). The Epstein–Barr virus (EBV) BZLF2 gene product associates with the gH and gL homologs of EBV and carries an epitope critical to infection of B cells but not of epithelial cells. J.Virol. *69*, 3987-3994.

Maruo, S.,Yang, L., andTakada, K. (2001). Roles of Epstein-Barr virus glycoproteins gp350 and gp25 in the infection of human epithelial cells. J. Gen. Virol *82*,2373-2383.

Miller, C.L., Lee, J.H., Kieff, E., and Longnecker, R. (1994). An integral membrane protein (LMP2) blocks reactivation of Epstein-Barr virus from latency following surface immunoglobulin cross-linking. Proc. Natl. Acad. Sci. USA. *91*, 772-776.

Molesworth, S.J., Lake, C.M., Borza, C.M., Turk, S.M., and Hutt-Fletcher, L.M. (2000). Epstein-Barr virus gH is essential for penetration of B cells but also plays a role in attachment of virus to epithelial cells. J. Virol. *74*, 6324-6332.

Morrison, J.A., Klingelhutz, A.J., and Raab-Traub, N. (2003). Epstein-Barr virus latent membrane protein 2A activates beta-catenin signaling in epithelial cells. J. Virol. *77*, 12276-12284.

Nakamura, S., Ueki, T., Yao, T., Ueyama, T., and Tsuneyoshi, M. (1994). Epstein-Barr virus in gastric carcinoma with lymphoid stroma. Special references to its detection by the polymerase chain reaction and *in situ* hybridization in 99 tumors, including a morphologic analysis. Cancer *73*, 2239–2249.

Nanbo, A., Inoue, K., Adachi-Takasawa, K., and Takada, K. (2002). Epstein-Barr virus RNA confers resistance to interferon-a-induced apoptosis in Burkitt's lymphoma. EMBO J. *5*, 954-965.

Nishikawa, J., Imai, S., Oda, T,. Kojima, T., Okita, K., and Takada, K. (1999). Epstein- Barr virus promotes epithelial cell growth in the absence of EBNA2 and LMP1 expression. J. Virol. *73*,1286–1292.

Nourad, M., Anwar, N., Magrath, I., Franklin, J., Kingma, D.W., and Jatle, E.S. (1994). Epstein-Barr virus associated gastric adenocarcinomas among Egyptians. Cancer. Mol. Biol. Res. *1*, 283–288.

Oda K, Tamaru J, Takenouchi T., Mikata, A., Nunomura, M., Saitoh, N., Sarashina, H., and Nakajima, N. (1993). Association of Epstein-Barr virus with gastric carcinoma with lymphoid stroma. Am. J. Pathol. *143*, 1063–1071.

Oda, T., Imai, S,. Chiba, S., and Takada, K. (2000). Epstein-Barr virus lacking glycoprotein gp85 cannot infect B cells and epithelial cells. Virology *276*, 52-58.

Oh, S.T., Seo, J.S., Moon, U.Y., Kang, K.H., Shin, D-J., Yoon, S.K., Kim, W.H., Park, J-G., and Lee, S.K. (2004). A naturally derived gastric cancer cell line shows latency I Epstein–Barr virus infection closely resembling EBV-associated gastric cancer. Virology *320*, 330-336.

Ohfuji, S,, Osaki, M,, Tsujitani, S,, Ikeguchi, M., Sairenji, T., and Itoh, H. (1996). Low frequency of apoptosis in Epstein-Barr virus-associated gastric carcinoma with lymphoid stroma. Int. J. Cancer *68*, 710–715.

Ojima H, Fukuda T, Nakajima T, and Nagamachi, Y. (1997). Infrequent overexpression of p53 protein in Epstein-Barr virus-associated gastric carcinomas. Jpn. J. Cancer. Res. *88*, 262–266.

Ott, G., Kirchner, T., and Muller-Hermelink, H.K. (1994). Monoclonal Epstein-Barr virus genomes but lack of EBV-related protein expression in different types of gastric carcinoma. Histopathology *25*, 323–329.

Parsonnet, J., Friedman, G.D., Vandersteen, D.P., Chang, Y., Vogelman, J.H., Orentreich, N., and Sibley, R.K. (1991). Helicobactor pylori infection and the risk of gastric carcinoma. N. Engl. J. Med. *325*, 1127–1131.

Qiu, K., Tomita ,Y., Hashimoto, M., Ohsawa, M., Kawano, K., Wu, D.M., and Aozasa, K. (1997). Epstein-Barr virus in gastric carcinoma in Suzhou, China and Osaka, Japan: association with clinico-pathologic factors and HLA-subtype. Int. J. Cancer *71*,155–158.

Rickinson, A.B. and Kieff, E. (2001) Epstein-Barr virus In Virology 4th Ed.: B. N. Fields, D.M. Knipe, P.M. Howley, eds. (Philadelphia Lippincott-Raven), pp. 2575-2628.

Rowlands, D.C., Ito, M., Mangham, D.C., Reynolds, G., Herbst, H., Hallissey, M.T., Fielding, J.W., Newbold, K.M., Jones, E.L., Young, L.S., and Niedobitek, G. (1993). Epstein-Barr virus and carcinomas: rare association of the virus with gastric adenocarcinomas. Br. J. Cancer. *68*, 1014-1019.

Ruf, I.K., Rhyne, P.W., Yang, H., Borza, C.M., Hutt-Fletcher, L.M., Cleveland, J.L., and Sample, J.T. (1999). Epstein-Barr virus regulates c-MYC, apoptosis, and tumorigenicity in Burkitt lymphoma. Mol. Cell. Biol. *19*, 1651–1660.

Saiki, Y., Ohtani, H., Naito, Y., Miyazawa, M., and Nagura, H. (1996). Immunophenotypic characterization of Epstein-Barr virus-associated gastric carcinoma: massive infiltration by proliferating CD8+ T-lymphocytes. Lab. Invest. *75*, 67–76.

Scholle, F., Bendt K.M., and Raab-Traub, N. Epstein-Barr virus LMP2A transforms epithelial cells, inhibits cell differentiation, and activates Akt. J. Virol. *74*, 10681-10689.

Shiao, Y., Rugge, M., Correa, P., Lehmann, H.P., and Scheer, W.D. (1994). p53 alteration in gastric precancerous lesions. Am. J. Pathol. *144*, 511–517.

Shibata, D., Hawes, D., Stemmermann, G.N., and Weiss, L.M. (1993). Epstein-Barr virus-associated gastric adenocarcinoma in Hawaii Japanese. Cancer Epidemiol. Biomarkers Prev. *2*, 213–217.

Shibata, D., Tokunaga, M., Uemura ,Y., Sato, E., Tanaka, S., and Weiss, L.M. (1991). Association of Epstein-Barr virus with undifferentiated gastric carcinomas with intense lymphoid infiltration. Am. J. Pathol. *139*, 469–744.

Shibata, D. and Weiss, L.M. (1992). Epstein-Barr virus-associated gastric adenocarcinoma. Am. J. Pathol. *140*, 769–774.

Shimizu, N., Tanabe-Tochikura, A., Kuroiwa, Y., and Takada, K. (1994). Isolation of Epstein-Barr virus (EBV)-negative cell clones from the EBV-positive Burkitt's lymphoma (BL) line Akata: malignant phenotypes of BL cells are dependent on EBV. J.Virol. *68*, 6069–6073.

Shimizu N, Yoshiyama H, and Takada K. (1996). Clonal propagation of Epstein-Barr virus (EBV) recombinants in EBV-negative Akata cells. J.Virol. *70*, 7260–7263.

Shin, W.S., Kang, M.W., Kang, J.H., Choi, M.K., Ahn, B.M., Kim, J.K., Sun, H.S., and Min, K.W. (1996). Epstein-Barr virus-associated gastric adenocarcinomas among Koreans. Am. J. Clin. Pathol. *105*, 174–181.

Sixbey, J.W. and Yao, Q. (1992). Immunoglobulin A-induced shift of Epstein-Barr virus tissue tropism. Science *255*, 1578–1580.

Sugiura, M., Imai, S., Tokunaga, M., Koizumi, S., Uchizawa, M., Okamoto, K., and Osato, T. (1996). Transcriptional analysis of Epstein-Barr virus gene expression in EBV-positive gastric carcinoma; unique viral latency in the tumour cells. Br. J. Cancer *74*, 625–631.

Tajima, M., Komura, M., and Okinaga, K. (1998). Establishment of Epstein-Barr virus-positive human gastric epithelial cell lines. Jpn. J. Cancer Res. *89*, 262–268.

Takada, K. (2000). Epstein–Barr virus and gastric carcinoma. Mol. Pathol. *53*, pp. 255–261.

Takano, Y., Kato, Y., and Sugano, H. (1994). Epstein-Barr virus-associated medullary carcinomas with lymphoid infiltration of the stomach. Cancer Res. Clin. Oncol. *120*, 303–308.

Takasaka, N., Tajima, M., Okinaga, K., Satoh, Y., Hoshikawa, Y., Katsumoto, T., Kurata, T., and Sairenji, T. (1998). Productive infection of Epstein-Barr virus (EBV) in EBV-genome-positive epithelial cell lines (GT38 and GT39) derived from gastric tissues. Virology *247*, 152–159.

Tashiro, Y., Arikawa, J., Itoh, T., and Tokunaga, M. (1998). Clinico-pathological findings of Epstein-Barr virus-related gastric cancer. In: T. Osato, K. Takada, M. M. Tokunaga, eds. Epstein-Barr virus and human cancer. Japanese Cancer Association Gann Monograph on Cancer Research No. 45. Basel: (Karger), 87–97.

Tokunaga, M., Uemura, Y., Tokudome, T., Ishidate, T., Masuda, H., Okazaki, E., Kaneko, K., Naoe, S., Ito, M., and Okamura, A. (1993). Epstein-Barr virus related gastric cancer in Japan: a molecular patho-epidemiological study. Acta. Pathol. Jpn. *43*, 574–581.

Tokunaga, M., Land, C.E., Uemura, Y., Tokudome, T., Tanaka, S., and Sato, E. (1993). Epstein-Barr virus in gastric carcinoma. Am. J. Pathol. *143*, 1250–1254.

Tugizov, S.M., Berline, J.W., and Palefsky, J.M. (2003).Epstein-Barr virus infection of polarized tongue and nasopharyngeal epithelial cells. Nat. Med *9,* 307-314.

Wang, X. and Hutt-Fletcher, L. M. (1998). Epstein–Barr virus lacking glycoprotein gp42 can bind to B cells but is not able to infect. J.Virol. *72,* 158-163.

Wang, X., Kenyon, W.J., Li, Q., Mullberg, J., and Hutt-Fletcher, L.M. (1998). Epstein-Barr virus uses different complexes of glycoproteins gH and gL to infect B lymphocytes and epithelial cells. J. Virol. *72,* 5552-5558.

Wei, M.X. and Ooka, T. (1989). A transforming function of the BARF1 gene encoded by Epstein-Barr virus. EMBO J. *8,* 2897–2903.

Wen, S., Shimizu, N., Yoshiyama, H., Mizugaki, Y., Shinozaki, F., and Takada, K. (1996). Association of Epstein-Barr virus (EBV) with Sjogren's syndrome: differential EBV expression between epithelial cells and lymphocytes in salivary glands. Am. J. Pathol. *149,* 1511–1517.

Wu, M.S., Shun, C.T., Wu, C.C., Hsu, T.Y., Lin, M.T., Chang, M.C., Wang, H.P., and Lin, J.T. (2000). Epstein-Barr virus-associated gastric carcinomas: relation to H. pylori infection and genetic alterations. Gastroenterology *118,* 1031-1038.

Yanai, H., Takada, K., Shimizu, N., Mizugaki, Y., Tada, M., and Okita, K. (1997). Epstein-Barr virus infection in non-carcinomatous gastric epithelium. J. Pathol. *183,* 293-298.

Yanai, H., Murakami, T., Yoshiyama, H., Takeuchi, H., Nishikawa, J., Nakamura, H., Okita, K., Miura, O., Shimizu, N., and Takada, K. (1999). Epstein-Barr virus-associated gastric carcinoma and atrophic gastritis. J. Clin. Gastroenterol. *29,* 39–43.

Yoshiyama, H., Imai, S., Shimizu, N., and Takada, K. (1997). Epstein-Barr virus infection of human gastric carcinoma cells: implication of the existence of a new virus receptor different from CD21. J. Virol. *71,* 5688–5691.

Yoshiyama, H., Shimizu, N., and Takada, K. (1995). Persistent Epstein-Barr virus Infection in a human T-cell line: unique program of latent virus expression. EMBO J. *14,* 3706–3711.

Yuen, S.T., Chung, L.P,. Leung, S.Y., Luk, I.S., Chan, S.Y., and Ho, J. (1994). In situ detection of Epstein-Barr virus in gastric and colorectal adenocarcinomas. Am. J. Surg. Pathol. *18,* 1158–1163.

Zur Hausen, A, Brink, A.A., Craanen, M.E., Middeldorp,.J.M., Meijer, C.J., and van den Brule, A.J. (2000). Unique transcription pattern of Epstein-Barr virus (EBV) in EBV-carrying gastric adenocarcinomas: expression of the transforming BARF1 gene. Cancer Res. *60,* 2745–2748.

Chapter 12

Epstein-Barr Virus Invasion and Metastasis

*Tomokazu Yoshizaki, Naohiro Wakisaka
and Joseph S. Pagano**

ABSTRACT

In addition to its central role in transformation of cells by EBV, the LMP-1 protein is able also to induce directly a comprehensive set of factors that promote tumor progression. These are cellular factors that mediate steps in the process of invasion and metastasis of tumor cells as well as angiogenesis. Principal functions and factors identified to date, most induced by LMP-1, include cell detachment (MUC1); penetration of basement membrane (MMP-9 and MMP-1); and increase in cell motility (α v integrins). Other viral genes may contribute to cell mobility (LMP-2A) and counteraction of the metastatic suppressor protein, Nm23-H1 (EBNA-3C). Neoangiogenesis is also essential for invasion by tumor cells both locally and at distant sites. LMP-1 induces key factors that mediate this complex process including IL-8, FGF-2, COX-2, HIF-1α and VEGF. Both NPC (Type II latency) and B-lymphoproliferative diseases (Type III latency) are highly invasive, and expression of LMP-1 *in vivo* correlates with invasive indices for NPC. Thus EBV may as an addition or alternative to an etiologic relation with its associated tumors exert the capacity to alter later stages of malignant progression of existing tumor cells.

INTRODUCTION

The oncogenic properties of human tumor viruses including human papillomavirus, Kaposi's sarcoma herpesvirus, SV_{40} and the human papovaviruses, human T-cell leukemia virus, hepatitis B and C viruses as well as EBV are well characterized (Pagano et al., 2004). However the conceptual approach to the relation between these viruses and the neoplasms linked to them has been confined largely to establishing etiology. Whether and how tumor viruses bring about malignant transformation of cells has been addressed repeatedly over the past 30 years. That tumor viruses might have a relation other than etiologic to virus-associated tumors has received little attention. The idea that a tumor virus might instead modify behavior of an already existing tumor cell is a relatively new concept.

*For correspondence email joseph_pagano@med.unc.edu

1. Initiation

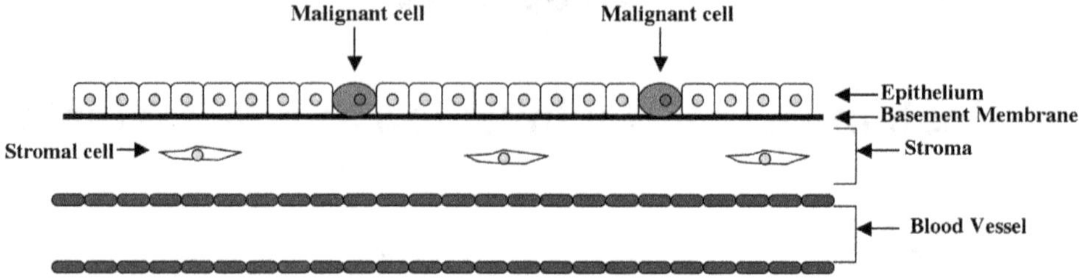

2. Growth

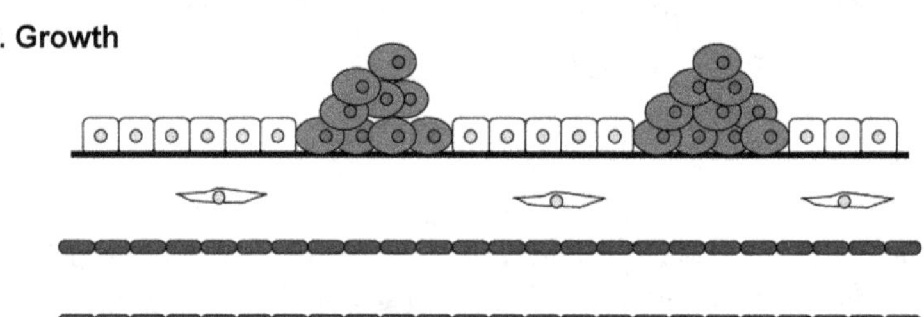

3. Angiogenesis and Invasion (*1*)

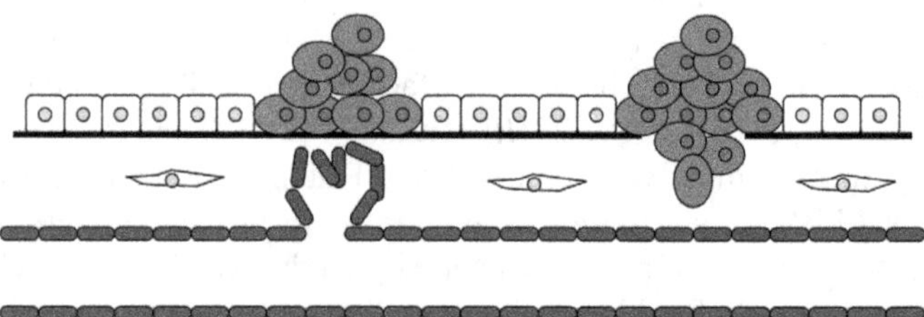

4. Angiogenesis and Invasion (*2*)

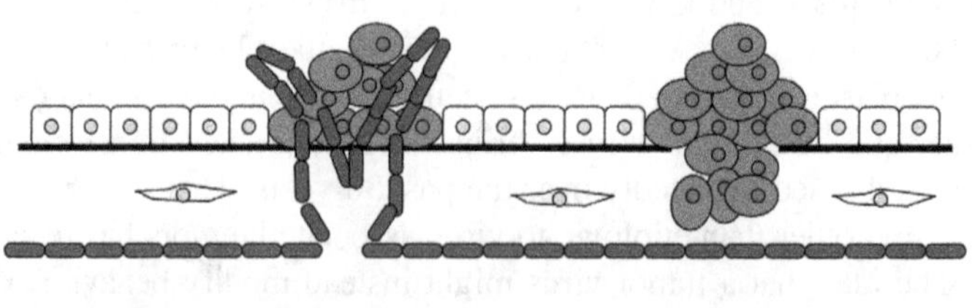

5. Intravasation

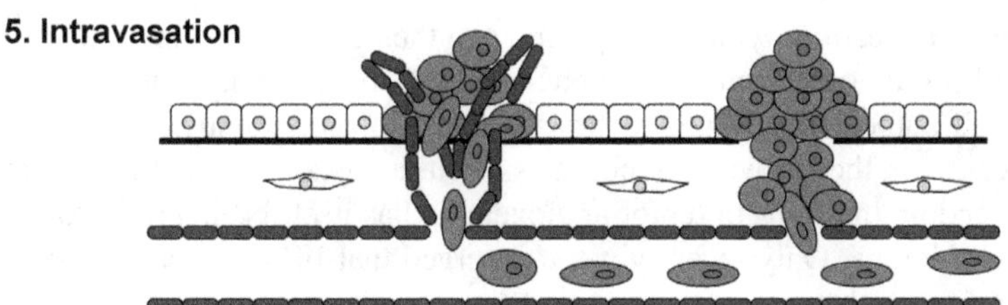

6. Extravasation

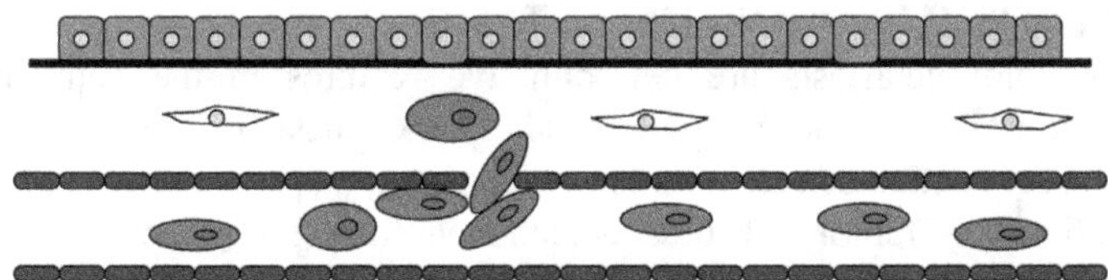

Figure 1. Stages of tumor progression.1. *Initiation*. When a tumor cell acquires the ability to invade and destroy normal tissue, it is termed malignant. The ability to form tumors at a distant site (metastasis) is a characteristic of highly malignant generally fatal cancers. 2. *Growth*. A single cell becomes modified to exhibit a growth advantage over the surrounding tissue. 3. *Angiogenesis and Invasion (1)*. (i) At some point, tumor cells require new vessels to supply nutrients for further growth. (ii) As the tumor becomes malignant, it acquires the ability to invade surrounding normal tissue. 4. *Angiogenesis and Invasion (2)*. (i) New vasculature reaches the nest of tumor cells. (ii) During invasion, the tumor cells penetrate the basement membrane underlying the tumor. 5. *Intravasation*. The tumor cells then move through extracellular matrix to reach the circulation through the vasculature. The process of entering the blood stream is termed intravasation. Intravasation occurs when the tumor cells cross the basement membrane and enter the circulation. 6. *Extravasation*. Following survival of tumor cells in the circulation, extravasation occurs when the tumor cells leave the circulation and penetrate the host tissue, again crossing through basement membrane. Metastasis occurs if the tumor cells can establish and grow at the secondary site.

This chapter assembles the evidence that an oncogenic virus might contribute to the metastatic properties of tumor cells either transformed or simply infected by the virus. EBV takes the lead in this area, and it also illustrates that a tumorigenic agent may be capable of both types of relations, perhaps independently. As the evidence accumulates it is becoming clear that ideas about the oncogenic properties of tumor viruses, at least EBV, need to be broadened to include the final and fatal steps of the malignant process, those involved in invasion and metastasis.

The chapter starts with a review of the steps in the complex cellular processes leading to invasion and metastasis by tumor cells, about which much is known at the molecular pathogenetic level. These processes involve detachment of cells from the tumor, invasion through the noncellular and cellular layers surrounding the tumor including a critical step, penetration of basement membrane; during these events tumor cells increase their motility. These steps result in localized invasion by the tumor. They also set the stage for the finale, namely, invasion of tumor at a distant site, for which penetration of vascular endothelium, circulation through the blood stream or lymphatic system, and then exodus and invasion into parenchyma of distant organs are critical phenomena. Crucial for the autonomy of these tumor-cell outposts is neoangiogenesis, needed to supply nutrients through the bloodstream for the growing tumor mass—either locally or at the distant sites. Understanding the steps in tumor angiogenesis has itself been a remarkable odyssey in cell biology. Only recently has it emerged that EBV triggers or alters critical molecular touchstones from the earliest to the concluding stages of these processes, as will be summarized here. Finally, there is a consideration of whether any of these discoveries might pertain to other tumor viruses.

STEPS IN INVASION AND METASTASIS

Invasion and metastasis are determinative features in the pathogenesis and progression of malignant neoplasms. The pathogenesis of metastasis consists of multiple, sequential, selective and interdependent steps (Figure 1). To establish a metastatic focus, tumor cells must detach from the primary tumor (suppression of cell-to-cell and cell-matrix adhesion), degrade and invade the extracellular matrix (ECM), increase in cell motility and enter the circulation, arrest in a capillary bed, gain entrance into organ parenchyma, proliferate and induce angiogenesis (Ellis et al., 1996). It is now well established that the processes of invasion and angiogenesis are essential for the growth and metastasis of both primary and metastatic tumors (Folkman et al., 1971).

Matrix metalloproteinases (MMPs) are a family of structurally related zinc-dependent endopeptidases collectively capable of degrading essentially all components of ECM. Based on their structure and substrate specificity, MMPs are classified into subgroups of collagenases: stromelysins and stromelysin-like MMPs and other MMPs (Shapiro et al., 1998; Nagase et al., 1999; Westermarck et al., 1999; Vihinen et al., 2002). MMPs play an important role in the physiologic degradation of ECM, *e.g.,* in tissue morphogenesis, tissue repair and in angiogenesis. MMPs also have important functions in pathologic conditions characterized by excessive degradation of ECM, such as rheumatoid arthritis, osteoarthritis, periodontitis, autoimmune blistering disorders of the skin, as well as in tumor invasion and metastasis (Shapiro et al., 1998; Nagase et al., 1999; Westermarck et al., 1999; Vihinen et al., 2002). MMP-1 (collagenase-1) cleaves fibrillar collagens with preference for type III collagen, which denatures into gelatin and is further degraded by other MMPs, such as gelatinases (Nagase et al., 1999; Vihinen et al., 2002). MMP-2 (gelatinase-A) and MMP-9 (gelatinase-B) can both degrade the type IV collagen of basement membranes, the first barrier to tumor invasion. MMPs facilitate tumor cell invasion and metastasis by at least three distinct mechanisms (Kleiner et al., 1999). First, proteinase action removes physical barriers to invasion through degradation of ECM macromolecules such as collagens, laminins, and proteoglycans. Second, MMPs have the ability to modulate cell adhesion. For cells to move through ECM, they must be able to form new cell-matrix and cell-cell attachments and break existing ones. Finally, MMPs may act on ECM components or other proteins to uncover hidden biologic activities.

High expression levels of certain MMPs are related to tumor invasive capacity *in vivo*. The expression of MMPs is primarily regulated at the level of transcription, and their proteolytic activity requires zymogen activation. Recent observations show that insertion of G nucleotide at -1607 bp in the MMP-1 promoter generates a new ETs-binding site and increases the transcription of the MMP-1 gene (Rutter et al., 1998). The proteolytic activity of MMPs is specifically inhibited by tissue inhibitors of metalloproteinases (TIMPs). The tumor stroma is made up primarily of fibroblasts, infiltrating immune cells and endothelial cells. All these different types of cells are capable of expressing various MMPs. They also produce cytokines, which enhance expression of MMPs by tumor and stromal cells. It is likely that MMPs form a network in which a single MMP cleaves certain native or partially degraded matrix components and activates other latent MMPs (Vihinen et al., 2002; Biswas et al., 1995; Westermarck et al., 2000).

The epithelial to mesenchymal transition (EMT) is characterized by the loss of epithelial characteristics and the gain of mesenchymal attributes in epithelial cells. It has been associated with physiological and pathological processes requiring epithelial cell migration and invasion (Gilles et al., 2005). In addition, evidence is mounting to support an important role of EMT pathways in the progression of carcinoma to metastasis by providing epithelial tumor cells with the ability to migrate, invade the surrounding stroma and disseminate in secondary organs (Birchmeier et al., 1996; Thiery et al., 2003; Gotzmann et al., 2004). Intermediate filament proteins provide a convenient and abundant marker, with keratins indicating epithelium, and vimentin a mesenchymal phenotype (Hay et al., 1995). Another commonly employed index for the epithelial state is the presence and junctional localization of the classically epithelial homotypic cell-adhesion molecule E-cadherin and associated catenins, forming the adherence junctions (Gilles et al., 2005). These cell-cell adhesion-related criteria are almost exclusively absent in mesenchymal cells (Birchmeier et al., 1996). In cancer, the maintenance of epithelia is lost, and dissociation of cells is associated with metastatic dissemination. Dissociation of cells can occur through a decrease in the local expression level of E-cadherin. EMT appears as a sequence of changes that lead to the expression of migratory and invasive properties by epithelial cells, and the expression of MMPs (and particularly of MT1-MMP) is clearly part of EMT processes (Gilles et al., 2005). Several MMPs can cleave E-cadherin, thereby inducing E-cadherin complex fragility and EMT changes (Gilles et al., 2005). The E-cadherin-mediated cell-adhesion system is also modulated by the mucin-like glycoprotein mucin 1 (MUC1), one of the mucin protein family. In breast-cancer cells, the cytoplasmic domain of MUC1 binds β-catenin, leading to a decrease of E-cadherin-β-catenin complex and resulting in an anti-adhesive effect (Li et al., 1998).

Angiogenesis, the growth of new capillaries from pre-existing blood vessels, is essential for cancers to grow beyond minimal size (Folkman, 1971; 1986; 1990; 1992). Angiogenesis is a complex multi-step process involving extravasation of plasma protein, degradation of extracellular matrix, endothelial migration and proliferation, and capillary tube formation. Tumor angiogenesis is induced by increased secretion of angiogenic factors and by down-regulation of angiogenic inhibitors (Detmar, 2003). Sustained angiogenesis is characteristic of diseases such as diabetes, psoriasis, and rheumatoid arthritis; it is essential for tumor growth and metastasis (Folkman, 1990). To initiate neovascularization, a tumor must switch to an angiogenic phenotype. Although this state may be a result of genetic changes,

Table 1. Effects of EBV gene products on invasion, metastasis and angiogenic factors

EBV gene product	Factor	Mechanism of induction	Reference
LMP-1	MMP-9	induces transcription	Yoshizaki et al., 1998; Takeshita et al., 1999; Horikawa et al., 2000
LMP-1	COX-2	induces transcription	Moruno et al., 2001
LMP-1	VEGF	induces transcription	Murono et al., 2001; Wakisaka et al., 2004
LMP-1	FGF-2	induces transcription induces release of protein	Wakisaka et al., 2002; Ceccarelli et al., in press
LMP-1	HIF-1α	induces transcription, synthesis & stabilization*	Wakaisaka et al., 2004
LMP-1	MUC-1	induces transcription	Kondo et al., 2005
LMP-1	MMP-1	induces transcription	Kondo et al., 2005
LMP-1	IL-8	induces transcription	Yoshizaki et al., 2001
BZLF-1	MMP-9	induces transcription	Yoshizaki et al., 1999
EBNA-3C	Nm23-H1	interacts with tumor suppressor protein, Nm23-H1	Subramanian et al., 2001
LMP-2A	mTOR ERK	increases protein translation of growth-ralated genes and down-regulates epithelial differentiation and intercellular contact and promotes motility of cells	Scholle et al., 2000 Chen et al., 2002 Moody et al., 2005

*Unpublished data, S. Kondo and J. Pagano.

responses to local stresses including hypoxia also have significant roles (Fox et al., 2001). The induction of angiogenesis is mediated by multiple regulatory molecules released by both tumor and host cells (Liotta et al., 1991; Fidler et al., 1994). Among these molecules are fibroblast growth factor (FGF) -2, vascular endothelial growth factor (VEGF) and interleukin (IL) -8. TNF-α increases the expression of VEGF and its receptors, as well as IL-8 and FGF-2, by endothelial cells, thus explaining its angiogenic properties *in vivo* (Yoshida et al., 1997; Giraudo et al., 1998). FGF-2 is one of the first angiogenic factors to be characterized and has been studied extensively. FGF-2 shows angiogenic activity *in vivo* and induces cell proliferation, chemotaxis, and plasminogen-activator production in cultured endothelial cells (Presta et al., 1986). Also, FGF-2 was found to induce capillary tube formation in collagen gels (Bussolino et al., 1996). VEGF has been shown to regulate vascular permeability, which is important for the initiation of angiogenesis (Dvorak et al., 1995). Recently, it has been suggested that cyclooxygenase (COX)-2 and hypoxia-inducible factor (HIF)-1 play key roles in the induction of VEGF and finally of angiogenesis (Murono et al., 2001; Wakisaka et al., 2004).

Recent findings extend the role for MMPs during multiple stages of tumor progression to include other functions such as tumor cell-growth and migration and angiogenesis. The complex process of tumorigenesis requires the coordinated regulation of multiple events including the formation of tumor vasculature. As vascular endothelial cells must penetrate the basement membrane in order to lay down new blood vessels, MMP activity might be important for endothelial invasion during neovascularization. Thus, modulation of MMP activity appears to modulate angiogenesis. The requirement for MMP activity during neovascularization has implications for the growth of both the initial tumor and distant metastasis into clinically detectable lesions (Stetler-Stevenson et al., 1999; McCawley et al., 2000).

Until recently, the role of tumorviruses in such processes has remained unexplored (Yoshizaki et al., 1998; Murono et al., 2000; Murono et al., 2001; Wakisaka et al., 2002). Although a number of viruses including human papilloma virus, Kaposi's sarcoma virus, SV-40 and the human polyomaviruses, human T-cell leukemia virus, hepatitis B and C viruses as well as EBV, are accepted as tumor viruses in human beings and are well characterized (Pagano et al., 2004), the contribution if any of oncogenic viruses to metastatic properties of the tumor cells transformed or infected by the virus had not been considered. This field of investigation has now opened with a series of studies that implicate EBV in the induction of invasive properties of cells transformed by EBV as well as in metastasis of NPC.

BIOLOGICAL EFFECTS OF EBV INFECTION ON CELLULAR INVASIVENESS
Cell motility and invasion
EBV infection of cells causes an increase in their ability to transmigrate across a Matrigel barrier, which correlates with the increased motility of EBV-infected cells (Kassis et al., 2002). In a membrane-invasion culture system to study the ability of EBV to enhance invasion and migration of nasopharyngeal carcinoma (NPC) cells (Wu et al., 2003), the number of invasive cells was higher in EBV-positive cells than in EBV-negative cells, and expression of MMP-2, MMP-9, and heparanase gene expression was moderately up-regulated. Also, VEGF was slightly upregulated. In an animal model, EBV-positive NPC tumors grew faster and larger than EBV-negative NPC tumors (Lin et al., 1997). With the use of an EBV-negative NPC cell line, TW03, established from an initially EBV-positive NPC that was reinfected with EBV by cocultivation with irradiated Akata cells carrying recombinant EBV containing a neomycin-resistant gene (Teramoto et al., 2000), an *in vitro* invasion assay showed that EBV-positive TW03 cells had greater

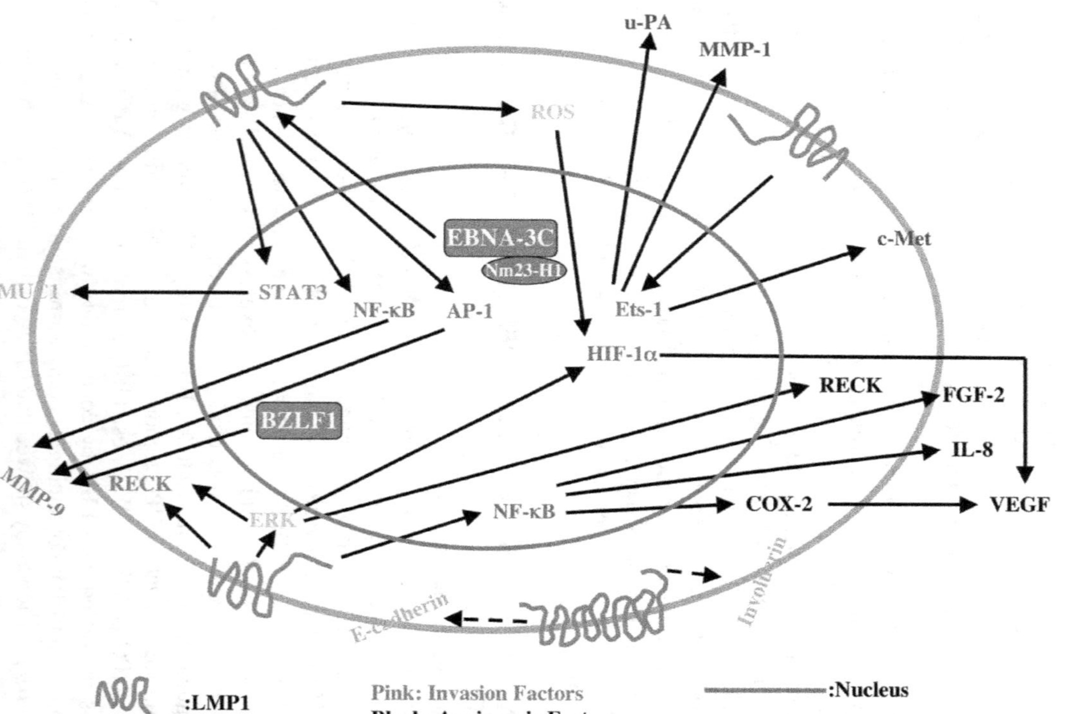

Figure 2. EBV Latent membrane protein 1 (LMP-1) induces invasiveness, angiogenic, cell motility, and transcriptional factors. MMP-9, matrix metalloproteinase-9; COX-2, cyclooxygenase-2; FGF-2, fibroblast growth factor-2; IL-8, interleukin-8; ROS, reactive oxygen species; VEGF, vascular endothelial growth factor; MMP-1, matrix metalloproteinase-1; u-PA, urokinase-type plasminogen activator; AP-1, activator protein-1; ERK, extracellular signal-regulated kinase; HIF-1α, hypoxia-inducible factor-1α; NF-κB, nuclear factor-κB; MUC1, mucin 1. See this figure in colour in the Colour Plate Section at the back of the book

invasive activity than EBV-negative TW03 cells. EBV-positive TW03 cells also exhibited greater tumorigenicity in nude mice than the EBV-negative cells.

αv Integrins play an important role in epithelial-derived cell migration, cell growth and tumor invasion and metastasis. EBV infection increases αv, β3 and β5 integrin subunit mRNAs as well as upregulates the expression of the αvβ3 integrin protein in human B cells (Huang et al., 2000). Treatment of EBV-transformed B-lymphocytes with αv antisense oligonucleotides specifically reduced cell-surface expression of αv integrins, inhibited cell growth in low serum, reduced cell invasion in Matrigel and decreased expression of metalloproteinase, MMP-9. These studies indicate that αv integrins play a significant role in EBV-induced B-lymphocyte proliferation and invasion.

These accumulating data indicate that EBV promotes the tumorigenicity of cells and enhances their invasive activity and help to explain why the plasma and serum levels of EBV DNA in patients with NPC have prognostic implications (Lo et al., 2000) both for recurrence and metastasis and for ultimate survival from the disease.

Down-regulation of intercellular adhesion by LMP-1

The intercellular adhesion system in epithelial cells is composed of tight junctions, adherens junctions, and desmosomes. One of the main adhesion molecules that plays a major role is E-cadherin. E-cadherin is a homophylic connector protein that is localized in adherens junctions. Its function is supported by β-catenin that binds to the cytoplasmic domain of E-cadherin. Reduction of E-cadherin expression or function, due to abnormal functioning of β-catenin, is frequently a feature of various cancers and has been reported to correlate with metastatic properties of tumors (Paciucci et al., 1998; Groden, 2000; Jiang et al., 2000). Expression of LMP-1 in human epithelial and murine adenocarcinoma cells down-regulates expression of E-cadherin. Moreover, the LMP-1-expressing cells had increased invasive capacity (Fahraeus et al., 1992; Farwell et al., 1999) (Figure 2). These reports underscore the suppressive effect of LMP-1 on the E-cadherin-mediated cell-adhesion system in epithelial cells. Expression of E-cadherin/β-catenin is also downregulated in NPC (Lou et al., 1999; Zheng et al., 1999). However, the mechanism of downregulation of the E-cadherin/β-catenin system by LMP-1 has not been clarified.

Degradation of ECM by LMP-1:MMP-9

Destruction of extracellular matrix (ECM) is also an essential step for tumor invasion and metastasis. The basement membrane is an important barrier, and thus the degree of basement-membrane destruction is one of the major predictors of tumor metastatic ability. Type IV collagen is a major component of the basement membrane, and expression of two type IV collagenases, MMP-2 and -9, is closely associated with the metastatic status of the tumor (Stetler-Stevenson, 2001). Direct evidence that MMP-9 has a role in tumor metastasis was presented in 1994. MMP-9 expression was induced in tumorigenic but nonmetastatic rat cells by transfection with an MMP-9 expression vector, which then conferred metastatic capacity on the nonmetastatic cells (Bernhard et al., 1994).

Expression of MMP-9 protein is closely correlated with expression of LMP-1 protein in latently EBV-infected cell lines. Moreover, transfection of LMP-

1-expression vector into the human epithelial cell line C33A induces expression of MMP-9. Electrophoretic mobility-shift assays reveal that cellular transcription factors bind to NF-κB and AP-1 DNA-binding sequences in the MMP-9 promoter. Furthermore, mutation of either the NF-κB or AP-1 binding regions dramatically decreases transactivation of the promoter by LMP-1 (Yoshizaki et al., 1998). Thus, LMP-1 transactivates MMP-9 production through the activation of NF-κB and AP-1-signaling pathways.

LMP-1 mutants lacking either of the protein's C-terminal activation regions, CTAR-1 or CTAR-2, can transactivate the MMP-9 promoter and induce MMP-9 independently and additively, although both have weaker MMP-9-inducing ability than wild-type LMP-1. Moreover, both CTAR-1 and CTAR-2 can activate both AP-1 and NF-κB, which suggests mediation of signaling by TRAF2 (Takeshita et al., 1999).

LMP-1 and MMP-9 protein expression studied immunohistochemically in NPC tissue discloses a significant correlation between expression of the two proteins. Furthermore, the expression of MMP-9 was significantly correlated with cervical lymph-node metastasis. Patients with NPC-expressing LMP-1 showed a tendency to cervical lymph-node metastasis, although not to a statistically significant level. These observations suggest a close association of MMP-9 with spread of tumor to cervical lymph nodes (Horikawa et al., 2000). In Hodgkin's disease, the other type II latency EBV-associated malignant disease, MMP-9 expression was constitutively detected in Reed-Sternberg cells regardless of either EBV or LMP-1 status. There was not a significant correlation between EBV-status and expression (Flavell et al., 2000).

Urokinase–type plasminogen activator (u-PA)

Urokinase-type plasminogen activator (u-PA) is a serine proteinase that also degrades ECM. Thus, its expression is generally correlated with tumor metastasis (Bindal et al., 1994). LMP-1 also induces u-PA when expressed in Madin-Darby canine kidney (MDCK) cells. Expression of a dominant-negative type of Ets-1 in LMP-1 transformed cells downregulated u-PA expression. Deletion of CTAR-1 domain abolished u-PA inducing ability, but a deletion mutant lacking the CTAR-2 domain still retained u-PA inducing ability (Kim et al., 2000). Plasminogen activator inhibitor (PAI)-1 is an inhibitor of u-PA, but paradoxically, expression of PAI-1 is also correlated with tumor metastasis (Sappino et al., 1991; Foekens et al., 1994). PAI-1 mRNA is also induced by transfecting LMP expression vector in MDCK cells (Kim et al., 2000). However, the clinical implication of either u-PA or PAI-1 clinically has not been clarified in EBV-associated cancers.

MMP-1

Although type IV collagen is an important component of ECM in ECM as a whole the most abundant component is interstitial collagen (type I collagen). MMP-1 is one of the most prevalent interstitial collagenases. MMP-1 mRNA is overexpressed in NPC tissue compared with other head-and-neck cancer tissues . In addition, LMP-1 induces expression of mRNA and protein of MMP-1 in the human keratinocyte line, RHEK-1 (Lu et al., 2003).

RECK

The RECK gene, a metastasis-suppressor gene, encodes a membrane glycoprotein that can negatively regulate MMP-2 and MMP-9 activity and inhibit tumor angiogenesis and metastasis (Takahashi et al., 1998; Oh et al., 2001). Whereas RECK mRNA is highly expressed in most human tissues and nontransformed cells, it is undetectable in many tumor-cell lines or in cells expressing active oncogenes (Takahashi et al., 1998). LMP-1 induces downregulation of RECK by transcriptional repression *via* the extracellular signal-regulated kinase (ERK) /Sp1 signaling pathway in an EBV-negative NPC line, TW04 cells (Liu et al., 2003). Biologically, the effect of expression of LMP-1 is to increase the invasive ability of TW04 cells. Restoration of RECK expression by PD98059, an inhibitor of ERK, reduced LMP-1 induced release of active MMP-9. Suppression of PD98059-induced RECK expression by siRNA abolished the inhibitory action of PD98059 on LMP-1 induced invasiveness. Coexpression of RECK with LMP-1 in TW04 cells effectively suppressed cell invasiveness induced by LMP-1. These data indicate that inhibition of RECK is a critical step in LMP-1 induced invasion of basement membrane leading to tumor metastasis.

Upregulation of cell motility by LMP-1

Tumor cells that do not migrate miss the chance of entering the lymphatic system and microvessels. Thus, motility is also an important factor that influences invasion and metastatic ability of tumor cells. The correlation of lymphatic and hematogenous metastasis with expression of cell motility and up-regulating factors such as Ets-1, autocrine motility factor receptor, and c-Met has been reported for various cancers (Nakayama et al., 1996; Nakada et al., 1999; Nakajima et al., 1999).

MDCK cells when transfected with LMP-1 lose cell-to-cell contacts and show morphological changes from a cobblestone appearance to a long spindle-shape. The enhanced motility of LMP-1 transformants is also detectable by scrape-wound migration assays. Differential display analysis of mRNA from control MDCK cells and from LMP-1-transformed MDCK cells revealed up-regulation of Ets-1. Expression of a dominant-negative form of Ets-1 (Ets-DN) greatly reduced motility of LMP-1-transfected MDCK cells, which however still showed reduced cell-to-cell adhesion (Kim et al., 2001). These results suggest an important role for Ets-1 in LMP-1 downstream signaling cascades for enhancement of cell motility.

LMP-1 has two cytoplasmic domains (CTAR-1 and CTAR-2). Expression of an LMP-1 mutant lacking the CTAR-1 domain failed to transform MDCK cells and thereby lost LMP-1's up-regulating effect on cellular motility. However, MDCK cells with an LMP-1 mutant lacking CTAR-2 exhibited morphological changes and high motility in the scrape-wound migration assay (Kim et al., 2001). Therefore, CTAR-1 may be the domain responsible for up-regulation of cell motility.

Expression of Ets-1 and c-Met in NPC

In vitro analyses suggest that Ets-1 plays an important role in LMP-1-mediated up-regulation of cell motility. Meanwhile, there has been increasing evidence that expression of c-Met, a hepatocyte growth factor (HGF) receptor, is highly associated with cell motility (Nakajima et al., 1999; Cortesina et al., 2000; Comoglio et al., 2001). Immunohistochemical analysis of expression of LMP-1,

Ets-1 and c-Met revealed close associations among these factors in the primary tumor tissue of patients with NPC. Moreover, both Ets-1 and c-Met expression significantly correlated with cervical lymph-node metastasis status. Finally, Ets-1 also modulates induction of c-Met by LMP-1 in MDCK cells (Horikawa et al., 2001). c-Met is localized at the cell membrane, and binding of its ligand, hepatocyte growth factor (HGF), to c-Met induces autophosphorylation of the intracellular portion of c-Met, which triggers activation of various cellular signal-transduction factors such as Rho, and as a result, enhances cell motility (Bardelli et al., 1997; Comoglio et al., 2001). Although enhancement of motility by LMP-1 is HGF-independent, induction of c-Met by LMP-1 may have additional roles in up-regulation of cell motility in the presence of HGF.

Pathways for promotion of angiogenesis by LMP-1

IL-8 is thought to contribute to angiogenesis as shown by a corneal pocket assay (Strieter et al., 1992). Characteristic of NPC is the high level of IL-8 detected in the tumor (Yoshizaki et al., 2001; Ren et al., 2004). Expression of LMP-1 is the probable explanation for this observation, in that transfection with LMP-1 expression plasmid induces IL-8 protein expression in C33A cells. The expression of LMP-1 transactivates the IL-8 promoter, as demonstrated by IL-8 promoter luciferase-reporter assays. Mutation of the NF-κB-responsive element in the IL-8 promoter region completely abolishes transactivation by LMP-1. These results suggest that LMP-1 induces expression of IL-8 through the NF-κB binding site, and IL-8 may contribute in part to angiogenesis in NPC (Yoshizaki et al., 2001).

FGF-2, also known as basic FGF, belongs to a family of nineteen structurally related members and is a major angiogenic factor (Bikfalvi et al., 1997). The five FGF-2 isoforms result from a process of alternative initiation of translation at five in-frame codons (Prats et al., 1989; Arnaud et al., 1999). Even though all FGF-2 isoforms do not contain a hydrophobic signal-peptide sequence, FGF-2 can be released from FGF-2-producing cells (Florkiewicz et al., 1995). LMP-1 induces the expression of 18, 22-22.5, and 24-kDa isoforms of FGF-2 in Ad-AH cells, an EBV-negative nasopharyngeal epithelial cell line (Wakisaka et al., 2002).

Only the 18-kDa isoform of FGF-2 is detected in the culture medium. In contrast, although treatment of these cells with PMA induced a large amount of 18-kDa protein detected in the cell lysates, the medium itself contained virtually no FGF-2. Moreover the epithelial cell line 293 does not contain detectable endogenous FGF-2 protein, nor is its expression induced in these cells by LMP-1. Nevertheless LMP-1 induces extracellular release of transfected 18-kDa FLAG-tagged FGF-2. Correspondingly, FLAG-tagged FGF-2 levels remaining in the cells are reduced depending on the amount of the protein released. These results indicate that LMP-1 stimulates the release of 18-kDa FGF-2 protein independently of its ability to induce expression of the protein. Analysis of the mechanism whereby LMP-1 causes release of FGF-2 by study of the effect of ouabain, which inhibits the activity of the Na/K-ATPase transporter, α1 subunit isoform, and of Brefeldin A, which inhibits the classical ER/Golgi-dependent pathway, has been carried out (Florkiewicz et al., 1995). Recent results show that Na/K-ATPase participates in FGF-2 release (Dahl et al., 2000), and ouabain reduces its secretion from LMP-1-positive 293 cells. IκBα expression also strongly suppresses the extracellular release of FLAG-tagged FGF-2 protein. These data suggest that Na/K-ATPase and NF-κB signaling are involved in the LMP-1 induced extracellular release of FGF-2.

We analyzed further the mechanism of LMP-1 induced FGF-2 secretion by confocal immunofluorescence microscopy, which revealed that FGF-2 is not only diffusely located in the central cytosol, but also concentrated in small dot-like foci located in proximity to the plasma membrane and along cellular projections, where it co-localizes with LMP-1. These peripheral foci, which are doubly positive for LMP-1 and 18-kDa FGF-2, also stained positively for CD63 and cathepsin D, markers of late endosomes or multivesicular bodies (MVBs) and lysosomes. Finally, both biochemical analysis and immunoelectron microscopy of the exosomal fractions demonstrated both increased release of exosomes and the presence of both LMP-1 and FGF-2 in these structures. These unpublished data suggest that LMP-1 induced FGF-2-secretion depends on an endosomal/MVB/ exosomal pathway. Further clarification of the mechanism of FGF-2 secretion as induced by LMP-1 should also shed light on routes of secretion that can be used by FGF-2 in other contexts (Ceccarelli et al., 2005).

Cyclooxygenase

Cyclooxygenase (COX) is the key enzyme in the biosynthetic pathway of prostaglandins (PGs) and thromboxanes from arachidonic acid. There are at least two isoforms of cyclooxygenase, COX-1 and COX-2. COX-1 is considered a house-keeping protein because it is constitutively expressed. COX-2 is highly inducible by mitogens, cytokines, and tumor promoters (Fosslien et al., 2000). The potential role of tumor-associated COX-2 expression in angiogenesis has been demonstrated by the ability of COX-2-expressing tumor cells to provoke new vessel formation in tumor implants (Tsuji et al., 1998). Transfection of LMP-1 expression plasmid clearly induces COX-2 in 422F6 cells, an EBV-negative clone of NPC-KT cells, which is a cell line created by fusion of originally EBV-positive NPC tissue with Ad-AH cells, an EBV-negative nasopharyngeal epithelial line (Murono et al., 2001). Coexpression of IκBα completely abolishes LMP-1 induced COX-2 expression and also represses COX-2 promoter activity induced by LMP-1. These results suggest that NF-κB is essential for COX-2-induction by LMP-1. Finally, LMP-1 expressing 422F6 cells increase secretion of VEGF protein, which is partially suppressed by a COX-2 specific inhibitor, NS398. Thus, LMP-1 induces VEGF, at least in part, through a COX-2-dependent pathway.

Hypoxia-inducible factor 1

Hypoxia-inducible factor 1 (HIF-1) is a heterodimeric basic helix-loop-helix transcription factor composed of HIF-1α and HIF-1β that is the central regulator of responses to hypoxia (Wang et al., 1995). HIF-1α is induced exponentially in response to decreases in cellular O_2 concentration, but HIF-1β is not regulated by cellular oxygen tension (Jiang et al., 1996). The specific binding of HIF-1 to the hypoxic-response element (HRE) activates the transcription of genes whose products are required for tumor progression, including VEGF, glucose transporters, and insulin-like growth factor 2 (Semenza, 1999). HIF-1α protein undergoes rapid ubiquitination and degradation by proteasomes under normoxic conditions.

HIF-1 target genes encode proteins that increase O_2 delivery and mediate adaptive responses to O_2 deprivation. Thus, a single protein, HIF-1α, appears to determine the responses to hypoxic conditions of at least 40 genes (Semenza, 2000; Semenza, 2002). That LMP-1 induces synthesis of HIF-1α protein without increasing its mRNA levels was reported recently (Wakisaka et al., 2004). LMP-1 also appears to stabilize HIF-1α protein by a novel mechanism (Kondo et al., 2005). Also, the mitogen-activated protein kinase (MAPK) inhibitor PD98059 greatly reduces induction of HIF-1α by LMP-1. Catalase, an H_2O_2 scavenger, strongly suppresses LMP-1 induced production of H_2O_2, which results in decrease of the expression of HIF-1α by LMP-1. Inhibition of NF-κB, c-jun N-terminal kinase, p38 MAPK, and phosphatidylinositol 3-kinase pathways does not affect HIF-1α expression. Finally LMP-1 increases the appearance of VEGF protein in extracellular fluids, and induction of VEGF is suppressed by PD98059 or catalase. These results suggest that LMP-1 increases HIF-1 activity through both induction of expression and stabilization of HIF-1α protein, which is controlled by p42/p44 MAPK activity and H_2O_2.

Many biological correlations with the various laboratory findings have been deduced. Significantly there is increased microvessel density in the areas of most intense neovascularization, which is a significant indicator of lymph-node metastasis in NPC (Wakisaka et al., 1999; Yoshizaki et al., 2001). Moreover in an immunohistochemical study, the expression of LMP-1 correlated with microvessel count in NPC tissue, suggesting that LMP-1 may induce angiogenic factors, now demonstrated in several studies (Wakisaka et al., 1999; Yoshizaki et al., 2001).

Recently, the role of IκB kinase β (IKKβ) during hypoxia, a regulator of NF-κB, was characterized. IKKβ mediates cell survival during hypoxia and is induced in a variety of squamous cell carcinoma cell lines (Chen et al., 2004). There may be some involvement of NF-κB signaling pathway in hypoxia-related invasion and metastasis and tumor cell survival.

Role of LMP-1 in the progression of NPC

Generally, the acquisition of metastatic ability is thought to be a rather late event in the progression of malignant tumors. Although LMP-1 is detected in approximately 70% of NPC tissues, all preinvasive NPC tumor cells already expressed LMP-1 (Pathmanathan et al., 1995). Thus, LMP-1 expression at an early stage might contribute to the highly metastatic characteristics of NPC. Because each component of metastasis such as down-regulation of intracellular adhesion, ECM destruction, cell motility upregulation, and angiogenesis is essential for the tumor to metastasize, inhibition of any step in the sequential metastatic cascade markedly reduces metastatic ability of the tumor. Following positive results in animal models, clinical trials were undertaken with the oral MMP inhibitor Marimastat®. However, these studies did not show efficacy of any MMP inhibitors in series of patients with various cancers including NPC. Anti-angiogenic drugs are undergoing clinical studies (Zheng et al., 1999; Zucker et al., 2000). Recently, the dose-limiting factors for Marimastat® have been overcome, and a new phase I study revealed that an approximately 100-fold greater dose of the MMP inhibitor, BMS-275291, could be safely administered. Thus, newly developed MMP inhibitors could be effective medicines for the treatment of EBV-related malignant diseases (Rizvi et al., 2004).

In addition, aspirin suppresses the induction of MMP-9 by LMP-1 (Murono et al., 2000). Thus, prophylactic administration of MMP inhibitors or anti-angiogenic drugs to patients whose NPC expressed LMP-1 may improve prognosis.

LMP-2A

LMP-2A is widely expressed in both latently EBV-infected cells and EBV-associated malignancies (Longnecker et al., 2000). LMP-2A may exploit MAPK kinase and affect both the phosphorylation and stability of c-jun protein. Moreover, LMP-2A induces cell invasion that is inhibited in the presence of ERK pathway inhibitor. LMP-2A may thereby promote the mobility of the cells. In doing so, it may enhance the mobility of EBV-infected cells and contribute to the process of invasion and metastasis of malignant cells (Chen et al., 2002). In fact Raft cultures derived from LMP-2A expressing cells were hyperproliferative, and epithelial differentiation was inhibited. Phosphatidylinositol-3-kinase (PI3-kinase)-dependent activation of the Akt was detected in LMP-2A expressing cells and LMP-2A tumors. PI3-kinase inhibitor blocked growth in soft agar (Scholle et al., 2000). These data indicate that LMP-2A greatly affects cell growth and differentiation pathways in epithelial cells, in part through activation of the PI3-kinase-Akt pathway. In animals injected with LMP-2A expressing cells metastases developed frequently that predominantly involved lymphoid organs. Involucrin, a marker of epithelial differentiation, and E-cadherin, involved in the maintenance of intercellular contact, were downregulated in LMP-2A expressing tumors (Scholle et al., 2000). Thus, LMP-2A may contribute to the development of metastases by downregulating intercellular contact.

EBNA-3C

EBNA-3C, a transcription factor encoded by EBV, is a member of the EBNA-3 family of genes arranged in tandem on the viral genome. The proteins have similar structural motifs in that they have binding sites for RBP-Jκ, a leucine zipper, acidic domains, proline and glutamine-rich repeats and several arginine or lysine residues responsible for nuclear translocation (Robertson, 1997). Although EBNA-3C does not specifically bind DNA, it upregulates expression of CD21 and of LMP-1 in G1-arrested, EBV-infected Raji cells (Mitchell et al., 1989; Allday et al.,1994). EBNA-3C can interact with RBP-Jκ and modulate EBNA-2 transactivation of viral promoters, but also can repress transcription directly when targeted specifically to promoters in transient transfection assays (Robertson et al., 1996). Thus, this dual role exhibited by EBNA-3C in transcriptional activation and repression may be critical for transformation of B-lymphocytes.

EBNA-3C may also affect metastasis through nucleoside diphosphate (NDP) kinases, specifically, Nm23-H1 and -H2. Suppression of metastasis can be produced in tumor-cell lines transfected with Nm23/NDP kinase-A (Lim et al., 1998; Hartsough et al., 2000). Typically, there is an inverse relation between expression of Nm23-H1 and metastasis (de la Rossa et al., 1995). The carboxy-terminal region of EBNA-3C interacts specifically with the metastatic suppressor protein, Nm23-H1. Moreover, EBNA-3C reverses the ability of Nm23-H1 to suppress the migration of Burkitt-lymphoma cells and breast-carcinoma cells. The proposal is that EBNA-3C contributes to EBV-associated human cancers by targeting and altering the role of the metastasis-suppressor, Nm23-H1 (Subramanian et al., 2001).

EBV replication and metastasis

The influence of EBV lytic gene expression on invasion and metastasis has yet to be clarified. Recent advances in diagnostic molecular biology have established the diagnostic value of EBV DNA quantification in peripheral blood cells and in serum as a sensitive and specific marker for the status of EBV-associated diseases (Kondo et al., 2004; Lin et al., 2004). Although, the predictive value of the EBV DNA load in patients is not completely agreed on, detection of EBV DNA in the peripheral blood cells of patients with NPC correlates with a greater risk of distant metastasis (Lin et al., 2001). In addition, the proportion of patients with NPC with circulating EBV DNA is much greater among patients who have distant metastases compared with patients who have locoregional recurrence (Hong et al., 2004). Median concentrations of plasma EBV DNA of 681 copies/ml among 25 patients with stage III disease, 1703 copies/ml among 74 patients with stage IV disease, and 291,940 copies/ml among 19 control patients with distant metastasis (P<0.001) are reported (Lin et al., 2004). There seem to be two interpretations of the relation between circulating EBV DNA and metastasis. One is that the amount of circulating EBV DNA reflects the entire tumor burden of the patient because the sub-population of tumor cells undergoing lytic viral infection increases as the tumor becomes larger. The other interpretation is that the cellular environment which promotes EBV replication also works to promote metastasis. One example would be induction of MMP-9 by the immediate-early gene, BZLF1. Direct binding of BZLF1 protein to the AP-1 binding site of the MMP-9 promoter transactivates the promoter (Yoshizaki et al., 1999). Moreover, host immunosuppression is an important factor which permits EBV replication. Reactivation of EBV occurs in patients with AIDS and in recipients of organ transplants. Although it is not a usual feature in patients with NPC mild suppression of immune function may occur and induce an increase in EBV lytic infection in some cells and enhanced EBV gene expression that modulates cell functions in the other EBV-positive cells.

KSHV and invasion and metastasis

Kaposi's sarcoma-associated herpesvirus (KSHV/HHV-8) is linked to Kaposi's sarcoma, primary effusion lymphoma, and multicentric Castleman's disease (Chang et al., 1994; Cesarman et al.,1995; Soulier et al., 1995). In addition to endothelial cells and B-lymphocytes, KSHV also infects epithelial cells and keratinocytes (Bechtel et al., 2003). KSHV is a γ herpesvirus related to EBV, and its genome contains many homologs of EBV genes (Dourmishev et al., 2003).

The first open reading frame of KSHV encodes a transmembrane glycoprotein named K1. Its genomic position is equivalent to that of EBV LMP-1. K1 is a signaling protein capable of eliciting B-cell activation. KSHV K1 can also induce expression of VEGF in epithelial and endothelial cells (Wang et al., 2004), and K1 induces expression of MMP-9 in endothelial cells. These results indicate that K1 signaling may contribute to KSHV-associated pathogenesis through a paracrine mechanism by promoting the secretion of VEGF and MMP-9 into the surrounding matrix. To date, MMP-9 is the only MMP that has been identified to signal through the VEGF-VEGFR system and to play a key role in the angiogenic switch of early tumorigenesis. MMP-9 may regulate vascular invasion by releasing VEGF bound to the ECM, and this released VEGF is free to bind VEGF receptors on endothelial cells, leading to an angiogenic loop that eventually results in cell migration and proliferation (Bergers et al., 2000).

It was recently observed that a constitutively active G protein-coupled receptor (GPCR) encoded by KSHV is oncogenic and stimulates angiogenesis by increasing secretion of VEGF. KSHV GPCR enhances the expression of VEGF by stimulating the activity of the transcriptional factor HIF-1, which activates transcription from a hypoxia response element (HRE) within the 5'-flanking region of the VEGF promoter (Sodhi et al., 2000). Stimulation of HIF-1α by KSHV GPCR involves the phosphorylation of its regulatory/inhibitory domain by the p38 and MAPK signaling pathways, thereby enhancing its transcriptional activity. Thus, the KSHV GPCR oncogene subverts convergent physiological pathways, leading to angiogenesis, and provides insight into a mechanism whereby growth factors and oncogenes acting upstream from MAPK, as well as inflammatory cytokines and cellular stress that activate p38, can interact with the hypoxia-dependent machinery of angiogenesis (Sodhi et al., 2000). Expression of GPCR in transfected epithelial, monocytic, and T-cell lines induces constitutive activation of transcription factors NF-κB and AP-1. This process is associated with constitutive induction of the proinflammatory NF-κB-dependent cytokines IL-1β, IL-6, and TNF-α, and the chemokines, monocyte-chemoattracted protein-1 and IL-8, as well as AP-1-dependent FGF-2. The GPCR-induced proinflammatory cytokine and growth-factor gene expression are mediated by a signaling determinant within at least five amino acids of the C terminus, a domain that is also critical for direct cell transformation (Schwarz et al., 2001). Thus, this virally encoded GPCR acts as a potent angiogenic activator by inducing the expression and secretion of VEGF.

CONCLUSIONS AND BIOLOGIC SIGNIFICANCE: SPECULATIONS

Clearly one and probably at least 2 human tumor viruses can affect expression of an array of cellular factors involved in invasion and metastasis including neoangiogenesis. In particular EBV has been shown to induce major representatives of key factors that affect the step-wise processes that lead to the final stages of malignant progression. In the case of EBV the most common molecular mechanism is through induction of transcription of these factors with regulation at the level of their promoters. In each case, NF-κB, AP-1, Ets-1, HIF-1, and STAT3 are involved in control of the normal cellular signaling pathways that are activated. For angiogenesis, induction of several factors, COX-2, FGF-2 and HIF-1α, precede induction of more proximal factors in angiogenesis such as VEGF. In the case of FGF-2 a key post-translational step, namely, release of the protein into the extracellular fluid in exosomes, is also induced (Ceccarelli et al., 2005). A number of correlations between expression of these factors and tissue changes in tumors have been made (Liotta et al., 1991; Fidler et al., 1994; Yoshida et al., 1997; Giraudo et al., 1998; Kondo et al., 2005).

Most of these effects are centralized to functions of the LMP-1 oncoprotein. Genetic analysis of LMP-1 has disclosed that both of its c-terminal activation domains are required for induction of these factors. It is interesting that EBNA-2 by itself, although it is a potent transactivator of cellular genes involved in cellular proliferation and cell-surface activation markers, appears not to induce invasion and metastasis factors directly. EBNA-2 does of course transactivate the LMP promoter and hence synthesis of LMP-1 in EBV-infected cells. Therefore LMP-1 uses a viral pathway independent of EBNA-2, the other principal EBV oncogene, to stimulate invasion and metastasis. At the same time two other EBV genes that contribute to oncogenesis, LMP-2A and EBNA-3C, also may be able to contribute to tumor progression. Thus rather than serving as a tumor initiator there is the alternative that EBV may enter preexisting neoplastic lesions and drive malignant progression. Another example is provided by Moody et al. who have shown that LMP-2A activates the mammalian target of rapamycin (mTOR) that acts downstream of PI3-kinase/Akt to mediate cellular growth signals by increasing translation and synthesis of growth-related genes. While insufficient for tumorigenesis these effects are also thought possibly to promote tumor progression (Moody et al., 2005).

EBV latency type generally correlates with the biologic characteristics of the diseases in which the different types are exhibited. Both NPC (Type II latency) and B-lymphoproliferative diseases (Type III latency) are highly invasive early in the course of neoplastic growth, whereas BL (Type 1) generally is localized at least initially. Expression of Type III products, specifically LMP-1, correlates with indices of invasiveness of NPC (Horikawa et al., 2000). Whether a similar correlation can be adduced for Hodgkin's lymphoma remains to be seen.

Thus an increasing number of papers related to the general topic of this chapter suggest that EBV, even if it is not etiologic, may exert a major impact on a developing malignancy by altering its phenotype. The concept is emerging that a tumor virus need not have exclusively an etiologic relation to a tumor with which it is associated. Rather the virus may instead infect an already existing tumor cell or be expressed in it at a later stage of oncogenesis with the result that the tumor becomes more invasive and aggressive.

In some malignancies EBV is detected only in subsets of the tumors: ~ 15% of gastric carcinomas and ~ 50% of Hodgkin's lymphomas, taking the various types together. A generally accepted criterion for an etiologic relation between infectious agent and tumor is the consistent presence of the agent in all the tumors. However it is possible that the presence of EBV defines a biologic subset of gastric carcinomas, and perhaps HL as well, although the characteristics of possible subsets is not immediately obvious either clinically or histologically. In these cases the EBV genome seems to be detected in all the malignant cells of a given EBV-positive tumor (Reed-Sternberg cells in the case of HL), and LMP-1 is expressed both in HL and gastric lymphoepithelioma-like carcinoma (Osato et al., 1996). Analysis of whether the pathobiologic characteristics of these tumors, specifically propensity to invasiveness, differs in EBV-positive *vs* EBV-negative HL of a given histologic subtype, and perhaps in EBV-positive *vs* -negative gastric carcinoma as well, could be revealing. Whether EBV plays an etiologic or tumor-modifying role in these malignancies is unclear.

Another situation is breast cancer. EBV is reported to be associated with more aggressive cases of breast cancer as indicated by ER/PR-negativity and other clinico-histologic criteria (Bonnet et al., 1999). In cases of this carcinoma only some are EBV-positive, and the average EBV genome copy number per tumor cell is highly variable, but usually low (see Chapter 10 by Arbach and Joab). Moreover not every cell in a given tumor specimen contains viral genomes. Therefore the association of virus and tumor has been dismissed as not biologically meaningful.

However the possibility exists that the role of EBV in this malignancy is not in initiation of the malignancy, but in later infection of some of the tumor cells in the population, propelling them to a more invasive and aggressive phenotype.

Finally, do other tumor viruses have similar effects? The most likely candidate based on the emerging evidence is clearly KSHV. KSHV genome is detected in all KS cells and all PEL cells, so the situation is more like EBV and NPC or the B-cell lymphoproliferative diseases, in which the virus possibly bears both an etiologic relation and a tumor-modifying effect. This virus induces invasiveness and angiogenic factors, and both KS and PEL are characteristically invasive tumors. Could a similar case be made for HPV and cervical carcinoma? Cervical carcinoma certainly can become invasive, but in contrast to NPC this behavior usually occurs late in the course of disease. Progression of cervical carcinomas is characterized by pronounced genetic instability (Lazo et al., 1999), and its invasive properties are probably more likely to be due to mutations acquired during tumor progression and not as a result of viral induction. Nevertheless it would be interesting to test whether HPV genes can induce specific cellular invasiveness or angiogenic factors.

References

Allday, M.J., and Farrell, P.J. (1994). Epstein-Barr virus EBNA-3C/6 expression maintains the level of latent membrane protein 1 in G1-arrested cells. J. Virol. *68*, 3491-3498.

Arnaud, E., Touriol, C., Boutonnet, C., Gensac, M.C., Vagner, S., Prats, H. and Prats, A.C. (1999). A new 34-kilodaltonisoform of human fibroblast growth factor 2 is cap dependently synthesized by using a non-AUG start codon and behaves as a survival factor. Mol. Cell. Biol. *19*, 505-514.

Bardelli, A., Longati, P., Gramaglia, D., Stella, M.C., and Comoglio, P.M. (1997). Gab1 coupling to the HGF/Met receptor multifunctional docking site requires binding of Grb2 and correlates with the transforming potential. Oncogene *15*, 3103-3111.

Bechtel, J.T., Liang, Y., Hvidding, J., and Ganem, D. (2003). Host range of Kaposi's sarcoma-associated herpesvirus in cultured cells. J. Virol. *77*, 6474-6481.

Bergers, G., Brekken, R., McMahon, G., Vu, T.H., Itoh, T., Tamaki, K., Tanzawa, K., Thorpe, P., Itohara, S., Werb, Z., and Hanahan, D. (2000). Matrix metalloproteinase-9 triggers the angiogenic switch during carcinogenesis. Nat. Cell Biol. *2*, 737-744.

Bernhard, E.J., Gruber, S.B., and Muschel, R.J. (1994). Direct evidence linking expression of matrix metalloproteinase 9 (92-kDa gelatinase/collagenase) to the metastatic phenotype in transformed rat embryo cells.Proc. Natl. Acad. Sci. USA. *91*, 4293-7.

Bikfalvi, A., Klein, S., Pintucci, G., and Rafkin, D.B. (1997). Biological roles of fibroblast growth factor-2. Endocr. Rev. *18*, 26-45.

Bindal, A.K., Hammoud, M., Shi, W.M., Wu, S.Z., Sawaya, R., and Rao, J.S. (1994). Prognostic significance of proteolytic enzymes in human brain tumors.J. Neurooncol. *22*, 101-10.

Birchmeier, C., Birchmeier, W., and Brand-Saberi, B. (1996). Epithelial-mesenchymal transition in cancer progression. Acta Anat. *156*, 217-226.

Biswas, C., Zhang, Y., DeCastro, R., Guo, H., Nakamura, T., Kataoka, H., and Nabeshima, K. (1995). The human tumor cell-derived collagenase stimulatory factor (renamed EMMPRIN) is a member of the immunoglobulin superfamily. Cancer Res. *55*, 434-439.

Bonnet, M., Guinebretiere, J.M., Kremmer, E., Grunewald, V., Benhamou, E., Contesso, G., and Joab, I. (1999). Detection of Epstein-Barr virus in invasive breast cancers. J. Natl. Cancer Inst. *91*, 1376-1381.

Bussolino, F., Albini, A., Camussi, G., Presta, M., Viglietto, G., Ziche, M., and Persico, G. (1996). Role of a soluble mediators in angiogenesis. Eur. J. Cancer *32A*, 2401-2412.

Ceccarelli, S., Visco, V., Wakisaka, N., Pagano, J.S., and Torrisi, M.R. (2005). Epstein-Barr virus latent membrane protein 1 promotes secretion of fibroblast growth factor 2 by exosomes. In Press.

Cesarman, E., Chang, Y., Moore, P.S., Said, J.W., and Knowles, D.M. (1995). Kaposi's sarcoma-associated herpesvirus-like DNA sequences in AIDS-related body-cavity-based lymphomas. N. Engl. J. Med. *332*, 1186-1191.

Chang, Y., Cesarman, E., Pessin, M.S., Lee, F., Culpepper, J., Knowles, D.M., and Moore, P.S. (1994). Identification of herpesvirus-like DNA sequence in AIDS-associated Kaposi's sarcoma. Science *266*, 1865-1869.

Chen, S.Y., Lu, J., Shih, Y.C., and Tsai, C.H. (2002). Epstein-Barr virus latent membrane protein 2A regulates c-jun protein through extracellular signal-regulated kinase. J. Virol. *76*, 9556-9561.

Chen, Y., Shi, G., Xia, W., Kong, C., Zhao, S., Gaw, A., Chen, E., Yang, G., Giaccia, A., and Le, Q.T,. Koong, A. (2004). Identification of Hypoxia-Regulated Proteins in Head and Neck Cancer by Proteomic and Tissue Array Profiling. Cancer Res. *64*, 7302-7310.

Comoglio, P.M., and Boccaccio, C. (2001). Scatter factors and invasive growth. Semin. Cancer Biol. *11*, 153-165.

Cortesina, G., Martone, T., Galeazzi, E., Olivero, M., De Stefani, A., Bussi, M. Valente, G., Comoglio, P.M., and Di Renzo, M.F. (2000). Staging of head and neck squamous cell carcinoma using the MET oncogene product as marker of tumor cells in lymph node metastases. Int. J. Cancer *89*, 286-292.

Dahl, J.P., Binda, A., Canfield, V.A., and Levenson, R. (2000). Participation of Na, K-ATPase in FGF-2 secretion: rescue of ouabain-inhibitable FGF-2 secretion by ouabain-resistant Na, K-ATPase α subunits. Biochem. *39*, 14877-14883.

De La Rosa, A., Williams, R.L., and Steeg, P.S. (1995). Nm23/nucleoside diphosphate kinase: toward a structural and biochemical understanding of its biological functions. Bioessays *17*, 53-62.

Detmar, M. (2003). Tumor angiogenesis. J. Invest. Dermatol. Symp. Proc. *5*, 20-23.

Dourmishev, L.A., Dourmishev, A.L., Palmeri, D., Schwartz, R.A., and Lukac, D.M. (2003). Molecular genetics of Kaposi's sarcoma-associated herpesvirus (human herpesvirus 8) epidemiology and pathogenesis. Microbiol. Mol. Biol. Rev. *67*, 175-212.

Dvorak, H.F., Brown, L.F., Detmar, M., and Dvorak, A.M. (1995). Vascular permeability factor/vascular endothelial growth factor, microvascular hyperpermeability, and angiogenesis. Am. J. Pathol. *146*, 1029-1039.

Ellis, L.M., and Fidler, I.J. (1996). Angiogenesis and metastasis. Eur. J. Cancer *32A*, 2451-2460.

Fahraeus, R., Chen, W., Trivedi, P., Klein, G., and Obrink, B. (1992). Decreased expression of E-cadherin and increased invasive capacity in EBV-LMP-transfected human epithelial and murine adenocarcinoma cells. Int. J. Cancer *52*, 834-838.

Farwell, D.G., McDougall, J.K., and Coltrera, M.D. (1999). Expression of Epstein-Barr virus latent membrane proteins leads to changes in keratinocyte cell adhesion. Ann. Otol. Rhinol. Laryngol. *108*, 851-859.

Fidler, I.J., and Ellis, L.M.(1994). The implications of angiogenesis to the biology and therapy of cancer metastasis. Cell *79*, 185-188.

Flavell, J.R., Baumforth, K.R., Williams, D.M., Lukesova, M., Madarova, J., Noskova, V., Prochazkova, J., Lowe, D., Kolar, Z., Murray, P.G., and Nelson, P.N. (2000). Expression of the matrix metalloproteinase 9 in Hodgkin's disease is independent of EBV status. Mol. Pathol. *53*, 145-9.

Florkiewicz, R.Z., Majack, R.A., Buechler, R.D., and Florkiewictz, E. (1995). Quantitative export of FGF-2 occurs through an alternative, energy-dependent, non-ER/Golgi pathway. J. Cell Physiol. *162*, 388-399.

Foekens, J.A., Schmitt, M., van Putten, W.L., Peters, H.A., Kramer, M.D., Janicke, F., and Klijn, J.G. (1994). Plasminogen activator inhibitor-1 and prognosis in primary breast cancer. J Clin Oncol. *12*, 1648-58.

Folkman, J. (1971). Tumor angiogenesis: Therapeutic implications. N. Engl. J. Med. *285*, 1182-1186.

Folkman, J. (1986). How is blood vessel growth regulated in normal and neoplastic tissue? G.H.A. Clowes Memorial Award Lecture. Cancer Res. *46*, 467-473.

Folkman, J. (1990). What is the evidence that tumors are angiogenesis dependent? J. Natl. Cancer Inst. *82*, 4-6.

Folkman, J. (1992). The role of angiogenesis in tumor growth. Semin. Cacer Biol. *3*, 65-71.

Kleiner, D.E., and Stetler-Stevenson, W.G. (1999). Matrix metalloproteinases and metastasis. Cancer Chemother. Pharmacol. *43S*, S42-S51.

Fosslien, E. (2000). Molecular pathology of cyclooxygenase-2 in neoplasia. Ann. Clin. Lab. Sci. *30*, 3-21.

Fox, S.B., Gasparini, G., and Harris, A.L. (2001). Angiogenesis: pathological, prognostic, and growth-factor pathways and their link to trial design and anticancer drugs. Lancet Oncol. *2*, 278-289.

Gilles, C., Newgreen, D.F., Sato, H., and Thompson, E.W. (2005). Matrix metalloproteinases and epithelial-to mesenchymal transition: Implications for carcinoma metastasis. In press.

Giraudo, E., Primo, L., Audero, E., Gerber, H.P., Koolwijk, P., Soker, S., Klagsburn, M., Ferrara, N., and Bussolino, F. (1998). Tumor necrosis factor-α regulates expression of vascular endothelial growth factor receptor-2 and of its co-receptor neuropilin-1 in human vascular endothelial cells. J. Biol. Chem. *273*, 22128-22135.

Gotzmann, J., Mikula, M., Eger, A., Schulte-Hermann, R., Foisner, R., Beug, H., and Mikulits, W. (2004). Molecular aspects of epithelial cell plasticity: implications for local tumor invasion and metastasis. Mutat. Res. *566*, 9-20.

Groden J. (2000). Touch and go: mediating cell-to-cell interactions and Wnt signaling in gastrointestinal tumor formation. Gastroenterology *119*, 1161-1164.

Hartsough, M.T., and Steeg, P.S. (2000). Nm23/nucleoside diphosphate kinase in human cancers. J. Bioenerg. Biomembr. *32*, 301-308.

Hay, E.D. (1995). An overview of epithelio-mesenchymal transformation. Acta Anat. *154*, 8-20.

Hong, R.L., Lin, C.Y., Ting, L.L., Ko, J.Y., and Hsu, M.M. (2004). Comparison of clinical and molecular surveillance in patients with advanced nasopharyngeal carcinoma after primary therapy: the potential role of quantitative analysis of circulating Epstein-Barr virus DNA. Cancer *100*, 1429-1437.

Horikawa, T., Yoshizaki, T., Sheen, T.S., Lee, S.Y., and Furukawa, M. (2000). Association of latent membrane protein 1 and matrix metalloproteinase 9 with metastasis in nasopharyngeal carcinoma. Cancer *89*, 715-723.

Horikawa, T., Takeshita, H., Sheen, T.S., Sato, H., Furukawa, M., and Yoshizaki, T. (2001). c-Met proto-oncogene is induced by Epstein-Barr virus latent membrane protein-1 and correlates with cervical lymphnode metastasis of nasopharyngeal carcinoma. Am. J. Pathol. *159*, 27-33.

Huang, S., Stupack, D., Liu, A., Cheresh, D., and Nemerow, G.R. (2000). Cell growth and matrix invasion of EBV-immortalized human B lymphocyte is regulated by expression of αv integrins. Oncogene *19*, 1915-1923.

Jiang, B.H., Semenza, G.L., Bauer, C., and Marti, H.H. (1996). Hypoxia-inducible factor 1 levels vary exponentially over a physiologically relevant range of O_2 tension. Am. J. Physiol. *271*, C1172-C1180.

Jiang, W.G, and Mansel, R.E. (2000). E-cadherin complex and its abnormalities in human breast cancer. Surg. Oncol. *9*, 151-171.

Kassis, J., Maeda, A., Teramoto, N., Takada, K., Wu, C., Klein, G., and Wells, A. (2002). EBV-expressing AGS gastric carcinoma cell sublines present increased motility and invasiveness. Int. J. Cancer *99*, 644-651.

Kim, K.R., Yoshizaki, T., Miyamori, H., Hasegawa, K., Horikawa, T., Furukawa, M., Harada, S., Seiki, M., and Sato, H. (2000). Transformation of Madin-Darby canine kidney (MDCK) epithelial cells by Epstein-Barr virus latent membrane protein 1 (LMP-1) induces expression of Ets1 and invasive growth. Oncogene *19*, 1764-1771.

Kondo, S., Horikawa, T., Takeshita, H., Kanegane, C., Kasahara, Y., Sheen, T.S., Sato, H., Furukawa, M., and Yoshizaki, T. (2004). Diagnostic value of serum EBV-DNA quantification and antibody to viral capsid antigen in nasopharyngeal carcinoma patients. Cancer Sci. *95*, 508-513.

Kondo, S., Yoshizaki, T., Wakisaka, N., Murono, S., Jang, K.L., Furukawa, M., and Pagano, J.S. MUC1 induced by Epstein-Barr virus latent membrane protein 1 causes dissociation of cell-matrix interaction and cellular invasiveness via NF-κB and STAT signaling. In Press.

Lazo, P.A. (1999). The molecular genetics of cervical carcinoma. Br. J. Cancer *80*, 2008-2018.

Li, Y., Bharti, A., Chen, D., Gong, J., and Kufe, D. (1998). Interaction of glycogen synthase kinase 3 beta with DF3/MUC1 carcinoma-associated antigen and beta-catenin. Mol. Biol. Cell *18*, 7216-7224.

Lim, S., Lee, H.Y., and Lee, H. (1998). Inhibition of colonization and cell-matrix adhesion after nm23-H1 transfection of human prostate carcinoma cells. Cancer Lett. *133*, 143-149.

Liotta, L.A., Steeg, P.S., and Stetlet-Stevenson, W.G. (1991). Cancer metastasis and angiogenesis: an imbalance positive and negative regulation. Cell *64*, 327-336.

Lin, C.T., Lin, C.R., Tan, G.K., Chen, W., Dee, A.N., and Chan, W.Y. (1997). The mechanism of Epstein-Barr virus infection in nasopharyngeal carcinoma cells. Am. J. Pathol. *150*, 1745-1756.

Lin, C.T. (2003). Functional analysis of EBV in nasopharyngeal carcinoma cells. Lab. Invest. *83*, 797-812.

Lin, J.C., Chen, K.Y., Wang, W.Y., Lin, J.C., Chen, K.Y., and Wang, W.Y. (2001). Detection of Epstein-Barr virus DNA the peripheral-blood cells of patients with nasopharyngeal carcinoma: relationship to distant metastasis and survival. J. Clin. Oncol. *19*, 2607-2615.

Lin, J.C., Wang, W.Y., Chen, K.Y., Wei, Y.H., Liang, W.M., Jan, J.S., and Jiang, R.S. (2004). Quantification of plasma Epstein-Barr virus DNA in patients with advanced nasopharyngeal carcinoma. N. Engl. J. Med. *350*, 2461-70.

Liu, L.T., Peng, J.P., Chang, H.C., and Hung, W.C. (2003). RECK is a target of Epstein-Barr virus latent membrane protein 1. Oncogene *22*, 8263-8270.

Longnecker, R. (2000). Epstein-Barr virus latency: LMP2, a regulator or means for Epstein-Barr virus persistence? Adv. Cancer Res. *79*, 175-200.

Lou, P.J., Chen, W.P., Sheen, T., Ko, J., Hsu, M., and Wu, J. (1999). Expression of E-cadherin/catenin complex in nasopharyngeal carcinoma: correlation with clinicopathological parameters. Oncol Rep. *6*, 1065-1071.

Lu, J, Chua, H.H., Chen, S.Y., Chen, J.Y., and Tsai, C.H. (2003). Regulation of matrix metalloproteinase-1 by Epstein-Barr virus proteins. Cancer Res. *63*, 256-262.

McCawley, L.J., and Matrisian, L.M. (2000). Matrix metalloproteinases: multifunctional contributors to tumor progression. Mol. Med. Today *6*, 149-156.

Mitchell, P.J., and Tjian, R. (1989). Transcriptional regulation in mammalian cells by sequence-specific DNA binding proteins. Science *245*, 371-378.

Moody, C.A., Scott, R.S., Amirghahari, N., Nathan, C.A., Young, L.S., Dawson, C.W., and Sixbey JW (2005). Epstein-Barr virus-encoded LMP-2A signals through the cell growth regulator mTOR to enhance protein translation. J. Virol. *79*, 5499-5506.

Murono, S., Yoshizaki, T., Sato, H., Takeshita, H., Furukawa, M., and Pagano, J.S. (2000). Aspirin inhibits tumor cell invasiveness induced by Epstein-Barr virus latent membrane protein 1 through suppression of matrix metalloproteinase-9 expression. Cancer Res. *60*, 2555-2561.

Murono, S., Inoue, H., Tanabe, T., Joab, I., Yoshizaki, T., Furukawa, M., and Pagano, J.S. (2001). Induction of cyclooxygenase-2 by Epstein-Barr virus latent membrane protein 1 is involved in vascular endothelial growth factor production in nasopharyngeal carcinoma cells. Proc. Natl. Acad. Sci. USA. *98*, 6905-6910.

Nagase, H., and Woessner, J.F. Jr. (1999). Matrix metalloproteinases. J. Biol. Chem. *274*, 21491-21494.

Nakada, M., Yamashita, J., Okada, Y., and Sato, H. (1999). Ets-1 positively regulates expression of urokinase-type plasminogen activator (uPA) and invasiveness of astrocytic tumors. J. Neuropathol. Exp. Neurol. *58*, 329-334.

Nakajima, M., Sawada, H., Yamada, Y., Watanabe, A., Tatsumi, M., Yamashita, J., Matsuda, M., Sakaguchi, T., Hirao, T., and Nakano, H. (1999). The prognostic significance of amplification and overexpression of c-met and c-erb B-2 in human gastric carcinoma. Cancer *85*, 1894-1902.

Nakayama, T., Ito, M., Ohtsuru, A., Naito, S., Nakashima, M., Fagin, J.A., Yamashita, S., and Sekine, I. (1996). Expression of the Ets-1 proto-oncogene in human gastric carcinoma: correlation with tumor invasion. Am. J. Pathol. *149*, 1931-1939.

Nemunaitis, J., Poole, C., Primrose, J., Rosemurgy, A., Malfetano, J., Brown, P., Berrington, A., Cornish, A., Lynch, K., Rasmussen, H, Kerr, D., Cox, D., and Millar, A. (1998). Combined analysis of studies of the effects of the matrix metalloproteinase inhibitor marimastat on serum tumor markers in advanced cancer: selection of a biologically active and tolerable dose for longer-term studies. Clin Cancer Res. *4*, 1101-9.

Oh, J., Takahashi, R., Kondo, S., Mizoguchi, A., Adachi, E., Sasahara, R.M., Nishimura, S., Imamura, Y., Kitayama, H., Alexander, D.B., Ide, C., Horan, T.P., Arakawa, T., Yoshida, H., Nishikawa, S., Itoh, Y., Seiki, M., Itohara, S., Takahashi, C., and Noda, M. (2001). The membrane-anchored MMP inhibitor RECK is a key regulator of extracellular matrix integrity and angiogenesis. Cell *107*, 789-800.

Osato, T, and Imai, S. (1996). Epstein-Barr virus and gastric carcinoma. Semin. Cancer Biol. *7*, 175-182.

Prats, H., Kaghad, M., Prats, A. C., Klagsburn, M., Lelias, J. M., Liauzum, P.,Chalon, P., Tauber, J. P., Amarlic, F., Smith, J. A., and Caput, D. (1989). High molecular mass form of basic fibroblast growth factor are initiated by alternative CUG codons. Proc. Natl. Acad. Sci. USA. *86*, 1836-1840.

Paciucci, R., Vila, M.R., Adell, T., Diaz, V.M., Tora, M., Nakamura, T, and Real, F.X. (1998). Activation of the urokinase plasminogen activator/urokinase plasminogen activator receptor system and redistribution of E-cadherin are associated with hepatocyte growth factor-induced motility of pancreas tumor cells overexpressing Met. Am. J. Pathol. *153*, 201-212.

Pagano, J.S., Blaser, M., Buendia, M.A., Damania, B., Khalili, K., Raab-Traub, N., Roizman, B. (2004). Infectious agents and cancer: criteria for a causal relation. Semin. Cancer Biol. *48*, 453-471.

Pathmanathan, R., Prasad, U., Sadler, R., Flynn, K., and Raab-Traub, N. (1995). Clonal proliferation of cells infected with Epstein-Barr virus in preinvasive lesions related to nasopharyngeal carcinoma. N. Engl. J. Med. *333*, 693-698.

Presta, M., Moscatelli, D., Joseph-Silverstein, J., and Rifkin, D.B. (1986). Purification from a human hepatoma cell line of a basic fibroblast growth factor-like molecule that stimulates capillary endothelial cell plasminogen activator production, DNA synthesis, and migration. Mol. Cell. Biol. *6*, 4060-4066.

Ren, Q., Sato, H., Murono, S., Furukawa, M., and Yoshizaki, T. (2004). Epstein-Barr virus (EBV) latent membrane protein 1 induces interleukin-8 through the nuclear factor-κB signaling pathway in EBV-infected nasopharyngeal carcinoma cell line. Laryngoscope *114*, 855-859.

Rizvi, N.A., Humphrey, J.S., Ness, E.A., Johnson, M.D., Gupta, E., Williams, K., Daly, D.J., Sonnichsen, D., Conway, D., Marshall, J., Hurwitz, H. (2004). A phase I study of oral BMS-275291, a novel nonhydroxamate sheddase-sparing matrix metalloproteinase inhibitor, in patients with advanced or metastatic cancer. Clin. Cancer Res. *10*,1963-70.

Robertson, E.S., Lin, J., and Kieff, E. (1997). The amino-terminal domains of Epstein-Barr virus nuclear protein 3A, 3B, and 3C interact with RBPJ(κ). J. Virol. *69*, 3068-3074.

Robertson, E.S. (1997). The Epstein-Barr virus EBNA-3 family as regulators of transcription. Epstein-Barr Virus Report *4*, 143-150.

Rutter, J.L., Mitchell, T.I., Buttice, G., Meyers, J., Gusella, J.F., Ozelius, L.J., and Brinckerhoff, C.E. (1998). A single nucleotide polymorphism in the matrix metalloproteinase-1 promoter creates an Ets binding site and augments transcription. Cancer Res. *58*, 5321-5325.

Scholle, F., Bendt, K.M., and Raab-Traub, N. (2000). Epstein-Barr virus LMP2A transforms epithelial cells, inhibits cell differentiation, and activates Akt. J. Virol. *74*, 10681-10689.

Sappino, A.P., Belin, D., Huarte, J., Hirschel-Scholz, S., Saurat, J.H., and Vassalli, J.D. (1991). Differential protease expression by cutaneous squamous and basal cell carcinomas. J. Clin. Invest. *88*, 1073-9.

Schwarz, M., and Murphy, P.M. (2001). Kaposi's sarcoma-associated herpesvirus G protein-coupled receptor constitutively activated NF-κB and induces proinflammatory cytokine and chemokine production via a C-terminal signaling determinant. J. Immunol. *167*, 505-513.

Semenza, G.L. (1999). Regulation of mammalian O2 homeostasis by hypoxia-inducible factor 1. Annu. Rev. Cell Dev. Biol. *15*, 551-578.

Semenza, G.L. (2000). Hypoxia, clonal selection, and the role of HIF-1 in tumor progression. Crit. Rev. Biochem. Mol. Biol. *35*, 71-103.

Semenza, G.L. (2002). Signal transduction to hypoxia-inducible factor 1. Biochem. Pharmacol. *64*, 993-998.

Shapiro, S.D. (1998). Matrix metalloproteinase degradation of extracellular matrix: biological consequences. Curr. Opin. Cell Biol. *10*, 602-608.

Sodhi, A., Montaner, S., Patel, V., Zohar, M., Bais, C., Mesri, E.A., and Gutkind, J.S. (2000). The Kaposi's sarcoma-associated herpes virus G protein-coupled receptor up-regulates vascular endothelial growth factor expression and secretion through mitogen-activated protein kinase and p38 pathways acting on hypoxia-inducible factor 1α. Cancer Res. *60*, 4873-4880.

Soulier, J., Grollet, L., Oksenhendler, E., Cacoub, P., Cazals-Hatem, D., Babinet, P., d'Agay, M.F., Clauvel, J.P., Raphael, M., Degos, L., and Sigaux, F. (1995). Kaposi's sarcoma-associated herpesvirus-like DNA sequence in multicentric Castleman's disease. Blood *86*, 1276-1280.

Stetler-Stevenson, W.G. (1999). Matrix metalloproteinases in angiogenesis: a moving target for therapeutic intervention. J. Clin. Invest. *103*, 1237-1241.

Stetler-Stevenson W.G. (2001). The role of matrix metalloproteinases in tumor invasion, metastasis, and angiogenesis. Surg. Oncol. Clin. N. Am. *10*, 383-392.

Strieter, R.M., Kunkel, S.L., Elner, V.M., Martonyi, C.L., Koch, A.E., Polverini, P.J., and Elner, S.G. (1992). Interleukin-8. A corneal factor that induces neovascularization. Am. J. Pathol. *141*, 1279-1284.

Subramanian, C., Cotter II, M.A., and Robertson, E.S. (2001). Epstein-Barr virus nuclear protein EBNA-3C interacts with the human metastatic suppressor Nm23-H1: a molecular link to cancer metastasis. Nat. Med. *7,* 350-355.

Takahashi, C., Sheng, Z., Horan, T.P., Kitayama, H., Maki, M., Hitomi, K., Kitaura, Y., Takai, S., Sasahara, R.M., Horimoto, A., Ikawa, Y., Ratzkin, B.J., Arakawa, T., and Noda, M. (1998). Regulation of matrix metalloproteinase-9 and inhibition of tumor invasion by the membrane-anchored glycoprotein RECK. Proc. Natl. Acad. Sci. USA *95,* 13221-13226.

Takeshita, H., Yoshizaki, T., Miller, W.E., Sato, H., Furukawa, M., Pagano, J.S., and Raab-Traub, N. (1999). Matrix metalloproteinase 9 expression is induced by Epstein-Barr 1 C - terminal activation region 1 and 2. J. Virol. *73,* 5548-5555.

Teramoto, N., Maeda, A., Kobayashi, K., Hayashi, K., Oka, T., Takahashi, K., Takada, K., Klein, G., and Akagi, T. (2000). Epstein-Barr virus infection to Epstein-Barr virus-negative nasopharyngeal carcinoma cell line TW03 enhances its tumorigenicity. Lab. Invest. *80,* 303-312.

Thiery, J.P. (2003). Epithelial-mesenchymal transition in development and pathologies. Curr. Opin. Cell Biol. *15,* 740-746.

Tsuji, M., Kawano, S., Tsuji, S., Sawaoka, H., Hori, M., and DuBois, R.N. (1998). Cyclooxygenase regulates angiogenesis induced by colon cancer cells. Cell *93,* 705-716.

Vihinen, P., and Kahari, V.M. (2002). Matrix metalloproteinases in cancer: prognostic markers and therapeutic targets. Int. J. Cancer *99,* 157-166.

Wakisaka, N., Wen, Q.H., Yoshizaki, T., Nishimura, T., Furukawa, M., Kawahara, E., and Nakanishi, I. (1999). Association of vascular endothelial growth factor expression with angiogenesis and lymph node metastasis in nasopharyngeal carcinoma. Laryngosope *109,* 810-814.

Wakisaka, N., Murono, S., Yoshizaki, T., Furukawa, M., and Pagano, J.S. (2002). Epstein-Barr virus latent membrane protein 1 induces and causes release of fibroblast growth factor-2. Cancer Res. *62,* 6337-6344.

Wakisaka, N., Kondo, S., Yoshizaki, T., Murono, S., Furukawa, M., Pagano, J.S. (2004). Epstein-Barr virus latent membrane protein 1 induces synthesis of hypoxia-inducible factor 1α. Mol. Cell. Biol. *24,* 5223-5234.

Wang, G.L., Jiang, B.H., Rue, E.A., and Semenza, G.L. (1995). Hypoxia-inducible factor 1 is a basic-helix-loop-helix-PAS heterodimer regulated by cellular O_2 tension. Proc. Natl. Acad. Sci. USA. *92,* 5510-5514.

Wang, L., Wakisaka, N., Tomlinson, C. C., DeWire, S.M., Krall, S., Pagano, J.S., and Damania, B. (2004). The Kaposi's sarcoma-associated herpesvirus (KSHV/HHV-8) K1 protein induces expression of angiogenic and invasion factors. Cancer Res. *64,* 2774-2781.

Westermarck, J., and Kahari, V.M. (1999). Regulation of matrix metalloproteinase expression in tumor invasion. FASEB J. *13,* 781-792.

Westermarck, J., Jakkola, P., Kallunki, T., Grenman, R., Kahari, V.M. (2000). Activation of fibloblast collagenase-1 expression by tumor cells of squamous cell carcinomas is mediated by p38 mitogen-activated protein kinase and c-jun NH2-terminal kinase-2. Cancer Res. *60,* 7156-7162.

Yoshida, S., Ono, M., Shono, T., Izumi, H., Ishibashi, T., Suzuki, H., and Kuwano, M. (1997). Involvement of interleukin-8, vascular endothelial growth factor, and basic fibroblast growth factor in tumor necrosis factor alpha-dependent angiogenesis. Mol. Cell. Biol. *17,* 4015-4023.

Yoshizaki, T., Sato, H., Furukawa, M., and Pagano, J.S. (1998). The expression of matrix metalloproteinase 9 is enhanced by Epstein-Barr virus latent membrane protein 1. Proc. Natl. Acad. Sci. USA. *95,* 3621-3626.

Yoshizaki, T., Sato, H., Murono S., Pagano, J.S., and Furukawa, M. (1999). Matrix metalloproteinase 9 is induced by the Epstein-Barr virus BZLF1 transactivator. Clin. Exp. Metastasis. *17*, 431-6.

Yoshizaki, T., Horikawa, T., Qing-chun, R., Wakisaka, N., Takeshita, H., Sheen, T.S., Lee, S.Y., Sato, H., and Furukawa, M. (2001). Induction of interleukin-8 by Epstein-Barr virus latent membrane protein-1 and its correlation to angiogenesis in nasopharyngeal carcinoma. Clin. Cancer Res. *7*, 1946-1951.

Zheng, Z., Pan, J., Chu, B., Wong, Y.C., Cheung, A.L., and Tsao, S.W. (1999). Downregulation and abnormal expression of E-cadherin and beta-catenin in nasopharyngeal carcinoma: close association with advanced disease stage and lymph node metastasis. Hum Pathol. *30*, 458-466.

Zucker, S., Cao, J., Chen, W.T. (2000). Critical appraisal of the use of matrix metalloproteinase inhibitors in cancer treatment. Oncogene *19*, 6642-6650.

Chapter 13

Immune Response and Evasion in the Host-EBV Interaction

*Christian Münz**

ABSTRACT

Epstein-Barr virus (EBV) is a ubiquitous human tumor virus that has probably shaped the human immune system in its long co-evolution with man. Its persistent infection is immune controlled by strong cell-mediated adaptive immunity mainly via T cells. Initiation of this immune control involves cells of the innate immune system. Dendritic cells likely activate Natural Killer cells at the beginning of an EBV specific immune response and these then limit viral burden as well as assist in the polarization of the DC primed T cell response. The result is anti-viral Th1 immunity that keeps most EBV carriers free of EBV associated malignancies for their lifetime. This chapter will highlight the composition of EBV specific immune control and point out features of the human immune system that might have evolved for better immunity against EBV as well as strategies utilized by the virus to evade the host immune response.

INTRODUCTION

The immune system of a species develops during evolution in the face of pathogenic challenges. Some of these pathogens are species-specific and the evolutionary pressure they exert on the species probably shapes the immune system to diverge from immune systems of other species. The evolutionary pressure on the immune system to cope with a particular pathogen increases with the penetration the pathogen achieves in the species and the threat it poses for the survival of the host.

Epstein-Barr virus (EBV) is such a pathogen that probably shaped the human immune system. More than 90% of the human adult population is infected, and EBV has strong growth transforming capacities for B cells *in vivo* and *in vitro*. Therefore the survival of the majority of the human species depends on efficient immune control of EBV. While simian homologues of EBV exist in the Lymphocryptovirus genus of the γ-Herpesvirus subfamily, no members are known in rodents. Therefore, EBV immune control might reveal differences between the murine and human immune system and allow deeper insight into the workings of the latter.

*For correspondence email munzc@rockefeller.edu

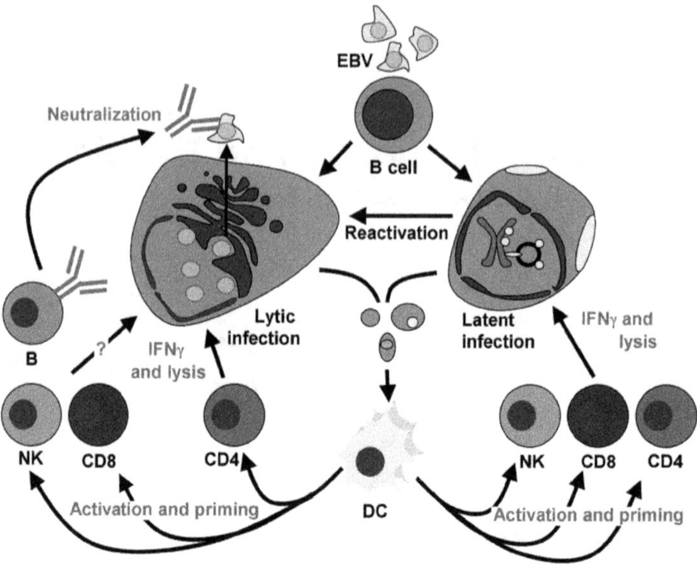

Figure 1. EBV specific immune control is mediated by innate and adaptive immunity. The initiation of EBV specific immune control is probably mediated by dendritic cells cross-presenting EBV antigens from infected B cells. NK cells, CD4$^+$ and CD8$^+$ T cells target EBV transformed B cells (latent infection). B cells are able to neutralize EBV particles via antibodies. While it has been demonstrated that EBV specific CD4$^+$ T cells can recognize B cells harboring lytic EBV infection, the role of NK cells and CD8$^+$ T cells for the control of lytic EBV infection remains unclear.

In this chapter, I will review the known components of EBV specific immune control that ensure disease-free coexistence of host and pathogen in the vast majority of EBV carriers. Furthermore, I will discuss how this comprehensive immune control might be initiated and point out differences between the human and murine immune system that are helpful in EBV specific immune control. These might have developed in the face of the evolutionary pressure exerted by EBV on the human species. Furthermore, I will describe mechanisms by which EBV counter-balances this immune control. Finally, I will present evidence how the comprehensive immune control fails in patients that develop EBV associated malignancies, like Hodgkin's disease, nasopharyngeal carcinoma and Burkitt's lymphoma.

INNATE IMMUNITY TO EBV

Both *in vitro* infection of peripheral blood mononuclear cells (PBMCs) with EBV and distinct stages of *in vivo* infection have been analyzed for EBV immunity. EBV readily transforms B cells *in vitro* and lymphoblastoid cell lines (LCL) can be established this way. In these cultures latent EBV infection predominates and the transformed cells carry in their majority all latent EBV antigens, the so-called latency III program. However, a low level (<5%) of lytic EBV infection is detectable too. *In vivo*, infectious mononucleosis (IM) patients are investigated for acute infection and healthy EBV carriers for chronic infection.

The earliest immune response to EBV infection is the production of type I IFNs (IFN α and β) (Figure 1). Since IM patients search medical help only after occurrence of clinical symptoms, the very early stages of EBV infection have not been investigated *in vivo*. *In vitro* experiments, however, have demonstrated that IFNα production peaks 24h after EBV infection of PBMCs (Kikuta et al., 1984). During this time window IFNα is mainly produced by B cells and Natural Killer cells (Lotz et al., 1986). Numerous protective effects have been attributed to this early type I IFN production. The most important role of IFNα is probably that it reduces B cell outgrowth after EBV transformation (Garner et al., 1984; Lotz et al., 1985). This reduction in EBV transformation is only achieved when IFNα is present during the first 24h of infection, while LCLs become resistant to antiviral effects of type I IFNs afterwards (Lotz et al., 1985). Several mechanisms by which IFNα prevents EBV transformed B cell outgrowth have been proposed: inhibition of EBV induced CD21 capping (Delcayre et al., 1993), repression of Epstein-Barr virus nuclear antigen 1 (EBNA1) transcription (Nonkwelo et al., 1997) and expression of the IR4 transcript modulating the double-stranded RNA-activated protein kinase (PKR), which prevents translation (Gao et al., 1999). While establishment of latent EBV infection takes 48h, the earliest latent EBV nuclear antigens EBNA2 and EBNA-LP reach steady-state levels at 24h (Kieff and Rickinson, 2001) and this coincides with resistance to type I IFNs. EBNA2 and EBNA-LP transfection also confers resistance to IFNα (Aman and von Gabain, 1990). In addition, the latent membrane protein 1 (LMP1) seems to mediate protection from IFNα induced apoptosis (Henderson et al., 1991). During latent EBV infections that lack EBNA2, EBNA-LP and LMP1 expression, like in Burkitt's lymphoma, resistance to IFNα is mediated via the non-translated EBV encoded RNAs, the EBERs (Nanbo et al., 2002). Therefore, the innate type I IFN response has only a short window of opportunity to prevent B cell transformation by EBV, namely the first 24h, and other innate immune responses have to protect the infected individual until after around 5 days the adaptive immune response takes over.

Natural Killer (NK) cells are a major component of innate immunity to herpesviruses (Biron et al., 1999). They counteract EBV transformation both *in vitro* (Masucci et al., 1983; Moretta et al., 1997; Thorley-Lawson et al., 1977; Wilson and Morgan, 2002) and *in vivo* (Baiocchi et al., 2001). Failure of NK cell responses during early EBV infection account in part for the inability to control EBV transformed B cells in males suffering from X-linked lymphoproliferative disease (Parolini et al., 2000). In these patients, the NK coreceptor 2B4 transmits inhibitory signals due to mutations in its adaptor molecule SAP, while 2B4 transduces a NK activating signal in healthy individuals. Since 2B4 binds CD48 on the surface of LCL (Brown et al., 1998), NK recognition of EBV transformed B cells is impaired (Parolini et al., 2000). Apart from activating coreceptors, NK cells use a limited number of main activating receptors to target transformed or virus-infected cells (Moretta et al., 2001). These are the Natural Cytotoxicity Receptors (NCRs) and the NKG2D receptor. It is not entirely clear which of these mediate recognition of EBV transformed B cells. For different cell lines different contributions of NCRs and NKG2D for NK cell recognition have been found (Pende et al., 2002). 2 LCLs and 2 Burkitt's lymphoma cell lines were previously analyzed. All of them expressed low levels or none of the NKG2D ligands. However, NK recognition of the LCL C1R and the Burkitt's lymphoma cell line DAUDI was mediated by both NCRs and NKG2D, while NK recognition of the LCL 721.221 and the Burkitt's lymphoma cell line RAJI was exclusively mediated by NCR recognition (Pende et al., 2002). Therefore, NCR mediated NK cell recognition plays a major role against EBV transformed B cells and might be one adaptation of the human immune system to the challenges posed by efficient EBV infection. Of the three NCRs in humans, NKp30, NKp44 and NKp46, only for NKp46 an ortholog has been found in mice (Moretta et al., 2001). It is conceivable that NKp30 and NKp44 have developed to strengthen the EBV specific NK response. With respect to the effector function, by which NK cells might inhibit EBV mediated B cell transformation, the initiation site of EBV specific immune control is of interest. During asymptomatic seroconversion after EBV infection the peripheral T cell repertoire is not perturbed (Silins et al., 2001). This indicates that EBV immune control is established at the primary site of EBV infection, in the tonsil. The secondary lymphoid organs tonsils and lymph nodes harbor a unique NK cell population that lacks cytolytic capacity, but promptly secretes cytokines upon activation (Fehniger et al., 2002; Ferlazzo and Münz, 2004; Ferlazzo et al., 2004). IFNγ is released by tonsillar NK cells upon activation (Ferlazzo et al., 2004) and IFNγ is 7-10 fold more potent than IFNα in the inhibition of B cell transformation by EBV (Lotz et al., 1985). In addition, IFNγ can suppress B cell transformation even when added 3-4 days after EBV infection, while the IFNα effect is limited to the first 24h (Lotz et al., 1985). Therefore, NK cells might limit EBV induced B cell transformation in tonsils via IFNγ secretion. These studies suggest that NK cells limit the burden of EBV transformed cells until the adaptive immune system becomes activated and establishes EBV specific immune control.

INITIATION OF EBV SPECIFIC IMMUNE CONTROL

Most reports on EBV specific immune control have characterized the already initiated or primed adaptive immunity during acute (infectious mononucleosis) or persistent infection in healthy EBV carriers. Little is known about the events involved in the initiation of EBV specific immune control itself. Meanwhile this topic is of interest, since especially EBV seronegative organ transplant recipients are at great risk of developing EBV associated post-transplant lymphoproliferative disease from EBV positive transplants (Boyle et al., 1997; Sokal et al., 1997) and it has proven difficult to prime their T cells for adoptive transfer *in vitro* (Metes et al., 2000; Savoldo et al., 2002). Moreover, patients with spontaneously arising EBV associated malignancies like Hodgkin's disease and nasopharyngeal carcinoma might lack essential components of T cell mediated EBV specific immune control, whose priming *ex vivo* for adoptive transfer or *in vivo* via vaccination would be of therapeutic benefit (Bickham and Münz, 2003).

The organs of initiation of EBV specific immune control are the tonsils and asymptomatic EBV infection seems to be entirely handled here (Silins et al., 2001). The tonsils are also the primary site of EBV infection and retain the highest concentration of EBV transformed B cells in the body (Laichalk et al., 2002). However, EBV transformed B cells are unable to prime protective EBV specific T cell responses in peripheral blood lymphocyte cultures from EBV seronegative donors (Calender et al., 1987; Hurley and Thorley-Lawson, 1988; Konttinen et al.,

1985; Nikiforow et al., 2001; Okano and Purtilo, 1995). Only when CD25[+] T cells were selected for LCL stimulation or the cultures were supplemented with IL-12 (Metes et al., 2000; Savoldo et al., 2002), could LCLs prime EBV specific CD4[+] and CD8[+] T cells. On the contrary, LCLs readily expand primed EBV specific T cells, which have been used to cure post-transplant lymphoproliferative disease (Rooney et al., 2001). Therefore, although EBV transformed B cells serve as targets of EBV specific T cell immunity and can recall memory T cell responses against EBV, they seem not to initiate the immune control of EBV (Bickham and Münz, 2003).

Instead, dendritic cells (DCs), when added to EBV infected lymphocyte cultures of EBV seronegative donors, can prevent the outgrowth of EBV transformed B cells and prime EBV specific T cell responses (Bickham et al., 2003). The regression of EBV transformed B cells in these cultures seems T cell mediated, because inhibition of T cell function eliminates protective immune control and the regression of EBV transformed B cells follows kinetics of a primary adaptive immune response (Bickham et al., 2003). The mechanism that DCs use for the priming of EBV specific T cells is probably cross-presentation of EBV antigens from EBV infected B cells, since DCs do not get infected by EBV (Bickham et al., 2003; Li et al., 2002). Cross-presentation of EBV antigens has been reported for the priming of naïve T cells (Bickham et al., 2003; Metes et al., 2001), expansion of memory T cells (Subklewe et al., 2001) and recognition of cross-presenting DCs by effector T cells (Blake et al., 2000; Münz et al., 2000). Both CD4[+] and CD8[+] T cells can be primed by cross-presentation of EBV antigens via DCs and mediate the protective immune control against EBV transformation of B cells *in vitro* (Bickham et al., 2003). Therefore, DCs and not EBV transformed B cells probably prime EBV specific T cells and initiate immune control of EBV. This might also explain the Th1 polarization of EBV specific T cell responses, because DCs are strong stimulators of this form of cell-mediated immunity (Cella et al., 1996; Koch et al., 1996; Macatonia et al., 1995).

Apart from priming EBV specific T cells, DCs might also be responsible to activate NK cells, which curb EBV mediated B cell transformation prior to the adaptive immune response. Myeloid DCs have been shown to stimulate NK cell proliferation, IFNγ secretion and an increase in NK cell cytotoxicity against, among other targets, a MHC class I negative EBV transformed lymphoblastoid cell line (LCL721.221) (Ferlazzo et al., 2002; Gerosa et al., 2002; Piccioli et al., 2002). Interestingly, DC-mediated NK cell activation mainly targets the NK cell subset enriched in human secondary lymphoid organs (Ferlazzo et al., 2004; Vitale et al., 2004), which secretes cytokines as its immediate effector function and only becomes cytolytic after prolonged activation (Ferlazzo and Münz, 2004). Furthermore, DCs and NK cells also colocalize in T cell areas of human secondary lymphoid tissues (Fehniger et al., 2002; Ferlazzo et al., 2004), and DCs induce NK cells from lymph nodes, tonsils and spleen to secrete the anti-viral cytokine IFNγ (Ferlazzo et al., 2004). Therefore, dendritic cells might activate tonsillar NK cells to limit EBV infection and these NK cells could assist in the polarization of EBV specific T cell immunity via cytokine secretion.

While DCs are able to cross-present antigens from EBV infected B cells, to prime EBV specific T cell responses and to activate NK cells, which limit EBV burden initially, the stimulus for DC activation by EBV infection remains unclear.

In this respect future studies are necessary to address which DC subpopulations encounter EBV infection in tonsils and how they detect EBV. DC activation by inflammation or by pathogen pattern recognition (Janeway and Medzhitov, 2002) might be the initial trigger to alarm the immune system against EBV.

ANTIBODY RESPONSES TO EBV

The reliability of serological responses to EBV has been the basis for the development of routine tests for the presence of EBV. EBV specific antibody responses can be divided into responses during acute versus persistent EBV infection, and into responses to lytic EBV antigens versus latent EBV antigens. Most data on antibody responses during acute EBV infection have been obtained by studying IM patients.

Most antibody responses against latent and lytic antigens peak during acute infection. These include IgM, IgA and IgG responses to virus-encoded nucleocapsid antigens and immediate-early and early lytic EBV antigens and IgG responses to the EBV latent antigen EBNA2 (Rickinson and Kieff, 2001). The expansion of these antibody responses probably follows the availability of their respective antigens. Interestingly, antibody responses against two EBV antigens, the latent EBV nuclear antigen EBNA1 and the most abundant viral envelope protein gp350, do not follow this pattern. Both reach their highest titer during convalescence from IM (Henle et al., 1987; Thorley-Lawson and Poodry, 1982). The contribution of these antibody responses to EBV specific immune control remains unclear and the EBNA antigens, which are only expressed intracellularly, are probably useless for protective humoral immunity. IgG responses against gp350, however, have proven to be neutralizing (Hoffman et al., 1980; Thorley-Lawson and Geilinger, 1980) and might contribute to the resolution of IM. IgA responses against this antigen are, on the contrary, associated with the EBV positive epithelial malignancy nasopharyngeal carcinoma (Henle and Henle, 1976) and gp350 specific IgA might facilitate EBV infection of epithelial cells (Sixbey and Yao, 1992).

During persistent EBV infection most healthy virus carriers maintain anti-EBNA1, anti-gp350 and anti-VCA IgG responses (Rickinson and Kieff, 2001). These responses are used diagnostically to assess if individuals have been infected with EBV. While any one of these responses can be absent, lack of all three antibody responses correlates well with lack of EBV infection. Similarly to antibody responses during acute infection, only the gp350 directed antibody response might contribute to EBV specific immune control.

CD8[+] T CELL RESPONSES TO EBV

The main immunological effector cells against EBV infection are probably CD8[+] T cells. They expand dramatically during acute infection with up to 25-50% of all CD8[+] T cells directed against individual EBV lytic antigens in certain infectious mononucleosis patients (Callan et al., 1998).

Latent and lytic EBV antigen specific CD8[+] T cells are differently regulated during primary EBV infection. Lytic EBV antigen specific CD8[+] T cells constitute the majority of expanded T cells in infectious mononucleosis patients with individual antigen specificities sometimes exceeding 5% of the total CD8[+] T cells (Callan et al., 1996; Callan et al., 1998). These cells cycle rapidly during acute EBV infection

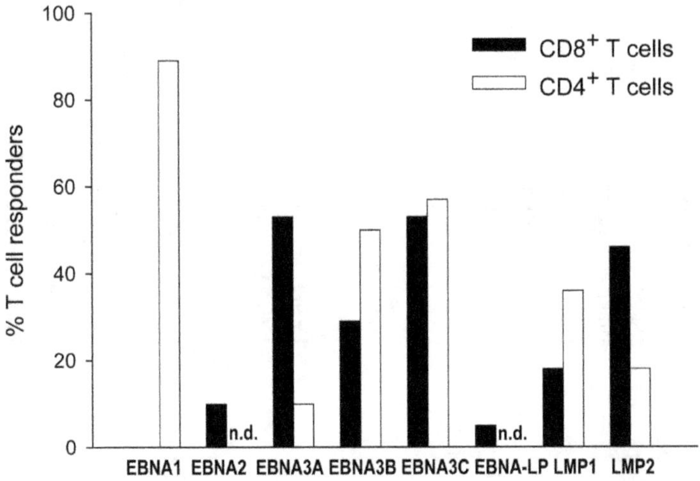

Figure 2. CD4+ and CD8+ T cell responses focus on different latent EBV antigens. Latent EBV antigen specific CD4+ (white bars) and CD8+ (black bars) T cells have been analyzed in healthy EBV carriers without selection for specific HLA haplotypes (Bickham et al., 2001; Duraiswamy et al., 2003; Khanna et al., 1992; Leen et al., 2001; Marshall et al., 2003; Meij et al., 2002; Münz et al., 2000; Murray et al., 1992; Rajnavolgyi et al., 2000; Steigerwald-Mullen et al., 2000; Steven et al., 1996). The percentages of healthy EBV carriers with specific T cell responses against latent EBV antigens are shown.

(Macallan et al., 2003). Interestingly, CD8+ T cells specific for lytic EBV antigens contract dramatically at the resolution of primary EBV infection, while latent EBV antigen specific CD8+ T cells sometimes reach their maximal expansion at this timepoint (Callan et al., 1998; Hislop et al., 2002; Hoshino et al., 1999). Although latent EBV antigen specific CD8+ T cells never reach the high frequencies of lytic EBV antigen specific CD8+ T cells, they peak during convalescence and it is tempting to speculate that they, therefore, mediate the resolution of primary EBV infection. This hypothesis is supported by the fact that entry into lytic EBV replication down-regulates MHC class I molecules dramatically, while MHC class II molecules are decreased, but maintained at detectable levels (Keating et al., 2002). This might indicate that CD8+ T cells cannot target infected B cells undergoing lytic EBV replication.

During persistent EBV infection both lytic and latent EBV antigen specific CD8+ T cells can be maintained at frequencies up to 1-5% of peripheral blood CD8+ T cells (Tan et al., 1999). Most latent EBV antigen specific CD8+ T cells are of the early CD28+/CD27+ differentiation phenotype (Appay et al., 2002) with lytic EBV antigen specific CD8+ T cells being sometimes of the late CD28-/CD27- differentiation phenotype (Hislop et al., 2001). The early differentiation phenotype, found for latent EBV antigen specific CD8+ T cells, includes the frequent expression of CCR7, a homing marker for secondary lymphoid organs (Appay et al., 2002; Hislop et al., 2001). Homing to these sites should be crucial since reactivation of EBV antigen expression occurs at these sites, for example in the tonsils (Thorley-Lawson, 2001).

Table 1. T cell epitopes from latent EBV antigens with known HLA restriction

EBV antigen	Position	Sequence	T cell specificity	HLA restriction	Reference
EBNA1	72–80	RPQKRPSCI	CD8	HLA-B7	Blake et al., 2000
	407–415	HPVGEADYF	CD8	HLA-B53	Blake et al., 2000
	407–417	HPVGEADYFEY	CD8	HLA-B35	Blake et al., 1997
	518–526	YNLRRGTAL	CD8	HLA-B8	Voo et al., 2004
	528–536	IPQCRLTPL	CD8	HLA-B7	Blake et al., 2000
	574–582	VLKDAIKDL	CD8	HLA-A2	Blake et al., 1997
	403–417	RPFFHPVGEADYFEY	CD4	HLA-DQ2	Leen et al., 2001
	475–489	NPKFENIAEGLRALL	CD4	HLA-DR11	Leen et al., 2001
	481–496	AEGLRALLARSHVER	CD4	HLA-DQ2/DQ3	Paludan et al., 2002
	515–528	TSLYNLRRGTALA	CD4	HLA-DR1	Khanna et al., 1995
	518–530	YNLRRGTALAIPQ	CD4	HLA-DP3	Voo et al., 2002
	551–570	PGPLRESIVCYFMVFLQTHI	CD4	HLA-DR7	Paludan et al., 2002
	563–577	MVFLQTHIFAEVLKD	CD4	HLA-DR15	Leen et al., 2001
	589–608	PTCNIKATVCSFDDGVDLPP	CD4	HLA-DR1	Leen et al., 2001
EBNA2	42–51	DTPLIPLTIF	CD8	HLA-A2/B51	Schmidt et al., 1991
	280–290	TVFYNIPPMPL	CD4	HLA-DQ2/DQ7	Khanna et al., 1997b
EBNA3A	158–166	QAKWRLQTL	CD8	HLA-B8	Burrows et al., 1994
	176–184	AYSSWMYSY	CD8	HLA-A30	Steven et al., 1996
	246–253	RYSIFFDY	CD8	HLA-A24	Burrows et al., 1994
	325–333	FLRGRAYGL	CD8	HLA-B8	Burrows et al., 1990b
	379–387	RPPIFIRRL	CD8	HLA-B7	Hill et al., 1995a
	406–414	LEKARGSTY	CD8	HLA-B62	Rickinson and Moss, 1997

	458-466	YPLHEQHGM	CD8	HLA-B35	Burrows et al., 1994
	491-499	VFSDGRVAC	CD8	HLA-B29	Rickinson and Moss, 1997
	502-510	VPAPAGPIV	CD8	HLA-B7	Rickinson and Moss, 1997
	596-604	SVRDRLARL	CD8	HLA-A2	Burrows et al., 1994
	603-611	RLRAEAQVK	CD8	HLA-A3	Hill et al., 1995b
EBNA3B	101-115	NPTQAPVIQLVHAVY	CD8	HLA-A11	Gavioli et al., 1993
	149-157	HRCQAIRKK	CD8	HLA-B27	Rickinson and Moss, 1997
	217-225	TYSAGIVQI	CD8	HLA-A24	Rickinson and Moss, 1997
	244-254	RRARSLSAERY	CD8	HLA-B27	Rickinson and Moss, 1997
	399-408	AVFDRKSDAK	CD8	HLA-A11	Gavioli et al., 1993
	416-424	IVTDFSVIK	CD8	HLA-A11	Gavioli et al., 1993
	481-495	LPGPQVTAVLLHESS	CD8	HLA-A11	Gavioli et al., 1993
	488-496	AVLLHEESM	CD8	HLA-B35	Rickinson and Moss, 1997
	551-563	DEPASTEPVHDQLL	CD8	HLA-A11	Gavioli et al., 1993
	657-666	VEITPYKPTW	CD8	HLA-B44	Rickinson and Moss, 1997
	831-839	GQGGSPTAM	CD8	HLA-B62	Rickinson and Moss, 1997
EBNA3C	163-171	EGGVGWRHW	CD8	HLA-B44	Morgan et al., 1996
	213-222	QNGALAINTF	CD8	HLA-B62	Kerr et al., 1996
	249-258	LRGKWQRRYR	CD8	HLA-B27	Brooks et al., 1993
	258-266	RRIYDLIEL	CD8	HLA-B27	Brooks et al., 1993
	271-278	HHIWQNLL	CD8	HLA-B39	Rickinson and Moss, 1997
	281-290	EENLLDFVRF	CD8	HLA-B44	Burrows et al., 1990a
	284-293	LLDFVRFMGV	CD8	HLA-A2	Kerr et al., 1996

Table 1, continued

EBV antigen	Position	Sequence	T cell specificity	HLA restriction	Reference
	285-293	LDFVRFMGV	CD8	HLA-B37	Khanna and Burrows, 2000
	335-343	KEHVIQNAF	CD8	HLA-B44	Khanna et al., 1992
	343-351	FRKAQIQGL	CD8	HLA-B27	Rickinson and Moss, 1997
	881-889	QPRAPIRPI	CD8	HLA-B7	Hill et al., 1995a
	386-400	SDDELPYIDPNMEPV	CD4	HLA-DR1	Leen et al., 2001
	741-760	PAPQAPYQGYQEPPAPQAPY	CD4	HLA-DR1/DR4	Rajnavolgyi et al., 2000
	916-930	PSMPFASDYSQGAFT	CD4	HLA-DR4	Leen et al., 2001
LMP1	51-60	ALLVLYSFAL	CD8	HLA-A2	Duraiswamy et al., 2003
	54-62	VLYSFALML	CD8	HLA-A2	Duraiswamy et al., 2003
	92-100	LLLIALWNL	CD8	HLA-A2	Khanna et al., 1998a
	125-133	YLLEMLWRL	CD8	HLA-A2	Khanna et al., 1998a
	156-164	IALYLQQNW	CD8	HLA-B57/B58	Duraiswamy et al., 2003
	159-167	YLQQNWWTL	CD8	HLA-A2	Khanna et al., 1998a
	166-174	TLLVDLLWL	CD8	HLA-A2	Khanna et al., 1998a
	167-175	LLVDLLWLL	CD8	HLA-A2	Khanna et al., 1998a
	212-226	SGHESDSNSNEGRHH	CD4	HLA-DQ2	Leen et al., 2001
LMP2	131-139	PYLFWLAAI	CD8	HLA-A23	Khanna et al., 1996
	200-208	IEDPPFNSL	CD8	HLA-B60	Lee et al., 1997
	236-244	RRRWRRLTV	CD8	HLA-B27	Brooks et al., 1993
	329-337	LLWTLVVLL	CD8	HLA-A2	Lee et al., 1997
	340-350	SSCSSCPLSKI	CD8	HLA-A11	Lee et al., 1997

	345-352	CPLSKILL	CD8	HLA-B8	unpublished
	356-364	FLYALALLL	CD8	HLA-A2	Lautscham et al., 2003
	419-427	TYGPVFMCL	CD8	HLA-A24	Lee et al., 1997
	426-434	CLGGLLTMV	CD8	HLA-A2	Lee et al., 1993a
	442-451	VMSNTLLSAW	CD8	HLA-A25	Rickinson and Moss, 1997
	453-461	LTAGFLIFL	CD8	HLA-A2	Lee et al., 1997
BARF0	n.d.	LLWAARPRL	CD8	HLA-A2	Kienzle et al., 1998

Distinct latent and lytic EBV antigens are consistently recognized by CD8[+] T cells in healthy EBV carriers (Paludan and Münz, 2003). The two immediate early lytic EBV antigens BZLF1 and BRLF1 and the three early lytic EBV antigens BMLF1, BMRF1 and BALF2 were found to be targeted in infectious mononucleosis patients and healthy EBV carriers (Steven et al., 1997) and especially the BZLF1 specific CD8[+] T cell response is maintained at high frequencies (5%) in HLA-B8 positive EBV carriers (Tan et al., 1999). There is very little information on late lytic EBV antigen specific CD8[+] T cell responses, but these responses seem to be less frequent than immediate early and early EBV antigen specific CD8[+] T cell responses (Khanna et al., 1999; Saulquin et al., 2001). Since, however, not even a third of all lytic EBV antigens has been evaluated for CD8[+] T cell recognition, there might still be dominant CD8[+] T cell antigens hidden within the lytic EBV antigens. The EBNA3 proteins, EBNA3A, EBNA3B and EBNA3C, are the dominant latent EBV antigens recognized by CD8[+] T cells (Figure 2) (Khanna et al., 1992; Murray et al., 1992; Steven et al., 1996). Again, as for lytic EBV antigens, an HLA-B8 restricted CD8[+] T cell response seems to be one of the most dominant and is specific for EBNA3A with a frequency of 1% of CD8[+] T cells in healthy EBV carriers (Burrows et al., 1990b; Tan et al., 1999). Apart from the EBNA antigens LMP2 specific CD8[+] T cell responses have received a lot of attention. Although less frequent, they target an antigen that is expressed in most spontaneously arising EBV associated malignancies like Hodgkin's lymphoma and a subset of nasopharyngeal carcinomas (Rickinson and Kieff, 2001). Eleven CD8[+] T cell epitopes from LMP2 have been identified (Table 1) (Lautscham et al., 2001) with the HLA-A2 restricted CD8[+] T cell responses being the strongest (Khanna et al., 1992; Murray et al., 1992; Steven et al., 1996). LMP2 specific immunization has therefore been tried for immunotherapy of nasopharyngeal carcinoma, but with modest success (Lin et al., 2002).

Latent EBV antigens have also been used to elucidate the MHC class I processing pathways for CD8[+] T cell recognition of EBV transformed B cells and epithelial cells. Most MHC class I ligands are generated by the proteasome and bind after import into the endoplasmic reticulum (ER) via the transporter associated with antigen processing (TAP) to newly synthesized MHC class I molecules (Pamer and Cresswell, 1998). LMP2 is of particular interest for MHC class I processing, since this transmembrane protein gives rise to a number of very hydrophobic CD8[+] T cell epitopes that are favored by some HLA haplotypes like HLA-A2 (Rammensee et al., 1999). These LMP2 derived epitopes are generated by the proteasome, but a large number of them enters the ER in a TAP independent manner (Lautscham et al., 2001). From these studies it was suggested that hydrophobic peptides, like the LMP2 derived CD8[+] T cell epitopes, can gain access to MHC class I presentation without TAP mediated transport. Another interesting latent EBV antigen with respect to MHC class I processing is EBNA1. It was one of the first viral immune escape proteins reported in the literature (Levitskaya et al., 1995). Full-length EBNA1 prevents its own degradation by the proteasome with its Gly-Ala repeat region (GA domain) (Levitskaya et al., 1997). The same GA domain also inhibits EBNA1 mRNA translation, thereby lowering EBNA1 protein levels and making it harder to detect for the immune system (Yin et al., 2003). Despite this clever strategy, EBNA1 does not fully escape MHC class I presentation and CD8[+] T cell detection on EBV transformed B cells. It has been noted that defective ribosomal products (DRiPs) are preferentially degraded by the proteasome and DRiP derived peptides are the major TAP substrates (Reits et al., 2000; Schubert et al., 2000). These observations suggest that a substantial amount of MHC class I ligands are derived from DRiPs (Princiotta et al., 2003) and EBNA1 is an excellent example for DRiP processing leading to MHC class I presentation (Münz, 2004). Although EBNA1 specific CD8[+] T cells have not been detected in the initial screens for EBV specific CD8[+] T cell antigens (Khanna et al., 1992; Murray et al., 1992; Steven et al., 1996), they exist at substantial frequencies in individuals with certain HLA haplotypes like HLA-B35 and HLA-B7 (Blake et al., 2000; Blake et al., 1997). These CD8[+] T cells target EBV transformed B cells and, although full-length EBNA1 is resistant to proteasome degradation, proteasome inhibition abolishes LCL recognition by EBNA1 specific CD8[+] T cells (Lee et al., 2004; Tellam et al., 2004; Voo et al., 2004). In addition, inhibition of protein biosynthesis blocks EBNA1 specific CD8[+] T cell responses to LCLs long before full-length EBNA1 degradation can be detected (Voo et al., 2004). These studies indicate that EBNA1 in EBV transformed B cells is detected by CD8[+] T cells exclusively after proteasomal processing of EBNA1 derived DRiPs. Unfortunately, EBNA1 specific CD8[+] T cells are not able to target Burkitt's lymphoma cells, since these not only down-regulate all EBV latent antigens except for EBNA1, but in addition decrease expression of essential components of the MHC class I processing machinery (Lee et al., 2004). Low levels of TAP and immunoproteasomes have been implicated in the lack of MHC class I antigen processing by Burkitt's lymphoma cells (Henriquez et al., 1999; Khanna et al., 1997c).

CD8[+] T cells can control EBV infection via two main mechanisms. They command cytolytic mechanisms as well as anti-viral cytokine secretion. In the earlier studies on EBV specific CD8[+] T cell immunity the emphasis was laid on perforin mediated cytotoxicity against lymphoblastoid cells lines (Khanna et al., 1992; Murray et al., 1992; Steven et al., 1996). However, some studies uncovered that CD8[+] T cells stimulated with EBV transformed B cells or EBV derived CD8[+] T cell epitopes showed weak to undetectable LCL lysis (Hill et al., 1995b; Shi et al., 1997). Especially EBNA1 specific CD8[+] T cells were found to have low cytolytic potential against LCLs (Blake et al., 2000; Blake et al., 1997). Since then more attention has been paid to the capacity of CD8[+] T cells to secrete the anti-viral cytokine IFNγ. It was found that EBNA3C specific CD8[+] T cells can control the outgrowth of freshly EBV infected B cells via IFNγ secretion (Shi and Lutz, 2002). Similarly, EBNA1 specific CD8[+] T cell clones with low killing activity against LCLs, produce IFNγ upon LCL recognition and can control LCL outgrowth *in vitro* (Lee et al., 2004). Upon deletion of the GA domain from EBNA1 the same EBNA1 specific T cell clones secrete around 4 fold more IFNγ and can now kill LCLs, which have been transformed with EBNA1ΔGA containing EBV (Lee et al., 2004). Therefore, EBV specific CD8[+] T cells might control EBV mediated transformation of B cells and epithelial cells without much tissue damage by IFNγ, similarly to non-cytolytic Hepatitis B virus clearance in the liver (Guidotti et al., 1999).

Table 2. T cell epitopes from lytic EBV antigens with known HLA restriction.

EBV antigen	Position	Sequence	T cell specificity	HLA restriction	Reference
BCRF1	3-11	RRLVVTLQC	CD8	HLA-B27	Saulquin et al., 2001
BHRF1	45-56	TVVLRYHLLEEI	CD4	HLA-DR4	Khanna and Burrows, 2000}
	122-133	PYYVVDLSVRGM	CD4	HLA-DR4	Landais et al., 2004
	171-189	AGLTLSLLVICSYLFISRG	CD4	HLA-DR2	White et al., 1996
BMLF1	280-288	GLCTLVAML	CD8	HLA-A2	Steven et al., 1997
	397-405	DEVEFLGHY	CD8	HLA-B18	Steven et al., 1997
BMRF1	86-100	FRNLAYGRTCVLGKE	CD8	HLA-C3	Steven et al., 1997
	268-276	YRSGIIAVV	CD8	HLA-C6	Steven et al., 1997
BRLF1	28-37	DYCNVLNKEF	CD8	HLA-A24	Pepperl et al., 1998
	25-39	LVSDYCNVLNKEFT	CD8	HLA-B18	Pepperl et al., 1998
	134-142	ATIGTAMYK	CD8	HLA-A11	Pepperl et al., 1998
	148-156	RVRAYTYSK	CD8	HLA-A3	Benninger-Doring et al., 1999
	225-239	RALIKTLPRASYSSH	CD8	HLA-A2	Pepperl et al., 1998
	393-407	ERPIFPHPSKPTFLP	CD8	HLA-Cw4	Pepperl et al., 1998
	529-543	QKEEAAICGQMDLS	CD8	HLA-B61	Pepperl et al., 1998
BZLF1	190-197	RAKFKQLL	CD8	HLA-B8	Bogedain et al., 1995
	186-201	RKCCRAKFKQLLQHYR	CD8	HLA-C6	Bogedain et al., 1995
	209-217	SENDRLRLL	CD8	HLA-B40	Scotet et al., 1996
gp85	225-233	SLVIVTTFV	CD8	HLA-A2	Khanna et al., 1999
	420-428	TLFIGSHVV	CD8	HLA-A2	Khanna et al., 1999
	542-550	LMIIPLINV	CD8	HLA-A2	Khanna et al., 1999
gp110	106-114	ILIYNGWYA	CD8	HLA-A2	Khanna and Burrows, 2000
	190-198	APGWLIWTY	CD8	HLA-B35	Khanna and Burrows, 2000
	544-552	VPGSETMCY	CD8	HLA-B35	Khanna and Burrows, 2000
gp350	863-871	VLQWASLAV	CD8	HLA-A2	Khanna et al., 1999
	65-75	FGQLTPHTKAV	CD4	HLA-DR2	Wallace et al., 1991
	169-183	NITAVVRAQGLDVTL	CD4	HLA-DR5	Wallace et al., 1991

CD4⁺ T CELL RESPONSES TO EBV

CD4⁺ T cells are thought to orchestrate virus specific immune responses and are crucial for the priming as well as maintenance of CD8⁺ T cells (Bennett et al., 1998; Cardin et al., 1996; Ridge et al., 1998; Schoenberger et al., 1998; Zajac et al., 1998). Polarization of virus specific CD4⁺ T cells decides if efficient humoral or cell-mediated immune responses can be raised. Th1 responses, which are characterized by the secretion of the anti-viral cytokine INFγ, are more protective against viral infections and support the generation of virus specific CD8⁺ T cells, which are the effectors of cell-mediated adaptive immunity (Maloy et al., 2000; Rentenaar et al., 2000). In contrast, Th2 responses, characterized by IL-4, IL-5 and IL-6 production, are thought to be supportive of humoral antibody-mediated immunity (Constant and Bottomly, 1997; Kapsenberg, 2003). Despite their crucial function CD4⁺ T cell responses have not been studied in EBV infection until recently.

One reason for this is probably that EBV specific CD4⁺ T cells never reach the high frequencies of EBV specific CD8⁺ T cells in peripheral blood. Even during primary infection in infectious mononucleosis patients, EBV specific CD4⁺ T cells, secreting IFNγ in response to all EBV antigens-containing LCL-lysates, reach with 1.4% on average only frequencies of individual CD8⁺ T cell specificities during persistent infections (Amyes et al., 2003). Individual latent and lytic EBV antigens like EBNA1, EBNA3A, BZLF1 and BMLF1 stimulate around 0.3% of antigen-experienced CD4⁺ T cells in infectious mononucleosis patients (Precopio et al., 2003). During persistent EBV infection, CD4⁺ T cell responses to LCL lysates drop to 0.34% on average with frequencies for individual antigens of no more than 0.1% (Amyes et al., 2003; K.N. Heller and C. Münz, unpublished observation). Therefore, EBV specific CD4⁺ T cells reach only one tenth of the frequency of their EBV specific CD8⁺ T cell counterparts during primary and persistent infection.

While initially only CD4⁺ T cell clones against EBNA1, EBNA2 and gp350 were decribed (Khanna et al., 1997a; Khanna et al., 1995; 1997b; Lee et al., 1993b; Wallace et al., 1991), more systematic studies revealed that CD4⁺ T cells target a different set of latent EBV antigens than CD8⁺ T cells (Figure 2). Surprisingly, EBNA1 was consistently recognized by CD4⁺ T cells of healthy EBV carriers and evoked responses more frequently than any other latent EBV antigen. Consistent EBNA1 specific responses were found, when CD4⁺ T cells were restimulated with panels of latent EBV antigens expressed with recombinant vaccinia viruses in dendritic cells (Münz et al., 2000) or added as overlapping peptide libraries (Leen et al., 2001) or recombinant proteins (Mautner et al., 2004; Steigerwald-Mullen et al., 2000). In addition, EBNA3B and EBNA3C were also recognized in around half of the healthy EBV carriers (Paludan and Münz, 2003). In contrast, CD4⁺ T cells rarely recognize the prominent CD8⁺ T cell antigen EBNA3A and the promising CD8⁺ T cell target LMP2. Vice versa, the LMP1 antigen is rarely detected by EBV specific CD8⁺ T cells (Meij et al., 2002), but frequently by CD4⁺ T cells of healthy EBV carriers (Marshall et al., 2003; Münz et al., 2000). Most of these LMP1 specific CD4⁺ T cells seem not to be of the protective Th1 type, but secrete immunosuppressive IL-10 (Marshall et al., 2003). In contrast to CD4⁺ T cell recognition of latent EBV antigens, still very little is known about CD4⁺ T cell responses to lytic EBV antigens. Apart from gp350 specific responses, BZLF1 specific, BMLF1 specific and HLA-DR4 restricted BHRF1 specific CD4$^+$ T cells have recently been identified (Landais et al., 2004; Precopio et al., 2003) (Table 2). However, a comparison of lytic EBV antigen panels remains to be performed to determine the dominant and most consistent CD4$^+$ T cell antigen of lytic EBV infection.

With the exception of IL-10 producing LMP1 specific CD4$^+$ T cells (Marshall et al., 2003), EBV specific CD4$^+$ T cells seem biased to secrete Th1 cytokines and mainly IFNγ (Amyes et al., 2003). For two EBV antigens, EBNA1 and EBNA3C, Th polarization of the respective CD4$^+$ T cell responses has been investigated in more detail (Bickham et al., 2001; Steigerwald-Mullen et al., 2000). CD4$^+$ T cells against both antigens were found to secrete mainly IFNγ *ex vivo* (Bickham et al., 2001; Leen et al., 2001). In contrast to EBNA3C specific CD4$^+$ T cell responses, EBNA1 specific CD4$^+$ T cells contain a subdominant population of Th2 type CD4$^+$ T cells, which can be isolated by means of their IL-4 production in response to EBNA1 (Bickham et al., 2001). This subpopulation also grows out after multiple *in vitro* restimulations (Steigerwald-Mullen et al., 2000). The function of this small EBNA1 specific Th2 polarized CD4$^+$ T cell population is unclear. The predominant Th1 polarization of EBNA1 specific CD4$^+$ T cells, however, is also reflected *in vivo* in the EBNA1 specific antibody response. EBNA1 specific antibodies were found to be mainly of the IgG1 isotype, associated with Th1 responses in humans (Bickham et al., 2001). These data indicate that EBV specific CD4$^+$ T cell responses are mainly of Th1 polarization, which have been shown to mediate more efficient protection against viral infections in mice and man (Maloy et al., 2000; Rentenaar et al., 2000).

Apart from orchestrating anti-viral Th1 responses, EBV specific CD4$^+$ T cells also target EBV associated tumor cells directly. Already in early studies on T cell responses against LCLs, immunologists found cytolytic EBV specific CD4$^+$ T cells (Misko et al., 1984; Misko et al., 1991). Both *in vitro* expanded and *in vitro* primed T cell lines were found to lyse LCLs by EBV specific CD4$^+$ T cells via Fas/FasL dependent and independent mechanisms (Savoldo et al., 2002; Sun et al., 2002; Wilson et al., 1998). The latent EBV antigens EBNA1, EBNA2 and LMP2 as well as the lytic EBV antigen BHRF1 have been identified as targets of cytolytic CD4$^+$ T cells (Bickham et al., 2001; Khanna et al., 1997b; Landais et al., 2004; Nikiforow et al., 2003; Omiya et al., 2002; Paludan et al., 2002; Su et al., 2001). For EBNA1, only Th1 polarized, but not Th2 polarized CD4$^+$ T cells were able to kill LCLs (Bickham et al., 2001). When the killing mechanism was analyzed in more detail, EBNA1 specific CD4$^+$ T cells were found to lyse EBV positive tumor cells by apoptosis induction via Fas/FasL (Nikiforow et al., 2003; Paludan et al., 2002), while LMP2 specific CD4$^+$ T cells killed by a Fas/FasL independent mechanism (Su et al., 2002). Interestingly, Burkitt's lymphoma cells, that evade CD8$^+$ T cell recognition by avoiding the expression of all EBV proteins except EBNA1 and by down-regulating the MHC class I antigen processing machinery (Henriquez et al., 1999; Khanna et al., 1997c; Rowe et al., 1995), can be targeted by EBV specific cytolytic CD4$^+$ T cells (Khanna et al., 1997b; Paludan et al., 2002). EBNA1 specific CD4$^+$ T cells seem to be the only T cell reactivity that can target Burkitt's lymphoma so far (Paludan et al., 2002; Voo et al., 2002). Apart from targeting EBV associated tumor cell lines, EBV specific CD4$^+$ T cells can also prevent B cell transformation by EBV *in vitro* (Bickham et al., 2003; Nikiforow et al., 2001). Both EBNA1 and EBNA2 specific CD4$^+$ T cells have been shown to cause regression of freshly EBV infected B cells (Nikiforow et al., 2003; Omiya et al., 2002). Similar to the prevention of EBV mediated B cell transformation by CD8$^+$ T cells, probably both cytolytic mechanisms and IFNγ secretion contribute to the immune control of freshly EBV infected B cells by CD4$^+$ T cells *in vitro*. *In vivo*, priming of EBV specific MHC class II restricted T cells was found to correlate with the recovery of cottontop tamarins (*Saguinus oedipus oedipus*) from EBV induced lymphoproliferative disease and these cells were able to inhibit EBV transformation *in vitro* (Cleary et al., 1985; Finerty et al., 1988; Wilson et al., 1996). In addition, EBNA1 specific CD4$^+$ T cells could control mouse Burkitt's lymphoma *in vivo* (Fu et al., 2004). These studies indicate that CD4$^+$ T cells exert EBV specific immune control as effectors in their own right and this might be an adaptation of the human immune system to resist pathogens like EBV with tropism for MHC class II positive host cells.

In contrast to mice, MHC class II expression is indeed less restricted to professional antigen-presenting cells in humans and can be up-regulated on nearly all tissues after inflammation or activation. Therefore, EBV antigens could be presented to CD4$^+$ T cells by MHC class II on inflamed epithelial tissues, activated lymphocytes as well as professional antigen presenting cells (Reith and Mach, 2001). Therefore, EBV specific CD4$^+$ T cells should be able to target most EBV associated malignancies, like epithelial cancers, NK/T cell lymphomas and B cell lymphomas. Accordingly, both nasopharyngeal carcinoma and gastric carcinoma, EBV associated tumors of epithelial cell origin, are MHC class II positive (Khanna et al., 1998b; Lee et al., 2000; Saiki et al., 1996). It is conceivable that the more promiscuous MHC class II expression is an adaptation of the human immune system towards pathogens like EBV, allowing CD4$^+$ T cell mediated immune control of EBV transformed epithelial, NK, T and B cells.

IMMUNE ESCAPE BY EBV

In order to counterbalance the comprehensive immune control of EBV by the human immune system the virus has developed mechanisms to avoid clearance from its human host. While EBV achieves long-term persistence in memory B cells without detectable antigen expression (Babcock et al., 1998), reactivation of lytic and latent EBV infection regularly occurs in healthy EBV carriers despite comprehensive immune control. These two programs of EBV infection face distinct challenges with respect to escape from eradication by T cells.

In latent EBV infection, the probably most efficient immune escape mechanism is the small number of EBV antigens that are expressed only at low copy numbers per cell. This low antigen level per EBV transformed cell leads to low CD8$^+$ T cell epitope presentation with some epitopes like RRIYDLIEL from EBNA3C being as infrequent as less than 1 epitope per cell (Crotzer et al., 2000). This results in low and sometimes hard to detect recognition of EBV transformed B cells by T cell clones of excellent peptide affinity (Hill et al., 1995b; Shi et al., 1997). In addition, EBNA1 contains a GA-repeat domain that further down-regulates translation

(Yin et al., 2003) and prevents proteasomal processing of the full-length protein (Levitskaya et al., 1995; 1997). The decreased translation of EBNA1 is thought to prevent formation of defective ribosomal products (DRiPs) of EBNA1, the major source for CD8$^+$ T cell epitopes from this antigen (Münz, 2004). Inhibition of proteasomal degradation by the GA-repeat domain probably prevents full-length EBNA1 from being processed for MHC class I presentation. Deletion of the GA domain from EBNA1 (EBNA1ΔGA) increases LCL recognition by EBNA1 specific CD8$^+$ T cells, leading to around fourfold increase in IFNγ secretion (Lee et al., 2004) and cytolytic recognition of EBNA1ΔGA expressing LCLs (Blake et al., 2000; 1997). Therefore, EBNA1 contains the cis-acting GA-repeat domain, which further limits the antigenic amount that can be processed onto MHC class I and EBNA1 specific CD8$^+$ T cells have to rely on the degradation of EBNA1 derived DRiPs for the recognition of EBV transformed cells. In addition to the low levels and low numbers of latent EBV antigens in LCLs, expression is further down-regulated dependent on the differentiation stage of the EBV infected B cell with only three antigens (EBNA1, LMP1 and LMP2) being expressed in EBV$^+$ germinal center B cells (Babcock et al., 2000) and no EBV antigen expression in the majority of EBV$^+$ peripheral blood memory B cells (Hochberg et al., 2004). EBV associated malignancies like Hodgkin's disease, nasopharyngeal carcinoma and Burkitt's lymphoma, which arise in otherwise immunocompetent individuals, express also only subsets of latent EBV antigens with EBNA1 being the only antigen on protein level in all these tumors (Rickinson and Kieff, 2001). While most of these malignancies have intact MHC class I and II processing pathways (Khanna et al., 1998b; Lee et al., 2000; 1998), Burkitt's lymphoma displays deficiencies in MHC class I restricted antigen presentation, in addition to expressing EBNA1 as the sole latent EBV protein. The c-myc overexpression by translocation in Burkitt's lymphoma cells imposes this less immunogenic phenotype of the lymphoma cells with down-regulation of both the transporter associated with antigen processing (TAP) and the immunoproteasome (Staege et al., 2002). The deficiency in MHC class I processing prevents recognition of Burkitt's lymphoma cells by CD8$^+$ T cells (Lee et al., 2004; Rooney et al., 1985). In contrast, Burkitt's lymphoma cell lines can be targeted by EBV specific CD4$^+$ T cell responses (Khanna et al., 1997b; Paludan et al., 2002). All these findings underscore the notion that EBV latent antigens are expressed at very low levels, barely detectable for human T cells. Nevertheless, the human immune system is able to target these low amounts of latent EBV antigens to prevent EBV induced tumorigenesis in most infected individuals, even so the recognition is not sensitive enough to suppress latent EBV infection entirely in healthy EBV carriers, and enough EBV infected cells reach the peripheral memory B cell pool for life-long persistence of the virus in its human host.

Lytic EBV infection on the contrary involves the expression of more than 60 viral proteins that are expressed at fairly high copy numbers per cell (Kieff and Rickinson, 2001). In order for lytically EBV infected cells to produce EBV particles for transmission of EBV from carriers with established EBV specific T cell immunity, these cells probably require immune escape mechanisms to avoid MHC class I and II restricted recognition long enough to complete lytic replication.

Indeed, it was found in LCL cultures that cells with lytic EBV replication, as visualized by BZLF1 expression, had a four-to fivefold down-regulation of MHC class I and a twofold down-regulation of MHC class II compared to cells with latent EBV infection (Keating et al., 2002). This MHC down-regulation seemed to be an early event in lytic EBV infection, since it also occurred when progression through the late lytic cycle was inhibited by acyclovir. The mechanism of MHC down-regulation in early lytic EBV infection is unknown and only during late lytic EBV infection distinct EBV antigens have been identified that compromise MHC restricted recognition of EBV transformed B cells. One of these late EBV antigens that inhibits the MHC class I processing pathway is the viral IL-10 homologue encoded by the BCRF1 gene (Liu et al., 1997). Like human IL-10 it was found to mediate lmp2 and TAP1 down-regulation, thereby compromising peptide generation by the immunoproteasome and peptide import into the ER for MHC class I loading, respectively (Zeidler et al., 1997). This mechanism might contribute to the inhibition of T cell mediated immune control against EBV induced B cell transformation *in vitro* by viral IL-10 (Bejarano and Masucci, 1998). The other late lytic EBV antigen that was shown to compromise antigen presentation on MHC class II to T cells is the gp42 protein, encoded by the BZLF2 gene (Ressing et al., 2003). It was shown that gp42 does not inhibit MHC class II maturation and surface expression, but sterically inhibits MHC class II interaction with the T cell receptor, thereby blocking CD4[+] T cell recognition. Therefore, two late lytic EBV antigens, viral IL-10 and gp42 target MHC class I and II presentation, respectively. Furthermore, the early lytic EBV antigen BARF1 constitutes a soluble receptor for colony-stimulating factor 1 (CSF-1) (Strockbine et al., 1998). It was shown to inhibit CSF-1 mediated IFNα secretion of PBMCs after EBV infection (Cohen and Lekstrom, 1999). This could dampen innate anti-viral responses mediated by type I Interferons during the first 24h after infection as well as compromise IFNα mediated NK cell activation. Finally, EBV was shown to inhibit dendritic cell (DC) differentiation from monocytes via the late lytic EBV antigens of its fusion complex gH/gL/gp42 (BXLF2/BKRF2/ BZLF2) (Li et al., 2002). Apoptosis induction in developing DCs was more efficient when EBV particles, shed from epithelial cells, were used (Guerreiro-Cacais et al., 2004). Since these viruses contain more gp42 in their fusion complexes (Borza and Hutt-Fletcher, 2002), these data indicate that gp42 is an important component for the interaction/infection of developing DCs with EBV and seems required for efficient inhibition of DC differentiation. This mechanism might inhibit DC mediated T cell priming of lytic EBV antigen specific T cells and dampen NK cell activation in secondary lymphoid organs with high lytic EBV replication. However, it remains to be established that inhibition of DC differentiation by EBV applies also to DC differentation as it occurs *in vivo*. In this respect, it will be crucial to demonstrate that inflammatory CD16[+] monocytes, which preferentially develop into DCs upon activation *in vivo* (Geissmann et al., 2003; Randolph et al., 2002), are susceptible to EBV mediated apoptosis induction.

In summary, EBV expresses antigens at the detection limit for MHC restricted T cell recognition during latent infection, while various immune escape mechanisms, that have yet to be fully characterized, seem to operate during lytic infection to inhibit lytic EBV antigen recognition by the immune system.

FAILURE OF EBV SPECIFIC IMMUNE CONTROL IN PATIENTS WITH EBV ASSOCIATED MALIGNANCIES

In contrast to EBV associated lymphoproliferative disease in immunocompromised individuals, EBV expresses only a limited set of latent EBV antigens in spontaneously arising EBV$^+$ tumors in immunocompetent patients. EBNA1, LMP1 and LMP2 are maintained in most of these EBV associated malignancies like Hodgkin's disease and nasopharyngeal carcinoma. In Burkitt's lymphoma, EBV latent antigen expression is further restricted to EBNA1 only (Rickinson and Kieff, 2001). These restricted EBV expression patterns have also been identified in healthy EBV carriers, Namely, germinal center B cells express only EBNA1, LMP1 and LMP2 (Babcock et al., 2000), and 1% of homeostatically proliferating EBV$^+$ memory B cells expresses the EBNA1 only program (Hochberg et al., 2004). Therefore, it is tempting to speculate that a selective failure of EBV specific immune control against this restricted antigen set contributes to the development of Hodgkin's lymphoma, nasopharyngeal carcinoma and Burkitt's lymphoma (Bickham and Münz, 2003).

Hodgkin's lymphoma is characterized by the presence of a small number (approximately 1% of total tumor cells) of large, multinucleated cells of primarily B cell origin, termed Reed-Sternberg cells, which are embedded in a large number of T cells, eosinophils, B cells, plasma cells and histiocytes (Kuppers and Rajewsky, 1998). Reed-Sternberg cells produce the cytokine, IL-13, *in vivo* (Kapp et al., 1999) which promotes immunoglobulin class switching to IgE and IgG4 (Zurawski and de Vries, 1994) and may play a role in the priming of Th2 polarized CD4$^+$ T cells (Murphy et al., 2000). This might explain the Th2 polarized CD4$^+$ T cells and the eosinophils found to surround Reed-Sternberg cells in Hodgkin's disease biopsy samples. Patients whose tumor specimens reveal an abundance of eosinophils (von Wasielewski et al., 2000) or elevated IL-10 levels in the tumor (Herbst et al., 1996) have a poor prognosis, highlighting the clinical relevance of the T cell polarization in the microenvironment of the tumor.

Systemic effects of cytokine dysregulation are apparent in Hodgkin's disease patients as well. They have a constellation of symptoms, which includes weight loss, night sweats and fever. These symptoms are thought to be associated with increased levels of cytokines in the peripheral blood. Elevations of Th2 cytokines in the serum, such as IL-6 (Kurzrock, 1997) and IL-10 (Sarris et al., 1999; Bohlen et al., 2000; Kurzrock, 1997; Viviani et al., 2000), are known to be poor prognostic indicators in patients. Though this Th2 cytokine profile is present in both EBV positive and negative cases of Hodgkin's lymphoma, it suggests a loss of efficient antiviral Th1 polarization in EBV specific immune control. This loss could result from the absence of protective, cytotoxic Th1 responses or the acquisition of permissive Th2 polarized responses to the EBV latent antigens expressed in Hodgkin's disease. Consequently, the presence of a Th2 cytokine milieu both systemically and in the tumor's microenvironment may well facilitate the escape of Hodgkin's tumor cells from immune surveillance.

A link between the epithelial cell cancer nasopharyngeal carcinoma (NPC) and EBV was first postulated because of unusually high EBV IgG antibody titers in NPC patients compared to controls (Henle et al., 1970; de Schryver et al., 1969). Confirming EBV association with NPC, it was found that all nonkeratinizing nasopharyngeal carcinomas carry EBV (Wolf et al., 1973; zur Hausen et al., 1970) and are derived from a single EBV positive progenitor cell (zur Hausen et al., 1970). Upon further investigation, it was shown that the presence of IgA antibodies to viral capsid antigen (VCA) was so peculiar to patients with NPC (Henle and Henle, 1976) that they are now used for the diagnosis of pre-clinical malignancy. Elevations in VCA IgA antibody titers can be seen as early as 8-30 months prior to the development of NPC and are therefore extremely useful in the detection of early stage disease (Zeng, 1985). Furthermore, levels of antibody titers are correlated with tumor burden and can be used to monitor for recurrence of disease (Sam et al., 1989). Because antibody production is a prominent feature in a Th2 polarized immune response and antibodies of IgA isotype are preferentially induced by TGF-β and IL-5 of Th2 polarized T cells, the appearance of these antibodies in NPC patients could mark a shift from the normal protective Th1 phenotype of EBV specific immune control to a permissive Th2 response allowing for the outgrowth of tumor cells.

Moreover, EBV induced B cell transformation occurs uncontrolled upon infection of PBMCs from NPC patients demonstrating a defect in EBV specific immune control. In a study comparing 54 NPC patients with 38 healthy controls, only 6% of cultures from NPC patients showed complete inhibition of EBV induced B cell transformation compared to 34% of controls (Chan and Chew, 1981). Further studies revealed that especially newly diagnosed NPC patients were significantly impaired in their ability to control outgrowth of LCL by their PBMCs, while long-term survivors showed more protective EBV specific immunity *in vitro*. This lack of regression was noted most consistently in newly diagnosed patients with high IgA antibody titers (Moss et al., 1983). Interestingly, the T cell deficiency that fails to cause LCL regression in the newly diagnosed patients appears to be rectified with the resolution of disease, as is seen in long-term survivors, thus arguing for a tumor-derived factor responsible for this phenomenon. The increase in EBV antibody titers of IgA isotype and the loss of cytolytic activity against EBV transformed B cells suggest that EBV specific immune control is polarized towards a Th2 phenotype in NPC patients, which is insufficient to control this EBV associated malignancy.

Endemic Burkitt's lymphoma is a B cell lymphoma and develops mainly in children of holoendemic malaria regions (Burkitt, 1962; de-Thé et al., 1978). In these children, the immune responses to both pathogens have to develop concurrently. Continuous exposure to *Plasmodium falciparum* and the cytokine milieu of the active immune defense against malaria most likely influence the priming of EBV immunity. The clearance of the erythrocytic stage of the parasite is mediated by a humoral, Th2 based immune response (Good and Doolan, 1999; Riley, 1999). It is tempting to speculate that the Th2 cytokine milieu created by the systemic malaria-specific immune response polarizes or diverts EBV immunity that is just emerging to Th2. The outcome is a decrease in Th1 type, EBV-specific immune control in the malaria patients (Whittle et al., 1984).

This failure in the establishment of EBV immunity might originate from the inhibition of DCs by *Plasmodium falciparum* (Urban et al., 1999). Erythrocytes are the main host cell during systemic *Plasmodium falciparum* infection. Malaria-infected red blood cells adhere to DCs, prevent their maturation and subsequently reduce their capacity to stimulate T cells (Urban et al., 1999). Instead of DCs, B cells are required for protection against systemic malaria infection, probably inducing a Th2 like immune response (Langhorne et al., 1998). A lack of EBV antigen presentation by Th1 inducing DCs, combined with direct presentation by infected B cells, might lead to less efficient priming and a more Th2-polarized EBV specific immune response. This immunity, less efficient against latently EBV infected B cells (Whittle et al., 1984), might be incapable of controlling Burkitt's lymphoma and, therefore, could at least in part be responsible for the outgrowth of this tumor in the affected children.

Therefore, in three EBV associated malignancies, Hodgkin's disease, nasopharyngeal carcinoma and Burkitt's lymphoma, EBV specific T cell immunity seems to be compromised and diverted from protective antiviral Th1 immunity to Th2 polarized T cell responses. Although it is unclear if the failure of EBV specific immune control in EBV associated malignancies contributes to the development of these malignancies or results from the tumor's subversion of the immune system, establishing Th1 polarized immune responses against the latent EBV antigens of these cancers should be the goal of immunotherapy against EBV associated malignancies like Hodgkin's disease, nasopharyngeal carcinoma and Burkitt's lymphoma.

SUMMARY

Immune control of EBV infection involves comprehensive immune responses by NK cells, T cells and B cells. The Th1 pattern of protective immunity against EBV is probably initiated by DCs and centers around strong memory $CD4^+$ and $CD8^+$ T cell responses, whereby the $CD4^+$ T cells maintain EBV specific Th1 immunity, and both $CD4^+$ and $CD8^+$ T cells target EBV infected cells directly. IFNγ secretion might be more relevant than cytolytic function in the control of EBV transformed cells by NK cells, $CD4^+$ and $CD8^+$ T cells. The challenge for the future is to design a vaccine that will harness all these immune system compartments and reestablish EBV specific immune control in patients with EBV associated malignancies.

References

Aman, P., and von Gabain, A. (1990). An Epstein-Barr virus immortalization associated gene segment interferes specifically with the IFN-induced anti-proliferative response in human B-lymphoid cell lines. EMBO J. *9*, 147-152.

Amyes, E., Hatton, C., Montamat-Sicotte, D., Gudgeon, N., Rickinson, A. B., McMichael, A. J., and Callan, M. F. (2003). Characterization of the $CD4^+$ T cell response to Epstein-Barr virus during primary and persistent infection. J. Exp. Med. *198*, 903-911.

Appay, V., Dunbar, P. R., Callan, M., Klenerman, P., Gillespie, G. M., Papagno, L., Ogg, G. S., King, A., Lechner, F., Spina, C. A., et al. (2002). Memory $CD8^+$ T cells vary in differentiation phenotype in different persistent virus infections. Nat. Med. *8*, 379-385.

Babcock, G. J., Decker, L. L., Volk, M., and Thorley-Lawson, D. A. (1998). EBV persistence in memory B cells *in vivo*. Immunity *9*, 395-404.

Babcock, J. G., Hochberg, D., and Thorley-Lawson, A. D. (2000). The expression pattern of Epstein-Barr virus latent genes *in vivo* is dependent upon the differentiation stage of the infected B cell. Immunity *13*, 497-506.

Baiocchi, R. A., Ward, J. S., Carrodeguas, L., Eisenbeis, C. F., Peng, R., Roychowdhury, S., Vourganti, S., Sekula, T., O'Brien, M., Moeschberger, M., and Caligiuri, M. A. (2001).

GM-CSF and IL-2 induce specific cellular immunity and provide protection against Epstein-Barr virus lymphoproliferative disorder. J. Clin. Invest *108*, 887-894.

Bejarano, M. T., and Masucci, M. G. (1998). Interleukin-10 abrogates the inhibition of Epstein-Barr virus-induced B-cell transformation by memory T-cell responses. Blood *92*, 4256-4262.

Bennett, S. R., Carbone, F. R., Karamalis, F., Flavell, R. A., Miller, J. F., and Heath, W. R. (1998). Help for cytotoxic-T-cell responses is mediated by CD40 signalling. Nature *393*, 478-480.

Benninger-Doring, G., Pepperl, S., Deml, L., Modrow, S., Wolf, H., and Jilg, W. (1999). Frequency of CD8[+] T lymphocytes specific for lytic and latent antigens of Epstein-Barr virus in healthy virus carriers. Virology *264*, 289-297.

Bickham, K., Goodman, K., Paludan, C., Nikiforow, S., Tsang, M. L., Steinman, R. M., and Münz, C. (2003). Dendritic cells initiate immune control of Epstein-Barr virus transformation of B lymphocytes *in vitro*. J. Exp. Med. *198*, 1653-1663.

Bickham, K., and Münz, C. (2003). Contrasting roles of dendritic cells and B cells in the immune control of Epstein-Barr virus. Curr. Top Microbiol. Immunol. *276*, 55-76.

Bickham, K., Münz, C., Tsang, M. L., Larsson, M., Fonteneau, J. F., Bhardwaj, N., and Steinman, R. (2001). EBNA1-specific CD4[+] T cells in healthy carriers of Epstein-Barr virus are primarily Th1 in function. J. Clin. Invest. *107*, 121-130.

Biron, C. A., Nguyen, K. B., Pien, G. C., Cousens, L. P., and Salazar-Mather, T. P. (1999). Natural killer cells in antiviral defense: function and regulation by innate cytokines. Annu. Rev. Immunol. *17*, 189-220.

Blake, N., Haigh, T., Shaka'a, G., Croom-Carter, D., and Rickinson, A. (2000). The importance of exogenous antigen in priming the human CD8[+] T cell response: lessons from the EBV nuclear antigen EBNA1. J. Immunol. *165*, 7078-7087.

Blake, N., Lee, S., Redchenko, I., Thomas, W., Steven, N., Leese, A., Steigerwald-Mullen, P., Kurilla, M. G., Frappier, L., and Rickinson, A. (1997). Human CD8[+] T cell responses to EBV EBNA1: HLA class I presentation of the (Gly-Ala)-containing protein requires exogenous processing. Immunity *7*, 791-802.

Bogedain, C., Wolf, H., Modrow, S., Stuber, G., and Jilg, W. (1995). Specific cytotoxic T lymphocytes recognize the immediate-early transactivator Zta of Epstein-Barr virus. J. Virol. *69*, 4872-4879.

Bohlen, H., Kessler, M., Sextro, M., Diehl, V., and Tesch, H. (2000). Poor clinical outcome of patients with Hodgkin's disease and elevated interleukin-10 serum levels. Clinical significance of interleukin-10 serum levels for Hodgkin's disease. Ann. Hematol *79*, 110-113.

Borza, C. M., and Hutt-Fletcher, L. M. (2002). Alternate replication in B cells and epithelial cells switches tropism of Epstein-Barr virus. Nat. Med. *8*, 594-599.

Boyle, G. J., Michaels, M. G., Webber, S. A., Knisely, A. S., Kurland, G., Cipriani, L. A., Griffith, B. P., and Fricker, F. J. (1997). Posttransplantation lymphoproliferative disorders in pediatric thoracic organ recipients. J. Pediatr. *131*, 309-313.

Brooks, J. M., Murray, R. J., Thomas, W. A., Kurilla, M. G., and Rickinson, A. B. (1993). Different HLA-B27 subtypes present the same immunodominant Epstein-Barr virus peptide. J. Exp. Med. *178*, 879-887.

Brown, M. H., Boles, K., van der Merwe, P. A., Kumar, V., Mathew, P. A., and Barclay, A. N. (1998). 2B4, the natural killer and T cell immunoglobulin superfamily surface protein, is a ligand for CD48. J. Exp. Med. *188*, 2083-2090.

Burkitt, D. (1962). A children's cancer dependent on climatic factors. Nature *194*, 232-234.

Burrows, S. R., Gardner, J., Khanna, R., Steward, T., Moss, D. J., Rodda, S., and Suhrbier, A. (1994). Five new cytotoxic T cell epitopes identified within Epstein-Barr virus nuclear antigen 3. J. Gen. Virol. *75 (Pt 9)*, 2489-2493.

Burrows, S. R., Misko, I. S., Sculley, T. B., Schmidt, C., and Moss, D. J. (1990a). An Epstein-Barr virus-specific cytotoxic T-cell epitope present on A- and B-type transformants. J. Virol. *64*, 3974-3976.

Burrows, S. R., Sculley, T. B., Misko, I. S., Schmidt, C., and Moss, D. J. (1990b). An Epstein-Barr virus-specific cytotoxic T cell epitope in EBV nuclear antigen 3 (EBNA 3). J. Exp. Med. *171*, 345-349.

Calender, A., Billaud, M., Aubry, J. P., Banchereau, J., Vuillaume, M., and Lenoir, G. M. (1987). Epstein-Barr virus (EBV) induces expression of B-cell activation markers on *in vitro* infection of EBV-negative B-lymphoma cells. Proc. Natl. Acad. Sci. USA. *84*, 8060-8064.

Callan, M. F., Steven, N., Krausa, P., Wilson, J. D., Moss, P. A., Gillespie, G. M., Bell, J. I., Rickinson, A. B., and McMichael, A. J. (1996). Large clonal expansions of CD8+ T cells in acute infectious mononucleosis. Nat. Med. *2*, 906-911.

Callan, M. F., Tan, L., Annels, N., Ogg, G. S., Wilson, J. D., O'Callaghan, C. A., Steven, N., McMichael, A. J., and Rickinson, A. B. (1998). Direct visualization of antigen-specific CD8+ T cells during the primary immune response to Epstein-Barr virus *In vivo*. J. Exp. Med. *187*, 1395-1402.

Cardin, R. D., Brooks, J. W., Sarawar, S. R., and Doherty, P. C. (1996). Progressive loss of CD8+ T cell-mediated control of a gamma-herpesvirus in the absence of CD4+ T cells. J. Exp. Med. *184*, 863-871.

Cella, M., Scheidegger, D., Palmer-Lehmann, K., Lane, P., Lanzavecchia, A., and Alber, G. (1996). Ligation of CD40 on dendritic cells triggers production of high levels of interleukin-12 and enhances T cell stimulatory capacity: T-T help via APC activation. J. Exp. Med. *184*, 747-752.

Chan, S. H., and Chew, T. S. (1981). Lack of regression in Epstein-Barr virus infected leucocyte cultures of nasopharyngeal carcinoma patients. Lancet *2*, 1353.

Cleary, M. L., Epstein, M. A., Finerty, S., Dorfman, R. F., Bornkamm, G. W., Kirkwood, J. K., Morgan, A. J., and Sklar, J. (1985). Individual tumors of multifocal EB virus-induced malignant lymphomas in tamarins arise from different B-cell clones. Science *228*, 722-724.

Cohen, J. I., and Lekstrom, K. (1999). Epstein-Barr virus BARF1 protein is dispensable for B-cell transformation and inhibits alpha interferon secretion from mononuclear cells. J. Virol. *73*, 7627-7632.

Constant, S. L., and Bottomly, K. (1997). Induction of Th1 and Th2 CD4+ T cell responses: the alternative approaches. Annu. Rev. Immunol. *15*, 297-322.

Crotzer, V. L., Christian, R. E., Brooks, J. M., Shabanowitz, J., Settlage, R. E., Marto, J. A., White, F. M., Rickinson, A. B., Hunt, D. F., and Engelhard, V. H. (2000). Immunodominance among EBV-derived epitopes restricted by HLA-B27 does not correlate with epitope abundance in EBV-transformed B-lymphoblastoid cell lines. J. Immunol. *164*, 6120-6129.

de Schryver, A., Friberg, S., Jr., Klein, G., Henle, W., Henle, G., De-The, G., Clifford, P., and Ho, H. C. (1969). Epstein-Barr virus-associated antibody patterns in carcinoma of the post-nasal space. Clin. Exp. Immunol. *5*, 443-459.

Delcayre, A. X., Lotz, M., and Lernhardt, W. (1993). Inhibition of Epstein-Barr virus-mediated capping of CD21/CR2 by alpha interferon (IFN-alpha): immediate antiviral activity of IFN-alpha during the early phase of infection. J. Virol. *67*, 2918-2921.

de-The, G., Geser, A., Day, N. E., Tukei, P. M., Williams, E. H., Beri, D. P., Smith, P. G., Dean, A. G., Bronkamm, G. W., Feorino, P., and Henle, W. (1978). Epidemiological evidence for causal relationship between Epstein-Barr virus and Burkitt's lymphoma from Ugandan prospective study. Nature *274*, 756-761.

Duraiswamy, J., Burrows, J. M., Bharadwaj, M., Burrows, S. R., Cooper, L., Pimtanothai, N., and Khanna, R. (2003). *Ex vivo* analysis of T-cell responses to Epstein-Barr virus-encoded oncogene latent membrane protein 1 reveals highly conserved epitope sequences in virus isolates from diverse geographic regions. J. Virol. *77*, 7401-7410.

Fehniger, T. A., Cooper, M. A., Nuovo, G. J., Cella, M., Facchetti, F., Colonna, M., and Caligiuri, M. A. (2002). CD56[bright] Natural Killer Cells are Present in Human Lymph Nodes and are Activated by T cell Derived IL-2: a Potential New Link between Adaptive and Innate Immunity. Blood *12*, 12.

Ferlazzo, G., and Münz, C. (2004). Natural killer cell compartments and their activation by dendritic cells. Journal of Immunology *172*, 1333-1339.

Ferlazzo, G., Thomas, D., Lin, S. L., Goodman, K., Morandi, B., Muller, W. A., Moretta, A., and Münz, C. (2004). The abundant NK cells in human lymphoid tissues require activation to express killer cell Ig-like receptors and become cytolytic. Journal of Immunology *172*, 1455-1462.

Ferlazzo, G., Thomas, D., Pack, M., Paludan, C., Schmid, D., Strowig, T., Muller, W. A., Moretta, L., and Münz, C. 2004. Distinct roles of IL-12 and IL-15 in human natural killer cell activation by dendritic cells from secondary lymphoid organs. Proc. Natl. Acad. Sci. USA *101*, 16606-16611.

Ferlazzo, G., Tsang, M. L., Moretta, L., Melioli, G., Steinman, R. M., and Münz, C. (2002). Human dendritic cells activate resting NK cells and are recognized via the NKp30 receptor by activated NK cells. J. Exp. Med. *195*, 343-351.

Finerty, S., Scullion, F. T., and Morgan, A. J. (1988). Demonstration *in vitro* of cell mediated immunity to Epstein-Barr virus in cotton-top tamarins. Clin. Exp. Immunol. *73*, 181-185.

Fu, T., Voo, K. S., and Wang, R. F. (2004). Critical role of EBNA1-specific CD4[+] T cells in the control of mouse Burkitt lymphoma *in vivo*. J. Clin. Invest *114*, 542-550.

Gao, Y., Xue, S., and Griffin, B. E. (1999). Sensitivity of an Epstein-Barr virus-positive tumor line, Daudi, to alpha interferon correlates with expression of a GC-rich viral transcript. Mol. Cell Biol. *19*, 7305-7313.

Garner, J. G., Hirsch, M. S., and Schooley, R. T. (1984). Prevention of Epstein-Barr virus-induced B-cell outgrowth by interferon alpha. Infect. Immun. *43*, 920-924.

Gavioli, R., Kurilla, M. G., de Campos-Lima, P. O., Wallace, L. E., Dolcetti, R., Murray, R. J., Rickinson, A. B., and Masucci, M. G. (1993). Multiple HLA A11-restricted cytotoxic T-lymphocyte epitopes of different immunogenicities in the Epstein-Barr virus-encoded nuclear antigen 4. J. Virol. *67*, 1572-1578.

Geissmann, F., Jung, S., and Littman, D. R. (2003). Blood monocytes consist of two principal subsets with distinct migratory properties. Immunity *19*, 71-82.

Gerosa, F., Baldani-Guerra, B., Nisii, C., Marchesini, V., Carra, G., and Trinchieri, G. (2002). Reciprocal activating interaction between natural killer cells and dendritic cells. J. Exp. Med. *195*, 327-333.

Good, M. F., and Doolan, D. L. (1999). Immune effector mechanisms in malaria. Curr. Opin. Immunol. *11*, 412-419.

Guerreiro-Cacais, A. O., Li, L., Donati, D., Bejarano, M. T., Morgan, A., Masucci, M. G., Hutt-Fletcher, L., and Levitsky, V. (2004). Capacity of Epstein-Barr virus to infect monocytes and inhibit their development into dendritic cells is affected by the cell type supporting virus replication. J. Gen. Virol. *85*, 2767-2778.

Guidotti, L. G., Rochford, R., Chung, J., Shapiro, M., Purcell, R., and Chisari, F. V. (1999). Viral clearance without destruction of infected cells during acute HBV infection. Science *284*, 825-829.

Henderson, S., Rowe, M., Gregory, C., Croom-Carter, D., Wang, F., Longnecker, R., Kieff, E., and Rickinson, A. (1991). Induction of bcl-2 expression by Epstein-Barr virus latent membrane protein 1 protects infected B cells from programmed cell death. Cell *65*, 1107-1115.

Henle, G., and Henle, W. (1976). Epstein-Barr virus-specific IgA serum antibodies as an outstanding feature of nasopharyngeal carcinoma. Int. J. Cancer *17*, 1-7.

Henle, W., Henle, G., Andersson, J., Ernberg, I., Klein, G., Horwitz, C. A., Marklund, G., Rymo, L., Wellinder, C., and Straus, S. E. (1987). Antibody responses to Epstein-Barr virus-determined nuclear antigen (EBNA)-1 and EBNA-2 in acute and chronic Epstein-Barr virus infection. Proc. Natl. Acad. Sci. USA. *84*, 570-574.

Henle, W., Henle, G., Ho, H. C., Burtin, P., Cachin, Y., Clifford, P., de Schryver, A., de-The, G., Diehl, V., and Klein, G. (1970). Antibodies to Epstein-Barr virus in nasopharyngeal carcinoma, other head and neck neoplasms, and control groups. J. Natl. Cancer Inst *44*, 225-231.

Henriquez, N. V., Floettmann, E., Salmon, M., Rowe, M., and Rickinson, A. B. (1999). Differential responses to CD40 ligation among Burkitt lymphoma lines that are uniformly responsive to Epstein-Barr virus latent membrane protein 1 [In Process Citation]. J. Immunol. *162*, 3298-3307.

Herbst, H., Foss, H. D., Samol, J., Araujo, I., Klotzbach, H., Krause, H., Agathanggelou, A., Niedobitek, G., and Stein, H. (1996). Frequent expression of interleukin-10 by Epstein-Barr virus-harboring tumor cells of Hodgkin's disease. Blood *87*, 2918-2929.

Hill, A., Worth, A., Elliott, T., Rowland-Jones, S., Brooks, J., Rickinson, A., and McMichael, A. (1995a). Characterization of two Epstein-Barr virus epitopes restricted by HLA-B7. Eur. J. Immunol. *25*, 18-24.

Hill, A. B., Lee, S. P., Haurum, J. S., Murray, N., Yao, Q. Y., Rowe, M., Signoret, N., Rickinson, A. B., and McMichael, A. J. (1995b). Class I major histocompatibility complex-restricted cytotoxic T lymphocytes specific for Epstein-Barr virus (EBV)-transformed B lymphoblastoid cell lines against which they were raised. J. Exp. Med. *181*, 2221-2228.

Hislop, A. D., Annels, N. E., Gudgeon, N. H., Leese, A. M., and Rickinson, A. B. (2002). Epitope-specific Evolution of Human CD8+ T Cell Responses from Primary to Persistent Phases of Epstein-Barr Virus Infection. J. Exp. Med. *195*, 893-905.

Hislop, A. D., Gudgeon, N. H., Callan, M. F., Fazou, C., Hasegawa, H., Salmon, M., and Rickinson, A. B. (2001). EBV-specific CD8+ T cell memory: relationships between epitope specificity, cell phenotype, and immediate effector function. J. Immunol. *167*, 2019-2029.

Hochberg, D., Middeldorp, J. M., Catalina, M., Sullivan, J. L., Luzuriaga, K., and Thorley-Lawson, D. A. (2004). Demonstration of the Burkitt's lymphoma Epstein-Barr virus phenotype in dividing latently infected memory cells *in vivo*. Proc. Natl. Acad. Sci. USA. *101*, 239-244.

Hoffman, G. J., Lazarowitz, S. G., and Hayward, S. D. (1980). Monoclonal antibody against a 250,000-dalton glycoprotein of Epstein-Barr virus identifies a membrane antigen and a neutralizing antigen. Proc. Natl. Acad. Sci. USA. *77*, 2979-2983.

Hoshino, Y., Morishima, T., Kimura, H., Nishikawa, K., Tsurumi, T., and Kuzushima, K. (1999). Antigen-driven expansion and contraction of CD8+-activated T cells in primary EBV infection. J. Immunol. *163*, 5735-5740.

Hurley, E. A., and Thorley-Lawson, D. A. (1988). B cell activation and the establishment of Epstein-Barr virus latency. J. Exp. Med. *168*, 2059-2075.

Janeway, C. A., Jr., and Medzhitov, R. (2002). Innate immune recognition. Annu. Rev. Immunol. *20*, 197-216.

Kapp, U., Yeh, W. C., Patterson, B., Elia, A. J., Kagi, D., Ho, A., Hessel, A., Tipsword, M., Williams, A., Mirtsos, C., et al. (1999). Interleukin 13 is secreted by and stimulates the growth of hodgkin and reed-sternberg cells. J. Exp. Med. *189*, 1939-1946.

Kapsenberg, M. L. (2003). Dendritic-cell control of pathogen-driven T-cell polarization. Nat. Rev. Immunol. *3*, 984-993.

Keating, S., Prince, S., Jones, M., and Rowe, M. (2002). The lytic cycle of Epstein-Barr virus is associated with decreased expression of cell surface major histocompatibility complex class I and class II molecules. J. Virol. *76*, 8179-8188.

Kerr, B. M., Kienzle, N., Burrows, J. M., Cross, S., Silins, S. L., Buck, M., Benson, E. M., Coupar, B., Moss, D. J., and Sculley, T. B. (1996). Identification of type B-specific and cross-reactive cytotoxic T-lymphocyte responses to Epstein-Barr virus. J. Virol. *70*, 8858-8864.

Khanna, R., and Burrows, S. R. (2000). Role of Cytotoxic T Lymphocytes in Epstein-Barr Virus-Associated Diseases. Annu. Rev. Microbiol. *54*, 19-48.

Khanna, R., Burrows, S. R., Kurilla, M. G., Jacob, C. A., Misko, I. S., Sculley, T. B., Kieff, E., and Moss, D. J. (1992). Localization of Epstein-Barr virus cytotoxic T cell epitopes using recombinant vaccinia: implications for vaccine development. J. Exp. Med. *176*, 169-176.

Khanna, R., Burrows, S. R., Moss, D. J., and Silins, S. L. (1996). Peptide transporter (TAP-1 and TAP-2)-independent endogenous processing of Epstein-Barr virus (EBV) latent membrane protein 2A: implications for cytotoxic T-lymphocyte control of EBV-associated malignancies. J. Virol. *70*, 5357-5362.

Khanna, R., Burrows, S. R., Nicholls, J., and Poulsen, L. M. (1998a). Identification of cytotoxic T cell epitopes within Epstein-Barr virus (EBV) oncogene latent membrane protein 1 (LMP1): evidence for HLA A2 supertype-restricted immune recognition of EBV-infected cells by LMP1- specific cytotoxic T lymphocytes. Eur. J. Immunol. *28*, 451-458.

Khanna, R., Burrows, S. R., Steigerwald-Mullen, P. M., Moss, D. J., Kurilla, M. G., and Cooper, L. (1997a). Targeting Epstein-Barr virus nuclear antigen 1 (EBNA1) through the class II pathway restores immune recognition by EBNA1-specific cytotoxic T lymphocytes: evidence for HLA-DM-independent processing. Int. Immunol. *9*, 1537-1543.

Khanna, R., Burrows, S. R., Steigerwald-Mullen, P. M., Thomson, S. A., Kurilla, M. G., and Moss, D. J. (1995). Isolation of cytotoxic T lymphocytes from healthy seropositive individuals specific for peptide epitopes from Epstein-Barr virus nuclear antigen 1: implications for viral persistence and tumor surveillance. Virology *214*, 633-637.

Khanna, R., Burrows, S. R., Thomson, S. A., Moss, D. J., Cresswell, P., Poulsen, L. M., and Cooper, L. (1997b). Class I processing-defective Burkitt's lymphoma cells are recognized efficiently by CD4+ EBV-specific CTLs. J. Immunol. *158*, 3619-3625.

Khanna, R., Busson, P., Burrows, S. R., Raffoux, C., Moss, D. J., Nicholls, J. M., and Cooper, L. (1998b). Molecular characterization of antigen-processing function in nasopharyngeal carcinoma (NPC): evidence for efficient presentation of Epstein-Barr virus cytotoxic T-cell epitopes by NPC cells. Cancer Res. *58*, 310-314.

Khanna, R., Cooper, L., Kienzle, N., Moss, D. J., Burrows, S. R., and Khanna, K. K. (1997c). Engagement of CD40 antigen with soluble CD40 ligand up-regulates peptide transporter expression and restores endogenous processing function in Burkitt's lymphoma cells. J. Immunol. *159*, 5782-5785.

Khanna, R., Sherritt, M., and Burrows, S. R. (1999). EBV structural antigens, gp350 and gp85, as targets for *ex vivo* virus-specific CTL during acute infectious mononucleosis: potential use of gp350/gp85 CTL epitopes for vaccine design. J. Immunol. *162*, 3063-3069.

Kieff, E., and Rickinson, A. (2001). Epstein-Barr Virus and its replication. In Fields Virology, D. M. Knipe, and P. M. Howley, eds. (Philadelphia, Lippincott-Raven), pp. 2511-2573.

Kienzle, N., Sculley, T. B., Poulsen, L., Buck, M., Cross, S., Raab-Traub, N., and Khanna, R. (1998). Identification of a cytotoxic T-lymphocyte response to the novel BARF0 protein of Epstein-Barr virus: a critical role for antigen expression. J. Virol. *72*, 6614-6620.

Kikuta, H., Mizuno, F., Yano, S., and Osato, T. (1984). Interferon production by Epstein-Barr virus in human mononuclear leukocytes. J. Gen. Virol. *65 (Pt 4)*, 837-841.

Koch, F., Stanzl, U., Jennewein, P., Janke, K., Heufler, C., Kampgen, E., Romani, N., and Schuler, G. (1996). High level IL-12 production by murine dendritic cells: upregulation via MHC class II and CD40 molecules and downregulation by IL-4 and IL-10. J. Exp. Med. *184*, 741-746.

Konttinen, Y. T., Bluestein, H. G., and Zvaifler, N. J. (1985). Regulation of the growth of Epstein-Barr virus-infected B cells. I. Growth regression by E rosetting cells from VCA-positive donors is a combined effect of autologous mixed leukocyte reaction and activation of T8$^+$ memory cells. J. Immunol. *134*, 2287-2293.

Kuppers, R., and Rajewsky, K. (1998). The origin of Hodgkin and Reed/Sternberg cells in Hodgkin's disease. Annu. Rev. Immunol. *16*, 471-493.

Kurzrock, R. (1997). Cytokine deregulation in hematological malignancies: clinical and biological implications. Clin. Cancer Res. *3*, 2581-2584.

Laichalk, L. L., Hochberg, D., Babcock, G. J., Freeman, R. B., and Thorley-Lawson, D. A. (2002). The dispersal of mucosal memory B cells: evidence from persistent EBV infection. Immunity *16*, 745-754.

Landais, E., Saulquin, X., Scotet, E., Trautmann, L., Peyrat, M. A., Yates, J. L., Kwok, W. W., Bonneville, M., and Houssaint, E. (2004). Direct killing of Epstein-Barr virus (EBV)-infected B cells by CD4 T cells directed against the EBV lytic protein BHRF1. Blood *103*, 1408-1416.

Langhorne, J., Cross, C., Seixas, E., Li, C., and von der Weid, T. (1998). A role for B cells in the development of T cell helper function in a malaria infection in mice. Proc. Natl. Acad. Sci. USA. *95*, 1730-1734.

Lautscham, G., Haigh, T., Mayrhofer, S., Taylor, G., Croom-Carter, D., Leese, A., Gadola, S., Cerundolo, V., Rickinson, A., and Blake, N. (2003). Identification of a TAP-independent, immunoproteasome-dependent CD8$^+$ T-cell epitope in Epstein-Barr virus latent membrane protein 2. J. Virol. *77*, 2757-2761.

Lautscham, G., Mayrhofer, S., Taylor, G., Haigh, T., Leese, A., Rickinson, A., and Blake, N. (2001). Processing of a multiple membrane spanning Epstein-Barr virus protein for CD8$^+$ T cell recognition reveals a proteasome-dependent, transporter associated with antigen processing-independent pathway. J. Exp. Med. *194*, 1053-1068.

Lee, S. P., Brooks, J. M., Al-Jarrah, H., Thomas, W. A., Haigh, T. A., Taylor, G. S., Humme, S., Schepers, A., Hammerschmidt, W., Yates, J. L., et al. (2004). CD8 T cell recognition of endogenously expressed Epstein-Barr virus nuclear antigen 1. J. Exp. Med. *199*, 1409-1420.

Lee, S. P., Chan, A. T., Cheung, S. T., Thomas, W. A., CroomCarter, D., Dawson, C. W., Tsai, C. H., Leung, S. F., Johnson, P. J., and Huang, D. P. (2000). CTL control of EBV in nasopharyngeal carcinoma (NPC): EBV-specific CTL responses in the blood and tumors of NPC patients and the antigen- processing function of the tumor cells. J. Immunol. *165*, 573-582.

Lee, S. P., Constandinou, C. M., Thomas, W. A., Croom-Carter, D., Blake, N. W., Murray, P. G., Crocker, J., and Rickinson, A. B. (1998). Antigen presenting phenotype of Hodgkin Reed-Sternberg cells: analysis of the HLA class I processing pathway and the effects of interleukin-10 on epstein-barr virus-specific cytotoxic T-cell recognition. Blood *92*, 1020-1030.

Lee, S. P., Thomas, W. A., Murray, R. J., Khanim, F., Kaur, S., Young, L. S., Rowe, M., Kurilla, M., and Rickinson, A. B. (1993a). HLA A2.1-restricted cytotoxic T cells recognizing a range of Epstein-Barr virus isolates through a defined epitope in latent membrane protein LMP2. J. Virol. *67*, 7428-7435.

Lee, S. P., Tierney, R. J., Thomas, W. A., Brooks, J. M., and Rickinson, A. B. (1997). Conserved CTL epitopes within EBV latent membrane protein 2: a potential target for CTL-based tumor therapy. J. Immunol. *158*, 3325-3334.

Lee, S. P., Wallace, L. E., Mackett, M., Arrand, J. R., Searle, P. F., Rowe, M., and Rickinson, A. B. (1993b). MHC class II-restricted presentation of endogenously synthesized antigen: Epstein-Barr virus transformed B cell lines can present the viral glycoprotein gp340 by two distinct pathways. Int. Immunol. *5*, 451-460.

Leen, A., Meij, P., Redchenko, I., Middeldorp, J., Bloemena, E., Rickinson, A., and Blake, N. (2001). Differential immunogenicity of Epstein-Barr virus latent-cycle proteins for human CD4+ T-helper 1 responses. J. Virol. *75*, 8649-8659.

Levitskaya, J., Coram, M., Levitsky, V., Imreh, S., Steigerwald-Mullen, P. M., Klein, G., Kurilla, M. G., and Masucci, M. G. (1995). Inhibition of antigen processing by the internal repeat region of the Epstein-Barr virus nuclear antigen-1. Nature *375*, 685-688.

Levitskaya, J., Sharipo, A., Leonchiks, A., Ciechanover, A., and Masucci, M. G. (1997). Inhibition of ubiquitin/proteasome-dependent protein degradation by the Gly-Ala repeat domain of the Epstein-Barr virus nuclear antigen 1. Proc. Natl. Acad. Sci. USA. *94*, 12616-12621.

Li, L., Liu, D., Hutt-Fletcher, L., Morgan, A., Masucci, M. G., and Levitsky, V. (2002). Epstein-Barr virus inhibits the development of dendritic cells by promoting apoptosis of their monocyte precursors in the presence of granulocyte macrophage-colony-stimulating factor and interleukin-4. Blood *99*, 3725-3734.

Lin, C. L., Lo, W. F., Lee, T. H., Ren, Y., Hwang, S. L., Cheng, Y. F., Chen, C. L., Chang, Y. S., Lee, S. P., Rickinson, A. B., and Tam, P. K. (2002). Immunization with Epstein-Barr Virus (EBV) peptide-pulsed dendritic cells induces functional CD8+ T-cell immunity and may lead to tumor regression in patients with EBV-positive nasopharyngeal carcinoma. Cancer Res. *62*, 6952-6958.

Liu, Y., de Waal Malefyt, R., Briere, F., Parham, C., Bridon, J. M., Banchereau, J., Moore, K. W., and Xu, J. (1997). The EBV IL-10 homologue is a selective agonist with impaired binding to the IL-10 receptor. J. Immunol. *158*, 604-613.

Lotz, M., Tsoukas, C. D., Fong, S., Carson, D. A., and Vaughan, J. H. (1985). Regulation of Epstein-Barr virus infection by recombinant interferons. Selected sensitivity to interferon-gamma. Eur. J. Immunol. *15*, 520-525.

Lotz, M., Tsoukas, C. D., Fong, S., Dinarello, C. A., Carson, D. A., and Vaughan, J. H. (1986). Release of lymphokines after infection with Epstein Barr virus *in vitro*. II. A monocyte-dependent inhibitor of interleukin 1 downregulates the production of interleukin 2 and interferon-gamma in rheumatoid arthritis. J. Immunol. *136*, 3643-3648.

Macallan, D. C., Wallace, D. L., Irvine, A. J., Asquith, B., Worth, A., Ghattas, H., Zhang, Y., Griffin, G. E., Tough, D. F., and Beverley, P. C. (2003). Rapid turnover of T cells in acute infectious mononucleosis. Eur. J. Immunol. *33*, 2655-2665.

Macatonia, S. E., Hosken, N. A., Litton, M., Vieira, P., Hsieh, C. S., Culpepper, J. A., Wysocka, M., Trinchieri, G., Murphy, K. M., and O'Garra, A. (1995). Dendritic cells produce IL-12 and direct the development of Th1 cells from naive CD4+ T cells. J. Immunol. *154*, 5071-5079.

Maloy, K. J., Burkhart, C., Junt, T. M., Odermatt, B., Oxenius, A., Piali, L., Zinkernagel, R. M., and Hengartner, H. (2000). CD4+ T cell subsets during virus infection. Protective capacity depends on effector cytokine secretion and on migratory capability. J. Exp. Med. *191*, 2159-2170.

Marshall, N. A., Vickers, M. A., and Barker, R. N. (2003). Regulatory T cells secreting IL-10 dominate the immune response to EBV latent membrane protein 1. J. Immunol. *170*, 6183-6189.

Masucci, M. G., Bejarano, M. T., Masucci, G., and Klein, E. (1983). Large granular lymphocytes inhibit the *in vitro* growth of autologous Epstein-Barr virus-infected B cells. Cell Immunol. *76*, 311-321.

Mautner, J., Pich, D., Nimmerjahn, F., Milosevic, S., Adhikary, D., Christoph, H., Witter, K., Bornkamm, G. W., Hammerschmidt, W., and Behrends, U. (2004). Epstein-Barr virus nuclear antigen 1 evades direct immune recognition by CD4[+] T helper cells. Eur. J. Immunol. *34*, 2500-2509.

Meij, P., Leen, A., Rickinson, A. B., Verkoeijen, S., Vervoort, M. B., Bloemena, E., and Middeldorp, J. M. (2002). Identification and prevalence of CD8[+] T-cell responses directed against Epstein-Barr virus-encoded latent membrane protein 1 and latent membrane protein 2. Int. J. Cancer *99*, 93-99.

Metes, D., Storkus, W., Zeevi, A., Patterson, K., Logar, A., Rowe, D., Nalesnik, M. A., Fung, J. J., and Rao, A. S. (2000). *Ex vivo* generation of effective Epstein-Barr virus (EBV)-specific CD8[+] cytotoxic T lymphocytes from the peripheral blood of immunocompetent Epstein Barr virus-seronegative individuals. Transplantation *70*, 1507-1515.

Metes, D., Storkus, W. J., Zeevi, A., Watkins, S., Patterson, K., Nellis, J., Logar, A., Fung, J. J., and Rao, A. S. (2001). Use of autologous dendritic cells loaded with apoptotic LCL for *ex vivo* generation of specific CTL from the PBMC of EBV(-) individuals. Transplant. Proc. *33*, 441.

Misko, I. S., Pope, J. H., Hutter, R., Soszynski, T. D., and Kane, R. G. (1984). HLA-DR-antigen-associated restriction of EBV-specific cytotoxic T-cell colonies. Int. J. Cancer *33*, 239-243.

Misko, I. S., Sculley, T. B., Schmidt, C., Moss, D. J., Soszynski, T., and Burman, K. (1991). Composite response of naive T cells to stimulation with the autologous lymphoblastoid cell line is mediated by CD4 cytotoxic T cell clones and includes an Epstein-Barr virus-specific component. Cell Immunol. *132*, 295-307.

Moretta, A., Bottino, C., Vitale, M., Pende, D., Cantoni, C., Mingari, M. C., Biassoni, R., and Moretta, L. (2001). Activating receptors and coreceptors involved in human natural killer cell-mediated cytolysis. Annu. Rev. Immunol. *19*, 197-223.

Moretta, A., Comoli, P., Montagna, D., Gasparoni, A., Percivalle, E., Carena, I., Revello, M. G., Gerna, G., Mingrat, G., Locatelli, F., et al. (1997). High frequency of Epstein-Barr virus (EBV) lymphoblastoid cell line-reactive lymphocytes in cord blood: evaluation of cytolytic activity and IL-2 production. Clin. Exp. Immunol. *107*, 312-320.

Morgan, S. M., Wilkinson, G. W., Floettmann, E., Blake, N., and Rickinson, A. B. (1996). A recombinant adenovirus expressing an Epstein-Barr virus (EBV) target antigen can selectively reactivate rare components of EBV cytotoxic T-lymphocyte memory *in vitro*. J. Virol. *70*, 2394-2402.

Moss, D. J., Chan, S. H., Burrows, S. R., Chew, T. S., Kane, R. G., Staples, J. A., and Kunaratnam, N. (1983). Epstein-Barr virus specific T-cell response in nasopharyngeal carcinoma patients. Int. J. Cancer *32*, 301-305.

Münz, C. (2004). Epstein-Barr virus nuclear antigen 1: from immunologically invisible to a promising T cell target. J. Exp. Med. *199*, 1301-1304.

Münz, C., Bickham, K. L., Subklewe, M., Tsang, M. L., Chahroudi, A., Kurilla, M. G., Zhang, D., O'Donnell, M., and Steinman, R. M. (2000). Human CD4[+] T lymphocytes consistently respond to the latent Epstein-Barr virus nuclear antigen EBNA1. J. Exp. Med. *191*, 1649-1660.

Murphy, K. M., Ouyang, W., Farrar, J. D., Yang, J., Ranganath, S., Asnagli, H., Afkarian, M., and Murphy, T. L. (2000). Signaling and transcription in T helper development. Annu. Rev. Immunol. *18*, 451-494.

Murray, R. J., Kurilla, M. G., Brooks, J. M., Thomas, W. A., Rowe, M., Kieff, E., and Rickinson, A. B. (1992). Identification of target antigens for the human cytotoxic T cell response to Epstein-Barr virus (EBV): implications for the immune control of EBV-positive malignancies. J. Exp. Med. *176*, 157-168.

Nanbo, A., Inoue, K., Adachi-Takasawa, K., and Takada, K. (2002). Epstein-Barr virus RNA confers resistance to interferon-alpha-induced apoptosis in Burkitt's lymphoma. Embo J. *21*, 954-965.

Nikiforow, S., Bottomly, K., and Miller, G. (2001). CD4$^+$ T-cell effectors inhibit Epstein-Barr virus-induced B-cell proliferation. J. Virol. *75*, 3740-3752.

Nikiforow, S., Bottomly, K., Miller, G., and Münz, C. (2003). Cytolytic CD4$^+$-T-Cell Clones Reactive to EBNA1 Inhibit Epstein-Barr Virus-Induced B-Cell Proliferation. J. Virol. *77*, 12088-12104.

Nonkwelo, C., Ruf, I. K., and Sample, J. (1997). Interferon-independent and -induced regulation of Epstein-Barr virus EBNA-1 gene transcription in Burkitt lymphoma. J. Virol. *71*, 6887-6897.

Okano, M., and Purtilo, D. T. (1995). Simple assay for evaluation of Epstein-Barr virus specific cytotoxic T lymphocytes. J. Immunol. Methods *184*, 149-152.

Omiya, R., Buteau, C., Kobayashi, H., Paya, C. V., and Celis, E. (2002). Inhibition of EBV-induced lymphoproliferation by CD4$^+$ T cells specific for an MHC class II promiscuous epitope. J. Immunol. *169*, 2172-2179.

Paludan, C., Bickham, K., Nikiforow, S., Tsang, M. L., Goodman, K., Hanekom, W. A., Fonteneau, J. F., Stevanovic, S., and Münz, C. (2002). EBNA1 specific CD4$^+$ Th1 cells kill Burkitt's lymphoma cells. J. Immunol. *169*, 1593-1603.

Paludan, C., and Münz, C. (2003). CD4$^+$ T cell responses in the immune control against latent infection by Epstein-Barr virus. Curr. Mol. Med. *3*, 341-347.

Pamer, E., and Cresswell, P. (1998). Mechanisms of MHC class I-restricted antigen processing. Annu. Rev. Immunol. *16*, 323-358.

Parolini, S., Bottino, C., Falco, M., Augugliaro, R., Giliani, S., Franceschini, R., Ochs, H. D., Wolf, H., Bonnefoy, J. Y., Biassoni, R., et al. (2000). X-linked lymphoproliferative disease. 2B4 molecules displaying inhibitory rather than activating function are responsible for the inability of natural killer cells to kill Epstein-Barr virus-infected cells. J. Exp. Med. *192*, 337-346.

Pende, D., Rivera, P., Marcenaro, S., Chang, C. C., Biassoni, R., Conte, R., Kubin, M., Cosman, D., Ferrone, S., Moretta, L., and Moretta, A. (2002). Major histocompatibility complex class I-related chain a and UL16-binding protein expression on tumor cell lines of different histotypes: Analysis of tumor susceptibility to NKG2D-dependent natural killer cell cytotoxicity. Cancer Research *62*, 6178-6186.

Pepperl, S., Benninger-Doring, G., Modrow, S., Wolf, H., and Jilg, W. (1998). Immediate-early transactivator Rta of Epstein-Barr virus (EBV) shows multiple epitopes recognized by EBV-specific cytotoxic T lymphocytes. J. Virol. *72*, 8644-8649.

Piccioli, D., Sbrana, S., Melandri, E., and Valiante, N. M. (2002). Contact-dependent stimulation and inhibition of dendritic cells by natural killer cells. J. Exp. Med. *195*, 335-341.

Precopio, M. L., Sullivan, J. L., Willard, C., Somasundaran, M., and Luzuriaga, K. (2003). Differential kinetics and specificity of EBV-specific CD4$^+$ and CD8$^+$ T cells during primary infection. J. Immunol. *170*, 2590-2598.

Princiotta, M. F., Finzi, D., Qian, S. B., Gibbs, J., Schuchmann, S., Buttgereit, F., Bennink, J. R., and Yewdell, J. W. (2003). Quantitating protein synthesis, degradation, and endogenous antigen processing. Immunity *18*, 343-354.

Rajnavolgyi, E., Nagy, N., Thuresson, B., Dosztanyi, Z., Simon, A., Simon, I., Karr, R. W., Ernberg, I., Klein, E., and Falk, K. I. (2000). A repetitive sequence of Epstein-Barr virus nuclear antigen 6 comprises overlapping T cell epitopes which induce HLA-DR-restricted CD4$^+$ T lymphocytes. Int. Immunol. *12*, 281-293.

Rammensee, H., Bachmann, J., Emmerich, N. P., Bachor, O. A., and Stevanovic, S. (1999). SYFPEITHI: database for MHC ligands and peptide motifs. Immunogenetics *50*, 213-219.

Randolph, G. J., Sanchez-Schmitz, G., Liebman, R. M., and Schakel, K. (2002). The CD16$^+$ (FcgammaRIII$^+$) subset of human monocytes preferentially becomes migratory dendritic cells in a model tissue setting. J. Exp. Med. *196*, 517-527.

Reith, W., and Mach, B. (2001). The bare lymphocyte syndrome and the regulation of MHC expression. Annu. Rev. Immunol. *19*, 331-373.

Reits, E. A., Vos, J. C., Gromme, M., and Neefjes, J. (2000). The major substrates for TAP *in vivo* are derived from newly synthesized proteins. Nature *404*, 774-778.

Rentenaar, R. J., Gamadia, L. E., van DerHoek, N., van Diepen, F. N., Boom, R., Weel, J. F., Wertheim-van Dillen, P. M., van Lier, R. A., and ten Berge, I. J. (2000). Development of virus-specific CD4$^+$ T cells during primary cytomegalovirus infection. J. Clin. Invest *105*, 541-548.

Ressing, M. E., van Leeuwen, D., Verreck, F. A., Gomez, R., Heemskerk, B., Toebes, M., Mullen, M. M., Jardetzky, T. S., Longnecker, R., Schilham, M. W., et al. (2003). Interference with T cell receptor-HLA-DR interactions by Epstein-Barr virus gp42 results in reduced T helper cell recognition. Proc. Natl. Acad. Sci. USA. *100*, 11583-11588.

Rickinson, A. B., and Kieff, E. (2001). Epstein-Barr Virus. In Fields Virology, D. M. Knipe, and P. M. Howley, eds. (Philadelphia, Lippincott-Raven), pp. 2575-2627.

Rickinson, A. B., and Moss, D. J. (1997). Human cytotoxic T lymphocyte responses to Epstein-Barr virus infection. Annu. Rev. Immunol. *15*, 405-431.

Ridge, J. P., Di Rosa, F., and Matzinger, P. (1998). A conditioned dendritic cell can be a temporal bridge between a CD4$^+$ T- helper and a T-killer cell. Nature *393*, 474-478.

Riley, E. M. (1999). Is T-cell priming required for initiation of pathology in malaria infections? Immunol. Today *20*, 228-233.

Rooney, C. M., Aguilar, L. K., Huls, M. H., Brenner, M. K., and Heslop, H. E. (2001). Adoptive immunotherapy of EBV-associated malignancies with EBV-specific cytotoxic T-cell lines. Curr. Top Microbiol. Immunol. *258*, 221-229.

Rooney, C. M., Rowe, M., Wallace, L. E., and Rickinson, A. B. (1985). Epstein-Barr virus-positive Burkitt's lymphoma cells not recognized by virus-specific T-cell surveillance. Nature *317*, 629-631.

Rowe, M., Khanna, R., Jacob, C. A., Argaet, V., Kelly, A., Powis, S., Belich, M., Croom-Carter, D., Lee, S., Burrows, S. R., and et al. (1995). Restoration of endogenous antigen processing in Burkitt's lymphoma cells by Epstein-Barr virus latent membrane protein-1: coordinate up-regulation of peptide transporters and HLA-class I antigen expression. Eur. J. Immunol. *25*, 1374-1384.

Saiki, Y., Ohtani, H., Naito, Y., Miyazawa, M., and Nagura, H. (1996). Immunophenotypic characterization of Epstein-Barr virus-associated gastric carcinoma: massive infiltration by proliferating CD8$^+$ T-lymphocytes. Lab. Invest *75*, 67-76.

Sam, C. K., Prasad, U., and Pathmanathan, R. (1989). Serological markers in the diagnosis of histopathological types of nasopharyngeal carcinoma. Eur. J. Surg. Oncol. *15*, 357-360.

Sarris, A. H., Kliche, K. O., Pethambaram, P., Preti, A., Tucker, S., Jackow, C., Messina, O., Pugh, W., Hagemeister, F. B., McLaughlin, P., et al. (1999). Interleukin-10 levels are often elevated in serum of adults with Hodgkin's disease and are associated with inferior failure-free survival. Ann. Oncol. *10*, 433-440.

Saulquin, X., Bodinier, M., Peyrat, M. A., Hislop, A., Scotet, E., Lang, F., Bonneville, M., and Houssaint, E. (2001). Frequent recognition of BCRF1, a late lytic cycle protein of Epstein-Barr virus, in the HLA-B*2705 context: evidence for a TAP-independent processing. Eur. J. Immunol. *31*, 708-715.

Savoldo, B., Cubbage, M. L., Durett, A. G., Goss, J., Huls, M. H., Liu, Z., Teresita, L., Gee, A. P., Ling, P. D., Brenner, M. K., et al. (2002). Generation of EBV-specific CD4$^+$ cytotoxic T cells from virus naive individuals. J. Immunol. *168*, 909-918.

Schmidt, C., Burrows, S. R., Sculley, T. B., Moss, D. J., and Misko, I. S. (1991). Nonresponsiveness to an immunodominant Epstein-Barr virus-encoded cytotoxic T-lymphocyte epitope in nuclear antigen 3A: implications for vaccine strategies. Proc. Natl. Acad. Sci. USA. *88*, 9478-9482.

Schoenberger, S. P., Toes, R. E., van der Voort, E. I., Offringa, R., and Melief, C. J. (1998). T-cell help for cytotoxic T lymphocytes is mediated by CD40-CD40L interactions. Nature *393*, 480-483.

Schubert, U., Anton, L. C., Gibbs, J., Norbury, C. C., Yewdell, J. W., and Bennink, J. R. (2000). Rapid degradation of a large fraction of newly synthesized proteins by proteasomes. Nature *404*, 770-774.

Scotet, E., David-Ameline, J., Peyrat, M. A., Moreau-Aubry, A., Pinczon, D., Lim, A., Even, J., Semana, G., Berthelot, J. M., Breathnach, R., et al. (1996). T cell response to Epstein-Barr virus transactivators in chronic rheumatoid arthritis. J. Exp. Med. *184*, 1791-1800.

Shi, Y., and Lutz, C. T. (2002). Interferon--gamma control of EBV-transformed B cells: a role for CD8$^+$ T cells that poorly kill EBV-infected cells. Viral Immunol. *15*, 213-225.

Shi, Y., Smith, K. D., Kurilla, M. G., and Lutz, C. T. (1997). Cytotoxic CD8$^+$ T cells recognize EBV antigen but poorly kill autologous EBV-infected B lymphoblasts: immunodominance is elicited by a peptide epitope that is presented at low levels *in vitro*. J. Immunol. *159*, 1844-1852.

Silins, S. L., Sherritt, M. A., Silleri, J. M., Cross, S. M., Elliott, S. L., Bharadwaj, M., Le, T. T., Morrison, L. E., Khanna, R., Moss, D. J., et al. (2001). Asymptomatic primary Epstein-Barr virus infection occurs in the absence of blood T-cell repertoire perturbations despite high levels of systemic viral load. Blood *98*, 3739-3744.

Sixbey, J. W., and Yao, Q. Y. (1992). Immunoglobulin A-induced shift of Epstein-Barr virus tissue tropism. Science *255*, 1578-1580.

Sokal, E. M., Antunes, H., Beguin, C., Bodeus, M., Wallemacq, P., de Ville de Goyet, J., Reding, R., Janssen, M., Buts, J. P., and Otte, J. B. (1997). Early signs and risk factors for the increased incidence of Epstein- Barr virus-related posttransplant lymphoproliferative diseases in pediatric liver transplant recipients treated with tacrolimus. Transplantation *64*, 1438-1442.

Staege, M. S., Lee, S. P., Frisan, T., Mautner, J., Scholz, S., Pajic, A., Rickinson, A. B., Masucci, M. G., Polack, A., and Bornkamm, G. W. (2002). MYC overexpression imposes a nonimmunogenic phenotype on Epstein-Barr virus-infected B cells. Proc. Natl. Acad. Sci. USA. *99*, 4550-4555.

Steigerwald-Mullen, P., Kurilla, M. G., and Braciale, T. J. (2000). Type 2 cytokines predominate in the human CD4$^+$ T-lymphocyte response to Epstein-Barr virus nuclear antigen 1. J. Virol. *74*, 6748-6759.

Steven, N. M., Annels, N. E., Kumar, A., Leese, A. M., Kurilla, M. G., and Rickinson, A. B. (1997). Immediate early and early lytic cycle proteins are frequent targets of the Epstein-Barr virus-induced cytotoxic T cell response. J. Exp. Med. *185*, 1605-1617.

Steven, N. M., Leese, A. M., Annels, N. E., Lee, S. P., and Rickinson, A. B. (1996). Epitope focusing in the primary cytotoxic T cell response to Epstein- Barr virus and its relationship to T cell memory. J. Exp. Med. *184*, 1801-1813.

Strockbine, L. D., Cohen, J. I., Farrah, T., Lyman, S. D., Wagener, F., DuBose, R. F., Armitage, R. J., and Spriggs, M. K. (1998). The Epstein-Barr virus BARF1 gene encodes a novel, soluble colony-stimulating factor-1 receptor. J. Virol. *72*, 4015-4021.

Su, Z., Peluso, M. V., Raffegerst, S. H., Schendel, D. J., and Roskrow, M. A. (2001). The generation of LMP2a-specific cytotoxic T lymphocytes for the treatment of patients with Epstein-Barr virus-positive Hodgkin disease. Eur. J. Immunol. *31*, 947-958.

Su, Z., Peluso, M. V., Raffegerst, S. H., Schendel, D. J., and Roskrow, M. A. (2002). Antigen presenting cells transfected with LMP2a RNA induce CD4$^+$ LMP2a-specific cytotoxic T lymphocytes which kill via a Fas-independent mechanism. Leuk Lymphoma *43*, 1651-1662.

Subklewe, M., Paludan, C., Tsang, M. L., Mahnke, K., Steinman, R. M., and Münz, C. (2001). Dendritic cells cross-present latency gene products from Epstein-Barr virus-transformed B cells and expand tumor-reactive CD8+ killer T cells. J. Exp. Med. *193*, 405-411.

Sun, Q., Burton, R. L., and Lucas, K. G. (2002). Cytokine production and cytolytic mechanism of CD4+ cytotoxic T lymphocytes in *ex vivo* expanded therapeutic Epstein-Barr virus-specific T-cell cultures. Blood *99*, 3302-3309.

Tan, L. C., Gudgeon, N., Annels, N. E., Hansasuta, P., O'Callaghan, C. A., Rowland-Jones, S., McMichael, A. J., Rickinson, A. B., and Callan, M. F. (1999). A re-evaluation of the frequency of CD8+ T cells specific for EBV in healthy virus carriers. J. Immunol. *162*, 1827-1835.

Tellam, J., Connolly, G., Green, K. J., Miles, J. J., Moss, D. J., Burrows, S. R., and Khanna, R. (2004). Endogenous presentation of CD8+ T cell epitopes from Epstein-Barr virus nuclear antigen 1. J. Exp. Med. *199*, 1421-1431.

Thorley-Lawson, D. A. (2001). Epstein-Barr virus: exploiting the immune system. Nature Reviews Immunology *1*, 75-82.

Thorley-Lawson, D. A., Chess, L., and Strominger, J. L. (1977). Suppression of *in vitro* Epstein-Barr virus infection. A new role for adult human T lymphocytes. J. Exp. Med. *146*, 495-508.

Thorley-Lawson, D. A., and Geilinger, K. (1980). Monoclonal antibodies against the major glycoprotein (gp350/220) of Epstein-Barr virus neutralize infectivity. Proc. Natl. Acad. Sci. USA. *77*, 5307-5311.

Thorley-Lawson, D. A., and Poodry, C. A. (1982). Identification and isolation of the main component (gp350-gp220) of Epstein-Barr virus responsible for generating neutralizing antibodies *in vivo*. J. Virol. *43*, 730-736.

Urban, B. C., Ferguson, D. J., Pain, A., Willcox, N., Plebanski, M., Austyn, J. M., and Roberts, D. J. (1999). Plasmodium falciparum-infected erythrocytes modulate the maturation of dendritic cells. Nature *400*, 73-77.

Vitale, M., Della Chiesa, M., Carlomagno, S., Romagnani, C., Thiel, A., Moretta, L., and Moretta, A. (2004). The small subset of CD56^{bright}CD16- natural killer cells is selectively responsible for both cell proliferation and interferon-gamma production upon interaction with dendritic cells. Eur. J. Immunol. *34*, 1715-1722.

Viviani, S., Notti, P., Bonfante, V., Verderio, P., Valagussa, P., and Bonadonna, G. (2000). Elevated pretreatment serum levels of Il-10 are associated with a poor prognosis in Hodgkin's disease, the milan cancer institute experience. Med. Oncol *17*, 59-63.

von Wasielewski, R., Seth, S., Franklin, J., Fischer, R., Hubner, K., Hansmann, M. L., Diehl, V., and Georgii, A. (2000). Tissue eosinophilia correlates strongly with poor prognosis in nodular sclerosing Hodgkin's disease, allowing for known prognostic factors. Blood *95*, 1207-1213.

Voo, K. S., Fu, T., Heslop, H. E., Brenner, M. K., Rooney, C. M., and Wang, R. F. (2002). Identification of HLA-DP3-restricted peptides from EBNA1 recognized by CD4+ T cells. Cancer Res. *62*, 7195-7199.

Voo, K. S., Fu, T., Wang, H. Y., Tellam, J., Heslop, H. E., Brenner, M. K., Rooney, C. M., and Wang, R. F. (2004). Evidence for the Presentation of Major Histocompatibility Complex Class I-restricted Epstein-Barr Virus Nuclear Antigen 1 Peptides to CD8+ T Lymphocytes. J. Exp. Med. *199*, 459-470.

Wallace, L. E., Wright, J., Ulaeto, D. O., Morgan, A. J., and Rickinson, A. B. (1991). Identification of two T-cell epitopes on the candidate Epstein-Barr virus vaccine glycoprotein gp340 recognized by CD4+ T-cell clones. J. Virol. *65*, 3821-3828.

White, C. A., Cross, S. M., Kurilla, M. G., Kerr, B. M., Schmidt, C., Misko, I. S., Khanna, R., and Moss, D. J. (1996). Recruitment during infectious mononucleosis of CD3+CD4+CD8+ virus-specific cytotoxic T cells which recognise Epstein-Barr virus lytic antigen BHRF1. Virology *219*, 489-492.

Whittle, H. C., Brown, J., Marsh, K., Greenwood, B. M., Seidelin, P., Tighe, H., and Wedderburn, L. (1984). T-cell control of Epstein-Barr virus-infected B cells is lost during P. falciparum malaria. Nature *312*, 449-450.

Wilson, A. D., and Morgan, A. J. (2002). Primary immune responses by cord blood CD4$^+$ T cells and NK cells inhibit Epstein-Barr virus B-cell transformation *in vitro*. J. Virol. *76*, 5071-5081.

Wilson, A. D., Redchenko, I., Williams, N. A., and Morgan, A. J. (1998). CD4$^+$ T cells inhibit growth of Epstein-Barr virus-transformed B cells through CD95-CD95 ligand-mediated apoptosis. Int. Immunol. *10*, 1149-1157.

Wilson, A. D., Shooshstari, M., Finerty, S., Watkins, P., and Morgan, A. J. (1996). Virus-specific cytotoxic T cell responses are associated with immunity of the cottontop tamarin to Epstein-Barr virus (EBV). Clin. Exp. Immunol. *103*, 199-205.

Wolf, H., zur Hausen, H., and Becker, V. (1973). EB viral genomes in epithelial nasopharyngeal carcinoma cells. Nat. New Biol. *244*, 245-247.

Yin, Y., Manoury, B., and Fahraeus, R. (2003). Self-inhibition of synthesis and antigen presentation by Epstein-Barr virus-encoded EBNA1. Science *301*, 1371-1374.

Zajac, A. J., Blattman, J. N., Murali-Krishna, K., Sourdive, D. J., Suresh, M., Altman, J. D., and Ahmed, R. (1998). Viral immune evasion due to persistence of activated T cells without effector function. J. Exp. Med. *188*, 2205-2213.

Zeidler, R., Eissner, G., Meissner, P., Uebel, S., Tampe, R., Lazis, S., and Hammerschmidt, W. (1997). Downregulation of TAP1 in B lymphocytes by cellular and Epstein-Barr virus-encoded interleukin-10. Blood *90*, 2390-2397.

Zeng, Y. (1985). Seroepidemiological studies on nasopharyngeal carcinoma in China. Adv. Cancer Res. *44*, 121-138.

zur Hausen, H., Schulte-Holthausen, H., Klein, G., Henle, W., Henle, G., Clifford, P., and Santesson, L. (1970). EBV DNA in biopsies of Burkitt tumours and anaplastic carcinomas of the nasopharynx. Nature *228*, 1056-1058.

Zurawski, G., and de Vries, J. E. (1994). Interleukin 13, an interleukin 4-like cytokine that acts on monocytes and B cells, but not on T cells. Immunol. Today *15*, 19-26.

Chapter 14

EBV Induces an Endogenous Superantigen: Implications for Pathogenesis

*Natalie Sutkowski and Brigitte T. Huber**

ABSTRACT

Superantigens are microbial proteins that strongly stimulate T cells. A human endogenous retroviral superantigen is expressed after Epstein-Barr virus (EBV) infection. The superantigen gene is located on chromosome 1, and is encoded by the envelope (*env*) gene of human endogenous retrovirus HERV-K18. The EBV latent membrane proteins are sufficient for transactivation of HERV-K18 *env*. In this chapter, we review the literature on viral superantigens, and summarize the evidence leading to the discovery of the EBV associated superantigen. In light of these findings, we discuss a possible role for the superantigen associated T cell activation in EBV biology, and postulate how this strong immune activation might at times enhance EBV's ability to cause disease.

Abbreviations

BCR, B cell receptor; BL, Burkitt's lymphoma; CMV, cytomegalovirus; CTL, cytotoxic T cell; EBERs, EBV encoded RNA transcripts; EBV, Epstein-Barr virus; Env, envelope; HERV, human endogenous retrovirus; HSV, herpes simplex virus; HVS, herpesvirus saimiri; IDDM, insulin dependent diabetes mellitus; IDDMK, insulin dependent diabetes mellitus K virus; IDRP, insertion/deletion-related polymorphism; HIV, human immunodeficiency virus; Hu PBL, human peripheral blood leukocytes; IFN-α, interferon-α; IFN-γ, interferon- γ; IM, infectious mononucleosis; ITAM, immunoreceptor tyrosine-based activation motifs; JAK, Janus kinase; JRA, juvenile rheumatoid arthritis; LCL, lymphoblastoid cell line; LMP, latent membrane protein; LTR, long terminal repeat; MAPK, mitogen-activated protein kinase; MHC, major histocompatibility complex; MHV-68, murine gammaherpesvirus-68; MMTV, mouse mammary tumor virus; MS, multiple sclerosis; MSRV, multiple sclerosis associated retroviral virions; NFKB, nuclear factor kappa beta; NPC, nasopharyngeal carcinoma; Orf, open reading frame; PBMC, peripheral blood mononuclear cells; PKC, protein kinase C; PTLD, post-transplant lymphoproliferative disease; RA, rheumatoid arthritis; SAP, SLAM-associated protein; SLAM, signaling lymphocytic activation marker; SCID, severe combined immunodeficiency; SLE, systemic lupus erythematosus; STAT, signal transducer and activator of transcription; SU, subunit; TCR, T cell receptor; TCRBV, T cell receptor variable region beta chain; TM, transmembrane; TNF-α, tumor necrosis factor-α; TRAF, TNF receptor associated factor; XLP, X-linked lymphoproliferative disorder

*For correspondence email brigitte.huber@tufts.edu

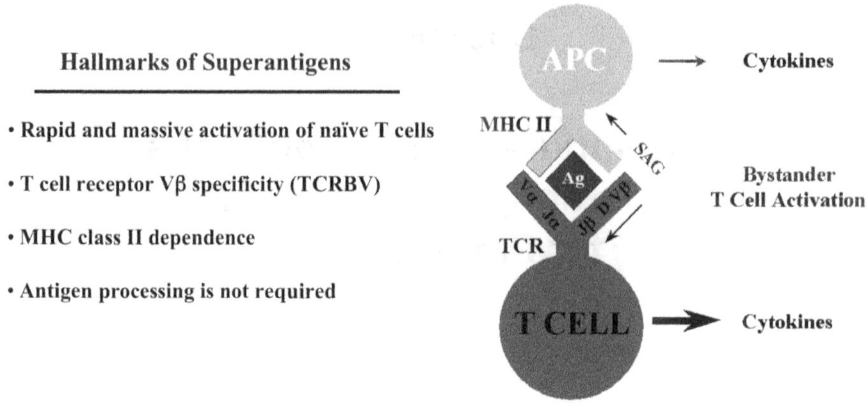

Hallmarks of Superantigens

• **Rapid and massive activation of naïve T cells**

• **T cell receptor Vβ specificity (TCRBV)**

• **MHC class II dependence**

• **Antigen processing is not required**

Figure 1. Superantigens cause massive T cell activation. Superantigens (SAG) bind in unprocessed form to the TCR, outside of the peptide-binding cleft used by conventional Ags. They bind solely to the TCRBV portion of the TCR, forming a bridge between MHC class II molecules and the TCR, which activates T cells to secrete cytokines. The hallmarks of a superantigen response are listed.

INTRODUCTION TO VIRAL SUPERANTIGENS
What are superantigens?

Superantigens are microbial pathogen-derived proteins that elicit a vigorous T cell response from the host (Kappler et al., 1989; 1988) and reviewed in Herman et al. (1991) and Leung et al. (1997). It is thought that the strong T cell proliferation, instead of limiting the pathogen, is actually in some way beneficial for the pathogen, because superantigens are evolutionarily conserved within a particular species. Superantigens differ from conventional antigens in their mode of binding to the T cell receptor (TCR, see Figure 1). Conventional antigens are processed by antigen presenting cells into peptides, and are presented to T cells in the context of major histocompatibility (MHC) class I or II proteins. T cells recognize these peptide/MHC complexes through the groove formed by the recombined TCR variable gene products, consisting of the beta chain V, D, and J segments and alpha chain V and J segments. Due to the diversity generated by this recombination process, relatively few naive T cells are capable of binding to a particular peptide antigen. In contrast, superantigens associate in unprocessed form with MHC class II molecules and are recognized by T cells through interaction with the TCR beta chain V gene product (TCRBV), forming a bridge between the antigen presenting cell and the T cell (Dellabona et al., 1990; Pullen et al., 1990). Since there are a limited number of TCRBV genes in the genome, *e.g.*, in the human genome there are 52 known functional TCRBV genes (Manavalan et al., 2004; Wei et al., 1994), this means that on average approximately 1/52 naive T cells, or roughly 2 %, would be capable of binding a particular superantigen. In actuality, the response is often even greater, because of the redundancy in the superantigen-binding site of TCRBV gene products. Superantigen-activated T cells proliferate, resulting in expansion and eventually in activation induced cell death (White et al., 1989). In addition, these T cells secrete cytokines that can further the response by stimulating bystander T cells (Mitchell et al., 1999; Zhang et al., 1996). Thus, the hallmarks of superantigens are strong and rapid proliferation of naive T cells, which is TCRBV restricted, MHC class II dependent, although not restricted to a particular MHC haplotype, and antigen processing into peptides is not required (Figure 1).

Microbial superantigens

There are two types of microbial superantigens: bacterial and viral (Kappler et al., 1989; 1988; Marrack et al., 1993; Scherer et al., 1993). The former are enterotoxins that are secreted and bind exogenously to MHC class II molecules for presentation to T cells (Choi et al., 1989; Kappler et al., 1989; Lavoie et al., 1999; Marrack et al., 1990; Marrack and Kappler, 1990; White et al., 1989). The latter consist of viral glycoproteins that are endogenously produced in infected antigen presenting cells (Beutner et al., 1992b; Choi et al., 1992; Hsu et al., 2001; Winslow et al., 1994; 1992). So far, two families of viruses have been shown to encode superantigens: retroviruses and rhabdoviruses.

Retroviral superantigens
MMTV

Superantigens were first described in the mouse mammary tumor virus (MMTV), a polymorphic B-type retrovirus that is either contained in the murine genome as an endogenous provirus or transmitted as infectious virus through the milk from mother to offspring (Acha-Orbea et al., 1991; Choi et al., 1991; Frankel et al., 1991; Marrack et al., 1991). An open reading frame (Orf) in the 3′ long terminal repeat (LT R) segment of MMTV encodes the superantigen (Beutner et al., 1992a; Choi et al., 1991). The endogenous *Mtv-7* superantigen, Mls-1, is considered the prototypic viral superantigen. It has been documented that the superantigen-mediated T cell activation is absolutely required for transmission of exogenous MMTV, which infects B cells associated with the gut of the newborn mouse (Beutner et al., 1994; 1996; Golovkina et al., 1992; Held et al., 1993). These infected B cells have to travel from the digestive tract to the mammary tissue, where epithelial cells are infected, resulting in viral production in the lactating mammary glands and transmission of infectious virus to the next generation of mice. In the absence of superantigen-mediated T cell activation, there is no expansion of the infected B cells, and consequently no infection of the mammary tissue; transmission of exogenous MMTV is halted. Expression of endogenous superantigens leads to thymic deletion of the responding T cells due to self-tolerance induction (Acha-Orbea et al., 1993; Huber, 1992; Kappler et al., 1988; Pullen et al., 1991; 1989; Scherer et al., 1995; Subramanyam et al., 1993). Thus, endogenous MMTVs shape the murine T cell repertoire by the deletion of TCRBV subsets that are specific for their superantigens.

HERV-K18

The first human retrovirus identified with superantigen activity was the human endogenous provirus, HERV-K18 (Conrad et al., 1997; Stauffer et al., 2001; Sutkowski et al., 2001). This virus was originally cloned by Conrad et al. and was called IDDMK$_{1,2}$22, for the insulin dependent diabetes mellitus associated K virus. These investigators showed that the envelope (*env*) gene of IDDMK$_{1,2}$22 encodes the superantigen (Conrad et al., 1997). Subsequently, two groups reported that IDDMK$_{1,2}$22, is an allelic variant of HERV-K18, located on chromosome one in the first intron of the CD48 gene (Hasuike et al., 1999; Tonjes et al., 1999). Two other alleles were identified in the Caucasian population, and the new terminology HERV-K18.1, -K18.2 and -K18.3 was coined (Stauffer et al., 2001). The HERV-K18 superantigen preferentially activates human TCRBV13 T cells (Sutkowski et al., 2001), and probably to a lesser extent TCRBV7 T cells (Conrad et al., 1997; Stauffer et al., 2001). The HERV-K18 provirus resides in a dormant state, but its *env* expression can be induced by the latent membrane proteins (LMPs) of the Epstein-Barr virus (EBV) and by the cytokine interferon-α (IFN-α) (Stauffer et al., 2001; Sutkowski et al., 2004) (see below). Since IFN-α is produced as a general defense mechanism in response to various viral infections, this endogenous superantigen may be expressed periodically after infections with viruses other than EBV.

HERV-W

Recently, a member of the HERV-W family was reported to have possible superantigen activity (Perron et al., 2001). Multiple sclerosis associated retroviral virions (MSRVs) are extracellular retroviral particles that are produced in cells derived from multiple sclerosis (MS) patients, but not from healthy individuals (Komurian-Pradel et al., 1999). Recombinant MSRVs were shown to preferentially stimulate human TCRBV16 T cells, but there was no demonstration that the response was MHC class II dependent; thus, it is as yet unclear whether these virion proteins are indeed superantigens (Perron et al., 2001). MSRV has homology to HERV-W Env protein (Blond et al., 1999), causing the investigators to speculate that the "superantigen-like" activity might be induced in neuronal cells in response to herpes simplex virus (HSV) infection. They have shown that the HERV-W Env protein is upregulated in neuroblastoma cells by HSV-1 (Lafon et al., 2002). When purified MSRVs were introduced into humanized SCID (severe combined immunodeficient) mice, the animals succumbed to lethal brain hemorrhage, which was aggravated by the presence of human T cells (Firouzi et al., 2003). Interestingly, HERV-W Env also causes syncytia formation; this viral protein is expressed in placenta during fetal development, and a role in placental morphogenesis is implicated (Mi et al., 2000).

Rhabdovirus superantigen

The N protein, a component of the rabies virus nucleocapsid, has been shown to be a superantigen that stimulates human TCRBV8 T cells (Lafon et al., 1992). When this purified nucleocapsid protein was injected into BALB/c mice, TCRBV6 T cells were preferentially expanded and then deleted (Lafon et al., 1994). The authors have speculated that some of the pathology associated with rabies infection might be due to the inflammatory superantigen response. In support of their theory, immune related paralysis as a result of rabies virus infection was decreased in BALB/c mice lacking TCRBV6 T cells, and upon adoptive transfer of that T cell subset, the paralysis was exacerbated (Lafon et al., 1994). While these studies are compelling, they have not been confirmed by any independent researchers, and no new information is available to date on the function of this superantigen.

Herpesvirus associated superantigens

Several members of the herpesvirus family are associated with superantigens or "superantigen-like" activities, although so far, none of the herpesviruses have been shown to encode a superantigen.

HVS

The herpesvirus saimiri (HVS) Orf14 protein (strain 295C) dramatically stimulates T cells in a polyclonal fashion, but this T cell activation is not TCRBV restricted, the primary hallmark of superantigens (Yao et al., 1996). The activity is, therefore, termed "superantigen-like". Orf14 is a highly glycosylated secreted protein with > 40 % sequence identity to the MMTV superantigen in restricted portions of the molecule. Recombinant Orf14 - Fc fusion protein strongly binds MHC class II molecules in an unrestricted fashion, signifying that the T cell activation is MHC class II dependent (Yao et al., 1996). Interestingly, deletion of Orf14 from saimiri results in a virus that is defective in its ability to transform T cells *in vitro* and *in vivo* in common marmosets, providing suggestive evidence that the T cell stimulatory activity is important for saimiri infection and/or oncogenesis (Duboise et al., 1998).

CMV

Human cytomegalovirus (CMV) was associated with a TCRBV12-specific superantigen activity (Dobrescu et al., 1995b). CMV infection of the MHC class II$^+$ monocyte cell line, U937, results in expression of this T cell stimulatory molecule. The activity could not be recapitulated by transfection of the CMV immediate early genes, *ie1* and *ie2,* into the same cell line (Dobrescu et al., 1995b). According to the authors, the TCRBV12 T cell subset becomes a reservoir for HIV infection in patients that are co-infected with both viruses, since HIV$^+$ TCRBV12 CD4 T cells were expanded in these subjects (Dobrescu et al., 1995a; 1995b; Posnett et al., 1995). While the CMV associated T cell stimulatory activity has all of the hallmarks of a superantigen, no CMV gene encoding the activity has been identified so far, leading to the postulate that CMV might be transactivating a human endogenous superantigen yet to be discovered, as is the case with EBV.

EBV

The ubiquitous herpesvirus EBV was first reported to be associated with superantigen activity in 1996, consisting of the preferential activation of TCRBV13 T cells (Sutkowski et al., 1996). Later, it was shown that the superantigen activity was not encoded by EBV itself, but instead was due to EBV's ability to transactivate HERV-K18, whose *env* gene encodes the superantigen (Sutkowski et al., 2001). This constituted the first report that a pathogen can induce a host-encoded superantigen (see below).

MHV-68

A "superantigen-like" activity is also associated with MHV-68 (Doherty et al., 1997; Tripp et al., 1997), a γ-herpesvirus alleged to be the murine counterpart of EBV, because it causes an "infectious mononucleosis (IM)-like" disease similar to EBV (Doherty et al., 1997; Tripp et al., 1997). Intranasal exposure to this virus induces an acute phase lytic type response in the lungs within the first 2 weeks of infection, which is followed by splenomegaly that coincides with latent infection of splenic B cells. The infection then resolves slowly over several weeks to months. A vigorous TCRBV4 T cell expansion, mainly in the CD8 T cell compartment, is seen > 3 weeks post-infection and is thought to arise as an offshoot of viral latency (Hardy et al., 2000; Stevenson et al., 1999; Tripp et al., 1997). The T cell expansion does not occur in MHC class II$^{-/-}$ mice, an indication that the T cell activation is MHC class II dependent (Brooks et al., 1999). However, because TCRBV4$^+$CD8$^+$ T cell hybridomas, derived from MHV-68 infected mice, were stimulated by virus infected spleen cells in the absence of class I or class II molecules, the classical superantigen presenting pathway does not seem to be involved (Coppola et al., 1999). No gene encoding the T cell stimulatory activity has been identified in MHV-68 so far, and it is possible that, like EBV's induction of HERV-K18 (see below), MHV-68 transactivates an as yet unidentified murine endogenous superantigen.

THE HERV-K18 SUPERANTIGEN

HERV sequences comprise up to 8 % of the human genome, with proviral copies integrated throughout the genome (Lander et al., 2001). The HERVs contain *gag, pol* and *env* genes that are similar to exogenous retroviruses, although most are defective, rendering the proviruses replication incompetent (Barbulescu et al., 1999; Ono, 1986; Tonjes et al., 1999; Turner et al., 2001). They are derived from infectious retroviruses that over various time points in evolutionary history integrated into germ cells. Once a provirus enters the germline, it is passed on from generation to generation in a Mendelian fashion (reviewed in (Boeke and Stoye, 1997; Urnovitz and Murphy, 1996)).

The HERV-K family

Retroviruses of the HERV-K family are defined by their use of the amino acid lysine (K) as primer-binding site during reverse transcription. There are estimated to be between 50 - 170 HERV-K proviruses, which are part of the C-type retrovirus family (Barbulescu et al., 1999; Ono, 1986; Ono et al., 1986; Tristem, 2000); however, they are closely related to the B-type retroviruses, such as MMTVs, based on amino acid similarity in the reverse transcriptase gene (Ono et al., 1986). This is particularly interesting, because MMTV represents the first group of viruses discovered to encode superantigens (Acha-Orbea et al., 1992; Beutner et al., 1992a; Frankel et al., 1991; Marrack et al., 1991). HERV-K18 is a relatively recent integrant in the genome, as it is found in Old World primates, but not New World primates, indicating that it was acquired subsequent to the evolutionary divergence of these species (Hughes and Coffin, 2001). Thus, HERV-K18 is not present in the marmoset genome, providing an explanation why the B95-8 EBV producer marmoset cell line does not stimulate T cells (see below).

HERV-K18 Env, the EBV associated superantigen
EBV associated superantigen activity
It had been postulated for years that EBV infection is associated with a superantigen, because EBV is the causative agent of IM, a disease characterized by massive lymphoproliferation of both B and T cells, which resolves spontaneously, *i.e.*, it

is self-limiting. IM has many of the predicted features of a superantigen-mediated illness, including the atypical lymphocytosis (Wood and Frenkel, 1967). Thus, several investigators analyzed the TCRBV repertoire in the peripheral blood (PBMC) of IM patients; however, inconsistent results were obtained. Instead, a superantigen activity associated with EBV was first identified with the invention of an *in vitro* model of acute infection (Sutkowski et al., 1996). The idea was to replicate the initial events that occur when T cells come in contact with EBV infected B cells. To do this, peripheral blood B cells from healthy adult volunteers were transformed with a common laboratory strain of EBV, B95-8, and with an IM-derived virus, IM-1. The resulting B lymphoblastoid cell lines (LCL) served as superantigen presenting cells in a classical proliferation assay, using autologous T cells as responders. After 48 h, a vigorous T cell proliferation, equivalent in magnitude and kinetics to a mitogen response, was observed. To demonstrate the EBV-dependence of this T cell stimulatory activity, the experiments were repeated, using EBV⁻ Burkitt's lymphoma (BL) and their counterpart, EBV⁺ BL (Calender et al., 1987), as antigen presenting cells. Only the latter induced strong T cell activation. Furthermore, this T cell proliferation was MHC class II dependent, because it could be blocked with antibodies specific for HLA.DR. The response of cord blood T cells was identical to that of PBMC from adult donors, indicating that this stimulation is not due to an anamnestic EBV response, since cord blood is by definition EBV⁻. The main characteristic of a superantigen is the TCRBV restriction. The T cell stimulation in this model of acute EBV infection was shown to be TCRBV13-specific, because the early activation antigen, CD69, was preferentially upregulated on TCRBV13⁺ peripheral blood T cells, but not on T cells bearing other TCRBVs. This TCRBV13 restriction was confirmed by the preferential activation of TCRBV13⁺ T cell hybridomas, but not other T hybridomas (Sutkowski et al., 1996). Thus, all of the characteristics of a superantigen were identified in association with EBV infection: rapid and massive proliferation of primary T cells, MHC class II dependence and TCRBV13 restriction.

EBV transactivation of HERV-K18 env

Despite intense investigations that ensued, no EBV gene could be identified that encoded the superantigen activity. The lack of T cell stimulation associated with the B95-8 producer marmoset B cell line (Miller and Lipman, 1973) provided the first confounding evidence that EBV did not encode the superantigen (Sutkowski et al., 2001). This was a particularly surprising finding, since B95-8 EBV transformed human B cells and B95-8 infected BL cells caused strong T cell activation (Sutkowski et al., 2001; 1996). The major difference between these cells is that the former is not of human origin, leading to the postulate that EBV is transactivating a human endogenous superantigen. This hypothesis gained further support in 1999, when two different laboratories localized the human endogenous retrovirus, HERV-K18, to chromosome 1 (Hasuike et al., 1999; Tonjes et al., 1999), where it was found to reside in the first intron of *CD48*, in opposite orientation to *CD48* (Hasuike et al., 1999) (see Figure 2). This observation was notable for two reasons: 1) HERV-K18 was shown to be an allelic variant of IDDMK$_{1,2}$22, the first HERV identified with superantigen activity (Conrad et al., 1997). 2) *CD48* is one of the

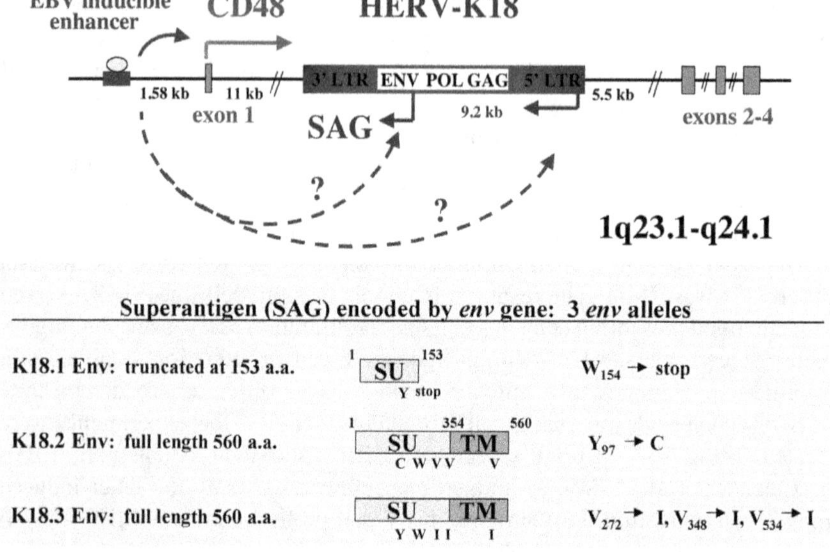

Figure 2. HERV-K18 resides in the first intron of *CD48*, an EBV inducible gene. HERV-K18 is an endogenous provirus located on chromosome 1q23.1-q24.1 in the first intron of *CD48*. There is an EBV inducible enhancer in the promoter region of *CD48,* possibly responsible for EBV's transactivation of HERV-K18 Env. The three known alleles of HERV-K18 *env* encode retroviral Env proteins that differ significantly. The HERV-K18.1 protein is a truncated envelope subunit (SU), while K18.2 and K18.3 proteins have complete SU and transmembrane (TM) domains that have conservative amino acid changes at 4 residues.

first cellular genes that are upregulated in B cells upon EBV infection (Staunton and Thorley-Lawson, 1987; Thorley-Lawson et al., 1993; 1982; Yokoyama et al., 1991). An EBV inducible enhancer had been mapped approximately 1.6 kb upstream of the *CD48* start site (Klaman and Thorley-Lawson, 1995). Taken together, these findings pointed to the possibility that EBV might be capable of enhancing transcription of a HERV with superantigen activity. Transactivation by EBV was not only plausible, but was readily detected by RNase protection assay for HERV-K18 *env* transcripts in EBV transformed LCL and EBV⁺, but not EBV⁻, BL (Sutkowski et al., 2001). The HERV-K18 Env protein was indeed stimulatory for T cells, since mouse B cells transfected with the *env* gene stimulated human peripheral blood T cells to proliferate, with magnitude and kinetics similar to a mitogen response. Furthermore, the *env* transfectants preferentially activated TCRBV13 T cell hybridomas, but not other T cell hybrids, and the activation was inhibited with antibodies specific for mouse class II molecules, indicating that it was MHC class II dependent. Interestingly, this activity required pretreatment of the *env*-transfected cells with phorbol ester, a requirement not seen with other viral superantigens, such as Mls-1. The ultimate proof of EBV's transactivation of a host-encoded superantigen was provided by the fact that the previously identified EBV-associated superantigen activity could be blocked by antiserum raised against HERV-K18 Env (Sutkowski et al., 2001).

EBV LMPs mediate transactivation

In an attempt to map the EBV gene(s) responsible for inducing HERV-K18 expression, various EBV deletion mutants were tested for superantigen activity (Sutkowski et al., 2004). Viral constructs deleted for the majority of lytic genes (Kempkes et al., 1995; Robertson and Kieff, 1995) were stimulatory for T cells, while the well-characterized P3HR1 virus (Hammerschmidt and Sugden, 1989), which lacks EBNA2, was non-stimulatory. Since EBNA2 is a major transactivator of the latent genes, this cell line is latent gene defective. The transcriptional activation of HERV-K18 *env* RNA was also shown to be EBNA2 dependent, confirming the functional data (Sutkowski et al., 2004). These studies led to the conclusion that a latent EBV gene product was responsible for the transactivation of HERV-K18. Infection of an EBV⁻ BL with vaccinia vectors containing the various EBV latent genes demonstrated that LMP-2A transactivates HERV-K18. Confirmation of LMP-2A's role in transactivation was provided with adenoviral vectors containing LMP-2A. In this manner it was shown that LMP-2A was sufficient for transactivation of the HERV-K18 superantigen, and LMP-1 could transactivate to a lesser extent, while both induced functional superantigen activity (Sutkowski et al., 2004).

RNase protection assays documented that, aside from EBV, the type 1 interferon, IFN-α, also transactivates the HERV-K18 superantigen (Stauffer et al., 2001). IFN-α signaling occurs mainly through the JAK/STAT (Janus kinase/ signal transducer and activator of transcription) pathway; in particular, STAT1 and STAT2 are generally activated, but activation of other STATs also occurs. The EBV latent membrane proteins, LMP-1 and LMP-2A, share some signaling elements with IFN-α, but the three signal transduction pathways are very different overall (see Figure 3). LMP-2A, a functional mimic of the B cell receptor (BCR) (Caldwell et al., 1998; Merchant et al., 2001; Portis et al., 2002), is a transmembrane protein with intracellular domains containing immunoreceptor tyrosine-based activation motifs (ITAMs) that play a role in B cell survival (Merchant et al., 2000). In EBV-infected B cells, the tyrosine kinases Syk and Lyn bind to the phosphorylated ITAMs on LMP-2A through their SH2 domains. In this manner, they are sequestered away from the BCR, resulting in phospholipase C-γ2 activation (Caldwell et al., 2000; Caldwell et al., 1998; Longnecker and Miller, 1996; Merchant et al., 2000; Miller et al., 1994) and constitutive activation of the PKC pathway. It has been reported that IFN-α binding to its receptor activates the JAK Tyk-2, signaling Lyn to bind to the phosphorylated ITAM on Tyk-2 through its SH2 domain (Uddin et al., 1998). Thus, Lyn appears in the IFN-α receptor pathway, as well as the LMP-2A and BCR signaling pathways, suggesting a possible link. Furthermore, the type 1 IFNs are reported to enhance B cell responses to BCR ligation, by lowering the threshold for stimulation (Braun et al., 2002). Moreover, like LMP-2A, type 1 IFNs reportedly increase B cell survival (Braun et al., 2002).

LMP-1, on the other hand, is a functional mimic of CD40 and binds multiple TNF receptor associated factor (TRAF) family members. LMP-1 is known to activate the mitogen-activated protein kinase (MAPK) and nuclear factor kappa B (NFKB) pathways (reviewed in (Eliopoulos and Young, 2001)). Interestingly, there is an NFKB site in the designated EBV-inducible enhancer sequence upstream of the *CD48* promoter on chromosome 1 (Klaman and Thorley-Lawson, 1995), within easy distance of HERV-K18, which resides in the first *CD48* intron (Hasuike

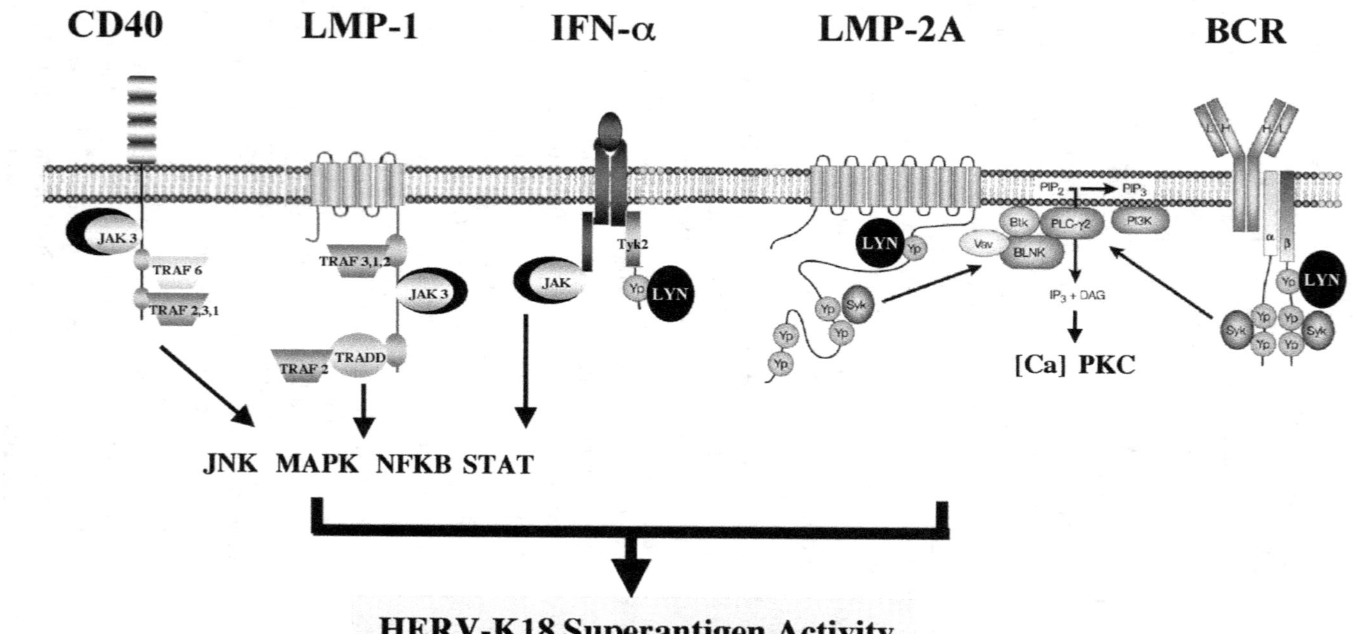

Figure 3. Activation pathways of LMP1, LMP2A and IFN-α. The EBV latent membrane proteins and the type 1 interferon, IFN-α, are each capable of transactivating HERV-K18 *env*. They share particular adaptor molecules, and have some downstream transcriptional activators in common. The EBV membrane proteins, LMP-1 and LMP-2A, are functional mimics of important B cell signaling molecules, CD40 and the BCR, respectively. Since the membrane proteins activate many of the same pathways as their cellular homologs, we are currently investigating whether crosslinking the BCR or CD40 ligation results in HERV-K18 superantigen activity.

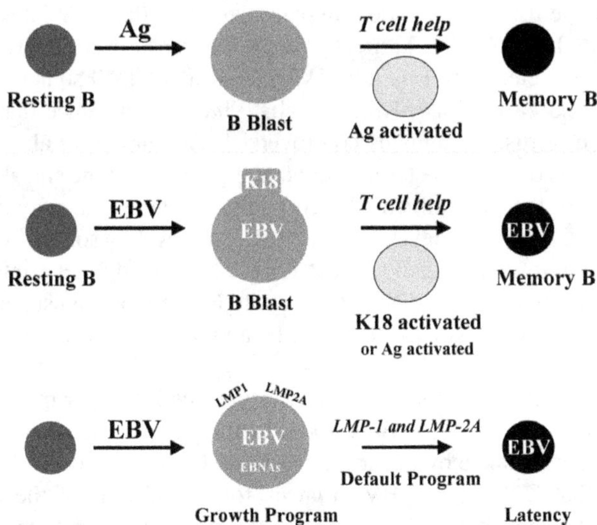

Figure 4. Potential role for the HERV-K18 superantigen in EBV biology. A leading theory of EBV infection is that EBV mimics the antigen activated B cell differentiation pathway in order to gain access to the memory cell compartment, the site of long-term viral persistence (Thorley-Lawson and Gross, 2004). Like antigen, the EBV latent membrane proteins cause the resting B cell to blast. For antigen activated B cells to differentiate into memory cells, T cell help from antigen activated T cells is required. For EBV[+] B cell blasts expressing the growth program to differentiate into memory cells, HERV-K18 superantigen-activated T cell might efficiently supply the help. In addition, antigen activated T cells responding to exogenous antigens, and fueled by superantigen-mediated cytokine production, could also provide help. Alternatively, it has been suggested that the EBV latent membrane proteins alone (default program) might be capable of inducing B cell differentiation, in the absence of T cells. It is likely that EBV employs a combination, or even all, of these mechanisms at times for gaining access to the memory compartment, where viral gene expression is down-regulated, and the virus can persist masked from the immune system.

et al., 1999). It is possible that the *CD48* enhancer boosts transcription of HERV-K18 (Sutkowski et al., 2001) through this NFKB site. Moreover, recent evidence suggests that, like IFN-α, LMP-1 also signals through the JAK - STAT pathway, phosphorylating both STAT3 and STAT5 in different cell types (Chen et al., 2003; 2001).

Possible role of the HERV-K18 superantigen in EBV biology

It has been postulated that EBV mimics the antigen activated B cell differentiation pathway in order to gain access to long-term viral latency in memory B cells (Babcock et al., 1998; Thorley-Lawson, 2001; Thorley-Lawson and Gross, 2004) (see Figure 4). EBV infects resting B cells, and like antigen, the EBV latent proteins cause the B cells to blast. Banchereau and his colleagues have provided convincing evidence that T cell help, in the form of cellular interaction through CD40/ CD40L signaling, (Liu and Banchereau, 1997), is required for antigen activated B cells to differentiate into long-term memory B cells. Thus, we postulate that HERV-K18 Env activated T cells provide the requisite signals for EBV[+] lymphoblasts to differentiate into memory cells, the site of long-term EBV persistence (Babcock et al., 1999). While it has been widely hypothesized that the EBV latent membrane proteins, LMP-1 and LMP-2A, are sufficient for transmitting the differentiation signals to the infected B cell by mimicking BCR and CD40 signals (Caldwell et al., 1998; Casola et al., 2004; Kilger et al., 1998), the evidence that this occurs in the absence of disease remains controversial (Brauninger et al., 2001). It has also been proposed that conventional antigen might provide the signal required for differentiation of the EBV infected B cell into the long-term latency compartment (Babcock et al., 1998; Thorley-Lawson, 2001; Thorley-Lawson and Gross, 2004). This would require that antigen specific for the BCR of the EBV infected B cell is coincidentally present at the site of infection. Expression of a superantigen that activates T cells to express CD40L would be a more efficient means of achieving the same end.

Once EBV gains access to the memory B cell compartment, viral gene expression is down-regulated (Babcock et al., 2000), and EBV remains hidden from immunosurveillance by cytotoxic T cells (CTL) of the host. It has been well documented that CTLs keep EBV in check for the lifetime of the healthy host (reviewed in (Khanna and Burrows, 2000)). We hypothesize that the superantigen-activated CD4$^+$ T cells, favoring viral persistence, provide the counterpart to the CTLs. Furthermore, since CD40 - CD40L signaling causes viral reactivation by induction of BZLF1, the major transactivator of the lytic cycle (Fu and Cannon, 2000), superantigen-activated T cells might play a role in viral reactivation. It is possible that during times of immunosuppression, when EBV specific CTLs are compromised, the superantigen activated T cells could tilt the balance towards disease, as will be discussed in detail in the next section.

THE HERV-K18 SUPERANTIGEN AND DISEASE
Potential role of superantigen activated T cells in EBV associated diseases

It has been speculated for years that HERVs are associated with pathogenesis, and studies have implicated the endogenous retroviruses in various diseases, including, but not limited to, IDDM (Conrad et al., 1994; 1997), systemic lupus erythematosus (SLE) (Adelman and Marchalonis, 2002; Muir et al., 1999), schizophrenia (Karlsson et al., 2001; Nakamura et al., 2003), MS (Perron et al., 2001) and cancer (Goedert et al., 1999; Ono et al., 1987; Schulte et al., 1996; Wang-Johanning et al., 2003). To date, these studies remain mainly descriptive, providing a possible link with disease.

Several HERV families have conserved gene products that have functional effects in humans. Most notably, the fusogenic HERV-W *syncytin* gene product was proposed to have a role in trophoblast differentiation (Mi et al., 2000; Stoye and Coffin, 2000). The placenta specific transcription factor, GcMA, appears to regulate expression of HERV-W (Yu et al., 2002), leading to the hypothesis that aberrant expression of *syncytin* may have a role in preeclampsia (Knerr et al., 2002; Kudo et al., 2003; Lee et al., 2001). Interestingly, a recent report suggests that this HERV is functionally and positionally conserved in the human genome, as well as in orthologous loci in the genome of chimpanzees, gorillas, orangutans and gibbons, indicating a conserved role throughout recent evolution (Blaise et al., 2003).

HERV-K18 superantigen, IM and XLP

IM, a frequent outcome of acute EBV infection in adolescents and young adults, is characterized by an atypical lymphocytosis (Wood and Frenkel, 1967). The clinical manifestations of IM are suggestive of superantigen stimulation, providing the original rationale for looking for an EBV associated superantigen. EBV specific cytotoxic T cells (CTL) are activated during IM in high numbers and seem to persist for years thereafter (Callan et al., 1998). Large clonal expansions are seen in the $CD8^+$ T cell compartment, consisting mainly of class I restricted CTL that are specific for the EBV lytic proteins (Callan, 2003; Callan et al., 1998; Tomkinson et al., 1989). While $CD4^+$ T cells are also activated during IM (Tomkinson et al., 1987), clonal expansion of the $CD4^+$ cells is undetectable; instead, the $CD4^+$ compartment remains polyclonal (Maini et al., 2000), as is seen with superantigen activation. The strong CTL response differs significantly from a superantigen response in that it is peptide specific and MHC class I restricted. Superantigen responses are instead MHC class II dependent and unrestricted. We propose that the superantigen-activated CD4 T cells in IM enhance the viral-specific CTL response. A similar process is known to occur with bacterial superantigens, which can directly activate influenza virus specific $CD8^+$ CTL *in vivo,* resulting in vigorous proliferation of the CTL, even after elimination of the virus (Coppola and Blackman, 1997). It has also been shown in mice that, conversely, vaccinia virus infection prolongs the T cell activation induced by bacterial superantigen (Mitchell et al., 1999). Furthermore, lymphocytic choriomeningitis virus infection primes T cells such that later bacterial superantigen stimulation results in lethal cytokine production (Sarawar et al., 1994). A synergistic lethal effect in response to bacterial superantigen was also seen after influenza virus infection (Zhang et al., 1996).

In X-linked lymphoproliferative disease (XLP), a genetic disorder resulting from mutation of the *SAP* (signaling lymphocytic activation molecule (SLAM)-associated protein) gene (Sayos et al., 1998), EBV infection is associated with fulminant IM that is often fatal. SAP, a small cytosolic adaptor protein with a single SH2 domain, is mainly expressed in T cells and NK cells (Sayos et al., 2000). It binds to the phosphorylated ITAM motifs of SLAM and other SLAM family members (Poy et al., 1999) in activated lymphocytes, preventing binding of other SH2-containing molecules, such as the SHP proteases. Loss of SAP results in aberrant lymphocyte activation (reviewed in Engel et al., 2003). It is unknown why EBV infection is so devastating to XLP patients, whose response to other viral infections is generally normal. However, it is tempting to speculate that the abnormal lymphoproliferative response is aggravated by HERV-K18 superantigen expression in the EBV infected cells of these patients.

HERV-K18 superantigen and EBV malignancies

A role for T cells in EBV associated cancers has been inferred from the presence of large T cell infiltrates that are frequently associated with various malignancies, such as Hodgkin's lymphoma, post-transplant lymphoproliferative disorder (PTLD) and nasopharyngeal carcinoma (NPC). Interestingly, the EBV latent proteins, LMP-2A and/or LMP-1, are expressed in these diseases (Delecluse et al., 1995; Herbst et al., 1991; Liebowitz, 1998; Niedobitek et al., 1997; Rea et al., 1994), suggesting the presence of the HERV-K18 superantigen.

Hodgkin's lymphoma

LMP-2A has been implicated in the pathogenesis of Hodgkin's lymphoma (Kuppers et al., 2002), because the characteristic Reed-Sternberg cells seen in this disease have a molecular phenotype resembling that of LMP-2A expressing B cells, in which B cell specific gene expression is down-regulated (Hertel et al., 2002; Portis et al., 2003; Portis and Longnecker, 2003; Schwering et al., 2003). It is tempting to speculate that the infiltrating T cells found within Hodgkin's lymphomas are activated in response to the HERV-K18 superantigen that is induced by LMP-2A in the tumor cells. It has been suggested that T cell derived cytokines and chemokines alter the microenvironment of the Hodgkin's Reed Sternberg cells, such that the tumor cells escape CTL detection (Beck et al., 2001).

NPC

EBV$^+$ NPCs commonly express LMP-2A (Busson et al., 1992; Heussinger et al., 2004; Lennette et al., 1995), frequently in association with LMP-1 (Brooks et al., 1992). Likewise, these epithelial cell tumors are often infiltrated with T cells (Lee, 2002; Thomas et al., 1984), which secrete cytokines and chemokines that are predicted to affect metastasis of the carcinoma (Agathanggelou et al., 1995; Niedobitek, 2000; Niedobitek et al., 1996). HERV-K18 *env* transcripts are present in all primary NPC samples tested so far (Sutkowski and Huber, unpublished observation). Hence, it is possible that HERV-K18 superantigen activated T cells influence the associated lymphoid stroma and play a role in metastasis of EBV$^+$ NPC.

PTLD

In an animal disease model resembling human PTLD, the Hu PBL SCID mouse, it is well documented that the development of EBV$^+$ tumors is T cell dependent (Amadori et al., 1992; Johannessen et al., 2000; Johannessen and Crawford, 1999; Veronese et al., 1992). Mosier and his co-workers discovered that injection of human PBMC from EBV seropositive healthy donors into SCID mice results in a large proportion of animals spontaneously succumbing to EBV$^+$ B cell lymphomas within 2 months (Mosier et al., 1988). However, when T cells were depleted from the PBMC inoculate, no tumors developed (Amadori et al., 1992; Johannessen et al., 2000; Veronese et al., 1992). It was subsequently shown that direct T cell - B cell interaction is required, because antibodies that inhibit CD40/CD40L interaction blocked lymphomagenesis (Murphy et al., 1995). Furthermore, cytokine production does not seem to be responsible for the enhanced B cell transformation, since injection of various T cell derived cytokines into Hu PBL SCID mice lacking T cells did not enhance tumor formation, while conversely, in the presence of T cells, injection of various anti-cytokine antibodies did not prevent lymphomas from developing (Veronesi et al., 1994). We have recently shown that the HERV-K18 superantigen is functionally expressed in Hu PBL SCID mice (Sutkowski and Huber, manuscript submitted), suggesting a possible role for superantigen activated T cells in the process of lymphoma development. *In vitro* studies have demonstrated that T cell signaling through CD40L induces viral reactivation through BZLF1, the major transactivator of the lytic cycle (Fu and Cannon, 2000). Increased viral reactivation might contribute to the enhanced lymphoma development in the mice.

Furthermore, we have observed that the EBV infected B cells do not survive in the SCID mouse in the absence of T cells, clearly demonstrating a direct role for T cells in the maintenance of the infected cells. Thus, these results suggest that superantigen activated T cells enhance EBV oncogenesis in two ways: 1) by increasing viral reactivation, and 2) by aiding survival of infected B cells.

HERV-K18 superantigen and autoimmunity

It is a long-standing hypothesis that the immune response to microbial pathogens can lead to the onset of autoimmunity by the activation of T cell clones that react with self (Fraser et al., 2000; Kotzin et al., 1993; Macphail, 1999; Portis, 2002; Torres and Johnson, 1998). The observation that IFN-α induces HERV-K18 expression lends a mechanism for how infection with various microbes might result in activation of autoimmune T cells, *e.g.*, through their elicitation of IFN-α in response to infection (Stauffer et al., 2001). EBV not only elicits IFN-α, but also directly transactivates the HERV-K18 superantigen (Sutkowski et al., 2001). For years, researchers have looked for links between EBV and various autoimmune phenomena (Balandraud et al., 2003; Ferrell et al., 1981; Huggins et al., 2003; James et al., 2001; Kang et al., 2004; Koide et al., 1997; Niedobitek et al., 2000; Saal

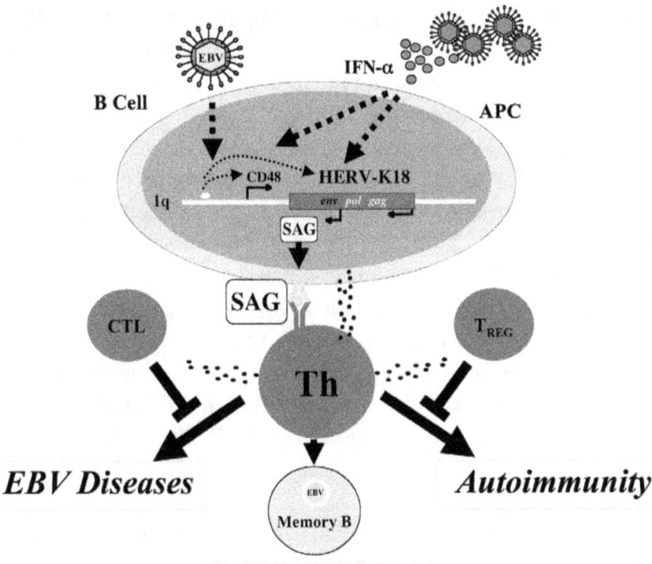

Figure 5. Model of superantigen in EBV diseases and autoimmunity. EBV infection elicits HERV-K18 superantigen expression in B cells, resulting in localized T cell stimulation. In a healthy host, the resulting T cell activation is self-limiting once EBV gains access to the EBV reservoir, the memory B cell compartment. In an immunosuppressed host, where virus specific memory cytotoxic T cells are suppressed, superantigen-activated T cells and cytokine production further perturb the balance. Alternatively, HERV-K18 superantigen expression in response to EBV, or other viruses by their elicitation of IFN-α production, could play a role in the activation of autoimmunity in susceptible individuals. By directly activating autoimmune T cells, or by cytokine-mediated disruption of regulatory T cells that control disease, the induction of autoimmunity is possible.

et al., 1999; Tosato et al., 1984; Venables et al., 1985). The strong T cell activation induced by the superantigen might be directly responsible for the activation of autoimmune T cells in susceptible individuals (Figure 5). In addition, the cytokine microenvironment resulting from localized HERV-K18 superantigen expression might influence the activation of autoimmune T cells. Alternatively, superantigen activated cytokine production might affect the growth of regulatory T cells that control autoreactive T cells (see Figure 5).

JRA vs. pediatric SLE

Many researchers have postulated a role for EBV in rheumatoid arthritis (RA) (Alspaugh et al., 1983; Balandraud et al., 2003; Billings et al., 1983; Ferrell et al., 1981; Fox et al., 1992; Kosaka, 1979; Venables et al., 1985), and numerous reports claim that EBV genomes are increased in the blood and joints of RA patients (Balandraud et al., 2003; Blaschke et al., 2000; Newkirk et al., 1994; Tosato et al., 1984; Tsai et al., 1995; Yao et al., 1986). In addition, there are reports of HERV up-regulation in RA synovial tissue (Brand et al., 1999; Gaudin et al., 2000; Griffiths et al., 1999; Nakagawa et al., 1997; Neidhart et al., 2000; Ogasawara et al., 2001; Takeuchi et al., 1995). In support of these findings, we have observed that HERV-K18 transcripts are significantly up-regulated in the peripheral blood of juvenile RA (JRA) patients, but not pediatric systemic Lupus Erythematosus (SLE) patients, compared with healthy child volunteers (Sicat et al., 2005). On the other hand, we could not find any correlation between the high level of HERV-K18 *env* expressed in PBMC or lymphocytes from the joint and known modifiers of HERV-K18 expression, including EBV seropositivity and serum IFN-α levels. While it has been reported that SLE patients have higher circulating levels of IFN-α, and that serum IFN-α levels correlate with disease activity (Bennett et al., 2003; Blanco et al., 2001; Pascual et al., 2003), HERV-K18 transcription was not significantly elevated in the pediatric Lupus patients compared with healthy children. However, as the SLE patients were all treated with steroids, a negative regulatory effect of the drug therapy cannot be ruled out. Interestingly, since the use of recombinant IFN-α in clinical trials as an anti-viral/anti-tumor agent, drug-related autoimmune manifestations, including thyroiditis, polyarthritis, arthralgias, myalgias and autoantibody production have been reported (Taki, 2002).

IDDM

As discussed above in the section entitled "Introduction to Viral Superantigens", an allelic variant of the HERV-K18 virus was first cloned from the peripheral blood of IDDM patients and, therefore, was originally called IDDMK$_{1,2}$22 (Conrad et al., 1997; Stauffer et al., 2001). Conrad and his colleagues had earlier discovered retroviral particles budding from the pancreatic islets of these patients (Conrad et al., 1994). The islets were simultaneously infiltrated with expanded populations of TCRBV7 T cells. The isolation of the HERV superantigen led to the hypothesis that superantigen activated T cells destroy the pancreatic islets in these patients (Conrad et al., 1997; Stauffer et al., 2001).

Sjogren's syndrome

Sjogren's syndrome is an autoimmune disease characterized by keratoconjunctivitis and sialadenitis, associated with glandular B and T cell infiltration, and the presence of antinuclear autoantibodies, anti-Ro and anti-La. The La antigen is a complex involving a 48 kD phosphoprotein bound to nascent RNA polymerase III transcripts, or alternatively bound to EBV encoded RNA transcripts (EBER1 and EBER2). The autoantibodies are characteristic of Sjogren's syndrome patients, but are also periodically found in SLE patients, leading to the hypothesis that EBV plays an etiologic role in both diseases (Lerner et al., 1981; McNeilage et al., 1984; Rosa et al., 1981; Whittingham et al., 1985). Because TCRBV13 T cells are frequently expanded in lymphocytic infiltrates found in the affected Sjogren's salivary glands (Legras et al., 1994; Smith et al., 1994; Sumida et al., 1992; Yonaha et al., 1992), it is possible that the HERV-K18 superantigen is involved.

Interestingly, there is a significant linkage association with homozygosity of the *TCRBV13S2*2* locus and disease (Kay et al., 1995). An insertion/deletion-related polymorphism (IDRP) at this locus was identified, the insertion resulting in a duplication of the *TCRBV13S2*1* allele and insertion of a *TCRBV7S2* gene, while the deletion results in a single copy of a *TCRBV13S2* allele and complete deletion of *TCRBV7S2*. Individuals that are homozygous for the deleted *TCRBV13S2*2* allele have no TCRBV7S2$^+$ T cells, whereas those carrying an inserted allele have double the TCRBV13S2$^+$ and TCRBV7$^+$ T cells (Manavalan et al., 2004), the exact subsets that respond to HERV-K18 Env. Thus, the prediction is that individuals carrying the inserted allele are more susceptible to Sjogren's syndrome; however, a recent report concludes that in fact it is the absence of the TCRBV7S2 subset that correlates with Sjogren's syndrome; the presence of *TCRBV13S2*2* is merely a marker of deletion (Manavalan et al., 2004; Zhao et al., 1994).

HERV-K18 env alleles: how genotype might influence acquisition of disease

Three alleles of HERV-K18 *env* have been observed in the Caucasian population, called K18.1, K18.2 and K18.3 *env* (Stauffer et al., 2001) (see Figure 2). HERV-K18.1 *env* was the first gene cloned; it has a stop codon at position 154, which results in a truncation of the full-length 560 a.a. retroviral Env protein to a 153 a.a. short form that lacks the C-terminal transmembrane domain. This allele is found in approximately 45 % of Caucasian chromosomes. HERV-K18.2 and HERV-K18.3 are both full-length Env proteins with 5 and 4 mutations, respectively, compared with HERV-K18.1 *env*. The HERV-K18.2 allele is present on approximately 45 % of Caucasian chromosomes, while HERV-K18.3 allele has a much lower distribution of about 10 % (Stauffer et al., 2001). All three of the proteins have functional superantigen activity and strongly stimulate T cells, as demonstrated in transfection experiments (Sutkowski et al., 2001). While HERV-K18.2 and K18.3 alleles proved to be highly toxic when expressed under strong exogenous promoters, the truncated HERV-K18.1 Env protein was comparatively less toxic upon over-expression (Sutkowski et al., 2001). Because the truncated HERV-K18.1 protein lacks the transmembrane domain, it might be secreted, although no evident secretory signal is predicted from the coding sequence (Accession AL121985). It is possible that the full-length HERV-K18 Env transmembrane proteins disrupt the integrity of the cell membrane, rendering them comparatively more toxic. The allelic changes might, thus, result in differences in the amount of Env protein expressed and/or in the stability of the expressed proteins. If this were indeed the case, individuals carrying the less toxic, *i.e.*, relatively more stable, HERV-K18.1 *env* allele could have higher superantigen expression than individuals carrying HERV-K18.2 and K18.3 alleles. As a functional consequence, this would result in relatively higher levels of superantigen-mediated T cell activation, possibly affecting disease progression. In addition, expression of the different proteins in the thymus during neonatal development might affect deletion of particular T cell clones through tolerance induction. Again, toxicity and stability of various alleles might alter thymic selection. Shaping of the T cell repertoire could have effects on the selection or deletion of T cells that would be capable of mounting anti-viral responses (Huang et al., 2002), or alternatively, of responding to self (Manavalan et al., 2004). Therefore, the HERV-K18 genotype might affect later acquisition of EBV associated diseases and/or the eventual development of autoimmunity.

Cases of BL in Africa are virtually 100% EBV⁺, while BLs in the United States are rarely positive for this herpesvirus (Crawford, 2001; Epstein et al., 1964). In addition, BLs are endemic in East Africa, the so-called lymphoma belt that comprises Kenya, Uganda and Tanzania, but are infrequently seen in other parts of Africa. It has been proposed that malaria is a co-factor in the etiology of BL, because this mosquito-born disease is also endemic in these same regions and induces general immunosuppression in the afflicted individuals (Morrow, 1985). However, there are many other areas in Africa where Malaria is endemic, but the frequency of BL is low. Thus, this cannot serve as the main explanation for the susceptibility to BL. More likely, a genetic trait unique to the East-African population is a deciding factor in BL-development. An intriguing possibility is that different alleles of HERV-K18 Env, which may vary in their superantigenicity, play a role in susceptibility to EBV⁺ African BL.

Why are endogenous superantigens maintained in the host genome?

Multiple endogenous MMTV superantigens have been cloned with various TCRBV specificities. The different proteins are functional, and the superantigenic portions of the molecules are conserved. The hypothesis is that endogenous MMTV superantigens are maintained in the mouse genome, because they protect the mice from highly oncogenic infectious MMTVs with superantigens of similar specificity. As the name implies, MMTVs cause mammary tumors, and some viral strains cause particularly aggressive tumors that are quite deadly (Acha-Orbea et al., 1993). Endogenous MMTVs expressed in the thymus during neonatal development result in deletion of particular TCRBV T cell subsets. Since transmission of infectious virus is dependent upon superantigen activated T cells amplifying the infected B cell reservoir, transmission of the infectious virus is effectively halted if the responding TCRBV specific T cell subsets have been deleted (Golovkina et al., 1992; Held et al., 1993). Thus, thymic deletion in response to endogenous superantigens is a protective mechanism that the murine host has acquired for preventing transmission of oncogenic viruses. It is as yet unclear why the human genome has maintained so many retroviruses, which are not known to produce functional viral particles. We postulate that these proviruses play a similar beneficial role as the MMTVs that are maintained in the murine genome.

References

Acha-Orbea, H., Held, W., Waanders, G. A., Shakhov, A. N., Scarpellino, L., Lees, R. K., and MacDonald, H. R. (1993). Exogenous and endogenous mouse mammary tumor virus superantigens. Immunol. Rev. *131*, 5-25.

Acha-Orbea, H., Scarpellino, L., Shakhov, A. N., Held, W., and MacDonald, H. R. (1992). Inhibition of mouse mammary tumor virus-induced T cell responses *in vivo* by antibodies to an open reading frame protein. J. Exp. Med. *176*, 1769-1772.

Acha-Orbea, H., Shakhov, A. N., Scarpellino, L., Kolb, E., Muller, V., Vessaz, S. A., Fuchs, R., Blochlinger, K., Rollini, P., Billotte, J., and et, a. l. (1991). Clonal deletion of V beta 14-bearing T cells in mice transgenic for mammary tumour virus. Nature *350*, 207-211.

Adelman, M. K., and Marchalonis, J. J. (2002). Endogenous retroviruses in systemic lupus erythematosus: candidate lupus viruses. Clin. Immunol. *102*, 107-116.

Agathanggelou, A., Niedobitek, G., Chen, R., Nicholls, J., Yin, W., and Young, L. S. (1995). Expression of immune regulatory molecules in Epstein-Barr virus-associated nasopharyngeal carcinomas with prominent lymphoid stroma. Evidence for a functional interaction between epithelial tumor cells and infiltrating lymphoid cells. Amer. J. Pathol. *147*, 1152-1160.

Alspaugh, M. A., Shoji, H., and Nonoyama, M. (1983). A search for rheumatoid arthritis-associated nuclear antigen and Epstein-Barr virus specific antigens or genomes in tissues and cells from patients with rheumatoid arthritis. Arthritis Rheum. *26*, 712-720.

Amadori, A., Veronese, M. L., Mazza, M. R., Zamarchi, R., Mion, M., D'Andrea, E., Menin, C., Del Mistro, A., Leszl, A., Calderazzo, F., and et al. (1992). Lymphoma induction by human B cells in scid mice. Leukemia *6*, 23S-25S.

Babcock, G. J., Decker, L. L., Freeman, R. B., and Thorley-Lawson, D. A. (1999). Epstein-barr virus-infected resting memory B cells, not proliferating lymphoblasts, accumulate in the peripheral blood of immunosuppressed patients. J. Exp. Med. *190*, 567-576.

Babcock, G. J., Decker, L. L., Volk, M., and Thorley-Lawson, D. A. (1998). EBV persistence in memory B cells *in vivo*. Immunity *9*, 395-404.

Babcock, G. J., Hochberg, D., and Thorley-Lawson, A. D. (2000). The expression pattern of Epstein-Barr virus latent genes *in vivo* is dependent upon the differentiation stage of the infected B cell. Immunity *13*, 497-506.

Balandraud, N., Meynard, J. B., Auger, I., Sovran, H., Mugnier, B., Reviron, D., Roudier, J., and Roudier, C. (2003). Epstein-Barr virus load in the peripheral blood of patients with rheumatoid arthritis: accurate quantification using real-time polymerase chain reaction. Arthritis Rheum. *48*, 1223-1228.

Barbulescu, M., Turner, G., Seaman, M. I., Deinard, A. S., Kidd, K. K., and Lenz, J. (1999). Many human endogenous retrovirus K (HERV-K) proviruses are unique to humans. Curr. Biol. *9*, 861-868.

Beck, A., Pazolt, D., Grabenbauer, G. G., Nicholls, J. M., Herbst, H., Young, L. S., and Niedobitek, G. (2001). Expression of cytokine and chemokine genes in Epstein-Barr virus-associated nasopharyngeal carcinoma: comparison with Hodgkin's disease. J. Pathol. *194*, 145-151.

Bennett, L., Palucka, A. K., Arce, E., Cantrell, V., Borvak, J., Banchereau, J., and Pascual, V. (2003). Interferon and granulopoiesis signatures in systemic lupus erythematosus blood. J. Exp. Med. *197*, 711-723.

Beutner, U., Frankel, W. N., Cote, M. S., Coffin, J. M., and Huber, B. T. (1992a). Mls-1 is encoded by the long terminal repeat open reading frame of the mouse mammary tumor provirus Mtv-7. Proc. Natl. Acad. Sci. USA. *89*, 5432-5436.

Beutner, U., Kraus, E., Kitamura, D., Rajewsky, K., and Huber, B. T. (1994). B cells are essential for murine mammary tumor virus transmission, but not for presentation of endogenous superantigens. J. Exp. Med. *179*, 1457-1466.

Beutner, U., McLellan, B., Kraus, E., and Huber, B. T. (1996). Lack of MMTV superantigen presentation in MHC class II-deficient mice. Cell Immunol. *168*, 141-147.

Beutner, U., Rudy, C., and Huber, B. T. (1992b). Molecular characterization of Mls-1. Int. Rev. Immunol. *8*, 279-288.

Billings, P. B., Hoch, S. O., White, P. J., Carson, D. A., and Vaughan, J. H. (1983). Antibodies to the Epstein-Barr virus nuclear antigen and to rheumatoid arthritis nuclear antigen identify the same polypeptide. Proc. Natl. Acad. Sci. USA. *80*, 7104-7108.

Blaise, S., de Parseval, N., Benit, L., and Heidmann, T. (2003). Genomewide screening for fusogenic human endogenous retrovirus envelopes identifies syncytin 2, a gene conserved on primate evolution. Proc. Natl. Acad. Sci. USA. *100*, 13013-13018.

Blanco, P., Palucka, A. K., Gill, M., Pascual, V., and Banchereau, J. (2001). Induction of dendritic cell differentiation by IFN-alpha in systemic lupus erythematosus. Science *294*, 1540-1543.

Blaschke, S., Schwarz, G., Moneke, D., Binder, L., Muller, G., and Reuss-Borst, M. (2000). Epstein-Barr virus infection in peripheral blood mononuclear cells, synovial fluid cells, and synovial membranes of patients with rheumatoid arthritis. J. Rheumatol *27*, 866-873.

Blond, J. L., Beseme, F., Duret, L., Bouton, O., Bedin, F., Perron, H., Mandrand, B., and Mallet, F. (1999). Molecular characterization and placental expression of HERV-W, a new human endogenous retrovirus family. J. Virol. *73*, 1175-1185.

Boeke, J. D., and Stoye, J. P. (1997). Retrotransposons, endogenous retroviruses and the evolution of retroelements. In Retroviruses, J. M. Coffin, S. H. Hughes, and H. E. Varmus, eds. (Cold Spring Harbor, NY, Cold Spring Harbor Press), pp. 343-435.

Brand, A., Griffiths, D. J., Herve, C., Mallon, E., and Venables, P. J. (1999). Human retrovirus-5 in rheumatic disease. J. Autoimmun *13*, 149-154.

Braun, D., Caramalho, I., and Demengeot, J. (2002). IFN-alpha/beta enhances BCR-dependent B cell responses. Int. Immunol. *14*, 411-419.

Brauninger, A., Spieker, T., Willenbrock, K., Gaulard, P., Wacker, H. H., Rajewsky, K., Hansmann, M. L., and Kuppers, R. (2001). Survival and clonal expansion of mutating "forbidden" (immunoglobulin receptor-deficient) epstein-barr virus-infected b cells in angioimmunoblastic t cell lymphoma. J. Exp. Med. *194*, 927-940.

Brooks, J. W., Hamilton-Easton, A. M., Christensen, J. P., Cardin, R. D., Hardy, C. L., and Doherty, P. C. (1999). Requirement for CD40 ligand, CD4(+) T cells, and B cells in an infectious mononucleosis-like syndrome. J. Virol. *73*, 9650-9654.

Brooks, L., Yao, Q. Y., Rickinson, A. B., and Young, L. S. (1992). Epstein-Barr virus latent gene transcription in nasopharyngeal carcinoma cells: coexpression of EBNA1, LMP1, and LMP2 transcripts. J. Virol. *66*, 2689-2697.

Busson, P., McCoy, R., Sadler, R., Gilligan, K., Tursz, T., and Raab-Traub, N. (1992). Consistent transcription of the Epstein-Barr virus LMP2 gene in nasopharyngeal carcinoma. J. Virol. *66*, 3257-3262.

Caldwell, R. G., Brown, R. C., and Longnecker, R. (2000). Epstein-Barr virus LMP2A-induced B-cell survival in two unique classes of EmuLMP2A transgenic mice. J. Virol. *74*, 1101-1113.

Caldwell, R. G., Wilson, J. B., Anderson, S. J., and Longnecker, R. (1998). Epstein-Barr virus LMP2A drives B cell development and survival in the absence of normal B cell receptor signals. Immunity *9*, 405-411.

Calender, A., Billaud, M., Aubry, J. P., Banchereau, J., Vuillaume, M., and Lenoir, G. M. (1987). Epstein-Barr virus (EBV) induces expression of B-cell activation markers on *in vitro* infection of EBV-negative B-lymphoma cells. Proc. Natl. Acad. Sci. USA. *84*, 8060-8064.

Callan, M. F. (2003). The evolution of antigen-specific CD8+ T cell responses after natural primary infection of humans with Epstein-Barr virus. Viral Immunol. *16*, 3-16.

Callan, M. F., Tan, L., Annels, N., Ogg, G. S., Wilson, J. D., O'Callaghan, C. A., Steven, N., McMichael, A. J., and Rickinson, A. B. (1998). Direct visualization of antigen-specific CD8+ T cells during the primary immune response to Epstein-Barr virus In vivo. J. Exp. Med. *187*, 1395-1402.

Casola, S., Otipoby, K. L., Alimzhanov, M., Humme, S., Uyttersprot, N., Kutok, J. L., Carroll, M. C., and Rajewsky, K. (2004). B cell receptor signal strength determines B cell fate. Nat. Immunol. *5*, 317-327.

Chen, H., Hutt-Fletcher, L., Cao, L., and Hayward, S. D. (2003). A positive autoregulatory loop of LMP1 expression and STAT activation in epithelial cells latently infected with Epstein-Barr virus. J. Virol. *77*, 4139-4148.

Chen, H., Lee, J. M., Zong, Y., Borowitz, M., Ng, M. H., Ambinder, R. F., and Hayward, S. D. (2001). Linkage between STAT regulation and Epstein-Barr virus gene expression in tumors. J. Virol. *75*, 2929-2937.

Choi, Y., Kappler, J. W., and Marrack, P. (1991). A superantigen encoded in the open reading frame of the 3' long terminal repeat of mouse mammary tumour virus. Nature *350*, 203-207.

Choi, Y., Marrack, P., and Kappler, J. W. (1992). Structural analysis of a mouse mammary tumor virus superantigen. J. Exp. Med. *175*, 847-852.

Choi, Y. W., Kotzin, B., Herron, L., Callahan, J., Marrack, P., and Kappler, J. (1989). Interaction of Staphylococcus aureus toxin "superantigens" with human T cells. Proc. Natl. Acad. Sci. USA. *86*, 8941-8945.

Conrad, B., Weidmann, E., Trucco, G., Rudert, W. A., Behboo, R., Ricordi, C., Rodriquez-Rilo, H., Finegold, D., and Trucco, M. (1994). Evidence for superantigen involvement in insulin-dependent diabetes mellitus aetiology [see comments]. Nature *371*, 351-355.

Conrad, B., Weissmahr, R. N., Boni, J., Arcari, R., Schupbach, J., and Mach, B. (1997). A human endogenous retroviral superantigen as candidate autoimmune gene in type I diabetes. Cell *90*, 303-313.

Coppola, M. A., and Blackman, M. A. (1997). Bacterial superantigens reactivate antigen-specific CD8+ memory T cells. Int. Immunol. *9*, 1393-1403.

Coppola, M. A., Flano, E., Nguyen, P., Hardy, C. L., Cardin, R. D., Shastri, N., Woodland, D. L., and Blackman, M. A. (1999). Apparent MHC-independent stimulation of CD8+ T cells *in vivo* during latent murine gammaherpesvirus infection. J. Immunol. *163*, 1481-1489.

Crawford, D. H. (2001). Biology and disease associations of Epstein-Barr virus. Philos Trans R Soc. Lond B Biol. Sci. *356*, 461-473.

Delecluse, H. J., Kremmer, E., Rouault, J. P., Cour, C., Bornkamm, G., and Berger, F. (1995). The expression of Epstein-Barr virus latent proteins is related to the pathological features of post-transplant lymphoproliferative disorders. Amer. J. Pathol. *146*, 1113-1120.

Dellabona, P., Peccoud, J., Kappler, J., Marrack, P., Benoist, C., and Mathis, D. (1990). Superantigens interact with MHC class II molecules outside of the antigen groove. Cell *62*, 1115-1121.

Dobrescu, D., Kabak, S., Mehta, K., Suh, C. H., Asch, A., Cameron, P. U., Hodtsev, A. S., and Posnett, D. N. (1995a). Human immunodeficiency virus 1 reservoir in CD4+ T cells is restricted to certain V beta subsets. Proc. Natl. Acad. Sci. USA. *92*, 5563-5567.

Dobrescu, D., Ursea, B., Pope, M., Asch, A. S., and Posnett, D. N. (1995b). Enhanced HIV-1 replication in V beta 12 T cells due to human cytomegalovirus in monocytes: evidence for a putative herpesvirus superantigen. Cell *82*, 753-763.

Doherty, P. C., Tripp, R. A., Hamilton-Easton, A. M., Cardin, R. D., Woodland, D. L., and Blackman, M. A. (1997). Tuning into immunological dissonance: an experimental model for infectious mononucleosis. Curr. Opin. Immunol. *9*, 477-483.

Duboise, M., Guo, J., Czajak, S., Lee, H., Veazey, R., Desrosiers, R. C., and Jung, J. U. (1998). A role for herpesvirus saimiri orf14 in transformation and persistent infection. J. Virol. *72*, 6770-6776.

Eliopoulos, A. G., and Young, L. S. (2001). LMP1 structure and signal transduction. Semin Cancer Biol. *11*, 435-444.

Engel, P., Eck, M. J., and Terhorst, C. (2003). The SAP and SLAM families in immune responses and X-linked lymphoproliferative disease. Nat. Rev. Immunol. *3*, 813-821.

Epstein, M. A., Achong, B. G., and Barr, Y. M. (1964). Virus Particles in Cultured Lymphoblasts from Burkitt's Lymphoma. Lancet *15*, 702-703.

Ferrell, P. B., Aitcheson, C. T., Pearson, G. R., and Tan, E. M. (1981). Seroepidemiological study of relationships between Epstein-Barr virus and rheumatoid arthritis. J. Clin. Invest *67*, 681-687.

Firouzi, R., Rolland, A., Michel, M., Jouvin-Marche, E., Hauw, J. J., Malcus-Vocanson, C., Lazarini, F., Gebuhrer, L., Seigneurin, J. M., Touraine, J. L., et al. (2003). Multiple sclerosis-associated retrovirus particles cause T lymphocyte-dependent death with brain hemorrhage in humanized SCID mice model. J. Neurovirol *9*, 79-93.

Fox, R. I., Luppi, M., Pisa, P., and Kang, H. I. (1992). Potential role of Epstein-Barr virus in Sjogren's syndrome and rheumatoid arthritis. J. Rheumatol Suppl *32*, 18-24.

Frankel, W. N., Rudy, C., Coffin, J. M., and Huber, B. T. (1991). Linkage of Mls genes to endogenous mammary tumour viruses of inbred mice. Nature *349*, 526-528.

Fraser, J., Arcus, V., Kong, P., Baker, E., and Proft, T. (2000). Superantigens - powerful modifiers of the immune system. Mol. Med. Today *6*, 125-132.

Fu, Z., and Cannon, M. J. (2000). Functional analysis of the CD4(+) T-cell response to Epstein-Barr virus: T-cell-mediated activation of resting B cells and induction of viral BZLF1 expression. J. Virol. *74*, 6675-6679.

Gaudin, P., Ijaz, S., Tuke, P. W., Marcel, F., Paraz, A., Seigneurin, J. M., Mandrand, B., Perron, H., and Garson, J. A. (2000). Infrequency of detection of particle-associated MSRV/HERV-W RNA in the synovial fluid of patients with rheumatoid arthritis. Rheumatology (Oxford) *39*, 950-954.

Goedert, J. J., Sauter, M. E., Jacobson, L. P., Vessella, R. L., Hilgartner, M. W., Leitman, S. F., Fraser, M. C., and Mueller-Lantzsch, N. G. (1999). High prevalence of antibodies against HERV-K10 in patients with testicular cancer but not with AIDS. Cancer Epidemiol. Biomarkers Prev *8*, 293-296.

Golovkina, T. V., Chervonsky, A., Dudley, J. P., and Ross, S. R. (1992). Transgenic mouse mammary tumor virus superantigen expression prevents viral infection. Cell *69*, 637-645.

Griffiths, D. J., Cooke, S. P., Herve, C., Rigby, S. P., Mallon, E., Hajeer, A., Lock, M., Emery, V., Taylor, P., Pantelidis, P., et al. (1999). Detection of human retrovirus 5 in patients with arthritis and systemic lupus erythematosus. Arthritis Rheum. *42*, 448-454.

Hammerschmidt, W., and Sugden, B. (1989). Genetic analysis of immortalizing functions of Epstein-Barr virus in human B lymphocytes. Nature *340*, 393-397.

Hardy, C. L., Silins, S. L., Woodland, D. L., and Blackman, M. A. (2000). Murine gamma-herpesvirus infection causes V(beta)4-specific CDR3- restricted clonal expansions within CD8(+) peripheral blood T lymphocytes. Int. Immunol. *12*, 1193-1204.

Hasuike, S., Miura, K., Miyoshi, O., Miyamoto, T., Niikawa, N., Jinno, Y., and Ishikawa, M. (1999). Isolation and localization of an IDDMK1,2-22-related human endogenous retroviral gene, and identification of a CA repeat marker at its locus. J. Hum. Genet. *44*, 343-347.

Held, W., Wanders, G. A., Shakhow, A. N., Scarpellino, L., Acha, O. H., and MacDonald, H. R. (1993). Superantigen-induced immune stimulation amplifies mouse mammary tumor virus infection and allows virus transmission. Cell *74*, 529-540.

Herbst, H., Dallenbach, F., Hummel, M., Niedobitek, G., Pileri, S., Muller-Lantzsch, N., and Stein, H. (1991). Epstein-Barr virus latent membrane protein expression in Hodgkin and Reed-Sternberg cells. Proc. Natl. Acad. Sci. USA. *88*, 4766-4770.

Herman, A., Kappler, J. W., Marrack, P., and Pullen, A. M. (1991). Superantigens: mechanism of T-cell stimulation and role in immune responses. Annu. Rev. Immunol. *9*, 745-772.

Hertel, C. B., Zhou, X. G., Hamilton-Dutoit, S. J., and Junker, S. (2002). Loss of B cell identity correlates with loss of B cell-specific transcription factors in Hodgkin/Reed-Sternberg cells of classical Hodgkin lymphoma. Oncogene *21*, 4908-4920.

Heussinger, N., Buttner, M., Ott, G., Brachtel, E., Pilch, B. Z., Kremmer, E., and Niedobitek, G. (2004). Expression of the Epstein-Barr virus (EBV)-encoded latent membrane protein 2A (LMP2A) in EBV-associated nasopharyngeal carcinoma. J. Pathol. *203*, 696-699.

Hsu, P. N., Wolf Bryant, P., Sutkowski, N., McLellan, B., Ploegh, H. L., and Huber, B. T. (2001). Association of mouse mammary tumor virus superantigen with MHC class II during biosynthesis. J. Immunol. *166*, 3309-3314.

Huang, C. C., Shah, S., Nguyen, P., Altman, J. D., and Blackman, M. A. (2002). Bacterial Superantigen Exposure after Resolution of Influenza Virus Infection Perturbs the Virus-Specific Memory CD8(+)-T-Cell Repertoire. J. Virol. *76*, 6852-6856.

Huber, B. T. (1992). Mls superantigens: how retroviruses influence the expressed T cell receptor repertoire. Semin Immunol. *4*, 313-318.

Huggins, M. L., Todd, I., and Powell, R. J. (2003). Reactivation of Epstein-Barr virus in patients with systemic lupus erythematosus. Rheumatol Int.

Hughes, J. F., and Coffin, J. M. (2001). Evidence for genomic rearrangements mediated by human endogenous retroviruses during primate evolution. Nat. Genet. *29*, 487-489.

James, J. A., Neas, B. R., Moser, K. L., Hall, T., Bruner, G. R., Sestak, A. L., and Harley, J. B. (2001). Systemic lupus erythematosus in adults is associated with previous Epstein-Barr virus exposure. Arthritis Rheum. *44*, 1122-1126.

Johannessen, I., Asghar, M., and Crawford, D. H. (2000). Essential role for T cells in human B-cell lymphoproliferative disease development in severe combined immunodeficient mice. Br. J. Haematol. *109*, 600-610.

Johannessen, I., and Crawford, D. H. (1999). In vivo models for Epstein-Barr virus (EBV)-associated B cell lymphoproliferative disease (BLPD). Rev. Med. Virol. *9*, 263-277.

Kang, I., Quan, T., Nolasco, H., Park, S. H., Hong, M. S., Crouch, J., Pamer, E. G., Howe, J. G., and Craft, J. (2004). Defective Control of Latent Epstein-Barr Virus Infection in Systemic Lupus Erythematosus. J. Immunol. *172*, 1287-1294.

Kappler, J., Kotzin, B., Herron, L., Gelfand, E. W., Bigler, R. D., Boylston, A., Carrel, S., Posnett, D. N., Choi, Y., and Marrack, P. (1989). V beta-specific stimulation of human T cells by staphylococcal toxins. Science *244*, 811-813.

Kappler, J. W., Staerz, U., White, J., and Marrack, P. C. (1988). Self-tolerance eliminates T cells specific for Mls-modified products of the major histocompatibility complex. Nature *332*, 35-40.

Karlsson, H., Bachmann, S., Schroder, J., McArthur, J., Torrey, E. F., and Yolken, R. H. (2001). Retroviral RNA identified in the cerebrospinal fluids and brains of individuals with schizophrenia. Proc. Natl. Acad. Sci. USA. *98*, 4634-4639.

Kay, R. A., Hutchings, C. J., and Ollier, W. E. (1995). A subset of Sjogren's syndrome associates with the TCRBV13S2 locus but not the TCRBV2S1 locus. Hum. Immunol. *42*, 328-330.

Kempkes, B., Pich, D., Zeidler, R., and Hammerschmidt, W. (1995). Immortalization of human primary B lymphocytes *in vitro* with DNA. Proc. Natl. Acad. Sci. USA. *92*, 5875-5879.

Khanna, R., and Burrows, S. R. (2000). Role of cytotoxic T lymphocytes in Epstein-Barr virus-associated diseases. Annu. Rev. Microbiol. *54*, 19-48.

Kilger, E., Kieser, A., Baumann, M., and Hammerschmidt, W. (1998). Epstein-Barr virus-mediated B-cell proliferation is dependent upon latent membrane protein 1, which simulates an activated CD40 receptor. Embo J. *17*, 1700-1709.

Klaman, L. D., and Thorley-Lawson, D. A. (1995). Characterization of the CD48 gene demonstrates a positive element that is specific to Epstein-Barr virus-immortalized B-cell lines and contains an essential NF-kappa B site. J. Virol. *69*, 871-881.

Knerr, I., Beinder, E., and Rascher, W. (2002). Syncytin, a novel human endogenous retroviral gene in human placenta: evidence for its dysregulation in preeclampsia and HELLP syndrome. Amer. J. Obstet. Gynecol. *186*, 210-213.

Koide, J., Takada, K., Sugiura, M., Sekine, H., Ito, T., Saito, K., Mori, S., Takeuchi, T., Uchida, S., and Abe, T. (1997). Spontaneous establishment of an Epstein-Barr virus-infected fibroblast line from the synovial tissue of a rheumatoid arthritis patient. J. Virol. *71*, 2478-2481.

Komurian-Pradel, F., Paranhos-Baccala, G., Bedin, F., Ounanian-Paraz, A., Sodoyer, M., Ott, C., Rajoharison, A., Garcia, E., Mallet, F., Mandrand, B., and Perron, H. (1999). Molecular cloning and characterization of MSRV-related sequences associated with retrovirus-like particles. Virology *260*, 1-9.

Kosaka, S. (1979). Detection of antibody to a new antigen induced by Epstein-Barr virus in rheumatoid arthritis. Tohoku J. Exp. Med. *127*, 157-160.

Kotzin, B. L., Leung, D. Y., Kappler, J., and Marrack, P. (1993). Superantigens and their potential role in human disease. Adv. Immunol. *54*, 99-166.

Kudo, Y., Boyd, C. A., Sargent, I. L., and Redman, C. W. (2003). Hypoxia alters expression and function of syncytin and its receptor during trophoblast cell fusion of human placental BeWo cells: implications for impaired trophoblast syncytialisation in pre-eclampsia. Biochim Biophys Acta *1638*, 63-71.

Kuppers, R., Schwering, I., Brauninger, A., Rajewsky, K., and Hansmann, M. L. (2002). Biology of Hodgkin's lymphoma. Ann. Oncol *13 Suppl 1*, 11-18.

Lafon, M., Jouvin-Marche, E., Marche, P. N., and Perron, H. (2002). Human viral superantigens: to be or not to be transactivated? Trends Immunol. *23*, 238-239; author reply 239.

Lafon, M., Lafage, M., Martinez-Arends, A., Ramirez, R., Vuillier, F., Charron, D., Lotteau, V., and Scott-Algara, D. (1992). Evidence for a viral superantigen in humans. Nature *358*, 507-510.

Lafon, M., Scott-Algara, D., Marche, P. N., Cazenave, P. A., and Jouvin-Marche, E. (1994). Neonatal deletion and selective expansion of mouse T cells by exposure to rabies virus nucleocapsid superantigen. J. Exp. Med. *180*, 1207-1215.

Lander, E. S., Linton, L. M., Birren, B., Nusbaum, C., Zody, M. C., Baldwin, J., Devon, K., Dewar, K., Doyle, M., FitzHugh, W., et al. (2001). Initial sequencing and analysis of the human genome. Nature *409*, 860-921.

Lavoie, P. M., Thibodeau, J., Erard, F., and Sekaly, R. P. (1999). Understanding the mechanism of action of bacterial superantigens from a decade of research. Immunol. Rev. *168*, 257-269.

Lee, S. P. (2002). Nasopharyngeal carcinoma and the EBV-specific T cell response: prospects for immunotherapy. Semin Cancer Biol. *12*, 463-471.

Lee, X., Keith, J. C., Jr., Stumm, N., Moutsatsos, I., McCoy, J. M., Crum, C. P., Genest, D., Chin, D., Ehrenfels, C., Pijnenborg, R., et al. (2001). Downregulation of placental syncytin expression and abnormal protein localization in pre-eclampsia. Placenta *22*, 808-812.

Legras, F., Martin, T., Knapp, A. M., and Pasquali, J. L. (1994). Infiltrating T cells from patients with primary Sjogren's syndrome express restricted or unrestricted T cell receptor V beta regions depending on the stage of the disease. Eur. J. Immunol. *24*, 181-185.

Lennette, E. T., Winberg, G., Yadav, M., Enblad, G., and Klein, G. (1995). Antibodies to LMP2A/2B in EBV-carrying malignancies. Eur. J. Cancer *31A*, 1875-1878.

Lerner, M. R., Andrews, N. C., Miller, G., and Steitz, J. A. (1981). Two small RNAs encoded by Epstein-Barr virus and complexed with protein are precipitated by antibodies from patients with systemic lupus erythematosus. Proc. Natl. Acad. Sci. USA. *78*, 805-809.

Leung, D. Y., Huber, B. T., and Schlievert, P. M. (1997). Superantigens: Molecular Biology, Immunology and Relevance to Human Disease (New York, NY, Marcel Dekker, Inc.).

Liebowitz, D. (1998). Epstein-Barr virus and a cellular signaling pathway in lymphomas from immunosuppressed patients. N. Engl. J. Med. *338*, 1413-1421.

Liu, Y. J., and Banchereau, J. (1997). Regulation of B-cell commitment to plasma cells or to memory B cells. Semin Immunol. *9*, 235-240.

Longnecker, R., and Miller, C. L. (1996). Regulation of Epstein-Barr virus latency by latent membrane protein 2. Trends Microbiol. *4*, 38-42.

Macphail, S. (1999). Superantigens: mechanisms by which they may induce, exacerbate and control autoimmune diseases. Int. Rev. Immunol. *18*, 141-180.

Maini, M. K., Gudgeon, N., Wedderburn, L. R., Rickinson, A. B., and Beverley, P. C. (2000). Clonal expansions in acute EBV infection are detectable in the CD8 and not the CD4 subset and persist with a variable CD45 phenotype. J. Immunol. *165*, 5729-5737.

Manavalan, S. J., Valiando, J. R., Reeves, W. H., Arnett, F. C., Necker, A., Simantov, R., Lyons, R., Satoh, M., and Posnett, D. N. (2004). Genomic absence of the gene encoding T cell receptor Vbeta7.2 is linked to the presence of autoantibodies in Sjogren's syndrome. Arthritis Rheum. *50*, 187-198.

Marrack, P., Blackman, M., Kushnir, E., and Kappler, J. (1990). The toxicity of staphylococcal enterotoxin B in mice is mediated by T cells. J. Exp. Med. *171*, 455-464.

Marrack, P., and Kappler, J. (1990). The staphylococcal enterotoxins and their relatives. Science *248*, 1066.

Marrack, P., Kushnir, E., and Kappler, J. (1991). A maternally inherited superantigen encoded by a mammary tumour virus. Nature *349*, 524-526.

Marrack, P., Winslow, G. M., Choi, Y., Scherer, M., Pullen, A., White, J., and Kappler, J. W. (1993). The bacterial and mouse mammary tumor virus superantigens; two different families of proteins with the same functions. Immunol. Rev. *131*, 79-92.

McNeilage, L. J., Whittingham, S., and Mackay, I. R. (1984). Autoantibodies reactive with small ribonucleoprotein antigens: a convergence of molecular biology and clinical immunology. J. Clin. Lab. Immunol. *15*, 1-17.

Merchant, M., Caldwell, R. G., and Longnecker, R. (2000). The LMP2A ITAM is essential for providing B cells with development and survival signals *in vivo*. J. Virol. *74*, 9115-9124.

Merchant, M., Swart, R., Katzman, R. B., Ikeda, M., Ikeda, A., Longnecker, R., Dykstra, M. L., and Pierce, S. K. (2001). The effects of the Epstein-Barr virus latent membrane protein 2A on B cell function. Int. Rev. Immunol. *20*, 805-835.

Mi, S., Lee, X., Li, X., Veldman, G. M., Finnerty, H., Racie, L., LaVallie, E., Tang, X. Y., Edouard, P., Howes, S., et al. (2000). Syncytin is a captive retroviral envelope protein involved in human placental morphogenesis. Nature *403*, 785-789.

Miller, C. L., Lee, J. H., Kieff, E., Burkhardt, A. L., Bolen, J. B., and Longnecker, R. (1994). Epstein-Barr virus protein LMP2A regulates reactivation from latency by negatively regulating tyrosine kinases involved in sIg-mediated signal transduction. Infect. Agents Dis. *3*, 128-136.

Miller, G., and Lipman, M. (1973). Release of infectious Epstein-Barr virus by transformed marmoset leukocytes. Proc. Natl. Acad. Sci. USA. *70*, 190-194.

Mitchell, T., Kappler, J., and Marrack, P. (1999). Bystander virus infection prolongs activated T cell survival. J. Immunol. *162*, 4527-4535.

Morrow, R. H., Jr. (1985). Epidemiological evidence for the role of falciparum malaria in the pathogenesis of Burkitt's lymphoma. IARC Sci. Publ, 177-186.

Mosier, D. E., Gulizia, R. J., Baird, S. M., and Wilson, D. B. (1988). Transfer of a functional human immune system to mice with severe combined immunodeficiency. Nature *335*, 256-259.

Muir, A., Ruan, Q. G., Marron, M. P., and She, J. X. (1999). The IDDMK(1,2)22 retrovirus is not detectable in either mRNA or genomic DNA from patients with type 1 diabetes. Diabetes *48*, 219-222.

Murphy, W. J., Funakoshi, S., Beckwith, M., Rushing, S. E., Conley, D. K., Armitage, R. J., Fanslow, W. C., Rager, H. C., Taub, D. D., Ruscetti, F. W., and et al. (1995). Antibodies to CD40 prevent Epstein-Barr virus-mediated human B-cell lymphomagenesis in severe combined immune deficient mice given human peripheral blood lymphocytes. Blood *86*, 1946-1953.

Nakagawa, K., Brusic, V., McColl, G., and Harrison, L. C. (1997). Direct evidence for the expression of multiple endogenous retroviruses in the synovial compartment in rheumatoid arthritis. Arthritis Rheum. *40*, 627-638.

Nakamura, A., Okazaki, Y., Sugimoto, J., Oda, T., and Jinno, Y. (2003). Human endogenous retroviruses with transcriptional potential in the brain. J. Hum. Genet. *48*, 575-581.

Neidhart, M., Rethage, J., Kuchen, S., Kunzler, P., Crowl, R. M., Billingham, M. E., Gay, R. E., and Gay, S. (2000). Retrotransposable L1 elements expressed in rheumatoid arthritis synovial tissue: association with genomic DNA hypomethylation and influence on gene expression. Arthritis Rheum. *43*, 2634-2647.

Newkirk, M. M., Watanabe Duffy, K. N., Leclerc, J., Lambert, N., and Shiroky, J. B. (1994). Detection of cytomegalovirus, Epstein-Barr virus and herpes virus-6 in patients with rheumatoid arthritis with or without Sjogren's syndrome. Br. J. Rheumatol *33*, 317-322.

Niedobitek, G. (2000). Epstein-Barr virus infection in the pathogenesis of nasopharyngeal carcinoma. Mol. Pathol. *53*, 248-254.

Niedobitek, G., Agathanggelou, A., and Nicholls, J. M. (1996). Epstein-Barr virus infection and the pathogenesis of nasopharyngeal carcinoma: viral gene expression, tumour cell phenotype, and the role of the lymphoid stroma. Semin Cancer Biol. *7*, 165-174.

Niedobitek, G., Kremmer, E., Herbst, H., Whitehead, L., Dawson, C. W., Niedobitek, E., von Ostau, C., Rooney, N., Grasser, F. A., and Young, L. S. (1997). Immunohistochemical detection of the Epstein-Barr virus-encoded latent membrane protein 2A in Hodgkin's disease and infectious mononucleosis. Blood *90*, 1664-1672.

Niedobitek, G., Lisner, R., Swoboda, B., Rooney, N., Fassbender, H. G., Kirchner, T., Aigner, T., and Herbst, H. (2000). Lack of evidence for an involvement of Epstein-Barr virus infection of synovial membranes in the pathogenesis of rheumatoid arthritis. Arthritis Rheum. *43*, 151-154.

Ogasawara, H., Okada, M., Kaneko, H., Hishikawa, T., Sekigawa, I., Iida, N., Maruyama, N., Yamamoto, N., and Hashimoto, H. (2001). Quantitative comparison of human endogenous retrovirus mRNA between SLE and rheumatoid arthritis. Lupus *10*, 517-518.

Ono, M. (1986). Molecular cloning and long terminal repeat sequences of human endogenous retrovirus genes related to types A and B retrovirus genes. J. Virol. *58*, 937-944.

Ono, M., Kawakami, M., and Ushikubo, H. (1987). Stimulation of expression of the human endogenous retrovirus genome by female steroid hormones in human breast cancer cell line T47D. J. Virol. *61*, 2059-2062.

Ono, M., Yasunaga, T., Miyata, T., and Ushikubo, H. (1986). Nucleotide sequence of human endogenous retrovirus genome related to the mouse mammary tumor virus genome. J. Virol. *60*, 589-598.

Pascual, V., Banchereau, J., and Palucka, A. K. (2003). The central role of dendritic cells and interferon-alpha in SLE. Curr. Opin. Rheumatol *15*, 548-556.

Perron, H., Jouvin-Marche, E., Michel, M., Ounanian-Paraz, A., Camelo, S., Dumon, A., Jolivet-Reynaud, C., Marcel, F., Souillet, Y., Borel, E., et al. (2001). Multiple sclerosis retrovirus particles and recombinant envelope trigger an abnormal immune response *in vitro*, by inducing polyclonal Vbeta16 T-lymphocyte activation. Virology *287*, 321-332.

Portis, J. L. (2002). Perspectives on the role of endogenous human retroviruses in autoimmune diseases. Virology *296*, 1-5.

Portis, T., Cooper, L., Dennis, P., and Longnecker, R. (2002). The LMP2A signalosome--a therapeutic target for Epstein-Barr virus latency and associated disease. Front Biosci *7*, d414-426.

Portis, T., Dyck, P., and Longnecker, R. (2003). Epstein-Barr virus (EBV) LMP2A induces alterations in gene transcription similar to those observed in Reed-Sternberg cells of Hodgkin's lymphoma. Blood.

Portis, T., and Longnecker, R. (2003). Epstein-Barr virus LMP2A interferes with global transcription factor regulation when expressed during B-lymphocyte development. J. Virol. *77*, 105-114.

Posnett, D. N., Kabak, S., Dobrescu, D., and Hodtsev, A. S. (1995). The HIV-1 reservoir in distinct V beta subsets of CD4 T cells: evidence for a putative superantigen. J. Clin. Immunol. *15*, 18S-21S.

Poy, F., Yaffe, M. B., Sayos, J., Saxena, K., Morra, M., Sumegi, J., Cantley, L. C., Terhorst, C., and Eck, M. J. (1999). Crystal structures of the XLP protein SAP reveal a class of SH2 domains with extended, phosphotyrosine-independent sequence recognition. Mol. Cell *4*, 555-561.

Pullen, A. M., Bill, J., Kubo, R. T., Marrack, P., and Kappler, J. W. (1991). Analysis of the interaction site for the self superantigen Mls-1a on T cell receptor V beta. J. Exp. Med. *173*, 1183-1192.

Pullen, A. M., Kappler, J. W., and Marrack, P. (1989). Tolerance to self antigens shapes the T-cell repertoire. Immunol. Rev. *107*, 125-139.

Pullen, A. M., Wade, T., Marrack, P., and Kappler, J. W. (1990). Identification of the region of T cell receptor beta chain that interacts with the self-superantigen Mls-1a. Cell *61*, 1365-1374.

Rea, D., Fourcade, C., Leblond, V., Rowe, M., Joab, I., Edelman, L., Bitker, M. O., Gandjbakhch, I., Suberbielle, C., Farcet, J. P., and et al. (1994). Patterns of Epstein-Barr virus latent and replicative gene expression in Epstein-Barr virus B cell lymphoproliferative disorders after organ transplantation. Transplantation *58*, 317-324.

Robertson, E., and Kieff, E. (1995). Reducing the complexity of the transforming Epstein-Barr virus genome to 64 kilobase pairs. J. Virol. *69*, 983-993.

Rosa, M. D., Gottlieb, E., Lerner, M. R., and Steitz, J. A. (1981). Striking similarities are exhibited by two small Epstein-Barr virus-encoded ribonucleic acids and the adenovirus-associated ribonucleic acids VAI and VAII. Mol. Cell Biol. *1*, 785-796.

Saal, J. G., Krimmel, M., Steidle, M., Gerneth, F., Wagner, S., Fritz, P., Koch, S., Zacher, J., Sell, S., Einsele, H., and Muller, C. A. (1999). Synovial Epstein-Barr virus infection increases the risk of rheumatoid arthritis in individuals with the shared HLA-DR4 epitope. Arthritis Rheum. *42*, 1485-1496.

Sarawar, S. R., Blackman, M. A., and Doherty, P. C. (1994). Superantigen shock in mice with an inapparent viral infection. J. Infect. Dis. *170*, 1189-1194.

Sayos, J., Nguyen, K. B., Wu, C., Stepp, S. E., Howie, D., Schatzle, J. D., Kumar, V., Biron, C. A., and Terhorst, C. (2000). Potential pathways for regulation of NK and T cell responses: differential X-linked lymphoproliferative syndrome gene product SAP interactions with SLAM and 2B4. Int. Immunol. *12*, 1749-1757.

Sayos, J., Wu, C., Morra, M., Wang, N., Zhang, X., Allen, D., van Schaik, S., Notarangelo, L., Geha, R., Roncarolo, M. G., et al. (1998). The X-linked lymphoproliferative-disease gene product SAP regulates signals induced through the co-receptor SLAM. Nature *395*, 462-469.

Scherer, M. T., Ignatowicz, L., Pullen, A., Kappler, J., and Marrack, P. (1995). The use of mammary tumor virus (Mtv)-negative and single-Mtv mice to evaluate the effects of endogenous viral superantigens on the T cell repertoire. J. Exp. Med. *182*, 1493-1504.

Scherer, M. T., Ignatowicz, L., Winslow, G. M., Kappler, J. W., and Marrack, P. (1993). Superantigens: bacterial and viral proteins that manipulate the immune system. Annu. Rev. Cell Biol. *9*, 101-128.

Schulte, A. M., Lai, S., Kurtz, A., Czubayko, F., Riegel, A. T., and Wellstein, A. (1996). Human trophoblast and choriocarcinoma expression of the growth factor pleiotrophin attributable to germ-line insertion of an endogenous retrovirus. Proc. Natl. Acad. Sci. USA. *93*, 14759-14764.

Schwering, I., Brauninger, A., Klein, U., Jungnickel, B., Tinguely, M., Diehl, V., Hansmann, M. L., Dalla-Favera, R., Rajewsky, K., and Kuppers, R. (2003). Loss of the B-lineage-specific gene expression program in Hodgkin and Reed-Sternberg cells of Hodgkin lymphoma. Blood *101*, 1505-1512.

Sicat, J., Sutkowski, N., and Huber, B.T. (2005). Expression of the human endogenous retrovirus HERV-K18 superantigen is elavated in juvenile rheumatoid arthritis. J. Rheumatol. In press.

Smith, M. D., Lamour, A., Boylston, A., Lancaster, F. C., Pennec, Y. L., van Agthoven, A., Rook, G. A., Roncin, S., Lydyard, P. M., and Youinou, P. Y. (1994). Selective expression of V beta families by T cells in the blood and salivary gland infiltrate of patients with primary Sjogren's syndrome. J. Rheumatol *21*, 1832-1837.

Stauffer, Y., Marguerat, S., Meylan, F., Ucla, C., Sutkowski, N., Huber, B., Pelet, T., and Conrad, B. (2001). Interferon-alpha-induced endogenous superantigen. a model linking environment and autoimmunity. Immunity *15*, 591-601.

Staunton, D. E., and Thorley-Lawson, D. A. (1987). Molecular cloning of the lymphocyte activation marker Blast-1. Embo J. *6*, 3695-3701.

Stevenson, P. G., Belz, G. T., Altman, J. D., and Doherty, P. C. (1999). Changing patterns of dominance in the CD8+ T cell response during acute and persistent murine gamma-herpesvirus infection. Eur. J. Immunol. *29*, 1059-1067.

Stoye, J. P., and Coffin, J. M. (2000). A provirus put to work. Nature *403*, 715, 717.

Subramanyam, M., Mohan, N., Mottershead, D., Beutner, U., McLellan, B., Kraus, E., and Huber, B. T. (1993). Mls-1 superantigen: molecular characterization and functional analysis. Immunol. Rev. *131*, 117-130.

Sumida, T., Yonaha, F., Maeda, T., Tanabe, E., Koike, T., Tomioka, H., and Yoshida, S. (1992). T cell receptor repertoire of infiltrating T cells in lips of Sjogren's syndrome patients. J. Clin. Invest *89*, 681-685.

Sutkowski, N., Chen, G., Calderon, G., and Huber, B. T. (2004). Epstein-Barr virus latent membrane protein LMP-2A is sufficient for transactivation of the human endogenous retrovirus HERV-K18 superantigen. J. Virol. *78*, 7852-7860.

Sutkowski, N., Conrad, B., Thorley-Lawson, D. A., and Huber, B. T. (2001). Epstein-Barr virus transactivates the human endogenous retrovirus HERV- K18 that encodes a superantigen. Immunity *15*, 579-589.

Sutkowski, N., Palkama, T., Ciurli, C., Sekaly, R. P., Thorley-Lawson, D. A., and Huber, B. T. (1996). An Epstein-Barr virus-associated superantigen. J. Exp. Med. *184*, 971-980.

Takeuchi, K., Katsumata, K., Ikeda, H., Minami, M., Wakisaka, A., and Yoshiki, T. (1995). Expression of endogenous retroviruses, ERV3 and lambda 4-1, in synovial tissues from patients with rheumatoid arthritis. Clin. Exp. Immunol. *99*, 338-344.

Taki, S. (2002). Type I interferons and autoimmunity: lessons from the clinic and from IRF-2-deficient mice. Cytokine Growth Factor Rev. *13*, 379-391.

Thomas, J. A., Iliescu, V., Crawford, D. H., Ellouz, R., Cammoun, M., and de-The, G. (1984). Expression of HLA-DR antigens in nasopharyngeal carcinoma: an immunohistological analysis of the tumour cells and infiltrating lymphocytes. Int. J. Cancer *33*, 813-819.

Thorley-Lawson, D. A. (2001). Epstein-Barr virus: exploiting the immune system. Nat. Rev. Immunol. *1*, 75-82.

Thorley-Lawson, D. A., and Gross, A. (2004). Persistence of the Epstein-Barr virus and the origins of associated lymphomas. N. Engl. J. Med. *350*, 1328-1337.

Thorley-Lawson, D. A., Ianelli, C., Klaman, L. D., Staunton, D., and Yokoyama, S. (1993). Function of CD48 and its regulation by Epstein-Barr virus. Biochem. Soc. Trans *21*, 976-980.

Thorley-Lawson, D. A., Schooley, R. T., Bhan, A. K., and Nadler, L. M. (1982). Epstein-Barr virus superinduces a new human B cell differentiation antigen (B-LAST 1) expressed on transformed lymphoblasts. Cell *30*, 415-425.

Tomkinson, B. E., Maziarz, R., and Sullivan, J. L. (1989). Characterization of the T cell-mediated cellular cytotoxicity during acute infectious mononucleosis. J. Immunol. *143*, 660-670.

Tomkinson, B. E., Wagner, D. K., Nelson, D. L., and Sullivan, J. L. (1987). Activated lymphocytes during acute Epstein-Barr virus infection. J. Immunol. *139*, 3802-3807.

Tonjes, R. R., Czauderna, F., and Kurth, R. (1999). Genome-wide screening, cloning, chromosomal assignment, and expression of full-length human endogenous retrovirus type K. J. Virol. *73*, 9187-9195.

Torres, B. A., and Johnson, H. M. (1998). Modulation of disease by superantigens. Curr. Opin. Immunol. *10*, 465-470.

Tosato, G., Steinberg, A. D., Yarchoan, R., Heilman, C. A., Pike, S. E., De Seau, V., and Blaese, R. M. (1984). Abnormally elevated frequency of Epstein-Barr virus-infected B cells in the blood of patients with rheumatoid arthritis. J. Clin. Invest *73*, 1789-1795.

Tripp, R. A., Hamilton-Easton, A. M., Cardin, R. D., Nguyen, P., Behm, F. G., Woodland, D. L., Doherty, P. C., and Blackman, M. A. (1997). Pathogenesis of an infectious mononucleosis-like disease induced by a murine gamma-herpesvirus: role for a viral superantigen? J. Exp. Med. *185*, 1641-1650.

Tristem, M. (2000). Identification and characterization of novel human endogenous retrovirus families by phylogenetic screening of the Human Genome Mapping Project Database. J. Virol. *74*, 3715-3730.

Tsai, Y. T., Chiang, B. L., Kao, Y. F., and Hsieh, K. H. (1995). Detection of Epstein-Barr virus and cytomegalovirus genome in white blood cells from patients with juvenile rheumatoid arthritis and childhood systemic lupus erythematosus. Int. Arch. Allergy Immunol. *106*, 235-240.

Turner, G., Barbulescu, M., Su, M., Jensen-Seaman, M. I., Kidd, K. K., and Lenz, J. (2001). Insertional polymorphisms of full-length endogenous retroviruses in humans. Curr. Biol. *11*, 1531-1535.

Uddin, S., Grumbach, I. M., Yi, T., Colamonici, O. R., and Platanias, L. C. (1998). Interferon alpha activates the tyrosine kinase Lyn in haemopoietic cells. Br. J. Haematol. *101*, 446-449.

Urnovitz, H. B., and Murphy, W. H. (1996). Human endogenous retroviruses: nature, occurrence, and clinical implications in human disease. Clin. Microbiol. Rev. *9*, 72-99.

Venables, P. J., Ross, M. G., Charles, P. J., Melsom, R. D., Griffiths, P. D., and Maini, R. N. (1985). A seroepidemiological study of cytomegalovirus and Epstein-Barr virus in rheumatoid arthritis and sicca syndrome. Ann. Rheum. Dis. *44*, 742-746.

Veronese, M. L., Veronesi, A., D'Andrea, E., Del Mistro, A., Indraccolo, S., Mazza, M. R., Mion, M., Zamarchi, R., Menin, C., Panozzo, M., and et al. (1992). Lymphoproliferative disease in human peripheral blood mononuclear cell- injected SCID mice. I. T lymphocyte requirement for B cell tumor generation. J. Exp. Med. *176*, 1763-1767.

Veronesi, A., Coppola, V., Veronese, M. L., Menin, C., Bruni, L., D'Andrea, E., Mion, M., Amadori, A., and Chieco-Bianchi, L. (1994). Lymphoproliferative disease in human peripheral-blood-mononuclear-cell- injected scid mice. II. Role of host and donor factors in tumor generation. Int. J. Cancer *59*, 676-683.

Wang-Johanning, F., Frost, A. R., Jian, B., Epp, L., Lu, D. W., and Johanning, G. L. (2003). Quantitation of HERV-K env gene expression and splicing in human breast cancer. Oncogene *22*, 1528-1535.

Wei, S., Charmley, P., Robinson, M. A., and Concannon, P. (1994). The extent of the human germline T-cell receptor V beta gene segment repertoire. Immunogenetics *40*, 27-36.

White, J., Herman, A., Pullen, A. M., Kubo, R., Kappler, J. W., and Marrack, P. (1989). The V beta-specific superantigen staphylococcal enterotoxin B: stimulation of mature T cells and clonal deletion in neonatal mice. Cell *56*, 27-35.

Whittingham, S., McNeilage, J., and Mackay, I. R. (1985). Primary Sjogren's syndrome after infectious mononucleosis. Ann. Intern. Med. *102*, 490-493.

Winslow, G. M., Marrack, P., and Kappler, J. W. (1994). Processing and major histocompatibility complex binding of the MTV7 superantigen. Immunity *1*, 23-33.

Winslow, G. M., Scherer, M. T., Kappler, J. W., and Marrack, P. (1992). Detection and biochemical characterization of the mouse mammary tumor virus 7 superantigen (Mls-1a). Cell *71*, 719-730.

Wood, T. A., and Frenkel, E. P. (1967). The atypical lymphocyte. Amer. J. Med. *42*, 923-936.

Yao, Q. Y., Rickinson, A. B., Gaston, J. S., and Epstein, M. A. (1986). Disturbance of the Epstein-Barr virus-host balance in rheumatoid arthritis patients: a quantitative study. Clin. Exp. Immunol. *64*, 302-310.

Yao, Z., Maraskovsky, E., Spriggs, M. K., Cohen, J. I., Armitage, R. J., and Alderson, M. R. (1996). Herpesvirus saimiri open reading frame 14, a protein encoded by T lymphotropic herpesvirus, binds to MHC class II molecules and stimulates T cell proliferation. J. Immunol. *156*, 3260-3266.

Yokoyama, S., Staunton, D., Fisher, R., Amiot, M., Fortin, J. J., and Thorley-Lawson, D. A. (1991). Expression of the Blast-1 activation/adhesion molecule and its identification as CD48. J. Immunol. *146*, 2192-2200.

Yonaha, F., Sumida, T., Maeda, T., Tomioka, H., Koike, T., and Yoshida, S. (1992). Restricted junctional usage of T cell receptor V beta 2 and V beta 13 genes, which are overrepresented on infiltrating T cells in the lips of patients with Sjogren's syndrome. Arthritis Rheum. *35*, 1362-1367.

Yu, C., Shen, K., Lin, M., Chen, P., Lin, C., Chang, G. D., and Chen, H. (2002). GCMa regulates the syncytin-mediated trophoblastic fusion. J. Biol. Chem. *277*, 50062-50068.

Zhang, W. J., Sarawar, S., Nguyen, P., Daly, K., Rehg, J. E., Doherty, P. C., Woodland, D. L., and Blackman, M. A. (1996). Lethal synergism between influenza infection and staphylococcal enterotoxin B in mice. J. Immunol. *157*, 5049-5060.

Zhao, T. M., Whitaker, S. E., and Robinson, M. A. (1994). A genetically determined insertion/deletion related polymorphism in human T cell receptor beta chain (TCRB) includes functional variable gene segments. J. Exp. Med. *180*, 1405-1414.

Chapter 15

Epstein-Barr Virus Genome

*Paul J. Farrell**

ABSTRACT

A genetic map has been compiled for Epstein-Barr virus using published features annotated to the EBVwt sequence. EBVwt was assembled from the B95-8 and Raji sequences, which were determined experimentally. The detailed annotation for the summary map illustrated in this chapter is available from the EMBL or Genbank databases under accession number AJ507799.

INTRODUCTION

Epstein-Barr virus (EBV) is one of the eight known human herpesviruses. Its systematic name is human herpesvirus 4 (HHV4). Its genome is composed of ds DNA, about 170kb in length, and the DNA is linear in the virus particle. A repeated DNA sequence (the terminal repeat) present at the ends of the linear form (Figure 1) mediates circularisation in the infected cell so that latently infected cells contain the genome as a circular plasmid in the nucleus, with usually about $10-20$ copies per infected cell. EBV was the first of the herpesviruses to be completely sequenced (Baer et al., 1984). Recently the sequences of relevant parts of the B95-8 and Raji strains of EBV have been used to create a composite sequence, known as EBVwt, which is now used as the reference sequence for the wild type EBV genome (de Jesus et al., 2003). EBV is a member of the gamma herpesviruses; the other human virus in this family is KSHV (HHV8).

REPEAT SEQUENCES AND GENOME STRUCTURE

The gamma herpesviruses typically have terminal repeats (TR) at the ends of the linear genome and a few other small repeat sequence arrays within the genome. EBV happens also to contain an unusually large tandemly repeated DNA sequence within the genome, known as the major internal repeat or IR1. It typically comprises between 5 and 10 copies of a sequence of 3072 bp and is functionally significant because it contains the Wp promoter for the EBNA genes that is active when the virus first infects B lymphocytes. IR1 is conventionally considered to divide the EBV genome (Figure 1) into the long and short unique sequences (U_L and U_S), the unique sequences being almost completely filled with closely packed genes.

*For correspondence email p.farrell@imperial.ac.uk

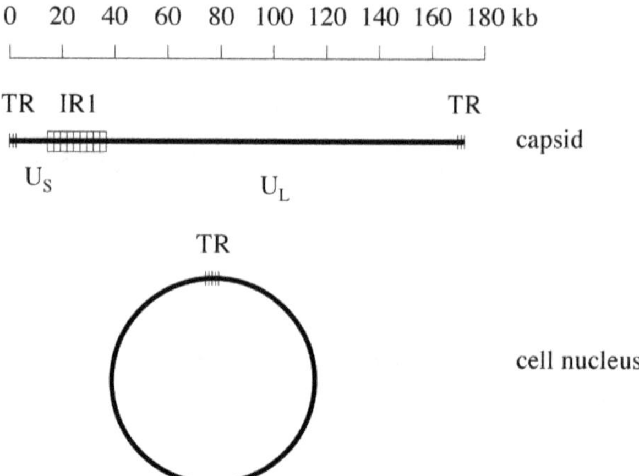

Figure 1. Diagram of linear and circular EBV genome structures, TR terminal repeat, U_L Unique long, U_S Unique short, IR1 internal repeat 1. The linear form in the capsid is circularised in the infected cell by joining at the terminal repeats.

Although similar names are used to describe the U_S and U_L regions of the alpha herpesviruses (eg Herpes Simplex Virus, HSV), the internal repeat sequences of HSV are functionally quite different from those of EBV. EBV does not have the genome isomerisation that is present in the alpha herpesviruses – there is a single arrangement of U_S -IR1- U_L in EBV.

Historical note on gene nomenclature

In other herpesviruses the genes are simply numbered according to their positions in the U_L or U_S region but in EBV the genes are systematically named according to the *Bam*HI restriction fragment in which they start. The *Bam*HI restriction fragments are conventionally named according to their size, A being the largest. So in EBV, for example, BARF1 is the first rightward reading frame starting in BamHI A, BXLF2 is the second leftward reading frame starting in Bam HI X. The systematic names may become replaced with more meaningful functional names as the genes become identified but the systematic names are still useful for clarity or where a function is not known.

The different gene nomeclature system arose because EBV was the first herpesvirus to be sequenced; some parts of the EBV genome were published before the entire sequence was completed so a nomenclature that was not dependent on the complete sequence was required. It was the first time that gene prediction solely from DNA sequence had been done on a large scale and there was some uncertainty about whether the gene predictions would be reliable; in fact the gene predictions proved to be remarkably accurate.

Evolutionary relationship between herpesviruses

The herpesvirus strategy involves infection at an external site on the host, usually a brief replication and then movement of the virus to another location and cell type in the host, where latent persistence occurs. Intermittent reactivation from that site and shedding at the exterior completes the infectious cycle for transmission to an additional host. The latent persistence is normally life-long. The different herpesviruses found in nature all represent different evolutionary solutions to this overall scheme and become latent in different cell niches. There is a general similarity in the capsid structure so the genes involved in lytic DNA replication, packaging and capsid structure are quite similar in all the herpesviruses. Although blocks of these genes may have become rearranged in evolution, there are several sections of the genome where lytic cycle genes are colinear in most herpesviruses. Alignment of these blocks has been described for all the sequenced gamma herpesviruses - EBV, KSHV, herpesvirus saimiri and murine herpesvirus 68 (Virgin et al., 1997). In contrast, the genes associated with latency, initial reactivation from latency and some of the surface proteins involved in cell type specific infection are generally unique to each virus, consistent with its unique site of latent persistence.

EBV STRAINS

Many strains of EBV have undergone simple characterisation at the DNA level by restriction mapping and Southern blotting the fragments. Isolates from people with infectious mononucleosis, Burkitt's lymphoma and nasopharyngeal carcinoma have been described in this way (Kieff and Rickinson, 2001). Curiously, no EBV strain isolated directly from a normal healthy carrier has been fully studied at the genome level. The genome structures from the different isolates have all been very similar, some isolates having characteristic deletions, which may have occurred in cell culture. There is significant variation in the copy numbers of several of the repeat arrays between strains and within any strain population so there is no such thing as a unique wild type EBV sequence.

The prototype strain B95-8, which has been completely sequenced (Baer et al., 1984), is convenient to work with because the virus can be produced relatively efficiently and immortalises B lymphocytes well. However, B95-8 has a large deletion relative to other strains and is also known to have some relatively unusual sequence variation in the LMP1 gene.

The main strain variation that has been recognised is the classification into A and B types (also known as type I and type II) based on the EBNA2 sequence (Kieff and Rickinson, 2001). A and B type EBNA2 can be distinguished serologically (there are some sera that specifically Western blot each type and some EBNA2 sera that recognise both forms). The EBNA2 gene sequence can also be distinguished at the DNA level using certain DNA probes that do not cross hybridise on Southern blots or specific PCR primers (Sample et al., 1990). There is additional co-variation of the EBNA3 family of genes with EBNA2 in the A and B type classification (Sample et al., 1990). A type strains produce LCLs more efficiently and the LCLs usually proliferate more rapidly than B type strains. Most Western isolates are A type but there is a significant proportion of B type EBV strains present in African populations. B95-8 and Akata are common examples of A type strains and AG876 and Jijoye are B type.

Several other patterns of sequence variation have been described, usually from sequencing single genes or other short segments of the viral DNA. LMP1 (Sung et al., 1998), LMP2 (Busson et al., 1992), EBNA1 (Wang et al., 2002) and BZLF1 (Packham et al., 1993) have all been studied in this way. Most work of this type has been done on LMP1, trying to determine whether there could be functionally different EBV types that could be relevant to the occurrence of EBV related cancers, for example (Edwards et al., 2004). At present there is no simple correlation of such viral sequence variation with disease but the possibility of a link remains open (Jenkins and Farrell, 1996). At present there seems to be no simple linkage between the A and B type variation and the other patterns of variation.

The selective forces that may result in the EBV sequence variation that is observed in human populations are not well understood and have been controversial, particularly whether immune surveillance may select some variants. There is some recent evidence consistent with immune selection acting on epitopes in some of the latent cycle genes depending on MHC type of the infected host (Midgley et al., 2003a; Midgley et al., 2003b)

REARRANGED DEFECTIVE VIRAL GENOMES

The clearest example of rearranged defective viral genomes in EBV is the het virus that has been described in P3HR1 EBV. By cloning P3HR1 Burkitt's lymphoma cells, lines which contained cells spontaneously in the viral lytic cycle could be separated from lines that had only latent infection. The lytic cycle induction correlated with the presence of additional het (heterogeneous) viral DNA (Miller et al., 1984). Sequence analysis of this showed that it comprises a rearranged and defective EBV genome that has the immediate early BZLF1 gene placed under the control of part of the constitutive Wp promoter so that BZLF1 is always expressed (Jenson et al., 1986), consequently inducing the lytic cycle. Cell cultures with het virus would presumably have standard P3HR1 in all the cells and some of the cells would also have the het virus. The replication caused by the het virus and infection of the other cells would tend to keep present it the population. A separate analysis of rearranged EBV DNA in P3HR1 came up with a similar but slightly different rearrangement (Cho et al., 1984) in which both BZLF1 and BRLF1 could be activated by Wp, consistent with the transcription analysis of the viral gene expression that occurs when an uncloned population of EBV from P3HR1 cells was superinfected into Raji cells (Biggin et al., 1987).

TRANSCRIPTION MAP OF EBV

Most of the gene expression analysis of EBV has been performed with the B95-8 strain. Major open reading frames were predicted from the DNA sequence to give the first genetic map and this provided the basis for transcript mapping. Data from many sources was used to compile the reference map of latent and lytic cycle gene expression (Appendix 1 [at the end of this chapter]). It is important to appreciate that EBV has several different patterns of gene expression in lymphocytes and epithelial cells. In lymphocytes the main patterns of latent gene expression have been described as Latency I, II and III (Rowe et al., 1987); there is some deviation in certain cell lines from these standard patterns but they are a good general description. EBV latent transcription in epithelial cells has not been analysed in such detail but latency I and latency II are observed. The map includes all of these patterns and should thus be used as a reference from which the relevant subsets of transcripts can be drawn.

Most of the lytic cycle mapping has been based on TPA induced B95-8 cells (or Raji cells for the B95-8 deletion region). PAA prevents EBV lytic DNA replication and consequently prevents late gene expression; this was used to distinguish late from early lytic cycle genes. So far, little attention has been given to relative timing of expression of early and late genes because of the experimental difficulty of studying a synchronous lytic cycle. TPA treatment probably activates the lytic cycle in a relatively non-specific way and the reactivation of the lytic cycle in Akata cells in response to cross linking the B cell receptor with anti-Ig (Takada and Ono, 1989) appears to be the most rapid, synchronous and biologically relevant system for studying the lytic cycle available at present. The first gene to be activated in the lytic cycle in lymphocytes is BZLF1. The BZLF1 protein is a transcription factor that activates BRLF1 and then these two proteins cooperate to activate most of the viral early genes. Induction of the Zp promoter, which gives rise to the mRNA for BZLF1, is the key step in the transition between latency and the EBV lytic cycle (Binné et al., 2002; Bryant and Farrell, 2002).

ORGANISATION OF THE TRANSCRIPTION MAP

The organisation of some blocks of early and late lytic cycle genes is recognisably conserved between herpesviruses but there is no simple early or late region, the lytic cycle genes of different expression classes are interspersed on the genome. There are many examples where a common poly A site is used for several mRNAs (Appendix 1), the longer mRNAs in such groups thus contain additional downstream open reading frames. Assuming EBV mRNAs are capped, usually only the most 5′ reading frame in each mRNA would be translated.

Latent cycle genes were discovered through a mixture of RNA and protein analysis. EBNA1, 2 and the EBNA3 family proteins were identified initially using human sera and the corresponding genes were subsequently assigned. The EBNA-LP, the EBER RNAs and the LMP genes were identified first by RNA analysis. Production of the mRNAs for the all latent cycle genes involves RNA splicing but this is rare in the lytic cycle genes.

In latency I the only EBNA to be expressed is EBNA1 and the mRNA is transcribed from the Qp promoter. In latency III all the EBNA mRNAs are produced from the same large transcription unit, alternate splicing giving rise to the mRNAs for the individual EBNAs. During initial infection of B cells, the latency III transcription inititiates at the Wp promoter in IR1, which is a very strong promoter. As a proportion of the infected cells grow out as immortalised cell lines, Wp activity subsides and Cp usually becomes the main promoter expressing the EBNA RNAs in latency III. Perhaps the purpose of Wp is to allow strong initial expression of EBNA proteins as EBV infects the resting B cell and this effect might be enhanced by the multiple copies of Wp present in the multiple IR1 repeats. As infection becomes established, the lower activity of Cp may be sufficient for expression of the correct EBNA protein levels. The very large transcription unit (100kb) for the EBNA genes, covering more than half the viral genome, might help to avoid accumulation of deleted viral episomes that could potentially accumulate during the long viral persistence *in vivo*.

There has been a tendency for the search for EBV genes to be directed by the open reading frame map so the search has not been completely systematic. It remains likely that there are a few additional genes to be discovered. Most of the genome is almost completely filled with genetic function and it seems implausible that other currently empty regions will be without function. This is particularly noticeable in the region around 148,000 on the map where no functions or significant open reading frames have been identified. The poorly understood BART genes are being studied in this respect but currently identified RNA structures such as A73 and RPMS1 (Smith et al., 2000) do not explain the empty region adequately. BART RNAs have been reported in all three latency states observed in cell culture but mostly using RT-PCR with little consideration of whether the levels detected are quantitatively significant. Novel and unexpected genetic functions continue to be found, for example, EBV encoded micro RNAs were recently described (Pfeffer et al., 2004). The possible functions of these in RNA silencing remain to be determined but this is the first example of such microRNAs to be found in a human virus.

By far the most abundant transcripts from the EBV genome in latent infection are the EBER RNAs. These are not translated into protein and have been proposed to function by preventing the activity of dsRNA dependent enzymes that can be induced by interferons (Clemens, 1994). More recent studies suggest a role for the EBER RNAs in tumourigenicity of Burkitt's Lymphoma cells by conferring reistance to interferon mediated apoptosis(Nanbo et al., 2002). EBER RNA can also bind to ribosomal protein L22 (Toczyski et al., 1994) and with the increasing appreciation of the role of small RNAs in regulating gene expression, it seems likely that additional functions of the EBER RNAs remain to be discovered.

MODIFICATION OF THE EBV GENOME AND CPG SUPPRESSION

Many studies have shown the accumulation of methylated CpG dinucleotides in EBV DNA in latently infected cell lines. Investigation of the Wp to Cp switch during early infection indicated that inactivation of Wp was associated with CpG methylation of the Wp promoter, the incoming viral DNA being unmethylated in this region. Intracellular DNA that has been replicated in the lytic cycle is usually unmethylated at CpG but one report has suggested a substantial level of cytosine methylation in EBV virion DNA (Diala and Hoffman, 1983). One remarkable feature of the EBV genome is the extent of CpG suppression in the sequence. The EBV DNA sequence is relatively GC rich (about 60% GC) but the CpG dinucleotide occurs much less frequently than would be predicted in random sequence of this GC content. Most likely this has occurred through deamidation of methylated cytosine residues in the genome over evolutionary time. Since EBV DNA spends most of its time in latency, this implies that the EBV genome is methylated in latency *in vivo*.

The 60% GC DNA content and CpG suppression results in an interesting codon usage bias in the genes of EBV. The excess GC is accommodated within the coding requirements of the viral proteins by tending to be in the third position of codons, where it does not affect the amino acid coded. Some genes of EBV have up to 80% GC in the third position of their codons. This constraint only applies if the sequence is actually expressed as a protein and this type of bioinformatic analysis has been used as part of the justification of open reading frames as real genes in EBV.

CONCLUSIONS

EBV was the first large virus genome to be sequenced and subsequent comparison of sequenced herpesviruses has provided a great insight into the evolutionary relationships and biology of the viruses. Continuous refinement of the annotation of the EBV sequence has provided a valuable reference map of the viral genome and the precise numbering allows convenient description of features at the sequence level. However, when using the map, it is important to remember that there is considerable natural variation of repeat copy numbers and some strain variation that is not yet fully understood.

References

Baer, R., Bankier, A. T., Biggin, M. D., Deininger, P. L., Farrell, P. J., Gibson, T. J., Hatfull, G., Hudson, G. S., Satchwell, S. C., Seguin, C., et al. (1984). DNA sequence and expression of the B95-8 Epstein-Barr virus genome. Nature *310*, 207-211.

Biggin, M., Bodescot, M., Perricaudet, M., and Farrell, P. (1987). Epstein-Barr virus gene expression in P3HR1-superinfected Raji cells. J. Virol. *61*, 3120-3132.

Binné, U., Amon, W., and Farrell, P. (2002). Promoter sequences required for reactivation of Epstein-Barr Virus from Latency. J. Virol. *76*, 10282-10289.

Bryant, H., and Farrell, P. (2002). Signal transduction and transcription factor modification during reactivation of Epstein-Barr virus from latency. J. Virol. *76*, 10290-10298.

Busson, P., McCoy, R., Sadler, R., Gilligan, K., Tursz, T., and Raab-Traub, N. (1992). Consistent transcription of the Epstein-Barr virus LMP2 gene in nasopharyngeal carcinoma. J. Virol. *66*, 3257-3262.

Cho, M. S., Gissmann, L., and Hayward, S. D. (1984). Epstein-Barr virus (P3HR-1) defective DNA codes for components of both the early antigen and viral capsid antigen complexes. Virology *137*, 9-19.

Clemens, M. J. (1994). Functional Significance of the Epstein-Barr Virus-encoded Small RNAs. Epstein-Barr Virus Report *1*, 107-111.

de Jesus, O., PR, S., Spender, L., Elgueta Karstegl, C., Niller, H., Huang, D., and PJ, F. (2003). Updated Epstein-Barr virus (EBV) DNA sequence and analysis of a promoter for the BART (CST, BARF0) RNAs of EBV. J. Gen. Virol. *84*, 1443-1450.

Diala, E. S., and Hoffman, R. M. (1983). Epstein-Barr HR-1 virion DNA is very highly methylated. J. Virol. *45*, 482-483.

Edwards, R. H., Sitki-Green, D., Moore, D. T., and Raab-Traub, N. (2004). Potential selection of LMP1 variants in nasopharyngeal carcinoma. J. Virol. *78*, 868-881.

Jenkins, P., and Farrell, P. (1996). Are particular Epstein-Barr virus strains linked to disease? Seminars in Cancer Biology *7*, 209-215.

Jenson, H. B., Rabson, M. S., and Miller, G. (1986). Palindromic structure and polypeptide expression of 36 kilobase pairs of heterogeneous Epstein-Barr virus (P3HR-1) DNA. J. Virol. *58*, 475-486.

Kieff, E., and Rickinson, A. (2001). Epstein-Barr virus and its replication. In Fields Virology, D. Knipe, and P. Howley, eds. (Philadelphia, Lippincott), pp. 2511-2573.

Midgley, R. S., Bell, A. I., McGeoch, D. J., and Rickinson, A. B. (2003a). Latent gene sequencing reveals familial relationships among Chinese Epstein-Barr virus strains and evidence for positive selection of A11 epitope changes. J. Virol. *77*, 11517-11530.

Midgley, R. S., Bell, A. I., Yao, Q. Y., Croom-Carter, D., Hislop, A. D., Whitney, B. M., Chan, A. T., Johnson, P. J., and Rickinson, A. B. (2003b). HLA-A11-restricted epitope polymorphism among Epstein-Barr virus strains in the highly HLA-A11-positive Chinese population: incidence and immunogenicity of variant epitope sequences. J. Virol. *77*, 11507-11516.

Miller, G., Rabson, M., and Heston, L. (1984). Epstein-Barr virus with heterogeneous DNA disrupts latency. J. Virol. *50*, 174-182.

Nanbo, A., Inoue, K., Adachi-Takasawa, K., and Takada, K. (2002). Epstein-Barr virus RNA confers resistance to interferon-alpha-induced apoptosis in Burkitt's lymphoma. Embo J. *21*, 954-965.

Packham, G., Brimmell, M., Cook, D., Sinclair, A., and Farrell, P. (1993). Strain variation in Epstein-Barr virus immediate early genes. Virology *192*, 541-550.

Pfeffer, S., Zavolan, M., Grasser, F. A., Chien, M., Russo, J. J., Ju, J., John, B., Enright, A. J., Marks, D., Sander, C., and Tuschl, T. (2004). Identification of virus-encoded microRNAs. Science *304*, 734-736.

Rowe, M., Rowe, D. T., Gregory, C. D., Young, L. S., Farrell, P. J., Rupani, H., and Rickinson, A. B. (1987). Differences in B cell growth phenotype reflect novel patterns of Epstein-Barr virus latent gene expression in Burkitt's lymphoma cells. EMBO J. *6*, 2743-2751.

Sample, J., Young, L., Martin, B., Chatman, T., Kieff, E., Rickinson, A., and Kieff, E. (1990). Epstein-Barr virus types 1 and 2 differ in their EBNA-3A, EBNA-3B, and EBNA-3C genes. J. Virol. *64*, 4084-4092.

Smith, P., de Jesus, O., Turner, D., Hollyoake, M., Elgueta Karstegl, C., Griffin, B., Karran, L., Wang, Y., Hayward, S., and Farrell, P. (2000). Structure and coding content of CST (BART) family RNAs of Epstein-Barr virus. Journal of Virology *74*, 3082-3092.

Sung, N. S., Edwards, R. H., Seillier-Moiseiwitsch, F., Perkins, A. G., Zeng, Y., and Raab-Traub, N. (1998). Epstein-Barr virus strain variation in nasopharyngeal carcinoma from the endemic and non-endemic regions of China. Int. J. Cancer *76*, 207-215.

Takada, K., and Ono, Y. (1989). Synchronous and sequential activation of latently infected Epstein-Barr virus genomes. J. Virol. *63*, 445-449.

Toczyski, D. P., Matera, A. G., Ward, D. C., and Steitz, J. A. (1994). The Epstein-Barr virus (EBV) small RNA EBER1 binds and relocalizes ribosomal protein L22 in EBV-infected human B lymphocytes. Proc. Natl. Acad. Sci. USA. *91*, 3463-3467.

Virgin, H. W. t., Latreille, P., Wamsley, P., Hallsworth, K., Weck, K. E., Dal Canto, A. J., and Speck, S. H. (1997). Complete sequence and genomic analysis of murine gammaherpesvirus 68. J. Virol. *71*, 5894-5904.

Wang, W. Y., Chien, Y. C., Jan, J. S., Chueh, C. M., and Lin, J. C. (2002). Consistent sequence variation of Epstein-Barr virus nuclear antigen 1 in primary tumor and peripheral blood cells of patients with nasopharyngeal carcinoma. Clin. Cancer Res. *8*, 2586-2590.

Appendix 1. Map of EBVwt. Apperndix 2 gives the coordinates of selected features shown on the map. The symbols on the map are explained in the panel.

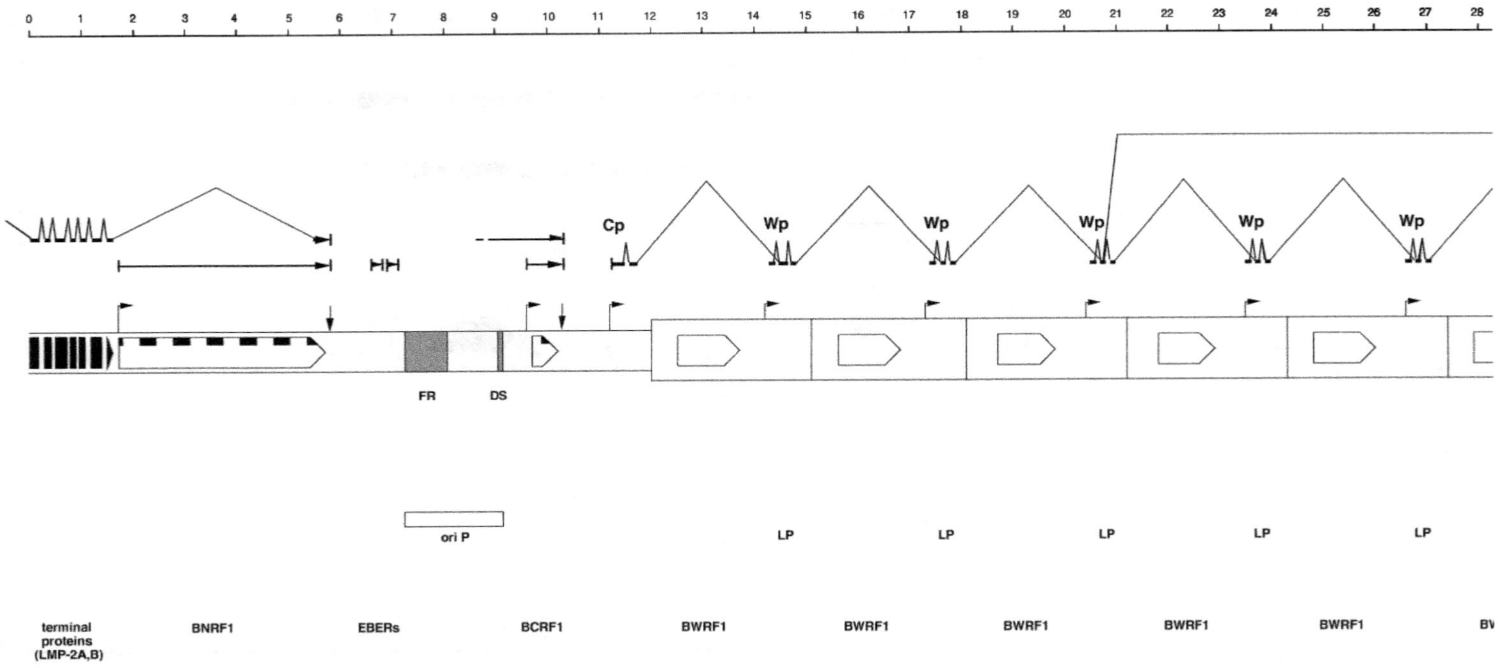

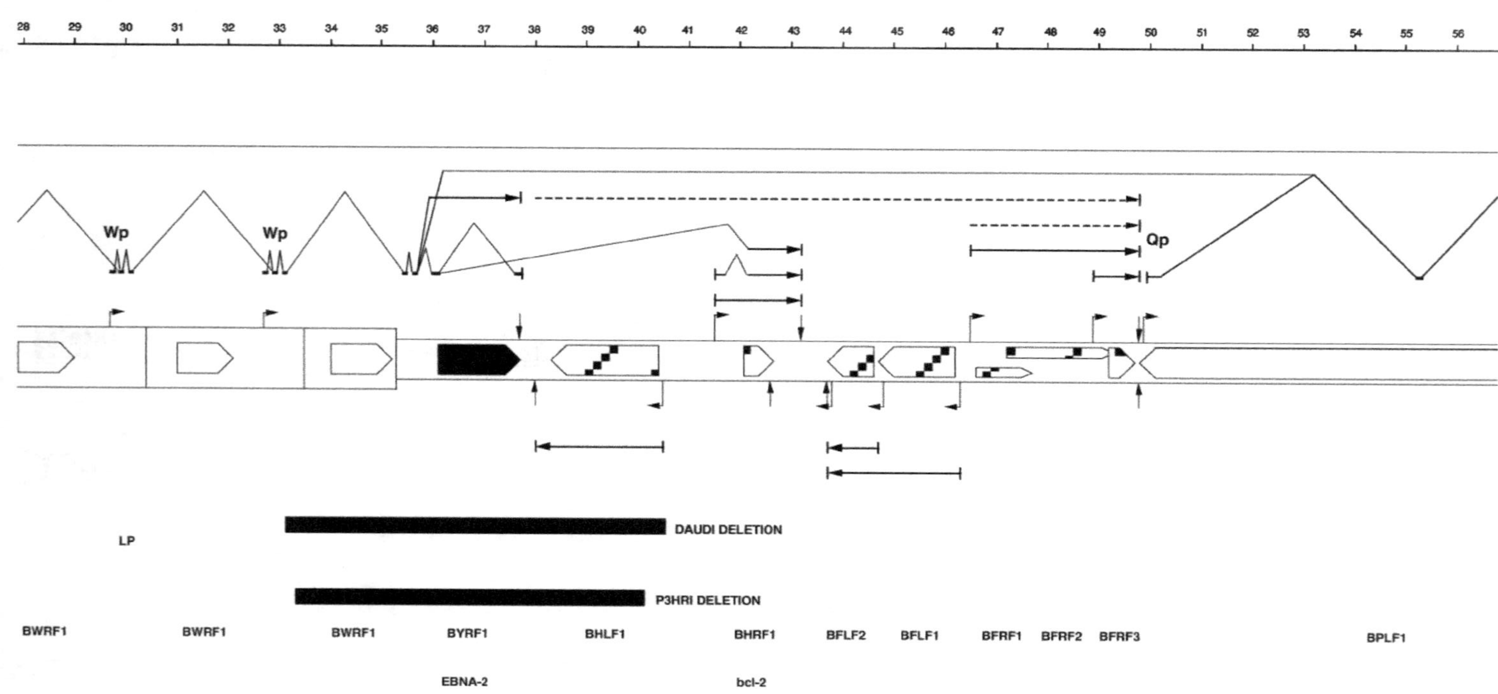

28 29 30 31 32 33 34 35 36 37 38 39 40 41 42 43 44 45 46 47 48 49 50 51 52 53 54 55 56
Wp
Wp
Qp
LP
DAUDI DELETION
P3HRI DELETION
BWRF1
BWRF1
BWRF1
BYRF1
BHLF1
BHRF1
BFLF2
BFLF1
BFRF1
BFRF2
BFRF3
BPLF1
EBNA-2
bcl-2

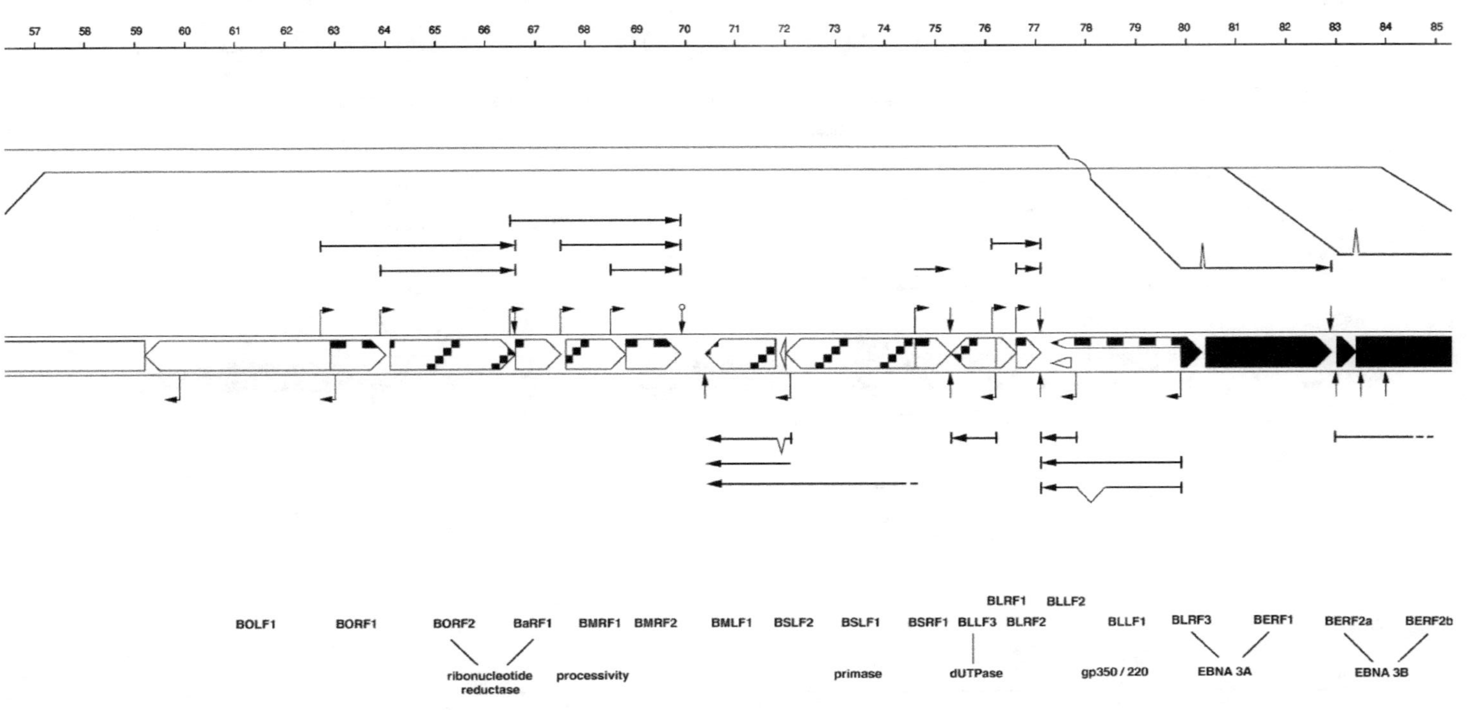

57 58 59 60 61 62 63 64 65 66 67 68 69 70 71 72 73 74 75 76 77 78 79 80 81 82 83 84 85
BOLF1
BORF1
BORF2
BaRF1
BMRF1
BMRF2
BMLF1
BSLF2
BSLF1
BSRF1
BLLF3
BLRF2
BLRF1
BLLF2
BLLF1
BLRF3
BERF1
BERF2a
BERF2b
ribonucleotide reductase
processivity
primase
dUTPase
gp350 / 220
EBNA 3A
EBNA 3B

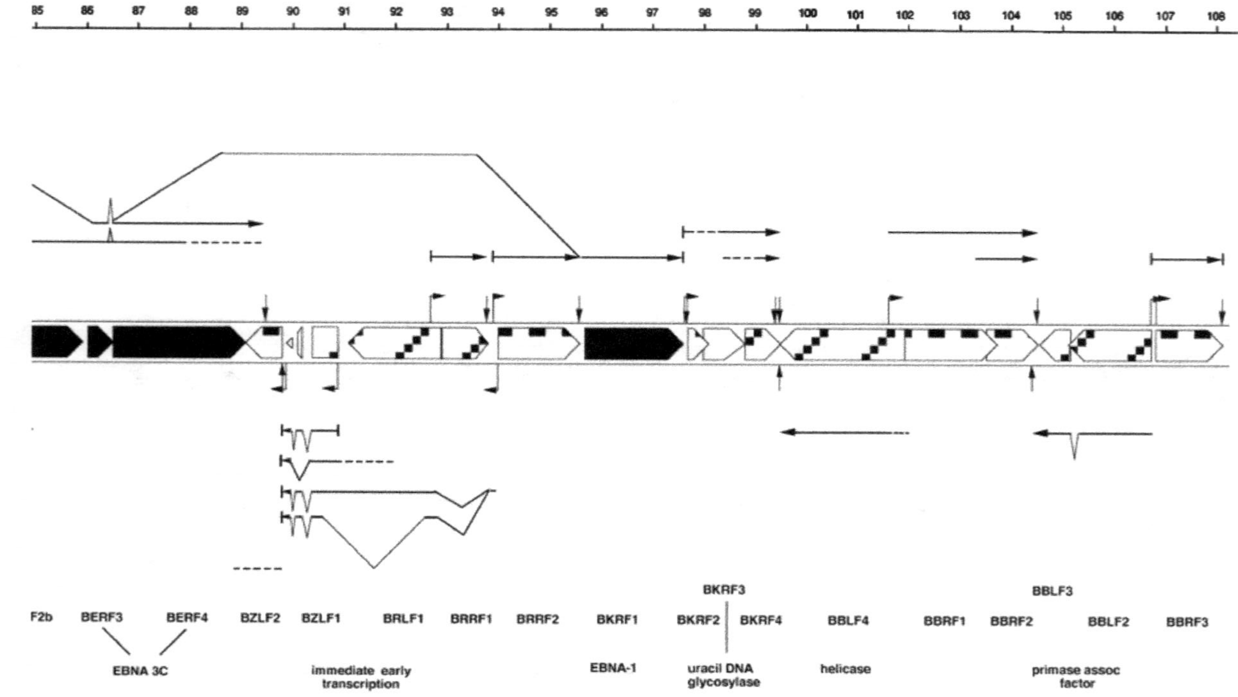

85 86 87 88 89 90 91 92 93 94 95 96 97 98 99 100 101 102 103 104 105 106 107 108
F2b BERF3 BERF4 BZLF2 BZLF1 BRLF1 BRRF1 BRRF2 BKRF1 BKRF2 BKRF4 BBLF4 BBRF1 BBRF2 BBLF2 BBRF3
BKRF3
BBLF3
EBNA 3C
immediate early transcription
EBNA-1
uracil DNA glycosylase
helicase
primase assoc factor

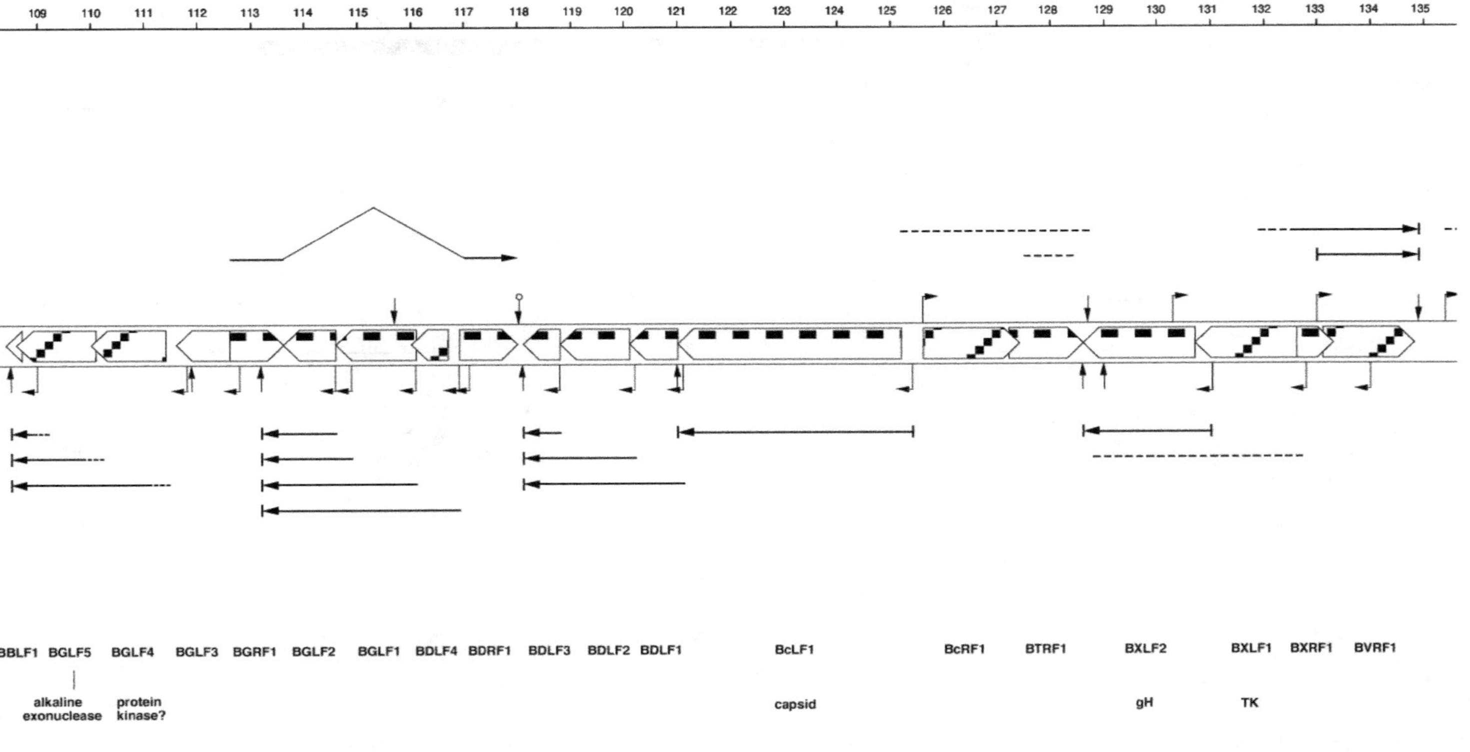

109 110 111 112 113 114 115 116 117 118 119 120 121 122 123 124 125 126 127 128 129 130 131 132 133 134 135
BBLF1 BGLF5 BGLF4 BGLF3 BGRF1 BGLF2 BGLF1 BDLF4 BDRF1 BDLF3 BDLF2 BDLF1 BcLF1 BcRF1 BTRF1 BXLF2 BXLF1 BXRF1 BVRF1
alkaline exonuclease
protein kinase?
capsid
gH
TK

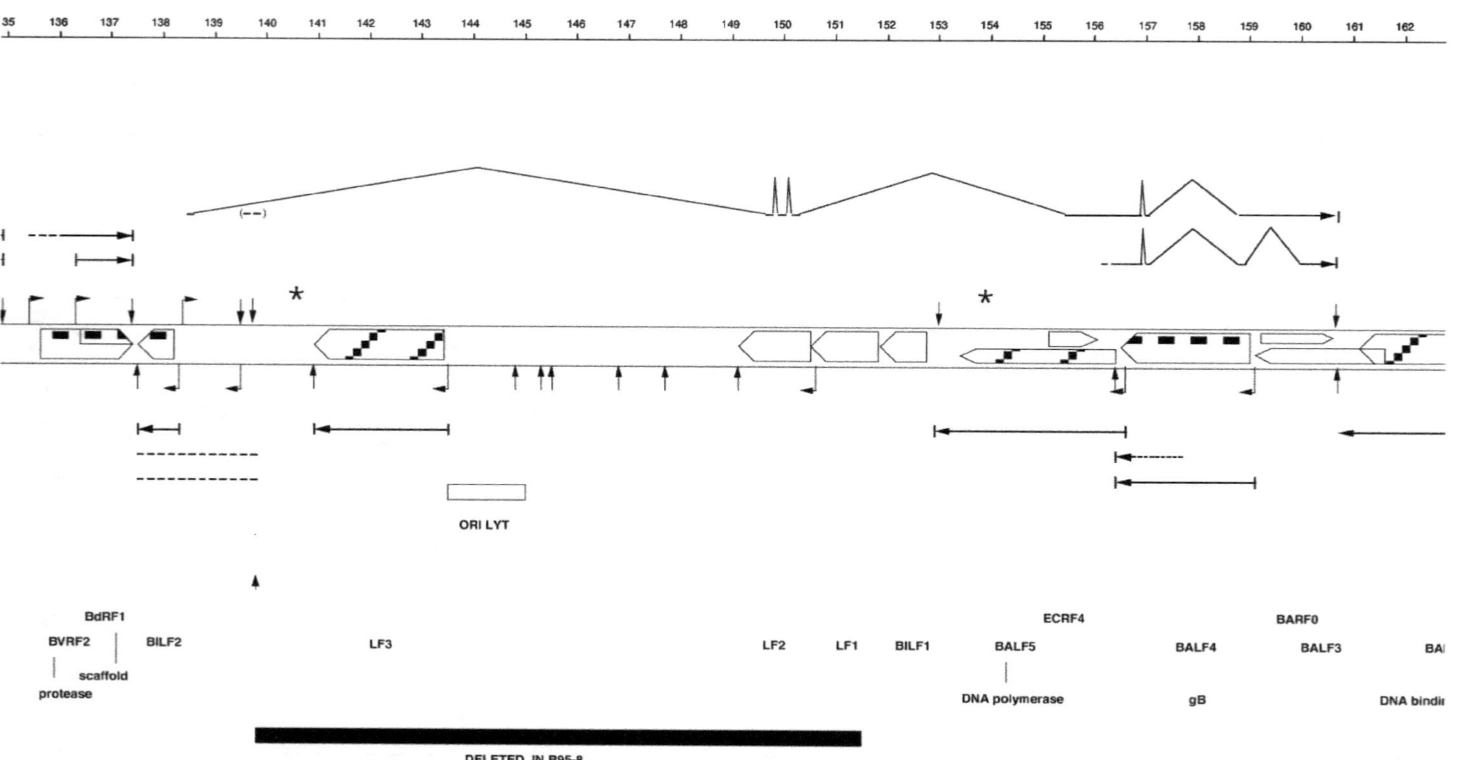

35 136 137 138 139 140 141 142 143 144 145 146 147 148 149 150 151 152 153 154 155 156 157 158 159 160 161 162
(--)
ORI LYT
BdRF1
BVRF2 BILF2 LF3 LF2 LF1 BILF1 ECRF4 BALF5 BALF4 BARF0 BALF3 BA
scaffold
protease DNA polymerase gB DNA bindi
DELETED IN B95-8

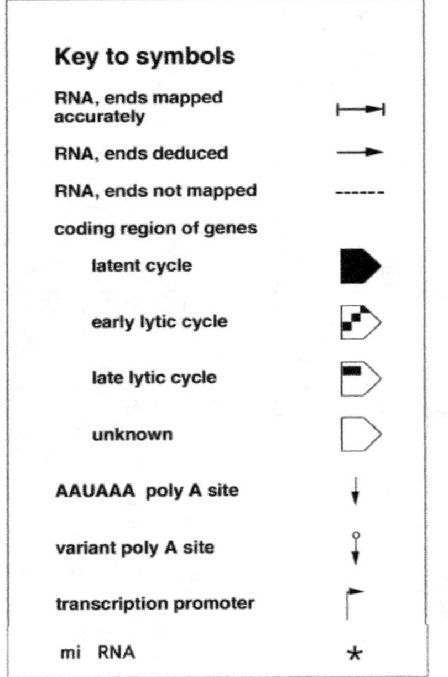
Key to symbols
RNA, ends mapped accurately
RNA, ends deduced
RNA, ends not mapped
coding region of genes
latent cycle
early lytic cycle
late lytic cycle
unknown
AAUAAA poly A site
variant poly A site
transcription promoter
mi RNA

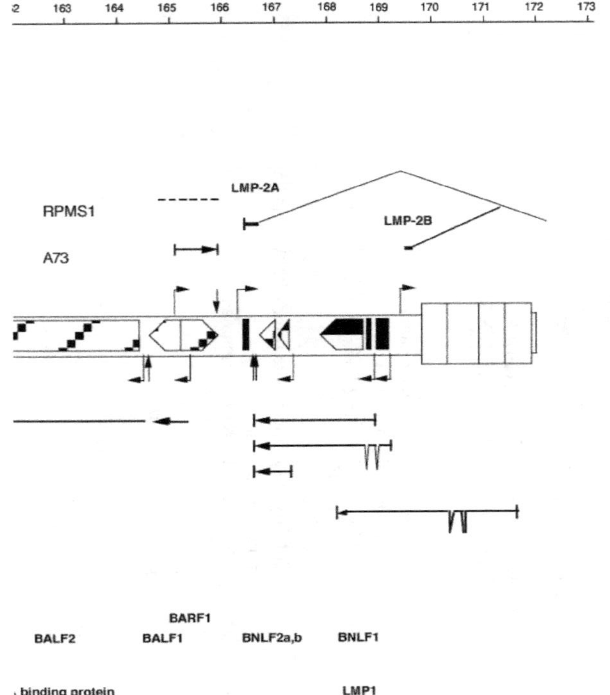
163 164 165 166 167 168 169 170 171 172 173
RPMS1
LMP-2A
LMP-2B
A73
BALF2
BALF1
BARF1
BNLF2a,b
BNLF1
binding protein
LMP1

Appendix 2. Simplified Feature Table for the map shown in Appendix 1.The full Feature Table which is periodically updated describes the coordinates of all the features illustrated and the protein sequences of all the identified open reading frames. It is available in accession number AJ507799 from the EMBL or Genbank data libraries.

FEATURES	LOCATION/QUALIFIERS
source	join (1..139723,151555..171823)
	/organism="Human herpesvirus 4" /strain="B95-8"
source	139724..151554
	/organism="Human herpesvirus 4" /strain="Raji"
CDS	join (166103..166458,58..272,360..458,540..788, 871..951,1026..1196,
	1280..1495,1574..1680)
	/product="terminal protein LMP2A" /gene="LMP2"
CDS	join (59..272,360..458,540..788,871..951,1026..1196,1280..1495,1574..1680)
	/product="terminal protein LMP2B" /gene="LMP2"
exon	58..272 /number="2" /gene="LMP2"
exon	360..458 /number="3" /gene="LMP2"
exon	540..788 /number="4" /gene="LMP2"
exon	871..951 /number="5" /gene="LMP2"
exon	1026..1196 /number="6" /gene="LMP2"
exon	1280..1495 /number="7" /gene="LMP2"
exon	1574..1682 /number="8" /gene="LMP2"
exon	5408..5856 /number="9" /gene="LMP2"
TATA_signal	1691..1695 /gene="BNRF1"
CDS	1736..5692 /gene="BNRF1"
	/product="putative nmembrane antigen p140"
misc_feature	3955..3955 /note="BAM: Bam H1 Nhet/h"
misc_feature	3994..3994 /note="BAM: Bam H1 h/C"
polyA_signal	5841..5846
misc_feature	6629..6795 /note="Pol III RNA EBER 1"
misc_feature	6956..7128 /note="Pol III RNA EBER 2"
rep_origin	7315..9312 /note="ori P"
repeat_region	7421..8042
	/note="binding sites for EBNA-1 (site I) "
	/note="tandem repeat part of oriP"
	/note="also functions as a cell type specific enhancer"
	/rpt_unit="7421..7450"
misc_feature	9021..9133
	/note="HPN: dyad symmetry, site II for EBNA-1 binding;
	dyad symmetry part of oriP"
TATA_signal	9631..9636 /note="BC-R1 late promoter"
CDS	9675..10187 /gene="BCRF1"
polyA_signal	10257..10262 /gene="BCRF1"
TATA_signal	11305..11308 /gene="BCR2"
exon	11336..11480 /number="C1" /gene="EBNA"
exon	11626..11657 /number="C2"
repeat_region	12001..15072 /note="3072 repeat 1"
CDS	<12541..13692 /note="reading frame 1"
	/gene="BCRF2"
	/product="hypothetical protein BCRF2"
misc_feature	13215..13215 /note="BAM: BamH1 C/W"
TATA_signal	14352..14352 /gene="EBNA-LP"
exon	14384..14410 /number="W0" /gene="EBNA-LP"
CDS	join (14409..14410,14559..14619,14701..14832, 17626..17691,17773..17904,
	20698..20763, 20845..20976,23770..23835,23917..24048, 26842..26907,
	26989..27120,29914..29979, 30061..30192,32986..33051,33133..33264,
	35473..35505,35590..35694)

	/note="can start at any W0/W1' or C1/W1'"
	/product="EBNA-LP protein" /gene="EBNA-LP"
exon	14554..14619 /number="W1" /gene="EBNA-LP"
exon	14559..14619 /number="W1'" /gene="EBNA-LP"
exon	14701..14832 /number="W2" /gene="EBNA-LP"
repeat_region	15073..18144 /note="3072 repeat 2"
CDS	<15613..16764 /note="reading frame 2"
	/product="hypothetical protein BWRF1"
	/gene="BWRF1"
misc_feature	16287..16287 /note="BAM: BamH1 W/W"
TATA_signal	17424..17424 /gene="EBNA-LP"
exon	17626..17691 /number="W1" /gene="EBNA-LP"
exon	17773..17904 /number="W2" /gene="EBNA-LP"
repeat_region	18145..21216 /note="3072 repeat 3"
CDS	<18685..19836 /note="reading frame 3"
	/product="hypothetical protein BWRF1"
	/gene="BWRF1"
misc_feature	19359..19359 /note="BAM: BamH1 W/W"
TATA_signal	20496..20496 /gene="EBNA-LP"
exon	20698..20763 /number="W1" /gene="EBNA-LP"
exon	20845..20976 /number="W2" /gene="EBNA-LP"
repeat_region	21217..24288 /note="3072 repeat 4"
CDS	<21757..22908 /note="reading frame 4"
	/product="hypothetical protein BWRF1"
	/gene="BWRF1"
misc_feature	22431..22431 /note="BAM: BamH1 W/W"
TATA_signal	23568..23568 /gene="EBNA-LP"
exon	23770..23835 /number="W1" /gene="EBNA-LP"
exon	23917..24048 /number="W2" /gene="EBNA-LP"
repeat_region	24289..27360 /note="3072 repeat 5"
CDS	<24829..25980 /note="reading frame 5"
	/product="hypothetical protein BWRF1"
	/gene="BWRF1"
misc_feature	25503..25503 /note="BAM: BamH1 W/W"
TATA_signal	26640..26640 /gene="EBNA-LP"
exon	26842..26907 /number="W1" /gene="EBNA-LP"
exon	26989..27120 /number="W2" /gene="EBNA-LP"
repeat_region	27361..30432 /note="3072 repeat 6"
CDS	<27901..29052 /note="reading frame 6"
	/product="hypothetical protein BWRF1"
	/gene="BWRF1"
misc_feature	28575..28575 /note="BAM: BamH1 W/W"
TATA_signal	29712..29712 /gene="EBNA-LP"
exon	29914..29979 /number="W1" /gene="EBNA-LP"
exon	30061..30192 /number="W2" /gene="EBNA-LP"
repeat_region	30433..33504 /note="3072 repeat 7"
CDS	<30973..32124 /note="reading frame 7"
	/product="hypothetical protein BWRF1"
	/gene="BWRF1"
misc_feature	31647..31647 /note="BAM: BamH1 W/W"
TATA_signal	32784..32784 /gene="EBNA-LP"
exon	32986..33051 /number="W1"
	/gene="EBNA-LP"
variation	33126..40537 /note="DAUDI deletion"
	/replace="gt"
exon	33133..33264 /number="W2" /gene="EBNA-LP"
variation	33355..40163 /note="P3HR1 deletion"
	/replace="gg"
repeat_region	33505..35355

```
CDS              <34045..35196 /note="reading frame 8"
                 /product="hypothetical protein BWRF1"
                 /gene="BWRF1"
misc_feature     34719..34719  /note="BAM: BamH1 W/Y"
exon             35473..35505 /number="Y1" /gene="EBNA-LP"
exon             35590..35711  /number="Y2" /gene="EBNA-LP"
exon             36098..37739  /number="YH"  /gene="EBNA-2"
exon             36098..36156 /gene="EBNA-LP"
CDS              36216..37679 /gene="BYRF1"
                 /product="EBNA-2 nuclear protein"
repeat_region    36390..36512  /rpt_unit="36390..36398"
misc_feature     36560..36560 /note="BAM: BamH1 Y/H"
repeat_region    37237..37290 /rpt_unit="37237..37242"
exon             37564..37744  /gene="EBNA-LP"
polyA_signal     37715..37720 /gene="EBNA-2"
polyA_signal     complement (38024..38029) /gene="BHLF1"
repeat_region    38290..39827 /rpt_unit="38290..38414"
CDS              complement (38287..40269)
                 /product="BHLF1 early reading frame"
rep_origin       40301..41293
                 /note="OriLyt, lytic origin of replication"
rep_origin       40301..40656
                 /note="part of OriLyt, upstream element"
misc_feature     40366..41409
                 /note="DRleft, similar to 143272..144328"
TATA_signal      complement (40524..40529) /gene="BHLF1"
TATA_signal      41471..41476
                 /note="likely promoter for class III and IV early RNAs encoding BHRF1"
                 /gene="BHF1"
misc_RNA         41474..41495  /note="miRNA"
CDS              42088..42663 /gene="BHRF1"
misc_feature     42565..42565 /note="BAM: BamH1 H/F"
polyA_signal     complement (42636..42641) /note="unknown gene"
misc_RNA         42853..42874  /note="miRNA"
misc_RNA         42888..42910  /note="miRNA, abundant"
misc_RNA         42968..42990 /note="miRNA, abundant"
polyA_signal     43230..43235  /gene="BHRF1"
CDS              complement (43691..44647) /gene="BFLF2"
                 /product="BFLF2 protein"
polyA_signal     complement (43697..43702) /gene="BFLF2"
CDS              complement (44660..46237) /gene="BFLF1"
TATA_signal       complement (44788..44793) /gene="BFLF2"
TATA_signal      complement (46275..46280) /gene="BFLF1"
TATA_signal      46544..46549  /gene="BFRF1"
CDS              46603..47613  /gene="BFRF1"
                 /product="BFRF1 protein"
CDS              47520..49295 /gene="BFRF2"
TATA_signal       49056..49061 /gene="BFRF3"
CDS              49219..49749 /gene="BFRF3"
                 /product="capsid protein VP26"
polyA_signal     complement (49775..49780) /gene="BPLF1"
polyA_signal     49781..49786 /gene="BFRF3"
CDS              complement (49790..59239) /gene="BPLF1"
                 /product="large tegument protein"
misc_feature     49961..49961  /note="BAM: BamH1 F/Q"
misc_feature     50142..50189
                 /note="Site III for EBNA-1 binding"
misc_feature     53833..53833 /note="BAM: BamH1 Q/U"
```

exon	55189..55361 /gene="EBNA-1"
misc_feature	57122..57122 /note="BAM: BamH1 U/P"
repeat_region	57396..57642 /rpt_unit="57396..57446"
repeat_region	58099..58233 /rpt_unit="58099..58113"
CDS	complement (59232..62951) /gene="BOLF1"
	/product="capsid assembly protein"
TATA_signal	complement (59899..59904) /gene="BPLF1"
misc_feature	61180..61180 /note="BAM: BamH1 P/O"
TATA_signal	62729..62729 /gene="BORF1"
CDS	62950..64044 /gene="BORF1"
	/product="putative capsid protein VP19C"
TATA_signal	complement (63029..63034) /gene="BOLF1"
TATA_signal	63881..63881 /gene="BORF2"
CDS	64119..66599 /gene="BORF2"
	/product="ribonucleoside-diphosphate reductase
large chain"	
misc_feature	65547..65552 /note="BAM: Bam H1 O/a"
TATA_signal	66516..66521 /gene="BaRF1"
polyA_signal	66595..66600 /gene="BORF2"
CDS	66612..67520 /gene="BaRF1"
	/product="ribonucleoside-diphosphate reductase small chain"
misc_feature	67249..67254 /note="BAM: Bam H1 a/M"
TATA_signal	67552..67557 /gene="BMRF1"
CDS	67611..68825 /gene="BMRF1"
	/product="early antigen protein D"
TATA_signal	68491..68496 /gene="BMRF2"
TATA_signal	68544..68549 /gene="BMRF2"
CDS	68830..69903 /gene="BMRF2"
	/product="BMRF2 protein"
polyA_signal	69892..69897 /gene="BMRF2"
repeat_region	70031..70173 /rpt_unit="70031..70101"
CDS	complement (70455..71771) /gene="BMLF1"
	/product="BMLF1 protein"
CDS	join (complement (71940..72000) , complement (70455..71833)
	/gene="SM, spliced BSLF2+BMLF1"
	/product="SM protein" /product="ICP27 homolog"
polyA_signal	complement (70454..70459) /gene="SM"
repeat_region	71352..71441 /rpt_unit="71352..71360"
CDS	complement (71881..72000) /gene="BSLF2"
	/product="BSLF2 protein"
misc_feature	71945..71945 /note="BAM: Bam H1 M/S"
CDS	complement (71969..74593) /gene="BSLF1"
	/product="helicase/primase complex protein"
TATA_signal	complement (72063..72068) /gene="SM"
TATA_signal	74594..74599 /gene="BSRF1"
CDS	74636..75292 /gene="BSRF1"
	/product="BSRF1 protein"
polyA_signal	75311..75316 /gene="BSRF1"
polyA_signal	complement (75320..75325) /gene="BLL3"
CDS	complement (75350..76186) /gene="BLLF3"
	/product="deoxyuridine 5'-triphosphate nucleotidohydrolase"
misc_feature	75362..75362 /note="BAM: Bam H1 S/L"
TATA_signal	76219..76224 /gene="BLRF1"
TATA_signal	complement (76226..76231) /gene="BLLF3"
CDS	76259..76567 /gene="BLRF1"
	/product="putative membrane protein BLRF1"
TATA_signal	76575..76580 /gene="BLRF2"
CDS	76637..77125 /gene="BLRF2"
	/product="putative BLRF2 protein"

polyA_signal	77124..77129 /gene="BLRF1"
polyA_signal	complement (77132..77137) /gene="BLLF1"
CDS	complement (77142..79865) /gene="BLLF1"
	/product="envelope glycoprotein gp350"
CDS	complement (77279..77725) /gene="BLLF2"
	/product="putative BLLF2 protein"
TATA_signal	complement (77758..77763) /gene="BLLF2"
	intron complement (77774..78364) /gene="gp220 form of gp350"
repeat_region	77889..78351 /rpt_unit="77889..77919"
TATA_signal	complement (79899..79904) /gene="BLLF2"
mRNA	join (79950..80293,80382..82960)
	/product="EBNA3A nuclear protein"
CDS	join (79955..80293,80382..82877)
	/product="EBNA3A nuclear protein"
misc_feature	80415..80415 /note="BAM: Bam H1 L/E"
repeat_region	81920..81989 /rpt_family="type A"
repeat_region	81993..82018 /rpt_family="type B"
repeat_region	82019..82093 /rpt_family="type C"
repeat_region	82098..82123 /rpt_family="type B"
repeat_region	82124..82201 /rpt_family="type C"
repeat_region	82202..82272 /rpt_family="type A"
repeat_region	82283..82360 /rpt_family="type C"
repeat_region	82361..82431 /rpt_family="type A"
repeat_region	82608..82694 /rpt_family="type D"
repeat_region	82695..82781 /rpt_family="type D"
polyA_signal	82933..82938 /gene="EBNA3A"
polyA_signal	complement (82979..82984) /note="unknown gene"
mRNA	join (83065..83421,83500..89502)
	/gene="EBNA3B"
CDS	join (83065..83421,83500..85959)
	/gene="EBNA3B"
	/product="EBNA-3B nuclear protein"
polyA_signal	complement (83526..83531) /note="unknown gene"
polyA_signal	complement (83983..83988) /note="unknown gene"
repeat_region	85234..85410 /rpt_unit="85234..85293"
mat_peptide	86035..86481 /gene="BERF3"
exon	86076..86442 /gene="EBNA1 leader sequence"
mRNA	join (86076..86442,86517..89502)
	/gene="EBNA3C"
CDS	join (86083..86442,86517..89135)
	/product="EBNA3C latent protein"
	/gene="EBNA3C"
variation	86838..89830 /note="=deletion in Raji"
	/replace="gc"
repeat_region	87834..88016 /rpt_unit="87834..87848"
misc_feature	88325..88325 /note="BAM: Bam H1 E/e1"
repeat_region	88377..88493 /rpt_unit="88377..88415"
misc_feature	88631..88631 /note="BAM: Bam H1 e1/e2"
misc_feature	89138..89138 /note="BAM: Bam H1 e2/e3"
CDS	complement (89157..89828) /gene="BZLF2"
	/product="putative BZLF2 protein"
polyA_signal	89477..89482 /gene="EBNA-3C"
misc_feature	89659..89659 /note="BAM: Bam H1 e3/Z"
mRNA	join (complement (90367..90906) ,
	complement (90138..90242) ,
	complement (89838..90053))
	/gene="BZLF1"
mRNA	join (complement (93838..93893) ,

	complement (90367..92897) ,
	complement (90138..90242) ,
	complement (89838..90053))
	/gene="BRLF1"
polyA_signal	complement (89863..89868) /gene="BZLF1"
TATA_signal	complement (89867..89872) /gene="BZLF2"
CDS	join (complement (90367..90867) ,
	complement (90138..90242) ,
	complement (89922..90053))
	/gene="BZLF1" /function="transcription activator"
repeat_region	90293..90364
TATA_signal	complement (90938..90943) /gene="BZLF1"
misc_feature	complement (90968..91023)
	/note="Upstream of Zp, homology to complement (93900 - 93955) "
CDS	complement (91078..92895) /gene="BRLF1"
	/function="transcription activator"
misc_feature	91453..91453 /note="BAM: Bam H1 Z/g"
misc_feature	91528..91528 /note="BAM: Bam H1 g/R"
exon	complement (92638..92897) /gene="RZ fusion"
TATA_signal	92728..92733 /gene="BRRF1"
CDS	92894..93826 /gene="BRRF1"
	/product="putative BRRF1 protein"
polyA_signal	93822..93827 /gene="BRRF1"
TATA_signal	complement (93920..93925) /gene="BRLF1"
TATA_signal	93955..93960 /gene="BRRF2"
CDS	94014..95627 /gene="BRRF2"
	/product="putative BRRF2 protein"
TATA_signal	complement (94092..94097) /note="unknown gene"
misc_feature	95169..95169 /note="BAM: Bam H1 R/f"
misc_feature	95277..95277 /note="BAM: Bam H1 f/K"
polyA_signal	95626..95631 /gene="BBRF2"
CDS	95662..97587 /gene="BKRF1"
	/product="EBNA-1 protein"
repeat_region	95929..96636
	/rpt_family="EBNA triplet repeat GGA,GCA,GGG"
TATA_signal	97617..97621 /gene="BKRF2"
polyA_signal	97649..97654 /gene="EBNA-1"
CDS	97670..98083 /gene="BKRF2"
	/product="glycoprotein L precursor"
CDS	98065..98832 /gene="BKRF3"
	/product="uracil-DNA glycosylase"
CDS	98846..99499 /gene="BKRF4"
	/product="putative BKRF4 protein"
polyA_signal	99431..99436 /gene="BKRF3"
polyA_signal	99499..99504
CDS	complement (99542..101971) /gene="BBLF4"
	/product="putative helicase"
polyA_signal	complement (99537..99542) /gene="BBLF4"
misc_feature	100332..100332 /note="BAM: Bam H1 K/B"
TATA_signal	101588..101593 /gene="BBRF1"
CDS	101916..103757 /note="BBRF1"
	/product="virion protein BBRF1"
CDS	103660..104496 /gene="BBRF2"
	/product="BBRF2 protein"
polyA_signal	complement (104408..104408) /gene="BBLF2-BBLF3"
CDS	join (complement (105227..>106792) ,
	complement (104493..>105098))
	/gene="BBLF2-BBLF3"
	/function="primase associated factor"

	/product="putative BBLF3 protein"
polyA_signal	104497..104502 /gene="BBRF1"
intron	complement (105098..105227) /note="intron spliced out in RNA linking BBLF2 and BBLF3"
TATA_signal	106693..106698 /gene="BBRF3"
TATA_signal	106810..106815 /gene="BBRF3"
CDS	106849..108066 /gene="BBRF3"
	/product="glycoprotein M"
polyA_signal	108070..108075 /gene="BBRF3"
CDS	complement (108459..108686) /note="BBLF1"
	/product="putative protein BBLF1"
polyA_signal	complement (108471..108476) /gene="BGLF5"
CDS	complement (108641..110053)
	/note="BGLF5" /product="alkaline exonuclease"
TATA_signal	complement (109038..109043) /gene="BBLF1"
misc_feature	110025..110025 /note="BAM: Bam H1 B/G"
CDS	complement (110040..111404) /gene="BGLF4"
	/product="putative serine/threonine-protein kinase"
CDS	complement (111653..112651)
	/gene="BGLF3" /product="BGLF3 protein"
TATA_signal	complement (111824..111829) /gene="BGLF4"
polyA_signal	complement (111926..111931) /gene="BGLF3"
CDS	join (112650..113585,116927..118063)
	/note="spliced BGRF1/BDRF1 reading frame"
	/product="BGRF1-BDRF1 protein"
	/product="putative DNA packaging protein"
TATA_signal	complement (112819..112825) /gene="BGLF3"
polyA_signal	complement (113196..113201) /gene="BGLF2"
CDS	complement (113575..114585) /gene="BGLF2"
	/product="BGLF2 protein"
CDS	complement (114563..116086) /gene="BGLF1"
	/product="BGLF1 protein"
TATA_signal	complement (114636..114641) /gene="BGLF2"
TATA_signal	complement (114944..114949) /gene="BGLF2"
polyA_signal	115741..115746/note="unknown gene"
CDS	complement (116056..116733) /gene="BDLF4"
	/product="BDLF4 protein"
TATA_signal	complement (116139..116144) /gene="BGLF1"
misc_feature	116560..116560/note="BAM: Bam H1 G/D"
TATA_signal	complement (116761..116766) /gene="BDLF4"
mat_peptide	116900..118060 /gene="BDRF1"
TATA_signal	complement (117086..117091) /gene="BDLF4"
polyA_signal	118059..118064 /gene="BGRF1-BDRF1"
polyA_signal	complement (118066..118071) /gene="BDLF3"
CDS	complement (118074..118778) /gene="BDLF3"
	/product="putative membrane antigen gp85"
TATA_signal	complement (118811..118816) /gene="BDLF3"
CDS	complement (118839..120101) /gene="BDLF2"
	/product="BDLF2 protein"
CDS	complement (120112..121017) /gene="BDLF1"
	/product="putative capsid protein vp23"
TATA_signal	complement (120183..120188) /gene="BDLF2"
polyA_signal	complement (121019..121024) /gene="BcLF1"
CDS	complement (121033..125178) /gene="BcLF1"
	/product="major capsid protein"
TATA_signal	complement (121059..121064) /gene="BDLF1"
TATA_signal	complement (121093..121098) /gene="BDLF1"
misc_feature	124580..124580 /note="BAM: Bam H1 D/c"

TATA_signal	complement (125417..125422) /gene="BcLF1"
TATA_signal	125569..125574 /gene="BcRF1"
CDS	125177..127429/gene="BcRF1"
	/product="BcRF1 protein "
misc_feature	125730..125730/note="BAM: Bam H1 c/b"
misc_feature	127063..127063/note="BAM: Bam H1 b/T"
CDS	<127353..>128627/gene="BTRF1"
	/product="BTRF1 protein"
polyA_signal	complement (128608..128613) /gene="BXLF2"
CDS	complement (128627..130747) /gene="BXLF2"
	/product="glycoprotein gp85 precursor"
	/product="gp85 protein"
polyA_signal	128681..128686 /gene="BcRF1"
polyA_signal	complement (128992..128997) /gene="BXLF2"
TATA_signal	130300..130305 /note="unknown gene"
misc_feature	130451..130451 /note="BAM: Bam H1 T/X"
CDS	complement (130749..132572) /gene="BXLF1"
	/product="thymidine kinase"
TATA_signal	complement (131016..131021) /gene="BXLF2"
CDS	132571..133317 /gene="BXRF1"
	/product="BXRF1 protein"
misc_feature	132573..132573 /note="BAM: Bam H1 X/V"
TATA_signal	complement (132839..132846) /gene="BXLF1"
TATA_signal	133013..133017 /gene="BVRF1"
CDS	133127..134839 /gene="BVRF1"
	/product="virion protein BVRF1"
polyA_signal	134881..134886 /gene="BVRF1"
TATA_signal	135432..135436 /gene="BVRF2"
CDS	135638..137455 /gene="BVRF2"
	/product="capsid protein P40"
misc_feature	135718..135718 /note="BAM: Bam H1 V/d"
TATA_signal	136331..136336 /gene="BdRF1"
CDS	136418..137455 /gene="BdRF1"
	/product="scaffold protein"
misc_feature	136826..136826 /note="BAM: Bam H1 d/I"
polyA_signal	137438..137443 /gene="BdRF1"
polyA_signal	complement (137464..137469) /gene="BILF2"
CDS	complement (137490..138236) /gene="BILF2"
	/product="probable membrane glycoprotein"
TATA_signal	complement (138277..138282) /gene="BILF2"
mRNA	join (138352..138480,149581..149712,149923..150077,
	150237..150348,155267..156737,156846..156928,
	158625..160531) /gene="RPMS1"
repeat_region	138947..139329
precursor_RNA	139348..139410 /note="BART1 miRNA precursor"
misc_RNA	139351..139371 /note="miRNA"
exon	139447..139552 /note="Exon IA of BART RNAs"
polyA_signal	139478..139483 /note="unknown gene"
TATA_signal	complement (139486..139491)
	/gene="possible start for 2kb RNAs crossing BILF2"
misc_feature	139723..>151555
	/note="DEL: B95-8 deletion with respect to Raji "
misc_feature	139724..151554
	/note="5-11835 of Raji DNA restoring the B95-8
deletion sequence"	
CDS	complement (140570..143344) /gene="LF3"
	/product="LF3 protein"
repeat_region	140765..143281 /rpt_family="PstI repeats"

rep_origin	143207..144444
	/note="OriLyt, lytic origin of replication"
misc_feature	143272..144328
	/note="DR right, similar to 40366..41409, ori lyt"
CDS	complement (149035..150324) /gene="LF2"
	/product="LF2 protein"
CDS	complement (150285..151694) /gene="LF1"
	/product="LF1 protein"
CDS	join (150323..150348,155267..155552)
	/gene="RPMS1" /product="RPMS1 protein"
CDS	complement (151703..152641)
	/gene="BILF1" /product="membrane protein"
misc_RNA	152747..152768 /note="BART2 miRNA"
precursor_RNA	152747..152804 /note="BART2 miRNA precursor"
polyA_signal	152796..152801 /gene="BALF5"
misc_feature	153179..153179 /note="HPN: 22bp 2-fold symmetric"
CDS	complement (153241..156288) /gene="BALF5"
	/product="DNA polymerase"
misc_feature	154289..154289 /note="BAM: Bam H1 I/A"
mRNA	join (155549..156737,156846..156928,158625..158751,
	159781..160531) /gene="A73"
polyA_signal	complement (156244..156249) /gene="BALF4"
CDS	complement (156291..158864) /gene="BALF4"
	/product="glycoprotein gp110 precursor"
CDS	join (156652..156737,156846..156928,158625..158751,
	159781..159865) /product="A73 protein"
	/gene="A73"
CDS	complement (158851..>161220) /gene="BALF3"
	/product="BALF3 protein"
TATA_signal	complement (158907..158912) /gene="BALF4"
CDS	<159121..160536 /gene="BARF0"
	/product="putative BARF0 protein"
polyA_signal	160508..160513 /gene="RPMS1 and A73 RNAs"
polyA_signal	complement (160550..160555) /gene="BALF2"
CDS	complement (160926..164312) /gene="BALF2"
	/product="major DNA-binding protein"
variation	163519..166178 /note="deletion in Raji"
	/replace="gg"
TATA_signal	complement (164351..164356) /gene="BALF2"
polyA_signal	complement (164388..164393) /gene="BALF1"
CDS	complement (164397..165059) /gene="BALF1"
	/product="hypothetical BALF1 protein"
TATA_signal	164979..164984 /gene="BALF2"
TATA_signal	165008..165013 /gene="BARF1"
CDS	165046..165711 /gene="BARF1"
TATA_signal	complement (165250..165255) /gene="BALF1"
polyA_signal	165707..165712 /gene="BARF1"
TATA_signal	166011..166016 /gene="LMP2A"
exon	166040..166458 /number="1"
	/product="LMP2A RNA" /gene="LMP2A"
misc_feature	166156..166156 /note="BAM: Bam H1 A/Nhet"
polyA_signal	complement (166483..166488) /gene="LMP1"
polyA_signal	complement (166487..166492) /gene="LMP1"
CDS	complement (166540..>166845) /gene="BNLF2b"
	/product="hypothetical BNLF2B protein"
CDS	complement (166846..167028) /gene="BNLF2a"
	/product="hypothetical BNLF2A protein"
TATA_signal	complement (167062..167067) /gene="BNLF2a"

	/gene="BNLFb"
exon	complement (167705..168507) /number="c"
	/gene="BNLF1"
CDS	join (complement (168749..169016) , complement (168584..168670) ,
	complement (167702..168507))
	/gene="LMP1" /function="latent membrane protein"
repeat_region	167941..168116 /rpt_unit="167941..167973"
intron	complement (168508..168583) /number="b"
	/gene="BNLF1"
exon	complement (168584..168670) /number="b"
	/gene="BNLF1"
intron	complement (168671..168748) /number="a"
	/gene="BNLF1"
TATA_signal	complement (168738..168743) /gene="lytic LMP1"
exon	complement (168749..169016) /number="a" /gene="BNLF1"
TATA_signal	complement (169083..169088) /gene="LMP1"
exon	169294..169448 /number="1" /product="LMP2B RNA" /gene="LMP2B"
repeat_region	169636..170173 /rpt_type="TERMINAL"
promoter	complement (170113..170121) / gene="LMP1"
	/product="3.5kb LMP1 mRNA
promoter	complement (170184) / gene="LMP1" /product="3.5kb LMP1 mRNA
repeat_region	170174..170696 /rpt_type="TERMINAL"
repeat_region	170697..171234 /rpt_type="TERMINAL"
repeat_region	171235..171773 /rpt_type="TERMINAL "

Chapter 16

Strategies for Reverse Genetic Analysis of EBV

*Kenneth M. Izumi**

ABSTRACT

Epstein-Barr virus (EBV) infects B-lymphocytes and then persists as an incomplete or "latent" virus in order to evade immune recognition while it expresses twelve gene products that efficiently transform cell growth into long-term proliferating lymphoblastoid cell lines. In healthy individuals, EBV's oncogenic strategy for persistence is restricted by adaptive immune responses, but the viral infection is not completely cleared. Thus, most adults are life-long asymptomatic carriers who routinely shed virus in saliva. In immuno-compromised individuals, EBV infected B-lymphocytes may progress to a malignant lymphoproliferative disease. EBV's strong association with the development of this and other malignancies has sparked considerable interest in understanding the molecular basis of transformation mediated by latent EBV gene expression. This article reviews the techniques that have been developed to expand our understanding of the molecular pathogenesis of EBV-mediated transformation.

INTRODUCTION

Epstein-Barr virus (EBV) infection of resting primary human B-lymphocytes typically results in the efficient transformation of cell growth into long-term proliferating lymphoblastoid cell lines (LCLs) while in a low percentage of cells EBV lytically infects resulting in virion production and ultimately cell death. In the growth-transformed cells, a structurally incomplete or "latent" EBV persists in the cell nucleus as a non-integrated circular DNA plasmid. While latent EBV DNA is genetically intact and competent for transcription, lytic infection genes are not usually expressed. Instead, the infected cell transcription and translation machinery is predisposed and subsequently altered such that six EBV nuclear antigen proteins (EBNA-1, -2, -3A, -3B, -3C, or -LP), three latent infection membrane proteins (LMP-1, -2A, or -2B), two EBV encoded RNAs (EBER-1 and -2), and a family of BamHI A rightward transcripts (BARTs) or complementary strand transcripts (CSTs) may be expressed (Figure 1). These EBV gene products are hypothesized to usurp the functions of host proteins in order to alter gene expression further so

*For correspondence email izumi@uthscsa.edu

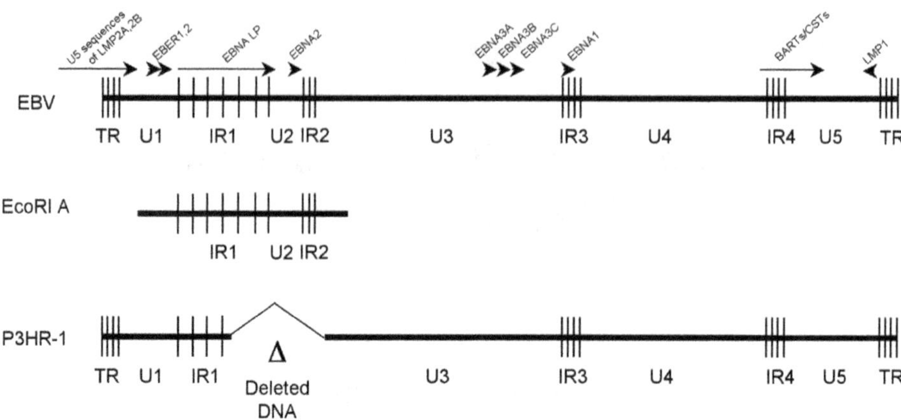

Figure 1. (Top line) Linear map of Epstein-Barr virus with the exons for latent gene expression depicted as arrows. Vertical lines denote terminal repeats (TR) or internal repeats (IR1 to IR4). Unique DNA sequences (U1 to U5) are labeled. Latent EBV DNA is a covalently closed circular plasmid. LMP2A/2B transcripts initiate within U5 DNA, cross over the TR, and terminate within U1 DNA. (Middle line) The *Eco*RI A restriction endonuclease fragment is shown in relationship to EBV and variant P3HR-1. (Bottom line) The P3HR-1 EBV genome and deleted DNA segment are shown.

that resting primary B-lymphocytes are transformed into long-term proliferating lymphoblastoid cell lines (LCLs). Latent EBV expressed proteins also parasitize the function of DNA replication enzymes so that viral DNA is synthesized alongside of cell chromatin DNA during S phase of the cell cycle and then distribute nascent viral genomes between the daughter cells during mitosis (Kieff et al., 2001; Rickinson et al., 2001).

Based upon the recovery of virus or detection of viral gene expression, EBV is strongly associated with the development of lymphoproliferative disease, a malignant B-lymphocyte tumor of immuno-compromised individuals such as patients with advanced AIDS or organ transplant recipients. In otherwise healthy individuals, EBV is etiologically associated with the development of several malignancies including Hodgkin's lymphoma, Burkitt's lymphoma, and nasopharyngeal carcinoma. Because latent EBV gene expression provides a plausible explanation for alterations in cell growth that enable progression to malignancy, reverse genetic techniques have been developed to identify the genes that are required for B–lymphocyte transformation. Further genetic and biochemical analyses used in conjunction with these assays have increased our understanding of the mechanisms by which these gene products usurp cell growth control. While additional investigation is needed to reveal fully the molecular aspects of transformation, the prospects for developing specific molecular inhibitors that can be used to prevent or treat these tumors appear to be excellent (Kieff et al., 2001; Rickinson et al., 2001).

REVERSE GENETICS TECHNIQUES

The early methods used to derive specifically mutated EBV recombinants are similar in principle to those used for other large DNA viruses. In brief, viral genomic DNA is maintained in *E. coli* as a library of restriction endonuclease fragments inserted into plasmid or cosmid cloning vectors. A specific mutation may be introduced into a gene of interest or the cloned DNA is derived from a virus strain that has a specific phenotype of interest. The cloned DNA is then transfected into cells that are lytically infected with a virus. During the lytic infection stage when viral DNA is undergoing replication, the cloned DNA recombines homologously with the replicating viral DNA. The resulting virus progeny is a mixture consisting mainly of unmodified or parental-type viruses and a minority of recombinant viruses that have incorporated the cloned DNA. Naïve cells are infected with the virus mixture, and recombinants are recovered using selectable markers or biological phenotypes. Recombinant viruses are then analyzed to delineate the effect of the mutation upon the phenotype of interest.

Lytic infection plays an important role in the derivation of specifically modified viral genomes because recombination occurs more efficiently with replicating viral DNA. This has made reverse genetic analysis of EBV more difficult because there is no cell or cell line that is truly permissive for lytic EBV infection. However, delineation of EBV BZLF1 protein as a major immediate early transactivator of lytic infection gene expression and the discovery of phorbol esters or butyric acid as chemical inducers of lytic infection were important steps forward in the development of reverse genetic techniques and in understanding EBV latent and lytic infection (Countryman et al., 1987; Hudewentz et al., 1980; Luka et al., 1979). While transfection of BZLF1 vector DNA into latent EBV infected cells is less efficient than chemical inducers at activating virus lytic infection, co-transfection of BZLF1 vector DNA with cloned DNA increases the probability of deriving EBV recombinants because the DNAs are likely to be introduced into the same cell. Thus, lytic infection activated by BZLF1 vector occurs in the cells that contain the cloned DNA.

The resulting virus mixture is next used to infect naïve cells. Resting primary B-lymphocytes derived from the peripheral blood of normal healthy individuals, placental cord blood, or routine tonsillectomy are typically used because they are efficiently infected. Most other cell types or cell lines are more difficult to infect likely due to under expression of CD21 and/or co-receptors such as MHC class II (Imai et al., 1998; Yoshiyama et al., 1997). Primary B-lymphocyte growth transformation is a significant biological marker that can be used to recover recombinant viruses whereas recovery of recombinants in continuous B-lymphoma cell lines requires installation of a selectable marker on the cloned DNA (Wang et al., 1991). Selectable markers typically have strong transcriptional promoters or enhancers that may affect nearby viral gene expression and possibly the phenotype of the recombinant virus. While infection of B-lymphoma cell lines is usually less efficient than primary B-lymphocytes and newly derived EBV-infected B-lymphoma cells may be difficult to induce to lytic infection (Marchini et al., 1991), an important consideration is that EBV recombinants that are deleted in essential transforming gene functions can be clonally recovered in B-lymphoma cells. In

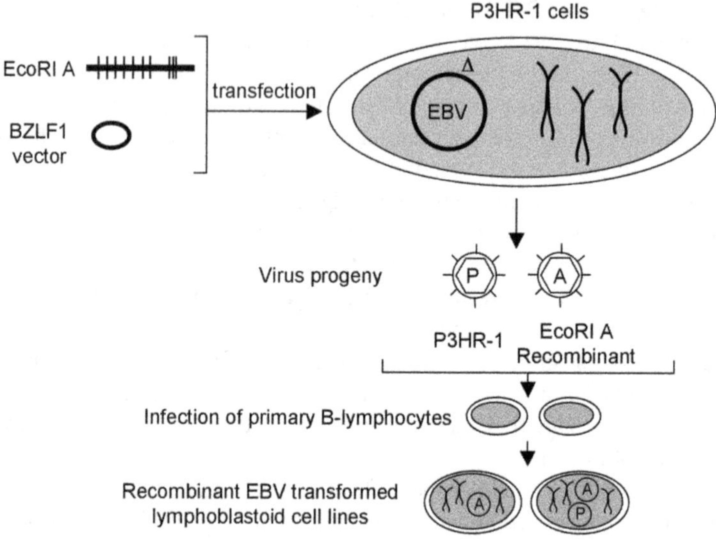

Figure 2. Marker rescue of P3HR-1 EBV. BZLF1 vector and EcoRI A DNA are co-transfected into P3HR-1 cells, which contain an EBV variant that is deleted (Δ) of DNA that encodes EBNA2 and part of EBNA LP. BZLF1 activates lytic virus replication, and EcoRI A recombines with the replicating DNA. The virus progeny consists of parental P3HR-1 EBV (P) and EcoRI A recombinants (A). Primary B-lymphocytes are infected with the progeny. Transformed B-lymphoblastoid cell lines (LCLs) are PCR analyzed for recombinant virus and P3HR-1 DNA. LCLs containing only recombinant virus DNA without co-infecting P3HR-1 DNA indicate the gene(s) on the EcoRI A DNA are critical for transformation.

primary B-lymphocytes an expressed copy of the complementing gene must be provided in trans (Lee et al., 1992a; Marchini et al., 1993; Wang et al., 1991).

RECOMBINANT EBV GENETIC ANALYSIS OF EBNA2 AND LP

The first delineation of an EBV latency gene product that is essential for B-lymphocyte transformation was based upon analyses of the EBV-positive B-lymphoma cell line P3HR-1. This continuous cell line, which was subcloned from the EBV-positive Jijoye B-lymphoma line, is latently infected with an EBV variant. P3HR-1 EBV differs from most other EBV including Jijoye because it is deleted of a segment of DNA that comprises all of the exons of EBNA2 and two C-terminal exons of EBNA LP (Figure 1). EBV P3HR1 is competent for latent infection and can be reactivated to lytic infection by transfecting P3HR-1 cells with BZLF1 vector DNA or treating cells with chemical inducers. While EBV P3HR-1 virions efficiently infect primary B-lymphocytes, none of the infected cells is growth transformed into LCLs (Miller et al., 1974; 1975). These results were early indications of a likely transforming function within the deleted segment.

Attempts were next made at marker rescuing P3HR-1 to determine whether the EBNA2 gene or the EBNA LP exons within the deleted DNA segment are responsible for the non-transforming phenotype. First, genomic DNA from a fully transforming EBV strain was cloned and propagated in *E. coli* as a library of restriction fragments. The approximately 55 kb EcoRI A DNA fragment, which spans the deleted DNA segment and contains substantial flanking DNA, was mutated to delete EBNA2 expression while not altering nearby EBNA-LP exons (Figure 1). P3HR-1 cells were co-transfected by electroporation with BZLF1 vector DNA and EcoRI A or, in the control transfection, with BZLF1 vector with EBNA2 deleted EcoRI A DNA (Figure 2). In the transfected cells, BZLF1 induces EBV P3HR-1 to lytic infection, and replicating viral DNA may recombine with the co-transfected cloned DNA. The viral progeny, which consists of about 10^7 unmodified P3HR-1 EBV and about 10^3 recombinant viruses, is used to infect primary B-lymphocytes. Infected cells are then seeded into 96-well plates and cultured *in vitro*. One to two months later, transformed cells were recovered from cultures infected with EcoRI A recombinants whereas none of the lymphocytes infected with the EBNA2 deleted EcoRI A recombinants was transformed. Homologous recombination of P3HR-1 genomes with EcoRI A DNA was confirmed by analyzing DNA structure or gene expression. Since the long-term proliferation of LCLs infected with nascent EcoRI A recombinants was similar to LCLs transformed with wild type EBV, these results clearly demonstrated an essential role for EBNA2 in B-lymphocyte transformation while establishing the methods used to derive recombinant viruses and test their transforming abilities (Cohen et al., 1989; Hammerschmidt et al., 1989).

More detailed reverse genetic analyses of EBNA2 codons in conjunction with biochemical assays have revealed critical functional domains. In brief, EBNA2 is a transcriptional activator that interacts with host proteins that recognize specific DNA sequences within gene promoters. Tethered to DNA, EBNA2 further interacts with transcription factors associated with RNA polymerase II to mediate activating effects on transcription (Aman et al., 1990; Knutson, 1990; Sung et al., 1991; Wang et al., 1987). EBNA2 tethered to DNA also displaces transcriptional repressor complexes from promoters (Sjoblom-Hallen et al., 1999). Because mutations in EBNA2 that disrupt protein interactions or transcriptional effects also abrogate the ability of EBV recombinants to transform B-lymphocyte growth (Cohen et al., 1991; Yalamanchili et al., 1994), an essential EBNA2 transforming function is to regulate, at the level of transcription, both viral and cell gene expression that enables long-term cell proliferation.

Recombinants with wild type EBNA2 genes but mutated in the C-terminal exons of EBNA-LP were derived by recombining modified DNA clones that span the deletion in P3HR-1 as described above (Mannick et al., 1991). LCLs containing only mutated EBNA LP recombinants were recovered, but the replication of the transformed cells was significantly impaired compared with recombinants with wild type EBNA LP and EBNA2 genes. While these results indicate an important role for EBNA LP in transformation, an essential role has not yet been formally excluded. Recent evidence reveals EBNA 2 mediated transactivation of viral genes such as LMP1 is co-activated by EBNA LP, and that co-activation activity maps to the repetitive exons that are 5′ of the deleted segment in P3HR-1 (Harada et al., 1997; Nitsche et al., 1997). Determination of an essential role in transformation and the contribution of the co-activating 5′ exon residues of EBNA LP would be illuminated by deleting its expression. Derivation of an EBNA LP knockout is difficult, however, because the 5′ exons are highly repeated and multiply spliced, and the EBNA LP promoter is a shared component of a complex transcription unit that codes for all EBNAs including the essential EBNA2 oncoprotein.

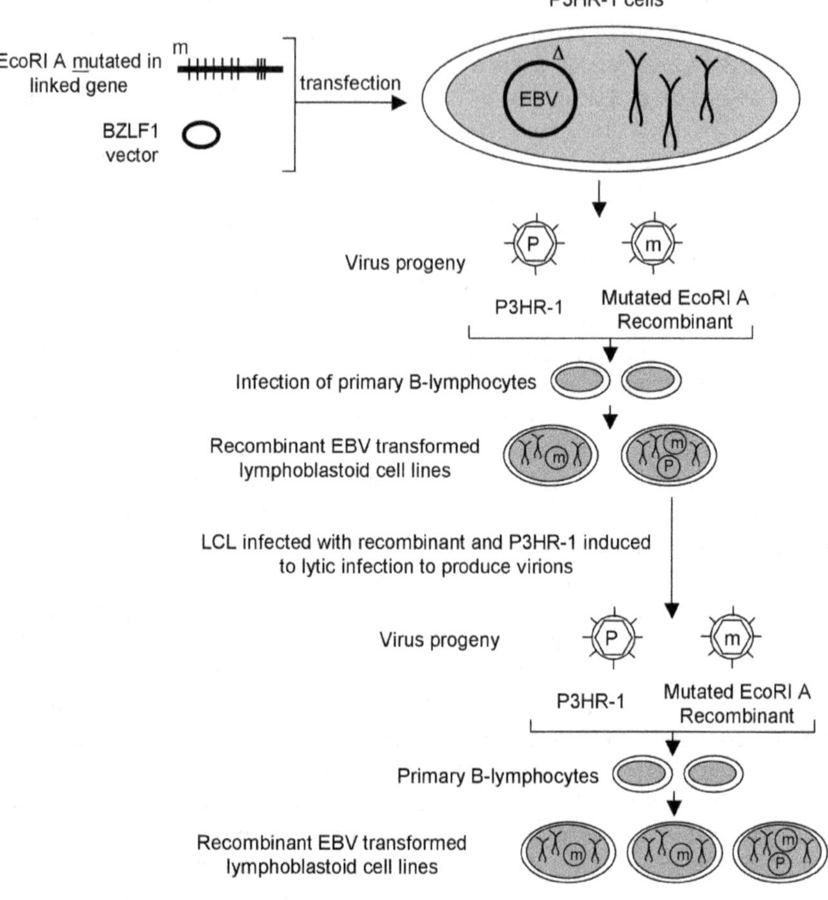

Figure 3. Derivation of EBV recombinants mutated in genes near the deleted DNA of P3HR-1. EcoRI A DNA is mutated (m) in a gene such as EBER1 and EBER2, BCRF1, or BHRF1. P3HR1 cells are transfected with BZLF1 vector and mutated EcoRI A, and replicating viral DNA recombines with the mutated EcoRI A DNA. The virus progeny consists mainly of P3HR-1 virus (P) and recombinants with the mutation (m). B-lymphocytes are infected with the virus progeny, and nascent LCLs are PCR analyzed for recombinant DNA. Typically, these LCLs contain both virus genomes (P and m). To delineate the transformation phenotype of the recombinant, co-infected LCLs are induced to lytic infection. The virus progeny is serially diluted and used to infect primary B-lymphocytes. LCLs are analyzed for viral DNA. Segregation of mutated recombinant from P3HR-1 reveals the mutation does not abrogate EBV transforming abilities.

REVERSE GENETIC ANALYSES OF GENES ADJACENT TO EBNA2

Because of the success in EBNA2 genetic analyses, genes near EBNA2 were targeted next. These recombinants were derived by recombining a single cloned DNA into P3HR-1 EBV (Figure 3). The DNA clones that were tested contain wild type EBNA2 and EBNA LP DNA linked to a mutation that deletes expression of the EBER1 and EBER2 latency transcripts, viral Bcl-2 encoded by the BHRF1 gene, or viral IL-10 encoded by the BCRF1 gene (Lee et al., 1992b; Marchini et al., 1991; Swaminathan et al., 1993; 1991). Next, P3HR-1 cells were co-transfected with a mutated DNA clone and BZLF1 vector DNA. In the co-transfected cells induced to lytic infection by BZLF1 vector expression, replicating viral DNA may recombine completely with the mutated DNA clone thereby rescuing the deleted DNA while incorporating the mutated gene into viral genomic DNA. Recombinants are recovered from the resulting viral progeny, which consists of about 10^7 unmodified P3HR-1 and 10^3 recombinants, by infecting primary B-lymphocytes. Since P3HR-1 EBV is non-transforming, LCLs that grow out in tissue culture are recombinants with EBNA2 and EBNA-LP DNA and likely, though not necessarily, with the mutated gene. PCR is used to analyze DNA from LCLs and identify those infected with recombinants that have the mutated gene of interest.

Several LCLs were recovered that contain recombinants mutated in the gene of interest but did not contain P3HR-1 DNA. Since the continued growth of these LCLs was similar to LCLs transformed with recombinants that have a wild type copy of the gene of interest, these data indicate that the mutated gene is not required for maintaining transformed cell growth.

While initial recovery of LCLs infected with recombinant EBV but lacking P3HR-1 EBV provides substantial evidence that the mutated gene is not required for transformation, more typically LCLs that contain the recombinant are co-infected with P3HR-1 EBV. This result is typical because the virus progeny used to infect primary B-lymphocytes consists of about 10^7 unmodified P3HR-1 and 10^3 recombinants or about 10,000-times more P3HR-1 than recombinant. While P3HR-1 EBV is competent for latent and lytic infection, critically important for analyses of essential transforming genes, P3HR-1 EBV provides a wild type copy of the gene of interest in trans. To delineate the transforming abilities of the recombinant, LCLs containing recombinant and P3HR-1 genomes are transfected with BZLF1 vector and/or treated with chemical inducers to activate lytic infection. Primary B-lymphocytes are infected with serial dilutions of the nascent progeny virus and then are seeded into 96-well plates. LCLs that grow out in tissue culture are analyzed by PCR for the mutated gene of interest and the wild type gene from P3HR-1. Successful segregation of recombinant virus from P3HR-1 and robust, long-term LCL proliferation provide clear evidence that the mutation does not adversely alter lytic infection or latent EBV growth transforming infection of B-lymphocytes.

EBV recombinants specifically deleted of EBER1 and EBER2, BCRF1, or BHRF1 expression were derived and tested for transforming abilities. EBER1 and EBER2 are non-coding RNA polymerase III transcripts that accumulate to about 10^4 to10^5 copies per transformed B cell. EBERs up-regulate *bcl-2* expression (Komano et al., 1999), inhibit protein kinase R (Clarke et al., 1991), or activate $2'$-$5'$ oligo-adenylate synthetase and therefore could have a role in transformation (Komano et al., 1998; Sharp et al., 1999). While BCRF1 protein is expressed mainly during lytic infection, BCRF1 is partially homologous with human IL-10, can induce B-cell proliferation, and regulate some effects on T-lymphocytes mediated by γ-interferon (Hsu et al., 1990). Thus, an accessory role in transformation for BCRF1, possibly critical *in vivo*, was hypothesized. BHRF1 protein is also expressed mainly during the early phase of lytic infection, but LCLs have low-levels of BHRF1 transcripts

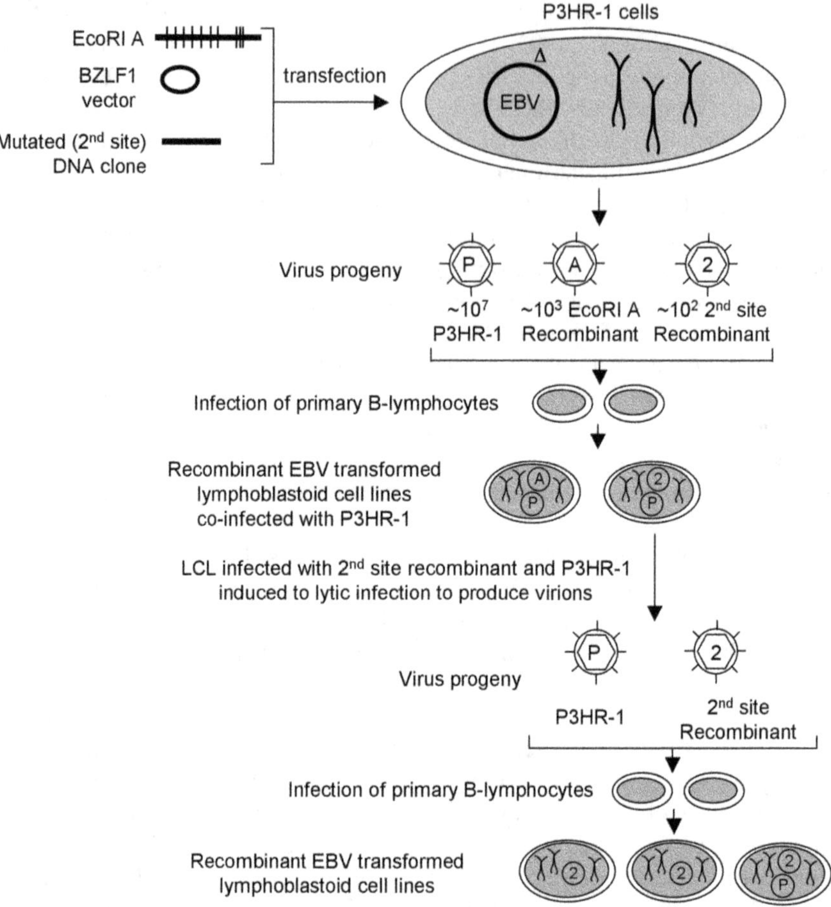

Figure 4. Derivation of EBV recombinants by second site homologous recombination method. P3HR1 cells are transfected with BZLF1 vector, EcoRI A DNA, and a mutated EBV DNA clone. In lytically infected cells, replicating EBV DNA recombines with EcoRI A DNA and mutated EBV DNA. B-lymphocytes are infected with the virus progeny, which consists of ~10^7 P3HR1 virus (P), ~10^3 EcoRI A recombinants (A), and ~10^2 EcoRI A recombinants with the second site mutation (2). Nascent LCLs are PCR analyzed for the second site recombinant, but most are co-infected with P3HR1 virus. To delineate the transformation phenotype of the recombinants with the second site mutation, co-infected LCLs are induced to lytic infection. The virus progeny is serially diluted and used to infect primary B-lymphocytes. LCLs are analyzed for viral DNA. Segregation of recombinant with the second-site mutation (2) from P3HR-1 reveals the mutation does not abrogate EBV transforming abilities.

(Lee et al., 1992b; Marchini et al., 1991). There was considerable interest in knocking out the BHRF1 gene because the protein is partially homologous with human *bcl-2*, an anti-apoptotic protein (Henderson et al., 1993). Comparison of EBER1 and EBER2, BCRF1, or BHRF1 deletion recombinant EBV with recombinants that have the wild type copy of the gene of interest revealed no significant difference in mediating B-lymphocyte growth transformation or lytic infection. Further, the long-term proliferation of LCLs infected with deletion recombinant or wild type gene recombinant was similar (Lee et al., 1992b; Marchini et al., 1991; Swaminathan et al., 1993). Whereas these results indicate that none of these genes is essential for transformation, an accessory role in EBV pathogenesis *in vivo* has not been excluded. More importantly, the successful recovery of these recombinant viruses affirmed the validity of the method of recombinant virus derivation and analysis of transformation. Further, these results confirmed the hypothesis that some viral gene products are not required for transformation of B-lymphocytes or do not have a phenotype that is evident *in vitro*.

SECOND-SITE HOMOLOGOUS RECOMBINATION
Many genes expressed by EBV are located relative to the deleted DNA of P3HR-1 such that a single cosmid cloned DNA cannot contain EBNA2/EBNA LP DNA and the mutated gene of interest. In order to derive P3HR-1 based recombinants that are rescued at the primary site for EBNA2 and EBNA LP DNA and mutated at a second site, P3HR-1 cells are co-transfected with BZLF1 vector, EcoRI A DNA, and an unlinked DNA clone that is mutated in a gene of interest (Figure 4). In cells induced to lytic infection, replicating viral DNA may recombine with EcoRI A DNA and with the cloned DNA containing the second-site mutation. The resulting virus progeny is a mixture consisting of about 10^7 unmodified P3HR-1 EBV, 10^3 EcoRI A recombinants, and 10^2 EcoRI A recombinants with the second-site mutation. Primary B-lymphocytes are infected with the virus mixture and seeded into 96-well microtiter plates. Next, LCLs that grow out in tissue culture are analyzed for the second-site mutated DNA.

Nearly all of the LCLs that contain recombinants with the second-site mutated DNA are co-infected with P3HR-1 EBV. P3HR-1 co-infection is typical because the virus mixture used to infect primary B-lymphocytes contains about 100,000-times more P3HR-1 EBV than second-site recombinant. Because P3HR-1 EBV provides a wild type copy of the mutated gene in trans, delineation of the transformation phenotype of the second-site recombinant requires segregation of the two genomes. Thus, LCLs are induced to lytic infection by transfection with BZLF1 vector and/ or treatment with chemical inducers of lytic infection, and primary B-lymphocytes are infected with serial dilutions of the resulting virus progeny. LCLs that grow out in tissue are PCR analyzed for the second-site mutated DNA and for P3HR-1 DNA.

The recovery of LCLs with the second-site mutated DNA but lacking P3HR-1 DNA provides substantial evidence that the mutated gene is not essential for EBV to mediate B-lymphocyte growth transformation. Alternatively, when all LCLs that have the second site mutated DNA also contain the wild type gene provided by P3HR-1, the evidence indicates the mutated gene is required for EBV to mediate cell transformation. Investigations using second-site homologous recombination technique have thus revealed the essential role of LMP1, EBNA3A, and EBNA3C in transformation while LMP2A, LMP2B, and EBNA3B are dispensable (Kaye et al., 1993; Longnecker et al., 1992; Tomkinson et al., 1992; 1993a).

Further recombinant virus genetic analyses combined with biochemical assays have been used to develop the current working model of LMP1. In brief, LMP1 is a constitutively activated, plasma membrane receptor that interacts with cell proteins

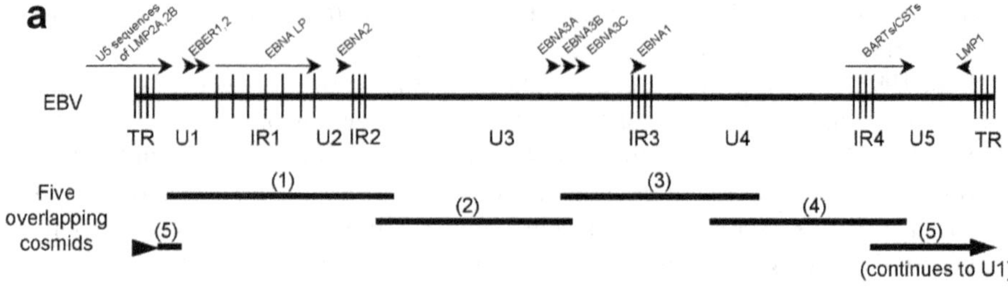

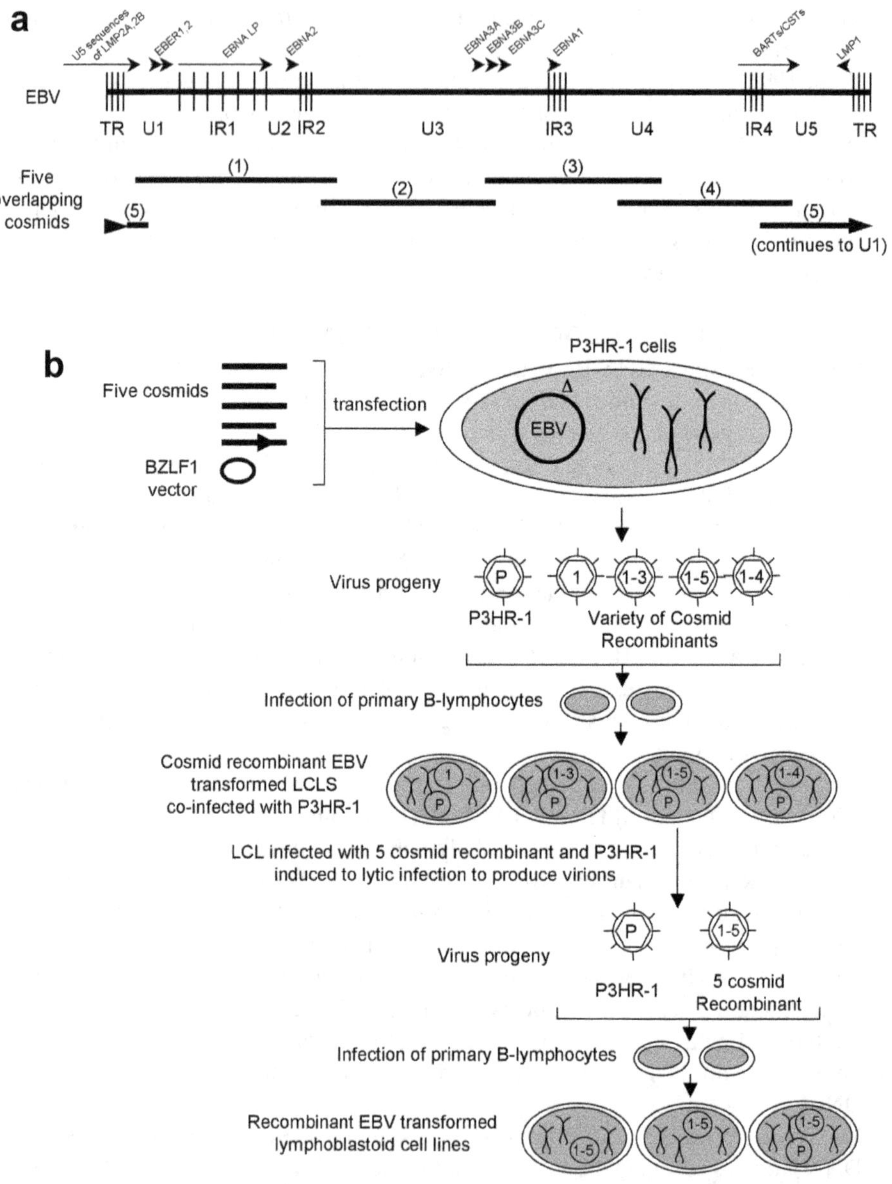

Figure 5. Derivation of EBV from five subgenomic DNA clones. (a) The EBV genome and latency gene transcripts shown are described in figure 1. Beneath this, the five DNA sequences cloned into cosmid vectors are represented as horizontal lines. (b) Five subgenomic DNA clones and BZLF1 vector are transfected into P3HR-1 cells. In lytically infected cells, the five DNAs recombine with replicating P3HR-1 DNA and with each other. The resulting virus progeny consists of P3HR-1 and several types of recombinant viruses (1, 1-3, 1-5, 1-4, etc). Primary B-lymphocytes are infected with the mixture, and nascent LCLs are PCR analyzed for recombinant virus DNA. LCLs containing recombinant genomes derived from all five subgenomic DNAs (1-5) usually contain P3HR-1. To delineate the phenotype of the

recombinant virus, co-infected LCLs are induced to lytic infection. The virus progeny is serially diluted and used to infect primary B-lymphocytes. LCLs that develop are analyzed for viral DNA. Successful segregation of recombinant EBV derived from the five cloned DNAs (1-5) reveals transforming EBV can be derived by this method.

in order to usurp the tumor necrosis factor receptor signal transduction pathway. While LMP1 does not require a ligand in order to aggregate in the plasma membrane, LMP1's ability to aggregate constitutively enables two sites in the LMP1 carboxyl cytoplasmic tail to transduce signals that alter gene expression through NF-κB or mitogen activated protein kinases. Altered cell or viral gene expression is the likely molecular basis of LMP1 transforming effects B-lymphocyte growth (Gires et al., 1999; Izumi et al., 1997a; 1997b; 1999; Mosialos et al., 1995).

Genetic and biochemical analyses of EBNA 3A or EBNA 3C have been used to develop the model that these structurally related nuclear proteins alter cell or viral gene expression by appropriating the function of the Notch family of proteins (Cotter et al., 2000a; Robertson et al., 1996). EBNA 3A or EBNA 3C regulate promoters that are transactivated by EBNA 2 through interaction with specific DNA binding proteins such as RBP-Jκ (Robertson et al., 1996). EBNA 3C further alters transcription through interaction with prothymosin α to regulate p300/CBP histone acetyl-transferase activity and with Nm23H1 to reverse its metastasis suppressor activity (Cotter et al., 2000b; Subramanian et al., 2001). EBNA3A, 3B, and 3C are complex proteins of about 800 to 1000 residues in length that are structurally related but have non-overlapping transforming functions. Further analyses are needed to delineate the functional domains of EBNA3A, 3B, and 3C.

EBV RECONSTRUCTED FROM CLONED SUB-GENOMIC DNAS

A more efficient technique to derive specifically mutated recombinant EBV is to co-transfect P3HR-1 cells with BZLF1 vector and five cosmid cloned EBV DNAs (Tomkinson et al., 1993b). These five DNAs collectively constitute the entire genome and, when arranged in viral genomic order, overlap at their ends with the adjacent clone (Figure 5a). In the transfected cells, lytic infection gene expression enables the cosmid clones to recombine with each other and with replicating P3HR-1 DNA (Figure 5b). The resulting EBV genomic DNAs are then incorporated into virions. Primary B-lymphocytes are infected with the virus progeny, and LCLs that grow out in tissue are analyzed by PCR. In a direct comparison, 3% of LCLs contain recombinant EBV that are derived by second-site homologous recombination technique while 26% of LCLs are transformed by recombinant EBV derived by the five-overlapping cosmid DNA approach.

As an extension of this approach, attempts were made to reduce the complexity of a transforming EBV genome since large segments of EBV code for genes that are involved in lytic infection or are essentially silent for transcription in latent EBV growth transformed LCLs. To test the hypothesis that these segments do not contain genes required for transformation, P3HR-1 cells were co-transfected with BZLF1 vector and three cosmid cloned DNAs (Figure 6a). Of the previous five clones, one cosmid was eliminated and the two clones that overlap with and are located 5′ or 3′ of the eliminated cosmid were endonuclease cut to delete more DNA sequences. The residual 5′ segment of the 5′ cosmid was ligated to the residual

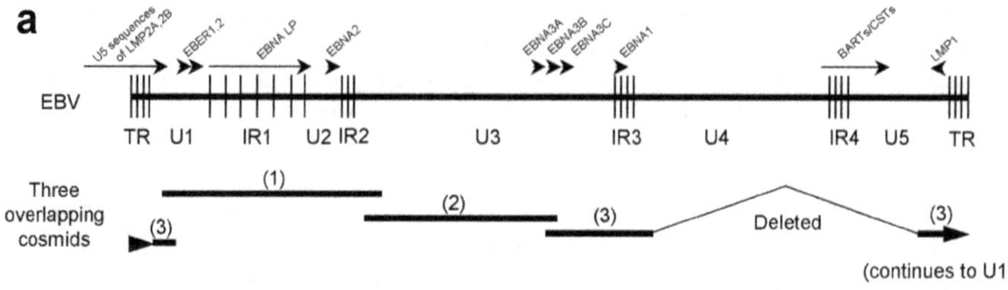

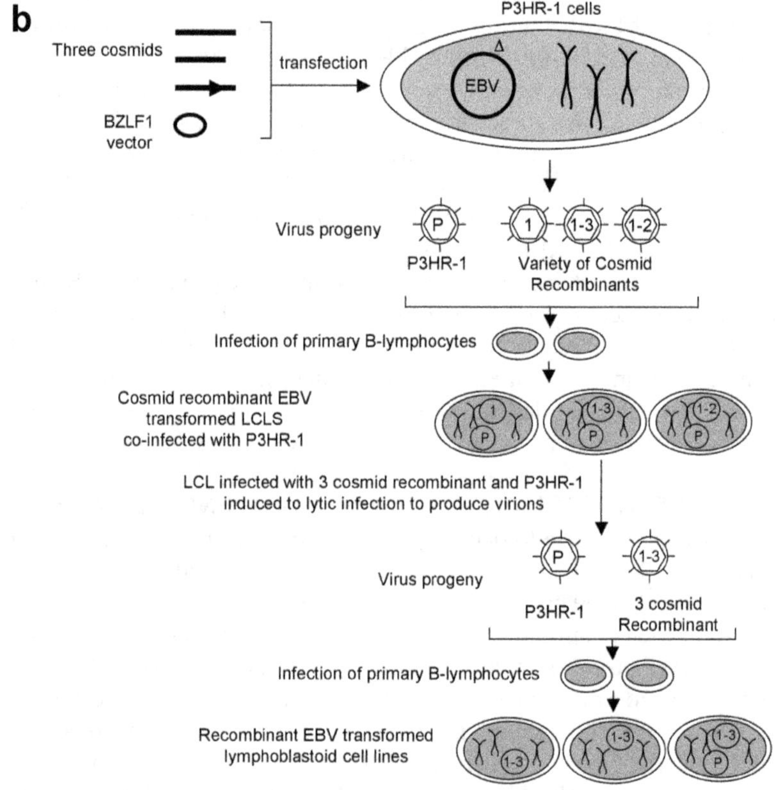

Figure 6. Derivation of EBV from three subgenomic DNA clones. (a) The EBV genome and latency gene transcripts shown are described in figure 1. Beneath this, the three DNA sequences cloned into cosmid vectors are represented as horizontal lines. Clone 3 is deleted of substantial EBV DNA including the BARTs/CSTs. (b) Three-subgenomic DNA clones and BZLF1 vector are transfected into P3HR-1 cells. In lytically infected cells, the three DNAs recombine with P3HR-1 DNA and with each other. The resulting virus progeny consists of P3HR-1 and several types of recombinant viruses (1, 1-3, 1-2, etc). Primary B-lymphocytes are infected, and nascent LCLs are PCR analyzed for recombinant virus DNA. LCLs containing mini-EBV genomes derived from the three subgenomic DNAs (1-3) usually contain P3HR-1. To delineate the phenotype of the mini-EBV genome, co-infected LCLs are induced to lytic infection. The virus progeny is serially diluted and used to infect primary B-lymphocytes. LCLs that develop are analyzed for viral DNA. Successful segregation of mini-EBV (1-3) from P3HR-1 reveals transforming EBV can be derived by this method.

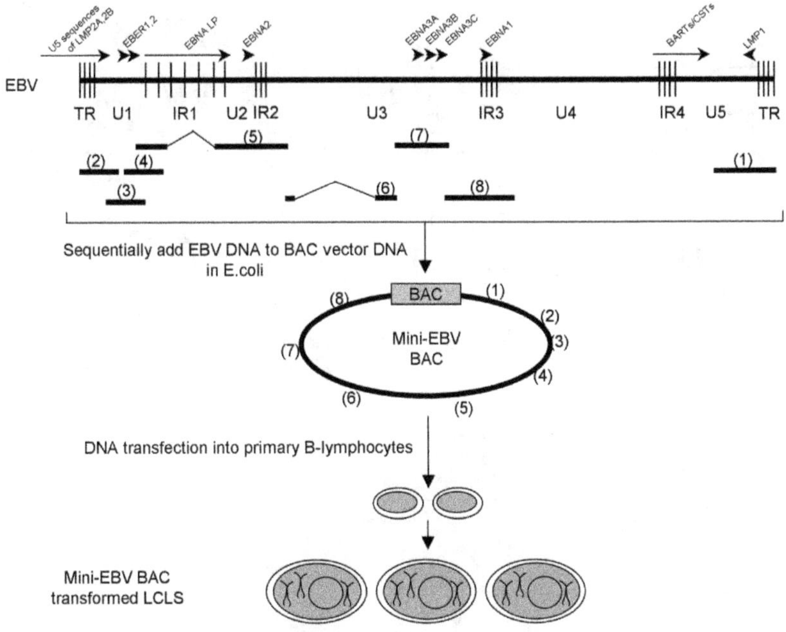

Figure 7. Mini-EBV bacterial artificial chromosome (BAC) genomes are transforming. (Top line) The EBV genome and latency gene transcripts shown are described in figure 1. Beneath this, the eight DNA sequences cloned into vectors are represented as horizontal lines. A mini-EBV BAC DNA was cloned by sequentially adding DNA sequences derived from eight cloned DNAs onto a BAC vector. The locations of the eight cloned DNAs are shown relative to the linear EBV genome (top line). Once the mini-EBV BAC cloned DNAs had been assembled, purified DNA was transfected directly into primary B-lymphocytes. LCLs that contained mini-EBV BAC DNA were derived.

3′ segment of the 3′ cosmid to derive a novel cosmid clone that essentially deletes over 50 kbp from EBV. In the co-transfected cells, lytic infection gene products expressed by replicating P3HR-1 enables the three cosmid clones to recombine with each other to form mini-EBV genomes that are packaged into virions (Figure 6b). The resulting virus progeny is then used to infect primary B-lymphocytes, and LCLs that grow out in tissue culture are analyzed by PCR for the recombinant virus genome (Robertson et al., 1995).

Initial LCLs containing recombinant mini-EBV genomes were co-infected with P3HR-1 as expected. Attempts were then made to segregate recombinant mini-EBV from P3HR-1. Thus, co-infected LCLs were induced to lytic infection by transfecting cells with BZLF1 vector and treating cells with chemical inducers. Primary B-lymphocytes were infected with serial dilutions of the resulting virus progeny, and LCLs that grew out were PCR or Southern blot analyzed for P3HR-1 and recombinant mini-EBV DNA. LCLs containing recombinant mini-EBV genomes without P3HR-1 were recovered, confirming the hypothesis that large segments of EBV are not required for transformation or efficient long-term LCL proliferation. The deleted DNA segments includes the BamHI A rightward

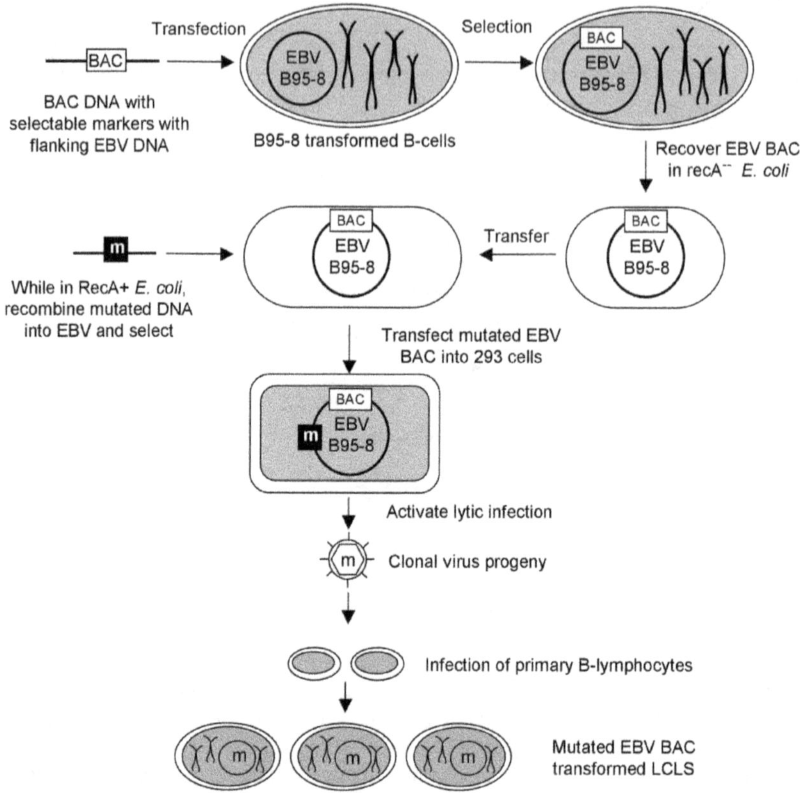

Figure 8. Genetic analysis using a complete EBV genome bacterial artificial chromosome (BAC). DNA containing BAC sequences, selectable markers, and flanking EBV DNA to target a specific site in the EBV B95-8 genome was cloned and transfected into B95-8 cells. This DNA recombines with the circular EBV DNA, and cells with the EBV BAC DNA are selected. EBV BAC DNA is recovered in *E. coli* and analyzed for precise recombination of BAC DNA into targeted EBV DNA. Mutated EBV DNA is recombined into the EBV BAC DNA by transforming recombinase positive *E. coli* with EBV BAC DNA and cloned mutated DNA, and then selecting for recombinant EBV BAC genomes. Recombinant BAC genomes are transformed into eukaryotic 293 cells and further analyzed for mutated EBV DNA. Lytic infection is activated in recombinant EBV BAC infected cells, and the clonal virus progeny is then used to infect primary B-lymphocytes. LCLs containing the mutated EBV recombinant BAC reveal the mutation does not abrogate viral transforming abilities.

transcripts (BARTs) or complementary strand transcripts (CSTs), which are expressed in latent EBV transformed LCLs (Robertson et al., 1995). These results indicate that these genes are not required for B cell transformation.

AN EBV BACTERIAL ARTIFICIAL CHROMOSOME

The most recent advancement in the derivation of recombinant EBV was to clone mini-EBV DNA or complete EBV DNA genomes as a bacterial artificial chromosome (BAC). Using this method, viral DNA is modified using simpler *E. coli* genetic systems that are well suited to large scale screening methods.

The mini-EBV BAC was cloned using DNA that contains an F-factor origin of replication and selectable markers for recovering recombinants in prokaryotic and eukaryotic cells (Figure 7). Large segments of cloned EBV DNA containing latency genes were added sequentially, and then the mini-EBV BAC was transfected into primary B-lymphocytes. Even though lytic infection or virions were not involved, long-term proliferating LCLs were derived that contain 71 kbp EBV plasmids. These mini-EBV BAC DNAs were deleted of DNA that codes for the BARTs/CSTs, thus demonstrating that these gene products are not required for transformation (Kempkes et al., 1995).

In order to derive a complete genomic EBV BAC, prokaryotic and eukaryotic cell selectable markers and F-factor DNA first were inserted into a cloned fragment of EBV DNA (Figure 8). EBV positive B95-8 cells were transfected with this cloned DNA, and viral genomes that had recombined homologously with the transfected DNA were selected. EBV BAC DNA was purified from these cells and transferred into *E. coli* for amplification. To mutate a gene of interest, EBV BAC DNA purified from *E. coli* was co-transformed with a cloned DNA mutated in a gene of interest into a recombinase-positive *E. coli*. In these cells, the mutated DNA is incorporated into EBV BAC DNA via the bacterial general recombination machinery, and recombinant EBV BACs are selected. Mutated BAC DNA is transferred back into recombinase-negative *E. coli* to analyze for the precise mutation. Mutated BAC DNA is stably transfected and selected in 293 human embryonic kidney cells, which are then induced to lytic infection by BZLF1 vector transfection. Primary B-lymphocytes are infected with the resulting recombinant virus clones to analyze the effect of the mutation. EBV BAC technology has been used to demonstrate the critical role of the terminal repeats in genomic DNA packaging during lytic infection, the individual and overlapping functions of immediate early activators of lytic infection BZLF1 and BRLF1, and the role of envelope glycoproteins in viral infection (Delecluse et al., 1999; Feederle et al., 2000; Janz et al., 2000). EBV BAC technology has been used to demonstrate EBNA1 is not required for B-cell transformation and to further delineate the functional domains of the plasma membrane oncogene product LMP1 (Dirmeier et al., 2003; Humme et al., 2003).

Summary

Recombinant virus genetic and biochemical analyses of Epstein-Barr virus mediated B-lymphocyte growth transformation have revealed the essential gene products and some of their growth altering functions that enable long-term cell proliferation. By adapting established techniques used for other DNA viruses while applying information regarding EBV to develop novel methodologies, the lack of a truly permissive cell line for lytic infection has been largely overcome. These efforts have been rewarded with a deeper understanding of the molecular mechanisms of EBV-mediated B-cell transformation. While new challenges have emerged, further application and refinement of these techniques will enable investigators to resolve those questions. The results of such analyses will foster progress towards development of specific therapies that can be used to treat or prevent malignancies associated with EBV infection.

Acknowledgments

This work was supported by institutional grants from the University of Texas Health Science Center at San Antonio, the Children's Cancer Research Foundation, and the Howard Hughes Medical Institute.

References

Aman, P., Rowe, M., Kai, C., Finke, J., Rymo, L., Klein, E., and Klein, G. (1990). Effect of the EBNA-2 gene on the surface antigen phenotype of transfected EBV-negative B-lymphoma lines. Int. J. Cancer *45*, 77-82.

Clarke, P.A., Schwemmle, M., Schickinger, J., Hilse, K., and Clemens, M.J. (1991). Binding of Epstein-Barr virus small RNA EBER-1 to the double-stranded RNA-activated protein kinase DAI. Nucleic Acids Res. *19*, 243-248.

Cohen, J.I., and Kieff, E. (1991). An Epstein-Barr virus nuclear protein 2 domain essential for transformation is a direct transcriptional activator. J. Virol. *65*, 5880-5885.

Cohen, J.I., Wang, F., Mannick, J., and Kieff, E. (1989). Epstein-Barr virus nuclear protein 2 is a key determinant of lymphocyte transformation. Proc. Natl. Acad. Sci. USA *86*, 9558-9562.

Cotter, M., Callahan, J., Aster, J., and Robertson, E. (2000a). Intracellular forms of human NOTCH1 functionally activate essential Epstein-Barr virus major latent promoters in the Burkitt's lymphoma BJAB cell line but repress these promoters in Jurkat cells. J. Virol. *74*, 1486-1494.

Cotter, M.A., 2nd, and Robertson, E.S. (2000b). Modulation of histone acetyltransferase activity through interaction of Epstein-Barr nuclear antigen 3C with prothymosin alpha. Mol. Cell. Biol. *20*, 5722-5735.

Countryman, J., Jenson, H., Seibl, R., Wolf, H., and Miller, G. (1987). Polymorphic proteins encoded within BZLF1 of defective and standard Epstein-Barr viruses disrupt latency. J. Virol. *61*, 3672-3679.

Delecluse, H.J., Pich, D., Hilsendegen, T., Baum, C., and Hammerschmidt, W. (1999). A first-generation packaging cell line for Epstein-Barr virus-derived vectors. Proc. Natl. Acad. Sci. USA *96*, 5188-5193.

Dirmeier, U., Neuhierl, B., Kilger, E., Reisbach, G., Sandberg, M.L., and Hammerschmidt, W. (2003). Latent membrane protein 1 is critical for efficient growth transformation of human B cells by epstein-barr virus. Cancer Res. *63*, 2982-2989.

Feederle, R., Kost, M., Baumann, M., Janz, A., Drouet, E., Hammerschmidt, W., and Delecluse, H.J. (2000). The Epstein-Barr virus lytic program is controlled by the co-operative functions of two transactivators. Embo J. *19*, 3080-3089.

Gires, O., Kohlhuber, F., Kilger, E., Baumann, M., Kieser, A., Kaiser, C., Zeidler, R., Scheffer, B., Ueffing, M., and Hammerschmidt, W. (1999). Latent membrane protein 1 of Epstein-Barr virus interacts with JAK3 and activates STAT proteins. Embo J. *18*, 3064-3073.

Hammerschmidt, W., and Sugden, B. (1989). Genetic analysis of immortalizing functions of Epstein-Barr virus in human B lymphocytes. Nature *340*, 393-397.

Harada, S., and Kieff, E. (1997). Epstein-Barr virus nuclear protein LP stimulates EBNA-2 acidic domain- mediated transcriptional activation. J. Virol. *71*, 6611-6618.

Henderson, S., Huen, D., Rowe, M., Dawson, C., Johnson, G., and Rickinson, A. (1993). Epstein-Barr virus-coded BHRF1 protein, a viral homologue of Bcl-2, protects human B cells from programmed cell death. Proc. Natl. Acad. Sci. USA *90*, 8479-8483.

Hsu, D.H., de Waal Malefyt, R., Fiorentino, D.F., Dang, M.N., Vieira, P., de Vries, J., Spits, H., Mosmann, T.R., and Moore, K.W. (1990). Expression of interleukin-10 activity by Epstein-Barr virus protein BCRF1. Science *250*, 830-832.

Hudewentz, J., Bornkamm, G.W., and Zur Hausen, H. (1980). Effect of the diterpene ester TPA on Epstein-Barr virus antigen- and DNA synthesis in producer and nonproducer cell lines. Virology *100*, 175-178.

Humme, S., Reisbach, G., Feederle, R., Delecluse, H.J., Bousset, K., Hammerschmidt, W., and Schepers, A. (2003). The EBV nuclear antigen 1 (EBNA1) enhances B cell immortalization several thousandfold. Proc. Natl. Acad. Sci. USA *100*, 10989-10994. Epub 12003 Aug 10928.

Imai, S., Nishikawa, J., and Takada, K. (1998). Cell-to-cell contact as an efficient mode of Epstein-Barr virus infection of diverse human epithelial cells. J. Virol. *72*, 4371-4378.

Izumi, K.M., Kaye, K.M., and Kieff, E.D. (1997a). The Epstein-Barr virus LMP1 amino acid sequence that engages tumor necrosis factor receptor associated factors is critical for primary B lymphocyte growth transformation. Proc. Natl. Acad. Sci. USA *94*, 1447-1452.

Izumi, K.M., and Kieff, E.D. (1997b). The Epstein-Barr virus oncogene product latent membrane protein 1 engages the tumor necrosis factor receptor-associated death domain protein to mediate B lymphocyte growth transformation and activate NF- kappaB. Proc. Natl. Acad. Sci. USA *94*, 12592-12597.

Izumi, K.M., McFarland, E.C., Ting, A.T., Riley, E.A., Seed, B., and Kieff, E.D. (1999). The Epstein-Barr virus oncoprotein latent membrane protein 1 engages the tumor necrosis factor receptor-associated proteins TRADD and receptor-interacting protein (RIP) but does not induce apoptosis or require RIP for NF-kappaB activation. Mol. Cell. Biol. *19*, 5759-5767.

Janz, A., Oezel, M., Kurzeder, C., Mautner, J., Pich, D., Kost, M., Hammerschmidt, W., and Delecluse, H.J. (2000). Infectious epstein-barr virus lacking major glycoprotein BLLF1 (gp350/220) demonstrates the existence of additional viral ligands. J. Virol. *74*, 10142-10152.

Kaye, K.M., Izumi, K.M., and Kieff, E. (1993). Epstein-Barr virus latent membrane protein 1 is essential for B- lymphocyte growth transformation. Proc. Natl. Acad. Sci. USA *90*, 9150-9154.

Kempkes, B., Pich, D., Zeidler, R., Sugden, B., and Hammerschmidt, W. (1995). Immortalization of human B lymphocytes by a plasmid containing 71 kilobase pairs of Epstein-Barr virus DNA. J. Virol. *69*, 231-238.

Kieff, E., and Rickinson, A. (2001). Epstein-Barr virus and its replication. In Virology, D.M. Knipe, P.M. Howley, D.E. Griffen, R.A. Lamb, M.A. Martin, B. Roizman, and S.E. Straus, eds. (Philadelphia, PA, Lippincott, Williams, and Wilkins), pp. 2511-2573.

Knutson, J.C. (1990). The level of c-fgr RNA is increased by EBNA-2, an Epstein-Barr virus gene required for B-cell immortalization. J. Virol. *64*, 2530-2536.

Komano, J., Maruo, S., Kurozumi, K., Oda, T., and Takada, K. (1999). Oncogenic role of Epstein-Barr virus-encoded RNAs in Burkitt's lymphoma cell line Akata. J. Virol. *73*, 9827-9831.

Komano, J., Sugiura, M., and Takada, K. (1998). Epstein-Barr virus contributes to the malignant phenotype and to apoptosis resistance in Burkitt's lymphoma cell line Akata. J. Virol. *72*, 9150-9156.

Lee, M.A., Kim, O.J., and Yates, J.L. (1992a). Targeted gene disruption in Epstein-Barr virus. Virology *189*, 253-265.

Lee, M.A., and Yates, J.L. (1992b). BHRF1 of Epstein-Barr virus, which is homologous to human proto- oncogene bcl2, is not essential for transformation of B cells or for virus replication *in vitro*. J. Virol. *66*, 1899-1906.

Longnecker, R., Miller, C.L., Miao, X.Q., Marchini, A., and Kieff, E. (1992). The only domain which distinguishes Epstein-Barr virus latent membrane protein 2A (LMP2A) from LMP2B is dispensable for lymphocyte infection and growth transformation *in vitro*; LMP2A is therefore nonessential. J. Virol. *66*, 6461-6469.

Luka, J., Kallin, B., and Klein, G. (1979). Induction of the Epstein-Barr virus (EBV) cycle in latently infected cells by n-butyrate. Virology *94*, 228-231.

Mannick, J.B., Cohen, J.I., Birkenbach, M., Marchini, A., and Kieff, E. (1991). The Epstein-Barr virus nuclear protein encoded by the leader of the EBNA RNAs is important in B-lymphocyte transformation. J. Virol. *65*, 6826-6837.

Marchini, A., Kieff, E., and Longnecker, R. (1993). Marker rescue of a transformation-negative Epstein-Barr virus recombinant from an infected Burkitt lymphoma cell line: a method useful for analysis of genes essential for transformation. J. Virol. *67*, 606-609.

Marchini, A., Tomkinson, B., Cohen, J.I., and Kieff, E. (1991). BHRF1, the Epstein-Barr virus gene with homology to Bcl2, is dispensable for B-lymphocyte transformation and virus replication. J. Virol. *65*, 5991-6000.

Miller, G., Robinson, J., Heston, L., and Lipman, M. (1974). Differences between laboratory strains of Epstein-Barr virus based on immortalization, abortive infection, and interference. Proc. Natl. Acad. Sci. USA *71*, 4006-4010.

Miller, G., Robinson, J., Heston, L., and Lipman, M. (1975). Differences between laboratory strains of Epstein-Barr virus based on immortalization, abortive infection and interference. IARC Sci. Publ. *11*, 395-408.

Mosialos, G., Birkenbach, M., Yalamanchili, R., VanArsdale, T., Ware, C., and Kieff, E. (1995). The Epstein-Barr virus transforming protein LMP1 engages signaling proteins for the tumor necrosis factor receptor family. Cell *80*, 389-399.

Nitsche, F., Bell, A., and Rickinson, A. (1997). Epstein-Barr virus leader protein enhances EBNA-2-mediated transactivation of latent membrane protein 1 expression: a role for the W1W2 repeat domain. J. Virol. *71*, 6619-6628.

Rickinson, A., and Kieff, E. (2001). Epstein-Barr virus. In Virology, D.M. Knipe, P.M. Howley, D.E. Griffen, R.A. Lamb, M.A. Martin, B. Roizman, and S.E. Straus, eds. (Philadelphia, PA, Lippincott, Williams, and Wilkins), pp. 2573-2627.

Robertson, E., and Kieff, E. (1995). Reducing the complexity of the transforming Epstein-Barr virus genome to 64 kilobase pairs. J. Virol. *69*, 983-993.

Robertson, E.S., Lin, J., and Kieff, E. (1996). The amino-terminal domains of Epstein-Barr virus nuclear proteins 3A, 3B, and 3C interact with RBPJ(kappa). J. Virol. *70*, 3068-3074.

Sharp, T.V., Raine, D.A., Gewert, D.R., Joshi, B., Jagus, R., and Clemens, M.J. (1999). Activation of the interferon-inducible (2'-5') oligoadenylate synthetase by the Epstein-Barr virus RNA, EBER-1. Virology *257*, 303-313.

Sjoblom-Hallen, A., Yang, W., Jansson, A., and Rymo, L. (1999). Silencing of the Epstein-Barr virus latent membrane protein 1 gene by the Max-Mad1-mSin3A modulator of chromatin structure. J. Virol. *73*, 2983-2993.

Subramanian, C., Cotter, M.A., and Robertson, E.S. (2001). Epstein-Barr virus nuclear protein EBNA-3C interacts with the human metastatic suppressor Nm23-H1: A molecular link to cancer metastasis. Nat. Med. *7*, 350-355.

Sung, N.S., Kenney, S., Gutsch, D., and Pagano, J.S. (1991). EBNA-2 transactivates a lymphoid-specific enhancer in the BamHI C promoter of Epstein-Barr virus. J. Virol. *65*, 2164-2169.

Swaminathan, S., Hesselton, R., Sullivan, J., and Kieff, E. (1993). Epstein-Barr virus recombinants with specifically mutated BCRF1 genes. J. Virol. *67*, 7406-7413.

Swaminathan, S., Tomkinson, B., and Kieff, E. (1991). Recombinant Epstein-Barr virus with small RNA (EBER) genes deleted transforms lymphocytes and replicates *in vitro*. Proc. Natl. Acad. Sci. USA *88*, 1546-1550.

Tomkinson, B., and Kieff, E. (1992). Use of second-site homologous recombination to demonstrate that Epstein-Barr virus nuclear protein 3B is not important for lymphocyte infection or growth transformation *in vitro*. J. Virol. *66*, 2893-2903.

Tomkinson, B., Robertson, E., and Kieff, E. (1993a). Epstein-Barr virus nuclear proteins EBNA-3A and EBNA-3C are essential for B-lymphocyte growth transformation. J. Virol. *67*, 2014-2025.

Tomkinson, B., Robertson, E., Yalamanchili, R., Longnecker, R., and Kieff, E. (1993b). Epstein-Barr virus recombinants from overlapping cosmid fragments. J. Virol. *67*, 7298-7306.

Wang, F., Gregory, C.D., Rowe, M., Rickinson, A.B., Wang, D., Birkenbach, M., Kikutani, H., Kishimoto, T., and Kieff, E. (1987). Epstein-Barr virus nuclear antigen 2 specifically induces expression of the B-cell activation antigen CD23. Proc. Natl. Acad. Sci. USA *84*, 3452-3456.

Wang, F., Marchini, A., and Kieff, E. (1991). Epstein-Barr virus (EBV) recombinants: use of positive selection markers to rescue mutants in EBV-negative B-lymphoma cells. J. Virol. *65*, 1701-1709.

Yalamanchili, R., Tong, X., Grossman, S., Johannsen, E., Mosialos, G., and Kieff, E. (1994). Genetic and biochemical evidence that EBNA 2 interaction with a 63-kDa cellular GTG-binding protein is essential for B lymphocyte growth transformation by EBV. Virology *204*, 634-641.

Yoshiyama, H., Imai, S., Shimizu, N., and Takada, K. (1997). Epstein-Barr virus infection of human gastric carcinoma cells: implication of the existence of a new virus receptor different from CD21. J. Virol. *71*, 5688-5691.

From: Epstein-Barr Virus. Edited by: Erle S. Robertson

Chapter 17

EBV Persistence and Latent Infection *In Vivo*

*David A. Thorley-Lawson**

ABSTRACT

This chapter will cover the current state of knowledge about how Epstein-Barr virus establishes and maintains persistent infection *in vivo*. EBV is a B lymphotrophic virus whose biology is closely entwined with the normal biology of B lymphocytes. Recent studies have led to a model of EBV persistence based on the idea that the virus uses all aspects of mature B cell biology to establish (activation and germinal center differentiation) and maintain (memory) persistent infection and to replicate (plasma cells). To achieve this, the virus uses different transcription programs in different cellular backgrounds. The virus activates and drives the growth of newly infected B cells, using the growth program, so that the cell can then differentiate, using the default program, via the germinal center reaction into the memory compartment. EBV persists within the long lived memory B cell compartment (latency and EBNA1 only programs) where it shuts down all viral protein expression, so it is neither pathogenic to the host nor detectable by the immune response. If the infected memory cell differentiates into a plasma cell the virus is released for further infectious spread. This model provides an explanation for the disparate behavior of EBV in different cell types and suggests that the EBV associated lymphomas arise from specific stages in the life cycle of EBV.

INTRODUCTION

Viral replication can be studied with any virus. The interesting thing about herpesviruses is that they establish latent infections that persist for the life time of the host, often without major pathological consequences. How they do this still remains to be fully elucidated, but amongst the human herpesviruses, persistent latent infection is best understood for Epstein-Bar virus.

The conundrum of EBV circa 1999

Epstein-Barr virus was discovered in 1964 as a consequence of efforts to identify human tumor viruses (Epstein et al., 1964). It was originally found in cultured tumor cells from patients with the endemic form of Burkitt's lymphoma. The irony

*For correspondence email david.thorley-lawson@tufts.edu

Table 1. EBV uses five transcription programs to establish and maintain persistent infection

Transcription program	Genes Expressed[2]	Infected B Cell Type[1]	Function
Growth	EBNA1,2, 3a, 3b, 3c, LP, LMP1, LMP2a and LMP2b	Naïve	Activate B Cell
Default	EBNA1,LMP1 and LMP2a	Germinal Center	Differentiate activated B cell into memory
Latency	None	Peripheral Memory	Allow life time persistence
EBNA-1 only	EBNA1	Dividing Peripheral Memory	Allow virus in latency program cell to divide
Lytic	All lytic genes	Plasma Cell	Replicate the virus in plasma cell

1- Except where indicated the cell types are primarily restricted to the lymphoid tissue of Waldeyer's ring.
2- Does not include the non-coding EBER and BART RNA's that are assumed to be ubiquitous but have not been rigorously identified in all of the infected subtypes.

is that 40 years later we have a large volume of information about EBV's molecular and cellular biology, immunology, virology, epidemiology, clinical manifestations and disease associations (Kieff and Rickinson, 2001; Rickinson and Kieff, 2001; Thorley-Lawson, 2001a and this text book) and we still don't know whether, let alone how, EBV plays a role in BL pathogenesis.

By 1999 we had already learned a great deal about EBV, but this large body of work existed as a series of independent pieces of information that did not hang together in a consistent and convincing conceptual frame work of viral biology and persistence. Amongst other things, we knew:

1. That EBV is capable of infecting normal resting B lymphocytes in culture and transforming them into proliferating lymphoblasts. This dramatic event was discovered early on (Henle et al., 1967; Pope et al., 1968) and has continued to be the main reason for believing that EBV could play a role in neoplastic disease. It is achieved through the concerted action of nine latent proteins (reviewed previously in (Kieff and Rickinson, 2001) but see other chapters in this book for the most recent advances). Phenotypically the cells look remarkably like antigen activated B cell blasts (Nilsson, 1979; Thorley-Lawson et al., 1985; 1982). This is the lymphoblastoid form of latency now called the "growth program" (Thorley-Lawson, 2001b; Thorley-Lawson and Gross, 2004) for obvious reasons (Table 1). Why would EBV cause infected cells to proliferate when this puts the host at risk for developing neoplastic disease? The host in which it establishes a life long persistent infection.

2. That EBV is found in several kinds of tumors especially B cell lymphomas. These include the immunoblastic lymphomas (IL) that occur in immunosuppressed individuals (Ho et al., 1985; Hopwood and Crawford, 2000), Hodgkin's disease (HD) (Glaser et al., 1997; Weiss et al., 1987), and Burkitt's lymphoma (BL) (Epstein et al., 1964). Curiously EBV expresses a different pattern of latent proteins in all three tumors. In IL the growth program is used (Hopwood and Crawford, 2000; Thomas et al., 1990). In HD, EBV only expresses three of the latent proteins (Deacon et al., 1993), now referred to as the "default program" (Table 1) (Thorley-Lawson, 2001b). One of the proteins is EBNA1 (Oudejans et al., 1996) which is required to replicate the viral DNA (Yates et al., 1985), and the other two are the latent membrane proteins LMP1 (Herbst et al., 1991) and LMP2a (Niedobitek et al., 1997). Initial skepticism that this was truly a distinct transcription program subsided when it was found that EBNA1 was expressed in the default program from a different promoter (Qp) than is used in the growth program (Nonkwelo et al., 1996; Schaefer et al., 1995). In BL only EBNA1 is expressed (Gregory et al., 1990), again from this novel promoter. Why would EBV have these different latent gene transcription patterns and why do different tumors express different patterns of viral latent genes?

3. That EBV infects and persists in >90% of the adult human population (Henle and Henle, 1979) almost always benignly. Despite its ability to make cells grow, EBV associated tumors are extremely rare compared to the number of people infected.

4. *In vitro* molecular signaling studies revealed that some of the EBV latent proteins, specifically LMP1 and LMP2a, had signaling properties remarkably similar to those involved in B cell differentiation and survival. In particular, it

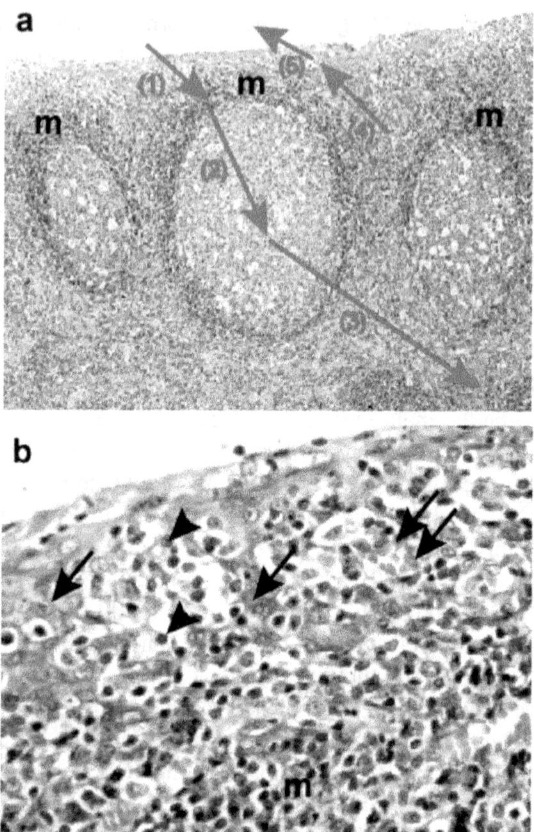

Figure 1. The lymphoepithelium of the palatine tonsils from Waldeyer's ring. The tonsil consists of a highly involuted surface creating a large surface area with deep invaginations called crypts. The surface of the crypt lumen is at the top of both micrographs. a. Numerous follicles containing B cells are arranged parallel to the crypt surface. The mantle zone (m), containing naive B cells, is always facing the crypt and is continuous with the reticulated epithelium. b. A higher magnification reveals the sponge like structure of the epithelial cells in the lymphopepithelium (black arrows) that create spaces extending all the way to the mantle zone and which are filled with infiltrating lymphocytes (arrow heads) such that there is frequently only a single epithelial cell between the outer surface and the lymphocytes. The infiltrating lymphocytes include memory B cells, plasma cells and T cells. (The micrograpahs were provided by Dr. Marta Perry). The colored arrows indicate the pathway of the antigen response that is reiterated by the virus with the difference that the virus provides most if not all of the signals required for the process through the activity of its latent proteins. The parallel signaling steps are detailed diagrammatically in Figure 2. Antigen and EBV both enter through saliva and cross the epithelial barrier (1). Antigen binds to and EBV infects naïve B cells in the mantle zone causing them to become B cell blasts that migrate into the germinal center (2). Here the B cells receive rescue and survival signals that allow them to exit as memory cells into the peripheral circulation through the efferent lymphatics. Memory cells reenter the lymphoepithelium by passing through the walls of small venules termed HEV (4) and occasionally differentiate into plasma cells that migrate into the epithelium where they release antibody onto the mucosal surface (5). If the plasma cell contains EBV the virus will be triggered to replicate and infectious virus will be shed into the saliva (5). For more details see Figure 2. See this figure in colour in the Colour Plate Section at the back of the book

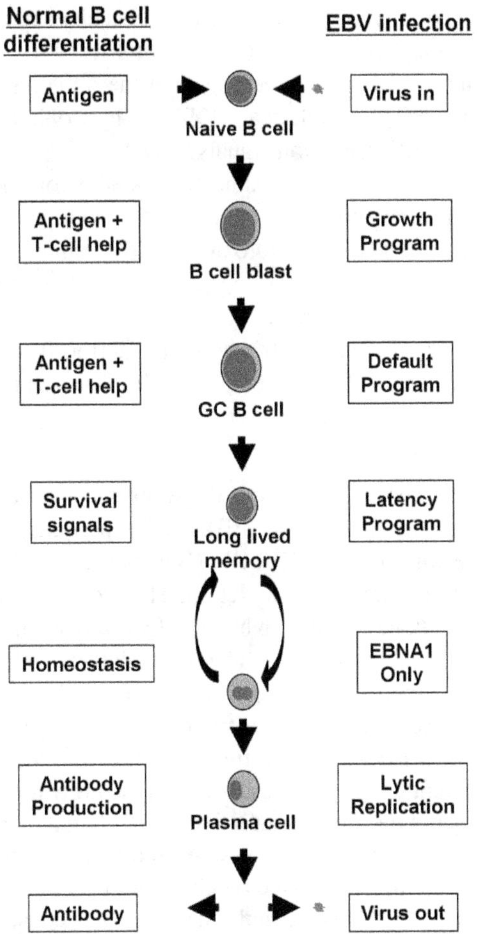

Figure 2. The parallels between normal mature B cell biology and the pathway of establishment, maintenance and replication of EBV. The virus infects resting naïve B cells in the lymphoid tissue of Waldeyer's ring and uses the growth program to activate these cells to become proliferating blasts. This parallels the process by which a naïve B cell becomes activated upon exposure to an antigen. The antigen activated B blast migrates into the follicle to undergo the germinal center reaction where it is selected for high affinity antibody production. The cell is ultimately rescued into the memory pool by receiving signals from antigen and antigen specific T helper cells. The virus infected B cell blast follows the same pathway. Once in the germinal center it switches to the default program to deliver the necessary rescue signals. The memory cells then exit the cell cycle and enter the peripheral circulation. For the virus infected cells this results in the shut down of all the protein encoding genes – the latency program. Memory B cells occasionally divide in the peripheral circulation as part of the homeostatic mechanism for maintaining stable numbers of cells. When this happens to a cell carrying EBV the virus expresses EBNA1 alone to allow the viral genome to divide with the cell. Memory cells reenter the lymphoepithelium and in response to unknown signals (perhaps bystander T cell help or cognate antigen) occasionally differentiate into plasma cells and migrate into the epithelium where they release antibody onto the mucosal surface. If the plasma cell contains EBV the virus will be triggered to replicate and infectious virus will be shed into the saliva.

was shown that LMP1 signaling paralleled that usually supplied by interaction with antigen specific T helper cells (Th cells) (Gires et al., 1997; Mosialos et al., 1995) and that LMP2a had at least some of the signaling capabilities of the antigen receptor itself (Caldwell et al., 1998). Why would EBV be mimicking B cell survival and differentiation signals?

5. One quirky piece of information came from studies on the immunology of EBV infection. Investigators have mapped the epitopes in different latent proteins that are recognized by cytotoxic T cells (Khanna et al., 1992; Murray et al., 1992). Usually it would be expected that these particular sequences would vary more rapidly than flanking sequences because they would be under selection pressure exerted by the CTL. Curiously in EBV the epitope sequences are conserved (Khanna et al., 1997). It was as though there was a selective advantage to the virus in having virus infected cells killed! What was the reason for this suicidal tendency?

The observation that brought all of this disparate and apparently irreconcilable behavior together was the finding that EBV in the peripheral blood of healthy carriers resides in resting, memory B cells (Babcock et al., 1998; Miyashita et al., 1997). This was the ultimate contradiction. The most widely known property of EBV, as mentioned above, was that when it infects any resting B cell in culture it invariably and efficiently causes them to become latently infected proliferating blasts. Proliferation was the *sine qua non* of latent EBV infection. Yet in the blood the virus was not in all cell types, but was tightly restricted to memory B cells and these were not proliferating blasts but resting cells that were subsequently found to express none of the latent proteins that are found in the growth program (Hochberg et al., 2004a). How could the virus sustain two such contradictory behaviors? The answer lay in recognizing the parallel between the behavior of virus infected B cells and that of normal antigen activated B cells (Thorley-Lawson, 2001b; Thorley-Lawson and Babcock, 1999). This recognition allowed the construction of a model for the establishment and maintenance of EBV persistence that could account for virtually every aspect of its known behavior. A description of this "grand unified theory" of EBV persistence and an update on its current status will be the subject of this chapter. In the first section I will describe the essential features of the model paralleling the behavior of the virus with that of normal B cell biology. In the subsequent section I will address the explanations that the model provides for the 5 questions listed above. In the final section I will discuss the insights that the model provides into the origins of EBV associated neoplastic disease particularly the lymphomas.

EBV INFECTION IN THE HEALTHY HOST

The sea change in understanding EBV was the recognition that under normal conditions it should not be thought of as an oncogenic virus. This despite its discovery in and association with tumors and its ability to transform B cells in culture. The essence of its biological behavior is that it does not aberrantly deregulate the behavior of infected B cells *in vivo*. It initiates, establishes and maintains persistent infection by subtly using various aspects of normal B cell biology. Ultimately this allows the virus to persist within memory B cells for the life time of the host in a fashion that is non-pathogenic. Not only is EBV not a natural tumor virus it has developed strategies to minimize its pathogenic potential to the host. A summary of normal mature B cell biology and the proposed parallels with EBV is given in Figures 1 and 2. A complete flow chart of all the steps so far known in EBV persistence (based on direct or indirect evidence) and described in this chapter is shown in Figure 3.

Salivary antigen and infectious EBV – crossing the epithelial barrier

It is generally believed that EBV is spread through salivary contact (Hoagland, 1955) and that the virus enters through the epithelium that lines the nasopharynx. The lymphoid system that surrounds the nasopharyngeal region includes the adenoids and tonsils and is called Waldeyer's ring. Together with the overlying epithelium it forms a continuous structure referred to as the lymphoepithelium (Figure 1) (Perry and Whyte, 1998). The epithelium is invaginated to form crypts below which resides the lymphoid tissue (Perry, 1994; Tang et al., 1995). Deep in the crypts, the epithelium can be only a single epithelial cell in thickness. Environmental antigens are sampled directly through the epithelium (Brandtzaeg et al., 1999a; 1999b; Perry and Whyte, 1998). The involuted nature of the crypts allows for a massive surface area for detecting antigens as they come in with food. The cellular structures involved in sampling antigens are not well characterized. The M cells which sample antigens from the lumen of the gut and are typically found in the Peyers patches of gut lymphoepithelium are not obviously present in Waldeyer's ring. It is assumed that similarly specialized epithelial cells in the crypts perform this function. The involuted surface area of the tonsil which is exposed to EBV bearing saliva, provides a large target for EBV infection. It is likely that the virus, in saliva, enters the crypts and crosses the thin layer of epithelial cells to infect B cells that reside below. How the virus crosses the epithelial barrier is not known. It has been speculated that the virus first infects the epithelial cells, replicates and then is released to infect B cells in the underlying areas, but there is no direct evidence for this and epithelial cells appear to be resistant to infection from the apical (i.e. mucosal) side (Tugizov et al., 2003). Other possibilities are that the virus is actively sampled and transported across the epithelium or that it passively migrates through gaps in the epithelium.

Intraepithelial B cells in the tonsil are a heterogeneous mixture of cells (Brandtzaeg et al., 1999a; 1999b; Dono et al., 1996a;1996b; Perry and Whyte, 1998). Plasma cells and resting IgM memory cells comprise the major components that lie between and immediately below the epithelium comprising the so called marginal zone (Spencer et al., 1985; Spencer et al., 1998). Immediately below, are the resting naive B cells that constitute the mantle zone. However, the mantle zone is not de-lineated from the epithelium so naïve B cells are also present throughout the lymphoepithelium. The sponge like nature and deep invaginations of the crypts ensure that all of these cell types are close to the surface and when EBV crosses the epithelium the target B cell it first encounters will almost certainly be in a resting state.

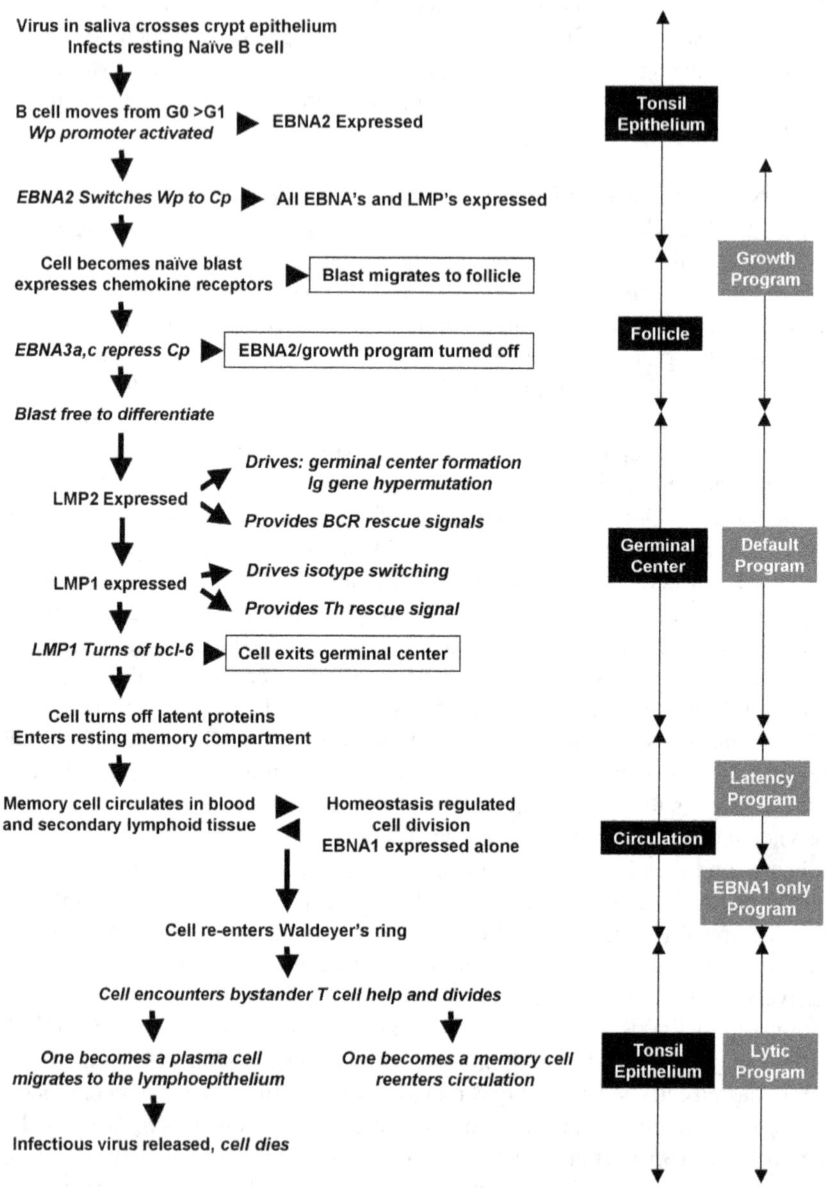

Figure 3. A flow chart summarizing the known and putative events in the establishment, maintenance and replication of EBV. The steps for which there is significant supportive evidence are shown. Italics denote steps for which there is good *in vitro* evidence and the boxes represent speculations based on *in vivo* observations. For details see the text.

Activating the naïve B cell: the response to antigen and the establishment of latency

What happens once EBV crosses the epithelial barrier is assumed to recapitulate what happens with an antigen. Naive B cells continuously re-circulate throughout the body. They extravasate from the peripheral circulation into secondary lymphoid structures such as the tonsils through specialized structures called high endothelial vesicles (HEV) which reside in the lymphoepithelium. The B cells migrate to the follicles where they remain for a few days and then they re-enter the circulation (Brandtzaeg et al., 1999a). If they encounter antigen presented by dendritic cells and antigen specific T helper cells the B cells will be activated to become proliferating blasts.

As far as we know whenever EBV encounters and infects a resting B cells it always drives that cell to become a proliferating lymphoblast (Thorley-Lawson and Mann, 1985). Therefore, if EBV encounters a naive B cell in the lymphoepithelium of Waldeyer's ring the result will be remarkably similar to an antigen driven response, however in this case the B cell becomes an activated blast not through interaction with antigen and T cell help but through the activity of the latent proteins encoded by the growth program (Kempkes et al., 1995). The only population that expresses the growth program in the tonsils of healthy carriers of the virus are naive B cells (Babcock et al., 2000; Joseph et al., 2000a), therefore the target of the incoming virus is the resting naive B cell that reside in the follicles immediately below the epithelium. This is the first example we will encounter of a latent gene transcription program, the growth program (Table 1, Figure 2), known to be used in lymphoma, in this case immunoblastic lymphoma (Figure 4), being found in a normal B cell counterpart, the naïve B cell in the tonsil (for a more detailed discussion of this issue see below).

Because the target for EBV infection is a resting cell, the virus must initiate latent gene transcription in a quiescent environment. It infects cells through the interaction of the viral glycoproteins gp350/220 with CD21 on the B cell (Fingeroth et al., 1984; Nemerow et al., 1985) and gp42/gH/gL with MHC class II on the B cell (Li et al., 1997). CD21 is the receptor for the C3d component of complement and forms part of a multimeric signal transduction complex with CD19, CD81 (TAPA-1), and Leu-13 (Matsumoto et al., 1993). The high density of gp350/220 on the virion (Thorley-Lawson and Poodry, 1982) ensures that the binding of viral particles will cause extensive cross-linking of the CD21 signaling complex. Cross-linking of surface molecules by the viral glycoproteins provides the signal (Sinclair and Farrell, 1995) to begin moving the resting B cells from G0 into the G1 phase of the cell cycle. During this time, the earliest expressed latent protein (EBNA2) is detected (Allday et al., 1989; Rooney et al., 1989). This protein is expressed from a promoter (Wp) that is present in multiple copies in the viral genome and may be designed to function in the transcriptionally sparse environment of a resting B cell (Woisetschlaeger et al., 1990). EBNA2 drives the cells through the first G1 (Sinclair et al., 1994). EBNA2 is a transcription factor that activates the promoters necessary to produce all nine of the latent proteins expressed in the growth program (reviewed in (Kieff and Rickinson, 2001) and this text book) (Table 1). At this point transcription also switches from Wp to Cp (Woisetschlaeger et al., 1990), a promoter that works optimally in B lymphoblasts and allows expression of all of

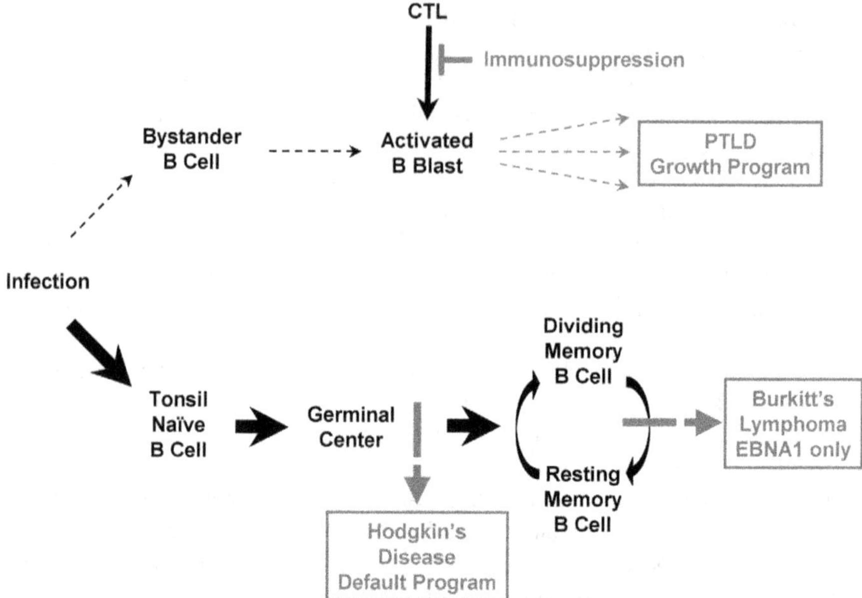

Figure 4. The putative check points in the EBV life cycle that might give rise to lymphoma. The events that occur normally in healthy carriers are denoted in black and the postulated events leading to lymphoma in grey. EBV normally infects naïve B cells in Waldeyer's ring and these cells can differentiate into memory cells and out of the cell cycle (thick arrows) so they are not pathogenic.Hodgkin's disease arises from an EBV-infected cell that is blocked at the germinal center cell stage. This results in constitutive expression of the default program.Burkitt's lymphoma arises from a germinal center cell that is entering the memory compartment but is stuck proliferating because of the activated c-myc oncogene. Consequently, the cell expresses EBNA1 only. For both of these tumors the critical event may be a cellular mutation that occurs during the immunological disturbance associated with acute EBV infection. Because the levels of infected cells are so high at this time there is a reasonable possibility that the cell undergoing mutation will have EBV in it by chance. PTLD arises if another cell other than the naïve B cell in Waldeyer's ring becomes infected and expresses the growth program. This cell will continue to proliferate because it can not differentiate out of the cell cycle. This is a very rare event, high lighting how carefully controlled EBV infection is. Normally these bystander B cell blasts would be destroyed by CTL but if the CTL response is suppressed then they can grow into PTLD lymphoma. Note shown: by stander blasts could also be produced if latently infected germinal center or memory cells fortuitously switched to the growth program.

the EBNA proteins. The result is that infected normal B cells become activated lymphoblasts and begin to proliferate in response to the actions of viral latent proteins. Although they should not be thought of as classically transformed cells, such as are obtained with other DNA tumor viruses (e.g. SV40, papillomavirus and adenovirus) (Allday et al., 1995), EBV driven cells are not completely normal as evidenced by deregulation of their cell cycle control resulting in immortal growth in culture (O'Nions and Allday, 2003; Wade and Allday, 2000) (reviewed in (O'Nions and Allday, 2004)).

The nine latent proteins of the growth program include six nuclear proteins (EBNA's - Epstein-Barr virus nuclear antigens – 1, 2, 3a, 3b, 3c and LP) and three membrane proteins (LMP's - latent membrane proteins) (reviewed in (Kieff and Rickinson, 2001) but see also this text book for recent updates). The presence of EBNA2 is diagnostic for the growth program (Hochberg and Thorley-Lawson, 2005). Several of the latent proteins have potent growth promoting activity and can act as oncogenes. These include EBNA2 (Kempkes et al., 1995), EBNA3a (Hickabottom et al., 2002), EBNA3c (Parker et al., 1996), and LMP1 (Wang et al., 1985).

In addition to the nine latent proteins, EBV infected lymphoblasts express two small non-polyadenylated RNA's, termed EBER1 and EBER2 (Arrand and Rymo, 1982), and a set of transcripts, derived from the Bam A fragment of the genome, called BARF0 or BART (Smith et al., 1993). The function of these genes is unclear. It has been suggested that the EBERs may interact with the interferon system (Sharp et al., 1999) and BARF0 may interact with Notch (Kusano and Raab-Traub, 2001). EBERs are the only viral transcripts expressed in the latency program (see below) suggesting that they may play a role in the maintenance of latency in memory cells (Sousa and Thorley-Lawson unpublished). The EBER transcripts are important because they are by far the most abundant RNA's found in latently infected cells and are routinely used as targets for *in situ* hybridization studies to identify EBV infected cells in normal tissues and tumors.

The latent genes are transcribed from the viral genome which exists as a covalently closed episomal circle (Adams and Lindahl, 1975). The linear genome forms this circle when the infected cell begins proliferating (Hurley and Thorley-Lawson, 1988). The status of the viral genome in a tissue provides a considerable amount of useful information. Linear genomes indicate viral replication. Circularized genomes, in the absence of linear genomes, are indicative of latently infected cells that have proliferated at some time in their history (Decker et al., 2001; Decker et al., 1996; Raab-Traub and Flynn, 1986). The number of discrete joint fragments is a measure of how many independent clones are present in the tissue. This latter technique is used to establish the clonal origin of the virus in tumors (Raab-Traub and Flynn, 1986).

EBV infected, proliferating lymphoblasts resemble, morphologically (Nilsson, 1979) and phenotypically (Thorley-Lawson and Mann, 1985), normal B cells that have been activated to become B blasts by antigen. This was first noted during studies that sought to identify EBV specific proteins on the surface of latently infected B cells. The first two candidate molecules (Blast 1 and Blast 2, now CD48 and CD23) turned out not to be EBV specific, but the first identified activation associated markers for B cells (Thorley-Lawson et al., 1985; 1982). These studies were subsequently extended to show that EBV latent genes, particularly EBNA2 and LMP1, could induce the expression of a large range of surface markers, including activation and adhesion molecules, characteristically expressed by B lymphoblasts (Wang et al., 1990a). As discussed above, normal naïve B cells need antigen, T cell help and lymphokines to achieve this activated, proliferating state, however EBV is able to perform the task alone. Therefore, the complex machinery of viral latent gene expression found in the growth program has evolved to drive the proliferation of new latently infected human B cells by inducing a normal, activated phenotype. It achieves this, not through some rare random event, such as the integration of the viral genome and disruption of cellular genes employed by retroviruses, but by a highly intricate transcriptional program. This mechanism regulates the expression of the latent genes which are uniquely designed to control the growth of human B cells. This ensures that EBV will efficiently and predictably establish latency and initiate cell growth whenever it encounters a resting naive B cell in the lymphoepithelium of the nasopharynx. This growth promoting transcription program puts the host, in which the virus intends to persist, at risk for developing neoplastic disease. Therefore the virus must use this program because it is essential. A central tenet of the current model of EBV persistence is that the virus needs the growth program because it has to drive the newly infected cell to become a proliferating blast so that it can then differentiate into a resting memory B cell. Once there the virus can shut down, become non-pathogenic and persist for the life of the host. How does an antigen activated B blast and, by analogy, the EBV infected B blast become a resting memory B cell?

The making of an antigen specific or a latently infected memory B cell

To understand how latently infected, naive B lymphoblasts expressing the growth program, can become resting memory B cells, with no viral gene expression, it is first necessary to describe how a normal naive B cell becomes a memory cell. Naïve B cells express surface IgM and IgD. When they become blasts, activated by antigen, they migrate into the follicles cease expression of surface immunoglobulin (Ig) and form a germinal center which consists of rapidly dividing antigen specific B cells. (Figure 1 and 2) (reviewed in (Liu and Arpin, 1997; MacLennan, 1994)). During this phase of rapid proliferation the cells are referred to as centroblasts. As centroblasts the cells undergo Ig class switching to express a single isotype which can be IgM, IgG, IgA or IgE. IgD+ memory cells also exist but probably do not arise through the germinal center reaction (Weller et al., 2003). During this time the B cells also randomly mutate their immunoglobulin genes. After several divisions the cells rest and re-express the isotype switched mutated immunoglobulin on the surface. These cells are referred to as centrocytes and they compete for antigen. If the mutation process has increased the affinity of the immunoglobulin expressed by the B cell it will successfully compete for antigen and survive. If not, the cell will die (Liu et al., 1989). Thus, B cells expressing the highest affinity B cell receptor (BCR) are selected. The cells that survive ultimately have two fates depending on the length and type of exposure to T helper (Th) cells and specific lymphokines (Banchereau et al., 1994). They can either terminally differentiate into antibody secreting plasma cells or enter the long lived memory compartment as resting memory B cells (Arpin et al., 1997; 1995). Both cell types ultimately home to the epithelium (Brandtzaeg et al., 1999a; 1999b; Dono et al., 1996a; 1996b; Liu et al., 1995). Memory B cells, as the names implies, carry immunological memory and are responsible for a heightened secondary response upon re-exposure to the specific antigen.

The events involved in the making of a plasma cell will be discussed in more detail later in the context of viral replication. The events involved in producing a resting memory cell from a germinal center cell are not fully understood, but what is clear is that in order for this to happen the B cell needs to receive two signals, one from antigen specific Th cells and one through the B cell receptor for antigen (BCR).

From the preceding discussion we can conclude that three events must occur if an EBV infected naive B lymphoblasts, expressing the growth program is to become a memory cell. First, the blast must migrate to the follicle, second the latent genes that drive proliferation must be turned off and third, the cells must receive the requisite survival signals.

a. Migration to the follicle

Migration of antigen activated B cell blasts into the follicle to form a germinal center is regulated by chemokines. Antigen bearing, activated, mature, interdigitating dendritic cells in the follicles produce the cytokine CCL19 (Ngo et al., 1998). Activation of B cells by antigen leads to expression of the receptor, CCR7 for this chemokine which is believed to cause the activated B cells to migrate into the follicles. When EBV activates the newly infected naïve B cell with the growth program one of the phenotypic changes it causes involves induction of CCR7 (Birkenbach et al., 1993) by EBNA2 (Burgstahler et al., 1995) which would allow the EBV activated blast to migrate into the follicles.

b. Turning off the growth program

Experimentally it has been shown *in vitro* that when EBNA2 is turned off in the presence of activated c-myc, which is characteristically expressed in germinal center cells (Martinez-Valdez et al., 1996), the B cell blasts down regulate surface markers characteristic of B blasts, such as CD23, and acquire germinal center specific markers, such as CD10 (Polack et al., 1996). Therefore, the infected lymphoblast appears free to acquire a germinal center phenotype once the differentiation block imposed by EBNA2, is removed. One of the direct targets of EBNA2 is c-myc, a known regulator of cell growth and apoptosis (Kaiser et al., 1999). c-myc is believed to be a major mediator of EBNA2 induced proliferation. We can assume therefore that upon arrival in the follicle the EBV lymphoblast receives a signal that turns EBNA2 and the growth program off whilst allowing c-myc expression to continue. What this signal is remains unknown but probably involves the Notch signaling pathway. EBNA2 does not bind directly to DNA but interacts with a cellular DNA binding protein referred to as RBPJκ or CBP1 (Ling et al., 1994). This suggests a regulatory mechanism for EBNA2 based on the known role of RBPJk/CBP1 in the Notch signaling pathway in developmental biology (reviewed in (Artavanis-Tsakonas et al., 1995)). In Drosophila development ligand interaction with the Notch receptor causes proteolytic cleavage of Notch and release of the intracellular domain Notch[IC]. Notch[IC] binds to the transcription factor Suppressor of Hairless (Su (H)) to act as a transcriptional co-activator resulting in direct activation of Notch target genes leading to cellular growth and the inhibition of differentiation. Hairless is a potent antagonist of this pathway which is believed to directly displace Su (H) in its interaction with Notch[IC]. The result is transcriptional repression, through interaction with the long range co-repressor protein Groucho and the short range co-repressor protein CtBP (Barolo et al., 2002; Morel et al., 2001). The latter interaction subsequently silences genes through recruitment of histone deacetlyases and histone methyltransferases (Shi et al., 2003). Su (H) is RBPJκ/CBP1 and EBNA2 is a functional homologue of Notch[IC] (Hsieh et al., 1996; Zimber-Strobl et al., 1994). It is has also been proposed that EBNA3a and 3c are functional homologues of Hairless (Zimber-Strobl and Strobl, 2001) first because they repress EBNA2 mediated transcriptional activation by acting as antagonists of EBNA2's interaction with RBPJκ/CBP1/Su (H) (Le Roux et al., 1994; Radkov et al., 1997) and second because they directly recruit CtBP (Hickabottom et al., 2002; Touitou et al., 2001) and histone deacetylases (Radkov et al., 1999).

Taken together these similarities to Notch signaling suggest a regulatory pathway for EBV driven proliferation and differentiation. Infection of the B cell leads to the lymphoblastoid phenotype and migration into the follicle. Expression of EBNA2, mimicking Notch[IC], blocks differentiation and allows cellular proliferation by interaction with RBPJκ/CBP1/Su (H). At the same time EBNA2 through its interaction with RBPJκ/CBP1/Su (H) activates the major EBV latent promoter Cp which leads directly to expression of the EBNA3 proteins including 3a and 3c. These, by analogy with Hairless, bind RBPJκ/CBP1/Su (H) and displace EBNA2 leading to growth arrest. At the same time they recruit CtBP leading to inactivation of Cp, blocking further production of EBNA2. CtBP recruitment leads to deacetylation and methylation of histones assembled into chromatin in and around the Cp promoter leading to its stable silencing. The silent chromatin state would probably trigger DNA methylation (reviewed in (Jaenisch and Bird, 2003)) and spread downstream engulfing Wp as well, until an "insulator" sequence was reached. This fits well with the documented observation that in the latency program Cp itself is active (Tierney et al., 1994) but transcription is abruptly aborted by downstream DNA methylation that extends through Wp (Paulson and Speck, 1999). Ablation of EBNA2 activity allows the cells to assume a germinal center phenotype and express the default program. The mechanisms that allow LMP1 and LMP2 expression, in the absence of EBNA2, *in vivo* are unknown at this point.

In this model growth regulation by EBV is a self regulating balance between EBNA2 and the EBNA3's that normally leads to rapid extinction of proliferation. It follows that large expansions of EBV infected cells would not be expected in germinal centers. It also suggests that the *in vitro* phenomenon of immortalization is a biological artifact where the balance has been shifted slightly in favor of EBNA2 presumably by the powerful selection pressure of *in vitro* growth.

c. *Providing the requisite survival signals*

Once the growth program is turned off we have good evidence that EBV is capable of providing the necessary signals to rescue the latently infected germinal center cell into memory. It has been shown that latently infected germinal center cells in the tonsil express the default program not the growth program (Table 1) (Babcock et al., 2000). The default program involves expression of three of the nine latent proteins, EBNA1, LMP1 and LMP2a (Babcock et al., 2000; Thorley-Lawson, 2001b). EBNA1 is expressed in the default program because it is essential for retaining the viral genome by tethering it to cellular DNA and allowing it to be replicated (Yates et al., 1985). For EBNA1 to be expressed without the other EBNA's the Q promoter is employed (Nonkwelo et al., 1996; Schaefer et al., 1995;

Tsai et al., 1995). EBNA1 transcription from Qp, in the absence of EBNA2 and the presence of LMP1and LMP2a, is diagnostic for the default program (Hochberg and Thorley-Lawson, 2005). LMP1 and LMP2a are expressed because together they are capable of delivering the two signals, T cell help and BCR, that rescue the germinal center cell into memory.

LMP1

LMP1 is a membrane protein that acts as a ligand independent, constitutively activated receptor (Gires et al., 1997). It does this by engaging a family of signaling molecules such as TRAF and TRADD (Izumi and Kieff, 1997; Mosialos et al., 1995) which normally transmit signals when CD40 engages its ligand (reviewed in (Lam and Sugden, 2003)). The interaction of CD40 ligand on Th cells with CD40 on B cells is the mechanism by which the Th cell delivers its rescue signal to the B cell (Banchereau et al., 1994). Indeed the cytoplasmic tails (the signaling end) of CD40 and LMP1 appear to be interchangeable (Gires et al., 1997). Thus, LMP1 is able to deliver a Th cell signal in the absence of either antigen or Th cells.

Since the first description of the model of EBV persistence several studies pertinent to the model have been performed on LMP1. It has been shown *in vitro* and *in vivo* that the parallel between CD40 based Th and LMP1 signaling extends to the ability of LMP1 to drive immunoglobulin class switching (He et al., 2003; Uchida et al., 1999). LMP1, like CD40, also turns off expression of bcl-6 the transcription factor believed to be the master regulator of germinal center function. bcl-6 is expressed in germinal center cells (Cattoretti et al., 1995), is essential for germinal center production (Ye et al., 1997) and, when turned off, causes the B cells to leave the germinal center and differentiate (Calame et al., 2003). Through its ability to turn off bcl-6, LMP1 (Carbone et al., 1998) almost certainly plays a role in driving the latently infected B cell to leave the germinal center and differentiate into a memory cell.

Studies with a transgenic mouse constitutively expressing LMP1 have suggested some modifications to the model (Uchida et al., 1999). These mice were unable to form germinal centers but in a CD40 knock out background the B cells were able to undergo class switching. This raises the possibility that EBV may exploit extrafollicular differentiation rather than the germinal center (see below). However, these mice also developed tumors. The problem with these studies is that LMP1 is not expressed from a constitutive promoter in the virus (Fahraeus et al., 1990; Wang et al., 1990b), is not oncogenic in healthy carriers, is not expressed alone in germinal center cells, but in the presence of LMP2 (Babcock et al., 2000). Most importantly the phenotype of these mice is an artifact of constutive expression of LMP1 because this prevents B cells from expressing bcl-6 and thereby makes it impossible for them to form germinal centers. It follows that, *in vivo*, LMP1 signaling, just like T cell help, only occurs once the infected B cell is in the germinal center.

LMP2a

LMP2a is also a membrane protein but it delivers a constitutive, ligand independent, BCR signal (Caldwell et al., 1998). LMP2a contains the same signaling motifs (ITAM's) (Beaufils et al., 1993), as the α and β chains of the BCR. These motifs allow it to engage members of the src family of non-receptor protein tyrosine kinases such as lyn and other signaling molecules usually employed by the BCR when it signals after binding cognate antigen (Kurosaki, 1999; Miller et al., 1995). The BCR produces two types of signals (MacLennan, 1998). One type is required simply to ensure the survival of resting B cells (Lam et al., 1997; Maruyama et al., 2000). The other leads to cellular activation, proliferation and ultimately differentiation into immunoglobulin secreting plasma cells (Liu and Arpin, 1997; MacLennan, 1994). The two signals, therefore, may be referred to as survival and activating respectively. LMP2a is able to provide the survival but not the activating signal.

Recent studies on conditional LMP2a transgenic mice have provided important support for the model (Casola et al., 2004). It was shown that LMP2a, in the absence of a BCR, was capable of driving germinal center formation but only in mucosal tissue. Furthermore these germinal center cells showed evidence of having undergone mutation of their immunoglobulin genes.

By expressing LMP1 and LMP2a the virus is providing the latently infected B cell with a whole range of signals associated with germinal center development. LMP2a ensures that the virus infected cells form germinal centers in mucosal epithelium; LMP1 drives class switching and LMP2a hyper-mutation, the hall marks of the germinal center reaction; LMP1 and LMP2a together provide the requisite survival signals to rescue a germinal center cell into memory and LMP1 ensures exit from the germinal center and terminal differentiation by switching off bcl6. Thus, EBV ensures that the latently infected B cell will receive the signals necessary for it to survive a germinal center and exit as a resting memory cell into the peripheral circulation (Babcock et al., 1998). This is the second example we will encounter of a latent gene transcription program, the default program (Table 1), known to be used in lymphoma, in this case Hodgkin's disease (Figure 4), being found in a normal B cell counterpart, the infected germinal center B cell in the tonsil (Figure 2) (for a more detailed discussion of this issue see below).

One critical remaining question is: does EBV do it all? Are the persistently infected memory cells produced entirely through the action of viral latent proteins or do these proteins act to augment the functions of a normal immune response? One very important unresolved issue is what role does antigen play in the production of the latently infected memory cells? Are they *bona fide* antigen specific memory cells co-opted by EBV for persistent infection or does LMP2a do all the work in rescuing the cell into memory. One way to distinguish these possibilities would be to see whether or not the latently infected memory cells show hypermutation patterns characteristic of antigen driven selection. Preliminary evidence suggests that EBV does not do it all. It has been observed that infection of B cells is much more efficient when the CD40 molecule on the surface of the target cell is engaged with CD40 ligand (i.e. receives help from Th cells) (Imadome et al., 2003). Also the latently infected memory cells in the periphery do not express LMP2a, therefore they must express an intact, functional and presumably antigen elected, immunoglobulin on their surface.

The other options: extrafollicular maturation or direct infection of memory cells

The preponderance of evidence to date favors the hypothesis that latently infected memory cells in the peripheral blood are derived through germinal center maturation. One of the key observations in support of this hypothesis is that the infected cells from the tonsils of healthy carriers which bear germinal center markers also express the default program. However, this result was obtained by cell fractionation (Babcock et al., 2000). Although CD10 and CD77 (the markers used) are considered highly specific for germinal center cells (Uckun, 1990) it has not been directly demonstrated that the infected cells bearing these markers actually reside in a follicle. There are two other possibilities that can not be excluded at this time. These are; that the newly infected naïve B cells undergo extrafollicular maturation, or that the virus does not use the differentiation process at all but directly infects memory cells.

Extrafollicular maturation

There is increasing evidence that B cells can undergo class switching, hypermutation and differentiation without forming a germinal center (Fagarasan and Honjo, 2000). This is termed extrafollicular maturation. The best evidence for this in humans comes from individuals with hyper IgM syndrome (Ramesh et al., 1998). They lack the ligand on Th cells for CD40 and therefore are unable to provide T cell help. They do not make germinal centers and lack B cells that are isotype switched and hypermutated. This makes them particularly vulnerable to bacterial infections. They are able to develop one class of memory cells, however, the IgD+ memory cell (Agematsu et al., 1998; Weller et al., 2001). This has led to the proposal that IgD+ memory is produced by a separate pathway that does not require classical germinal center development (Weller et al., 2003). As mentioned above, the inability of LMP1 transgenic mice to produce germinal centers has been used as an argument for the extrafollicular development of EBV infected memory cells. B cells in these mice are able to class switch to IgG (Uchida et al., 1999) and LMP1 is also capable of inducing switching to IgG *in vitro* (He et al., 2003). As argued above however the transgenic results should be interpreted with caution. One important reason for believing that EBV preferentially enters memory through the germinal center route is that the virus is found in all subsets of isotype switched memory cells with the single striking exception that it is not present in the IgD+ memory cells (Joseph et al., 2000b). This observation suggests that under normal circumstances EBV does not use the extrafollicular route identified in hyper IgM syndrome and provides further evidence that the IgD+ memory subset is different from the other memory cells in that it can not sustain persistent latent EBV infection.

Direct Infection of memory cells

Based on studies with tonsils from individuals acutely infected with EBV i.e. with infectious mononucleosis it has been proposed that EBV-infected cells do not participate in the germinal center reaction, but rather that EBV directly infects memory B cells (Kurth et al., 2003; 2000). This idea is based on the observation that EBV-infected cells, undergoing clonal expansion, were identified within the germinal centers of these tonsils. These cells expressed EBNA2, characteristic of the growth program, and were not mutating their immunoglobulin genes which, as we have seen, are a defining property of germinal center B cells. The authors interpreted this observation to mean that when EBV was present in a germinal center cell it did not express the default program but simply drove cellular expansion using the growth program. There are a number of difficulties with this study. The first is that it does not explain the presence of cells in healthy carriers that possess a germinal center phenotype and express the default program. Indeed the techniques used in the infectious mononucleosis study were unable to detect such cells yet we know they exist (Babcock et al., 2000). In fact in the memory cell scenario there is no apparent need of or explanation for the default program which appears specifically evolved to rescue a latently infected germinal center cell. A second problem is that the studies were non-quantitative, relying on immunohistochemical approaches to find infected cells. While this approach is able to identify infected populations it will naturally focus on the dominant population and, since the lower limit of sensitivity of the technique is not known, may allow important infected cells to go undetected. Infectious mononucleosis is a highly debilitating disease associated with major disorganization of the tonsil architecture. Infection in the tonsil will therefore be deregulated, inappropriate cells may become infected and the predominant infected population will reflect on the pathogenic nature of the infection but may not be biologically relevant.

One surprising thing about these studies is that the authors failed to mention that their results, rather than contradicting the model, are equally consistent with and predicted by the model. The model holds that latently infected naïve B cells rapidly differentiate out of cell cycle via the default program, to become resting memory cells. Indeed if the EBNA2/EBNA3 feed back model of growth program regulation described above holds true then the cells may only divide at most a few times. Therefore infected naive cells expressing the growth program and germinal center cells expressing the default program will be short lived. Because of the high level of viremia and disruption of lymphoid tissue in infectious mononucleosis, germinal center or memory B cells may incidentally become directly infected These cells should then express the growth program (EBNA2 positive), not the default program a prediction that has been confirmed *in vitro* (Babcock et al., 2000). They will expand rapidly (Figure 4) because they cannot differentiate out of the cell cycle and will therefore become the dominant infected populations found in tonsils during infectious mononucleosis i.e. these are the cells that will be seen by immunohistochemical approaches. Eventually, the CTL response destroys them; leaving behind only the small numbers of infected naive cells expressing the growth program and germinal center cells expressing the default program which are seen in healthy carriers of the virus.

EBV persistence in the peripheral memory B cell compartment

EBV, in the peripheral blood, is found only in B cells (Miyashita et al., 1995) that have the cell surface phenotype expected of a long lived resting memory cell that has undergone germinal center differentiation i.e. classical memory B cells (Babcock et al., 1998; Joseph et al., 2000b). They express the classical memory B cell marker CD27 (Joseph et al., 2000b; Klein et al., 1998), they do not express IgD, but can express any of the other immunoglobulin isotypes (Babcock et al., 1998;

Joseph et al., 2000b). Indeed the virus appears to be evenly distributed among B cells carrying the different isotypes. Claims that EBV resides preferentially in IgA bearing B cells (Ehlin-Henriksson et al., 1999) have not been substantiated (for a detailed discussion of the issues see (Joseph et al., 2000b)). Restriction to the memory compartment is so tight that in a recent quantitative study it was shown that less than 1 in 10,000 latently infected cells in the blood are in the naïve compartment (Hochberg et al., 2004b). The cells are CD23 and CD80 (B cell activation markers) negative (Miyashita et al., 1995), characteristic of resting memory B cells, and >90% are in the G0 stage of the cell cycle (Hochberg et al., 2004b; Miyashita et al., 1997). The cells do not express CD5 and therefore are not B1 cells (Joseph et al., 2000b). B1 cells are a long lived compartment of B cells distinct from classical memory cells (Kantor, 1991; Youinou et al., 1999). They do not develop through germinal centers and frequently have specificity for poly-antigens such as bacterial cell wall components. The absence of EBV from this subset supports a requirement for the germinal center in the production of latently infected memory B cells.

Memory cells latently infected with EBV in the peripheral blood do not express any of the known latent proteins (Hochberg et al., 2004a). This is an important point to stress. Several studies have identified EBV latent gene expression in the peripheral blood however; they have all been performed on bulk preparations of B cells frequently from acutely infected individuals (Babcock et al., 1999; Chen et al., 1995; Qu and Rowe, 1992; Tierney et al., 1994). Because the RTPCR assays used are so sensitive it is impossible to know if the signals are from rare infected cells expressing the transcript or are representative of the whole infected population of cells. It turns out the former is true. By performing a limiting dilution RTPCR analysis ((Hochberg et al., 2004a; Hochberg and Thorley-Lawson, 2005) it was possible to show that >99% of the infected cells do not express transcripts for any of the known latent proteins. Indeed the single cell analysis afforded by this approach revealed that when latent gene transcripts were found they were not part of any known transcription programs indicating that they almost certainly are residual transcripts of no biological significance.

These observations have led to the proposal that the memory B cell is the site of long term viral persistence. Here it can persist for the life time of the host because immunological memory persists for life, but the virus is no longer pathogenic to the host because the genes that drive cellular proliferation and threaten neoplastic disease are turned off. Similarly the virus is safe from immunosurveillance because no viral proteins are expressed to act as targets of the immune system. The transcription program used in these cells, where no viral proteins are expressed, is called the latency program (Hochberg et al., 2004a; Table 1) reflecting its role at the site of latent persistence.

The frequency of infected cells for an individual healthy carrier is relatively stable over time although this has not been studied rigorously with precise quantitative methods (Khan et al., 1996). However, the level of infected cells in a population ranges widely from 5-3,000 for every 10^7 memory B cells both in the peripheral blood (mean $110/10^7$) and in Waldeyer's ring (mean $175/10^7$) (the virus is evenly distributed throughout the ring) (Laichalk et al., 2002). Early estimates of total body load of infected cells, made by direct extrapolation of this number (Khan et al., 1996), turn out to be incorrect. This is because the level of infected cells is similar between peripheral blood and Waldeyer's ring but at least 20 fold lower in the other lymph nodes tested (spleen and mesenteric lymph node) (Laichalk et al., 2002). Based on these measurements the total body load calculates to 10^4-10^7 (mean 0.5×10^6) infected cells per person. Only ~1% of these cells reside in the peripheral blood raising the concern that, numerically, this site is not important for persistent infection. The arguments against this concern rests on the specificity of the infection for memory cells in the blood and the lack of latent gene expression suggesting that this is a very small, stable and, most critically, "safe" pool of infected cells that guarantees long term persistence. It is also likely that a significant fraction of the latently infected memory cells reside in the lymphoid tissue at any given time so measurement on the peripheral blood alone will be substantial under estimates.

These studies also provide insights into the behavior of memory B cells. It is generally held that memory B cells stay in the lymph node of origin and, unlike naive B cells, do not recirculate (Gray et al., 1982). Nevertheless, memory cells are found in the peripheral circulation (Klein et al., 1994). In addition, EBV infected memory B cells are produced in the tonsil yet are also found in the peripheral circulation and throughout the secondary lymphoid tissue albeit at a 20 fold lower level. This led to the suggestion that EBV may be viewed as a marker for memory B cell migration i.e. memory cells can leave the site of production enter the circulation and return. With some diminished frequency they can also enter distant sites in the secondary lymphoid tissue. This makes sense because it would ensure that memory B cells preferentially colonize the draining lymph nodes at the site of infection but also populate, to a lesser extent, the other lymph nodes in the event that the infection enters the body elsewhere. The distribution of EBV infected memory cells can be seen therefore as simply representing the circulation and distribution of normal memory B cells.

Memory B cell homeostasis - maintenance of long term memory and persistent infection

The maintenance of long term memory B cells requires the expression of a functional BCR (Maruyama et al., 2000). This function can be completely replaced by LMP2 in transgenic mouse experiments (Caldwell et al., 1998), raising the possibility that the persistently infected cells could be BCR independent. However, the best evidence suggests that the infected cells in the periphery do not usually express LMP2a (Hochberg et al., 2004a). This means that they must express a functional, presumably, antigen selected BCR.

The other mechanism that maintains stable levels of memory B cells is homeostasis. It has been estimated that human memory B cells divide about once a month presumably to replace cells lost through attrition and maintain their levels (Derek Macallan et al., 2005). Memory B cells in the periphery of adult humans are >90% in a resting state but at any given time around 2-3% of the cells are undergoing cell division based on DNA staining (Hochberg et al., 2004b; Miyashita et al., 1997) or deuterated glucose labeling (Derek Macallan et al., 2005) consistent with an average division time of the order of a month. Analysis of the latently infected memory B cells in the blood reveals that they undergo a similar rate of cell division (Hochberg et al., 2004b; Miyashita et al., 1997). Since these cells express no latent proteins this observation argues that the pool of latently infected cells is indistinguishable to the host from normal memory B cells and hence are maintained by the normal mechanisms of homeostasis. It appears though that when the cells divide the rule of no viral gene expression is broken. This is because the viral genome can not replicate along with the cell unless the viral EBNA1 protein is expressed (Yates et al., 1985). Predictably therefore the latently infected memory cells in the periphery express only EBNA1 when they undergo cell division (Hochberg et al., 2004a). This is the third example we will encounter of a latent gene transcription program, the EBNA1 only program (Table 1), known to be used in a lymphoma, in this case Burkitt's lymphoma (Figure 4), being found in a normal B cell counterpart, the dividing memory B cell in the blood (Figure 2) (for a more detailed discussion of this issue see below). EBNA1 expression during cell division is the only potential point of attack for the immune system against the pool of latently infected cells. It is perhaps not surprising therefore that EBNA1 has evolved so as not to be processed and presented efficiently to the immune system (Levitskaya et al., 1995; Levitskaya et al., 1997) thus protecting the only possible point of attack.

Terminal differentiation - maintenance of stable antibody production and viral shedding

The last component of persistent infection to be discussed is that infectious virus is continuously shed into the saliva (Golden et al., 1973). Memory B cells, transiting the nasopharyngeal lymphoid tissue, must occasionally initiate virus replication and release the virus back into the crypts. B cells replicating the virus in the lymphoepithelium have been shown to have the characteristics of plasma cells, the antibody secreting cells, both by morphological criteria (Anagnostopoulos et al., 1995; Niedobitek et al., 2000) and from cell surface phentoyping of fractionated tonsil cells (Laichalk and Thorley-Lawson, 2004). This suggests that, in some circumstances and at a given rate, the latently infected, memory B cells differentiate into plasma cells and release virus. Since the circulating memory cells express no viral proteins the signal that causes them to undergo terminal differentiation *in vivo* is probably host derived. An explanation for how this could work is offered by recent studies suggesting that immunological B cell memory does not require antigen but is sustained by polyclonal stimuli such as bystander T cell help (Bernasconi et al., 2002). The consequence is that a memory B cell transiting through a lymph node will, when it encounters bystander T cell help, undergo a cell division that will generate one memory cell and one plasma cell. The memory cell replaces the original memory cell and the plasma cell will go on to produce antibody. Recurrence of these events over time ensures the stability of the memory pool while a continuous supply of plasma cells is produced that will guarantee stable production of antibody over the life time of the host. Applied to EBV this would explain, since EBV replication is initiated by plasma cell differentiation, how the population of latently infected memory cells could be maintained for years while, through the generation of plasma cells, virus can also be continuously produced.

An alternate hypothesis is that the generation of plasma cells replicating EBV is stimulated by cognate antigen and T cell help. This would mean that EBV becomes reactivated when the latently infected memory cell undergoes an antigen specific secondary response. This hypothesis has the attractive feature that latency would need to be established in antigen specific memory cells in the tonsil. These cells would then enter the peripheral circulation where they would maintain persistent infection. As these cells reenter secondary lymphoid tissue, the site where they would most likely reencounter cognate antigen would be the tonsil. This would provide a mechanism for preferential homing and reactivation of the latently infected memory cells in the tonsil compared to other lymph nodes. *In vitro* studies with the Akata cell line appear at first glance to support this idea. Cross linking of the antigen receptor on these cells leads to their synchronous entry into the lytic cycle (Takada and Ono, 1989). However, a number of properties of this system suggest it may not be representative of *in vivo* replication. First and most striking is that activation of the BZLF1 promoter in Akata cells occurs within minutes of the antigen receptor signal whereas differentiation into plasma cells takes days. The response in Akata cells is too acute to be the same as the *in vivo* mechanism and not surprisingly the Akata cells do not undergo plasma cell differentiation prior to viral replication. The Akata system may be more closely related to the spontaneous reactivation that occurs when latently infected peripheral memory cells are placed into culture (Rickinson et al., 1977). This is an acute reactivation, presumably in response to the stress induced upon being placed in culture and also occurs too rapidly to be associated with terminal differentiation. It is also important to note that the signals that induce viral replication in cell lines, including surface Ig cross linking of Akata, are associated with the induction of apoptosis in the cells (Inman et al., 2001). However, expression of lytic EBV genes protects these cells from apoptosis (Inman et al., 2001) suggesting that EBV has acquired specific genes to save a stressed cell from death while the virus replicates. Indeed it is known that EBV encodes for homologues of the anti-apoptotic gene bcl-2 that are expressed during viral replication *in vitro* (Henderson et al., 1993). This discussion raises the possibility that there may be two mechanisms for reactivating EBV *in vivo*. One may be an acute reactivation in resting memory cells in response to stress that allows the virus to escape quickly before the cell dies. The other allows replication of EBV in plasma cells located in the epithelium of the tonsil producing high levels of virions that may infect new naïve B cells or be shed directly into saliva.

Quantitative estimates suggest that somewhere in the region of 5-1,000 cells initiate replication in Waldeyer's ring at any one time (Laichalk and Thorley-Lawson, 2004). However, sequentially fewer cells express the immediate early, early and then late lytic antigens such that it appears that only ~10% of the cells complete the replicative cycle. This suggests that fewer than one hundred cells are actually releasing virus in Waldeyer's ring at any given time. This may account for the relatively low titers of EBV reported in the saliva of most healthy carriers (Yao et al., 1985). The sequential diminution in the numbers of cells replicating the virus as they proceed through the cycle may indicate that replication is frequently abortive or may be the result of aggressive immunosurveillance by CTL against the immediate early proteins of the lytic cycle.

Herpesviruses like HSV encode immediate early proteins that specifically blunt CTL responses by down regulating MHC (Tortorella et al., 2000). Although MHC Class I is down regulated during acute reactivation *in vitro* (Keating et al., 2002) this does not occur in plasma cells *in vivo* (Laichalk and Thorley-Lawson, 2004) yet again high-lighting the differences between *in vitro* and *in vivo* replication. The failure of EBV to down regulate MHC Class I may reflect different strategies for these viruses. For HSV the strategy of the virus is to sporadically reactivate and produce infectious virus for a brief period of time before viral replication is shut down by the immune response. EBV on the other hand is chronically shed into the saliva so transient delay of the immune response would not be useful. Rather it appears that EBV is able to continue generating replicative cells, a few of which complete the cycle, despite the cellular immune response. This suggests that EBV may have developed a mechanism to constitutively reduce the local effectiveness of the CTL response, perhaps through the production of viral IL10 during the lytic cycle (Bejarano and Masucci, 1998; Matsuda et al., 1994).

The role of epithelial cells in viral persistence

The possibility that epithelial cells may play a role in viral persistence has had a checkered history. There are several reasons to think that epithelial cells might play a role. EBV is found in several carcinomas most notably nasopharyngeal carcinoma (Henle and Henle, 1979) where 100% of tumors are EBV positive (Andersson-Anvret et al., 1977?) usually expressing the default program (Brooks et al., 1992; Fahraeus et al., 1988; Young et al., 1988). This implies that EBV can establish latent infection in epithelial cells. EBV is also found in the disease oral hairy leukoplakia which represents a lesion where EBV is actively replicating in the epithelium of the tongue (Greenspan et al., 1985). This indicates that EBV can also replicate in epithelial cells. Lastly the presumed site of entry and exit of EBV is the lymphoepithelium of the organs comprising Waldeyer's ring. In order for the virus to pass between saliva and the lymphoid tissue it needs to traverse the epithelial cell lining. The simplest model would be that EBV infects epithelial cells and replicates both on the way in and on the way out. This would provide a way to transport the infectious virus across the epithelium while amplifying the titre, thus increasing the chances of hitting a B cell in the underling lymphoid tissue on the way in and increasing the amount of virus shed into saliva on the way out. The role of epithelium in persistence was first suggested based on a series of studies claiming to find EBV infected, exfoliated epithelial cells in the throats of acute infectious mononucleosis patients (Rickinson, 1984; Sixbey et al., 1984; 1983; Young and Sixbey, 1988). However, these studies were subsequently discredited (Karajannis et al., 1997; Niedobitek et al., 2000). Identifying infected epithelial cells in the lymphoepithelium of the tonsil by immunohistochemical approaches has been complicated by the fact that there are abundant infected B cells in this region during acute infection. The role of the epithelium in persistence was further discredited based on studies with two groups of patients. The first demonstrated that patients undergoing complete bone marrow ablation as part of a bone marrow transplant either lost their EBV, if the donor was EBV negative, or retained the donor EBV strain over that of the recipient if the donor was EBV positive (Gratama et al., 1988). Second it was observed that patients with XLA, an X linked genetic disorder where the patients lack B cell, also showed no signs of being able to sustain persistent EBV infection (Faulkner et al., 1999). Taken together these results indicate that an intact lymphoid system, specifically the presence of B cells, is required for viral persistence. Perhaps because of an over enthusiastic back lash against epithelial cells these results have been over interpreted to imply that epithelial cells do not play a role in persistence whereas all they actually show is that B cells are essential. They do not eliminate the possibility that epithelial cell are also essential but not sufficient.

More recently, interest in epithelial cells has again begun to increase based on three separate observations. It has been shown that:

1. The glycoprotein patterns on the virus differ depending on whether the virus emerges from a B cell or an epithelial cell (Borza and Hutt-Fletcher, 2002). This happens in such a way that the virus bears an epithelial tropic pattern of viral glycoproteins when it emerges from B cells and a B lymphotrophic pattern when it emerges from epithelial cells. These results imply that the virus has evolved to efficiently shuttle back and forth between epithelial and B cells. This behavior further suggests that EBV only uses epithelial cells in one direction e.g. if it passes from B cell to epithelial cells on the way out the emerging virus would be B lymphotrophic and therefore would preferentially infect B cells not epithelial cells on the way in or vise versa. Similarly, the newly emerging virus from the plasma cells (e.g. in Figures 2 and 5) would be predicted to pass through an intermediate epithelial replicative stage before infecting new B cells in the lymphoepithelium.
2. A unique receptor, $\alpha5\beta1$ integrin, for EBV is expressed on epithelial cells that allows infection only on the basolateral surface. As with the glycoprotein patterns this implies that epithelial cell infection by EBV is only used in one direction, in this case specifically restricted to exit by the virus.
3. Cell cultures of primary epithelial cells from tonsils carry already infected cells that are both latently infected and replicating the virus (Pegtel et al., 2004).

Taking these three pieces of evidence together presents a strong circumstantial argument that normal nasopharyngeal epithelium is actively infected with EBV as an ongoing part of normal viral persistence and provides an explanation for the presence of the virus in the associated diseases of epithelial cells.

THE MODEL – A SUMMARY

In summary EBV can be seen as a virus that elegantly exploits all aspects of mature B cell biology to establish persistent infection (B cell activation with the growth program and germinal center differentiation with the default program) maintain persistent infection (latency in the long lived memory pool, maintained and regulated through the processes of homeostasis) and replicate for reinfection and infectious spread (reactivation of viral replication in response to terminal differentiation into plasma cells) (Figures 2 and 5).

Thus persistence can be seen as a self renewing circle of infection, differentiation, persistent infection, reactivation and reinfection (Figure 5). The stability of persistent infection depends on a pool of latently infected memory cells where the virus is quiescent and the cells behave as normal memory B cells being

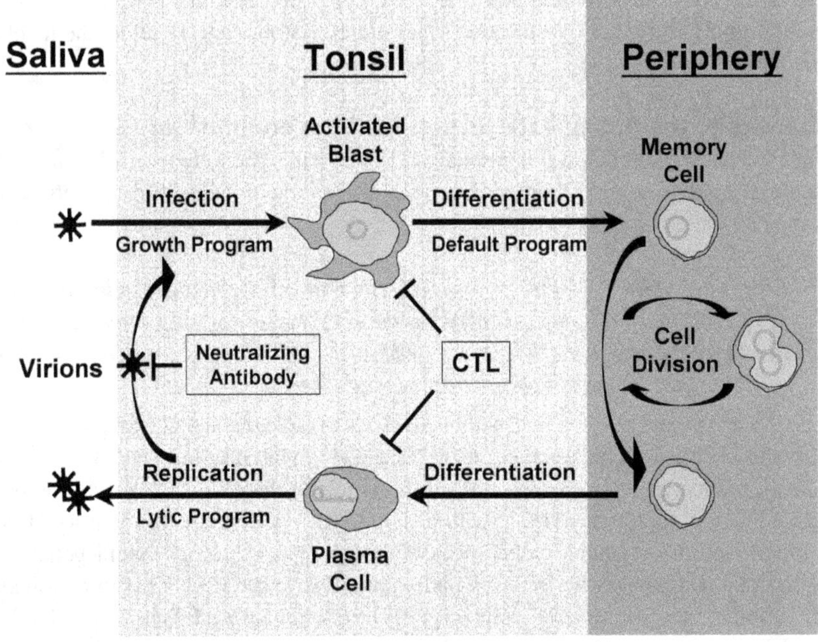

Figure 5. The Circle of Events in Persistent EBV Infection. EBV infects naive B cells that differentiate through the germinal center reaction into the memory compartment. These cells circulate in the periphery then upon return to the tonsil occasionally reactivate and release infectious virus to renew the cycle. In the absence of immune regulation this cycle spins out of control and the memory compartment fills up with latently infected cells (Hochberg et al., 2004b). However, the overall load of virus and infected cells is regulated by the immune system which recognizes every step in the pathway with the exception of the resting memory cells where the virus is quiescent. CTL recognize all the other infected cells types and antibody neutralizes the virus. The immunologically invisible pool of latently infected memory cells is believed to be critical for the stable maintenance of persistent infection.

regulated by homeostasis. However, the absolute levels of infection are defined by the level of the immune response against viral proteins. This includes CTL against cells expressing latent and lytic proteins and neutralizing antibody against infectious virus. The significance of the immune system in moderating the overall viral load can be seen from studies of immunosuppressed individuals where the levels of virus infected cells increases on average 50 fold (Babcock et al., 1999). However, even in the absence of an effective immune response regulation of viral persistence is intact and the virus in the blood remains restricted to resting memory B cells. This means the immune response per se plays no role in regulating the mechanisms of viral persistence it only regulates absolute levels of the infection.

That the system is a simple circle is amply demonstrated by studies on acutely infected individuals. Here the system is allowed to run unchecked until the immune response is activated. In this case as many as a staggering 50% of all memory cells may become latently infected with EBV until the immune system begins to reduce the overall load of infection (Hochberg et al., 2004). Impressively the regulation still holds and the virus remains restricted to resting memory cells in the blood again high lighting that the immune system only functions to regulate the level of infection.

Answers to the five questions/points of the conundrum

Once the basic ideas behind the model are in place it is possible to resolve all of the major contradictions and confusions enunciated by the five questions/points posed at the beginning of this Chapter.

1. *EBV is capable of infecting normal B lymphocytes and transforming them into proliferating lymphoblasts. Why would EBV cause latently infected cells to proliferate if this puts the host, in which it establishes a persistent life long infection, at risk for developing neoplastic disease?*

It is now apparent from the model that EBV has to have the growth program, which activates B cells and makes them grow, because the latently infected B cell has to transition through the activated state so it can differentiate into a resting memory cell. EBV has to activate the B cell so it can gain access to the memory B cell compartment. Once there it can shut off the growth promoting latent genes and persist in a state that is no longer a pathogenic risk to the host. This infection and activation of newly infected B cells is highly efficient because this is the mechanism that EBV uses to establish persistent infection.

2. EBV *is found in Burkitt's lymphoma, Hodgkin's disease and immunoblastic lymphomas and uses a different transcription program in each one. Why?*

EBV has different transcription programs because it has to affect different functions depending on the B cell in which it is found. The newly infected cell has to use the growth program to activate the cell. The germinal center cell has to express the default program so that it can be rescued from the germinal center into long term memory. The latently infected memory cell expresses the latency program where no viral proteins are expressed to ensure that the cell is not recognized and eliminated by the immune response, that the cell is non-pathogenic and therefore not a threat to the host and so the infected cells can be maintained by the same homeostatic mechanisms that maintain normal memory cells. When the latently infected memory cells in the periphery divides it has to make EBNA1 alone to ensure that the viral genome also divides.

The existence of tumors expressing all three of the active transcription programs associated with division of infected cells (grow program, default program and EBNA1 only program) suggest that multiple sites in the EBV life cycle are vulnerable to deregulation leading to lymphoma. It follows that the pattern of genes expressed by the different lymphomas may be indicative of the infected cell of origin (see below).

3. *EBV infects and persists in >90% of the adult human population almost always benignly despite its ability to make cells grow.*

The virus does not use the growth program to maintain persistent infection only to establish it. The proliferating cells are short lived and the virus persists in non-proliferating cells. This is because, unlike in tissue culture, the activation of newly infected B cells in the tonsil is a transient event on the way to becoming a resting memory cell. The proliferating blasts are not normally a pathogenic threat because the virus is programmed to ensure that they rapidly differentiate into resting memory cells.

4. *LMP1 and LMP2a, have signaling properties analogous to T cell help and the BCR respectively. Why would two EBV latent proteins mimic B cell survival and differentiation signals?*

LMP1 and LMP2a have these properties because they are replicating the signals that are normally used to rescue and differentiate normal germinal center B cells into memory. These molecules rescue the EBV blast into memory by mimicking the germinal center process. Hence LMP1 and LMP2a are the only viral regulatory proteins expressed in infected germinal center cells.

5. *The epitopes on the latent proteins recognized by cytotoxic T cells are conserved. What is the reason for this suicidal tendency?*

It is apparent that EBV must use the potentially threatening growth program because it is an essential step in the process of producing a latently infected memory cell. However, once the virus has colonized the memory compartment the growth program is no longer essential. Newly infected naïve B cells will swiftly differentiate out of the cell cycle and into a resting memory state. However, any infected naive B cell that continues to proliferate or any other type of B cell that gets infected fortuitously (bystander infection in Figure 4) will continue to proliferate and be a potential threat to the host in which the virus wants to persist for as long as possible. Therefore the virus ensures that any cell, continuing to proliferate due to the growth program, will be eliminated by conserving the targets of EBV specific CTL.

INFECTIOUS MONONUCLEOSIS - ACUTE INFECTION IN THE CONTEXT OF THE MODEL

Most people in the world become infected with EBV at a very early age, usually in the first years of life (Henle and Henle, 1979). Primary infection of children is rarely recognized because it is usually asymptomatic. In Western and economically advanced countries, infection is frequently delayed into adolescence or adulthood when it can cause IM (Niederman et al., 1968). IM is exclusively a disease of EBV negative adolescents and adults and in this group 20-50% of primary EBV infections result in IM (Niederman et al., 1968). The intensity of the disease varies but can last for weeks or months before finally resolving (Hoagland, 1967).

IM is characterized by the production of IgM heterophile and auto-antibodies with no specificity for EBV and a lymphocytosis (Wood and Frenkel, 1967) due to the appearance of large numbers of "atypical" lymphocytes which are predominantly CD8+ T cells, representing a vigorous CTL response to the virus (Callan et al., 1998b; Strang and Rickinson, 1987). This response can, during the height of the disease, involve up to half of all the T cells in the body and is predominantly directed against cells replicating the virus (Steven et al., 1997) although cells expressing latent genes are also targeted (Callan et al., 1998b). Presumably, this explosion of T cells is responsible for resolving the infection, but the ensuing destruction of infected tissue leads to the fever and malaise characteristic of IM. At the same time as the cellular immune response is given over so massively to responding to the virus infected cells, the invasion by EBV of the B cell system is similarly massive. As many as 50% of all memory cells may be latently infected with EBV at the time of the first visit to the clinic (Hochberg et al., 2004b). Despite this overwhelming invasion, the regulation still holds and the virus remains restricted to resting memory B cells. This acute infection is eventually resolved by the immune response but at an excruciatingly slow rate. Symptoms usually disappear after a few weeks or months, but numbers of latently infected cells continue to drop even after one year.

A likely scenario for IM is that when the virus initially infects there is nothing to stop the cycle of infection, latency, reactivation and reinfection, shown in Figure 5. Consequently the memory compartment begins to fill up with latently infected cells until the immune system is triggered. This limits the output of infectious virus and the input of newly latently infected cells, however, the latently infected memory cells lack viral gene expression so they can not be removed by the immune system. Presumably their levels decrease simply by attrition as they initiate viral replication at some rate and then die. This may explain the parallel decrease in the number of latently infected memory B cells and the number of CTL recognizing cells replicating the virus (Catalina et al., 2001; Hochberg et al., 2004b). The slow rate of disappearance of latently infected memory cells may simply be a consequence of the average life time of a normal memory cell which has been estimated to be ~26 days (Derek Macallan et al., 2005).

Over the course of a year or more the symptoms of IM regress, fewer viruses is shed into the saliva and infected B cells disappear from the peripheral blood. The ensuing events, however, are strikingly different from infection with most other viruses. Shedding of EBV does not stop, but stabilizes and continues for life at a lower, steady rate (Yao et al., 1985). B cells expressing the growth and default programs persist in the nasopharyngeal lymph nodes (Babcock et al., 2000) and latently infected memory cells, expressing the latency program, remain in the blood for life (Babcock et al., 1998; Khan et al., 1996). At the same time, levels of neutralizing antibodies (Henle and Henle, 1979) and CTL also stabilize (Callan et al., 1998a; Steven et al., 1996) and continue at significant levels for the life time of the host. It is apparent therefore that the immune response ameliorates but never clears the infection. Equilibrium is established between the immune response and the various states of viral latency allowing the virus to persist and be shed at stable levels, without causing significant impairment to the host

The normal course of EBV infection involves the precisely regulated expression of transcription programs that allow latent EBV to gain access to and persist within the long lived memory compartment. This process is normally asymptomatic, as observed in the primary infection of young children (Joncas et al., 1974). Why adolescence and adults get IM is not clear. It is likely that IM is immunopathologic in nature, meaning the disease symptoms are caused by the inflammatory response of the immune system rather than the virus itself. The age dependence of symptoms has led to the suggestion that IM is a disease of a mature immune response. What this means mechanistically is less clear. Another explanation has been suggested based on animal studies on the memory CTL response (Selin et al., 1998). Early in life the CTL response to new infections is by naive T cells. These produce high affinity CTL that efficiently clear the virus infection and then become memory CTL. As the organism ages, memory CTL accumulate to a variety of pathogens. Exposure to a novel infection later in life, is more likely to reactivate cross reacting memory CTL than naive CTL. These CTL are of lower affinity and may produce an ineffectual response allowing extensive viral replication and spread resulting in a massive inflammatory response that lasts until the virus is finally brought under control.

THE OTHER EBV ASSOCIATED DISEASES IN THE CONTEXT OF THE MODEL

EBV has also been associated with a number of diseases that generally fall into two distinct categories. These are autoimmune disease and cancer.

Autoimmune disease

EBV has variously been linked with a number of autoimmune diseases including SLE (James et al., 1997), rheumatoid arthritis (Lotz and Roudier, 1989) and Sjorgens syndrome (Fox et al., 1987). The driving force behind attempts to find such associations is the knowledge that EBV can cause the proliferation of B cells in an antigen independent fashion and that such cells are immortal in tissue culture. This raises the possibility that EBV infection could allow the survival of forbidden clones of B cells producing autoimmune antibodies. However, EBV does not persist by immortalizing B cells *in vivo* but by differentiating the infected cells into a resting memory state. Consequently the fundamental rationale for the association would have to be modified to suggest that the EBV latent genes may rescue a forbidden clone from a germinal center into the memory compartment. The cell itself could not produce antibodies because upon plasma cell differentiation it would produce infectious virus and die. However, theoretically such a cell could present autoimmune antigens and perhaps break tolerance. To know how viable this possibility may be it is again important to discover the relative contribution of viral latent proteins and antigen to the selection of memory cells carrying latent EBV.

Demonstrating experimentally a causative role for EBV in autoimmune disease has been difficult. This is because infection usually occurs early in life and by adulthood >90% of the population world wide is infected. EBV then stays with the host for the rest of their lives. This begs the question if almost everybody is infected with the virus why does it only cause autoimmune disease in a few? Analysis is further complicated by the fact that EBV is carried in the peripheral circulation by infected memory B cells. This means that infected cells are naturally distributed throughout the tissues and lymphoid system so sensitive tests will detect EBV in any inflamed tissue regardless of the virus's roll in causing the inflammation. The model of EBV persistence adds another layer of complication which is that EBV uses virtually every aspect of mature B cell biology to establish and maintain persistent infection and virus shedding. The corollary is that EBV is exquisitely sensitive to changes in the immune system. Any disease that affects the immune system will have an impact on the regulation of EBV persistence. This could result in an increase in the numbers of infected cells in the blood (peripheral blood burden) and/or an increase in virus shedding. Thus, changes in virological or immunological parameters of EBV infection are much more likely to be an indirect effect of a compromised immune system caused by the auto-immune disease rather than being evidence that the autoimmune disease is caused by aberrant virus infection.

Cancer

The motivating force behind associating EBV with cancer is obvious. Since it is well established that EBV has latent proteins that can drive cellular proliferation, at least in B lymphocytes, it is highly likely that inappropriate or deregulated expression of these genes could play a causative role in tumor development. EBV associated cancers fall into three discrete groups:

1. Those for which there are claims that remain to be substantiated, this would include, but not be limited to, breast (Bonnet et al., 1999) and hepatocellular carcinoma (Sugawara et al., 1999). In these cases the doubt usually exists through the inability of investigators to reproducibly detect the virus in the tumor cells. Since the assays used are usually based on either PCR or immunohistochemistry they are subject to the vagaries of those techniques which include a high level of false positives and dependence on technical skill to perform well controlled studies. Thus it becomes difficult to resolve if conflicting results are caused by false positive artifacts or technical inconsistencies producing false negative results.

2. Those for which there is strong supportive evidence but the tumors arise in cell types for which no biologically relevant role has been established. Such tumors may arise through accidental infection leading to inappropriate viral latent gene expression. This includes such tumors as nasopharyngeal (Raab-Traub, 2002), gastric (Shibata et al., 1991; Shibata and Weiss, 1992) and salivary gland (Raab-Traub et al., 1991) carcinomas, leimyosarcoma (Timmons et al., 1995; van Gelder et al., 1995) and T and NK lymphoma (Chiang et al., 1996; Tao et al., 1995). In these cases there is a high degree of correlation between the disease and EBV (e.g. 100% of undifferentiated NPC contain EBV), clear and reproducible detection of the virus in the tumor cells and frequent expression of the latent gene LMP1 which is known to be highly oncogenic when constitutively expressed. This type of tumor is relevant to the model of viral persistence. The model posits that EBV latent gene expression patterns have evolved to be regulated in concert with normal B cell activation and differentiation with the ultimate goal of establishing persistent infection in a memory B cell where the latent genes are no longer expressed. It follows that if EBV fortuitously gains access to a cell type which is not a natural target of infection i.e. a non-B cell, this could lead to aberrant latent gene expression that would not be regulated appropriately. This could result in constitutive expression of LMP1 which is known to be oncogenic from *in vitro* (Baichwal and Sugden, 1988; Moorthy and Thorley-Lawson, 1992; Nicholson et al., 1997; Wang et al., 1985) and *in vivo* (Uchida et al., 1999) studies.

3. The third group is EBV positive lymphomas that occur in the normal infected counterparts postulated in the model of B cell persistence. This includes immunoblastic lymphomas that arise in the immunosuppressed, Hodgkin's disease, Burkitt's lymphoma and possibly XLP syndrome.

Lymphoma in the immunosuppressed

This lymphoma occurs in the immunosuppressed who are at risk for B cell lymphoproliferative diseases such as the post-transplant lymphoproliferative diseases (PTLD) in organ transplant patients and immunoblastic lymphoma in AIDS patients. These are a heterogeneous collection of disorders (reviewed in (Hopwood and Crawford, 2000)) that usually carry the virus and express the growth program (Thomas et al., 1990). The obvious explanation for PTLD is that immunosuppression of the CTL allows uninhibited growth of EBV infected cells; however it is not that simple. If EBV was able to freely drive cell growth in the absence of a CTL response, there would be several consequences. First, all immunosuppressed patients who are EBV infected should develop the disease. Second, it should be a polyclonal and disseminated disease since EBV would drive the growth of many, infected cells throughout the body. Finally, the lymphomas would be expected to arise most frequently in the places where infected cells are known to express the growth program - Waldeyer's ring (Joseph et al., 2000a). In reality the disease has none of these features. Only a small fraction of patients (0.1-10% depending on the setting) develop the disease, it is usually an oligoclonal disease arising from one or a few infected cells; and it often occurs in extra nodal sites such as the brain and gut (Hopwood and Crawford, 2000; Penn, 1998). This indicates that the disease is not simple EBV driven growth but involves a rare event where EBV infection has gone wrong.

The model of EBV persistence states that the growth program of EBV is used specifically to activate B cells so they can then differentiate into resting memory B cells. If the cell fails to do this it will be rapidly eliminated by CTL due to expression of conserved CTL targets from the latent proteins. In follows that for a lymphoblastoid cell, proliferating due to the growth program, to survive and evolve into a lymphoma two events must occur. First, the CTL response must be crippled so that these lymphoblasts are not killed and the EBV infected cell must be unable to exit the cell cycle and become a resting memory cell.

The infected naïve B cells in Waldeyer's ring expressing the growth program will not give rise to lymphoma because they are able to differentiate out of the cell cycle and become resting memory cells (Figure 4 thick arrows). This regulation occurs irrespective of the status of the immune system as is apparent in the immunosuppressed where the tumors that arise are not of naïve B cells in the Waldeyer's ring. The problem may start when some other type of B cell becomes infected and expresses the growth program (bystander cells in Figure 4). Alternatively a latently infected germinal center or memory cell could fortuitously receive a signal that activates the growth program. None of these cells can exit the growth program and would continue to grow if they were not normally eliminated by CTL. However, in the immunosuppressed they will continue to proliferate due to the absence of effective T cell immunity. This explains the heterogeneity of the disease, which comes from a mixture of B cell types (Timms et al., 2003) consistent with the disease not arising from a specific subset of infected cells but from a variety of bystander B cells that get infected by chance.

This concept of bystander infection is illustrated by studies of IM patients. In IM, the extraordinarily high viral load results in the infection of bystander cells. Analysis of B cell populations from the tonsil of IM patients detects clonal expansion of both germinal center and memory cells expressing the growth program (Kurth et al., 2003; 2000). This means the virus has infected these cells directly and the inappropriate direct infection of these bystander cells results in EBV getting stuck expressing the growth program and driving clonal expansion (Figure 4). This differs from the normal process of infection in Waldeyer's ring where infected germinal center cells expressing the default program are not derived by direct infection but from directly infected naïve B cells that undergo a maturation process that allows them to transition from growth program to default program. Clonal expansions of memory and germinal center cells driven by EBNA2 are only seen in acute infection because they can only occur before the primary immune response is activated and the infected cells eliminated. Thus the clonal expansions observed in IM do not develop into full blown immunoblastic lymphoma because the CTL response eventually eliminates any cells persistently expressing the growth program. Even in immunosuppressed individuals the events leading to lymphoma must be rare because only one or two of the millions of infected cells found *in vivo* end up as tumors. That every infected B cell does not simply proliferate out of control further attests to how tightly regulated EBV driven proliferation is *in vivo*, even in the presence of a crippled immune response. The event is probably even rarer in healthy carriers because the burden of infected cells and virus is much lower than in the immunosuppressed. It is not surprising though, that EBV negative organ recipients are at the highest risk for developing PTLD (Hopwood and Crawford, 2000). Upon primary infection of these patients, often by the organ, the lack of any kind of primed immune response allows high viral loads and infection of bystander cells with no possibility to develop a CTL response to control them because of immunosuppression.

Immunoblastic lymphoma in the immunosuppressed is therefore a consequence of two events. The rare event is the expression of the growth program in a cell that can not exit the cell cycle. The global event is immunosuppression that prevents the elimination of these rare cells. At this stage of the disease the tumor cells are still susceptible to immunosurveillance and regression can be achieved by reducing immunosuppression (Starzl et al., 1984) or by treatment with autologous CTL (Rooney et al., 1995). However, in the absence of T cell immunity, the proliferating cells acquire additional genetic damage and become selected for the appearance of a more malignant clone (Knowles et al., 1995). These cells ultimately become unresponsive to reduced immunosuppression or immunotherapy and are usually fatal.

Hodgkin's disease (HD)

HD is characterized by the unusual Hodgkin's Reed-Sternberg (HRS) tumor cells. The other place where HRS cells are found is in IM (Anagnostopoulos et al., 1995) which, along with elevated antibody titers to the virus, is a risk factor for HD (Henle and Henle, 1979). This was the first evidence linking EBV with HD. Subsequently, up to 40% of the tumors were found to contain EBV (Glaser et al., 1997). The virus is clonal and the tumor cells express the default transcription program (Deacon et al., 1993; Herbst et al., 1991; Niedobitek et al., 1997; Oudejans et al., 1996), the same transcription program used by latently infected germinal center B cells (Babcock et al., 2000). Previously it had been shown that the immunoglobulin genes of HRS cells are hypermutated (Kuppers and Rajewsky, 1998). This means that HD is a tumor of a germinal center or post-germinal center i.e. memory, B cell. Thus, the immunoglobulin mutations and the viral gene expression data agree that HD arises from an EBV infected germinal center B cell (Figure 4).

A significant fraction of HD tumors are EBV negative but there are no obvious differences between EBV positive and negative HD to indicate that they might be distinct. Therefore HD can arise with or without the virus. The presence of EBV in ~40% of the tumors though, would seem to rule out a chance association. But this does not take in to account that the levels of EBV infected cells reaches extremely high levels during IM [frequently 10-50 % (Hochberg et al., 2004b)]. Therefore if the immunologic disruption of IM rather than EBV itself is the risk factor, there is a very high probability (as high as 50%) that the pre-malignant germinal center cell will have EBV in it by chance.

A common lesion found in HD that may supercede other differences, including the presence or absence of EBV, is the deregulated expression of NFkB frequently through inaction of the inhibitory IkB component of the complex (Cabannes et al., 1999; Staudt, 2000). NFkB is a multipotent transcription factor (Ghosh et al., 1998) that is tightly regulated and controls the expression of many important genes related to B cell function and growth. The over activity of the immune system during IM provides large populations of activated and proliferating lymphocytes that could acquire genetic mutations. If such a mutation results in constitutive expression of NFkB in a germinal center cell it could block differentiation and prevent the cell from exiting the cell cycle. As a result, if the tumor cell happened to carry EBV, it would continue to express the default program which could provide survival signals through LMP1 and/or LMP2a (Caldwell et al., 1998; Kaye et al., 1993; Zimber-Strobl et al., 1996) that favor the development of the tumor. However, it is conceivable that the viral latent proteins play no role in tumor development at all. They may be expressed in HD simply because the virus is trapped in a tumor that continues to express the default program because it is derived from an infected germinal center cell (Figure 4).

Burkitt's lymphoma (BL)

BL has the pedigree of being the tumor in which EBV was originally discovered, but its contribution to BL remains enigmatic. The defining genetic lesion in BL is deregulated activation of the c-myc oncogene due to reciprocal translocation with one of the immunoglobulin genes (Klein, 1983; Leder, 1985; Manolov and Manolova, 1972). Expression in transgenic mice of c-myc, in the context of the immunoglobulin translocation, has been shown to produce Burkitt's lymphoma like tumors (Kovalchuk et al., 2000) suggesting that deregulated c-myc is sufficient to produce the tumors. This raises the question as to what role EBV may play. The most compelling evidence of EBV's involvement in BL is the high frequency of tumors carrying the virus (de-The, 1985). In the most common form of BL, endemic

BL in Africa (eBL), EBV DNA is present in >95% of tumors. The frequency is lower (15-85%) in the sporadic form of the tumor (sBL). The presence of clonal EBV in all of the tumor cells (Gulley et al., 1992) has also been interpreted as evidence of EBV's role but in actuality it only means that EBV was present prior to the last transformation event that produced the tumor. Curiously, the only latent protein expressed in the tumor cells is EBNA1 (Gregory et al., 1990) none of the viral growth promoting genes are expressed. EBNA1 has to be expressed because tumor cells by definition are proliferating and EBNA1 is essential for duplication of the viral genome. Without it the genome would be lost. Therefore, transcriptionally this looks like a situation where the virus is just along for the ride.

Currently, there is no satisfactory model explaining how EBV might play a role in BL pathogenesis. Diametrically opposite models have been proposed. In one, EBV driven growth predisposes to a subsequent c-myc translocation (Klein, 1987; Klein, 1994) and in the other the c-myc translocation becomes stabilized by subsequent EBV infection (Lenoir and Bornkamm, 1987). No compelling evidence has been presented that favors one model over the other. Evidence has been presented that EBNA1 (Wilson et al., 1996) or the small untranslated RNA EBER (Komano et al., 1999) may be oncogenic. However, these claims require additional confirmation. It has also been suggested that the growth program drives proliferation early in BL development and is subsequently suppressed by some unknown mechanism leaving only EBNA1 expression (Kelly et al., 2002).

What is missing from current models of how EBV may predispose to BL is the lack of any explanation for why only EBNA1 is expressed. Earlier on it was proposed that this was part of an immune evasion tactic since EBNA1 is poorly recognized by CTL (Khanna et al., 1992; Levitskaya et al., 1995). However, BL tumor cells expressing CTL targets are poorly recognized by CTL irrespective of the presence or absence of EBV (see Rickinson and Kieff, 1996) for a more detailed discussion of this issue) and HD NPC and other EBV containing tumors survive in the face of an intact immune system despite expressing LMP1 and LMP2, so why only EBNA1 in BL? In the case of HD we have suggested that the default program is expressed because the tumor derives from an EBV infected germinal center cell and the default program is what the virus naturally expresses in a germinal center cell. In effect the viral gene expression pattern in HD is not selected by tumor growth but is a natural consequence of the cellular origin of the tumor. Applying this thinking to BL, there is currently only one way known to produce the "EBNA1 only" phenotype of BL in a non-tumor cell. This is when a latently infected memory cell expressing the latency program i.e. no latent proteins, divides, as part of normal B cell homeostasis (Figures 2 and 5) and turns on expression of "EBNA1 only" to ensure replication of the viral DNA with the cell. BL has the same pattern of immunoglobulin gene hypermutation as memory B cells (Klein et al., 1995) therefore a consistent scenario is that BL derives from a latently infected memory cell that expresses EBNA1 because it is continuously in cell cycle due to deregulated c-myc expression (Figure 4). Discordant with this idea is that the cellular phenotype of BL is that of a germinal center cell (Gregory et al., 1987). Thus, BL may be derived from a germinal center cell on the way to becoming a resting memory cell that is stuck proliferating, because of the activated c-myc, and therefore constitutively expresses EBNA1 only. Both versions would account for the presence of clonal EBV in the tumors because the virus would already be present as a passenger when the major transformation event, c-myc translocation occurs.

Not every BL tumor carries EBV, but the c-myc translocation is universal, superceding the presence or absence of EBV and the distinction between eBL and sBL. As discussed above, EBV infected cells can be present in infected subsets at staggeringly high levels during acute infection. If the c-myc translocation occurs because of the severe disruption of the immune response at the time of acute EBV infection then there is a good chance that the emergent tumor cell will fortuitously contain latent EBV. Given these considerations it remains a possibility that EBV may play no role in the tumor but simply be a fortuitous passenger.

X-linked proliferative syndrome

XLP is a rare X-linked immunodeficiency (Purtilo et al., 1975; Seemayer et al., 1995) which frequently results in lymphoma or fulminating IM (reviewed in Purtilo et al., (1975) and Seemayer et al. (1995)). It usually results in the death of afflicted boys within one month of primary EBV infection. Upon infection with EBV the boys typically mount what has been described as a "dysregulated exuberant response" (Seemayer et al., 1995) to the virus. EBV infected B cells, CD8+ T cells and macrophages enter tissues throughout the body causing massive tissue damage, particularly in the liver and the bone marrow, ultimately resulting in hepatitis and hypoplasia. A common feature of XLP is a pronounced virus associated hemophagocytic syndrome (VAHS) characterized by widely disseminated histiocytes filled with erythrocyte and nuclear debris. Death is usually associated early on with the accumulation of EBV infected B cells expressing the growth program in tissues such as the liver or subsequently by the widespread tissue damage of the VAHS. XLP has been fatal in 75% of cases. Surviving boys typically have severely disrupted immune systems resulting in varying degrees of hypogammaglobulinaemia. This process suggests the normal regulation of EBV infection has gone awry. Identification of the responsible gene was expected to shed important light on to the mechanism of EBV persistence. Sadly this was not to be. The XLP gene itself SH2D1A or SAP (Coffey et al., 1998; Sayos et al., 1998), encodes for a small signaling molecule of 128 amino-acids that consists essentially of a single SH2 domain with a small C-terminal extension. Prior to the onset of IM, boys carrying mutations in the XLP gene are not obviously immunodeficient, though several boys have manifested XLP without detectable EBV infection (Strahm et al., 2000). This implies that EBV precipitates an incipient predisposition. What this predisposition may be and how mutations in the XLP gene causes it remain unknown. SAP is an important regulator of T and NK cell interactions and activation. This has encouraged speculation that the XLP defect may interfere with NK activity or result in a failure to dampen T cell responses resulting in extensive tissue damage. Why these effects would be so specifically precipitated by EBV is unclear. Studies in SAP knockout mice indicate that they have defects in long term B cell memory (Crotty et al., 2003). This may be a clue that the true defect in XLP may be the inability to efficiently move EBV infected lymphoblasts into the memory compartment. Based on the model described above this would result in the infected cells being permanently stuck in the proliferative phase driven by the growth program. It may be that in the absence of the ability to move most of these cells into a resting state the immune system is overwhelmed and unable to clear the proliferating cells. This would explain both the appearance of overwhelming numbers of infected lymphoblasts and T cells trying to control them in these patients.

CONCLUSIONS

The model of EBV persistence has provided a framework for uniting all of the disparate observations on EBV. Although the details are likely to change and much is still to be learned it seems certain that an ultimate understanding of EBV infection will involve a model whereby EBV uses the normal biology of mature B lymphocytes to establish and maintain persistent infection.

The model has also allowed some possible insights into the origin of EBV associated disease particularly the B cell lymphomas and has provided some insight into B cell biology. The failure of EBV to persist in the IgD+ memory subset was one of the first pieces of evidence that they may be a separate, germinal center independent, lineage of memory cells. Similarly, studies on the distribution of latently infected memory B cells provide the first evidence of the circulation and distribution of memory B cells in man.

One of the most interesting unanswered questions about EBV persistence is why the virus encodes for a BCR surrogate if it is persisting in B cells with an apparently normal BCR. What is the relative contribution of viral latent proteins (especially LMP1 and LMP2a) and physiologic signals (Th and BCR) to the production of these latently infected memory cells? Could the requirement for both be providing us new insights into the complexities involved in producing and maintaining immunological memory?

Acknowledgements

I would like to thank Prof Martin Allday for critical reading of the manuscript and especially for his help in developing and describing the EBNA2/EBNA3 feed back model. I am indebted to Dr. Marta Perry who provided copious material from her own files from which I could select the micrographs shown in Figure 1. Her clearly written and informative review articles on the tonsil have been the major source of my education in the ways of this organ. DTL is supported by Public Health Service Grants AI 18757, and CA 65883.

References

Adams, A., and Lindahl, T. (1975). Epstein-Barr virus genomes with properties of circular DNA molecules in carrier cells. Proc. Natl. Acad. Sci. USA *72*, 1477-1481.

Agematsu, K., Nagumo, H., Shinozaki, K., Hokibara, S., Yasui, K., Terada, K., Kawamura, N., Toba, T., Nonoyama, S., Ochs, H. D., and Komiyama, A. (1998). Absence of IgD-CD27 (+) memory B cell population in X-linked hyper-IgM syndrome. J. Clin. Invest. *102*, 853-860.

Allday, M. J., Crawford, D. H., and Griffin, B. E. (1989). Epstein-Barr virus latent gene expression during the initiation of B cell immortalization. J. Gen. Virol. *70*, 1755-1764.

Allday, M. J., Sinclair, A., Parker, G., Crawford, D. H., and Farrell, P. J. (1995). Epstein-Barr virus efficiently immortalizes human B cells without neutralizing the function of p53. EMBO J. *14*, 1382-1391.

Anagnostopoulos, I., Hummel, M., Kreschel, C., and Stein, H. (1995). Morphology, immunophenotype, and distribution of latently and/or productively Epstein-Barr virus-infected cells in acute infectious mononucleosis: implications for the interindividual infection route of Epstein-Barr virus. Blood *85*, 744-750.

Andersson-Anvret, M., Forsby, N., Klein, G., and Henle, W. (1977). Relationship between the Epstein-Barr virus and undifferentiated nasopharyngeal carcinoma: correlated nucleic acid hybridization and histopathological examination. Int. J. Cancer *20*, 486-494.

Arpin, C., Banchereau, J., and Liu, Y. J. (1997). Memory B cells are biased towards terminal differentiation: a strategy that may prevent repertoire freezing. J. Exp. Med. *186*, 931-940.

Arpin, C., Dechanet, J., Van, K. C., Merville, P., Grouard, G., Briere, F., Banchereau, J., and Liu, Y. J. (1995). Generation of memory B cells and plasma cells *in vitro*. Science *268*, 720-722.

Arrand, J. J., and Rymo, L. (1982). Characterisation of the major Epstein-Barr virus specific RNA in Burkitt lymphoma derived cells. J. Virol. *41*, 376-389.

Artavanis-Tsakonas, S., Matsuno, K., and Fortini, M. E. (1995). Notch signaling. Science *268*, 225-232.

Babcock, G. J., Decker, L. L., Freeman, R. B., and Thorley-Lawson, D. A. (1999). Epstein-barr virus-infected resting memory B cells, not proliferating lymphoblasts, accumulate in the peripheral blood of immunosuppressed patients. J. Exp. Med. *190*, 567-576.

Babcock, G. J., Decker, L. L., Volk, M., and Thorley-Lawson, D. A. (1998). EBV persistence in memory B cells *in vivo*. Immunity *9*, 395-404.

Babcock, G. J., Hochberg, D., and Thorley-Lawson, A. D. (2000). The expression pattern of Epstein-Barr virus latent genes *in vivo* is dependent upon the differentiation stage of the infected B cell. Immunity *13*, 497-506.

Baichwal, V. R., and Sugden, B. (1988). Transformation of Balb 3T3 cells by the BNLF-1 gene of Epstein-Barr virus. Oncogene *2*, 461-467.

Banchereau, J., Bazan, F., Blanchard, D., Briere, F., Galizzi, J. P., van, K. C., Liu, Y. J., Rousset, F., and Saeland, S. (1994). The CD40 antigen and its ligand. Ann. Rev. Immunol. *12*, 881-922.

Barolo, S., Stone, T., Bang, A. G., and Posakony, J. W. (2002). Default repression and Notch signaling: Hairless acts as an adaptor to recruit the corepressors Groucho and dCtBP to Suppressor of Hairless. Genes Dev. *16*, 1964-1976.

Beaufils, P., Choquet, D., Mamoun, R. Z., and Malissen, B. (1993). The (YXXL/I)2 signalling motif found in the cytoplasmic segments of the bovine leukaemia virus envelope protein and Epstein-Barr virus latent membrane protein 2A can elicit early and late lymphocyte activation events. EMBO J. *12*, 5105-5112.

Bejarano, M. T., and Masucci, M. G. (1998). Interleukin-10 abrogates the inhibition of Epstein-Barr virus-induced B-cell transformation by memory T-cell responses. Blood *92*, 4256-4262.

Bernasconi, N. L., Traggiai, E., and Lanzavecchia, A. (2002). Maintenance of serological memory by polyclonal activation of human memory B cells. Science *298*, 2199-2202.

Birkenbach, M., Josefsen, K., Yalamanchili, R., Lenoir, G., and Kieff, E. (1993). Epstein-Barr virus-induced genes: first lymphocyte-specific G protein-coupled peptide receptors. J. Virol. *67*, 2209-2220.

Bonnet, M., Guinebretiere, J. M., Kremmer, E., Grunewald, V., Benhamou, E., Contesso, G., and Joab, I. (1999). Detection of Epstein-Barr virus in invasive breast cancers. J. Natl. Cancer Inst. *91*, 1376-1381.

Borza, C. M., and Hutt-Fletcher, L. M. (2002). Alternate replication in B cells and epithelial cells switches tropism of Epstein-Barr virus. Nat. Med. *8*, 594-599.

Brandtzaeg, P., Baekkevold, E. S., Farstad, I. N., Jahnsen, F. L., Johansen, F. E., Nilsen, E. M., and Yamanaka, T. (1999a). Regional specialization in the mucosal immune system: what happens in the microcompartments? Immunol. Today *20*, 141-151.

Brandtzaeg, P., Farstad, I. N., and Haraldsen, G. (1999b). Regional specialization in the mucosal immune system: primed cells do not always home along the same track. Immunol. Today *20*, 267-277.

Brooks, L., Yao, Q. Y., Rickinson, A. B., and Young, L. S. (1992). Epstein-Barr virus latent gene transcription in nasopharyngeal carcinoma cells: coexpression of EBNA1, LMP1, and LMP2 transcripts. J. Virol. *66*, 2689-2697.

Burgstahler, R., Kempkes, B., Steube, K., and Lipp, M. (1995). Expression of the chemokine receptor BLR2/EBI1 is specifically transactivated by Epstein-Barr virus nuclear antigen 2. Biochem Biophys Res Commun *215*, 737-743.

Cabannes, E., Khan, G., Aillet, F., Jarrett, R. F., and Hay, R. T. (1999). Mutations in the IkBa gene in Hodgkin's disease suggest a tumour suppressor role for IkappaBalpha. Oncogene *18*, 3063-3070.

Calame, K. L., Lin, K. I., and Tunyaplin, C. (2003). Regulatory mechanisms that determine the development and function of plasma cells. Annu. Rev. Immunol. *21*, 205-230. Epub 2001 Dec 2019.

Caldwell, R. G., Wilson, J. B., Anderson, S. J., and Longnecker, R. (1998). Epstein-Barr virus LMP2A drives B cell development and survival in the absence of normal B cell receptor signals. Immunity *9*, 405-411.

Callan, M. F., Annels, N., Steven, N., Tan, L., Wilson, J., McMichael, A. J., and Rickinson, A. B. (1998a). T cell selection during the evolution of CD8+ T cell memory *in vivo*. Eur J Immunol *28*, 4382-4390.

Callan, M. F., Tan, L., Annels, N., Ogg, G. S., Wilson, J. D., O'Callaghan, C. A., Steven, N., McMichael, A. J., and Rickinson, A. B. (1998b). Direct visualization of antigen-specific CD8+ T cells during the primary immune response to Epstein-Barr virus *in vivo*. J. Exp. Med. *187*, 1395-1402.

Carbone, A., Gaidano, G., Gloghini, A., Larocca, L. M., Capello, D., Canzonieri, V., Antinori, A., Tirelli, U., Falini, B., and Dalla-Favera, R. (1998). Differential expression of BCL-6, CD138/syndecan-1, and Epstein-Barr virus-encoded latent membrane protein-1 identifies distinct histogenetic subsets of acquired immunodeficiency syndrome-related non-Hodgkin's lymphomas. Blood *91*, 747-755.

Casola, S., Otipoby, K. L., Alimzhanov, M., Humme, S., Uyttersprot, N., Kutok, J. L., Carroll, M. C., and Rajewsky, K. (2004). B cell receptor signal strength determines B cell fate. Nat. Immunol. *5*, 317-327.

Catalina, M. D., Sullivan, J. L., Bak, K. R., and Luzuriaga, K. (2001). Differential evolution and stability of epitope-specific CD8 (+) T cell responses in EBV infection. J. Immunol. *167*, 4450-4457.

Cattoretti, G., Chang, C. C., Cechova, K., Zhang, J., Ye, B. H., Falini, B., Louie, D. C., Offit, K., Chaganti, R. S., and Dalla-Favera, R. (1995). BCL-6 protein is expressed in germinal-center B cells. Blood *86*, 45-53.

Chen, F., Zou, J. Z., di, R. L., Winberg, G., Hu, L. F., Klein, E., Klein, G., and Ernberg, I. (1995). A subpopulation of normal B cells latently infected with Epstein-Barr virus resembles Burkitt lymphoma cells in expressing EBNA-1 but not EBNA-2 or LMP1. J. Virol. *69*, 3752-3758.

Chiang, A. K., Tao, Q., Srivastava, G., and Ho, F. C. (1996). Nasal NK- and T-cell lymphomas share the same type of Epstein-Barr virus latency as nasopharyngeal carcinoma and Hodgkin's disease. Int. J. Cancer *68*, 285-290.

Coffey, A. J., Brooksbank, R. A., Brandau, O., Oohashi, T., Howell, G. R., Bye, J. M., Cahn, A. P., Durham, J., Heath, P., Wray, P., *et al.* (1998). Host response to EBV infection in X-linked lymphoproliferative disease results from mutations in an SH2-domain encoding gene. Nat. Genet. *20*, 129-135.

Crotty, S., Kersh, E. N., Cannons, J., Schwartzberg, P. L., and Ahmed, R. (2003). SAP is required for generating long-term humoral immunity. Nature *421*, 282-287.

Deacon, E. M., Pallesen, G., Niedobitek, G., Crocker, J., Brooks, L., Rickinson, A. B., and Young, L. S. (1993). Epstein-Barr virus and Hodgkin's disease: transcriptional analysis of virus latency in the malignant cells. J. Exp. Med. *177*, 339-349.

Decker, L. L., Babcock, G. J., and Thorley-Lawson, D. A. (2001). Detection and discrimination of latent and replicative herpesvirus infection at the single cell level *in vivo*. Methods Mol Biol *174*, 111-116.

Decker, L. L., Klaman, L. D., and Thorley-Lawson, D. A. (1996). Detection of the latent form of Epstein-Barr virus DNA in the peripheral blood of healthy individuals. J. Virol. *70*, 3286-3289.

de-The, G. (1985). Epstein-Barr virus and Burkitt's lymphoma worldwide: the causal relationship revisited. In Burkitt's Lymphoma: a human cancer model, C. L. M. Olweny, ed. (New York, Oxford University Press), pp. 165-176.

Dono, M., Burgio, V. L., Tacchetti, C., Favre, A., Augliera, A., Zupo, S., Taborelli, G., Chiorazzi, N., Grossi, C. E., and Ferrarini, M. (1996a). Subepithelial B cells in the human palatine tonsil. I. Morphologic, cytochemical and phenotypic characterization. Eur. J. Immunol. *26*, 2035-2042.

Dono, M., Zupo, S., Augliera, A., Burgio, V. L., Massara, R., Melagrana, A., Costa, M., Grossi, C. E., Chiorazzi, N., and Ferrarini, M. (1996b). Subepithelial B cells in the human palatine tonsil. II. Functional characterization. Eur. J. Immunol. *26*, 2043-2049.

Ehlin-Henriksson, B., Zou, J. Z., Klein, G., and Ernberg, I. (1999). Epstein-Barr virus genomes are found predominantly in IgA-positive B cells in the blood of healthy carriers. Int. J. Cancer *83*, 50-54.

Epstein, M. A., Achong, B. G., and Barr, Y. M. (1964). Virus particles in cultured lymphoblasts from Burkitt's lymphoma. Lancet *1*, 702-703.

Fagarasan, S., and Honjo, T. (2000). T-Independent immune response: new aspects of B cell biology. Science *290*, 89-92.

Fahraeus, R., Fu, H. L., Ernberg, I., Finke, J., Rowe, M., Klein, G., Falk, K., Nilsson, E., Yadav, M., Busson, P., and et al. (1988). Expression of Epstein-Barr virus-encoded proteins in nasopharyngeal carcinoma. Int. J. Cancer *42*, 329-338.

Fahraeus, R., Jansson, A., Ricksten, A., Sjoblom, A., and Rymo, L. (1990). Epstein-Barr virus-encoded nuclear antigen 2 activates the viral latent membrane protein promoter by modulating the activity of a negative regulatory element. Proc. Natl. Acad. Sci. USA *87*, 7390-7394.

Faulkner, G. C., Burrows, S. R., Khanna, R., Moss, D. J., Bird, A. G., and Crawford, D. H. (1999). X-Linked agammaglobulinemia patients are not infected with Epstein-Barr virus: implications for the biology of the virus. J. Virol. *73*, 1555-1564.

Fingeroth, J. D., Weis, J. J., Tedder, T. F., Strominger, J. L., Biro, P. A., and Fearon, D. T. (1984). Epstein-Barr virus receptor of human B lymphocytes is the C3d receptor CR2. Proc. Natl. Acad. Sci. USA *81*, 4510-4514.

Fox, R. I., Chilton, T., Scott, S., Benton, L., Howell, F. V., and Vaughan, J. H. (1987). Potential role of Epstein-Barr virus in Sjogren's syndrome. Rheum. Dis. Clin. North Am. *13*, 275-292.

Ghosh, S., May, M. J., and Kopp, E. B. (1998). NF-kappa B and Rel proteins: evolutionarily conserved mediators of immune responses. Annu. Rev. Immunol. *16*, 225-260.

Gires, O., Zimber-Strobl, U., Gonnella, R., Ueffing, M., Marschall, G., Zeidler, R., Pich, D., and Hammerschmidt, W. (1997). Latent membrane protein 1 of Epstein-Barr virus mimics a constitutively active receptor molecule. Embo J. *16*, 6131-6140.

Glaser, S. L., Lin, R. J., Stewart, S. L., Ambinder, R. F., Jarrett, R. F., Brousset, P., Pallesen, G., Gulley, M. L., Khan, G., O'Grady, J., *et al.* (1997). Epstein-Barr virus-associated Hodgkin's disease: epidemiologic characteristics in international data. Int. J. Cancer *70*, 375-382.

Golden, H. D., Chang, R. S., Prescott, W., Simpson, E., and Cooper, T. Y. (1973). Leukocyte-transforming agent: prolonged excretion by patients with mononucleosis and excretion by normal individuals. J. Infect. Dis. *127*, 471-473.

Gratama, J. W., Oosterveer, M. A., Zwaan, F. E., Lepoutre, J., Klein, G., and Ernberg, I. (1988). Eradication of Epstein-Barr virus by allogeneic bone marrow transplantation: implications for sites of viral latency. Proc. Natl. Acad. Sci. USA *85*, 8693-8696.

Gray, D., MacLennan, I. C., Bazin, H., and Khan, M. (1982). Migrant mu+ delta+ and static mu+ delta- B lymphocyte subsets. Eur. J. Immunol. *12*, 564-569.

Greenspan, J. S., Greenspan, D., Lennette, E. T., Abrams, D. I., Conant, M. A., Petersen, V., and Freese, U. K. (1985). Replication of Epstein-Barr virus within the epithelial cells of oral "hairy" leukoplakia, an AIDS-associated lesion. N. Engl. J. Med. *313*, 1564-1571.

Gregory, C. D., Rowe, M., and Rickinson, A. B. (1990). Different Epstein-Barr virus-B cell interactions in phenotypically distinct clones of a Burkitt's lymphoma cell line. J. Gen. Virol. *71*, 1481-1495.

Gregory, C. D., Tursz, T., Edwards, C. F., Tetaud, C., Talbot, M., Caillou, B., Rickinson, A. B., and Lipinski, M. (1987). Identification of a subset of normal B cells with a Burkitt's lymphoma (BL)-like phenotype. J. Immunol. *139*, 313-318.

Gulley, M. L., Raphael, M., Lutz, C. T., Ross, D. W., and Raab-Traub, N. (1992). Epstein-Barr virus integration in human lymphomas and lymphoid cell lines. Cancer *70*, 185-191.

He, B., Raab-Traub, N., Casali, P., and Cerutti, A. (2003). EBV-encoded latent membrane protein 1 cooperates with BAFF/BLyS and APRIL to induce T cell-independent Ig heavy chain class switching. J. Immunol. *171*, 5215-5224.

Henderson, S., Huen, D., Rowe, M., Dawson, C., Johnson, G., and Rickinson, A. (1993). Epstein-Barr virus-coded BHRF1 protein, a viral homologue of Bcl-2, protects human B cells from programmed cell death. Proc. Natl. Acad. Sci. USA *90*, 8479-8483.

Henle, W., Diehl, V., Kohn, G., Zur Hausen, H., and Henle, G. (1967). Herpes-type virus and chromosome marker in normal leukocytes after growth with irradiated Burkitt cells. Science *157*, 1064-1065.

Henle, W., and Henle, G. (1979). Seroepidemiology of the virus. In The Epstein-Barr virus, B. G. Achong, ed. (Berlin, Springer-Verlag), pp. 61-78.

Herbst, H., Dallenbach, F., Hummel, M., Niedobitek, G., Pileri, S., Muller, L. N., and Stein, H. (1991). Epstein-Barr virus latent membrane protein expression in Hodgkin and Reed-Sternberg cells. Proc. Natl. Acad. Sci. USA *88*, 4766-4770.

Hickabottom, M., Parker, G. A., Freemont, P., Crook, T., and Allday, M. J. (2002). Two nonconsensus sites in the Epstein-Barr virus oncoprotein EBNA3A cooperate to bind the co-repressor carboxyl-terminal-binding protein (CtBP). J. Biol. Chem. *277*, 47197-47204. Epub 42002 Oct 47197.

Ho, M., Miller, G., Atchison, R. W., Breinig, M. K., Dummer, J. S., Andiman, W., Starzl, T. E., Eastman, R., Griffith, B. P., Hardesty, R. L., and et al. (1985). Epstein-Barr virus infections and DNA hybridization studies in posttransplantation lymphoma and lymphoproliferative lesions: the role of primary infection. J. Infect. Dis. *152*, 876-886.

Hoagland, R. J. (1955). The transmission of infectious mononucleosis. Amer. J. Med. Sci. *229*, 262-272.

Hoagland, R. J. (1967). Infectious mononucleosis. In Infectious mononucleosis (New York and London, Grune and Stratton Inc.).

Hochberg, D., Middeldorp, J. M., Catalina, M., Sullivan, J. L., Luzuriaga, K., and Thorley-Lawson, D. A. (2004a). Demonstration of the Burkitt's lymphoma Epstein-Barr virus phenotype in dividing latently infected memory cells *in vivo*. Proc. Natl. Acad. Sci. USA *101*, 239-244.

Hochberg, D., Souza, T., Catalina, M., Sullivan, J. L., Luzuriaga, K., and Thorley-Lawson, D. A. (2004b). Acute infection with Epstein-Bar virus targets and overwhelms the peripheral memory B cell compartment with resting, latently infected cells. J. Virol. *78*, 5194-5204.

Hochberg, D. and Thorley-Lawson, D. A. (2005). Q uantative detection of viral gene expression in populations of Epstein-Bar infected cells *in vivo*. Methods Mol. Biol. *292*, 39-56.

Hopwood, P., and Crawford, D. H. (2000). The role of EBV in post-transplant malignancies: a review. J. Clin. Pathol. *53*, 248-254.

Hsieh, J. J., Henkel, T., Salmon, P., Robey, E., Peterson, M. G., and Hayward, S. D. (1996). Truncated mammalian Notch1 activates CBF1/RBPJk-repressed genes by a mechanism resembling that of Epstein-Barr virus EBNA2. Mol. Cell. Biol. *16*, 952-959.

Hurley, E. A., and Thorley-Lawson, D. A. (1988). B cell activation and the establishment of Epstein-Barr virus latency. J. Exp. Med. *168*, 2059-2075.

Imadome, K., Shirakata, M., Shimizu, N., Nonoyama, S., and Yamanashi, Y. (2003). CD40 ligand is a critical effector of Epstein-Barr virus in host cell survival and transformation. Proc. Natl. Acad. Sci. USA *100*, 7836-7840. Epub 2003 Jun 7812.

Inman, G. J., Binne, U. K., Parker, G. A., Farrell, P. J., and Allday, M. J. (2001). Activators of the Epstein-Barr virus lytic program concomitantly induce apoptosis, but lytic gene expression protects from cell death. J. Virol. *75*, 2400-2410.

Izumi, K. M., and Kieff, E. D. (1997). The Epstein-Barr virus oncogene product latent membrane protein 1 engages the tumor necrosis factor receptor-associated death domain protein to mediate B lymphocyte growth transformation and activate NF- kappaB. Proc. Natl. Acad. Sci. USA *94*, 12592-12597.

Jaenisch, R., and Bird, A. (2003). Epigenetic regulation of gene expression: how the genome integrates intrinsic and environmental signals. Nat. Genet. *33*, 245-254.

James, J. A., Kaufman, K. M., Farris, A. D., Taylor-Albert, E., Lehman, T. J., and Harley, J. B. (1997). An increased prevalence of Epstein-Barr virus infection in young patients suggests a possible etiology for systemic lupus erythematosus. J. Clin. Invest. *100*, 3019-3026.

Joncas, J., Boucher, J., Granger-Julien, M., and Filion, C. (1974). Epstein-Barr virus infection in the neonatal period and in childhood. Can Med Assoc J *110*, 33-37.

Joseph, A. M., Babcock, G. J., and Thorley-Lawson, D. A. (2000a). Cells expressing the Epstein-Barr virus growth program are present in and restricted to the naive B-cell subset of healthy tonsils. J. Virol. *74*, 9964-9971.

Joseph, A. M., Babcock, G. J., and Thorley-Lawson, D. A. (2000b). EBV persistence involves strict selection of latently infected B cells. J. Immunol. *165*, 2975-2981.

Kaiser, C., Laux, G., Eick, D., Jochner, N., Bornkamm, G. W., and Kempkes, B. (1999). The proto-oncogene c-myc is a direct target gene of Epstein-Barr virus nuclear antigen 2. J. Virol. *73*, 4481-4484.

Kantor, A. B. (1991). The development and repertoire of B-1 cells (CD5 B cells). Immunol. Today *12*, 389-391.

Karajannis, M. A., Hummel, M., Anagnostopoulos, I., and Stein, H. (1997). Strict lymphotropism of Epstein-Barr virus during acute infectious mononucleosis in nonimmunocompromised individuals. Blood *89*, 2856-2862.

Kaye, K. M., Izumi, K. M., and Kieff, E. (1993). Epstein-Barr virus latent membrane protein 1 is essential for B-lymphocyte growth transformation. Proc. Natl. Acad. Sci. USA *90*, 9150-9154.

Keating, S., Prince, S., Jones, M., and Rowe, M. (2002). The lytic cycle of Epstein-Barr virus is associated with decreased expression of cell surface major histocompatibility complex class I and class II molecules. J. Virol. *76*, 8179-8188.

Kelly, G., Bell, A., and Rickinson, A. (2002). Epstein-Barr virus-associated Burkitt lymphomagenesis selects for downregulation of the nuclear antigen EBNA2. Nat. Med. *8*, 1098-1104.

Kempkes, B., Spitkovsky, D., Jansen-Durr, P., Ellwart, J. W., Kremmer, E., Delecluse, H. J., Rottenberger, C., Bornkamm, G. W., and Hammerschmidt, W. (1995). B-cell proliferation and induction of early G1-regulating proteins by Epstein-Barr virus mutants conditional for EBNA2. Embo J. *14*, 88-96.

Khan, G., Miyashita, E. M., Yang, B., Babcock, G. J., and Thorley-Lawson, D. A. (1996). Is EBV persistence *in vivo* a model for B cell homeostasis? Immunity *5*, 173-179.

Khanna, R., Burrows, S. R., Kurilla, M. G., Jacob, C. A., Misko, I. S., Sculley, T. B., Kieff, E., and Moss, D. J. (1992). Localization of Epstein-Barr virus cytotoxic T cell epitopes using recombinant vaccinia: implications for vaccine development. J. Exp. Med. *176*, 169-176.

Khanna, R., Slade, R. W., Poulsen, L., Moss, D. J., Burrows, S. R., Nicholls, J., and Burrows, J. M. (1997). Evolutionary dynamics of genetic variation in Epstein-Barr virus isolates of diverse geographical origins: evidence for immune pressure- independent genetic drift. J. Virol. *71*, 8340-8346.

Kieff, E., and Rickinson, A. B. (2001). Epstein-Barr Virus and Its Replication. In Virology, D. M. Knipe, and P. M. Howley, eds. (New York, Lippincott Williams and Wilkins), pp. 2511-2574.

Klein, G. (1983). Specific chromosomal translocations and the genesis of B-cell-derived tumors in mice and men. Cell *32*, 311-315.

Klein, G. (1987). In defense of the "old" Burkitt lymphoma scenario. In Advances in Viral Oncology, G. Klein, ed. (New York, Raven Press), pp. 207-211.

Klein, G. (1994). Epstein-Barr Virus Strategy in Normal and Neoplastic B-cells. Cell *77*, 791-793.

Klein, U., Klein, G., Ehlin-Henriksson, B., Rajewsky, K., and Kuppers, R. (1995). Burkitt's lymphoma is a malignancy of mature B cells expressing somatically mutated V region genes. Mol. Med. *1*, 495-505.

Klein, U., Kuppers, R., and Rajewsky, K. (1994). Variable region gene analysis of B cell subsets derived from a 4-year-old child: somatically mutated memory B cells accumulate in the peripheral blood already at young age. J. Exp. Med. *180*, 1383-1393.

Klein, U., Rajewsky, K., and Kuppers, R. (1998). Human immunoglobulin (Ig)M+IgD+ peripheral blood B cells expressing the CD27 cell surface antigen carry somatically mutated variable region genes: CD27 as a general marker for somatically mutated (memory) B cells. J. Exp. Med. *188*, 1679-1689.

Knowles, D. M., Cesarman, E., Chadburn, A., Frizzera, G., Chen, J., Rose, E. A., and Michler, R. E. (1995). Correlative morphologic and molecular genetic analysis demonstrates three distinct categories of posttransplantation lymphoproliferative disorders. Blood *85*, 552-565.

Komano, J., Maruo, S., Kurozumi, K., Oda, T., and Takada, K. (1999). Oncogenic role of Epstein-Barr virus-encoded RNAs in Burkitt's lymphoma cell line Akata. J. Virol. *73*, 9827-9831.

Kovalchuk, A. L., Qi, C. F., Torrey, T. A., Taddesse-Heath, L., Feigenbaum, L., Park, S. S., Gerbitz, A., Klobeck, G., Hoertnagel, K., Polack, A., *et al.* (2000). Burkitt lymphoma in the mouse. J. Exp. Med. *192*, 1183-1190.

Kuppers, R., and Rajewsky, K. (1998). The origin of Hodgkin and Reed/Sternberg cells in Hodgkin's disease. Annu. Rev. Immunol. *16*, 471-493.

Kurosaki, T. (1999). Genetic analysis of B cell antigen receptor signaling. Annu. Rev. Immunol. *17*, 555-592.

Kurth, J., Hansmann, M. L., Rajewsky, K., and Kuppers, R. (2003). Epstein-Barr virus-infected B cells expanding in germinal centers of infectious mononucleosis patients do not participate in the germinal center reaction. Proc. Natl. Acad. Sci. USA *100*, 4730-4735.

Kurth, J., Spieker, T., Wustrow, J., Strickler, G. J., Hansmann, L. M., Rajewsky, K., and Kuppers, R. (2000). EBV-infected B cells in infectious mononucleosis: viral strategies for spreading in the B cell compartment and establishing latency. Immunity *13*, 485-495.

Kusano, S., and Raab-Traub, N. (2001). An Epstein-Barr virus protein interacts with Notch. J. Virol. *75*, 384-395.

Laichalk, L. L., Hochberg, D., Babcock, G. J., Freeman, R. B., and Thorley-Lawson, D. A. (2002). The dispersal of mucosal memory B cells: evidence from persistent EBV infection. Immunity *16*, 745-754.

Laichalk, L. L., and Thorley-Lawson, D. A. **(2005)**. Terminal Differentiation into Plasma Cells Initiates the Replicative Cycle of Epstein-Barr Virus *in vivo*. J. Virol. *79*, 1296-1307.

Lam, K. P., Kuhn, R., and Rajewsky, K. (1997). *In vivo* ablation of surface immunoglobulin on mature B cells by inducible gene targeting results in rapid cell death. Cell *90*, 1073-1083.

Lam, N., and Sugden, B. (2003). CD40 and its viral mimic, LMP1: similar means to different ends. Cell Signal *15*, 9-16.

Le Roux, A., Kerdiles, B., Walls, D., Dedieu, J. F., and Perricaudet, M. (1994). The Epstein-Barr virus determined nuclear antigens EBNA-3A, -3B, and -3C repress EBNA-2-mediated transactivation of the viral terminal protein 1 gene promoter. Virology *205*, 596-602.

Leder, P. (1985). Translocations among antibody genes in human cancer. In Burkitt's Lymphoma: a human cancer model, G. M. Lenoir, G. T. O'Conor, and C. L. M. Olweny, eds. (New York, Oxford University Press), pp. 341-371.

Lenoir, G. M., and Bornkamm, G. W. (1987). Burkitt's lymphoma, a human cancer model for the study of the multistep development of cancer: Proposal for a new scenario. Adv. Viral Oncol. *7*, 173-206.

Levitskaya, J., Coram, M., Levitsky, V., Imreh, S., Steigerwald, M. P., Klein, G., Kurilla, M. G., and Masucci, M. G. (1995). Inhibition of antigen processing by the internal repeat region of the Epstein-Barr virus nuclear antigen-1. Nature *375*, 685-688.

Levitskaya, J., Sharipo, A., Leonchiks, A., Ciechanover, A., and Masucci, M. G. (1997). Inhibition of ubiquitin/proteasome-dependent protein degradation by the Gly-Ala repeat domain of the Epstein-Barr virus nuclear antigen 1. Proc. Natl. Acad. Sci. USA *94*, 12616-12621.

Li, Q., Spriggs, M. K., Kovats, S., Turk, S. M., Comeau, M. R., Nepom, B., and Hutt-Fletcher, L. M. (1997). Epstein-Barr virus uses HLA class II as a cofactor for infection of B lymphocytes. J. Virol. *71*, 4657-4662.

Ling, P. D., Hsieh, J. J., Ruf, I. K., Rawlins, D. R., and Hayward, S. D. (1994). EBNA-2 upregulation of Epstein-Barr virus latency promoters and the cellular CD23 promoter utilizes a common targeting intermediate, CBF1. J. Virol. *68*, 5375-5383.

Liu, Y. J., and Arpin, C. (1997). Germinal center development. Immunol. Rev. *156*, 111-126.

Liu, Y. J., Barthelemy, C., de Bouteiller, O., Arpin, C., Durand, I., and Banchereau, J. (1995). Memory B cells from human tonsils colonize mucosal epithelium and directly present antigen to T cells by rapid up-regulation of B7-1 and B7-2. Immunity *2*, 239-248.

Liu, Y. J., Joshua, D. E., Williams, G. T., Smith, C. A., Gordon, J., and MacLennan, I. C. (1989). Mechanism of antigen-driven selection in germinal centres. Nature *342*, 929-931.

Lotz, M., and Roudier, J. (1989). Epstein-Barr virus and rheumatoid arthritis: cellular and molecular aspects. Rheumatol. Int. *9*, 147-152.

Macallan, D.C., Wallace, D.L., Zhang, Y., Ghattas, H., Asquith, B., de Lara, C., Worth, A., Panayiotakopoulos, G., Griffin, G.E., Tough, D.F. and Beverley P.C. (2005). B-cell kinetics in humans: rapid turnover of peripheral blood memory cells. Blood 105, 3633-3640.

MacLennan, I. C. (1994). Germinal centers. Ann. Rev.Immunol. *12*, 117-139.

MacLennan, I. C. (1998). B-cell receptor regulation of peripheral B cells. Curr. Opin. Immunol. *10*, 220-225.

Manolov, G., and Manolova, Y. (1972). Marker band in one chromosome 14 from Burkitt lymphomas. Nature *237*, 33-34.

Martinez-Valdez, H., Guret, C., de Bouteiller, O., Fugier, I., Banchereau, J., and Liu, Y. J. (1996). Human germinal center B cells express the apoptosis-inducing genes Fas, c-myc, P53, and Bax but not the survival gene bcl-2. J. Exp. Med. *183*, 971-977.

Maruyama, M., Lam, K. P., and Rajewsky, K. (2000). Memory B-cell persistence is independent of persisting immunizing antigen. Nature *407*, 636-642.

Matsuda, M., Salazar, F., Petersson, M., Masucci, G., Hansson, J., Pisa, P., Zhang, Q. J., Masucci, M. G., and Kiessling, R. (1994). Interleukin 10 pretreatment protects target cells from tumor- and allo-specific cytotoxic T cells and downregulates HLA class I expression. J. Exp. Med. *180*, 2371-2376.

Matsumoto, A. K., Martin, D. R., Carter, R. H., Klickstein, L. B., Ahearn, J. M., and Fearon, D. T. (1993). Functional dissection of the CD21/CD19/TAPA-1/Leu-13 complex of B lymphocytes. J. Exp. Med. *178*, 1407-1417.

Miller, C. L., Burkhardt, A. L., Lee, J. H., Stealey, B., Longnecker, R., Bolen, J. B., and Kieff, E. (1995). Integral membrane protein 2 of Epstein-Barr virus regulates reactivation from latency through dominant negative effects on protein-tyrosine kinases. Immunity *2*, 155-166.

Miyashita, E. M., Yang, B., Babcock, G. J., and Thorley-Lawson, D. A. (1997). Identification of the site of Epstein-Barr virus persistence *in vivo* as a resting B cell. J. Virol. *71*, 4882-4891.

Miyashita, E. M., Yang, B., Lam, K. M., Crawford, D. H., and Thorley-Lawson, D. A. (1995). A novel form of Epstein-Barr virus latency in normal B cells *in vivo*. Cell *80*, 593-601.

Moorthy, R., and Thorley-Lawson, D. A. (1992). Mutational analysis of the transforming function of the EBV encoded LMP-1. Curr. Top. Microbiol. Immunol. *182*, 359-365.

Morel, V., Lecourtois, M., Massiani, O., Maier, D., Preiss, A., and Schweisguth, F. (2001). Transcriptional repression by suppressor of hairless involves the binding of a hairless-dCtBP complex in Drosophila. Curr. Biol. *11*, 789-792.

Mosialos, G., Birkenbach, M., Yalamanchili, R., VanArsdale, T., Ware, C., and Kieff, E. (1995). The Epstein-Barr virus transforming protein LMP1 engages signaling proteins for the tumor necrosis factor receptor family. Cell *80*, 389-399.

Murray, R. J., Kurilla, M. G., Brooks, J. M., Thomas, W. A., Rowe, M., Kieff, E., and Rickinson, A. B. (1992). Identification of target antigens for the human cytotoxic T cell response to Epstein-Barr virus (EBV): implications for the immune control of EBV-positive malignancies. J. Exp. Med. *176*, 157-168.

Nemerow, G. R., Wolfert, R., McNaughton, M. E., and Cooper, N. R. (1985). Identification and characterization of the Epstein-Barr virus receptor on human B lymphocytes and its relationship to the C3d complement receptor (CR2). J. Virol. *55*, 347-351.

Ngo, V. N., Tang, H. L., and Cyster, J. G. (1998). Epstein-Barr virus-induced molecule 1 ligand chemokine is expressed by dendritic cells in lymphoid tissues and strongly attracts naive T cells and activated B cells. J. Exp. Med. *188*, 181-191.

Nicholson, L. J., Hopwood, P., Johannessen, I., Salisbury, J. R., Codd, J., Thorley-Lawson, D., and Crawford, D. H. (1997). Epstein-Barr virus latent membrane protein does not inhibit differentiation and induces tumorigenicity of human epithelial cells. Oncogene *15*, 275-283.

Niederman, J. C., McCollum, R. W., Henle, G., and Henle, W. (1968). Infectious mononucleosis. Clinical manifestations in relation to EB virus antibodies. JAMA *203*, 205-209.

Niedobitek, G., Agathanggelou, A., Steven, N., and Young, L. S. (2000). Epstein-Barr virus (EBV) in infectious mononucleosis: detection of the virus in tonsillar B lymphocytes but not in desquamated oropharyngeal epithelial cells. Mol. Pathol. *53*, 37-42.

Niedobitek, G., Kremmer, E., Herbst, H., Whitehead, L., Dawson, C. W., Niedobitek, E., von Ostau, C., Rooney, N., Grasser, F. A., and Young, L. S. (1997). Immunohistochemical detection of the Epstein-Barr virus-encoded latent membrane protein 2A in Hodgkin's disease and infectious mononucleosis. Blood *90*, 1664-1672.

Nilsson, K. (1979). The nature of lymphoid cell lines and their relationship to the virus. In The Epstein-Barr virus, M. A. Epstein, and B. G. Achong, eds. (Berlin, Springer-Verlag), pp. 225-281.

Nonkwelo, C., Skinner, J., Bell, A., Rickinson, A., and Sample, J. (1996). Transcription start sites downstream of the Epstein-Barr virus (EBV) Fp promoter in early-passage Burkitt lymphoma cells define a fourth promoter for expression of the EBV EBNA-1 protein. J. Virol. *70*, 623-627.

O'Nions, J., and Allday, M. J. (2003). Epstein-Barr virus can inhibit genotoxin-induced G1 arrest downstream of p53 by preventing the inactivation of CDK2. Oncogene *22*, 7181-7191.

O'Nions, J., and Allday, M. J. (2004). Deregulation of the Cell Cycle by the Epstein-Barr Virus. In Advances in Cancer Research, G. F. Vande Woude, and G. Klein, eds. (New York, Elsevier), pp. 119-178.

Oudejans, J. J., Dukers, D. F., Jiwa, N. M., van den Brule, A. J., Grasser, F. A., de Bruin, P. C., Horstman, A., Vos, W., van Gorp, J., Middeldorp, J. M., and Meijer, C. J. (1996). Expression of epstein-barr virus encoded nuclear antigen 1 in benign and malignant tissues harbouring EBV. J. Clin. Pathol. *49*, 897-902.

Parker, G. A., Crook, T., Bain, M., Sara, E. A., Farrell, P. J., and Allday, M. J. (1996). Epstein-Barr virus nuclear antigen (EBNA)3C is an immortalizing oncoprotein with similar properties to adenovirus E1A and papillomavirus E7. Oncogene *13*, 2541-2549.

Paulson, E. J., and Speck, S. H. (1999). Differential methylation of Epstein-Barr virus latency promoters facilitates viral persistence in healthy seropositive individuals. J. Virol. *73*, 9959-9968.

Pegtel, D.M., Middeldorp, J., and Thorley-Lawson DA. (2005). Epstein-Barr virus infection in *ex vivo* tonsil epithelial cell cultures of asymptomatic carriers. J Virol. *78*, 12613-12624.

Penn, I. (1998). The role of immunosuppression in lymphoma formation. Springer Semin Immunopathol *20*, 343-355.

Perry, M., and Whyte, A. (1998). Immunology of the tonsils. Immunol. Today *19*, 414-421.

Perry, M. E. (1994). The specialised structure of crypt epithelium in the human palatine tonsil and its functional significance. J. Anat. *185 (Pt 1)*, 111-127.

Polack, A., Hortnagel, K., Pajic, A., Christoph, B., Baier, B., Falk, M., Mautner, J., Geltinger, C., Bornkamm, G. W., and Kempkes, B. (1996). c-myc activation renders proliferation of Epstein-Barr virus (EBV)- transformed cells independent of EBV nuclear antigen 2 and latent membrane protein 1. Proc. Natl. Acad. Sci. USA *93*, 10411-10416.

Pope, J. H., Horne, M. K., and Scott, W. (1968). Transformation of foetal human keukocytes *in vitro* by filtrates of a human leukaemic cell line containing herpes-like virus. Int. J. Cancer *3*, 857-866.

Purtilo, D. T., Cassel, C. K., Yang, J. P., and Harper, R. (1975). X-linked recessive progressive combined variable immunodeficiency (Duncan's disease). Lancet *1*, 935-940.

Qu, L., and Rowe, D. T. (1992). Epstein-Barr virus latent gene expression in uncultured peripheral blood lymphocytes. J. Virol. *66*, 3715-3724.

Raab-Traub, N. (2002). Epstein-Barr virus in the pathogenesis of NPC. Semin. Cancer Biol. *12*, 431-441.

Raab-Traub, N., and Flynn, K. (1986). The structure of the termini of the Epstein-Barr virus as a marker of clonal cellular proliferation. Cell *47*, 883-889.

Raab-Traub, N., Rajadurai, P., Flynn, K., and Lanier, A. P. (1991). Epstein-Barr virus infection in carcinoma of the salivary gland. J. Virol. *65*, 7032-7036.

Radkov, S. A., Bain, M., Farrell, P. J., West, M., Rowe, M., and Allday, M. J. (1997). Epstein-Barr virus EBNA3C represses Cp, the major promoter for EBNA expression, but has no effect on the promoter of the cell gene CD21. J. Virol. *71*, 8552-8562.

Radkov, S. A., Touitou, R., Brehm, A., Rowe, M., West, M., Kouzarides, T., and Allday, M. J. (1999). Epstein-Barr virus nuclear antigen 3C interacts with histone deacetylase to repress transcription. J. Virol. *73*, 5688-5697.

Ramesh, N., Seki, M., Notarangelo, L. D., and Geha, R. S. (1998). The hyper-IgM (HIM) syndrome. Springer Semin. Immunopathol. *19*, 383-399.

Rickinson, A. B. (1984). Epstein-Barr virus in epithelium. Nature *310*, 99-100.

Rickinson, A. B., Finerty, S., and Epstein, M. A. (1977). Mechanism of the establishment of Epstein-Barr virus genome-containing lymphoid cell lines from infectious mononucleosis patients: studies with phosphonoacetate. Int. J. Cancer *20*, 861-868.

Rickinson, A. B., and Kieff, E. (1996). Epstein-Barr virus. In Virology, B. N. Fields, D. M. Knipe, P. M. Howley, and e. al, eds. (New York, Raven Press), pp. 2397-2446.

Rickinson, A. B., and Kieff, E. (2001). Epstein-Barr Virus. In Virology, D. M. Knipe, and P. M. Howley, eds. (New York, Lippincott Williams and Wilkins), pp. 2575-2628.

Rooney, C., Howe, J. G., Speck, S. H., and Miller, G. (1989). Influence of Burkitt's lymphoma and primary B cells on latent gene expression by the nonimmortalizing P3J-HR-1 strain of Epstein-Barr virus. J. Virol. *63*, 1531-1539.

Rooney, C. M., Smith, C. A., Ng, C. Y., Loftin, S., Li, C., Krance, R. A., Brenner, M. K., and Heslop, H. E. (1995). Use of gene-modified virus-specific T lymphocytes to control Epstein-Barr-virus-related lymphoproliferation. Lancet *345*, 9-13.

Sayos, J., Wu, C., Morra, M., Wang, N., Zhang, X., Allen, D., van Schaik, S., Notarangelo, L., Geha, R., Roncarolo, M. G., *et al.* (1998). The X-linked lymphoproliferative-disease gene product SAP regulates signals induced through the co-receptor SLAM. Nature *395*, 462-469.

Schaefer, B. C., Strominger, J. L., and Speck, S. H. (1995). Redefining the Epstein-Barr virus-encoded nuclear antigen EBNA-1 gene promoter and transcription initiation site in group I Burkitt lymphoma cell lines. Proc. Natl. Acad. Sci. USA *92*, 10565-10569.

Seemayer, T. A., Gross, T. G., Egeler, R. M., Pirruccello, S. J., Davis, J. R., Kelly, C. M., Okano, M., Lanyi, A., and Sumegi, J. (1995). X-linked lymphoproliferative disease: twenty-five years after the discovery. Pediatr. Res. *38*, 471-478.

Selin, L. K., Varga, S. M., Wong, I. C., and Welsh, R. M. (1998). Protective heterologous antiviral immunity and enhanced immunopathogenesis mediated by memory T cell populations. J. Exp. Med. *188*, 1705-1715.

Sharp, T. V., Raine, D. A., Gewert, D. R., Joshi, B., Jagus, R., and Clemens, M. J. (1999). Activation of the interferon-inducible (2'-5') oligoadenylate synthetase by the Epstein-Barr virus RNA, EBER-1. Virology *257*, 303-313.

Shi, Y., Sawada, J., Sui, G., Affar el, B., Whetstine, J. R., Lan, F., Ogawa, H., Luke, M. P., and Nakatani, Y. (2003). Coordinated histone modifications mediated by a CtBP co-repressor complex. Nature *422*, 735-738.

Shibata, D., Tokunaga, M., Uemura, Y., Sato, E., Tanaka, S., and Weiss, L. M. (1991). Association of Epstein-Barr virus with undifferentiated gastric carcinomas with intense lymphoid infiltration. Lymphoepithelioma-like carcinoma. Am. J. Pathol. *139*, 469-474.

Shibata, D., and Weiss, L. M. (1992). Epstein-Barr virus-associated gastric adenocarcinoma. Am. J. Pathol. *140*, 769-774.

Sinclair, A. J., and Farrell, P. J. (1995). Host cell requirements for efficient infection of quiescent primary B lymphocytes by Epstein-Barr virus. J. Virol. *69*, 5461-5468.

Sinclair, A. J., Palmero, I., Peters, G., and Farrell, P. J. (1994). EBNA-2 and EBNA-LP cooperate to cause G0 to G1 transition during immortalization of resting human B lymphocytes by Epstein-Barr virus. EMBO J. *13*, 3321-3328.

Sixbey, J. W., Nedrud, J. G., Raab, T. N., Hanes, R. A., and Pagano, J. S. (1984). Epstein-Barr virus replication in oropharyngeal epithelial cells. N. Engl. J. Med. *310*, 1225-1230.

Sixbey, J. W., Vesterinen, E. H., Nedrud, J. G., Raab, T. N., Walton, L. A., and Pagano, J. S. (1983). Replication of Epstein-Barr virus in human epithelial cells infected *in vitro*. Nature *306*, 480-483.

Smith, P. R., Gao, Y., Karran, L., Jones, M. D., Snudden, D., and Griffin, B. E. (1993). Complex nature of the major viral polyadenylated transcripts in Epstein-Barr virus associated tumors. J. Virol. *67*, 3217-3225.

Spencer, J., Finn, T., Pulford, K. A., Mason, D. Y., and Isaacson, P. G. (1985). The human gut contains a novel population of B lymphocytes which resemble marginal zone cells. Clin. Exp. Immunol. *62*, 607-612.

Spencer, J., Perry, M. E., and Dunn-Walters, D. K. (1998). Human marginal-zone B cells. Immunol. Today *19*, 421-426.

Starzl, T. E., Nalesnik, M. A., Porter, K. A., Ho, M., Iwatsuki, S., Griffith, B. P., Rosenthal, J. T., Hakala, T. R., Shaw Jr, B. W., Hardesty, R. L., and et, a. l. (1984). Reversibility of lymphomas and lymphoproliferative lesions developing under cyclosporin-steroid therapy. Lancet *1*, 583-587.

Staudt, L. M. (2000). The molecular and cellular origins of Hodgkin's disease. J. Exp. Med. *191*, 207-212.

Steven, N. M., Annels, N. E., Kumar, A., Leese, A. M., Kurilla, M. G., and Rickinson, A. B. (1997). Immediate early and early lytic cycle proteins are frequent targets of the Epstein-Barr virus-induced cytotoxic T cell response. J. Exp. Med. *185*, 1605-1617.

Steven, N. M., Leese, A. M., Annels, N. E., Lee, S. P., and Rickinson, A. B. (1996). Epitope focusing in the primary cytotoxic T cell response to Epstein-Barr virus and its relationship to T cell memory. J. Exp. Med. *184*, 1801-1813.

Strahm, B., Rittweiler, K., Duffner, U., Brandau, O., Orlowska-Volk, M., Karajannis, M. A., Stadt, U., Tiemann, M., Reiter, A., Brandis, M., *et al.* (2000). Recurrent B-cell non-Hodgkin's lymphoma in two brothers with X-linked lymphoproliferative disease without evidence for Epstein-Barr virus infection. Br. J. Haematol. *108*, 377-382.

Strang, G., and Rickinson, A. B. (1987). Multiple HLA class I-dependent cytotoxicities constitute the "non-HLA-restricted" response in infectious mononucleosis. Eur. J. Immunol. *17*, 1007-1013.

Sugawara, Y., Mizugaki, Y., Uchida, T., Torii, T., Imai, S., Makuuchi, M., and Takada, K. (1999). Detection of Epstein-Barr virus (EBV) in hepatocellular carcinoma tissue: a novel EBV latency characterized by the absence of EBV-encoded small RNA expression. Virology *256*, 196-202.

Takada, K., and Ono, Y. (1989). Synchronous and sequential activation of latently infected Epstein-Barr virus genomes. J. Virol. *63*, 445-449.

Tang, X., Hori, S., Osamura, R. Y., and Tsutsumi, Y. (1995). Reticular crypt epithelium and intra-epithelial lymphoid cells in the hyperplastic human palatine tonsil: an immunohistochemical analysis. Pathol. Int. *45*, 34-44.

Tao, Q., Ho, F. C., Loke, S. L., and Srivastava, G. (1995). Epstein-Barr virus is localized in the tumour cells of nasal lymphomas of NK, T or B cell type. Int. J. Cancer *60*, 315-320.

Thomas, J. A., Hotchin, N. A., Allday, M. J., Amlot, P., Rose, M., Yacoub, M., and Crawford, D. H. (1990). Immunohistology of Epstein-Barr virus-associated antigens in B cell disorders from immunocompromised individuals. Transplantation *49*, 944-953.

Thorley-Lawson, D. A. (2001a). Epstein-Barr virus. In Sampter's Immunologic Diseases, K. F. Austen, M. M. Frank, J. P. Atkinson, and H. Cantor, eds. (New York, Williams and Wilkins), pp. 970-985.

Thorley-Lawson, D. A. (2001b). Epstein-Barr virus: exploiting the immune system. Nat. Rev. Immunol. *1*, 75-82.

Thorley-Lawson, D. A., and Babcock, G. J. (1999). A model for persistent infection with Epstein-Barr virus: the stealth virus of human B cells. Life Sci. *65*, 1433-1453.

Thorley-Lawson, D. A., and Gross, A. (2004). Persistence of the Epstein-Barr virus and the origins of associated lymphomas. N. Engl. J. Med. *350*, 1328-1337.

Thorley-Lawson, D. A., and Mann, K. P. (1985). Early events in Epstein-Barr virus infection provide a model for B cell activation. J. Exp. Med. *162*, 45-59.

Thorley-Lawson, D. A., Nadler, L. M., Bhan, A. K., and Schooley, R. T. (1985). BLAST-2 [EBVCS], an early cell surface marker of human B cell activation, is superinduced by Epstein Barr virus. J. Immunol. *134*, 3007-3012.

Thorley-Lawson, D. A., and Poodry, C. A. (1982). Identification and isolation of the main component (gp350-gp220) of Epstein-Barr virus responsible for generating neutralizing antibodies *in vivo*. J. Virol. *43*, 730-736.

Thorley-Lawson, D. A., Schooley, R. T., Bhan, A. K., and Nadler, L. M. (1982). Epstein-Barr virus superinduces a new human B cell differentiation antigen (B-LAST 1) expressed on transformed lymphoblasts. Cell *30*, 415-425.

Tierney, R. J., Steven, N., Young, L. S., and Rickinson, A. B. (1994). Epstein-Barr virus latency in blood mononuclear cells: analysis of viral gene transcription during primary infection and in the carrier state. J. Virol. *68*, 7374-7385.

Timmons, C. F., Dawson, D. B., Richards, C. S., Andrews, W. S., and Katz, J. A. (1995). Epstein-Barr virus-associated leiomyosarcomas in liver transplantation recipients. Origin from either donor or recipient tissue. Cancer *76*, 1481-1489.

Timms, J. M., Bell, A., Flavell, J. R., Murray, P. G., Rickinson, A. B., Traverse-Glehen, A., Berger, F., and Delecluse, H. J. (2003). Target cells of Epstein-Barr-virus (EBV)-positive post-transplant lymphoproliferative disease: similarities to EBV-positive Hodgkin's lymphoma. Lancet *361*, 217-223.

Tortorella, D., Gewurz, B. E., Furman, M. H., Schust, D. J., and Ploegh, H. L. (2000). Viral subversion of the immune system. Annu. Rev. Immunol. *18*, 861-926.

Touitou, R., Hickabottom, M., Parker, G., Crook, T., and Allday, M. J. (2001). Physical and functional interactions between the corepressor CtBP and the Epstein-Barr virus nuclear antigen EBNA3C. J. Virol. *75*, 7749-7755.

Tsai, C. N., Liu, S. T., and Chang, Y. S. (1995). Identification of a novel promoter located within the Bam HI Q region of the Epstein-Barr virus genome for the EBNA 1 gene. DNA & Cell Biol. *14*, 767-776.

Tugizov, S. M., Berline, J. W., and Palefsky, J. M. (2003). Epstein-Barr virus infection of polarized tongue and nasopharyngeal epithelial cells. Nat. Med. *9*, 307-314.

Uchida, J., Yasui, T., Takaoka-Shichijo, Y., Muraoka, M., Kulwichit, W., Raab-Traub, N., and Kikutani, H. (1999). Mimicry of CD40 signals by Epstein-Barr virus LMP1 in B lymphocyte responses. Science *286*, 300-303.

Uckun, F. M. (1990). Regulation of human B-cell ontogeny. Blood *76*, 1908-1923.

van Gelder, T., Vuzevski, V. D., and Weimar, W. (1995). Epstein-Barr virus in smooth-muscle tumors. N. Engl. J. Med. *332*, 1719.

Wade, M., and Allday, M. J. (2000). Epstein-Barr virus suppresses a G (2)/M checkpoint activated by genotoxins. Mol. Cell. Biol. *20*, 1344-1360.

Wang, D., Liebowitz, D., and Kieff, E. (1985). An EBV membrane protein expressed in immortalized lymphocytes transforms established rodent cells. Cell *43*, 831-840.

Wang, F., Gregory, C., Sample, C., Rowe, M., Liebowitz, D., Murray, R., Rickinson, A., and Kieff, E. (1990a). Epstein-Barr virus latent membrane protein (LMP1) and nuclear proteins 2 and 3C are effectors of phenotypic changes in B lymphocytes: EBNA-2 and LMP1 cooperatively induce CD23. J. Virol. *64*, 2309-2318.

Wang, F., Tsang, S. F., Kurilla, M. G., Cohen, J. I., and Kieff, E. (1990b). Epstein-Barr virus nuclear antigen 2 transactivates latent membrane protein LMP1. J. Virol. *64*, 3407-3416.

Weiss, L. M., Strickler, J. G., Warnke, R. A., Purtilo, D. T., and Sklar, J. (1987). Epstein-Barr viral DNA in tissues of Hodgkin's disease. Am. J. Pathol. *129*, 86-91.

Weller, S., Faili, A., Aoufouchi, S., Gueranger, Q., Braun, M., Reynaud, C. A., and Weill, J. C. (2003). Hypermutation in Human B Cells *in vivo* and *in vitro*. Ann N Y Acad Sci *987*, 158-165.

Weller, S., Faili, A., Garcia, C., Braun, M. C., Le Deist, F. F., de Saint Basile, G. G., Hermine, O., Fischer, A., Reynaud, C. A., and Weill, J. C. (2001). CD40-CD40L independent Ig gene hypermutation suggests a second B cell diversification pathway in humans. Proc. Natl. Acad. Sci. USA *98*, 1166-1170.

Wilson, J. B., Bell, J. L., and Levine, A. J. (1996). Expression of Epstein-Barr virus nuclear antigen-1 induces B cell neoplasia in transgenic mice. Embo J. *15*, 3117-3126.

Woisetschlaeger, M., Yandava, C. N., Furmanski, L. A., Strominger, J. L., and Speck, S. H. (1990). Promoter switching in Epstein-Barr virus during the initial stages of infection of B lymphocytes. Proc. Natl. Acad. Sci. USA *87*, 1725-1729.

Wood, T. A., and Frenkel, E. P. (1967). The atypical lymphocyte. Am J Med *42*, 923-936.

Yao, Q. Y., Rickinson, A. B., and Epstein, M. A. (1985). A re-examination of the Epstein-Barr virus carrier state in healthy seropositive individuals. Int. J. Cancer *35*, 35-42.

Yates, J. L., Warren, N., and Sugden, B. (1985). Stable replication of plasmids derived from Epstein-Barr virus in various mammalian cells. Nature *313*, 812-815.

Ye, B. H., Cattoretti, G., Shen, Q., Zhang, J., Hawe, N., de Waard, R., Leung, C., Nouri-Shirazi, M., Orazi, A., Chaganti, R. S., *et al.* (1997). The BCL-6 proto-oncogene controls germinal-centre formation and Th2-type inflammation. Nat. Genet. *16*, 161-170.

Youinou, P., Jamin, C., and Lydyard, P. M. (1999). CD5 expression in human B-cell populations. Immunol. Today *20*, 312-316.

Young, L. S., Dawson, C. W., Clark, D., Rupani, H., Busson, P., Tursz, T., Johnson, A., and Rickinson, A. B. (1988). Epstein-Barr virus gene expression in nasopharyngeal carcinoma. J. Gen. Virol. *69*, 1051-1065.

Young, L. S., and Sixbey, J. W. (1988). Epstein-Barr virus and epithelial cells: a possible role for the virus in the development of cervical carcinoma. Cancer Surv. *7*, 507-518.

Zimber-Strobl, U., Kempkes, B., Marschall, G., Zeidler, R., Van Kooten, C., Banchereau, J., Bornkamm, G. W., and Hammerschmidt, W. (1996). Epstein-Barr virus latent membrane protein (LMP1) is not sufficient to maintain proliferation of B cells but both it and activated CD40 can prolong their survival. Embo J. *15*, 7070-7078.

Zimber-Strobl, U., and Strobl, L. J. (2001). EBNA2 and Notch signalling in Epstein-Barr virus mediated immortalization of B lymphocytes. Semin. Cancer Biol. *11*, 423-434.

Zimber-Strobl, U., Strobl, L. J., Meitinger, C., Hinrichs, R., Sakai, T., Furukawa, T., Honjo, T., and Bornkamm, G. W. (1994). Epstein-Barr virus nuclear antigen 2 exerts its transactivating function through interaction with recombination signal binding protein RBP-J kappa, the homologue of Drosophila Suppressor of Hairless. EMBO J. *13*, 4973-4982.

From: Epstein-Barr Virus. Edited by: Erle S. Robertson

Chapter 18

EBV Entry and Epithelial Infection

*Lindsey Hutt-Fletcher**

ABSTRACT

Epstein-Barr virus (EBV) is a highly, but not exclusively lymphotropic virus. It has variously been reported to infect macrophages, smooth muscle cells, T cells, and natural killer cells. However, the second major target of the virus is generally accepted to be the epithelial cell. Epithelial cell malignancies comprise a large part of the burden of EBV-associated disease worldwide and epithelial cells may play an important role in the spread of virus both between and within hosts. Our understanding of how virus initiates infection of either a B cell or an epithelial cell is incomplete, but several of the key virus and cell proteins involved in B cell infection have been identified. Less is clear about the important players in epithelial cell infection, where a very different picture is gradually emerging. This chapter summarizes what we currently know about early events in epithelial infection and contrasts it with the models that have been derived from the more extensive and comprehensive work done on B cells. It also explores how replication of virus in each cell type may influence virus trafficking, transmission and persistence.

INTRODUCTION

Defining the role that epithelial cells play in establishment and persistence of infection with Epstein-Barr virus (EBV) has been among the most contentious issues in the field. Opinion has veered from one extreme to the other for much of the time that the virus has been studied. There was never any real doubt that EBV could infect an epithelial cell. The association of the virus with nasopharyngeal carcinoma (NPC) was revealed only a few years after the virus was discovered (Henle and Henle, 1979; Old et al., 1968) and the presence of EBV DNA in the epithelial cells of the tumor was confirmed not much later (Wolf et al., 1973; zur Hausen et al., 1970). The issue that has been in play for so long is whether or not infection of epithelial cells is a common event that plays a role in infection of normal healthy individuals, or whether it occurs only in the context of malignancy or immunosuppression.

*For correspondence email *lhuttf@lsuhsc.edu*

THE ROLE OF EPITHELIAL CELLS IN COLONIZATION OF THE HOST

Replication in epithelial cells during primary infection was originally explored because of the difficulties experienced in obtaining virus from B cell lines *in vitro*. Since cell free virus was clearly found in the saliva of patients with infectious mononucleosis the hypothesis that there was another cell type responsible for support of lytic replication was an attractive one. Many herpesviruses replicate in epithelial cells and the model of Marek's disease virus, which transforms lymphocytes and replicates productively in epithelial cells of the feather follicles of the birds that it infects, provided a plausible analogy. *In situ* cytohybridization analyses of cells collected from throat washings then revealed levels of EBV DNA in desquamating epithelial cells that were consistent with productive replication (Lemon et al., 1977; Sixbey et al., 1983). Evidence for infection of exfoliating cervical epithelium was also obtained (Sixbey et al., 1986) and expression of lytic cycle antigens in epithelium of salivary gland biopsies was reported (Venables et al., 1989). It remained difficult to demonstrate efficient infection of epithelial cells *in vitro* (Sixbey et al., 1984), but this was thought to represent a failure to establish an appropriate culture system.

By this time it had been recognized that oral hairy leukoplakia, (OHL) seen in end-stage AIDS patients, is a lesion that is driven by lytic replication of EBV in epithelial cells of the oral mucosa (Friedman-Kien, 1986; Greenspan et al., 1985). Replication occurs in the more differentiated upper layers of stratified epithelium (Gilligan et al., 1990; Young et al., 1991) and the hypothesis that emerged was that EBV persists in a latent form in basal cells, as in NPC, and that cells become permissive for replication as they differentiate. Inefficient replication in cells in tissue culture was explained in terms of an inability to reproduce the appropriate state of differentiation, much in the same way that completion of the papilloma virus replication cycle is only possible in differentiating cells. Latent infection of a basal cell, established during primary infection, was seen as providing a self-renewing population that serves as the reservoir of virus in a persistently infected individual. In this model, virus made in an epithelial cell is responsible for maintaining the small number of infected B cells that can always be detected in the peripheral circulation. EBV infected B cells are clearly susceptible to immune destruction *in vitro* and without continuous reseeding by replicating virus were expected to be cleared rapidly from the persistently infected host (Allday and Crawford, 1988; Rickinson et al., 1985; Sixbey et al., 1987). B cell infection was seen briefly as a secondary consequence of expression of a shared receptor on B cells and epithelial cells.

The central role that was assigned to epithelial cells in this model was, however, soon challenged. EBV-seropositive individuals treated with acyclovir for up to four weeks showed no change in frequencies of EBV-positive B cells in peripheral blood (Yao et al., 1989). This implied that continuous reactivation of virus from latently infected epithelial cells is not required, at least in the short term, for maintaining a population of infected B cells. It also redirected attention to the possibility, now thought to be correct (Thorley-Lawson and Gross, 2004), that infected B cells can survive in circulation because virus gene expression is more restricted *in vivo* than it is in transformed B cells *in vitro* (Crawford et al., 1978). Two patients who received radiation therapy that ablated their lymphoid systems, but left their epithelial cells relatively intact, lost all evidence of infection with EBV (Gratama et al., 1988) further suggesting that epithelial cells are not the primary reservoir of infection *in vivo*. The model reverted to one in which virus replication in epithelial cells occurs during primary infection, facilitating access to B cells, but in which B cells are the primary source of persistent virus (Rickinson, 1990). A more extreme view held that not only do epithelial cells play no role as a reservoir of latent infection, but also that infection of an epithelial cell never occurs at all in a normal individual (Thorley-Lawson et al., 1996). This was based on the difficulty that some had in finding infected epithelial cells in normal individuals (Anagnastopoulos et al., 1995; Karjannis et al., 1997; Niedobitek et al., 1989; Tao et al., 1995) and, later, on a study of six patients with agammaglobulinemia, lacking mature B cells, in whom no evidence for infection with EBV could be found (Faulkner et al., 1999). More recently, however, as described below, a possible explanation for the difficulty in finding virus in epithelial cells *in vivo* has been proposed (Borza and Hutt-Fletcher, 2002), efficient replication in epithelial cells has been achieved *in vitro* (Tugizov et al., 2003) and also new evidence of infection of normal epithelium *in vivo* has been obtained (Pegtel et al., 2004). The model that in healthy individuals epithelial cells are transient players that serve to facilitate access to B cells and amplify or maintain the B cell reservoir would seem to be back in play.

INITIATION OF INFECTION
EBV is typically transmitted in saliva which contains both cell-free virus and infected cells, either of which might potentially be the source of virus for establishing infection of a new host. Early models proposed that infection began as epithelial cells were infected when virus producing B cells fused with their cell membranes (Bayliss and Wolf, 1980), but no subsequent studies have been done either to confirm or refute this hypothesis. Infection of epithelial cells in close contact with virus-producing B cells is more efficient than infection by cell free virus (Imai et al., 1998; Tugizov et al., 2003). However, investigation of virus and cell proteins that might be involved in the initiation of infection has been more easily and directly addressed by examining the behavior of cell free virus.

Entry of cell free virus into B cell and epithelial cells has long been known to occur by different routes. Virus enters B cells by endocytosis (Nemerow and Cooper, 1984) where it fuses with the endosomal membrane at low pH (Miller and Hutt-Fletcher, 1992). Endocytosis is required, but low pH is not. In contrast, virus does not require endocytosis for entry into an epithelial cell and fusion takes place at neutral pH, probably at the cell surface. Beyond this, however, the virus and cell proteins that mediate attachment to and penetration of the cell membrane are either different, or behave in different ways during infection of B cells and epithelial cells. These differences, while incompletely understood, have profound implications for the ability of the virus to traffic between them and maintain or amplify the virus load.

Attachment
B cells
Attachment to B cells is mediated by a high affinity (Moore et al., 1989) protein-protein interaction between the virus envelope glycoprotein gp350/220 (Table 1 and Figure 1a) and the complement receptor type 2, CR2 or CD21 (Fingeroth et al.,

Table 1. EBV membrane proteins currently implicated in virus entry[1]

Open reading frame	Protein[2]	Type	Role
BLLF1a/BLLF1b	gp350/220	single pass type1	attachment to CR2
BXLF2	gH/gp85	single pass type 1	fusion; attachment to epithelial cell receptor/coreceptor; oligomerizes with gL and gp42
BKRF2	gL/gp25	single pass type 2	fusion; attachment to epithelial cell receptor/coreceptor; oligomerizes with gH and gp42, chaperone for gH
BZLF2	gp42	single pass type 2	interaction with B cell coreceptor HLA class II
BALF4	gB/gp110	single pass (?) type 1	fusion
BLRF1	gN	single pass type 1	co-dependent for expression; possibly involved in post-fusion events
BBRF3	gM	multi-spanning	
BMRF2	BMRF2	multi-spanning	binds integrins; important to infection of polarized epithelial cells
BDLF3	gp150	single pass type 1	non-essential for infection *in vitro;* virus lacking gp150 is about two fold more infectious for epithelial cells

[1]References are cited in the text.
[2]EBV glycoproteins are named for their apparent mass, by a letter designation if they have homologs in the alpha and betaherpesviruses, or by their open reading frame if they have not been extensively characterized.

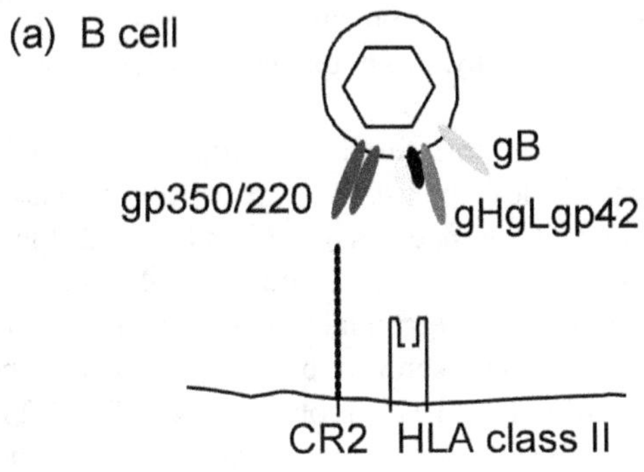

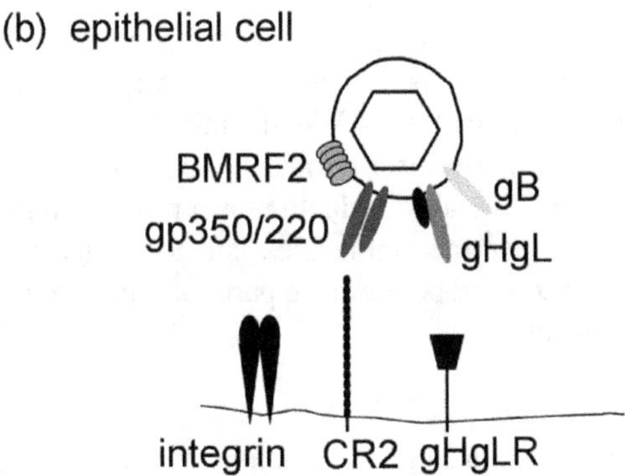

Figure 1. Virus envelope glycoproteins and cell surface proteins that have been implicated as playing major roles in entry of virus into B lymphocytes (a) or epithelial cells (b). On a B cell gp350/220 interacts with CR2, gp42, one of the three proteins in the gHgLgp42 complex required for virus fusion, interacts with HLA class II. On some epithelial cells gp350/220 interacts with CR2, gHgL, as a two part complex required for fusion, interacts with a novel gHgL receptor (gHgLR) and BMRF2 interacts with intergrins. Glycoprotein gB is necessary for fusion with both B cells and epithelial cells. The symbols used in this figure are used throughout.

1984; Frade et al., 1985; Nemerow et al., 1987; 1985; Tanner et al., 1987; 1988). gp350/220 is an abundant, highly glycosylated 907 residue protein that has a dual nomenclature because its gene is expressed in two alternatively spliced forms with masses of approximately 350 and 220 kDa (Beisel et al., 1985; Hummel et al., 1984). The splice maintains the reading frame of the protein and the CR2 binding domain at the amino terminus of the molecule. The biologic significance, if any, of the expression of two forms of the protein is unknown, but the initial interaction with CR2 apparently separates the virus envelope and the cell membrane by approximately 50 nm (Nemerow and Cooper, 1984), a considerable distance. One possibility is that exchange of the larger for the smaller form might bring the virus a little closer. The region of gp350 that is lost as a result of the splice includes three repeats of a 21 amino acid motif with amphipathic characteristics and it has also been suggested that this may be membrane interactive (Tanner et al., 1988). However, what advantage, if any, might accrue by deletion of such a structure is unclear.

CR2 is a member of a large family of proteins involved in tissue repair, inflammation and the immune response. Its entire extracellular domain is composed of at least 15 discrete structural modules, 60-75 amino acids long, known as short consensus repeats (SCRs) (Ahearn and Fearon, 1989; Moore et al., 1987; Weis et al., 1988)that may provide some segmental flexibility to the molecule (Moore et al., 1989). The EBV binding site has been very precisely mapped to the amino terminal SCR-1 and SCR-2 (Martin et al., 1991; Prota et al., 2002). CR2 is part of a signal transduction complex on B cells (Fearon and Carter, 1995) and the interaction with gp350/220 does more than simply tether virus to the cell surface. Cross-linking of CR2 triggers endocytosis of the virus (Tanner et al., 1987) and activates NFκB (Sinclair and Farrell, 1995; Sugano et al., 1997). Downstream effects include induction of IL-6 via a protein kinase C pathway (D'Addario et al., 2001; Tanner et al., 1996), transcriptional activation of Wp, the initial viral latent gene promoter (Sugano et al., 1997), and posttranscriptional control of viral gene expression (Sinclair and Farrell, 1995). Signal transduction as a result of interactions of viral envelope proteins with cell surface molecules is a topic that is ripe for further exploration and may, as discussed below, have particular importance for successful infection of an epithelial cell.

Epithelial cells

The ligands and receptors responsible for attachment of virus to an epithelial cell are much less clear. One possibility is that CR2 plays a role in epithelial as well as B cell infection (Figure 1b). It is clear that many epithelial tumor cell lines express at least low levels of CR2 (Fingeroth et al., 1999; Imai et al., 1998) and stable expression of CR2 in an epithelial cell from a transfected cDNA clone facilitates a high level of infection (Borza and Hutt-Fletcher, 2002; Li et al., 1992). However, the relevance of these observations to infection *in vivo* remains uncertain. Identification of CR2-expressing epithelial cells *in vivo* has been confounded by the fact that the monoclonal antibodies used in initial studies cross-reacted with an unrelated epithelial cell protein; these studies have not since been repeated (Young et al., 1986; 1989). Should CR2 be used for infection of at least a subpopulation of epithelial cells it is of interest to note that in this context, as well as in B cells, it may play a role as a signal transducer, although it may play no role in endocytosis (Borza et al., 2004). EBV-mediated cross-linking of CR2 on epithelial cells, which lack the other components of the CR2 signaling complex found on B cells, stimulates relocalization and clustering of CR2 and the formin homolog overexpressed in spleen (FHOS/FHOD1). Formins are molecular scaffolds that nucleate actin, linking signal transduction to actin reorganization and gene transcription (Gill et al., 2004).

Epithelial cells that carry the polymorphic IgA receptor can be infected *in vitro* with virus that is coated with IgA specific for gp350/220 (Sixbey and Yao, 1992). This may be particularly relevant to infection of cells at the basolateral surface in an immune host, although the efficiency of infection is not very high. However, in polarized cells virus is transported intact from the basolateral to the apical surface

(Gan et al., 1997), so the process, if it occurs *in vivo*, may be more important to trans epithelial transport than direct infection of an epithelial cell, unless loss of intrinsic polarization occurs as the result of trauma, or perhaps malignant conversion. Epithelial lines also express an as yet unidentified molecule that facilitates virus binding via a complex of two EBV proteins, gH and gL (Molesworth et al., 2000; Oda et al., 2000). Virus lacking gHgL loses the ability to bind to these lines and soluble forms of gHgL can be shown to attach specifically (Borza et al., 2004). However, again infection rates are low and, as will be discussed below, it is possible that, in the absence of a primary attachment receptor, this interaction represents inefficient use of a gHgL receptor (gHgLR) whose major role is as a coreceptor mediating post attachment events.

Finally, one of the most intriguing recent findings is that the multi-span EBV envelope glycoprotein, BMRF2, interacts with the $\alpha5\beta1$ integrin, possibly via an RGD motif found in one of its extracellular loops (Tugizov et al., 2003). Antibodies to integrins and to a BMRF2 fusion protein partially blocked binding to polarized epithelial cells and had a more significant impact on infection via the basolateral surface of the polarized monolayer. Infection was efficient, at least at the fairly high multiplicities of infection that were used, and both lytic cycle and latent cycle proteins were expressed. These observations are particularly interesting since polarized epithelial cells are probably closer to the environment encountered by virus *in vivo*. Further work will be necessary to determine whether the interaction with integrins is really most relevant to attachment, to penetration or perhaps, again, to triggering of signal transduction pathways essential to post entry events.

Penetration

Penetration of any enveloped virus into a cell involves fusion of the virion envelope with the membrane of the cell either at the cell surface or after endocytosis. The best understood paradigm for virus cell fusion is provided by the RNA viruses such as the human immunodeficiency virus, which fuses at the cell surface, and influenza virus, which fuses with the endocytic vesicle. The virus glycoproteins that mediate fusion are made as single type 1 membrane proteins, but are cleaved during processing to create two species which reassociate in a metastable state (Colman and Lawrence, 2003). The fragment that retains the transmembrane domain includes a hydrophobic sequence or "fusion peptide" that can be triggered by conformational changes to insert into an opposing cell membrane and initiate formation of a fusion pore. The conformational change in the human immunodeficiency virus is triggered by interaction with coreceptors; the conformational change in the influenza virus fusion protein is triggered by exposure to the low pH of the endosome. However, no clear cut paradigm has yet been identified for EBV or any other herpesvirus. Fusion appears to require cooperation between several unique protein species, none of which includes a readily identifiable "fusion peptide". As might be expected for a virus that does not use exposure to low pH to trigger the fusion event, it also requires the presence of a coreceptor.

All known herpesviruses express homologs of three essential glycoproteins, gB, gH and gL that are involved in virus-cell fusion. EBV gB is a 857 residue protein that shares some structural, although little sequence homology with its counterparts in other herpes viruses (Gong et al., 1987). Like some, but not all, gB homologs it undergoes cleavage to produce two polypeptides of approximately 56 and 80 kDa that are linked by disulfide bonds (C. Lake and L. Hutt-Fletcher, unpublished). Alpha and betaherpesvirus gB homologs are relatively abundant in virions and the roles so far ascribed to them all involve interactions with the cell surface. They include attachment to heparan sulphate or other cell surface molecules, virus cell fusion and triggering of signaling responses (Akula et al., 2002; Cohen and Straus, 2001; Mocarski and Courcelle, 2001; Roizman and Knipe, 2001). In contrast, the EBV gB homolog, gp110, is present predominantly in the membranes of the nucleus and endoplasmic reticulum rather than the plasma membrane (Emini et al., 1987; Gong and Kieff, 1990; Gong et al., 1987) and carries mostly the high mannose sugars that are consistent with a lack of processing in the Golgi apparatus (Gong and Kieff, 1990; Lee, 1999; Papworth et al., 1997). A string of four arginine residues in the cytoplasmic tail of gB functions as an ER retention motif (Lee, 1999). The importance of a nuclear localization is supported by the phenotype of a recombinant virus lacking gB. Lymphoblastoid cell lines transformed with this virus and induced into the lytic cycle contained none of the capsids and enveloped virions typical of cells transformed with wild type virus, but instead contained in the nucleus what were described as morphologically abnormal structures resembling partially assembled nucleocapsids. No virus was released from cells harboring this recombinant unless it was complemented with wild type gB provided *in trans* (Lee and Longnecker, 1997).

The essential role that gB plays in virus assembly and egress initially precluded an analysis of its potential role in fusion. More recently, however, Haan and colleagues (Haan et al., 2001) provided compelling evidence that EBV gB has retained the fusogenic function of its alpha and betaherpesvirus homologs. These authors applied a quantitative assay to determine the minimal complement of EBV glycoproteins necessary for cell/cell fusion and found that gB was one of the essential players. This suggests that the small amounts of a more completely processed form of gB, gp125, that some reports indicated were present in the virus (Emini et al., 1987; Gong and Kieff, 1990; Gong et al., 1987), are probably important to virus entry. An important role for gB in virus entry is further supported by observations that increasing gB expression in a B95-8 BAC, to levels consistent with those found in virus strains that carry more gB in the virion, increases efficiency of infection of non-B cells (Neuhierl et al., 2002).

B cells

Fusion of EBV with a B cell requires three glycoproteins, gH, gL, and gp42 in addition to gB. Liposomes that contain all virus envelope proteins, except gH, gL and gp42, bind to receptor positive cells but fail to fuse (Haddad and Hutt-Fletcher, 1989) and recombinant virus that lacks all three can bind to but cannot penetrate B cells (Molesworth et al., 2000). The proteins form a non-covalently linked complex in the virion. Glycoprotein gH, the largest of the three is a 708 residue type 1 membrane protein with 5 potential N-linked glycosylation sites and it carries about 10 kDa of N-linked sugar (Baer et al., 1984; Heineman et al., 1988; Oba and Hutt-Fletcher, 1988). As its homologs in all other herpesviruses, gH is dependent on the smaller membrane protein, gL, for folding and transport through the cell. The EBV gL is a 137 residue glycoprotein of approximately 25 kDa (Li et

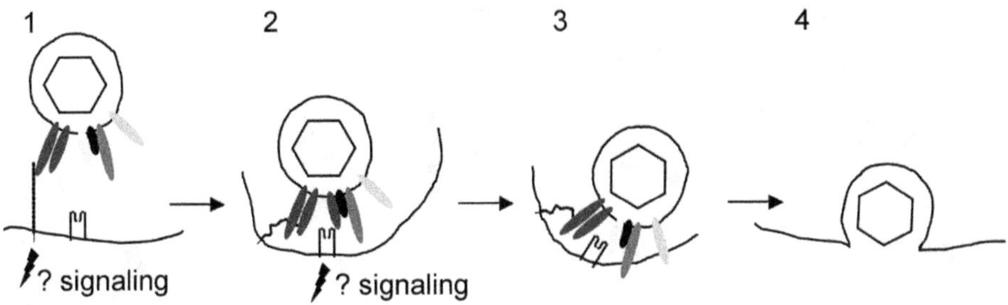

Figure 2. A putative model of the steps involved in entry of virus into a B lymphocyte. 1) Virus binds to CR2 via gp350, possibly initiating signaling events and triggering endocytosis. 2) CR2 may now bind to gp220 rather than gp350 and this, combined with the potential flexibility of CR2 may allow virus to approach closer to the cell membrane where gp42 can interact with HLA class II. 3) Interaction of gp42 with HLA class II triggers the interaction of the core fusion machinery, gHgL and gB, with the endosomal membrane and may also initiate further signaling events. 4) Virus and endosomal membranes fuse allowing entry of the tegumented capsid into the cytoplasm.

al., 1995; Yaswen et al., 1993). Cell surface labeling indicates that it that remains anchored in the envelope by an uncleaved signal sequence (Li et al., 1995) and no soluble form of the protein is detectable in culture medium (S. Turk and L. Hutt-Fletcher, unpublished). Because of the dependence of gH on gL and because EBV in which expression of the gH gene is interrupted loses expression of gL as well (Molesworth et al., 2000), there are only few instances in which the functions of the two proteins can be separated. However, the third protein, gp42, plays no known role in gHgL maturation and its behavior is unique among human herpesviruses. It interacts with the variable domain of the beta chain of HLA class II (Mullen et al., 2002; Spriggs et al., 1996), which functions as an essential coreceptor for B cell infection (Haan et al., 2000; Li et al., 1997). A soluble form of gp42 competes with gp42 in virus for binding to HLA class II and blocks infection. A monoclonal antibody to gp42 (originally incorrectly mapped to gH) that blocks the interaction with HLA class II inhibits virus cell fusion and a monoclonal antibody to HLA class II that blocks gp42 binding neutralizes virus infection (Li et al., 1995; Miller and Hutt-Fletcher, 1988). In further support of a critical role for gp42 in B cell infection, a virus that lacks gp42 fails to infect B cells unless cells and bound virus are fused with polyethylene glycol (Wang and Hutt-Fletcher, 1998) or a small amount of soluble gp42, which lacks a transmembrane domain but which retains a gHgL binding domain, is added *in trans* (Wang et al., 1998). The model proposed for B cell infection is that following attachment of gp350/220 to CR2, gp42 interacts with HLA class II and that this interaction is critical to activation of the core fusion machinery comprised of gHgL and gB (Figure 2). Interestingly, the interaction between gp42 and HLA class II has allelic specificity (Haan and Longnecker, 2000), which may have implications for the efficiency of infection and virus load. Whether or not additional signaling events may be initiated as a result of a gp42 HLA class II interaction is not yet known, but since signaling via HLA class II molecules is well documented under other circumstances, this remains a reasonable possibility (Watt, 1997).

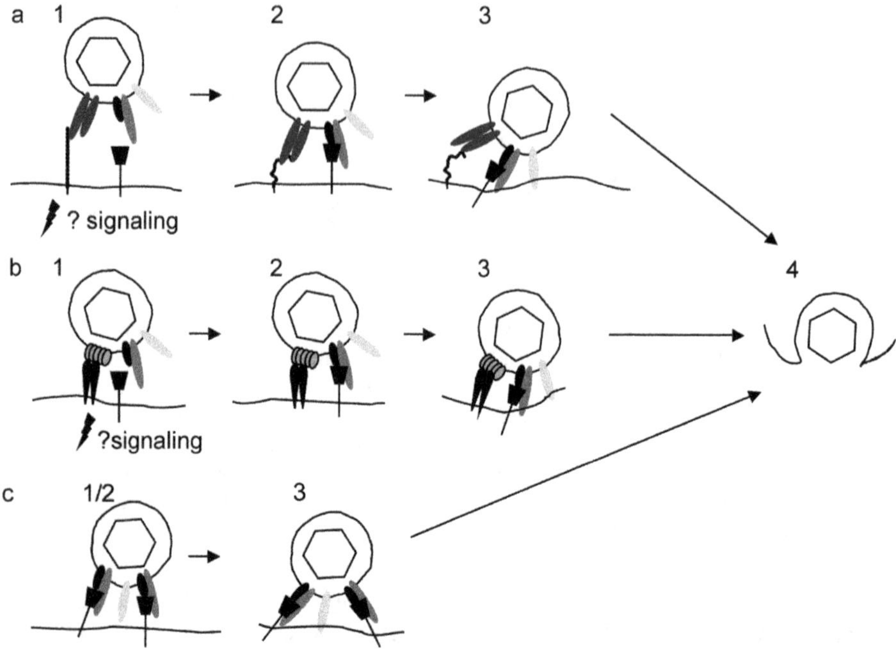

Figure 3. Putative models of the steps involved in entry of virus into epithelial cells after attachment via different cell surface proteins. a. Entry via CR2 and gHgLR: 1) virus binds to CR2 via gp350, signaling may occur but endocytosis is not triggered 2) a switch to interaction with gp220 combined with the potential flexibility of CR2 may allow virus to approach closer to the cell membrane where gHgL can interact with gHgLR. 3) interaction of gHgL with gHgLR triggers the interaction of core fusion machinery, gHgL and gB, with the cell membrane. b. Entry via integrins and gHgLR: 1) BMRF2 interacts with $\alpha5\beta1$ integrins, possibly initiating signaling events. 2) gHgL interacts with gHgLR. 3) interaction of gHgL with gHgLR triggers the interaction of core fusion machinery, gHgL and gB, with the cell membrane. c. Entry via gHgLR alone (which is less efficient): 1) there is no separate attachment event 2) gHgL interacts directly with gHgLR 3)) interaction of gHgL with gHgLR triggers the interaction of core fusion machinery, gHgL and gB, with the cell membrane. Step 4 is the same for all routes, virus and cell membrane fuse allowing entry of the tegumented capsid into the cytoplasm.

Epithelial cells

Fusion with an epithelial cell is intriguingly different. It requires gB, gH and gL (L. Wu and L Hutt-Fletcher, unpublished), but in contrast to the essential role that gp42 plays in B cell infection, the protein is not only completely dispensable for entry into epithelial cells that do not constitutively express HLA class II, but its presence can also be inhibitory (Wang et al., 1998). Stoichiometric analysis of wild type virus revealed the presence of much larger amounts of gHgL than gp42 in the virion, implying that some complexes naturally lack gp42. Conversion of these two part gHgL complexes to three part gHgLgp42 complexes by addition of soluble gp42 *in trans* blocked epithelial infection. In addition, infection of epithelial cells, but not B cells, could be blocked by antibodies that interact with gH or gHgL. These findings were interpreted to mean that a gHgL receptor, gHgLR, functions as a coreceptor on epithelial cells, substituting for HLA class II. gHgL interacts with this receptor only in the absence of gp42. The neutralizing antibodies to gH or gHgL or the presence of gp42 are thought to impede interaction with this novel coreceptor. Entry into an epithelial cell could then involve a variety of attachment receptors that function in conjunction with gHgLR (Figure 3).

As mentioned above, increased levels of gB in virus have been implicated as increasing the ability of attached virus to infect an epithelial cell, a phenomenon that may represent a requirement for higher levels of gB for fusion with an epithelial cell than for fusion with a B cell. In contrast, presence of another mucin-like glycoprotein, gp150, influences epithelial cell infection in a negative way. Glycoprotein gp150 is a 234 amino acid protein, rich in serine and threonine residues, and at least 70% of its apparent mass is accounted for by N-linked and O-linked sugar (Kurilla et al., 1995; Nolan and Morgan, 1995). A recombinant gp150 null virus infects B cells with the same efficiency as wild type virus. However, it reproducibly infects an epithelial cell, either one that expresses both CR2 and gHgLR, or one that expresses only gHgLR, twice as well as wild type virus (Borza and Hutt-Fletcher, 1998). The explanation for this phenotype, which probably reflects a secondary consequence of another undiscovered function of gp150, remains unknown. Loss of gp150 does not appear to alter virus binding, although subtle changes in binding affinity would not have been discovered by the techniques used for its analysis. One possibility is that the charge of a virus particle that lacks the large amount of sugar carried by gp150 is presumably very different and perhaps alters constraints on conformational changes in other proteins that are required for penetration of epithelial cells, but not B cells.

Shuttling between B cells and epithelial cells

The exclusive use of three part gHgLgp42 complexes for B cell infection and two part gHgL complexes for epithelial infection suggested that changes in the stoichiometry of the complexes might influence tropism of EBV for the two cell types. Comparisons of virus made in HLA class II positive B cells and HLA class II negative epithelial cells supported the hypothesis that such changes might occur *in vivo* (Borza and Hutt-Fletcher, 2002). Virus made in HLA class II-negative epithelial cells was as much as two logs more infectious for B lymphocytes than the same amount of virus produced by an HLA class II-positive B cell. Virus originating from either cell type bound equally well to CR2 on the B cell surface, but virus made in the B cell entered less efficiently. This appeared to reflect the fact that in a class II-positive virus-producing cell some complexes containing gp42 interacted with class II during biosynthesis and were targeted to the class II trafficking pathway where they are vulnerable to degradation. The resulting loss of three-part complexes from virus reduced the efficiency of class II-dependent entry. Such a loss did not occur in a class II-negative epithelial cell where virus had a relative increase in gp42 and an increased efficiency for class II-dependent entry. The levels of gp42 in virus also impacted infection of epithelial cells via the class II-independent pathway. B cell virus is on average fivefold better at infecting epithelial cells than epithelial virus. These findings suggested that gp42 may function as a switch of virus tropism that might be relevant to spread of virus between tissues *in vivo* (Borza and Hutt-Fletcher, 2002). The model proposed is that virus shed from

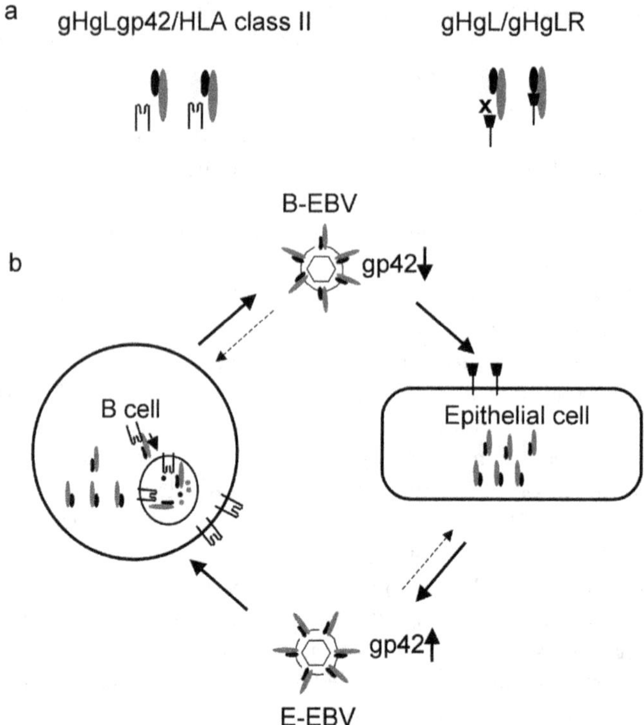

Figure 4. Glycoprotein gp42 as a switch of cell tropism. a) B cell infection requires an interaction between gHgLgp42 and HLA class II whereas epithelial cell infection requires an interaction between gHgL and gHgLR; complexes lacking gp42 cannot bind to HLA class II and complexes containing gp42 cannot bind to gHgLR. b) gHgLgp42 complexes can interact with HLA class II in the endoplasmic reticulum of a B cell. They may then be targeted to endosomal vesicles rich in proteases where incoming proteins are digested into peptides for loading into the peptide binding groove of HLA class II before it continues on to the cell surface. The result is a loss of gHgLgp42 complexes to degradation and a relative decrease in gHgLgp42 complexes in virus produced from a B cell or a plasmablast (B-EBV). Complexes in an epithelial cell are not lost to this degradative pathway and virus produced from an epithelial cell (E-EBV) is richer in gp42. E-EBV high in gp42 is better able to infect a B cell and B-EBV low in g42 is better able to infect an epithelial cell.

reactivating B cells or plasmablasts into saliva is low in gp42 and thus prepared for entry into an HLA class II-negative epithelial cell in a new host, whereas any virus emerging from an epithelial cell is higher in gp42 and strongly targeted to an HLA class II-positive B cell (Figure 4). This would support and provide an explanation for a relatively transient infection of epithelial cells that would serve primarily to facilitate access to B cells. Use of a coreceptor that targets proteins to a degradative pathway in B cells has a significant negative trade off for re-infection of a B cell, for which only replication in an HLA class II negative cell can compensate.

The effects of the stoichiometry of gHgL and gp42 on virus penetration described above can account entirely for the striking difference in infectivity of B cell derived EBV and epithelial cell-derived EBV for B cells expressing CR2 since both viruses attach equally well via gp350/220. However, the effects of infection of epithelial cells that lack CR2, where virus may use gHgL for attachment as well as for fusion, is not only less striking, but also more complicated. Virus produced from a B cell binds to gHgLR almost as well as it binds to CR2 and binds as much as 2 logs better than does virus from an epithelial cell which has a higher gp42 content. However, this translates into only about a fivefold higher level of infectivity of B cell virus than epithelial virus for gHgLR positive, CR2 negative cells. Binding efficiently to gHgLR comes at the expense of the ability to mediate efficient entry and it is proposed that this reflects the fact that virus entry is efficient if gHgLR is used as a coreceptor, but not if it is used efficiently as an attachment receptor as well (Borza et al., 2004). Engagement of a large number of gHgL complexes for attachment may not be compatible with fusion, or at least with expansion of a fusion pore. An enhanced ability to bind to gHgLR may, however, be important to ensure that free virus in saliva is not immediately washed unproductively across a mucosal surface, or perhaps to tether any virus-producing B cells in close contact with the epithelial surface. Efficient infection with cell free virus clearly requires expression of a second receptor such as CR2 or an integrin. It will be important to determine if this requirement reflects not only the need for attachment and fusion to be mediated by separate virus glycoproteins, but also a requirement for signal transduction to occur concurrent with entry.

Post entry events

After release of virus to the cytoplasm, the capsid and the associated tegument proteins must move away from the cell membrane in a reversal of the budding process that is required to produce enveloped virus in the first place. What triggers this reversal is not known. Two proteins, however, are known to be essential to envelopment, glycoproteins gN and gM. Glycoprotein gN is a 102 residue type 1 membrane proteins that carries only O-linked sugar. It is incompletely processed in the absence of gM, with which it forms a non-covalently linked complex (Lake et al., 1998). Glycoprotein gM is a multispan membrane protein of 405 residues. A recombinant virus in which the gN open reading frame is interrupted is phenotypically null for both gN and gM. Only very small amounts of enveloped recombinant virus are made and what little there is that binds specifically to CR2 expressing cells appears to have a post entry defect. It has been proposed the long cytoplasmic tail of gM, which is rich in proline and arginine residues and potential phosphorylation sites, is important to association of the tegumented capsid and the cell membrane from which the envelope is derived and to the subsequent dissociation required for virus to continue its movement into the cell (Lake and Hutt-Fletcher, 2000). No test of this hypothesis has yet been provided.

Little is known also about how deenveloped EBV is transported through the cytoplasm, a process that is rapid. Transmission electron microscopy of B cells fixed at 10 minutes after initiation of entry reveals capsids already at the nucleopore (L. Hutt-Fletcher, unpublished). Analogy with other herpesviruses and other large DNA viruses that replicate in the nucleus would suggest that transport probably requires interaction with the cellular cytoskeleton (Smith and Enquist, 2002; Sodeik, 2000). Recent work with Kaposi's sarcoma herpesvirus suggests that signaling events at the cell surface that lead to cytoskeletal rearrangements may influence its efficiency (Sharma-Walia et al., 2004). In this respect, the effects of EBV-mediated cross-linking of CR2 on actin rearrangements are of interest and the newly discovered interaction with integrins, which may impact focal adhesion kinases and microtubule function, are of great interest and will no doubt be the object of further study.

SUMMARY

Our understanding of the early events in infection of either B cells or epithelial cells is sadly incomplete. Much remains to be learned about the dynamic molecular interactions that are involved in entry, and of course in events that immediately follow. What we know to this point, however, reveals some fascinating differences between what occurs in the two cell types. Some of these differences, such as the different requirements for gHgL and gp42 may complement each other and have relevance to trafficking between cells. If the current cyclical models of persistence are correct and if, as seems increasingly likely, replication in epithelial cells plays a role in initiating or maintaining the cycle, this provides a fascinating example of adaptive evolution. The most recent discovery of a role for integrins in epithelial infection indicates that more cell type specific interactions are likely to be found. This work also emphasizes the ongoing need to move studies as much as possible into model systems that more closely resemble the *in vivo* environment. Eventually it will be necessary to devise more creative ways in which to study epithelial infections *in vivo*. Some interesting beginnings have been made in this direction by investigators who study epithelial infection in AIDS patients, where viral loads are very high (Walling et al., 2003; Walling et al., 2001), and the role of epithelial cell infection in normal individuals is beginning to be explored with the very sensitive techniques that are now available (Pegtel et al., 2004). Some of the fundamental processes in initiation of infection of B cell and epithelial cells are undoubtedly the same and many of the findings that are made with B cells, which are much easier to study, will provide insight into infection of either cell type. However, since epithelial cells make such a significant contribution to the burden of EBV-associated disease particular attention must be paid not only to the features that they share in common with B cells, but also to parallel events that are mediated in unique ways.

References

Ahearn, J., M,, and Fearon, D. T. (1989). Structure and function of the complement receptors, CR1 (CD35) and CR2 (CD21). Adv. Immunol. *46*, 183-219.

Akula, S. M., Pramod, N. P., Wang, F. Z., and Chandran, B. (2002). Integrin a3b1 (CD49c/29) is a cellular receptor for Kaposi's sarcoma-associated herpesvirus (KSHV/HHV8) entry into target cells. Cell *108*, 407-419.

Allday, M. J., and Crawford, D. H. (1988). Role of epithelium in EBV persistence and pathogenesis of B cell tumours. Lancet *1*, 855-856.

Anagnastopoulos, I., M., H., Kreschel, C., and Stein, H. (1995). Morphology, immunophenotype, and distribution of latently and/or productively Epstein-Barr virus-infected cells in acute infectious mononucleosis: implications for the interindividual infection route of Epstein-Barr virus. Blood *85*, 744-750.

Baer, R., Bankier, A. T., Biggin, M. D., Deininger, P. L., Farrell, P. J., Gibson, T. J., Hatfull, G., Hudson, G. S., Satchwell, S. C., Seguin, C., et al. (1984). DNA sequence and expression of the B95-8 Epstein-Barr virus genome. Nature *310*, 207-211.

Bayliss, G. J., and Wolf, H. (1980). Epstein-Barr virus induced cell fusion. Nature *287*, 164-165.

Beisel, C., Tanner, J., Matsuo, T., Thorley-Lawson, D., Kezdy, F., and Kieff, E. (1985). Two major outer envelope glycoproteins of Epstein-Barr virus are encoded by the same gene. J. Virol. *54*, 665-674.

Borza, C., and Hutt-Fletcher, L. M. (1998). Epstein-Barr virus recombinant lacking expression of glycoprotein gp150 infects B cells normally but is enhanced for infection of the epithelial line SVKCR2. J. Virol. *72*, 7577-7582.

Borza, C. M., and Hutt-Fletcher, L. M. (2002). Alternate replication in B cells and epithelial cells switches tropism of Epstein-Barr virus. Nature Med. *8*, 594-599.

Borza, C. M., Morgan, A. J., Turk, S. M., and Hutt-Fletcher, L. M. (2004). Use of gHgL for attachment of Epstein-Barr virus to epithelial cells compromises infection. J. Virol. *78*, 5007-5014.

Cohen, J. I., and Straus, S. E. (2001). Varicella-zoster virus and its replication. In Fields Virolggy, D. M. Knipe, and P. M. Howley, eds. (Philadelphia, Lippincott WIlliams and Wilkins), pp. 2707-2767.

Colman, P. M., and Lawrence, M. C. (2003). The structural biology of type 1 viral membrane fusion. Nature Rev. Mol. Cell Biol. *4*, 309-319.

Crawford, D. H., Rickinson, A. B., Finerty, S., and Epstein, M. A. (1978). Epstein-Barr (EB) virus genome-containing, EB nuclear antigen negative B lymphocyte populations in acute infectious mononucleosis. J. Gen. Virol. *38*, 449-460.

D'Addario, M., Libermann, T. A., Xu, J., Ahmad, A., and Menezes, J. (2001). Epstein-Barr virus and its glycoprotein-350 upregulate IL-6 in human B cells via CD21, involving activation of NF-kB and different signaling pathways. J. Mol. Biol. *308*, 501-514.

Emini, E. A., Luka, J., Armstrong, M. E., Keller, P. M., Ellis, R. W., and Pearson, G. R. (1987). Identification of an Epstein-Barr virus glycoprotein which is antigenically homologous to the varicella-zoster glycoprotein II and the herpes simplex virus glycoprotein B. Virology *157*, 552-555.

Faulkner, G. C., Burrows, S. R., Khanna, R., Moss, D. J., Bird, A. G., and Crawford, D. H. (1999). X-linked agammaglobulinemia patients are not infected with Epstein-Barr virus: implications for the biology of the virus. J. Virol. *73*, 1555-1564.

Fearon, D. T., and Carter, R. H. (1995). The CD19/CR2/TAPA-1 complex of B lymphocytes: linking natural to acquired immunity. Ann. Rev. Immunol. *13*, 127-149.

Fingeroth, J. D., Diamond, M. E., Sage, D. R., Hayman, J., and Yates, J. L. (1999). CD-21 dependent infection of an epithelial cell line, 293, by Epstein-Barr virus. J. Virol. *73*, 2115-2125.

Fingeroth, J. D., Weis, J. J., Tedder , T. F., Strominger, J. L., Biro, P. A., and Fearon, D. T. (1984). Epstein-Barr virus receptor of human B lymphocytes is the C3d complement CR2. Proc. Natl. Acad. Sci. USA. *81*, 4510-4516.

Frade, R., Barel, M., Ehlin-Henricksson, B., and Klein, G. (1985). gp140 the C3d receptor of human B lymphocytes is also the Epstein-Barr virus receptor. Proc. Natl. Acad. Sci. USA. *82*, 1490-1493.

Friedman-Kien, A.E. (1986). Viral origin of oral hairy leukoplakia. Lancet *ii*, 694-695.

Gan, Y., Chodosh, J., Morgan, A., and Sixbey, J. W. (1997). Epithelial cell polarization is a determinant in the infectious outcome of immunoglobulin A-mediated entry by Epstein-Barr virus. J. Virol. *71*, 519-526.

Gill, M. B., Roecklein-Canfield, J., Sage, D. R., Zambela-Soediono, M., Longtine, N., Uknis, M., and Fingeroth, J. D. (2004). EBV attachment stimulates FHOF/FHOD1 redistribution and co-aggregation with CD21:formin interactins with the cytoplasmic domain of human CD21. J. Cell Sci. *117*, 2709-2720.

Gilligan, K., Rajadurai, P., Resnick, L., and Raab-Traub, N. (1990). Epstein-Barr virus small nuclear RNAs are not expressed in permissively infected cells in AIDS-associated leukoplakia. Proc. Natl. Acad. Sci. USA. *87*, 8790-8794.

Gong, M., and Kieff, E. (1990). Intracellular trafficking of two major Epstein-Barr virus glycoproteins, gp350/220 and gp110. J. Virol. *64*, 1507-1516.

Gong, M., Ooka, T., Matsuo, T., and Kieff, E. (1987). Epstein-Barr virus glycoprotein homologous to herpes simplex virus gB. J. Virol. *61*, 499-508.

Gratama, J. W., Oosterveer, M. A. P., Zwaan, F. E., Lepoutre, J., Klein, G., and Ernberg, I. (1988). Eradication of Epstein-Barr virus by allogeneic bone marrow transplantation: implication for sites of latency. Proc. Natl. Acad. Sci. USA. *85*, 8693-8696.

Greenspan, J. S., Greenspan, D., Lennette, E. T., Abrams, D. I., Conant, M. A., Petersen, V., and Freese, U. K. (1985). Replication of Epstein-Barr virus within the epithelial cells of oral "hairy" leukoplakia, an AIDS-associated lesion. New Engl. J. Med. *313*, 1564-1571.

Haan, K. M., Kwok, W. W., Longnecker, R., and Speck, P. (2000). Epstein-Barr virus entry utilizing HLA-DP or HLA-DQ as a coreceptor. J. Virol. *74*, 2451-2454.

Haan, K. M., Lee, S. K., and Longnecker, R. (2001). Different functional domains in the cytoplasmic tail of glycoprotein gB are involved in Epstein-Barr virus induced membrane fusion. Virology *290*, 106-114.

Haan, K. M., and Longnecker, R. (2000). Coreceptor restriction within the HLA-DQ locus for Epstein-Barr virus infection. Proc. Natl. Acad. Sci. USA. *97*, 9252-9257.

Haddad, R. S., and Hutt-Fletcher, L. M. (1989). Depletion of glycoprotein gp85 from virosomes made with Epstein-Barr virus proteins abolishes their ability to fuse with virus receptor-bearing cells. J. Virol. *63*, 4998-5005.

Heineman, T., Gong, M., Sample, J., and Kieff, E. (1988). Identification of the Epstein-Barr virus gp85 gene. J. Virol. *62*, 1101-1107.

Henle, W., and Henle, G. (1979). Seroepidemiology of the virus. In The Epstein-Barr virus, M. A. Epstein, and B. G. Achong, eds. (New York, Springer-Verlag), pp. 61-78.

Hummel, M., Thorley-Lawson, D., and Kieff, E. (1984). An Epstein-Barr virus DNA fragment encodes messages for the two major envelope glycoproteins (gp350/300 and gp220/200). J. Virol. *49*, 413-417.

Imai, S., Nishikawa, J., and Takada, K. (1998). Cell-to-cell contact as an efficient mode of Epstein-Barr virus infection of diverse human epithelial cells. J. Virol. *72*, 4371-4378.

Karjannis, M. A., Hummel, M., Anagnostopoulos, I., and Stein, H. (1997). Strict lymphotropism of Epstein-Barr virus during acute infectious mononucleosis in nonimmunocompromised individuals. Blood *89*, 2856-2862.

Kurilla, M. G., Heineman, T., Davenport, L. C., Kieff, E., and Hutt-Fletcher, L. M. (1995). A novel Epstein-Barr virus glycoprotein gp150 expressed from the BDLF3 open reading frame. Virology *209*, 108-121.

Lake, C. M., and Hutt-Fletcher, L. M. (2000). Epstein-Barr virus that lacks glycoprotein gN is impaired in assembly and infection. J. Virol. *74*, 11162-11172.

Lake, C. M., Molesworth, S. J., and Hutt-Fletcher, L. M. (1998). The Epstein-Barr virus (EBV) gN homolog BLRF1 encodes a 15 kilodalton glycoprotein that cannot be authentically processed unless it is co-expressed with the EBV gM homolog BBRF3. J. Virol. *72*, 5559-5564.

Lee, S. K. (1999). Four consecutive arginine residues at positions 836-839 of EBV gp110 determine intracellular localization of gp110. Virology *264*, 350-358.

Lee, S. K., and Longnecker, R. (1997). The Epstein-Barr virus glycoprotein 110 carboxy-terminal tail domain is essential for lytic virus replication. J. Virol. *71*, 4092-4097.

Lemon, S. M., Hutt, L. M., Shaw, J. E., Li, J.-L. H., and Pagano, J. S. (1977). Replication of EBV in epithelial cells during infectious mononucleosis. Nature *268*, 268-270.

Li, Q. X., Spriggs, M. K., Kovats, S., Turk, S. M., Comeau, M. R., Nepom, B., and Hutt-Fletcher, L. M. (1997). Epstein-Barr virus uses HLA class II as a cofactor for infection of B lymphocytes. J. Virol. *71*, 4657-4662.

Li, Q. X., Turk, S. M., and Hutt-Fletcher, L. M. (1995). The Epstein-Barr virus (EBV) BZLF2 gene product associates with the gH and gL homologs of EBV and carries an epitope critical to infection of B cells but not of epithelial cells. J. Virol. *69*, 3987-3994.

Li, Q. X., Young, L. S., Niedobitek, G., Dawson, C. W., Birkenbach, M., Wang, F., and Rickinson, A. B. (1992). Epstein-Barr virus infection and replication in a human epithelial system. Nature *356*, 347-350.

Martin, D. R., Yuryev, A., Kalli, K. R., Fearon, D. T., and Ahearn, J. M. (1991). Determination of the structural basis for selective binding of Epstein-Barr virus to human complement receptor type 2. J. Exp. Med. *174*, 1299-1311.

Miller, N., and Hutt-Fletcher, L. M. (1988). A monoclonal antibody to glycoprotein gp85 inhibits fusion but not attachment of Epstein-Barr virus. J. Virol. *62*, 2366-2372.

Miller, N., and Hutt-Fletcher, L. M. (1992). Epstein-Barr virus enters B cells and epithelial cells by different routes. J. Virol. *66*, 3409-3414.

Mocarski, E. S., and Courcelle, C. T. (2001). Cytomegaloviruses and their replication. In Fields Virology, D. M. Knipe, and P. M. Howley, eds. (Philadelpia, Lippincott Williams and WIlkins), pp. 2629-2673.

Molesworth, S. J., Lake, C. M., Borza, C. M., Turk, S. M., and Hutt-Fletcher, L. M. (2000). Epstein-Barr virus gH is essential for penetration of B cell but also plays a role in attachment of virus to epithelial cells. J. Virol. *74*, 6324-6332.

Moore, M. D., Cooper, N. R., Tack, B. F., and Nemerow, G. R. (1987). Molecular cloning of the cDNA encoding the Epstein-Barr virus/C3d receptor (complement receptor type 2) of human B lymphocytes. Proc. Natl. Acad. Sci. USA. *84*, 9194-9198.

Moore, M. D., DiScipio, R. G., Cooper, N. R., and Nemerow, G. R. (1989). Hydrodynamic, electron microscopic and ligand binding analysis of the Epstein-Barr virus/C3dg receptor (CR2). J. Biol. Chem. *34*, 20576-20582.

Mullen, M. M., Haan, K. M., Longnecker, R., and Jardetzky, T. S. (2002). Structure of the Epstein-Barr virus gp42 protein bound to the MHC class II receptor HLA-DR1. Mol. Cell *9*, 375-385.

Nemerow, G. R., and Cooper, N. R. (1984). Early events in the infection of human B lymphocytes by Epstein-Barr virus. Virology *132*, 186-198.

Nemerow, G. R., Mold, C., Schwend, V. K., Tollefson, V., and Cooper, N. R. (1987). Identification of gp350 as the viral glycoprotein mediating attachment of Epstein-Barr virus (EBV) to the EBV/C3d receptor of B cells: sequence homology of gp350 and C3 complement fragment C3d. J. Virol. *61*, 1416-1420.

Nemerow, G. R., Wolfert, R., McNaughton, M., and Cooper, N. R. (1985). Identification and characterization of the Epstein-Barr virus receptor on human B lymphocytes and its relationship to the C3d complement receptor (CR2). J. Virol. *55*, 347-351.

Neuhierl, B., Feederle, R., W., H., and Delecluse, H. J. (2002). Glycoprotein gp110 of Epstein-Barr virus determines viral tropism and efficiencey of infection. Proc. Natl. Acad. Sci. USA. *99*, 15036-15041.

Niedobitek, G., Hamilton-Dutoit, S., Herbst, H., Finn, T., Vetner, M., Pallesen, G., and Stein, H. (1989). Identification of Epstein-Barr virus-infected cells in tonsils of acute infectious mononucleosis by in situ hybridization. Hum. Pathol. *20*, 796-799.

Nolan, L. A., and Morgan, A. J. (1995). The Epstein-Barr virus open reading frame BDLF3 codes for a 100-150 kDa glycoprotein. J. Gen. Virol. *76*, 1381-1392.

Oba, D. E., and Hutt-Fletcher, L. M. (1988). Induction of antibodies to the Epstein-Barr virus glycoprotein gp85 with a synthetic peptide corresponding to a sequence in the BXLF2 open reading frame. J. Virol. *62*, 1108-1114.

Oda, T., Imai, S., Chiba, S., and Takada, K. (2000). Epstein-Barr virus lacking glycoprotein gp85 cannot infect B cells and epithelial cells. Virology *276*, 52-58.

Old, L. J., Boyse, E. A., Geering, G., and Oettgen, H. F. (1968). Serologic approaches to the study of cancer in animals and man. Cancer Res. *28*, 1288-1299.

Papworth, M. A., Van Dijk, A. A., Benyon, G. R., Allen, T. D., Arrand, J. R., and Mackett, M. (1997). The processing, transport and heterologous expression of Epstein-Barr virus gp110. J. Gen. Virol. *78*, 2179-2189.

Pegtel, D. M., Middeldorp, J., and Thorley-Lawson, D. A. (2004). Epstein-Barr virus infection in ex-vivo tonsil epithelial cell cultures of asymptomatic carriers. J. Virol. *78*, 12613-12624.

Prota, A. E., Sage, D. R., Stehle, T., and Fingeroth, J. D. (2002). The crystal structure of human CD21: implications for Epstein-Barr virus and C3d binding. Proc. Natl. Acad. Sci. USA. *99*, 10641-10646.

Rickinson, A. B. (1990). On the biology of Epstein-Barr virus: a reappraisal. Adv. Exp. Med. Biol. *278*, 137-146.

Rickinson, A. B., Yao, Q.-Y., and Wallace, L. E. (1985). The Epstein-Barr virus as a model of virus-host interactions. Brit. Med. Bull *41*, 75-79.

Roizman, B., and Knipe, D. M. (2001). Herpes simplex viruses and their replication. In Fields Virology, D. M. Knipe, and P. M. Howley, eds. (Philadelphia, Lippincott Williams and Wilkins), pp. 2399-2459.

Sharma-Walia, N., Naranatt, P., Krishnan, H. H., Zeng, L., and Chandran, B. (2004). Kaposi's sarcoma-associated herpesvirus/human herpesvirus 8 envelope glycoprotein gB induces the integrin-dependent focal adhesion kinase-src-phophatidylinositol 3-kinase-rho GTPase signal pathways and cytoskeletal rearrangements. J. Virol. *78*, 4207-4223.

Sinclair, A. J., and Farrell, P. J. (1995). Host cell requirements for efficient infection of quiescent primary B lymphocytes by Epstein-Barr virus. J. Virol. *69*, 5461-5468.

Sixbey, J. W., Davis, D. S., Young, L. S., Hutt-Fletcher, L., Tedder, T. F., and Rickinson, A. B. (1987). Human epithelial cell expression of an Epstein-Barr virus receptor. J. Gen. Virol. *68*, 805-811.

Sixbey, J. W., Lemon, S. M., and Pagano, J. S. (1986). A second site for Epstein-Barr virus shedding: The uterine cervix. Lancet *2*, 1122-1124.

Sixbey, J. W., Nedrud, J. G., Raab-Traub, N., Hanes, R. A., and Pagano, J. S. (1984). Epstein-Barr virus replication in oropharyngeal epithelial cells. New Engl. J. Med. *310*, 1225-1230.

Sixbey, J. W., Vesterinen, E. H., Nedrud, J. G., Raab-Traub, N., Walton, L. A., and Pagano, J. S. (1983). Replication of Epstein-Barr virus in human epithelial cells infected *in vitro*. Nature *306*, 480-483.

Sixbey, J. W., and Yao, Q.-Y. (1992). Immunoglobulin A-induced shift of Epstein-barr virus tissue tropism. Science *255*, 1578-1580.

Smith, G. A., and Enquist, L. W. (2002). Break ins and breakouts: viral interactions with the cytoskeleton of mammalian cells. Ann. Rev. Cell Dev. Biol. *18*, 135-161.

Sodeik, B. (2000). Mechanisms of viral transport in the cytoplasm. Trends Microbiol. *8*, 465-472.

Spriggs, M. K., Armitage, R. J., Comeau, M. R., Strockbine, L., Farrah, T., MacDuff, B., Ulrich, D., Alderson, M. R., Mullberg, J., and Cohen, J. I. (1996). The extracellular domain of the Epstein-Barr virus BZLF2 protein binds the HLA-DR beta chain and inhibits antigen presentation. J. Virol. *70*, 5557-5563.

Sugano, N., Chen, W., Roberts, M. L., and Cooper, N. R. (1997). Epstein-Barr virus binding to CD21 activates the initial viral promoter via NFkB induction. J. Exp. Med. *186*, 731-737.

Tanner, J., Weis, J., Fearon, D., Whang , Y., and Kieff, E. (1987). Epstein-Barr virus gp350/220 binding to the B lymphocyte C3d receptor mediates adsorption, capping and endocytosis. Cell *50*, 203-213.

Tanner, J., Whang, Y., Sample, J., Sears, A., and Keiff, E. (1988). Soluble gp350/220 and deletion mutant glycoproteins block Epstein-Barr virus adsorption to lymphocytes. J. Virol. *62*, 4452-4464.

Tanner, J. E., Alfieri, C., Chatila, T. A., and Diaz-Mitoma, F. (1996). Induction of interleukin-6 after stimulation of human B-cell CD21 by Epstein-Barr virus glycoproteins gp350 and gp220. J. Virol. *70*, 570-575.

Tao, Q., Srivastava, G., Chan, A. C. L., Chung, L. P., S.L., L., and Ho, F. C. S. (1995). Evidence for lytic infection by Epstein-Barr virus in mucosal lymphocytes instead of nasopharyngeal epithelial cells in normal individuals. J. Med. Virol. *45*, 71-77.

Thorley-Lawson, D. A., and Gross, A. (2004). Persistence of Epstein-Barr virus and the origins of associated lymphomas. New Engl. J. Med. *350*, 1328-1337.

Thorley-Lawson, D. A., Miyashita, E. M., and Khan, G. (1996). Epstein-Barr virus and the B cell: that's all it takes. Trends Microbiol. *4*, 204-208.

Tugizov, S. M., Berline, J. W., and Palefsky, J. M. (2003). Epstein-Barr virus infection of polarized tongue and nasopharyngeal epithelial cells. Nature Med. *9*, 307-314.

Venables, P. J. W., Teo, C. G., Baboonian, C., Griffin, B. E., and Hughes, R. A. (1989). Persistence of Epstein-Barr virus in salivary gland biopsies from healthy individuals and patients with Sjogren's syndrome. Clin. Exp. Immunol. *75*, 359-364.

Walling, D. M., Flaitz, C. M., Adler-Storthz, K., and Nichols, C. M. (2003). A non-invasive technique for studying oral epithelial Epstein-Barr virus infection and disease. Oral Oncol *39*, 436-444.

Walling, D. M., Flaitz, C. M., Nichols, C. M., Hudnall, S. D., and Adler-Storthz, K. (2001). Persistent productive Epstein-Barr virus replication in normal epithelial cells *in vivo*. J. Inf Dis. *184*, 1499-1507.

Wang, X., and Hutt-Fletcher, L. M. (1998). Epstein-Barr virus lacking glycoprotein gp42 can bind to B cells but is not able to infect. J. Virol. *72*, 158-163.

Wang, X., Kenyon, W. J., Li, Q. X., Mullberg, J., and Hutt-Fletcher, L. M. (1998). Epstein-Barr virus uses different complexes of glycoproteins gH and gL to infect B lymphocytes and epithelial cells. J. Virol. *72*, 5552-5558.

Watt, T. H. (1997). Signaling via MHC class II molecules. In Lymphocyte Signaling: Mechanisms, subversion and manipulation, M. Harnete, and K. P. Rigley, eds. (New York, John Wiley and Sons Ltd), pp. 141-161.

Weis, J. J., Toothaker, L. E., Smith, J. A., Weis, J. H., and D.T., F. (1988). Structure of the human B lymphocyte receptor for C3d and the Epstein-Barr virus and relatedness to other members of the family of C3/C4 binding proteins. J. Exp. Med. *167*, 1047-1066.

Wolf, H., zur Hausen, H., and Becker, V. (1973). EB viral genome in epithelial nasopharyngeal carcinoma cells. Nature New Biol. *244*, 245-247.

Yao, Q. Y., Ogan, P., Rowe, M., Wood, M., and Rickinson, A. B. (1989). Epstein-Barr virus-infected B cells persist in the circulation of acyclovir-treated virus carriers. Int. J. Cancer, 67-71.

Yaswen, L. R., Stephens, E. B., Davenport, L. C., and Hutt-Fletcher, L. M. (1993). Epstein-Barr virus glycoprotein gp85 associates with the BKRF2 gene product and is incompletely processed as a recombinant protein. Virology *195*, 387-396.

Young, L. S., Clark, D., Sixbey, J. W., and Rickinson, A. B. (1986). Epstein-Barr virus receptors on human pharyngeal epithelium. Lancet *1*, 240-242.

Young, L. S., Dawson, C. W., Brown, K. W., and Rickinson, A. B. (1989). Identification of a human epithelial cell surface protein sharing an epitope with the C3d/Epstein-Barr virus receptor molecule of B lymphocytes. Int. J. Cancer *43*, 786-794.

Young, L. S., Lau, R., Rowe, M., Niedobitek, G., Packham, G., Shanahan, F., Rowe, D. T., Greenspan, D., Greenspan, J. S., Rickinson, A. B., and Farrell, P. J. (1991). Differentiation-associated expression of the Epstein-Barr virus BZLF1 transactivator protein in oral hairy leukoplakia. J. Virol. *65*, 2868-2874.

zur Hausen, H., Schulte-Holthauzen, H., Klein, G., Clifford, P., and Santesson, L. (1970). EBV DNA in biopsies of Burkitt tumours and anaplastic carcinomas of the nasopharynx. Nature *228*, 1056-1058.

From: Epstein-Barr Virus. Edited by: Erle S. Robertson

Chapter 19

The Plasmid Replicon of EBV: An Element That Underlies EBV's Transforming Functions

*Amanda A. Mack and Bill Sugden**

ABSTRACT

Epstein-Barr virus (EBV) is a particularly successful human parasite; it successfully establishes life-long latent infections in a large majority of all people. EBV is also a strikingly successful cellular parasite that in some cases contributes to its host cell's survival. In this chapter we describe how EBV's plasmid replicon underlies all viral contributions to the latent infection of cells. These contributions provide the infected cells advantages that ensure the maintenance of EBV even in a proliferating population of cells and thus also can underlie EBV's contributions to human cancers.

INTRODUCTION: EBV, A PARASITE THAT PROVIDES FOR ITS HOST

Viruses are obligate cellular parasites. In this chapter, we consider the ways Epstein-Barr virus reverses this relationship and provides its host cell a selective advantage such that it is dependent on the virus. For example, EBV can foster the proliferation and survival of infected cells and, thereby, promote tumorigenesis. We first discuss EBV's plasmid replicon as a genetic element essential for all of EBV's functions. This element is not perfectly stable in proliferating cells so that if it is lost from a population of cells, its loss indicates that whatever EBV's functions were in the parental cells, they failed to provide the parental cells a sufficient advantage to outgrow the daughters that had lost EBV. EBV's overcoming such a loss reflects advantages it affords infected cells. We then describe both the advantages EBV brings to its host cell using genetic and biochemical evidence for its contributions to the proliferation and/or survival of these cells and the viral oncogenes that encode these advantages. Lastly, we consider two EBV-associated cancers in order to illustrate how the retention or loss of EBV's replicon can help identify viral contributions to these cancers. It is important to note that infection with EBV only

*For correspondence email sugden@oncology.wisc.edu

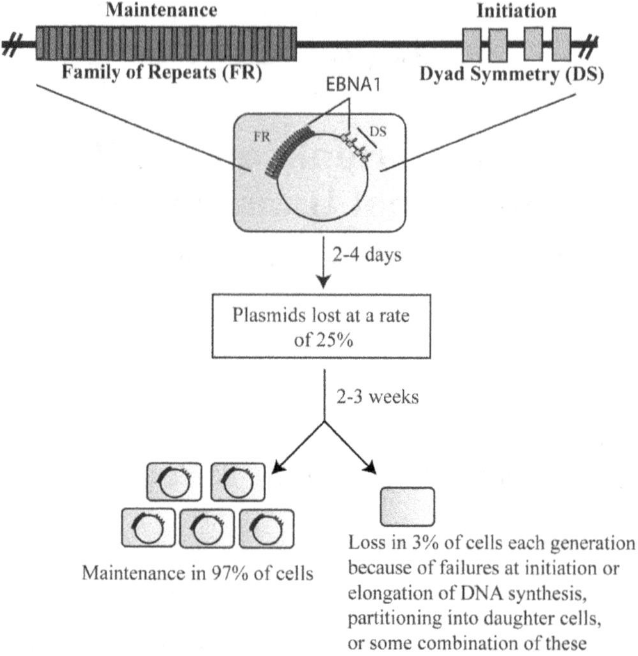

Figure 1. The family of repeats (FR) and the Dyad Symmetry (DS) are *cis*-acting elements within *oriP* that are required to support extrachromosomal replication and are dependent on their binding EBNA1. FR contains 20 EBNA1-binding sites that are required for maintenance of the viral plasmid during replication. DS contains two pairs of EBNA1-binding sites that are required for initiation of DNA synthesis. Dimers of EBNA1 bind the sites within both FR and DS. Viral plasmids are lost at a rate of 25% 2-4 days following introduction into cells (Kirchmaier and Sugden, 1995; Leight and Sugden, 2001). After multiple rounds of replication (2-3 weeks), plasmids are lost at a rate of only 3% and if the plasmids encode a selective advantage such as resistance to a drug, cells that maintain the plasmids will overgrow those that have lost them. Vertical dark grey bars, EBNA1 binding sites within FR; Vertical light grey bars, EBNA1 binding sites within DS; ellipses, EBNA1 dimers.

rarely leads to oncogenesis; in the vast majority of people, EBV cohabits with its human host successfully without causing life-threatening diseases.

THE LOSS OF EBV FROM PROLIFERATING CELLS

EBV is not an infallible parasite; it is lost from cells and this loss can be detected under multiple circumstances. An early recognition of this failing came from analysis of EBV's plasmid replicon. Approximately, 2 to 3 weeks following the introduction into cells of a replicon derived from EBV, the replicon becomes established within them (Figure 1) (Leight and Sugden, 2001). By established, we mean that its DNA synthesis and partitioning occur at a constant rate for as long as has been measured. However, this established plasmid replicon is still lost at a rate of 3% each cell cycle (Figure 1) (Kirchmaier and Sugden, 1995). That is, in only 97% of the cell cycles the plasmid replicon is inherited by daughter cells; this rate is independent of cell type. The loss of intact EBV from cells can be detected too, but only when the cells from which it is lost survive to be measured. For example, passaging of cells derived from nasopharyngeal carcinomas (NPC) in culture usually results in the loss of EBV while passaging a Burkitt's lymphoma (BL) in culture rarely results in loss of EBV, too (Huang et al., 1980; Lin et al., 1994). In particular, Kenzo Takada and his colleagues have demonstrated that the Akata cell line derived from an EBV-positive BL loses EBV following serial passaging (Shimizu et al., 1994). In these cases, the loss of EBV yields cells that still survive albeit *in vitro* but apparently lack an advantage that EBV afforded them *in vivo* (Komano et al., 1999). Furthermore, infection of EBV-negative BL cells *in vitro* has yielded EBV-positive cells; however, the EBV DNA is often integrated within the chromosomal DNA of these cells (Gulley et al., 1992; Hurley et al., 1991). Clearly, these parental tumor cell lines proliferate and survive in culture without EBV; infection with EBV fails to provide these tumor cells a selective advantage sufficient to maintain EBV extrachromosomally. All of these observations indicate that when EBV infects cells latently and the cells proliferate, the viral plasmid will be lost from cells, probably at a rate of 3% or more per cell cycle. If the cells that maintain EBV have an adequate selective advantage, they will overgrow those that lose it. Those rare cells in which EBV DNA is integrated will obviously also maintain it (Razzouk et al., 1996).

Two examples that contrast with the loss of EBV plasmids from proliferating host cells help illustrate EBV's distinctive means of maintaining its genome extrachromosomally. Multiple DNA tumor viruses such as human papilloma viruses (HPV) are usually maintained extrachromosomally during all phases of their viral life cycle. However, the cervical carcinomas that arise following infection with papilloma viruses have viral DNA integrated into the host cell DNA such that the viral genome is specifically disrupted to foster expression of two viral oncogenes (Jeon and Lambert, 1995). Thus, integration of papilloma viral DNA affects not only the maintenance of the viral DNA but also regulation of viral genes some time during the course of tumorigenesis. In contrast to HPV, EBV must rely on its own imperfect replication coupled with any selective advantages it provides to be retained in proliferating cells. Those herpesviruses that do not promote proliferation of a particular, infected host cell provide a second example. Herpes simplex virus types I and II, for example, infect neurons latently and encode no means to maintain themselves in proliferating cells; they do not have to because these host cells do not proliferate. EBV itself is maintained in non-proliferating, small, dense, memory B-cells in the periphery (Babcock et al., 1998; Kurth et al., 2000; Miyashita et al., 1997). Here EBV's loss from proliferating cells is irrelevant to its life-long maintenance in the human host. As with HSV-infected neurons, EBV-positive memory B-cells are non-proliferating but long-lived.

EBV'S PLASMID REPLICON

The loss of EBV from proliferating cells is an intrinsic feature of its plasmid replicon which is composed of an origin of DNA synthesis, *oriP*, and one viral protein, EBNA1 (Figure 1). All other factors required by this replicon are provided by the host cell. *OriP* was identified as the only element within EBV that could support extrachromosomal replication of a shuttle vector in EBV-positive cells (Yates et al., 1984). *OriP* is composed of two *cis*-acting sequences approximately 1 kilobase pair

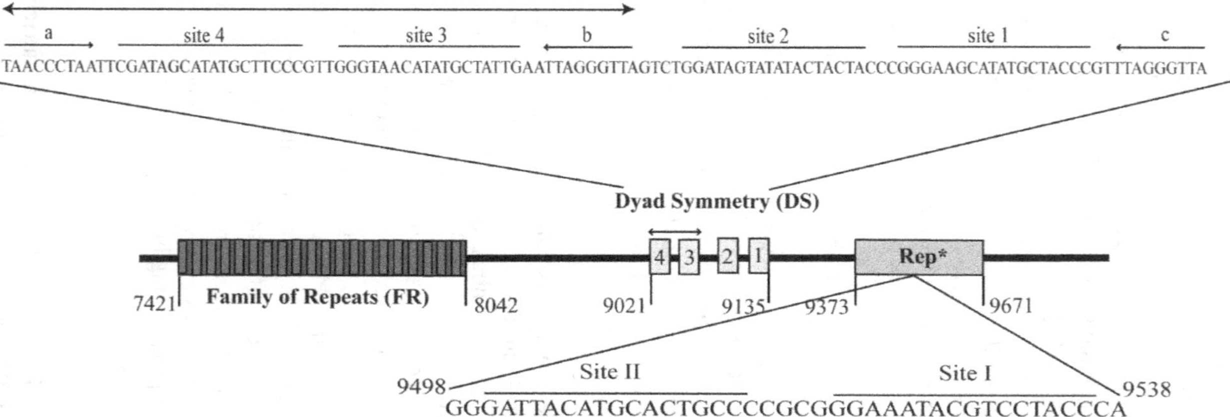

Figure 2. Schematic representation of *oriP* and Rep*. The Dyad Symmetry (DS) consists of 4 EBNA1-binding sites that function in pairs (4,3 and 2,1). EBNA1 binds to a pair of sites which are spaced 21 base pairs apart center-to-center to facilitate DNA synthesis. A set of nonamer repeats (single arrowheads) are located at three sites within DS and yield modest reductions in replication when mutated. The actual DS element, for which this region was termed, is indicated by double arrowheads. Rep* is a second sequence that can functionally substitute for DS albeit inefficiently. It also has a pair of EBNA1-binding sites spaced 21 base pairs apart center-to-center required for DNA synthesis.

apart known as the family of repeats (FR) and the dyad symmetry (DS) (Figures 1 and 2) (Reisman et al., 1985). The initiation of DNA synthesis occurs at or near DS as determined by 2-dimensional gel analysis; and, as might be expected, plasmids lacking DS do not support DNA synthesis (Gahn and Schildkraut, 1989; Reisman et al., 1985; Yates et al., 1984).

The elements within DS required for it to function as an origin of DNA synthesis have been identified first by its extensive mutational analyses and second by comparing and contrasting its features with those of a related, newly discovered origin within EBV, Rep* (Kirchmaier and Sugden, 1998). One assay underlies much of the functional characterization of DS and its derivatives. Plasmids containing DS or its engineered derivatives are propagated in *E. coli* carrying dam methylase which methylates adenines in the sequence "GATC". These methylated DNAs are susceptible to digestion with the endonuclease *DpnI*; DNAs that have been synthesized for two rounds or more in mammalian cells will have lost these methyl residues and be resistant to digestion with *DpnI*. The DNAs to be assayed are isolated from *E. coli* and introduced into an appropriate recipient, mammalian cell by transfection or electroporation. The recipient cells are propagated for 2 days to 4 weeks, plasmid DNAs are isolated from them by selective extraction, digested with *DpnI*, and the resistant DNAs detected by Southern blotting (Reisman et al., 1985). Two additional features of *oriP* need to be considered in these assays for the origin function of DS. First, while DS supports the initiation of DNA synthesis, FR bound by EBNA1 is essential to maintain *oriP*-containing plasmids in proliferating cells (Figure 1). This requirement means that DS can be assayed in the absence of FR for short times, 2-4 days, but for longer times the absence of FR leads to a rapid loss of detection of potentially replicated DNAs. Second, plasmids containing wild type *oriP* on being introduced into cells replicate in the presence of EBNA1 but are initially lost precipitously at rates exceeding 25% per cell generation (Figure 1) (Leight and Sugden, 2001). Some epigenetic change occurs during the first two weeks following their introduction such that *oriP*-plasmids become established and now are lost at a constant rate of only 3% per cell generation. While mutations in DS that reduce the levels of replicated (*DpnI*-resistant) DNA detected at a given time point are usually interpreted to affect a loss in origin function, it is often possible that these mutations could also affect the maintenance or rates of establishment of the mutated plasmids.

DS was defined initially in assays measuring resistance to digestion with *DpnI* by its being an essential element of *oriP* (Reisman et al., 1985) which could substitute, when multimerized, for FR while FR could not substitute functionally for it (Wysokenski and Yates, 1989). Footprinting experiments demonstrated that EBNA1 binds four sites within DS (Rawlins et al., 1985) and multiple studies indicated that these four sites are distributed within DS as two functional pairs each separated by 21 base pairs center-to-center (Figure 2) (Harrison et al., 1994; Koons et al., 2001; Yates et al., 2000). One pair of EBNA1-binding sites with this exact spacing is essential for DS to function minimally as an origin (Bashaw and Yates, 2001). However, separate mutations in each EBNA1-binding site can reduce the overall efficiency with which *oriP*-derivatives yield stably replicating plasmids (Koons et al., 2001). A set of nonamer repeats positioned at three sites in DS have also been identified; footprinting indicates that these sites are likely occupied *in vivo* (labeled a, b, and c in Figure 2) (Niller et al., 1995). Mutations in these nonamer binding sites yield modest reductions in the replication of derivatives of DS measured over long times (Koons et al., 2001; Yates et al., 2000). Finally, the structure for which DS is named, the dyad symmetry element, encompasses two EBNA1-binding sites (3 and 4) and two nonamer binding sites (a and b) but does not extend to the other pair of EBNA1-binding sites (1 and 2) (Figure 2). These latter two sites can support origin function albeit less efficiently than the dyad pair and less efficiently than intact DS (Harrison et al., 1994; Koons et al., 2001; Yates et al., 2000).

The many characterizations of DS have provided exacting measurements of the functional defects of derivatives of DS relative to their wild type parent. These characterizations have been made more informative by comparisons with a newly identified, second EBNA1-dependent origin in EBV. Rep* was identified serendipitously as a sequence that can functionally substitute for DS inefficiently (Kirchmaier and Sugden, 1998). It is located several hundred base pairs away from DS in the viral genome (Figure 2). Eight copies of Rep* (8xRep*) substituted for DS in *oriP* yields a plasmid that supports replication in the presence of EBNA1 as efficiently as does wild type *oriP* (Wang, L.S.E., Leight, E.R. and Sugden, B., personal communication). Analysis using 2-dimensional gel electrophoresis indicates that DNA synthesis initiates at or near Rep* and density shift experiments with BrdU demonstrate that Rep* plasmids replicate once per cell cycle as does *oriP* (*ibid*). DS and 8xRep* thus appear to provide identical origin functions in the context of FR *in cis* and EBNA1 *in trans*. Analyses of the sequences in Rep* have revealed that it too has a pair of binding sites for EBNA1 spaced 21 base pairs apart center-to-center and no other elements *in cis* are required for its origin function (Figure 2) (*ibid*). Apparently, appropriately spaced, binding sites for EBNA1 are the essential elements for origin function for EBV's plasmid replicon.

FR is the second *cis*-acting element essential for *oriP* function and is an enigma in mammalian biology. FR consists of 20 binding sites for EBNA1 and all current data indicate that FR's functions require dimers of EBNA1 being bound to it (Figure 1 and 2). Multiple observations indicate that FR bound by EBNA1 maintains EBV's plasmid replicons in proliferating cells. That is, in some limited sense, it provides these plasmids a centromere-like function although, clearly, it is less complex and lacks some critical features of centromeres (Sugden and Warren, 1988). The data supporting FR's acting to maintain plasmids in proliferating cells is derived from two kinds of experiments. First, DNAs containing FR are retained longer in cells which express EBNA1 whether or not these DNAs are competent to support replication (Krysan et al., 1989; Middleton and Sugden, 1994; Wade-Martins et al., 1999). Second, DNAs containing different origins including DS, Rep*, and cellular sequences are maintained in proliferating cells when FR is present *in cis* and EBNA1 *in trans* but not when FR is absent (Kirchmaier and Sugden, 1995; Krysan et al., 1989; Reisman et al., 1985; Wang, L.S.E., Leight, E.R. and Sugden, B., personal communication).

FR has a second associated function. It can enhance transcription from heterologous promoters and from EBV's adjacent promoters (Gahn and Sugden, 1995; Puglielli et al., 1996; Reisman and Sugden, 1986; Sugden and Warren, 1989). The mechanism of FR's serving as a transcriptional enhancer has been

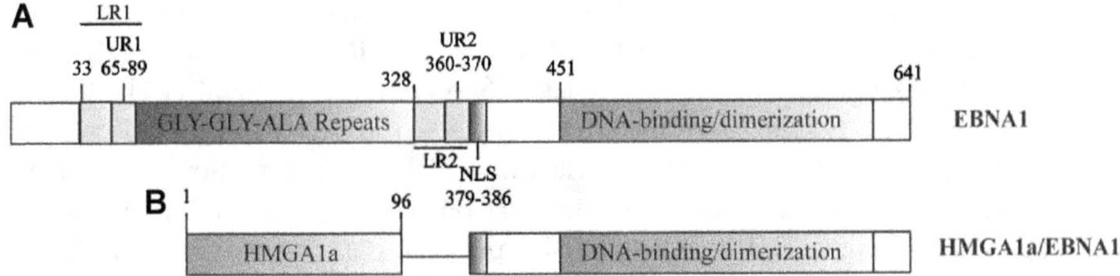

Figure 3. EBNA1 is required for replication and maintenance of EBV. EBNA1 is composed of 641 amino acids that include linking region 1 and 2 (LR1, LR2), two unique regions (UR1, UR2), a nuclear localization signal (NLS), and a DNA-binding/dimerization domain. Substitution of the amino-terminal half of EBNA1 with full-length, human HMGA1a supports replication of *oriP* to levels equal to wild type EBNA1.

difficult to define as a result of the types of experiments used to measure it. In general, DNA templates plus or minus FR have been introduced into cells with or without EBNA1 and transcription from the templates measured either directly or via a surrogate reporter gene. EBNA1 on binding FR in newly transfected DNAs appears to foster nuclear translocation of these DNAs (Langle-Rouault et al., 1998) thereby potentially obscuring whether FR bound by EBNA1 promotes transcription of adjacent promoters already in the nucleus. EBNA1's role could be only to conduit bound DNAs into the nucleus. Recent work has shown, however, that a template containing FR after integration into a host cell's chromosomes enhances transcription of an adjacent promoter dependent upon EBNA1 (Kennedy and Sugden, 2003).

EBNA1 *in trans* provides all viral contributions to all known functions of *oriP*. EBNA1 is a stable protein of 641 residues when encoded by the B95-8 laboratory strain of EBV (Figure 3) (Baer et al., 1984). It is expressed in all known cell types infected with EBV with the possible exception of resting, memory B-cells, the likely host for long-time, viral persistence *in vivo* (Hochberg et al., 2004; Miyashita et al., 1997; Qu and Rowe, 1992; Rowe et al., 1987). EBNA1 has been studied genetically and biochemically revealing that, like many viral proteins, it can carry out multiple, complex functions.

EBNA1 is essential for the replication of plasmids containing *oriP* (Lupton and Levine, 1985; Yates et al., 1985). Its carboxy-terminal DNA-binding domain acts as a dimer to bind DNA at defined sites within FR, DS, and Rep* (Figure 3) (Polvino-Bodnar and Schaffer, 1992; Rawlins et al., 1985; Wang, L.S.E., Leight, E.R. and Sugden, B., personal communication). The structure of this moiety of EBNA1 bound to a consensus DNA sequence has been solved by X-ray crystallography (Bochkarev et al., 1996). This carboxy-terminal DNA-binding domain is not sufficient to support EBNA1's contributions to *oriP*; in fact, it behaves as a dominant negative derivative inhibiting all of these functions (Kirchmaier and Sugden, 1997; 1998). EBNA1 has been purified from insect and monkey cells infected with replicating vectors and found to lack enzymatic activities such as a helicase or DNA-dependent ATPase (Frappier and O'Donnell, 1991; Middleton and Sugden, 1992). A role for it in promoting the initiation of DNA synthesis at or near DS although likely has been obscure. For example, other viral origin-binding proteins such as the T-antigen of Simian Virus 40 and the E1 protein of Bovine Papilloma Virus encode helicase activities involved in the unwinding of DNA to promote RNA-priming and DNA synthesis (Seo et al., 1993; Stahl et al., 1986; Yang et al., 1993). On the other hand, these papova viral plasmid replicons can be replicated multiple times in one S-phase while EBV plasmid replicons replicate once per cell cycle (Adams, 1987; Gilbert and Cohen, 1987; Yates and Guan, 1991; Wang, L.S.E., Leight, E.R. and Sugden, B., personal communication). EBNA1's contribution to initiating DNA synthesis at or near DS has been in part resolved by chromatin-immunoprecipitation with antibodies to EBNA1 and to ORC (origin recognition complex). In these experiments, precipitation of EBNA1 from cells carrying EBV or EBV-derived plasmid replicons, co-precipitates DS sequences and FR sequences; precipitation with antibodies to ORC co-precipitates DS but not FR (Chaudhuri et al., 2001; Dhar et al., 2001; Schepers et al., 2001). Thus, EBNA1 on binding DS but not FR serves as a platform to bind ORC. ORC is required to assemble the presumptive replicative helicase complex, MCM (minichromosome maintenance) for cellular origins of DNA synthesis and chromatin immunoprecipitation with antibodies to members of the MCM complex co-immunoprecipitate DS, too (Chaudhuri et al., 2001). We hypothesize that the structure EBNA1 forms with DS or with Rep* is recognized by ORC. It is not known what structures are recognized in the cell to mediate the association of ORC with origins of DNA synthesis. Comparative studies of EBNA1 binding to DS and Rep* have identified a second, shared feature of these viral origins which may be important for their recognition by ORC. EBNA1 bends each DNA on binding; we hypothesize that this bound, bent DNA forms a structure that ORC can bind (Bashaw and Yates, 2001; Wang, L.S.E., Leight, E.R. and Sugden, B., personal communication). It appears that once ORC binds a structure it can bind the proteins Cdt1 and Cdc6 which recruit MCM to form a pre-replication complex (Cook et al., 2004; Maiorano et al., 2000). It is reasonable to propose that ORC on binding to EBNA1 can also bind Cdt1 and Cdc6 to recruit MCM and initiate DNA synthesis although this proposition is yet to be tested.

EBNA1 in association with FR is essential for maintaining EBV-derived plasmid replicons in proliferating cells because replicons lacking FR replicate but are lost from cells in days while plasmids with intact *oriP* fail to replicate in the absence of EBNA1 (Reisman et al., 1985; Yates et al., 2000; Wang, L.S.E., Leight, E.R. and Sugden, B., personal communication). The exact mechanism by which EBNA1 binding FR supports maintenance of plasmids is still a mystery. Studies on populations of cells indicate that FR-containing plasmids with different origins of synthesis are lost in the absence of selection at rates of approximately 3% per generation (Kirchmaier and Sugden, 1995; Sugden and Warren, 1988; Yates et al., 1984; Wang, L.S.E., Leight, E.R. and Sugden, B., personal communication) and, importantly, this rate is independent of the average number of replicons per cell (Kirchmaier and Sugden, 1995). This latter finding likely means that the partitioning of EBV-derived plasmids is likely to be non-random. Were loss to be random, the rates at which *oriP*-negative cells arise at low numbers of plasmids per cell (*i.e.* 3-5) should be greater than at high numbers (>10-15). However, all of the biochemical data provides support only for mechanisms of maintenance of EBV-derived plasmids by EBNA1 that are random. In particular, Marechal and his colleagues have shown that two regions of EBNA1 within the amino-terminal half of EBNA1, often termed linking region 1 and 2, when fused to GFP mediate binding of GFP to mitotic chromosomes (Figure 3A) (Marechal et al., 1999). This observation has lead to a model whereby the carboxy-terminal DNA-binding domain of EBNA1 binds *oriP* within EBV site-specifically while the amino-terminal linking regions facilitate the association of this protein:DNA complex to mitotic chromosomes. This association would allow EBV-derived replicons to piggy-back on chromosomes to be distributed to daughter cells at the end of mitosis. This model lacks a mechanism to ensure that sister molecules of *oriP*-containing plasmids are separately piggy-backed on sister chromatids and therefore currently is consistent only with a random partitioning of these plasmids. Whether EBV-derived plasmids in fact are positioned randomly or by a directional mechanism is yet to be resolved.

The means by which the linking regions of EBNA1 bind metaphase chromosomes is now being defined. These two regions of EBNA1 link the DNAs to which these proteins are bound (Frappier and O'Donnell, 1991; Laine and Frappier, 1995; Mackey et al., 1995; Middleton and Sugden, 1992; Su et al., 1991) even when fused to the yeast Gal4 DNA-binding domain (Mackey et al., 1995; Mackey and Sugden, 1997). This linking could mediate the association of EBNA1 to mitotic chromosomes. Linking region 2 has also been found to bind a cellular protein EBP2 (EBV-binding protein 2) in yeast two-hybrid assays and data in yeast supports this binding to be a necessary bridge for associating EBNA1 with mitotic chromosomes and contributing to plasmid maintenance (Kapoor and Frappier, 2003; Kapoor et al., 2001; Shire et al., 1999; Wu et al., 2000). On the other hand, genetic analyses of EBNA1 indicate that multiple copies of linking region 1 can substitute for linking region 2 to support wild type levels of *oriP* plasmids which are maintained as stably as is wild type (Sears et al., 2004). These derivatives do not bind EBP2 (*ibid*). Both linking regions in EBNA1 are rich in glycine-arginine repeats, a motif common to an element termed an A-T hook, which binds A-T-rich DNAs (Figure 3B) (Reeves and Nissen, 1990; Sears et al., 2004). The linking regions of EBNA1 do act as A-T hooks (Sears et al., 2004) and, importantly, substitution of the amino-terminal half of EBNA1 with the cellular protein HMGI/Y, also termed HMGA1a, which also has A-T hook activity yields a fusion protein that supports *oriP* replication with wild type efficiency (Figure 3) (Hung et al., 2001; Reeves and Beckerbauer, 2001; Sears et al., 2003; 2004). HMGI/Y provides another parallel with EBNA1 in that it too can link DNAs (Bagga et al., 2000). It thus appears likely that the A-T hook activity of EBNA1's linking regions mediate the linking of complexes of EBNA1 and DNAs to which it binds site-specifically to mitotic chromosomes.

The mechanism by which EBNA1 positively regulates transcription of nearby promoters on binding FR is not known. A derivative of EBNA1 in which a unique sequence of 25 residues located in linking region 1 (UR1) is deleted fails to enhance this transcription (Figure 3A) (Kennedy and Sugden, 2003). It therefore seems likely that one or more cellular transcription factors require this unique sequence to associate with EBNA1 to foster transcription, although candidate factors have yet to be identified. The mechanism by which EBNA1 itself or EBNA1 bound to a DNA template partitions into the nucleus has been illuminated with yeast two-hybrid assays which have identified multiple nuclear import factors which EBNA1 can bind (Fischer et al., 1997; Kim et al., 1997). It seems likely that these cellular binding partners not only direct EBNA1 into the nucleus but also can facilitate a complex of EBNA1 bound to DNA entering the nucleus.

EBV's plasmid replicon in a proliferating host cell is successfully passed to both daughter cells in 97% of all mitoses. The 3% failure for established plasmids is independent of the origin in that plasmids with FR plus DS or FR plus 8xRep* are lost from proliferating cells at the same rate (Kirchmaier and Sugden, 1995; Wang, L.S.E., Leight, E.R. and Sugden, B., personal communication). We favor a model in which the failure arises at synthesis, not at partitioning of the replicons. Experiments in which selection for an *oriP* replicon has been removed for 20, 40, and 60 generations and then reapplied have shown that the surviving, resistant cells retain the same average number of plasmids as did the initial parent population of cells prior to removing selection (Kirchmaier and Sugden, 1995). A failure of synthesis is consistent with this finding. A failure of partitioning would be expected in these experiments to yield cells that retain plasmids with higher numbers of plasmids than present in the parental population. Whatever the mechanism of this loss is, it is clear that once EBV's plasmid replicon is lost from a cell, the advantages encoded by the replicon are lost too. Only if those advantages allow cells that retain the replicons to overgrow those that lose it would EBV be found in most cells of a proliferating population.

EBV's plasmid replicon is the vehicle that delivers and maintains all viral functions in infected cells. Among these functions are those encoded by viral oncogenes that provide selective advantages to infected cells and can also contribute to viral pathogenesis.

THE GENETIC IDENTIFICATION OF EBV'S ONCOGENES

Multiple genetic analyses have identified viral genes required for efficient initiation and/or maintenance of proliferation of EBV-infected primary B-lymphocytes. Driving cells to proliferate can allow them to overgrow an uninfected, quiescent population. These analyses have used recombination to generate deletions and tested a population of derived virus particles for whether infected, proliferating B-lymphoblasts can be recovered containing EBV with only the engineered deletion. Construction of these deletions often began with the P3HR1 strain of EBV; P3HR1 DNA has a deletion of the EBNA2 gene (Jeang and Hayward, 1983) shown early to be required for initiation and/or maintenance of proliferation of infected cells (Hammerschmidt and Sugden, 1989; Jeang and Hayward, 1983). Recombination restoring EBNA2 to the P3HR1 strain enriches for additional recombinations that substitute one stretch of DNA for another elsewhere in the viral genome (Longnecker et al., 1992). This general approach has identified EBNA2, EBNA3a, EBNA3c, and LMP1 as essential transforming genes (Hammerschmidt and Sugden, 1989; Kaye et al., 1993; Tomkinson et al., 1993). Although this analysis has been quite informative, it still has limitations. In addition to the evident difficulty of interpreting negative results, three problems have arisen in interpreting these genetic analyses based on deletions of putative transforming genes. First, because the stocks of recombinant viruses could not be purified to yield individual recombinants, it was not possible to measure the contributions of individual genes quantitatively. This limitation led to an under appreciation of two transforming genes, EBERs and LMP2a (Brielmeier et al., 1996; Takada, K. personal communication). Second, the read-out of these genetic assays is based on the induced and maintained proliferation of infected cells; failure at either step yields no growing cells. This read-out did not allow the identification of transforming genes as being active in only one step of this process. EBNA3a, for example, apparently contributes an essential activity to the early stages of infection but is not required for maintaining proliferation of infected B-cells (Kempkes et al., 1995). Thirdly, these genetic analyses focused on peripheral B-lymphocytes as targets for infection. It is now clear that this practical choice prevented an appreciation of the contributions of LMP2a to infected B-cells differentiating *in vivo* (Caldwell et al., 1998). These genetic analyses based on deletional studies have been augmented by additional genetic and transgene approaches to address their limitations.

EBV DNA has been cloned into *E.coli* as an F-plasmid, transfected into 293 cells, and propagated under drug selection in these established, adherent cells (Delecluse et al., 1998). These cells can be induced to support the lytic phase of EBV's life-cycle thus allowing biological purification and rescue of mutant derivatives of EBV. The 293 cells can also be engineered to express viral genes so that they complement mutations introduced into the F-plasmid. EBNA1, for example, has been found to contribute approximately a 10,000-fold increase in the induction and maintenance of proliferation of infected cells through studies of a derivative constructed in an F-plasmid and rescued from a complementing 293 host cell (Humme et al., 2003). These F-plasmids have also been engineered to express potential, transforming genes conditionally. In an earlier study, LMP1 was expressed from a tetracycline-dependent promoter on an F-plasmid introduced into and packaged as a viral particle in P3HR1 cells. Infection of primary, tonsil-derived B-lymphocytes with the recombinant virus in the presence of tetracycline yielded proliferating cells whose continued growth depended on tetracycline (Kilger et al., 1998). Genetic analysis using bacterial artificial chromosomes (BACs) have also been extended to a second strain of EBV, Akata, allowing complementary studies to those being done with the B95-8 strain in F-plasmids (Kanda et al., 2004).

To overcome the limited host cells available for infection, individual transforming genes of EBV have been studied as transgenes in mice. These studies have revealed an important role for LMP2a in mimicking activation signals from the B-cell receptor (BCR) to foster survival during differentiation of the host cell (Caldwell et al., 1998; Portis and Longnecker, 2003). The varied genetic and transgene analyses have identified LMP1, LMP2a, EBERs, EBNA1, EBNA2, EBNA3a, and EBNA3c as being important for the induction and/or maintenance of proliferation of B-cells in culture and LMP2a as being important for survival of developing B-cells *in vivo*. These contributions likely are important for infections *in vivo*, too.

EBV'S ONCOGENES

We shall briefly discuss functional properties of four oncogenes of EBV: LMP1, LMP2a, EBERs, and EBNA1, which continue to be expressed in many EBV-positive tumors (Crawford, 2001). More detailed descriptions of these viral gene products can be found in other chapters in this text. These four genes of EBV apparently

Table 1. Properties encoded by viral products likely to provide a selective advantage in established tumor cells

	Functional properties	**NPC**	**BL**
EBNA1	Required for replication and extrachromosomal maintenance of the viral plasmid and may regulate transcription and survival	+	+
EBNA2	Required for initiation and/or maintenance of proliferatioin. (Hammerschmidt and Sugden, 1989; Jeang and Hayward, 1983)		
LMP1	Induces signaling through NF-kB, AP-1, JAK/STAT to promote proliferation. (Kilger et al., 1998; Mosiales et al., 1995)	+	
LMP2a	Inhibits signaling through the BCR and promotes survival during B-cell development. (Caldwell et al., 1998; Portis and Longnecker, 2003)	+	
EBERs	Small, non-polyadenylated RNAs that inhibit interferon-α induced apoptosis, stimulate IL-10 production, promote proliferation. (Kitagawa et al., 2000; Komano et al., 1999; Nanbo et al., 2002)	+	+

NPC, nasopharyngeal carcinoma
PTLD, post-transplant lymphoproliferative disease
HD, Hodgkin's disease
BL, Burkitt's lymphoma
+, viral products detected in the indicated malignancy

can provide selective advantages to the tumors in which they are expressed *in vivo* (Table 1). Another family of transcripts, the BARTs (BamH1 A rightward transcripts), is often expressed in EBV-positive tumors but its functions are only now being elucidated (Crawford, 2001; de Jesus et al., 2003). An appreciation of the functions encoded by these viral genes serves as a background to consider the selective advantages afforded to normal and tumorigenic host cells by EBV. It is apparent that these advantages in established tumors overcome any immune responses of the host to the viral products providing these advantages. Conditions in which EBV's plasmid replicon is lost from cells are those in which the viral oncogenes expressed fail to provide the selective advantages needed for viral-positive cells to outgrow viral-negative progeny.

LMP1

Genetic analyses have shown LMP1 to be critical for inducing and maintaining proliferation of infected B-cells (Kilger et al., 1998). Biochemical studies have identified multiple activities that make LMP1 an important regulator of its host cell proliferation. LMP1 signals through NF-κB by associating with cellular TRAF (tumor necrosis factor receptor-associated factor) molecules (Devergne et al., 1996; Mosialos et al., 1995; Sandberg et al., 1997). It also can signal through AP-1 (activating protein 1) and through the JAK (Janus kinase)/STAT (signal transducers and activators of transcription) pathways through its binding to cellular TRADD (TRAF-associated death domain) and JAK3 (Eliopoulos and Young, 1998; Gires et al., 1999; Izumi et al., 1997; Kieser et al., 1997). LMP1 shares much of its signaling capacity with CD40, a cellular receptor that can regulate B-cell proliferation and differentiation (Busch and Bishop, 1999; Kilger et al., 1998; Uchida et al., 1999). LMP1 and CD40 share some, but not all functions. Activities of CD40 and LMP1 other than those mediated through their directly binding TRAF molecules contribute also to their promoting proliferation of B-cells. LMP1's non-canonical binding of TRADD or its activation of the JAK/STAT pathway may contribute to its support of proliferation, for example (Eliopoulos et al., 1999; Gires et al., 1999; Izumi et al., 1999; Kieser et al., 1999; Schultheiss et al., 2001).

One potential activity of LMP1 which could provide its cells a selective advantage is its inhibiting apoptosis. Despite some previous conflicting observations (Henderson et al., 1991; Kilger et al., 1998; Martin et al., 1993), more recent analyses of genes directly regulated by LMP1 in EBV-infected B-cells may have resolved these conundrums. LMP1 coordinately regulates both pro- and anti-apoptotic genes as it controls proliferation such that the balance of induced gene expression supports cell survival (Dirmeier et al., submitted). No one pro- or anti-apoptotic gene alone dictates the outcome of LMP1's contributions to survival. The introduction of LMP1 into epithelial cell lines can induce the anti-apoptotic gene, A20, and therefore could lead to an inhibition of apoptosis in them (Fries et al., 1996; Laherty et al., 1992). However, it appears likely that LMP1 will regulate a balance of pro- and anti-apoptotic genes in primary epithelial cells, too. EBV-positive NPC cells usually lose EBV upon being explanted into cell culture indicating that neither LMP1 nor any EBV gene is required for their survival and proliferation in cell culture (Huang et al., 1980; Kilger et al., 1998; Lin et al., 1994). This latter finding underscores how important it is for our understanding of EBV's tumorigenesis that we consider the significance of the spontaneous loss of EBV's plasmid replicon. This consideration allows us to conclude that the advantages EBV affords mature NPC tumors in general act *in vivo*, not *in vitro*. We need to search for functions of EBV's transforming genes that have such a special phenotype.

LMP2a

LMP2a is a remarkable protein, cobbled together by EBV during its evolution, to foster EBV's survival in B-cells *in vivo*. It has been examined biochemically, genetically, and as a transgene in mice. LMP2a resides in the plasma membrane of B-cells infected *in vitro*. Its amino-terminus is on the cytoplasmic face of the plasma membrane and encodes an ITAM (immunoreceptor tyrosine activation motif) (Longnecker and Kieff, 1990). This motif can bind cellular protein kinases and is present within the B-cell receptor (BCR) as well. Activation of the BCR is dependent on the ITAM for signaling. LMP2a inhibits both signaling mediated by the BCR and the induction of EBV's lytic cycle mediated by the activation of the BCR *in vitro* (Burkhardt et al., 1992; Fruehling and Longnecker, 1997; Miller et al., 1994). LMP2a's role in EBV's life-cycle *in vivo* has been illuminated by transgenic studies. Mice that express LMP2a from a promoter derived from the immunoglobulin (Ig) heavy chain along with an intronic enhancer express LMP2a in their spleen and bone marrow but do not undergo normal rearrangement of Ig heavy chains during their development (Caldwell et al., 1998). These IgM⁻, LMP2a⁺ B-cells develop even in mice genetically incapable of supporting immunoglobulin rearrangement and demonstrate that LMP2a can substitute for the developmental signals that the correct rearrangement and expression of a heavy chain provide precursor B-cells (*ibid*). In fact, different levels of expression of LMP2a as transgenes in mice support the development of different subsets of mature B-cells (Casola et al., 2004). Bone marrow and splenic B-cells from normal and transgenic mice also express different sets of transcription factors consistent with the notion that LMP2a signals to regulate B-cell development (Portis and Longnecker, 2003).

EBERs

The EBERs are two small, non-coding, non-polyadenylated RNAs which are present in infected cells as ribonucleoprotein complexes and expressed abundantly (Arrand and Rymo, 1982; Rosa et al., 1981). The abundant expression of EBERs has allowed their ready detection by *in situ* hybridization and their use as a surrogate marker in individual tumors for the presence of EBV (Barletta et al., 1993; Howe and Steitz, 1986; Wu et al., 1991). Complexes containing the EBERs can be precipitated with antibodies from some patients with autoimmune diseases (Lerner et al., 1981). The EBERs have also been found to associate with the ribosomal protein L22 and with PKR apparently at an overlapping site on one of these two viral RNAs (Clarke et al., 1991; Elia et al., 2004; Toczyski et al., 1994). These properties of the EBERs or others yet to be identified underlie complex phenotypes associated with these viral RNAs. They can inhibit apoptosis induced by treating B-cells with interferon-α through binding PKR, stimulate synthesis and secretion of IL-10, and promote tumorigenic phenotypes in derivatives of Burkitt's lymphoma cells (Kitagawa et al., 2000; Komano et al., 1999; Nanbo et al., 2002). Whether all of their mechanisms of action are mediated through their binding PKR is not known.

Early genetic analysis of EBERs had found no role for them during infection (Swaminathan et al., 1991). The absence of a role for EBERs is puzzling given their potential anti-apoptotic and growth-promoting roles detected in derivatives of Burkitt's lymphoma cells (Kitagawa et al., 2000; Komano et al., 1999; Nanbo et al., 2002). These early genetic analyses were performed with a derivative of P3HR1 under conditions in which pure stocks of EBERs-null recombinants could not be assayed quantitatively (Swaminathan et al., 1991). However, the engineering of pure stocks of EBERs-null EBV and its reconstituted wild type derivative in the Akata strain of EBV has allowed the measurement of a 10 to100-fold difference in the efficiencies with which these recombinant viruses induce and/or maintain proliferation of infected primary B-cells (Takada, K., personal communication). Despite their deceptive size, the EBERs' role in proliferation and/or cell survival is clearly evident.

EBNA1

The central role EBNA1 provides EBV's plasmid replicon underlies all of EBV's possible contributions to its associated tumors and has been described earlier. However, a specific role for EBNA1 in EBV's tumorigenesis has been revealed recently in experiments designed to inhibit EBNA1 in normal and tumor B-cells. Retroviral vectors encoding dominant-negative derivatives of EBNA1 were used to inhibit EBNA1 and surprisingly the retroviral-infected cells died prior to loss of EBV DNA (Kennedy et al., 2003). These EBV-positive normal and tumor cells died apoptotically. In reconstruction experiments it appeared that wild type EBNA1 in selected EBV-negative cell lines could inhibit specifically induced apoptosis making it likely that EBNA1 encodes an anti-apoptotic function (*ibid*). Genetic analyses indicated that the unique sequence in linking region 1 of EBNA1 is required for its inhibition of apoptosis consistent with a model that EBNA1's ability to regulate transcription mediates this inhibition.

EBV'S ONCOGENESIS

EBV's contributions to its associated cancers depend on the types of cancer considered. It is a risk factor for the development of both carcinomas and lymphomas (Crawford, 2001). We shall contrast NPC with BL to illustrate that loss or retention of EBV's plasmid replicon needs to be considered to identify those selective advantages encoded by EBV that it provides its associated cancers. These two EBV-positive tumors are fully discussed in later chapters.

Nasopharyngeal carcinoma

Infection with EBV is a risk factor for developing NPC (Zeng et al., 1985). All NPC tumors are EBV-positive; that is, within the limits of the assays used, all of the tumor cells in NPC biopsies express one or more viral products identified either immunohistochemically or by *in situ* hybridization (Lo and Huang, 2002). Two observations indicate, however, that EBV's contributions to the development of NPC are enigmatic. First, EBV appears to infect epithelial cells that have already accumulated at least one predisposing mutation (Chan et al., 2000). Dolly Huang and her colleagues have found that nasopharyngeal tissue exhibits loss of heterozygosity of chromosome 3p prior to infection with EBV and prior to tumor development in people with a high risk for developing NPC (*ibid*). Greater than 80% of NPC tumors exhibit a deletion of this locus and all are EBV-positive making it likely that any role for EBV is not exerted at the stage of initiation of tumorigenesis (*ibid*). Second, EBV is usually not maintained in NPC biopsies placed into cell culture (Huang et al., 1980). This second observation indicates that whatever selective advantages EBV confers on NPC tumor cells, these advantages do not extend to cells in culture. What then might be EBV's oncogenic contributions to NPC? Four viral products are known to be expressed in NPC: LMP1, LMP2a, EBERs, and EBNA1 (Table 1) (Crawford, 2001). It is not known what role EBNA1 might play in the development of NPC other than to allow viral plasmid replication and maintenance. Given that NPC cells survive and proliferate in culture following the loss of the viral replicon it seems unlikely that EBNA1's inhibition of apoptosis is necessary for tumor survival *in vivo*. The EBERs have been found to increase expression of insulin-like growth factor 1 in a transfected gastric carcinoma which is consistent with a role for EBERs in promoting proliferation (Iwakiri et al., 2003). It is not known what LMP2a could contribute to the development of NPC, and it is clear that not all EBERs-positive NPCs express LMP2a or LMP1. However, LMP2a can be a target for the host's immune response so that its expression in NPC tumors makes it likely that it does provide the tumor cell a selective advantage. LMP1 has been found to stimulate STAT activation in epithelial cells, to increase expression of the EGF receptor in them and, perhaps, to stimulate expression of metalloproteinase 1 in them, too (Chan et al., 2000; Lu et al., 2003; Miller et al., 1995). Of all of these activities only the last, activation of an enzyme involved potentially in tissue remodeling, would be likely to provide an oncogenic activity *in vivo*, not detected in cell culture. LMP1 is also a target of the host's immune response indicating that it too likely provides the tumor cell that expresses it some advantage *in vivo*. EBV-positive gastric carcinomas usually lose EBV replicons on propagation in cell culture, too (Takada, 2000) . This parallel supports EBV providing carcinomas some phenotype peculiarly important for tumor growth and/ or survival *in vivo*. What is clear is that more research is needed to unravel the mechanisms of EBV's oncogenesis in NPC and in carcinomas in general.

Burkitt's lymphoma

Where BL is endemic, it is often correlated with the presence of EBV, malaria, and translocation of the *c-myc* locus. Approximately 95% of tumor cells derived from African BL are associated with EBV (Crawford, 2001; Delecluse et al., 1993). EBV is maintained extrachromosomally in cells of these tumors. Not surprisingly, EBNA1 has been consistently detected in these cells in addition to EBERs (Table 1).

Elegant studies by Lam et al. of children from the Gambia found a strong correlation between the development of BL and high levels of EBV infection (Lam et al., 1991). They also demonstrated that higher viral loads corresponded to peak incidence in malaria compared to periods when malaria was at its lowest level. Malaria suppresses the immune response and, therefore, likely facilitates expansion of EBV-infected cells. 95% of endemic BLs are monoclonal with each tumor cell exhibiting a translocation of the *c-myc* locus to any of three immunoglobulin loci (Taub et al., 1982). Ig genes are constitutively expressed in B-cells so that the juxtaposition of them next to *c-myc* leads to its expression and promotion of proliferation in the absence of EBNA2 and LMP1 (Polack et al., 1996). This property of BL tumor cells might mean that EBV does not provide a selective advantage to fully developed BL cells; however, 95% of the biopsies derived from BLs are EBV-positive.

EBNA1 is expressed in BL tumor cells and must be expressed if EBV is to be maintained as a plasmid. Any information that EBV might provide a BL tumor cell would be dependent on maintenance of the viral DNA and thus on EBNA1. EBNA1 has been inhibited functionally with two of its dominantly-acting derivatives which compete for EBNA1's binding site-specifically to DNA and inhibit wild type EBNA1's transcriptional activation. Inhibition of EBNA1 in BL cell lines by either of these derivatives inhibited survival of BL cells (Kennedy et al., 2003). Many of these BL cells died apoptotically indicating that EBNA1 directly or indirectly inhibits apoptosis in EBV-positive normal and BL cells (*ibid*). For a few cell types in which p53 induces apoptosis, the introduction of EBNA1 in the absence of other viral genes, inhibited apoptosis. Thus, EBNA1 may be promoting survival of BL tumors by directly inhibiting apoptosis in them (*ibid*).

EBERs are also expressed in BL cells; these small RNAs inhibit interferon-α-induced apoptosis by inhibiting PKR (Nanbo et al., 2002). The EBERs thus may also contribute to the survival of BL cells. The EBERs stimulate the synthesis and secretion of IL-10 so they may contribute to the proliferation of BL cells, too (Kitagawa et al., 2000).

EBV is neither lost from BL cells *in vivo* nor on propagating them *in vitro* indicating that EBNA1 and/or the EBERs encode phenotypes providing selective advantages to BL cells *in vivo* and *in vitro*. Survival mediated by the inhibition of apoptosis is obviously one such advantage. It is also clear that for the uncommon BL cells that can shed EBV and still survive in culture, one or more cellular mutations must override the apoptotic pathway and any other advantages afforded these tumor cells by retaining the viral plasmid replicon.

SUMMA VIROLOGICA

EBV is an obligate cellular parasite that under rare circumstances *in vivo* provides an infected host cell uncontrolled proliferative and survival functions. These functions are a selective advantage for the infected cell but a drastic disadvantage for its human host. Even in the presence of an immune response, which in most infected people is robust, these infected cells can evolve oncogenically to avoid the immune response and become life-threatening to the human host while remaining dependent on one or more viral functions. The selective advantage EBV affords its tumor cells require the viral replicon to express the oncogenes encoding them. Identifying conditions in which these replicons are lost from tumor cells will help delineate the kinds of selective advantages EBV provides. Forcing the cells *in vivo* to lose their viral replicons is a reasonable means to treat these tumors.

References

Adams, A. (1987). Replication of latent Epstein-Barr virus genomes in Raji cells. J. Virol. *61*, 1743-1746.

Arrand, J. R., and Rymo, L. (1982). Characterization of the major Epstein-Barr virus-specific RNA in Burkitt lymphoma-derived cells. J. Virol. *41*, 376-389.

Babcock, G. J., Decker, L. L., Volk, M., and Thorley-Lawson, D. A. (1998). EBV persistence in memory B cells *in vivo*. Immunity *9*, 395-404.

Baer, R., Bankier, A. T., Biggin, M. D., Deininger, P. L., Farrell, P. J., Gibson, T. J., Hatfull, G., Hudson, G. S., Satchwell, S. C., and Seguin, C. (1984). DNA sequence and expression of the B95-8 Epstein-Barr virus genome. Nature *310*, 207-211.

Bagga, R., Michalowski, S., Sabnis, R., Griffith, J. D., and Emerson, B. M. (2000). HMG I/Y regulates long-range enhancer-dependent transcription on DNA and chromatin by changes in DNA topology. Nucleic Acids Res. *28*, 2541-2550.

Barletta, J. M., Kingma, D. W., Ling, Y., Charache, P., Mann, R. B., and Ambinder, R. F. (1993). Rapid in situ hybridization for the diagnosis of latent Epstein-Barr virus infection. Mol. Cell Probes *7*, 105-109.

Bashaw, J. M., and Yates, J. L. (2001). Replication from oriP of Epstein-Barr virus requires exact spacing of two bound dimers of EBNA1 which bend DNA. J. Virol. *75*, 10603-10611.

Bochkarev, A., Barwell, J. A., Pfuetzner, R. A., Bochkareva, E., Frappier, L., and Edwards, A. M. (1996). Crystal structure of the DNA-binding domain of the Epstein-Barr virus origin-binding protein, EBNA1, bound to DNA. Cell *84*, 791-800.

Brielmeier, M., Mautner, J., Laux, G., and Hammerschmidt, W. (1996). The latent membrane protein 2 gene of Epstein-Barr virus is important for efficient B cell immortalization. J. Gen. Virol. *77 (Pt 11)*, 2807-2818.

Burkhardt, A. L., Bolen, J. B., Kieff, E., and Longnecker, R. (1992). An Epstein-Barr virus transformation-associated membrane protein interacts with src family tyrosine kinases. J. Virol. *66*, 5161-5167.

Busch, L. K., and Bishop, G. A. (1999). The EBV transforming protein, latent membrane protein 1, mimics and cooperates with CD40 signaling in B lymphocytes. J. Immunol. *162*, 2555-2561.

Caldwell, R. G., Wilson, J. B., Anderson, S. J., and Longnecker, R. (1998). Epstein-Barr virus LMP2A drives B cell development and survival in the absence of normal B cell receptor signals. Immunity *9*, 405-411.

Casola, S., Otipoby, K. L., Alimzhanov, M., Humme, S., Uyttersprot, N., Kutok, J. L., Carroll, M. C., and Rajewsky, K. (2004). B cell receptor signal strength determines B cell fate. Nat. Immunol. *5*, 317-327.

Chan, A. S., To, K. F., Lo, K. W., Mak, K. F., Pak, W., Chiu, B., Tse, G. M., Ding, M., Li, X., Lee, J. C., and Huang, D. P. (2000). High frequency of chromosome 3p deletion in histologically normal nasopharyngeal epithelia from southern Chinese. Cancer Res. *60*, 5365-5370.

Chaudhuri, B., Xu, H., Todorov, I., Dutta, A., and Yates, J. L. (2001). Human DNA replication initiation factors, ORC and MCM, associate with oriP of Epstein-Barr virus. Proc. Natl. Acad. Sci. USA. *98*, 10085-10089.

Clarke, P. A., Schwemmle, M., Schickinger, J., Hilse, K., and Clemens, M. J. (1991). Binding of Epstein-Barr virus small RNA EBER-1 to the double-stranded RNA-activated protein kinase DAI. Nucleic Acids Res. *19*, 243-248.

Cook, J. G., Chasse, D. A., and Nevins, J. R. (2004). The regulated association of Cdt1 with minichromosome maintenance proteins and Cdc6 in mammalian cells. J. Biol. Chem. *279*, 9625-9633.

Crawford, D. H. (2001). Biology and disease associations of Epstein-Barr virus. Philos Trans R Soc. Lond B Biol. Sci. *356*, 461-473.

de Jesus, O., Smith, P. R., Spender, L. C., Elgueta Karstegl, C., Niller, H. H., Huang, D., and Farrell, P. J. (2003). Updated Epstein-Barr virus (EBV) DNA sequence and analysis of a promoter for the BART (CST, BARF0) RNAs of EBV. J. Gen. Virol. *84*, 1443-1450.

Delecluse, H. J., Bartnizke, S., Hammerschmidt, W., Bullerdiek, J., and Bornkamm, G. W. (1993). Episomal and integrated copies of Epstein-Barr virus coexist in Burkitt lymphoma cell lines. J. Virol. *67*, 1292-1299.

Delecluse, H. J., Hilsendegen, T., Pich, D., Zeidler, R., and Hammerschmidt, W. (1998). Propagation and recovery of intact, infectious Epstein-Barr virus from prokaryotic to human cells. Proc. Natl. Acad. Sci. USA. *95*, 8245-8250.

Devergne, O., Hatzivassiliou, E., Izumi, K. M., Kaye, K. M., Kleijnen, M. F., Kieff, E., and Mosialos, G. (1996). Association of TRAF1, TRAF2, and TRAF3 with an Epstein-Barr virus LMP1 domain important for B-lymphocyte transformation: role in NF-kappaB activation. Mol. Cell Biol. *16*, 7098-7108.

Dhar, S. K., Yoshida, K., Machida, Y., Khaira, P., Chaudhuri, B., Wohlschlegel, J. A., Leffak, M., Yates, J., and Dutta, A. (2001). Replication from oriP of Epstein-Barr virus requires human ORC and is inhibited by geminin. Cell *106*, 287-296.

Elia, A., Vyas, J., Laing, K. G., and Clemens, M. J. (2004). Ribosomal protein L22 inhibits regulation of cellular activities by the Epstein-Barr virus small RNA EBER-1. Eur. J. Biochem. *271*, 1895-1905.

Eliopoulos, A. G., Blake, S. M., Floettmann, J. E., Rowe, M., and Young, L. S. (1999). Epstein-Barr virus-encoded latent membrane protein 1 activates the JNK pathway through its extreme C terminus via a mechanism involving TRADD and TRAF2. J. Virol. *73*, 1023-1035.

Eliopoulos, A. G., and Young, L. S. (1998). Activation of the cJun N-terminal kinase (JNK) pathway by the Epstein-Barr virus-encoded latent membrane protein 1 (LMP1). Oncogene *16*, 1731-1742.

Fischer, N., Kremmer, E., Lautscham, G., Mueller-Lantzsch, N., and Grasser, F. A. (1997). Epstein-Barr virus nuclear antigen 1 forms a complex with the nuclear transporter karyopherin alpha2. J. Biol. Chem. *272*, 3999-4005.

Frappier, L., and O'Donnell, M. (1991). Epstein-Barr nuclear antigen 1 mediates a DNA loop within the latent replication origin of Epstein-Barr virus. Proc. Natl. Acad. Sci. USA. *88*, 10875-10879.

Fries, K. L., Miller, W. E., and Raab-Traub, N. (1996). Epstein-Barr virus latent membrane protein 1 blocks p53-mediated apoptosis through the induction of the A20 gene. J. Virol. *70*, 8653-8659.

Fruehling, S., and Longnecker, R. (1997). The immunoreceptor tyrosine-based activation motif of Epstein-Barr virus LMP2A is essential for blocking BCR-mediated signal transduction. Virology *235*, 241-251.

Gahn, T. A., and Schildkraut, C. L. (1989). The Epstein-Barr virus origin of plasmid replication, oriP, contains both the initiation and termination sites of DNA replication. Cell *58*, 527-535.

Gahn, T. A., and Sugden, B. (1995). An EBNA-1-dependent enhancer acts from a distance of 10 kilobase pairs to increase expression of the Epstein-Barr virus LMP gene. J. Virol. *69*, 2633-2636.

Gilbert, D. M., and Cohen, S. N. (1987). Bovine papilloma virus plasmids replicate randomly in mouse fibroblasts throughout S phase of the cell cycle. Cell *50*, 59-68.

Gires, O., Kohlhuber, F., Kilger, E., Baumann, M., Kieser, A., Kaiser, C., Zeidler, R., Scheffer, B., Ueffing, M., and Hammerschmidt, W. (1999). Latent membrane protein 1 of Epstein-Barr virus interacts with JAK3 and activates STAT proteins. Embo J. *18*, 3064-3073.

Gulley, M. L., Raphael, M., Lutz, C. T., Ross, D. W., and Raab-Traub, N. (1992). Epstein-Barr virus integration in human lymphomas and lymphoid cell lines. Cancer *70*, 185-191.

Hammerschmidt, W., and Sugden, B. (1989). Genetic analysis of immortalizing functions of Epstein-Barr virus in human B lymphocytes. Nature *340*, 393-397.

Harrison, S., Fisenne, K., and Hearing, J. (1994). Sequence requirements of the Epstein-Barr virus latent origin of DNA replication. J. Virol. *68*, 1913-1925.

Henderson, S., Rowe, M., Gregory, C., Croom-Carter, D., Wang, F., Longnecker, R., Kieff, E., and Rickinson, A. (1991). Induction of bcl-2 expression by Epstein-Barr virus latent membrane protein 1 protects infected B cells from programmed cell death. Cell *65*, 1107-1115.

Hochberg, D., Souza, T., Catalina, M., Sullivan, J. L., Luzuriaga, K., and Thorley-Lawson, D. A. (2004). Acute infection with Epstein-Barr virus targets and overwhelms the peripheral memory B-cell compartment with resting, latently infected cells. J. Virol. *78*, 5194-5204.

Howe, J. G., and Steitz, J. A. (1986). Localization of Epstein-Barr virus-encoded small RNAs by in situ hybridization. Proc. Natl. Acad. Sci. USA. *83*, 9006-9010.

Huang, D. P., Ho, J. H., Poon, Y. F., Chew, E. C., Saw, D., Lui, M., Li, C. L., Mak, L. S., Lai, S. H., and Lau, W. H. (1980). Establishment of a cell line (NPC/HK1) from a differentiated squamous carcinoma of the nasopharynx. Int. J. Cancer *26*, 127-132.

Humme, S., Reisbach, G., Feederle, R., Delecluse, H. J., Bousset, K., Hammerschmidt, W., and Schepers, A. (2003). The EBV nuclear antigen 1 (EBNA1) enhances B cell immortalization several thousandfold. Proc. Natl. Acad. Sci. USA. *100*, 10989-10994.

Hung, S. C., Kang, M. S., and Kieff, E. (2001). Maintenance of Epstein-Barr virus (EBV) oriP-based episomes requires EBV-encoded nuclear antigen-1 chromosome-binding domains, which can be replaced by high-mobility group-I or histone H1. Proc. Natl. Acad. Sci. USA. *98*, 1865-1870.

Hurley, E. A., Agger, S., McNeil, J. A., Lawrence, J. B., Calendar, A., Lenoir, G., and Thorley-Lawson, D. A. (1991). When Epstein-Barr virus persistently infects B-cell lines, it frequently integrates. J. Virol. *65*, 1245-1254.

Iwakiri, D., Eizuru, Y., Tokunaga, M., and Takada, K. (2003). Autocrine growth of Epstein-Barr virus-positive gastric carcinoma cells mediated by an Epstein-Barr virus-encoded small RNA. Cancer Res. *63*, 7062-7067.

Izumi, K. M., Cahir McFarland, E. D., Ting, A. T., Riley, E. A., Seed, B., and Kieff, E. D. (1999). The Epstein-Barr virus oncoprotein latent membrane protein 1 engages the tumor necrosis factor receptor-associated proteins TRADD and receptor-interacting protein (RIP) but does not induce apoptosis or require RIP for NF-kappaB activation. Mol. Cell Biol. *19*, 5759-5767.

Izumi, K. M., Kaye, K. M., and Kieff, E. D. (1997). The Epstein-Barr virus LMP1 amino acid sequence that engages tumor necrosis factor receptor associated factors is critical for primary B lymphocyte growth transformation. Proc. Natl. Acad. Sci. USA. *94*, 1447-1452.

Jeang, K. T., and Hayward, S. D. (1983). Organization of the Epstein-Barr virus DNA molecule. III. Location of the P3HR-1 deletion junction and characterization of the NotI repeat units that form part of the template for an abundant 12-O-tetradecanoylphorbol-13-acetate-induced mRNA transcript. J. Virol. *48*, 135-148.

Jeon, S., and Lambert, P. F. (1995). Integration of human papillomavirus type 16 DNA into the human genome leads to increased stability of E6 and E7 mRNAs: implications for cervical carcinogenesis. Proc. Natl. Acad. Sci. USA. *92*, 1654-1658.

Kanda, T., Yajima, M., Ahsan, N., Tanaka, M., and Takada, K. (2004). Production of high-titer Epstein-Barr virus recombinants derived from Akata cells by using a bacterial artificial chromosome system. J. Virol. *78*, 7004-7015.

Kapoor, P., and Frappier, L. (2003). EBNA1 partitions Epstein-Barr virus plasmids in yeast cells by attaching to human EBNA1-binding protein 2 on mitotic chromosomes. J. Virol. *77*, 6946-6956.

Kapoor, P., Shire, K., and Frappier, L. (2001). Reconstitution of Epstein-Barr virus-based plasmid partitioning in budding yeast. Embo J. *20*, 222-230.

Kaye, K. M., Izumi, K. M., and Kieff, E. (1993). Epstein-Barr virus latent membrane protein 1 is essential for B-lymphocyte growth transformation. Proc. Natl. Acad. Sci. USA. *90*, 9150-9154.

Kempkes, B., Pich, D., Zeidler, R., Sugden, B., and Hammerschmidt, W. (1995). Immortalization of human B lymphocytes by a plasmid containing 71 kilobase pairs of Epstein-Barr virus DNA. J. Virol. *69*, 231-238.

Kennedy, G., Komano, J., and Sugden, B. (2003). Epstein-Barr virus provides a survival factor to Burkitt's lymphomas. Proc. Natl. Acad. Sci. USA. *100*, 14269-14274.

Kennedy, G., and Sugden, B. (2003). EBNA-1, a bifunctional transcriptional activator. Mol. Cell Biol. *23*, 6901-6908.

Kieser, A., Kaiser, C., and Hammerschmidt, W. (1999). LMP1 signal transduction differs substantially from TNF receptor 1 signaling in the molecular functions of TRADD and TRAF2. Embo J. *18*, 2511-2521.

Kieser, A., Kilger, E., Gires, O., Ueffing, M., Kolch, W., and Hammerschmidt, W. (1997). Epstein-Barr virus latent membrane protein-1 triggers AP-1 activity via the c-Jun N-terminal kinase cascade. Embo J. *16*, 6478-6485.

Kilger, E., Kieser, A., Baumann, M., and Hammerschmidt, W. (1998). Epstein-Barr virus-mediated B-cell proliferation is dependent upon latent membrane protein 1, which simulates an activated CD40 receptor. Embo J. *17*, 1700-1709.

Kim, A. L., Maher, M., Hayman, J. B., Ozer, J., Zerby, D., Yates, J. L., and Lieberman, P. M. (1997). An imperfect correlation between DNA replication activity of Epstein-Barr virus nuclear antigen 1 (EBNA1) and binding to the nuclear import receptor, Rch1/importin alpha. Virology *239*, 340-351.

Kirchmaier, A. L., and Sugden, B. (1995). Plasmid maintenance of derivatives of oriP of Epstein-Barr virus. J. Virol. *69*, 1280-1283.

Kirchmaier, A. L., and Sugden, B. (1997). Dominant-negative inhibitors of EBNA-1 of Epstein-Barr virus. J. Virol. *71*, 1766-1775.

Kirchmaier, A. L., and Sugden, B. (1998). Rep*: a viral element that can partially replace the origin of plasmid DNA synthesis of Epstein-Barr virus. J. Virol. *72*, 4657-4666.

Kitagawa, N., Goto, M., Kurozumi, K., Maruo, S., Fukayama, M., Naoe, T., Yasukawa, M., Hino, K., Suzuki, T., Todo, S., and Takada, K. (2000). Epstein-Barr virus-encoded poly(A)(-) RNA supports Burkitt's lymphoma growth through interleukin-10 induction. Embo J. *19*, 6742-6750.

Komano, J., Maruo, S., Kurozumi, K., Oda, T., and Takada, K. (1999). Oncogenic role of Epstein-Barr virus-encoded RNAs in Burkitt's lymphoma cell line Akata. J. Virol. *73*, 9827-9831.

Koons, M. D., Van Scoy, S., and Hearing, J. (2001). The replicator of the Epstein-Barr virus latent cycle origin of DNA replication, oriP, is composed of multiple functional elements. J. Virol. *75*, 10582-10592.

Krysan, P. J., Haase, S. B., and Calos, M. P. (1989). Isolation of human sequences that replicate autonomously in human cells. Mol. Cell Biol. *9*, 1026-1033.

Kurth, J., Spieker, T., Wustrow, J., Strickler, G. J., Hansmann, L. M., Rajewsky, K., and Kuppers, R. (2000). EBV-infected B cells in infectious mononucleosis: viral strategies for spreading in the B cell compartment and establishing latency. Immunity *13*, 485-495.

Laherty, C. D., Hu, H. M., Opipari, A. W., Wang, F., and Dixit, V. M. (1992). The Epstein-Barr virus LMP1 gene product induces A20 zinc finger protein expression by activating nuclear factor kappa B. J. Biol. Chem. *267*, 24157-24160.

Laine, A., and Frappier, L. (1995). Identification of Epstein-Barr virus nuclear antigen 1 protein domains that direct interactions at a distance between DNA-bound proteins. J. Biol. Chem. *270*, 30914-30918.

Lam, K. M., Syed, N., Whittle, H., and Crawford, D. H. (1991). Circulating Epstein-Barr virus-carrying B cells in acute malaria. Lancet *337*, 876-878.

Langle-Rouault, F., Patzel, V., Benavente, A., Taillez, M., Silvestre, N., Bompard, A., Sczakiel, G., Jacobs, E., and Rittner, K. (1998). Up to 100-fold increase of apparent gene expression in the presence of Epstein-Barr virus oriP sequences and EBNA1: implications of the nuclear import of plasmids. J. Virol. *72*, 6181-6185.

Leight, E. R., and Sugden, B. (2001). Establishment of an oriP replicon is dependent upon an infrequent, epigenetic event. Mol. Cell Biol. *21*, 4149-4161.

Lerner, M. R., Andrews, N. C., Miller, G., and Steitz, J. A. (1981). Two small RNAs encoded by Epstein-Barr virus and complexed with protein are precipitated by antibodies from patients with systemic lupus erythematosus. Proc. Natl. Acad. Sci. USA. *78*, 805-809.

Lin, C. T., Dee, A. N., Chen, W., and Chan, W. Y. (1994). Association of Epstein-Barr virus, human papilloma virus, and cytomegalovirus with nine nasopharyngeal carcinoma cell lines. Lab. Invest *71*, 731-736.

Lo, K. W., and Huang, D. P. (2002). Genetic and epigenetic changes in nasopharyngeal carcinoma. Semin Cancer Biol. *12*, 451-462.

Longnecker, R., and Kieff, E. (1990). A second Epstein-Barr virus membrane protein (LMP2) is expressed in latent infection and colocalizes with LMP1. J. Virol. *64*, 2319-2326.

Longnecker, R., Miller, C. L., Miao, X. Q., Marchini, A., and Kieff, E. (1992). The only domain which distinguishes Epstein-Barr virus latent membrane protein 2A (LMP2A) from LMP2B is dispensable for lymphocyte infection and growth transformation *in vitro*; LMP2A is therefore nonessential. J. Virol. *66*, 6461-6469.

Lu, J., Chua, H. H., Chen, S. Y., Chen, J. Y., and Tsai, C. H. (2003). Regulation of matrix metalloproteinase-1 by Epstein-Barr virus proteins. Cancer Res. *63*, 256-262.

Lupton, S., and Levine, A. J. (1985). Mapping genetic elements of Epstein-Barr virus that facilitate extrachromosomal persistence of Epstein-Barr virus-derived plasmids in human cells. Mol. Cell Biol. *5*, 2533-2542.

Mackey, D., Middleton, T., and Sugden, B. (1995). Multiple regions within EBNA1 can link DNAs. J. Virol. *69*, 6199-6208.

Mackey, D., and Sugden, B. (1997). Studies on the mechanism of DNA linking by Epstein-Barr virus nuclear antigen 1. J. Biol. Chem. *272*, 29873-29879.

Maiorano, D., Moreau, J., and Mechali, M. (2000). XCDT1 is required for the assembly of pre-replicative complexes in Xenopus laevis. Nature *404*, 622-625.

Marechal, V., Dehee, A., Chikhi-Brachet, R., Piolot, T., Coppey-Moisan, M., and Nicolas, J. C. (1999). Mapping EBNA-1 domains involved in binding to metaphase chromosomes. J. Virol. *73*, 4385-4392.

Martin, J. M., Veis, D., Korsmeyer, S. J., and Sugden, B. (1993). Latent membrane protein of Epstein-Barr virus induces cellular phenotypes independently of expression of Bcl-2. J. Virol. *67*, 5269-5278.

Middleton, T., and Sugden, B. (1992). A chimera of EBNA1 and the estrogen receptor activates transcription but not replication. J. Virol. *66*, 1795-1798.

Middleton, T., and Sugden, B. (1994). Retention of plasmid DNA in mammalian cells is enhanced by binding of the Epstein-Barr virus replication protein EBNA1. J. Virol. *68*, 4067-4071.

Miller, C. L., Lee, J. H., Kieff, E., and Longnecker, R. (1994). An integral membrane protein (LMP2) blocks reactivation of Epstein-Barr virus from latency following surface immunoglobulin crosslinking. Proc. Natl. Acad. Sci. USA. *91*, 772-776.

Miller, W. E., Earp, H. S., and Raab-Traub, N. (1995). The Epstein-Barr virus latent membrane protein 1 induces expression of the epidermal growth factor receptor. J. Virol. *69*, 4390-4398.

Miyashita, E. M., Yang, B., Babcock, G. J., and Thorley-Lawson, D. A. (1997). Identification of the site of Epstein-Barr virus persistence *in vivo* as a resting B cell. J. Virol. *71*, 4882-4891.

Mosialos, G., Birkenbach, M., Yalamanchili, R., VanArsdale, T., Ware, C., and Kieff, E. (1995). The Epstein-Barr virus transforming protein LMP1 engages signaling proteins for the tumor necrosis factor receptor family. Cell *80*, 389-399.

Nanbo, A., Inoue, K., Adachi-Takasawa, K., and Takada, K. (2002). Epstein-Barr virus RNA confers resistance to interferon-alpha-induced apoptosis in Burkitt's lymphoma. Embo J. *21*, 954-965.

Niller, H. H., Glaser, G., Knuchel, R., and Wolf, H. (1995). Nucleoprotein complexes and DNA 5'-ends at oriP of Epstein-Barr virus. J. Biol. Chem. *270*, 12864-12868.

Polack, A., Hortnagel, K., Pajic, A., Christoph, B., Baier, B., Falk, M., Mautner, J., Geltinger, C., Bornkamm, G. W., and Kempkes, B. (1996). c-myc activation renders proliferation of Epstein-Barr virus (EBV)-transformed cells independent of EBV nuclear antigen 2 and latent membrane protein 1. Proc. Natl. Acad. Sci. USA. *93*, 10411-10416.

Polvino-Bodnar, M., and Schaffer, P. A. (1992). DNA binding activity is required for EBNA 1-dependent transcriptional activation and DNA replication. Virology *187*, 591-603.

Portis, T., and Longnecker, R. (2003). Epstein-Barr virus LMP2A interferes with global transcription factor regulation when expressed during B-lymphocyte development. J. Virol. *77*, 105-114.

Puglielli, M. T., Woisetschlaeger, M., and Speck, S. H. (1996). oriP is essential for EBNA gene promoter activity in Epstein-Barr virus-immortalized lymphoblastoid cell lines. J. Virol. *70*, 5758-5768.

Qu, L., and Rowe, D. T. (1992). Epstein-Barr virus latent gene expression in uncultured peripheral blood lymphocytes. J. Virol. *66*, 3715-3724.

Rawlins, D. R., Milman, G., Hayward, S. D., and Hayward, G. S. (1985). Sequence-specific DNA binding of the Epstein-Barr virus nuclear antigen (EBNA-1) to clustered sites in the plasmid maintenance region. Cell *42*, 859-868.

Razzouk, B. I., Srinivas, S., Sample, C. E., Singh, V., and Sixbey, J. W. (1996). Epstein-Barr Virus DNA recombination and loss in sporadic Burkitt's lymphoma. J. Infect. Dis. *173*, 529-535.

Reeves, R., and Beckerbauer, L. (2001). HMGI/Y proteins: flexible regulators of transcription and chromatin structure. Biochim Biophys Acta *1519*, 13-29.

Reeves, R., and Nissen, M. S. (1990). The A.T-DNA-binding domain of mammalian high mobility group I chromosomal proteins. A novel peptide motif for recognizing DNA structure. J. Biol. Chem. *265*, 8573-8582.

Reisman, D., and Sugden, B. (1986). trans activation of an Epstein-Barr viral transcriptional enhancer by the Epstein-Barr viral nuclear antigen 1. Mol. Cell Biol. *6*, 3838-3846.

Reisman, D., Yates, J., and Sugden, B. (1985). A putative origin of replication of plasmids derived from Epstein-Barr virus is composed of two cis-acting components. Mol. Cell Biol. *5*, 1822-1832.

Rosa, M. D., Gottlieb, E., Lerner, M. R., and Steitz, J. A. (1981). Striking similarities are exhibited by two small Epstein-Barr virus-encoded ribonucleic acids and the adenovirus-associated ribonucleic acids VAI and VAII. Mol. Cell Biol. *1*, 785-796.

Rowe, M., Rowe, D. T., Gregory, C. D., Young, L. S., Farrell, P. J., Rupani, H., and Rickinson, A. B. (1987). Differences in B cell growth phenotype reflect novel patterns of Epstein-Barr virus latent gene expression in Burkitt's lymphoma cells. Embo J. *6*, 2743-2751.

Sandberg, M., Hammerschmidt, W., and Sugden, B. (1997). Characterization of LMP-1's association with TRAF1, TRAF2, and TRAF3. J. Virol. *71*, 4649-4656.

Schepers, A., Ritzi, M., Bousset, K., Kremmer, E., Yates, J. L., Harwood, J., Diffley, J. F., and Hammerschmidt, W. (2001). Human origin recognition complex binds to the region of the latent origin of DNA replication of Epstein-Barr virus. Embo J. *20*, 4588-4602.

Schultheiss, U., Puschner, S., Kremmer, E., Mak, T. W., Engelmann, H., Hammerschmidt, W., and Kieser, A. (2001). TRAF6 is a critical mediator of signal transduction by the viral oncogene latent membrane protein 1. Embo J. *20*, 5678-5691.

Sears, J., Kolman, J., Wahl, G. M., and Aiyar, A. (2003). Metaphase chromosome tethering is necessary for the DNA synthesis and maintenance of oriP plasmids but is insufficient for transcription activation by Epstein-Barr nuclear antigen 1. J. Virol. *77*, 11767-11780.

Sears, J., Ujihara, M., Wong, S., Ott, C., Middeldorp, J., and Aiyar, A. (2004). The Amino Terminus of Epstein-Barr Virus (EBV) Nuclear Antigen 1 Contains AT Hooks That Facilitate the Replication and Partitioning of Latent EBV Genomes by Tethering Them to Cellular Chromosomes. J. Virol. *78*, 11487-11505.

Seo, Y. S., Muller, F., Lusky, M., Gibbs, E., Kim, H. Y., Phillips, B., and Hurwitz, J. (1993). Bovine papilloma virus (BPV)-encoded E2 protein enhances binding of E1 protein to the BPV replication origin. Proc. Natl. Acad. Sci. USA. *90*, 2865-2869.

Shimizu, N., Tanabe-Tochikura, A., Kuroiwa, Y., and Takada, K. (1994). Isolation of Epstein-Barr virus (EBV)-negative cell clones from the EBV-positive Burkitt's lymphoma (BL) line Akata: malignant phenotypes of BL cells are dependent on EBV. J. Virol. *68*, 6069-6073.

Shire, K., Ceccarelli, D. F., Avolio-Hunter, T. M., and Frappier, L. (1999). EBP2, a human protein that interacts with sequences of the Epstein-Barr virus nuclear antigen 1 important for plasmid maintenance. J. Virol. *73*, 2587-2595.

Stahl, H., Droge, P., and Knippers, R. (1986). DNA helicase activity of SV40 large tumor antigen. Embo J. *5*, 1939-1944.

Su, W., Middleton, T., Sugden, B., and Echols, H. (1991). DNA looping between the origin of replication of Epstein-Barr virus and its enhancer site: stabilization of an origin complex with Epstein-Barr nuclear antigen 1. Proc. Natl. Acad. Sci. USA. *88*, 10870-10874.

Sugden, B., and Warren, N. (1988). Plasmid origin of replication of Epstein-Barr virus, oriP, does not limit replication in cis. Mol. Biol. Med. *5*, 85-94.

Sugden, B., and Warren, N. (1989). A promoter of Epstein-Barr virus that can function during latent infection can be transactivated by EBNA-1, a viral protein required for viral DNA replication during latent infection. J. Virol. *63*, 2644-2649.

Swaminathan, S., Tomkinson, B., and Kieff, E. (1991). Recombinant Epstein-Barr virus with small RNA (EBER) genes deleted transforms lymphocytes and replicates *in vitro*. Proc. Natl. Acad. Sci. USA. *88*, 1546-1550.

Taub, R., Kirsch, I., Morton, C., Lenoir, G., Swan, D., Tronick, S., Aaronson, S., and Leder, P. (1982). Translocation of the c-myc gene into the immunoglobulin heavy chain locus in human Burkitt lymphoma and murine plasmacytoma cells. Proc. Natl. Acad. Sci. USA. *79*, 7837-7841.

Toczyski, D. P., Matera, A. G., Ward, D. C., and Steitz, J. A. (1994). The Epstein-Barr virus (EBV) small RNA EBER1 binds and relocalizes ribosomal protein L22 in EBV-infected human B lymphocytes. Proc. Natl. Acad. Sci. USA. *91*, 3463-3467.

Tomkinson, B., Robertson, E., and Kieff, E. (1993). Epstein-Barr virus nuclear proteins EBNA-3A and EBNA-3C are essential for B-lymphocyte growth transformation. J. Virol. *67*, 2014-2025.

Uchida, J., Yasui, T., Takaoka-Shichijo, Y., Muraoka, M., Kulwichit, W., Raab-Traub, N., and Kikutani, H. (1999). Mimicry of CD40 signals by Epstein-Barr virus LMP1 in B lymphocyte responses. Science *286*, 300-303.

Wade-Martins, R., Frampton, J., and James, M. R. (1999). Long-term stability of large insert genomic DNA episomal shuttle vectors in human cells. Nucleic Acids Res. *27*, 1674-1682.

Wu, H., Ceccarelli, D. F., and Frappier, L. (2000). The DNA segregation mechanism of Epstein-Barr virus nuclear antigen 1. EMBO Rep. *1*, 140-144.

Wu, T. C., Mann, R. B., Epstein, J. I., MacMahon, E., Lee, W. A., Charache, P., Hayward, S. D., Kurman, R. J., Hayward, G. S., and Ambinder, R. F. (1991). Abundant expression of EBER1 small nuclear RNA in nasopharyngeal carcinoma. A morphologically distinctive target for detection of Epstein-Barr virus in formalin-fixed paraffin-embedded carcinoma specimens. Amer. J. Pathol. *138*, 1461-1469.

Wysokenski, D. A., and Yates, J. L. (1989). Multiple EBNA1-binding sites are required to form an EBNA1-dependent enhancer and to activate a minimal replicative origin within oriP of Epstein-Barr virus. J. Virol. *63*, 2657-2666.

Yang, L., Mohr, I., Fouts, E., Lim, D. A., Nohaile, M., and Botchan, M. (1993). The E1 protein of bovine papilloma virus 1 is an ATP-dependent DNA helicase. Proc. Natl. Acad. Sci. USA. *90*, 5086-5090.

Yates, J., Warren, N., Reisman, D., and Sugden, B. (1984). A cis-acting element from the Epstein-Barr viral genome that permits stable replication of recombinant plasmids in latently infected cells. Proc. Natl. Acad. Sci. USA. *81*, 3806-3810.

Yates, J. L., Camiolo, S. M., and Bashaw, J. M. (2000). The minimal replicator of Epstein-Barr virus oriP. J. Virol. *74*, 4512-4522.

Yates, J. L., and Guan, N. (1991). Epstein-Barr virus-derived plasmids replicate only once per cell cycle and are not amplified after entry into cells. J. Virol. *65*, 483-488.

Yates, J. L., Warren, N., and Sugden, B. (1985). Stable replication of plasmids derived from Epstein-Barr virus in various mammalian cells. Nature *313*, 812-815.

Zeng, Y., Zhang, L. G., Wu, Y. C., Huang, Y. S., Huang, N. Q., Li, J. Y., Wang, Y. B., Jiang, M. K., Fang, Z., and Meng, N. N. (1985). Prospective studies on nasopharyngeal carcinoma in Epstein-Barr virus IgA/VCA antibody-positive persons in Wuzhou City, China. Int. J. Cancer *36*, 545-547.

From: Epstein-Barr Virus. Edited by: Erle S. Robertson

Chapter 20

Regulation of EBV Latency-Associated Gene Expression

*Samuel H. Speck**

ABSTRACT

The association of Epstein-Barr virus (EBV) with a number of cancers, and the ability of EBV to latently infect and immortalize B cells *in vitro*, has led to an intense interest in identifying the mechanisms by which EBV is able to transform B cells. To identify viral genes involved in B cell growth transformation, an extensive effort was mounted to identify transcriptionally active regions of the viral genome in various lymphoblastoid cell lines (LCLs) and Burkitt lymphoma (BL) cell lines. These studies ultimately revealed that transcription of the genes encoding the EBV latency-associated antigens is singularly complex and intriguing. This Chapter will focus on what we have learned about transcription of the Epstein-Barr nuclear antigen (EBNA) genes and the latency-associated membrane protein (LMP) genes, as well as speculate about mechanisms that may be involved in regulating EBV gene expression.

INTRODUCTION
EBV latency - a complex relationship between virus and host cell
EBV immortalized B lymphoblasts express 9 virus-encoded antigens, 6 nuclear antigens (Epstein-Barr nuclear antigens, EBNAs) and 3 membrane proteins (latency-associated membrane proteins, LMPs), as well as 2 small pol III transcripts referred to as the EBERs (Epstein-Barr encoded RNAs) (latency III program (Figure 1)) (Kieff, 2001). The functions of these viral antigens is discussed elsewhere in this book (see Chapters 15, 17, 26, 19, 21, 23, and 25), and the organization of these genes and their transcription is discussed below. Because of the ability of EBV to growth transform B cells, for many years most investigators in the EBV field tacitly assumed that this growth transforming latency program was involved in driving proliferation of EBV-associated tumors. This view was, in large part, fueled by early characterizations of viral gene expression in cell lines established from Burkitt's lymphoma (BL) tumors which showed that, in general, these tumor cell

*For correspondence email ssspeck@rmy.emory.edu

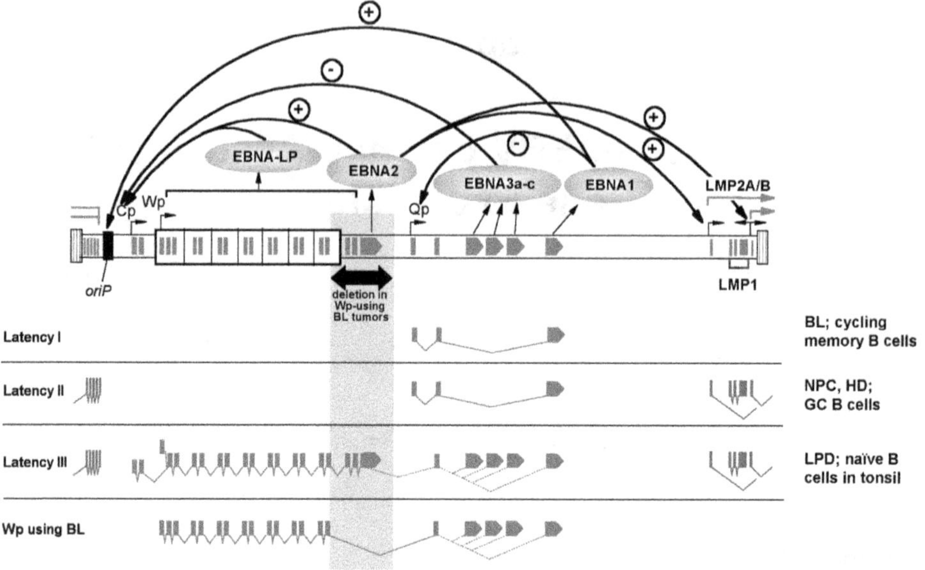

Figure 1. Diagram depicting the locations within the 172 Kb EBV genome of the EBNA gene promoters Cp, Wp, and Qp, and the LMP gene promoters, and illustrating the structures of the EBNA gene and LMP transcripts observed during the distinct latency programs. It should be noted that the EBV genome exists as a covalently closed episome in the nucleus of infected cells, and that transcription of the LMP2A and LMP2B genes spans the fused terminal repeats in the viral episome. In the Latency I and Latency II programs the only EBNA gene expressed is EBNA 1, which transcribed from the Qp promoter. In Latency II, in addition to EBNA1, the LMP genes are also expressed. The Latency III program involves expression of all six EBNAs, generated by alternative 3′ processing coupled with alternative splicing of long primary transcripts that initiate at either Cp or Wp. Note that since Wp is encoded within the major internal repeat of the virus there are multiple copies of Wp in the EBV genome (only the first Wp promoter is indicated in the diagram). In addition, the LMP genes are expressed in the Latency III program. Transcription of the EBNA and LMP genes is autoregulated by the EBNA gene products, as indicated. EBNA1 binds to multiple sites within the viral latency-associated origin, oriP, and enhances transcription from Cp, Wp and the LMP gene promoters. In contrast, EBNA1 inhibits Qp activity through binding to two low affinity sites downstream of the Qp transcription initiation site. EBNA2 activates transcription from Cp and the LMP promoters through interaction with the cellular factor CBF1/RBPJk, which is bound to sites within enhancers located upstream of each of these promoters. EBNA-LP serves to augment EBNA2 activation of Cp and the LMP promoters, while the EBNA 3 gene products have all been shown to be capable of counteracting EBNA2 activation of viral gene expression. For the sake of clarity, the actions of EBNA-LP and EBNA3a-c are only indicated on Cp, but also occur at the LMP promoters. The region deleted in the Wp-using BL tumors described by Kelley et al. is indicated, as is the EBV transcription program observed in this subset of BL tumors (Kelly et al., 2002).

lines phenotypically resembled EBV immortalized B blasts (Gregory et al., 1990; Rowe et al., 1986). However, subsequent analyses of fresh BL biopsies, as well as low passage cell lines established from BL tumors, demonstrated that BL tumors have a poorly differentiated resting B cell phenotype (Rowe et al., 1986; 1987b).

In addition, BL tumor biopsies express only a single EBNA gene product, EBNA1, demonstrating the establishment of a restricted form of EBV latency (Rowe et al., 1986; Rowe et al., 1987b). Subsequent analyses of BL tumors in culture demonstrated that many of these tumor lines "drift" from a resting B cell phenotype to an activated B blast phenotype in which the cells ultimately express the full range of EBV latency antigens detected in EBV immortalized LCLs. Characterization of other EBV-associated neoplasms that arise in immunocompetent individuals (*e.g.*, nasopharyngeal carcinoma and certain subtypes of Hodgkin's disease) has also revealed restricted EBNA gene expression, along with variable expression of the LMPs (Rickinson, 2001).

Building a lasting memory

More recently, analyses of EBV infection of specific B cell populations in seropositive individuals has revealed that EBV encodes at least 3 distinct latency programs that are tightly coupled to the differentiation state of the infected B cell (Thorley-Lawson, 2001). Examination of peripheral blood recovered from seropositive individuals has demonstrated the presence of EBV only in resting memory B cells (Babcock et al., 1998). However, analysis of EBV infection in tonsils demonstrated the presence of virus in naive, germinal center and memory B cells (Babcock and Thorley-Lawson, 2000; Joseph et al., 2000). Furthermore, analysis of EBV gene expression in the B cell subsets revealed different patterns of viral gene expression (Thorley-Lawson, 2001). In infected naive B cells (CD19+, IgD+), the full pattern of EBNA and LMP gene expression was observed (latency III program; see Figure 1), while in germinal center B cells only EBNA1 and the LMP gene were expressed (latency II program; see Figure 1). In contrast, the majority of latently infected memory B cells in the periphery do not appear to express any viral antigens (Hochberg et al., 2004). However, it was recently shown that those memory B cells that have recently divided express EBNA1 (latency I program; see Figure 1) (Hochberg et al., 2004), consistent with a role for EBNA1 in maintenance of the viral episome in this population of latently infected cells. Thus, the restricted patterns of EBV gene expression observed in various EBV-associated tumors are also seen in specific populations of infected B cells in normal seropositive individuals (Thorley-Lawson, 2001).

The existence of different EBV latency programs raises the question of their role in chronic infection. The most compelling model put forward to date argues that EBV has evolved to usurp normal B cell differentiation to ensure maintenance of chronic infection in the face of the host's adaptive immune response (Thorley-Lawson, 2001). As discussed elsewhere in this book (see Chapter 17), the EBV latency III program (also referred to as the growth program) appears to have evolved to drive differentiation naive B cells, leading to germinal center formation and, ultimately, latently infected memory B cells. Whether there is any role for antigen in this process is unclear. Importantly, accessing the memory B cell reservoir with concomitant down-regulation of antigenic EBNA and LMP gene expression would appear to allow EBV to persist in the infected host without detection by the host immune system.

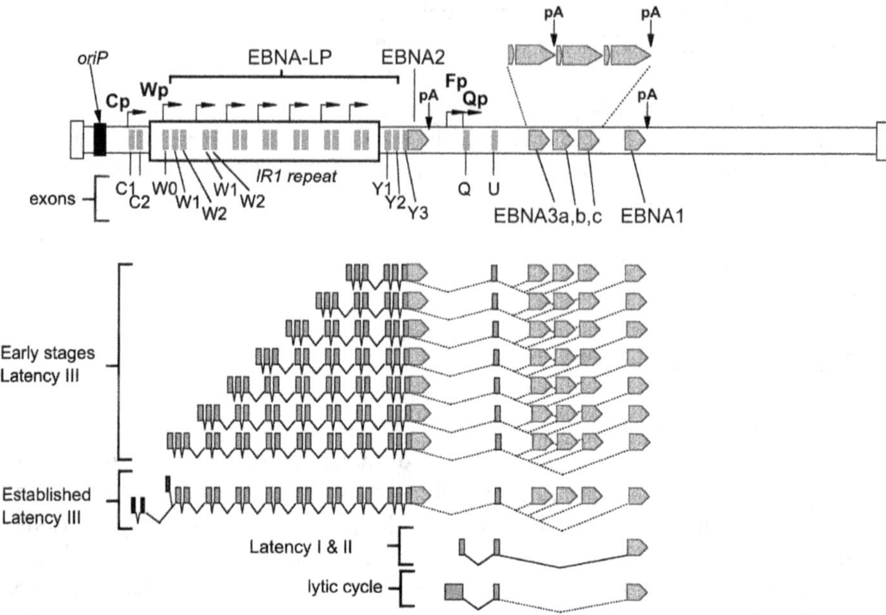

Figure 2. Exon organization of EBNA gene transcripts. Transcription of the EBNA genes in the latency III program is driven from either Cp or Wp, while transcription of the EBNA1 gene during type I and type II latency is driven from Qp. During the establishment of the latency III program multiple size EBNA-LP species are observed and, as depicted, are thought to arise from initiation of EBNA-LP gene transcription at different copies of Wp present within the IR1 repeat. Fp-initiated transcripts are detected during the viral lytic cycle, and a small percentage of these transcripts splice to the EBNA1 coding exon. As shown in the enlargement of the EBNA3 coding region, the EBNA 3a, 3b and 3c coding exons are composed of a short coding exon which splices to a long coding exon located just downstream of the short coding exon. The locations of consensus polyadenylation signals (pA) closely linked to EBNA coding exons are also shown.

ORGANIZATION OF THE EBNA GENES AND STRUCTURE OF EBNA GENE TRANSCRIPTS IN EBV IMMORTALIZED LYMPHOBLASTOID CELLS

The early studies detecting EBNA expression in latently infected cell lines relied heavily on EBV-positive human serum (Reedman and Klein, 1973; Reedman et al., 1974; Rowe et al., 1987a). Northern blots using RNA from LCLs and BL cell lines revealed regions of the EBV genome that are transcriptionally active during latent infection, and subsequent studies concentrated on these regions of the viral genome (Baer et al., 1984; Heller et al., 1982; van Santen et al., 1983; 1981; Weigel and Miller, 1983; 1985). The first EBNA to be described, EBNA1, was shown to be encoded in the BamHI K region of the genome; upon transfection of murine fibroblasts with a plasmid containing the EBV BamHI K fragment, the cells were rendered EBNA positive when screened with EBV-positive human serum (Summers et al., 1982). Three years later, a cDNA containing the 5′ end of the EBNA1 open reading frame was cloned from the JY LCL using an EBNA1 specific primer to initiate cDNA synthesis (Speck and Strominger, 1985). This cDNA demonstrated that the EBNA1 transcript is highly spliced, containing a number of small exons upstream of the EBNA1 coding exon, and originated from a transcriptional unit spanning over 70 kb. Earlier studies had shown that extensively spliced transcripts from the region of the major internal repeat (IR1; BamHI W repeats) were present in BL cells and LCLs (Bodescot et al., 1984; King et al., 1980), but this was the first time that an extensively spliced cDNA was shown to encode an EBNA (Speck and Strominger, 1985). The EBNA1 cDNA contained exons encoded within the IR1 repeat, as well as from the BamHI Y, U, E, and K regions of the viral genome (Figure 2). This initial report was followed by several other analyses of EBNA cDNA clones (Bodescot et al., 1986; Bodescot and Perricaudet, 1986; Sample et al., 1986; Speck et al., 1986), which confirmed the prediction that all the EBNA gene transcripts share common 5′ exons encoded by the major internal repeat of EBV (IR1) that are alternatively spliced to 3′ exons encoding the various EBNA gene products (Speck and Strominger, 1985).

For reasons that are not completely apparent, EBNA gene transcripts are quite difficult to clone. Nevertheless, a limited number of cDNAs have been characterized and these have provided a fairly detailed picture of the structure of EBNA gene transcripts (Figure 2). From the analysis of cDNA clones it became clear that EBNA gene transcripts initiate from one of two promoters; Cp, located in the unique region just upstream of IR1 (Bodescot et al., 1987; Woisetschlaeger et al., 1989), or Wp, encoded within the major internal repeat of the virus (IR1) and thus present in multiple copies (Figure 2) (Sample et al., 1986; Speck et al., 1986; Woisetschlaeger et al., 1989). Cp initiated transcripts contain two unique noncoding exons, C1 and C2, at their 5′ end. Wp initiated transcripts contain a single copy of the noncoding W0 exon at their 5′ end. Both the C2 and W0 exons splice to the first copy of the W1 repeat exon (Figure 2). As discussed below, based on cell lines that exclusively utilize Cp or Wp to drive EBNA gene transcription (Woisetschlaeger et al., 1989), it appears that both promoters are capable of driving expression of all 6 EBNA genes. Every EBNA transcript analyzed contains multiple copies of the W1 and W2 exons, which in most cases splice to 2 or 3 small exons (Y1, Y2 and Y3) encoded within the unique region just downstream of IR1 (Figure 2). Alternative splicing from the Y2 and Y3 exons, and from a downstream exon (U exon), dictate which 3′ coding exons are present in the mature EBNA gene transcripts (Figure 2).

While 5 of the EBNA genes are encoded by discrete exons located at the 3′ end of their respect transcripts, EBNA-LP (also referred to as EBNA4 in some early reports) is encoded by the repeat exons W1 and W2 and the unique exons Y1 and Y2 present at the 5′ end of all EBNA gene transcripts (Dillner et al., 1986; Rowe et al., 1987a; Sample et al., 1986; Speck et al., 1986; Speck and Strominger, 1985). Notably, the presence of the EBNA-LP translation initiation codon is dictated by alternative splicing since the AUG translation initiation codon is generated by the splice junction between the C2 and W1 exons or the W0 and W1 exons (Figure 3) (Rogers et al., 1990). Splicing to the first copy of the W1 exon in an EBNA transcript may utilize one of two possible splice acceptor sites; with the upstream (and more frequently used) splice acceptor failing to generate the translation initiation codon. Importantly, there are no other possible sites of translation initiation within the

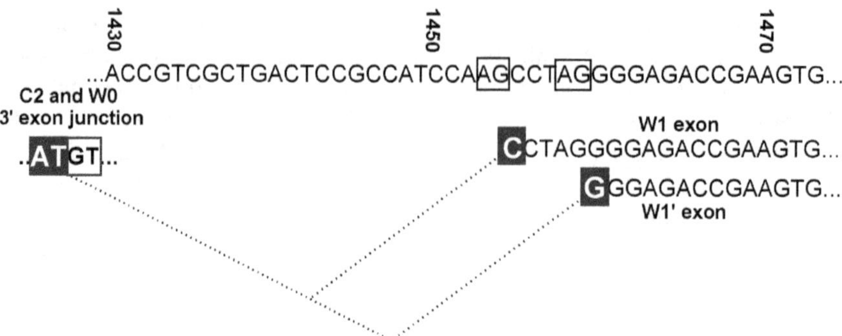

Figure 3. Alternative splicing for the C2 or W0 exons to the first W1 exon. Shown are the two splice acceptor sites present at the 5′ end of the W1 exon. Splicing to the upstream splice acceptor fails to generate the EBNA-LP translation initiation codon, while splicing to the downstream splice acceptor generates the EBNA-LP translation initiation codon. Notably, there are no other possible translation initiation sites within the EBNA-LP coding sequence.

EBNA-LP coding sequence. Thus, it appears that this mechanism has evolved to allow expression of EBNA-LP without interfering with efficient translation of the other EBNA gene products.

Alternative splicing and 3′ processing - existence of a splicing gradient?

With the exception of the sequences encoding EBNA3b and EBNA-LP, the EBNA coding exons are closely linked to a 3′ consensus polyadenylation signal (Baer et al., 1984). Since all of the EBNAs are transcribed from a single long transcriptional unit, the question arises as to how splicing to the downstream EBNAs is efficiently achieved. It appears that there is significant read-through transcription which ignores proximal polyadenylation signals. Studies of eukaryotic mRNA transcription and processing have increasingly supported the idea that transcription elongation, capping, RNA splicing, cleavage, and polyadenylation are all coupled to each other (Minvielle-Sebastia and Keller, 1999). For instance, analyses of the RNA polymerase II transcription complex have shown that the phosphorylated carboxy-terminal domain of RNA pol II associates with a number of factors involved in splicing and 3′ end processing, including SR proteins, the cleavage and polyadenylation specificity factor (CPSF), and the cleavage stimulation factor (CstF) (Minvielle-Sebastia and Keller, 1999). It is therefore possible that during elongation of the primary EBNA transcripts splicing 'enhancer' sequences in the downstream EBNA coding exons, or else favored polyadenylation signals for downstream EBNA genes, may increase the frequency of transcripts that splice to the downstream EBNA gene coding exons.

Suppression of spontaneous reactivation - antisense transcription?

It is curious that the EBNA genes are located such large distances from each other (spread over ~100 kb of the viral genome), and yet utilize the same promoters, Cp and Wp. There are at least two obvious hypotheses for the existence of such an extensive transcription and alternative splicing strategy in EBV latently infected cells. First, it is possible that by placing the promoters near the latency origin of replication (*oriP*), the virus is better able to regulate transcription of latency associated genes. *OriP* has been shown to function as an EBNA1-dependent enhancer of Cp- and Wp-initiated transcription (Puglielli et al., 1996; Sugden and Warren, 1989). Indeed, reporter gene assays in LCLs has demonstrated that *oriP* is essential for Cp and Wp activity (Puglielli et al., 1996). Consistent with the hypothesis that *oriP* serves as a major regulator of latency-associated gene transcription, the latent membrane protein-1 (LMP-1) gene promoter (which is also located near oriP) has been shown to be upregulated by EBNA1 binding to oriP (Gahn and Sugden, 1995). The role of viral origins of replication as enhancers of viral gene transcription is a common theme in many DNA viruses (Chiang et al., 1992; Wu and Carstens, 1996; Yang and Botchan, 1990). Notably, concentrating the latency promoters around *oriP*, and thus segregating them from lytic cycle-associated viral genes, may diminish spurious transcription of lytic genes.

Second, it is notable that the viral genes, BRLF1 and BZLF1, involved in triggering reactivation of latent EBV (Speck et al., 1997), are positioned within the EBNA1 gene primary transcriptional unit, but encoded on the opposite strand. Thus, transcription of EBNA1 is antisense to the BRLF1 and BZLF1 genes and this may reduce the basal level transcription of these genes. Some support for this hypothesis originated from a study showing that double stranded hybrid RNAs encoding the BZLF1 transcript and the noncoding EBNA1 intron are present in the P3HR1 BL cell line, and over-expression of a mutated (noncoding) BZLF1 transcript could saturate this mechanism and trigger induction of the lytic cycle (Prang et al., 1999). Further evidence for a linkage between the levels of EBNA gene transcripts and entry into the lytic cycle come from studies in the Jijoye BL cell line in which it was demonstrated that removal of glucocorticoids from the culture media resulted in diminished levels of Cp-initiated EBNA gene transcripts and an increase in the percentage of cells spontaneously entering the lytic cycle (Sinclair et al., 1992). Notably, a similar scenario has been postulated for neurons latently infected with herpes simplex virus (HSV) where the latency associated transcripts (LATs) are antisense to the immediate early lytic gene ICP0 (Stevens et al., 1987). Although this theory is hotly debated for HSV, it remains an intriguing explanation for EBNA gene transcript structure.

REGULATION OF EBNA GENE TRANSCRIPTION FROM Cp AND Wp

Significant effort has gone into identifying cis-elements that regulate Cp and Wp activity (Figure 4). As previously mentioned, *oriP* functions as an important enhancer element for both Cp and Wp (Puglielli et al., 1996; Sugden and Warren, 1989; Woisetschlaeger et al., 1989). Since EBNA1 is transcribed relatively early after infection of primary B cells (Alfieri et al., 1991; Schlager et al., 1996), it is possible that upregulation of Cp by EBNA1/*oriP* plays a major role during the process of immortalization. Other cis-regulatory elements that have been identified as involved in modulating Cp-initiated transcription (using transient transfection reporter gene assays) include the glucocorticoid response elements (GREs) upstream of Cp (Kupfer and Summers, 1990), two CAAT boxes located upstream of Cp (Puglielli et al., 1996), and an EBNA2-responsive enhancer (Sung et al.,

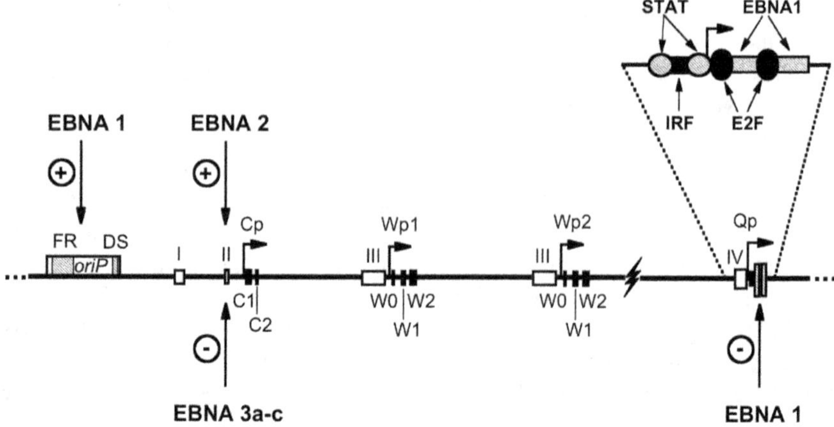

Figure 4. Locations of the major cis-elements regulating EBNA gene transcription during latency. I, glucocorticoid response element; II, EBNA2-dependent enhancer; III, enhancer shown to upregulate transcription from Cp and Wp; IV, IRF binding site which may also bind members of the STAT transcription factor family. In addition, EBNA1 binding to oriP has been shown to strongly upregulated transcription initiation from Cp/Wp. EBNA1 binding to two low affinity sites immediately downstream of the Qp transcription initiation site inhibits Qp activity.

1991; Woisetschlaeger et al., 1991). The EBNA2-responsive enhancer contains binding sites for the cellular transcription factors CBF1 and CBF2 (Jin and Speck, 1992; Ling et al., 1993). CBF1, also known as RBP-Jκ, is a ubiquitous cellular DNA binding factor shown to tether EBNA2 to DNA (Grossman et al., 1994; Henkel et al., 1994; Waltzer et al., 1994). The function of EBNA2 and the role of CBF1 in regulating viral latency-associated gene transcription are discussed in detailed elsewhere in this book (see Chapters 22 and 23).

An enhancer located upstream of Wp has been shown to upregulate both Cp and Wp activity (Bell et al., 1998; Puglielli et al., 1997; Ricksten et al., 1988; Walls and Perricaudet, 1991). Subsequent analyses of this region has led to the identification of 3 domains (termed UAS1, 2 and 3) that play an important role in regulating Wp activity, as assessed employing Wp fragments containing relatively limited upstream and downstream sequences (Bell et al., 1998; Kirby et al., 2000; Tierney et al., 2000a). Binding sites for YY1 and CREB/ATF factors have been identified and shown to be functionally important (Bell et al., 1998; Kirby et al., 2000). Notably, one of these domains, UAS1, appears to confer B cell-specificity to Wp. Binding of RFX family members (RFX1 and RFX3) and the B cell specific activator protein BSAP/Pax5 to UAS1 regulatory sequences has been shown (Tierney et al., 2000a). Whether Wp function is limited to B cells is unclear. A potential shortcoming of the latter analyses is that they were based on transient transfection of reporter constructs containing limited viral sequences. As such, distal cis-elements which may be critical in regulating Wp activity, were not present and may influence the activity of Wp in various cell types. Thus, some caution should be taken in extrapolating results obtained with such reporter constructs to the outcome of virus infection in non-B cell lines.

Autoregulation of Cp activity

It is increasingly clear that Cp is tightly autoregulated by the EBNA gene products. While EBNA1 and EBNA2 upregulate Cp-initiated transcription (Jin and Speck, 1992; Puglielli et al., 1996; Rooney et al., 1992; Schlager et al., 1996; Sung et al., 1991; Woisetschlaeger et al., 1991), the EBNA3 family are able to interfere with EBNA2-mediated transactivation through CBF1 (Krauer et al., 1996; Marshall and Sample, 1995; Robertson et al., 1995a). EBNA-LP can augment EBNA2 transcriptional activation through CBF1, which is mediated through the C-terminal transactivation domain of EBNA 2 (Harada and Kieff, 1997; McCann et al., 2001; Peng et al., 2000). While EBNA2 and EBNA-LP cooperate to activate target promoters, the EBNA3 proteins have been shown to bind CBF1 resulting in destabilization of CBF1 DNA binding and subsequent down-regulation of transcription. Thus, the Cp EBNA2-dependent enhancer appears to serve as an auto-regulated rheostat, through the opposing actions of the EBNA3 proteins and EBNA2, which adjusts the levels of EBNA gene transcription. Notably, as discussed below, Wp activity does not appear to be dependent on the Cp EBNA2 enhancer (Yoo and Speck, 2000; Yoo et al., 1997).

Analysis of EBV mutants with alterations in cis-elements involved in regulating Cp activity

Although the sequences surrounding Cp and Wp have been extensively mapped by transient reporter gene assays, very few promoter mutations have been made in the context of the whole virus due to the technical difficulty in generating mutant strains of EBV. Nevertheless, several LCLs with mutations in the viral genome that are relevant to Cp and Wp regulation have been generated. One of these viral mutants contains a deletion of Cp and over 3 kb of flanking genome, including the GRE, the EBNA2 responsive enhancer, and the C1 and C2 exons (Swaminathan, 1996). Notably, this mutant virus was fully capable of transforming B cells with normal efficiency and had normal EBNA and LMP1 expression levels. This demonstrated that Cp is not required for immortalization of B cells and that Wp is apparently able to fully compensate for Cp's transcriptional functions *in vitro*. The latter result is consistent with the previous identification of LCLs that harbor viral genomes in which Cp has been deleted (Woisetschlaeger et al., 1990; Yandava and Speck, 1992).

A mutant virus lacking only the GRE elements upstream of Cp has also been reported (Evans et al., 1996). This virus exhibited increased Cp activity compared to wild-type strains. This was a somewhat surprising result, since ablation of the GRE was predicted to downregulate Cp, based on the data from transient transfection assays (Gahn and Sugden, 1995; Sinclair et al., 1992). Finally, two distinct mutations of the EBNA 2-dependent enhancer upstream of Cp have been reported (Evans et al., 1996; Yoo et al., 1997). The first reported EBNA2 enhancer mutant virus involved the introduction of point mutations into the CBF1 binding site in the Cp EBNA2 responsive enhancer (Evans et al., 1996). The LCLs generated with this mutant virus were apparently normal with respect to Cp and Wp activity, in contrast to the results obtained by transient transfection of reporter constructs which suggested that ablation of the CBF1 site should lead to diminished Cp activity. However, a distinct phenotype was observed with a mutant virus in which a 200 bp deletion, encompassing the Cp EBNA2 responsive enhancer, including the CBF1 and CBF2 binding sites, was generated (Yoo et al., 1997). LCLs generated with the EBNA2 enhancer deletion mutant exhibited greatly reduced levels of Cp-initiated transcripts and a commensurate increase in Wp-initiated transcripts. The difference in the behavior of these two EBNA2 enhancer mutant viruses may reflect residual enhancer activity present in the recombinant virus containing point mutations in the CBF1 binding site.

Finally, EBV mutants were analyzed that deleted one copy of the enhancer located upstream of Wp (Yoo et al, 2002). Recombinant viruses were generated in which a deletion extending from -169 to -369bp upstream of the Wp transcription initiation site was introduced into one the copies of Wp in the virus. EBV immortalized lymphoblastoid cell lines were generated and analyzed for the site of deletion. Notably, of 10 recombinant viruses analyzed, only a single lymphoblastoid cell line contained this deletion upstream of the first copy of Wp (Wp1) - which lies immediately proximal to Cp. This result was in stark contrast to the analysis of a control targeting vector in which the Wp regulatory sequences were intact, but which contained a sequence tag within the W0 exon. With the control targeting vector, of 5 recombinants analyzed, 4 had recombined the sequence tag into Wp1. Thus, it appears that while recombination favors insertion into Wp1, deletion of the enhancer sequences upstream of Wp1 is unfavorable for immortalization of B cells. Analysis of the lone lymphoblastoid cell line harboring the deletion upstream of Wp1 revealed: (i) the level of Cp-initiated transcription was significantly diminished compared to that of wt virus immortalized lymphoblastoid cell lines; (ii) the decreased Cp-initiated transcription was not efficiently compensated by transcription initiation from Wp1; and (iii) transcription initiation from downstream Wp promoters was detectable, most likely from Wp2. The latter observation was the first documentation of transcription initiation from a Wp promoter downstream of Wp1 in an established lymphoblastoid cell line. Overall, this analysis demonstrated the critical importance of the first copy of the Wp enhancer in regulating both Wp- and Cp-initiated transcription.

Role of Cp and Wp?

The analysis of mutant viruses has clearly demonstrated that Cp is dispensable for EBV-driven B cell immortalization. This would suggest that Cp is not important for establishment of a latent infection by EBV, at least *in vitro*, and raises the question of the importance of Cp during virus infection. The major indication that Cp is important lies in the fact that Cp and associated regulatory cis-elements are very well conserved in several lymphocryptoviruses (EBV, rhesus EBV and herpesvirus papio), consistent with Cp playing an important role *in vivo* (Fuentes-Panana et al., 1999). Thus, the latter data argues that *in vitro* B cell immortalization may not accurately recapitulate the selection pressures that EBV encounters *in vivo*.

Early efforts to define the roles of Cp and Wp led to the observation that Wp is active during the initial stages of B cell infection, followed by upregulation of transcription from Cp at later times (Alfieri et al., 1991; Woisetschlaeger et al., 1991; 1990). The presence of multiple copies of Wp (8-10 copies on average per genome) may facilitate initial EBNA gene transcription in resting B cells. Indeed, as depicted in Figure 2, the ladder of EBNA-LP polypeptides observed during the initial stages of infection is consistent with transcription initiation from multiple copies of Wp (Rooney et al., 1989). Synthesis of sufficient levels of EBNA1 and EBNA2 leads to upregulation of Cp-initiated transcription (Schlager et al., 1996). Thus, while Wp may have evolved to be responsive to the transcriptional milieu in quiescent, resting B cells, Cp has clearly evolved to be exquisitely autoregulated by viral latency gene products, namely all six EBNAs. By encoding multiple copies of Wp, the virus helps ensure that viral gene transcription gets "jump started" in a resting B cell.

During the initial stages of infection in primary B cells, Wp-initiated transcripts first become detectable as early as 6 hours post-infection and Cp-initiated transcripts are activated between 48 and 72 hours post-infection (Schlager et al., 1996; Woisetschlaeger et al., 1990). Around the time that Cp gets upregulated, Wp activity begins to decline (Woisetschlaeger et al., 1990). It has been shown that the upregulation of Cp is dependent on EBNA2 and/or EBNA-LP, since infection of B cells with the EBNA2/EBNA-LP negative P3HR-1 virus fails to result in Cp upregulation (Woisetschlaeger et al., 1991). Furthermore, Cp does not get upregulated in B cells infected with an EBV strain lacking the Cp EBNA2-responsive enhancer (Yoo and Speck, 2000).

While Wp activity is detectable at very early times post-infection, the EBNA genes are not all immediately detectable. EBNA-LP and EBNA2 are the earliest EBNA genes transcribed, consistent with their location near the 5′ end of the long EBNA gene transcriptional unit (Figure 2) (Alfieri et al., 1991; Schlager et al., 1996). Expression of EBNA2 and/or EBNA-LP may be important for transcription of EBNA1, since infection of primary B cells with P3HR-1 virus results in reduced EBNA1 gene transcription (Schlager et al., 1996). This suggests that EBNA2 or EBNA-LP may affect transcriptional elongation through to the EBNA1 coding exon, or alternatively, may affect splicing. Once EBNA1 is produced, it binds to oriP and upregulates transcription from Cp.

Utilization of Cp and/or Wp in established LCLs - transcriptional interference

Early analyses of Cp and Wp activity in established LCLs and BL cell lines demonstrated that the activities of these promoters is generally mutually exclusive (Altiok et al., 1992; Woisetschlaeger et al., 1989). The majority of cell lines exclusively used Cp, but four cell lines were identified that exclusively used Wp. Interestingly, of the four cell lines examined which were Wp users, the viral genomes in two (X50-7 and IB4) were subsequently shown to have deletions that encompassed Cp (Woisetschlaeger et al., 1990; Yandava and Speck, 1992), suggesting that Cp is usually the dominant promoter when present.

More recent analyses of newly established LCLs generated by immortalization of primary B cells with either B95.8 or Akata strain viruses has demonstrated that while LCLs are generally biased toward Cp usage, they also exhibited significant levels of Wp-initiated transcripts (Yoo et al., 1997). It now appears that Cp and Wp activities are only mutually exclusive in extensively passaged LCLs and not in newly established (low passage) LCLs. A possible explanation for this observation lies in the fact that the steady-state level of Cp-initiated EBNA gene transcripts appears to be significantly higher in long-established LCLs than in newly established LCLs.

What is the relationship between levels of Cp activity and Wp activity? Since Cp is distal to Wp, transcription initiation from Cp may interfere with transcription initiation from Wp (transcriptional interference). Several lines of evidence support the transcriptional interference model. First, transient transfection of Cp/Wp driven reporter constructs demonstrates that in LCLs Cp is the dominant promoter, and that inversion of Cp leads to a dramatic increase in Wp activity without affecting Cp activity (Puglielli et al., 1997; 1996). Second, as discussed above, LCLs established with Cp EBNA2-enhancer mutant virus exhibit low level Cp activity and a compensatory increase in Wp-initiated transcripts (Yoo et al., 1997). Third, titration of EBNA2 function in an LCL harboring an EBNA2 conditional mutant (EBNA2-estrogen receptor ligand binding domain fusion) demonstrated that decreased levels of functional EBNA 2 resulted in a drop in the levels of Cp-initiated transcripts and a significant increase in the levels of Wp-initiated transcripts (Yoo and Speck, 2000). The latter observation is consistent with the phenotype of the Cp EBNA2 enhancer mutant virus, and strongly argues that there is an inverse relationship between the levels of Cp-initiated and Wp-initiated transcripts. Finally, during the establishment of EBV latency in B cells the appearance of Cp-initiated transcripts correlates with the fall in Wp-initiated transcripts (Woisetschlaeger et al., 1990).

More recently, DNA methylation of Wp regulatory sequences has been suggested as a mechanism for regulating the switch from Wp to Cp-driven EBNA gene transcription (Tierney et al., 2000b). It was shown, during the first 21 days post-infection of purified B cells in tissue culture, that there is progressive methylation of Wp regulatory sequences but not Cp regulatory sequences. Indeed, analyses of Wp regulatory sequences by day 21 post-infection revealed hypermethylation of Wp. However, subsequent examination of low passage LCLs, while demonstrating variable methylation of Wp regulatory sequences, also revealed the presence of hypomethylated Wp regulatory sequences in all LCLs examined (Elliot et al., 2004). The latter observation suggests that either, (i) Wp methylation is lost during establishment and growth of LCLs or (ii) those EBV infected B cells in which all copies of Wp become hypermethylated are unable to grow out as LCLs.

EBNA1 GENE TRANSCRIPTION DURING RESTRICTED VIRAL LATENCY IN EBV-ASSOCIATED TUMORS.
Identification of Qp

Based on our understanding of EBNA gene expression in LCLs, how is it possible to express only EBNA1 in EBV-associated tumors? Characterization of EBNA1 gene transcription in BL tumor cell lines that retain the restricted latency phenotype revealed that Cp- and Wp-initiated transcripts were not detectable (Sample et al., 1991; Schaefer et al., 1991). Initial attempts to clone the 5′ end of the EBNA1 gene transcript from BL cell lines exhibiting restricted latency, by rapid amplification of cDNA ends (RACE), identified a candidate promoter, Fp, mapping near the junction between the BamHI F and Q fragments (Figure 2) (Sample et al., 1991; Schaefer et al., 1991). However, subsequent analyses of Fp revealed that this is a promoter that is only active during lytic virus replication (Sample et al., 1991; Schaefer et al., 1991), and Fp-initiated transcripts do not account for EBNA1 gene expression during restricted viral latency (Schaefer et al., 1995a; Schaefer et al., 1995b).

Further RACE analyses identified the true 5' end of EBNA1 transcripts in BL cell lines exhibiting the restricted latency phenotype, which led to the identification of Qp (Nonkwelo et al., 1996; Schaefer et al., 1995b; Tsai et al., 1995). Qp-initiated EBNA1 gene transcripts contain a short 5' exon, Q, which splices to the U exon (also present in EBNA1 transcripts in LCLs), followed by splicing to the EBNA1 coding exon (Figure 2). Unlike EBNA gene transcription in LCLs, there is no detectable splicing from the U exon to any of the EBNA3 coding exons in these BL cell lines. This suggests that 5' transcript structure may play a role in dictating alternative splicing from the U exon.

Notably, Fp- and Qp-initiated transcripts share common 5' sequences, with the Q exon being entirely contained within the first exon, FQ, of Fp-initiated transcripts (Figure 2). In addition, Fp-initiated transcripts splice to the U exon, although only a very small percentage of these transcript splice to the EBNA1 coding exon; the majority of Fp-initiated transcripts fail to splice at the U exon splice donor site, and have an undefined 3' structure (Schaefer et al., 1995a). Thus, the initial identification of Fp as the EBNA1 gene promoter in type I BL cell lines most likely reflected amplification of Fp-initiated transcripts from a small percentage of cells that had entered the lytic cycle.

Cis-elements regulating Qp activity

Qp is a TATA-less promoter, and Qp-driven reporter constructs are active in a wide variety of cell types, which is reminiscent of eukaryotic housekeeping gene promoters (Schaefer et al., 1995b). The cis-elements involved in regulating Qp activity appear to be clustered in a ca. 80 bp region surrounding the site of transcription initiation. Immediately upstream of the site of transcription initiation is a site that binds members of the interferon regulatory factor (IRF) family of transcription factors (Nonkwelo et al., 1997b; Schaefer et al., 1997a; Zhang and Pagano, 1997; 1999). In BL cell lines retaining the latency I program the predominant factor binding this site is IRF2 (low level IRF1 binding can also be detected) (Schaefer et al., 1997a). In contrast to other promoters where IRF-2 binding appears to be involved in repressing promoter activity, IRF-2 binding to Qp appears to activate transcription (Schaefer et al., 1997a), although in some cell lines over-expression of IRF-2 can modestly inhibit Qp activity (Zhang and Pagano, 1999). Mutations in the Qp IRF binding site significantly inhibit, but do not ablate, Qp activity (Nonkwelo et al., 1997b; Schaefer et al., 1997a). In LCLs, where Qp is not active, it has been reported that binding of IRF-7 serves to inhibit Qp activity (Zhang and Pagano, 1997). Consistent with this hypothesis, IRF-7 levels are generally higher in LCLs than in latency I BL cell lines (Zhang and Pagano, 1997). However, over-expression of IRF-7 in two EBV-negative BL cell lines failed to repress a Qp-driven reporter construct (Nonkwelo et al., 1997b). The latter result suggests that IRF-7 modulation of Qp activity may be influenced by multiple factors. Thus, while it is clear that several members of the IRF family of transcription factors can bind to the Qp IRF site, questions remain about the role of individual IRF family members in the regulation of Qp-initiated transcription.

Qp is also subject to potent negative autoregulation by EBNA1, which is mediated through EBNA1 binding to two low affinity sites located just downstream of the Qp transcription initiation site (Sample et al., 1992; Schaefer et al., 1997b).

Notably, these are the only known EBNA1 binding sites outside of *oriP* in the EBV genome, and were identified through a general screen for EBNA1 binding sites prior to the identification of Qp (Rawlins et al., 1985). Two E2F binding sites, overlapping the low affinity EBNA1 binding sites, have also been reported (Sung et al., 1994). Consistent with the binding of an E2F family member(s) to this site, Qp-initiated transcription has been shown to be cell cycle regulated in the Akata BL cell line (Davenport and Pagano, 1999). Based on these observations, it has been argued that induction of a E2F family member during entry into the cell cycle promotes dissociation of EBNA1 from the low affinity binding sites, and induction of Qp-initiated transcription (Davenport and Pagano, 1999). Mutations in the putative E2F binding have been shown to diminish constitutive Qp-driven reporter gene activity in the absence of EBNA1, suggesting that binding of an E2F family member is involved in Qp activation independent of its postulated role in modulating EBNA1 binding (Nonkwelo et al., 1997a). However, as in the case of IRF family members binding to the Qp IRF site, there remain questions about the role of E2F family members in regulating Qp-initiated transcription (Schaefer et al., 1997a).

Finally, two STAT binding sites have been reported adjacent to the Qp IRF binding site (Chen et al., 1999). Coexpression of either Jak1 or STAT1 can activate a Qp-driven CAT reporter construct, and purified STAT4 can bind to these sites (Chen et al., 1999). In addition, mutation of either putative STAT binding site resulted in diminished response to JAK1, although the impact of these mutations on IRF binding was not assessed (Chen et al., 1999). Thus, STAT activation may play an important role in regulating Qp activity. Precisely which STATs are involved remains unclear, although there is evidence that during the latency II program STAT3 is activated via LMP1 induction of IL6 expression (see discussion below of LMP gene regulation during restricted EBV latency) (Chen et al., 2003; Chen et al., 2001). However, while LMP1 induction of STAT3 activation could nicely account for STAT activation during the latency II program, mechanisms that might be involved in generating activated STAT proteins during the latency I program are unknown.

Does regulation of Qp-initiated transcription need to be this complex?

Upon the initial identification and characterization of Qp, it appeared to be a relatively simple promoter; it is a TATA-less promoter that is active in a wide variety of cell types, and only requires sequences to -35 bp for maximal activity in reporter gene assays (Schaefer et al., 1995b). As such it was proposed to be a default promoter that ensured production of sufficient levels of EBNA1 to maintain replication of the viral episome during latent infection in settings where Cp and Wp are inactive (Schaefer et al., 1995b). Functional characterization of cis-elements in Qp has placed significant emphasis on understanding why Qp is quiescent during the latency III program, and how it is activated during entry into the latency I or II programs. Of the available data on the regulation of Qp, the most compelling is the strong negative regulation of Qp by EBNA1 binding to the low affinity sites adjacent the Qp transcription initiation site (Sample et al., 1992; Schaefer et al., 1997b). Since the Qp EBNA1 binding sites are of lower affinity than the EBNA1 binding sites in oriP, this negative autoregulation would be expected to maintain a sufficient supply of EBNA1 to fill the binding sites in *oriP*.

Based on negative autoregulation of Qp by EBNA1, a relatively simple model can be proposed in which Qp levels drop when the concentration of EBNA1 exceeds that required to fill the Qp EBNA1 binding sites. If this occurs in a quiescent memory B cell, the long term reservoir for latent EBV (Babcock et al., 1998; Chen et al., 1995; Miyashita et al., 1995), one would not expect to detect Qp-initiated EBNA1 gene transcripts in these cells. This is indeed what is observed in the peripheral blood of healthy seropositive individuals, where Qp-initiated transcripts can only sporadically be detected (Babcock and Thorley-Lawson, 2000; Chen et al., 1995; Hochberg et al., 2004; Qu and Rowe, 1992). However, when these cells traffic through secondary lymphoid tissue and are transiently stimulated to proliferate (Babcock and Thorley-Lawson, 2000), EBNA1 levels would be expected to fall thereby releasing EBNA1-mediated repression of Qp-initiated transcription. In this situation, basal levels of IRF-1 and IRF-2 alone may be sufficient to drive Qp activity upon release from EBNA1-mediated repression, or other factors (*e.g.*, STATs and/or E2F) may be required. Support for the role of other factors comes from the observation, in transient reporter gene assays, that mutation of the IRF site does not appear to completely ablate Qp activity (Schaefer et al., 1997a). Based on this simple model, it is not readily apparent why Qp needs to be cell cycle regulated. Alternatively, it could be argued that since the putative E2F sites appear to play an important role in upregulating Qp-initiated transcription (Nonkwelo et al., 1997a), that displacement of these factors by EBNA1 binding to the low affinity Qp EBNA1 sites may facilitate EBNA1-mediated repression of Qp.

Differential methylation of the EBV genome in tumors and in latently infected memory B cells isolated from seropositive individuals

While Qp is negatively autoregulated by EBNA1, transcription initiation from Cp and Wp is strongly upregulated by EBNA1 (Puglielli et al., 1996; Schaefer et al., 1997b). The latter point raises the issue of how restricted latency (Qp-driven EBNA1 expression) is maintained. Extensive analyses of cell lines exhibiting restricted EBV latency have repeatedly demonstrated the absence of Cp/Wp-initiated transcripts. The lack of Cp/Wp activity could reflect: (i) the absence of a necessary transcription factor(s); (ii) the presence of a Cp/Wp transcriptional repressor; (iii) mutation of Cp/Wp in the viral genome; or (iv) an epigenetic phenomenon (*e.g.*, cytosine methylation). Several lines of data argue against the first three possibilities. Introduction of reporter constructs [either by transient transfection Paulson et al. (2002), Paulson and Speck (1999), Schaefer et al. (1997b), or stably Paulson and Speck (unpublished data)], into cell lines exhibiting restricted latency readily demonstrated the presence of Cp/Wp-initiated transcripts arising from the reporter construct. These analyses demonstrated that the transcription factors required for Cp/Wp activity are present in cell lines exhibiting restricted viral latency. Also, since these reporter constructs contained extensive sequences upstream and downstream of Cp and Wp (from just upstream of *oriP* through the second W1 exon (fused to a reporter gene), a 10 kb region of the viral genome), it seems unlikely that Cp/Wp is repressed by a cellular factor. Furthermore, generation of LCLs using virus recovered from BL cell lines exhibiting restricted viral latency demonstrated that Cp and Wp in the endogenous viral genomes are functional. This raises the possibility that an epigenetic phenomenon, such as methylation of the viral genome, is involved in the establishment and maintenance of restricted EBV latency. Indeed, the suppressed CpG dinucleotide frequency in EBV suggests that this virus has been subjected to methylation during its evolution (Baer et al., 1984).

Methylation of genomic DNA is observed in many species across the phylogenetic spectrum, from prokaryotes to humans. While the principle function of methylation of bacterial genomes is the protection of self DNA from cleavage by endogenous restriction endonucleases, most evidence suggests that the primary function of genome methylation in higher eukaryotes is regulation of gene expression. It has been repeatedly demonstrated that cytosine methylation represses gene expression by inducing changes in chromatin structure which make methylated promoters inaccessible to the transcriptional machinery (reviewed in (Bestor, 1998)).

Analyses of methylation of the EBV genome has consistently demonstrated that the viral genome in LCLs is hypomethylated, while it is extensively methylated in cell lines that exhibit restricted viral latency (Altiok et al., 1992; Ernberg et al., 1989; Jansson et al., 1992; Masucci et al., 1989; Minarovits et al., 1994; Paulson et al., 2002; Paulson and Speck, 1999; Robertson and Ambinder, 1997a; Robertson and Ambinder, 1997b; Robertson et al., 1995b; Schaefer et al., 1997b; 1991). Detailed analyses of methylation around Cp, Wp and Qp have further revealed that while Cp and Wp are hypermethylated in tumor cell lines, the region around Qp is not methylated (Jansson et al., 1992; Paulson et al., 2002; Paulson and Speck, 1999; Robertson and Ambinder, 1997a; Robertson et al., 1995b; Schaefer et al., 1997b; 1991). These analyses have been extended to show that Cp and Wp are methylated in the long term latency reservoir in healthy seropositive individuals, while Qp is protected from methylation (Paulson and Speck, 1999). Importantly, the latter observations demonstrate that methylation of the EBV genome is a normal aspect of the life cycle of EBV *in vivo* and is not restricted to EBV-associated tumors. Currently, the major challenges with respect to the role of methylation in regulating EBNA gene expression are to determine: (i) how specific regions of the viral genome are targeted or protected from methylation; (ii) at what stage of B cell development the viral genome becomes methylated; and (iii) whether methylation is the primary mechanism involved in the switch from the latency III/growth program to the more restricted latency I and II programs, or secondary to an independent switch mechanism. The observation that, at least in some cell lines, the shift from restricted viral latency to the latency III program can be accomplished by addition of the methyltransferase inhibitor 5-azacytidine, indicates that repression of Cp and Wp requires maintenance of viral genome methylation.

REGULATION OF LMP GENE EXPRESSION
LMP gene expression in LCLs
While EBNA gene transcription is quite complex, the organization of the LMP genes is relatively simple. There are 3 LMP gene products expressed in the latency II and III programs - LMP1, LMP2A and LMP2B. LMP1 is encoded by 3 closely spaced exons located in the unique sequence at the righthand end of the viral

genome near the terminal repeats (see Figure 1). As discussed elsewhere in this book (see Chapter 25), LMP2A and LMP2B are encoded by transcripts driven by independent promoters located in the unique region of the EBV genome near the righthand end of the viral genome (see Figure 1). As such, the LMP2A and LMP2B transcripts each encode a unique 5′ exon that is spliced to common 3′ exons which are located in the unique region at the left end of the viral genome. Thus, expression of the LMP2 gene products requires circularization of the viral genome, which occurs at an early stage of infection, since transcription of these genes spans the fused terminal repeats. Notably, transcription of the LMP1 and LMP2B genes in LCLs is driven by a bi-directional promoter (reviewed in (Kieff, 2001)).

Like Cp, the promoters regulating LMP1/LMP2B and LMP2A expression are highly dependent on EBNA2 regulation through CBF1 binding sites in these promoters (reviewed in (Kieff, 2001)). The LMP1/LMP2B divergent promoter also contains a site for the B cell-specific transcription factor PU.1, which also appears to play a role in recruiting EBNA2 (Johannsen et al., 1995). Two independent lines of evidence have indicated the importance of EBNA2 for LMP gene expression in LCL and BL exhibiting the latency III pattern of EBNA gene expression. In BL lines in which the EBNA2 gene has been deleted there is little expression of the LMP genes (Abbot, 1990; Murray, 1988; Zimber-Strobl, 1990; 1991). Indeed, recent data has revealed a class of BL tumors in which the EBNA 2 coding exon and a portion of the EBNA-LP coding sequence have been deleted (Kelly et al., 2002; Speck, 2002). In these BL tumors EBNA gene transcription is not driven from Qp, but rather is driven from Wp, and there is little or no expression of the LMP genes (Kelly et al., 2002). This observation appears to underscore the importance of EBNA2 (and perhaps EBNA-LP) for both Cp-initiated EBNA gene transcription and LMP gene expression. Consistent with these observations, analysis of viral gene expression in an EBNA2 conditional LCL has demonstrated that functional EBNA2 expression is required to maintain LMP gene expression.

LMP gene expression during restricted latency

Since expression of the LMP genes in LCLs and latency III BL cell lines is highly dependent on the expression of EBNA2, this raises the question of how LMP gene expression arises during the latency II program (*i.e.*, in the absence of EBNA2 expression). While the promoters regulating LMP1/LMP2B and LMP2A expression during the latency III program have been clearly defined (reviewed in (Kieff, 2001)), significantly less is known about regulation of LMP gene expression during the latency II program. Currently there is no insight into the mechanisms involved in regulating LMP2 gene expression during the latency II program. More is known about LMP1 expression during the latency II program. In NPC tumors LMP1 gene expression is driven from a TATA-less GC box containing promoter (referred to as TR-L1 and ED-L1E) that is located in the terminal repeat of the EBV genome ca. 600bp upstream of the promoter (ED-L1) used to drive LMP1 gene transcription during the latency III program in LCLs (Gilligan et al., 1990; Kieff, 1995; Sadler and Raab-Traub, 1995; Tsai et al., 1999). The GC box present in this promoter has been shown to bind both Sp1 and Sp3, and in reporter gene assays both Sp1 and Sp3 can activate transcription from this promoter (Tsai et al., 1999). In addition, recent data has indicated a role for STAT proteins in the activation of transcription from both the ED-L1 and TR-L1 LMP gene promoters (Chen et al., 2001). Furthermore, it appears that LMP1, through induction of IL6, reinforces JAK-STAT signaling largely through activation of STAT3 (Chen et al., 2003; Chen et al., 2001). This auto-regulatory loop may thus serve to drive both LMP1 gene expression from TR-L1, as well as Qp-driven EBNA1 gene expression (see discussion of Qp regulation above) during the latency II program. Consistent with this hypothesis, immunohistochemistry has shown the presence of activated STAT3 and STAT5 in NPC and activated STAT 3 and STAT6 in Hodgkin's disease biopsies (Chen et al., 2001). Whether the TR-L1 promoter is also used to drive LMP1 expression during the latency II program in germinal center B cells in normal seropositive individuals is unclear, although it seems likely since: (i) a similar transcript has been observed in the EBV immortalized marmoset LCL B95-8, and (ii) the observation of activated STAT3 in Hodgkin's disease tissue.

CONCLUSIONS

EBV has evolved remarkably sophisticated and elegant transcriptional programs to regulate the expression of the EBNA and LMP genes. A central theme, from the analysis of EBV latency-associated viral gene expression, is the localization of EBNA and LMP gene transcription initiation to the region of the viral genome near the terminal repeats and oriP, even though the exons encoding EBNA 1, 3a, 3b and 3c lie far downstream of Cp and Wp. The only exception to this clustering of latency-associated promoters is Qp, which drives EBNA1 gene transcription during the latency I and II programs. Overall, this organization of viral transcription (limiting the regions of the viral genome that are actively initiating transcription during latency) may have evolved as a mechanism to avoid spurious transcription of lytic genes and activation of virus replication. It is clear that there are still gaps in our understanding, most notably the regulation of LMP2 gene expression during the latency II program. In addition, the mechanisms involved in extinguishing LMP gene expression in the transition to latently infected resting memory B cell are also unresolved. Finally, the precise role of DNA methylation in regulating the transition between the latency III program and the more restricted latency I and II programs remains an open question.

References

Abbot, S.D., Rowe, M., Cadwallader, K., Ricksten, A., Gordon, J., Wang, F., Rymo, L., Rickinson, A.B. (1990) Epstein-Barr virus nuclear antigen 2 induces expression of the virus-encoded latent membrane protein. J. Virol. *64*, 2126-2134.

Alfieri, C., Birkenbach, M., and Kieff, E. (1991). Early events in Epstein-Barr virus infection of human B lymphocytes. Virology *181*, 595-608.

Altiok, E., Minarovits, J., Hu, L. F., Contreras-Brodin, B., Klein, G., and Ernberg, I. (1992). Host-cell-phenotype-dependent control of the BCR2/BWR1 promoter complex regulates the expression of Epstein-Barr virus nuclear antigens 2-6. Proc. Natl. Acad. Sci. USA. *89*, 905-909.

Babcock, G. J., Decker, L. L., Volk, M., and Thorley-Lawson, D. A. (1998). EBV persistence in memory B cells *in vivo*. Immunity *9*, 395-404.

Babcock, G. J., and Thorley-Lawson, D. A. (2000). Tonsillar memory B cells, latently infected with Epstein-Barr virus, express the restricted pattern of latent genes previously found only in Epstein-Barr virus-associated tumors. Proc. Natl. Acad. Sci. USA. *97*, 12250-12255.

Baer, R., Bankier, A. T., Biggin, M. D., Deininger, P. L., Farrell, P. J., Gibson, T. J., Hatfull, G., Hudson, G. S., Satchwell, S. C., Seguin, C., and et al. (1984). DNA sequence and expression of the B95-8 Epstein-Barr virus genome. Nature *310*, 207-211.

Bell, A., Skinner, J., Kirby, H., and Rickinson, A. (1998). Characterisation of regulatory sequences at the Epstein-Barr virus BamHI W promoter. Virology *252*, 149-161.

Bestor, T. H. (1998). Gene silencing. Methylation meets acetylation. Nature *393*, 311-312.

Bodescot, M., Brison, O., and Perricaudet, M. (1986). An Epstein-Barr virus transcription unit is at least 84 kilobases long. Nucleic Acids Res. *14*, 2611-2620.

Bodescot, M., Chambraud, B., Farrell, P., and Perricaudet, M. (1984). Spliced RNA from the IR1-U2 region of Epstein-Barr virus: presence of an open reading frame for a repetitive polypeptide. Embo J. *3*, 1913-1917.

Bodescot, M., and Perricaudet, M. (1986). Epstein-Barr virus mRNAs produced by alternative splicing. Nucleic Acids Res. *14*, 7103-7114.

Bodescot, M., Perricaudet, M., and Farrell, P. J. (1987). A promoter for the highly spliced EBNA family of RNAs of Epstein-Barr virus. J. Virol. *61*, 3424-3430.

Chen, F., Zou, J. Z., di Renzo, L., Winberg, G., Hu, L. F., Klein, E., Klein, G., and Ernberg, I. (1995). A subpopulation of normal B cells latently infected with Epstein-Barr virus resembles Burkitt lymphoma cells in expressing EBNA-1 but not EBNA-2 or LMP1. J. Virol. *69*, 3752-3758.

Chen, H., Hutt-Fletcher, L., Cao, L., and Hayward, S. D. (2003). A positive autoregulatory loop of LMP1 expression and STAT activation in epithelial cells latently infected with Epstein-Barr virus. J. Virol. *77*, 4139-4148.

Chen, H., Lee, J. M., Wang, Y., Huang, D. P., Ambinder, R. F., and Hayward, S. D. (1999). The Epstein-Barr virus latency BamHI-Q promoter is positively regulated by STATs and Zta interference with JAK/STAT activation leads to loss of BamHI-Q promoter activity. Proc. Natl. Acad. Sci. USA. *96*, 9339-9344.

Chen, H., Lee, J. M., Zong, Y., Borowitz, M., Ng, M. H., Ambinder, R. F., and Hayward, S. D. (2001). Linkage between STAT regulation and Epstein-Barr virus gene expression in tumors. J. Virol. *75*, 2929-2937.

Chiang, C. M., Dong, G., Broker, T. R., and Chow, L. T. (1992). Control of human papillomavirus type 11 origin of replication by the E2 family of transcription regulatory proteins. J. Virol. *66*, 5224-5231.

Davenport, M. G., and Pagano, J. S. (1999). Expression of EBNA-1 mRNA is regulated by cell cycle during Epstein-Barr virus type I latency. J. Virol. *73*, 3154-3161.

Dillner, J., Kallin, B., Alexander, H., Ernberg, I., Uno, M., Ono, Y., Klein, G., and Lerner, R. A. (1986). An Epstein-Barr virus (EBV)-determined nuclear antigen (EBNA5) partly encoded by the transformation-associated Bam WYH region of EBV DNA: preferential expression in lymphoblastoid cell lines. Proc. Natl. Acad. Sci. USA. *83*, 6641-6645.

Elliot, J., Goodhew, B., Krug, L.T., Shakhnovsky, N., Yoo, L., and Speck, S.H. (2004). Variable methylation of the Epstein-Barr virus Wp EBNA gene promoter in B-lymphoblastoid cell lines. J. Virol. *78*, 14062-14065.

Ernberg, I., Falk, K., Minarovits, J., Busson, P., Tursz, T., Masucci, M. G., and Klein, G. (1989). The role of methylation in the phenotype-dependent modulation of Epstein-Barr nuclear antigen 2 and latent membrane protein genes in cells latently infected with Epstein-Barr virus. J. Gen. Virol. *70 (Pt 11)*, 2989-3002.

Evans, T. J., Farrell, P. J., and Swaminathan, S. (1996). Molecular genetic analysis of Epstein-Barr virus Cp promoter function. J. Virol. *70*, 1695-1705.

Fuentes-Panana, E. M., Swaminathan, S., and Ling, P. D. (1999). Transcriptional activation signals found in the Epstein-Barr virus (EBV) latency C promoter are conserved in the latency C promoter sequences from baboon and Rhesus monkey EBV-like lymphocryptoviruses (cercopithicine herpesviruses 12 and 15). J. Virol. *73*, 826-833.

Gahn, T. A., and Sugden, B. (1995). An EBNA-1-dependent enhancer acts from a distance of 10 kilobase pairs to increase expression of the Epstein-Barr virus LMP gene. J. Virol. *69*, 2633-2636.

Gilligan, K., Sato, H., Rajadurai, P., Busson, P., Young, L., Rickinson, A., Tursz, T., and Raab-Traub, N. (1990). Novel transcription from the Epstein-Barr virus terminal EcoRI fragment, DIJhet, in a nasopharyngeal carcinoma. J. Virol. *64*, 4948-4956.

Gregory, C. D., Rowe, M., and Rickinson, A. B. (1990). Different Epstein-Barr virus-B cell interactions in phenotypically distinct clones of a Burkitt's lymphoma cell line. J. Gen. Virol. *71 (Pt 7)*, 1481-1495.

Grossman, S. R., Johannsen, E., Tong, X., Yalamanchili, R., and Kieff, E. (1994). The Epstein-Barr virus nuclear antigen 2 transactivator is directed to response elements by the J. kappa recombination signal binding protein. Proc. Natl. Acad. Sci. USA. *91*, 7568-7572.

Harada, S., and Kieff, E. (1997). Epstein-Barr virus nuclear protein LP stimulates EBNA-2 acidic domain-mediated transcriptional activation. J. Virol. *71*, 6611-6618.

Heller, M., van Santen, V., and Kieff, E. (1982). Simple repeat sequence in Epstein-Barr virus DNA is transcribed in latent and productive infections. J. Virol. *44*, 311-320.

Henkel, T., Ling, P. D., Hayward, S. D., and Peterson, M. G. (1994). Mediation of Epstein-Barr virus EBNA2 transactivation by recombination signal-binding protein J. kappa. Science *265*, 92-95.

Hochberg, D., Middeldorp, J. M., Catalina, M., Sullivan, J. L., Luzuriaga, K., and Thorley-Lawson, D. A. (2004). Demonstration of the Burkitt's lymphoma Epstein-Barr virus phenotype in dividing latently infected memory cells *in vivo*. Proc. Natl. Acad. Sci. USA. *101*, 239-244.

Jansson, A., Masucci, M., and Rymo, L. (1992). Methylation of discrete sites within the enhancer region regulates the activity of the Epstein-Barr virus BamHI W promoter in Burkitt lymphoma lines. J. Virol. *66*, 62-69.

Jin, X. W., and Speck, S. H. (1992). Identification of critical cis elements involved in mediating Epstein-Barr virus nuclear antigen 2-dependent activity of an enhancer located upstream of the viral BamHI C promoter. J. Virol. *66*, 2846-2852.

Johannsen, E., Koh, E., Mosialos, G., Tong, X., Kieff, E., and Grossman, S. R. (1995). Epstein-Barr virus nuclear protein 2 transactivation of the latent membrane protein 1 promoter is mediated by J. kappa and PU.1. J. Virol. *69*, 253-262.

Joseph, A. M., Babcock, G. J., and Thorley-Lawson, D. A. (2000). Cells expressing the Epstein-Barr virus growth program are present in and restricted to the naive B-cell subset of healthy tonsils. J. Virol. *74*, 9964-9971.

Kelly, G., Bell, A., and Rickinson, A. (2002). Epstein-Barr virus-associated Burkitt lymphomagenesis selects for downregulation of the nuclear antigen EBNA2. Nat. Med. *8*, 1098-1104.

Kieff, E. (1995). Epstein-Barr virus--increasing evidence of a link to carcinoma. N. Engl. J. Med. *333*, 724-726.

Kieff, E. R., A.B. (2001). Epstein-Barr virus and its replication. In Fields Virology, D. M. H. Knipe, P.M., ed. (Philidelphia, Lippincott Williams and Wilkins), pp. 2511-2573.

King, W., Thomas-Powell, A. L., Raab-Traub, N., Hawke, M., and Kieff, E. (1980). Epstein-Barr virus RNA. V. Viral RNA in a restringently infected, growth-transformed cell line. J. Virol. *36*, 506-518.

Kirby, H., Rickinson, A., and Bell, A. (2000). The activity of the Epstein-Barr virus BamHI W promoter in B cells is dependent on the binding of CREB/ATF factors. J. Gen. Virol. *81*, 1057-1066.

Krauer, K. G., Kienzle, N., Young, D. B., and Sculley, T. B. (1996). Epstein-Barr nuclear antigen-3 and -4 interact with RBP-2N, a major isoform of RBP-J kappa in B lymphocytes. Virology *226*, 346-353.

Kupfer, S. R., and Summers, W. C. (1990). Identification of a glucocorticoid-responsive element in Epstein-Barr virus. J. Virol. *64*, 1984-1990.

Ling, P. D., Rawlins, D. R., and Hayward, S. D. (1993). The Epstein-Barr virus immortalizing protein EBNA-2 is targeted to DNA by a cellular enhancer-binding protein. Proc. Natl. Acad. Sci. USA. *90*, 9237-9241.

Marshall, D., and Sample, C. (1995). Epstein-Barr virus nuclear antigen 3C is a transcriptional regulator. J. Virol. *69*, 3624-3630.

Masucci, M. G., Contreras-Salazar, B., Ragnar, E., Falk, K., Minarovits, J., Ernberg, I., and Klein, G. (1989). 5-Azacytidine up regulates the expression of Epstein-Barr virus nuclear antigen 2 (EBNA-2) through EBNA-6 and latent membrane protein in the Burkitt's lymphoma line rael. J. Virol. *63*, 3135-3141.

McCann, E. M., Kelly, G. L., Rickinson, A. B., and Bell, A. I. (2001). Genetic analysis of the Epstein-Barr virus-coded leader protein EBNA-LP as a co-activator of EBNA2 function. J. Gen. Virol. *82*, 3067-3079.

Minarovits, J., Hu, L. F., Minarovits-Kormuta, S., Klein, G., and Ernberg, I. (1994). Sequence-specific methylation inhibits the activity of the Epstein-Barr virus LMP 1 and BCR2 enhancer-promoter regions. Virology *200*, 661-667.

Minvielle-Sebastia, L., and Keller, W. (1999). mRNA polyadenylation and its coupling to other RNA processing reactions and to transcription. Curr. Opin. Cell Biol. *11*, 352-357.

Miyashita, E. M., Yang, B., Lam, K. M., Crawford, D. H., and Thorley-Lawson, D. A. (1995). A novel form of Epstein-Barr virus latency in normal B cells *in vivo*. Cell *80*, 593-601.

Murray, R. J., Young, L. S., Calender, A., Gregory, C. D., Rowe, M., Lenoir, G. M., Rickinson, A. B. (1988) Different patterns of Epstein-Barr virus gene expression and of cytotoxic T-cell recognition in B-cell lines infected with transforming (B95.8) or nontransforming (P3HR1) virus strains. J. Virol. *62*, 849-901.

Nonkwelo, C., Ruf, I. K., and Sample, J. (1997a). The Epstein-Barr virus EBNA-1 promoter Qp requires an initiator-like element. J. Virol. *71*, 354-361.

Nonkwelo, C., Ruf, I. K., and Sample, J. (1997b). Interferon-independent and -induced regulation of Epstein-Barr virus EBNA-1 gene transcription in Burkitt lymphoma. J. Virol. *71*, 6887-6897.

Nonkwelo, C., Skinner, J., Bell, A., Rickinson, A., and Sample, J. (1996). Transcription start sites downstream of the Epstein-Barr virus (EBV) Fp promoter in early-passage Burkitt lymphoma cells define a fourth promoter for expression of the EBV EBNA-1 protein. J. Virol. *70*, 623-627.

Paulson, E. J., Fingeroth, J. D., Yates, J. L., and Speck, S. H. (2002). Methylation of the EBV genome and establishment of restricted latency in low-passage EBV-infected 293 epithelial cells. Virology *299*, 109-121.

Paulson, E. J., and Speck, S. H. (1999). Differential methylation of Epstein-Barr virus latency promoters facilitates viral persistence in healthy seropositive individuals. J. Virol. *73*, 9959-9968.

Peng, R., Tan, J., and Ling, P. D. (2000). Conserved regions in the Epstein-Barr virus leader protein define distinct domains required for nuclear localization and transcriptional cooperation with EBNA2. J. Virol. *74*, 9953-9963.

Prang, N., Wolf, H., and Schwarzmann, F. (1999). Latency of Epstein-Barr virus is stabilized by antisense-mediated control of the viral immediate-early gene BZLF-1. J. Med. Virol. *59*, 512-519.

Puglielli, M. T., Desai, N., and Speck, S. H. (1997). Regulation of EBNA gene transcription in lymphoblastoid cell lines: characterization of sequences downstream of BCR2 (Cp). J. Virol. *71*, 120-128.

Puglielli, M. T., Woisetschlaeger, M., and Speck, S. H. (1996). oriP is essential for EBNA gene promoter activity in Epstein-Barr virus-immortalized lymphoblastoid cell lines. J. Virol. *70*, 5758-5768.

Qu, L., and Rowe, D. T. (1992). Epstein-Barr virus latent gene expression in uncultured peripheral blood lymphocytes. J. Virol. *66*, 3715-3724.

Rawlins, D. R., Milman, G., Hayward, S. D., and Hayward, G. S. (1985). Sequence-specific DNA binding of the Epstein-Barr virus nuclear antigen (EBNA-1) to clustered sites in the plasmid maintenance region. Cell *42*, 859-868.

Reedman, B. M., and Klein, G. (1973). Cellular localization of an Epstein-Barr virus (EBV)-associated complement-fixing antigen in producer and non-producer lymphoblastoid cell lines. Int. J. Cancer *11*, 499-520.

Reedman, B. M., Klein, G., Pope, J. H., Walters, M. K., Hilgers, J., Singh, S., and Johansson, B. (1974). Epstein-Barr virus-associated complement-fixing and nuclear antigens in Burkitt lymphoma biopsies. Int. J. Cancer *13*, 755-763.

Rickinson, A. B. K., E. (2001). Epstein-Barr virus. In Fields Virology, D. M. H. Knipe, P.M., ed. (Philadelphia, Lippincott, Williams and Wilkins), pp. 2575-2628.

Ricksten, A., Olsson, A., Andersson, T., and Rymo, L. (1988). The 5' flanking region of the gene for the Epstein-Barr virus-encoded nuclear antigen 2 contains a cell type specific cis-acting regulatory element that activates transcription in transfected B-cells. Nucleic Acids Res. *16*, 8391-8410.

Robertson, E. S., Grossman, S., Johannsen, E., Miller, C., Lin, J., Tomkinson, B., and Kieff, E. (1995a). Epstein-Barr virus nuclear protein 3C modulates transcription through interaction with the sequence-specific DNA-binding protein J. kappa. J. Virol. *69*, 3108-3116.

Robertson, K. D., and Ambinder, R. F. (1997a). Mapping promoter regions that are hypersensitive to methylation-mediated inhibition of transcription: application of the methylation cassette assay to the Epstein-Barr virus major latency promoter. J. Virol. *71*, 6445-6454.

Robertson, K. D., and Ambinder, R. F. (1997b). Methylation of the Epstein-Barr virus genome in normal lymphocytes. Blood *90*, 4480-4484.

Robertson, K. D., Hayward, S. D., Ling, P. D., Samid, D., and Ambinder, R. F. (1995b). Transcriptional activation of the Epstein-Barr virus latency C promoter after 5-azacytidine treatment: evidence that demethylation at a single CpG site is crucial. Mol. Cell Biol. *15*, 6150-6159.

Rogers, R. P., Woisetschlaeger, M., and Speck, S. H. (1990). Alternative splicing dictates translational start in Epstein-Barr virus transcripts. Embo J. *9*, 2273-2277.

Rooney, C., Howe, J. G., Speck, S. H., and Miller, G. (1989). Influence of Burkitt's lymphoma and primary B cells on latent gene expression by the nonimmortalizing P3J-HR-1 strain of Epstein-Barr virus. J. Virol. *63*, 1531-1539.

Rooney, C. M., Brimmell, M., Buschle, M., Allan, G., Farrell, P. J., and Kolman, J. L. (1992). Host cell and EBNA-2 regulation of Epstein-Barr virus latent-cycle promoter activity in B lymphocytes. J. Virol. *66*, 496-504.

Rowe, D. T., Farrell, P. J., and Miller, G. (1987a). Novel nuclear antigens recognized by human sera in lymphocytes latently infected by Epstein-Barr virus. Virology *156*, 153-162.

Rowe, D. T., Rowe, M., Evan, G. I., Wallace, L. E., Farrell, P. J., and Rickinson, A. B. (1986). Restricted expression of EBV latent genes and T-lymphocyte-detected membrane antigen in Burkitt's lymphoma cells. Embo J. *5*, 2599-2607.

Rowe, M., Rowe, D. T., Gregory, C. D., Young, L. S., Farrell, P. J., Rupani, H., and Rickinson, A. B. (1987b). Differences in B cell growth phenotype reflect novel patterns of Epstein-Barr virus latent gene expression in Burkitt's lymphoma cells. Embo J. *6*, 2743-2751.

Sadler, R. H., and Raab-Traub, N. (1995). The Epstein-Barr virus 3.5-kilobase latent membrane protein 1 mRNA initiates from a TATA-Less promoter within the first terminal repeat. J. Virol. *69*, 4577-4581.

Sample, J., Brooks, L., Sample, C., Young, L., Rowe, M., Gregory, C., Rickinson, A., and Kieff, E. (1991). Restricted Epstein-Barr virus protein expression in Burkitt lymphoma is due to a different Epstein-Barr nuclear antigen 1 transcriptional initiation site. Proc. Natl. Acad. Sci. USA. *88*, 6343-6347.

Sample, J., Henson, E. B., and Sample, C. (1992). The Epstein-Barr virus nuclear protein 1 promoter active in type I latency is autoregulated. J. Virol. *66*, 4654-4661.

Sample, J., Hummel, M., Braun, D., Birkenbach, M., and Kieff, E. (1986). Nucleotide sequences of mRNAs encoding Epstein-Barr virus nuclear proteins: a probable transcriptional initiation site. Proc. Natl. Acad. Sci. USA. *83*, 5096-5100.

Schaefer, B. C., Paulson, E., Strominger, J. L., and Speck, S. H. (1997a). Constitutive activation of Epstein-Barr virus (EBV) nuclear antigen 1 gene transcription by IRF1 and IRF2 during restricted EBV latency. Mol. Cell Biol. *17*, 873-886.

Schaefer, B. C., Strominger, J. L., and Speck, S. H. (1995a). The Epstein-Barr virus BamHI F promoter is an early lytic promoter: lack of correlation with EBNA 1 gene transcription in group 1 Burkitt's lymphoma cell lines. J. Virol. *69*, 5039-5047.

Schaefer, B. C., Strominger, J. L., and Speck, S. H. (1995b). Redefining the Epstein-Barr virus-encoded nuclear antigen EBNA-1 gene promoter and transcription initiation site in group I Burkitt lymphoma cell lines. Proc. Natl. Acad. Sci. USA. *92*, 10565-10569.

Schaefer, B. C., Strominger, J. L., and Speck, S. H. (1997b). Host-cell-determined methylation of specific Epstein-Barr virus promoters regulates the choice between distinct viral latency programs. Mol. Cell Biol. *17*, 364-377.

Schaefer, B. C., Woisetschlaeger, M., Strominger, J. L., and Speck, S. H. (1991). Exclusive expression of Epstein-Barr virus nuclear antigen 1 in Burkitt lymphoma arises from a third promoter, distinct from the promoters used in latently infected lymphocytes. Proc. Natl. Acad. Sci. USA. *88*, 6550-6554.

Schlager, S., Speck, S. H., and Woisetschlager, M. (1996). Transcription of the Epstein-Barr virus nuclear antigen 1 (EBNA1) gene occurs before induction of the BCR2 (Cp) EBNA gene promoter during the initial stages of infection in B cells. J. Virol. *70*, 3561-3570.

Sinclair, A. J., Brimmell, M., and Farrell, P. J. (1992). Reciprocal antagonism of steroid hormones and BZLF1 in switch between Epstein-Barr virus latent and productive cycle gene expression. J. Virol. *66*, 70-77.

Speck, S. H. (2002). EBV framed in Burkitt lymphoma. Nat. Med. *8*, 1086-1087.

Speck, S. H., Chatila, T., and Flemington, E. (1997). Reactivation of Epstein-Barr virus: regulation and function of the BZLF1 gene. Trends Microbiol. *5*, 399-405.

Speck, S. H., Pfitzner, A., and Strominger, J. L. (1986). An Epstein-Barr virus transcript from a latently infected, growth-transformed B-cell line encodes a highly repetitive polypeptide. Proc. Natl. Acad. Sci. USA. *83*, 9298-9302.

Speck, S. H., and Strominger, J. L. (1985). Analysis of the transcript encoding the latent Epstein-Barr virus nuclear antigen I: a potentially polycistronic message generated by long-range splicing of several exons. Proc. Natl. Acad. Sci. USA. *82*, 8305-8309.

Stevens, J. G., Wagner, E. K., Devi-Rao, G. B., Cook, M. L., and Feldman, L. T. (1987). RNA complementary to a herpesvirus alpha gene mRNA is prominent in latently infected neurons. Science *235*, 1056-1059.

Sugden, B., and Warren, N. (1989). A promoter of Epstein-Barr virus that can function during latent infection can be transactivated by EBNA-1, a viral protein required for viral DNA replication during latent infection. J. Virol. *63*, 2644-2649.

Summers, W. P., Grogan, E. A., Shedd, D., Robert, M., Liu, C. R., and Miller, G. (1982). Stable expression in mouse cells of nuclear neoantigen after transfer of a 3.4-megadalton cloned fragment of Epstein-Barr virus DNA. Proc. Natl. Acad. Sci. USA. *79*, 5688-5692.

Sung, N. S., Kenney, S., Gutsch, D., and Pagano, J. S. (1991). EBNA-2 transactivates a lymphoid-specific enhancer in the BamHI C promoter of Epstein-Barr virus. J. Virol. *65*, 2164-2169.

Sung, N. S., Wilson, J., Davenport, M., Sista, N. D., and Pagano, J. S. (1994). Reciprocal regulation of the Epstein-Barr virus BamHI-F promoter by EBNA-1 and an E2F transcription factor. Mol. Cell Biol. *14*, 7144-7152.

Swaminathan, S. (1996). Characterization of Epstein-Barr virus recombinants with deletions of the BamHI C promoter. Virology *217*, 532-541.

Thorley-Lawson, D. A. (2001). Epstein-Barr virus: exploiting the immune system. Nat. Rev. Immunol. *1*, 75-82.

Tierney, R., Kirby, H., Nagra, J., Rickinson, A., and Bell, A. (2000a). The Epstein-Barr virus promoter initiating B-cell transformation is activated by RFX proteins and the B-cell-specific activator protein BSAP/Pax5. J. Virol. *74*, 10458-10467.

Tierney, R. J., Kirby, H. E., Nagra, J. K., Desmond, J., Bell, A. I., and Rickinson, A. B. (2000b). Methylation of transcription factor binding sites in the Epstein-Barr virus latent cycle promoter Wp coincides with promoter down-regulation during virus-induced B-cell transformation. J. Virol. *74*, 10468-10479.

Tsai, C. N., Lee, C. M., Chien, C. K., Kuo, S. C., and Chang, Y. S. (1999). Additive effect of Sp1 and Sp3 in regulation of the ED-L1E promoter of the EBV LMP 1 gene in human epithelial cells. Virology *261*, 288-294.

Tsai, C. N., Liu, S. T., and Chang, Y. S. (1995). Identification of a novel promoter located within the Bam HI Q region of the Epstein-Barr virus genome for the EBNA 1 gene. DNA Cell Biol. *14*, 767-776.

van Santen, V., Cheung, A., Hummel, M., and Kieff, E. (1983). RNA encoded by the IR1-U2 region of Epstein-Barr virus DNA in latently infected, growth-transformed cells. J. Virol. *46*, 424-433.

van Santen, V., Cheung, A., and Kieff, E. (1981). Epstein-Barr virus RNA VII: size and direction of transcription of virus-specified cytoplasmic RNAs in a transformed cell line. Proc. Natl. Acad. Sci. USA. *78*, 1930-1934.

Walls, D., and Perricaudet, M. (1991). Novel downstream elements upregulate transcription initiated from an Epstein-Barr virus latent promoter. Embo J. *10*, 143-151.

Waltzer, L., Logeat, F., Brou, C., Israel, A., Sergeant, A., and Manet, E. (1994). The human J. kappa recombination signal sequence binding protein (RBP-J kappa) targets the Epstein-Barr virus EBNA2 protein to its DNA responsive elements. Embo J. *13*, 5633-5638.

Weigel, R., and Miller, G. (1983). Major EB virus-specific cytoplasmic transcripts in a cellular clone of the HR-1 Burkitt lymphoma line during latency and after induction of viral replicative cycle by phorbol esters. Virology *125*, 287-298.

Weigel, R., and Miller, G. (1985). Latent and viral replicative transcription *in vivo* from the BamHI K fragment of Epstein-Barr virus DNA. J. Virol. *54*, 501-508.

Woisetschlaeger, M., Jin, X. W., Yandava, C. N., Furmanski, L. A., Strominger, J. L., and Speck, S. H. (1991). Role for the Epstein-Barr virus nuclear antigen 2 in viral promoter switching during initial stages of infection. Proc. Natl. Acad. Sci. USA. *88*, 3942-3946.

Woisetschlaeger, M., Strominger, J. L., and Speck, S. H. (1989). Mutually exclusive use of viral promoters in Epstein-Barr virus latently infected lymphocytes. Proc. Natl. Acad. Sci. USA. *86*, 6498-6502.

Woisetschlaeger, M., Yandava, C. N., Furmanski, L. A., Strominger, J. L., and Speck, S. H. (1990). Promoter switching in Epstein-Barr virus during the initial stages of infection of B lymphocytes. Proc. Natl. Acad. Sci. USA. *87*, 1725-1729.

Wu, Y., and Carstens, E. B. (1996). Initiation of baculovirus DNA replication: early promoter regions can function as infection-dependent replicating sequences in a plasmid-based replication assay. J. Virol. *70*, 6967-6972.

Yandava, C. N., and Speck, S. H. (1992). Characterization of the deletion and rearrangement in the BamHI C region of the X50-7 Epstein-Barr virus genome, a mutant viral strain which exhibits constitutive BamHI W promoter activity. J. Virol. *66*, 5646-5650.

Yang, L., and Botchan, M. (1990). Replication of bovine papillomavirus type 1 DNA initiates within an E2-responsive enhancer element. J. Virol. *64*, 5903-5911.

Yoo, L., and Speck, S. H. (2000). Determining the role of the Epstein-Barr virus Cp EBNA2-dependent enhancer during the establishment of latency by using mutant and wild-type viruses recovered from cottontop marmoset lymphoblastoid cell lines. J. Virol. *74*, 11115-11120.

Yoo, L. I., Mooney, M., Puglielli, M. T., and Speck, S. H. (1997). B-cell lines immortalized with an Epstein-Barr virus mutant lacking the Cp EBNA2 enhancer are biased toward utilization of the oriP-proximal EBNA gene promoter Wp1. J. Virol. *71*, 9134-9142.

Yoo, L.I., Woloszynek, J., Templeton, S., and Speck, S.H. (2002). Deletion of Epstein-Barr virus regulatory sequences upstream of the EBNA gene promoter Wp1 is unfavorable for B-cell immortalization. J. Virol. 76, 11763-11769.

Zhang, L., and Pagano, J. S. (1997). IRF-7, a new interferon regulatory factor associated with Epstein-Barr virus latency. Mol. Cell Biol. *17*, 5748-5757.

Zhang, L., and Pagano, J. S. (1999). Interferon regulatory factor 2 represses the Epstein-Barr virus BamHI Q latency promoter in type III latency. Mol. Cell Biol. *19*, 3216-3223.

Zimber-Strobl, U., Suentzenich, K., Falk, M., Laux, G., Cordier, M., Calender, A., Billaud, M., Lenoir, G. M., Bornkamm, G. W. (1990) Epstein-Barr virus terminal protein gene transcription is dependent on EBNA2 expression and provides evidence for viral integration into the host genome. Curr. Top. Microbiol. Immunol. *166,* 359-366.

Zimber-Strobl, U., Suentzenich, K., Laux, G., Erik, D., Cordier, M., Calender, A., Billaud, M., Lenoir, G. M., Bornkamm, G. W. (1991) Epstein-Barr virus nuclear antigen 2 activates transcription of the terminal protein gene. J. Virol. *65,* 415-423.

Chapter 21

Role of EBERs

*Kenzo Takada**

ABSTRACT

Epstein-Barr virus (EBV)-encoded small nonpolyadenylated RNAs (EBERs) are the most abundant viral transcripts in latently EBV-infected cells. However, their roles in viral infection were long totally unknown. Recently, several reports demonstrated that EBERs play a key role in oncogenesis. EBERs confer resistance to interferon α-induced apoptosis by directly binding to double-stranded RNA-activated protein kinase and inhibiting its phosphorylation. Alternatively, EBERs can also induce the expression of cellular growth factors. EBERs induce the expression of interleukin-10 in B cells, interleukin-9 in T cells, and insulin-like growth factor-1 in epithelial cells, each of which acts as an autocrine growth factor. These studies open the way toward supporting a new concept that RNA molecules can contribute to the oncogenic process.

INTRODUCTION

Epstein-Barr virus (EBV) encodes small nonpolyadenylated RNAs termed EBERs. EBERs, EBER1 and EBER2 (Lerner et al., 1981), are the most abundant viral transcripts in latently EBV-infected cells (Rymo, 1979), 167 and 172 nucleotides long, respectively, and transcribed by RNA polymerase III (Rosa et al., 1981). Their abundance allows using them as target molecules for detection of EBV-infected cells in tissues by *in situ* hybridization (ISH) (Chang et al., 1992), and their existence is considered a reliable marker of the existence of EBV. EBERs were reported to bind some cellular proteins (Lerner et al., 1981; Toczyski and Steitz, 1991; Clarke et al., 1991); however, their roles in viral infection were long totally unknown. Recently, it was found that EBERs play a key role in the maintenance of malignant phenotypes of Burkitt's lymphoma (BL) cells (Komano et al., 1999). They confer clonability in soft agarose, tumorigenicity in mice, and resistance to apoptosis against various stimuli in BL. Furthermore, EBERs induce transcription of interleukin (IL)-10, which acts as an autocrine growth factor of BL (Kitagawa et al., 2000). These studies open the way toward the new concept that RNA molecules can act in oncogenesis.

*For correspondence email kentaka@igm.hokudai.ac.jp

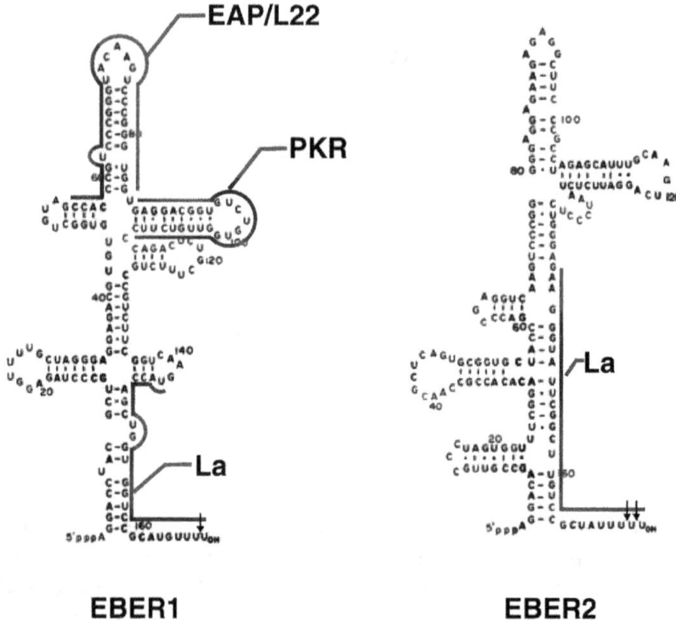

Figure 1. Potential secondary structures of EBERs. Reproduced from Rosa et al. (1981).

STRUCTURE OF EBERS

EBERs are encoded by the right-hand 1,000 base pairs of the EcoRI J fragment of the EBV genome. EBER1 is 166 nucleotides long and EBER2 is 172 nucleotides long (Rosa et al., 1981). The EBER genes are separated by 161 base pairs and are transcribed from left to right on the EBV map. Both EBER genes carry intragenic transcription control regions for RNA polymerase III, and can be transcribed by it. The primary sequence similarity between EBER1 and EBER2 is only 54%, but both EBER1 and EBER2 show striking similarity in their secondary structures with extensively base-paired structures containing a number of short stem loops (Figure 1). Striking similarities in the secondary structures also exist between EBERs and adenovirus-associated RNAs (VAs), which are small nonpolyadenylated RNAs like EBERs (Rosa et al., 1981).

The primary sequences of the EBERs are strongly conserved among a number of EBV strains (Arrand et al., 1989). Within the 1 kilobase EBER region, ten single base changes which group the strains into two families (1 and 2) have been identified. The EBER1 sequence is completely conserved, two base changes are within the EBER2-coding sequence and eight are outside the coding regions. Family 1 and family 2 parallel type A and type B EBVs (also called type I and type II EBVs) as defined by variations in EBNA2 and EBNA3; however, some isolates appear to be intertypic recombinants, which are often observed in isolates from HIV patients (Yao et al., 1996). The overall high conservation of the sequence suggests that EBERs are important for the virus life cycle.

EXPRESSION OF EBERS

EBERs are transcribed by RNA polymerase III (Rosa et al., 1981). Class III promoters are characterized by their intragenic location and, in the case of EBERs, these sequences are nearly identical to the consensus sequences derived from boxes A and B. In addition, they contain three upstream elements that together stimulate *in vivo* expression 50-fold and contain a TATA box and ATF- and Sp1-like promoter elements, which resemble sites associated with typical class II promoters (Howe and Shu, 1989). However, it is not known whether EBERs are transcribed by RNA polymerase II.

It appears that the copy number of EBERs is related to the copy number of EBV DNA molecules in each cell type (Lerner et al., 1981). EBER plasmids that contain 10 tandem repeats of EBER genes give higher EBER expression in transfected B lymphoma cells compared with EBER plasmid containing a single copy of the EBER gene (Komano et al., 1999). Following EBV infection of primary B lymphocytes, EBV-determined nuclear antigen 2 (EBNA2) appears first at 6 h after infection, followed by other EBNAs, latent membrane proteins (LMPs) and EBERs. On the other hand, the non-transforming P3HR-1 strain, from which the EBNA2 gene is deleted, expresses only EBNA-leader protein (EBNA-LP) and trace amounts of EBER1 in primary B lymphocytes, while the same virus can express EBNA1, EBNA3, EBNA-LP and EBERs in EBV-genome-negative BL cell lines (Rooney et al., 1989). These findings suggest that EBER expression is dependent on the host cell, perhaps through products specific for the cell cycle or the state of B-cell differentiation.

State of viral life cycle seems to influence EBER expression. Nuclear run-on assays showed downregulation of EBER transcription during the switch from latent infection to lytic replication of the virus (Greifenegger et al., 1998). In contrast, the amounts of EBERs remain unaltered within 72 h after induction of lytic replication. Although both EBERs are transcribed at approximately equal rates, the steady-state level of EBER1 is 10-fold greater than that of EBER2. This is due to a much faster rate of turnover of EBER2. In the presence of actinomycin D, the half-lives of EBER1 and EBER2 are 8 to 9 h and 45 min, respectively (Clarke et al., 1992).

EBERs are not expressed in the tissue of oral hairy leukoplakia, Sjogren's syndrome, salivary gland lymphoma or oral papilloma, where EBV actively replicates (Gilligan et al., 1990; Wen et al., 1996; 1997; Mizugaki et al., 1998). It is believed that EBERs are not expressed in permissively infected cells, but are consistently expressed in latently infected cells. Therefore, an *in situ* hybridization targeted EBER1 has been extensively used as a reliable marker of EBV infection in tissue specimens (Chang et al., 1992). Sugawara et al. (1999) have demonstrated that this is not always correct. They have shown that there is a high EBV load in the tissue of hepatocellular carcinoma and that EBV-infected carcinoma cells express EBNA1 and transcripts from the *Bam*HI A region (BARF0), but are completely negative for EBER expression, even when using a very sensitive RT-PCR assay. A similar pattern of EBV infection is also observed in the tissue of breast carcinoma (Bonnet et al., 1999). Yao et al. (2000) have reported heterogeneity of EBER expression in individual tumor cells of NPC; some cells express high levels of EBER, yet adjacent tumor cells express very little or none. However, in this case, it must be clarified whether EBER-negative cells are positive for the EBV genome. In conclusion, EBER-negative latent infection does exist, and by using the EBER ISH, a number of EBV-infected cells could remain undetected.

ASSOCIATION OF EBERS WITH CELLULAR PROTEINS

EBERs are located in the nucleus as evidenced by intense nuclear staining by EBER ISH (Chang et al., 1992; Howe and Steitz, 1986). However, high resolution ISH using confocal laser scanning microscopy has revealed that EBERs are found in cytoplasm as well as in the nuclei of interphase cells (Schwemmle et al., 1992). The cytoplasmic staining is not homogenous, with the perinuclear region being preferentially stained, which corresponds to the location of the rough ER and Golgi apparatus.

EBERs exist in nuclear ribonucleoprotein (RNP) complexes that are precipitated by anti-La antibodies associated with systemic lupus erythematosus (SLE) (Lerner et al., 1981). In these complexes, La binds the oligouridylate stretch at the 3' termini of all mammalian RNA polymerase III transcripts, transiently for most RNAs but stably in the case of the EBERs (Howe and Shu, 1988), though the significance of their interactions is not known.

Adenovirus-associated RNAs, VA1 and VA2, are small RNAs transcribed by RNA polymerase III (Akusjarvi et al., 1980). Although there is no striking nucleotide sequence homology between EBERs and VAs, similarities exist in their size, degree of secondary structure and genomic organization (Bhat and Thimmappaya, 1983). Like VA RNAs, EBERs bind double-stranded (ds) RNA-activated protein kinase (PKR), which is a key mediator of the antiviral effect of interferon (Clarke et al., 1991; Meurs et al., 1990; Sharp et al., 1993). Once activated by dsRNA, PKR phosphorylates the α subunit of protein synthesis initiation factor eIF2, causing inhibition of translation at the level of initiation. *In vitro* assays have demonstrated that EBERs can inhibit PKR activation and block phosphorylation of eIF2α, thus resulting in the blockage of inhibition of protein synthesis by eIF2α (Sharp et al., 1993; Clarke et al., 1990; Katze et al, 1991). More recently, Nanbo et al. (2002) demonstrated that in BL cells, EBERs confer resistance to interferon α-induced apoptosis by directly binding to PKR and inhibiting its phosphorylation.

It has been reported that expression of a functionally defective mutant of PKR in NIH3T3 cells results in malignant transformation, suggesting that PKR may function as a tumor suppressor (Koromilas et al., 1992; Meurs et al., 1993). There is a report from the proceedings of a meeting that the stable expression of EBER1 in NIH3T3 cells results in a morphologically transformed phenotype and confers on these cells the ability to form colonies and grow in soft agarose (Laing et al., 1995). However, the problem with that report is the absence of report that succeeds in repeating these experiments.

A second highly abundant protein designated EAP (EBER-associated protein) was identified in La-containing RNP complexes (Toczyski and Steitz, 1991). EAP was subsequently shown to be the ribosomal protein L22 (Toczyski et al., 1994). A model was proposed, in which L22 was captured by EBER in the nucleoplasm before it was assembled into the nucleolus. L22 binds a stem-loop structure (stem-loop 3) of EBER1. A possible role for the EBER-L22 interaction is to sequester the cellular L22 molecules. The gene encoding L22 has been identified as the target of a chromosomal translocation in certain patients with leukemia (Liu et al., 1993; Nucifosa et al., 1993), suggesting that L22 levels may be a determinant in cell transformation. Recently, it was reported that L22 interferes with the ability of the EBERs to inhibit the activation of PKR by dsRNA (Elia at al., 2004). Transient expression of EBER1 in murine embryonic fibroblasts stimulates reporter gene expression and partially reverses the inhibitory effect of PKR. However, EBER1 is also stimulatory when transfected into PKR knockout cells, suggesting an additional, PKR-independent, mode of action of EBERs. Expression of L22 prevents both the PKR-dependent and –independent effects of EBER1 *in vivo*. These results suggest that the association of L22 with EBER1 in EBV-infected cells can attenuate the biological effect of the viral RNA. Such effects include both the inhibition of PKR and additional mechanism(s) by which EBER1 stimulates gene expression.

ROLE OF EBERS IN ONCOGENESIS

The lack of a suitable *in vitro* system that represents BL-type EBV infection, which is characterized by expression of a limited number of viral genes (termed type I latency), including EBNA1, EBERs and BARF0 (Rickinson and Kieff, 2002), has hampered study of the role of EBV in the genesis of BL. The Japanese BL-derived Akata cell line (Takada, 1984; Takada and Ono, 1989; Takada et al., 1991) is unique in that it retains the *in vivo* phenotype of EBV expression even after long-term culture *in vitro* (Shimizu et al., 1994). Isolation of EBV-negative cell clones from the parental Akata cell culture allowed for more systematic studies of the role of EBV in BL (Shimizu et al., 1994). Comparison of EBV-positive and negative cell clones revealed that the presence of EBV in Akata cells was required for the cells to be more malignant and apoptosis resistant (Shimizu et al., 1994; Chodosh et al., 1998; Komano et al., 1998; Ruf et al., 2000), which underlined the oncogenic role of EBV in the genesis of BL. Subsequent studies revealed that EBERs were responsible for these phenotypes (Komano et al., 1999). Transfection of the EBER genes into EBV-negative Akata clones restored the capacity for growth in soft agar, tumorigenicity in SCID mice, resistance to apoptotic inducers, and upregulated expression of bcl-2 that was originally retained in parental EBV-positive Akata cells and lost in EBV-negative subclones. After their report, other studies presented essentially similar results (Ruf et al., 2000; Yamamoto et al., 2000; Maruo et al., 2001).

More recently, it was demonstrated that EBERs induce human IL-10 expression in BL cells (Kitagawa et al., 2000). EBV-negative subclones were isolated from BL-derived Akata and Mutu cell lines, which retained BL-type EBV expression. It was found that EBV-positive Akata and Mutu cell clones expressed higher levels of IL-10 than their EBV-negative subclones at the transcriptional level. Transfection of an individual EBV latent gene into EBV-negative Akata cells revealed that EBERs were responsible for IL-10 induction. Recombinant IL-10 enabled EBV-negative Akata cells to grow in low (0.1%) serum conditions. On the other hand, growth of EBV-positive Akata cells was blocked by treatment either with an anti-IL-10 antibody or antisense oligonucleotide against IL-10. EBV-positive BL biopsies consistently expressed IL-10, but EBV-negative BL biopsies did not. These results suggest that IL-10 induced by EBERs acts as an autocrine growth factor for BL.

EBV associates with various T-cell-proliferating diseases such as chronic active EBV infection and nasal lymphoma. A human T-cell line, MT-2, was susceptible to EBV infection, and EBV-infected cell clones showed type II EBV latency, which was identical with those seen in EBV-infected T-cells *in vivo* (Yoshiyama et al., 1995). EBV-infected MT-2 cell clones had shorter cell doubling times and higher saturation density than EBV-uninfected counterparts. It was found that EBV-positive MT-2 cells expressed higher levels of IL-9 than EBV-negative MT-2 cells at the transcriptional level (Yang et al., 2004). It was also demonstrated that EBERs were responsible for IL-9 expression. Addition of recombinant IL-9 accelerated the growth of MT-2 cells, whereas growth of the EBV-converted MT-2 cells was blocked by treatment with an anti-IL-9 antibody. These results suggest that IL-9 induced by EBERs acts as an autocrine growth factor for EBV-infected T-cells. Analysis of nasal lymphoma biopsies indicated that three of four specimens expressed IL-9. These findings suggest that EBERs directly affect the pathogenesis of EBV-associated T-cell diseases.

About 5 to 10% of gastric carcinoma (GC) cases worldwide are associated with EBV (Takada, 2000). Iwakiri et al. (2003) reported that EBV infection induces expression of insulin-like growth factor 1 (IGF1) in the GC-derived EBV-negative cell line NU-GC-3, and that the secreted IGF1 acts as an autocrine growth factor. Transfection of individual EBV latent genes into NU-GC-3 cells revealed that the EBERs were responsible for IGF1 expression. Addition of recombinant IGF1 accelerated growth of NU-GC-3 cells, while growth of the EBV-converted NU-GC-3 cells was blocked by treatment with an anti-IGF1 antibody. These results suggest that IGF1 induced by EBERs acts as an autocrine growth factor for EBV-positive GC. These findings seem to be operative *in vivo*, as EBV-positive GC biopsies consistently express IGF1, while EBV-negative GC biopsies do not. These findings strongly suggest that EBERs directly affect the pathogenesis of EBV-positive GC

EBERs induce transcription of three different growth factors in three different cell types and make key contributions to both lymphoid and epithelioid carcinogenesis. Studies on dominant-negative PKR and the PKR inhibitor suggested that PKR inhibition was not involved in transcriptional activation of these growth factors. It remains to be clarified whether binding of EBER with other cellular factors, including La and EAP/L22, plays some role in activation of the growth factors. It would also be interesting to examine whether EBER is cleaved to small RNA and has activity as a small interfering RNA.

Regarding the role of EBERs in the process of EBV-induced B cell transformation, Swaminathan et al. (1991) demonstrated that EBERs were not essential for the immortalization of B lymphocytes or for the replication of the virus. They restored the transformation-defect of the P3HR-1 strain EBV, having a deletion of the essential-transforming gene EBNA-2, using homologous recombination between the P3HR-1 deleted genome and a EBER-deleted EBV DNA fragment spanning the EBNA-2 locus. Their attempt resulted in obtaining lymphoblastoid cell lines (LCLs) harboring only EBER-deleted recombinant viruses, indicating that EBERs are dispensable for B cell transformation. However, they failed to produce a large quantity of pure EBER-deleted EBV. Instead, a co-cultivation method was used to passage the EBER-deleted EBV from primary LCLs to secondary LCLs. Therefore, the transforming titer of EBER-deleted EBV has never been determined by using a pure recombinant virus. Recently, Yajima et al. (2005) re-visited this issue by producing a large quantity of EBER-deleted

EBV using an Akata cell system. Although the EBER-deleted virus efficiently infected B lymphocytes, its 50% transforming dose was approximately 100-fold less than that of the EBER-positive EBV. They then engineered the genome of EBER-deleted virus and generated a recombinant virus with the EBER genes reconstituted at their native locus. The resultant EBER-reconstituted EBV exhibited restored transforming ability. In addition, LCLs established with the EBER-deleted EBV grew significantly slower than those established with wild-type or EBER-reconstituted EBV, and the difference of growth rates was especially highlighted when the cells were plated at low cell densities. These results clearly demonstrate that EBERs significantly contribute to efficient growth transformation of B lymphocytes by enhancing the growth potential of transformed lymphocytes.

References

Akusjarvi, G., Mathews, M.B., Andersson, P., Vennstrom, B., and Pettersson, U. (1980) Structure of genes for virus-associated RNAI and RNAII of adenovirus type 2. Proc. Natl. Acad. Sci. USA. *77*, 2424-2428.

Arrand, J.R., Young, L.S., and Tugwood, J.D. (1989) Two families of sequences in the small RNA-encoding region of Epstein- Barr virus (EBV) correlate with EBV types A and B. J. Virol. *63*, 983-986.

Bhat, R.A., and Thimmappaya, B. (1983) Two small RNAs encoded by Epstein-Barr virus can functionally substitute for the virus-associated RNAs in the lytic growth of adenovirus 5. Proc. Natl. Acad. Sci. USA. *80*, 4789-4793.

Bonnet, M., Guinebretiere, J.M., Kremmer, E., Grunewald, V., Benhamou, E., Contesso, G., and Joab, I. (1999) Detection of Epstein-Barr virus in invasive breast cancers. J. Natl. Cancer Inst. *91*, 1376-1381.

Chang, K.L., Chen, Y.Y., Shibata, D., and Weiss, L.M. (1992) Description of an in situ hybridization methodology for detection of Epstein-Barr virus RNA in paraffin-embedded tissues, with a survey of normal and neoplastic tissues. Diagn. Mol. Pathol. *1*, 246-255.

Chodosh, J., Holder, V.P., Gan, Y.J., Belgaumi, A., Sample, J., and Sixbey, J.W. (1998) Eradication of latent Epstein-Barr virus by hydroxyurea alters the growth-transformed cell phenotype. J. Infect. Dis. *177*, 1194-1201.

Clarke, P.A., Schwemmle, M., Schickinger, J., Hilse, K., and Clemens, M.J. (1991) Binding of Epstein-Barr virus small RNA EBER-1 to the double-stranded RNA-activated protein kinase DAI. Nucleic. Acids. Res. *19*, 243-248.

Clarke, P.A., Sharp, N.A., and Clemens, M.J. (1990) Translational control by the Epstein-Barr virus small RNA EBER-1. Reversal of the double-stranded RNA-induced inhibition of protein synthesis in reticulocyte lysates. Eur. J. Biochem. *193*, 635-641.

Clarke, P.A., Sharp, N.A., and Clemens, M.J. (1992) Expression of genes for the Epstein-Barr virus small RNAs EBER-1 and EBER-2 in Daudi Burkitt's lymphoma cells: effects of interferon treatment. J. Gen. Virol. *73*, 3169-3175.

Elia, A., Vyas, J., Laing, K.G. and Clemens, M.J. (2004) Ribosomal protein L22 inhibits regulation of cellular activities by the Epstein-Barr virus small RNA EBER-1. Eur. J. Biochem. *271*, 1895-1905.

Gilligan, K., Rajadurai, P., Resnick, L., and Raab-Traub, N. (1990) Epstein-Barr virus small nuclear RNAs are not expressed in permissively infected cells in AIDS-associated leukoplakia. Proc. Natl. Acad. Sci. USA. *87*, 8790-8794.

Greifenegger, N., Jager, M., Kunz-Schughart, L.A., Wolf, H., and Schwarzmann, F. (1998) Epstein-Barr virus small RNA (EBER) genes: differential regulation during lytic viral replication. J. Virol. *72*, 9323-9328.

Howe, J.G., and Shu, M.D. (1988) Isolation and characterization of the genes for two small RNAs of herpesvirus papio and their comparison with Epstein-Barr virus-encoded EBER RNAs. J. Virol. *62*, 2790-2798.

Howe, J.G., and Shu, M.D. (1989) Epstein-Barr virus small RNA (EBER) genes: unique transcription units that combine RNA polymerase II and III promoter elements. Cell. *57*, 825-834.

Howe, J.G., and Steitz, J.A. (1986) Localization of Epstein-Barr virus-encoded small RNAs by in situ hybridization. Proc. Natl. Acad. Sci. USA. *83*, 9006-9010.

Iwakiri, D., Eizuru, Y., Tokunaga, M., and Takada, K. (2003). Advances in brief: Autocrine growth of Epstein-Barr virus positive gastric carcinoma mediated by an Epstein-Barr virus encoded small RNA. Cancer Res. *63*, 7062-7067.

Katze, M.G., Wambach, M., Wong, M.L., Garfinkel, M., Meurs, E., Chong, K., Williams, B.R., Hovanessian, A.G., and Barber, G.N. (1991) Functional expression and RNA binding analysis of the interferon-induced, double-stranded RNA-activated, 68,000-Mr protein kinase in a cell-free system. Mol. Cell Biol. *11*, 5497-5505.

Kitagawa, N., Goto, M., Kurozumi, K., Maruo, S., Fukayama, M., Naoe, T., Yasukawa, M., Hino, K., Suzuki, T., Todo, S., and Takada, K. (2000) Epstein-Barr virus-encoded poly(A)(-) RNA supports Burkitt's lymphoma growth through interleukin-10 induction. EMBO J. *19*, 6742-6750.

Komano, J., Maruo, S., Kurozumi, K., Oda, T., and Takada, K. (1999) Oncogenic role of Epstein-Barr virus-encoded RNAs in Burkitt's lymphoma cell line Akata. J. Virol. *73*, 9827-9831.

Komano, J., Sugiura, M., and Takada, K. (1998) Epstein-Barr virus contributes to the malignant phenotype and to apoptosis resistance in Burkitt's lymphoma cell line Akata. J. Virol. *72*, 9150-9156.

Koromilas, A.E., Roy, S., Barber, G.N., Katze, M.G., and Sonenberg, N. (1992) Malignant transformation by a mutant of the IFN-inducible dsRNA-dependent protein kinase. Science. *257*, 1685-1689.

Laing, K.G., Matys, V., Clemens, M.J. (1995) Effects of expression of the Epstein-Barr virus small RNA EBER-1 in heterologous cells on protein synthesis and cell growth. Biochem. Soc. Trans. *23*, 311S.

Lerner, M.R., Andrews, N.C., Miller, G., and Steitz, J.A. (1981) Two small RNAs encoded by Epstein-Barr virus and complexed with protein are precipitated by antibodies from patients with systemic lupus erythematosus. Proc. Natl. Acad. Sci. USA. *78*, 805-809.

Liu, P., Tarle, S.A., Hajra, A., Claxton, D.F., Marlton, P., Freedman, M., Siciliano, M.J., and Collins, F.S. (1993) Fusion between transcription factor CBF beta/PEBP2 beta and a myosin heavy chain in acute myeloid leukemia. Science. *261*, 1041-1044.

Maruo, S., Nanbo, A. and Takada, K. (2001) Replacement of the Epstein-Barr virus plasmid with the EBER plasmid in Burkitt's lymphoma cells. J. Virol. *75*, 9977-9982.

Meurs, E., Chong, K., Galabru, J., Thomas, N.S., Kerr, I.M., Williams, B.R., and Hovanessian, A.G. (1990) Molecular cloning and characterization of the human double-stranded RNA- activated protein kinase induced by interferon. Cell. *62*, 379-390.

Meurs, E.F., Galabru, J., Barber, G.N., Katze, M.G., and Hovanessian, A.G. (1993) Tumor suppressor function of the interferon-induced double-stranded RNA-activated protein kinase. Proc. Natl. Acad. Sci. USA. *90*, 232-236.

Mizugaki, Y., Sugawara, Y., Shinozaki, F., and Takada, K. (1998) Detection of Epstein-Barr virus in oral papilloma. Jpn. J. Cancer Res. 89, 604-607.

Nanbo, A., Inoue, K., Adachi-Takasawa, K. and Takada, K. (2002) Epstein-Barr virus RNA confers resistance to interferon-α-induced apoptosis in Burkitt's lymphoma. EMBO J. *21*, 954-965.

Nucifora, G., Begy, C.R., Erickson, P., Drabkin, H.A., and Rowley, J.D. (1993) The 3;21 translocation in myelodysplasia results in a fusion transcript between the AML1 gene and the gene for EAP, a highly conserved protein associated with the Epstein-Barr virus small RNA EBER 1. Proc. Natl. Acad. Sci. USA. *90*, 7784-7788.

Rickinson, A.B., and Kieff, E. (2002) Epstein-Barr virus, in Fields Virology, Fields B.N., Knipe D.M., and Hoeley, P.M., eds. (Lippincott- Raven Publishers, Philadelphia) pp 2575-2627.

Rooney, C., Howe, J.G., Speck, S.H., and Miller, G. (1989) Influence of Burkitt's lymphoma and primary B cells on latent gene expression by the nonimmortalizing P3J-HR-1 strain of Epstein-Barr virus. J. Virol. *63*, 1531-1539.

Rosa, M.D., Gottlieb, E., Lerner, M.R., and Steitz, J.A. (1981) Striking similarities are exhibited by two small Epstein-Barr virus- encoded ribonucleic acids and the adenovirus-associated ribonucleic acids VAI and VAII. Mol. Cell. Biol. *1*, 785-796.

Ruf, I.K., Rhyne, P.W., Yang, C., Cleveland, J.L., and Sample, J.T. (2000) Epstein-Barr virus small RNAs potentiate tumorigenicity of Burkitt lymphoma cells independently of an effect on apoptosis. J. Virol. *74*, 10223-10228.

Ruf, I.K., Rhyne, P.W., Yang, H., Borza, C.M., Hutt-Fletcher, L.M., Cleveland, J.L., and Sample, J.T. (1999) Epstein-Barr virus regulates c-MYC, apoptosis, and tumorigenicity in Burkitt lymphoma. Mol. Cell. Biol. *19*, 1651-1660.

Rymo, L. (1979) Identification of transcribed regions of Epstein-Barr virus DNA in Burkitt lymphoma-derived cells. J. Virol. *32*, 8-18.

Schwemmle, M., Clemens, M.J., Hilse, K., Pfeifer, K., Troster, H., Muller, W.E., and Bachmann, M. (1992) Localization of Epstein-Barr virus-encoded RNAs EBER-1 and EBER-2 in interphase and mitotic Burkitt lymphoma cells. Proc. Natl. Acad. Sci. USA. *89*, 10292-10296.

Sharp, T.V., Schwemmle, M., Jeffrey, I., Laing, K., Mellor, H., Proud, C.G., Hilse, K., and Clemens, M.J. (1993) Comparative analysis of the regulation of the interferon-inducible protein kinase PKR by Epstein-Barr virus RNAs EBER-1 and EBER-2 and adenovirus VAI RNA. Nucleic. Acids. Res. *21*, 4483-4490.

Shimizu, N., Tanabe-Tochikura, A., Kuroiwa, Y., and Takada, K. (1994) Isolation of Epstein-Barr virus (EBV)-negative cell clones from the EBV- positive Burkitt's lymphoma (BL) line Akata: malignant phenotypes of BL cells are dependent on EBV. J. Virol. *68*, 6069-6073.

Sugawara, Y., Mizugaki, Y., Uchida, T., Torii, T., Imai, S., Makuuchi, M., and Takada, K. (1999) Detection of Epstein-Barr virus (EBV) in hepatocellular carcinoma tissue: a novel EBV latency characterized by the absence of EBV-encoded small RNA expression. Virology. *256*, 196-202.

Swaminathan, S., Tomkinson, B., and Kieff, E. (1991) Recombinant Epstein-Barr virus with small RNAs (EBERs) genes deleted transforms lymphocytes and replicates *in vitro*. J. Virol. *66*, 5133-5136.

Takada, K. (1984) Cross-linking of cell surface immunoglobulins induces Epstein-Barr virus in Burkitt lymphoma lines. Int. J. Cancer. *33*, 27-32.

Takada, K. (2000) Epstein-Barr virus and gastric carcinoma. Mol. Pathol. *53*, 255-261.

Takada, K., Horinouchi, K., Ono, Y., Aya, T., Osato, T., Takahashi, M., and Hayasaka, S. (1991) An Epstein-Barr virus-producer line Akata: establishment of the cell line and analysis of viral DNA. Virus Genes. *5*, 147-156.

Takada, K., and Ono, Y. (1989) Synchronous and sequential activation of latently infected Epstein-Barr virus genomes. J. Virol. *63*, 445-449.

Toczyski, D.P., Matera, A.G., Ward, D.C., and Steitz, J.A. (1994) The Epstein-Barr virus (EBV) small RNA EBER1 binds and relocalizes ribosomal protein L22 in EBV-infected human B lymphocytes. Proc. Natl. Acad. Sci. USA. *91*, 3463-3467.

Toczyski, D.P., and Steitz, J.A. (1991) EAP, a highly conserved cellular protein associated with Epstein-Barr virus small RNAs (EBERs). EMBO J. *10*, 459-466.

Wen, S., Mizugaki, Y., Shinozaki, F., and Takada, K. (1997) Epstein-Barr virus (EBV) infection in salivary gland tumors: lytic EBV infection in nonmalignant epithelial cells surrounded by EBV-positive T- lymphoma cells. Virology *227,* 484-487.

Wen, S., Shimizu, N., Yoshiyama, H., Mizugaki, Y., Shinozaki, F., and Takada, K. (1996) Association of Epstein-Barr virus (EBV) with Sjogren's syndrome: differential EBV expression between epithelial cells and lymphocytes in salivary glands. Am. J. Pathol. *149,* 1511-1517.

Yajima, M., Kanda, T. and Takada, K. (2005). Role of Epstein-Barr virus (EBV)-encoded RNA (EBER) in efficient EBV-induced B-lymphocyte growth transformation. J. Virol. *79,* 4298-4307.

Yamamoto, N., Takizawa, T., Iwanaga, Y., Shimizu, N., and Yamamoto, N. (2000) Malignant transformation of B lymphoma cell line BJAB by Epstein-Barr virus-encoded small RNAs. FEBS Lett. *484,* 153-158.

Yang, L., Aozasa, K., Oshimi, K. and Takada, K. (2004). Epstein-Barr virus (EBV)-encoded RNA promotes growth of EBV-infected T-cells through interleukin-9 induction. Cancer R. *64,* 5332-5337.

Yao, Q-Y., Tierney, R.J., Croom-Carter, D., Cooper, G.M., Ellis, C.J., Rowe, M., and Rickinson, A.B. (1996) Isolation of intertypic recombinants of Epstein-Barr virus from T-cell immunocompromised individuals. J. Virol. *70,* 4895-4903.

Yao, Y., Minter, H.A., Chen, X., Reynolds, G.M., Bromley, M., and Arrand, J.R. (2000) Heterogeneity of HLA and EBER expression in Epstein-Barr virus-associated nasopharyngeal carcinoma. Int. J. Cancer. *88,* 949-955.

Yoshiyama, H., Shimizu, N., and Takada, K. (1995) Persistent Epstein-Barr virus infection in a human T-cell line: unique program of latent virus expression. EMBO J. *14,*3706-3711.

Chapter 22

EBNA2 Transcription Regulation in EBV Latency

*Henrik Zetterberg and Lars Rymo**

ABSTRACT

In vitro and upon primary infection *in vivo*, Epstein-Barr virus (EBV) infects resting B-cells and transforms them into proliferating lymphoblasts with a remarkable efficiency. Nine virally encoded proteins and a limited number of cellular proteins are expressed under the control of a master transcription factor, the EBV nuclear antigen 2 (EBNA2), and play a decisive role in the transformation process. While much is known at the molecular level about the EBNA2-dependent expression program of EBV genes, the pattern of cellular target genes remains poorly defined. However, it is known that EBNA2 specifically up-regulates expression of the B-cell activation marker CD23, the complement and EBV receptor CD21, the chemokine receptor BLR2/EBI1 and the AP-1 family member BATF, and activates the proto-oncogenes *c-fgr* and *c-myc*, the latter of which seems to be the most prominent cellular target essential for the ability of EBNA2 to transform B-cells. Here, we review current knowledge of the molecular mechanisms by which EBNA2 controls viral and cellular gene expression in order to promote lymphoproliferation.

INTRODUCTION: EBNA2 – STRUCTURE AND FUNCTION

EBNA2 is a viral transcription factor that does not bind DNA directly but is tethered to the EBNA2-responsive promoters by interacting with the cellular repressor recombination signal binding protein J kappa (RBP-Jκ) and several other cellular DNA-binding proteins. It is a phosphoprotein (Grässer et al., 1992; Grässer et al., 1991) that localizes to various compartments of the nucleus, including the nucleoplasm, the chromatin fraction, and the nuclear matrix (Petti et al., 1990; Sauter and Mueller-Lantzsch, 1987). EBNA2 does not bear any significant sequence homology with cellular proteins. Instead, much information about structure-function relationships has been derived from deletion mutation analyses of the *EBNA2* gene (Cohen and Kieff, 1991; Cohen et al., 1991; Ling et al., 1994; Sjöblom et al., 1995b; Tong et al., 1994; Yalamanchili et al., 1994; Zimber-Strobl et al., 1994) and comparative analysis of evolutionary conserved regions of EBNA2-equivalent proteins in isolates from different primate lymphocryptoviruses (Ling

*For correspondence email lars.rymo@clinchem.gu.se

et al., 1993b). Altogether, nine conserved regions have been identified (CR1 to CR9) and characterized functionally. CR1 to CR4 contain two self-association domains that mediate homotypic interaction (dimerization or oligomerization), which is important for the ability of EBNA2 to transform B-cells and to activate the *LMP1* promoter in reporter assays (Harada et al., 1998; 2001; Tsui and Schubach, 1994; Yalamanchili et al., 1996). CR1 seems to be especially critical for the enhancement of *LMP1* promoter activity (Gordadze et al., 2004; Sjöblom et al., 1995b). Residues between CR4 and CR5 are referred to as the divergent region because of very low sequence homology between the two major variants of EBNA2 (-2A and -2B). Despite the diversity, this region has been implicated in interactions with cellular proteins, such as the RNA helicase DP103, the RNA processing complex associated with the survival motor neuron protein, and the SWI/SNF complex, involved in the mediation of EBNA2 function during B-cell immortalization (Grundhoff et al., 1999; Voss et al., 2001; Wu et al., 1996; 2000). CR5 interacts with SKIP (Zhou et al., 2000), which facilitates the functionally important CR6-mediated interaction with RBP-Jκ (Grossman et al., 1994; Henkel et al., 1994; Ling et al., 1993a; Waltzer et al., 1995; Zimber-Strobl et al., 1994). CR8 contains an acidic activation domain and is important for interaction with several components of the basal transcription machinery and co-activators, including TFIIH, TAF40, TFIIB, and p100 that interacts with TFIIE, and the p300, CBP and PCAF histone acetyltransferases (Tong et al., 1995a; 1995b; 1995c; Wang et al., 2000). CR9 contains a nuclear localization signal and a conserved arginine-glycine rich region (Cohen et al., 1991; Ling et al., 1993b).

TRANSCRIPTIONAL CONTROL – AN OVERVIEW

The EBV genome is packaged into chromatin with an overall structure resembling that of the host cell chromatin (Dyson and Farrell, 1985). The packaging of DNA into chromatin restricts the access of the many regulatory proteins required for essential biological processes including transcription factors and the basal transcriptional machinery. In a simplified model, at least two distinct events are needed for initiation of transcription to occur: (i) relief by modifying enzymes and remodeling factors of nucleosome-mediated repression of the DNA template, and (ii) assembly of the RNA polymerase II transcription-initiation complex at the promoter in the exposed DNA (Kornberg and Lorch, 2002). As will become evident below, EBNA2 plays important roles during both of these events.

The basic structural unit of chromatin is the nucleosome, which is composed of 146 bp of DNA wrapped around a disc-shaped octamer of histone proteins. There are at least three fundamental ways by which histones contribute to the dynamics of chromosome function. Firstly, chromatin remodeling machines, such as the SWI/SNF complex, alter the structure of the nucleosome in an ATP-dependent manner, presumably by modifying the histone-DNA interface, and often cause nucleosome sliding (Peterson, 2002a; 2002b). Secondly, the composition of nucleosomes can be modulated by the replacement of the major histone proteins with specialized variants (Smith, 2002). Finally, the histones are subject to a rich variety of covalent modifications that regulate their function. Known modifications include co-translational acetylation of the N-terminus, post-translational lysine acetylation, lysine and arginine methylations, serine phosphorylation, and lysine ubiquitination

(Berger, 2002). In contrast to the topological and structural changes induced by SWI/SNF-like complexes, the covalent modification of histones does not cause dramatic changes in histone-DNA interactions or in nucleosome positioning. Rather, it has been proposed that distinct patterns of covalent histone modification function as a recognition code (histone code) for the recruitment of chromatin remodeling complexes and regulatory factors (Strahl and Allis, 2000).

DNA methyltransferases catalyze methylation of cytosines in the dinucleotide sequence CpG in DNA. In transcription regulatory regions, this modification often results in repression of transcription through at least two different mechanisms. Firstly, DNA methylation inhibits binding of certain transcription factors to regulatory elements in the DNA. Secondly, DNA methylation attracts methyl-CpG binding domain proteins, which interact with co-repressors with histone deacetylase (HDAC) activity that modify the nucleosomes establishing a condensed, transcriptionally inactive chromatin state. Thus, there is a relationship between DNA methylation and histone modification in the regulation of gene expression. Generally, DNA methylation is considered a less dynamic event compared with histone modifications. In fact, DNA methylation seems to be responsible for the stable maintenance of a particular gene expression pattern through mitotic cell division, and highly regulated DNA methylation is essential during cell differentiation and embryonic development (Jaenisch and Bird, 2003).

Gene activation: Relief of repression by chromatin structure

Local modification of nucleosomes on enhancers and promoters is required to activate gene repression. Transcription factors that bind nucleosome-free regions of DNA or to DNA within nucleosomes recruit two types of enzymatic activities that modify the surrounding chromatin architecture and function as co-activators of transcription. The first includes the SWI/SNF activity and the related RSC, NuRD, NURF, Mi-2, CHRAC and other complexes that utilize ATP to alter the nucleosomal structure and/or arrangement. The second includes histone acetyltransferases (HATs) that acetylate lysine residues in the N-terminal tails of histones. It is postulated that remodeling is accomplished either by nucleosome sliding along the DNA (Langst et al., 1999; Lorch et al., 2001; Whitehouse et al., 1999) or by inducing a continuous ATP-dependent DNA twist that provides the force for creating DNA sites accessible for regulatory factors even in the absence of histone movement (Gavin et al., 2001; Havas et al., 2000). It is possible that both mechanisms are linked in such a way that changes of the DNA structure might facilitate subsequent nucleosome sliding. However, the nature of chromatin alterations during the course of transcriptional switches is still not fully understood. Recent studies have elucidated the temporal sequence in which co-activators of transcription such as chromatin remodeling complexes and histone acetylase complexes are recruited to promoters *in vivo* and how their enzymatic properties contribute to gene activation (Featherstone, 2002). The best-characterized example in mammals is provided by the human β-interferon (*IFN-β*) gene. The gene is switched on by three transcription factors and an architectural protein which bind cooperatively to the nucleosome-free enhancer DNA to form an enhanceosome (Merika and Thanos, 2001). The enhanceosome targets the modification and repositioning of a nucleosome that blocks the formation of a transcriptional preinitiation complex on the *IFN*-β promoter. This is accomplished by the ordered recruitment of HATs, SWI/SNF, and basal transcription factors. The bending of DNA induced upon TFIID binding to the promoter causes sliding of the SWI/SNF-modified nucleosome to a new position 36 bp downstream, thus allowing the initiation of transcription (Lomvardas and Thanos, 2001). This ordered recruitment and nucleosome sliding is consistent with the view that histone acetylation sets the stage for ATP-dependent remodeling by establishing a recognition surface for the bromodomains present in the SWI/SNF-like remodeling machines. However, it is also well established that ATP-dependent chromatin remodeling might precede histone acetylation. Thus, in the case of mitotically expressed yeast genes, histone acetylation requires prior chromatin remodeling (Cosma et al., 1999).

The *cis*-acting regulatory elements in the DNA sequence that regulate transcription are designated enhancers and promoters. These are composed of short sequence motifs on which the *trans*-acting regulatory apparatus (transcription factors) binds, and by doing so, affects the level of transcription of the nearby gene, as well as the transcription start site. The transcription factors may be positive (activators) or negative (repressors) regulators of transcription. The terms co-activator and co-repressor are here used for factors that do not necessarily bind to regulatory DNA-sites directly but affect the interaction between DNA-binding activators and repressors with the basal transcription factors. It is the unique combination of transcription factors bound on enhancer DNA that specifies the expression program of a given gene at a given time. Transcription factors affect transcription not only by modulating the recruitment of the basal transcriptional machinery to a nearby promoter (Ptashne and Gann, 1997) but also by recruiting chromatin-modifying activities. The transcriptional machinery is composed of the large RNA polymerase II transcription-initiation complex. RNA polymerase II is by itself composed of 10 polypeptides and requires several general transcription factors (GTFs) to locate the proper start site in a DNA template and initiate transcription. These include TFIID, which binds to a TATA box through its TATA box-binding subunit, TBP. Two GTF subunits have helicase activity that separates the template strands during transcription. Furthermore, RNA polymerase II requires the highly conserved multiprotein mediator complex that associates with RNA polymerase II at the promoter, forming the large holoenzyme complex that also includes most of the GTFs. The mediator plays an essential role in transcriptional regulation by acting as an interface between RNA polymerase II and transcription factors bound to the enhancers and the promoter-proximal elements of the promoter. When completely assembled the active RNA polymerase II transcription-initiation complex comprises as many as 60-70 proteins with a total mass of 3 MDa (Kornberg, 2001). The fact that the basal transcriptional machinery is unable to assemble on condensed promoter structures ensures that transcription will not occur spontaneously at inappropriate times and places. Some transcription factors can bind to sites that are part of a nucleosome and subsequently recruit chromatin modulating enzymatic activities (histone acetyltransferases and SWI/SNF-related ATPases) (Di Croce et al., 1999). In other cases, the critical regulatory *cis* elements are already exposed since they are positioned between nucleosomes, and therefore the initial access of these regulators to the DNA is not an issue (Agalioti et al., 2000). However, chromatin rearrangement triggered by these chromatin-bound

Table 1. EBNA2-responsive viral and cellular genes and their possible functions

Gene	Function	EBNA2 effect
LMP1	Viral oncogene. A constitutively active CD40 homologue that induces B-cell proliferation and protects from apoptosis.	Up-regulation
LMP2	Inhibits virus lytic cycle. A constitutively active B-cell receptor homologue that protects from apoptosis.	Up-regulation
EBNA1-6	Maintenance of viral genome (EBNA1) and transactivation of viral and cellular genes involved in lymphoproliferation (EBNA1-6).	Up-regulation
CD21	The physiological C3d receptor that binds EBV.	Up-regulation
CD23	B-cell activation markers implicated in cell cycle progression.	Up-regulation
BLR2/ EBI1	G-protein coupled receptor that may play a role in lymphocyte trafficking.	Up-regulation
BATF	Negative regulator of AP-1 function. Inhibits lytic cycle.	Up-regulation
c-fgr	Cellular proto-oncogene. Encodes a tyrosine kinase that promotes cell proliferation.	Up-regulation
c-myc	Cellular proto-oncogene and transcription factor. Activates gene sets involved in cell cycle progression.	Up-regulation
Ig-μ	Encodes IgM. The biological relevance of the down-regulation is uncertain.	Down-regulation

factors allows binding of additional regulators to nearby nucleosomes and/or assembly of the basal transcriptional machinery, which then initiates transcription (Agalioti et al., 2000; Gregory et al., 1998; Venter et al., 1994).

EBNA2 AND TRANSCRIPTION REGULATION

EBNA2 and EBNA5 (also designated EBNA-leader protein; EBNA-LP) are the first viral gene products to be expressed after EBV infection and EBNA2 plays a crucial role in the process by activating transcription of viral genes encoding EBNA1-6, LMP1 and LMP2 (the growth program, also designated latency III). In addition to viral targets, EBNA2 can enhance the transcription of cellular genes encoding proteins that contribute to the cellular growth associated with EBV infection (Table 1). Since EBNA2 lacks sequence-specific DNA-binding activity it has to be recruited to the promoters of cellular and viral target genes by other DNA-binding factors. It appears that the interaction with the DNA-binding protein RBP-Jκ is critically important to direct EBNA2 to most EBNA2-responsive promoters. In fact, EBV carrying a mutated EBNA2 unable to bind RBP-Jκ is incapable of immortalizing B-cells (Yalamanchili et al., 1994). RBP-Jκ binds to the conserved core sequence GTGGGAA that is present in most EBNA2-responsive promoters and is involved in both gene activation and repression depending on the recruitment of distinct co-repressor and co-activator complexes. RBP-Jκ functions as a repressor by interfering with the interaction between the GTFs TFIIA and TFIID (Olave et al., 1998) and by recruiting histone deacetylase (HDAC) co-repressor complexes to the promoter. These co-repressor complexes may include SMRT and HDAC1 (Kao et al., 1998; Waltzer et al., 1995), CIR, SAP30 and HDAC2 (Hsieh et al., 1999), and SKIP (Zhou et al., 2000). Binding of EBNA2 to RBP-Jκ is believed to displace the co-repressor complex from RBP-Jκ (Kao et al., 1998). On the other hand, it is clearly established that EBNA2 also can activate cellular and viral promoters in an RBP-Jκ-independent manner (Fåhraeus et al., 1990; 1993; Johannsen et al., 1995; Ling et al., 1993a; Sjöblom et al., 1995a; 995b), suggesting that other transcription factors may target EBNA2 to responsive promoters. EBNA2 can interact with several other cellular transcription factors such as the Ets-related protein PU.1 (Johannsen et al., 1995; Laux et al., 1994a), the POU domain protein (Sjöblom et al., 1995a), members of the ATF/CREB protein family (Sjöblom et al., 1998), AUF1 (Fuentes-Pananá et al., 2000) and DP103 (Voss et al., 2001) and these cellular proteins are all possible candidates that may contribute to EBNA2 positioning at EBNA2-responsive promoters in an RBP-Jκ-independent way. Furthermore, EBNA2 interacts with components of the basal transcription machinery (Tong et al., 1995a; 1995b; 1995c), the co-repressor BS69 (Ansieau and Leutz, 2002), the co-activators CBP, p300 and PCAF (Wang et al., 2000) and the SWI-SNF (Wu et al., 1996) and SKIP (Zhou et al., 2000) complexes. These latter interactions imply that promoter transactivation by EBNA2 is in part due to its ability to mediate modulation of the chromatin structure at the promoter. EBNA2-mediated transcription activation is also strongly and specifically potentiated by direct interaction with EBNA5 (Harada and Kieff, 1997; Nitsche et al., 1997; Peng et al., 2004) and this effect is important for inducing the transition of the infected cells from G0 to G1 (Sinclair et al., 1994). Moreover, EBNA6 up-regulates EBNA2-mediated gene activation by binding to a SUMO-activated repressor and inhibiting its repressive effects on CBP, p300 and PCAF and possibly other transcription factors (Rosendorff et al., 2004). It should be noted that, in the absence of other EBNA proteins, EBNA2 expression might retard cell growth (Lin et al., 2000).

The EBNA2-induced proliferation program is believed to be required for the multiplication of virus-infected cells in the host *in vivo* and the efficient establishment of viral latency in a high number of cells (Thorley-Lawson and Gross, 2004). However, the EBNA2-6 proteins are highly immunogenic and their expression has to be down-regulated if T-cell recognition is to be avoided in an immunocompetent host. It is so far unknown how EBNA2 expression is down-regulated *in vivo*. The simplest model would involve a direct reversal of the activation mechanism to be discussed below. A physiological B-cell differentiation signal might transform the host cell into a phenotype in which the cellular repertoire of transcription regulatory factors does not support the synthesis of all factors necessary for maintaining C promoter (Cp) activity and EBNA2 expression. The process might involve remodeling of the promoter chromatin followed by the recruitment of repressive HDAC-containing complexes to the promoter. It is also well established that hypermethylation of Cp locks the promoter in an inactive state in the restricted latency types found in latently infected B-cells in healthy carriers and in BL and NPC tumor biopsies (latencies 0-II). Immunological selection for cells that do not express EBNA2 or the other EBNAs might contribute to the enrichment of cells that express these restricted latency programs. Interestingly, it was recently reported that EBNA2 is hyperphosphorylated in a mitosis-specific manner, and that the cyclin-dependent kinase p34cdc2 is likely involved in the hyperphosphorylation. The modification appears to suppress the ability of EBNA2 to activate the *LMP1* promoter (Yue et al., 2004). Possibly, this is a general effect that suppresses all other EBNA2-responsive genes during M phase in type III latency, which may be important for accurate division of EBV episomes into daughter cells. It might also be part of a posttranslational mechanism by which the ability of EBNA2 to activate the growth program can be controlled in a proliferating cell population. These results might at a first glance seem difficult to reconcile with those of Schubach and co-workers who reported that a phosphorylated form of EBNA2 interacts with SWI/SNF. Phosphorylation of a consensus CKII site at residue 469 in the C-terminus of EBNA2 is important for the binding of SWI/ SNF and for B-cell growth transformation and immortalization (Kwiatkowski et al., 2004; Wu et al., 1996; 2000). Previous work by Fåhraeus et al. suggests that EBNA2 inhibits protein phosphatase 1-like activity in EBV-transformed cells leading to increased phosphorylation of transcription factors and increased activity of the *LMP1* promoter (Fåhraeus et al., 1994). Taken together, these observations might, however, be interpreted as a manifestation of the fact that site-specific phosphorylation of EBNA2 by different kinases modulates its function and that the function may vary during the cell cycle in a strictly regulated manner.

EBNA2-RESPONSIVE VIRAL GENES
Latent Membrane Protein 1

The *LMP1* oncogene is repressed in human B-cells in the absence of an inducer, and EBNA2 seems to be required for relief of this repression in most B-cell lines (Abbot et al., 1990; Fåhraeus et al., 1990; Wang et al., 1990b). In contrast, LMP1 is expressed independently of EBNA2 in epithelial cells and in the EBV-positive human malignancies NPC, Hodgkin's disease and T-cell lymphomas demonstrating that other cellular and/or viral factors can substitute for EBNA2 functions under certain circumstances (Rickinson and Kieff, 2001). The *LMP1* gene can be transcribed from different promoters identified in different cell lines. In B-cells the EcoRI leftward promoter (ED-L1, also designated LMP1p) is the major promoter initiating *LMP1* gene transcripts (Farrell et al., 1983) and this promoter is EBNA2-responsive (Abbot et al., 1990; Fåhraeus et al., 1990). Another promoter designated ED-L1A gives rise to a truncated version of LMP1 during lytic infection of B-cells (Hudson et al., 1985). In NPC cells, ED-L1 and ED-L1A appear to direct *LMP1* gene transcription together with a third promoter, ED-L1B (Chen et al., 1995). Yet another *LMP1* gene promoter in NPC has been identified in the first terminal repeat of the viral genome (Sadler and Raab-Traub, 1995) and all these different promoters may contribute to the independence of EBNA2 for *LMP1* gene expression in NPC.

A sequence spanning the region between positions −634 and +40 relative to the ED-L1 promoter initiation site has been defined as the *LMP1* transcription regulatory sequence (LRS) (Fåhraeus et al., 1990). This sequence contains both positive and negative transcriptional *cis* elements. Hypermethylation of the *LMP1* promoter region is regarded as one important factor underlying the repression in B-cell lines that do not express EBNA2, since treatment with the nucleotide analogue 5-azacytidine that inhibits methylation up-regulates *LMP1* gene expression (Masucci et al., 1989). In agreement with this view, there is a correlation between unmethylated DNA at the *LMP1* promoter and expression of LMP1 in various EBV-infected cell lines of different latency types and in NPC tumor biopsies (Ernberg et al., 1989; Salamon et al., 2001; Takacs et al., 2001). In addition, histone deacetylation appears to be an important mechanism for maintaining the *LMP1* promoter in a repressed state. In the absence of EBNA2, HDAC-containing large complexes are targeted to at least two different positions in LRS: the RBP-Jκ site at position –223/-217 by the RBP-Jκ co-repressor (Johannsen et al., 1995; Laux et al., 1994a; 1994b), and the E-box at position –56/-51 by the Max-Mad1 repressor proteins interacting with mSin3-HDAC-containing complexes (Sjöblom-Hallén et al., 1999).

The *LMP1* promoter can be activated by interferon regulatory factor 7 (Ning et al., 2003) and by signals transmitted via the cellular protein kinase A (PKA) (Fåhraeus et al., 1994) and PKC (Rowe et al., 1992) pathways. The promoter is also activated by EBNA1 (Gahn and Sugden, 1995), EBNA6 (also referred to as EBNA3C) (Allday et al., 1993) and EBNA2, respectively, the latter acting alone (Fåhraeus et al., 1990) or in concert with EBNA5 (also designated EBNA-LP) (Harada and Kieff, 1997; Nitsche et al., 1997). Three regions of LRS contribute to EBNA2 responsiveness independently of each other: the -106/+40, -176/-136 and -234/-205 regions (Fåhraeus et al., 1990; 1993; Ghosh and Kieff, 1990; Sjöblom et al., 1995a; Tsang et al., 1991). Subsequent investigations have identified individual regulatory *cis* elements in these regions. There are two binding sites for RBP-Jκ in LRS but only the most promoter-proximal seems to be implicated in transcriptional activation (Johannsen et al., 1995; Laux et al., 1994a; 1994b) and its role in EBNA2 transactivation of ED-L1 is complex. Reporter constructs that lack the RBP-Jκ site can still be activated by EBNA2 (Fåhraeus et al., 1990; Fåhraeus et al., 1993; Johannsen et al., 1995; Sjöblom et al., 1995a; 1995b) and a reporter construct that contains four tandem RBP-Jκ sites is not EBNA2-responsive (Ling et al., 1993a). Furthermore, activated Notch, which is capable of activating other RBP-Jκ site containing promoters, cannot replace EBNA2 in continuous *LMP1* gene activation (Gordadze et al., 2001; Höfelmayr et al., 2001). Four promoter elements down-stream of the RBP-Jκ site seem to contribute to the EBNA2-mediated transactivation of ED-L1: a purine rich sequence (PU-box) at position –171/-155 binding PU.1/Spi-1 and Spi-B (Johannsen et al., 1995; Laux et al., 1994a; Sjöblom et al., 1995a; 1995b), an octamer motif at position –147/-139 binding a POU-domain protein (Sjöblom et al., 1995a), an AP-2 consensus site at position –103/-95 binding AP-2 factors (L. Rymo, unpublished), and an ATF/CRE site at position –44/-38 that binds members of the ATF/CREB family (Fåhraeus et al., 1994; Sjöblom et al., 1995a; 1998). These elements may play an important role in EBNA2-dependent activation of cellular promoters that do not contain an RBP-Jκ site and in EBNA2-independent activation of LMP1p. Other positive transcriptional *cis* elements in LRS do not seem to mediate EBNA2 responsiveness but might still be a functional part of both EBNA2-independent and -dependent transactivation of the promoter

(Johannsen et al., 1995; Sjöblom et al., 1995a; 1995b). In addition to the interaction with the above mentioned transcription factors, EBNA2 contributes to *LMP1* gene expression via interaction through its activation domain with components of the RNA polymerase II transcription complex (Tong et al., 1995a; 1995c), with a co-activator called p100 (Tong et al., 1995b), and with the p300, CBP and PCAF histone acetyltransferases (Wang et al., 2000). These data are in agreement with the recent demonstration by chromatin immunoprecipitation analysis that recruitment of EBNA2 to ED-L1 results in hyperacetylation of the promoter regions, which correlates with transcriptional activation (Alazard et al., 2003).

Latent Membrane Protein 2

The *LMP2* gene encodes two related proteins, LMP2A and -2B, which share their 12 transmembrane domains and the C-terminal tail. LMP2A carries an additional N-terminal domain of 119 amino acids that resembles the cytoplasmic domain of the normal B-cell receptor (Longnecker and Kieff, 1990). Neither LMP2A nor LMP2B are essential for B-cell transformation *in vitro* but LMP2A is known to inhibit lytic viral reactivation from latency by interfering with normal B-cell signal transduction processes and in doing so may also provide a survival signal that could be important for viral persistence *in vivo* (Longnecker, 2000; Scholle et al., 2000). Much less is known about the function of LMP2B. The presence of this protein has been predicted on the basis of DNA sequence and RNA expression data, but it has never been visualized due to the absence of the appropriate reagents. LMP2A and –2B expression has been analyzed at the RNA level and found to be regulated either directly or indirectly by EBNA2 (Zimber-Strobl et al., 1991). The mechanism behind this regulation remains to be elucidated. It should be noted, however, that the *LMP2* gene regulatory sequence shares its EBNA2 response elements (binding sites for RBP-Jκ and PU.1/Spi-1) with the ED-L1 promoter (Laux et al., 1994a; 1994b), through which EBNA2 might recruit the SWI/SNF complex and the CBP and p300 co-activators to activate LMP2 expression (Wu et al., 2000).

EBNA1-6

After *in vitro* infection of B-lymphocytes, transcription is initiated from the W promoter (Wp) in the BamHI W repeat region of the EBV genome, leading to the expression of EBNA2 and EBNA5 (Alfieri et al., 1991; Allday et al., 1989; Ricksten et al., 1988; Rooney et al., 1989; Woisetschlaeger et al., 1990). Within 48 h there is a switch in promoter usage from Wp to the upstream C promoter (Cp) in the *Bam*HI C region leading to the expression of EBNA1-6 that promote B-cell proliferation (Woisetschlaeger et al., 1991; 1990). The mechanism behind this switch is not fully understood at the molecular level, but clearly EBNA2 plays an important role (Woisetschlaeger et al., 1991; Yoo and Speck, 2000). It should also be noted that the EBV antigens EBNA3, EBNA4, EBNA5 and EBNA6 are also referred to as EBNA3A, EBNA3B, EBNALP and EBNA3C, respectively.

The regulation of Cp has been the subject of several investigations, and a number of positive *cis*-acting transcription regulatory elements have been identified in the regions up- and downstream of the promoter. EBNA1 homodimers activate Cp by binding to a sub-element of the latency origin of replication (oriPI, also designated the family of repeats, FR), which functions as an EBNA1-dependent enhancer of

Cp (Nilsson et al., 1993; Puglielli et al., 1996; Reisman and Sugden, 1986; Sugden and Warren, 1989). A glucocorticoid-responsive element has been identified in the Cp upstream region (Kupfer and Summers, 1990). A third *cis* element identified upstream of Cp is the EBNA2-responsive enhancer (E2RE) (Jin and Speck, 1992; Sung et al., 1991; Yoo et al., 1997). This element contains at least two binding sites; one for RBP-Jκ and one for AUF1, and both these cellular factors are believed to mediate EBNA2 transactivation of Cp (Fuentes-Pananá et al., 2000; Grossman et al., 1994; Henkel et al., 1994; Ling et al., 1993a). EBNA6 acts as a repressor of Cp possibly via interaction with RBP-Jκ (Radkov et al., 1997). Cp also requires a promoter-proximal CCAAT box that binds nuclear factor Y (NF-Y) (Boreström et al., 2003; Nilsson et al., 2001; Puglielli et al., 1996), promoter-proximal GC-rich sequences that bind Sp1 (Boreström et al., 2003; Nilsson et al., 2001) and a C/EBP-like site that binds different isoforms of C/EBP (Nilsson et al., 2001) for activity in lymphoid cell lines. A weak positive *cis* element has been identified in the region between +2680 and +2880 relative to the Cp transcription initiation site (Puglielli et al., 1997).

In transient transfection experiments, Cp can be activated independently of EBNA2. However, in the endogenous viral genome, Cp activity correlates completely with EBNA2 expression (Boreström et al., 2003; Yoo and Speck, 2000). This correlation may be coupled to the ability of EBNA2 to interact with co-activators possessing ATP-dependent chromatin remodeling (SWI/SNF) and intrinsic HAT (p300, CBP and PCAF) activities (Wang et al., 2000). Assuming that the viral promoter sequence is stabilized in a condensed chromatin conformation in the silent state, a conceivable role of EBNA2 would be to recruit these co-activators to Cp. This might lead to modulation of chromatin structure allowing the recruitment of C/EBP, Sp1, NF-Y, the oriPI-EBNA1 complex and additional factors to the promoter and the subsequent activation of transcription. Recent experimental data by Alazard et al. showing a correlation between EBNA2 recruitment to Cp and increased levels of histone acetylation at the promoter supports this view (Alazard et al., 2003).

EBNA2-RESPONSIVE CELLULAR GENES
CD21 and *CD23*

Conversion of EBV-negative BL cell lines by the immortalizing viral strain B95-8 induces high-level expression of the EBV-receptor CD21 (Fingeroth et al., 1984; Frade et al., 1985) and the B-cell activation antigen CD23 (Bonnefoy et al., 1989; Thorley-Lawson et al., 1985; Yukawa et al., 1987). CD21 and CD23 are normal constituents of the B-cell membrane, expressed early during B-cell activation and implicated in the regulation of the cell cycle. In fact, CD23 seems to function as an autocrine growth factor (Wang et al., 1990a). Stable expression of EBNA2 in P3HR-1-infected BL cell lines or conditional EBNA2 expression in the EREB2-5 cell line specifically activates expression of both CD21 and CD23 and the promoters of these genes have been claimed to be direct targets of EBNA2 (Cordier et al., 1990; Kempkes et al., 1995a; Wang et al., 1987), possibly via RBP-Jκ interaction at RBP-Jκ consensus sites present in the regulatory sequences of both promoters (Makar et al., 1998; 2001; Wang et al., 1991). Recently, chromatin immunoprecipitation experiments provided evidence that EBNA2 recruits the SWI/SNF complex to the *CD23* gene regulatory region, which underscores the importance of EBNA2-induced chromatin modification as a mechanism for the activation of cellular target genes (Wu et al., 2000).

BATF

The *BATF* gene was originally isolated from a cDNA library prepared from EBV stimulated human B-cells (Dorsey et al., 1995). The gene encodes a 125 amino acid nuclear protein possessing a basic leucine zipper domain that is most similar to the basic leucine zipper found in AP-1 family members (Dorsey et al., 1995), and functions as a negative regulator of AP-1 activity (Echlin et al., 2000). The induction of the *BATF* gene by EBV is mediated by EBNA2 as shown in infection experiments of B-cells with either wild-type or EBNA2-deleted viruses (Johansen et al., 2003). Interestingly, BATF induction by EBNA2 is a B-cell specific process that can be duplicated by Notch, suggesting that RBP-Jκ is involved and that additional B-cell transcription factors or signaling events may be needed for *BATF* gene expression. The documented role of AP-1 in the transactivation of the lytic switch-activating *BZLF1* gene suggests that BATF may have a role in repression of the EBV lytic cycle. In line with this view, it was recently shown that BATF negatively impacts the expression of a *BZLF1* reporter gene and reduces the efficiency of lytic cycle-induction (Johansen et al., 2003).

BLR2/EBI1

The discovery of this gene was made by subtractive hybridization analysis of cDNA clones from BL cells after EBV-infection *in vitro*. The results revealed EBV-induced up-regulation of two novel genes, EBV-induced genes 1 and 2 (*EBI1* and *EBI2*), that encode proteins belonging to the same family as BLR1 and LCR1, and the receptors for IL-8 and MIP1a (Birkenbach et al., 1993). These proteins are G-protein coupled chemokine receptors predominantly expressed in lymphocytes, and may play a role in lymphocyte trafficking into secondary lymphatic organs (Springer, 1994). The results were extended by Burgstahler et al., who showed that *EBI1*, also denoted *BLR2*, is specifically and rapidly transactivated by EBNA2 without the necessity of other viral proteins (Burgstahler et al., 1995). This effect may be biologically important during dissemination of EBV infection in naïve individuals upon primary infection, although no direct evidence that BLR2/EBI1 participates in this process has been presented as yet.

c-fgr

The cellular proto-oncogene *c-fgr* encodes a protein tyrosine kinase and is a member of the src gene family. The gene was first identified in the oncogenic retrovirus Gardner-Rasheed feline sarcoma virus (*v-fgr*) (Rasheed et al., 1982) but equivalent transcripts have also been found in normal and neoplastic human cells. Especially high expression levels have been detected in EBV-positive lymphomas and also upon infection of EBV-negative BL cells and primary B-cells *in vitro* (Cheah et al., 1986; Klein et al., 1988). Three lines of evidence show that EBNA2 mediates this effect. Firstly, acute infection of EBV-negative BJAB and Ramos cells by the EBNA2 mutant strain P3HR-1 does not affect *c-fgr* RNA levels. Secondly, cell lines constitutively expressing only the EBNA2 gene display increased *c-fgr* RNA levels (Knutson, 1990). Thirdly, nuclear run-on assays show that this induction results from an increased rate of transcription of the *c-fgr* gene in response to EBNA2 (Patel et al., 1990). Notably, the 5′ untranslated region of the *c-fgr* gene contains a number of RBP-Jκ-like sites that may play a role in this process (Gutkind et al., 1991; Link et al., 1992)

c-myc

The *c-myc* gene belongs to the group of proto-oncogenes that encodes transcription factors. Overwhelming evidence has shown that c-myc both positively and negatively regulates distinct sets of genes, most of which are involved in cell growth and proliferation, through binding to its consensus sequence, CACGTG (Amati et al., 2001; Zeller et al., 2003). Furthermore, its interactions with chromatin remodeling complexes and with histone acetyltransferases establish that role (Amati et al., 2001). In general, c-myc appears to be a relatively weak transactivator of a broad array of genes that are rate-limiting steps in nearly all aspects of normal cellular proliferation. *c-myc* gene expression is deregulated in approximately 70% of human cancers. The deregulation occurs through diverse mechanisms, which may include translocations (as in Burkitt's lymphoma), amplifications, enhanced translation or protein stability. However, most often *c-myc* expression is activated indirectly through alterations in signaling pathways that induce *c-myc* transcription (Nilsson and Cleveland, 2003). By using conditionally EBNA2-expressing cells, it was demonstrated that EBNA2 induces *c-myc* gene expression by enhancement of transcription initiation, and that this effect is direct and not requiring *de novo* protein synthesis (Kaiser et al., 1999a). Although the precise mechanism by which *c-myc* gene expression is up-regulated by EBNA2 remains to be elucidated, some data indicate that an RBP-Jκ site in the first intron of *c-myc* may be important for mediating the EBNA2 induction (Cooper et al., 2003). Interestingly, overexpression of c-myc to a much higher level than what is normally observed in EBV-immortalized cells is sufficient to support B-cell proliferation independently of EBNA2 and LMP1 (Polack et al., 1996). The proliferation program imposed by enforced overexpression of c-myc, however, differs substantially from that activated by EBV as revealed by the fact that EBV-immortalized cells grow in clumps, express a variety of activation markers and adhesion molecules, and are resistant to apoptosis, while c-myc-driven cells grow as single cells, do not express the EBV-induced surface molecules and are highly sensitive to apoptosis (Polack et al., 1996). These distinct patterns imply that there must be substantial differences in target genes between EBNA2 and c-myc, a notion that was recently verified by proteome analysis of EBNA2- and c-myc-conditional cell lines (Schlee et al., 2004).

The immunoglobulin μ gene

Activation of a conditional mutant of EBNA2 results in down-regulation of cell surface IgM and *Ig–*μ steady-state RNA expression (Jochner et al., 1996) but the biological relevance of this finding is still unclear. In LCLs, activation of EBNA2 is required for maintaining proliferation, whereas in BL cell lines with t(8;14) translocations, activation of EBNA2 induces growth arrest. In these cells, Northern and nuclear run-on analyses have revealed rapid simultaneous repression of *Ig*-μ and *c-myc* transcription as early as 30 min after activation of EBNA2. Since *c-myc* expression is under the control of the Ig heavy chain locus in BL cell lines with a t(8;14) translocation, *Ig*-μ and *c-myc* may be down-regulated by EBNA2 through a common mechanism (Jochner et al., 1996).

Possible EBNA2 targets among cellular genes

Most likely, EBNA2 affects the transcription of several additional targets among cellular genes. One such possible target is the anti-apoptotic *bcl-2* oncogene, since transfection of B-cells with an EBNA2-carrying plasmid up-regulates *bcl-2* expression. The molecular mechanism behind this up-regulation remains unknown (Finke et al., 1992). Some information on possible EBNA2-responsive genes has been derived from large-scale studies of the cellular transcriptome and proteome following EBV-infection and/or activation of conditionally expressed EBNA2 (Kaiser et al., 1999b; Schlee et al., 2004; Schuhmacher et al., 2001; Spender et al., 2002). EBNA2 induces and/or modifies cell cycle-regulating proteins, such as cdc2, cyclin D2, E, A, E2F-1, cdk2 and cdk4, but whether these effects are direct or indirect are unknown (Kempkes et al., 1995b; Spender et al., 2001). Additional possible EBNA2 targets are the genes encoding interleukin-16 and the transcription factor AML-2, as revealed in microarray expression profiles of cell lines after activation of conditional EBNA2 activity (Spender et al., 2002). A proteome analysis of [^{35}S]methionine- and [^{35}S]cysteine-labelled EBNA2-conditional or myc-conditional cells and 2D gel electrophoresis revealed six proteins specifically regulated by EBNA2 but not c-myc (Bid, IgE-HRF, Annexin IV, γ-actin, GMFγ, and AF103803; the first two being up-regulated and the last four being down-regulated by EBNA2) (Schlee et al., 2004). However, again it is yet to be determined if the changes in expression are direct consequences of EBNA2 regulation or reflect some general change in cellular gene expression, and if they are biologically relevant. A third approach that has verified some of the above-mentioned findings is the recently reported construction of a fusion peptide between the 10 amino acid peptide from the RBP-Jκ interaction domain of EBNA2 and the protein transduction domain of HIV-1 TAT (Farrell et al., 2004). This fusion peptide effectively blocks EBNA2-RBP-Jκ interaction *in vitro* and inhibits EBV-mediated lymphoproliferation and down-regulates the genes encoding LMP1, LMP2, CD23, BATF, p34^{cdc2} and ICAM-1. It should be useful in the identification of additional genes that are regulated by EBNA2 in an RBP-Jκ-dependent way.

CONCLUDING REMARKS

Clearly, EBNA2 controls viral and cellular gene expression during the EBV-driven growth program in a highly complicated manner with a host of cellular and viral transcription factors as interaction partners. EBNA2 affects both the chromatin structure and the recruitment of transcription factors and the RNA polymerase II holoenzyme to the promoters of target genes, but the sequential events underlying EBNA2-induced transcription activation remain to be elucidated. In latently infected B-lymphocytes, the EBV genome exists as a non-integrated nuclear episome with nucleosome phasing indistinguishable from that of cellular chromatin. Generally, initiation sites for transcription are specified by sequence-specific DNA binding proteins that are subject to multiple levels of regulation through signaling pathways that respond to changes in intra- and extracellular conditions. Higher order chromatin structures prohibit sequence-specific transcription factors from binding to their recognition sites. Precisely how the sequence-specific factors can access their binding sites in these regions of the genome, recruit chromatin-modifying and remodeling activities, and initiate the process of transcription is poorly understood. With regard to EBV, evidence is, however, accumulating indicating that chromatin-based repression is one important component of maintaining the transcriptional silence of lytic cycle genes during EBV latency. Thus, some EBV-encoded transcription factors must possess the ability to disrupt the chromatin-repressed state of a subset of latent viral and cellular genes to initiate the transcription programs associated with infection and cellular growth transformation.

The molecular mechanism underlying EBNA2 induction of *LMP1* promoter activity has been studied in some detail and may serve as a model for how EBNA2 may activate other viral and cellular target genes. Hypothetically, the first step in LMP1 transcription activation is the targeting of EBNA2 to the promoter. As EBNA2 does not bind directly to DNA it has to be recruited by interaction with DNA-binding proteins. There are several candidates for this task including the ATF/CRE-binding factors, the AP-2 factors, the PU.1 factor and RBP-Jκ. Notably, only one or two of these factor-binding sites has to be present in *LMP1* promoter-containing reporter plasmids to confer EBNA2 inducibility to the construct. Presumably, the critical binding sites are already exposed and positioned between nucleosomes. EBNA2 binds to the factor and a program is initiated involving the EBNA2-mediated displacement of deacetylating activities (SIN3, NuRD) and the recruitment of components of acetylating (CBP, p300, GCN5/PCAF) and chromatin remodeling (SWI/SNF) complexes. The temporal order of these events remains to be clarified. The program culminates with the binding of the RNA polymerase II holoenzyme and TFIID, a possible sliding of the nucleosome adjacent to the TATA box, and the initiation of transcription.

This model suggests that a significant part of the ability of EBNA2 to activate transcription from cellular and viral genes to promote lymphoproliferation and establish EBV infection in the host relies upon its ability to interact with different transcription factors and recruit chromatin-modifying activities to the promoter. The EBNA2-recruiting factor may be different between different target promoters and cell types depending on the element composition of the promoter and the factors present in the cell. The general outcome, however, is the same: the formation of a transcriptionally competent promoter structure that allows assembly of the basal transcriptional machinery at the promoter and initiation of transcription.

Acknowledgments

This study was supported by grants from the Swedish Medical Research Council (project 5667), the Swedish Cancer Society, the Sahlgrenska University Hospital and the Göteborg Medical Society.

References

Abbot, S.D., Rowe, M., Cadwallader, K., Ricksten, A., Gordon, J., Wang, F., Rymo, L., and Rickinson, A.B. (1990). Epstein-Barr virus nuclear antigen 2 induces expression of the virus-encoded latent membrane protein. J. Virol. *64*, 2126-2134.

Agalioti, T., Lomvardas, S., Parekh, B., Yie, J., Maniatis, T., and Thanos, D. (2000). Ordered recruitment of chromatin modifying and general transcription factors to the IFN-beta promoter. Cell *103*, 667-678.

Alazard, N., Gruffat, H., Hiriart, E., Sergeant, A., and Manet, E. (2003). Differential hyperacetylation of histones H3 and H4 upon promoter-specific recruitment of EBNA2 in Epstein-Barr virus chromatin. J. Virol. *77*, 8166-8172.

Alfieri, C., Birkenbach, M., and Kieff, E. (1991). Early events in Epstein-Barr virus infection of human B lymphocytes. Virol. *181*, 595-608.

Allday, M.J., Crawford, D.H., and Griffin, B.E. (1989). Epstein-Barr virus latent gene expression during the initiation of B cell immortalization. J. Gen. Virol. *70*, 1755-1764.

Allday, M.J., Crawford, D.H., and Thomas, J.A. (1993). Epstein-Barr virus (EBV) nuclear antigen 6 induces expression of the EBV latent membrane protein and an activated phenotype in Raji cells. J. Gen. Virol. *74*, 361-369.

Amati, B., Frank, S.R., Donjerkovic, D., and Taubert, S. (2001). Function of the c-Myc oncoprotein in chromatin remodeling and transcription. Biochim. Biophys. Acta *1471*, M135-145.

Ansieau, S., and Leutz, A. (2002). The conserved Mynd domain of BS69 binds cellular and oncoviral proteins through a common PXLXP motif. J. Biol. Chem. *277*, 4906-4910.

Berger, S.L. (2002). Histone modifications in transcriptional regulation. Curr. Opin. Genet. Dev. *12*, 142-148.

Birkenbach, M., Josefsen, K., Yalamanchili, R., Lenoir, G., and Kieff, E. (1993). Epstein-Barr virus-induced genes: first lymphocyte-specific G protein-coupled peptide receptors. J. Virol. *67*, 2209-2220.

Bonnefoy, J.Y., Denoroy, M.C., Guillot, O., Martens, C.L., and Banchereau, J. (1989). Activation of normal human B cells through their antigen receptor induces membrane expression of IL-1 alpha and secretion of IL-1 beta. J. Immunol. *143*, 864-869.

Boreström, C., Zetterberg, H., Liff, K., and Rymo, L. (2003). Functional interaction of nuclear factor Y and Spl is required for activation of the Epstein-Barr virus C promoter. J. Virol. *77*, 821-829.

Burgstahler, R., Kempkes, B., Steube, K., and Lipp, M. (1995). Expression of the chemokine receptor BLR2/EBI1 is specifically transactivated by Epstein-Barr virus nuclear antigen 2. Biochem. Biophys. Res. Commun. *215*, 737-743.

Cheah, M.S., Ley, T.J., Tronick, S.R., and Robbins, K.C. (1986). fgr proto-oncogene mRNA induced in B lymphocytes by Epstein-Barr virus infection. Nature *319*, 238-240.

Chen, M.L., Hsu, N.C., Liu, S.T., and Chang, Y.S. (1995). Identification of an internal promoter of the latent membrane protein 1 gene of Epstein-Barr virus. DNA Cell. Biol. *14*, 205-211.

Cohen, J.I., and Kieff, E. (1991). An Epstein-Barr virus nuclear protein 2 domain essential for transformation is a direct transcriptional activator. J. Virol. *65*, 5880-5885.

Cohen, J.I., Wang, F., and Kieff, E. (1991). Epstein-Barr virus nuclear protein 2 mutations define essential domains for transformation and transactivation. J. Virol. *65*, 2545-2554.

Cooper, A., Johannsen, E., Maruo, S., Cahir-McFarland, E., Illanes, D., Davidson, D., and Kieff, E. (2003). EBNA3A association with RBP-Jkappa down-regulates *c-myc* and Epstein-Barr virus-transformed lymphoblast growth. J. Virol. *77*, 999-1010.

Cordier, M., Calender, A., Billaud, M., Zimber, U., Rousselet, G., Pavlish, O., Banchereau, J., Tursz, T., Bornkamm, G., and Lenoir, G. M. (1990). Stable transfection of Epstein-Barr virus (EBV) nuclear antigen 2 in lymphoma cells containing the EBV P3HR1 genome induces expression of B-cell activation molecules CD21 and CD23. J. Virol. *64*, 1002-1013.

Cosma, M. P., Tanaka, T., and Nasmyth, K. (1999). Ordered recruitment of transcription and chromatin remodeling factors to a cell cycle- and developmentally regulated promoter. Cell *97*, 299-311.

Di Croce, L., Koop, R., Venditti, P., Westphal, H.M., Nightingale, K.P., Corona, D.F., Becker, P.B., and Beato, M. (1999). Two-step synergism between the progesterone receptor and the DNA-binding domain of nuclear factor 1 on MMTV minichromosomes. Mol. Cell *4*, 45-54.

Dorsey, M.J., Tae, H.J., Sollenberger, K.G., Mascarenhas, N.T., Johansen, L.M., and Taparowsky, E.J. (1995). B-ATF: a novel human bZIP protein that associates with members of the AP-1 transcription factor family. Oncogene *11*, 2255-2265.

Dyson, P.J., and Farrell, P.J. (1985). Chromatin structure of Epstein-Barr virus. J. Gen. Virol. *66*, 1931-1940.

Echlin, D.R., Tae, H.J., Mitin, N., and Taparowsky, E.J. (2000). B-ATF functions as a negative regulator of AP-1 mediated transcription and blocks cellular transformation by Ras and Fos. Oncogene *19*, 1752-1763.

Ernberg, I., Falk, K., Minarovits, J., Busson, P., Tursz, T., Masucci, M.G., and Klein, G. (1989). The role of methylation in the phenotype-dependent modulation of Epstein-Barr nuclear antigen 2 and latent membrane protein genes in cells latently infected with Epstein-Barr virus. J. Gen. Virol. *70*, 2989-3002.

Farrell, C.J., Lee, J.M., Shin, E.C., Cebrat, M., Cole, P.A., and Hayward, S.D. (2004). Inhibition of Epstein-Barr virus-induced growth proliferation by a nuclear antigen EBNA2-TAT peptide. Proc. Natl. Acad. Sci. U.S.A. *101*, 4625-4630.

Farrell, P.J., Bankier, A., Seguin, C., Deininger, P., and Barrell, B.G. (1983). Latent and lytic cycle promoters of Epstein-Barr virus. EMBO J. *2*, 1331-1338.

Featherstone, M. (2002). Coactivators in transcription initiation: here are your orders. Curr. Opin. Genet. Dev. *12*, 149-155.

Fingeroth, J.D., Weis, J.J., Tedder, T.F., Strominger, J.L., Biro, P.A., and Fearon, D.T. (1984). Epstein-Barr virus receptor of human B lymphocytes is the C3d receptor CR2. Proc. Natl. Acad. Sci. U.S.A. *81*, 4510-4514.

Finke, J., Fritzen, R., Ternes, P., Trivedi, P., Bross, K.J., Lange, W., Mertelsmann, R., and Dolken, G. (1992). Expression of *bcl-2* in Burkitt's lymphoma cell lines: induction by latent Epstein-Barr virus genes. Blood *80*, 459-469.

Frade, R., Myones, B.L., Barel, M., Krikorian, L., Charriaut, C., and Ross, G.D. (1985). gp140, a C3b-binding membrane component of lymphocytes, is the B cell C3dg/C3d receptor (CR2) and is distinct from the neutrophil C3dg receptor (CR4). Eur. J. Immunol. *15*, 1192-1197.

Fuentes-Pananá, E.M., Peng, R., Brewer, G., Tan, J., and Ling, P.D. (2000). Regulation of the Epstein-Barr virus C promoter by AUF1 and the cyclic AMP/protein kinase A signaling pathway. J. Virol. *74*, 8166-8175.

Fåhraeus, R., Jansson, A., Ricksten, A., Sjöblom, A., and Rymo, L. (1990). Epstein-Barr virus-encoded nuclear antigen 2 activates the viral latent membrane protein promoter by modulating the activity of a negative regulatory element. Proc. Natl. Acad. Sci. U.S.A. *87*, 7390-7394.

Fåhraeus, R., Jansson, A., Sjöblom, A., Nilsson, T., Klein, G., and Rymo, L. (1993). Cell phenotype-dependent control of Epstein-Barr virus latent membrane protein 1 gene regulatory sequences. Virology *195*, 71-80.

Fåhraeus, R., Palmqvist, L., Nerdstedt, A., Farzad, S., Rymo, L., and Lain, S. (1994). Response to cAMP levels of the Epstein-Barr virus EBNA2-inducible LMP1 oncogene and EBNA2 inhibition of a PP1-like activity. EMBO J. *13*, 6041-6051.

Gahn, T.A., and Sugden, B. (1995). An EBNA-1-dependent enhancer acts from a distance of 10 kilobase pairs to increase expression of the Epstein-Barr virus *LMP* gene. J. Virol. *69*, 2633-2636.

Gavin, I., Horn, P.J., and Peterson, C.L. (2001). SWI/SNF chromatin remodeling requires changes in DNA topology. Mol. Cell *7*, 97-104.

Ghosh, D., and Kieff, E. (1990). *cis*-acting regulatory elements near the Epstein-Barr virus latent-infection membrane protein transcriptional start site. J. Virol. *64*, 1855-1858.

Gordadze, A.V., Onunwor, C.W., Peng, R., Poston, D., Kremmer, E., and Ling, P.D. (2004). EBNA2 amino acids 3 to 30 are required for induction of LMP-1 and immortalization maintenance. J. Virol. *78*, 3919-3929.

Gordadze, A.V., Peng, R., Tan, J., Liu, G., Sutton, R., Kempkes, B., Bornkamm, G.W., and Ling, P.D. (2001). Notch1IC partially replaces EBNA2 function in B cells immortalized by Epstein-Barr virus. J. Virol. *75*, 5899-5912.

Grässer, F.A., Gottel, S., Haiss, P., Boldyreff, B., Issinger, O.G., and Mueller-Lantzsch, N. (1992). Phosphorylation of the Epstein-Barr virus nuclear antigen 2. Biochem. Biophys. Res. Commun. *186*, 1694-1701.

Grässer, F.A., Haiss, P., Gottel, S., and Mueller-Lantzsch, N. (1991). Biochemical characterization of Epstein-Barr virus nuclear antigen 2A. J. Virol. *65*, 3779-3788.

Gregory, P.D., Schmid, A., Zavari, M., Lui, L., Berger, S.L., and Horz, W. (1998). Absence of Gcn5 HAT activity defines a novel state in the opening of chromatin at the PHO5 promoter in yeast. Mol. Cell *1*, 495-505.

Grossman, S.R., Johannsen, E., Tong, X., Yalamanchili, R., and Kieff, E. (1994). The Epstein-Barr virus nuclear antigen 2 transactivator is directed to response elements by the J kappa recombination signal binding protein. Proc. Natl. Acad. Sci. U.S.A. *91*, 7568-7572.

Grundhoff, A.T., Kremmer, E., Tureci, O., Glieden, A., Gindorf, C., Atz, J., Mueller-Lantzsch, N., Schubach, W.H., and Grässer, F.A. (1999). Characterization of DP103, a novel DEAD box protein that binds to the Epstein-Barr virus nuclear proteins EBNA2 and EBNA3C. J. Biol. Chem. *274*, 19136-19144.

Gutkind, J.S., Link, D.C., Katamine, S., Lacal, P., Miki, T., Ley, T.J., and Robbins, K.C. (1991). A novel *c-fgr* exon utilized in Epstein-Barr virus-infected B lymphocytes but not in normal monocytes. Mol. Cell. Biol. *11*, 1500-1507.

Harada, S., and Kieff, E. (1997). Epstein-Barr virus nuclear protein LP stimulates EBNA-2 acidic domain-mediated transcriptional activation. J. Virol. *71*, 6611-6618.

Harada, S., Yalamanchili, R., and Kieff, E. (1998). Residues 231 to 280 of the Epstein-Barr virus nuclear protein 2 are not essential for primary B-lymphocyte growth transformation. J. Virol. *72*, 9948-9954.

Harada, S., Yalamanchili, R., and Kieff, E. (2001). Epstein-Barr virus nuclear protein 2 has at least two N-terminal domains that mediate self-association. J. Virol. 75, 2482-2487.

Havas, K., Flaus, A., Phelan, M., Kingston, R., Wade, P.A., Lilley, D.M., and Owen-Hughes, T. (2000). Generation of superhelical torsion by ATP-dependent chromatin remodeling activities. Cell *103*, 1133-1142.

Henkel, T., Ling, P.D., Hayward, S.D., and Peterson, M.G. (1994). Mediation of Epstein-Barr virus EBNA2 transactivation by recombination signal-binding protein J kappa. Science *265*, 92-95.

Hsieh, J.J., Zhou, S., Chen, L., Young, D.B., and Hayward, S.D. (1999). CIR, a corepressor linking the DNA binding factor CBF1 to the histone deacetylase complex. Proc. Natl. Acad. Sci. U.S.A. *96*, 23-28.

Hudson, G.S., Farrell, P.J., and Barrell, B.G. (1985). Two related but differentially expressed potential membrane proteins encoded by the EcoRI Dhet region of Epstein-Barr virus B95-8. J. Virol. *53*, 528-535.

Höfelmayr, H., Strobl, L.J., Marschall, G., Bornkamm, G.W., and Zimber-Strobl, U. (2001). Activated Notch1 can transiently substitute for EBNA2 in the maintenance of proliferation of LMP1-expressing immortalized B cells. J. Virol. *75*, 2033-2040.

Jaenisch, R., and Bird, A. (2003). Epigenetic regulation of gene expression: how the genome integrates intrinsic and environmental signals. Nat. Genet. *33* Suppl, 245-254.

Jin, X.W., and Speck, S.H. (1992). Identification of critical *cis* elements involved in mediating Epstein- Barr virus nuclear antigen 2-dependent activity of an enhancer located upstream of the viral BamHI C promoter. J. Virol. *66*, 2846-2852.

Jochner, N., Eick, D., Zimber-Strobl, U., Pawlita, M., Bornkamm, G.W., and Kempkes, B. (1996). Epstein-Barr virus nuclear antigen 2 is a transcriptional suppressor of the immunoglobulin mu gene: implications for the expression of the translocated *c-myc* gene in Burkitt's lymphoma cells. EMBO J. *15*, 375-382.

Johannsen, E., Koh, E., Mosialos, G., Tong, X., Kieff, E., and Grossman, S.R. (1995). Epstein-Barr virus nuclear protein 2 transactivation of the latent membrane protein 1 promoter is mediated by J kappa and PU.1. J. Virol. *69*, 253-262.

Johansen, L.M., Deppmann, C.D., Erickson, K.D., Coffin, W.F., 3rd, Thornton, T.M., Humphrey, S.E., Martin, J.M., and Taparowsky, E.J. (2003). EBNA2 and activated Notch induce expression of BATF. J. Virol. *77*, 6029-6040.

Kaiser, C., Laux, G., Eick, D., Jochner, N., Bornkamm, G.W., and Kempkes, B. (1999a). The proto-oncogene *c-myc* is a direct target gene of Epstein-Barr virus nuclear antigen 2. J. Virol. *73*, 4481-4484.

Kaiser, C., von Stein, O., Laux, G., and Hoffmann, M. (1999b). Functional genomics in cancer research: identification of target genes of the Epstein-Barr virus nuclear antigen 2 by subtractive cDNA cloning and high-throughput differential screening using high-density agarose gels. Electrophoresis *20*, 261-268.

Kao, H.Y., Ordentlich, P., Koyano-Nakagawa, N., Tang, Z., Downes, M., Kintner, C.R., Evans, R.M., and Kadesch, T. (1998). A histone deacetylase corepressor complex regulates the Notch signal transduction pathway. Genes Dev. *12*, 2269-2277.

Kempkes, B., Pawlita, M., Zimber-Strobl, U., Eissner, G., Laux, G., and Bornkamm, G.W. (1995a). Epstein-Barr virus nuclear antigen 2-estrogen receptor fusion proteins transactivate viral and cellular genes and interact with RBP-J kappa in a conditional fashion. Virology *214*, 675-679.

Kempkes, B., Spitkovsky, D., Jansen-Dürr, P., Ellwart, J.W., Kremmer, E., Delecluse, H.J., Rottenberger, C., Bornkamm, G.W., and Hammerschmidt, W. (1995b). B-cell proliferation and induction of early G1-regulating proteins by Epstein-Barr virus mutants conditional for EBNA2. EMBO J. *14*, 88-96.

Klein, C., Busson, P., Tursz, T., Young, L.S., and Raab-Traub, N. (1988). Expression of the *c-fgr* related transcripts in Epstein-Barr virus-associated malignancies. Int. J. Cancer *42*, 29-35.

Knutson, J.C. (1990). The level of *c-fgr* RNA is increased by EBNA-2, an Epstein-Barr virus gene required for B-cell immortalization. J. Virol. *64*, 2530-2536.

Kornberg, R.D. (2001). The eukaryotic gene transcription machinery. Biol. Chem. *382*, 1103-1107.

Kornberg, R.D., and Lorch, Y. (2002). Chromatin and transcription: where do we go from here? Curr. Opin. Genet. Dev. *12*, 249-251.

Kupfer, S.R., and Summers, W.C. (1990). Identification of a glucocorticoid-responsive element in Epstein-Barr virus. J. Virol. *64*, 1984-1990.

Kwiatkowski, B., Chen, S.Y., and Schubach W.H. (2004). CKII site in Epstein-Barr virus nuclear protein 2 controls binding to hSNF5/Ini1 and is important for growth transformation. J. Virol. *78*, 6067-6072.

Langst, G., Bonte, E.J., Corona, D.F., and Becker, P.B. (1999). Nucleosome movement by CHRAC and ISWI without disruption or *trans*-displacement of the histone octamer. Cell *97*, 843-852.

Laux, G., Adam, B., Strobl, L.J., and Moreau-Gachelin, F. (1994a). The Spi-1/PU.1 and Spi-B ets family transcription factors and the recombination signal binding protein RBP-

J kappa interact with an Epstein-Barr virus nuclear antigen 2 responsive *cis*-element. EMBO J. *13*, 5624-5632.

Laux, G., Dugrillon, F., Eckert, C., Adam, B., Zimber-Strobl, U., and Bornkamm, G.W. (1994b). Identification and characterization of an Epstein-Barr virus nuclear antigen 2-responsive *cis* element in the bidirectional promoter region of latent membrane protein and terminal protein 2 genes. J. Virol. *68*, 6947-6958.

Lin, C.S., Kuo, H.H., Chen, J.Y., Yang, C.S., and Wang W.B. (2000). Epstein-Barr virus nuclear antigen 2 retards cell growth, induces p21(WAF1) expression, and modulates p53 activity post-translationally. Virology *303*, 7-23.

Ling, P.D., Hsieh, J.J., Ruf, I.K., Rawlins, D.R., and Hayward, S.D. (1994). EBNA-2 upregulation of Epstein-Barr virus latency promoters and the cellular CD23 promoter utilizes a common targeting intermediate, CBF1. J. Virol. *68*, 5375-5383.

Ling, P.D., Rawlins, D.R., and Hayward, S.D. (1993a). The Epstein-Barr virus immortalizing protein EBNA-2 is targeted to DNA by a cellular enhancer-binding protein. Proc. Natl. Acad. Sci. U.S.A. *90*, 9237-9241.

Ling, P.D., Ryon, J.J., and Hayward, S.D. (1993b). EBNA-2 of herpesvirus papio diverges significantly from the type A and type B EBNA-2 proteins of Epstein-Barr virus but retains an efficient transactivation domain with a conserved hydrophobic motif. J. Virol. *67*, 2990-3003.

Link, D.C., Gutkind, S.J., Robbins, K.C., and Ley, T.J. (1992). Characterization of the 5′ untranslated region of the human *c-fgr* gene and identification of the major myelomonocytic *c-fgr* promoter. Oncogene *7*, 877-884.

Lomvardas, S., and Thanos, D. (2001). Nucleosome sliding via TBP DNA binding *in vivo*. Cell *106*, 685-696.

Longnecker, R. (2000). Epstein-Barr virus latency: LMP2, a regulator or means for Epstein-Barr virus persistence? Adv. Cancer Res. *79*, 175-200.

Longnecker, R., and Kieff, E. (1990). A second Epstein-Barr virus membrane protein (LMP2) is expressed in latent infection and colocalizes with LMP1. J. Virol. *64*, 2319-2326.

Lorch, Y., Zhang, M., and Kornberg, R.D. (2001). RSC unravels the nucleosome. Mol. Cell *7*, 89-95.

Makar, K.W., Pham, C.T., Dehoff, M.H., O'Connor, S.M., Jacobi, S.M., and Holers, V.M. (1998). An intronic silencer regulates B lymphocyte cell- and stage-specific expression of the human complement receptor type 2 (CR2, CD21) gene. J. Immunol. *160*, 1268-1278.

Makar, K.W., Ulgiati, D., Hagman, J., and Holers, V.M. (2001). A site in the complement receptor 2 (CR2/CD21) silencer is necessary for lineage specific transcriptional regulation. Int. Immunol. *13*, 657-664.

Masucci, M.G., Contreras-Salazar, B., Ragnar, E., Falk, K., Minarovits, J., Ernberg, I., and Klein, G. (1989). 5-Azacytidine up regulates the expression of Epstein-Barr virus nuclear antigen 2 (EBNA-2) through EBNA-6 and latent membrane protein in the Burkitt's lymphoma line rael. J. Virol. *63*, 3135-3141.

Merika, M., and Thanos, D. (2001). Enhanceosomes. Curr. Opin. Genet. Dev. 11, 205-208.

Ning, S., Hahn, A.M., Huye, L.E., and Pagano, J.S. (2003). Interferon regulatory factor 7 regulates expression of Epstein-Barr virus latent membrane protein 1: a regulatory circuit. J. Virol. *77*, 9359-9368.

Nilsson, J.A., and Cleveland, J.L. (2003). Myc pathways provoking cell suicide and cancer. Oncogene *22*, 9007-9021.

Nilsson, T., Sjöblom, A., Masucci, M.G., and Rymo, L. (1993). Viral and cellular factors influence the activity of the Epstein-Barr virus BCR2 and BWR1 promoters in cells of different phenotype. Virology *193*, 774-785.

Nilsson, T., Zetterberg, H., Wang, Y.C., and Rymo, L. (2001). Promoter-proximal regulatory elements involved in oriP-EBNA1-independent and -dependent activation of the Epstein-Barr virus C promoter in B-lymphoid cell lines. J. Virol. *75*, 5796-5811.

Nitsche, F., Bell, A., and Rickinson, A. (1997). Epstein-Barr virus leader protein enhances EBNA-2-mediated transactivation of latent membrane protein 1 expression: a role for the W1W2 repeat domain. J. Virol. *71*, 6619-6628.

Olave, I., Reinberg, D., and Vales, L.D. (1998). The mammalian transcriptional repressor RBP (CBF1) targets TFIID and TFIIA to prevent activated transcription. Genes Dev. *12*, 1621-1637.

Patel, M., Leevers, S.J., and Brickell, P.M. (1990). Regulation of *c-fgr* proto-oncogene expression in Epstein-Barr virus infected B-cell lines. Int. J. Cancer *45*, 342-346.

Peng, C.W., Xue, Y., Zhao, B., Johannsen, E., Kieff, E., and Harada, S. (2004). Direct interactions between Epstein-Barr virus leader protein LP and the EBNA2 acidic domain underlie coordinate transcriptional regulation. Proc. Natl. Acad. Sci. U.S.A. *101*, 1033-1038.

Peterson, C.L. (2002a). Chromatin remodeling enzymes: taming the machines. EMBO Rep. *3*, 319-322.

Peterson, C.L. (2002b). Chromatin remodeling: nucleosomes bulging at the seams. Curr. Biol. *12*, R245-247.

Petti, L., Sample, C., and Kieff, E. (1990). Subnuclear localization and phosphorylation of Epstein-Barr virus latent infection nuclear proteins. Virology *176*, 563-574.

Polack, A., Hortnagel, K., Pajic, A., Christoph, B., Baier, B., Falk, M., Mautner, J., Geltinger, C., Bornkamm, G.W., and Kempkes, B. (1996). *c-myc* activation renders proliferation of Epstein-Barr virus (EBV)-transformed cells independent of EBV nuclear antigen 2 and latent membrane protein 1. Proc. Natl. Acad. Sci. U.S.A. *93*, 10411-10416.

Ptashne, M., and Gann, A. (1997). Transcriptional activation by recruitment. Nature *386*, 569-577.

Puglielli, M.T., Desai, N., and Speck, S.H. (1997). Regulation of EBNA gene transcription in lymphoblastoid cell lines: characterization of sequences downstream of BCR2 (Cp). J. Virol. *71*, 120-128.

Puglielli, M.T., Woisetschlaeger, M., and Speck, S.H. (1996). oriP is essential for EBNA gene promoter activity in Epstein-Barr virus- immortalized lymphoblastoid cell lines. J. Virol. *70*, 5758-5768.

Radkov, S.A., Bain, M., Farrell, P.J., West, M., Rowe, M., and Allday, M.J. (1997). Epstein-Barr virus EBNA3C represses Cp, the major promoter for EBNA expression, but has no effect on the promoter of the cell gene CD21. J. Virol. *71*, 8552-8562.

Rasheed, S., Barbacid, M., Aaronson, S., and Gardner, M.B. (1982). Origin and biological properties of a new feline sarcoma virus. Virology *117*, 238-244.

Reisman, D., and Sugden, B. (1986). *trans* activation of an Epstein-Barr viral transcriptional enhancer by the Epstein-Barr viral nuclear antigen 1. Mol. Cell. Biol. *6*, 3838-3846.

Rickinson, A.B., and Kieff, E. (2001). Epstein-Barr virus. In Fields Virology, D.M. Knipe, and P.M. Howley, eds. (Philadelphia, Lippincott-Raven Publishers), pp. 2575-2627.

Ricksten, A., Olsson, A., Andersson, T., and Rymo, L. (1988). The 5′ flanking region of the gene for the Epstein-Barr virus-encoded nuclear antigen 2 contains a cell type specific *cis*-acting regulatory element that activates transcription in transfected B-cells. Nucleic Acids Res. *16*, 8391-8410.

Rooney, C., Howe, J.G., Speck, S.H., and Miller, G. (1989). Influence of Burkitt's lymphoma and primary B cells on latent gene expression by the nonimmortalizing P3J-HR-1 strain of Epstein-Barr virus. J. Virol. *63*, 1531-1539.

Rosendorff, A., Illanes, D., David, G., Lin, J., Kieff, E., and Johannsen, E. (2004). EBNA3C coactivation with EBNA2 requires a SUMO homology domain. J. Virol. *78*, 367-377.

Rowe, M., Lear, A.L., Croom-Carter, D., Davies, A.H., and Rickinson, A.B. (1992). Three pathways of Epstein-Barr virus gene activation from EBNA1-positive latency in B lymphocytes. J. Virol. *66,* 122-131.

Sadler, R.H., and Raab-Traub, N. (1995). The Epstein-Barr virus 3.5-kilobase latent membrane protein 1 mRNA initiates from a TATA-Less promoter within the first terminal repeat. J. Virol. *69,* 4577-4581.

Salamon, D., Takacs, M., Ujvari, D., Uhlig, J., Wolf, H., Minarovits, J., and Niller, H.H. (2001). Protein-DNA binding and CpG methylation at nucleotide resolution of latency-associated promoters Qp, Cp, and LMP1p of Epstein-Barr virus. J. Virol. *75,* 2584-2596.

Sauter, M., and Mueller-Lantzsch, N. (1987). Characterization of an Epstein-Barr virus nuclear antigen 2 variant (EBNA 2B) by specific sera. Virus Res. *8,* 141-152.

Schlee, M., Krug, T., Gires, O., Zeidler, R., Hammerschmidt, W., Mailhammer, R., Laux, G., Sauer, G., Lovric, J., and Bornkamm, G.W. (2004). Identification of Epstein-Barr virus (EBV) nuclear antigen 2 (EBNA2) target proteins by proteome analysis: activation of EBNA2 in conditionally immortalized B cells reflects early events after infection of primary B cells by EBV. J. Virol. *78,* 3941-3952.

Scholle, F., Bendt, K.M., and Raab-Traub, N. (2000). Epstein-Barr virus LMP2A transforms epithelial cells, inhibits cell differentiation, and activates Akt. J. Virol. *74,* 10681-10689.

Schuhmacher, M., Kohlhuber, F., Holzel, M., Kaiser, C., Burtscher, H., Jarsch, M., Bornkamm, G.W., Laux, G., Polack, A., Weidle, U.H., and Eick, D. (2001). The transcriptional program of a human B cell line in response to Myc. Nucleic Acids Res. *29,* 397-406.

Sinclair, A.J., Palmero, I., Peters, G., and Farrell, P.J. (1994). EBNA-2 and EBNA-LP cooperate to cause G0 to G1 transition during immortalization of resting human B lymphocytes by Epstein-Barr virus. EMBO J. *13,* 3321-3328.

Sjöblom, A., Jansson, A., Yang, W., Lain, S., Nilsson, T., and Rymo, L. (1995a). PU box-binding transcription factors and a POU domain protein cooperate in the Epstein-Barr virus (EBV) nuclear antigen 2-induced transactivation of the EBV latent membrane protein 1 promoter. J. Gen. Virol. *76,* 2679-2692.

Sjöblom, A., Nerstedt, A., Jansson, A., and Rymo, L. (1995b). Domains of the Epstein-Barr virus nuclear antigen 2 (EBNA2) involved in the transactivation of the latent membrane protein 1 and the EBNA Cp promoters. J. Gen. Virol. *76,* 2669-2678.

Sjöblom, A., Yang, W., Palmqvist, L., Jansson, A., and Rymo, L. (1998). An ATF/CRE element mediates both EBNA2-dependent and EBNA2-independent activation of the Epstein-Barr virus *LMP1* gene promoter. J. Virol. *72,* 1365-1376.

Sjöblom-Hallén, A., Yang, W., Jansson, A., and Rymo, L. (1999). Silencing of the Epstein-Barr virus latent membrane protein 1 gene by the Max-Mad1-mSin3A modulator of chromatin structure. J. Virol. *73,* 2983-2993.

Smith, M.M. (2002). Centromeres and variant histones: what, where, when and why? Curr. Opin. Cell. Biol. *14,* 279-285.

Spender, L.C., Cornish, G.H., Rowland, B., Kempkes, B., and Farrell, P.J. (2001). Direct and indirect regulation of cytokine and cell cycle proteins by EBNA-2 during Epstein-Barr virus infection. J. Virol. *75,* 3537-3546.

Spender, L.C., Cornish, G.H., Sullivan, A., and Farrell, P.J. (2002). Expression of transcription factor AML-2 (RUNX3, CBF(alpha)-3) is induced by Epstein-Barr virus EBNA-2 and correlates with the B-cell activation phenotype. J. Virol. *76,* 4919-4927.

Springer, T.A. (1994). Traffic signals for lymphocyte recirculation and leukocyte emigration: the multistep paradigm. Cell *76,* 301-314.

Strahl, B.D., and Allis, C.D. (2000). The language of covalent histone modifications. Nature *403,* 41-45.

Sugden, B., and Warren, N. (1989). A promoter of Epstein-Barr virus that can function during latent infection can be transactivated by EBNA-1, a viral protein required for viral DNA replication during latent infection. J. Virol. *63*, 2644-2649.

Sung, N.S., Kenney, S., Gutsch, D., and Pagano, J.S. (1991). EBNA-2 transactivates a lymphoid-specific enhancer in the BamHI C promoter of Epstein-Barr virus. J. Virol. *65*, 2164-2169.

Takacs, M., Salamon, D., Myohanen, S., Li, H., Segesdi, J., Ujvari, D., Uhlig, J., Niller, H.H., Wolf, H., Berencsi, G., and Minarovits, J. (2001). Epigenetics of latent Epstein-Barr virus genomes: high resolution methylation analysis of the bidirectional promoter region of latent membrane protein 1 and 2B genes. Biol. Chem. *382*, 699-705.

Thorley-Lawson, D.A., and Gross, A. (2004). Persistence of the Epstein-Barr virus and the origins of associated lymphomas. N. Engl. J. Med. *350*, 1328-1337.

Thorley-Lawson, D.A., Nadler, L.M., Bhan, A.K., and Schooley, R.T. (1985). BLAST-2 [EBVCS], an early cell surface marker of human B cell activation, is superinduced by Epstein Barr virus. J. Immunol. *134*, 3007-3012.

Tong, X., Drapkin, R., Reinberg, D., and Kieff, E. (1995a). The 62- and 80-kDa subunits of transcription factor IIH mediate the interaction with Epstein-Barr virus nuclear protein 2. Proc. Natl. Acad. Sci. U.S.A. *92*, 3259-3263.

Tong, X., Drapkin, R., Yalamanchili, R., Mosialos, G., and Kieff, E. (1995b). The Epstein-Barr virus nuclear protein 2 acidic domain forms a complex with a novel cellular coactivator that can interact with TFIIE. Mol. Cell. Biol. *15*, 4735-4744.

Tong, X., Wang, F., Thut, C.J., and Kieff, E. (1995c). The Epstein-Barr virus nuclear protein 2 acidic domain can interact with TFIIB, TAF40, and RPA70 but not with TATA-binding protein. J. Virol. *69*, 585-588.

Tong, X., Yalamanchili, R., Harada, S., and Kieff, E. (1994). The EBNA-2 arginine-glycine domain is critical but not essential for B-lymphocyte growth transformation; the rest of region 3 lacks essential interactive domains. J. Virol. *68*, 6188-6197.

Tsang, S.F., Wang, F., Izumi, K.M., and Kieff, E. (1991). Delineation of the *cis*-acting element mediating EBNA-2 transactivation of latent infection membrane protein expression. J. Virol. *65*, 6765-6771.

Tsui, S., and Schubach, W.H. (1994). Epstein-Barr virus nuclear protein 2A forms oligomers *in vitro* and *in vivo* through a region required for B-cell transformation. J. Virol. *68*, 4287-4294.

Waltzer, L., Bourillot, P.Y., Sergeant, A., and Manet, E. (1995). RBP-J kappa repression activity is mediated by a co-repressor and antagonized by the Epstein-Barr virus transcription factor EBNA2. Nucleic Acids Res. *23*, 4939-4945.

Wang, F., Gregory, C., Sample, C., Rowe, M., Liebowitz, D., Murray, R., Rickinson, A., and Kieff, E. (1990a). Epstein-Barr virus latent membrane protein (LMP1) and nuclear proteins 2 and 3C are effectors of phenotypic changes in B lymphocytes: EBNA-2 and LMP1 cooperatively induce CD23. J. Virol. *64*, 2309-2318.

Wang, F., Gregory, C.D., Rowe, M., Rickinson, A.B., Wang, D., Birkenbach, M., Kikutani, H., Kishimoto, T., and Kieff, E. (1987). Epstein-Barr virus nuclear antigen 2 specifically induces expression of the B-cell activation antigen CD23. Proc. Natl. Acad. Sci. U.S.A. *84*, 3452-3456.

Wang, F., Kikutani, H., Tsang, S.F., Kishimoto, T., and Kieff, E. (1991). Epstein-Barr virus nuclear protein 2 transactivates a *cis*-acting CD23 DNA element. J. Virol. *65*, 4101-4106.

Wang, F., Tsang, S.F., Kurilla, M.G., Cohen, J.I., and Kieff, E. (1990b). Epstein-Barr virus nuclear antigen 2 transactivates latent membrane protein LMP1. J. Virol. *64*, 3407-3416.

Wang, L., Grossman, S.R., and Kieff, E. (2000). Epstein-Barr virus nuclear protein 2 interacts with p300, CBP, and PCAF histone acetyltransferases in activation of the *LMP1* promoter. Proc. Natl. Acad. Sci. U.S.A. *97*, 430-435.

Venter, U., Svaren, J., Schmitz, J., Schmid, A., and Horz, W. (1994). A nucleosome precludes binding of the transcription factor Pho4 *in vivo* to a critical target site in the PHO5 promoter. EMBO J. *13*, 4848-4855.

Whitehouse, I., Flaus, A., Cairns, B.R., White, M.F., Workman, J.L., and Owen-Hughes, T. (1999). Nucleosome mobilization catalysed by the yeast SWI/SNF complex. Nature *400*, 784-787.

Woisetschlaeger, M., Jin, X.W., Yandava, C.N., Furmanski, L.A., Strominger, J.L., and Speck, S.H. (1991). Role for the Epstein-Barr virus nuclear antigen 2 in viral promoter switching during initial stages of infection. Proc. Natl. Acad. Sci. U.S.A. *88*, 3942-3946.

Woisetschlaeger, M., Yandava, C.N., Furmanski, L.A., Strominger, J.L., and Speck, S.H. (1990). Promoter switching in Epstein-Barr virus during the initial stages of infection of B lymphocytes. Proc. Natl. Acad. Sci. U.S.A. *87*, 1725-1729.

Voss, M.D., Hille, A., Barth, S., Spurk, A., Hennrich, F., Holzer, D., Mueller-Lantzsch, N., Kremmer, E., and Grässer, F.A. (2001). Functional cooperation of Epstein-Barr virus nuclear antigen 2 and the survival motor neuron protein in transactivation of the viral *LMP1* promoter. J. Virol. *75*, 11781-11790.

Wu, D.Y., Kalpana, G.V., Goff, S.P., and Schubach, W.H. (1996). Epstein-Barr virus nuclear protein 2 (EBNA2) binds to a component of the human SNF-SWI complex, hSNF5/ Ini1. J. Virol. *70*, 6020-6028.

Wu, D.Y., Krumm, A., and Schubach, W.H. (2000). Promoter-specific targeting of human SWI-SNF complex by Epstein-Barr virus nuclear protein 2. J. Virol. *74*, 8893-8903.

Yalamanchili, R., Harada, S., and Kieff, E. (1996). The N-terminal half of EBNA2, except for seven prolines, is not essential for primary B-lymphocyte growth transformation. J. Virol. *70*, 2468-2473.

Yalamanchili, R., Tong, X., Grossman, S., Johannsen, E., Mosialos, G., and Kieff, E. (1994). Genetic and biochemical evidence that EBNA 2 interaction with a 63-kDa cellular GTG-binding protein is essential for B lymphocyte growth transformation by EBV. Virology *204*, 634-641.

Yoo, L., and Speck, S.H. (2000). Determining the role of the Epstein-Barr virus Cp EBNA2-dependent enhancer during the establishment of latency by using mutant and wild-type viruses recovered from cottontop marmoset lymphoblastoid cell lines. J. Virol. *74*, 11115-11120.

Yoo, L.I., Mooney, M., Puglielli, M.T., and Speck, S.H. (1997). B-cell lines immortalized with an Epstein-Barr virus mutant lacking the Cp EBNA2 enhancer are biased toward utilization of the oriP-proximal EBNA gene promoter Wp1. J. Virol. *71*, 9134-9142.

Yue, W., Davenport, M.G., Shackelford, J., and Pagano, J.S. (2004). Mitosis-specific hyperphosphorylation of Epstein-Barr virus nuclear antigen 2 suppresses its function. J. Virol. *78*, 3542-3552.

Yukawa, K., Kikutani, H., Owaki, H., Yamasaki, K., Yokota, A., Nakamura, H., Barsumian, E.L., Hardy, R.R., Suemura, M., and Kishimoto, T. (1987). A B cell-specific differentiation antigen, CD23, is a receptor for IgE (Fc epsilon R) on lymphocytes. J. Immunol. *138*, 2576-2580.

Zeller, K.I., Jegga, A.G., Aronow, B.J., O'Donnell, K.A., and Dang, C.V. (2003). An integrated database of genes responsive to the Myc oncogenic transcription factor: identification of direct genomic targets. Genome Biol. *4*, R69.

Zhou, S., Fujimuro, M., Hsieh, J.J., Chen, L., and Hayward, S.D. (2000). A role for SKIP in EBNA2 activation of CBF1-repressed promoters. J. Virol. *74*, 1939-1947.

Zimber-Strobl, U., Strobl, L.J., Meitinger, C., Hinrichs, R., Sakai, T., Furukawa, T., Honjo, T., and Bornkamm, G.W. (1994). Epstein-Barr virus nuclear antigen 2 exerts its transactivating function through interaction with recombination signal binding protein RBP-J kappa, the homologue of Drosophila Suppressor of Hairless. EMBO J. *13*, 4973-4982.

Zimber-Strobl, U., Suentzenich, K.O., Laux, G., Eick, D., Cordier, M., Calender, A., Billaud, M., Lenoir, G.M., and Bornkamm, G.W. (1991). Epstein-Barr virus nuclear antigen 2 activates transcription of the terminal protein gene. J. Virol. *65*, 415-423.

Chapter 23

EBNA-2 and Notch Signalling

Bettina Kempkes, Lothar J. Strobl, Georg W. Bornkamm, and Ursula Zimber-Strobl*

ABSTRACT

Epstein-Barr virus nuclear antigen-2 (EBNA-2) plays a key role in B cell growth transformation by initiating and maintaining the proliferation of B cells upon EBV infection *in vitro*. EBNA-2 is one of the first viral genes expressed after virus infection. By activating viral as well as cellular target genes EBNA-2 initiates the transcription of a cascade of primary and secondary target genes, which eventually govern the activation of the resting B cell, cell cycle entry and proliferation of the growth transformed cells. The growth transformed B cells exhibit a phenotype reminiscent of antigen activated B cells. In addition, EBNA-2's anti-apoptotic activities protect the infected B cell. The multiple mechanisms by which EBNA-2 exerts its function are reflected by the association of EBNA-2 with several cellular and viral proteins as well as a spectrum of activated target genes. The recent finding that cellular effector proteins of EBNA-2 are also part of the signalling pathways activated by the Notch transmembrane receptor has suggested that EBNA-2 might be the viral functional analogue of the Notch signalling cascade. This review describes the common denominators of EBNA-2 and Notch signalling pathways and will try to integrate our current knowledge on the physiological role of Notch activation in lymphocyte development and differentiation as well as on the role of activated Notch in lymphocyte malignancies.

INTRODUCTION: EBNA-2, A KEY REGULATOR OF EBV GROWTH TRANSFORMATION

EBNA-2 expression

In vivo EBNA-2 is expressed during a short time window immediately after infection of tonsillar B cells of healthy individuals as well as in tonsillar EBV infected B cells in patients suffering from infectious mononucleosis (Thorley-Lawson, 2001; Kurth et al., 2003). Expression of EBNA-2 in EBV related malignant diseases is found in immunodeficient patients, which lack efficient T-cell immunosurveillance following immunosuppressive drug treatment post transplantation or HIV infection. Since the viral expression pattern in these situations closely resembles

*For correspondence email *kempkes@gsf.de*

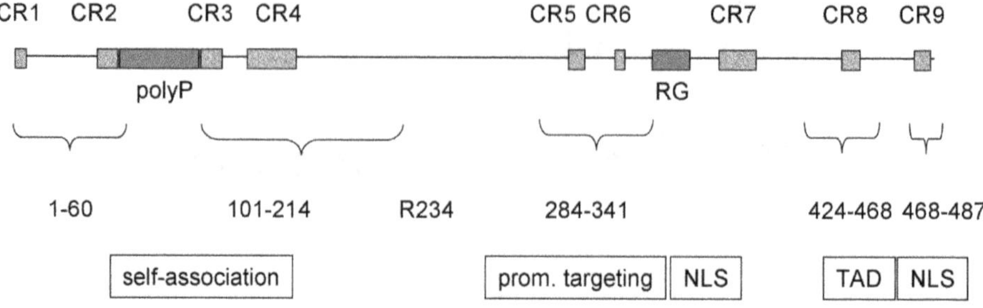

Figure 1. Schematic illustration showing the primary structure of EBNA-2. Characteristic features of EBNA-2 are a poly-proline (polyP) and a poly-arginine-glycine (RG) stretch and conserved regions (CR1-9), which have been defined by comparison of EBV strain type 1 and 2, baboon and rhesus macaque lymphocryptoviruses. Regions of EBNA-2, which mediate self-association are found in the amino terminus as well as in the divergent region between CR4 and CR5 adjacent to amino acid position R234. CR5 and CR6 lie within the region, which mediates promoter targeting by association with RBP-Jκ or PU.1. The transactivation domain (TAD) and the nuclear localization signals (NLS) are found at the carboxyl terminus. Amino acid numbering refers to the B95-8 primary structure of EBNA-2.

the pattern seen in immortalized B cells *in vitro*, EBNA-2 is likely to drive the proliferation of these highly malignant cells. EBNA-2 is not expressed in latently infected memory B cells of healthy individuals or in EBV associated malignancies of immunocompetent people, like Burkitt's lymphoma or Hodgkin's disease (for review see: Bornkamm and Hammerschmidt, 2001; Macsween and Crawford, 2003).

The EBNA-2 protein

Within the nucleus EBNA-2 forms granules and localizes to the nucleoplasm, the chromatin fraction and the nuclear matrix. The EBNA-2 open reading frame of EBV type 1 strain B95-8 encodes a protein of 487 amino acids[1]. EBNA-2, isolated from cellular extracts, has an apparent molecular weight of approximately 84 kDa as judged by SDS-PAGE. Phosphorylation of EBNA-2 only partially accounts for the discrepancy between the expected molecular weight and the electrophoretic mobility of the protein (Grasser et al., 1991). Notably, EBNA-2 proteins from distinct nuclear compartments exhibit differential modification patterns and these phosphorylation patterns vary during the cell cycle (Petti et al., 1990; Yue et al., 2004). Based on sequence identity and similarities, conserved regions (CR1-9) in the primary structure of EBNA-2 derived from EBV Type 1 or 2, baboon and rhesus lymphocryptoviruses have been defined (Peng et al., 2000). CR1/2 and CR3/4 are separated by a poly-proline repeat of variable sizes; CR6 and CR7 are separated by a poly-arginine-glycine motif, which is less conserved in the EBNA-2 protein of lymphocryptoviruses and functions as a nuclear localization motif in cooperation with a stretch of 20 amino acids at the extreme carboxyl terminus (Cohen and Kieff, 1991) (Figure 1).

[1]In this review we will refer to the primary structure of EBNA-2 using the Swiss-Prot data entry:P12978[2]

Table 1. EBNA-2 and Notch associated protein

Associated Protein	Reference for EBNA-2 binding	Reference for Notch binding
RBP-Jκ	Grossman et al., 1994 Henkel et al., 1994 Waltzer et al., 1994 Zimber-Strobl et al., 1994	Tamura et al., 1995 Jarriault et al., 1995
PU.1/Spi-1	Johannsen et al., 1995 Laux et al., 1994a	n.e.*
TFIIB	Tong et al., 1995c	n.e.
TAF40	Tong et al., 1995c	n.e.
CBP/p300	Wang et al., 2000	Oswald et al., 2001
PCAF/GCN5	Wang et al., 2000	Kurooka et al., 1998
P100	Tong et al., 1995b	n.e.
Nur77	Lee et al., 2002	Jehn et al., 1999
SKIP	Zhou et al., 2000a	Zhou et al., 2000b
DP103	Grundhoff et al., 1999	n.e.
SMN	Barth et al., 2003	n.e.
hSNF5/Ini1	Wu et al., 1996	n.e.
BS69	Ansieau and Leutz, 2002	n.e.
p34 cdc2	Yue et al., 2004	n.e.
ATF-2/c-Jun	Sjoblom et al., 1998	n.e.

* no evidence n.e.

Two amino terminal regions (1-60 and 92-206) mediate homotypic association (Harada et al., 2001). A third region, the large divergent stretch between CR4 and CR5, supports self-association (Tsui and Schubach, 1994). Recombinant EBNA-2 protein purified from insect cells can form oligomers which consist of up to eight monomers, while in cellular extracts high molecular weight complexes of up to 1 MDa are present (Grasser et al., 1991; Tsui and Schubach, 1994; Wu et al., 1996; Wu et al., 2000a).

EBNA-2 carries the characteristic features common to all transcription factors: A transactivation domain, nuclear localization motifs and a region which mediates promoter contact. Since EBNA-2 cannot bind to DNA directly it gains access to the promoters of its target genes by using adaptor proteins, which confer DNA contact and thereby recruit the transcriptional machinery, histone acetylase and chromatin remodelling activity to specific sites in the viral and cellular genome.

EBNA-2 associated proteins, which mediate promoter targeting

So far, the best studied cellular DNA adaptor protein of EBNA-2 is the DNA binding protein RBP-Jκ, which was first identified as a downstream effector molecule of EBNA-2 in the context of viral promoter activation (Table 1). RBP-Jκ is a ubiquitously expressed protein and belongs to the group of CSL proteins (CBF1 in humans for C-promoter binding protein, $\underline{S}$u(H) in *Drosophila melanogaster*, $\underline{L}$ag1 in *Caenorhabditis elegans* also known as recombination binding protein-Jκ in mice). The minimal domain of EBNA-2 which mediates RBP-Jκ binding has been mapped to the EBNA-2 CR6 region (Ling and Hayward, 1995). RBP-Jκ is a sequence specific DNA binding protein, which in the absence of EBNA-2 recruits a co-repressor complex to the promoter of target genes. Constituents of this co-repressor complex are SMRT/N-CoR, CIR, SKIP, Sin3A, SAP30 and HDAC1, which either directly or indirectly interfere with histone acetylation of target gene chromatin, thereby repressing transcription (reviewed in Lai (2002)). Binding of EBNA-2 relieves this repression by competition with co-repressor binding as well as the recruitment of co-activators by virtue of an intrinsic transactivation domain (Hsieh and Hayward, 1995).

Since RBP-Jκ is also an important downstream element of the cellular Notch signal transduction pathway, the discovery of RBP-Jκ in the context of the viral protein EBNA-2 has provoked an intense search for potential parallels of Notch and EBNA-2 signalling as will be discussed in the second part of this review.

In EBV infected B cells the viral C-promoter (Cp) as well as the promoters of the viral LMP-1 (LMP-1p), LMP-2A (LMP-2Ap) and LMP-2B (LMP-2Bp) genes are activated by EBNA-2 (Abbot et al., 1990; Fahraeus et al., 1990; Ghosh and Kieff, 1990; Wang et al., 1990b; Sung et al., 1991; Jin and Speck, 1992). By deletion analysis of promoter reporter constructs as well as gel retardation assays, EBNA-2 responsive elements (EBNA-2RE), which share functional binding motifs for RBP-Jκ as a common denominator, have been identified within all these promoters (Zimber-Strobl et al., 1991; Ling et al., 1993; Zimber-Strobl et al., 1993; Grossman et al., 1994; Henkel et al., 1994; Laux et al., 1994b; Waltzer et al., 1994; Zimber-Strobl et al., 1994). The detailed analysis of the viral promoters finally elucidated the molecular mechanism by which EBNA-2 exerts its function. The high affinity RBP-Jκ binding sites within the EBNA-2REs of Cp and LMP-2Ap were instrumental for the purification and identification of RBP-Jκ by four independent groups in 1994 (Grossman et al., 1994; Henkel et al., 1994; Waltzer et al., 1994; Zimber-Strobl et al., 1994).

All these RBP-Jκ binding motifs are flanked by diverse transcription factor binding sites, some of which contribute directly to the EBNA-2 responsiveness of the individual viral promoters, while others contribute to basal activity rather than modifying the EBNA-2 response of these promoters. Early after B cell infection EBNA-2 expression is primarily initiated from the W-promoter. Initiation then switches to Cp, a central promoter, which drives expression of all EBNA genes in EBV growth transformed established B cell lines (Woisetschlaeger et al., 1991; Puglielli et al., 1996; Schlager et al., 1996; Puglielli et al., 1997). Within the EBNA-2RE of Cp a single RBP-Jκ binding site is flanked by a binding site for the cyclic AMP responsive AUF1/hnRNP D protein, also called CBF2. The binding sites for both factors are evolutionary conserved as shown by sequence comparison of EBV like lymphocryptoviruses found in baboon and rhesus macaques (Fuentes-Panana and Ling, 1998; Fuentes-Panana et al., 1999; 2000). Binding of RBP-Jκ to the C-promoter is absolutely critical for promoter activation in transient reporter assays, but both factors contribute to stimulation of Cp by EBNA-2. Regions distal to the EBNA-2RE, both upstream and downstream, which bind additional cellular factors like SP1, Egr-1, NF-Y or the viral EBNA-1/oriP complex modulate the basal activity of the C-promoter in an EBNA-2 independent fashion (Puglielli et al., 1996; Borestrom et al., 2003).

Since EBNA-2 drives the expression of all EBNA genes in established growth transformed B cells by activating Cp, the relevance of the EBNA-2 responsive element in the context of the viral genome for the growth transformation process is crucial for the understanding of EBNA-2 function and has been studied by genetic experiments using viral mutants. While inactivation of the RBP-Jκ binding site partially inactivated Cp, deletion of the binding sites for RBP-Jκ as well as AUF1/hnRNP D resulted in significant loss of Cp activity and switch to Wp usage. However, the growth transformation capacities of the mutant viruses were not significantly reduced (Evans et al., 1996; Yoo et al., 1997; Yoo and Speck, 2000). Two functional binding sites for RBP-Jκ have been mapped to a 54 bps core sequence within the EBNA-2-RE of the viral LMP-2A promoter. Multimerization of this 54 bps core sequence can confer EBNA-2 responsivenes to heterologous promoters. An adjacent 30 bps region is required for full EBNA-2 activation of a singular 54 bps promoter fragment, suggesting that additional yet unknown cellular factors might enhance the interaction of EBNA-2 and RBP-Jκ (Zimber-Strobl et al., 1993; Meitinger et al., 1994; Zimber-Strobl et al., 1994).

Analysis of LMP-1p has revealed a complex pattern of transcription factor binding sites. Both binding of PU.1/Spi-1 as well as RBP-Jκ to a *cis*-regulatory region of LMP-1p are critical for EBNA-2 function. In addition, ATF-2/c-Jun heterodimers enhance EBNA-2 effects (Laux et al., 1994a; 1994b; Johannsen et al., 1995; Sjoblom et al., 1998). Further transcription factor binding sites like an interferon stimulated response element, an Sp1 binding site and a yet undefined POU-Box protein contribute to LMP-1p activity (Sjoblom et al., 1995a; 1995b).

In summary, it appears that EBNA-2 requires at least two DNA binding factors to target viral promoters. We would like to stress though, that the relative contribution of these binding factors to EBNA-2 activities in the context of the entire viral genome has so far only been studied for Cp. Thus, our recent understanding of viral promoter activation by EBNA-2 might be biased by the reporter gene studies that have been performed by us and others. An additional layer of complexity to this problem is the capacity of EBNA-2 to functionally interact with the viral proteins EBNA-LP and EBNA-3 proteins, which fine tunes the EBNA-2 responses. EBNA-LP directly interacts with the transactivation domain of EBNA-2 and enhances EBNA-2 transcriptional activation by a mechanism that is irrespective of the promoter context (Harada and Kieff, 1997; Nitsche et al., 1997; Peng et al., 2004). All EBNA-3 proteins can interfere with EBNA-2 transactivation of RBP-Jκ dependent promoters (Le Roux et al., 1994). EBNA-3C, however, can cooperate with EBNA-2 to activate PU.1 dependent transcription from the LMP-1 promoter (Marshall and Sample, 1995; Zhao and Sample, 2000). In addition, RPMS1, a viral protein expressed in all EBV infected B cells, counteracts RBP-Jκ dependent EBNA-2 activation (Zhang et al., 2001). Thus, the biological activity of EBNA-2 is significantly modulated by the viral context.

EBNA-2 associated proteins

The transactivation domain of EBNA-2 is located in the carboxyl terminus of the protein (AA: 422-466) and shows structural similarities with the transactivation domain of herpes simplex viral protein VP16. An EBNA-2 amino acid replacement mutant (W458A) inactivates this transactivation domain (Cohen and Kieff, 1991; Cohen, 1992). The core fragment of the transactivation domain (AA: 453-466) can be replaced by an acidic fragment of the VP16 transactivation domain indicating that EBNA-2 and VP16 share functional similarities (Cohen, 1992). In fact both transactivation domains bind to TFIIB and TAF40, components of the transcription initiation complex, and TFIIH, a factor involved in promoter clearance. In addition, both bind to RPA70, the replication protein A (Tong et al., 1995a; 1995c). In contrast to VP16, EBNA-2 does not bind to TBP (Tong et al., 1995c). Both transactivation domains also recruit histone acetyltransferase activity by interacting with CBP, p300 and PCAF (Wang et al., 2000). Chromatin immunoprecipitations using EBNA-2, histone H3 and H4 specific antibodies recently proved that LMP-1p and Cp are differentially acetylated in the presence of EBNA-2 (Alazard et al., 2003). In addition, EBNA-2 forms a complex with a novel cellular coactivator, p100, which can bind to the general transcription factor TFIIE and thereby bridges STAT6/RNA polymerase II interactions (Tong et al., 1995b; Yang et al., 2002). All these interactions within the transactivation domain rely on the integrity of position W458 of the transactivation domain of EBNA-2. Two potential scenarios could account for this result. Either the W458 mutation alters the structure of the EBNA-2 protein at secondary disparate binding sites and thereby simultaneously abolishes several interactions. Alternatively, by oligomerization EBNA-2 could easily recruit multiple effector molecules to the same multiprotein complex by using a single interaction site.

Apart from recruiting HAT activity and general transcription factors, phosphorylated EBNA-2 also interacts with hSNF5/Ini, a component of the hSWI/SNF chromatin remodelling complex and potential tumour suppressor gene. EBNA-2 recruits this protein to target promoters. This interaction is conferred by less conserved regions of EBNA-2 flanking CR5 and depends on the integrity of IPP285 and DQQ111 as well as phosphorylation of SS469 adjacent to the transactivation domain of EBNA-2 (Wu et al., 1996; 2000a; Kwiatkowski et al., 2004). The EBNA-2 interaction with hSNF5/Ini1 could potentially serve a second function. It might interfere with the growth suppressing activities of hSNF5/Ini1 in heterotrimeric complexes with GADD34 and PP-1 (Wu et al., 2002a).

The carboxyl terminus, CR7 and CR8, can be further targeted by the Mynd domain protein and corepressor BS69, a protein which was first described as an E1A associated protein (Hateboer et al., 1995). It was shown that BS69 can inhibit transcriptional activation by EBNA-2, but the functional implications of these interactions have not yet been analyzed in B cells (Ansieau and Leutz, 2002).

A potential function of EBNA-2 in RNA processing has been suggested by the identification of the DEAD box protein DP103 which binds to EBNA-2 CR4 as well as to CR4 flanking regions (AA:121-213) (Grundhoff et al., 1999). The RG repeat region of EBNA-2 is methylated at arginine residues and recruits the survival motor neuron (SMN) (Barth et al., 2003). SMN, a protein involved in

RNA splicing, directly interacts with DP103 and can enhance LMP-1 promoter activation by EBNA-2 (Voss et al., 2001).

The Ski-interacting protein (SKIP) is a multifunctional protein, which is a component of the spliceosome, a co-activator or co-repressor of transcription and an RB and E7 binding protein (reviewed in (Folk et al., 2004). SKIP binds to RBP-Jκ and either facilitates binding of the SMRT, CIR, Sin3A and HDAC2 co-repressor complex or potentiates binding of EBNA-2 to RBP-Jκ by interacting with CR5 (Zhou et al., 2000a).

Recently, a novel anti-apoptotic function of EBNA-2, based on the finding that EBNA-2 binds to Nur77 (TR3, NGFI-B), has been described (Lee et al., 2002). Nur77 is an orphan member of the nuclear hormone receptor superfamily and a bifunctional molecule. Nur77, a nuclear protein, either acts as a transcription factor or can be translocated from the nucleus into the cytoplasm and trigger cytochrom c release in response to apoptotic stimuli (Philips et al., 1997; Li et al., 2000). EBNA-2 can protect cells from apoptotic cell death by retaining Nur77 in the nucleus upon apoptotic stimuli (Lee et al., 2002). Whether EBNA-2 also modulates functions of Nur77 related to its role as transcription factor has not been analyzed to date.

Regions of EBNA-2 critical or essential for growth transformation

The functional and genetic analysis of the role of EBNA-2 in B cell growth transformation by EBV has been facilitated by the discovery of the EBV strain P3HR-1, a laboratory strain, which is infectious and induces early antigen expression upon infection of Raji cells but which fails to immortalize primary B cells (Miller et al., 1974). The EBV-P3HR-1 strain is named after the Burkitt's lymphoma cell line P3HR-1 that produces the virus. The P3HR-1 Burkitt's lymphoma cell line has been generated by single cell cloning of the Burkitt's lymphoma line Jijoye by Hinuma and coworkers in 1967 (Hinuma et al., 1967). Remarkably, the virus produced by the parental Jijoye cell line is transformation competent and carries a functional EBNA-2B gene of EBV type 2 (Bornkamm et al., 1982; King et al., 1982; Zimber et al., 1986). The Burkitt's lymphoma line Raji carries a replication-deficient viral genome (Zur Hausen and Schulte-Holthausen, 1970) with two large deletions (Polack et al., 1984). Treatment of Raji cells with the phorbolester tetradecanoyl-phorbole-13 acetate (TPA) and/or sodium butyrate induces early antigen expression but does not lead to viral replication (Hudewentz et al., 1980). Infection of Raji cells with P3HR1 virus, in contrast, induces not only early antigen expression but also viral replication and release of infectious virus. As shown by the elegant work of the late Karl-Otto Fresen, supernatants of Raji cells, that had been superinfected with P3HR1 virus, have immortalizing capacity for primary B cells (Fresen et al., 1978). Analysis of the viral genome of cells immortalized by such supernatants revealed that extensive recombination had taken place between the Raji- and the incoming P3HR1-viral genomes and that these recombination events have led to the formation of immortalizing virus in which not only the region deleted in P3HR-1 virus had been invariably restored but also the two regions deleted in the viral genome of Raji cells (Skare et al., 1985). With the advent of recombinant DNA technology these early classical genetic experiments were confirmed and extended by *cis* and *trans* complementation experiments using the cloned EBNA-2 gene. These experiments directly proved that EBNA-2 plays a pivotal and essential role in B cell growth transformation (Cohen et al., 1989; Hammerschmidt and Sugden, 1989). In addition, they allowed a detailed functional analysis of EBNA-2 mutants in the context of the viral genome. Cosmid clones comprising the EBNA-2 open reading frame and flanking regions were introduced into the P3HR-1 genome by homologous recombination and marker rescue. By this technique, a comprehensive set of EBNA-2 deletion as well as amino acid replacement mutants were tested for their capacity to complement the EBNA-2 defect in P3HR-1 upon primary B cell infection (Cohen et al., 1991). More recently, an alternate transcomplementation assay was developed, which uses a lentiviral transduction system to express EBNA-2 mutants in a lymphoblastoid cell line conditional for EBNA-2 and was used to test whether specific EBNA-2 mutants can maintain B cell proliferation (Gordadze et al., 2001).

Importantly, all experiments conclusively showed that regions of EBNA-2 which mediate RBP-Jκ (CR6) or SKIP (CR5) binding as well as the transactivation domain of EBNA-2 are absolutely essential for B cell growth transformation (Cohen and Kieff, 1991; Cohen et al., 1991; Cohen, 1992). Deletion of the poly RG repeat leads to loss of EBNA-2 chromatin association and reduces the growth transformation efficiency but is compatible with B cell growth transformation (Tong et al., 1994). Using the two systems, conflicting results were reported concerning the relative importance of the amino terminus or the poly-proline region of EBNA-2. While deletion of CR1 (AA: 3-30) was not compatible with B cell proliferation in the transcomplementation system, lymphoblastoid cell lines carrying deletions of the amino terminus comprising CR1 and CR2 plus most of the poly-proline region could be generated in the ciscomplementation system. In contrast, the poly-proline region scored as dispensable in the transcomplementation system while a minimal number of seven prolines was critical for growth transformation after complementation in *cis* (Yalamanchili et al., 1996; Gordadze et al., 2002; 2004). Smaller deletions (AA: 101-126 or 116-145 or 146-231 or 235-284) within the central part of EBNA-2 affected growth transformation efficiencies. Deletion of the entire region (AA: 114-292) was absolutely incompatible with growth transformation (Cohen et al., 1991; Yalamanchili et al., 1994; Harada et al., 1998). Which of the EBNA-2 interactions, that have been assigned to this region *e.g.* Nur77, DP103 and SNF5/Ini1 binding, are relevant for B cell growth transformation will need to be explored further.

Target genes of EBNA-2

EBNA-2 not only initiates a cascade of events leading to cell cycle entry of EBV infected B cells, it also modulates the phenotype of the B cell towards an activated B cell blast, which is a potent antigen presenting cell. Since the viral EBNA-2 targets LMP-1 and LMP-2A induce expression of cellular target genes on their own, direct cellular target genes of EBNA-2 are difficult to define. In addition EBNA-2 in cooperation with EBNA-LP has a profound impact on cellular target gene expression, which in turn modulates the expression of secondary target genes. This strong impact on cellular target gene expression can be exemplified by induction of cyclin D2 upon EBNA-2 and EBNA-LP expression in primary human B cells, indicating that both viral proteins cooperatively induce G0/G1 transition in the absence of any further viral proteins (Sinclair et al., 1994). Later it was shown, that even these early events most likely require the contribution of additional cellular processes for activation (Kaiser et al., 1999; Spender et al., 2001).

Table 2. EBNA-2 and Notch activated cellular genes

	Target gene	Reference for activation by EBNA-2	Reference for activation by Notch
Transcription factors	Hes1	Sakai et al., 1998	Jarriault et al., 1995
	Hes5	n.e.	Nishimura et al., 1998
	Hes7	n.e.	Bessho et al., 2001
	HERP1/Hey2/Hesr2/ HRT2/CHF1/Gridlock	n.e.	Iso et al., 2003
	HERP2/Hey1/Hesr1/ HRT1/CHF2	n.e.	Maier and Gessler, 2000
	HERP3/HeyL/Hesr3/ HRT3	n.e.	Nakagawa et al., 2000
	AML-2/RUNX3/ CBFα-3/PEBP3α-C	Spender et al., 2002	n.e.
	BATF	Johansen et al., 2003	Johansen et al., 2003
	c-myc	Kaiser et al., 1999	Gordadze et al., 2001
	NFκB2	n.e.	Oswald et al., 1998
	PU.1	n.e.	Schroeder et al., 2003
Receptors	CCR7/EBI-1/BLR2	Burgstahler et al., 1995	n.e.
	CD21	Cordier et al., 1990	Strobl et al., 2000
	CD23	Wang et al., 1987	Gordadze et al., 2001 Hubmann et al., 2002
	erbB-2	n.e.	Chen et al., 1997
	PTCRA	n.e.	Reizis and Leder, 2002
Modulators of Notch signalling	Deltex1	n.e.	Deftos et al., 2000
	NRARP	n.e.	Krebs et al., 2001 Lamar et al., 2001
Modulator of NFκB signalling	IκBα	n.e.	Oakley et al., 2003
Cell cycle regulation	CyclinD1	n.e.	Ronchini and Capobianco, 2001
GTPase	p21	n.e.	Rangarajan et al., 2001
	Nodal	n.e.	Krebs et al., 2003a

* no evidence n.e.

To date, CD23, CD21, CCR7 (BLR2/EBI1), Hes-1, the proto-oncogene c-*myc*, AML-2 (RUNX3, CBFα-3) and BATF have been defined as direct or primary target genes (Table 2) (Wang et al., 1987; Calender et al., 1990; Cordier et al., 1990; Burgstahler et al., 1995; Sakai et al., 1998; Kaiser et al., 1999; Spender et al., 2001; 2002; Johansen et al., 2003). Two complementary approaches have been used to identify cellular target genes of EBNA-2: Expression of EBNA-2 in established

EBV negative B cell lines or expression of a conditional EBNA-2 protein in the context of an EBV immortalized lymphoblastoid cell line.

In order to study the impact of EBNA-2 on cellular gene expression in the absence of viral proteins, EBNA-2 was expressed in established EBV negative B cell lines and the phenotype of the transfectants was screened for potential alterations (Wang et al., 1987; Calender et al., 1990; Cordier et al., 1990; Burgstahler et al., 1995; Johansen et al., 2003). CD23 is the human low affinity Fc epsilon receptor II. Transcription of CD23 can be initiated from two alternative promoters designated type a or b, the former one being predominantly used in B cells (Yokota et al., 1988). An EBNA-2 responsive fragment could be delineated in the CD23 type a promoter and subsequently was shown to carry a functional RBP-Jκ binding site (Wang et al., 1990a; 1991; Ling et al., 1994).

CD21, the complement receptor 2, is part of the B cell receptor complex and a ligand for CD23 (Fearon and Carroll, 2000). CD21 promoter reporter constructs are activated by EBNA-2, but the relevant EBNA-2 response elements within this promoter have not been identified to date (Radkov et al., 1997). The first intron of the human CD21 gene carries an intronic silencer, which binds RBP-Jκ (Makar et al., 1998; 2001). Whether EBNA-2 can bind to this RBP-Jκ site has not been described yet.

The chemokine receptor CCR7 (EBI-1/BLR2) is a G-coupled receptor which is involved in lymphocyte homing of B- and T-cells to secondary lymphoid organs, which express the CCR7 ligands CCL19 and CCL21 (Muller et al., 2003). CCR7 is induced upon B cell activation.

BATF is a member of the AP-1/ATF transcription factor superfamily, which heterodimerizes with Jun and antagonizes AP-1 activity of Fos:Jun complexes. Since AP-1 is a potent inducer of BZLF-1, a protein which governs lytic cycle entry, in the context of EBV, BATF might control viral latency by repressing BZLF1 expression (Johansen et al., 2003). The EBNA-2 responsive elements within the CCR7 or BATF genes have not been defined.

In order to recapitulate the events which follow EBNA-2 expression after primary B cell infection a cellular system, ER/EB, has been established that is conditional for EBNA-2. By fusing the open reading frame of the estrogen receptor hormone binding domain to the open reading frame of EBNA-2 (ER/EBNA-2) EBNA-2 function was rendered strictly dependent on estrogen in the cell culture medium (Kempkes et al., 1995a). This ER/EBNA-2 protein was used to complement the EBNA-2 deletion in the P3HR-1 virus strain. By infection of primary B cells a lymphoblastoid cell line (ER/EB) was established, which reversibly growth arrests and re-enters the cell cycle upon estrogen starvation and induction thus proving that EBNA-2 is required for initiation and maintenance of EBV driven proliferation (Kempkes et al., 1995b). Cell cycle arrest of ER/EB cells could be rescued by high level Myc expression, indicating that EBNA-2 induced Myc expression is rate limiting for the proliferation of EBV growth transformed B cells (Polack et al., 1996; Pajic et al., 2000). Cellular target genes of EBNA-2 can either be directly activated by EBNA-2 or may be activated by the cascade of events, initiated by viral or primary cellular EBNA-2 target genes. Since in ER/EB cells EBNA-2 activation is independent of *de novo* protein synthesis, this cellular system has been used to dissect primary direct and secondary indirect target genes by studying gene activation in the presence of protein synthesis inhibitors. Importantly, by this criteria the proto-oncogene *c-myc* was defined as a direct target gene, while induction of early cell cycle effectors like cyclinD2 and cdk4 appeared to rely on additional cellular or viral functions (Kaiser et al., 1999; Spender et al., 2001). A screen for EBNA-2 target genes using this system also led to the identification of AML-2 as a direct target gene of EBNA-2. AML-2 is a member of the "Runt family" of transcription factors and is specifically expressed in group III Burkitt's lymphoma cell lines, which express a transcriptional program of viral and cellular genes similar to the program used by growth transformed B cells (Spender et al., 2002). Thus, the transcription factor AML-2 could have a potential functional impact on the gene expression program expressed by lymphoblastoid cell lines. Most recently, the ER/EB system has been used to identify 20 novel EBNA-2 target genes by a proteomics approach and thereby underscored the value of EREB2-5 cells as an appropriate model system for the analysis of early events in the process of EBV-mediated B-cell immortalization (Schlee et al., 2004).

EBNA-2 not only activates transcription of target genes but also actively represses transcription of IgM by a so far unknown mechanism. In the context of Burkitt's lymphoma cell lines, which carry a chromosomal translocation that juxtaposes the IgM and the *c-myc* gene locus on chromosome 8, repression of IgM coincides with *c-myc* repression and a potent growth inhibitory activity of EBNA-2 (Jochner et al., 1996; Kempkes et al., 1996).

These findings strongly suggest that EBNA2- and c-myc-driven proliferation programs are incompatible with each other (Pajic et al., 2001). In fact, EBNA2 downregulation is selected for in Burkitt's lymphoma cells *in vivo* for at least two reasons: firstly, it downregulates a translocated c-myc gene (Jochner et al., 1996), and secondly, it drives expression of LMP-1 and T-helper-specific chemokines thus rendering the cells strongly immunogenic (Kelly et al., 2002).

In summary, compared to our detailed understanding of the EBNA-2 responsive viral promoters, the mechanisms by which EBNA-2 activates cellular target genes are less well understood. This is exemplified by the fact, that the relative contribution of the RBP-Jκ site has been well established for CD23 promoter activation, but not for any of the other cellular target genes. In addition, the cellular context significantly influenced phenotypic alterations in response to EBNA-2 (Wang et al., 1990a; Cordier-Bussat et al., 1993). Thus, further unknown cellular factors, which are not ubiquitously expressed in every B cell line, are rate limiting for EBNA-2 dependent activation of CD23. It has been suggested, that EBNA-2 might not only contact RBP-Jκ bound to promoter but also to intronic sites. The first intron of CD21 and c-myc carry an intronic RBP-Jκ binding motif, which in the case of the CD21 gene has been proved to be a functional intronic silencer (Makar et al., 2001). Still, it is an open question, whether EBNA-2 contacts such intronic silencing elements and how this might induce target gene activation.

Notably, EBNA-2 target genes and target genes of the viral LMP-1 protein are not mutually exclusive. There might be a significant overlap of EBNA-2 and LMP-1 cellular target genes as indicated by activation of CD23 and suppression of IgM expression by both proteins independently (Wang et al., 1990a; Floettmann et al., 1996). This finding could suggest a molecular effector mechanism common to both viral proteins. However, a common cellular pathway downstream of EBNA-2 and LMP-1 has not yet been described, suggesting that EBV has evolved to mimic B cell activation by inducing complementary and perhaps redundant pathways. EBNA-2 acts as a key determinant of the activated phenotype of EBV infected B cells. By activating BATF it might contribute to control the balance of the viral lytic and latent life cycle. Activation of *c-myc* by EBNA-2 is likely to be the major rate limiting step for initiation and maintenance of proliferation. Whether activation of further cellular target genes like CD21, CD23 or CCR7 is crucial for the growth transformation process is currently unknown.

FUNCTIONAL EQUIVALENCE OF EBNA-2 AND ACTIVATED NOTCH

Signal transduction through the classical Notch signalling pathway is initiated by binding of the Notch extracellular domain to a member of two conserved families of ligands (Delta and Serrate/Jagged) and involves a regulated set of proteolytic events leading to nuclear translocation of the Notch intracellular fragment (Notch-IC) and the subsequent binding to the transcription factor RBP-Jκ. The binding of Notch-IC to RBP-Jκ converts it from a transcriptional repressor into a transcriptional activator that leads to the activation of Notch target genes. Since both EBNA-2 and activated Notch (Notch-IC) interact with RBP-Jκ to activate gene expression, it is tempting to speculate that EBNA-2 activates the Notch signalling pathway. The discovery of RBP-Jκ as a common interaction partner for Notch-IC and EBNA-2 has provoked an intensive search for potential parallels of EBNA-2 and Notch signalling. We will summarize what is known about the mechanism of Notch mediated target gene transactivation in mammals and will discuss the common and distinct features of Notch and EBNA-2.

Primary structure of Notch

Mammals have four Notch receptors encoded by four different genes (Notch1–4). Notch receptors are large single transmembrane spanning proteins composed of a series of well-defined structural motifs (Figure 2) (for review: Radtke et al., 2004b). The extracellular part of Notch contains epidermal growth factor-like repeats that bind Notch ligands, and LIN12/Notch repeats that prevent inappropriate Notch receptor activation prior to ligand binding. The intracellular Notch (Notch-IC) contains several functional domains which mediate Notch signal transduction. These include the RAM domain and ankyrin repeats that interact with downstream effector proteins, nuclear localization sequences, a transactivation domain (only for Notch1 and 2) and a C-terminal PEST (proline-, glutamate-, serine-, threonine-rich) sequence, which regulates protein stability and contributes to the transactivation domain of Notch (Dumont et al., 2000).

Notch receptors have at least three sites (S1 to S3), at which they are proteolytically cleaved to gain effector function. During maturation, the precursor protein (300kDa in the case of Notch 1) is cleaved by a furine-like convertase at S1, generating an aminoterminal fragment (180kDa) containing the majority of the extracellular domain, and a carboxyterminal fragment (120kDa) containing a short extracellular part, the transmembrane domain and the intracellular part (Notch-IC) (Blaumueller et al., 1997; Logeat et al., 1998). The two fragments are reassembled non-covalently in the Golgi apparatus to form a heterodimer, which can then be expressed on the cell surface (Rand et al., 2000).

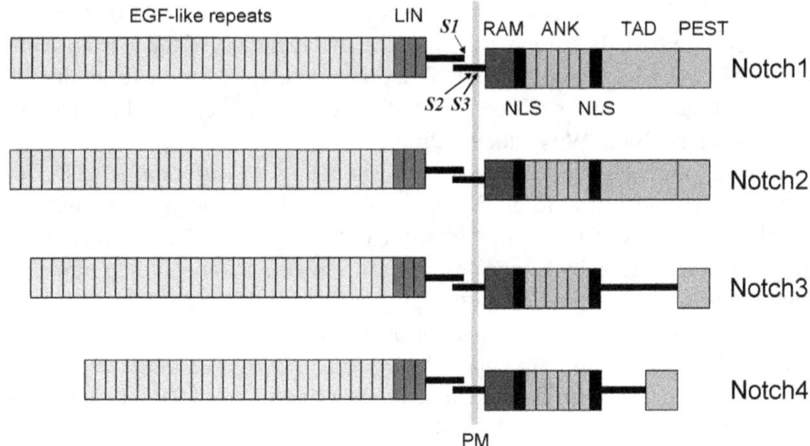

Figure 2. Pimary structure of the mammalian Notch receptors. The extra cellular parts contain 29 to 36 EGF-like repeats associated with ligand binding and three LIN repeats that prevent ligand independent activation. The intracellular parts consist of the RAM domain and six ankyrin repeats (ANK) which both are necessary for protein-protein interaction, two nuclear localization signals (NLS) a transactivation domain (TAD), which is missing in Notch 3 and 4 and a PEST sequence.

Signalling and interaction partners

Notch activation converts a transmembrane receptor into a transcriptional coactivator in the nucleus. This multistep process is initiated by receptor ligand interactions between adjacent cells. At present, two Notch ligand families, Delta and Jagged/Serrate, which comprise at least five different members (Delta-like 1, 3, 4 and Jagged 1, 2) in mammals, have been described. Like Notch, these ligands are transmembrane proteins and their extracellular part is largely composed of EGF-like repeats and a unique N-terminal DSL domain. An expression pattern of the different ligands is described for the hematopoetic system in detail in (Radtke et al., 2004b). If the different ligands exert distinct biological effects is still in discussion.

Upon ligand binding, the Notch receptor undergoes a conformational change that allows the cleavage at the site S2 (Figure 2) close to the transmembrane domain through the action of TNF-α-converting enzyme (TACE), a metalloprotease (Brou et al., 2000; Mumm et al., 2000). The membrane tethered product of the S2 cleavage is itself a substrate for the proteolytic activity of the γ-secretase complex that cleaves Notch at the S3 site within the transmembrane domain (De Strooper et al., 1999; Huppert et al., 2000; Okochi et al., 2002), to release soluble Notch-IC (Schroeter et al., 1998). The latter product is translocated to the nucleus to activate genes by interaction with the DNA binding protein RBP-Jκ, which is acting as a transcriptional repressor in the absence of Notch-IC.

Activated Notch (Notch-IC) relieves this repression in two ways: by binding RBP-Jκ through RAM and ankyrin repeat domains, thereby displacing corepressors, and by recruiting through its C-terminal TAD the histone acetyltransferases (HATs) PCAF/GCN5 (Kurooka et al., 1998) or p300/CBP (Oswald et al., 2001). Other components of this coactivator complex are SKIP (Zhou et al., 2000b), which is also part of the RBP-Jκ corepressor complex, and members of the mastermind-like family (MAML1-3), which appear to function as a scaffold for the formation of the large coactivator complex (Wu et al., 2000b; Kitagawa et al., 2001; Jeffries et al., 2002; Lin et al., 2002; Wu et al., 2002b).

Several modulators regulate Notch signalling at different levels. During maturation in the trans-Golgi, Notch is also modified by glycosyltransferases, including members of the Fringe family (Radical Fringe, Manic Fringe, Lunatic Fringe). Glycosylation by Fringe proteins restricts activation of Notch signalling to Delta ligands, whereas Jagged induced signalling is inhibited (Hicks et al., 2000). In the cytoplasm Deltex1 is a regulator of Notch signalling. It can antagonize the function of Notch1 in the thymus (Izon et al., 2002), but its mode of interaction with Notch has still to be clarified. The nuclear protein NRARP interacts through its ankyrin repeats with Notch-IC and RBP-Jκ, thereby inhibiting Notch signalling (Yun and Bevan, 2003). Another nuclear protein called MINT in mice and SHARP in humans competes with Notch-IC for interaction with RBP-Jκ, also resulting in a negative regulation of Notch signalling (Oswald et al., 2002; Kuroda et al., 2003). Ubiquitin ligase and proteolytic steps are involved in the final down-regulation of the Notch signal. Mammalian homologues of the *C. elegans* protein Sel-10, which is a substrate-targeting component of an E3 ubiquitin ligase, can stimulate phosphorylation dependent ubiquitination of Notch-IC and trigger its proteosome dependent degradation (Gupta-Rossi et al., 2001; Oberg et al., 2001). This pathway is dependent on the C-terminal region of Notch, which includes the PEST domain. Another E3 ubiquitin ligase implicated in Notch pathway down-regulation is Itch, which associates with Notch and promotes PEST domain-independent ubiquitination (Qiu et al., 2000).

Target genes of Notch-IC

The direct effect of Notch activation is increased transcription of Notch target genes. During the last 10 years, several Notch target genes in vertebrates have been identified. Different experimental approaches have suggested that HES1,5,7, and HERP1,2,3 (Hes related protein, also named CHF/gridlock/Hesr/Hey/HRT) are target genes of Notch. It is assumed that these homologues of the Drosophila enhancer of split [E(spl)] family act as the main primary effectors of Notch signalling. Their promoters are activated by constitutively active Notch-IC in transiently transfected reporter gene assays (Jarriault et al., 1995; Nishimura et al., 1998; Maier and Gessler, 2000; Nakagawa et al., 2000; Bessho et al., 2001; Iso et al., 2002). In addition, endogenous HERP1 and -2 mRNA is induced after transfection of Notch-IC in several different cell lines (Iso et al., 2001). By using more physiological systems (coculture approaches with Notch ligand-expressing cells, which generate much lower amounts of endogenous Notch-IC in the nucleus) HES1, HERP1, and HERP2 could be confirmed as target genes of Notch (Kuroda et al., 1999; Iso et al., 2001; 2002). Beside these transcriptional repressor proteins of the basic helix-loop-helix (bHLH) type in the meantime several other cellular Notch target genes which exert their function as transcription factors, receptors or modulators of Notch and NFκB signalling have been described (Table 2).

Comparison of EBNA-2 and Notch-IC

Although both EBNA-2 and Notch-IC interact with RBP-Jκ leading to gene activation there is no obvious sequence and structural homology between both proteins. Binding of Notch-IC and EBNA-2 is mutually exclusive since both proteins bind to a central region using a limited number of amino acids within the central region of the RBP-Jκ protein (Chung et al., 1994; Sakai et al., 1998; Callahan et al., 2000; Fuchs et al., 2001). As described above Notch signalling is regulated before and after ligand activation at different levels, which might allow a transient well controlled signal. In contrast, EBNA-2 seems to circumvent all negative regulations by cellular factors, leading to a constitutive active signal. EBNA-2 is acting ligand independently, and there is no evidence that it interacts with any of the negative modulators like NRARP or SHARP. Accordingly, the cell does not have the ability to counterregulate EBNA-2 function. Instead, EBNA-2 activity is tightly controlled by viral proteins as discussed above. The transactivation mechanism of Notch-IC and EBNA-2 seems to be very similar since both molecules recruit the same set of coactivators in the region of Notch/EBNA-2-regulated promoters. Both interact with histone acetyltransferases (HATs) PCAF/GCN5, p300/CBP and SKIP (Figure 3). In addition both EBNA-2 and Notch-IC interact with Nur77, thereby protecting the cell from apoptosis (Jehn et al., 1999;

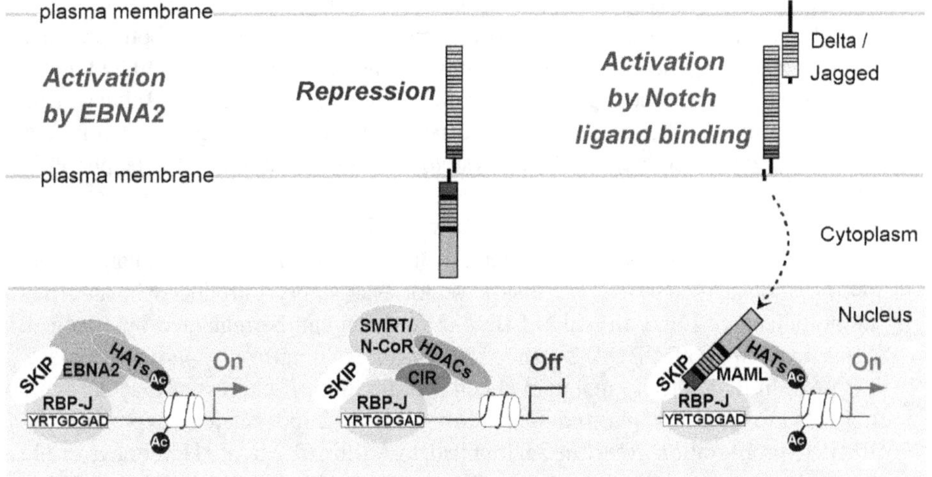

Figure 3. Comparison of EBNA-2 and Notch-IC mediated transactivation of genes. EBNA-2 as well as Notch-IC transactivate genes mainly by recruiting histone acetyl transferase (HAT) activity to the promoters of their target genes. At first interaction of EBNA-2 or Notch-IC with the DNA binding protein RBP-Jκ dislocates the associated corepressor complex containing histone deacetylases (HDACs) from RBP-Jκ. In a second step EBNA-2 and Notch-IC interact via their TADs with HATs. The accessory protein SKIP has been described as a component of the corepressor complex as well as a part of the coactivator complexes of EBNA-2 and Notch. Until yet an interaction of MAML has only been described for the Notch coactivator complex. Transactivation of genes by Notch-IC is dependent on a previous activation of the Notch receptor by a ligand, whereas EBNA-2 transactivation is autonomous.

Lee et al., 2002). These common interactions could point to common pathways activated by EBNA-2 and Notch. The interaction partners of Notch-IC and EBNA-2 are compared in Table1. To study whether EBNA-2 and Notch have common target genes we and others have transfected EBNA-2 and Notch-IC expression plasmids into different cell lines.

In transient transfection assays it could be shown that the viral EBNA-2 regulated promoters Cp, LMP-1p and LMP-2p can be activated by Notch1-IC (Hofelmayr et al., 1999; Cotter et al., 2000). Mutations or deletions of the EBNA-2-responsive regions revealed that the RBP-Jκ/EBNA-2 and RBP-Jκ/Notch1-IC interaction is essential but not sufficient to transactivate EBNA-2-responsive promoters. Additional cellular factors binding within the EBNA-2-responsive regions appeared to play an important role. Binding of AUF1/hnRNP D in the Cp, PU.1 in LMP-1p and L2BF2 in the LMP-2Ap turned out to be essential for both EBNA-2 and Notch1-IC mediated transactivation. By stably transfecting different B cell lines with Notch1-IC or Notch2-IC or by stimulating the endogenous Notch with ligand the known EBNA-2 target genes CD21, CD23, LMP-1 and LMP-2A appeared to be activated by Notch signalling (Bash et al., 1999; Strobl et al., 2000; Gordadze et al., 2001; Hubmann et al., 2002). Interestingly, overexpression of Notch-IC in Burkitt's Lymphoma cells resulted in the downregulation of IgM and c-myc (Strobl et al., 2000), as had been described for EBNA-2. Although the mechanism leading to the EBNA-2 or Notch-IC mediated downregulation of genes is still unclear, it seems likely that a common pathway is used by both proteins to downregulate gene expression. In summary, as far as it has been studied, most target genes seem to be regulated by both, EBNA-2 and Notch-IC (Table2). But, since for most of these genes data had been obtained by using a constitutively active overexpressed Notch-IC the physiological relevance has still to be proven.

Can Notch-IC and EBNA-2 substitute each other?

To study, if Notch-IC is able to replace EBNA-2 in its ability to maintain B cell proliferation a Notch1-IC expression vector was stably introduced in an EBV-immortalized cell line, in which EBNA-2 function can be regulated by estrogen. After switching off EBNA-2's function, Notch-IC could only partially replace EBNA-2 in its ability to maintain B cell proliferation. If expressed together with a LMP-1 expression plasmid that allows EBNA-2-independent expression of LMP-1, proliferation could be maintained to a limited extent (Hofelmayr et al., 2001). Rescue of proliferation in the absence of EBNA-2 and constitutive LMP-1 could only be achieved if extremely high expression levels of Notch1-IC could be selected by using a bicistronic expression system. Only cells with high levels of Notch1-IC expressed the target genes c-myc and LMP-1 (Gordadze et al., 2001). These data stress both, the similarities and the differences between EBNA-2 and Notch1-IC. They underline that Notch1-IC and EBNA-2 indeed utilize the same or similar cellular pathways, but they also indicate that both proteins exhibit pronounced differences in their efficiency to activate EBNA-2 target genes, like *c-myc* and LMP-1. Vice versa, it was studied if EBNA-2 could replace Notch1-IC in its function to suppress the differentiation of C2C12 myoblast progenitor cells. EBNA-2 indeed was able to induce Hes-1 expression and to suppress myoblast differentiation in C2C12 cells in a similar fashion as Notch1-IC (Sakai et al., 1998).

THE ROLE OF NOTCH SIGNALLING DURING HAEMATOPOIESIS

The data summarized above support the notion that EBNA-2 uses the Notch signalling pathway to induce B cell immortalisation. It has been shown in mice that constitutively active Notch (Notch-IC) leads to proliferation of T cells, but up to now there is no experimental proof that Notch-IC is also able to drive B cell proliferation *in vivo*. We will discuss now what is known about the function of Notch during lymphocyte development.

T/B lineage decision in common lymphoid progenitor cells (CLP)

The best characterized function of Notch signalling in haematopoiesis is its role in the development of T cells (Figure 4). Both gain and loss of function experiments show that Notch1 plays a critical role in the cell fate decision between the B and T cell lineages. In transgenic mice inducible deletion of Notch1 or RBP-Jκ in common lymphoid progenitor cells led to a complete block in T cell development at a very early stage. Instead of T cells, early stages of B cells appeared in the thymus (Radtke et al., 1999; Wilson et al., 2001; Han et al., 2002). In contrast to Notch1, neither Notch2 nor Notch3 deficiency impaired T cell development (Krebs et al., 2003b; Saito et al., 2003). Gain of function experiments using mice reconstituted with bone marrow (BM) expressing constitutive active Notch1 or the Notch ligand Dll4 resulted in the reciprocal phenotype. B cell development was blocked in the BM, instead, T cells which were mainly CD4+ and CD8+ double positive (DP), developed thymus independently in the BM (Pui et al., 1999; Yan et al., 2001).

These results indicate that Notch1 is controlling the fate of a bipotential B/T precursor. Notch1 signalling has to be suppressed in the BM to allow B cell development and has to be activated in lymphoid precursor cells to generate T cells. Whether the commitment to T cells occurs in the BM or as soon as the progenitor cell colonizes the thymus is still unclear. In the thymus a Notch1 signal prevents common lymphoid precursor cells from adopting a B cell fate.

Since Notch signals promote T-cell development and block B cell development, Notch signalling must be absent or negatively regulated during B cell development in the BM. Notch1 and Notch2 are expressed on progenitor cells and the Notch ligands jagged1, jagged2, Dll1, Dll3 and Dll4 on BM stromal cells. It is therefore likely that Notch signals in common lymphoid progenitor cells have to be negatively regulated to suppress T-cell development in the BM. Two molecules, which have been shown to negatively regulate Notch signalling in lymphoid progenitor cells, are Deltex-1 and NRARP (Izon et al., 2002; Yun and Bevan, 2003). But since these results are based on gain of function studies, the physiological role of these negative regulators has still to be shown. Another molecule suspected to antagonize Notch1 is Pax-5. Pax5 drives B cell development in the BM by transcriptionally downregulating Notch1 expression in lymphoid progenitor cells (Souabni et al., 2002).

Notch in T cells

After commitment to T cells the early T cell development takes place in the thymus. T cells develop from the DN (double negative) stage (CD4-, CD8-) to the DP (double positive) - (CD4+, CD8+) and finally to the SP (single positive) stage

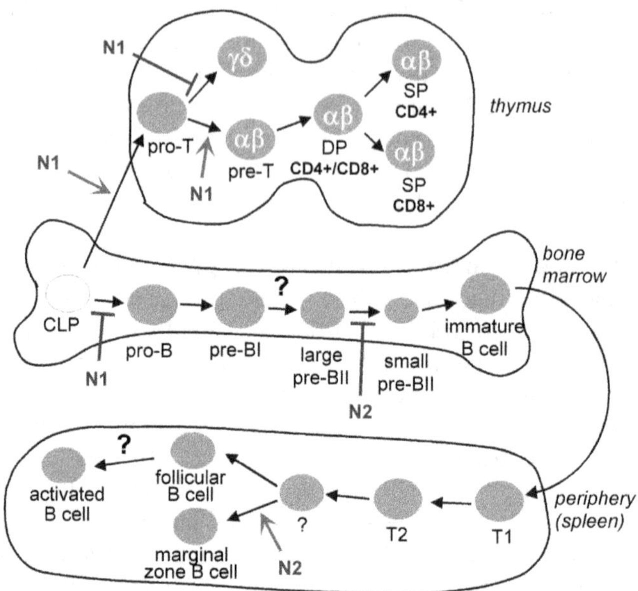

Figure 4. Notch in lymphocyte development. Notch1 signalling (N1) induces the differentiation of common lymphoid progenitor cells (CLP) to pro T cells. Notch1 signals in pro T cells inhibit the generation and emigration of γ/δ T cells and induce further differentiation of pre T cells of the α/β lineage. Double positive (DP) T cells, expressing CD4 and CD8, mature to CD4 or CD8 single positive (SP) T cells, which migrate to the periphery. Notch signals have to be absent in the bone marrow allowing CLPs to differentiate to the B cell lineage. Rearrangement of the immunoglobulin heavy chain starts in pro B cells. Large cycling pre BII cells express the pre B cell receptor on the surface. In small resting pre BII cells the pre BCR is down regulated and the immunoglobulin light chains are rearranged. Overexpression of Notch2-IC (N2) blocks B cell development at the large pre-BII stage. Immature B cells, which have rearranged the heavy and light chains and express IgM on the surface, leave the bone marrow and migrate as so-called transitional B cells (T1, T2) to the spleen, where they differentiate to follicular or MZB cells. In the spleen Notch2 signals induce the differentiation of marginal zone B cells. The progenitor cell receiving the Notch signal is unknown (?). In addition it is not known whether Notch signals play a role during early B cell development in the bone marrow or in the B cell activation process (?). Thus, there are still gaps in our detailed understanding of the physiological role of Notch signalling in early and late B cell development.

(CD4+ or CD8+). During the DN stage the decision is taken to enter the α/β or γ/δ lineage. Some data suggest the involvement of Notch in the α/β versus γ/δ lineage decision. Hemizygous Notch1+/- BM precursors give rise to relatively more γ/δ than α/β cells after transplantation in irradiated mice (Washburn et al., 1997). The group of Honjo recently reported that Notch signalling inhibits the generation and emigration of γ/δ T cells (Tanigaki et al., 2004). Cells, which are committed to the α/β lineage depend for further differentiation again on Notch signalling. Notch1-deficient T cells can no longer rearrange the TCR-β chain and therefore fail to express a pre-TCR on the cell surface and to undergo further differentiation

(Wolfer et al., 2002). The last step during T cell development is the differentiation from the DP- to the SP-stage, where a cell has to adopt a CD4+ or a CD8+ SP cell fate. Notch does not seem to be involved in this decision, although there has been an ongoing discussion on this issue for several years (Wolfer et al., 2001; Tanigaki et al., 2004). Several functions of Notch have been discussed in differentiation of peripheral T cells, which are beyond the scope of this review. We would like to refer to other reviews for this topic (Maillard et al., 2003; Radtke et al., 2004a).

Notch in B cells

In contrast to the function of Notch in T cell development, contribution of Notch to B cell development is only poorly understood (Figure 4). B cell development can be subdivided in early B cell development, which takes place in the bone marrow (BM), and late B cell development, taking place in the secondary lymphoid organs.

In the BM B cells differentiate from pro-B to pre-B and finally to immature B cells. The rearrangement of the immunoglobulin heavy chain (IgH) genes starts in late pro-B cells. Pre-B cells, which have successfully rearranged the IgH locus, express the pre-B cell receptor consisting of a heavy chain and the surrogate light chain. The pre-B cells are subdivided in two stages, the so-called early large cycling pre-B cells and late small resting pre-B cells, which start to rearrange the immunoglobulin light chain (IgL) genes. Immature B cells, which have successfully rearranged the immunoglobulin light chain genes and express surface IgM, exit the BM and migrate as so-called transitional B cells to the spleen. In the spleen the transitional B cells differentiate to follicular B cells (FoB) or marginal zone B cells (MZB) (Figure 4). FoB cells recirculate between the blood, secondary lymphoid organs, and BM. They respond to T-cell dependent antigens and form the germinal centres, where they undergo class switching and hypermutation to differentiate into plasma cells secreting high affinity antibodies or into memory B cells. MZB cells are localized in the so-called marginal zone surrounding the white pulp in the spleen. MZB cells are more responsive to LPS and differentiate more rapidly to antibody secreting cells than FoB cells. MZB cells are mainly involved in the T cell independent response to antigen and seem to play a crucial role in the defence against blood borne bacteria (for review (Martin and Kearney, 2002).

Notch function in early B cells

Several groups have studied the expression of the Notch genes in the BM. Even though the data describing the expression levels of Notch1 and Notch2 in the course of early B cell differentiation are contradictory, all investigators agree that Notch1 and Notch2 are expressed on early B cell subsets (Walker et al., 2001; Saito et al., 2003; Witt et al., 2003). Only Bertrand and coauthors who studied Notch expression not in murine but human B cell development claimed that Notch1 is expressed during all stages, while Notch2 has been detected only in a very sharp window in pre-B cells (Bertrand et al., 2000).

Although Notch1 and Notch2 are expressed on early B cell subsets and the Notch ligands jagged1, jagged2 and Dll1 are detected on BM-stromal cells, no clear function could be assigned to Notch signalling during early B cell development up to now. Neither inactivation of Notch1, Notch2 nor RBP-Jκ impaired B cell development in the BM. Gain of function studies revealed that Notch1-IC and Notch2-IC block B cell development at two different stages. Whereas expression of Notch1-IC resulted in a very early block of B cell differentiation before the pro-B cell stage, Notch2-IC blocked B cell development in large pre-B cells (Radtke et al., 1999; Witt et al., 2003). Whether this difference is due to different expression levels of Notch1-IC and Notch2-IC after retroviral delivery, or to functional differences between these two molecules, is still an open question. It has been proposed that large pre-B cells have to exit from the cell cycle to progress from the large pre-B to the small pre-B stage. This could imply that Notch signalling has to be down modulated between the large and small pre-B cell stage to allow further differentiation.

Notch function in late B cells

B cell specific inactivation of Notch2 and RBP-Jκ results in the loss of MZB cells implying that Notch2 signalling is involved in the generation of MZB cells. (Tanigaki et al., 2002; Saito et al., 2003). Very recently, it was shown that inducible inactivation of the Notch ligand Dll1 in all cell types of the hematopoetic system also results in the loss of MZB cells, whereas the development of T cells is not impaired (Hozumi et al., 2004). This suggests that a specific receptor ligand interaction between Notch2 and Dll1 determines MZB development and that Dll1 cannot be replaced by other ligands which have been shown to be expressed in the spleen such as Dll4 and jagged1 (Bash et al., 1999; Kuroda et al., 2003). B cell specific inactivation of Dll1 did not result in the disappearance of MZB cells indicating that ligand-receptor interaction is taking place between MZB precursor cells and non-B cells, potentially dendritic cells which have been shown to express Dll1 and Dll4. Inactivation of MINT, which interacts with RBP-Jκ thereby displacing Notch-IC from the complex, resulted in an increased number of MZBs (Kuroda et al., 2003). This finding implies that MINT acts as negative regulator of Notch signalling during the development of MZB cells. Summarizing these results, the only well-characterized function of Notch in B cells up to now is its contribution to MZB cell development. It remains to be demonstrated, whether gain of function proteins like Notch-IC and EBNA-2 are driving B cells to the marginal zone.

THE ROLE OF NOTCH IN LEUKEMO- AND LYMPHOMAGENESIS

The oncogenic potential of Notch-IC is clearly defined in T cell acute lymphoblastic lymphoma/leukaemia, but a direct oncogenic effect of Notch-IC in B cells, reminiscent of the function of EBNA-2 during B cell immortalization, has not been described so far. An overview of tumors induced by an activated Notch is given in Table 3.

Overexpression of Notch-IC in the development of T cell leukemias and lymphomas

The human Notch1 gene was first described in the context of a chromosomal t(7;9) translocation in a case of acute T cell leukemia (T-ALL) (Ellisen et al., 1991). Through this chromosomal translocation the active part of Notch1 is fused to the TCR-β locus resulting in overexpression of a constitutive active form of Notch1 (Ellisen et al., 1991). Initially, Notch1-IC had been called TAN-1 for translocation

Table 3. Involvement of Notch in tumor development

Activation of Notch	Induced tumors	Reference
Activation of Notch1-IC by chromosomal translocation TAN-1 t(7;9)	Acute T cell leukaemia in humans	Ellisen et al., 1991
Retroviral transduction of TAN-1 in BM cells	T cell leukaemia/ lymphoma in mice	Pear et al., 1996
Retroviral transduction of Notch1-IC in BM-cells	T cell leukaemia/ lymphoma in mice	Aster et al., 2000
Feline leukaemia virus carrying Notch2-IC	Thymic lymphomas in cats	Rohn et al., 1996
Transgenic mice expressing Notch3-IC under the control of the lck-promoter	T cell-lymphomas in mice	Bellavia et al., 2000
Activation of Notch4 by retroviral insertion	Mammary adenocarcinoma in mice	Gallahan and Callahan, 1997
Retroviral transduction of the Notch-ligand Dll-4 in BM cells	T cell leukaemia/ lymphoma in mice	Yan et al., 2001; Dorsch et al., 2002

associated Notch homolog. Experimental proof that TAN-1 is the causative agent of the leukaemia was provided by Pear et al., who reconstituted lethally irradiated mice with progenitor cells that had been retrovirally transduced with TAN-1 (Pear et al., 1996). Overexpression of truncated forms of Notch3 can also cause acute T cell leukaemias/lymphomas (Bellavia et al., 2000). Furthermore, the murine Notch4 gene has been cloned as a gene activated by retroviral insertion in mammary adenocarcinomas in mice (Gallahan and Callahan, 1997). Notch2 is also involved in lymphomagenesis, as a Feline Leukemia virus carrying a Notch2-IC gene has been isolated from a thymic lymphoma in cats (Rohn et al., 1996). Recently, it was reported that retroviral transduction of the Notch ligand Dll4 also results in the development of lymphoproliferative disorder or the development of T cell leukaemia/lymphoma (Yan et al., 2001; Dorsch et al., 2002).

To learn more about the mechanism, by which Notch-IC contributes to tumour development, different deletion mutants of Notch1-IC have been tested for their ability to induce T cell leukaemia after transplantation of retrovirally transduced BM (Aster et al., 2000). The transforming activity of Notch1-IC appears to correlate with the transactivation ability. Only mutants which could efficiently transactivate the Hes-1 promoter were able to cause leukaemia. The minimal region of Notch1-IC required to induce T cell leukaemogenesis, was defined as the ankyrin repeats and the transactivation domain. This indicates that recruitment of transcriptional coactivators to the promoters of Notch regulated genes plays an essential role in T-cell leukaemogenesis. In a T-ALL cell line carrying a t(7;9) translocation, which leads to the overexpression of Notch1-IC a high molecular weight complex has recently been described consisting of Notch-IC, MAML1 and RBP-Jκ. Dominant negative forms of MAML1, which inhibit the recruitment of coactivators in this complex induced G1 arrest and apoptosis in human and murine T-ALL supporting the notion that expression of Notch regulated genes plays an essential role in the development of T-cell leukaemia (Jeffries et al., 2002; Weng et al., 2003).

It is striking that transplantation experiments with BM cells that have been retrovirally transduced with Notch-IC and Dll4 resulted exclusively in the development of T-cell leukaemia and T cell lymphoma, whereas tumours of B lymphoid or myeloid origin have never been reported. The proliferating T cells mainly have a CD4+, CD8+ (DP) phenotype and only transgenic mice expressing Notch3-IC under the control of the lck promoter developed tumours with the phenotype of late immature DN T cells (Pear et al., 1996; Bellavia et al., 2000). It is tempting to speculate that Notch-IC upregulates a set of transforming target genes in T cells or cooperates with some T cell specific genes in inducing T cell leukaemia. It has been shown recently that expression of the pre-TCR is required for T cell leukemogenesis (Allman et al., 2001; Bellavia et al., 2002). These results imply that the transforming capacity of Notch-IC in T cells is limited to immature T cells which have rearranged the TCR-β chain. It may be speculated that activated Notch cooperates with a preTCR signal for expansion and transformation of T cells.

Notch and B cell neoplasias

In contrast to T cells, translocations leading to the expression of constitutive active Notch molecules have not been described in B cells. This could imply that overexpression of Notch-IC might be disadvantageous for B cells at some developmental stages. The risk of B cells to acquire translocations is high in BM progenitor cells, where the recombination machinery for immunoglobulin rearrangement is active, and in germinal centre cells, where hypermutation and class switching take place. It has been shown that overexpression of Notch1-IC and Notch2-IC in BM progenitor cells results in a block of B cell development. B cells carrying a translocation that activates Notch might thus be eliminated in the BM. The impact of a constitutive active Notch signal in germinal centre cells has not been studied so far. However, as EBNA-2 is usually not expressed in germinal centre cells, it is conceivable that constitutive Notch signalling may be deleterious for B cells at this differentiation stage.

Although chromosomal translocations leading to a constitutive active Notch signal have not been described in B cells, two studies have emphasized that Notch signalling might play a role in the development of some B cell malignancies. Hubmann et al. reported that Notch2 is highly expressed in B cell chronic lymphocytic leukaemia (B-CLL). In gel retardation assays they were able to detect Notch2-IC in a complex at the CD23a promoter suggesting that Notch2 is responsible for high expression of CD23 in CLL (Hubmann et al., 2002). Jundt et al. suggested that Notch plays an important role in the pathogenesis of Hodgkin's Disease (HD)(Jundt et al., 2002). They showed that Notch1 as well as Notch2 are highly expressed in B cell derived Hodgkin's lymphoma. They could further show the expression of the Notch ligand jagged1 on bystander cells surrounding the tumour. Activation of Notch by the ligand jagged1 induces proliferation and survival of cultured HD cells (Jundt et al., 2002). These two reports provide some evidence that dysregulation of Notch expression may be involved in the development of some B cell lymphomas. However, it is still unclear, whether the high Notch expression level is involved in tumour development or if it is only a marker for HD and B-CLL.

Open questions

No obvious phenotype has been observed so far in RBP-Jκ deficient B cells during B cell activation and germinal centre differentiation. Likewise, no change in immunoglobulin production has been noted in response to T cell dependent and independent antigens after B cell specific inactivation of RBP-Jκ (Tanigaki et al., 2002). It has therefore remained an enigma why EBNA-2 usurps a cellular pathway for B cell activation and induction of proliferation that does not seem to play a major role in B cell physiology. One possible explanation for this discrepancy may be that EBNA-2 is not causing B cell activation and proliferation on its own. EBNA-2 induces the viral oncogenes LMP-1 and LMP-2 that mimic T cell help by an activated CD40 receptor and an antigen-activated B cell receptor, respectively. EBNA-2 thus induces a network of genes in concert with its downstream viral target genes LMP-1 and LMP-2 to mediate B cell immortalization. The obvious question arises whether the role of EBNA-2 in B cell immortalisation is restricted to the induction of LMP-1 and LMP-2, or whether EBNA-2 contributes additionally by activating a number of important cellular target genes or by inactivation of growth inhibitory pathways. We have shown that constitutive expression of LMP-1 is unable to rescue B cell immortalisation in the absence of EBNA-2 (Zimber-Strobl et al., 1996), but the question has not been addressed whether LMP-1 and LMP-2 together are able to rescue B cell proliferation in the absence of functional EBNA-2.

It will also be important to better understand the impact of EBNA-2 and activated Notch on B cell differentiation and activation in the absence and presence of the other viral oncogenes in transgenic mice. Notch signalling has recently been shown also to be involved in T-cell receptor (TCR) mediated activation of peripheral T cells and to be required for TCR-dependent proliferation (Palaga et al., 2003). It will be interesting to see whether the same holds true also for B cells.

It is another open question, whether endogenous Notch signalling may substitute for EBNA-2 at some stages of the viral infection *in vivo*. All viral genes expressed in immortalized B cells are under control of promoters containing RBP-Jκ binding sites, suggesting that they may be regulated by EBNA-2 as well as Notch. Although definitive experimental proof that the viral promoters can be activated by ligand induced Notch signalling is still missing, it is tempting to speculate that the endogenous Notch pathway may be used to activate viral promoters in the absence of EBNA-2. *In vivo*, proliferation of EBV-infected B cells is limited, because the cells are rapidly attacked and eliminated by cytotoxic T cells that recognize the viral proteins expressed in EBV-immortalized B cells. Only some cells, which shut down the expression of viral proteins, can escape the immune response and establish latency in memory B cells. According to the model proposed by the group of Thorley-Lawson (Babcock et al., 2000), EBV-infected B cells use the normal differentiation program of B cells to shut down the viral proteins and to get access to the memory B cell compartment. Thorley-Lawson and collaborators have studied EBV-gene expression in B cells with germinal centre phenotype and could show that viral gene expression is restricted to EBNA-1, LMP-1 and LMP-2A. One may speculate that an activated Notch signal might upregulate LMP-1 and LMP-2 in these cells.

In addition, it is conceivable that Notch signalling might also be responsible for induction of LMP-1 and LMP-2A expression in EBNA-2-negative tumour cells. Usually EBNA-2 is not expressed in tumours of immunocompetent hosts like Hodgkin's Disease and Naspopharyngeal carcinomas. For example in Hodgkin's lymphoma, 40% of the cases are EBV positive, yet EBNA-1, LMP-1 and LMP-2A are expressed independently of EBNA-2. For EBNA-1 an alternative EBNA-2 independent promoter has been described. It is, however, still elusive how LMP-1 and LMP-2A are expressed in these cells. Since Notch1 and Notch2 appear to be highly expressed in Hodgkin's lymphoma, it is conceivable that Notch signalling takes over the function of EBNA-2 in upregulating LMP-1 and LMP-2A.

Acknowledgments

We thank Nathalie Uyttersprot, Hella Kohlhof and Sabine Maier for critical reading of the manuscript. This work was supported by the Deutsche Forschungsgemeinschaft through SFB455 and through Grant STR 461/3-2 and by the Deutsche Krebshilfe (10-1963-Ke I) and Wilhelm Sander-Stiftung (2003.143.1).

References

Abbot, S. D., Rowe, M., Cadwallader, K., Ricksten, A., Gordon, J., Wang, F., Rymo, L., and Rickinson, A. B. (1990). Epstein-Barr virus nuclear antigen 2 induces expression of the virus-encoded latent membrane protein. J. Virol. *64*, 2126-2134.

Alazard, N., Gruffat, H., Hiriart, E., Sergeant, A., and Manet, E. (2003). Differential hyperacetylation of histones H3 and H4 upon promoter-specific recruitment of EBNA2 in Epstein-Barr virus chromatin. J. Virol. *77*, 8166-8172.

Allman, D., Karnell, F. G., Punt, J. A., Bakkour, S., Xu, L., Myung, P., Koretzky, G. A., Pui, J. C., Aster, J. C., and Pear, W. S. (2001). Separation of Notch1 promoted lineage commitment and expansion/transformation in developing T cells. J. Exp. Med. *194*, 99-106.

Ansieau, S., and Leutz, A. (2002). The conserved Mynd domain of BS69 binds cellular and oncoviral proteins through a common PXLXP motif. J. Biol. Chem. *277*, 4906-4910.

Aster, J. C., Xu, L., Karnell, F. G., Patriub, V., Pui, J. C., and Pear, W. S. (2000). Essential roles for ankyrin repeat and transactivation domains in induction of T-cell leukemia by notch1. Mol. Cell Biol. *20*, 7505-7515.

Babcock, G. J., Hochberg, D., and Thorley-Lawson, A. D. (2000). The expression pattern of Epstein-Barr virus latent genes *in vivo* is dependent upon the differentiation stage of the infected B cell. Immunity *13*, 497-506.

Barth, S., Liss, M., Voss, M. D., Dobner, T., Fischer, U., Meister, G., and Grasser, F. A. (2003). Epstein-Barr virus nuclear antigen 2 binds via its methylated arginine-glycine repeat to the survival motor neuron protein. J. Virol. *77*, 5008-5013.

Bash, J., Zong, W. X., Banga, S., Rivera, A., Ballard, D. W., Ron, Y., and Gelinas, C. (1999). Rel/NF-kappaB can trigger the Notch signaling pathway by inducing the expression of Jagged1, a ligand for Notch receptors. Embo J. *18*, 2803-2811.

Bellavia, D., Campese, A. F., Alesse, E., Vacca, A., Felli, M. P., Balestri, A., Stoppacciaro, A., Tiveron, C., Tatangelo, L., Giovarelli, M., et al. (2000). Constitutive activation of NF-kappaB and T-cell leukemia/lymphoma in Notch3 transgenic mice. Embo J. *19*, 3337-3348.

Bellavia, D., Campese, A. F., Checquolo, S., Balestri, A., Biondi, A., Cazzaniga, G., Lendahl, U., Fehling, H. J., Hayday, A. C., Frati, L., et al. (2002). Combined expression of pTalpha and Notch3 in T cell leukemia identifies the requirement of preTCR for leukemogenesis. Proc. Natl. Acad. Sci. USA. *99*, 3788-3793.

Bertrand, F. E., Eckfeldt, C. E., Lysholm, A. S., and LeBien, T. W. (2000). Notch-1 and Notch-2 exhibit unique patterns of expression in human B-lineage cells. Leukemia *14*, 2095-2102.

Bessho, Y., Miyoshi, G., Sakata, R., and Kageyama, R. (2001). Hes7: a bHLH-type repressor gene regulated by Notch and expressed in the presomitic mesoderm. Genes Cells *6*, 175-185.

Blaumueller, C. M., Qi, H., Zagouras, P., and Artavanis-Tsakonas, S. (1997). Intracellular cleavage of Notch leads to a heterodimeric receptor on the plasma membrane. Cell *90*, 281-291.

Borestrom, C., Zetterberg, H., Liff, K., and Rymo, L. (2003). Functional interaction of nuclear factor y and sp1 is required for activation of the epstein-barr virus C promoter. J. Virol. *77*, 821-829.

Bornkamm, G. W., and Hammerschmidt, W. (2001). Molecular virology of Epstein-Barr virus. Philos Trans R Soc. Lond B Biol. Sci. *356*, 437-459.

Bornkamm, G. W., Hudewentz, J., Freese, U. K., and Zimber, U. (1982). Deletion of the nontransforming Epstein-Barr virus strain P3HR-1 causes fusion of the large internal repeat to the DSL region. J. Virol. *43*, 952-968.

Brou, C., Logeat, F., Gupta, N., Bessia, C., LeBail, O., Doedens, J. R., Cumano, A., Roux, P., Black, R. A., and Israel, A. (2000). A novel proteolytic cleavage involved in Notch signaling: the role of the disintegrin-metalloprotease TACE. Mol. Cell *5*, 207-216.

Burgstahler, R., Kempkes, B., Steube, K., and Lipp, M. (1995). Expression of the chemokine receptor BLR2/EBI1 is specifically transactivated by Epstein-Barr virus nuclear antigen 2. Biochem. Biophys Res. Commun. *215*, 737-743.

Calender, A., Cordier, M., Billaud, M., and Lenoir, G. M. (1990). Modulation of cellular gene expression in B lymphoma cells following *in vitro* infection by Epstein-Barr virus (EBV). Int. J. Cancer *46*, 658-663.

Callahan, J., Aster, J., Sklar, J., Kieff, E., and Robertson, E. S. (2000). Intracellular forms of human NOTCH1 interact at distinctly different levels with RBP-jkappa in human B and T cells. Leukemia *14*, 84-92.

Chen, Y., Fischer, W. H., and Gill, G. N. (1997). Regulation of the ERBB-2 promoter by RBPJkappa and NOTCH. J. Biol. Chem. *272*, 14110-14114.

Chung, C. N., Hamaguchi, Y., Honjo, T., and Kawaichi, M. (1994). Site-directed mutagenesis study on DNA binding regions of the mouse homologue of Suppressor of Hairless, RBP-J kappa. Nucleic Acids Res. *22*, 2938-2944.

Cohen, J. I. (1992). A region of herpes simplex virus VP16 can substitute for a transforming domain of Epstein-Barr virus nuclear protein 2. Proc. Natl. Acad. Sci. USA. *89*, 8030-8034.

Cohen, J. I., and Kieff, E. (1991). An Epstein-Barr virus nuclear protein 2 domain essential for transformation is a direct transcriptional activator. J. Virol. *65*, 5880-5885.

Cohen, J. I., Wang, F., and Kieff, E. (1991). Epstein-Barr virus nuclear protein 2 mutations define essential domains for transformation and transactivation. J. Virol. *65*, 2545-2554.

Cohen, J. I., Wang, F., Mannick, J., and Kieff, E. (1989). Epstein-Barr virus nuclear protein 2 is a key determinant of lymphocyte transformation. Proc. Natl. Acad. Sci. USA. *86*, 9558-9562.

Cordier, M., Calender, A., Billaud, M., Zimber, U., Rousselet, G., Pavlish, O., Banchereau, J., Tursz, T., Bornkamm, G., and Lenoir, G. M. (1990). Stable transfection of Epstein-Barr virus (EBV) nuclear antigen 2 in lymphoma cells containing the EBV P3HR1 genome induces expression of B-cell activation molecules CD21 and CD23. J. Virol. *64*, 1002-1013.

Cordier-Bussat, M., Billaud, M., Calender, A., and Lenoir, G. M. (1993). Epstein-Barr virus (EBV) nuclear-antigen-2-induced up-regulation of CD21 and CD23 molecules is dependent on a permissive cellular context. Int. J. Cancer *53*, 153-160.

Cotter, M., Callahan, J., Aster, J., and Robertson, E. (2000). Intracellular forms of human NOTCH1 functionally activate essential Epstein-Barr virus major latent promoters in the Burkitt's lymphoma BJAB cell line but repress these promoters in Jurkat cells. J. Virol. *74*, 1486-1494.

De Strooper, B., Annaert, W., Cupers, P., Saftig, P., Craessaerts, K., Mumm, J. S., Schroeter, E. H., Schrijvers, V., Wolfe, M. S., Ray, W. J., et al. (1999). A presenilin-1-dependent gamma-secretase-like protease mediates release of Notch intracellular domain. Nature *398*, 518-522.

Deftos, M. L., Huang, E., Ojala, E. W., Forbush, K. A., and Bevan, M. J. (2000). Notch1 signaling promotes the maturation of CD4 and CD8 SP thymocytes. Immunity *13*, 73-84.

Dorsch, M., Zheng, G., Yowe, D., Rao, P., Wang, Y., Shen, Q., Murphy, C., Xiong, X., Shi, Q., Gutierrez-Ramos, J. C., et al. (2002). Ectopic expression of Delta4 impairs hematopoietic development and leads to lymphoproliferative disease. Blood *100*, 2046-2055.

Dumont, E., Fuchs, K. P., Bommer, G., Christoph, B., Kremmer, E., and Kempkes, B. (2000). Neoplastic transformation by Notch is independent of transcriptional activation by RBP-J signalling. Oncogene *19*, 556-561.

Ellisen, L. W., Bird, J., West, D. C., Soreng, A. L., Reynolds, T. C., Smith, S. D., and Sklar, J. (1991). TAN-1, the human homolog of the Drosophila notch gene, is broken by chromosomal translocations in T lymphoblastic neoplasms. Cell *66*, 649-661.

Evans, T. J., Farrell, P. J., and Swaminathan, S. (1996). Molecular genetic analysis of Epstein-Barr virus Cp promoter function. J. Virol. *70*, 1695-1705.

Fahraeus, R., Jansson, A., Ricksten, A., Sjoblom, A., and Rymo, L. (1990). Epstein-Barr virus-encoded nuclear antigen 2 activates the viral latent membrane protein promoter by modulating the activity of a negative regulatory element. Proc. Natl. Acad. Sci. USA. *87*, 7390-7394.

Fearon, D. T., and Carroll, M. C. (2000). Regulation of B lymphocyte responses to foreign and self-antigens by the CD19/CD21 complex. Annu. Rev. Immunol. *18*, 393-422.

Floettmann, J. E., Ward, K., Rickinson, A. B., and Rowe, M. (1996). Cytostatic effect of Epstein-Barr virus latent membrane protein-1 analyzed using tetracycline-regulated expression in B cell lines. Virology *223*, 29-40.

Folk, P., Puta, F., and Skruzny, M. (2004). Transcriptional coregulator SNW/SKIP: the concealed tie of dissimilar pathways. Cell Mol. Life Sci. *61*, 629-640.

Fuchs, K. P., Bommer, G., Dumont, E., Christoph, B., Vidal, M., Kremmer, E., and Kempkes, B. (2001). Mutational analysis of the J. recombination signal sequence binding protein (RBP-J)/Epstein-Barr virus nuclear antigen 2 (EBNA2) and RBP-J/Notch interaction. Eur. J. Biochem. *268*, 4639-4646.

Fuentes-Panana, E. M., and Ling, P. D. (1998). Characterization of the CBF2 binding site within the Epstein-Barr virus latency C promoter and its role in modulating EBNA2-mediated transactivation. J. Virol. *72*, 693-700.

Fuentes-Panana, E. M., Peng, R., Brewer, G., Tan, J., and Ling, P. D. (2000). Regulation of the Epstein-Barr virus C promoter by AUF1 and the cyclic AMP/protein kinase A signaling pathway. J. Virol. *74*, 8166-8175.

Fuentes-Panana, E. M., Swaminathan, S., and Ling, P. D. (1999). Transcriptional activation signals found in the Epstein-Barr virus (EBV) latency C promoter are conserved in the latency C promoter sequences from baboon and Rhesus monkey EBV-like lymphocryptoviruses (cercopithicine herpesviruses 12 and 15). J. Virol. *73*, 826-833.

Gallahan, D., and Callahan, R. (1997). The mouse mammary tumor associated gene INT3 is a unique member of the NOTCH gene family (NOTCH4). Oncogene *14*, 1883-1890.

Ghosh, D., and Kieff, E. (1990). cis-acting regulatory elements near the Epstein-Barr virus latent-infection membrane protein transcriptional start site. J. Virol. *64*, 1855-1858.

Gordadze, A. V., Onunwor, C. W., Peng, R., Poston, D., Kremmer, E., and Ling, P. D. (2004). EBNA2 amino acids 3 to 30 are required for induction of LMP-1 and immortalization maintenance. J. Virol. *78*, 3919-3929.

Gordadze, A. V., Peng, R., Tan, J., Liu, G., Sutton, R., Kempkes, B., Bornkamm, G. W., and Ling, P. D. (2001). Notch1IC partially replaces EBNA2 function in B cells immortalized by Epstein-Barr virus. J. Virol. *75*, 5899-5912.

Gordadze, A. V., Poston, D., and Ling, P. D. (2002). The EBNA2 polyproline region is dispensable for Epstein-Barr virus-mediated immortalization maintenance. J. Virol. *76*, 7349-7355.

Grasser, F. A., Haiss, P., Gottel, S., and Mueller-Lantzsch, N. (1991). Biochemical characterization of Epstein-Barr virus nuclear antigen 2A. J. Virol. *65*, 3779-3788.

Grossman, S. R., Johannsen, E., Tong, X., Yalamanchili, R., and Kieff, E. (1994). The Epstein-Barr virus nuclear antigen 2 transactivator is directed to response elements by the J. kappa recombination signal binding protein. Proc. Natl. Acad. Sci. USA. *91*, 7568-7572.

Grundhoff, A. T., Kremmer, E., Tureci, O., Glieden, A., Gindorf, C., Atz, J., Mueller-Lantzsch, N., Schubach, W. H., and Grasser, F. A. (1999). Characterization of DP103, a novel DEAD box protein that binds to the Epstein-Barr virus nuclear proteins EBNA2 and EBNA3C. J. Biol. Chem. *274*, 19136-19144.

Gupta-Rossi, N., Le Bail, O., Gonen, H., Brou, C., Logeat, F., Six, E., Ciechanover, A., and Israel, A. (2001). Functional interaction between SEL-10, an F-box protein, and the nuclear form of activated Notch1 receptor. J. Biol. Chem. *276*, 34371-34378.

Hammerschmidt, W., and Sugden, B. (1989). Genetic analysis of immortalizing functions of Epstein-Barr virus in human B lymphocytes. Nature *340*, 393-397.

Han, H., Tanigaki, K., Yamamoto, N., Kuroda, K., Yoshimoto, M., Nakahata, T., Ikuta, K., and Honjo, T. (2002). Inducible gene knockout of transcription factor recombination signal binding protein-J reveals its essential role in T versus B lineage decision. Int. Immunol. *14*, 637-645.

Harada, S., and Kieff, E. (1997). Epstein-Barr virus nuclear protein LP stimulates EBNA-2 acidic domain-mediated transcriptional activation. J. Virol. *71*, 6611-6618.

Harada, S., Yalamanchili, R., and Kieff, E. (1998). Residues 231 to 280 of the Epstein-Barr virus nuclear protein 2 are not essential for primary B-lymphocyte growth transformation. J. Virol. *72*, 9948-9954.

Harada, S., Yalamanchili, R., and Kieff, E. (2001). Epstein-Barr virus nuclear protein 2 has at least two N-terminal domains that mediate self-association. J. Virol. *75*, 2482-2487.

Hateboer, G., Gennissen, A., Ramos, Y. F., Kerkhoven, R. M., Sonntag-Buck, V., Stunnenberg, H. G., and Bernards, R. (1995). BS69, a novel adenovirus E1A-associated protein that inhibits E1A transactivation. Embo J. *14*, 3159-3169.

Henkel, T., Ling, P. D., Hayward, S. D., and Peterson, M. G. (1994). Mediation of Epstein-Barr virus EBNA2 transactivation by recombination signal-binding protein J. kappa. Science *265*, 92-95.

Hicks, C., Johnston, S. H., diSibio, G., Collazo, A., Vogt, T. F., and Weinmaster, G. (2000). Fringe differentially modulates Jagged1 and Delta1 signalling through Notch1 and Notch2. Nat. Cell Biol. *2*, 515-520.

Hinuma, Y., Konn, M., Yamaguchi, J., Wudarski, D. J., Blakeslee, J. R., Jr., and Grace, J. T., Jr. (1967). Immunofluorescence and herpes-type virus particles in the P3HR-1 Burkitt lymphoma cell line. J. Virol. *1*, 1045-1051.

Hofelmayr, H., Strobl, L. J., Marschall, G., Bornkamm, G. W., and Zimber-Strobl, U. (2001). Activated Notch1 can transiently substitute for EBNA2 in the maintenance of proliferation of LMP1-expressing immortalized B cells. J. Virol. *75*, 2033-2040.

Hofelmayr, H., Strobl, L. J., Stein, C., Laux, G., Marschall, G., Bornkamm, G. W., and Zimber-Strobl, U. (1999). Activated mouse Notch1 transactivates Epstein-Barr virus nuclear antigen 2-regulated viral promoters. J. Virol. *73*, 2770-2780.

Hozumi, K., Negishi, N., Suzuki, D., Abe, N., Sotomaru, Y., Tamaoki, N., Mailhos, C., Ish-Horowicz, D., Habu, S., and Owen, M. J. (2004). Delta-like 1 is necessary for the generation of marginal zone B cells but not T cells *in vivo*. Nat. Immunol. *5*, 638-644.

Hsieh, J. J., and Hayward, S. D. (1995). Masking of the CBF1/RBPJ kappa transcriptional repression domain by Epstein-Barr virus EBNA2. Science *268*, 560-563.

Hubmann, R., Schwarzmeier, J. D., Shehata, M., Hilgarth, M., Duechler, M., Dettke, M., and Berger, R. (2002). Notch2 is involved in the overexpression of CD23 in B-cell chronic lymphocytic leukemia. Blood *99*, 3742-3747.

Hudewentz, J., Bornkamm, G. W., and Zur Hausen, H. (1980). Effect of the diterpene ester TPA on Epstein-Barr virus antigen- and DNA synthesis in producer and nonproducer cell lines. Virology *100*, 175-178.

Huppert, S. S., Le, A., Schroeter, E. H., Mumm, J. S., Saxena, M. T., Milner, L. A., and Kopan, R. (2000). Embryonic lethality in mice homozygous for a processing-deficient allele of Notch1. Nature *405*, 966-970.

Iso, T., Chung, G., Hamamori, Y., and Kedes, L. (2002). HERP1 is a cell type-specific primary target of Notch. J. Biol. Chem. *277*, 6598-6607.

Iso, T., Kedes, L., and Hamamori, Y. (2003). HES and HERP families: multiple effectors of the Notch signaling pathway. J. Cell Physiol *194*, 237-255.

Iso, T., Sartorelli, V., Chung, G., Shichinohe, T., Kedes, L., and Hamamori, Y. (2001). HERP, a new primary target of Notch regulated by ligand binding. Mol. Cell Biol. *21*, 6071-6079.

Izon, D. J., Aster, J. C., He, Y., Weng, A., Karnell, F. G., Patriub, V., Xu, L., Bakkour, S., Rodriguez, C., Allman, D., and Pear, W. S. (2002). Deltex1 redirects lymphoid progenitors to the B cell lineage by antagonizing Notch1. Immunity *16*, 231-243.

Jarriault, S., Brou, C., Logeat, F., Schroeter, E. H., Kopan, R., and Israel, A. (1995). Signalling downstream of activated mammalian Notch. Nature *377*, 355-358.

Jeffries, S., Robbins, D. J., and Capobianco, A. J. (2002). Characterization of a high-molecular-weight Notch complex in the nucleus of Notch(ic)-transformed RKE cells and in a human T-cell leukemia cell line. Mol. Cell Biol. *22*, 3927-3941.

Jehn, B. M., Bielke, W., Pear, W. S., and Osborne, B. A. (1999). Cutting edge: protective effects of notch-1 on TCR-induced apoptosis. J. Immunol. *162*, 635-638.

Jin, X. W., and Speck, S. H. (1992). Identification of critical cis elements involved in mediating Epstein-Barr virus nuclear antigen 2-dependent activity of an enhancer located upstream of the viral BamHI C promoter. J. Virol. *66*, 2846-2852.

Jochner, N., Eick, D., Zimber-Strobl, U., Pawlita, M., Bornkamm, G. W., and Kempkes, B. (1996). Epstein-Barr virus nuclear antigen 2 is a transcriptional suppressor of the immunoglobulin mu gene: implications for the expression of the translocated c-myc gene in Burkitt's lymphoma cells. Embo J. *15*, 375-382.

Johannsen, E., Koh, E., Mosialos, G., Tong, X., Kieff, E., and Grossman, S. R. (1995). Epstein-Barr virus nuclear protein 2 transactivation of the latent membrane protein 1 promoter is mediated by J. kappa and PU.1. J. Virol. *69*, 253-262.

Johansen, L. M., Deppmann, C. D., Erickson, K. D., Coffin, W. F., 3rd, Thornton, T. M., Humphrey, S. E., Martin, J. M., and Taparowsky, E. J. (2003). EBNA2 and activated Notch induce expression of BATF. J. Virol. *77*, 6029-6040.

Jundt, F., Anagnostopoulos, I., Forster, R., Mathas, S., Stein, H., and Dorken, B. (2002). Activated Notch1 signaling promotes tumor cell proliferation and survival in Hodgkin and anaplastic large cell lymphoma. Blood *99*, 3398-3403.

Kaiser, C., Laux, G., Eick, D., Jochner, N., Bornkamm, G. W., and Kempkes, B. (1999). The proto-oncogene c-myc is a direct target gene of Epstein-Barr virus nuclear antigen 2. J. Virol. *73*, 4481-4484.

Kelly, G., Bell, A., and Rickinson, A. (2002). Epstein-Barr virus-associated Burkitt lymphomagenesis selects for downregulation of the nuclear antigen EBNA2. Nat. Med. *8*, 1098-1104.

Kempkes, B., Pawlita, M., Zimber-Strobl, U., Eissner, G., Laux, G., and Bornkamm, G. W. (1995a). Epstein-Barr virus nuclear antigen 2-estrogen receptor fusion proteins transactivate viral and cellular genes and interact with RBP-J kappa in a conditional fashion. Virology *214*, 675-679.

Kempkes, B., Spitkovsky, D., Jansen-Durr, P., Ellwart, J. W., Kremmer, E., Delecluse, H. J., Rottenberger, C., Bornkamm, G. W., and Hammerschmidt, W. (1995b). B-cell proliferation and induction of early G1-regulating proteins by Epstein-Barr virus mutants conditional for EBNA2. Embo J. *14*, 88-96.

Kempkes, B., Zimber-Strobl, U., Eissner, G., Pawlita, M., Falk, M., Hammerschmidt, W., and Bornkamm, G. W. (1996). Epstein-Barr virus nuclear antigen 2 (EBNA2)-oestrogen receptor fusion proteins complement the EBNA2-deficient Epstein-Barr virus strain P3HR1 in transformation of primary B cells but suppress growth of human B cell lymphoma lines. J. Gen. Virol. *77*, 227-237.

King, W., Dambaugh, T., Heller, M., Dowling, J., and Kieff, E. (1982). Epstein-Barr virus DNA XII. A variable region of the Epstein-Barr virus genome is included in the P3HR-1 deletion. J. Virol. *43*, 979-986.

Kitagawa, M., Oyama, T., Kawashima, T., Yedvobnick, B., Kumar, A., Matsuno, K., and Harigaya, K. (2001). A human protein with sequence similarity to Drosophila mastermind coordinates the nuclear form of notch and a CSL protein to build a transcriptional activator complex on target promoters. Mol. Cell Biol. *21*, 4337-4346.

Krebs, L. T., Deftos, M. L., Bevan, M. J., and Gridley, T. (2001). The Nrarp gene encodes an ankyrin-repeat protein that is transcriptionally regulated by the notch signaling pathway. Dev. Biol. *238*, 110-119.

Krebs, L. T., Iwai, N., Nonaka, S., Welsh, I. C., Lan, Y., Jiang, R., Saijoh, Y., O'Brien, T. P., Hamada, H., and Gridley, T. (2003a). Notch signaling regulates left-right asymmetry determination by inducing Nodal expression. Genes Dev. *17*, 1207-1212.

Krebs, L. T., Xue, Y., Norton, C. R., Sundberg, J. P., Beatus, P., Lendahl, U., Joutel, A., and Gridley, T. (2003b). Characterization of Notch3-deficient mice: normal embryonic development and absence of genetic interactions with a Notch1 mutation. Genesis *37*, 139-143.

Kuroda, K., Han, H., Tani, S., Tanigaki, K., Tun, T., Furukawa, T., Taniguchi, Y., Kurooka, H., Hamada, Y., Toyokuni, S., and Honjo, T. (2003). Regulation of marginal zone B cell development by MINT, a suppressor of Notch/RBP-J signaling pathway. Immunity *18*, 301-312.

Kuroda, K., Tani, S., Tamura, K., Minoguchi, S., Kurooka, H., and Honjo, T. (1999). Delta-induced Notch signaling mediated by RBP-J inhibits MyoD expression and myogenesis. J. Biol. Chem. *274*, 7238-7244.

Kurooka, H., Kuroda, K., and Honjo, T. (1998). Roles of the ankyrin repeats and C-terminal region of the mouse notch1 intracellular region. Nucleic Acids Res. *26*, 5448-5455.

Kurth, J., Hansmann, M. L., Rajewsky, K., and Kuppers, R. (2003). Epstein-Barr virus-infected B cells expanding in germinal centers of infectious mononucleosis patients do not participate in the germinal center reaction. Proc. Natl. Acad. Sci. USA. *100*, 4730-4735.

Kwiatkowski, B., Chen, S. Y., and Schubach, W. H. (2004). CKII site in Epstein-Barr virus nuclear protein 2 controls binding to hSNF5/Ini1 and is important for growth transformation. J. Virol. *78*, 6067-6072.

Lai, E. C. (2002). Keeping a good pathway down: transcriptional repression of Notch pathway target genes by CSL proteins. EMBO Rep. *3*, 840-845.

Lamar, E., Deblandre, G., Wettstein, D., Gawantka, V., Pollet, N., Niehrs, C., and Kintner, C. (2001). Nrarp is a novel intracellular component of the Notch signaling pathway. Genes Dev. *15*, 1885-1899.

Laux, G., Adam, B., Strobl, L. J., and Moreau-Gachelin, F. (1994a). The Spi-1/PU.1 and Spi-B ets family transcription factors and the recombination signal binding protein RBP-J kappa interact with an Epstein-Barr virus nuclear antigen 2 responsive cis-element. Embo J. *13*, 5624-5632.

Laux, G., Dugrillon, F., Eckert, C., Adam, B., Zimber-Strobl, U., and Bornkamm, G. W. (1994b). Identification and characterization of an Epstein-Barr virus nuclear antigen 2-responsive cis element in the bidirectional promoter region of latent membrane protein and terminal protein 2 genes. J. Virol. *68*, 6947-6958.

Le Roux, A., Kerdiles, B., Walls, D., Dedieu, J. F., and Perricaudet, M. (1994). The Epstein-Barr virus determined nuclear antigens EBNA-3A, -3B, and -3C repress EBNA-2-mediated transactivation of the viral terminal protein 1 gene promoter. Virology *205*, 596-602.

Lee, J. M., Lee, K. H., Weidner, M., Osborne, B. A., and Hayward, S. D. (2002). Epstein-Barr virus EBNA2 blocks Nur77- mediated apoptosis. Proc. Natl. Acad. Sci. USA. *99*, 11878-11883. Epub 12002 Aug 11823.

Li, H., Kolluri, S. K., Gu, J., Dawson, M. I., Cao, X., Hobbs, P. D., Lin, B., Chen, G., Lu, J., Lin, F., et al. (2000). Cytochrome c release and apoptosis induced by mitochondrial targeting of nuclear orphan receptor TR3. Science *289*, 1159-1164.

Lin, S. E., Oyama, T., Nagase, T., Harigaya, K., and Kitagawa, M. (2002). Identification of new human mastermind proteins defines a family that consists of positive regulators for notch signaling. J. Biol. Chem. *277*, 50612-50620.

Ling, P. D., and Hayward, S. D. (1995). Contribution of conserved amino acids in mediating the interaction between EBNA2 and CBF1/RBPJk. J. Virol. *69*, 1944-1950.

Ling, P. D., Hsieh, J. J., Ruf, I. K., Rawlins, D. R., and Hayward, S. D. (1994). EBNA-2 upregulation of Epstein-Barr virus latency promoters and the cellular CD23 promoter utilizes a common targeting intermediate, CBF1. J. Virol. *68*, 5375-5383.

Ling, P. D., Rawlins, D. R., and Hayward, S. D. (1993). The Epstein-Barr virus immortalizing protein EBNA-2 is targeted to DNA by a cellular enhancer-binding protein. Proc. Natl. Acad. Sci. USA. *90*, 9237-9241.

Logeat, F., Bessia, C., Brou, C., LeBail, O., Jarriault, S., Seidah, N. G., and Israel, A. (1998). The Notch1 receptor is cleaved constitutively by a furin-like convertase. Proc. Natl. Acad. Sci. USA. *95*, 8108-8112.

Macsween, K. F., and Crawford, D. H. (2003). Epstein-Barr virus-recent advances. Lancet Infect. Dis. *3*, 131-140.

Maier, M. M., and Gessler, M. (2000). Comparative analysis of the human and mouse Hey1 promoter: Hey genes are new Notch target genes. Biochem. Biophys Res. Commun. *275*, 652-660.

Maillard, I., Adler, S. H., and Pear, W. S. (2003). Notch and the immune system. Immunity *19*, 781-791.

Makar, K. W., Pham, C. T., Dehoff, M. H., O'Connor, S. M., Jacobi, S. M., and Holers, V. M. (1998). An intronic silencer regulates B lymphocyte cell- and stage-specific expression of the human complement receptor type 2 (CR2, CD21) gene. J. Immunol. *160*, 1268-1278.

Makar, K. W., Ulgiati, D., Hagman, J., and Holers, V. M. (2001). A site in the complement receptor 2 (CR2/CD21) silencer is necessary for lineage specific transcriptional regulation. Int. Immunol. *13*, 657-664.

Marshall, D., and Sample, C. (1995). Epstein-Barr virus nuclear antigen 3C is a transcriptional regulator. J. Virol. *69*, 3624-3630.

Martin, F., and Kearney, J. F. (2002). Marginal-zone B cells. Nat. Rev. Immunol. *2*, 323-335.

Meitinger, C., Strobl, L. J., Marschall, G., Bornkamm, G. W., and Zimber-Strobl, U. (1994). Crucial sequences within the Epstein-Barr virus TP1 promoter for EBNA2-mediated transactivation and interaction of EBNA2 with its responsive element. J. Virol. *68*, 7497-7506.

Miller, G., Robinson, J., Heston, L., and Lipman, M. (1974). Differences between laboratory strains of Epstein-Barr virus based on immortalization, abortive infection, and interference. Proc. Natl. Acad. Sci. USA. *71*, 4006-4010.

Muller, G., Hopken, U. E., and Lipp, M. (2003). The impact of CCR7 and CXCR5 on lymphoid organ development and systemic immunity. Immunol. Rev. *195*, 117-135.

Mumm, J. S., Schroeter, E. H., Saxena, M. T., Griesemer, A., Tian, X., Pan, D. J., Ray, W. J., and Kopan, R. (2000). A ligand-induced extracellular cleavage regulates gamma-secretase-like proteolytic activation of Notch1. Mol. Cell *5*, 197-206.

Nakagawa, O., McFadden, D. G., Nakagawa, M., Yanagisawa, H., Hu, T., Srivastava, D., and Olson, E. N. (2000). Members of the HRT family of basic helix-loop-helix proteins act as transcriptional repressors downstream of Notch signaling. Proc. Natl. Acad. Sci. USA. *97*, 13655-13660.

Nishimura, M., Isaka, F., Ishibashi, M., Tomita, K., Tsuda, H., Nakanishi, S., and Kageyama, R. (1998). Structure, chromosomal locus, and promoter of mouse Hes2 gene, a homologue of Drosophila hairy and Enhancer of split. Genomics *49*, 69-75.

Nitsche, F., Bell, A., and Rickinson, A. (1997). Epstein-Barr virus leader protein enhances EBNA-2-mediated transactivation of latent membrane protein 1 expression: a role for the W1W2 repeat domain. J. Virol. *71*, 6619-6628.

Oakley, F., Mann, J., Ruddell, R. G., Pickford, J., Weinmaster, G., and Mann, D. A. (2003). Basal expression of IkappaBalpha is controlled by the mammalian transcriptional repressor RBP-J (CBF1) and its activator Notch1. J. Biol. Chem. *278*, 24359-24370.

Oberg, C., Li, J., Pauley, A., Wolf, E., Gurney, M., and Lendahl, U. (2001). The Notch intracellular domain is ubiquitinated and negatively regulated by the mammalian Sel-10 homolog. J. Biol. Chem. *276*, 35847-35853.

Okochi, M., Steiner, H., Fukumori, A., Tanii, H., Tomita, T., Tanaka, T., Iwatsubo, T., Kudo, T., Takeda, M., and Haass, C. (2002). Presenilins mediate a dual intramembranous gamma-secretase cleavage of Notch-1. Embo J. *21*, 5408-5416.

Oswald, F., Kostezka, U., Astrahantseff, K., Bourteele, S., Dillinger, K., Zechner, U., Ludwig, L., Wilda, M., Hameister, H., Knochel, W., et al. (2002). SHARP is a novel component of the Notch/RBP-Jkappa signalling pathway. Embo J. *21*, 5417-5426.

Oswald, F., Liptay, S., Adler, G., and Schmid, R. M. (1998). NF-kappaB2 is a putative target gene of activated Notch-1 via RBP-Jkappa. Mol. Cell Biol. *18*, 2077-2088.

Oswald, F., Tauber, B., Dobner, T., Bourteele, S., Kostezka, U., Adler, G., Liptay, S., and Schmid, R. M. (2001). p300 acts as a transcriptional coactivator for mammalian Notch-1. Mol. Cell Biol. *21*, 7761-7774.

Pajic, A., Spitkovsky, D., Christoph, B., Kempkes, B., Schuhmacher, M., Staege, M. S., Brielmeier, M., Ellwart, J., Kohlhuber, F., Bornkamm, G. W., et al. (2000). Cell cycle activation by c-myc in a Burkitt lymphoma model cell line. Int. J. Cancer *87*, 787-793.

Pajic, A., Staege, M. S., Dudziak, D., Schuhmacher, M., Spitkovsky, D., Eissner, G., Brielmeier, M., Polack, A., and Bornkamm, G. W. (2001). Antagonistic effects of c-myc and Epstein-Barr virus latent genes on the phenotype of human B cells. Int. J. Cancer *93*, 810-816.

Palaga, T., Miele, L., Golde, T. E., and Osborne, B. A. (2003). TCR-mediated Notch signaling regulates proliferation and IFN-gamma production in peripheral T cells. J. Immunol. *171*, 3019-3024.

Pear, W. S., Aster, J. C., Scott, M. L., Hasserjian, R. P., Soffer, B., Sklar, J., and Baltimore, D. (1996). Exclusive development of T cell neoplasms in mice transplanted with bone marrow expressing activated Notch alleles. J. Exp. Med. *183*, 2283-2291.

Peng, C. W., Xue, Y., Zhao, B., Johannsen, E., Kieff, E., and Harada, S. (2004). Direct interactions between Epstein-Barr virus leader protein LP and the EBNA2 acidic domain underlie coordinate transcriptional regulation. Proc. Natl. Acad. Sci. USA. *101*, 1033-1038.

Peng, R., Gordadze, A. V., Fuentes Panana, E. M., Wang, F., Zong, J., Hayward, G. S., Tan, J., and Ling, P. D. (2000). Sequence and functional analysis of EBNA-LP and EBNA2 proteins from nonhuman primate lymphocryptoviruses. J. Virol. *74*, 379-389.

Petti, L., Sample, C., and Kieff, E. (1990). Subnuclear localization and phosphorylation of Epstein-Barr virus latent infection nuclear proteins. Virology *176*, 563-574.

Philips, A., Lesage, S., Gingras, R., Maira, M. H., Gauthier, Y., Hugo, P., and Drouin, J. (1997). Novel dimeric Nur77 signaling mechanism in endocrine and lymphoid cells. Mol. Cell Biol. *17*, 5946-5951.

Polack, A., Delius, H., Zimber, U., and Bornkamm, G. W. (1984). Two deletions in the Epstein-Barr virus genome of the Burkitt lymphoma nonproducer line Raji. Virology *133*, 146-157.

Polack, A., Hortnagel, K., Pajic, A., Christoph, B., Baier, B., Falk, M., Mautner, J., Geltinger, C., Bornkamm, G. W., and Kempkes, B. (1996). c-myc activation renders proliferation of Epstein-Barr virus (EBV)-transformed cells independent of EBV nuclear antigen 2 and latent membrane protein 1. Proc. Natl. Acad. Sci. USA. *93*, 10411-10416.

Puglielli, M. T., Desai, N., and Speck, S. H. (1997). Regulation of EBNA gene transcription in lymphoblastoid cell lines: characterization of sequences downstream of BCR2 (Cp). J. Virol. *71*, 120-128.

Puglielli, M. T., Woisetschlaeger, M., and Speck, S. H. (1996). oriP is essential for EBNA gene promoter activity in Epstein-Barr virus-immortalized lymphoblastoid cell lines. J. Virol. *70*, 5758-5768.

Pui, J. C., Allman, D., Xu, L., DeRocco, S., Karnell, F. G., Bakkour, S., Lee, J. Y., Kadesch, T., Hardy, R. R., Aster, J. C., and Pear, W. S. (1999). Notch1 expression in early lymphopoiesis influences B versus T lineage determination. Immunity *11*, 299-308.

Qiu, L., Joazeiro, C., Fang, N., Wang, H. Y., Elly, C., Altman, Y., Fang, D., Hunter, T., and Liu, Y. C. (2000). Recognition and ubiquitination of Notch by Itch, a hect-type E3 ubiquitin ligase. J. Biol. Chem. *275*, 35734-35737.

Radkov, S. A., Bain, M., Farrell, P. J., West, M., Rowe, M., and Allday, M. J. (1997). Epstein-Barr virus EBNA3C represses Cp, the major promoter for EBNA expression, but has no effect on the promoter of the cell gene CD21. J. Virol. *71*, 8552-8562.

Radtke, F., Wilson, A., and MacDonald, H. R. (2004a). Notch signaling in T- and B-cell development. Curr. Opin. Immunol. *16*, 174-179.

Radtke, F., Wilson, A., Mancini, S. J., and MacDonald, H. R. (2004b). Notch regulation of lymphocyte development and function. Nat. Immunol. *5*, 247-253.

Radtke, F., Wilson, A., Stark, G., Bauer, M., van Meerwijk, J., MacDonald, H. R., and Aguet, M. (1999). Deficient T cell fate specification in mice with an induced inactivation of Notch1. Immunity *10*, 547-558.

Rand, M. D., Grimm, L. M., Artavanis-Tsakonas, S., Patriub, V., Blacklow, S. C., Sklar, J., and Aster, J. C. (2000). Calcium depletion dissociates and activates heterodimeric notch receptors. Mol. Cell Biol. *20*, 1825-1835.

Rangarajan, A., Talora, C., Okuyama, R., Nicolas, M., Mammucari, C., Oh, H., Aster, J. C., Krishna, S., Metzger, D., Chambon, P., et al. (2001). Notch signaling is a direct determinant of keratinocyte growth arrest and entry into differentiation. Embo J. *20*, 3427-3436.

Reizis, B., and Leder, P. (2002). Direct induction of T lymphocyte-specific gene expression by the mammalian Notch signaling pathway. Genes Dev. *16*, 295-300.

Rohn, J. L., Lauring, A. S., Linenberger, M. L., and Overbaugh, J. (1996). Transduction of Notch2 in feline leukemia virus-induced thymic lymphoma. J. Virol. *70*, 8071-8080.

Ronchini, C., and Capobianco, A. J. (2001). Induction of cyclin D1 transcription and CDK2 activity by Notch(ic): implication for cell cycle disruption in transformation by Notch(ic). Mol. Cell Biol. *21*, 5925-5934.

Saito, T., Chiba, S., Ichikawa, M., Kunisato, A., Asai, T., Shimizu, K., Yamaguchi, T., Yamamoto, G., Seo, S., Kumano, K., et al. (2003). Notch2 is preferentially expressed in mature B cells and indispensable for marginal zone B lineage development. Immunity *18*, 675-685.

Sakai, T., Taniguchi, Y., Tamura, K., Minoguchi, S., Fukuhara, T., Strobl, L. J., Zimber-Strobl, U., Bornkamm, G. W., and Honjo, T. (1998). Functional replacement of the intracellular region of the Notch1 receptor by Epstein-Barr virus nuclear antigen 2. J. Virol. *72*, 6034-6039.

Schlager, S., Speck, S. H., and Woisetschlager, M. (1996). Transcription of the Epstein-Barr virus nuclear antigen 1 (EBNA1) gene occurs before induction of the BCR2 (Cp) EBNA gene promoter during the initial stages of infection in B cells. J. Virol. *70*, 3561-3570.

Schlee, M., Krug, T., Gires, O., Zeidler, R., Hammerschmidt, W., Mailhammer, R., Laux, G., Sauer, G., Lovric, J., and Bornkamm, G. W. (2004). Identification of Epstein-Barr virus (EBV) nuclear antigen 2 (EBNA2) target proteins by proteome analysis: activation of EBNA2 in conditionally immortalized B cells reflects early events after infection of primary B cells by EBV. J. Virol. *78*, 3941-3952.

Schroeder, T., Kohlhof, H., Rieber, N., and Just, U. (2003). Notch signaling induces multilineage myeloid differentiation and up-regulates PU.1 expression. J. Immunol. *170*, 5538-5548.

Schroeter, E. H., Kisslinger, J. A., and Kopan, R. (1998). Notch-1 signalling requires ligand-induced proteolytic release of intracellular domain. Nature *393*, 382-386.

Sinclair, A. J., Palmero, I., Peters, G., and Farrell, P. J. (1994). EBNA-2 and EBNA-LP cooperate to cause G0 to G1 transition during immortalization of resting human B lymphocytes by Epstein-Barr virus. Embo J. *13*, 3321-3328.

Sjoblom, A., Jansson, A., Yang, W., Lain, S., Nilsson, T., and Rymo, L. (1995a). PU box-binding transcription factors and a POU domain protein cooperate in the Epstein-Barr virus (EBV) nuclear antigen 2-induced transactivation of the EBV latent membrane protein 1 promoter. J. Gen. Virol. *76*, 2679-2692.

Sjoblom, A., Nerstedt, A., Jansson, A., and Rymo, L. (1995b). Domains of the Epstein-Barr virus nuclear antigen 2 (EBNA2) involved in the transactivation of the latent membrane protein 1 and the EBNA Cp promoters. J. Gen. Virol. *76*, 2669-2678.

Sjoblom, A., Yang, W., Palmqvist, L., Jansson, A., and Rymo, L. (1998). An ATF/CRE element mediates both EBNA2-dependent and EBNA2-independent activation of the Epstein-Barr virus LMP1 gene promoter. J. Virol. *72*, 1365-1376.

Skare, J., Farley, J., Strominger, J. L., Fresen, K. O., Cho, M. S., and zur Hausen, H. (1985). Transformation by Epstein-Barr virus requires DNA sequences in the region of BamHI fragments Y and H. J. Virol. *55*, 286-297.

Souabni, A., Cobaleda, C., Schebesta, M., and Busslinger, M. (2002). Pax5 promotes B lymphopoiesis and blocks T cell development by repressing Notch1. Immunity *17*, 781-793.

Spender, L. C., Cornish, G. H., Rowland, B., Kempkes, B., and Farrell, P. J. (2001). Direct and indirect regulation of cytokine and cell cycle proteins by ebna-2 during epstein-barr virus infection. J. Virol. *75*, 3537-3546.

Spender, L. C., Cornish, G. H., Sullivan, A., and Farrell, P. J. (2002). Expression of transcription factor AML-2 (RUNX3, CBF(alpha)-3) is induced by Epstein-Barr virus EBNA-2 and correlates with the B-cell activation phenotype. J. Virol. *76*, 4919-4927.

Strobl, L. J., Hofelmayr, H., Marschall, G., Brielmeier, M., Bornkamm, G. W., and Zimber-Strobl, U. (2000). Activated Notch1 modulates gene expression in B cells similarly to Epstein-Barr viral nuclear antigen 2. J. Virol. *74*, 1727-1735.

Sung, N. S., Kenney, S., Gutsch, D., and Pagano, J. S. (1991). EBNA-2 transactivates a lymphoid-specific enhancer in the BamHI C promoter of Epstein-Barr virus. J. Virol. *65*, 2164-2169.

Tamura, K., Taniguchi, Y., Minoguchi, S., Sakai, T., Tun, T., Furukawa, T., and Honjo, T. (1995). Physical interaction between a novel domain of the receptor Notch and the transcription factor RBP-J kappa/Su(H). Curr. Biol. *5*, 1416-1423.

Tanigaki, K., Han, H., Yamamoto, N., Tashiro, K., Ikegawa, M., Kuroda, K., Suzuki, A., Nakano, T., and Honjo, T. (2002). Notch-RBP-J signaling is involved in cell fate determination of marginal zone B cells. Nat. Immunol. *3*, 443-450.

Tanigaki, K., Tsuji, M., Yamamoto, N., Han, H., Tsukada, J., Inoue, H., Kubo, M., and Honjo, T. (2004). Regulation of alphabeta/gammadelta T cell lineage commitment and peripheral T cell responses by Notch/RBP-J signaling. Immunity *20*, 611-622.

Thorley-Lawson, D. A. (2001). Epstein-Barr virus: exploiting the immune system. Nature Rev. Immunol. *1*, 75-82.

Tong, X., Drapkin, R., Reinberg, D., and Kieff, E. (1995a). The 62- and 80-kDa subunits of transcription factor IIH mediate the interaction with Epstein-Barr virus nuclear protein 2. Proc. Natl. Acad. Sci. USA. *92*, 3259-3263.

Tong, X., Drapkin, R., Yalamanchili, R., Mosialos, G., and Kieff, E. (1995b). The Epstein-Barr virus nuclear protein 2 acidic domain forms a complex with a novel cellular coactivator that can interact with TFIIE. Mol. Cell Biol. *15*, 4735-4744.

Tong, X., Wang, F., Thut, C. J., and Kieff, E. (1995c). The Epstein-Barr virus nuclear protein 2 acidic domain can interact with TFIIB, TAF40, and RPA70 but not with TATA-binding protein. J. Virol. *69*, 585-588.

Tong, X., Yalamanchili, R., Harada, S., and Kieff, E. (1994). The EBNA-2 arginine-glycine domain is critical but not essential for B-lymphocyte growth transformation; the rest of region 3 lacks essential interactive domains. J. Virol. *68*, 6188-6197.

Tsui, S., and Schubach, W. H. (1994). Epstein-Barr virus nuclear protein 2A forms oligomers *in vitro* and *in vivo* through a region required for B-cell transformation. J. Virol. *68*, 4287-4294.

Voss, M. D., Hille, A., Barth, S., Spurk, A., Hennrich, F., Holzer, D., Mueller-Lantzsch, N., Kremmer, E., and Grasser, F. A. (2001). Functional cooperation of Epstein-Barr virus nuclear antigen 2 and the survival motor neuron protein in transactivation of the viral LMP1 promoter. J. Virol. *75*, 11781-11790.

Walker, L., Carlson, A., Tan-Pertel, H. T., Weinmaster, G., and Gasson, J. (2001). The notch receptor and its ligands are selectively expressed during hematopoietic development in the mouse. Stem Cells *19*, 543-552.

Waltzer, L., Logeat, F., Brou, C., Israel, A., Sergeant, A., and Manet, E. (1994). The human J. kappa recombination signal sequence binding protein (RBP-J kappa) targets the Epstein-Barr virus EBNA2 protein to its DNA responsive elements. Embo J. *13*, 5633-5638.

Wang, F., Gregory, C., Sample, C., Rowe, M., Liebowitz, D., Murray, R., Rickinson, A., and Kieff, E. (1990a). Epstein-Barr virus latent membrane protein (LMP1) and nuclear proteins 2 and 3C are effectors of phenotypic changes in B lymphocytes: EBNA-2 and LMP1 cooperatively induce CD23. J. Virol. *64*, 2309-2318.

Wang, F., Gregory, C. D., Rowe, M., Rickinson, A. B., Wang, D., Birkenbach, M., Kikutani, H., Kishimoto, T., and Kieff, E. (1987). Epstein-Barr virus nuclear antigen 2 specifically induces expression of the B-cell activation antigen CD23. Proc. Natl. Acad. Sci. USA. *84*, 3452-3456.

Wang, F., Kikutani, H., Tsang, S. F., Kishimoto, T., and Kieff, E. (1991). Epstein-Barr virus nuclear protein 2 transactivates a cis-acting CD23 DNA element. J. Virol. *65*, 4101-4106.

Wang, F., Tsang, S. F., Kurilla, M. G., Cohen, J. I., and Kieff, E. (1990b). Epstein-Barr virus nuclear antigen 2 transactivates latent membrane protein LMP1. J. Virol. *64*, 3407-3416.

Wang, L., Grossman, S. R., and Kieff, E. (2000). Epstein-Barr virus nuclear protein 2 interacts with p300, CBP, and PCAF histone acetyltransferases in activation of the LMP1 promoter. Proc. Natl. Acad. Sci. USA. *97*, 430-435.

Washburn, T., Schweighoffer, E., Gridley, T., Chang, D., Fowlkes, B. J., Cado, D., and Robey, E. (1997). Notch activity influences the alphabeta versus gammadelta T cell lineage decision. Cell *88*, 833-843.

Weng, A. P., Nam, Y., Wolfe, M. S., Pear, W. S., Griffin, J. D., Blacklow, S. C., and Aster, J. C. (2003). Growth suppression of pre-T acute lymphoblastic leukemia cells by inhibition of notch signaling. Mol. Cell Biol. *23*, 655-664.

Wilson, A., MacDonald, H. R., and Radtke, F. (2001). Notch 1-deficient common lymphoid precursors adopt a B cell fate in the thymus. J. Exp. Med. *194*, 1003-1012.

Witt, C. M., Hurez, V., Swindle, C. S., Hamada, Y., and Klug, C. A. (2003). Activated Notch2 potentiates CD8 lineage maturation and promotes the selective development of B1 B cells. Mol. Cell Biol. *23*, 8637-8650.

Woisetschlaeger, M., Jin, X. W., Yandava, C. N., Furmanski, L. A., Strominger, J. L., and Speck, S. H. (1991). Role for the Epstein-Barr virus nuclear antigen 2 in viral promoter switching during initial stages of infection. Proc. Natl. Acad. Sci. USA. *88*, 3942-3946.

Wolfer, A., Bakker, T., Wilson, A., Nicolas, M., Ioannidis, V., Littman, D. R., Lee, P. P., Wilson, C. B., Held, W., MacDonald, H. R., and Radtke, F. (2001). Inactivation of Notch 1 in immature thymocytes does not perturb CD4 or CD8T cell development. Nat. Immunol. *2*, 235-241.

Wolfer, A., Wilson, A., Nemir, M., MacDonald, H. R., and Radtke, F. (2002). Inactivation of Notch1 impairs VDJbeta rearrangement and allows pre-TCR-independent survival of early alpha beta Lineage Thymocytes. Immunity *16*, 869-879.

Wu, D. Y., Kalpana, G. V., Goff, S. P., and Schubach, W. H. (1996). Epstein-Barr virus nuclear protein 2 (EBNA2) binds to a component of the human SNF-SWI complex, hSNF5/Ini1. J. Virol. *70*, 6020-6028.

Wu, D. Y., Krumm, A., and Schubach, W. H. (2000a). Promoter-specific targeting of human SWI-SNF complex by Epstein-Barr virus nuclear protein 2. J. Virol. *74*, 8893-8903.

Wu, D. Y., Tkachuck, D. C., Roberson, R. S., and Schubach, W. H. (2002a). The human SNF5/INI1 protein facilitates the function of the growth arrest and DNA damage-inducible protein (GADD34) and modulates GADD34-bound protein phosphatase-1 activity. J. Biol. Chem. *277*, 27706-27715.

Wu, L., Aster, J. C., Blacklow, S. C., Lake, R., Artavanis-Tsakonas, S., and Griffin, J. D. (2000b). MAML1, a human homologue of Drosophila mastermind, is a transcriptional co-activator for NOTCH receptors. Nat. Genet. *26*, 484-489.

Wu, L., Sun, T., Kobayashi, K., Gao, P., and Griffin, J. D. (2002b). Identification of a family of mastermind-like transcriptional coactivators for mammalian notch receptors. Mol. Cell Biol. *22*, 7688-7700.

Yalamanchili, R., Harada, S., and Kieff, E. (1996). The N-terminal half of EBNA2, except for seven prolines, is not essential for primary B-lymphocyte growth transformation. J. Virol. *70*, 2468-2473.

Yalamanchili, R., Tong, X., Grossman, S., Johannsen, E., Mosialos, G., and Kieff, E. (1994). Genetic and biochemical evidence that EBNA 2 interaction with a 63-kDa cellular

GTG-binding protein is essential for B lymphocyte growth transformation by EBV. Virology *204*, 634-641.

Yan, X. Q., Sarmiento, U., Sun, Y., Huang, G., Guo, J., Juan, T., Van, G., Qi, M. Y., Scully, S., Senaldi, G., and Fletcher, F. A. (2001). A novel Notch ligand, Dll4, induces T-cell leukemia/lymphoma when overexpressed in mice by retroviral-mediated gene transfer. Blood *98*, 3793-3799.

Yang, J., Aittomaki, S., Pesu, M., Carter, K., Saarinen, J., Kalkkinen, N., Kieff, E., and Silvennoinen, O. (2002). Identification of p100 as a coactivator for STAT6 that bridges STAT6 with RNA polymerase II. Embo J. *21*, 4950-4958.

Yokota, A., Kikutani, H., Tanaka, T., Sato, R., Barsumian, E. L., Suemura, M., and Kishimoto, T. (1988). Two species of human Fc epsilon receptor II (Fc epsilon RII/CD23): tissue-specific and IL-4-specific regulation of gene expression. Cell *55*, 611-618.

Yoo, L., and Speck, S. H. (2000). Determining the role of the Epstein-Barr virus Cp EBNA2-dependent enhancer during the establishment of latency by using mutant and wild-type viruses recovered from cottontop marmoset lymphoblastoid cell lines. J. Virol. *74*, 11115-11120.

Yoo, L. I., Mooney, M., Puglielli, M. T., and Speck, S. H. (1997). B-cell lines immortalized with an Epstein-Barr virus mutant lacking the Cp EBNA2 enhancer are biased toward utilization of the oriP-proximal EBNA gene promoter Wp1. J. Virol. *71*, 9134-9142.

Yue, W., Davenport, M. G., Shackelford, J., and Pagano, J. S. (2004). Mitosis-specific hyperphosphorylation of Epstein-Barr virus nuclear antigen 2 suppresses its function. J. Virol. *78*, 3542-3552.

Yun, T. J., and Bevan, M. J. (2003). Notch-regulated ankyrin-repeat protein inhibits Notch1 signaling: multiple Notch1 signaling pathways involved in T cell development. J. Immunol. *170*, 5834-5841.

Zhang, J., Chen, H., Weinmaster, G., and Hayward, S. D. (2001). Epstein-Barr virus BamHi-a rightward transcript-encoded RPMS protein interacts with the CBF1-associated corepressor CIR to negatively regulate the activity of EBNA2 and NotchIC. J. Virol. *75*, 2946-2956.

Zhao, B., and Sample, C. E. (2000). Epstein-barr virus nuclear antigen 3C activates the latent membrane protein 1 promoter in the presence of Epstein-Barr virus nuclear antigen 2 through sequences encompassing an spi-1/Spi-B binding site. J. Virol. *74*, 5151-5160.

Zhou, S., Fujimuro, M., Hsieh, J. J., Chen, L., and Hayward, S. D. (2000a). A role for SKIP in EBNA2 activation of CBF1-repressed promoters. J. Virol. *74*, 1939-1947.

Zhou, S., Fujimuro, M., Hsieh, J. J., Chen, L., Miyamoto, A., Weinmaster, G., and Hayward, S. D. (2000b). SKIP, a CBF1-associated protein, interacts with the ankyrin repeat domain of NotchIC To facilitate NotchIC function. Mol. Cell Biol. *20*, 2400-2410.

Zimber, U., Adldinger, H. K., Lenoir, G. M., Vuillaume, M., Knebel-Doeberitz, M. V., Laux, G., Desgranges, C., Wittmann, P., Freese, U. K., and Schneider, U. (1986). Geographical prevalence of two types of Epstein-Barr virus. Virology *154*, 56-66.

Zimber-Strobl, U., Kempkes, B., Marschall, G., Zeidler, R., Van Kooten, C., Banchereau, J., Bornkamm, G. W., and Hammerschmidt, W. (1996). Epstein-Barr virus latent membrane protein (LMP1) is not sufficient to maintain proliferation of B cells but both it and activated CD40 can prolong their survival. Embo J. *15*, 7070-7078.

Zimber-Strobl, U., Kremmer, E., Grasser, F., Marschall, G., Laux, G., and Bornkamm, G. W. (1993). The Epstein-Barr virus nuclear antigen 2 interacts with an EBNA2 responsive cis-element of the terminal protein 1 gene promoter. Embo J. *12*, 167-175.

Zimber-Strobl, U., Strobl, L. J., Meitinger, C., Hinrichs, R., Sakai, T., Furukawa, T., Honjo, T., and Bornkamm, G. W. (1994). Epstein-Barr virus nuclear antigen 2 exerts its transactivating function through interaction with recombination signal binding protein RBP-J kappa, the homologue of Drosophila Suppressor of Hairless. Embo J. *13*, 4973-4982.

Zimber-Strobl, U., Suentzenich, K. O., Laux, G., Eick, D., Cordier, M., Calender, A., Billaud, M., Lenoir, G. M., and Bornkamm, G. W. (1991). Epstein-Barr virus nuclear antigen 2 activates transcription of the terminal protein gene. J. Virol. *65*, 415-423.

Zur Hausen, H., and Schulte-Holthausen, H. (1970). Presence of EB virus nucleic acid homology in a "virus-free" line of Burkitt tumour cells. Nature *227*, 245-248.

Chapter 24

Epstein-Barr Virus and the Cell Cycle

*Jason S. Knight and Erle S. Robertson**

ABSTRACT

Epstein-Barr virus (EBV) is the causative agent of post-transplant lymphoproliferative disease (PTLD) and is strongly associated with other human malignancies including endemic Burkitt's lymphoma, nasopharyngeal carcinoma, and some subtypes of Hodgkin's disease. Through the expression of nine latency proteins, a subset of which are absolutely essential, EBV activates and drives the proliferation of B-lymphocytes *in vitro* resulting in indefinitely proliferating lymphoblastoid cell lines (LCLs). The small DNA tumor viruses have classically been linked to cell cycle deregulation, presumably a necessary step in cell immortalization/transformation. Specifically, adenovirus, the polyomavirus SV40, and human papillomavirus express specific viral antigens that bind and modulate the function of critical cell cycle gatekeepers such as the retinoblastoma protein and p53. By comparison, the links between EBV latency antigens and cell cycle regulators have remained tenuous. Here, we will examine current evidence implicating EBV latency antigens in cell cycle deregulation. This review focuses on four latency proteins EBNA2, EBNA-LP, EBNA3C, and LMP1. EBNA2 in cooperation with EBNA-LP facilitates cell cycle entry and regulates the cyclin D2 promoter. EBNA3C also facilitates passage through the G1/S restriction point, and additionally targets S-phase cyclins. LMP1 clearly provides anti-apoptotic signals to the cell and potentially also targets the cyclin D2 promoter. We will attempt to assimilate these and other single gene studies into a model that explains how EBV utilizes its latency proteins to fulfil unique requirements in the stimulation of B cell proliferation and how this may distance EBV from other DNA tumor viruses.

INTRODUCTION TO EPSTEIN-BARR VIRUS AND ITS LATENT ANTIGENS

Epstein-Barr virus (EBV) was the first described oncogenic virus in humans. EBV was discovered in 1962 by electron microscopy of cells cultured from Burkitt's lymphoma, an endemic and aggressive malignancy predominantly seen in equatorial Africa (Epstein et al., 1964). Subsequent epidemiologic and cell biology studies implicated EBV as the primary causative agent of infectious

*For correspondence email erle@mail.med.upenn.edu.

mononucleosis (IM) (Diehl et al., 1968; Kieff and Rickinson, 2002; Rickinson and Kieff, 2002). While IM is not a malignancy, the ability of EBV to infect, activate, and expand B-lymphocytes is fundamental to the etiology of this disease. The clinical presentation of IM results from an overwhelming induction of cytotoxic T-lymphocytes reactive to the expanded pool of EBV-infected B-lymphocytes. More recently, EBV was identified as the causative agent of post-transplant lymphoproliferative disease (PTLD) and the majority of cases of AIDS-associated lymphoma (Kieff and Rickinson, 2002; Rickinson and Kieff, 2002). EBV has also been frequently and clonally detected in nasopharyngeal carcinomas especially of Eastern Asia and some subtypes of Hodgkin's disease (Kieff and Rickinson, 2002; Rickinson and Kieff, 2002); as for Burkitt's lymphoma, a likely causative, and minimally contributory, role for EBV in these malignancies is suspected although not yet formally proven. EBV has also been linked to gastric carcinoma and more controversially metastatic breast carcinoma (Kieff and Rickinson, 2002; Rickinson and Kieff, 2002; Subramanian et al., 2001).

Early studies of EBV noted its potent ability to transform resting primary B-lymphocytes in cell culture (Nilsson et al., 1971; Pope et al., 1968). Indeed, as long as donor cytotoxic T lymphocytes are inhibited, EBV infection results in the formation of lymphoblastoid cell lines (LCLs) with a surface phenotype strongly resembling activated B-lymphocytes *in vivo* (Thorley-Lawson et al., 1985; 1982). These LCLs can be propagated indefinitely in culture, underscoring the remarkable ability of the virus to dominate and drive the B-lymphocyte cell division cycle. At least four latency transcription programs for EBV have been described (Kieff and Rickinson, 2002; Rickinson and Kieff, 2002); type III latency, sometimes referred to as the growth program (Thorley-Lawson, 2001), is dominant in the immortalization of B-lymphocytes *in vitro* and in lymphomas of the immunocompromised host.

Type III latency results in the expression of all known EBV latency proteins, and not coincidentally only dominates in contexts where the T-cell immune response does not. In the normal EBV life cycle (which does not typically result in lymphoma), this latency program may serve to activate and expand initially infected naïve B-lymphocytes in the tonsillar region (Babcock et al., 2000; Joseph et al., 2000). Viral infection of these naïve cells potentially initiates B-lymphocyte maturation events, generating memory B-cells through a germinal center-like reaction and down-regulating the Type III latency program (Babcock and Thorley-Lawson, 2000; Thorley-Lawson, 2001). It is in memory B-cells that the virus can be maintained episomally without detectable EBV antigen expression and, consequently, without immune recognition by the host (Babcock et al., 1998; Reedman and Klein, 1973). These cells are the site of long-term viral persistence in the infected host. There is also evidence that, in the context of the high viral loads of infectious mononucleosis, germinal center or memory B-lymphocytes may be directly infected by EBV leading to viral-driven expansion of these populations (Kurth et al., 2003; 2000); however, it is not clear whether these populations are then able to down-regulate the Type III latency program necessary for long-term viral persistence in the immunocompetent host. Importantly for the purposes of this chapter, the majority of data regarding EBV latency proteins and their role in EBV biology comes from the study of type III latency due to the accessibility of LCLs for study in the tissue culture setting.

As is mentioned above, all known EBV latency proteins are expressed in type III latency (Kieff and Rickinson, 2002; Rickinson and Kieff, 2002). These include the Epstein-Barr Nuclear Antigens (EBNAs 1, 2, 3A, 3B, 3C, and -LP) and the Latent Membrane Proteins (LMPs 1, 2A, and 2B). EBNA1 is the most promiscuously expressed of the EBV latency proteins as it is thought to be expressed in any dividing cell that harbors the viral genome (Reedman and Klein, 1973); EBNA1 is likely not expressed in circulating memory B-cells that have exited the cell division cycle (Reedman and Klein, 1973). EBNA1 binds the latent origin of episome replication OriP and facilitates both genome replication and the precise segregation of replicated genomes to daughter cells (Yates et al., 1984; 1985). EBNA2 (along with EBNA-LP) is the first latency protein detected after viral infection of B-lymphocytes (Alfieri et al., 1991) and is the master regulator of latent transcription. EBNA2 stimulates transcription from LMP promoters as well as the BamHI C promoter (Cp) which regulates the expression of all EBNAs (Abbot et al., 1990; Fahraeus et al., 1990; Ghosh and Kieff, 1990; Jin and Speck, 1992; Sung et al., 1991; Wang et al., 1990; Zimber-Strobl et al., 1993). EBNA2 does not directly bind DNA, but relies on cellular factors like RBP-Jκ and PU.1 to target it to specific viral and cellular promoters (Grossman et al., 1994; Henkel et al., 1994; Johannsen et al., 1995; 1996; Yalamanchili et al., 1994). EBNA-LP is also detected early in the course of the viral latency program and synergistically cooperates with EBNA2 in promoter regulation (Harada and Kieff, 1997; Nitsche et al., 1997).

The three EBNA3 proteins are arranged in a collinear fashion in the viral genome suggesting origin from an ancestral gene duplication event (Kieff and Rickinson, 2002; Rickinson and Kieff, 2002). The EBNA3 proteins have limited homology at their amino termini (approximately 30% at the amino acid level) and have some conserved functions (Kieff and Rickinson, 2002; Rickinson and Kieff, 2002); however, as will be discussed below, additional functions have also clearly diverged. Historically, the first clear role for the EBNA3 proteins was to downregulate EBNA2-driven transcription from RBP-Jκ-targeted promoters by competing with EBNA2 for RBP-Jκ binding (Johannsen et al., 1996; Krauer et al., 1996; Le Roux et al., 1994; Marshall and Sample, 1995; Robertson et al., 1995; Waltzer et al., 1996; Zhao et al., 1996). This effect is mediated by all three proteins to some extent, and is thought to provide negative feedback for Cp transcription (Robertson et al., 1996).

The LMP proteins possess multiple transmembrane domains and localize to cytoplasmic membranes. Clustering in these membranes is clearly important for LMP1 functions, one of which is to mimic constitutively-activated CD40 receptor (Liebowitz et al., 1986; Uchida et al., 1999). This is understood in some detail mechanistically as LMP1 regulates multiple members of the TRAF family and in doing so stimulates NF-κB transcriptional activity in the nucleus (Devergne et al., 1996; Kaye et al., 1996; Mosialos et al., 1995). LMP2 proteins, especially LMP2A, modulate B-cell receptor (BCR) signaling and, in doing so, may help to maintain the latent state by preventing BCR-dependent initiation of lytic replication (Miller et al., 1995; 1993).

As the molecular genetics of EBV matured in the late 1980s, a question of considerable interest was whether all nine latency proteins would prove to be essential for the immortalization of B-lymphocytes *in vitro*. An important event, as investigators began to address this question, was the description of a lymphoblastoid cell line containing a mutant virus, P3HR1, which was competent for lytic replication, but could not establish lymphoblastoid cell lines on fresh B-lymphocytes (Miller et al., 1975). A genome deletion in P3HR1 removes part of the EBNA-LP coding region and the entire EBNA2 coding region hinting that one or both of these proteins may be essential for B-lymphocyte immortalization (Dambaugh et al., 1984; Heller et al., 1981; Hennessy and Kieff, 1983; King et al., 1980; Raab-Traub et al., 1978). Indeed, cosmid delivery of intact B95-8 viral DNA overlapping the deletion rescues both EBNA expression and the immortalization phenotype (Cohen et al., 1989; Hammerschmidt and Sugden, 1989; Skare et al., 1985). Mutagenesis of cosmid DNA confirmed the absolute requirement of EBNA2 in this process (Cohen et al., 1989) (Hammerschmidt and Sugden, 1989), while also suggesting a supportive role for EBNA-LP, although not an absolute requirement (Mannick et al., 1991). Subsequent studies employing similar strategies demonstrated that, similar to EBNA2, LMP1 and EBNA3C are absolutely required for the establishment of lymphoblastoid cell lines (Kaye et al., 1993; Tomkinson et al., 1993). While rare LCLs with a limited growth phenotype could be obtained in the absence of EBNA3A, it is generally believed that EBNA3A, like EBNA-LP, strongly contributes to immortalization (Tomkinson et al., 1993). In contrast, EBNA3B, LMP2A, and LMP2B are dispensable for the *in vitro* transformation of B-lymphocytes although they almost certainly play a role in the *in vivo* biology of EBV (Kim and Yates, 1993; Longnecker et al., 1992; 1993a; 1993b; Tomkinson and Kieff, 1992). In summary, EBNA1 is essential for maintenance of the EBV genome in replicating cells, EBNA2, LMP1, and EBNA3C are absolutely required for *in vitro* B-lymphocyte immortalization, and EBNA3A and EBNA-LP also strongly contribute to transformation. These genetic studies are compatible with the data presented below which will argue that EBNA2 and EBNA3C are the strongest contributors to EBV regulation of the cell division cycle.

TUMOR VIRUSES AND THE CELL CYCLE

Progression through the cell cycle is regulated by many different proteins and is reviewed in detail elsewhere (Nevins et al., 1997; Sherr, 1996). The issues highlighted here are particularly relevant to the below discussion and are not meant to be comprehensive. The cell division cycle of eukaryotic cells is tightly regulated ensuring faithful replication of chromosomal DNA with equal partitioning into the resulting daughter cells. In eukaryotes four clear and distinct cell cycle phases have been defined, namely the G1, S, G2, and M phases. Together G1, S, and G2 represent classic microscopic interphase. During S-phase, DNA is replicated. In contrast, G1 and G2 are regulatory, ensuring sequential progression of phases and appropriate stockpiling of factors necessary for DNA synthesis and cell division. M-phase represents mitosis where DNA condenses and distributes into daughter cells. For multicellular organisms, the initiation and progression of this process is tightly regulated by external stimuli, ensuring that cell division happens only in the desired systems and compartments. An example of desired cell division is the antigenically-driven expansion of naïve B-lymphocytes.

Internal to the cell, the best understood regulatory force behind sequential cell cycle progression is phosphorylation by protein kinase complexes. These regulatory complexes consist of a cyclin regulatory subunit and a cyclin-dependent kinase (cdk) enzymatic subunit (Draetta et al., 1989). As expected, these complexes are themselves regulated at multiple levels. At least four families of cyclins, D, E, A, and B, play important roles in mammalian cells, each expressed with characteristic peak and valley during specific phases of the cell cycle. The expression of D-type cyclins is particularly responsive to growth factors and consequently plays a critical role in initiation of the cell cycle. D-type cyclins form active kinase complexes with both cdk4 and cdk6 (Matsushime et al., 1992; Xiong et al., 1992). Cyclin E complexes with cdk2 and plays a role in entry into, and progression through, S-phase (Sauer and Lehner, 1995). Cyclin A complexes with cdk2 in S-phase and cdk1 in G2-phase (Pagano et al., 1992). Heterologous expression of cyclin A drives cells into S-phase, similar to cyclin E, suggesting that they may share many of the same substrates (Resnitzky et al., 1995); however, during internally regulated cell cycle progression, cyclin E is clearly expressed and active first. Cyclin B, in complex with cdk1 is the classic regulator of mitotic function (Maller et al., 1989; Pines and Hunter, 1989). In addition to cyclin binding, cyclin-dependent kinase complexes are additionally regulated by a number of cyclin-dependent kinase inhibitors (CDKIs). CDKIs are primarily of two families. The INK4 family, including p16 and p15, interact with cdk4 and cdk6 directly and are though to abrogate kinase activity by competing with D-type cyclins for cdk binding (Serrano et al., 1993). The CIP/KIP family, including p21 and p27, have broader specificity targeting both cyclin D/cdk4,6 complexes as well as cyclin E/cdk2 and cyclin A/cdk2 (Polyak et al., 1994; Toyoshima and Hunter, 1994). p27 has been crystallized in complex with cdk2 and a truncated form of cyclin A (Russo et al., 1996). This structure suggests that p27 binds to both the cyclin and cdk subunit and interferes with normal ATP processing by the kinase.

While many of the critical substrates for these cyclin-dependent kinase activities remain to be fully elucidated, a well characterized example is the retinoblastoma tumor suppressor (Rb)-family of proteins, also called the "pocket"-domain proteins. Rb (p105), p107, and p130 bind and repress members of the E2F family of transcriptional activators (Helin et al., 1992). E2F-responsive promoters regulate genes essential for DNA synthesis and S-phase progression as well as cyclins E and A (Ohtani et al., 1995). The Rb-family may physically disrupt E2F activators from relevant promoters, although it is also clear that in some cases Rb can recruit histone deacetylase activity to E2F-responsive promoters thereby silencing transcriptional activity (Luo et al., 1998). The active form of Rb (similar to p107 and p130) is hypophosphorylated; however, upon stimulation of both cyclin D and cyclin E-associated kinase activities Rb becomes hyperphosphorylated abrogating its ability to repress E2F-dependent transcription (Buchkovich et al., 1989; Chen et al., 1989; DeCaprio et al., 1989). The importance of active Rb in blocking S-phase entry is underscored by the frequent disruption of this gene product in human cancer permitting cells to more freely enter S-phase without the strict need for external cues.

Classic work with the small DNA tumor viruses has demonstrated that these viruses drive cell proliferation and block apoptosis by specifically targeting cell cycle gatekeeper and checkpoint molecules (Debbas and White, 1993; Hickman et al., 1994; Howes et al., 1994; Lowe and Ruley, 1993; Pan and Griep, 1994; White et al., 1994). The adenovirus E1A protein, the human papillomavirus E7 protein, and the SV40 large T antigen promote DNA replication by inactivating the so-called "pocket"-domain proteins, the best studied member of which is Rb, as described above (DeCaprio et al., 1988; Dyson et al., 1989; Whyte et al., 1989). E1A, E7, and large-T share an LxCxE motif which allows physical interaction with Rb family members through the pocket motif. This interaction is thought to physically displace other Rb-binding proteins including members of the E2F and HDAC families. In the context of this chapter, it is important to consider that the "motivation" of the small DNA tumor viruses for stimulating S-phase entry may differ significantly from what we will consider below for EBV. These viruses are dependent on the cell's replication machinery to propagate their genomes and likely drive cells into S-phase to better access this machinery. In contrast, activation and proliferation of B-lymphocytes seems to be a fundamental part of EBV biology.

EBV's closest relative among the human herpesviruses is Kaposi's sarcoma-associated herpesvirus (KSHV). Like EBV, KSHV is strongly linked to human malignancies with latent KSHV infection detected in Kaposi's sarcoma, body-cavity based lymphomas in AIDS patients, and multicentric Castleman's disease. Importantly, KSHV has been shown to disrupt the function of Rb, apparently through multiple mechanisms. It is believed that the number of genes expressed during KSHV latent infection is more limited than for EBV, as the strongest latent promoter directs the expression of only three gene products, encoded by open-reading frames (ORFs) 71, 72, and 73, respectively. Early on in the study of KSHV, ORF 72 was shown to encode a cyclin D homologue, v-cyclin, which binds and constitutively activates cdk6, providing one potential mechanism whereby KSHV latent infection regulates Rb (Chang et al., 1996; Godden-Kent et al., 1997). More recently, the latency-associated nuclear antigen (LANA), encoded by ORF 73, has also been shown to target Rb (Radkov et al., 2000). Functionally homologous to EBNA1, LANA clearly plays a role in facilitating the segregation of viral episomes into daughter nuclei during mitosis. However, it has also been suggested, based on the ubiquitous detection of LANA in KSHV-associated malignancies, that LANA may play a role in KSHV-mediated oncogenesis. Indeed LANA was shown to transactivate E2F-regulated reporter genes, while physically associating with Rb both *in vitro* and *in vivo* (Radkov et al., 2000). LANA expression also gave phenotypes typical of Rb-antagonizing oncogenes including the ability to rescue an Rb-induced flat cell phenotype in Saos-2 cells and transformation of primary rat embryonic fibroblasts in cooperation with H-ras (Radkov et al., 2000). In support of the link between LANA and Rb, a recent study from our lab has suggested that a LANA homologue encoded by ORF73 of Herpesvirus Saimiri (a virus of the squirrel monkey more closely related to KSHV than EBV) similarly regulates Rb function. However, as the role of KSHV in viral-associated malignancies is still not well understood, and as no clear model system exists for KSHV-mediated oncogenesis, the role of ORF 72 and 73 in KSHV-associated tumors remains to be firmly established.

Table 1. EBV latent antigens associated with cell cycle regulation

Essential latency antigen	Reported cell cycle regulation	References
EBNA-LP	promotes G0/G1 transition and cyclin D2 induction in cooperation with EBNA2 in primary B cells effect may be limited to quiescent cells	Sinclair et al., 1994
	binds Rb and p53 *in vitro*, although regulation of these pathways has not been demonstrated *in vivo*	Inman and Farrell, 1995; Jiang et al., 1991; Szekely et al., 1993
	phosphorylated by cdk1 which may facilitate its co-activation of the LMP1 promoter along with EBNA2	Kato et al., 2003; Kitay and Rowe, 1996
EBNA2	promotes G0/G1 transition and cyclin D2 induction in cooperation with EBNA-LP in primary B-cells	Sinclair et al., 1994
	activates a reporter gene fused to the cyclin D2 promoter independent of EBNA-LP	Spender et al., 2001
	functional disruption induces G1 and G2 arrest in LCLs	Kempkes et al., 1995
	induces transcription of the c-*myc* gene	Kaiser et al., 1999
	binds and is phosphorylated by cdk1 which may abrogate activation of the LMP1 promoter	Yue et al., 2004
	may cause growth arrest in some transformed cell types	Lin et al., 2000
EBNA3A	induces G0/G1 growth arres with prolonged viability	Cooper et al., 2003
EBNA3C	rescues LMP1 levels during G1 arrest of Raji cells	Allday et al., 1993
	binds Rb *in vitro* and regulates Rb pathways in primary rat fibroblasts	Parker et al., 1996
	promotes entry into S-phase in spite of serum withdrawal in NIH-3T3 and U2OS cells	Parker et al., 2000
	promotes aberrant passage through mitotic spindle checkpoints in response to DNA damage	Parker et al., 2000
	binds cyclin A in LCLs and stimulates cyclin A-associated kinase activity	Knight and Robertson, 2004
	functionally disrupts the association between cyclin A and p27 in BJAB cells	Knight and Robertson, 2004
EBNA1	prolongs G2/M phase	Chuang et al., 2003
LMP1	promotes S-phase entry in primary B cells	Peng and Lundgren, 1992
	induces cyclin D2 transcription in EBV-negative BL cells	Arvanitakis et al., 1995

EPSTEIN - BARR VIRUS AND THE CELL CYCLE – A HISTORICAL PERSPECTIVE

In this chapter we will focus on those studies which address the role of type III latency gene products in cell cycle regulation (Table 1). This is justified based not only on the already firmly established link between type III latency and both *in vitro* and *in vivo* immortalization, but also on our emerging understanding of the contribution of type III latency to normal EBV biology, likely leading to the activation and expansion of infected naïve B-lymphocytes in the tonsillar lymph nodes. In spite of this focus, the role of individual gene products in the initiation of EBV-associated malignancies such as Burkitt's lymphoma and Hodgkin's disease should not be discounted. The initial events precipitated by viral infection that contribute to frank tumorigenesis remain poorly understood. Disruption of normal cell cycle gatekeeper and caretaker functions may contribute to genomic instability which could facilitate tumorigenesis. Recently identified links between DNA tumor viruses and genomic instability have been reviewed elsewhere (Lavia et al., 2003).

The prevailing model of EBV cell cycle regulation suggests that the virus utilizes its latency proteins to drive B-lymphocyte proliferation, inducing a phenotype that closely mimics antigen-driven B-lymphocyte activation and expansion. In the case of immortalization of B-lymphocytes *in vitro*, and likely for EBV-driven B-lymphocyte expansion *in vivo*, this process occurs in the absence of B-cell receptor (BCR) engagement and T-cell signalling. An open and important question is whether EBV latency antigens directly bind and modulate the function of cell cycle regulatory proteins as has been well described for the small DNA tumor virus proteins or whether EBV relies more heavily on transcriptional regulation to modify the cell cycle program. The first stories to emerge implicated the latter with roles suggested for EBNA2, EBNA-LP, and LMP1. However, more recent studies have suggested a role for EBNA3C in directly binding and regulating critical cell cycle proteins. Below we will further explore this data for individual latency gene products, but first we will consider studies that have more broadly addressed the pathways targeted by EBV latency.

An important study in 1993 by Palmero and colleagues identified cyclin D2 as the predomiant D-type cyclin in cell lines expressing EBV type III latency (Palmero et al., 1993). Their work began with the observation that the best studied D-type cyclin, cyclin D1, is not present in most B-lineage cell lines leading to the hypothesis that other related gene products may functionally substitute for cyclin D1 in these cells. A probe containing the coding region of mouse cyclin D1 was used to screen a Raji cDNA library, yielding two overlapping phage clones with sequence corresponding to human cyclin D2 (Palmero et al., 1993). A probe corresponding to cyclin D2 was then hybridized against numerous B cell lines; it was noted that cyclin D2 is expressed predominantly in EBV-positive Burkitt's lymphoma lines with a type III latency pattern of gene expression and in LCLs, but not in Burkitt's lymphoma lines with a type I latency pattern (where Cp is silent and only EBNA1 is expressed from an alternative promoter Qp) or in non-EBV infected lines (Palmero et al., 1993). This study clearly hinted at a link between the expression of EBV latency genes and that of cyclin D2. Subsequent studies have demonstrated that expression of cyclin D2 is dependent upon EBV gene expression as UV-irradiated virus does not induce cyclin D2 transcripts (Spender et al., 1999). In fact, specific EBV latency antigens have now been linked to cyclin D2 transcription as will be discussed below.

A follow up study demonstrated a role for methylation in silencing the cyclin D2 promoter in at least one Burkitt's lymphoma cell line with a type I latency pattern of expression (Sinclair et al., 1995). Treatment of Mutu Cl 179 Burkitt's lymphoma cells with the demethylating agent 5-azacytidine leads to changes in the methylation status of a CCGG restriction site near the transcription start site of the cyclin D2 gene, and, importantly, leads to expression of the cyclin D2 gene as detected by RT-PCR (Sinclair et al., 1995). While these data have not been further explored, they hint at an interesting role for methylation of the cyclin D2 promoter in transitioning between different types of EBV latency.

As is mentioned above, cyclin D2 is not typically expressed in Burkitt's lymphoma cell lines that are either EBV-positive with a type 1 latency pattern of gene expression or EBV-negative. In these lines, a third D-type cyclin, cyclin D3 is utilized (Palmero et al., 1993; Pokrovskaja et al., 1996). This pattern is consistent across numerous cell lines tested (Pokrovskaja et al., 1996). While the D-type cyclins all share similar functionality, cyclin D3 differs from D1 and D2 in that it lacks a serum-inducible box and certain negative regulatory elements in its promoter consistent with different regulatory roles for the cyclin D proteins (Herber et al., 1994). Interestingly, infection of EBV-negative Burkitt's lymphoma cell lines with EBV, while resulting in the expression of all EBNAs and LMPs, does not result in a switch from cyclin D3 to cyclin D2 usage (Pokrovskaja et al., 1996). The overall impression of these studies is that preferential D-type cyclin promoter usage is quite stably maintained and may be highly dependent on the properties of the B-lymphocytes from which these cell lines were derived. Further studies are clearly necessary to dissect this complex interplay between viral factors and the host cell in D-type cyclin regulation.

Since the initial characterization of EBV's putative role in the induction of cyclin D2, additional cell cycle regulatory proteins have been considered. One study compared transcript levels in primary B-lymphocytes isolated from peripheral blood to freshly established LCLs (Hollyoake et al., 1995). The authors found approximately 100-fold upregulation of not only cyclin D2 transcripts, but also cyclin E and cdc-2 (cdk1) as detected by RT-PCR (Hollyoake et al., 1995). To determine whether these genes were indeed early targets of EBV infection, freshly isolated B-lymphocytes were either infected with EBV or treated with a cocktail of anti-CD40, anti-IgM, and IL-4 to stimulate B-cell proliferation. After 4 days of treatment, cyclin D2, cyclin E, and cdk1 transcripts are clearly detectable in both EBV-infected and cocktail-treated cells, but not in control cells (Hollyoake et al., 1995). While this study did little to implicate specific latency proteins in these processes, it did suggest that infection with EBV may to some degree mimic antigen-dependent B-cell activation.

The same group elaborated on the above with a related study which employed western blotting to examine numerous cell cycle regulatory proteins including the pocket-family proteins, cyclins, cdks, and CDKIs (Cannell et al., 1996). For most experiments, primary B-lymphocytes from the peripheral blood were compared to a freshly-immortalized LCL. For the pocket-family proteins, p130 levels decrease, while p107 and Rb levels paradoxically increase (Cannell et al., 1996). However, the increase in p107 and Rb is accompanied by retarded mobility on SDS-PAGE consistent with a shift to the inactive, hyperphosphorylated forms of the proteins (Cannell et al., 1996). The Rb result was confirmed by monitoring fresh infection of B-lymphocytes by EBV. By 51 hours, Rb levels are significantly increased and a clear accumulation of the hyperphosphorylated species is present. By 144 hours, the Rb profile is essentially indistinguishable from that seen in fully established LCLs (Cannell et al., 1996). Additionally, cdk2, cdk4, cdk6, cyclin D2, and cyclin E are all strongly upregulated in the LCL compared to B-lymphocytes (Cannell et al., 1996). Considering the CDKIs, both p16 and p27 are strongly downregulated, p15 levels are essentially unchanged, and p21 is upregulated in the LCL (Cannell et al., 1996). In spite of p21 upregulation, functional cyclin E/cdk2 complexes are easily detected by immunoprecipiation from LCLs (Cannell et al., 1996). It is important to point out, that, with the exception of Rb for which an infection/ time-course analysis was utilized, this data was generated by simply comparing primary B-lymphocytes to a lymphoblastoid cell line. As such, some of the aforementioned changes are presumably more directly linked to EBV antigens than others. However, to our knowledge, this study represents the only global consideration of critical cell cycle regulatory molecules in EBV-immortalized cells. Such studies are challenging due, in part, to an inability to monitor individually infected cells until they have generated an obvious lymphoblastoid cell line in tissue culture. Further studies, perhaps with novel viral reagents that can be tracked during the immortalization process, are needed to confirm the above and better our understanding of the molecules most critically and directly targeted by EBV.

EBNA2 REGULATES EARLY EVENTS IN CELL CYCLE ENTRY

As is described in detail above, studies comparing D-type cyclin transcript levels in B-cell lines differentially infected with EBV suggest that cyclin D2 is preferentially expressed in the context of EBV type III latency (Palmero et al., 1993). This observation hinted that one of the type III latency antigens may be responsible for either direct or indirect upregulation of the cyclin D2 promoter. Sinclair and colleagues infected B-cells with B95-8 EBV and monitored viral gene expression by western blotting and cyclin D2 transcript levels by RNase protection assay (Sinclair et al., 1994). EBNA-LP and EBNA2 are both detectable by 24 hours, while LMP1 is not detected until 144 hours (Sinclair et al., 1994). The EBNA3 proteins and EBNA1 were not considered in this study. Similar to EBNA-LP and EBNA2, cyclin D2 mRNA is readily detectable by 24 hours post-infection and is dependent upon viral protein expression as either UV-irradiation of virus or treatment of cells with a protein synthesis inhibitor anisomycin readily blocks the effect (Sinclair et al., 1994). As was hinted above, the timing of cyclin D2 transcription suggests a potential role for EBNA-LP and EBNA2 in this process. Indeed, delivery of EBNA-LP and EBNA2 expression plasmids into primary B-lymphocytes by electroporation, in cooperation with treatment of the cells with a purified preparation of the EBV surface glycoprotein 340, upregulates cyclin D2 transcription (Sinclair et al., 1994). Neither EBNA2 nor EBNA-LP alone is sufficient to stimulate transcription (Sinclair et al., 1994). It should be pointed out that transcripts were detected by RT-PCR in this assay necessitated by low levels of induction. Additionally, EBNA protein levels were below the level detectable by western blotting due to limited expression of plasmids in primary B-lymphocytes even with gp340 stimulation. Also, other EBNAs or LMPs were not considered in the transfection protocol. Nevertheless, these data suggest that the expression of only two viral genes, EBNA2 and EBNA-LP, along with gp340 stimulation, may be sufficient to convert a resting B-lymphocyte into a B-lymphocyte that is primed for S-phase entry.

Another study further implicates EBNA2 in the early events of G1 regulation. Kempkes and colleagues constructed a system whereby EBNA2's transcription regulatory functions were estrogen-inducible (Kempkes et al., 1995). Sequence encoding the estrogen binding domain of the estrogen receptor (ER) was fused to either the 5' or 3' end of the EBNA2 gene such that in-frame fusion proteins, either ER/EBNA2 or EBNA2/ER, were expressed; fusion proteins were dependent on estrogen for activity as both transactivated the LMP1 promoter in the presence, but not absence, of estrogen (Kempkes et al., 1995). The inducible EBNA2 genes were cloned into a mini-EBV vector containing oriP, oriLyt, and TR; these elements permit latent replication, lytic replication, and packaging of the viral genome, respectively (Kempkes et al., 1995). As is described in detail above, P3HR1 cells are infected with a virus that replicates lytically but is deficient for EBNA2 and consequently cannot establish fresh LCLs. P3HR1 cells were transfected with the mini-EBV vector and induced for lytic replication. Supernatants, presumably containing both P3HR1 viral genomes and mini-EBV vectors, were then used to infect freshly-isolated primary B-lymphoctyes in the presence of estrogen.

Of eight lymphoblastoid cell lines analyzed, all contained both mini-EBV genomes (expressing chimeric EBNA2 protein) as autonomously replicating plasmids and P3HR1 viral genomes (Kempkes et al., 1995). Upon withdrawal of estrogen from the culture medium, approximately 50% of the cells remain viable for at least 5 days, but within this viable population the proportion of S-phase cells is severely decreased (Kempkes et al., 1995). In contrast, the distribution of cells in G1 and G2-phases is not significantly altered upon the removal of estrogen suggesting at least two blocks to cell cycle progression, one at G1 and one at G2 (Kempkes et al., 1995). Upon re-addition of estrogen, G1-arrested cells are able to begin DNA synthesis; however, as no mitotic cells are detected until after DNA synthesis is completed, the G2-arrested population apparently does not immediately recover in the presence of functional EBNA2 (Kempkes et al., 1995). As a whole, these data are supportive of a role for EBNA2 in early cell cycle events as G1-arrested cells accumulate in the absence of functional EBNA2 and then enter S-phase when functional EBNA2 is again present (although whether EBNA2 is directly regulating these events can again be questioned). It is interesting that cells also arrest in G2 with estrogen withdrawal suggesting a novel role for EBNA2, or perhaps another latency protein whose transcription is dependent on EBNA2, in G2 progression. The inability of EBNA2, upon estrogen re-addition, to immediately stimulate mitosis could be consistent with a role for another EBV antigen in this transition.

In a subsequent study, the same EBNA2 estrogen-responsive system was used to address whether EBNA2 is directly regulating the promoters of cell cycle regulatory genes or whether these effects are indirect and dependent upon de novo protein synthesis (Kaiser et al., 1999). As above, estrogen withdrawal was used to arrest cells. After re-addition of estrogen, transcript levels for cyclin D2, cdk4, c-*myc* and LMP1 were assayed both in the presence and absence of the protein synthesis inhibitors anisomycin and cycloheximide. EBNA2 induces c-*myc* and LMP1 transcription independent of de novo protein synthesis (Kaiser et al., 1999). For LMP1, the induction profiles are very similar in the presence and absence of cycloheximide; in contrast, while c-*myc* transcription is clearly induced in spite of cycloheximide treatment, it never reaches the levels seen for untreated samples suggesting that EBNA2 may regulate the c-*myc* promoter through both direct and indirect mechanisms (Kaiser et al., 1999). Indirect regulation is clearly suggested for both cyclin D2 and cdk4 as EBNA2-mediated induction of transcription is completely ablated in the presence of protein synthesis inhibitors (Kaiser et al., 1999).

While the aforementioned studies suggest a role for EBNA2 in induction of cyclin D2 transcription, this effect is apparently dependent upon de novo protein synthesis, hinting at a critical role for intermediary factors (Kaiser et al., 1999). The Myc proto-oncogene is known to directly induce the cyclin D2 promoter; specifically, the cyclin D2 promoter is repressed by Mad-Max complexes and de-repressed by Myc via a highly conserved E-box element (Bouchard et al., 1999). Myc has also been shown to regulate the proliferation of a Burkitt's lymphoma model cell line in which c-*myc* was placed under the control of a tetracycline-inducible promoter (Pajic et al., 2000). In this system, the G1/S transition and cyclin D2 expression are dependent upon the Myc protein (Pajic et al., 2000). As the c-*myc* gene can be directly regulated by EBNA2 (Kaiser et al., 1999), a question of considerable interest was whether most or all of EBNA2's cell cycle effects might be mediated through induction of c-*myc*.

This question has been partially addressed by luciferase report studies. EBNA2 activates a reporter containing 1624 basepairs of the cyclin D promoter approximately 10-fold in DG75 cells (Spender et al., 2001). Interestingly this activation is neither dependent on, nor enhanced by, expression of EBNA-LP conflicting with previous data suggesting that both EBNA-LP and EBNA2 are necessary to induce cyclin D2 transcription in primary B-cells (Sinclair et al., 1994). The authors suggest that EBNA-LP's role may be restricted to quiescent cells, thereby making it dispensable in the reporter assay (Spender et al., 2001). Importantly, truncation of the cyclin D2 promoter to -652, which removes the aforementioned E-box regulatory elements necessary for Myc activation, does not abrogate EBNA2 activation (Spender et al., 2001). Indeed, EBNA2 activates this truncation promoter approximately 20-fold (Spender et al., 2001). Further truncation mutants gradually reduce activity leaving no single element as the definitive regulator of the cyclin D2 promoter by EBNA2 (Spender et al., 2001). Additional work will be necessary to elucidate the factors required for regulation of the cyclin D2 promoter by EBNA2; however, at present it appears that factors other than Myc are mediating this effect.

More recently, EBNA2 has been linked to another cell cycle regulatory protein cdk1 (p34^{cdc2}) (Yue et al., 2004). Cdk1 forms complexes with both cyclin A and cyclin B and is a critical regulator of mitotic progression. EBNA2 and cdk1 co-immunoprecpitate from LCL lysates, while hyperphosphorylation of EBNA2 correlates temporally with increased cdk1 activity (Yue et al., 2004).

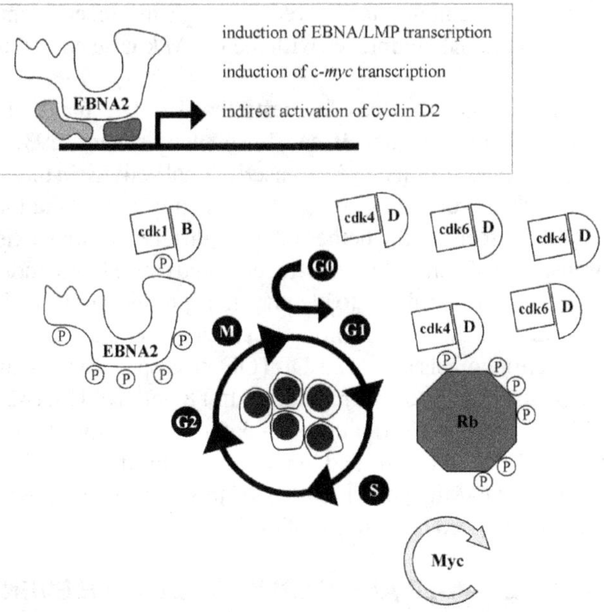

Figure 1. EBNA2 targets early events in B-lymphocyte activation and cell cycle progression. In addition to serving as the master regulator of EBV-latent transcription, EBNA2 also regulates cellular gene transcription, including direct induction of c-*myc* and indirect activation of cyclin D2. EBNA2 is also phosphorylated by cdk1 in M-phase. D, cyclin D2. B, cyclin B.

Interestingly, hyperphosphorylation of EBNA2 may suppress its ability to activate the LMP1 promoter based on comparative reporter assays in M-phase arrested and asynchronously-growing cells (Yue et al., 2004). Consistent with this result, PU.1, a critical factor responsible for targeting EBNA2 to the LMP1 promoter, co-immunoprecipitates less strongly with EBNA2 in M-phase cells (Yue et al., 2004). The advantages associated with downregulating EBNA2-dependent transcriptional activity during M-phase are not entirely clear at present, and further studies of additional EBNA2-regulated genes as well as mapping of specific phosphorylation sites are clearly necessary. Also, potential regulation of cdk1 activity by EBNA2 has not been explored.

In summary, EBNA2 is essential for primary B-lymphocyte immortalization by EBV. Cumulative data suggests a role for EBNA2 in early events in cell cycle progression (Figure 1). In cooperation with EBNA-LP, EBNA2 induces the G0 to G1 transition in primary B-lymphocytes as defined by expression of cyclin D2. Also, disruption of EBNA2 function in established LCLs causes a block to cell cycle progression in both G1 and G2 phases; restoration of EBNA2 promotes S-phase entry. Historically, cyclin D2 induction was the first cell cycle protein to be clearly regulated by EBNA2; however the data now suggests that this effect is indirect as de novo protein synthesis is required for promoter induction. Importantly, cyclin D2 induction is not completely dependent upon EBNA2's ability to activate the c-*myc* gene (a known regulator of cyclin D2), and consequently the identity of the critical intermediary factor or factors is not clear). Why disruption of EBNA2 causes arrest in G2 also remains unresolved although the recent demonstration that EBNA2 forms functional complexes with the G2/M kinase cdk1 may hint at a mechanism.

It should be noted that one study suggests that EBNA2 may actually have growth suppressive functions in non-B cell lines such as Vero, 293, and U2OS possibly due to its ability to stimulate p53 transcriptional activity in these cells (Lin et al., 2000). However, more studies are needed to better understand the cell-type specificity of EBNA2 effects, and whether inherent differences may exist between primary and transformed cells. Another recent study used the inducible ER/EBNA2 LCL system described above to identify gene products induced by EBNA2 following withdrawal and then re-addition of estrogen (Schlee et al., 2004). Two-dimensional gel electrophoresis and MALDI-TOF mass spectrometry identified at least 20 proteins as either up- or down-regulated after 8 hours of EBNA2 activation (Schlee et al., 2004). None were obvious regulators of cell cycle function, and, importantly, cyclin D2 was not identified in this screen. Further experimentation is needed to better understand the gene products of this screen and how they might contribute to EBNA2-induced cell proliferation.

A LINK BETWEEN EBNA-LP AND EARLY CELL CYCLE EVENTS.

As outlined above, EBNA-LP, is the first latency protein detected during de novo infection of primary B-lymphocytes by EBV with robust expression by 24 hours post-infection, timing that suggests an early role in cell activation and cell cycle progression. In addition, a more specific role for EBNA-LP in cell cycle regulation is indirectly suggested by transfection experiments where both EBNA-LP and EBNA2, but neither protein alone, induce expression of cyclin D2 transcripts in resting B-lymphocytes (Sinclair et al., 1994). Based on these clues, early attempts were made to link EBNA-LP directly to the critical cell cycle gatekeepers Rb and p53 (Jiang et al., 1991). Immunofluorescence localization studies employed a specific Rb monoclonal antibody in IB4 cells (essentially a long-term LCL); this antibody detects large intranuclear foci that also stain with the EBNA-LP monoclonal antibody JF186 (Jiang et al., 1991). However, another Rb monoclonal antibody stains the nuclei of IB4 cells diffusely, without distinct nuclear foci (Jiang et al., 1991).

The potential association between Rb and EBNA-LP was further developed in a subsequent study (Szekely et al., 1993). GST-Rb precipitates low levels of EBNA-LP from an EBNA-LP over-expressing cell line (Szekely et al., 1993). A mutant Rb (C706F) that is deficient for large T-antigen binding also precipitates EBNA-LP at a level similar to wild-type; however, in a completely *in vitro* binding assay, an E7-derived peptide competes with EBNA-LP for Rb binding making it unclear whether EBNA-LP is binding to the same region of Rb as these well-studied tumor virus antigens (Szekely et al., 1993). The authors additionally show that GST-p53 precipitates EBNA-LP from EBNA-LP over-expressing cells and that both Rb and p53 binding map to the W repeat at the amino-terminus of EBNA-LP (Szekely et al., 1993). Whether EBNA-LP disrupts either Rb or p53 functional pathways was not addressed in this study.

A subsequent report by Inman and colleagues questioned the functional relevance of the aforementioned *in vitro* association between EBNA-LP and the tumor suppressors Rb and p53 (Inman and Farrell, 1995). Using transfection reporter assays, the ability of EBNA-LP to regulate Rb and p53 function was compared to known viral oncoproteins SV40 large T-antigen and human papillomavirus E6. To test Rb function, the B-cell line DG-75 was transfected with a GAL4 chloramphenicol acetyl-transferase reporter and a GAL4-E2F-1 fusion protein. The reporter system is responsive to GAL4-E2F-1, an effect that is suppressed by both Rb and p107. While large T-antigen is able to rescue both Rb and p107 suppression, EBNA-LP rescues neither, despite extensive titration (Inman and Farrell, 1995). To test p53 function, Akata cells, a Group I Burkitt's lymphoma line null for p53, were transfected with an artificial p53 reporter and a wild-type p53 expression plasmid. EBNA-LP, again despite extensive titration of the expression plasmid, does not prevent p53-mediated transactivation of the reporter (Inman and Farrell, 1995). In contrast, the human papillomavirus E6 protein potently suppresses p53-mediated transactivation in this assay (Inman and Farrell, 1995). These results suggest that if EBNA-LP does target Rb and p53 *in vivo*, then it either affects other functions of the proteins or modulates their gene regulatory properties such that the effects are not detected by transfection into cycling, transformed cells.

Similar to EBNA2, there is data to indicate that EBNA-LP is phosphorylated in a cell cycle dependent manner. One study suggests that EBNA-LP in X50-7 cells is maximally phoshorylated when cells are arrested in G2/M by nocodazole treatment or when cells are synchronized in G1 by hydroxyurea and then released for progression into G2 (Kitay and Rowe, 1996). Amino acid analysis demonstrates phosphorylation on serine residues which is consistent with *in vitro* kinase assays demonstrating phosphorylation of EBNA-LP by both casein kinase 2 and cdk1 (Kitay and Rowe, 1996). Preliminary mapping experiments suggest that cdk1-mediated phosphorylation most likely occurs on the W2 exon-encoded region of EBNA-LP (Kitay and Rowe, 1996). A subsequent study by another group also indicates that cdk1 can phosphorylate EBNA-LP both *in vitro* and *in vivo*, implicating serine-35 in the W2 repeat region as a major site for this phosphorylation (Kato et al., 2003). Interestingly, it has been suggested that this serine is critical for regulation of the coactivator function of EBNA-LP as serine-35 mutants are impaired in their ability to activate the LMP1 promoter in concert with EBNA2 (Knight et al., 2003; Yokoyama et al., 2001). This is in contrast to data for EBNA2 suggesting that hyperphosphorylation of EBNA2 in M-phase disrupts PU.1 binding and LMP1 transactivation (Yue et al., 2004). The manner by which cdk1 regulates the transcriptional properties of EBNA2 and EBNA-LP, and ultimately activation of the LMP1 gene, appears complicated and will require further studies to understand the cell cycle-dependent interplay between these two phosphoproteins.

In summary, the expression pattern of EBNA-LP, the first latency protein detected during EBV infection of naïve B-lymphocytes *in vitro*, is perhaps the strongest evidence to suggest that EBNA-LP plays a critical role in early B-lymphocyte activation events. Indeed, in a system where EBNA-LP and EBNA2 are delivered by transfection into primary B-lymphocytes, EBNA-LP is necessary for EBNA2-mediated induction of cyclin D2 transcripts. However, reporter assays in asynchronously cycling cells suggest that EBNA-LP is dispensable for cyclin D2 induction. EBNA-LP binds both Rb and p53 *in vitro*, although the functional relevance of this observation is not clear as EBNA-LP does not functionally disrupt either Rb or p53 in transcriptional reporter assays. Similar to EBNA2, EBNA-LP functionally associates with cdk1 and is itself phosphorylated; interestingly, while phosphorylation of EBNA2 may antagonize its ability to activate the LMP1 promoter, phosphorylation of EBNA-LP stimulates activation. As LMP1 expression is regulated in a cell cycle-dependent fashion in some unique LCL backgrounds (Allday and Farrell, 1994), the role of phosphorylation in this process seems to warrant further study.

EBNA3C REGULATION OF CELL CYCLE EVENTS

EBNA3C of type 1 EBV is a 992 amino acid protein that localizes to the nucleus in either a diffuse or punctuate pattern depending on the cell type and expression level (Knight et al., 2003; Rosendorff et al., 2004; Subramanian et al., 2001). EBNA3C has been shown to play a regulatory role in the transcription of both viral and cellular genes. EBNA3C antagonizes EBNA2-mediated transactivation by competing with EBNA2 for RBP-Jκ binding (Johannsen et al., 1996; Radkov et al., 1997; Robertson et al., 1996), and, conversely, cooperates with EBNA2 in the upregulation of some promoters like the LMP1 promoter (Allday et al., 1993; Allday and Farrell, 1994; Zhao and Sample, 2000). Additionally, EBNA3C recruits both histone acetylase and deacetylase activities (Knight et al., 2003; Radkov et al., 1999; Subramanian et al., 2002), in part through its association with the small acidic nuclear protein Prothymosin-alpha (Cotter and Robertson, 2000).

EBNA3C was first linked to the cell cycle, albeit indirectly, through studies of LMP1 expression in the lymphoblastoid cell line Raji (Allday and Farrell, 1994). The Raji viral genome has a large deletion that removes essential lytic genes rendering the Raji virus deficient for lytic replication. Importantly for this study, the Raji deletion also removes the EBNA3C coding region although all other latency antigens are expressed. LMP1 is normally one of the last latent genes to be expressed, the likely result of its known dependence on other latent genes, specifically EBNA2, for its expression. However, expression of EBNA2 does not uniformly induce expression of LMP1. In fact, during early infection of resting B cells no LMP1 is detected for many hours, and in some cases days, after EBNA2 is expressed (Alfieri et al., 1991). It is now known that EBNA3C plays an important role in LMP1 regulation in cooperation with EBNA2, but at the time of this study it was quite surprising to find a phenotype apparently unique to Raji cells, namely that LMP1 expression is dependent on the proliferative state of the cells (Allday and Farrell, 1994). Shortly after feeding, LMP1 levels increase; however, 96 hours later the levels are almost undetectable when the majority of the cells are arrested in the G1 phase of the cell cycle (Allday and Farrell, 1994). Induction of LMP1 is independent of the presence of serum in the diluting medium strongly suggesting that LMP1 expression is not due to stimulation by serum factors (Allday and Farrell, 1994). If EBNA3C is expressed in trans in Raji cells, the level of LMP1 closely mimics that seen in LCLs expressing a full type III latency pattern of gene expression and is maintained through G1, implicating EBNA3C as a potential player in early EBV infection prior to progression of infected cells through the restriction point and on into S phase (Allday and Farrell, 1994).

Subsequent studies by Parker and colleagues further, and more specifically, implicated EBNA3C in cell cycle regulation (Parker et al., 1996). B-*myb* is a cellular gene that is normally repressed in G1 and is activated by the E2F family of transcription factors. EBNA3C activates the B-*myb* promoter by gene reporter assay in DG-75 cells (Parker et al., 1996). A promoter with a mutant E2F site was relatively non-responsive to EBNA3C suggesting specificity for E2F and potentially Rb/E2F restriction point complexes (Parker et al., 1996). To further dissect the role of EBNA3C in the regulation of fundamental Rb pathways, EBNA3C was tested against human papillomavirus E7 in primary rodent fibroblast transformation assays. Indeed, EBNA3C and H-ras induce transformed colonies with a similar efficiency as E7 and H-ras (Parker et al., 1996). To more directly address the transforming potential of EBNA3C as it relates to Rb, fibroblast transformation was assayed in the presence of the CDKI p16^{INK4a} known to bind and potently inhibit cdk4 and cdk6. p16 inhibits transformation of rat fibroblasts by H-ras in combination with oncogenes such as c-*myc* and highly-transforming p53 mutants that do not specifically target Rb (Parker et al., 1996). In contrast, viral oncoproteins such as E1A and E7 that antagonize Rb function are resistant to suppression of transformation by p16. In this study, both EBNA3C and E7, but not mutant p53, overcome suppression by p16 further implicating EBNA3C in the regulation of Rb pathways (Parker et al., 1996). The final link between EBNA3C and Rb was in the context of *in vitro* binding studies. Bacterially-expressed Rb binds *in vitro*-translated EBNA3C in a manner that is at least partially dependent upon the pocket domain of Rb (Parker et al., 1996). The pocket domain has classically been shown to interact with the consensus LxCxE motif present in SV40 large T-antigen, human papillomavirus E7, and adenovirus E1A. Although this consensus motif is not present in EBNA3C, the EBNA3C protein showed several-fold lower affinity for an Rb pocket domain mutant hinting that the Rb-EBNA3C interaction might functionally resemble the interaction between Rb and the aforementioned tumor virus antigens (Parker et al., 1996). It must be noted that despite this apparently specific *in vitro* interaction (and similar to EBNA-LP), there is no evidence showing that EBNA3C and Rb associate *in vivo*. As such it is left to be determined if these data can be substantiated with consistent *in vivo* experimental evidence.

Another study by the same group investigates more directly the ability of EBNA3C to regulate cell cycle progression (Parker et al., 2000). EBNA3C-positive and negative NIH3T3 cells show little difference in cell cycle profile when grown in 15% serum; however, serum-starved cells show increased DNA content in EBNA3C-positive versus negative populations. Specifically, after 48 hours in 0.5% serum, 32 percent of EBNA3C-posiitve cells showed 4N DNA content by propidium iodide staining, in contrast to 10 percent of EBNA3C-negative cells (Parker et al., 2000). Interestingly, re-gating of the same experiment demonstrated that a significant percentage of EBNA3C-positive cells had greater than 4N DNA content suggesting that DNA replication was proceeding in the absence of cell division (Parker et al., 2000). In support of this, cells containing from 1-7 nuclei were easily visualized.

To eliminate the possibility that long-term drug selection influences the cell cycle profile, further experiments were performed with U2OS cells manipulated to express EBNA3C upon Muristerone A induction. Serum-deprived, EBNA3C-expressing U2OS cells show a similar DNA profile to that observed previously, with a higher DNA content than EBNA3C-negative cells in 0.1% serum (Parker et al., 2000). Also, western blotting demonstrated that p27^{Kip1} protein levels dramatically deteriorate upon EBNA3C induction under reduced serum conditions (Parker et al., 2000). p27 is a CDKI known to potently bind G1 and S-phase cyclin/cdk complexes including cyclin E/cdk2 and cyclin A/cdk2. In general, the levels of p27 are highest in senescent cells and dramatically decrease upon cell cycle entry, therefore it is not clear from this study whether EBNA3C is directly regulating p27 or whether this is an indirect effect of EBNA3C-stimulated cell cycle progression. Surprisingly, in these studies Rb does not shift to the hyperphosphorylated state as would be expected in cycling cells, although total levels of the hypophosphorylated species do appear to decrease, suggesting that EBNA3C may specifically target and disable active, hypophosphorylated Rb repression complexes (Parker et al., 2000). However, it must be reiterated that a reproducible *in vivo* link between Rb and EBNA3C has yet to be demonstrated by co-immunoprecipitation or co-localization studies.

As EBNA3C expression is associated with multinucleated cells, the same authors tested whether EBNA3C overcomes mitotic spindle checkpoints induced by microtubule poisoning drugs. Indeed, EBNA3C-negative U2OS cells demonstrate a high mitotic index (80-90%) following treatment with nocodazole suggesting that the majority of cells have arrested by a mitotic spindle checkpoint (Parker et al., 2000). In contrast, EBNA3C-expressing cells have a dramatically lower mitotic index (10%) as these cells presumably bypass the aforementioned checkpoint and proceed through mitosis (Parker et al., 2000).

More recently, work from our lab has implicated EBNA3C in the regulation of cyclin/cdk complexes (Knight and Robertson, 2004). To identify EBNA3C protein binding partners, a truncated form of EBNA3C corresponding to amino acids 365-992 was fused in frame with the GAL4 DNA binding domain and tested against an LCL-derived cDNA library in a yeast two-hybrid screen. Partial sequences from two positive cDNA clones identified coding sequence within the amino-terminal one-third of the human cyclin A gene (Knight and Robertson, 2004). Indeed, cyclin A, an activator of S-phase progression, binds tightly to EBNA3C. EBNA3C binds cyclin A *in vitro* by pull-down experiments using a bacterially-expressed GST-cyclin A fusion protein and associates with cyclin A complexes in LCLs by co-immunoprecipitation studies (Knight and Robertson, 2004). As a measure of cyclin A-associated kinase activity, cyclin A complexes were immunoprecipitated from either EBNA3C-expressing or control U2OS cells and incubated with a known *in vitro* substrate for cdk activity, histone H1. While EBNA3C expression does not affect cyclin A protein levels, cyclin A-associated kinase complexes were at least five-fold more active in cells expressing EBNA3C (Knight and Robertson, 2004). A mechanism for this effect was suggested by the observation that EBNA3C potently rescues p27-mediated inhibition of cyclin A/cdk2 kinase in both U2OS and BJAB cells (Knight and Robertson, 2004). In BJAB, we also clearly see a decrease in the molecular association between cyclin A and p27 by co-immunoprecipitation experiments (Knight and Robertson, 2004). Interestingly, we do not observe potent downregulation of p27 protein levels as was described above when EBNA3C expression was induced under low serum conditions (Parker et al., 2000). However, our experiments were generally in the context of asynchronously cycling cells, not serum-starved, G1-arrested cells.

We additionally find that cyclin A is targeted by a region at the carboxy-terminus of EBNA3C. The extreme carboxy terminus of EBNA3C amino acids 957-992 shows some binding to cyclin A in an *in vitro* binding assay, and EBNA3C amino acids 621-992 replace the full-length molecule in rescuing p27 suppression of kinase activity (Knight and Robertson, 2004). However, in a subsequent report we find that although the carboxy terminus of EBNA3C predominantly regulates cyclin A-dependent kinase activity, the region of greatest affinity for cyclin A lies within the EBNA3 amino-terminal homology domain (Knight et al., 2004). Detailed mapping studies employing both *in vitro* binding assays and co-immunoprecipitation experiments implicate a small region of EBNA3C, amino acids 130-159, as having the greatest affinity for cyclin A. The EBNA3 homology domain (approximately amino acids 90-320) has the highest degree of amino acid similarity between the EBNA3 proteins, and, indeed, EBNA3B, but not EBNA3A, shows binding activity with cyclin A in an *in vitro* binding assay, although the relevance of this association *in vivo* has yet to be pursued. We also demonstrate that the amino terminus of EBNA3C binds to the α1 helix of the highly conserved mammalian cyclin box, with cyclin A amino acids 206-226 required for strong binding to EBNA3C. Interestingly, EBNA3C also binds human cyclins D1 and E *in vitro*, although the affinity was approximately 30% that seen for Cyclin A.

Surprisingly, the amino terminus of EBNA3C, with the highest binding affinity for cyclin A, is unable to rescue p27 suppression of kinase activity despite extensive titration of the expression levels of EBNA3C amino acids 1-365; this suggests that the carboxy terminus of EBNA3C is the predominant functional domain as discussed above. While the carboxy terminus of EBNA3C (amino acids 621-992) can regulate cyclin A kinase activity in the absence of the amino terminus in our assays, this is possibly an effect of the overexpression system employed. In LCLs, where EBNA3C levels are lower and EBNA3C protein may be localized differently, the amino terminus could play a critical role in recruiting cyclin/cdk complexes to EBNA3C. In support of this, delivery of EBNA3C amino acids 1-365, but not 1-129, into a recently established LCL by electroporation suppresses cyclin A-associated kinase activity. As a whole, these data suggest a complex interplay between different domains of EBNA3C in the regulation of cyclin/cdk activity which will require further experimentation and targeted mutation of the EBNA3C protein to clarify.

Another link between EBNA3C and the cell cycle, albeit indirect, can be inferred from its interaction with the small acidic nuclear protein Prothymosin α (Cotter and Robertson, 2000). A number of studies have indicated that Prothymosin α is associated with cellular proliferation events. For example, c-*myc* induction causes immediate transcriptional activation of Prothymosin α (Eilers et al., 1991; Vareli et al., 1995), and Prothymosin α antisense oligonucleotides inhibit myeloma cell division (Sburlati et al., 1991). Prothymosin α mRNA and protein accumulate maximally during S/G2 resembling Cyclin B expression (Vareli et al., 1996). While EBNA3C has yet to be directly linked to the cell cycle regulatory properties of Prothymosin α, EBNA3C and Prothymosin α do appear to interact at the level of chromatin to regulate both histone acetylase and deacetylase activities (Cotter and Robertson, 2000; Knight et al., 2003). Further studies will be needed to determine whether Prothymosin and EBNA3C regulate chromatin in a cell cycle-dependent fashion.

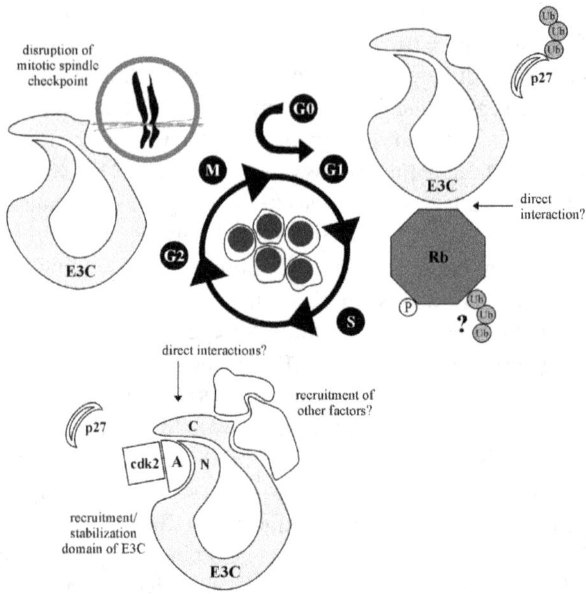

Figure 2. EBNA3C regulates multiple phases of the cell cycle, potentially though direct protein-protein interactions. EBNA3C regulates the G1/S transition by targeting Rb pathways and potentially by decreasing p27 stability. EBNA3C complexes with cyclin A/cdk2, and uses both N- and C-terminal domains to stimulate cyclin A-associated kinase activity; a mechanism may be the disruption of p27 from kinase complexes. EBNA3C also abrogates mitotic spindle checkpoints. Ub, ubiquitin. A, cyclin A. N, amino-terminus of EBNA3C. C, carboxy-terminus of EBNA3C.

In summary, EBNA3C is an essential latency antigen that was initially described as a negative regulator of EBNA2 transcriptional activity at the viral Cp promoter. Subsequent studies have, however, implicated EBNA3C in cell cycle regulation and B-lymphocyte proliferation (Figure 2). Similar to EBNA-LP, EBNA3C binds Rb *in vitro*. However, EBNA3C is also functionally linked to Rb pathways, inducing E2F-dependent promoters and stimulating primary rodent fibroblast transformation in cooperation with H-ras. Further, EBNA3C expression facilitates the progression of cells from G1 into later phases of the cell cycle and apparently promotes aberrant mitotic progression resulting in multinucleated cells. More recently, EBNA3C has been linked to cyclin/cdk complexes and specifically stimulates cyclin A/cdk2 kinase activity in some overexpression systems. EBNA3C may regulate cyclin/cdk activty by disrupting CDKIs such as p27 from these complexes. Multiple domains of EBNA3C have been implicated in this regulation and further studies are needed to understand their interplay.

LMP1 AND THE CYCLIN D2 PROMOTER

LMP1 has structural features and functions reminiscent of a constitutively active TNF family receptor. The signalling pathways targeted by this protein are myriad and have been reviewed in detail elsewhere (Eliopoulos and Young, 2001; Kieff and Rickinson, 2002). For the purposes of this chapter we will focus only on those topics directly related to cell cycle progression. In one study, LMP1 was delivered to primary B-lymphocytes by pretreating cells with UV-irradiated EBV and then delivering expression plasmids for LMP1 and a surface protein CD2 (Peng and Lundgren, 1992). CD2 was used as marker for transfected cells. Transient expression of the LMP1 gene induces cellular enlargement and up-regulation of activation markers such as CD23 and CD21. LMP1 also induces cells to enter S-phase (Peng and Lundgren, 1992). This is consistent with another study which demonstrates induction of cyclin D2 in LMP1-expressing, but not control, BL41 cells (an EBV-negative Burkitt's lymphoma cell line) (Arvanitakis et al., 1995). These same LMP1-expressing BL41 cells show interesting properties in response to TGF-β1 signalling. The TGF-β growth inhibition response is known to be dependent on the hypophosphorylated form of Rb and on inhibition of c-*myc* transcription (Arvanitakis et al., 1995). However, other mechanisms have also been suggested involving the regulation of cyclin/cdk complexes (Arvanitakis et al., 1995). With [^{3}H]thymidine incorporation as a measure of proliferation, control BL41 arrest when treated with TGF-β1; in contrast, LMP1-expressing cells continued to proliferate (Arvanitakis et al., 1995). The diffuse pathways regulated by TGF-β1 make interpretation of this result difficult, but it may be related to induction of cyclin D2 and resultant phosphorylation of the Rb protein.

In summary, the role of LMP1 in B-lymphocyte activation and immortalization is unquestioned. The potency of LMP1 in regulating such processes is underscored by its transforming ability in primary rodent fibroblast experiments. As LMP1 clearly provides anti-death signals to the cell, through NF-κB and other transcription factors, its role in B-cell immortalization is of fundamental importance. However, LMP1 has yet to be directly linked to the pathways and complexes that are thought to be most critical for classic cell cycle progression, namely Rb/E2F and cyclin/cdk. As the complex signalling cascades activated by LMP1 become even better elucidated, these links will very likely be demonstrated.

EPSTEIN-BARR VIRUS AND P53

Owing to the historical connection between Rb and p53 as critical factors targeted by tumor viruses, the regulation of p53 by EBV deserves brief mention here. The prevailing opinion is that EBV immortalizes B-lymphocytes without neutralizing the function of p53. An important study by Allday and colleagues demonstrates that, although EBV protects immortalized cells from apoptosis activated by the withdrawal of growth factors, EBV-immortalized B-lymphocytes, at least *in vitro*, remain responsive to DNA-damaging agents such as cisplatin which signal apoptosis through p53 (Allday et al., 1995). In fact, endogenous p53 levels are increased in LCLs relative to primary B-cells suggesting the virus is actually priming LCLs for apoptotic induction in response to some stimuli (Allday et al., 1995). These observations clearly suggest fundamental differences between EBV and the small DNA tumor viruses where abrogation of p53 function is essential for cell transformation and tumor formation. Recently KSHV LANA has also been shown to target p53 and disrupt p53 transcriptional effects (Friborg et al., 1999). These differences underscore the complexities of EBV-mediated immortalization and illustrate the fine symbiosis that has evolved between EBV and B-lymphocytes such that cell cycle progression is potently stimulated even in the presence of active p53.

EBV LYTIC REPLICATION AND THE CELL CYCLE

EBV-encoded genes expressed during viral latency have clear growth-promoting properties, and there is strong evidence that viral latency drives EBV-associated proliferative disorders and cancers. In contrast, there are few links between lytic viral replication and EBV-driven proliferation. Indeed, similar to Herpes Simplex virus and Cytomegalovirus, the EBV lytic program promotes cell cycle arrest in G1 which presumably facilitates lytic DNA replication (Flemington, 2001). To date, two specific EBV lytic gene products have been linked to cell cycle regulation, namely the immediate early viral transactivators, BZLF1 and BRLF1.

BZLF1 is sufficient for the initiation of lytic replication from latently infected cell lines and, with relevance to this discussion, induces G0/G1 cell cycle arrest when delivered into both EBV-positive and EBV-negative cell lines (Cayrol and Flemington, 1996b). This phenotype maps to a carboxyl-terminal region of Zta, excluding the primary transactivation domain, but including the bZIP domain (Cayrol and Flemington, 1996a). Expression of Zta results in the induction of the tumor suppressor protein, p53, and the cyclin-dependent kinase inhibitors, p21 and p27, and possibly in repression c-myc (Cayrol and Flemington, 1996b; Rodriguez et al., 2001; Zhang et al., 1994). It should, however, be noted that in some cell types such as primary keratinocytes and gastric carcinoma cells, BZLF1, may actually promote cell cycle progression (Mauser et al., 2002). A more thorough treatment of BZLF1 can be found in the chapter by Shannon Kenney.

The other EBV-encoded lytic transactivator, BRLF1, has been shown to interact with Rb and also to induce the expression of E2F1 using an adenovirus-BRLF1 transduction system (Swenson et al., 1999; Zacny et al., 1998). It remains to be fully resolved how these potential cell cycle-stimulating functions of BRLF1 synergize with BZLF1 to create what is presumably the optimal environment for viral DNA replication.

SYNTHESIS AND CONCLUSIONS

Does the initial activation of B-lymphocytes, which drives cells from quiescence into an activated and cycling state, require the same signals as are needed for the establishment of continuously proliferating, and potentially oncogenic, cell lines? The few studies which have ambitiously addressed the global impact of EBV on cell cycle regulators clearly suggest that EBV mimics normal B-lymphocyte activation upon de novo viral infection of resting primary B-lymphocytes. Indeed, EBNA2, EBNA-LP, and LMP1 may be the critical mediators of this activation phenotype providing signals similar to those seen with antigen-stimulated proliferation (Table 2). In general, studies of EBV-induced immortalization and cell cycle regulation have either considered the most initial stages of viral infection or the lymphoblastoid cell lines which are generated some weeks later. However, the critical period for understanding how EBV takes an activated lymphoblast and transforms it into a true immortalized cell line likely lies somewhere in between these two experimentally-accessible time points.

The studies examined above may lead the reader to believe that the EBNA3 proteins are dispensable for the earliest events in lymphoblast activation, and indeed they may be. However, EBNA3C, in particular, has properties that perhaps most closely resemble those of the well-described small-tumor-virus antigens in its

Table 2. Cell cycle transitions and the implicated EBV encoded antigens

Cell cycle transition	Implicated latency protein	Reference
G0/G1	EBNA2 +	Sinclair et al., 1994
	EBNA-LP +	Sinclair et al., 1994
	LMP1 +/-	Arvanitakis et al., 1995; Peng and Lundgren, 1992
G1/S	EBNA2 +	Sinclair et al., 1994; Spender et al., 2001; Kempkes et al., 1995; Kaiser et al., 1999
	LMP1 +	Arvanitakis et al., 1995
	EBNA3C +	Knight and Robertson, 2004; Parker et al., 1996; Parker et al., 2000
	EBNA-LP +/-	Sinclair et al., 1994
S/G2	EBNA3C +/-	Knight and Robertson, 2004
G2/M	EBNA2 +/-	Kempkes et al., 1995; Yue et al., 2004
	EBNA3C +/-	Parker et al., 2000

+ indicates a strong link; +/- indicates additional studies are needed

ability to interact directly with cell cycle proteins and in doing so modulate their function. Perhaps the critical role of the EBNA3 proteins will be found as we better understand the transition from lymphoblast to lymphoblastoid cell line. EBNA3C certainly appears to target multiple phases of the cell cycle and, as such, may be critical for facilitating immortal growth in culture (Table 2). Addressing this issue will require either inducible EBNA3 systems similar to the ER/EBNA2 system described above or the technology and knowledge to deliver dominant negative molecules into LCLs to assess the role of the EBNA3s in LCL maintenance.

Although clearly important, for the purposes of this chapter, we have generally not considered the role of cytokine-mediated signalling events as they relate to EBV-mediated B-lymphocyte proliferation and survival. An excellent review of this issue already exists (Brennan, 2001). To summarize the above, the role of the promiscuous transactivator EBNA2 and the EBV oncoprotein LMP1 in B-lymphocyte activation, proliferation, and survival is clearly paramount. However, the data generally suggests a diffuse picture where numerous intermediary molecules may be needed to facilitate these effects. EBNA-LP also clearly plays a supportive role in these processes and may yet be linked to a critical and very early event in B-lymphocyte activation. In contrast, our understanding of EBNA3C and its essential roles in B-lymphocyte transformation is more primitive. However, studies are beginning to point to a role for direct protein-protein interaction in the regulation B-lymphocyte proliferation, perhaps especially in the context of long-term LCL maintenance.

Acknowledgements

JSK is supported by the Lady Tata Memorial Trust. The project is supported by NIH grants NCI CA72150-07, NCI CA91792-01, NCI CA108461-01 and DCR DE14136-01 to ESR. ESR is a scholar of the Leukemia and Lymphoma Society of America.

References

Abbot, S. D., Rowe, M., Cadwallader, K., Ricksten, A., Gordon, J., Wang, F., Rymo, L., and Rickinson, A. B. (1990). Epstein-Barr virus nuclear antigen 2 induces expression of the virus- encoded latent membrane protein. J. Virol. *64*, 2126-2134.

Alfieri, C., Birkenbach, M., and Kieff, E. (1991). Early events in Epstein-Barr virus infection of human B lymphocytes [published erratum appears in Virology 1991 Dec;185(2):946]. Virology *181*, 595-608.

Allday, M. J., Crawford, D. H., and Thomas, J. A. (1993). Epstein-Barr virus (EBV) nuclear antigen 6 induces expression of the EBV latent membrane protein and an activated phenotype in Raji cells. J. Gen. Virol. *74*, 361-369.

Allday, M. J., and Farrell, P. J. (1994). Epstein-Barr virus nuclear antigen EBNA3C/6 expression maintains the level of latent membrane protein 1 in G1-arrested cells. J. Virol. *68*, 3491-3498.

Allday, M. J., Sinclair, A., Parker, G., Crawford, D. H., and Farrell, P. J. (1995). Epstein-Barr virus efficiently immortalizes human B cells without neutralizing the function of p53. Embo J. *14*, 1382-1391.

Arvanitakis, L., Yaseen, N., and Sharma, S. (1995). Latent membrane protein-1 induces cyclin D2 expression, pRb hyperphosphorylation, and loss of TGF-beta 1-mediated growth inhibition in EBV-positive B cells. J. Immunol. *155*, 1047-1056.

Babcock, G. J., Decker, L. L., Volk, M., and Thorley-Lawson, D. A. (1998). EBV persistence in memory B cells *in vivo*. Immunity *9*, 395-404.

Babcock, G. J., Hochberg, D., and Thorley-Lawson, A. D. (2000). The expression pattern of Epstein-Barr virus latent genes *in vivo* is dependent upon the differentiation stage of the infected B cell. Immunity *13*, 497-506.

Babcock, G. J., and Thorley-Lawson, D. A. (2000). Tonsillar memory B cells, latently infected with Epstein-Barr virus, express the restricted pattern of latent genes previously found only in Epstein-Barr virus-associated tumors. Proc. Natl. Acad. Sci. USA. *97*, 12250-12255.

Bouchard, C., Thieke, K., Maier, A., Saffrich, R., Hanley-Hyde, J., Ansorge, W., Reed, S., Sicinski, P., Bartek, J., and Eilers, M. (1999). Direct induction of cyclin D2 by Myc contributes to cell cycle progression and sequestration of p27. Embo J. *18*, 5321-5333.

Brennan, P. (2001). Signalling events regulating lymphoid growth and survival. Semin Cancer Biol. *11*, 415-421.

Buchkovich, K., Duffy, L. A., and Harlow, E. (1989). The retinoblastoma protein is phosphorylated during specific phases of the cell cycle. Cell *58*, 1097-1105.

Cannell, E. J., Farrell, P. J., and Sinclair, A. J. (1996). Epstein-Barr virus exploits the normal cell pathway to regulate Rb activity during the immortalisation of primary B-cells. Oncogene *13*, 1413-1421.

Cayrol, C., and Flemington, E. (1996a). G0/G1 growth arrest mediated by a region encompassing the basic leucine zipper (bZIP) domain of the Epstein-Barr virus transactivator Zta. J. Biol. Chem. *271*, 31799-31802.

Cayrol, C., and Flemington, E. K. (1996b). The Epstein-Barr virus bZIP transcription factor Zta causes G0/G1 cell cycle arrest through induction of cyclin-dependent kinase inhibitors. Embo J. *15*, 2748-2759.

Chang, Y., Moore, P. S., Talbot, S. J., Boshoff, C. H., Zarkowska, T., Godden, K., Paterson, H., Weiss, R. A., and Mittnacht, S. (1996). Cyclin encoded by KS herpesvirus. Nature *382*, 410.

Chen, P. L., Scully, P., Shew, J. Y., Wang, J. Y., and Lee, W. H. (1989). Phosphorylation of the retinoblastoma gene product is modulated during the cell cycle and cellular differentiation. Cell *58*, 1193-1198.

Chuang, T.C., Lee, Y.J., Liu, J.Y., Lin, Y.S., Li, J.W., Wang, V., Law, S.L., and Kao, M.C. (2003). EBNA1 may prolong G(2)/M phase and sensitize HER2/neu-overexpressing ovarian cancer cells to both topoisomerase II-targeting and paclitaxel drugs. Biochem. Biophys. Res. Commun. *307*, 653-659.

Cohen, J. I., Wang, F., Mannick, J., and Kieff, E. (1989). Epstein-Barr virus nuclear protein 2 is a key determinant of lymphocyte transformation. Proc. Natl. Acad. Sci. USA. *86*, 9558-9562.

Cooper, A., Johannsen, E., Maruo, S., Cahir-McFarland, E., Illanes, D., Davidson, D., and Kieff, E. (2003). EBNA3A association with RBP-Jkappa down-regulates c-myc and Epstein-Barr virus-transformed lymphoblast growth. J. Virol. *77*, 999-1010.

Cotter, M. A., 2nd, and Robertson, E. S. (2000). Modulation of histone acetyltransferase activity through interaction of epstein-barr nuclear antigen 3C with prothymosin alpha. Mol. Cell Biol. *20*, 5722-5735.

Dambaugh, T., Hennessy, K., Chamnankit, L., and Kieff, E. (1984). U2 region of Epstein-Barr virus DNA may encode Epstein-Barr nuclear antigen 2. Proc. Natl. Acad. Sci. USA. *81*, 7632-7636.

Debbas, M., and White, E. (1993). Wild-type p53 mediates apoptosis by E1A, which is inhibited by E1B. Genes Dev. *7*, 546-554.

DeCaprio, J. A., Ludlow, J. W., Figge, J., Shew, J. Y., Huang, C. M., Lee, W. H., Marsilio, E., Paucha, E., and Livingston, D. M. (1988). SV40 large tumor antigen forms a specific complex with the product of the retinoblastoma susceptibility gene. Cell *54*, 275-283.

DeCaprio, J. A., Ludlow, J. W., Lynch, D., Furukawa, Y., Griffin, J., Piwnica-Worms, H., Huang, C. M., and Livingston, D. M. (1989). The product of the retinoblastoma susceptibility gene has properties of a cell cycle regulatory element. Cell *58*, 1085-1095.

Devergne, O., Hatzivassiliou, E., Izumi, K. M., Kaye, K. M., Kleijnen, M. F., Kieff, E., and Mosialos, G. (1996). Association of TRAF1, TRAF2, and TRAF3 with an Epstein-Barr virus LMP1 domain important for B-lymphocyte transformation: role in NF-kappaB activation. Mol. Cell Biol. *16*, 7098-7108.

Diehl, V., Henle, G., W., H., and Kohn, G. (1968). Demonstration of a herpes group virus in cultures of peripheral leukocytes from patients with infectious mononucleosis. J. Virol. *2*, 663-669.

Draetta, G., Luca, F., Westendorf, J., Brizuela, L., Ruderman, J., and Beach, D. (1989). Cdc2 protein kinase is complexed with both cyclin A and B: evidence for proteolytic inactivation of MPF. Cell *56*, 829-838.

Dyson, N., Howley, P. M., Munger, K., and Harlow, E. (1989). The human papilloma virus-16 E7 oncoprotein is able to bind to the retinoblastoma gene product. Science *243*, 934-937.

Eilers, M., Schirm, S., and Bishop, J. M. (1991). The MYC protein activates transcription of the alpha-prothymosin gene. Embo J. *10*, 133-141.

Eliopoulos, A. G., and Young, L. S. (2001). LMP1 structure and signal transduction. Semin Cancer Biol. *11*, 435-444.

Epstein, M. A., Achong, B. G., and Barr, Y. M. (1964). Virus particles in cultured lymphoblasts from Burkitt's lymphoma. Lancet *1*, 702-703.

Fahraeus, R., Jansson, A., Ricksten, A., Sjoblom, A., and Rymo, L. (1990). Epstein-Barr virus-encoded nuclear antigen 2 activates the viral latent membrane protein promoter by modulating the activity of a negative regulatory element. Proc. Natl. Acad. Sci. USA. *87*, 7390-7394.

Flemington, E. K. (2001). Herpesvirus lytic replication and the cell cycle: arresting new developments. J. Virol. *75*, 4475-4481.

Friborg, J., Jr., Kong, W., Hottiger, M. O., and Nabel, G. J. (1999). p53 inhibition by the LANA protein of KSHV protects against cell death. Nature *402*, 889-894.

Ghosh, D., and Kieff, E. (1990). cis-acting regulatory elements near the Epstein-Barr virus latent- infection membrane protein transcriptional start site. J. Virol. *64*, 1855-1858.

Godden-Kent, D., Talbot, S. J., Boshoff, C., Chang, Y., Moore, P., Weiss, R. A., and Mittnacht, S. (1997). The cyclin encoded by Kaposi's sarcoma-associated herpesvirus stimulates cdk6 to phosphorylate the retinoblastoma protein and histone H1. J. Virol. *71*, 4193-4198.

Grossman, S. R., Johannsen, E., Tong, X., Yalamanchili, R., and Kieff, E. (1994). The Epstein-Barr virus nuclear antigen 2 transactivator is directed to response elements by the J. kappa recombination signal binding protein. Proc. Natl. Acad. Sci. USA. *91*, 7568-7572.

Hammerschmidt, W., and Sugden, B. (1989). Genetic analysis of immortalizing functions of Epstein-Barr virus in human B lymphocytes. Nature *340*, 393-397.

Harada, S., and Kieff, E. (1997). Epstein-Barr virus nuclear protein LP stimulates EBNA-2 acidic domain- mediated transcriptional activation. J. Virol. *71*, 6611-6618.

Helin, K., Lees, J. A., Vidal, M., Dyson, N., Harlow, E., and Fattaey, A. (1992). A cDNA encoding a pRB-binding protein with properties of the transcription factor E2F. Cell *70*, 337-350.

Heller, M., Dambaugh, T., and Kieff, E. (1981). Epstein-Barr virus DNA. IX. Variation among viral DNAs from producer and nonproducer infected cells. J. Virol. *38*, 632-648.

Henkel, T., Ling, P. D., Hayward, S. D., and Peterson, M. G. (1994). Mediation of Epstein-Barr virus EBNA2 transactivation by recombination signal-binding protein J. kappa. Science *265*, 92-95.

Hennessy, K., and Kieff, E. (1983). One of two Epstein-Barr virus nuclear antigens contains a glycine-alanine copolymer domain. Proc. Natl. Acad. Sci. USA. *80*, 5665-5669.

Herber, B., Truss, M., Beato, M., and Muller, R. (1994). Inducible regulatory elements in the human cyclin D1 promoter. Oncogene *9*, 2105-2107.

Hickman, E. S., Picksley, S. M., and Vousden, K. H. (1994). Cells expressing HPV16 E7 continue cell cycle progression following DNA damage induced p53 activation. Oncogene *9*, 2177-2181.

Hollyoake, M., Stuhler, A., Farrell, P., Gordon, J., and Sinclair, A. (1995). The normal cell cycle activation program is exploited during the infection of quiescent B lymphocytes by Epstein-Barr virus. Cancer Res. *55*, 4784-4787.

Howes, K. A., Ransom, N., Papermaster, D. S., Lasudry, J. G., Albert, D. M., and Windle, J. J. (1994). Apoptosis or retinoblastoma: alternative fates of photoreceptors expressing the HPV-16 E7 gene in the presence or absence of p53. Genes Dev. *8*, 1300-1310.

Inman, G. J., and Farrell, P. J. (1995). Epstein-Barr virus EBNA-LP and transcription regulation properties of pRB, p107 and p53 in transfection assays. J. Gen. Virol. *76 (Pt 9)*, 2141-2149.

Jiang, W. Q., Szekely, L., Wendel-Hansen, V., Ringertz, N., Klein, G., and Rosen, A. (1991). Co-localization of the retinoblastoma protein and the Epstein-Barr virus-encoded nuclear antigen EBNA-5. Exp. Cell Res. *197*, 314-318.

Jin, X. W., and Speck, S. H. (1992). Identification of critical cis elements involved in mediating Epstein-Barr virus nuclear antigen 2-dependent activity of an enhancer located upstream of the viral BamHI C promoter. J. Virol. *66*, 2846-2852.

Johannsen, E., Koh, E., Mosialos, G., Tong, X., Kieff, E., and Grossman, S. R. (1995). Epstein-Barr virus nuclear protein 2 transactivation of the latent membrane protein 1 promoter is mediated by J. kappa and PU.1. J. Virol. *69*, 253-262.

Johannsen, E., Miller, C. L., Grossman, S. R., and Kieff, E. (1996). EBNA-2 and EBNA-3C extensively and mutually exclusively associate with RBPJkappa in Epstein-Barr virus-transformed B lymphocytes. J. Virol. *70*, 4179-4183.

Joseph, A. M., Babcock, G. J., and Thorley-Lawson, D. A. (2000). Cells expressing the Epstein-Barr virus growth program are present in and restricted to the naive B-cell subset of healthy tonsils. J. Virol. *74*, 9964-9971.

Kaiser, C., Laux, G., Eick, D., Jochner, N., Bornkamm, G. W., and Kempkes, B. (1999). The proto-oncogene c-myc is a direct target gene of Epstein-Barr virus nuclear antigen 2. J. Virol. *73*, 4481-4484.

Kato, K., Yokoyama, A., Tohya, Y., Akashi, H., Nishiyama, Y., and Kawaguchi, Y. (2003). Identification of protein kinases responsible for phosphorylation of Epstein-Barr virus nuclear antigen leader protein at serine-35, which regulates its coactivator function. J. Gen. Virol. *84*, 3381-3392.

Kaye, K. M., Devergne, O., Harada, J. N., Izumi, K. M., Yalamanchili, R., Kieff, E., and Mosialos, G. (1996). Tumor necrosis factor receptor associated factor 2 is a mediator of NF- kappa B activation by latent infection membrane protein 1, the Epstein- Barr virus transforming protein. Proc. Natl. Acad. Sci. USA. *93*, 11085-11090.

Kaye, K. M., Izumi, K. M., and Kieff, E. (1993). Epstein-Barr virus latent membrane protein 1 is essential for B- lymphocyte growth transformation. Proc. Natl. Acad. Sci. USA. *90*, 9150-9154.

Kempkes, B., Spitkovsky, D., Jansen-Durr, P., Ellwart, J. W., Kremmer, E., Delecluse, H. J., Rottenberger, C., Bornkamm, G. W., and Hammerschmidt, W. (1995). B-cell proliferation and induction of early G1-regulating proteins by Epstein-Barr virus mutants conditional for EBNA2. Embo J. *14*, 88-96.

Kieff, E., and Rickinson, A. B. (2002). Epstein-Barr Virus and Its Replication. In Fields Virology, D. Knipe, and P. Howley, eds. (Lippincott Williams and Wilkins), pp. 2511-2573.

Kim, O. J., and Yates, J. L. (1993). Mutants of Epstein-Barr virus with a selective marker disrupting the TP gene transform B cells and replicate normally in culture. J. Virol. *67*, 7634-7640.

King, W., Thomas-Powell, A. L., Raab-Traub, N., Hawke, M., and Kieff, E. (1980). Epstein-Barr virus RNA. V. Viral RNA in a restringently infected, growth-transformed cell line. J. Virol. *36*, 506-518.

Kitay, M. K., and Rowe, D. T. (1996). Cell cycle stage-specific phosphorylation of the Epstein-Barr virus immortalization protein EBNA-LP. J. Virol. *70*, 7885-7893.

Knight, J. S., Lan, K., Subramanian, C., and Robertson, E. S. (2003). Epstein-Barr virus nuclear antigen 3C recruits histone deacetylase activity and associates with the corepressors mSin3A and NCoR in human B-cell lines. J. Virol. *77*, 4261-4272.

Knight, J. S., and Robertson, E. S. (2004). Epstein-Barr virus nuclear antigen 3C regulates cyclin A/p27 complexes and enhances cyclin A-dependent kinase activity. J. Virol. *78*, 1981-1991.

Knight, J. S., Sharma, N., Kalman, D. E., and Robertson, E. S. (2004). A cyclin-binding motif within the amino-terminal homology domain of EBNA3C binds cyclin A and modulates cyclin A-dependent kinase activity in EBV-infected cells. J. Virol. *78*, 12857-12867.

Krauer, K. G., Kienzle, N., Young, D. B., and Sculley, T. B. (1996). Epstein-Barr nuclear antigen-3 and -4 interact with RBP-2N, a major isoform of RBP-J kappa in B lymphocytes. Virology *226*, 346-353.

Kurth, J., Hansmann, M. L., Rajewsky, K., and Kuppers, R. (2003). Epstein-Barr virus-infected B cells expanding in germinal centers of infectious mononucleosis patients do not participate in the germinal center reaction. Proc. Natl. Acad. Sci. USA. *100*, 4730-4735.

Kurth, J., Spieker, T., Wustrow, J., Strickler, G. J., Hansmann, L. M., Rajewsky, K., and Kuppers, R. (2000). EBV-infected B cells in infectious mononucleosis: viral strategies for spreading in the B cell compartment and establishing latency. Immunity *13*, 485-495.

Lavia, P., Mileo, A. M., Giordano, A., and Paggi, M. G. (2003). Emerging roles of DNA tumor viruses in cell proliferation: new insights into genomic instability. Oncogene *22*, 6508-6516.

Le Roux, A., Kerdiles, B., Walls, D., Dedieu, J. F., and Perricaudet, M. (1994). The Epstein-Barr virus determined nuclear antigens EBNA-3A, -3B, and - 3C repress EBNA-2-mediated transactivation of the viral terminal protein 1 gene promoter. Virology *205*, 596-602.

Liebowitz, D., Wang, D., and Kieff, E. (1986). Orientation and patching of the latent infection membrane protein encoded by Epstein-Barr virus. J. Virol. *58*, 233-237.

Lin, C. S., Kuo, H. H., Chen, J. Y., Yang, C. S., and Wang, W. B. (2000). Epstein-barr virus nuclear antigen 2 retards cell growth, induces p21(WAF1) expression, and modulates p53 activity post-translationally. J. Mol. Biol. *303*, 7-23.

Longnecker, R., Miller, C. L., Miao, X. Q., Marchini, A., and Kieff, E. (1992). The only domain which distinguishes Epstein-Barr virus latent membrane protein 2A (LMP2A) from LMP2B is dispensable for lymphocyte infection and growth transformation *in vitro*; LMP2A is therefore nonessential. J. Virol. *66*, 6461-6469.

Longnecker, R., Miller, C. L., Miao, X. Q., Tomkinson, B., and Kieff, E. (1993a). The last seven transmembrane and carboxy-terminal cytoplasmic domains of Epstein-Barr virus latent membrane protein 2 (LMP2) are dispensable for lymphocyte infection and growth transformation *in vitro*. J. Virol. *67*, 2006-2013.

Longnecker, R., Miller, C. L., Tomkinson, B., Miao, X. Q., and Kieff, E. (1993b). Deletion of DNA encoding the first five transmembrane domains of Epstein-Barr virus latent membrane proteins 2A and 2B. J. Virol. *67*, 5068-5074.

Lowe, S. W., and Ruley, H. E. (1993). Stabilization of the p53 tumor suppressor is induced by adenovirus 5 E1A and accompanies apoptosis. Genes Dev. *7*, 535-545.

Luo, R. X., Postigo, A. A., and Dean, D. C. (1998). Rb interacts with histone deacetylase to repress transcription. Cell *92*, 463-473.

Maller, J., Gautier, J., Langan, T. A., Lohka, M. J., Shenoy, S., Shalloway, D., and Nurse, P. (1989). Maturation-promoting factor and the regulation of the cell cycle. J. Cell Sci. Suppl *12*, 53-63.

Mannick, J. B., Cohen, J. I., Birkenbach, M., Marchini, A., and Kieff, E. (1991). The Epstein-Barr virus nuclear protein encoded by the leader of the EBNA RNAs is important in B-lymphocyte transformation. J. Virol. *65*, 6826-6837.

Marshall, D., and Sample, C. (1995). Epstein-Barr virus nuclear antigen 3C is a transcriptional regulator. J. Virol. *69*, 3624-3630.

Matsushime, H., Ewen, M. E., Strom, D. K., Kato, J. Y., Hanks, S. K., Roussel, M. F., and Sherr, C. J. (1992). Identification and properties of an atypical catalytic subunit (p34PSK-J3/cdk4) for mammalian D type G1 cyclins. Cell *71*, 323-334.

Mauser, A., Holley-Guthrie, E., Zanation, A., Yarborough, W., Kaufmann, W., Klingelhutz, A., Seaman, W. T., and Kenney, S. (2002). The Epstein-Barr virus immediate-early protein BZLF1 induces expression of E2F-1 and other proteins involved in cell cycle progression in primary keratinocytes and gastric carcinoma cells. J. Virol. *76*, 12543-12552.

Miller, C. L., Burkhardt, A. L., Lee, J. H., Stealey, B., Longnecker, R., Bolen, J. B., and Kieff, E. (1995). Integral membrane protein 2 of Epstein-Barr virus regulates reactivation from latency through dominant negative effects on protein- tyrosine kinases. Immunity *2*, 155-166.

Miller, C. L., Longnecker, R., and Kieff, E. (1993). Epstein-Barr virus latent membrane protein 2A blocks calcium mobilization in B lymphocytes. J. Virol. *67*, 3087-3094.

Miller, G., Robinson, J., Heston, L., and Lipman, M. (1975). Differences between laboratory strains of Epstein-Barr virus based on immortalization, abortive infection and interference. IARC Sci. Publ *(11*, 395-408.

Mosialos, G., Birkenbach, M., Yalamanchili, R., VanArsdale, T., Ware, C., and Kieff, E. (1995). The Epstein-Barr virus transforming protein LMP1 engages signaling proteins for the tumor necrosis factor receptor family. Cell *80*, 389-399.

Nevins, J. R., Leone, G., DeGregori, J., and Jakoi, L. (1997). Role of the Rb/E2F pathway in cell growth control. J. Cell Physiol *173*, 233-236.

Nilsson, K., Klein, G., Henle, W., and Henle, G. (1971). The establishment of lymphoblastoid cell lines from adult and from foetal human lymphoid tissue and its dependence on EBV. Int. J. Cancer *8*, 443-450.

Nitsche, F., Bell, A., and Rickinson, A. (1997). Epstein-Barr virus leader protein enhances EBNA-2-mediated transactivation of latent membrane protein 1 expression: a role for the W1W2 repeat domain. J. Virol. *71*, 6619-6628.

Ohtani, K., DeGregori, J., and Nevins, J. R. (1995). Regulation of the cyclin E gene by transcription factor E2F1. Proc. Natl. Acad. Sci. USA. *92*, 12146-12150.

Pagano, M., Pepperkok, R., Verde, F., Ansorge, W., and Draetta, G. (1992). Cyclin A is required at two points in the human cell cycle. Embo J. *11*, 961-971.

Pajic, A., Spitkovsky, D., Christoph, B., Kempkes, B., Schuhmacher, M., Staege, M. S., Brielmeier, M., Ellwart, J., Kohlhuber, F., Bornkamm, G. W., et al. (2000). Cell cycle activation by c-myc in a burkitt lymphoma model cell line. Int. J. Cancer *87*, 787-793.

Palmero, I., Holder, A., Sinclair, A. J., Dickson, C., and Peters, G. (1993). Cyclins D1 and D2 are differentially expressed in human B-lymphoid cell lines. Oncogene *8*, 1049-1054.

Pan, H., and Griep, A. E. (1994). Altered cell cycle regulation in the lens of HPV-16 E6 or E7 transgenic mice: implications for tumor suppressor gene function in development. Genes Dev. *8*, 1285-1299.

Parker, G. A., Crook, T., Bain, M., Sara, E. A., Farrell, P. J., and Allday, M. J. (1996). Epstein-Barr virus nuclear antigen (EBNA)3C is an immortalizing oncoprotein with similar properties to adenovirus E1A and papillomavirus E7. Oncogene *13*, 2541-2549.

Parker, G. A., Touitou, R., and Allday, M. J. (2000). Epstein-Barr virus EBNA3C can disrupt multiple cell cycle checkpoints and induce nuclear division divorced from cytokinesis. Oncogene *19*, 700-709.

Peng, M., and Lundgren, E. (1992). Transient expression of the Epstein-Barr virus LMP1 gene in human primary B cells induces cellular activation and DNA synthesis. Oncogene *7*, 1775-1782.

Pines, J., and Hunter, T. (1989). Isolation of a human cyclin cDNA: evidence for cyclin mRNA and protein regulation in the cell cycle and for interaction with p34cdc2. Cell *58*, 833-846.

Pokrovskaja, K., Ehlin-Henriksson, B., Bartkova, J., Bartek, J., Scuderi, R., Szekely, L., Wiman, K. G., and Klein, G. (1996). Phenotype-related differences in the expression of D-type cyclins in human B cell-derived lines. Cell Growth Differ *7*, 1723-1732.

Polyak, K., Lee, M. H., Erdjument-Bromage, H., Koff, A., Roberts, J. M., Tempst, P., and Massague, J. (1994). Cloning of p27Kip1, a cyclin-dependent kinase inhibitor and a potential mediator of extracellular antimitogenic signals. Cell *78*, 59-66.

Pope, J. H., Horne, M. K., and Scott, W. (1968). Transformation of foetal human leukocytes *in vitro* by filtrates of a human leukaemic cell line containing herpes-like virus. Int. J. Cancer *3*, 857-866.

Raab-Traub, N., Pritchett, R., and Kieff, E. (1978). DNA of Epstein-Barr virus. III. Identification of restriction enzyme fragments that contain DNA sequences which differ among strains of Epstein-Barr virus. J. Virol. *27*, 388-398.

Radkov, S. A., Bain, M., Farrell, P. J., West, M., Rowe, M., and Allday, M. J. (1997). Epstein-Barr virus EBNA3C represses Cp, the major promoter for EBNA expression, but has no effect on the promoter of the cell gene CD21. J. Virol. *71*, 8552-8562.

Radkov, S. A., Kellam, P., and Boshoff, C. (2000). The latent nuclear antigen of Kaposi sarcoma-associated herpesvirus targets the retinoblastoma-E2F pathway and with the oncogene Hras transforms primary rat cells. Nat. Med. *6*, 1121-1127.

Radkov, S. A., Touitou, R., Brehm, A., Rowe, M., West, M., Kouzarides, T., and Allday, M. J. (1999). Epstein-Barr virus nuclear antigen 3C interacts with histone deacetylase to repress transcription. J. Virol. *73*, 5688-5697.

Reedman, B. M., and Klein, G. (1973). Cellular localization of an Epstein-Barr virus (EBV)-associated complement-fixing antigen in producer and non-producer lymphoblastoid cell lines. Int. J. Cancer *11*, 499-520.

Resnitzky, D., Hengst, L., and Reed, S. I. (1995). Cyclin A-associated kinase activity is rate limiting for entrance into S phase and is negatively regulated in G1 by p27Kip1. Mol. Cell Biol. *15*, 4347-4352.

Rickinson, A. B., and Kieff, E. (2002). Epstein-Barr Virus. In Fields Virology, D. Knipe, and P. Howley, eds. (Lippincott Williams and Wilkins), pp. 2575-2627.

Robertson, E. S., Grossman, S., Johannsen, E., Miller, C., Lin, J., Tomkinson, B., and Kieff, E. (1995). Epstein-Barr virus nuclear protein 3C modulates transcription through interaction with the sequence-specific DNA-binding protein J. kappa. J. Virol. *69*, 3108-3116.

Robertson, E. S., Lin, J., and Kieff, E. (1996). The amino-terminal domains of Epstein-Barr virus nuclear proteins 3A, 3B, and 3C interact with RBPJ(kappa). J. Virol. *70*, 3068-3074.

Rodriguez, A., Jung, E. J., Yin, Q., Cayrol, C., and Flemington, E. K. (2001). Role of c-myc regulation in Zta-mediated induction of the cyclin-dependent kinase inhibitors p21 and p27 and cell growth arrest. Virology *284*, 159-169.

Rosendorff, A., Illanes, D., David, G., Lin, J., Kieff, E., and Johannsen, E. (2004). EBNA3C Coactivation with EBNA2 Requires a SUMO Homology Domain. J. Virol. *78*, 367-377.

Russo, A. A., Jeffrey, P. D., Patten, A. K., Massague, J., and Pavletich, N. P. (1996). Crystal structure of the p27Kip1 cyclin-dependent-kinase inhibitor bound to the cyclin A-Cdk2 complex. Nature *382*, 325-331.

Sauer, K., and Lehner, C. F. (1995). The role of cyclin E in the regulation of entry into S phase. Prog Cell Cycle Res. *1*, 125-139.

Sburlati, A. R., Manrow, R. E., and Berger, S. L. (1991). Prothymosin alpha antisense oligomers inhibit myeloma cell division. Proc. Natl. Acad. Sci. USA. *88*, 253-257.

Schlee, M., Krug, T., Gires, O., Zeidler, R., Hammerschmidt, W., Mailhammer, R., Laux, G., Sauer, G., Lovric, J., and Bornkamm, G. W. (2004). Identification of Epstein-Barr virus (EBV) nuclear antigen 2 (EBNA2) target proteins by proteome analysis: activation of EBNA2 in conditionally immortalized B cells reflects early events after infection of primary B cells by EBV. J. Virol. *78*, 3941-3952.

Serrano, M., Hannon, G. J., and Beach, D. (1993). A new regulatory motif in cell-cycle control causing specific inhibition of cyclin D/CDK4. Nature *366*, 704-707.

Sherr, C. J. (1996). Cancer cell cycles. Science *274*, 1672-1677.

Sinclair, A. J., Palmero, I., Holder, A., Peters, G., and Farrell, P. J. (1995). Expression of cyclin D2 in Epstein-Barr virus-positive Burkitt's lymphoma cell lines is related to methylation status of the gene. J. Virol. *69*, 1292-1295.

Sinclair, A. J., Palmero, I., Peters, G., and Farrell, P. J. (1994). EBNA-2 and EBNA-LP cooperate to cause G0 to G1 transition during immortalization of resting human B lymphocytes by Epstein-Barr virus. Embo J. *13*, 3321-3328.

Skare, J., Farley, J., Strominger, J. L., Fresen, K. O., Cho, M. S., and zur Hausen, H. (1985). Transformation by Epstein-Barr virus requires DNA sequences in the region of BamHI fragments Y and H. J. Virol. *55*, 286-297.

Spender, L. C., Cannell, E. J., Hollyoake, M., Wensing, B., Gawn, J. M., Brimmell, M., Packham, G., and Farrell, P. J. (1999). Control of cell cycle entry and apoptosis in B lymphocytes infected by Epstein-Barr virus. J. Virol. *73*, 4678-4688.

Spender, L. C., Cornish, G. H., Rowland, B., Kempkes, B., and Farrell, P. J. (2001). Direct and indirect regulation of cytokine and cell cycle proteins by EBNA-2 during Epstein-Barr virus infection. J. Virol. *75*, 3537-3546.

Subramanian, C., Cotter, M. A., 2nd, and Robertson, E. S. (2001). Epstein-Barr virus nuclear protein EBNA-3C interacts with the human metastatic suppressor Nm23-H1: a molecular link to cancer metastasis. Nat. Med. *7*, 350-355.

Subramanian, C., Hasan, S., Rowe, M., Hottiger, M., Orre, R., and Robertson, E. S. (2002). Epstein-Barr virus nuclear antigen 3C and prothymosin alpha interact with the p300 transcriptional coactivator at the CH1 and CH3/HAT domains and cooperate in regulation of transcription and histone acetylation. J. Virol. *76*, 4699-4708.

Sung, N. S., Kenney, S., Gutsch, D., and Pagano, J. S. (1991). EBNA-2 transactivates a lymphoid-specific enhancer in the BamHI C promoter of Epstein-Barr virus. J. Virol. *65*, 2164-2169.

Swenson, J. J., Mauser, A. E., Kaufmann, W. K., and Kenney, S. C. (1999). The Epstein-Barr virus protein BRLF1 activates S phase entry through E2F1 induction. J. Virol. *73*, 6540-6550.

Szekely, L., Selivanova, G., Magnusson, K. P., Klein, G., and Wiman, K. G. (1993). EBNA-5, an Epstein-Barr virus-encoded nuclear antigen, binds to the retinoblastoma and p53 proteins. Proc. Natl. Acad. Sci. USA. *90*, 5455-5459.

Thorley-Lawson, D. A. (2001). Epstein-Barr virus: exploiting the immune system. Nat. Rev. Immunol. *1*, 75-82.

Thorley-Lawson, D. A., Nadler, L. M., Bhan, A. K., and Schooley, R. T. (1985). BLAST-2 [EBVCS], an early cell surface marker of human B cell activation, is superinduced by Epstein Barr virus. J. Immunol. *134*, 3007-3012.

Thorley-Lawson, D. A., Schooley, R. T., Bhan, A. K., and Nadler, L. M. (1982). Epstein-Barr virus superinduces a new human B cell differentiation antigen (B-LAST 1) expressed on transformed lymphoblasts. Cell *30*, 415-425.

Tomkinson, B., and Kieff, E. (1992). Use of second-site homologous recombination to demonstrate that Epstein- Barr virus nuclear protein 3B is not important for lymphocyte infection or growth transformation *in vitro*. J. Virol. *66*, 2893-2903.

Tomkinson, B., Robertson, E., and Kieff, E. (1993). Epstein-Barr virus nuclear proteins EBNA-3A and EBNA-3C are essential for B-lymphocyte growth transformation. J. Virol. *67*, 2014-2025.

Toyoshima, H., and Hunter, T. (1994). p27, a novel inhibitor of G1 cyclin-Cdk protein kinase activity, is related to p21. Cell *78*, 67-74.

Uchida, J., Yasui, T., Takaoka-Shichijo, Y., Muraoka, M., Kulwichit, W., Raab-Traub, N., and Kikutani, H. (1999). Mimicry of CD40 signals by Epstein-Barr virus LMP1 in B lymphocyte responses. Science *286*, 300-303.

Vareli, K., Frangou-Lazaridis, M., and Tsolas, O. (1995). Prothymosin alpha mRNA levels vary with c-myc expression during tissue proliferation, viral infection and heat shock. FEBS Lett. *371*, 337-340.

Vareli, K., Tsolas, O., and Frangou-Lazaridis, M. (1996). Regulation of prothymosin alpha during the cell cycle. Eur. J. Biochem. *238*, 799-806.

Waltzer, L., Perricaudet, M., Sergeant, A., and Manet, E. (1996). Epstein-Barr virus EBNA3A and EBNA3C proteins both repress RBP-J kappa- EBNA2-activated transcription by inhibiting the binding of RBP-J kappa to DNA. J. Virol. *70*, 5909-5915.

Wang, F., Tsang, S. F., Kurilla, M. G., Cohen, J. I., and Kieff, E. (1990). Epstein-Barr virus nuclear antigen 2 transactivates latent membrane protein LMP1. J. Virol. *64*, 3407-3416.

White, A. E., Livanos, E. M., and Tlsty, T. D. (1994). Differential disruption of genomic integrity and cell cycle regulation in normal human fibroblasts by the HPV oncoproteins. Genes Dev. *8*, 666-677.

Whyte, P., Williamson, N. M., and Harlow, E. (1989). Cellular targets for transformation by the adenovirus E1A proteins. Cell *56*, 67-75.

Xiong, Y., Zhang, H., and Beach, D. (1992). D type cyclins associate with multiple protein kinases and the DNA replication and repair factor PCNA. Cell *71*, 505-514.

Yalamanchili, R., Tong, X., Grossman, S., Johannsen, E., Mosialos, G., and Kieff, E. (1994). Genetic and biochemical evidence that EBNA 2 interaction with a 63-kDa cellular GTG-binding protein is essential for B lymphocyte growth transformation by EBV. Virology *204*, 634-641.

Yates, J., Warren, N., Reisman, D., and Sugden, B. (1984). A cis-acting element from the Epstein-Barr viral genome that permits stable replication of recombinant plasmids in latently infected cells. Proc. Natl. Acad. Sci. USA. *81*, 3806-3810.

Yates, J. L., Warren, N., and Sugden, B. (1985). Stable replication of plasmids derived from Epstein-Barr virus in various mammalian cells. Nature *313*, 812-815.

Yokoyama, A., Tanaka, M., Matsuda, G., Kato, K., Kanamori, M., Kawasaki, H., Hirano, H., Kitabayashi, I., Ohki, M., Hirai, K., and Kawaguchi, Y. (2001). Identification of major phosphorylation sites of Epstein-Barr virus nuclear antigen leader protein (EBNA-LP): ability of EBNA-LP to induce latent membrane protein 1 cooperatively with EBNA-2 is regulated by phosphorylation. J. Virol. *75*, 5119-5128.

Yue, W., Davenport, M. G., Shackelford, J., and Pagano, J. S. (2004). Mitosis-specific hyperphosphorylation of Epstein-Barr virus nuclear antigen 2 suppresses its function. J. Virol. *78*, 3542-3552.

Zacny, V. L., Wilson, J., and Pagano, J. S. (1998). The Epstein-Barr virus immediate-early gene product, BRLF1, interacts with the retinoblastoma protein during the viral lytic cycle. J. Virol. *72*, 8043-8051.

Zhang, Q., Gutsch, D., and Kenney, S. (1994). Functional and physical interaction between p53 and BZLF1: implications for Epstein-Barr virus latency. Mol. Cell Biol. *14*, 1929-1938.

Zhao, B., Marshall, D. R., and Sample, C. E. (1996). A conserved domain of the Epstein-Barr virus nuclear antigens 3A and 3C binds to a discrete domain of Jkappa. J. Virol. *70*, 4228-4236.

Zhao, B., and Sample, C. E. (2000). Epstein-barr virus nuclear antigen 3C activates the latent membrane protein 1 promoter in the presence of Epstein-Barr virus nuclear antigen 2 through sequences encompassing an spi-1/Spi-B binding site. J. Virol. *74*, 5151-5160.

Zimber-Strobl, U., Kremmer, E., Grasser, F., Marschall, G., Laux, G., and Bornkamm, G. W. (1993). The Epstein-Barr virus nuclear antigen 2 interacts with an EBNA2 responsive cis-element of the terminal protein 1 gene promoter. Embo J. *12*, 167-175.

Chapter 25

Function of Latent Membrane Protein 2A

*Masato Ikeda, Makoto Fukuda
and Richard Longnecker**

ABSTRACT

Latent membrane protein 2A (LMP2A) is expressed both in normal EBV latency and EBV-associated pathogenesis. This persistent expression indicates that LMP2A likely functions in key aspects of EBV latency and in disease manifestations in the human host. Functional studies of LMP2A has has revealed that LMP2A mimics BCR-mediated signal transduction and activates anti-apoptotic and cell-survival signals. To regulate its own function, LMP2A utilizes ubiquitin-dependent processes which may allow LMP2A to modulate B cell differentiation and establish latent infection in memory B cells. Recent studies have showed that, despite the dispensability of LMP2A in EBV-induced B cell transformation, LMP2A has potent transforming activities in epithelial cells. Moreover, DNA microarray analysis suggests an important role of LMP2A in the development and pathogenesis of Hodgkin's Lymphoma. In summary, it appears that LMP2A plays a key role in the establishment of EBV latent infection and the development of EBV-related pathogenesis such as nasopharyngeal carcinoma and Hodgkin's Lymphoma. This chapter will focus on what is known of LMP2A function in B cells and epithelial cells.

LMP2A INTRODUCTION

Epstein-Barr virus (EBV) infects B cells circulating through the oral epithelium and establishes a lifelong latent infection in a subset of resting memory B cells (Kieff and Rickinson, 2001; Longnecker, 1998; Thorley-Lawson, 2001). LMP2A is expressed in humans with EBV latent infections (Kieff and Rickinson, 2001). Due to this persistent expression, Latent membrane protein 2A (LMP2A) is thought to play a key role for the establishment of EBV latency and its maintenance by allowing EBV to gain access to a specific B cell compartment and providing long life to the infected lymphocytes (Longnecker, 1998; Thorley-Lawson, 2001). In its amino-terminal cytoplasmic domain, LMP2A contains an immunoreceptor tyrosine-based activation motif (ITAM) that is also present in the Ig-α and Ig-β chains of B cell antigen receptor (BCR). LMP2A interacts and activates protein tyrosine kinases

*For correspondence email r-longnecker@northwestern.edu

(PTKs) in a constitutive manner mimicking a BCR providing development and survival signals for B cells in the absence of corresponding antigens (Longnecker, 2000). LMP2A also blocks normal B cell signal transduction that induces EBV lytic replication, preventing switch from latent to lytic replication (Longnecker, 2000).

In addition to the establishment of viral latency, LMP2A is implicated in the development of EBV-associated malignancies. In biopsies from Hodgkin's disease (HD) and nasopharyngeal carcinoma (NPC), LMP2A expression is consistently detected (Brooks et al., 1992; Busson et al., 1992; Niedobitek et al., 1997). This type of EBV latency has been designated as type II latency or the default program. LMP2A is also expressed in type III latency or the growth program (Longnecker, 2000). This form of latency is characterized by EBV transformed B lymphocytes grown in culture termed lymphoblastoid cell lines (LCLs) due to the activated phenotype of these cells. LMP2A is also a prime candidate to be expressed in the latency program found in individuals with EBV latent infection (Chen et al., 1995; Qu and Rowe, 1992; Tierney et al., 1994). LMP2A is not expressed in type I latency as typified by Burkitt's lymphoma (Longnecker, 2000). Besides LMP1 that is a primary viral oncoprotein, LMP2A is thought to play an important role in EBV oncogenesis since LMP2A has an ability to activate signal pathways important for cell activation and growth (Longnecker, 2000). Despite this signaling potential, LMP2A is incapable of inducing tumorgenicity or affecting immortalization in most B cells (Longnecker et al., 1992). However, recent studies have shown that LMP2A can cause cell proliferation in epithelial cell lines by stabilizing β-catenin through phosphatidylinositol 3-kinase (PI3K) and Akt (Morrison et al., 2003). In addition, LMP2A has been shown to regulate c-Jun which induced cell migration (Chen et al., 2002). These studies suggest that LMP2A contributes to epithelial oncogenesis. Finally, recent studies have shown that LMP2A induces similar changes in gene expression when compared to cells from Hodgkin's lymphoma biopsies (Portis et al., 2003).

STRUCTURAL FEATURES OF LMP2A

The LMP2 genes are encoded near the termini of the linear EBV genome (Laux et al., 1989; Sample et al., 1989). Upon circulation and generation of EBV episome at the terminal-repeat regions of the EBV genome, the two LMP2 genes (LMP2A and LMP2B) are alternatively transcribed from two separately distinct promoters. The LMP2A transcription start site is approximately 3 kb from the bidirectional promoter responsible for the expression of LMP2B and LMP1. The transcription from the LMP2A and LMP2B/LMP1 promoters are regulated positively by EBNA2 through cellular transcription factors RBP-Jκ and PU.1 (Laux et al., 1994a; 1994b; Meitinger et al., 1994). The two transcripts are multiply spliced, and the resultant mRNAs differ only in their first exons. The LMP2A first exon contains an initiator of translation that results in a hydrophilic 119-amino-acid terminus. The LMP2B first exon lacks a methionine codon, and translation of LMP2B initiates at a methionine near the beginning of second exon, which is common to both mRNAs. This latter methionine codon immediately precedes the first of 12 transmembrane domains, each of which traverses the plasma membrane, and a 27-amino-acid cytoplasmic

C-terminal tail both of which are common to LMP2A and LMP2B. Thus, LMP2A and LMP2B sequences are in common except for the cytosolic amino-terminus of LMP2A. Although a specific function for LMP2B has not been described, strong conservation of both LMP2B expression and the bidirectional EBNA2 responsive nature of its promoter in the EBV-related rhesus monkey lymphocryptovirus (LCV) suggest that LMP2B plays an important role in EBV biology (Rivailler et al., 1999).

LMP2A is a membrane protein containing a cytoplasmic amino terminus, 12 transmembrane domains, and a cytoplasmic carboxyl terminus (Figure 1 and Figure 2A). The LMP2A amino terminal domain is constitutively tyrosine phosphorylated (Longnecker et al., 1991) and associated with the Src homology 2 (SH2) domains of Src-family PTKs, such as Lyn, and the Syk PTK (Fruehling and Longnecker, 1997; Fruehling et al., 1998). By forming constitutive membrane complexes, LMP2A is able to mimic an activated BCR complex. Initially, Lyn is recruited to the LMP2A complexes (Fruehling et al., 1998). Since both LMP2A and Lyn are constitutively colocalized in the lipid rafts in B cells, Lyn may associate with LMP2A independent of protein–protein interaction motifs (Dykstra et al., 2001). Alternatively, Lyn may be recruited to a LMP2A proline-rich region through the interaction of the Lyn SH3 domain or a DQSL sequence in LMP2A similar to the DCSM sequence that has been shown to be important for recruitment of Lyn to the Ig-α chain of the BCR complex (Pleiman et al., 1994). The recruitment of Lyn to LMP2A is followed by the phosphorylation of LMP2A and binding of Lyn via its SH2 domain to Y112 of LMP2A (Fruehling et al., 1998). Once bound to Y112, Lyn phosphorylates the remaining LMP2A tyrosines, allowing the recruitment of other SH2-containing proteins to LMP2A. Specifically, the Syk PTK binds to the phosphorylated LMP2A ITAM (Y74 and Y85) (Fruehling and Longnecker, 1997), resulting in the activation of downstream signal molecules including PI3K, Btk, BLNK, and Akt (Engels et al., 2001; Merchant and Longnecker, 2001; Miller et al., 1995; Swart et al., 2000). Mutational analysis in remaining amino-terminal LMP2A tyrosine residues indicates that Y23, Y31, Y60, Y64, and Y101 are unnecessary for the LMP2A function in blocking of B cell activation (Swart et al., 1999). Y23 and Y64 are not conserved in EBV clinical isolates (Berger et al., 1999; Busson et al., 1995; Tanaka et al., 1999) and EBV-related simian LCVs (Franken et al., 1995; Rivailler et al., 1999), and their surrounding sequences reveal no homology to the consensus sequence of any known SH2-binding domains (Longnecker, 2000), indicating that Y23 and Y64 may not be functional significant. There is a stronger case for the function of Y31, Y60, and Y101. Although they are nonessential for the block of B cell activation by LMP2A, these tyrosine residues are conserved in EBV clinical isolates and simian LCVs (Swart et al., 1999). Interestingly, in the simian LCVs, there is an additional tyrosine located upstream of Y60 in the EBV LMP2A (Figure 1). The sequences around the tyrosines in the simian LMP2As is quite well conserved being flanked by acidic amino acids which is also similar to the sequences surrounding Y60 in EBV LMP2A. Comparison with the consensus sequence of SH2-domain binding motifs reveals that Y31 shows best homology with Shc, PI3K, and PLCγ2; Y60 with Abl, Crk, and Nck; and Y101 with Csk (Longnecker, 2000). These consensus sequences are also conserved in simian

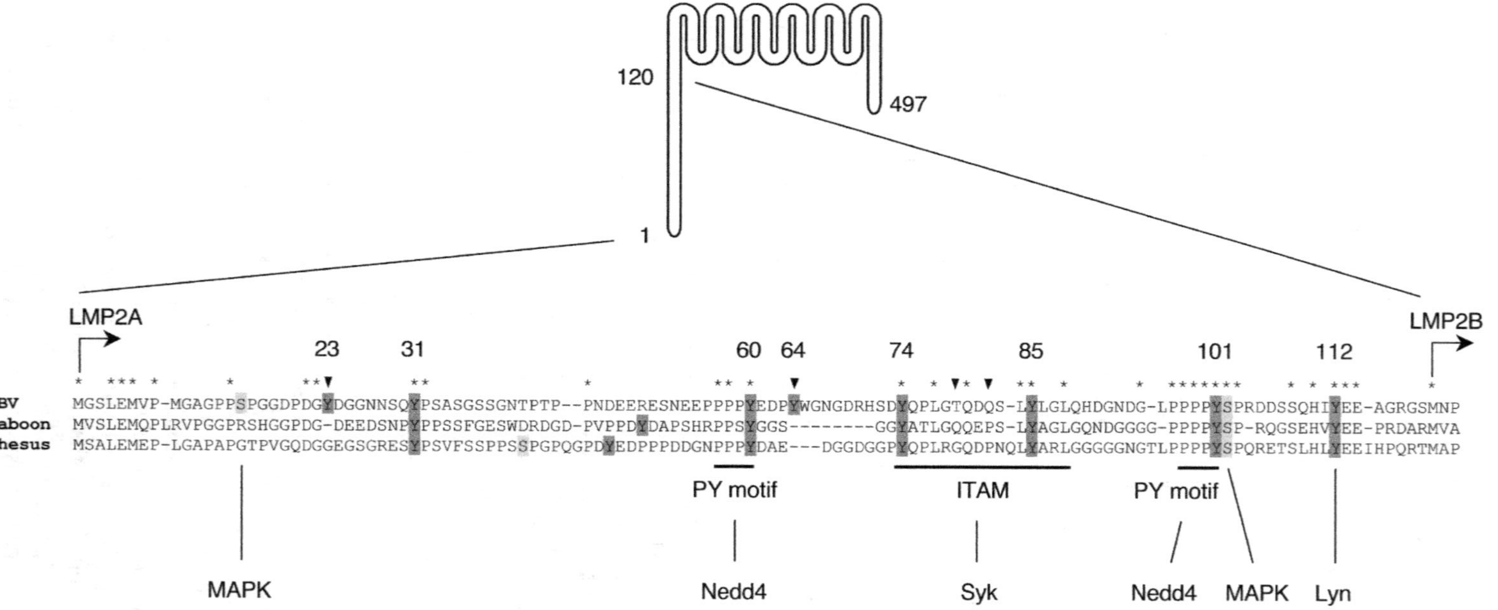

Figure 1. Amino acid sequence and functional domains of LMP2A amino terminus. Alignment of the amino terminal tail domains of LMP2A from EBV (Sample et al., 1989), baboon LCV (herpesvirus papio) (Franken et al., 1995), and rhesus LCV (Rivailler et al., 1999). The functional domains of EBV LMP2A are conserved in simian LCVs and clinical EBV isolates. Y112 binds Src-family PTKs such as Lyn. Y74 and Y85 form an ITAM and are required for association of LMP2A with Syk PTK. Two PY motifs bind to Nedd4-family ubiquitin ligases in a phosphorylation independent fashion. Tyrosine residues are indicated with dark shaded boxes, and residue numbers of EBV LMP2A tyrosines are indicated above the sequences. Light shaded boxes are serine residues phosphorylated by MAPK *in vitro*. Conserved residues in three LMP2As are indicated with asterisks. Arrowheads denote non-conservative changes detected in clinical isolates (Berger et al., 1999; Busson et al., 1995; Tanaka et al., 1999). Arrows indicate initiation codons for LMP2A and LMP2B transcripts.

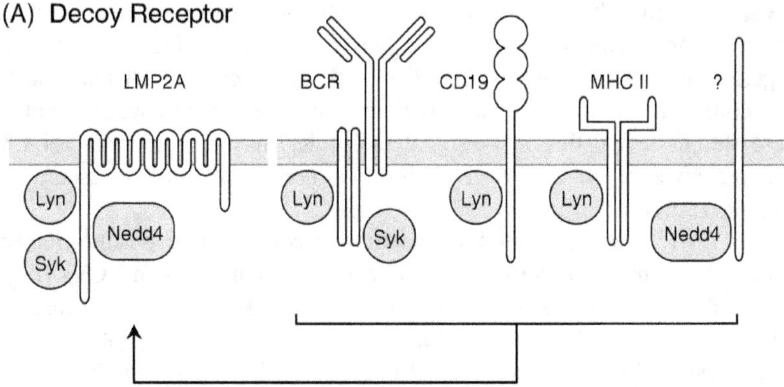

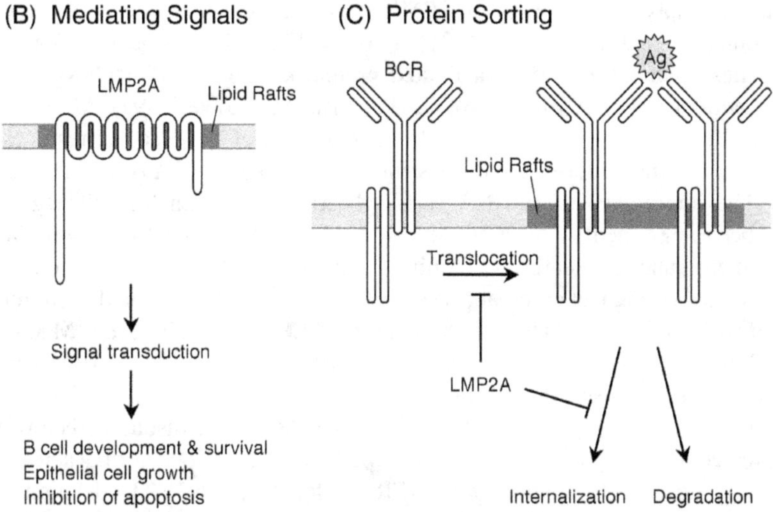

Figure 2. LMP2A functions. (A) LMP2A functions as a decoy receptor. LMP2A specifically associates with Lyn and Syk PTKs that are important for mediating signal transduction by BCR, CD19 or MHC Class II. Acting as a decoy receptor, LMP2A inhibits signal transduction through the BCR, CD19, and MHC Class II by recruitment of Lyn and Syk. LMP2A may inhibit other signal pathways in which Nedd4-family ubiquitin ligases play a critical role since LMP2A specifically binds to Nedd4-family ubiquitin ligases. (B) LMP2A mediated signals. LMP2A constitutively aggregates in lipid rafts and is able to mediate a variety of signals through Lyn, Syk, PI3K, or Akt. This LMP2A-mediated signals results in B cell development and survival, epithelial cell growth by inhibiting GSK-3β activity, and inhibition of TGF-β-associated apoptosis. (C) LMP2A regulates BCR sorting. The BCR is rapidly translocated into lipid rafts and therein activated by Src-family PTKs following activation by crosslinking. Once recruited into lipid rafts, the BCR is internalized to the early endosomes or MHC class II compartments, and a portion of BCR is further trafficked into lysosomes for the degradation. LMP2A is interferes with BCR translocation into lipd rafts and BCR internalization into endosomes.

LCVs, suggesting that these interactions are evolutionally important for LMP2A functions. LMP2A induces constitutive phosphorylation of PI3K and CrkL in B cells (Engels et al., 2001; Miller et al., 1995). Csk increases LMP2A phosphorylation when epithelial cells are stimulated with extra cellular matrix (Scholle et al., 1999). These studies reveal that these signal molecules may be important for LMP2A-mediated development and survival signals although they are not essential for the block of B cell activation.

The LMP2A amino-terminal domain also contains two proline-rich regions (PPPPY) that are conserved among clinical isolates and simian LCVs (Figure 1). These proline-rich regions are homologous to PY motifs (XPPXY) that are predicted to interact with WW domains (Sudol, 1996). LMP2A PY motifs specifically associate with WW-domains of Nedd4-family ubiquitin ligases, such as AIP4/Itchy, WWP2, Nedd4, and Nedd4-2/KIAA0439 (Ikeda et al., 2000; Winberg et al., 2000). This interaction results in the ubiquitination and degradation of LMP2A and its associated proteins such as Lyn, leading the downregulation of LMP2A-mediated signal transduction (Ikeda et al., 2001; 2002; Winberg et al., 2000). Peptide-mapping analysis has shown that LMP2A is also phosphorylated on serines and threonines (Longnecker et al., 1991). Two LMP2A serine residues, S15 and S102, are sites of *in vitro* mitogen-activated protein kinase (MAPK) phosphorylation of a purified LMP2A fusion protein (Panousis and Rowe, 1997). MAPK bound to LMP2A in LCLs, presumably at either or both of these serine residues. The functional significance of the MAPK interaction with LMP2A is not known.

The importance of LMP2A carboxyl-terminal domain for self-aggregation has been suggested by several studies. An LMP2A mutant lacking the last ten transmembrane and carboxyl-terminal domains localizes diffusely on plasma membrane (Longnecker et al., 1991). In addition, the carboxyl terminus has been suggested to mediate self-association of LMP2A in 293 cells (Matskova et al., 2001), although a deletion mutant of LMP2A that lacks the LMP2A carboxyl-terminus and last seven transmembrane domains is not totally deficient in the block of B cell activation (Miller et al., 1994). As will be discussed below, when expressed in B lymphocytes in culture, LMP2A is able to block normal activation of B cells following stimulation of the BCR (Miller et al., 1993). The deletion mutant lacking carboxyl-terminal and last ten transmembrane domains is still tyrosine phosphorylated (Longnecker et al., 1991). Although further analysis remains to be elucidated, these studies suggest that both or either of transmembrane domains and carboxyl-terminal domain is critical for LMP2A to self-cluster on the plasma membrane. Cysteine-rich regions of LMP2A carboxyl-terminal domain are the major sites of LMP2A palmitoylation in 293 cells (Matskova et al., 2001), whereas the domain in B cells does not appear as important for LMP2A palmitoylation (Katzman and Longnecker, 2004). Interestingly, LMP2A palmitoylation is nonessential for the localization of LMP2A in rafts or most LMP2A identified functions (Katzman and Longnecker, 2004).

LMP2A INHIBITS SIGNAL TRANSDUCTION AND LYTIC CYCLE ACTIVATION

LMP2A blocks B cell activation, including phosphorylation of PTKs, rapid calcium mobilization, and upregulation of gene transcription that follow the crosslinking of BCR, CD19 and MHC class II (Miller et al., 1993). As B cell activation induces the transcription of BZLF1 and following EBV lytic replication, LMP2A prevents EBV infected cells from entering into lytic cycle (Miller et al., 1994). The ability of LMP2A to block BCR activation is dependent on the association of LMP2A with Lyn and Syk, since LMP2A mutants that are deficient in the association of these PTKs are unable to block B cell activation in LCLs (Fruehling and Longnecker, 1997; Fruehling et al., 1998). One simple model for the LMP2A-mediate block in BCR signal transduction is that LMP2A recruits the Lyn and Syk PTKs that are essential for BCR-mediated signal which results in the depletion of these PTKs from BCR complex (Figure 2A). In essence, LMP2A acts as a decoy BCR. Alternatively, constitutive signals emanating from LMP2A through Lyn and Syk may induce negative-feedback signals and cause desensitization of normal B cell signal transduction (Figure 2B).

Other studies indicate a potential different mechanism of the LMP2A alteration of normal BCR signal transduction. LMP2A has been shown to block the translocation of the BCR into lipid rafts in LCLs (Dykstra et al., 2001). Lipid rafts are enriched in glycosphingolipids, cholestrol, and fatty acid modified proteins. One function of lipid rafts is to provide a sorting site for proteins involved in signal transduction. Some B cell signal proteins, such as Src-family PTKs, are constitutively detected in lipid rafts, whereas other proteins are recruited into lipid rafts following receptor activation, including the BCR itself, Syk, Btk, BLNK, Vav PLCγ2, SHIP, and PI3K (Pierce, 2002). Interestingly, LMP2A is constitutively expressed in lipid rafts on plasma membrane (Higuchi et al., 2001; Katzman and Longnecker, 2004; Matskova et al., 2001), allowing LMP2A to mimic an activated BCR in lipid rafts (Figure 2C). In LMP2A-expressing LCLs, however, the BCR is excluded from lipid rafts even after crosslinking (Dykstra et al., 2001). Although the translocation of BCR into lipid rafts is independent of the BCR signaling through Lyn, LMP2A excludes the BCR from entering rafts in a Lyn-dependent mechanism (Dykstra et al., 2001). LMP2A also blocks BCR trafficking and sorting using a different mechanism. Generally, the BCR targets bound antigen to the endocytic MHC class II peptide-loading compartments (Pierce, 2002). Once the BCR is translocated into lipid rafts, a portion of rafts is internalized along with the BCR and trafficked to the MHC class II compartments. Internalization also results in BCR degradation in lysosome and an end to signal transduction through the receptor. It is known that both BCR internalization and degradation require BCR signaling. However, LMP2A interferes with BCR internalization by a Lyn-independent mechanism, since mutants deficient in Lyn association with LMP2A that permit BCR signaling and raft translocation still blocks BCR trafficking (Dykstra et al., 2001). Further studies are required to refine the model for LMP2A function in BCR trafficking.

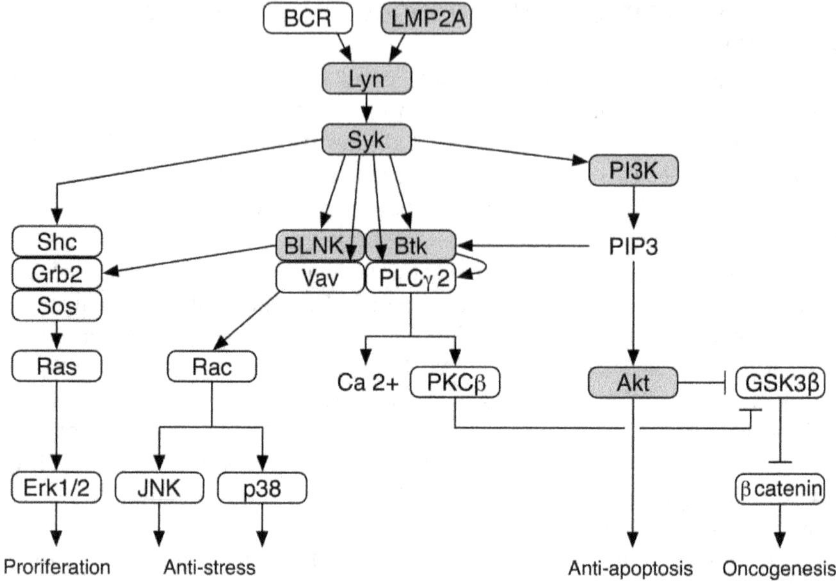

Figure 3. LMP2A signaling targets. As a functional BCR mimic, LMP2A activates B cell signaling molecules. LMP2A is constitutively phosphorylated and associated with Lyn and Syk, resulting in the constitutive activation of these PTKs and downstream signaling molecules including BLNK, Btk and PI3K. The activation of PI3K-Akt pathway causes cell survival in B cells and epithelial cells. Akt activation also results in the stablization of β-catenin by inhibiting GSK-3β activity in epithelial cells. LMP2A may be capable of activating other signal pathways although it has not been elucidated yet. Gray-shaded indicate proteins that are activated in LMP2A expressing cells.

SIGNAL TRANSDUCTION IN B CELLS

Since EBV can immortalize primary human B cells, a large portion of EBV research over the years has focused on the function of individual genes expressed in EBV transformed LCLs or by transfection into cell lines grown in culture. Despite the interaction of LMP2A with B cell PTKs, analysis of EBV recombinants indicates that LMP2A and LMP2B are not essential for EBV replication or immortalization of B cells *in vitro* (Kim and Yates, 1993; Longnecker et al., 1992; 1993a; 1993b). Only one report using mini-EBV plasmids suggests that LMP2A increases the efficiency of B cell immortalization (Brielmeier et al., 1996), but more recent studies suggests that LMP2A has no importance in B cell transformation by EBV (Speck et al., 1999).

Many studies have indicated that LMP2A is capable of inducing constitutive activation of signaling molecules in lymphocytes. LMP2A requires Syk, Btk and BLNK to mediate developmental and survival signals in mice (Engels et al., 2001; Merchant et al., 2000; Merchant and Longnecker, 2001). Chimeric receptors consisting of LMP2A N-terminal cytoplasmic and CD8 extracellular domains are capable of transducing signals including a calcium flux and IL-2 production in T cells (Beaufils et al., 1993). In LMP2A-expressing LCLs, constitutive tyrosine phosphorylation is observed for many signal transducers, including Lyn, Syk,

PI3K, BLNK, Btk and Akt (Fruehling et al., 1998; Ikeda et al., 2001; Merchant and Longnecker, 2001; Miller et al., 1995; Swart et al., 2000). These results clearly demonstrate that LMP2A triggers signal transduction in B cell lines. In BCR-mediated signal transduction, activated tyrosine kinases cause the phosphorylation of adaptor molecules and thereby induce the activation of downstream effectors including the important Ras–MAPK, PLCγ2–Ca^{2+}/PKC, and PI3K–Akt signaling pathways (Kurosaki, 2002). The phosphorylation profile in LMP2A-expressing LCLs suggests that PI3K–Akt pathway is constitutively activated in B cells (Swart et al., 2000). PI3K–Akt regulates numerous signaling pathways that promote cell survival, cell-cycle progression and cell growth. The constitutive Akt activation is consistent with the observation that LMP2A induces a survival signal in mice (Caldwell et al., 1998). In addition, LMP2A is able to mediate anti-apoptotic signals through PI3K–Akt pathway in B cell lines (Fukuda and Longnecker, 2004). Therefore, PI3K–Akt appears to be an important pathway targeted by LMP2A. Most recently, studies have shown that LMP2A also targets the Ras pathway and this is also important for EBV-mediated B cell survival (Portis and Longnecker, 2004). Due to the complexity of the signaling pathways that LMP2A modulates (Figure 3), further studies are required to fully elucidate signal transduction by LMP2A. This understanding will provide a significant insight for understanding how EBV maintains latency and contributes to proliferative disorders in humans with EBV latent infections.

DEVELOPMENT AND SURVIVAL SIGNALS

Latent EBV is found in one of 10^6 to 10^8 resting B cells in the peripheral blood of human hosts (Kieff and Rickinson, 2001; Thorley-Lawson, 2001). Although analysis of LMP2A associated proteins and LMP2A-activated signaling pathways in cell lines have provided many insights into how LMP2A helps maintain EBV latency, the paucity of these cells has prevented functional studies of EBV latent infection in healthy humans. In addition, the lack of an animal model that can be infected with EBV has made it difficult to draw conclusions about how LMP2A might function *in vivo*. However, studies using transgenic mice have proven informative in dissecting putative *in vivo* functions for LMP2A.

Transgenic mice have been generated expressing LMP2A downstream of the μ immunoglobulin (Ig) heavy chain enhancer and promoter (Eμ) in order to express LMP2A in B lineage cells (Caldwell et al., 1998). In wild type mice, competent BCR signaling is required in all stages of B cell development, from early B cell ontogeny to peripheral B cell maintenance (Figure 4). However, expression of the LMP2A transgene *in vivo* alters normal B cell development, allowing BCR-negative B cells to exit the bone marrow and survive in peripheral lymphoid organs. Thus, LMP2A acts as a surrogate BCR *in vivo* by providing constitutive signaling required for B cell development and survival. This LMP2A signal allows a bypass of normal Ig gene rearrangement resulting in an overall loss of BCR expression. Since LMP2A can provide a signal that mimics a normal BCR signal, this signal, along with the LMP1 signal that mimics CD40 may allow EBV to gain access to the resting memory B cell compartment without going through BCR-dependent germinal center formation. The signals that emanate from the BCR and CD40 provide important developmental and activation signals required for normal B cell

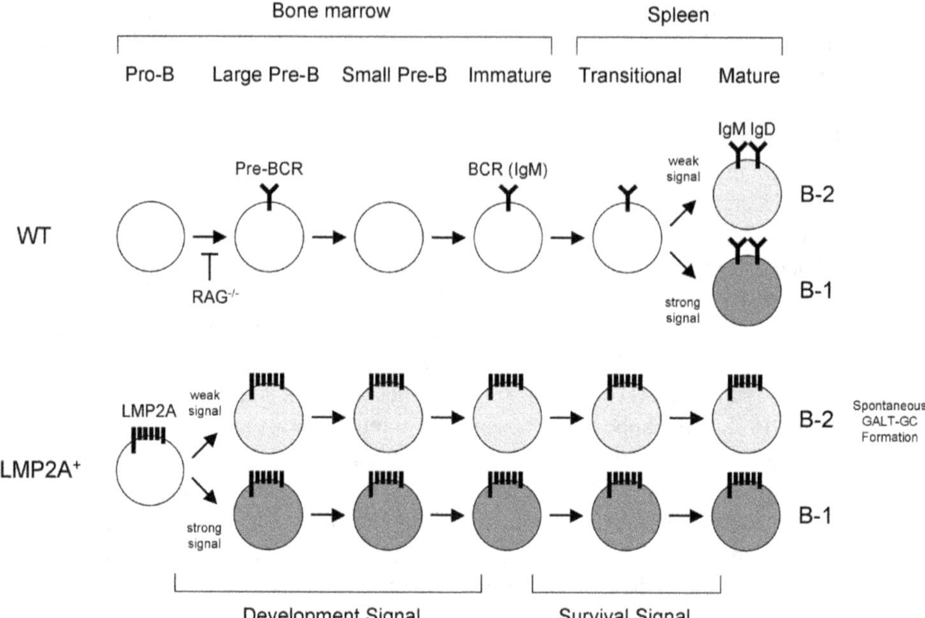

Figure 4. LMP2A *in vivo* phenotypes. In normal B cell development, expression of functional pre-BCR and BCR allow B cells to progress into pre-B and immature B cells, respectively. A deficiency in RAG genes causes the block of Ig gene rearrangement resulting in a block of normal B cell development. BCR expression is also required for B cells to exit bone marrow into peripheral organs. The signal from BCR is critical for B cells to survive in the periphery and escape from clonal deletion. It has been shown that the intensity of BCR signal determines the fate of B cell development. A strong signal from BCR at the transitional stage causes B cells develop into B-1 cells, whereas a relatively weak signal produces B-2 cells. LMP2A mimics the function of both pre-BCR and BCR and as a result LMP2A transgenic B cells display can bypass the requirement for Ig rearrangement resulting in the generation of BCR negative B cells. Similar to signals that emanate from a BCR in wild-type mice, a strong signal from LMP2A results in B-1 development exclusively.

development and survival (Banchereau et al., 1994). Both LMP1 and LMP2A have been shown to rescue B cells from apoptosis (Thorley-Lawson, 2001).

It has been shown that LMP2A utilizes B cell signal transduction pathways to mimic a normal BCR signaling. The Syk PTK is essential for the LMP2A mediated B cell development alteration as observed in the ΔITAM-LMP2A transgenic mice (Merchant et al., 2000). These murine transgenic lines express a mutant form of LMP2A in which the ITAM has been mutated. The ITAM motif in Ig-α and Ig-β is essential for coupling BCR signaling to antigen recognition (Kurosaki, 2002). In addition, LMP2A, like the BCR, requires Btk and BLNK, indicating similar signal requirements for both BCR and LMP2A development (Engels et al., 2001; Merchant and Longnecker, 2001). Interestingly, in addition to the Btk-dependent signals, LMP2A can also mediate Btk-independent signals although its pathway is still poorly understood (Merchant and Longnecker, 2001).

The expression level of LMP2A is also an important factor for determining the alteration in normal B cell development mediated by LMP2A observed in LMP2A transgenic mice. When expressed at high level, LMP2A allows the bypass of Ig gene rearrangement as described above. Low level expression of LMP2A is not sufficient to allow bypass of Ig gene rearrangement, instead B cells develop that express both a functional BCR and LMP2A (Caldwell et al., 2000). This result is likely due to the signal strength of LMP2A related to the absolute amount of LMP2A expressed. Previous studies examining BCR signal transduction have shown that signal strength through the BCR is an important determinate of B cell developmental outcomes (Hardy and Hayakawa, 2001). Further studies examining B cell development alterations in LMP2A transgenic mice have highlighted other interesting observations regarding developmental outcomes due to LMP2A. Recently it was shown that LMP2A can drive B cell development resulting in cells that represent CD5$^+$ B-1a cells (Ikeda et al., 2004). In wild type B cell differentiation, two distinct mature B cell populations, B-1 and B-2, are present in the mouse and human periphery (Hardy and Hayakawa, 2001). B-1 cells are more likely to produce autoantibodies and may be involved in T cell-independent responses to common environmental antigens. B-1 cells are found mainly in the peritoneal and pleuropericardial cavities, while B-2 cells are mainly found in spleen, lymph nodes and peripheral blood. In LMP2A transgenic mice expressing high levels of LMP2A, B-1 cells are the principal B cell population in bone marrow, spleen, and periphery (Ikeda et al., 2004). This switch to B-1 cells occurs at the pre-B stage in the bone marrow that is much earlier than appreciated for normal B-1 commitment (Ikeda et al., 2004). When expressed at low levels, LMP2A promotes follicular and marginal zone B cell development but not that of B-1 cells (Casola et al., 2004). Furthermore, spontaneous germinal centers (GC) are formed in gut-associated lymphoid tissue (GALT) of LMP2A transgenic mice, suggesting the ability of LMP2A to sustain a germinal center reaction (Casola et al., 2004).

These findings may provide further clues into EBV-related malignancies. Hodgkin and Reed-Sternberg (H-RS) cells arise from germinal center B cells and are positive for EBV in almost 50% of the cases of Hodgkin's Lymphoma (Rickinson and Kieff, 2001). H-RS cells have been known for quite sometime to express latency type II or the default program that includes both LMP1 and LMP2A. The ability of LMP2A to sustain a germinal center reaction is suggestive for the involvement of EBV infection in the Hodgkin's disease pathogenesis and recent studies do provide evidence that expression of LMP2A is responsible for the dramatic alteration in normal transcription that is observed in H-RS cells (Portis et al., 2003; Portis and Longnecker, 2003; 2004). For B-1 development, however, the involvement of EBV is unclear. Chronic lymphocytic leukemia (CLL) B cells are believed to be the transformed counterpart of normal B-1 cells (Caligaris-Cappio, 1996). Transformation into high-grade large cell Non-Hodgkin's lymphoma (NHL) often occurs in CLL patients and is clonally related to CLL in most cases. Although it has been shown that CLL is negative for EBV infection, CLL-related NHL and H-RS-like cells are often EBV positive (Ansell et al., 1999; Momose et al., 1992).

LMP2A FUNCTIONS IN EPITHELIAL CELLS

Early studies focused on the role of LMP2A in B cells, the primary compartment of EBV latency. However, LMP2 expression is found in NPC and nude mouse passaged NPC tumors suggesting that LMP2 contributes to the progression of this disease (Brooks et al., 1992; Busson et al., 1992). Recent studies on epithelial cells have indicated that LMP2A greatly affects cell proliferation and differentiation. Although LMP2A is constitutively phosphoryalted in B cells, LMP2A phosphorylation is induced by cell–extracellular matrix interactions in epithelial cells (Scholle et al., 1999). LMP2A interacts with the Src-family and Syk/ZAP-70-family PTKs and Csk, a negative Src regulator (Scholle et al., 1999; Scholle et al., 2001). In human keratinocyte cell line HaCaT, LMP2A expression induces activation of PI3K with subsequent phosphorylation and activation of Akt (Scholle et al., 2000). LMP2A-expressing HaCaT cells develop hyperproliferative raft cultures, form anchorage-independent colonies in soft agar, and induce the formation of aggressive metastatic tumors in nude mice (Scholle et al., 2000). This indicates that LMP2A has potent transforming abilities in epithelial cells. In addition, LMP2A activates and stabilizes β-catenin in epithelial cells through PI3K and Akt activation, which inhibits the Wnt signaling regulator glycogen synthase kinase-3β (GSK-3β) (Morrison et al., 2003). β-catenin is an important regulator of cell proliferation and differentiation during development and has been associated with oncogenic transformation (van de Wetering et al., 2002). The activity of β-catenin is tightly regulated by targeted ubiquitination and proteasomal degradation by GSK-3β (van de Wetering et al., 2002). A separate study shows that β-catenin is stablized in type III latency but degraded in type I latency (Shackelford et al., 2003). It is interesting to note that BCR activation also induces β-catenin activity by inhibiting GSK-3 similar to what is observed in epithelial cells (Christian et al., 2002). Other studies have shown that LMP2A induces MAPK cascades, a major proliferative signal including the phosphorylation of Erk and JNK in epithelial cell lines (Chen et al., 2002). LMP2A has also been shown to be a determinant of epithelial clonal emergence (Moody et al., 2003). In these studies, it was shown that the terminal repeat number varied inversely to the quantity of LMP2A transcripts whose unspliced precursors cross the joined genome terminal repeat regions. LMP2A may have two other roles important in epithelial cells. LMP2A has been shown to augment LMP1 signaling and inhibit TGF-β-associated apoptosis in epithelial cell lines (Dawson et al., 2001; Fukuda and Longnecker, 2004). Inhibition of TGF-β-associated apoptosis was also observed in B cells.

Despite the dramatic effects observed when LMP2A is expressed in epithelial cell lines, LMP2A transgenic mice with expression localized to the differentiating epithelia exhibited no dramatic alteration in normal epithelial phenotype (Longan and Longnecker, 2000). This suggests that LMP2A itself is insufficient to transform non-immortalized epithelial cells *in vivo*. It is thus likely that the activation of Akt by LMP2A, in combination with other genetic changes in immortalized epithelial cell lines, contributes to the malignant phenotype of NPC.

UBIQUITINATION AND TRAFFICKING

LMP2A signaling is regulated in B cells by association of LMP2A with members of the HECT-domain containing Nedd4-family ubiquitin protein ligases, including

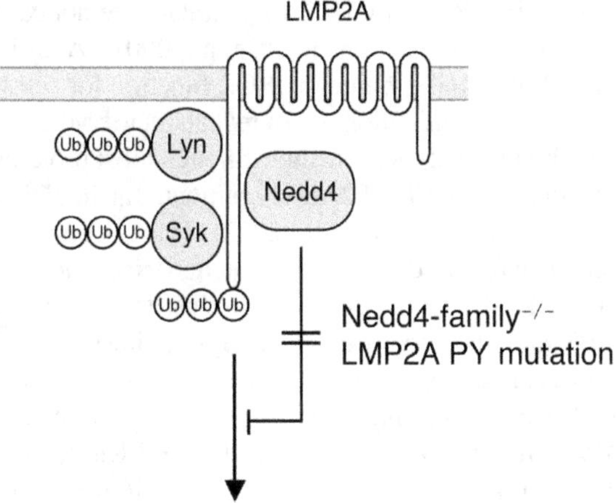

Figure 5. Nedd4-family ubiquitin ligases downregulate LMP2A signaling. LMP2A and its associated proteins such as Lyn and Syk are ubiquitinated by Nedd4-family ubiquitin ligases, resulting in the rapid degradation of LMP2A and LMP2A associated proteins. Lyn expression is severely reduced in LMP2A-expressing cells. When the association of LMP2A with Nedd4-family ubiquitin ligases is blocked by PY mutations in LMP2A or genetic disruption in Nedd4-family, Lyn is not degraded and remains associated with LMP2A in an activated form. This results in a stronger LMP2A signaling and elevated cellular tyrosine phosphorylation.

AIP4/Itchy, WWP2, Nedd4, and Nedd4-2/KIAA0439 (Ikeda et al., 2000; Winberg et al., 2000). Nedd4-family members bind specifically to PY motifs present in the cytoplasmic domain of LMP2A, resulting in the ubiquitination and degradation of LMP2A and its associated proteins such as the PTK Lyn (Figure 5). Negative regulation of LMP2A signal transduction by ubiquitination was confirmed by both *in vitro* mutational analysis and *in vivo* mouse genetic analysis. A LMP2A mutant in both of its PY motifs is incapable of associating with Nedd4-family ubiquitin ligases resulting in an absence of the degradation of LMP2A associated protein PTKs (Ikeda et al., 2001). Consequently, elevated cellular phosphorylation is observed in cells expressing this mutant. In addition, when LMP2A transgenic mice are crossed with Itchy mice that are deficient in AIP4, there is an increase in survival, signal transduction, and growth of LMP2A expressing B cells (Ikeda et al., 2003). This is compatible with previous results indicating that signal strength of LMP2A or relative expression of LMP2A is critical for determining B cell fate (Caldwell et al., 2000). It is likely that LMP2A utilizes ubiquitin-mediated degradation to regulate the strength of its own signal, which may allow LMP2A to modulate B cell processes, such as differentiation, activation, or survival. Nedd4-family ubiquitin ligases would appear to control LMP2A-mediated B-1/B-2 development by regulating LMP2A degradation, resulting in appropriate germinal center formation to allow infected B cells to differentiate into the resting memory B cells (Casola et al., 2004; Ikeda et al., 2004). However, LMP2A signaling itself may be insufficient or nonessential for B cell proliferation in germinal centers, since the augmented signals from LMP2A PY mutant are not capable of inducing cell growth and viral lytic replication (Ikeda et al., 2001). Additional signals may be required for LMP2A to fully mimic the BCR function for cell growth.

Results of *in vitro* ubiquitination assay reveals that LMP2A is modified with ubiquitin in a rare and unusual mechanism. Although most proteins are specifically ubiquitinated at lysine residues, LMP2A is ubiquitinated at the amino group of amino-terminal methionine residue (Ikeda et al., 2002). This uncommon amino-terminal ubiquitination has been reported in only six proteins (Ciechanover and Ben-Saadon, 2004). Interestingly, LMP1 is also included in this unusual degradation pathway indicating that EBV may specifically utilize this type of host cell modification during viral latency (Aviel et al., 2000). LMP1 aggregates together with LMP2A on plasma membrane in LCLs indicating that there may be some cooperativity of between LMP1 and LMP2A in viral latency (Longnecker and Kieff, 1990). Interestingly, LMP2A is mono-ubiquitinated (Ikeda et al., 2002). Ubiquitination has been shown to play a role in initiating and regulating endocytosis (Hicke, 2001). It has long been understood that conjugation of ubiquitin to cytoplasmic proteins can target them for degradation to the 26S proteasome. Recently, a large number of membranes proteins, including signaling receptors, have been reported to be ubiquitinated. In this case, ubiquitination is necessary for the first step of endocytosis. In contrast to proteasome degradation, which requires poly-ubiquitination, endocytosis requires only mono-ubiquitination (Hicke, 2001). Indeed, LMP2A expression is observed not only on plasma membrane but also in internal membrane organelles (Dawson et al., 2001; Longnecker and Kieff, 1990; Lynch et al., 2002). About 30% of LMP2A is associated with lipid rafts while the cytosolic soluble fraction also contains LMP2A (Higuchi et al., 2001). This suggests that LMP2A can be processed for degradation through ubiquitin-dependent endocytosis. LMP2A may also regulate the trafficking of other membrane proteins. LMP2A resides in lipid rafts and excludes the BCR from entering rafts, thus blocking both BCR signaling and antigen transport (Dykstra et al., 2001). Although it has not been demonstrated directly, it is likely that LMP2A utilizes ubiquitination to regulate the trafficking of BCR in order to maintain B cell survival during EBV latency.

LATENCY-SPECIFIC TRANSCRIPTION

Recent studies utilizing DNA microarray technology have linked LMP2A to the development and pathogenesis of Hodgkin's Lymphoma. Expresssion of LMP2A during B cell development *in vivo* causes reduced expression of several transcription factors critical for B cell development. Specifically, decreased expression of E2A, early B-cell factor (EBF), and Pax-5 correlate with a global down-regulation of gene transcription necessary for proper B cell development and function (Portis et al., 2003; Portis and Longnecker, 2003). A strikingly similar pattern of global down-regulation of B cell specific genes is observed in malignant H-RS cells of HD (Hertel et al., 2002; Schwering et al., 2003). H-RS cells are presumed to derive from germinal center B cells that have lost expression of Ig and other B cell-specific characteristics (Cossman et al., 1999; Kuppers, 2002). Thus, it is plausible that LMP2A is directly responsible for many of the transcriptional abnormalities identified in H-RS cells, which may be necessary for the cells to survive without a functional BCR. It is interesting to note that activated Notch signaling appears to play a significant role in the pathogenesis of HD (Jundt et al., 2002). Although direct activation of Notch by LMP2A has not yet been demonstrated, increased expression of RBP-Jk, together with reduced E2A activity, in LMP2A-expressing splenic B cells, suggests that LMP2A activates the Notch pathway *in vivo*. Future studies that address the mechanisms by which LMP2A-induced changes in transcription factor gene expression promote altered development and survival of B cells may provide better insight into how LMP2A maintains latency in EBV-infected B lymphocytes.

CONCLUSION

Although the exact function of LMP2A during EBV latency *in vivo* and EBV-associated pathogenesis is still unclear, studies discussed in this chapter provide important clues. By mimicking BCR signal transduction, LMP2A functions in regulating normal B cell processes such as differentiation and transcription factor expression that are imparted for BCR signal transduction. Signalling by LMP2A is fine tuned by ubiquitination resulting in lower levels of kinases important for normal BCR signal transduction and trafficking of the LMP2A complex. In regard to EBV-associated cancers, studies outlined in this chapter indicate that LMP2A likely plays an active role by providing inappropriate growth promoting or anti-apoptotic signals that contribute to cancers associated with EBV infection. Future studies that define LMP2A signalling will further delineate the role of LMP2A in human EBV infections and diseases associated with such infections.

Acknowledgements

R.L. is a Stohlman Scholar of the Leukemia and Lymphoma Society of America and supported by the Public Health Service grants CA62234, CA73507, and CA93444 from the National Cancer Institute and DE13127 from the National Institute of Dental and Craniofacial Research. M.I. is a Special Fellow of the Leukemia and Lymphoma Society of America. M.F. is a Philip Morris External Research Program Postdoctoral Fellow and supported by Philip Morris USA Inc. We would like to thank current and former members of the Longnecker Laboratory for their contribution to the work described, as well as our many colleagues throughout the world whose work we were unable to cite because of the limitation on number of citations.

References

Ansell, S. M., Li, C. Y., Lloyd, R. V., and Phyliky, R. L. (1999). Epstein-Barr virus infection in Richter's transformation. Amer. J. Hematol *60*, 99-104.

Aviel, S., Winberg, G., Massucci, M., and Ciechanover, A. (2000). Degradation of the epstein-barr virus latent membrane protein 1 (LMP1) by the ubiquitin-proteasome pathway. Targeting via ubiquitination of the N-terminal residue. J. Biol. Chem. *275*, 23491-23499.

Banchereau, J., Bazan, F., Blanchard, D., Briere, F., Galizzi, J. P., van Kooten, C., Liu, Y. J., Rousset, F., and Saeland, S. (1994). The CD40 antigen and its ligand. Annu. Rev. Immunol. *12*, 881-922.

Beaufils, P., Choquet, D., Mamoun, R. Z., and Malissen, B. (1993). The (YXXL/I)2 signalling motif found in the cytoplasmic segments of the bovine leukaemia virus envelope protein and Epstein-Barr virus latent membrane protein 2A can elicit early and late lymphocyte activation events. Embo J. *12*, 5105-5112.

Berger, C., Rothenberger, S., Bachmann, E., McQuain, C., Nadal, D., and Knecht, H. (1999). Sequence polymorphisms between latent membrane proteins LMP1 and LMP2A do not correlate in EBV-associated reactive and malignant lympho-proliferations. Int. J. Cancer *81*, 371-375.

Brielmeier, M., Mautner, J., Laux, G., and Hammerschmidt, W. (1996). The latent membrane protein 2 gene of Epstein-Barr virus is important for efficient B cell immortalization. J. Gen. Virol. *77*, 2807-2818.

Brooks, L., Yao, Q. Y., Rickinson, A. B., and Young, L. S. (1992). Epstein-Barr virus latent gene transcription in nasopharyngeal carcinoma cells: coexpression of EBNA1, LMP1, and LMP2 transcripts. J. Virol. *66*, 2689-2697.

Busson, P., Edwards, R. H., Tursz, T., and Raab-Traub, N. (1995). Sequence polymorphism in the Epstein-Barr virus latent membrane protein (LMP)-2 gene. J. Gen. Virol. *76*, 139-145.

Busson, P., McCoy, R., Sadler, R., Gilligan, K., Tursz, T., and Raab-Traub, N. (1992). Consistent transcription of the Epstein-Barr virus LMP2 gene in nasopharyngeal carcinoma. J. Virol. *66*, 3257-3262.

Caldwell, R. G., Brown, R. C., and Longnecker, R. (2000). Epstein-Barr virus LMP2A-induced B-cell survival in two unique classes of EmuLMP2A transgenic mice. J. Virol. *74*, 1101-1113.

Caldwell, R. G., Wilson, J. B., Anderson, S. J., and Longnecker, R. (1998). Epstein-Barr virus LMP2A drives B cell development and survival in the absence of normal B cell receptor signals. Immunity *9*, 405-411.

Caligaris-Cappio, F. (1996). B-chronic lymphocytic leukemia: a malignancy of anti-self B cells. Blood *87*, 2615-2620.

Casola, S., Otipoby, K. L., Alimzhanov, M., Humme, S., Uyttersprot, N., Kutok, J. L., Carroll, M. C., and Rajewsky, K. (2004). B cell receptor signal strength determines B cell fate. Nat. Immunol. *5*, 317-327. Epub 2004 Feb 2001.

Chen, F., Zou, J. Z., di Renzo, L., Winberg, G., Hu, L. F., Klein, E., Klein, G., and Ernberg, I. (1995). A subpopulation of normal B cells latently infected with Epstein-Barr virus resembles Burkitt lymphoma cells in expressing EBNA-1 but not EBNA-2 or LMP1. J. Virol. *69*, 3752-3758.

Chen, S. Y., Lu, J., Shih, Y. C., and Tsai, C. H. (2002). Epstein-Barr virus latent membrane protein 2A regulates c-Jun protein through extracellular signal-regulated kinase. J. Virol. *76*, 9556-9561.

Christian, S. L., Sims, P. V., and Gold, M. R. (2002). The B cell antigen receptor regulates the transcriptional activator beta-catenin via protein kinase C-mediated inhibition of glycogen synthase kinase-3. J. Immunol. *169*, 758-769.

Ciechanover, A., and Ben-Saadon, R. (2004). N-terminal ubiquitination: more protein substrates join in. Trends Cell Biol. *14*, 103-106.

Cossman, J., Annunziata, C. M., Barash, S., Staudt, L., Dillon, P., He, W.-W., Ricciardi-Castagnoli, P., Rosen, C. A., and Carter, K. C. (1999). Reed-Sternberg cell genome expression supports a B-cell lineage. Blood *94*, 411-416.

Dawson, C. W., George, J. H., Blake, S. M., Longnecker, R., and Young, L. S. (2001). The Epstein-Barr virus encoded latent membrane protein 2A augments signaling from latent membrane protein 1. Virology *289*, 192-207.

Dykstra, M. L., Longnecker, R., and Pierce, S. K. (2001). Epstein-Barr virus coopts lipid rafts to block the signaling and antigen transport functions of the BCR. Immunity *14*, 57-67.

Engels, N., Merchant, M., Pappu, R., Chan, A. C., Longnecker, R., and Wienands, J. (2001). Epstein-Barr virus latent membrane protein 2A (LMP2A) employs the SLP-65 signaling module. J. Exp. Med. *194*, 255-264.

Franken, M., Annis, B., Ali, A. N., and Wang, F. (1995). 5' Coding and regulatory region sequence divergence with conserved function of the Epstein-Barr virus LMP2A homolog in herpesvirus papio. J. Virol. *69*, 8011-8019.

Fruehling, S., and Longnecker, R. (1997). The immunoreceptor tyrosine-based activation motif of Epstein-Barr virus LMP2A is essential for blocking BCR-mediated signal transduction. Virology *235*, 241-251.

Fruehling, S., Swart, R., Dolwick, K. M., Kremmer, E., and Longnecker, R. (1998). Tyrosine 112 of latent membrane protein 2A is essential for protein tyrosine kinase loading and regulation of Epstein-Barr virus latency. J. Virol. *72*, 7796-7806.

Fukuda, M., and Longnecker, R. (2004). Latent membrane protein 2A inhibits transforming growth factor-beta 1-induced apoptosis through the phosphatidylinositol 3-kinase/Akt pathway. J. Virol. *78*, 1697-1705.

Hardy, R. R., and Hayakawa, K. (2001). B cell development pathways. Annu. Rev. Immunol. *19*, 595-621.

Hertel, C. B., Zhou, X., Hamilton-Dutoit, S. J., and Junker, S. (2002). Loss of B cell identity correlates with loss of B cell-specific transcription factors in Hodgkin/Reed-Sternberg cells of classical Hodgkin lymphoma. Oncogene *21*, 4908-4920.

Hicke, L. (2001). A new ticket for entry into budding vesicles-ubiquitin. Cell *106*, 527-530.

Higuchi, M., Izumi, K. M., and Kieff, E. (2001). Epstein-Barr virus latent-infection membrane proteins are palmitoylated and raft-associated: protein 1 binds to the cytoskeleton through TNF receptor cytoplasmic factors. Proc. Natl. Acad. Sci. USA. *98*, 4675-4680.

Ikeda, A., Caldwell, R. G., Longnecker, R., and Ikeda, M. (2003). Itchy, a Nedd4 ubiquitin ligase, downregulates latent membrane protein 2A activity in B-cell signaling. J. Virol. *77*, 5529-5534.

Ikeda, A., Merchant, M., Lev, L., Longnecker, R., and Ikeda, M. (2004). Latent membrane protein 2A, a viral B cell receptor homologue, induces CD5+ B-1 cell development. J. Immunol. *172*, 5329-5337.

Ikeda, M., Ikeda, A., Longan, L. C., and Longnecker, R. (2000). The Epstein-Barr virus latent membrane protein 2A PY motif recruits WW domain-containing ubiquitin-protein ligases. Virology *268*, 178-191.

Ikeda, M., Ikeda, A., and Longnecker, R. (2001). PY motifs of Epstein-Barr virus LMP2A regulate protein stability and phosphorylation of LMP2A-associated proteins. J. Virol. *75*, 5711-5718.

Ikeda, M., Ikeda, A., and Longnecker, R. (2002). Lysine-independent ubiquitination of Epstein-Barr virus LMP2A. Virology *300*, 153-159.

Jundt, F., Anagnostopoulos, I., Forster, R., Mathas, S., Stein, H., and Dorken, B. (2002). Activated Notch 1 signaling promotes tumor cell proliferation and survival in Hodgkin's and anaplastic large cell lymphoma. Blood *99*, 3398-3403.

Katzman, R. B., and Longnecker, R. (2004). LMP2A does not Require Palmitoylation to Localize to Buoyant Complexes or for Function. J. Virol. *78*, 10878-10887.

Kieff, E., and Rickinson, A. B. (2001). Epstein-Barrr virus and its replication. In Fields Virology, D. M. Knipe, and P. M. Howley, eds. (Philadelphia, PA, Lippincott-Raven Publishers), pp. 2511-2573.

Kim, O. J., and Yates, J. L. (1993). Mutants of Epstein-Barr virus with a selective marker disrupting the TP gene transform B cells and replicate normally in culture. J. Virol. *67*, 7634-7640.

Kuppers, R. (2002). Biology of Hogdkin's lymphoma. Ann. Oncology *13*, 11-18.

Kurosaki, T. (2002). Regulation of B-cell signal transduction by adaptor proteins. Nat. Rev. Immunol. *2*, 354-363.

Laux, G., Adam, B., Strobl, L. J., and Moreau-Gachelin, F. (1994a). The Spi-1/PU.1 and Spi-B ets family transcription factors and the recombination signal binding protein RBP-J kappa interact with an Epstein-Barr virus nuclear antigen 2 responsive cis-element. Embo J. *13*, 5624-5632.

Laux, G., Dugrillon, F., Eckert, C., Adam, B., Zimber-Strobl, U., and Bornkamm, G. W. (1994b). Identification and characterization of an Epstein-Barr virus nuclear antigen 2-responsive cis element in the bidirectional promoter region of latent membrane protein and terminal protein 2 genes. J. Virol. *68*, 6947-6958.

Laux, G., Economou, A., and Farrell, P. J. (1989). The terminal protein gene 2 of Epstein-Barr virus is transcribed from a bidirectional latent promoter region. J. Gen. Virol. *70*, 3079-3084.

Longan, L., and Longnecker, R. (2000). Epstein-Barr virus latent membrane protein 2A has no growth-altering effects when expressed in differentiating epithelia. J. Gen. Virol. *81*, 2245-2252.

Longnecker, R. (1998). Molecular Biology of Epstein-Barr Virus. In Human Tumor Viruses, D. J. McCance, ed. (Washington, D. C., American Society for Microbiology), pp. 133-172.

Longnecker, R. (2000). Epstein-Barr virus latency: LMP2, a regulator or means for Epstein-Barr virus persistence? Adv. Cancer Res. *79*, 175-200.

Longnecker, R., Druker, B., Roberts, T. M., and Kieff, E. (1991). An Epstein-Barr virus protein associated with cell growth transformation interacts with a tyrosine kinase. J. Virol. *65*, 3681-3692.

Longnecker, R., and Kieff, E. (1990). A second Epstein-Barr virus membrane protein (LMP2) is expressed in latent infection and colocalizes with LMP1. J. Virol. *64*, 2319-2326.

Longnecker, R., Miller, C. L., Miao, X. Q., Marchini, A., and Kieff, E. (1992). The only domain which distinguishes Epstein-Barr virus latent membrane protein 2A (LMP2A) from LMP2B is dispensable for lymphocyte infection and growth transformation *in vitro*; LMP2A is therefore nonessential. J. Virol. *66*, 6461-6469.

Longnecker, R., Miller, C. L., Miao, X. Q., Tomkinson, B., and Kieff, E. (1993a). The last seven transmembrane and carboxy-terminal cytoplasmic domains of Epstein-Barr virus latent membrane protein 2 (LMP2) are dispensable for lymphocyte infection and growth transformation *in vitro*. J. Virol. *67*, 2006-2013.

Longnecker, R., Miller, C. L., Tomkinson, B., Miao, X. Q., and Kieff, E. (1993b). Deletion of DNA encoding the first five transmembrane domains of Epstein-Barr virus latent membrane proteins 2A and 2B. J. Virol. *67*, 5068-5074.

Lynch, D. T., Zimmerman, J. S., and Rowe, D. T. (2002). Epstein-Barr virus latent membrane protein 2B (LMP2B) co-localizes with LMP2A in perinuclear regions in transiently transfected cells. J. Gen. Virol. *83*, 1025-1035.

Matskova, L., Ernberg, I., Pawson, T., and Winberg, G. (2001). C-terminal domain of the Epstein-Barr virus LMP2A membrane protein contains a clustering signal. J. Virol. *75*, 10941-10949.

Meitinger, C., Strobl, L. J., Marschall, G., Bornkamm, G. W., and Zimber-Strobl, U. (1994). Crucial sequences within the Epstein-Barr virus TP1 promoter for EBNA2-mediated transactivation and interaction of EBNA2 with its responsive element. J. Virol. *68*, 7497-7506.

Merchant, M., Caldwell, R. G., and Longnecker, R. (2000). The LMP2A ITAM is essential for providing B cells with development and survival signals *in vivo*. J. Virol. *74*, 9115-9124.

Merchant, M., and Longnecker, R. (2001). LMP2A survival and developmental signals are transmitted through Btk-dependent and Btk-independent pathways. Virology *291*, 46-54.

Miller, C. L., Burkhardt, A. L., Lee, J. H., Stealey, B., Longnecker, R., Bolen, J. B., and Kieff, E. (1995). Integral membrane protein 2 of Epstein-Barr virus regulates reactivation from latency through dominant negative effects on protein-tyrosine kinases. Immunity 2, 155-166.

Miller, C. L., Lee, J. H., Kieff, E., and Longnecker, R. (1994). An integral membrane protein (LMP2) blocks reactivation of Epstein-Barr virus from latency following surface immunoglobulin crosslinking. Proc. Natl. Acad. Sci. USA. *91*, 772-776.

Miller, C. L., Longnecker, R., and Kieff, E. (1993). Epstein-Barr virus latent membrane protein 2A blocks calcium mobilization in B lymphocytes. J. Virol. *67*, 3087-3094.

Momose, H., Jaffe, E. S., Shin, S. S., Chen, Y. Y., and Weiss, L. M. (1992). Chronic lymphocytic leukemia/small lymphocytic lymphoma with Reed-Sternberg-like cells and possible transformation to Hodgkin's disease. Mediation by Epstein-Barr virus. Amer. J. Surg Pathol. *16*, 859-867.

Moody, C. A., Scott, R. S., Su, T., and Sixbey, J. W. (2003). Length of Epstein-Barr virus termini as a determinant of epithelial cell clonal emergence. J. Virol. *77*, 8555-8561.

Morrison, J. A., Klingelhutz, A. J., and Raab-Traub, N. (2003). Epstein-Barr virus latent membrane protein 2A activates beta-catenin signaling in epithelial cells. J. Virol. *77*, 12276-12284.

Niedobitek, G., Kremmer, E., Herbst, H., Whitehead, L., Dawson, C. W., Niedobitek, E., von Ostau, C., Rooney, N., Grasser, F. A., and Young, L. S. (1997). Immunohistochemical detection of the Epstein-Barr virus-encoded latent membrane protein 2A in Hodgkin's disease and infectious mononucleosis. Blood *90*, 1664-1672.

Panousis, C. G., and Rowe, D. T. (1997). Epstein-Barr virus latent membrane protein 2 associates with and is a substrate for mitogen-activated protein kinase. J. Virol. *71*, 4752-4760.

Pierce, S. K. (2002). Lipid rafts and B-cell activation. Nat. Rev. Immunol. *2*, 96-105.

Pleiman, C. M., Abrams, C., Gauen, L. T., Bedzyk, W., Jongstra, J., Shaw, A. S., and Cambier, J. C. (1994). Distinct p53/56lyn and p59fyn domains associate with nonphosphorylated and phosphorylated Ig-alpha. Proc. Natl. Acad. Sci. USA. *91*, 4268-4272.

Portis, T., Dyck, P., and Longnecker, R. (2003). Epstein-Barr Virus (EBV) LMP2A induces alterations in gene transcription similar to those observed in Reed-Sternberg cells of Hodgkin lymphoma. Blood *102*, 4166-4178. Epub 2003 Aug 4167.

Portis, T., and Longnecker, R. (2003). Epstein-Barr virus LMP2A interferes with global transcription factor regulation when expressed during B-lymphocyte development. J. Virol. *77*, 105-114.

Portis, T., and Longnecker, R. (2004). Epstein-Barr virus (EBV) LMP2A alters normal transcriptional regulation following B-cell receptor activation. Virology *318*, 524-533.

Portis, T., and Longnecker, R. (2004). Epstein-Barr virus (EBV) LMP2A mediates B-lymphocyte survival through constitutive activation of the Ras/PI3K/Akt pathway. Oncogene. *23*, 8619-8628.

Qu, L., and Rowe, D. T. (1992). Epstein-Barr virus latent gene expression in uncultured peripheral blood lymphocytes. J. Virol. *66*, 3715-3724.

Rickinson, A. B., and Kieff, E. (2001). Epstein-Barrr virus. In Fields Virology, D. M. Knipe, and P. M. Howley, eds. (Philadelphia, PA, Lippincott-Raven Publishers).

Rivailler, P., Quink, C., and Wang, F. (1999). Strong selective pressure for evolution of an Epstein-Barr virus LMP2B homologue in the rhesus lymphocryptovirus. J. Virol. *73*, 8867-8872.

Sample, J., Liebowitz, D., and Kieff, E. (1989). Two related Epstein-Barr virus membrane proteins are encoded by separate genes. J. Virol. *63*, 933-937.

Scholle, F., Bendt, K. M., and Raab-Traub, N. (2000). Epstein-Barr virus LMP2A transforms epithelial cells, inhibits cell differentiation, and activates Akt. J. Virol. *74*, 10681-10689.

Scholle, F., Longnecker, R., and Raab-Traub, N. (1999). Epithelial cell adhesion to extracellular matrix proteins induces tyrosine phosphorylation of the Epstein-Barr virus latent membrane protein 2: a role for C-terminal Src kinase. J. Virol. *73*, 4767-4775.

Scholle, F., Longnecker, R., and Raab-Traub, N. (2001). Analysis of the phosphorylation status of Epstein-Barr virus LMP2A in epithelial cells. Virology *291*, 208-214.

Schwering, I., Brauninger, A., Klein, U., Jungnickel, B., Tinguely, M., Diehl, V., Hansmann, M.-L., Dalla-Favera, R., Rajewsky, K., and Kuppers, R. (2003). Loss of the B-lineage-specific gene expression program in Hodgkin and Reed-Sternberg cells of Hodgkin lymphoma. Blood *101*, 1505-1512.

Shackelford, J., Maier, C., and Pagano, J. S. (2003). Epstein-Barr virus activates β-catenin in type III latently infected B lymphocyte lines: association with deubiquitinating enzymes. Proc. Natl. Acad. Sci. USA. *100*, 15572-15576.

Speck, P., Kline, K. A., Cheresh, P., and Longnecker, R. (1999). Epstein-Barr virus lacking latent membrane protein 2 immortalizes B cells with efficiency indistinguishable from that of wild-type virus. J. Gen. Virol. *80*, 2193-2203.

Sudol, M. (1996). Structure and function of the WW domain. Prog Biophys Mol. Biol. *65*, 113-132.

Swart, R., Fruehling, S., and Longnecker, R. (1999). Tyrosines 60, 64, and 101 of Epstein-Barr virus LMP2A are not essential for blocking B cell signal transduction. Virology *263*, 485-495.

Swart, R., Ruf, I. K., Sample, J., and Longnecker, R. (2000). Latent membrane protein 2A-mediated effects on the phosphatidylinositol 3-Kinase/Akt pathway. J. Virol. *74*, 10838-10845.

Tanaka, M., Kawaguchi, Y., Yokofujita, J., Takagi, M., Eishi, Y., and Hirai, K. (1999). Sequence variations of Epstein-Barr virus LMP2A gene in gastric carcinoma in Japan. Virus Genes *19*, 103-111.

Thorley-Lawson, D. A. (2001). Epstein-Barr virus: exploiting the immune system. Nat. Rev. Immunol. *1*, 75-82.

Tierney, R. J., Steven, N., Young, L. S., and Rickinson, A. B. (1994). Epstein-Barr virus latency in blood mononuclear cells: analysis of viral gene transcription during primary infection and in the carrier state. J. Virol. *68*, 7374-7385.

van de Wetering, M., de Lau, W., and Clevers, H. (2002). WNT signaling and lymphocyte development. Cell *109*, S13-S19.

Winberg, G., Matskova, L., Chen, F., Plant, P., Rotin, D., Gish, G., Ingham, R., Ernberg, I., and Pawson, T. (2000). Latent membrane protein 2A of Epstein-Barr virus binds WW domain E3 protein-ubiquitin ligases that ubiquitinate B-cell tyrosine kinases. Mol. Cell Biol. *20*, 8526-8535.

From: Epstein-Barr Virus. Edited by: Erle S. Robertson

Chapter 26

Epstein-Barr Virus Latent Infection Membrane Protein One

*Ellen Cahir-McFarland and Elliott Kieff**

ABSTRACT

Epstein-Barr Virus (EBV) Latent Infection Membrane Protein 1 (LMP1) is expressed in latency III primary EBV infection of human B lymphoblasts, *in vivo*, in latency III post-transplant lymphoproliferative disease, in EBV latency III infected human B lymphocytes that have been converted into lymphoblasts capable of long term proliferation, *in vitro* (LCLs), in EBV associated Hodgkin's Disease, and in many nasopharyngeal cancers. LMP1 expression in human B lymphoma cells induces activation and adhesion molecule expression and cell clumping, which are characteristic of LCLs or of CD40 activated B lymphocytes. In immortalized fibroblasts, LMP1 mimics aspects of activated ras in enabling serum, contact, and anchorage independent growth. Reverse genetic analyses implicate the LMP1 6 transmembrane and 2 C-terminal cytoplasmic domains as the essential domains for LMP1 effects. The 6 transmembrane domains cause intermolecular interaction, whereas the C-terminal domains signal through tumor necrosis factor receptor (TNFR) associated factors or death domain proteins and activate NF-κB, JNK, and p38. Comparisons of LMP1 and TNFR signal transduction provides insights into the biology and chemistry of these pathways. Biochemical interactions among LMP1 transmembrane domains and lipid rafts are critical for signaling and NF-κB activation is essential for cell survival. Studies of the chemistry and biology of LMP1 effects are important for controlling EBV infected cell survival and growth.

INTRODUCTION

Epstein-Barr Virus (EBV) latent infection integral membrane protein one (LMP1) is frequently expressed in latent EBV infections associated with B lymphocyte proliferation and with Naso-Pharyngeal Cancer (NPC). LMP1 is uniformly expressed in latency III EBV infection of human B lymphocytes, *in vitro*, in resultant lymphoblastoid cell lines (LCLs), in early primary human infection associated latency III lymphocyte infection, *in vivo*, in early onset lymphoproliferative disease (LPD) in transplant recipients, and frequently in late onset LPDs in transplant recipients or HIV infected people. LMP1 is also expressed in latency II

*For correspondence email ekieff@rics.bwh.harvard.edu

EBV infection in Hodgkin's Disease (HD) B lymphocytes. In latency III infected B lymphocyte proliferation, LMP1 is expressed along with six EBV nuclear proteins (EBNAs), a second integral membrane protein (LMP2), two small RNAs (EBERs), and BamA rightwards transcripts (BARTs), whereas in latency II infection, LMP1 is expressed along with EBNA1, LMP2, EBERs, and BARTs. In NPC latency II, LMP1 is expressed along with EBNA1, EBERs, BARTs, and frequently LMP2 (Reviewed in (Rickinson and Kieff, 2001)).

LMP1 structure and gene expression

The LMP1 gene is adjacent to the EBV genome terminal repeat. The gene includes two ~80b introns. The first exon encodes 24 N-terminal cytoplasmic amino acids (aa), two hydrophobic transmembrane domains, RRD, and the beginning of a third hydrophobic transmembrane domain. The second exon encodes the rest of the third transmembrane domain and part of the fourth. The third exon encodes the rest of the 4th transmembrane domain, R, the 5th and 6th transmembrane domains, and 200 C-terminal cytoplasmic aa. The prototype LMP1 C-terminal cytoplasmic domain has 5 tandem 33 bp repeats (Fennewald et al., 1984). Other EBV LMP1 genes differ from the prototype in repeat number.

Three promoters can regulate LMP1 gene transcription. In initial latency III lymphocyte infection, LMP1 transcription is turned on by EBNA2 (Wang et al., 1990b) and EBNALP (Harada and Kieff, 1997) which also turn on EBNA1 and EBNA3A, 3B, and 3C, LMP2, and cell genes, including c-myc (Cooper et al., 2003; Kempkes et al., 1995). EBNA1 and EBNA3C partially regulate the latency III LMP1 promoter (Gahn and Sugden, 1995; Waltzer et al., 1996). LMP1 mRNA is among the most abundant EBV mRNAs in latency III infection. Despite a relatively short half life, LMP1 protein is readily detected in latency III infected cells. In latency II infected NPC cells, LMP1 transcription originates from a STAT regulated upstream promoter in the EBV terminal direct repeat (Sadler and Raab-Traub, 1995). In carcinoma cells, LMP2A expression can down-regulate NF-κB and IL-6 secretion and thereby down modulate STAT regulated LMP1 expression (Chen et al., 2003). A third promoter in the LMP1 first intron is activated late in lytic EBV replication and transcribes a 5' truncated D1LMP1 mRNA (Fennewald et al., 1984). D1LMP1 RNA encodes 4 N-terminal cytoplasmic residues, the 5th and 6th transmembrane domains, and the C-terminal cytoplasmic 200 aa.

LMP1 in LCLs or stably expressed at comparable levels in B lymphocytes, epithelial cells, or fibroblasts aggregates in characteristic patches in the cell plasma and cytoplasmic membranes. Chymotrypsin treatment of live latently infected lymphoblasts results in specific cleavage in 30% of LMP1 first outer reverse turns, indicating that at least 30% of LMP1 is in the plasma membrane (Liebowitz et al., 1986). N-terminal mutations that affect cytoplasmic anchoring of the first transmembrane domain result in more diffuse plasma membrane localization and are significantly compromised in biologic activity (Izumi et al., 1994; Liebowitz et al., 1992). Stable aggregation in plasma membranes may not be necessary for LMP1 signaling following 293 cell transfection; LMP1 can localize to other cytoplasmic membranes (Lam and Sugden, 2003). When expressed in cells following gene transfer, D1LMP1 accumulates at higher levels than LMP1, diffusely localizes to all cytoplasmic membranes, and has minimal signaling effects (Wang et al., 1985; 1988a; 1988b). Thus, the six LMP1 hydrophobic transmembrane domains are critical for aggregation in cytoplasmic membranes and aggregation correlates with LMP1 effects on lymphocyte or fibroblast growth.

LMP1 membrane localization in lipid rafts

Biochemical and reverse genetic analyses indicate that localization to cholesterol rich raft microdomains is mediated by the LMP1 transmembrane domains and is important for LMP1 effects. A significant fraction of LMP1 is raft associated (Ardila-Osorio et al., 1999; Higuchi et al., 2001). Raft association requires only the hydrophobic transmembrane domains and increases substantially with deletion of most of the C-terminal cytoplasmic (Higuchi et al., 2001; Yasui et al., 2004). The first two LMP1 transmembrane domains are particularly critical for LMP1 intermolecular aggregation and signaling. In transient transfection experiments, the LMP1 N-terminal cytoplasmic domain and first two transmembrane domains can enable ~40% of wild type LMP1 C-terminal cytoplasmic domain mediated activation effects; the N-terminal cytoplasmic domain and first 4 transmembrane domains can mediate ~70%, whereas the N-terminal cytoplasmic domain and transmembrane domains 3-6 have little effect in the absence of transmembrane domains 1 and 2. In the context of GST-LMP1 fusion protein expression in epithelial cells, transmembrane domains 1 and 2 can biochemically associate with transmembrane domains 3-6 on another molecule. This association is dependent on the FWLY sequence that is at the external end of the first transmembrane domain. FWLY is also critical for raft association and for LMP1 signaling, further linking these activities (Yasui et al., 2004). LMP1 transmembrane domains 1 and 2 can also interact with 1 and 2 on another LMP1 molecule and 3-6 can interact with 3-6; these interactions could also have roles in LMP1 effects. Full length LMP1 with FWLY mutated to AALA does not associate with rafts and does not signal. Further reverse genetic analyses may identify other specific transmembrane domain residues that effect intermolecular interactions between the transmembrane domains and enable signaling (Yasui et al., 2004).

LMP1 is palmitylated. However, mutations in the palmitylated cysteines that abrogate palmitylation do not affect stable raft association (Higuchi et al., 2001). Still, palmitylation could enhance the rate of raft association and could be significant for early latency III or later latency II EBV infection.

LMP1 associated with cytoskeleton

LMP1 is also significantly associated with the cytoskeleton; the association is only partially understood. Extraction of the cytoplamsic membranes with non-ionic detergents leaves most LMP1 strongly bound to the cytoskeleton (Liebowitz et al., 1987). Dounce homogenization or sonication can then fracture the cytoskeleton and release LMP1. The C-terminal cytoplasmic domain is required for cytosekeletal association. LMP1 strongly associates with vimentin and relocalization of vimentin during mitosis can relocalize LMP1 (Liebowitz et al., 1987). However, vimentin, per se, is not required for LMP1 effects on lymphocyte activation or adhesion; activation is similar in a lymphoblast that lacks vimentin (Liebowitz and Kieff, 1989). Perhaps, other intermediate filament proteins can substitute. The LMP1 C-terminal cytoplasmic domain also engages TNF receptor associated factors (TRAF) 3 and 2 (Mosialos et al., 1995). TRAF3 can stably associate with microtubules through MIP-T3 (Ling and Goeddel, 2000). Whether binding to TRAF3 can result in linkage to microtubules in unknown. CD40 activation detaches TRAF3 from MIP-T3 (Ling and Goeddel, 2000). Since LMP1 is similar to a constitutively active CD40 in biochemical and cell signaling effects, LMP1 might be expected to constitutively detach TRAF3 from microtubules. The LMP1 C-terminal cytoplasmic domain also binds TRAF2, albeit mostly through TRAF1 (Devergne et al., 1996; Mosialos et al., 1995). TRAF2 associates with filamin (Arron et al., 2002). TRAF2 association with filamin has been implicated in enabling downstream signaling effects. LMP1 appears to be stabilized by association with the cell cytoskeletal. In pulse labeling experiments, LMP1 moves from the non-ionic detergent soluble cell fraction to the insoluble cytoskeleton associated fraction, where the LMP1 half life is somewhat extended (Baichwal and Sugden, 1987; Liebowitz et al., 1987; Mann and Thorley, 1987; Moorthy and Thorley-Lawson, 1993b; Wang et al., 1988a). Cytoskeletal LMP1 is phosphorylated and eventually cleaved to generate a stable 25kDa C-terminal cytoplasmic domain fragment (Moorthy and Thorley-Lawson, 1993b).The temporal sequence of LMP1 association with rafts, cytoskeleton, and downstream signaling molecules has not as yet been synchronized with cytoplasmic effects.

LMP1 induces filopodia formation and actin-cytoskeleton rearrangements (Puls et al., 1999). Filopodia formation in Swiss 3T3s in independent of both the N- and C- terminal cytoplasmic domains may therefore likely mediated by the transmembrane domains. Dominant negative CDC42 blocks LMP1 mediated filopodia formation. LMP1, via NF-κB, induces CIP4, a CDC42 interacting protein that binds activated CDC42, WASP1 and LYN (Cahir-McFarland et al., 2004; Dombrosky-Ferlan et al., 2003). Thus LMP1 mediated CIP4 induction may be important in the transmembrane mediated activation of CDC42.

LMP1 induces a transformed phenotype

LMP1 expressed in immortalized fibroblasts following single gene transfer has some effects that are similar to those of activated ras. LMP1 enables NIH3T3 cells to grow in less serum, exhibit less contact inhibition, and adhere without actin cables. In Balb/c and Rat1 cells, LMP1 causes loss of anchorage dependence and growth into tumors after subcutaneous injection into nude mice (Baichwal and Sugden, 1988; 1989; Moorthy and Thorley-Lawson, 1993a; Wang et al., 1985). In non-EBV infected Burkitt tumor B lymphoblasts, LMP1 expression duplicates much of the phenotype of latency III EBV infection (Cahir-McFarland et al., 2004; Wang et al., 1988b; 1990a). LMP1 causes plasma membrane ruffling, increased adhesion and activation marker expression, and cell adhesion (Wang et al., 1988b; Wang et al., 1990a) ; effects that are also induced by CD40 ligand. In contrast, D1 LMP1, which lacks the N-terminal cytoplasmic and first 4 transmembrane domains, has almost no effects (Wang et al., 1988a).

LMP1 and CD40 ligand effects on B lymphocytes overlap extensively (Uchida et al., 1999). Both LMP1 and CD40 activate NF-kB, JNK and p38 (Gires et al., 1997; Hatzivassiliou et al., 1998; Zimber-Strobl et al., 1996). In a recombinant EBV infected LCL, where LMP1 expression can be conditionally regulated, CD40 ligand can maintain LCL proliferation in the absence of LMP1 expression (Kilger et al., 1998). Furthermore, an LMP1 N-terminal and transmembrane domain fusion to the CD40 CTD can replace LMP1 in LCL outgrowth assays (Dirmeier et al., 2003).

Transgenic LMP1 expression in mouse lymphocytes has been associated with abnormal lymphocyte proliferation. In three lineages of transgenic mice, LMP1 expression under control of the immunoglobulin heavy chain promoter and enhancer was associated with histologic evidence of lymphoma in 23-51% of the animals at 18 months versus a 12% prevalence of lymphoma histology in control animals (Kulwichit et al., 1998). Putative lymphomas frequently had clonal lymphocyte proliferations with over-expression of A20, Bcl2, or c-myc. Transgenic LMP1 mimicked the effects of activated CD40 on B lymphocytes and enhanced activation marker expression, spontaneously proliferation, *in vitro*, and foci of extra-follicular B cell differentiation. LMP1 also blocked germinal center formation, probably owing to constitutive signaling causing precocious stable localization before entry to lymphoid follicles (Uchida et al., 1999).

Direct comparison in CD40 (-/-) mice of transgenic CD40 or CD40 with an LMP1 C-terminal cytoplasmic domain revealed that LMP1 substituted for the CD40 cytoplasmic domain in creating normal lymphocyte subsets, antibody responses, class switching, and affinity maturation to the T cell dependent antigen TNP-KLH, but was also associated with hyper IL-6, splenomegaly, lymphadenopathy, expanded germinal center B cells, and auto-antibodies (Stunz et al., 2004). In surprising contrast, mice transgenic for LMP1 or LMP1 with the CD40 cytoplasmic domain resulted in LMP1CD40 having elevated numbers of marginal zone B cells with increased numbers of neutrophils (Panagopoulos et al., 2004; Uchida et al., 1999).

Transgenic LMP1 expression in mouse skin cells under control of the Polyoma virus enhancer resulted in thickened skin with increased epidermal cell mitotic index, expression of proliferation associated keratins, and no increase in carcinoma (Wilson et al., 1990). Topical treatment of LMP1 transgenic mice with carcinogens produced more small papillomas than in control animals, but progression to carcinoma was not enhanced. Crossing transgenic LMP1 into animals deleted for the INK4a locus and treatment with carcinogen resulted in an LMP1 related increase in carcinogen induced papilloma prevalence and consequently an increased carcinoma incidence (Macdiarmid et al., 2003). Overall, in transgenic mice, LMP1 appears to increase lymphocyte or epithelial cell growth or survival and thereby predispose to malignancy.

LMP1 is essential for B lymphocyte immortalization

Combined biochemical and reverse recombinant genetic studies of LMP1 mutations incorporated in the EBV genome and assayed for effects on human B lymphocyte conversion to LCLs indicate that LMP1 is critical for LCL outgrowth (Izumi et al., 1994; 1997; Kaye et al., 1993; 1995; 1999). Consequently, LCL outgrowth provided a definitive endpoint for biologically significant LMP1 effects. The LMP1 N-terminal cytoplasmic domain is R and P rich. The R-amino group probably interacts with phosphates on the cytoplasmic aspect of membrane lipids and anchors the N terminus of the first transmembrane domain to the membrane-cytoplasm interphase. Although LMP1 N-terminal Ps could by sequence comparison be a

SH3 interactive domain, associated proteins have not been identified. Furthermore, an EBV recombinant deleted for nearly all LMP1 N-terminal cytoplasmic domain residues, aa 6-24, is only slightly impaired in aggregation and can still transform B lymphocytes into LCLs with reduced efficiency (Izumi et al., 1994). Smaller deletions within 6-24 are close to wild type LMP1 in transforming efficiency, consistent with the notion that this R- and P- rich sequence does not mediate specific protein-protein interactions (Izumi et al., 1994). Moreover, LMP1 that is truncated for the entire N-terminus and most of the first transmembrane domain is diffusely distributed in the plasma membrane and is a null mutation for LCL outgrowth (Kaye et al., 1993).

The LMP1 transmembrane domains required for ligand independent aggregation

The LMP1 transmembrane domains confer ligand-independent aggregation in the plasma membrane, which then enables putative trimers of C-terminal cytoplasmic domains to stably engage cytoplasmic signaling molecules that are ordinarily engaged by CD40 and other TNFRs in response to TNF ligands. TNF ligands, TNFRs, and TNFR Associated cytoplasmic Factors (TRAFs) or Death Domain proteins share an intrinsic ability to trimerize (Ye et al., 1999). Associations with rafts may increase LMP1 local density and thereby enhance or stabilize oligomerization. Fusion of the LMP1 N terminus and transmembrane domains to the CD40 cytoplasmic domain results in constitutive CD40 C-terminal cytoplasmic domain signaling (Hatzivassiliou et al., 1998). Further, expression of a fusion of CD40 or other extracellular receptor domain to the C-terminal cytoplasmic signaling domain results in ligand or antibody dependent signaling from the LMP1 C-terminal cytoplasmic domain (Floettmann and Rowe, 1997).

The transmembrane domains may also affect signaling amplitude by altering the LMP1 half-life. Prototypical LMP1 has a 2-5 hour half life (Baichwal and Sugden, 1987; Liebowitz et al., 1987; Mann and Thorley, 1987; Moorthy and Thorley-Lawson, 1993b; Wang et al., 1988a), whereas the CAO EBV strain has a longer half life as well as higher biologic and chemical signaling activity (Blake et al., 2001; Chen et al., 1992; Mehl et al., 1998). The CAO LMP1 was identified in an Asian NPC, consistent with the notion that this EBV strain may have increased tumorigenic potential. However, CAO type EBV LMP1 is frequent in many Asian populations and similar EBV LMP1 genes are found world wide. CAO TMs differ from the prototype LMP1 in TMs2-6. Conversion of methionine 129 in B958 to the corresponding isoleucine found in CAO LMP1 increases LMP1 half-life and likely accounts for much of the biologic effect in cell based assays (Pandya and Walling, 2004). CAO LMP1 also differs from other LMP1s in a HOS binding site found adjacent to the TRAF binding site. HOS is a receptor for the SCFHOS/betaTrCP ubiquitin-protein isopeptide ligase and binding may affect LMP1 half-life although the significance of the binding is not clearly evident (Tang et al., 2003).

LMP1 carboxy terminal signaling

The LMP1 C-terminal cytoplasmic domain has two signaling components that are critical for LMP1 effects in latency III mediated LCL outgrowth; aa187-231, which engage TRAFs and aa351-386, which engage death domain proteins, such

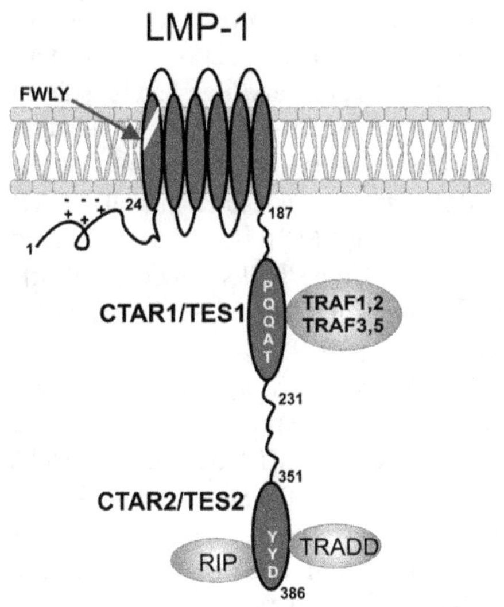

Figure 1. LMP1 has a cytoplasmic amino terminus, 6 transmembrane domains, and a 200 amino acid cytoplasmic domain. The amino terminus anchors LMP1 in the plasma membrane by electrostatic interactions. FWLY in first transmembrane domain mediates LMP1-LMP1 interactions as well as association with lipid rafts. The cytoplasmic C-terminus mediates binding to TRAFs through a PQQAT motif in CTAR1/TES1 and to TRADD or RIP through residues including YYD_{386} in CTAR2/TES2.

as TRADD and RIP (Devergne et al., 1996; Hammarskjold and Simurda, 1992; Izumi et al., 1999a; 1997; Izumi and Kieff, 1997; Mitchell and Sugden, 1995; Mosialos et al., 1995). These two domains have been referred to as **C-T**erminal **A**ctivation **R**egions or **T**ransformation **E**ffecter **S**ites 1 and 2 (CTAR1 and 2 or TES1 and 2) (Figure 1). Both sites independently activate NF-kB and are critical for LCL outgrowth. LMP1 CTAR1/TES1 $PQQAT_{208}$ binds to TRAFs 3, 5 and 1, 2 (Devergne et al., 1996). CTAR1/TES1 mutation $AQAAT_{208}$ abrogates TRAF binding and NF-κB activation from this domain (Devergne et al., 1996). Notably, LMP1 deleted for 186-210 cannot contribute to primary B lymphocyte conversion to LCLs (Izumi et al., 1997). Mutations $Y_{384}YD$ to ID or Y_{384} to G abrogate CTAR2/TES2 interaction with TRADD or RIP and NF-κB activation by this domain (Floettmann and Rowe, 1997; Izumi et al., 1999a; Izumi and Kieff, 1997). Nevertheless, LMP1 with these mutations or C-terminally truncated at aa232 can still contribute to initial LCL outgrowth (Kaye et al., 1999), although long term outgrowth is inefficient. LMP1 with both mutations does not activate NF-κB. LMP1 deleted for aa 232-351 or mutated to $F_{384}FD$ can fully enable LCL outgrowth, indicating that aa 232-351 and the terminal Ys are not critical for LMP1 growth or survival effects as measures in LCL outgrowth (Izumi et al., 1999b; 1999a).

Using fibroblast feeder layers to support the growth of lymphocytes infected with reverse genetically mutated EBV, EBV with either CTAR1/TES1 or CTAR2/TES2 mutations were 10 fold reduced in outgrowth, whereas EBV deletion for LMP1, the TMs, or the C-terminal cytoplasmic domain were 50-250 fold reduced in outgrowth efficiency (Dirmeier et al., 2003). Interestingly, once established, B cell lines infected with EBV having CTAR1/TES1 or CTAR2/TES2 mutations could be grown without feeders (Dirmeier et al., 2003) ; LCL cell to cell contact probably provides sufficient autocrine and cell to cell contact mediated signaling to replace fibroblast feeder support.

Since both CTAR1/TES1 and CTAR2/TES2 are required for efficient long term LCL outgrowth, both contribute to overall signaling. LMP1 CTAR2/TES2 weakly binds TRADD and RIP (Izumi et al., 1999a; Izumi and Kieff, 1997), without propagating a death signal, and substantially activates IKKβ phosphorylation of IkBα resulting in canonical NF-κB activation (Eliopoulos et al., 2003a; Saito et al., 2003), whereas CTAR1/TES1 strongly binds TRAFs at a very high level resulting in strong activation of NIK and IKKα phosphorylation of p100 NF-κB2, leading to p52/Rel B nuclear translocation (Atkinson et al., 2003; Eliopoulos et al., 2003a; Luftig et al., 2004; Saito et al., 2003). CTAR1 may also induce canonical NF-κB activation as LCLs transformed with virus expressing LMP1 1-231 have the same complement of nuclear NF-kB complexes as wild type LCLs. Nuclear NF-κB complexes include p52/RelA and /RelB heterodimers, p50/RelA, /RelB, and /c-Rel heterodimers, and p50/p50 homodimers (Kaye et al., 1999).

LMP1 CTAR1/TES1 activation of p52 complexes may account for CTAR1/TES1 preferential induction of TRAF1, EBI3 or EGFR expression and fibroblast independent initial LCL outgrowth. In epithelial cells, LMP1 induces p52/RelA and /RelB heterodimers, p50/p52 heterodimers and p50/p50 homodimers. CTAR1/TES1 null mutants induce p52/RelA and weakly induce p50/p50 homodimers, but do not induce p50/p52 or p52/RelB. The failure of CTAR1/TES1 mutants to induce p52/RelB and p50/p52 correlates with an inability of CTAR1/TES1 mutants to induce EGFR expression in these cells (Miller et al., 1998; Paine et al., 1995). In B cells, p52/RelB activation by CTAR1/ TES1 may be critical for TRAF1 induction and fibroblast independent initial LCL outgrowth (Atkinson et al., 2003; Devergne et al., 1998; Izumi et al., 1997; Luftig et al., 2004; Saito et al., 2003).

The role of individual TRAFs in LMP1 signaling is partially elucidated in knock out MEFs. Although TRAF1 and TRAF3 heterodimerize with TRAF2 and TRAF5, respectively, and TRAF 2 and 5 can activate NF-κB, LMP1 does not require TRAF2 or TRAF5 for NF-κB activation in MEFs (Luftig et al., 2003). LMP1 activates NF-κB to similar levels in TRAF2/5 double knockout and normal mice. Surprisingly, TRAF6 and IRAK1 are essential for LMP1 NF-κB and JNK activation in MEFs (Luftig et al., 2003; Schultheiss et al., 2001; Wan et al., 2004). However MEFs also do not express TRAF1 and this can increase LMP1 dependence on CTAR2/TES2 mediated signaling (Devergne et al., 1996; Eliopoulos et al., 2003b; Eliopoulos and Young, 1998). TRAF6 may not be as important for CTAR1/TES1 mediated NIK and IKKα activation of NF-κB, JNK and p38 activation.

Surprisingly, LMP1 activation of NF-κB, JNK and p38 in mouse lymphoma cell lines is almost completely TRAF3 dependent, placing TRAF3 in a central role for signaling from both domains (Xie and Bishop, 2004; Xie et al., 2004). However, analyses of LMP1 induced CD23, CD80, and CD95 in TRAF3 (-/-) B cells indicate a more complicated dependency on TRAF3 with CD80 induction being almost nil, whereas CD23 induction is ~40%, and CD95 induction is equivalent to that observed in TRAF3 (+/+) B cells (Xie et al., 2004). Further, LMP1 stimulated IgM secretion through CTAR1 is almost completely TRAF3 dependent, whereas CTAR2 is variably TRAF3 dependent (Xie and Bishop, 2004). Moreover, TRAFs 1 and 2 are not necessary for LMP1 mediated JNK activation in TRAF3 (+/+) mouse B cells (Xie et al., 2004). TRAF3 has a particularly significant role in directly binding NIK, causing NIK degradation (Liao et al., 2004). CD40 or BAFF, which activate NIK in a ligand dependent fashion, cause TRAF3 degradation, releasing NIK to bind to p100, recruit and activate IKKα, and cause NF-κB2/RelB nuclear translocation (Liao et al., 2004). LMP1 CTAR1 has the unique ability to aggregate much of the cell TRAF3 and this could enable NIK to assemble p100/NIK/IKKα complexes and activate NF-κB2/RelB nuclear translocation.

In IKKβ (-/-) MEFs, LMP1 does not activate an NF-κB reporter comprised of MHC class I sites (Luftig et al., 2003). Activation of a reporter comprised of NF-κB sites from the IgK locus in IKKβ (-/-) MEFs has been reported and would be most likely related to non-canonical NIK/IKKα activation (Saito et al., 2003). Surprisingly, LMP1 causes RelA nuclear translocation and activates both reporters in IKKγ (-/-) MEF. Thus, LMP1 activation of the canonical MHC class I KB subunits is atypical in not requiring IKKγ.

LMP1 activation of cytokine expression in knock out MEFs is also unusual in being almost entirely IKKβ dependent, frequently IKKγ independent, and frequently IKKα independent or hyper activated (Luftig et al., 2004). Using real-time PCR, LMP1 regulated cytokine genes fall into three groups. The first group includes MIP-2, which was hyper induced by LMP1 in IKKα (-/-) MEFs relative to wild type MEFs, very weakly induced in IKKβ (-/-) MEFs, and not induced in IKKγ (-/-) MEFs. Aside from hyper induction in IKKa (-/-) MEFs, this group is regulated by typical canonical NF-κB activation. The second group included MIG, which was weakly induced in IKKα (-/-) MEFs, not induced in IKKβ (-/-) MEFs, and induced to wild type levels in IKKγ (-/-) MEFs. This group is canonical with regard to IKKβ dependence, atypical in not requiring IKKγ for IKKβ activation, and non-canonical in requiring NIK/IKKα for full activation. The third group included I-Tac, which was hyper induced in IKKa (-/-) MEFs like group 1, was not induced in IKKβ (-/-) MEFs, and was wild type in IKKγ (-/-) MEFs. This group is canonical with regard to IKKβ dependence, atypical in not requiring IKKγ for IKKβ activation, and hyper activated in the absence of IKKα (Luftig et al., 2004). Overall, these data are consistent with a model that MEFs, under these assay conditions, have a low level of basal NF-κB activation and require canonical NF-κB activation for robust canonical or non-canonical NF-κB activation. Further, LMP1 is fundamentally atypical in activation of IKKβ in the absence of IKKγ. Since IKKβ activation following TNF, Toll, or IL1 receptor activation is IKKγ dependent (Yamaoka et al., 1998), LMP1 mediated IKKβ activation in the absence of IKKγ may be due to the unique LMP1 constitutive aggregation, which enables LMP1 to recruit and concentrate IKKβ in the absence of IKKγ.

LMP1 activation of JNK and p38 pathways is more CTAR2 dependent in epithelial cells but may be mediated by both CTAR1 and CTAR2 in lymphocytes (Eliopoulos and Young, 1998; Wan et al., 2004). Deletion of CTAR2 or mutation of CTAR2 Y to G, which abrogate TRADD binding, render LMP1 unable to activate JNK in epithelial cells (Eliopoulos and Young, 1998). However, epithelial cells do not express TRAF1 and TRAF1 expression enables CTAR1 to active JNK and NF-κB in epithelial cells (Eliopoulos et al., 2003b). In B cells and some epithelial cells, CTAR1 can induce TRAF1 and more effectively activate JNK and NF-κB (Devergne et al., 1996; Eliopoulos et al., 2003b). CTAR1 mediated JNK activation in B cells may be mediated by TRAF3 binding to T3JAM, a TRAF3 interacting protein that activates JNK after CD40 stimulation (Dadgostar et al., 2003). Interestingly, both CTAR1 and CTAR2 activate p38 in epithelial cells (Eliopoulos et al., 1999). Like NF-κB activation (see below), LMP1 mediated JNK activation is dependent on TRAF6 and is independent of IRAK4 (Wan et al., 2004). IRAK1 is required for LMP1 mediated NF-κB activation but is dispensable for JNK activity (Luftig et al., 2003; Wu et al., 2004). LMP1 mediated JNK activation also requires TAB2 and TAK1 and does not require MyD88, TRADD, RIP or TRAF2 (Wu et al., 2004). Since TRADD binding is essential for TES2 NF-κB activation, another adapter molecule, possibly BS69/BRAM1 (Chung et al., 2002), must interact with TES2 to activate the JNK pathway. These data indicate that LMP1 signaling can not be simply modeled after a TNF receptor since most TNFRs require TRAF2 for JNK activation or an IL-1/TLR receptor, since IL-1/TLRs require MyD88 and IRAK4 for signaling (Akira and Takeda, 2004). Rather LMP1 seems to function in a unique way that maximizes survival and growth signals, without the propensity to induce apoptosis (Figure 2).

Antiapoptotic functions of LMP1

LMP1 protects LCLs from apoptosis and can protect BL cells from growth factor withdrawal (Henderson et al., 1991; Rowe et al., 1994). Indeed, LMP1 activation of NF-κB pathway is critical for LCL survival and NF-κB inhibition induces LCL apoptosis without additional pro-apoptotic stimuli (Cahir-McFarland et al., 2004; 2000). LMP1 induces expression of many NF-κB dependent genes in LCLs important for survival. However, some may decrease at the protein level on a time scale consistent with a role in the ensuing apoptosis. These include c-FLIP, A20, Bfl-1 and c-IAPs (Cahir-McFarland et al., 2004; 2000). In LCLs, NF-κB inhibition also activates caspases, probably due to decreased c-IAP and c-FLIP expression. However, caspase activity is less central to NF-κB inhibition induced apoptosis in LCLs; caspase inhibitors reduce or eliminate caspase mediated cleavages, but do not prevent or delay apoptosis (Cahir-McFarland et al., 2000). Rather, BAX activation and the subsequent loss of mitochondrial membrane potential are critical events that temporally correlate with NF-κB inactivation and the onset of cell death (Cahir-McFarland et al., 2004; 2000). In this temporal cascade, NF-κB inhibition induced apoptosis in LCLs is phenotypically similar to enforced BAX dimerization mediated apoptosis (Gross et al., 1998). In both cases, caspase inhibition has little effect and Mitochondrial Membrane potential falls without cytochrome C release. Timely Bfl-1 mRNA decreases are consistent with Bfl-1 loss contributing to effective BAX activation.

LMP1 also provides growth signals to LCLs. In an LCL in which LMP1 expression can be regulated by tetracycline, LMP1 is required for proliferation (Kilger et al., 1998). When LMP1 expression is turned off, the cells exit cell cycle and survive up to 5 days. Hence, LMP1 is required for to maintain LCL proliferation. LMP1 growth effects are likely mostly mediated by JNK, since NF-κB inactivation has only modest affects on cell cycle progression and cells continued to undergo apoptosis as they progress through cycle (Cahir-McFarland et al., 2000).

In latency III infected LCLs or in latency II HD or NPC cells LMP1 co-localizes in the plasma membrane with LMP2 (Longnecker and Kieff, 1990). Biochemical interactions between these mimics of CD40 co-activation and B cell BCR activation likely contribute to subsequent B cell differentiation. Notably, latency III infected LCLs can also undergo class switching. Transcriptional profiling experiments suggest that LMP1 alone causes much of latency III effects on cell gene expression.

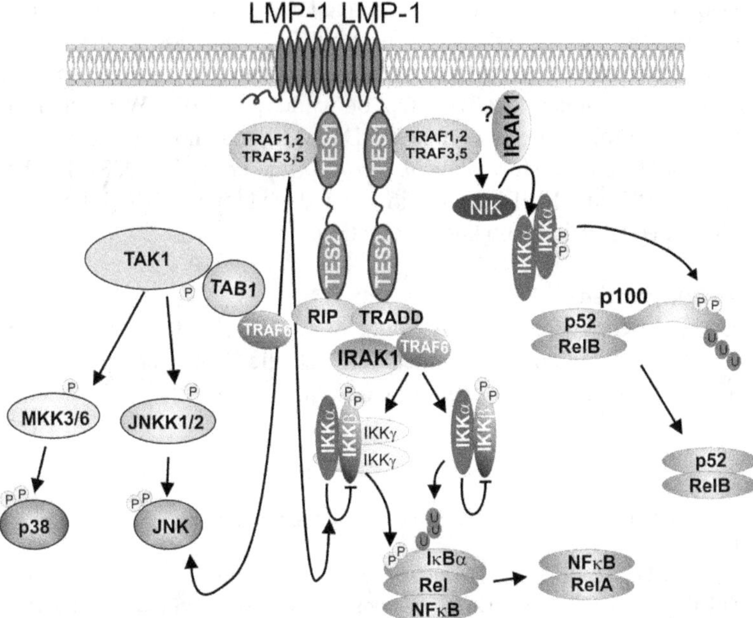

Figure 2. LMP1 functions as a constitutive TNFR family member by aggregation in the plasma membrane mediated through transmembrane interactions. The CTAR1/TES1 region binds TRAF1/2 heterdimers or TRAF3/5 heterodimers. TRAF binding mediates the activation of NIK and IKKα to induce the processing of p100 NFκB2 to p52 and nuclear translocation of p52/RelB complexes. CTAR2/TES2 binds TRADD or RIP. Death domain protein binding mediates canonical NF-kB activation by inducing the IKKβ mediated phosphorylation of IκBα. IKKβ activation is unusual in that IKKγ is not always required. LMP1 mediated NF-κB activation is hyperactive in IKKα deficient cells indicating that it negatively modulates IKKβ activity. TRAF6 is required for LMP1-mediated NF-κB activation and JNK activation whereas IRAK1 is required for LMP1 mediated NF-κB activation and dispensable for JNK activation. TAB1 and TAK1 are required for TES2-mediated JNK activation and have unknown roles in NF-κB activation. TES1 mediated JNK activation is dependent on TRAF1 expression in epithelial cells and TRAF3 in B cells but uncharacterized further.

Indeed, LMP1 expression in BL cells recapitulates most of EBV latency III effects in BLs (Cahir-McFarland et al., 2004). The highly evident LMP1 effects on transcription contrast with the more muted but significant LMP2 effects (Portis and Longnecker, 2004). Since LMP1 and LMP2 are both constitutively activated receptors that effectively recruit cytoplasmic signaling molecules, the less evident LMP2 effects are likely due to intrinsic differences in the regulation of tyrosine kinase and the NF-κB pathways.

The role of LMP1 in lytic replication is uncertain. In latency III infected cells, LMP1 activation of NF-κB contributes to high level Bcl-XL, Bcl-2, Bfl-1, and Mcl-1 expression and provides an anti-apoptotic environment (D'Souza et al., 2000; 2004; Finke et al., 1992; Henderson et al., 1991; Wang et al., 1996). Lytic infection in a latency III infected cell may not therefore require death protection from the EBV encoded Bcl-2 homologues, BHRF1 and BALF1. Further, induction of lytic infection in latency I infected lymphocytes can result in late lytic LMP1 expression (Chang et al., 2004; Rowe et al., 1992), which could still induce cell gene expression. The role of late lytic D1 LMP1 expression is less obvious. D1 LMP1 does not aggregate in the plasma membrane and does not interact with TRAFs and has minimal negative effect on wild type LMP1 (Wang et al., 1988a). Mutation of the initiator methionine for D1 LMP1 has no adverse effect on the efficiency of B cell transformation *in vitro* (Dirmeier et al., 2003). Furthermore, LMP1, D1LMP1, and LMP2 are not in purified enveloped EBV (Johannsen et al., 2004). Thus, D1LMP1 has no known function.

Acknowledgements

We are supported by grants #CA47006, #CA85180, and #CA87661 from the National Cancer Institute of the United States Public Heath Service. We thank our colleagues, Teruhito Yasui, Vishal Soni, and Jae Song for their critical reading of this manuscript.

References

Akira, S., and Takeda, K. (2004). Toll-like receptor signalling. Nat. Rev. Immunol. *4*, 499-511.

Ardila-Osorio, H., Clausse, B., Mishal, Z., Wiels, J., Tursz, T., and Busson, P. (1999). Evidence of LMP1-TRAF3 interactions in glycosphingolipid-rich complexes of lymphoblastoid and nasopharyngeal carcinoma cells. Int. J. Cancer *81*, 645-649.

Arron, J. R., Pewzner-Jung, Y., Walsh, M. C., Kobayashi, T., and Choi, Y. (2002). Regulation of the subcellular localization of tumor necrosis factor receptor-associated factor (TRAF) 2 by TRAF1 reveals mechanisms of TRAF2 signaling. J. Exp. Med. *196*, 923-934.

Atkinson, P. G., Coope, H. J., Rowe, M., and Ley, S. C. (2003). Latent membrane protein 1 of Epstein-Barr virus stimulates processing of NF-kappa B2 p100 to p52. J. Biol. Chem. *278*, 51134-51142.

Baichwal, V. R., and Sugden, B. (1987). Posttranslational processing of an Epstein-Barr virus-encoded membrane protein expressed in cells transformed by Epstein-Barr virus. J. Virol. *61*, 866-875.

Baichwal, V. R., and Sugden, B. (1988). Transformation of Balb 3T3 cells by the BNLF-1 gene of Epstein-Barr virus. Oncogene *2*, 461-467.

Baichwal, V. R., and Sugden, B. (1989). The multiple membrane-spanning segments of the BNLF-1 oncogene from Epstein-Barr virus are required for transformation. Oncogene *4*, 67-74.

Blake, S. M., Eliopoulos, A. G., Dawson, C. W., and Young, L. S. (2001). The transmembrane domains of the EBV-encoded latent membrane protein 1 (LMP1) variant CAO regulate enhanced signalling activity. Virology *282*, 278-287.

Cahir-McFarland, E. D., Carter, K., Rosenwald, A., Giltnane, J. M., Henrickson, S. E., Staudt, L. M., and Kieff, E. (2004). Role of NF-kappaB in Cell Survival and Transcription of Latent Membrane Protein 1-Expressing or Epstein-Barr Virus Latency III-Infected Cells. J. Virol. *78*, 4108-4119.

Cahir-McFarland, E. D., Davidson, D. M., Schauer, S. L., Duong, J., and Kieff, E. (2000). NF-kappa B inhibition causes spontaneous apoptosis in Epstein-Barr virus-transformed lymphoblastoid cells. Proc. Natl. Acad. Sci. USA. *97*, 6055-6060.

Chang, Y., Lee, H. H., Chang, S. S., Hsu, T. Y., Wang, P. W., Chang, Y. S., Takada, K., and Tsai, C. H. (2004). Induction of Epstein-Barr virus latent membrane protein 1 by a lytic transactivator Rta. J. Virol. *78*, 13028-13036.

Chen, H., Hutt-Fletcher, L., Cao, L., and Hayward, S. D. (2003). A positive autoregulatory loop of LMP1 expression and STAT activation in epithelial cells latently infected with Epstein-Barr virus. J. Virol. *77*, 4139-4148.

Chen, M. L., Tsai, C. N., Liang, C. L., Shu, C. H., Huang, C. R., Sulitzeanu, D., Liu, S. T., and Chang, Y. S. (1992). Cloning and characterization of the latent membrane protein (LMP) of a specific Epstein-Barr virus variant derived from the nasopharyngeal carcinoma in the Taiwanese population. Oncogene *7*, 2131-2140.

Chung, P. J., Chang, Y. S., Liang, C. L., and Meng, C. L. (2002). Negative regulation of Epstein-Barr virus latent membrane protein 1-mediated functions by the bone morphogenetic protein receptor IA-binding protein, BRAM1. J. Biol. Chem. *277*, 39850-39857.

Cooper, A., Johannsen, E., Maruo, S., Cahir-McFarland, E., Illanes, D., Davidson, D., and Kieff, E. (2003). EBNA3A association with RBP-Jkappa down-regulates c-myc and Epstein-Barr virus-transformed lymphoblast growth. J. Virol. *77*, 999-1010.

Dadgostar, H., Doyle, S. E., Shahangian, A., Garcia, D. E., and Cheng, G. (2003). T3JAM, a novel protein that specifically interacts with TRAF3 and promotes the activation of JNK (1). FEBS Lett. *553*, 403-407.

Devergne, O., Cahir McFarland, E. D., Mosialos, G., Izumi, K. M., Ware, C. F., and Kieff, E. (1998). Role of the TRAF binding site and NF-kappaB activation in Epstein-Barr virus latent membrane protein 1-induced cell gene expression. J. Virol. *72*, 7900-7908.

Devergne, O., Hatzivassiliou, E., Izumi, K. M., Kaye, K. M., Kleijnen, M. F., Kieff, E., and Mosialos, G. (1996). Association of TRAF1, TRAF2, and TRAF3 with an Epstein-Barr virus LMP1 domain important for B-lymphocyte transformation: role in NF-kappaB activation. Mol. Cell Biol. *16*, 7098-7108.

Dirmeier, U., Neuhierl, B., Kilger, E., Reisbach, G., Sandberg, M. L., and Hammerschmidt, W. (2003). Latent membrane protein 1 is critical for efficient growth transformation of human B cells by epstein-barr virus. Cancer Res. *63*, 2982-2989.

Dombrosky-Ferlan, P., Grishin, A., Botelho, R. J., Sampson, M., Wang, L., Rudert, W. A., Grinstein, S., and Corey, S. J. (2003). Felic (CIP4b), a novel binding partner with the Src kinase Lyn and Cdc42, localizes to the phagocytic cup. Blood *101*, 2804-2809.

D'Souza, B., Rowe, M., and Walls, D. (2000). The bfl-1 gene is transcriptionally upregulated by the Epstein-Barr virus LMP1, and its expression promotes the survival of a Burkitt's lymphoma cell line. J. Virol. *74*, 6652-6658.

D'Souza, B. N., Edelstein, L. C., Pegman, P. M., Smith, S. M., Loughran, S. T., Clarke, A., Mehl, A., Rowe, M., Gelinas, C., and Walls, D. (2004). Nuclear factor kappa B-dependent activation of the antiapoptotic bfl-1 gene by the Epstein-Barr virus latent membrane protein 1 and activated CD40 receptor. J. Virol. *78*, 1800-1816.

Eliopoulos, A. G., Caamano, J. H., Flavell, J., Reynolds, G. M., Murray, P. G., Poyet, J. L., and Young, L. S. (2003a). Epstein-Barr virus-encoded latent infection membrane protein 1 regulates the processing of p100 NF-kappaB2 to p52 via an IKKgamma/NEMO-independent signalling pathway. Oncogene *22*, 7557-7569.

Eliopoulos, A. G., Gallagher, N. J., Blake, S. M., Dawson, C. W., and Young, L. S. (1999). Activation of the p38 mitogen-activated protein kinase pathway by Epstein-Barr virus-encoded latent membrane protein 1 coregulates interleukin-6 and interleukin-8 production. J. Biol. Chem. *274*, 16085-16096.

Eliopoulos, A. G., Waites, E. R., Blake, S. M., Davies, C., Murray, P., and Young, L. S. (2003b). TRAF1 is a critical regulator of JNK signaling by the TRAF-binding domain of the Epstein-Barr virus-encoded latent infection membrane protein 1 but not CD40. J. Virol. *77*, 1316-1328.

Eliopoulos, A. G., and Young, L. S. (1998). Activation of the cJun N-terminal kinase (JNK) pathway by the Epstein- Barr virus-encoded latent membrane protein 1 (LMP1). Oncogene *16*, 1731-1742.

Fennewald, S., van Santen, V., and Kieff, E. (1984). Nucleotide sequence of an mRNA transcribed in latent growth- transforming virus infection indicates that it may encode a membrane protein. J. Virol. *51*, 411-419.

Finke, J., Fritzen, R., Ternes, P., Trivedi, P., Bross, K. J., Lange, W., Mertelsmann, R., and Dolken, G. (1992). Expression of bcl-2 in Burkitt's lymphoma cell lines: induction by latent Epstein-Barr virus genes. Blood *80*, 459-469.

Floettmann, J. E., and Rowe, M. (1997). Epstein-Barr virus latent membrane protein-1 (LMP1) C-terminus activation region 2 (CTAR2) maps to the far C-terminus and requires oligomerisation for NF-kappaB activation. Oncogene *15*, 1851-1858.

Gahn, T. A., and Sugden, B. (1995). An EBNA-1-dependent enhancer acts from a distance of 10 kilobase pairs to increase expression of the Epstein-Barr virus LMP gene. J. Virol. *69*, 2633-2636.

Gires, O., Zimber-Strobl, U., Gonnella, R., Ueffing, M., Marschall, G., Zeidler, R., Pich, D., and Hammerschmidt, W. (1997). Latent membrane protein 1 of Epstein-Barr virus mimics a constitutively active receptor molecule. Embo J. *16*, 6131-6140.

Gross, A., Jockel, J., Wei, M. C., and Korsmeyer, S. J. (1998). Enforced dimerization of BAX results in its translocation, mitochondrial dysfunction and apoptosis. Embo J. *17*, 3878-3885.

Hammarskjold, M. L., and Simurda, M. C. (1992). Epstein-Barr virus latent membrane protein transactivates the human immunodeficiency virus type 1 long terminal repeat through induction of NF-kappa B activity. J. Virol. *66*, 6496-6501.

Harada, S., and Kieff, E. (1997). Epstein-Barr virus nuclear protein LP stimulates EBNA-2 acidic domain-mediated transcriptional activation. J. Virol. *71*, 6611-6618.

Hatzivassiliou, E., Miller, W. E., Raab-Traub, N., Kieff, E., and Mosialos, G. (1998). A fusion of the EBV latent membrane protein-1 (LMP1) transmembrane domains to the CD40 cytoplasmic domain is similar to LMP1 in constitutive activation of epidermal growth factor receptor expression, nuclear factor-kappa B, and stress-activated protein kinase. J. Immunol. *160*, 1116-1121.

Henderson, S., Rowe, M., Gregory, C., Croom-Carter, D., Wang, F., Longnecker, R., Kieff, E., and Rickinson, A. (1991). Induction of bcl-2 expression by Epstein-Barr virus latent membrane protein 1 protects infected B cells from programmed cell death. Cell *65*, 1107-1115.

Higuchi, M., Izumi, K. M., and Kieff, E. (2001). Epstein-Barr virus latent-infection membrane proteins are palmitoylated and raft-associated: protein 1 binds to the cytoskeleton through TNF receptor cytoplasmic factors. Proc. Natl. Acad. Sci. USA. *98*, 4675-4680.

Izumi, K. M., Cahir McFarland, E. D., Ting, A. T., Riley, E. A., Seed, B., and Kieff, E. D. (1999a). The Epstein-Barr virus oncoprotein latent membrane protein 1 engages the tumor necrosis factor receptor-associated proteins TRADD and receptor-interacting protein (RIP) but does not induce apoptosis or require RIP for NF-kappaB activation. Mol. Cell Biol. *19*, 5759-5767.

Izumi, K. M., Cahir-McFarland, E., Riley, E. A., Rizzo, D., Chen, Y., and Kieff, E. (1999b). The residues between the two transformation effector sites of Epstein-Barr virus latent membrane protein 1 are not critical for B-lymphocyte growth transformation. J. Virol. *73*, 9908-9916.

Izumi, K. M., Kaye, K. M., and Kieff, E. D. (1994). Epstein-Barr virus recombinant molecular genetic analysis of the LMP1 amino-terminal cytoplasmic domain reveals a probable structural role, with no component essential for primary B-lymphocyte growth transformation. J. Virol. *68*, 4369-4376.

Izumi, K. M., Kaye, K. M., and Kieff, E. D. (1997). The Epstein-Barr virus LMP1 amino acid sequence that engages tumor necrosis factor receptor associated factors is critical for primary B lymphocyte growth transformation. Proc. Natl. Acad. Sci. USA. *94*, 1447-1452.

Izumi, K. M., and Kieff, E. D. (1997). The Epstein-Barr virus oncogene product latent membrane protein 1 engages the tumor necrosis factor receptor-associated death domain protein to mediate B lymphocyte growth transformation and activate NF- kappaB. Proc. Natl. Acad. Sci. USA. *94*, 12592-12597.

Johannsen, E., Luftig, M., Chase, M. R., Weicksel, S., Cahir-McFarland, E., Illanes, D., Sarracino, D., and Kieff, E. (2004). Proteins of purified Epstein-Barr virus. Proc. Natl. Acad. Sci. USA. *101*, 16286-16291.

Kaye, K. M., Izumi, K. M., and Kieff, E. (1993). Epstein-Barr virus latent membrane protein 1 is essential for B- lymphocyte growth transformation. Proc. Natl. Acad. Sci. USA. *90*, 9150-9154.

Kaye, K. M., Izumi, K. M., Li, H., Johannsen, E., Davidson, D., Longnecker, R., and Kieff, E. (1999). An Epstein-Barr virus that expresses only the first 231 LMP1 amino acids efficiently initiates primary B-lymphocyte growth transformation. J. Virol. *73*, 10525-10530.

Kaye, K. M., Izumi, K. M., Mosialos, G., and Kieff, E. (1995). The Epstein-Barr virus LMP1 cytoplasmic carboxy terminus is essential for B-lymphocyte transformation; fibroblast cocultivation complements a critical function within the terminal 155 residues. J. Virol. *69*, 675-683.

Kempkes, B., Spitkovsky, D., Jansen, D. P., Ellwart, J. W., Kremmer, E., Delecluse, H. J., Rottenberger, C., Bornkamm, G. W., and Hammerschmidt, W. (1995). B-cell proliferation and induction of early G1-regulating proteins by Epstein-Barr virus mutants conditional for EBNA2. Embo J. *14*, 88-96.

Kilger, E., Kieser, A., Baumann, M., and Hammerschmidt, W. (1998). Epstein-Barr virus-mediated B-cell proliferation is dependent upon latent membrane protein 1, which simulates an activated CD40 receptor. Embo J. *17*, 1700-1709.

Kulwichit, W., Edwards, R. H., Davenport, E. M., Baskar, J. F., Godfrey, V., and Raab-Traub, N. (1998). Expression of the Epstein-Barr virus latent membrane protein 1 induces B cell lymphoma in transgenic mice. Proc. Natl. Acad. Sci. USA. *95*, 11963-11968.

Lam, N., and Sugden, B. (2003). LMP1, a viral relative of the TNF receptor family, signals principally from intracellular compartments. Embo J. *22*, 3027-3038.

Liao, G., Zhang, M., Harhaj, E. W., and Sun, S. C. (2004). Regulation of the NF-kappaB-inducing kinase by tumor necrosis factor receptor-associated factor 3-induced degradation. J. Biol. Chem. *279*, 26243-26250.

Liebowitz, D., and Kieff, E. (1989). Epstein-Barr virus latent membrane protein: induction of B-cell activation antigens and membrane patch formation does not require vimentin. J. Virol. *63*, 4051-4054.

Liebowitz, D., Kopan, R., Fuchs, E., Sample, J., and Kieff, E. (1987). An Epstein-Barr virus transforming protein associates with vimentin in lymphocytes. Mol. Cell Biol. *7*, 2299-2308.

Liebowitz, D., Mannick, J., Takada, K., and Kieff, E. (1992). Phenotypes of Epstein-Barr virus LMP1 deletion mutants indicate transmembrane and amino-terminal cytoplasmic domains necessary for effects in B-lymphoma cells. J. Virol. *66*, 4612-4616.

Liebowitz, D., Wang, D., and Kieff, E. (1986). Orientation and patching of the latent infection membrane protein encoded by Epstein-Barr virus. J. Virol. *58*, 233-237.

Ling, L., and Goeddel, D. V. (2000). MIP-T3, a novel protein linking tumor necrosis factor receptor-associated factor 3 to the microtubule network. J. Biol. Chem. *275*, 23852-23860.

Longnecker, R., and Kieff, E. (1990). A second Epstein-Barr virus membrane protein (LMP2) is expressed in latent infection and colocalizes with LMP1. J. Virol. *64*, 2319-2326.

Luftig, M., Prinarakis, E., Yasui, T., Tsichritzis, T., Cahir-McFarland, E., Inoue, J., Nakano, H., Mak, T. W., Yeh, W. C., Li, X., et al. (2003). Epstein-Barr virus latent membrane protein 1 activation of NF-kappaB through IRAK1 and TRAF6. Proc. Natl. Acad. Sci. USA. *100*, 15595-15600.

Luftig, M., Yasui, T., Soni, V., Kang, M. S., Jacobson, N., Cahir-McFarland, E., Seed, B., and Kieff, E. (2004). Epstein-Barr virus latent infection membrane protein 1 TRAF-binding site induces NIK/IKK alpha-dependent noncanonical NF-kappaB activation. Proc. Natl. Acad. Sci. USA. *101*, 141-146.

Macdiarmid, J., Stevenson, D., Campbell, D. H., and Wilson, J. B. (2003). The latent membrane protein 1 of Epstein-Barr virus and loss of the INK4a locus: paradoxes resolve to cooperation in carcinogenesis *in vivo*. Carcinogenesis *24*, 1209-1218.

Mann, K. P., and Thorley, L. D. (1987). Posttranslational processing of the Epstein-Barr virus-encoded p63/LMP protein. J. Virol. *61*, 2100-2108.

Mehl, A. M., Fischer, N., Rowe, M., Hartmann, F., Daus, H., Trumper, L., Pfreundschuh, M., Muller-Lantzsch, N., and Grasser, F. A. (1998). Isolation and analysis of two strongly transforming isoforms of the Epstein-Barr-Virus (EBV) -encoded latent membrane protein-1 (LMP1) from a single Hodgkin's lymphoma. Int. J. Cancer *76*, 194-200.

Miller, W. E., Cheshire, J. L., and Raab-Traub, N. (1998). Interaction of tumor necrosis factor receptor-associated factor signaling proteins with the latent membrane protein 1 PXQXT motif is essential for induction of epidermal growth factor receptor expression. Mol. Cell Biol. *18*, 2835-2844.

Mitchell, T., and Sugden, B. (1995). Stimulation of NF-kappa B-mediated transcription by mutant derivatives of the latent membrane protein of Epstein-Barr virus. J. Virol. *69*, 2968-2976.

Moorthy, R. K., and Thorley-Lawson, D. A. (1993a). All three domains of the Epstein-Barr virus-encoded latent membrane protein LMP-1 are required for transformation of rat-1 fibroblasts. J. Virol. *67*, 1638-1646.

Moorthy, R. K., and Thorley-Lawson, D. A. (1993b). Biochemical, genetic, and functional analyses of the phosphorylation sites on the Epstein-Barr virus-encoded oncogenic latent membrane protein LMP-1. J. Virol. *67*, 2637-2645.

Mosialos, G., Birkenbach, M., Yalamanchili, R., VanArsdale, T., Ware, C., and Kieff, E. (1995). The Epstein-Barr virus transforming protein LMP1 engages signaling proteins for the tumor necrosis factor receptor family. Cell *80*, 389-399.

Paine, E., Scheinman, R. I., Baldwin, A. S., Jr., and Raab-Traub, N. (1995). Expression of LMP1 in epithelial cells leads to the activation of a select subset of NF-kappa B/Rel family proteins. J. Virol. *69*, 4572-4576.

Panagopoulos, D., Victoratos, P., Alexiou, M., Kollias, G., and Mosialos, G. (2004). Comparative analysis of signal transduction by CD40 and the Epstein-Barr virus oncoprotein LMP1 *in vivo*. J. Virol. *78*, 13253-13261.

Pandya, J., and Walling, D. M. (2004). Epstein-Barr virus latent membrane protein 1 (LMP-1) half-life in epithelial cells is down-regulated by lytic LMP-1. J. Virol. *78*, 8404-8410.

Portis, T., and Longnecker, R. (2004). Epstein-Barr virus (EBV) LMP2A alters normal transcriptional regulation following B-cell receptor activation. Virology *318*, 524-533.

Puls, A., Eliopoulos, A. G., Nobes, C. D., Bridges, T., Young, L. S., and Hall, A. (1999). Activation of the small GTPase Cdc42 by the inflammatory cytokines TNF (alpha) and IL-1, and by the Epstein-Barr virus transforming protein LMP1. J. Cell Sci. *112 (Pt 17)*, 2983-2992.

Rickinson, A., and Kieff, E. (2001). Epstein-Barr Virus. In Fields Virology, D. Knipe, P. Howley, D. Griffin, M. Martin, R. Lamb, B. Roizman, and S. Straus, eds. (Philadelphia, Lippincott Williams and Wilkins), pp. 2575-2628.

Rowe, M., Lear, A. L., Croom-Carter, D., Davies, A. H., and Rickinson, A. B. (1992). Three pathways of Epstein-Barr virus gene activation from EBNA1- positive latency in B lymphocytes. J. Virol. *66*, 122-131.

Rowe, M., Peng-Pilon, M., Huen, D. S., Hardy, R., Croom-Carter, D., Lundgren, E., and Rickinson, A. B. (1994). Upregulation of bcl-2 by the Epstein-Barr virus latent membrane protein LMP1: a B-cell-specific response that is delayed relative to NF-kappa B activation and to induction of cell surface markers. J. Virol. *68*, 5602-5612.

Sadler, R. H., and Raab-Traub, N. (1995). The Epstein-Barr virus 3.5-kilobase latent membrane protein 1 mRNA initiates from a TATA-Less promoter within the first terminal repeat. J. Virol. *69*, 4577-4581.

Saito, N., Courtois, G., Chiba, A., Yamamoto, N., Nitta, T., Hironaka, N., Rowe, M., and Yamaoka, S. (2003). Two carboxyl-terminal activation regions of Epstein-Barr virus latent membrane protein 1 activate NF-kappaB through distinct signaling pathways in fibroblast cell lines. J. Biol. Chem. *278*, 46565-46575.

Schultheiss, U., Puschner, S., Kremmer, E., Mak, T. W., Engelmann, H., Hammerschmidt, W., and Kieser, A. (2001). TRAF6 is a critical mediator of signal transduction by the viral oncogene latent membrane protein 1. Embo J. *20*, 5678-5691.

Stunz, L. L., Busch, L. K., Munroe, M. E., Sigmund, C. D., Tygrett, L. T., Waldschmidt, T. J., and Bishop, G. A. (2004). Expression of the cytoplasmic tail of LMP1 in mice induces hyperactivation of B lymphocytes and disordered lymphoid architecture. Immunity *21*, 255-266.

Tang, W., Pavlish, O. A., Spiegelman, V. S., Parkhitko, A. A., and Fuchs, S. Y. (2003). Interaction of Epstein-Barr virus latent membrane protein 1 with SCFHOS/beta-TrCP E3 ubiquitin ligase regulates extent of NF-kappaB activation. J. Biol. Chem. *278*, 48942-48949.

Uchida, J., Yasui, T., Takaoka-Shichijo, Y., Muraoka, M., Kulwichit, W., Raab-Traub, N., and Kikutani, H. (1999). Mimicry of CD40 signals by Epstein-Barr virus LMP1 in B lymphocyte responses. Science *286*, 300-303.

Waltzer, L., Perricaudet, M., Sergeant, A., and Manet, E. (1996). Epstein-Barr virus EBNA3A and EBNA3C proteins both repress RBP-J kappa- EBNA2-activated transcription by inhibiting the binding of RBP-J kappa to DNA. J. Virol. *70*, 5909-5915.

Wan, J., Sun, L., Mendoza, J. W., Chui, Y. L., Huang, D. P., Chen, Z. J., Suzuki, N., Suzuki, S., Yeh, W. C., Akira, S., et al. (2004). Elucidation of the c-Jun N-terminal kinase pathway mediated by Estein-Barr virus-encoded latent membrane protein 1. Mol. Cell Biol. *24*, 192-199.

Wang, D., Liebowitz, D., and Kieff, E. (1985). An EBV membrane protein expressed in immortalized lymphocytes transforms established rodent cells. Cell *43*, 831-840.

Wang, D., Liebowitz, D., and Kieff, E. (1988a). The truncated form of the Epstein-Barr virus latent-infection membrane protein expressed in virus replication does not transform rodent fibroblasts. J. Virol. *62*, 2337-2346.

Wang, D., Liebowitz, D., Wang, F., Gregory, C., Rickinson, A., Larson, R., Springer, T., and Kieff, E. (1988b). Epstein-Barr virus latent infection membrane protein alters the human B- lymphocyte phenotype: deletion of the amino terminus abolishes activity. J. Virol. *62*, 4173-4184.

Wang, F., Gregory, C., Sample, C., Rowe, M., Liebowitz, D., Murray, R., Rickinson, A., and Kieff, E. (1990a). Epstein-Barr virus latent membrane protein (LMP1) and nuclear proteins 2 and 3C are effectors of phenotypic changes in B lymphocytes: EBNA-2 and LMP1 cooperatively induce CD23. J. Virol. *64*, 2309-2318.

Wang, F., Tsang, S. F., Kurilla, M. G., Cohen, J. I., and Kieff, E. (1990b). Epstein-Barr virus nuclear antigen 2 transactivates latent membrane protein LMP1. J. Virol. *64*, 3407-3416.

Wang, S., Rowe, M., and Lundgren, E. (1996). Expression of the Epstein Barr virus transforming protein LMP1 causes a rapid and transient stimulation of the Bcl-2 homologue Mcl-1 levels in B-cell lines. Cancer Res. *56*, 4610-4613.

Wilson, J. B., Weinberg, W., Johnson, R., Yuspa, S., and Levine, A. J. (1990). Expression of the BNLF-1 oncogene of Epstein-Barr virus in the skin of transgenic mice induces hyperplasia and aberrant expression of keratin 6. Cell *61*, 1315-1327.

Wu, H. C., Lu, T. Y., Lee, J. J., Hwang, J. K., Lin, Y. J., Wang, C. K., and Lin, C. T. (2004). MDM2 expression in EBV-infected nasopharyngeal carcinoma cells. Lab. Invest *84*, 1547-1556.

Xie, P., and Bishop, G. A. (2004). Roles of TNF receptor-associated factor 3 in signaling to B lymphocytes by carboxyl-terminal activating regions 1 and 2 of the EBV-encoded oncoprotein latent membrane protein 1. J. Immunol. *173*, 5546-5555.

Xie, P., Hostager, B. S., and Bishop, G. A. (2004). Requirement for TRAF3 in signaling by LMP1 but not CD40 in B lymphocytes. J. Exp. Med. *199*, 661-671.

Yamaoka, S., Courtois, G., Bessia, C., Whiteside, S. T., Weil, R., Agou, F., Kirk, H. E., Kay, R. J., and Israel, A. (1998). Complementation cloning of NEMO, a component of the IkappaB kinase complex essential for NF-kappaB activation. Cell *93*, 1231-1240.

Yasui, T., Luftig, M., Soni, V., and Kieff, E. (2004). Latent infection membrane protein transmembrane FWLY is critical for intermolecular interaction, raft localization, and signaling. Proc. Natl. Acad. Sci. USA. *101*, 278-283.

Ye, H., Park, Y. C., Kreishman, M., Kieff, E., and Wu, H. (1999). The structural basis for the recognition of diverse receptor sequences by TRAF2. Mol. Cell *4*, 321-330.

Zimber-Strobl, U., Kempkes, B., Marschall, G., Zeidler, R., Van Kooten, C., Banchereau, J., Bornkamm, G. W., and Hammerschmidt, W. (1996). Epstein-Barr virus latent membrane protein (LMP1) is not sufficient to maintain proliferation of B cells but both it and activated CD40 can prolong their survival. Embo J. *15*, 7070-7078.

From: Epstein-Barr Virus. Edited by: Erle S. Robertson

Chapter 27

EBV Lytic Infection

*Bruce F. Israel and Shannon C. Kenney**

ABSTRACT

Reactivation of lytic EBV infection from latently infected cells is a highly regulated process that begins when expression of immediate-early (IE) viral genes is triggered by host cell transcription factors. The IE gene products are transcriptional transactivators that initiate a cascade of lytic viral gene expression. The early viral genes encode viral proteins that initiate replication of the viral genome from the *oriLyt* site. Late viral gene transcription occurs following viral replication, allowing expression of structural proteins. The viral genome is then packaged within a capsid to generate transmissible virions.

INTRODUCTION

In order to outlive its existing host and persist over time, a virus must be able to transmit its genome to new hosts. The herpesviruses all follow a strategy of life-long latency in the host, interrupted periodically by lytic reactivation in a fraction of the latent cellular reservoir to allow production of free virions for transmission between hosts. In lytic EBV infection a cascade of viral genes are activated and expressed which function to produce encapsidated, enveloped virions that can be transmitted to uninfected cells and new hosts. Although cell lines containing the latent form of EBV are readily grown in culture, a cell culture system that permits efficient primary lytic EBV infection (and hence serial passage of free virions) is not currently available. Thus, our understanding of lytic EBV replication derives largely from studies of cell lines with latent virus infection that are induced to reactivate to the lytic life cycle, using a variety of methodologies.

CLINICAL ASPECTS OF LYTIC INFECTION

Primary EBV infection in humans is often asymptomatic during childhood, but frequently causes the syndrome infectious mononucleosis in adolescents and adults. The infectious mononucleosis syndrome is characterized by fever, lymphadenopathy, pharyngitis, and increased atypical lymphocytes (Cohen, 2000; Jenson, 2000). Although EBV infects B cells and converts them to an activated phenotype, most of the atypical lymphocytes found in infectious mononucleosis are activated CD8+ cytotoxic T cells, directed especially against lytic EBV

*For correspondence email shann@med.unc.edu

antigens (Hislop et al., 2002; Steven et al., 1997). This vigorous immune response contributes to the symptoms of infectious mononucleosis but is also responsible for eventually eliminating most cells with lytic viral infection, as well as cells containing certain forms of latent infection.

The early events following primary EBV infection of humans have not been completely characterized, both because infection is frequently asymptomatic, and the incubation period between exposure and the development of infectious mononucleosis is at least four weeks. Initially, primary infection probably occurs in oropharyngeal epithelial cells, which are permissive for lytic infection (Greenspan et al., 1985; Herrmann et al., 2002; Walling et al., 2003b; 2001). This initial stage of infection appears to be asymptomatic. By the time that symptomatic disease (infectious mononucleosis) is present, lytic infection can be found in tonsilar lymphocytes, but has been mostly eliminated in the epithelial cells (Niedobitek et al., 1997; 2000; Sixbey et al., 1984). Thereafter, latent EBV persists for the life of the host in rare circulating B cells, and lytic viral replication in circulating B cells in the normal host is difficult to detect (Babcock et al., 1999; Decker et al., 1996; Herrmann et al., 2002). However, virus is frequently found in the saliva of normal hosts, presumably derived from aymptomatic viral reactivation in either oral epithelial cells or tonsillar B cells. Occasional patients with end-stage acquired immunodeficiency syndrome (AIDS) develop oral hairy leukoplakia, a chronic lytic EBV infection occurring in epithelial cells localized to the sides of the tongue (Greenspan et al., 1985; Lau et al., 1993; Resnick et al., 1990; Walling et al., 2003b; 2001). Anti-viral drugs that target lytic viral gene products, such as acyclovir, are usually effective at controlling this condition.

Following the establishment of latency in the B-cell reservoir, in the normal host approximately one in a million circulating B-cells contains latent episomal virus. The ease of recovering EBV from the saliva suggests that lytic replication occurs frequently throughout the life of the asymptomatic host, being readily detectable in saliva (Ikuta et al., 2000; Ling et al., 2003; Lucht et al., 1995; Shu et al., 1992; Walling et al., 2003a). EBV has also been reported in male and female genital secretions (Israele et al., 1991; Sixbey et al., 1986). The combination of life-long persistence and asymptomatic recurrent viral shedding in saliva is presumably responsible for the extremely high prevalence (over 90% in adults) of EBV infection in all populations throughout the world (Rickinson and Kieff, 2001).

Latent EBV infection is also rarely associated with the development of certain types of malignancies, including African Burkitt lymphoma, lymphoproliferative disorders in immunosuppressed patient, nasopharyngeal carcinoma (NPC), gastric carcinoma, and a subset of Hodgkin's disease. Interestingly, however, serological studies have suggested that onset of NPC, Burkitt lymphoma, and Hodgkin's disease are preceded by reactivation of lytic EBV replication (Chan et al., 1991; Chen et al., 1985; Mueller et al., 1989; Mueller et al., 1991). Therefore, the lytic form of EBV may conceivably contribute to the development of certain forms of EBV-associated malignancies.

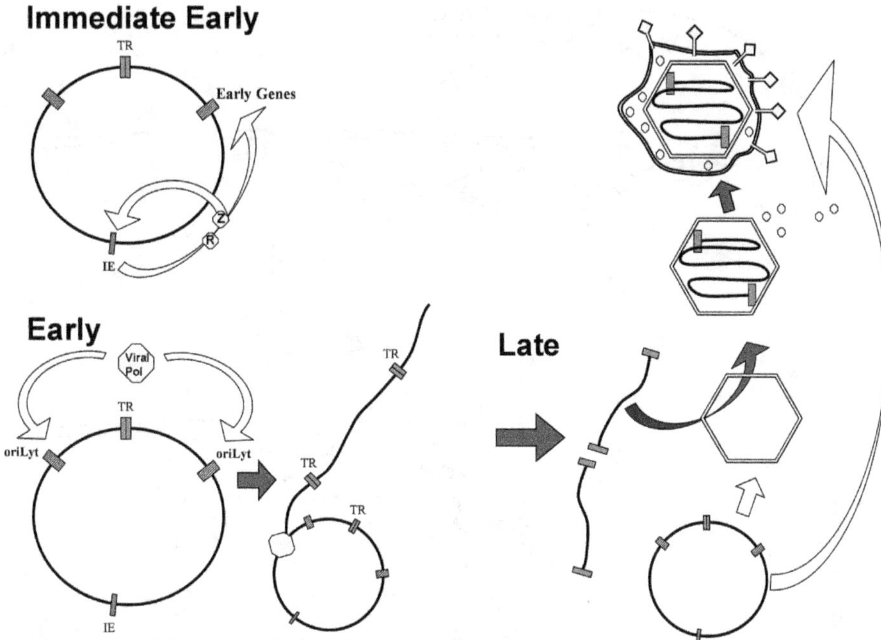

Figure 1. Lytic EBV infection proceeds through three sequential phases to generate virions. Initial activation of the immediate early (IE) genes leads to the production of viral proteins Z and R, which transactivate their own promoters and viral early genes. Early genes, located throughout the viral genome, encode viral proteins required for lytic viral replication, including the viral DNA polymerase (Viral Pol). Lytic replication initiates at one of two origins of lytic replication (oriLyt) and then proceeds repeatedly around the circular viral episome, producing long linear concatemers of the viral genome. The linear concatemers are clipped within the terminal repeat sequences (TR) to produce complete linear double-stranded viral genomes. Late genes which encode capsid proteins, viral tegument proteins, and viral membrane proteins, are transcribed following lytic viral replication, allowing the linear viral genome to be packaged.

LYTIC EBV REPLICATION: OVERVIEW

Lytic EBV replication proceeds through a sequential expression of viral gene categories shown in Figure 1, starting with expression the two immediate early (IE) genes, followed by expression of the early viral genes, and finally the late viral genes. The two IE genes, BZLF1 and BRLF1, are transcription factors that alter host cell cycle processes and transactivate viral early genes. Transcription of IE genes occurs even in the presence of protein synthesis inhibitors (Biggin et al., 1987; Flemington et al., 1991; Takada and Ono, 1989). The early genes are defined as those viral genes which are transcribed prior to lytic replication of the viral genome but are not transcribed in the presence of protein synthesis inhibitors. This category encodes viral proteins responsible for replication, including the viral DNA polymerase. Finally, following lytic viral replication, the late genes are transcribed, resulting in the production of the structural components found in free virions. Late gene expression can be largely blocked by inhibition of viral replication. In most instances, the host cell probably does not survive productive lytic EBV infection.

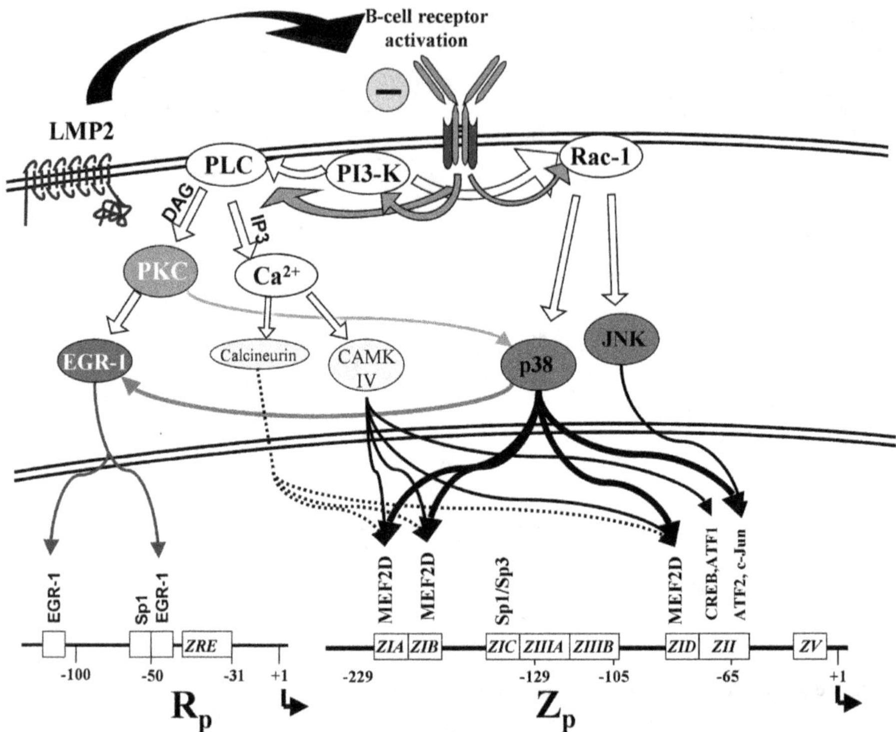

Figure 2. B-cell receptor activation induces several pathways that contribute to activation of the immediate early promoters. The R promoter and Z promoter (Rp and Zp), detailed in figure 4, are upregulated by several B-cell receptor (BCR)-associated pathways. Viral protein LMP2A inhibits BCR signaling and lytic promoter activation. BCR cross-linking leads to activation of phosphotidylinositol 3-kinase (PI3K) at the cytoplasmic membrane surface, which enhances recruitment of membrane-associated Ras-family GTPases including Rac1 and phospholipase Cγ-2 (PLC). Ras-family GTPases lead to activation of stress MAP kinases, p38 and c-jun N-terminal kinase (JNK), which phosphorylate (and upregulate) ATF-2 and c-jun, which bind at the ZII site of Zp. The p38 pathway also phosphorylates the transcriptional activator domain of MEF2 proteins, which bind to multiple ZI sites. PLC cleaves phosphotidylinositol diphosphate into diacylglycerol (DAG) and inositol triphosphate (IP3). DAG activates protein kinase C (PKC) which upregulates both EGR-1 and p38 pathways. Rp has two positive EGR-1 response elements. IP3 mobilizes intracellular calcium reserves and activates the calcium-dependent factors calcineurin and calcium/calmodulin-dependent kinase type IV/Gr (CAMK-IV/Gr). Calcineurin increases Zp activity through MEF2D sites, whereas CAMK-IV/Gr activates both MEF2D but also activates CREB.

REACTIVATION OF LYTIC INFECTION

Although most B-cells contain the latent form of EBV infection, it is clear that the virus can periodically reactivate to the lytic form in a small percentage of B-cells. Antigen-mediated activation of B-cells *in vivo* may be a physiologic stimulus for lytic reactivation of EBV in humans. Reactivation of some latently infected B-cell lines *in vitro* can be induced by B-cell receptor (BCR) stimulation through crosslinking surface immunoglobulin (Tovey et al., 1978). With the Burkitt lymphoma cell line Akata, this technique is especially effective, allowing up to 50% of cells to be induced to productive infection (Takada and Ono, 1989). However, most EBV-immortalized B cell lines (lymphoblastoid lines) cannot be efficiently reactivated by this method, due to ability of the EBV latent protein, LMP-2A, to suppress BCR-related signaling. Latent membrane protein 2A (LMP2), expressed in certain forms of viral latency and certain malignancies (Babcock and Thorley-Lawson, 2000; Rickinson and Kieff, 2001), competes with the BCR pathway at the cytosolic membrane as a dominant negative BCR analog, decreasing baseline Lyn protein-tyrosine kinase activity and reducing the activation of mediators downstream of the BCR. Thus lytic induction by BCR-dependent pathways is repressed (Miller et al., 1995; 1994a; 1994b).

The BCR-related signaling pathways which are thought to lead to lytic EBV infection are summarized in Figure 2 and reviewed in (Gold, 2002). BCR engagement in uninfected cells results in phosphorylation of BCR-associated proteins at immunoreceptor tyrosine-based activation motifs (ITAMs). ITAM phosphorylation enables the recruitment and activation of several signal transduction mediators, including membrane-bound phospholipase Cγ2 (PLC), phosphotidylinositol-3 kinase (PI3-Kinase), and Ras/Rac GTPases. PLC cleaves phosphatidylinositol diphosphate to generate two secondary messengers: diacylglycerol (which then activates protein kinase C (PKC)) and inositol triphosphate (IP3) (which then stimulates the release of Ca^{++}, enabling Ca^{++}-dependent pathways). PI3-Kinase generates phosphotidylinositol triphosphate, which enhances BCR-recruitment and activation of PLC and Rac-1, while Rac1/Ras activate cascades that upregulate c-jun N-terminal kinase and p38 MAPK (Gold, 2002).

Viral IE gene expression occurs within 30 minutes of BCR crosslinking, followed by early gene expression within a few hours (Flemington et al., 1991; Mellinghoff et al., 1991; Takada and Ono, 1989). The phospholipase Cγ2 (PLC) and Ras/Rac pathways probably both contribute to lytic viral induction. Two distinct calcium-dependent proteins that are activated through IP3, the calcium/calmodulin-dependent phosphatase, calcineurin, and the calcium/calmodulin-dependent protein kinase type IV/Gr, are required for lytic EBV induction by anti-IgG in Akata cells (Chatila et al., 1997; Goldfeld et al., 1995; Liu et al., 1997b). Inhibition of calcineurin by cyclosporin, an immunosuppressive drug in clinical use, prevents anti-IgG induction of lytic EBV (Bryant and Farrell, 2002; Goldfeld et al., 1995). BCR-mediated activation of PI3 kinase, PKC, and p38 kinase, have also been shown to be each essential for activation of lytic EBV (Adamson et al., 2000; Bryant and Farrell, 2002; Darr et al., 2001; Grab et al., 2004; Iwakiri and Takada, 2004; Krappmann et al., 2001; Mellinghoff et al., 1991). Together, the pathways induced by BCR activation ultimately collaborate to activate transcription of the two EBV promoters. The BZLF1 and BRLF1 gene products then further amplify the process by transcriptionally activating one another's promoters (Flemington et al., 1991).

In addition to antigen stimulation, other extracellular signals may be important to physiologic reactivation of lytic EBV expression in B-cells. TGF-beta can induce lytic viral infection in some Burkitt lymphoma lines *in vitro* via an ERK-dependent pathway (Adler et al., 2002; Fahmi et al., 2000). CD4+ T cells can interact with latent cells to induce lytic replication (Fu and Cannon, 2000).

Lytic EBV in oral epithelial cells in humans is probably induced by differentiation. In oral hairy leukoplakia lesions, viral DNA and lytic viral antigens are restricted to the more differentiated epithelial layers (Greenspan et al., 1985; Young et al., 1991). In paired differentiated and undifferentiated EBV-positive epithelial lines, initiation of lytic infection occurs preferentially in differentiated cells (Karimi et al., 1995; Li et al., 1992).

Whatever the stimulus, these distinct cues from the milieu produce nuclear signals, either unique to a pathway or convergent from different pathways, which are able to act upon the two viral immediate early promoters, Zp and Rp, to upregulate their activity (Feng et al., 2004; Flemington and Speck, 1990d; Shimizu and Takada, 1993; Zalani et al., 1995). The relative importance of particular signal transduction pathways may be influenced by the cell-type environment. In general, lytic inducing stimuli activate BZLF1 and BRLF1 transcription with similar kinetics (Biggin et al., 1987; Flemington et al., 1991; Takada and Ono, 1989). As each IE protein is able to transactivate its own promotor and that of the other (Adamson et al., 2000; Flemington et al., 1991; Liu and Speck, 2003; Ragoczy and Miller, 2001; Sinclair et al., 1991; Speck et al., 1997; Zalani et al., 1996), relatively weak inducing stimuli subsequently may be amplified by the IE proteins.

EBV IE GENE LOCUS

The IE gene locus encodes four potential gene products: the two immediate early genes, BZLF1 and BRLF1, an early gene, BRRF1, and a potential splice product, RAZ, as depicted in Figure 3. Transcription from Zp produces a message that produces the BZLF1 protein (Z). Transcription from Rp, which is upstream of BZLF1, can produce 3 alternately spliced mRNAs, of which the two larger messages could potentially produce both the BRLF1 protein (R) as well as Z (Manet et al., 1989). Translation of Z from the two larger bicistronic BRLF1 transcripts appears possible *in vitro*, and R may enhance this translation (Chang et al., 1998; Chang and Liu, 2001). However, the great majority of Z seems to be derived from the Zp promoter. The third Rp generated transcript produces a fusion product of R and Z called RAZ, which is only expressed later in lytic infection. RAZ may serve as a negative regulator of BZLF1 transcription (Furnari et al., 1994; Manet et al., 1989; Segouffin et al., 1996).

REGULATION OF THE ZP IE PROMOTER

A number of host cell factors have been shown to regulate the BZLF1 (Zp) and BRLF1 (Rp) promoters. Figure 4 depicts known transcription factor binding sites in the Zp and Rp promoters.

Previously, sequence motifs in Zp have been categorized into four groups: ZI, ZII, ZIII, and ZIV. The Zp region from -233 to $+12$ includes most of the major regulatory elements associated with lytic induction pathways. The four members of the ZI category (ZIA through ZID) represent A+T-rich motifs that are negative regulators of Zp in the absence of lytic stimuli. ZIA, ZIB, and ZID bind to MEF2D (Liu et al., 1997b). The ZIA, ZIC, and ZID motifs also weakly bind Sp1 and Sp3 (Liu et al., 1997a). In latently infected B-cells, MEF2D binding to the Zp ZI motifs recruits histone deacetylase complexes (HDACs), reducing the activity of Zp

EBV immediate-early gene region

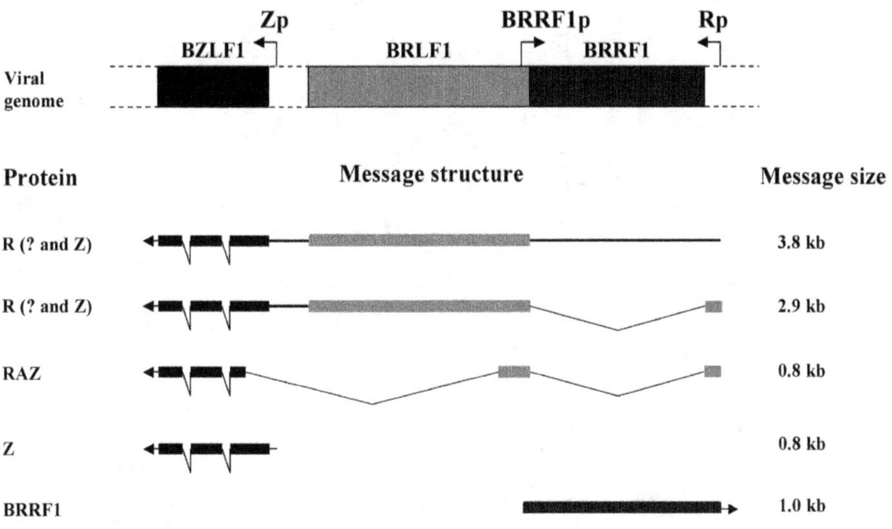

Figure 3. The immediate-early gene region generates several alternate spliced messages. The immediate-early (IE) gene region encodes two IE genes (BZLF1 and BRLF1) with two IE promoters (Zp and Rp) and one early gene (BRRF1) which is transcribed in the opposite direction from an independent promoter (BRRF1p). Rp directs transcription of a bicistronic message that encodes the upstream BRLF1 protein, R, and the downstream BZLF1 protein, Z. The two largest alternate splicings encode full length R and Z. The Rp-derived RAZ message may function as a negative regulator of Z. Zp has one message that encodes only Z. BRRF1p directs the BRRF1 message from the opposite DNA strand.

(Gruffat et al., 2002b). However, the ZI sites also act as positive regulators of BZLF1 transcription in the presence of certain inducing stimuli, including phorbol esters, calcium ionophores, and chemotherapy agents (Flemington and Speck, 1990d). Changes in MEF2D phosphorylation state, or other post-translational modifications, may alter its effect on Zp activity. BCR engagement leads to dephosphorylation of MEF2D, with cyclosporin abrogating this effect (Bryant and Farrell, 2002). Calmodulin-dependent protein kinase type IV/Gr may inhibit the MEF2D-HDAC association, thereby upregulating Zp (McKinsey et al., 2000). While phosphorylation of some MEF2D sites may be inhibitory, other phosphorylation sites may have the opposite effect (Li et al., 2001). MAP kinases directly activate MEF2D via phosphorylation of the transcriptional activator domain (McKinsey et al., 2000).

The Zp ZII motif is also required for most stimuli to successfully induce BZLF1 transcription (Binne et al., 2002; Chatila et al., 1997; Daibata et al., 1994; Feng et al., 2004; Flamand and Menezes, 1996; Flemington and Speck, 1990d). This site can bind multiple factors, including CREB, ATF-1, an ATF-2/c-jun heterodimer, and possibly the c-jun/c-fos heterodimer, also known as AP1 (Flamand and Menezes, 1996; Flemington and Speck, 1990d; Adamson et al., 2000; Liu et al.,

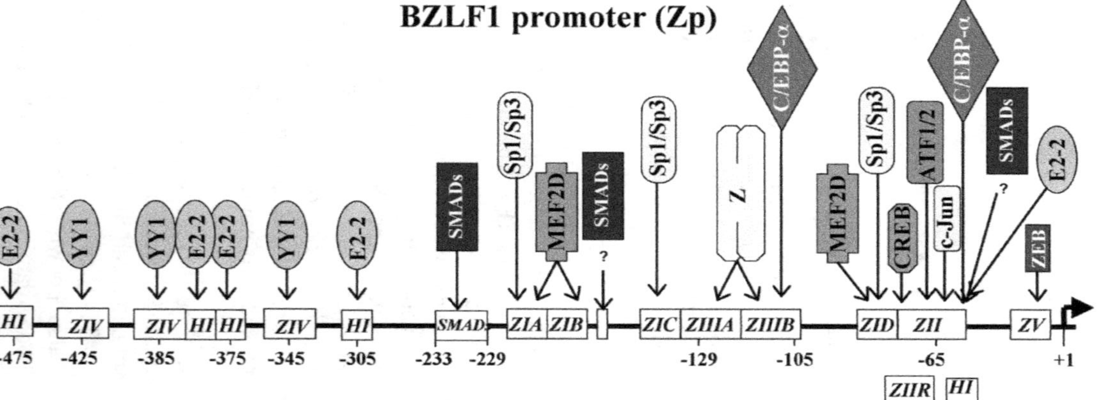

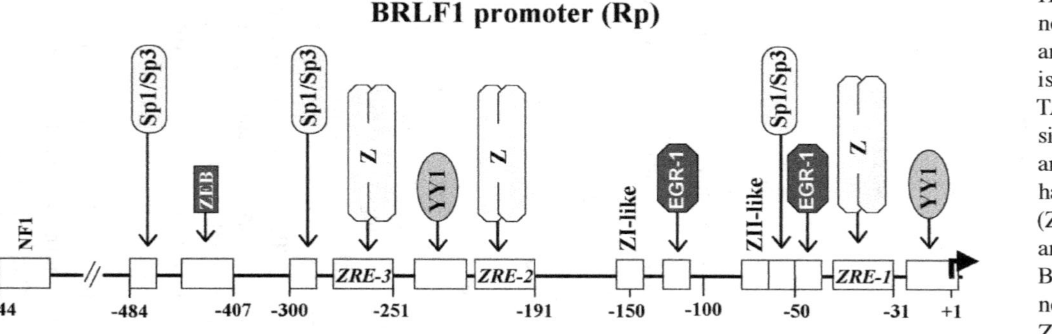

Figure 4. Transcription factors and cis-acting elements in the immediate-early promoters. BZLF1 promoter (Zp) ZI elements (ZIA, ZIB, and ZID) bind MEF2D, with overlapping binding sites for Sp1/Sp3 found in ZIA, ZIC, and ZID. ZII binds to CREB, ATF1/2, c-jun, and CAAT enhancer binding protein alpha (C/EBP-α). Two Z-binding sites (ZIIIA and ZIIIB) are required for positive Z-autoregulation of Zp, with ZIIIB also binding C/EBP-α. Multiple ZIV (bound by YY1) and H-box (bound by E2-2) function as negative regulatory elements. Finally, an important ZV negative element that is bound by ZEB is present near the TATA box. Potential SMAD binding sites which mediate TGF-β-activation are shown. The BRLF1 promoter (Rp) has multiple Z-responsive elements (ZRE-1, -2, and –3), two EGR-1 sites, and three Sp1/Sp3 sites that upregulate BRLF1p. YY1 binding sites that negatively regulate Rp, and a potential ZEB binding site, are also present.

1998; Wang et al., 1997). These transcription factors are commonly expressed across many cell types, but often require phosphorylation for activation. A number of different lytic-inducing stimuli, including BCR engagement, BRLF1 expression, and certain chemotherapy agents, can activate the stress MAP kinases, p38 and c-jun N-terminal kinase (JNK), which phosphorylate (and upregalate) ATF-2 and c-jun. Agents that inhibit JNK or p38 function are broadly inhibitory to activation of Zp by a variety of different stimuli (Adamson et al., 2000; Feng et al., 2004; 2002a). BCR-stimulation also results in phosphorylation of the ATF-1 and CREB transcription factors, mediated through calmodulin-dependent protein kinase IV/Gr (Bryant and Farrell, 2002; Matthews et al., 1994). The activated forms of CREB, ATF-1, ATF-2 and c-jun have all been shown to activate the BZLF1 promoter through the ZII motif in reporter-gene assays (Karimi et al., 1995; MacCallum et al., 1999). The ZII motif is also important, in conjunction with a Smad site, for TGF-beta-mediated activation of Zp in certain Burkitt lymphoma cell lines. C-jun binding to the ZII motif interacts directly with TGF-beta-triggered transcription factors (Smad3/Smad4) that directly bind Zp at –233 to –229 (Fahmi et al., 2000; Liang et al., 2002).

The two Zp ZIII sites (ZIIIA and ZIIIB) are Z-responsive elements (ZREs) that can be activated by direct binding of Z, and the ZIIIA site is required for efficient Zp activation in Akata cells (Binne et al., 2002; Flemington and Speck, 1990a). However, it remains controversial whether Z can activate its own promoter in the context of the intact latent viral genome (Le Roux et al., 1996; Yin et al., 2004; Zalani et al., 1996).

Interestingly, the ZII and ZIIIB sites in Zp were also recently shown to be bound by the cellular transcription factor, CAAT enhancer binding protein alpha (C/EBP-alpha), and it has been suggested that a heterodimer formed by Z and C/EBP-alpha may activate Zp through these sites (Wu et al., 2004).

Zp also contains several different elements that act to negatively regulate the promoter. Of these, the "ZV" motif present just upstream of the TATA box appears to be the most potent. ZV binds to ZEB, a zinc-finger protein (Kraus et al., 2001; Kraus et al., 2003), and deletion of this site results in dramatic enhancement of Zp activation following BCR cross-linking (Binne et al., 2002). The ZIV sites bind to the cellular YY-1 transcription factor, which downregulates Zp (Montalvo et al., 1995; Montalvo et al., 1991). In addition, a site named ZIIR overlaps the ZII and ZID motifs and is inhibitory for Zp, although the specific protein(s) binding to this site have not yet been identified (Liu et al., 1998). The ZI sites, in addition to MEF-2D, may also be bound directly by a cellular protein "smubp-2" to inhibit transcription (Zhang et al., 1999b). Finally, a number of "H-box" loci are scattered throughout Zp. Binding by the cellular E2-2 transcription factor to these motifs may serve to downregulate Zp in B-cells, while upregulating Zp in epithelial cells (Thomas et al., 2003).

REGULATION OF THE IE RP PROMOTER

Rp, which has not been characterized as extensively as Zp, has several defined and potential regulatory elements, with motifs that are common to Zp and some that are distinct. Sp1 and Sp3 may a more significant role for activation of Rp than Zp. Sp1/Sp3 binding sites are required for constitutive promoter activity, and the Sp1/Sp3 binding sites are also required for positive autoregulation of Rp by the R gene product (Liu and Speck, 2003; Ragoczy and Miller, 2001; Zalani et al., 1992). Rp also contains two EGR-1 (Zif268) sites, which have been shown to be important for induction of Rp activity (Feng et al., 2004; Zalani et al., 1995). BCR cross-linking, phorbol esthers, and chemotherapy all result in upregulation of EGR-1 (Krappmann et al., 2001). A binding site for NF1 has been reported to be a positive regulatory element in HeLa cells but not lymphoid cells (Glaser et al., 1998). There are three Z-responsive elements in Rp, and these elements mediate the ability of Z to upregulate Rp transcription (Bhende et al., 2004; Liu and Speck, 2003; Sinclair et al., 1991). Similar to Zp, Rp is downregulated through two YY1 binding sites (Zalani et al., 1997). Rp also has a potential ZEB site, although negative regulation through this site has not yet been shown "ZI"-like and "ZII"-like motifs are present, but their significance is unclear.

VIRAL AND HOST CELL FACTORS REGULATING INDUCTION OF LYTIC GENES

In latently infected B-cells, the promoters driving BZLF1 (Zp) and BRLF1 (Rp) transcription are not active, and virtually no lytic gene activity is present (Biggin et al., 1987; Flemington et al., 1991; Mellinghoff et al., 1991; Takada and Ono, 1989; Zalani et al., 1992). Epigenetic modifications of the viral genome, including deacetylation of the chromatin histone, and viral DNA methylation, inhibit Zp and Rp activity (Ambinder et al., 1999; Bhende et al., 2004; Falk and Ernberg, 1999; Gruffat et al., 2002b; Jenkins et al., 2000; Nonkwelo and Long, 1993; Paulson et al., 2002; Paulson and Speck, 1999; Szyf et al., 1985). Inhibiting histone deacetylation, using agents such as sodium butyrate and valproic acid, is sufficient to increase viral lytic gene expression in some, but not all, EBV-positive cell lines (Gradoville et al., 2002). Genes that respond to butyrate frequently have Sp1/Sp3 sites in their promoters (Davie, 2003), and this may play a role in the activation of Zp and Rp (Jenkins et al., 2000). Similarly, reversing methylation has been recently explored as a strategy to induce lytic infection. One demethylating agent, 5-azacytidine, has been shown to increase lytic viral gene expression in some cell lines but not others (Chan et al., 2004; Feng et al., 2004; Moore et al., 2001). Nevertheless, transfected Zp and Rp reporter gene constructs (lacking chromatin and DNA methylation) are not spontaneously active in most EBV-negative B-cell lines, but can be activated by various stimuli that induce lytic viral expression in EBV-positive cells, such as TPA and B-cell receptor cross linking (Bhende et al., 2004; Feng et al., 2004; Kenney et al., 1989b; Sinclair et al., 1991; Zalani et al., 1995). This finding emphasizes that activation of specific cellular transcription factors is required to begin the lytic cascade.

Several factors have been identified which repress induction of lytic viral expression. While Z and the BRLF1 protein (R) upregulate their own promoters, as discussed previously, some cellular factors can impede this positive feedback. Retinoic acid receptor and NF-KB both directly interact with Z and decrease Z transactivation (Gutsch et al., 1994; Morrison et al., 2004; Sista et al., 1993). LMP1 expression is associated with a decrease in BZLF1 induction by multiple pathways (Adler et al., 2002); this effect likely involves LMP-1 mediated activation of NF-κB. Nitric oxide can likewise down-regulate the lytic form of EBV infection (Gao et al., 1999; Kawanishi, 1995), although the mechanism for this effect is not clear.

VIRAL IMMEDIATE-EARLY PROTEINS

Translation of the BZLF1 and BRLF1 genes produces the proteins Z (also called ZEBRA, Zta, EB1) and R (Rta), respectively, which together are required for driving forward the program of lytic gene expression. High-level expression of Z is sufficient to induce the lytic form of infection in essentially all latently infected cell lines. Some, but not all, latently infected cell lines can also be switched to lytic infection by over-expression of R (Chevallier-Greco et al., 1986; Countryman and Miller, 1985; Ragoczy et al., 1998; Rooney et al., 1989; Takada et al., 1986; Westphal et al., 1999; Zalani et al., 1996). Thus, activation of BZLF1 expression (and hence regulation of the Zp promoter) may be relatively more important *in vivo* for inducing lytic EBV infection than activation of BRLF1 expression. Lytic EBV induction mediated by Z requires that Z initially activate Rp expression, as both Z and R together are required for activation of many early viral genes in the context of the intact viral genome. Conversely, R must first activate Zp expression in order to induce early lytic gene expression.

TRANSCRIPTIONAL EFFECTS OF Z

The Z protein is a member of the bZIP family of leucine-zipper transactivators, including the cellular transcription factors c-jun and c-fos (Farrell et al., 1989; Flemington and Speck, 1990c; Flemington et al., 1994; Kouzarides et al., 1991) (Figure 5). Z homodimerizes through a bZip domain at the carboxy-terminus of the protein, and binds to AP1-like sites known as Z-responsive elements (ZREs) through its basic DNA binding domain (Chang et al., 1990; Farrell et al., 1989;

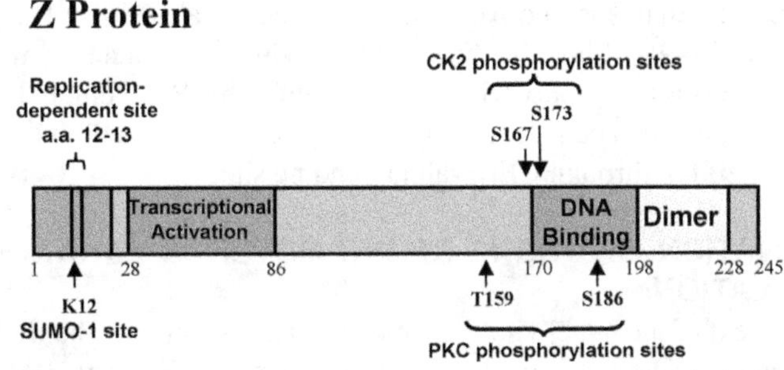

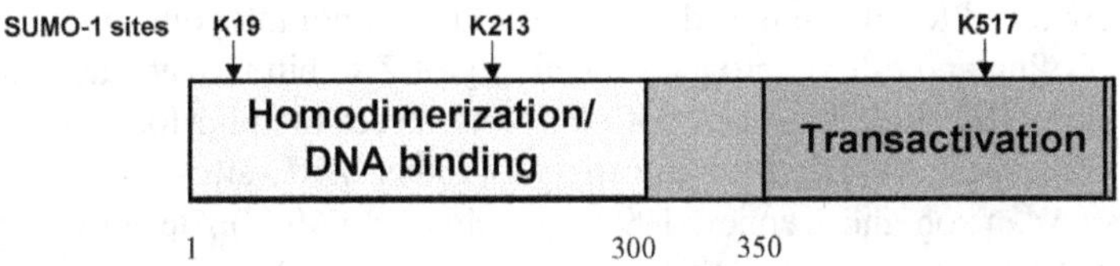

Figure 5. Immediate-early proteins. The protein domains that mediate dimerization, DNA-binding and transactivation functions for Z and R are shown, along with known sites of phosphorylation and sumo1 modification.

Flemington and Speck, 1990a; Lieberman and Berk, 1990; Lieberman et al., 1990; Packham et al., 1990). The amino-terminus of Z contains the transactivator domain (Deng et al., 2001; Flemington et al., 1992).

Z is phosphorylated *in vivo*, and *in vitro*, Z can be phosphorylated by protein kinase C (PKC) over threonine 159 and serine 186, and by casein kinase 2 (CK2) over serine 167 and serine 173. PKC-mediated phosphorylation of Z has been reported to be responsible for increased Z DNA binding activity in TPA-treated cells, as well as enhanced transcriptional function (Baumann et al., 1998). However, in the absence of PKC agonists, neither of the potential PKC phosphorylation sites in Z are phosphorylated *in vivo*, and PKC inhibitors do not reduce the ability of Z to disrupt viral latency (El-Guindy et al., 2002). The two CK2 sites in Z are constitutively phosphorylated *in vivo* (El-Guindy and Miller, 2004). Deletion of the CK2 sites does not inhibit the ability of Z to activate early lytic gene transcription, but reduces its ability to inhibit R-mediated activation of a viral late gene, BLRF2 (El-Guindy and Miller, 2004), suggesting that it may regulate temporal control of the EBV lytic life cycle.

ZRE motifs are found in many early viral promoters (Flemington and Speck, 1990a; Urier et al., 1989). The presence of two or more ZRE sites leads to synergistic transactivation by Z (Carey et al., 1992). Activation by Z requires that ZRE sites be located within a few hundred basepairs of the transcriptional start site. Z interacts directly with a number of basic transcription factors, including TFIID and TFIIA (Chi and Carey, 1993; Chi et al., 1995; Lieberman and Berk, 1991; Lieberman and Berk, 1994; Lieberman et al., 1997; Mikaelian et al., 1993). In addition, Z to interacts directly with histone acetylating complexes (including CBP and p300) results in acetylation of chromatin, converting it to a conformation favorable for transcription (Adamson and Kenney, 1999; Chen et al., 2001a; Deng et al., 2003; Zerby et al., 1999). Although Z does not heterodimerize efficiently with either c-fos or c-jun, it forms a complex with another cellular bZIP protein, C/EBP-alpha (Wu et al., 2003), and can activate at least some promoters, including its own promoter (Wu et al., 2004), through CEBP-alpha binding sites.

Z ACTIVATION OF THE IE RP PROMOTER IS ENHANCED BY ZRE METHYLATION

In the context of the intact viral genome, Z must first activate the BRLF1 promoter, Rp, in order to disrupt viral latency (Adamson and Kenney, 1998; Feederle et al., 2000). The mechanism by which Z activates the Rp promoter appears to be somewhat distinct from the mechanisms by which it activates the early promoters. For example, a Z mutant in which serine residue 186 in the DNA binding domain has been switched to an alanine (the residue found in homologous DNA binding domains of c-jun and c-fos) abrogates the ability of Z to bind to, and activate, the Rp promoter ZRE sites, but does not appear to affect binding to, or activation through, consensus AP-1sites or a variety of other ZRE sites found in early promoters (Adamson and Kenney, 1998; Francis et al., 1997). Because it cannot induce BRLF1 transcription from the latent viral genome, this mutant is unable to activate lytic EBV gene transcription, but can be rescued by co-transfection with an R expression vector (Adamson and Kenney, 1999; Francis et al., 1999).

Rp has three atypical ZRE sites, of which two contain CpG motifs (ZRE-2 and ZRE-3) and hence could be methylated in latently infected cells. Surprisingly, Z preferentially binds to the methylated form of the Rp ZRE-2 site (TGAGCGA), and binds only to the methylated form of the Rp ZRE-3 site (TTCGCGA) (Bhende et al.). The Rp ZRE-1 site (TGAGCCA) cannot be methylated. The crystal structure of a Fos-Jun heterodimer binding to a consensus AP-1 site indicates that the residues in the basic DNA binding domain of c-fos and c-jun which make direct contact with DNA are analogous to Z residues 182, 185, 186, 189, 190, and that Z residue 186 would directly contact the methylated cytosine in the Rp ZRE-2 motif (TGAGCGA) (Glover and Harrison, 1995). Interestingly, the residues that directly contact DNA are identical in the c-Fos, c-Jun, and Z proteins, with the exception of the Z serine residue at position 186, which is an alanine at the analogous position in c-Fos and c-Jun (Francis et al., 1997; Glover and Harrison, 1995).

The Rp promoter is highly methylated during the latent form of viral infection (Bhende et al.). Although DNA methylation of promoters generally acts as a potent inhibitor of cellular gene transcription, Z preferentially activates the methylated form of Rp in reporter-gene assays, and preferentially induces lytic EBV transcription from a methylated, versus unmethylated, viral genome (Bhende et al.). The unexpected ability of Z to activate methylated lytic viral promoters suggests a mechanism by which EBV evades the inhibitory effects of viral genome methylation. Z is the first transcription factor shown to preferentially activate the methylated form of a downstream target gene

TRANSCRIPTIONAL EFFECTS OF Z UPON CELLULAR GENES

Beyond viral gene activation, Z can activate certain cellular genes through AP-1 sites and/or ZRE sites, including c-fos (Flemington and Speck, 1990b), cellular IL-10 (Mahot et al., 2003), TGF-beta (Cayrol and Flemington, 1995), the tyrosine kinase TKT (Lu et al., 2000), and matrix metalloproteinases 1 and 9 (Lu et al., 2003; Yoshizaki et al., 1999). Z activation of human immunosuppressive cytokines such as IL-10 and TGF-beta may function to suppress responding host immune cells. The role of other Z-activated cellular genes is less clear for viral persistence or transmission, but intermittent lytic expression in malignant EBV disease could potentially contribute to oncogenesis, in that matrix metalloproteinases have been implicated in some malignancies as facilitating metastatic spread, while IL-10 could serve as a paracrine/autocrine growth factor for B cell tumors.

VIRAL REPLICATION FUNCTIONS OF Z

In addition to its essential role as a viral transcription factor, BZLF1 also plays a direct role in lytic viral replication. The EBV lytic origin of replication, *oriLyt*, contains a series of ZRE sites, and Z binding to these sites is required for *oriLyt* replication. (Fixman et al., 1992; 1995; Hammerschmidt and Sugden, 1988; Schepers et al., 1993a; 1993b). A Z mutant altered at amino acid residues 12 to 13 cannot replicate viral DNA, but is transcriptionally active (Sarisky et al., 1996). BZLF1 also interacts directly with some of the core viral replication proteins (Gao et al., 1998; Takagi et al., 1991; Zhang et al., 1996). Together, these results suggest that BZLF1 acts as an essential *oriLyt* binding protein during lytic EBV replication, and that this binding may promote formation of the initial replication complex.

A BZLF1-DELETED VIRUS CANNOT SUPPORT LYTIC EBV INFECTION

A BZLF1-deleted EBV was recently constructed (Feederle et al., 2000). Not surprisingly, this mutant cannot enter the lytic form of EBV infection unless Z is supplied in trans. In cells containing the Z-deleted virus, restoration of Z expression initially activates transcription of the other IE gene, BRLF1, and subsequently allows expression of the complete cascade of early and late lytic genes. Over-expression of R in cells containing the Z-deleted virus does not result in expression of the majority of early or late genes or lytic viral replication. The Z-deleted virus is not defective in immortalizing primary B cells. The studies with the Z-deleted virus indicate that both Z and R are required for the activation of many (but not all) lytic EBV genes in the context of the intact viral genome.

CELL-CYCLE EFFECTS OF Z

Herpesviruses commonly usurp the host cell cycle control mechanisms to assure adequate substrates for lytic viral DNA replication. Z has various effects upon cell cycle that are cell-type dependent. In some cell types, including fibroblasts, Z activates p21 expression (possibly as a heterodimer with C/EBP-α) and inhibits expression of cyclin A and c-myc. Together, these effects result in a severe G1/S block (Cayrol and Flemington, 1996a; 1996b; 1995; Mauser et al., 2002b; Rodriguez et al., 1999; 2001; Wu et al., 2003). HeLa cells respond to Z with decreased cyclin B, resulting in both a G2 block, and inhibition of chromosomal condensation that leads to mitotic block (Mauser et al., 2002a). In primary and telomerase-immortalized keratinocytes, which may more closely model the *in vivo* target that EBV might naturally infect in the oropharynx, Z activates S-phase associated proteins, E2F-1, cyclin E, and cyclin A (Mauser et al., 2002b). Less is known about the effect of Z on the cell cycle in lymphocytes. However, in the marmoset B95-8 B-cell line, Z expression results in upregulation of cyclin-dependent kinases, even though cellular DNA replication is arrested (Kudoh et al., 2003) in late G1/S phase. Most importantly, agents such as roscovitine that inhibit the activity of cyclin-dependent kinases also inhibit lytic EBV gene expression (Kudoh et al., 2004), although it is not currently understood why cyclin-dependent kinases are required for efficient lytic EBV replication. These results suggest that a "pseudo lateG1/S-phase" environment, in which certain late G1/S-phase restricted cellular proteins are expressed, but cellular DNA does not actually replicate, may be the ideal host cell environment for lytic EBV replication. Inhibition of cellular DNA replication presumably decreases competition between the virus and host cell DNA for limiting substrates involved in DNA replication, while the expression of certain G1/S-phase restricted cellular proteins, such as E2F-1, may be required for viral replication.

Z-MEDIATED DISPERSION OF PML BODIES

Promyelocytic leukemia (PML) bodies are structures in the nucleus. Although the exact function(s) of PML bodies is somewhat mysterious, the fact that the formation of these bodies is dramatically enhanced by both type I and type II interferons, and that a number of different viruses encode proteins capable of dispersing PML bodies, suggests that these bodies have an antiviral function (Bernardi and Pandolfi,

2003; Chee et al., 2003; Salomoni and Pandolfi, 2002). PML bodies always contain PML protein that have undergone post-translational modification by the small ubiquitin-related modifier (sumo-1), and may also include various cellular proteins such as CREB-binding protein (CBP), Rb, Sp100, and sumo-1 itself (Bernardi and Pandolfi, 2003; Salomoni and Pandolfi, 2002). A number of viruses have been found to encode proteins which disperse PML bodies (Chee et al., 2003), and it has been suggested that PML bodies serve an anti-viral function that viruses must overcome. Z expression in EBV-negative cells results in decreased PML modification by sumo-1 and dispersion of PML bodies (Adamson and Kenney, 2001; Bell et al., 2000). Interestingly, BZLF1 itself is efficiently modified by sumo-1 over lysine 12, and the ability of BZLF1 to inhibit PML protein sumo-1 modification may be at least partially due to competition between BZLF1 and PML for limiting amounts of sumo-1 in the host cell (Adamson and Kenney, 2001). A BZLF1 mutant altered at residues 12 and 13 is transcriptionally competent, but defective in mediating lytic replication (Sarisky et al., 1996). Thus, sumo-1 modification of BZLF1 may be primarily important for its replicative, rather than transcriptional, function.

Z EFFECTS UPON P53

Many viruses have evolved mechanisms to override cellular p53, a transcription factor which can respond to various markers of viral infection by activation of pro-apoptotic pathways. The effects of Z on p53 in the host cell are complex. Z expression increases the level of total p53, and induces a variety of post-translational modifications in p53 (including enhanced phosphorylation over multiple different residues, as well as enhanced acetylation) that would be expected to activate p53 function. Nevertheless, p53 transcriptional function is clearly decreased in the presence of Z. The inhibitory effect of Z on p53 function may reflect its ability to interact directly with p53 (Zhang et al., 1994). In addition, Z decreases the level of cellular TATA-binding protein, and reconstitution of TBP in Z-expressing cells partially restores p53 transcriptional function (Mauser et al., 2002c). The disruption of PML bodies by Z may also impair p53 function. Since p53 is an important mediator of cellular apoptosis, the ability of Z to inhibit p53 function likely plays a crucial role in protecting the virus from apoptosis during lytic infection.

Z-MEDIATED INHIBITION OF CELLULAR IMMUNITY

Z plays a critical role in impairing the host immune response to lytic EBV infection. First, Z minimizes immune presentation of lytic viral antigens by inhibiting expression of MHC class I on cells (Keating et al., 2002; Mahot et al., 2003). In addition, Z decreases the ability of the host cell to respond to interferon gamma by inhibiting transcription of the gene encoding the receptor for this cytokine (Morrison et al., 2001). Z inhibition of interferon gamma signaling is likely important not only for protecting the virus from the immunostimulatory effects of interferon gamma (including activation of MHC class II), but may also be required for epithelial cell-derived virus to efficiently infect B cells. Virus produced in cells expressing MHC class II preferentially infects epithelial cells, whereas virus produced in cells not expressing MHC class II preferentially infects B cells (Borza and Hutt-Fletcher, 2002).

Likewise, Z inhibits transcription of the gene encoding the major tumor necrosis factor alpha receptor (TNR-α-R1), and thus suppresses the response of the host cell to this antiviral cytokine. TNF-alpha not only activates expression of a number of important inflammatory genes (through its effects on NF-KB), but also induces cellular apoptosis. TNF-alpha cannot activate transcription of important downstream target genes such as ICAM-1 in BZLF1-expressing cells, and is also unable to induce cellular apoptosis (Morrison et al., 2004).

Z also interacts directly with the p65 subunit of NF-κB, and inhibits its transcriptional function (Gutsch et al., 1994; Hong et al., 1997; Keating et al., 2002; Morrison and Kenney, 2004). Z decreases NF-KB binding to promoters in the context of the intact cellular genome (Morrison and Kenney, 2004). Thus, the IL-1 cytokine cannot activate NF-KB-responsive cellular genes in the presence of BZLF1, even though the upstream components of the IL-1 signaling pathway appear to be unaffected by BZLF1 (Morrison and Kenney, 2004). Somewhat paradoxically, however, since one of the NF-KB responsive genes inhibited by BZLF1 is I-kappa B (IK-B) (Morrison and Kenney, 2004), and the IK-B protein normally acts to retain NF-KB in the cytoplasm in an inactive form, BZLF1-expressing cells actually have a very high level of nuclear NF-KB (Morrison and Kenney, 2004). This BZLF1-mediated translocation of NF-KB into the nucleus may act to negatively regulate BZLF1 transcriptional function and hence promote viral latency in situations where BZLF1 expression is limiting relative to the amount of nuclear NF-KB.

As discussed previously, Z stimulates expression of both TGF-beta, and IL-10 (Cayrol and Flemington, 1995). Both TGF-beta and IL-10 have immunomodulatory effects that would be expected to attenuate the cytotoxic T cell response directed against the virus. The immunomodulatory effects of BZLF1 are summarized in Figure 6.

TRANSCRIPTIONAL EFFECTS OF R

The R protein is also capable of inducing lytic EBV infection in a subset of latently infected cell lines (Feederle et al., 2000; Ragoczy et al., 1998; Westphal et al., 1999; Zalani et al., 1996). Interestingly, in the case of KSHV, only the R homologue (ORF50), and not the Z homologue (K8), can activate lytic viral gene expression. Although R expression alone is not sufficient for disrupting EBV latency in some cell lines (such as the Raji Burkitt lymphoma line), R expression is nevertheless essential (in addition to Z) in all EBV-infected lines for the activation of many early lytic genes, including BMRF1 (Feederle et al., 2000; Ragoczy and Miller, 1999). Like Z, R binds directly to the EBV *oriLyt* (Gruffat and Sergeant, 1994; Hammerschmidt and Sugden, 1988). However, R binding to *oriLyt* has not been shown to be critical for *oriLyt* replication (Fixman et al., 1992), at least in plasmid-based replication assays, although R binding to *oriLyt* may enhance the efficiency of replication in the context of the intact viral genome.

The DNA-binding domain and homodimerization domain (Manet et al., 1991) of R are contained within its amino-terminus, and a potent transcriptional activation domain is located with the carboxy-terminus (Hardwick et al., 1988; 1992). R interacts directly with TBP and TFIIB through its transcriptional activator domain (Manet et al., 1993). Like Z, R also interacts directly with the histone

cis-acting elements in *oriLyt*

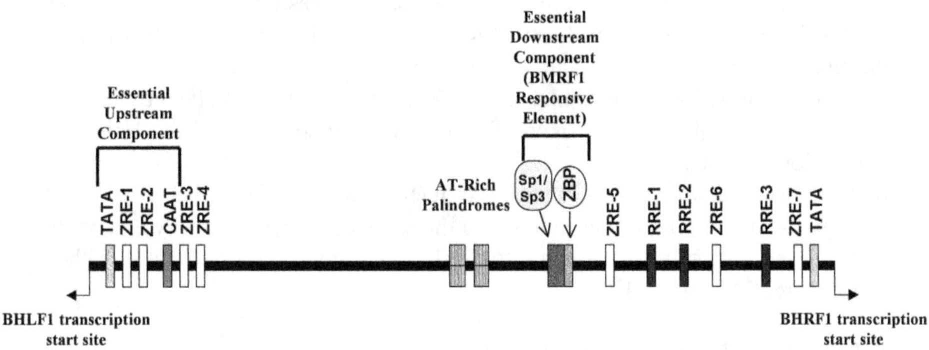

Figure 6. Cis-acting elements of oriLyt. The *oriLyt* locus (the origin of lytic replication) contains the promoters of two early genes, BHLF1 and BHRF1. The locations of Z and R binding sites, as well as the binding sites of cellular transcription factors ZBP-89, Sp1, and Sp3, are shown. The localization of two essential elements in *oriLyt* which are absolutely required for replication is indicated.

acetylase, CREB-binding protein (CBP)(Swenson et al., 2001). There is some evidence that R may be sumo-modified (Chang et al., 2004). Also, R appears to be phosphorylated, although the site(s) have not been characterized (Zacny et al., 1998).

R is primarily nuclear in its location, and activates lytic EBV promoters through at least two different mechanisms. R binds directly (as a homodimer) to a GC-rich motif (consensus $GGCCN_7GTGGTG$) present in the promoters of at least three early EBV genes, SM (formerly called BMLF1), BHRF1, and BMRF1 (Chevallier-Greco et al., 1989; Cox et al., 1990; Gruffat et al., 1992; 1990; Kenney et al., 1989a; Quinlivan et al., 1993). In contrast to ZRE sites, R-binding motifs often function as powerful enhancer elements in the presence of the R protein (Kenney et al., 1989a). R binding to the two RRE sites in the EBV *oriLyt* is enhanced by the cellular protein, HMGB1 (Mitsouras et al., 2002), apparently because HMGB1 induces bending of the DNA that promotes R binding. A number of early viral promoters, including the BMRF1 promoter, contain both Z and R binding motifs, and the combination of both Z and R is required to activate such promoters (at least in the context of the intact viral genome) (Holley-Guthrie et al., 1990; Quinlivan et al., 1993).

In contrast to the direct DNA binding mechanism described above, R activates its own promoter (Rp) and the BZLF1 promoter (Zp) through indirect mechanisms. R activates its own promoter at least partially through SP1 binding motifs (Liu and Speck, 2003; Ragoczy and Miller, 2001), although the mechanisms for this effect has not been defined. R activates the Zp promoter at least partially through the ZII motif (Adamson et al., 2000). This effect is mediated by R activation of the stress Map kinases (p38 and c-jun N-terminal kinase), which results in phosphorylation and activation of the c-jun and ATF-2 transcription factors (Adamson et al., 2000). R also activates the PI3 kinase pathway in cells (Darr et al., 2001). At least one early viral promoter, the DNA polymerase promoter, is also activated by R through an indirect mechanism, and in this case the effect involves USF and E2F-1 binding motifs (Liu et al., 1996). The ability of R to disrupt viral latency is abolished in the presence of either p38 stress Map kinase, or PI3 kinase, inhibitors, which prevent R from activating Zp (Adamson et al., 2000; Darr et al., 2001). In latently infected cells (such as Raji) where R expression by itself cannot induce fully lytic gene expression, this defect seems to be due to the inability of R to activate BZLF1 transcription (Zalani et al., 1996), and R efficiently activates the SM early promoter through the direct binding mechanism in the same cell lines (Ragoczy and Miller, 1999). The ability of R to activate Z transcription in some cell lines, but not others, is not currently explained.

R ACTIVATION OF CELLULAR GENES

R activates expression of at least two cellular genes: fatty acid synthase (FAS), and the c-mer tyrosine kinase (Li et al., 2004a; 2004b). High-level expression of FAS enzyme, which is essential for synthesis of lipids, including palmitate, is normally restricted to fat cells and liver cells. R activates FAS gene expression in host cells through a p38 kinase-dependent mechanism. Interestingly, specific inhibitors of the FAS enzyme (cerulenin and C75) also abrogate R-mediated induction of lytic EBV gene expression, as well as constitutively lytic EBV gene expression in AGS cells and anti-IgG mediated induction in Akata cells (Li et al., 2004b). Thus, activation of cellular FAS activity by R appears to be an essential aspect in the switch from latent to lytic infection. Why FAS activity is required is not clear, but is perhaps related to differences in the efficiency of palmitoylation of viral and/or cellular proteins (required for entry of some proteins into lipid rafts) in cells expressing low, versus high, levels of FAS.

R also activates expression of the c-mer tyrosine kinase in the host cell (Li et al., 2004a), and increases the ability of the cell to respond to the c-mer ligand, gas-6. The exact role of c-mer activation during lytic infection is not yet clear. Since c-mer over-expression has been suggested to play a role in the development of lymphoid malignancies, activation of c-mer expression by the R protein could potentially be relevant to the development of EBV-associated lymphomas.

BRLF1 cell cycle effects

R also regulates cell cycle progression in the host cell. R increases the number of cells in S-phase in both primary human fibroblasts and HeLa cells (Swenson et al., 1999). This effect is probably at least partially secondary to the ability of R to greatly increase E2F-1 expression in cells (Swenson et al., 1999). In addition, R interacts directly with the tumor suppressor protein, Rb (Zacny et al., 1998), suggesting another mechanism by which R activates cell cycle progression.

AN BRLF1-DELETED VIRUS CANNOT SUPPORT LYTIC EBV INFECTION

An R-deleted mutant virus has been constructed (Feederle et al., 2000). The R-deleted virus immortalizes primary B cells with similar efficiency as wild-type virus (Feederle et al., 2000). As expected, the R-deleted virus cannot activate most lytic viral genes, including the BZLF1 immediate-early gene, unless R is supplied in trans. These studies confirm that in the context of the intact virus, R is required for BZLF1 activation, and that the combination of Z and R together is required for activation of many early viral genes. However, as discussed later, this particular R-deleted virus was subsequently discovered to be unable to express an early EBV gene product, BRRF1, and thus the phenotype of this virus in fact is also partially due to the loss of BRRF1 expression (Hong et al., 2004).

REGULATION OF EARLY LYTIC EBV GENE TRANSCRIPTION

Z and R expression leads to activation of the viral early genes. Both the Z and R proteins are required for activation of many early viral promoters in the context of the intact viral genome (Cox et al., 1990; Feederle et al., 2000). Regulation of the early BMRF1 gene promoter has been studied most extensively. There are ZRE sites, as well as an R-binding motif, in the BMRF1 promoter (Holley-Guthrie et al., 1990; Quinlivan et al., 1993). Z alone efficiently activates the BMRF1 promoter in reporter gene assays, and this effect requires that both ZRE sites be present. While R by itself only modestly activates the BMRF1 promoter in reporter-gene assays, the combination of Z and R together synergistically activates the BMRF1 promoter in some cell types (Holley-Guthrie et al., 1990). The combination of both Z and R is required for significant activation of BMRF1 transcription from the intact viral genome, regardless of cell type (Adamson and Kenney, 1998; Feederle et al., 2000; Ragoczy and Miller, 1999; Zalani et al., 1996). However, some viral early promoters can be activated by expression of R alone in the context of the intact viral genome. For example, the early SM promoter, which contains a ZRE motif as well as an R-binding site, is activated by R in Raji cells in the absence of any concomitant Z expression (Ragoczy and Miller, 1999; Swenson et al., 2001).

OriLyt contains two divergent early promoters, BHLF1 and BHRF1, which have been extensively studied *in vitro*, although not in the context of the intact virus. There are three R binding sites in the overlapping BHLF1 and BHRF1 promoters (Gruffat and Sergeant, 1994, ADD Carey ref; Mitsouras et al., 2002). In reporter gene assays, R activates the BHRF1 promoter much more efficiently than the BHLF1 promoter (Cox et al., 1990; Hardwick et al., 1988). There are also a series of ZRE sites upstream of both the BHLF1 and BHRF1 promoters (Hardwick et al., 1988), and Z alone can activate both promoters. As was is the case for the BMRF1 promoter, the combination of Z and R together activates the *oriLyt* promoters more efficiently than either protein alone in reporter gene assays. Whether the combination of both IE proteins is required for activation of BHLF1 and/or BHRF1 transcription in the context of the intact virus has not been well studied.

EARLY LYTIC GENE PRODUCTS: PROTEINS WHICH REGULATE TRANSCRIPTION

At least two of the EBV early genes encode transcription factors. One of these is the BMRF1 gene product (Cho et al., 1985), which also has a separate function as the viral DNA polymerase-processivity factor (Tsurumi et al., 1993; 1994; Zhang et al., 1999a; 1997). BMRF1 activates the *oriLyt* promoter, BHLF1, as well as the cellular gastrin promoter (Holley-Guthrie et al., 2005). In both cases, the BMRF1-responsive region of these promoter contains a GC-rich domain that binds to the cellular transcription factors, ZBP-89 and SP1/SP3. BMRF1 interacts directly with both SP1 and ZBP-89, and enhances their transcriptional activity (Baumann et al., 1999; Holley-Guthrie et al., 2005). The transcriptional function of BMRF1 maps to its carboxy-terminal domain (Zhang et al., 1999a), whereas the DNA polymerase-processivity function requires only the first 300 amino acids of the protein (Kiehl and Dorsky, 1991; 1995). Interestingly, the region of *oriLyt* required BMRF1 to activate the BHLF1 promoter (the downstream essential domain) is also essential in cis for *oriLyt* replication (Zhang et al., 1997).

An early gene (BRRF1) that is encoded by the opposite strand of the first intron of the BRLF1 immediate-early gene (Segouffin-Cariou et al., 2000) also functions as a transcription factor (Hong et al., 2004). BRRF1 activates Zp expression in reporter gene assays, and is required for efficient R-mediated disruption of viral latency in certain cell types (Hong, 2004). BRRF1 activates Zp through the ZII motif, and enhances the phosphorylation, and transcriptional function, of c-jun (Hong et al., 2004). The mechanism for this BRRF1 effect has not been defined. The previously mentioned R-deleted virus, which was originally thought to specifically delete the BRLF1 gene product (Feederle et al., 2000), is also defective in BRRF1 transcription. The combination of R and BRRF1 is required to fully rescue the lytic phenotype of this mutant virus in some cell types (Hong et al., 2004).

EARLY LYTIC GENE PRODUCTS: PROTEINS WHICH REGULATE RNA TRANSPORT AND STABILITY

Many lytic EBV genes are lacking introns, and RNA derived from intronless genes is often unstable in cells. The early lytic protein, SM (previously known as BMLF1) (Cook et al., 1994), is an RNA-binding protein that enhances the stability of intronless lytic viral transcripts, and promotes the transport of such messages from the nucleus to the cytoplasm (Boyer et al., 2002; 1999; Buisson et al., 1999; 1989; Chen et al., 2001b; Farjot et al., 2000; Gruffat et al., 2002a; Hiriart et al., 2003a; 2003b; Kenney et al., 1989c; Ruvolo et al., 2001; 2004; 1998; Semmes et al., 1998). An SM-deleted virus was recently constructed and found to be deficient in viral replication and virion production (Gruffat et al., 2002a). In addition to its effects on viral mRNA, the SM protein also inhibits PKR activation (Poppers et al., 2003). PKR inhibition by SM may help the virus to escape the inhibitory effect of PKR on protein translation. In contrast, SM enhances STAT1 expression (Ruvolo et al., 2003); though a viral function for this has not been elucidated.

EARLY LYTIC GENE PRODUCTS: PROTEINS WHICH INHIBIT CELLULAR APOPTOSIS AND IMMUNE EVASION

At least two early EBV proteins may help the virus to cellular apoptosis during the lytic form of infection, and thus increase the efficiency of lytic EBV viral replication by preventing the death of the host cell. The early lytic protein BHRF1 is a homologue of the cellular anti-apoptosis gene, BCL-2, and prevents apoptosis mediated by a number of different agents (Henderson et al., 1993; Tarodi et al., 1994). The early lytic protein, BARF1, which encodes a soluble receptor for colony–stimulating factor 1 (Strockbine et al., 1998), likely helps to protect the virus from monocytes and macrophages. BARF1 decreases the ability of monocytes and macrophages to secrete alpha-interferon (Cohen and Lekstrom, 1999), and inhibits the differentiating and proliferative effects of this cytokine on monocytes/ macrophages (Strockbine et al., 1998).

EARLY LYTIC GENE PRODUCTS: A VIRAL KINASE

At least one early protein, encoded by the BGLF4 gene, functions as a viral kinase (EBV-PK). EBV-PK is homologous to the cytomegalovirus UL97 gene product. EBV-PK autophosphorylates itself, and phosphorylates at least two other EBV proteins, BMRF1 and the EBNA-Leader protein (Chen et al., 2000; Gershburg and Pagano, 2002; Kato et al., 2001; 2003). The effect of this phosphorylation on BMRF1 and EBNA Leader protein function is currently unknown. EBV-PK also phosphorylates the cellular protein, translation elongation factor 1 delta (Kato et al., 2001). EBV-PK phosphorylates protein motifs similar to those recognized by the cellular cdc2 kinase (Kawaguchi et al., 2003). The role(s) of EBV-PK in promoting lytic EBV replication has not yet been defined. The phosphorylation and activation of the antiviral nucleoside analogue, ganciclovir, that occurs in lytically infected EBV-positive cells is at least partially mediated by EBV-PK (Marschall et al., 2002). Virally mediated-activation of ganciclovir is required for its ability to inhibit lytic EBV replication.

VIRAL REPLICATION PROTEINS AND *ORILYT*

Lytic replication of the viral genome requires the lytic origin of replication (*oriLyt*) in cis, and the viral core replication proteins (BALF5, BMRF1, BBLF2, BBLF4, BSLF1, and BBLF2/3) in trans (Fixman et al., 1992; 1995). Lytic EBV replication involves the formation of head-to-tail concatamers and probably occurs through a rolling-circle mechanism. Following replication, the virus is clipped within the terminal repeat region of the genome into a linear, unit-length genome and packaged into virion particles. The viral proteins that mediate the cleavage and packaging functions have not been identified. The EBV terminal repeats are sufficient to allow packaging of plasmids in cis (Zimmermann and Hammerschmidt, 1995).

Most EBV strains contain two copies of *oriLyt*, although viral strains that contain only one copy of *oriLyt* (such as B95-8) seem to replicate equally well. *OriLyt* contains the divergent promoters of two EBV early genes, BHRF1 and BHLF1, as well as binding sites for both Z and R (Hammerschmidt and Sugden, 1988). Two domains in *oriLyt* are absolutely essential in cis for replication. The "upstream" domain contains two Z-responsive elements which must be bound by Z for successful replication (Schepers et al., 1993a; 1993b; 1996). The "downstream" domain is a GC-rich sequence which mediates responsiveness to the transcriptional effects of BMRF1 (Zhang et al., 1997), and binds to the cellular transcription factors Sp1, Sp3, and ZBP-89 (Baumann et al., 1999; Gruffat et al., 1995; Schepers et al., 1993b). The "downstream" domain can potentially form a triple helical DNA structure, which may play a role in lytic replication (Portes-Sentis et al., 1997). In addition, ZBP-89 binding to this domain may be required for *oriLyt* replication (Baumann et al., 1999).

The core viral replication proteins include the viral DNA polymerase (BALF5), the DNA polymerase processivity factor (BMRF1), helicase (BBLF4), primase (BSLF1), primase-associated protein (BBLF2/3), and the single-stranded DNA binding protein (BALF2) (Daibata and Sairenji, 1993; Fixman et al., 1992; Fujii et al., 2000; Gao et al., 1998; Liao et al., 2001). Several core replication proteins directly interact with one another, presumably allowing formation of a large replication initiation complex (Daibata and Sairenji, 1993; Fujii et al., 2000; Gao et al., 1998; Liao et al., 2001). Z is also required for lytic replication, and interacts directly with BMRF1 and BBLF4 (Liao et al., 2001; Zhang et al., 1996).

EBV DNA polymerase activity requires both the catalytic component of the enzyme (encoded by the BALF5 gene), as well as the polymerase processivity function (encoded by the BMRF1 gene). The catalytic component has $3'$-to-$5'$ proofreading exonuclease activity (Tsurumi et al., 1993), and in contrast to cellular polymerase, is active in high-salt (100 mM ammonium sulfate) (Tsurumi et al., 1993), The BMRF1 polymerase processivity factor assembles as a ring-like structure (Makhov et al., 2004) which presumably allows it to function as a sliding clamp during viral replication.

In addition to the core replication proteins, several viral early proteins encode enzymes involved in deoxynucleotide metabolism, regulating nucleoside substrates in the lytic host cell environment. These proteins include the viral thymidine kinase gene (BXLF1)(Littler et al., 1986), the viral dUTPase gene (Fleischmann et al., 2002) and viral ribonucleotide reductase (BORF2 and BARF1). The role of these EBV proteins may be to ensure efficient production of nucleotide substrates in non-replicating cells. The EBV thymidine kinase may also mediate phosphorylation of the nucleoside analog, acyclovir, in lyticly infected cells (Littler and Arrand, 1988), although this has been challenged (Gustafson et al., 1998).

LATE VIRAL GENE REGULATION

Regulation of late lytic EBV genes is less well understood than the regulation of IE and early genes. Late promoters are not characterized by Z-responsive or R-responsive elements. Late genes are traditionally defined as genes that are expressed after the onset of viral replication, and agents that prevent viral replication classically inhibit the expression of late genes. Studies performed in herpes simplex suggest that EBV late promoters may be primarily activated in cis by viral replication. Some late promoters in reporter gene assays can be activated in a replication-independent manner (Serio et al., 1998; 1997). More compellingly, studies performed with the Z and R-deleted viruses suggest that viral replication is not required for activation of all EBV late gene promoters. In particular, R induces expression of a subset of "late" genes in the Z-deleted virus in the absence of any viral replication (Feederle et al., 2000). R also activates some late genes in Raji cells in the absence of viral replication (Ragoczy and Miller, 1999). In reporter gene assays, certain late promoters are also activated in a replication-independent manner (Serio et al., 1998; 1997).

LATE VIRAL PROTEINS

The nucleocapsid proteins that make up the virion particle, including BNRF1, BFRF2/3, BLRF1/2/3, and BcLF1, are encoded by viral late genes. Viral membrane-associated glycoproteins which bind cellular receptor (gp350/220 (BLLF2), gp85 (BXLF2), gp42(BZLF2), gp25(BKRF2)) are also encoded by late genes. The role of these proteins in determining cellular tropism of EBV is discussed in the chapter by Hutt-Fletcher.

In addition, the late viral protein, vIL-10 (BCLF1), encodes a homologue of cellular IL-10 that is likely important for protecting the virus from the host immune response (Hsu et al., 1990), Viral IL-10 is secreted from lytically infected cells, shares with cellular IL-10 the ability to potently repress the cytotoxic T-cell response (Salek-Ardakani et al., 2002). Viral IL-10, like cellular IL-10, also functions as B-cell growth factor (Miyazaki et al., 1993; Rousset et al., 1992; Stuart et al., 1995). Secretion of vIL-10 from lytically-infected EBV-positive B cells could thus help support the growth of the latently infected cells through a paracrine mechanism. A vIL-10-deleted virus was not defective in B-cell transformation *in vitro* or lymphoma formation in mice (Swaminathan et al., 1993). The ability of Z to activate cellular IL-10 may compensate, at least partially, for the loss of viral IL-10. Finally, one or more late viral gene products also appear to inhibit apoptosis (Inman et al., 2001; Marshall et al., 1999), confirming that prevention of cellular apoptosis is required for efficient EBV replication throughout lytic replication.

VIRION MATURATION AND RELEASE

Mechanisms for EBV virion assembly and release have not been well documented and are largely presumed based upon studies performed with alpha herpeviruses. The mature EBV particle has a very similar structure to other herpesvirus virions, which consist of a nucleocapsid with 162 capsomeres, protein tegument, an envelope, and an outer envelope. The primary capsid proteins are homologous to those of herpes simplex, while the outer membrane glycoproteins are specific to EBV. EBV packaging of viral DNA within the capsid probably follows a very similar mechanism to HSV, which occurs within the nucleus followed by budding from the nuclear membrane. In HSV, varicella-zoster virus, and cytomegalovirus, this envelope fuses with the nuclear membrane releasing the encapsidated genome to the cytoplasm. From this point, it is thought that viral tegument material in the cytoplasm is packaged around the capsid and a second process of budding occurs to produce the final viral product (Mettenleiter, 2002). The validity of this model has yet to be substantiated for EBV specifically.

THERAPY FOR LYTIC EBV INFECTION

To date, there are no clinically approved therapies that target latent EBV products, but several nucleoside analogs that are effective for treating other herpesvirus lytic infections can be activated in cells undergoing lytic EBV infection. Acyclovir and ganciclovir are two nucleoside analogs that inhibit lytic EBV infection (Datta et al., 1980; Lin et al., 1987; Lin and Machida, 1988; Meerbach et al., 1998). In lytically infected EBV-positive cells, acyclovir is phosphorylated and activated by EBV thymidine kinase, while ganciclovir is likely primarily phosphorylated by the EBV BGLF4 gene product, EBV-PK (the functional homologue of CMV UL97) (Lin et al., 1986; Marschall et al., 2002; Moore et al., 2001). Monophosphorylated forms of acyclovir and ganciclovir are unable to diffuse back out of the cell, and are further phosphorylated by cellular enzymes to guanidine triphosphate analogs. Acyclovir triphosphate is preferentially utilized by viral DNA polymerase and reversibly impairs viral DNA replication. There is no significant benefit of using acyclovir to treat acute infectious mononucleosis (Torre and Tambini, 1999), in which symptoms may be largely attributable to the host immune response. Nevertheless, acyclovir is clearly effective for treating oral hairy leukoplakia. Ganciclovir acts similarly to acyclovir, but the incorporation of ganciclovir triphosphate into DNA irreversibly terminates the elongation of the DNA strand. Thus, while both drugs are preferentially activated to their toxic form in cells harboring lytic virus, ganciclovir is significantly more toxic, causing bone marrow suppression in particular. Foscarnet is another approved anti-viral drug which targets the DNA polymerase of all herpesviruses, and is active against EBV (Datta and Hood, 1981); however, its severe renal toxicity precludes routine use.

It is uncertain whether anti-viral drugs are useful in treating post-transplant polyclonal EBV-positive lymphoproliferative disease (Cohen, 2000; Oertel et al., 1999). Lytic EBV is not clearly implicated in establishing a malignant B-cell phenotype (Feederle et al., 2000), but nucleoside analog prophylaxis is associated with a reduced risk to developing such disease in transplant recipients (Darenkov et al., 1997; Farmer et al., 2002; Fong et al., 2000; Green et al., 2001; Malouf et al., 2002; McDiarmid et al., 1998). This may be a function of diminished intra-host lytic transmission, resulting in a smaller total reservoir of latently-infected cells.

THERAPEUTIC INDUCTION OF LYTIC EBV TO TREAT EBV-ASSOCIATED MALIGNANCIES

The mere presence of EBV in a wide array of malignancies and its absence in normal human tissue suggests the virus as a worthwhile candidate for targeted therapies. EBV-related malignancies grow with the latent form of replication, expressing a restricted set of viral proteins. "Therapeutic" induction of lytic EBV infection in EBV-positive tumor cells is actively being investigated as a novel way to selectively kill EBV-infected tumor cells (Gutierrez et al., 1996; Israel and Kenney, 2003; Westphal et al., 1999). Lytic induction could be theoretically achieved by either heterologous introduction of immediate early genes, or by providing stimuli that are sufficient to activate endogenous IE expression. Both approaches have been studied. Recombinant adenoviral vectors, which are capable of efficiently transducing various epithelial cell types, have been tested for delivery of Z or R *in vitro* and in a murine model of EBV-positive nasopharyngeal carcinoma. Expression of the IE genes sufficient to induce early gene activation was achieved in EBV-positive tumors, and resulted in inhibition of tumor growth (Feng et al., 2002b). Alternatively, many of the cytotoxic regimens used in cancer therapy, including gamma irradiation and some chemotherapy agents, can activate the signal transduction pathways, including p38 kinase, MAP kinase, PI3-kinase, and EGR-1 pathways (Sinclair et al. 1991), and result in lytic infection (Feng et al., 2004; Feng et al., 2002a; Roychowdhury et al., 2003; Westphal et al., 2000). Lytic induction by chemotherapy has been shown to depend upon these specific transduction pathways (Feng et al., 2004; 2002a). Agents that inhibit histone deacetylases (such as butyrate compounds) have also been shown to increase lytic EBV infection in some murine models of lymphoma (Westphal et al., 2000), though not all B-cell tumors respond this way (Feng et al., 2004).

Lytic induction strategies, especially when combined with the antiviral drug, ganciclovir (Faller et al., 2001; 2004; 2002a; Mentzer et al., 1998; Roychowdhury et al., 2003; Westphal et al., 2000), inhibit the growth of EBV-positive tumors.

As discussed previously, cells containing the lytic (but not latent) type of EBV infection express virally encoded kinases (EBV-PK and the viral thymidine kinase) that convert the nucleoside analogue, ganciclovir, to its active cytotoxic form (Moore et al., 2001). The activated form of ganciclovir inhibits the host cell DNA polymerase (in addition to the virally encoded DNA polymerase), and is thus cytotoxic. Most importantly, phosphorylated ganciclovir can also be transferred into nearby cells and thus induce "bystander" killing. As the currently available lytic inducing agents can only induce lytic EBV infection in only a portion of tumor cells, the combination of these agents with ganciclovir kills EBV-positive tumor cells in mouse models much more effectively than the inducing agents, or ganciclovir, alone (Feng et al., 2004; 2002a). The efficacy of ganciclovir, when combined with lytic induction strategies, in treatment of EBV-positive human tumors is currently being investigated (Faller et al., 2001; Mentzer et al., 2001).

References

Adamson, A. L., Darr, D., Holley-Guthrie, E., Johnson, R. A., Mauser, A., Swenson, J., and Kenney, S. (2000). Epstein-Barr virus immediate-early proteins BZLF1 and BRLF1 activate the ATF2 transcription factor by increasing the levels of phosphorylated p38 and c-Jun N-terminal kinases. J. Virol. *74*, 1224-1233.

Adamson, A. L., and Kenney, S. (1999). The Epstein-Barr virus BZLF1 protein interacts physically and functionally with the histone acetylase CREB-binding protein. J. Virol. *73*, 6551-6558.

Adamson, A. L., and Kenney, S. (2001). Epstein-barr virus immediate-early protein BZLF1 is SUMO-1 modified and disrupts promyelocytic leukemia bodies. J. Virol. *75*, 2388-2399.

Adamson, A. L., and Kenney, S. C. (1998). Rescue of the Epstein-Barr virus BZLF1 mutant, Z(S186A), early gene activation defect by the BRLF1 gene product. Virology *251*, 187-197.

Adler, B., Schaadt, E., Kempkes, B., Zimber-Strobl, U., Baier, B., and Bornkamm, G. W. (2002). Control of Epstein-Barr virus reactivation by activated CD40 and viral latent membrane protein 1. Proc. Natl. Acad. Sci. USA. *99*, 437-442.

Ambinder, R. F., Robertson, K. D., and Tao, Q. (1999). DNA methylation and the Epstein-Barr virus. Semin Cancer Biol. *9*, 369-375.

Babcock, G. J., Decker, L. L., Freeman, R. B., and Thorley-Lawson, D. A. (1999). Epstein-barr virus-infected resting memory B cells, not proliferating lymphoblasts, accumulate in the peripheral blood of immunosuppressed patients. J. Exp. Med. *190*, 567-576.

Babcock, G. J., and Thorley-Lawson, D. A. (2000). Tonsillar memory B cells, latently infected with Epstein-Barr virus, express the restricted pattern of latent genes previously found only in Epstein-Barr virus-associated tumors. Proc. Natl. Acad. Sci. USA. *97*, 12250-12255.

Baumann, M., Feederle, R., Kremmer, E., and Hammerschmidt, W. (1999). Cellular transcription factors recruit viral replication proteins to activate the Epstein-Barr virus origin of lytic DNA replication, *oriLyt*. Embo J. *18*, 6095-6105.

Baumann, M., Mischak, H., Dammeier, S., Kolch, W., Gires, O., Pich, D., Zeidler, R., Delecluse, H. J., and Hammerschmidt, W. (1998). Activation of the Epstein-Barr virus transcription factor BZLF1 by 12-O-tetradecanoylphorbol-13-acetate-induced phosphorylation. J. Virol. *72*, 8105-8114.

Bell, P., Lieberman, P. M., and Maul, G. G. (2000). Lytic but not latent replication of epstein-barr virus is associated with PML and induces sequential release of nuclear domain 10 proteins. J. Virol. *74*, 11800-11810.

Bernardi, R., and Pandolfi, P. P. (2003). Role of PML and the PML-nuclear body in the control of programmed cell death. Oncogene *22*, 9048-9057.

Bhende, P. M., Seaman, W. T., Delecluse, H.-J., and Kenney, S. C. (2004). The EBV lytic switch protein, Z, preferentially binds to and activates the methylated viral genome. Nat. Genet. *36*, 1099-1104.

Biggin, M., Bodescot, M., Perricaudet, M., and Farrell, P. (1987). Epstein-Barr virus gene expression in P3HR1-superinfected Raji cells. J. Virol. *61*, 3120-3132.

Binne, U. K., Amon, W., and Farrell, P. J. (2002). Promoter sequences required for reactivation of Epstein-Barr virus from latency. J. Virol. *76*, 10282-10289.

Borza, C. M., and Hutt-Fletcher, L. M. (2002). Alternate replication in B cells and epithelial cells switches tropism of Epstein-Barr virus. Nat. Med. *8*, 594-599.

Boyer, J. L., Swaminathan, S., and Silverstein, S. J. (2002). The Epstein-Barr virus SM protein is functionally similar to ICP27 from herpes simplex virus in viral infections. J. Virol. *76*, 9420-9433.

Boyle, S. M., Ruvolo, V., Gupta, A. K., and Swaminathan, S. (1999). Association with the cellular export receptor CRM 1 mediates function and intracellular localization of Epstein-Barr virus SM protein, a regulator of gene expression. J. Virol. *73*, 6872-6881.

Bryant, H., and Farrell, P. J. (2002). Signal Transduction and Transcription Factor Modification during Reactivation of Epstein-Barr Virus from Latency. J. Virol. *76*, 10290-10298.

Buisson, M., Hans, F., Kusters, I., Duran, N., and Sergeant, A. (1999). The C-terminal region but not the Arg-X-Pro repeat of Epstein-Barr virus protein EB2 is required for its effect on RNA splicing and transport. J. Virol. *73*, 4090-4100.

Buisson, M., Manet, E., Trescol-Biemont, M. C., Gruffat, H., Durand, B., and Sergeant, A. (1989). The Epstein-Barr virus (EBV) early protein EB2 is a posttranscriptional activator expressed under the control of EBV transcription factors EB1 and R. J. Virol. *63*, 5276-5284.

Carey, M., Kolman, J., Katz, D. A., Gradoville, L., Barberis, L., and Miller, G. (1992). Transcriptional synergy by the Epstein-Barr virus transactivator ZEBRA. J. Virol. *66*, 4803-4813.

Cayrol, C., and Flemington, E. (1996a). G0/G1 growth arrest mediated by a region encompassing the basic leucine zipper (bZIP) domain of the Epstein-Barr virus transactivator Zta. J. Biol. Chem. *271*, 31799-31802.

Cayrol, C., and Flemington, E. K. (1995). Identification of cellular target genes of the Epstein-Barr virus transactivator Zta: activation of transforming growth factor beta igh3 (TGF-beta igh3) and TGF-beta 1. J. Virol. *69*, 4206-4212.

Cayrol, C., and Flemington, E. K. (1996b). The Epstein-Barr virus bZIP transcription factor Zta causes G0/G1 cell cycle arrest through induction of cyclin-dependent kinase inhibitors. Embo J. *15*, 2748-2759.

Chan, A. T., Tao, Q., Robertson, K. D., Flinn, I. W., Mann, R. B., Klencke, B., Kwan, W. H., Leung, T. W., Johnson, P. J., and Ambinder, R. F. (2004). Azacitidine induces demethylation of the Epstein-Barr virus genome in tumors. J. Clin. Oncol *22*, 1373-1381.

Chan, C. K., Mueller, N., Evans, A., Harris, N. L., Comstock, G. W., Jellum, E., Magnus, K., Orentreich, N., Polk, B. F., and Vogelman, J. (1991). Epstein-Barr virus antibody patterns preceding the diagnosis of nasopharyngeal carcinoma. Cancer Causes Control *2*, 125-131.

Chang, L. K., Lee, Y. H., Cheng, T. S., Hong, Y. R., Lu, P. J., Wang, J. J., Wang, W. H., Kuo, C. W., Li, S. S., and Liu, S. T. (2004). Post-translational Modification of Rta of Epstein-Barr Virus by SUMO-1. J. Biol. Chem. *279*, 38803-38812.

Chang, P. J., Chang, Y. S., and Liu, S. T. (1998). Role of Rta in the translation of bicistronic BZLF1 of Epstein-Barr virus. J. Virol. *72*, 5128-5136.

Chang, P. J., and Liu, S. T. (2001). Function of the intercistronic region of BRLF1-BZLF1 bicistronic mRNA in translating the zta protein of Epstein-Barr virus. J. Virol. *75*, 1142-1151.

Chang, Y. N., Dong, D. L., Hayward, G. S., and Hayward, S. D. (1990). The Epstein-Barr virus Zta transactivator: a member of the bZIP family with unique DNA-binding specificity and a dimerization domain that lacks the characteristic heptad leucine zipper motif. J. Virol. *64*, 3358-3369.

Chatila, T., Ho, N., Liu, P., Liu, S., Mosialos, G., Kieff, E., and Speck, S. H. (1997). The Epstein-Barr virus-induced Ca2+/calmodulin-dependent kinase type IV/Gr promotes a Ca(2+)-dependent switch from latency to viral replication. J. Virol. *71*, 6560-6567.

Chee, A. V., Lopez, P., Pandolfi, P. P., and Roizman, B. (2003). Promyelocytic leukemia protein mediates interferon-based anti-herpes simplex virus 1 effects. J. Virol. *77*, 7101-7105.

Chen, C. J., Deng, Z., Kim, A. Y., Blobel, G. A., and Lieberman, P. M. (2001a). Stimulation of CREB binding protein nucleosomal histone acetyltransferase activity by a class of transcriptional activators. Mol. Cell Biol. *21*, 476-487.

Chen, J. Y., Hwang, L. Y., Beasley, R. P., Chien, C. S., and Yang, C. S. (1985). Antibody response to Epstein-Barr-virus-specific DNase in 13 patients with nasopharyngeal carcinoma in Taiwan: a retrospective study. J. Med. Virol. *16*, 99-105.

Chen, L., Liao, G., Fujimuro, M., Semmes, O. J., and Hayward, S. D. (2001b). Properties of two EBV Mta nuclear export signal sequences. Virology *288*, 119-128.

Chen, M. R., Chang, S. J., Huang, H., and Chen, J. Y. (2000). A protein kinase activity associated with Epstein-Barr virus BGLF4 phosphorylates the viral early antigen EA-D *in vitro*. J. Virol. *74*, 3093-3104.

Chevallier-Greco, A., Gruffat, H., Manet, E., Calender, A., and Sergeant, A. (1989). The Epstein-Barr virus (EBV) DR enhancer contains two functionally different domains: domain A is constitutive and cell specific, domain B is transactivated by the EBV early protein R. J. Virol. *63*, 615-623.

Chevallier-Greco, A., Manet, E., Chavrier, P., Mosnier, C., Daillie, J., and Sergeant, A. (1986). Both Epstein-Barr virus (EBV)-encoded trans-acting factors, EB1 and EB2, are required to activate transcription from an EBV early promoter. Embo J. *5*, 3243-3249.

Chi, T., and Carey, M. (1993). The ZEBRA activation domain: modular organization and mechanism of action. Mol. Cell Biol. *13*, 7045-7055.

Chi, T., Lieberman, P., Ellwood, K., and Carey, M. (1995). A general mechanism for transcriptional synergy by eukaryotic activators. Nature *377*, 254-257.

Cho, M. S., Milman, G., and Hayward, S. D. (1985). A second Epstein-Barr virus early antigen gene in BamHI fragment M encodes a 48- to 50-kilodalton nuclear protein. J. Virol. *56*, 860-866.

Cohen, J. I. (2000). Epstein-Barr virus infection. N. Engl. J. Med. *343*, 481-492.

Cohen, J. I., and Lekstrom, K. (1999). Epstein-Barr virus BARF1 protein is dispensable for B-cell transformation and inhibits alpha interferon secretion from mononuclear cells. J. Virol. *73*, 7627-7632.

Cook, I. D., Shanahan, F., and Farrell, P. J. (1994). Epstein-Barr virus SM protein. Virology *205*, 217-227.

Countryman, J., and Miller, G. (1985). Activation of expression of latent Epstein-Barr herpesvirus after gene transfer with a small cloned subfragment of heterogeneous viral DNA. Proc. Natl. Acad. Sci. USA. *82*, 4085-4089.

Cox, M. A., Leahy, J., and Hardwick, J. M. (1990). An enhancer within the divergent promoter of Epstein-Barr virus responds synergistically to the R and Z transactivators. J. Virol. *64*, 313-321.

Daibata, M., and Sairenji, T. (1993). Epstein-Barr virus (EBV) replication and expressions of EA-D (BMRF1 gene product), virus-specific deoxyribonuclease, and DNA polymerase in EBV-activated Akata cells. Virology *196*, 900-904.

Daibata, M., Speck, S. H., Mulder, C., and Sairenji, T. (1994). Regulation of the BZLF1 promoter of Epstein-Barr virus by second messengers in anti-immunoglobulin-treated B cells. Virology *198*, 446-454.

Darenkov, I. A., Marcarelli, M. A., Basadonna, G. P., Friedman, A. L., Lorber, K. M., Howe, J. G., Crouch, J., Bia, M. J., Kliger, A. S., and Lorber, M. I. (1997). Reduced incidence of Epstein-Barr virus-associated posttransplant lymphoproliferative disorder using preemptive antiviral therapy. Transplantation *64*, 848-852.

Darr, C. D., Mauser, A., and Kenney, S. (2001). Epstein-Barr virus immediate-early protein BRLF1 induces the lytic form of viral replication through a mechanism involving phosphatidylinositol-3 kinase activation. J. Virol. *75*, 6135-6142.

Datta, A. K., Colby, B. M., Shaw, J. E., and Pagano, J. S. (1980). Acyclovir inhibition of Epstein-Barr virus replication. Proc. Natl. Acad. Sci. USA. *77*, 5163-5166.

Datta, A. K., and Hood, R. E. (1981). Mechanism of inhibition of Epstein-Barr virus replication by phosphonoformic acid. Virology *114*, 52-59.

Davie, J. R. (2003). Inhibition of histone deacetylase activity by butyrate. J. Nutr. *133*, 2485S-2493S.

Decker, L. L., Klaman, L. D., and Thorley-Lawson, D. A. (1996). Detection of the latent form of Epstein-Barr virus DNA in the peripheral blood of healthy individuals. J. Virol. *70*, 3286-3289.

Deng, Z., Chen, C. J., Chamberlin, M., Lu, F., Blobel, G. A., Speicher, D., Cirillo, L. A., Zaret, K. S., and Lieberman, P. M. (2003). The CBP bromodomain and nucleosome targeting are required for Zta-directed nucleosome acetylation and transcription activation. Mol. Cell Biol. *23*, 2633-2644.

Deng, Z., Chen, C. J., Zerby, D., Delecluse, H. J., and Lieberman, P. M. (2001). Identification of acidic and aromatic residues in the Zta activation domain essential for Epstein-Barr virus reactivation. J. Virol. *75*, 10334-10347.

El-Guindy, A. S., Heston, L., Endo, Y., Cho, M. S., and Miller, G. (2002). Disruption of Epstein-Barr virus latency in the absence of phosphorylation of ZEBRA by protein kinase C. J. Virol. *76*, 11199-11208.

El-Guindy, A. S., and Miller, G. (2004). Phosphorylation of Epstein-Barr virus ZEBRA protein at its casein kinase 2 sites mediates its ability to repress activation of a viral lytic cycle late gene by Rta. J. Virol. *78*, 7634-7644.

Fahmi, H., Cochet, C., Hmama, Z., Opolon, P., and Joab, I. (2000). Transforming growth factor beta 1 stimulates expression of the Epstein-Barr virus BZLF1 immediate-early gene product ZEBRA by an indirect mechanism which requires the MAPK kinase pathway. J. Virol. *74*, 5810-5818.

Falk, K. I., and Ernberg, I. (1999). Demethylation of the Epstein-barr virus origin of lytic replication and of the immediate early gene BZLF1 is DNA replication independent. Brief report. Arch. Virol. *144*, 2219-2227.

Faller, D. V., Mentzer, S. J., and Perrine, S. P. (2001). Induction of the Epstein-Barr virus thymidine kinase gene with concomitant nucleoside antivirals as a therapeutic strategy for Epstein-Barr virus-associated malignancies. Curr. Opin. Oncol *13*, 360-367.

Farjot, G., Buisson, M., Duc Dodon, M., Gazzolo, L., Sergeant, A., and Mikaelian, I. (2000). Epstein-Barr virus EB2 protein exports unspliced RNA via a Crm-1-independent pathway. J. Virol. *74*, 6068-6076.

Farmer, D. G., McDiarmid, S. V., Winston, D., Yersiz, H., Cortina, G., Dry, S., Maxfield, A. J., Vandenbogaart, B., Correa, M., Kroeber, A., et al. (2002). Effectiveness of aggressive prophylatic and preemptive therapies targeted against cytomegaloviral and Epstein-Barr viral disease after human intestinal transplantation. Transplant. Proc. *34*, 948-949.

Farrell, P. J., Rowe, D. T., Rooney, C. M., and Kouzarides, T. (1989). Epstein-Barr virus BZLF1 trans-activator specifically binds to a consensus AP-1 site and is related to c-fos. Embo J. *8*, 127-132.

Feederle, R., Kost, M., Baumann, M., Janz, A., Drouet, E., Hammerschmidt, W., and Delecluse, H. J. (2000). The Epstein-Barr virus lytic program is controlled by the co-operative functions of two transactivators. Embo J. *19*, 3080-3089.

Feng, W. H., Hong, G., Delecluse, H. J., and Kenney, S. C. (2004). Lytic Induction Therapy for Epstein-Barr Virus-Positive B-Cell Lymphomas. J. Virol. *78*, 1893-1902.

Feng, W. H., Israel, B., Raab-Traub, N., Busson, P., and Kenney, S. C. (2002a). Chemotherapy induces lytic EBV replication and confers ganciclovir susceptibility to EBV-positive epithelial cell tumors. Cancer Res. *62*, 1920-1926.

Feng, W. H., Westphal, E., Mauser, A., Raab-Traub, N., Gulley, M. L., Busson, P., and Kenney, S. C. (2002b). Use of adenovirus vectors expressing Epstein-Barr virus (EBV) immediate-early protein BZLF1 or BRLF1 to treat EBV-positive tumors. J. Virol. *76*, 10951-10959.

Fixman, E. D., Hayward, G. S., and Hayward, S. D. (1992). trans-acting requirements for replication of Epstein-Barr virus ori-Lyt. J. Virol. *66*, 5030-5039.

Fixman, E. D., Hayward, G. S., and Hayward, S. D. (1995). Replication of Epstein-Barr virus *oriLyt*: lack of a dedicated virally encoded origin-binding protein and dependence on Zta in cotransfection assays. J. Virol. *69*, 2998-3006.

Flamand, L., and Menezes, J. (1996). Cyclic AMP-responsive element-dependent activation of Epstein-Barr virus zebra promoter by human herpesvirus 6. J. Virol. *70*, 1784-1791.

Fleischmann, J., Kremmer, E., Greenspan, J. S., Grasser, F. A., and Niedobitek, G. (2002). Expression of viral and human dUTPase in Epstein-Barr virus-associated diseases. J. Med. Virol. *68*, 568-573.

Flemington, E., and Speck, S. H. (1990a). Autoregulation of Epstein-Barr virus putative lytic switch gene BZLF1. J. Virol. *64*, 1227-1232.

Flemington, E., and Speck, S. H. (1990b). Epstein-Barr virus BZLF1 trans activator induces the promoter of a cellular cognate gene, c-fos. J. Virol. *64*, 4549-4552.

Flemington, E., and Speck, S. H. (1990c). Evidence for coiled-coil dimer formation by an Epstein-Barr virus transactivator that lacks a heptad repeat of leucine residues. Proc. Natl. Acad. Sci. USA. *87*, 9459-9463.

Flemington, E., and Speck, S. H. (1990d). Identification of phorbol ester response elements in the promoter of Epstein-Barr virus putative lytic switch gene BZLF1. J. Virol. *64*, 1217-1226.

Flemington, E. K., Borras, A. M., Lytle, J. P., and Speck, S. H. (1992). Characterization of the Epstein-Barr virus BZLF1 protein transactivation domain. J. Virol. *66*, 922-929.

Flemington, E. K., Goldfeld, A. E., and Speck, S. H. (1991). Efficient transcription of the Epstein-Barr virus immediate-early BZLF1 and BRLF1 genes requires protein synthesis. J. Virol. *65*, 7073-7077.

Flemington, E. K., Lytle, J. P., Cayrol, C., Borras, A. M., and Speck, S. H. (1994). DNA-binding-defective mutants of the Epstein-Barr virus lytic switch activator Zta transactivate with altered specificities. Mol. Cell Biol. *14*, 3041-3052.

Fong, I. W., Ho, J., Toy, C., Lo, B., and Fong, M. W. (2000). Value of long-term administration of acyclovir and similar agents for protecting against AIDS-related lymphoma: case-control and historical cohort studies. Clin. Infect. Dis. *30*, 757-761.

Francis, A., Ragoczy, T., Gradoville, L., Heston, L., El-Guindy, A., Endo, Y., and Miller, G. (1999). Amino acid substitutions reveal distinct functions of serine 186 of the ZEBRA protein in activation of early lytic cycle genes and synergy with the Epstein-Barr virus R transactivator. J. Virol. *73*, 4543-4551.

Francis, A. L., Gradoville, L., and Miller, G. (1997). Alteration of a single serine in the basic domain of the Epstein-Barr virus ZEBRA protein separates its functions of transcriptional activation and disruption of latency. J. Virol. *71*, 3054-3061.

Fu, Z., and Cannon, M. J. (2000). Functional analysis of the CD4(+) T-cell response to Epstein-Barr virus: T-cell-mediated activation of resting B cells and induction of viral BZLF1 expression. J. Virol. *74*, 6675-6679.

Fujii, K., Yokoyama, N., Kiyono, T., Kuzushima, K., Homma, M., Nishiyama, Y., Fujita, M., and Tsurumi, T. (2000). The Epstein-Barr virus pol catalytic subunit physically interacts with the BBLF4-BSLF1-BBLF2/3 complex. J. Virol. *74*, 2550-2557.

Furnari, F. B., Zacny, V., Quinlivan, E. B., Kenney, S., and Pagano, J. S. (1994). RAZ, an Epstein-Barr virus transdominant repressor that modulates the viral reactivation mechanism. J. Virol. *68*, 1827-1836.

Gao, X., Tajima, M., and Sairenji, T. (1999). Nitric oxide down-regulates Epstein-Barr virus reactivation in epithelial cell lines. Virology *258*, 375-381.

Gao, Z., Krithivas, A., Finan, J. E., Semmes, O. J., Zhou, S., Wang, Y., and Hayward, S. D. (1998). The Epstein-Barr virus lytic transactivator Zta interacts with the helicase-primase replication proteins. J. Virol. *72*, 8559-8567.

Gershburg, E., and Pagano, J. S. (2002). Phosphorylation of the Epstein-Barr virus (EBV) DNA polymerase processivity factor EA-D by the EBV-encoded protein kinase and effects of the L-riboside benzimidazole 1263W94. J. Virol. *76*, 998-1003.

Glaser, G., Vogel, M., Wolf, H., and Niller, H. H. (1998). Regulation of the Epstein-Barr viral immediate early BRLF1 promoter through a distal NF1 site. Arch. Virol. *143*, 1967-1983.

Glover, J. N., and Harrison, S. C. (1995). Crystal structure of the heterodimeric bZIP transcription factor c-Fos-c-Jun bound to DNA. Nature *373*, 257-261.

Gold, M. R. (2002). To make antibodies or not: signaling by the B-cell antigen receptor. Trends Pharmacol Sci. *23*, 316-324.

Goldfeld, A. E., Liu, P., Liu, S., Flemington, E. K., Strominger, J. L., and Speck, S. H. (1995). Cyclosporin A and FK506 block induction of the Epstein-Barr virus lytic cycle by anti-immunoglobulin. Virology *209*, 225-229.

Grab, L. T., Kearns, M. W., Morris, A. J., and Daniel, L. W. (2004). Differential role for phospholipase D1 and phospholipase D2 in 12-O-tetradecanoyl-13-phorbol acetate-stimulated MAPK activation, Cox-2 and IL-8 expression. Biochim Biophys Acta *1636*, 29-39.

Gradoville, L., Kwa, D., El-Guindy, A., and Miller, G. (2002). Protein kinase C-independent activation of the Epstein-Barr virus lytic cycle. J. Virol. *76*, 5612-5626.

Green, M., Reyes, J., Webber, S., and Rowe, D. (2001). The role of antiviral and immunoglobulin therapy in the prevention of Epstein-Barr virus infection and post-transplant lymphoproliferative disease following solid organ transplantation. Transpl Infect. Dis. *3*, 97-103.

Greenspan, J. S., Greenspan, D., Lennette, E. T., Abrams, D. I., Conant, M. A., Petersen, V., and Freese, U. K. (1985). Replication of Epstein-Barr virus within the epithelial cells of oral "hairy" leukoplakia, an AIDS-associated lesion. N. Engl. J. Med. *313*, 1564-1571.

Gruffat, H., Batisse, J., Pich, D., Neuhierl, B., Manet, E., Hammerschmidt, W., and Sergeant, A. (2002a). Epstein-Barr virus mRNA export factor EB2 is essential for production of infectious virus. J. Virol. *76*, 9635-9644.

Gruffat, H., Duran, N., Buisson, M., Wild, F., Buckland, R., and Sergeant, A. (1992). Characterization of an R-binding site mediating the R-induced activation of the Epstein-Barr virus BMLF1 promoter. J. Virol. *66*, 46-52.

Gruffat, H., Manet, E., Rigolet, A., and Sergeant, A. (1990). The enhancer factor R of Epstein-Barr virus (EBV) is a sequence-specific DNA binding protein. Nucleic Acids Res. *18*, 6835-6843.

Gruffat, H., Manet, E., and Sergeant, A. (2002b). MEF2-mediated recruitment of class II HDAC at the EBV immediate early gene BZLF1 links latency and chromatin remodeling. EMBO Rep. *3*, 141-146.

Gruffat, H., Renner, O., Pich, D., and Hammerschmidt, W. (1995). Cellular proteins bind to the downstream component of the lytic origin of DNA replication of Epstein-Barr virus. J. Virol. *69*, 1878-1886.

Gruffat, H., and Sergeant, A. (1994). Characterization of the DNA-binding site repertoire for the Epstein-Barr virus transcription factor R. Nucleic Acids Res. *22*, 1172-1178.

Gustafson, E. A., Chillemi, A. C., Sage, D. R., and Fingeroth, J. D. (1998). The Epstein-Barr virus thymidine kinase does not phosphorylate ganciclovir or acyclovir and demonstrates a narrow substrate specificity compared to the herpes simplex virus type 1 thymidine kinase. Antimicrob Agents Chemother *42*, 2923-2931.

Gutierrez, M. I., Judde, J. G., Magrath, I. T., and Bhatia, K. G. (1996). Switching viral latency to viral lysis: a novel therapeutic approach for Epstein-Barr virus-associated neoplasia. Cancer Res. *56*, 969-972.

Gutsch, D. E., Holley-Guthrie, E. A., Zhang, Q., Stein, B., Blanar, M. A., Baldwin, A. S., and Kenney, S. C. (1994). The bZIP transactivator of Epstein-Barr virus, BZLF1, functionally and physically interacts with the p65 subunit of NF-kappa B. Mol. Cell Biol. *14*, 1939-1948.

Hammerschmidt, W., and Sugden, B. (1988). Identification and characterization of *oriLyt*, a lytic origin of DNA replication of Epstein-Barr virus. Cell *55*, 427-433.

Hardwick, J. M., Lieberman, P. M., and Hayward, S. D. (1988). A new Epstein-Barr virus transactivator, R, induces expression of a cytoplasmic early antigen. J. Virol. *62*, 2274-2284.

Hardwick, J. M., Tse, L., Applegren, N., Nicholas, J., and Veliuona, M. A. (1992). The Epstein-Barr virus R transactivator (Rta) contains a complex, potent activation domain with properties different from those of VP16. J. Virol. *66*, 5500-5508.

Henderson, S., Huen, D., Rowe, M., Dawson, C., Johnson, G., and Rickinson, A. (1993). Epstein-Barr virus-coded BHRF1 protein, a viral homologue of Bcl-2, protects human B cells from programmed cell death. Proc. Natl. Acad. Sci. USA. *90*, 8479-8483.

Herrmann, K., Frangou, P., Middeldorp, J., and Niedobitek, G. (2002). Epstein-Barr virus replication in tongue epithelial cells. J. Gen. Virol. *83*, 2995-2998.

Hiriart, E., Bardouillet, L., Manet, E., Gruffat, H., Penin, F., Montserret, R., Farjot, G., and Sergeant, A. (2003a). A region of the Epstein-Barr virus (EBV) mRNA export factor EB2 containing an arginine-rich motif mediates direct binding to RNA. J. Biol. Chem. *278*, 37790-37798.

Hiriart, E., Farjot, G., Gruffat, H., Nguyen, M. V., Sergeant, A., and Manet, E. (2003b). A novel nuclear export signal and a REF interaction domain both promote mRNA export by the Epstein-Barr virus EB2 protein. J. Biol. Chem. *278*, 335-342.

Hislop, A. D., Annels, N. E., Gudgeon, N. H., Leese, A. M., and Rickinson, A. B. (2002). Epitope-specific evolution of human CD8(+) T cell responses from primary to persistent phases of Epstein-Barr virus infection. J. Exp. Med. *195*, 893-905.

Holley-Guthrie, E. A., Quinlivan, E. B., Mar, E. C., and Kenney, S. (1990). The Epstein-Barr virus (EBV) BMRF1 promoter for early antigen (EA-D) is regulated by the EBV transactivators, BRLF1 and BZLF1, in a cell-specific manner. J. Virol. *64*, 3753-3759.

Holley-Guthrie, E. A., Seaman, W. T., Bhende, P., Merchant, J. L., and Kenney, S. (2005). The Epstein-Barr virus protein, BMRF1, activates gastrin transcription. J. Virol. *79*, 745-755.

Hong, G. K., Delecluse, H.-J., Gruffat, H., Morrison, T. E., Feng, W., and Kenney, S. C. (2004). The BRRF1 early gene of Epstein-Barr virus encodes a transcription factor that enhances induction of lytic infection by BRLF1. J. Virol. *78*, 4983-4992.

Hong, Y., Holley-Guthrie, E., and Kenney, S. (1997). The bZip dimerization domain of the Epstein-Barr virus BZLF1 (Z) protein mediates lymphoid-specific negative regulation. Virology *229*, 36-48.

Hsu, D. H., de Waal Malefyt, R., Fiorentino, D. F., Dang, M. N., Vieira, P., de Vries, J., Spits, H., Mosmann, T. R., and Moore, K. W. (1990). Expression of interleukin-10 activity by Epstein-Barr virus protein BCRF1. Science *250*, 830-832.

Ikuta, K., Satoh, Y., Hoshikawa, Y., and Sairenji, T. (2000). Detection of Epstein-Barr virus in salivas and throat washings in healthy adults and children. Microbes Infect. *2*, 115-120.

Inman, G. J., Binne, U. K., Parker, G. A., Farrell, P. J., and Allday, M. J. (2001). Activators of the Epstein-Barr virus lytic program concomitantly induce apoptosis, but lytic gene expression protects from cell death. J. Virol. *75*, 2400-2410.

Israel, B. F., and Kenney, S. C. (2003). Virally targeted therapies for EBV-associated malignancies. Oncogene *22*, 5122-5130.

Israele, V., Shirley, P., and Sixbey, J. W. (1991). Excretion of the Epstein-Barr virus from the genital tract of men. J. Infect. Dis. *163*, 1341-1343.

Iwakiri, D., and Takada, K. (2004). Phosphatidylinositol 3-kinase is a determinant of responsiveness to B cell antigen receptor-mediated Epstein-Barr virus activation. J. Immunol. *172*, 1561-1566.

Jenkins, P. J., Binne, U. K., and Farrell, P. J. (2000). Histone acetylation and reactivation of Epstein-Barr virus from latency. J. Virol. *74*, 710-720.

Jenson, H. B. (2000). Acute complications of Epstein-Barr virus infectious mononucleosis. Curr. Opin. Pediatr. *12*, 263-268.

Karimi, L., Crawford, D. H., Speck, S., and Nicholson, L. J. (1995). Identification of an epithelial cell differentiation responsive region within the BZLF1 promoter of the Epstein-Barr virus. J. Gen. Virol. *76 (Pt 4)*, 759-765.

Kato, K., Kawaguchi, Y., Tanaka, M., Igarashi, M., Yokoyama, A., Matsuda, G., Kanamori, M., Nakajima, K., Nishimura, Y., Shimojima, M., et al. (2001). Epstein-Barr virus-encoded protein kinase BGLF4 mediates hyperphosphorylation of cellular elongation factor 1delta (EF-1delta): EF-1delta is universally modified by conserved protein kinases of herpesviruses in mammalian cells. J. Gen. Virol. *82*, 1457-1463.

Kato, K., Yokoyama, A., Tohya, Y., Akashi, H., Nishiyama, Y., and Kawaguchi, Y. (2003). Identification of protein kinases responsible for phosphorylation of Epstein-Barr virus nuclear antigen leader protein at serine-35, which regulates its coactivator function. J. Gen. Virol. *84*, 3381-3392.

Kawaguchi, Y., Kato, K., Tanaka, M., Kanamori, M., Nishiyama, Y., and Yamanashi, Y. (2003). Conserved protein kinases encoded by herpesviruses and cellular protein kinase cdc2 target the same phosphorylation site in eukaryotic elongation factor 1delta. J. Virol. *77*, 2359-2368.

Kawanishi, M. (1995). Nitric oxide inhibits Epstein-Barr virus DNA replication and activation of latent EBV. Intervirology *38*, 206-213.

Keating, S., Prince, S., Jones, M., and Rowe, M. (2002). The lytic cycle of Epstein-Barr virus is associated with decreased expression of cell surface major histocompatibility complex class I and class II molecules. J. Virol. *76*, 8179-8188.

Kenney, S., Holley-Guthrie, E., Mar, E. C., and Smith, M. (1989a). The Epstein-Barr virus BMLF1 promoter contains an enhancer element that is responsive to the BZLF1 and BRLF1 transactivators. J. Virol. *63*, 3878-3883.

Kenney, S., Kamine, J., Holley-Guthrie, E., Lin, J. C., Mar, E. C., and Pagano, J. (1989b). The Epstein-Barr virus (EBV) BZLF1 immediate-early gene product differentially affects latent versus productive EBV promoters. J. Virol. *63*, 1729-1736.

Kenney, S., Kamine, J., Holley-Guthrie, E., Mar, E. C., Lin, J. C., Markovitz, D., and Pagano, J. (1989c). The Epstein-Barr virus immediate-early gene product, BMLF1, acts in trans by a posttranscriptional mechanism which is reporter gene dependent. J. Virol. *63*, 3870-3877.

Kiehl, A., and Dorsky, D. I. (1991). Cooperation of EBV DNA polymerase and EA-D(BMRF1) *in vitro* and colocalization in nuclei of infected cells. Virology *184*, 330-340.

Kiehl, A., and Dorsky, D. I. (1995). Bipartite DNA-binding region of the Epstein-Barr virus BMRF1 product essential for DNA polymerase accessory function. J. Virol. *69*, 1669-1677.

Kouzarides, T., Packham, G., Cook, A., and Farrell, P. J. (1991). The BZLF1 protein of EBV has a coiled coil dimerisation domain without a heptad leucine repeat but with homology to the C/EBP leucine zipper. Oncogene *6*, 195-204.

Krappmann, D., Patke, A., Heissmeyer, V., and Scheidereit, C. (2001). B-cell receptor- and phorbol ester-induced NF-kappaB and c-Jun N-terminal kinase activation in B cells requires novel protein kinase C's. Mol. Cell Biol. *21*, 6640-6650.

Kraus, R. J., Mirocha, S. J., Stephany, H. M., Puchalski, J. R., and Mertz, J. E. (2001). Identification of a novel element involved in regulation of the lytic switch BZLF1 gene promoter of Epstein-Barr virus. J. Virol. *75*, 867-877.

Kraus, R. J., Perrigoue, J. G., and Mertz, J. E. (2003). ZEB negatively regulates the lytic-switch BZLF1 gene promoter of Epstein-Barr virus. J. Virol. *77*, 199-207.

Kudoh, A., Daikoku, T., Sugaya, Y., Isomura, H., Fujita, M., Kiyono, T., Nishiyama, Y., and Tsurumi, T. (2004). Inhibition of S-phase cyclin-dependent kinase activity blocks expression of Epstein-Barr virus immediate-early and early genes, preventing viral lytic replication. J. Virol. *78*, 104-115.

Kudoh, A., Fujita, M., Kiyono, T., Kuzushima, K., Sugaya, Y., Izuta, S., Nishiyama, Y., and Tsurumi, T. (2003). Reactivation of lytic replication from B cells latently infected with Epstein-Barr virus occurs with high S-phase cyclin-dependent kinase activity while inhibiting cellular DNA replication. J. Virol. *77*, 851-861.

Lau, R., Middeldorp, J., and Farrell, P. J. (1993). Epstein-Barr virus gene expression in oral hairy leukoplakia. Virology *195*, 463-474.

Le Roux, F., Sergeant, A., and Corbo, L. (1996). Epstein-Barr virus (EBV) EB1/Zta protein provided in trans and competent for the activation of productive cycle genes does not activate the BZLF1 gene in the EBV genome. J. Gen. Virol. *77 (Pt 3)*, 501-509.

Li, M., Linseman, D. A., Allen, M. P., Meintzer, M. K., Wang, X., Laessig, T., Wierman, M. E., and Heidenreich, K. A. (2001). Myocyte enhancer factor 2A and 2D undergo phosphorylation and caspase-mediated degradation during apoptosis of rat cerebellar granule neurons. J. Neurosci *21*, 6544-6552.

Li, Q. X., Young, L. S., Niedobitek, G., Dawson, C. W., Birkenbach, M., Wang, F., and Rickinson, A. B. (1992). Epstein-Barr virus infection and replication in a human epithelial cell system. Nature *356*, 347-350.

Li, Y., Mahajan, N., Webster-Cyriaque, J., Bhende, P., Hong, G., Earp, H. S., and Kenney, S. (2004a). C-mer is Induced by the Epstein-Barr Virus Immediate-early Protein BRLF1. J. Virol. *78*, 11778-11785.

Li, Y., Webster-Cyriaque, J., Tomlinson, C. C., Yohe, M., and Kenney, S. (2004b). Fatty acid synthase expression is induced by the Epstein-Barr virus immediate-early protein BRLF1 and required for lytic viral gene expression. J.Virol. *78*, 4197-4206.

Liang, C. L., Chen, J. L., Hsu, Y. P., Ou, J. T., and Chang, Y. S. (2002). Epstein-Barr virus BZLF1 gene is activated by transforming growth factor-beta through cooperativity of Smads and c-Jun/c-Fos proteins. J. Biol. Chem. *277*, 23345-23357.

Liao, G., Wu, F. Y., and Hayward, S. D. (2001). Interaction with the Epstein-Barr virus helicase targets Zta to DNA replication compartments. J. Virol. *75*, 8792-8802.

Lieberman, P. M., and Berk, A. J. (1990). In vitro transcriptional activation, dimerization, and DNA-binding specificity of the Epstein-Barr virus Zta protein. J. Virol. *64*, 2560-2568.

Lieberman, P. M., and Berk, A. J. (1991). The Zta trans-activator protein stabilizes TFIID association with promoter DNA by direct protein-protein interaction. Genes Dev. *5*, 2441-2454.

Lieberman, P. M., and Berk, A. J. (1994). A mechanism for TAFs in transcriptional activation: activation domain enhancement of TFIID-TFIIA--promoter DNA complex formation. Genes Dev. *8*, 995-1006.

Lieberman, P. M., Hardwick, J. M., Sample, J., Hayward, G. S., and Hayward, S. D. (1990). The zta transactivator involved in induction of lytic cycle gene expression in Epstein-Barr virus-infected lymphocytes binds to both AP-1 and ZRE sites in target promoter and enhancer regions. J. Virol. *64*, 1143-1155.

Lieberman, P. M., Ozer, J., and Gursel, D. B. (1997). Requirement for transcription factor IIA (TFIIA)-TFIID recruitment by an activator depends on promoter structure and template competition. Mol. Cell Biol. *17*, 6624-6632.

Lin, J. C., DeClercq, E., and Pagano, J. S. (1987). Novel acyclic adenosine analogs inhibit Epstein-Barr virus replication. Antimicrob Agents Chemother *31*, 1431-1433.

Lin, J. C., and Machida, H. (1988). Comparison of two bromovinyl nucleoside analogs, 1-beta-D-arabinofuranosyl-E-5-(2-bromovinyl)uracil and E-5-(2-bromovinyl)-2'-deoxyuridine, with acyclovir in inhibition of Epstein-Barr virus replication. Antimicrob Agents Chemother *32*, 1068-1072.

Lin, J. C., Nelson, D. J., Lambe, C. U., and Choi, E. I. (1986). Metabolic activation of 9([2-hydroxy-1-(hydroxymethyl)ethoxy]methyl)guanine in human lymphoblastoid cell lines infected with Epstein-Barr virus. J. Virol. *60*, 569-573.

Ling, P. D., Lednicky, J. A., Keitel, W. A., Poston, D. G., White, Z. S., Peng, R., Liu, Z., Mehta, S. K., Pierson, D. L., Rooney, C. M., et al. (2003). The dynamics of herpesvirus and polyomavirus reactivation and shedding in healthy adults: a 14-month longitudinal study. J. Infect. Dis. *187*, 1571-1580.

Littler, E., and Arrand, J. R. (1988). Characterization of the Epstein-Barr virus-encoded thymidine kinase expressed in heterologous eucaryotic and procaryotic systems. J. Virol. *62*, 3892-3895.

Littler, E., Zeuthen, J., McBride, A. A., Trost Sorensen, E., Powell, K. L., Walsh-Arrand, J. E., and Arrand, J. R. (1986). Identification of an Epstein-Barr virus-coded thymidine kinase. Embo J. *5*, 1959-1966.

Liu, C., Sista, N. D., and Pagano, J. S. (1996). Activation of the Epstein-Barr virus DNA polymerase promoter by the BRLF1 immediate-early protein is mediated through USF and E2F. J. Virol. *70*, 2545-2555.

Liu, P., Liu, S., and Speck, S. H. (1998). Identification of a negative cis element within the ZII domain of the Epstein-Barr virus lytic switch BZLF1 gene promoter. J. Virol. *72*, 8230-8239.

Liu, P., and Speck, S. H. (2003). Synergistic autoactivation of the Epstein-Barr virus immediate-early BRLF1 promoter by Rta and Zta. Virology *310*, 199-206.

Liu, S., Borras, A. M., Liu, P., Suske, G., and Speck, S. H. (1997a). Binding of the ubiquitous cellular transcription factors Sp1 and Sp3 to the ZI domains in the Epstein-Barr virus lytic switch BZLF1 gene promoter. Virology *228*, 11-18.

Liu, S., Liu, P., Borras, A., Chatila, T., and Speck, S. H. (1997b). Cyclosporin A-sensitive induction of the Epstein-Barr virus lytic switch is mediated via a novel pathway involving a MEF2 family member. Embo J. *16*, 143-153.

Lu, J., Chen, S. Y., Chua, H. H., Liu, Y. S., Huang, Y. T., Chang, Y., Chen, J. Y., Sheen, T. S., and Tsai, C. H. (2000). Upregulation of tyrosine kinase TKT by the Epstein-Barr virus transactivator Zta. J. Virol. *74*, 7391-7399.

Lu, J., Chua, H. H., Chen, S. Y., Chen, J. Y., and Tsai, C. H. (2003). Regulation of matrix metalloproteinase-1 by Epstein-Barr virus proteins. Cancer Res. *63*, 256-262.

Lucht, E., Biberfeld, P., and Linde, A. (1995). Epstein-Barr virus (EBV) DNA in saliva and EBV serology of HIV-1-infected persons with and without hairy leukoplakia. J. Infect. *31*, 189-194.

MacCallum, P., Karimi, L., and Nicholson, L. J. (1999). Definition of the transcription factors which bind the differentiation responsive element of the Epstein-Barr virus BZLF1 Z promoter in human epithelial cells. J. Gen. Virol. *80 (Pt 6)*, 1501-1512.

Mahot, S., Sergeant, A., Drouet, E., and Gruffat, H. (2003). A novel function for the Epstein-Barr virus transcription factor EB1/Zta: induction of transcription of the hIL-10 gene. J. Gen. Virol. *84*, 965-974.

Makhov, A. M., Subramanian, D., Holley-Guthrie, E. A., Kenney, S. C., and Griffith, J. D. (2004). The Epstein-Barr virus polymerase accessory factor, BMRF1 adopts a ring shaped structure as visualized by electron microscopy. J. Biol. Chem.

Malouf, M. A., Chhajed, P. N., Hopkins, P., Plit, M., Turner, J., and Glanville, A. R. (2002). Anti-viral prophylaxis reduces the incidence of lymphoproliferative disease in lung transplant recipients. J. Heart Lung Transplant. *21*, 547-554.

Manet, E., Allera, C., Gruffat, H., Mikaelian, I., Rigolet, A., and Sergeant, A. (1993). The acidic activation domain of the Epstein-Barr virus transcription factor R interacts *in vitro* with both TBP and TFIIB and is cell-specifically potentiated by a proline-rich region. Gene Expr *3*, 49-59.

Manet, E., Gruffat, H., Trescol-Biemont, M. C., Moreno, N., Chambard, P., Giot, J. F., and Sergeant, A. (1989). Epstein-Barr virus bicistronic mRNAs generated by facultative splicing code for two transcriptional trans-activators. Embo J. *8*, 1819-1826.

Manet, E., Rigolet, A., Gruffat, H., Giot, J. F., and Sergeant, A. (1991). Domains of the Epstein-Barr virus (EBV) transcription factor R required for dimerization, DNA binding and activation. Nucleic Acids Res. *19*, 2661-2667.

Marschall, M., Stein-Gerlach, M., Freitag, M., Kupfer, R., van den Bogaard, M., and Stamminger, T. (2002). Direct targeting of human cytomegalovirus protein kinase pUL97 by kinase inhibitors is a novel principle for antiviral therapy. J. Gen. Virol. *83*, 1013-1023.

Marshall, W. L., Yim, C., Gustafson, E., Graf, T., Sage, D. R., Hanify, K., Williams, L., Fingeroth, J., and Finberg, R. W. (1999). Epstein-Barr virus encodes a novel homolog of the bcl-2 oncogene that inhibits apoptosis and associates with Bax and Bak. J. Virol. *73*, 5181-5185.

Matthews, R. P., Guthrie, C. R., Wailes, L. M., Zhao, X., Means, A. R., and McKnight, G. S. (1994). Calcium/calmodulin-dependent protein kinase types II and IV differentially regulate CREB-dependent gene expression. Mol. Cell Biol. *14*, 6107-6116.

Mauser, A., Holley-Guthrie, E., Simpson, D., Kaufmann, W., and Kenney, S. (2002a). The Epstein-Barr virus immediate-early protein BZLF1 induces both a G(2) and a mitotic block. J. Virol. *76*, 10030-10037.

Mauser, A., Holley-Guthrie, E., Zanation, A., Yarborough, W., Kaufmann, W., Klingelhutz, A., Seaman, W. T., and Kenney, S. (2002b). The Epstein-Barr virus immediate-early protein BZLF1 induces expression of E2F-1 and other proteins involved in cell cycle progression in primary keratinocytes and gastric carcinoma cells. J. Virol. *76*, 12543-12552.

Mauser, A., Saito, S., Appella, E., Anderson, C. W., Seaman, W. T., and Kenney, S. (2002c). The Epstein-Barr virus immediate-early protein BZLF1 regulates p53 function through multiple mechanisms. J. Virol. *76*, 12503-12512.

McDiarmid, S. V., Jordan, S., Kim, G. S., Toyoda, M., Goss, J. A., Vargas, J. H., Martin, M. G., Bahar, R., Maxfield, A. L., Ament, M. E., et al. (1998). Prevention and preemptive therapy of posttransplant lymphoproliferative disease in pediatric liver recipients. Transplantation *66*, 1604-1611.

McKinsey, T. A., Zhang, C. L., Lu, J., and Olson, E. N. (2000). Signal-dependent nuclear export of a histone deacetylase regulates muscle differentiation. Nature *408*, 106-111.

Meerbach, A., Holy, A., Wutzler, P., De Clercq, E., and Neyts, J. (1998). Inhibitory effects of novel nucleoside and nucleotide analogues on Epstein-Barr virus replication. Antivir Chem. Chemother *9*, 275-282.

Mellinghoff, I., Daibata, M., Humphreys, R. E., Mulder, C., Takada, K., and Sairenji, T. (1991). Early events in Epstein-Barr virus genome expression after activation: regulation by second messengers of B cell activation. Virology *185*, 922-928.

Mentzer, S. J., Fingeroth, J., Reilly, J. J., Perrine, S. P., and Faller, D. V. (1998). Arginine butyrate-induced susceptibility to ganciclovir in an Epstein-Barr-virus-associated lymphoma. Blood Cells Mol. Dis. *24*, 114-123.

Mentzer, S. J., Perrine, S. P., and Faller, D. V. (2001). Epstein--Barr virus post-transplant lymphoproliferative disease and virus-specific therapy: pharmacological re-activation of viral target genes with arginine butyrate. Transpl Infect. Dis. *3*, 177-185.

Mikaelian, I., Manet, E., and Sergeant, A. (1993). The bZIP motif of the Epstein-Barr virus (EBV) transcription factor EB1 mediates a direct interaction with TBP. C R Acad. Sci. III *316*, 1424-1432.

Miller, C. L., Burkhardt, A. L., Lee, J. H., Stealey, B., Longnecker, R., Bolen, J. B., and Kieff, E. (1995). Integral membrane protein 2 of Epstein-Barr virus regulates reactivation from latency through dominant negative effects on protein-tyrosine kinases. Immunity *2*, 155-166.

Miller, C. L., Lee, J. H., Kieff, E., Burkhardt, A. L., Bolen, J. B., and Longnecker, R. (1994a). Epstein-Barr virus protein LMP2A regulates reactivation from latency by negatively regulating tyrosine kinases involved in sIg-mediated signal transduction. Infect. Agents Dis. *3*, 128-136.

Miller, C. L., Lee, J. H., Kieff, E., and Longnecker, R. (1994b). An integral membrane protein (LMP2) blocks reactivation of Epstein-Barr virus from latency following surface immunoglobulin crosslinking. Proc. Natl. Acad. Sci. USA. *91*, 772-776.

Mitsouras, K., Wong, B., Arayata, C., Johnson, R. C., and Carey, M. (2002). The DNA architectural protein HMGB1 displays two distinct modes of action that promote enhanceosome assembly. Mol. Cell Biol. *22*, 4390-4401.

Miyazaki, I., Cheung, R. K., and Dosch, H. M. (1993). Viral interleukin 10 is critical for the induction of B cell growth transformation by Epstein-Barr virus. J. Exp. Med. *178*, 439-447.

Montalvo, E. A., Cottam, M., Hill, S., and Wang, Y. J. (1995). YY1 binds to and regulates cis-acting negative elements in the Epstein-Barr virus BZLF1 promoter. J. Virol. *69*, 4158-4165.

Montalvo, E. A., Shi, Y., Shenk, T. E., and Levine, A. J. (1991). Negative regulation of the BZLF1 promoter of Epstein-Barr virus. J. Virol. *65*, 3647-3655.

Moore, S. M., Cannon, J. S., Tanhehco, Y. C., Hamzeh, F. M., and Ambinder, R. F. (2001). Induction of Epstein-Barr virus kinases to sensitize tumor cells to nucleoside analogues. Antimicrob Agents Chemother *45*, 2082-2091.

Morrison, T. E., and Kenney, S. C. (2004). BZLF1, an Epstein-Barr virus immediate-early protein induces p65 nuclear translocation while inhibiting p65 transcriptional function. Virol. *328*, 219-232.

Morrison, T. E., Mauser, A., Klingelhutz, A., and Kenney, S. C. (2004). Epstein-Barr virus immediate-early protein BZLF1 inhibits tumor necrosis factor alpha-induced signaling and apoptosis by downregulating tumor necrosis factor receptor 1. J. Virol. *78*, 544-549.

Morrison, T. E., Mauser, A., Wong, A., Ting, J. P., and Kenney, S. C. (2001). Inhibition of IFN-gamma signaling by an Epstein-Barr virus immediate-early protein. Immunity *15*, 787-799.

Mueller, N., Evans, A., Harris, N. L., Comstock, G. W., Jellum, E., Magnus, K., Orentreich, N., Polk, B. F., and Vogelman, J. (1989). Hodgkin's disease and Epstein-Barr virus. Altered antibody pattern before diagnosis. N. Engl. J. Med. *320*, 689-695.

Mueller, N., Mohar, A., Evans, A., Harris, N. L., Comstock, G. W., Jellum, E., Magnus, K., Orentreich, N., Polk, B. F., and Vogelman, J. (1991). Epstein-Barr virus antibody patterns preceding the diagnosis of non-Hodgkin's lymphoma. Int. J. Cancer *49*, 387-393.

Niedobitek, G., Agathanggelou, A., Herbst, H., Whitehead, L., Wright, D. H., and Young, L. S. (1997). Epstein-Barr virus (EBV) infection in infectious mononucleosis: virus latency, replication and phenotype of EBV-infected cells. J. Pathol. *182*, 151-159.

Niedobitek, G., Agathanggelou, A., Steven, N., and Young, L. S. (2000). Epstein-Barr virus (EBV) in infectious mononucleosis: detection of the virus in tonsillar B lymphocytes but not in desquamated oropharyngeal epithelial cells. Mol. Pathol. *53*, 37-42.

Nonkwelo, C. B., and Long, W. K. (1993). Regulation of Epstein-Barr virus BamHI-H divergent promoter by DNA methylation. Virology *197*, 205-215.

Oertel, S. H., Ruhnke, M. S., Anagnostopoulos, I., Kahl, A. A., Frewer, A. F., Bechstein, W. O., Hummel, M. W., and Riess, H. B. (1999). Treatment of Epstein-Barr virus-induced posttransplantation lymphoproliferative disorder with foscarnet alone in an adult after simultaneous heart and renal transplantation. Transplantation *67*, 765-767.

Packham, G., Economou, A., Rooney, C. M., Rowe, D. T., and Farrell, P. J. (1990). Structure and function of the Epstein-Barr virus BZLF1 protein. J. Virol. *64*, 2110-2116.

Paulson, E. J., Fingeroth, J. D., Yates, J. L., and Speck, S. H. (2002). Methylation of the EBV genome and establishment of restricted latency in low-passage EBV-infected 293 epithelial cells. Virology *299*, 109-121.

Paulson, E. J., and Speck, S. H. (1999). Differential methylation of Epstein-Barr virus latency promoters facilitates viral persistence in healthy seropositive individuals. J. Virol. *73*, 9959-9968.

Poppers, J., Mulvey, M., Perez, C., Khoo, D., and Mohr, I. (2003). Identification of a lytic-cycle Epstein-Barr virus gene product that can regulate PKR activation. J. Virol. *77*, 228-236.

Portes-Sentis, S., Sergeant, A., and Gruffat, H. (1997). A particular DNA structure is required for the function of a cis-acting component of the Epstein-Barr virus *OriLyt* origin of replication. Nucleic Acids Res. *25*, 1347-1354.

Quinlivan, E. B., Holley-Guthrie, E. A., Norris, M., Gutsch, D., Bachenheimer, S. L., and Kenney, S. C. (1993). Direct BRLF1 binding is required for cooperative BZLF1/BRLF1 activation of the Epstein-Barr virus early promoter, BMRF1. Nucleic Acids Res. *21*, 1999-2007.

Ragoczy, T., Heston, L., and Miller, G. (1998). The Epstein-Barr virus Rta protein activates lytic cycle genes and can disrupt latency in B lymphocytes. J. Virol. *72*, 7978-7984.

Ragoczy, T., and Miller, G. (1999). Role of the epstein-barr virus RTA protein in activation of distinct classes of viral lytic cycle genes. J. Virol. *73*, 9858-9866.

Ragoczy, T., and Miller, G. (2001). Autostimulation of the Epstein-Barr virus BRLF1 promoter is mediated through consensus Sp1 and Sp3 binding sites. J. Virol. *75*, 5240-5251.

Resnick, L., Herbst, J. S., and Raab-Traub, N. (1990). Oral hairy leukoplakia. J. Amer. Acad. Dermatol *22*, 1278-1282.

Rickinson, A., and Kieff, E. (2001). Epstein-Barr virus. In Fields virology, H. P, ed. (Philadelphia Pa, Lippincott-Raven publishers).

Rodriguez, A., Armstrong, M., Dwyer, D., and Flemington, E. (1999). Genetic dissection of cell growth arrest functions mediated by the Epstein-Barr virus lytic gene product, Zta. J. Virol. *73*, 9029-9038.

Rodriguez, A., Jung, E. J., and Flemington, E. K. (2001). Cell cycle analysis of Epstein-Barr virus-infected cells following treatment with lytic cycle-inducing agents. J. Virol. *75*, 4482-4489.

Rooney, C. M., Rowe, D. T., Ragot, T., and Farrell, P. J. (1989). The spliced BZLF1 gene of Epstein-Barr virus (EBV) transactivates an early EBV promoter and induces the virus productive cycle. J. Virol. *63*, 3109-3116.

Rousset, F., Garcia, E., Defrance, T., Peronne, C., Vezzio, N., Hsu, D. H., Kastelein, R., Moore, K. W., and Banchereau, J. (1992). Interleukin 10 is a potent growth and differentiation factor for activated human B lymphocytes. Proc. Natl. Acad. Sci. USA. *89*, 1890-1893.

Roychowdhury, S., Peng, R., Baiocchi, R. A., Bhatt, D., Vourganti, S., Grecula, J., Gupta, N., Eisenbeis, C. F., Nuovo, G. J., Yang, W., et al. (2003). Experimental treatment of Epstein-Barr virus-associated primary central nervous system lymphoma. Cancer Res. *63*, 965-971.

Ruvolo, V., Gupta, A. K., and Swaminathan, S. (2001). Epstein-Barr virus SM protein interacts with mRNA *in vivo* and mediates a gene-specific increase in cytoplasmic mRNA. J. Virol. *75*, 6033-6041.

Ruvolo, V., Navarro, L., Sample, C. E., David, M., Sung, S., and Swaminathan, S. (2003). The Epstein-Barr virus SM protein induces STAT1 and interferon-stimulated gene expression. J. Virol. *77*, 3690-3701.

Ruvolo, V., Sun, L., Howard, K., Sung, S., Delecluse, H. J., Hammerschmidt, W., and Swaminathan, S. (2004). Functional analysis of Epstein-Barr virus SM protein: identification of amino acids essential for structure, transactivation, splicing inhibition, and virion production. J. Virol. *78*, 340-352.

Ruvolo, V., Wang, E., Boyle, S., and Swaminathan, S. (1998). The Epstein-Barr virus nuclear protein SM is both a post-transcriptional inhibitor and activator of gene expression. Proc. Natl. Acad. Sci. USA. *95*, 8852-8857.

Salek-Ardakani, S., Arrand, J. R., and Mackett, M. (2002). Epstein-Barr virus encoded interleukin-10 inhibits HLA-class I, ICAM-1, and B7 expression on human monocytes: implications for immune evasion by EBV. Virology *304*, 342-351.

Salomoni, P., and Pandolfi, P. P. (2002). The role of PML in tumor suppression. Cell *108*, 165-170.

Sarisky, R. T., Gao, Z., Lieberman, P. M., Fixman, E. D., Hayward, G. S., and Hayward, S. D. (1996). A replication function associated with the activation domain of the Epstein-Barr virus Zta transactivator. J. Virol. *70*, 8340-8347.

Schepers, A., Pich, D., and Hammerschmidt, W. (1993a). A transcription factor with homology to the AP-1 family links RNA transcription and DNA replication in the lytic cycle of Epstein-Barr virus. Embo J. *12*, 3921-3929.

Schepers, A., Pich, D., and Hammerschmidt, W. (1996). Activation of *oriLyt*, the lytic origin of DNA replication of Epstein-Barr virus, by BZLF1. Virology *220*, 367-376.

Schepers, A., Pich, D., Mankertz, J., and Hammerschmidt, W. (1993b). cis-acting elements in the lytic origin of DNA replication of Epstein-Barr virus. J. Virol. *67*, 4237-4245.

Segouffin, C., Gruffat, H., and Sergeant, A. (1996). Repression by RAZ of Epstein-Barr virus bZIP transcription factor EB1 is dimerization independent. J. Gen. Virol. *77 (Pt 7)*, 1529-1536.

Segouffin-Cariou, C., Farjot, G., Sergeant, A., and Gruffat, H. (2000). Characterization of the epstein-barr virus BRRF1 gene, located between early genes BZLF1 and BRLF1. J. Gen. Virol. *81*, 1791-1799.

Semmes, O. J., Chen, L., Sarisky, R. T., Gao, Z., Zhong, L., and Hayward, S. D. (1998). Mta has properties of an RNA export protein and increases cytoplasmic accumulation of Epstein-Barr virus replication gene mRNA. J. Virol. *72*, 9526-9534.

Serio, T. R., Cahill, N., Prout, M. E., and Miller, G. (1998). A functionally distinct TATA box required for late progression through the Epstein-Barr virus life cycle. J. Virol. *72*, 8338-8343.

Serio, T. R., Kolman, J. L., and Miller, G. (1997). Late gene expression from the Epstein-Barr virus BcLF1 and BFRF3 promoters does not require DNA replication in cis. J. Virol. *71*, 8726-8734.

Shimizu, N., and Takada, K. (1993). Analysis of the BZLF1 promoter of Epstein-Barr virus: identification of an anti-immunoglobulin response sequence. J. Virol. *67*, 3240-3245.

Shu, C. H., Chang, Y. S., Liang, C. L., Liu, S. T., Lin, C. Z., and Chang, P. (1992). Distribution of type A and type B EBV in normal individuals and patients with head and neck carcinomas in Taiwan. J. Virol. Methods *38*, 123-130.

Sinclair, A. J., Brimmell, M., Shanahan, F., and Farrell, P. J. (1991). Pathways of activation of the Epstein-Barr virus productive cycle. J. Virol. *65*, 2237-2244.

Sista, N. D., Pagano, J. S., Liao, W., and Kenney, S. (1993). Retinoic acid is a negative regulator of the Epstein-Barr virus protein (BZLF1) that mediates disruption of latent infection. Proc. Natl. Acad. Sci. USA. *90*, 3894-3898.

Sixbey, J. W., Lemon, S. M., and Pagano, J. S. (1986). A second site for Epstein-Barr virus shedding: the uterine cervix. Lancet *2*, 1122-1124.

Sixbey, J. W., Nedrud, J. G., Raab-Traub, N., Hanes, R. A., and Pagano, J. S. (1984). Epstein-Barr virus replication in oropharyngeal epithelial cells. N. Engl. J. Med. *310*, 1225-1230.

Speck, S. H., Chatila, T., and Flemington, E. (1997). Reactivation of Epstein-Barr virus: regulation and function of the BZLF1 gene. Trends Microbiol. *5*, 399-405.

Steven, N. M., Annels, N. E., Kumar, A., Leese, A. M., Kurilla, M. G., and Rickinson, A. B. (1997). Immediate early and early lytic cycle proteins are frequent targets of the Epstein-Barr virus-induced cytotoxic T cell response. J. Exp. Med. *185*, 1605-1617.

Strockbine, L. D., Cohen, J. I., Farrah, T., Lyman, S. D., Wagener, F., DuBose, R. F., Armitage, R. J., and Spriggs, M. K. (1998). The Epstein-Barr virus BARF1 gene encodes a novel, soluble colony-stimulating factor-1 receptor. J. Virol. *72*, 4015-4021.

Stuart, A. D., Stewart, J. P., Arrand, J. R., and Mackett, M. (1995). The Epstein-Barr virus encoded cytokine viral interleukin-10 enhances transformation of human B lymphocytes. Oncogene *11*, 1711-1719.

Swaminathan, S., Hesselton, R., Sullivan, J., and Kieff, E. (1993). Epstein-Barr virus recombinants with specifically mutated BCRF1 genes. J. Virol. *67*, 7406-7413.

Swenson, J. J., Holley-Guthrie, E., and Kenney, S. C. (2001). Epstein-Barr virus immediate-early protein BRLF1 interacts with CBP, promoting enhanced BRLF1 transactivation. J. Virol. *75*, 6228-6234.

Swenson, J. J., Mauser, A. E., Kaufmann, W. K., and Kenney, S. C. (1999). The Epstein-Barr virus protein BRLF1 activates S phase entry through E2F1 induction. J. Virol. *73*, 6540-6550.

Szyf, M., Eliasson, L., Mann, V., Klein, G., and Razin, A. (1985). Cellular and viral DNA hypomethylation associated with induction of Epstein-Barr virus lytic cycle. Proc. Natl. Acad. Sci. USA. *82*, 8090-8094.

Takada, K., and Ono, Y. (1989). Synchronous and sequential activation of latently infected Epstein-Barr virus genomes. J. Virol. *63*, 445-449.

Takada, K., Shimizu, N., Sakuma, S., and Ono, Y. (1986). trans activation of the latent Epstein-Barr virus (EBV) genome after transfection of the EBV DNA fragment. J. Virol. *57*, 1016-1022.

Takagi, S., Takada, K., and Sairenji, T. (1991). Formation of intranuclear replication compartments of Epstein-Barr virus with redistribution of BZLF1 and BMRF1 gene products. Virology *185*, 309-315.

Tarodi, B., Subramanian, T., and Chinnadurai, G. (1994). Epstein-Barr virus BHRF1 protein protects against cell death induced by DNA-damaging agents and heterologous viral infection. Virology *201*, 404-407.

Thomas, C., Dankesreiter, A., Wolf, H., and Schwarzmann, F. (2003). The BZLF1 promoter of Epstein-Barr virus is controlled by E box-/HI-motif-binding factors during virus latency. J. Gen. Virol. *84*, 959-964.

Torre, D., and Tambini, R. (1999). Acyclovir for treatment of infectious mononucleosis: a meta-analysis. Scand. J. Infect. Dis. *31*, 543-547.

Tovey, M. G., Lenoir, G., and Begon-Lours, J. (1978). Activation of latent Epstein-Barr virus by antibody to human IgM. Nature *276*, 270-272.

Tsurumi, T., Daikoku, T., Kurachi, R., and Nishiyama, Y. (1993). Functional interaction between Epstein-Barr virus DNA polymerase catalytic subunit and its accessory subunit *in vitro*. J. Virol. *67*, 7648-7653.

Tsurumi, T., Daikoku, T., and Nishiyama, Y. (1994). Further characterization of the interaction between the Epstein-Barr virus DNA polymerase catalytic subunit and its accessory subunit with regard to the 3'-to-5' exonucleolytic activity and stability of initiation complex at primer terminus. J. Virol. *68*, 3354-3363.

Urier, G., Buisson, M., Chambard, P., and Sergeant, A. (1989). The Epstein-Barr virus early protein EB1 activates transcription from different responsive elements including AP-1 binding sites. Embo J. *8*, 1447-1453.

Walling, D. M., Brown, A. L., Etienne, W., Keitel, W. A., and Ling, P. D. (2003a). Multiple Epstein-Barr virus infections in healthy individuals. J. Virol. *77*, 6546-6550.

Walling, D. M., Flaitz, C. M., and Nichols, C. M. (2003b). Epstein-Barr virus replication in oral hairy leukoplakia: response, persistence, and resistance to treatment with valacyclovir. J. Infect. Dis. *188*, 883-890.

Walling, D. M., Flaitz, C. M., Nichols, C. M., Hudnall, S. D., and Adler-Storthz, K. (2001). Persistent productive Epstein-Barr virus replication in normal epithelial cells *in vivo*. J. Infect. Dis. *184*, 1499-1507.

Wang, Y. C., Huang, J. M., and Montalvo, E. A. (1997). Characterization of proteins binding to the ZII element in the Epstein-Barr virus BZLF1 promoter: transactivation by ATF1. Virology *227*, 323-330.

Westphal, E. M., Blackstock, W., Feng, W., Israel, B., and Kenney, S. C. (2000). Activation of lytic Epstein-Barr virus (EBV) infection by radiation and sodium butyrate *in vitro* and *in vivo*: a potential method for treating EBV-positive malignancies. Cancer Res. *60*, 5781-5788.

Westphal, E. M., Mauser, A., Swenson, J., Davis, M. G., Talarico, C. L., and Kenney, S. C. (1999). Induction of lytic Epstein-Barr virus (EBV) infection in EBV-associated malignancies using adenovirus vectors *in vitro* and *in vivo*. Cancer Res. *59*, 1485-1491.

Wu, F. Y., Chen, H., Wang, S. E., ApRhys, C. M., Liao, G., Fujimuro, M., Farrell, C. J., Huang, J., Hayward, S. D., and Hayward, G. S. (2003). CCAAT/enhancer binding protein alpha interacts with ZTA and mediates ZTA-induced p21(CIP-1) accumulation and G(1) cell cycle arrest during the Epstein-Barr virus lytic cycle. J. Virol. *77*, 1481-1500.

Wu, F. Y., Wang, S. E., Chen, H., Wang, L., Hayward, S. D., and Hayward, G. S. (2004). CCAAT/enhancer binding protein alpha binds to the Epstein-Barr virus (EBV) ZTA protein through oligomeric interactions and contributes to cooperative transcriptional activation of the ZTA promoter through direct binding to the ZII and ZIIIB motifs during induction of the EBV lytic cycle. J. Virol. *78*, 4847-4865.

Yin, Q., Jupiter, K., and Flemington, E. K. (2004). The Epstein-Barr virus transactivator Zta binds to its own promoter and is required for full promoter activity during anti-Ig- and TGF-beta1-mediated reactivation. Virology *327*, 134-143.

Yoshizaki, T., Sato, H., Murono, S., Pagano, J. S., and Furukawa, M. (1999). Matrix metalloproteinase 9 is induced by the Epstein-Barr virus BZLF1 transactivator. Clin. Exp. Metastasis *17*, 431-436.

Young, L. S., Lau, R., Rowe, M., Niedobitek, G., Packham, G., Shanahan, F., Rowe, D. T., Greenspan, D., Greenspan, J. S., Rickinson, A. B., and et al. (1991). Differentiation-associated expression of the Epstein-Barr virus BZLF1 transactivator protein in oral hairy leukoplakia. J. Virol. *65*, 2868-2874.

Zacny, V. L., Wilson, J., and Pagano, J. S. (1998). The Epstein-Barr virus immediate-early gene product, BRLF1, interacts with the retinoblastoma protein during the viral lytic cycle. J. Virol. *72*, 8043-8051.

Zalani, S., Coppage, A., Holley-Guthrie, E., and Kenney, S. (1997). The cellular YY1 transcription factor binds a cis-acting, negatively regulating element in the Epstein-Barr virus BRLF1 promoter. J. Virol. *71*, 3268-3274.

Zalani, S., Holley-Guthrie, E., and Kenney, S. (1995). The Zif268 cellular transcription factor activates expression of the Epstein-Barr virus immediate-early BRLF1 promoter. J. Virol. *69*, 3816-3823.

Zalani, S., Holley-Guthrie, E., and Kenney, S. (1996). Epstein-Barr viral latency is disrupted by the immediate-early BRLF1 protein through a cell-specific mechanism. Proc. Natl. Acad. Sci. USA. *93*, 9194-9199.

Zalani, S., Holley-Guthrie, E. A., Gutsch, D. E., and Kenney, S. C. (1992). The Epstein-Barr virus immediate-early promoter BRLF1 can be activated by the cellular Sp1 transcription factor. J. Virol. *66*, 7282-7292.

Zerby, D., Chen, C. J., Poon, E., Lee, D., Shiekhattar, R., and Lieberman, P. M. (1999). The amino-terminal C/H1 domain of CREB binding protein mediates zta transcriptional activation of latent Epstein-Barr virus. Mol. Cell Biol. *19*, 1617-1626.

Zhang, Q., Gutsch, D., and Kenney, S. (1994). Functional and physical interaction between p53 and BZLF1: implications for Epstein-Barr virus latency. Mol. Cell Biol. *14*, 1929-1938.

Zhang, Q., Holley-Guthrie, E., Dorsky, D., and Kenney, S. (1999a). Identification of transactivator and nuclear localization domains in the Epstein-Barr virus DNA polymerase accessory protein, BMRF1. J. Gen. Virol. *80 (Pt 1)*, 69-74.

Zhang, Q., Holley-Guthrie, E., Ge, J. Q., Dorsky, D., and Kenney, S. (1997). The Epstein-Barr virus (EBV) DNA polymerase accessory protein, BMRF1, activates the essential downstream component of the EBV *oriLyt*. Virology *230*, 22-34.

Zhang, Q., Hong, Y., Dorsky, D., Holley-Guthrie, E., Zalani, S., Elshiekh, N. A., Kiehl, A., Le, T., and Kenney, S. (1996). Functional and physical interactions between the Epstein-Barr virus (EBV) proteins BZLF1 and BMRF1: Effects on EBV transcription and lytic replication. J. Virol. *70*, 5131-5142.

Zhang, Q., Wang, Y. C., and Montalvo, E. A. (1999b). Smubp-2 represses the Epstein-Barr virus lytic switch promoter. Virology *255*, 160-170.

Zimmermann, J., and Hammerschmidt, W. (1995). Structure and role of the terminal repeats of Epstein-Barr virus in processing and packaging of virion DNA. J. Virol. *69*, 3147-3155.

Chapter 28

Biological Role of the *BARF1* Gene Encoded by Epstein-Barr Virus

*Tadamasa Ooka**

ABSTRACT

Of the approximately 90 genes encoded by the EBV genome, two viral oncogenes, *LMP1* and *BARF1,* were known to induce a malignant transformation when introduced into rodent fibroblasts. *LMP1* and *BARF1* are expressed in 50% and 90% respectively, of cases of invasive NPC. *BARF1* transcripts are detected in almost 100% of EBV-positive gastric carcinomas and in epithelial cells immortalized by NPC-derived EBV, however these cells are consistently negative for *LMP1*. The *BARF1* oncogene is capable of immortalizing a monkey kidney epithelial cell line *in vitro* and of inducing malignant transformation in several established cell lines including rodent fibroblast and human B cell lines. In these cells, the expression of cellular proteins like telomerase, Bcl2, c-myc, CD21, CD23 and CD71 was activated. N-terminal sequence (1 to 54 aa) of the BARF1 protein was sufficient to induce malignant transformation and to activate Bcl2 expression. Purified BARF1 protein was capable of activating the cell cycle of rodent fibroblasts, primary monkey epithelial cells and the human B cell line by an autocrine/paracrine mechanism suggesting that this protein can behave as a growth factor. On the other hand, BARF1 is involved in immunomodulation: secreted BARF1 protein formed a complex with purified CSF1 molecule, so that the activation of macrophage *in vitro* was inhibited. *BARF1* negative virus was capable of signicantly inhibiting secretion of interferon. BARF1 protein expressed in a human B cell line by transfection was recognized by Natural Killer cells (NK) in ADCC (Antibody Dependent Cell Cytotoxity) test. These observations suggest that BARF1 can intervene not only in immunomodulation, but also in the oncogenic process.

INTRODUCTION

The Epstein-Barr virus (EBV), the etiological agent of infectious mononucleosis, also associated with numerous human cancers including undifferentiated nasopharyngeal carcinoma (NPC), Burkitt's lymphoma, Hodgkin's lymphoma and Gastric carcinoma. NPC is a major health problem in South-East Asia and North

*For correspondence email ooka@sante.univ-lyon1.fr

Africa with an incidence of 5-40/100,000 individuals. Its association with EBV is constant except in a few cases of atypical highly differentiated cases. About 5-20% of gastric carcinoma, a widespread cancer in the world, is associated with EBV. All these observations indicate that EBV has clearly emerged as a virus with oncogenicity in humans. EBV is a lymphotropic and epitheliotropic DNA virus capable of immortalizing simian (Miller et al., 1972) and human B lymphocytes *in vitro* (Rickinson and Kieff, 1996). Recent data showed that EBV was capable also of immortalizing not only simian primary epithelial cells (Nishikawa et al., 1999), but also human gastric primary epithelial cells (Danve et al., 2001). *In vitro* immortalized B and epithelial cells are however not tumorigenic in nude mice. Among eleven latent viral genes, at least six genes (*EBNA1, EBNA2, EBNA3A, EBNA3C, LMP1* and EBERs) are indispensable for *in vitro* B cell immortalization (Kieff., 1996). The expression pattern of EBV genes was studied in EBV associated disorders and was classified to three type I, II and III. In lymphoblast cell lines (LCLs), all latent viral genes are expressed. This expression pattern is referred to as latency type III. *In vivo*, expression of the complete set of latent genes is only found in absence of an immune response, *i.e.* in the lymphomas of AIDS patients and of transplant recipients who received immunosuppressive therapy. Latency type I, which is characterized by restricted expression of *EBERs, EBNA1, BARF0* and three LMPs are expressed, was first detected in NPC but this is also the prevailing expression pattern in most other EBV positive tumors like Hodgkin's disease and non-Hodgkin's lymphoma. At present several other gene expression patterns are recognized that can not be grouped with known latency patterns. Gastric carcinoma and Nasopharyngeal carcinoma belong to this latter category.

In epithelial cells immortalized *in vitro* by EBV, at least four viral genes: *EBERs, EBNA1, LMP2A* and *BARF1* are expressed. In gastric carcinoma, at least five EBV genes were found to be expressed in tumor biopsies, *EBERs, EBNA1, LMP2, BARF0* and *BARF1*. Viral genes expressed in nasopharyngeal carcinoma are also different when compared to those expressed in EBV-carrying immortalized B cells among the 90 genes encoded by EBV genome (Baer et al., 1984). Some latent infection transcripts (*EBNA1, LMP1, EBERs, LMP2A* and *BARF0*) (Young et al., 1988; Fahraeus et al., 1988; Brooks et al., 1992; Busson et al., 1992; Raab-Traub et al., 1983; Hitt et al., 1989; Gilligan et al., 1991) and early genes (EA-D, DNAase, BZLF1 and *BARF1*) (Luka et al., 1988; Sbih-Lammali et al., 1996a; Cochet et al., 1993; Sbih-Lammali et al., 1996b) are expressed. Expression of some early genes in NPC biopsies is however likely to be limited to a small population of epithelial cells, while a large part of cell population expressed DNAase (Sbih-Lammali et al., 1996a) and *BARF1* (Decaussin et al., 2000). At least those 5 latent and early genes expressed in NPC biopsies might constitute the good candidates for the initiation, promotion and/or maintenance of tumoral transformation of epithelial cells.

Recent data using quantitative RT-PCR technique showed that among the early genes only *BARF1* was highly transcribed in NPC biopsies in the absence of expression of other EBV lytic genes (North-South Workshop of Nasopharyngeal carcinoma, December 2003, Villejuif, France). Taken together, these data show that EBV has a good tropism for epithelial cells and is capable of immortalizing primary epithelial cells. However, EBV-immortalized epithelial cells are not tumorigenic

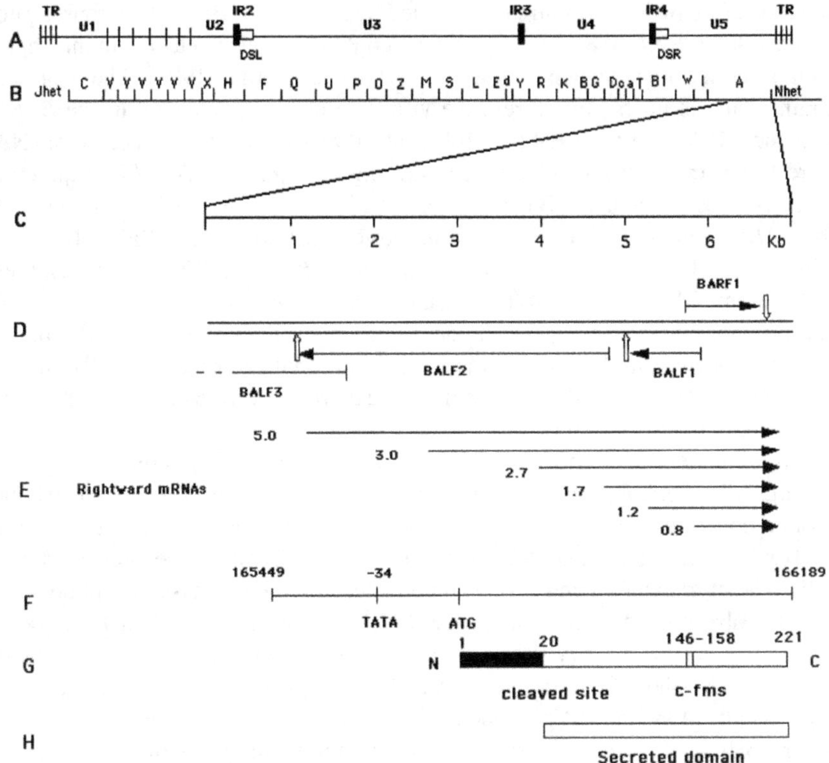

Figure 1. Transcription of *BARF1* in *Bam*H1-A fragment of EBV genome. A: Map of the EBV genome. B: its *Bam*H1 restriction sites. C: Genomic deletion in Raji strain. D: ORFs localized in the right-hand part of *Bam*H1-A. E: Rightward mRNAs transcription in *BARF1* region. F: *BARF1* cDNA sequence. G: BARF1 protein. H: secreted BARF1 protein domain

in nude mice. In this regard, interaction between EBV and cellular genes will be primordial for inducing malignant transformation of epithelial cells.

As the expression of *BARF1* gene was frequently found in NPC (Decaussin et al., 2000 ; Hayes et al., 1999) and gastric cancer (Zur Hausen et al., 2000), *BARF1* could play an important role in epithelial oncogenesis.

BARF1 gene and its protein

The *BARF1* gene was identified in a cDNA library constructed from P3HR1 cells treated with AraT (arabinofuranosyl thymine), TPA (Tetradecanoyl-13-Phorbolester) and SB (sodium butyrate) in which only latent and early viral phases are expressed. An isolated cDNA fragment of 0.74 kb (at position 165449-166189) localized in the right hand of *Bam*H1-A fragment (Zhang et al., 1988)(see Figure1). The cDNA contained, therefore, the total *BARF1* Open Reading Frame (ORF) plus 41 bp at the 5′ end including a promoter with the TATA sequence at -34 (Zhang et al., 1988) as well as a polyadenylation site sequence at the 3′ end and without splicing sequence (see Figure 1). An early promoter TATAAGA EDR1 at position

165466 before *BARF1* was already reported by Baer et al. (1984). The transcription of five other mRNA (0.8, 1.2, 2.7, 3.0 and 5.0 Kb) were also detected in this region with all co-terminal sequence (see Figure 1) (Zhang et al., 1988). The Raji EBV strain which is unable to express the viral late phase is defective in the *BARF1* sequence (Polack et al., 1994 ; Hatfull et al., 1988), but several truncated mRNAs were found in *BARF1* deleted region (Zhang et al., 1988). *BARF1* is therefore, unlikely to be essential for B cell immortalization. Raji cells were transfected with the *BALF2* gene (a major DNA binding protein also missing in Raji cells) along with a cosmid construct containing the essential *EBNA* and *EBNA3C* genes, these cells became EBV poducers following the chemical treatment [phorbol ester (TPA) and Sodium butyrate (SB)] (Decaussin et al., 1995; Robertson et al., 1996). The EBV produced from these cells was able to immortalize primary B cells but was inefficient and the LCLs derived from these experiments were difficult to maintain in culture in the long term (Robertson et al., 1996).

Total number of amino acids in the BARF1 protein is 221 (Figure 2). BARF1 is a highly hydrophobic protein and its pI is 6.2. Its half-life is greater than 20 hours. From a computer analysis, BARF1 has a predicted signal cleavage site at amino acid (aa) 20, so that aminoacid sequence of 21 to 221th could be secreted outside of cells. The myristylation and glycosylation sites as well as kinase phospholylation sites are also predicted in its sequence. Myristylation sites are at aa 18 (scvaa G qavta) , aa 26 (vtafl G ervtl) and aa 212 (keeah G vyvsg). The N-glycosylation site (vvtaa N ishdg) is found at aa 95. Three major phosphorylation sites were also found: cAMP/cGMP-dependent kinase phosphorylation site at position 39 (ywrrv S lgpei), Protein kinase C phosphorylation site at position 131 (tlsvh S ersqf) and Casein kinase II phosphorylation site at position 114 (getev T kqehl). A weak homology with c-fms protooncogene was also observed on the amino acid motif comprising residues 146 to 158 which was conserved between the BARF1 sequence and the receptor tyrosine kinases (hCSFR, hPDGFRfl, hVEGFR, hFGFR) (see Figure 2) (Strockbine et al., 1998). Some others sequence homologies with cellular genes were also found: B7-1 protein (of 54 to 214) with 26%, axl gene (of 122 to 182) with 26%, RET gene (of 104 to 138) with 26%, FGF-R (of 36 to 173) with 17%, and ICAM-1 (overall homologies) with 23.5%.

BARF1 encodes a p31-33 kDa early protein (Zhang et al., 1988 ; De Turenne-Tessier et al., 1997). This protein size was previously determined by immunoprecipitation of polypeptide synthesized *in vitro* by reticulocyte translation system with selected mRNA from P3HR1 cells (treated by araT, TPA and SB). *In vitro* synthesized polypeptide was immunoprecipitated with North African NPC serum containing a high titer of EA (1/1280) and VCA (1/1280)(21). Cellular localization of the BARF1 protein was initially demonstrated using NPC sera to be the cytoplasm and nuclei of *BARF1*-transfected rodent Balb/c3T3 fibroblasts (Zhang et al., 1988). However detection of the BARF1 protein in cell extracts, even those derived from *BARF1*-transfected cells, was difficult. The majority of BARF1 protein may be secreted as is predicted by computer analysis of the protein sequence. Secretion of BARF1 protein from B cell cultures was confirmed by Strockbine's group in Seatle (Strockbine et al., 1998). L-[^{35}S] Methionine/cysteine labelled BARF1 was found in the culture medium of *BARF1*-transfected CV1/

Protein structure predicted from the BARF1 ORF

Cleavage site

↓

1   MARFIAQLLLLASCVAAGQ   A VTAFLGEERVT

Transforming domaine

31   L TSY W RRVSLG RE I EVSWF KLGPGEEQVLI

cAMP PK

61   GRMHHDVI F IEWPFRGFFD I HRSANTFFLV

91   VTAA NISH DGNYLCRMKLGETEVTKQEHLS
       N-glyc   **CKII**

121   VVKPLTLSVH SER SQFPDFSVLTVT CTVNA
                   PKC

151   FPHPHVQW LMPEGVEPAPTAANGGVMKEKD
       c-fms receptor

181   GSLSVAVDLSLPKPWHLPVTCVGKNDKEEA

211   HGVYVSGYLSQ

Figure 2. The structure of the BARF1 protein predicted from BARF1 ORF encoded by EBV.

EBNA cells by immunoprecipitation method (Strockbine et al., 1998). Secreted BARF1 was also found in chemically (TPA and SB) induced EBV positive B95-8 B cells. This secreted protein could be complexed with CSF-1 protein (Strockbine et al., 1998). From our recent observations on human 293 epithelial cells, using BARF1 recombinant adenovirus or GFP/Flag BARF1 fusion system (with HaCaT cells), this protein is in fact secreted into culture medium of epithelial cells (Sall et al., 2004 ; de Turenne-Tessier et al., 2005; Liu and Ooka, in preparation). BARF1 protein was also found in cytoplasm, perinuclei and nuclei (Liu and Ooka, in preparation) of *BARF1*-transfected human HaCaT epithelial cells. Its localization in perinuclei and nuclei represents as a form of patch. The size of the secreted BARF1 protein was smaller (29 kDa) than that present in the cell (31 kDa). However two forms (29 and 31kDa) are found in HaCaT cells: the smaller one was present in the soluble cytoplasmic fraction and the larger one could be sedimented by ultracentrigugation at 105,000g for one hour. The p31 protein could be found in the cells which are stationary or cultured in serum free medium for several hours. The secreted 29kDa BARF1 protein (called as p29) from 293 and HaCAT cells has a similar size as those secreted from B cells (Strockbine et al., 1998). The difference in molecular weight comes from the presence of peptide signal present at 20th amino acid of p31 BARF1 protein (Sall et al., 2004; de Turenne-Tessier et al., 2005). Amino acid sequencing data showed that p29 protein produced from 293 cells by infection of adenovirus recombinant containing BARF1 sequence was the product of p31 protein cleaved at 20th a.a (Turenne-Tessier et al., 2005). Secreted p29 BARF1 protein has a molecular weight of around 180 kDa in non denaturing conditions suggesting it to be an oligomeric protein (Sall et al., 2004). Recent data obtained from crystalographic analysis suggest that the oligomeric BARF1 protein is likely to be in dimeric form, but this needs to be confirmed.

Oncogenic properties of the BARF1 gene

BARF1 not only has a malignant transforming activity, but also an immortalizing activity. Its oncogenic activity showed tropism for a large number of cell types: rodent fibroblasts, human B cell lines and monkey kidney epithelial cells (Figure 3).

BARF1 and cell immortalization

In general, rodent and human primary cells transfected by viral oncogene like papilloma *E7* gene are not tumorigenic when injected into mice (Zur Hausen et al., 2000 ; Mansure and Androphy. 1993). To induce a tumorigenic transformation, cellular genes such as *ras* were needed (Mansure et al., 1993). The major criteria that distinguish between immortalization and malignant (or tumoral) transformation are: Cell immortalization results in 1) morphological change, 2) continuous cell passage, 3) conservation of inhibition contact, 4) its inability to grow at high dilution of cell concentration and 5) the induction of telomerase activity. Malignant transformation results in 1) a loss of inhibition contact, 2) generally higher cell growth rate than immortalized cells, 4) serum starvation resistance, and 5) a tumor development when the cells are injected in immunodeficient mice, nude mice or SCID mice.

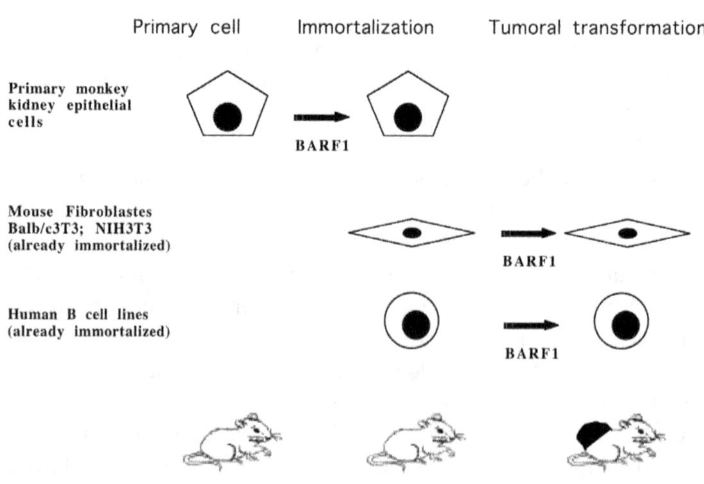

Figure 3. The schematic illustrates the oncogenic activity of the BARF1 protein.

In the case of SV40, a large T antigen alone was unable to immortalize the human cells and co-expression of telomerase gene (31) was needed. The additional introduction of a transforming gene like the mutated *ras*^{V12} oncogene in these immortalized cells (with SV40 large T and telomerase genes) was necessary for passage from the immortalized state to malignantly(ou tumoral) transformed state (Hahn et al., 1999).

All immortalized cells reported to date have been found to express significant telomerase activity (Counter et al., 1998; Meyerson et al., 1997). Therefore cell immortalization seems to be due to reactivation of telomerase, a ribonucleoprotein which possesses the enzymatic activity for synthesis of the telomeres. Telomerase activity was found in the majority of cancers and non in the normal cells excepting of germinal cells and some hematopoietic cells (Shay and Bucchetti. 1997). In cancer cells in which telomerase were not expressed, cells utilise a mechanism named ALT, which seems to stabilize the size of teromeres by chromosomal recombination (Bryan et al., 1995).

Regulation of telomerase activity is based on transcriptional and translational level. Most hTERT transcription was controled by the region of -200bp (named « Core promotor ») including several sites of fixation E-Box and SP1. Heterodimeric formation with Max, c-myc fixed on E-Box activate hTERT transcription. As Mad is an antagonist of c-myc protein, when heterodimeric Max/c-myc were replaced by Mad/Max, there is a diminution of activity of TERT promoter. Sp1 could play a role in the activation of hTERT promoter after binding on GC rich sites of core promoter. Several hormons were capable of activating hTERT transcription: estrogen active telomerase in cells expressing ER receptor (Estrogen Receptor). EGF (Epidemal Growth Factor) could also stimulate transcriptional expression of hTERT in cells expressing EGF-R. This activation comes from a promoter region – 22bp which is a sequence binding with Ets protein (Maida et al., 2002). Concerning post-translational regulation of hTERT, Ptotein kinase C (PKC) actives telomerase in catalizing the phosphorylation of hTERT protein, while the inhibitors of PKC abolished telomerase activity of NPC-derived epithelial cells (Ku et al., 1997). Hsp90 chaperon complex increased significantly telomerase activity in the cells expressing a low telomerase activity. The regulation of hTERT expression could come from post-translational modifications of histones as well as the inhibitors of HDAC (Histone Deacetylase) (Cong and Bucchetti., 2000).

In DNA viruses, the E6 protein of human papillomavirus type 16 was capable of activating telomerase activity in mammary epithelial cells and keratinocytes, but not in fibroblasts, indicating the regulation of hTERT expression is dependent on cell type (Klingelfutz et al., 1996). The activation of telomerase by E6 occurs through c-myc and the GC rich SP1 binding site (Oh et al., 2001), while the other reports showed that the complex of E6 and E6AP (E6-associated protein) would be needed for telomerase activation (Gewin and Galloway., 2001).

The introduction of the *BARF1* gene into primary monkey kidney epithelial cells led to morphological changes, continuous cell growth (over 100 passages) and the capacity to grow in highly diluted culture condition (Wei et al., 1997). However contact inhibition was conserved in these cells and no tumors were induced when injected in nude mice (Wei et al., 1997). *BARF1* by itself therefore was only able to immortalize these primary epithelial cells. In the earlier works done by Dr.

Griffin's group in England, a 40Kb EBV genomic fragment (called p31 localized of *Bam*H1-D to *Bam*H1-A) was capable of immortalizing *in vitro* the monkey primary epithelial cells (Griffin and Karrin, 1984). The *BARF1* gene is included in this large DNA fragment. Interestingly, the immortalized cells containing the 40Kb fragment expressed one major protein of p31 kDa (similar molecular weight as that of BARF1 protein) that was recognized by mouse serum obtained after injection of cells immortalized by a 40Kb fragment (Karran et al., 1990; Dr. Griffin : personal communication). Cell immortalization obtained by this large fragment is probably due to the presence of *BARF1* in this fragment, but we need further confirmation. For instance, the immortalization assay should be repeated using a 40 Kb fragment in which the *BARF1* gene is inactivated. More interesting results obtained from Dr. Griffin's laboratory are that *BARF1* was specifically localized in the double minute chromosome (DMs) when transfected in monkey epithelial cells (Gao et al., 2002). DMs have proved in other cases (*e.g.* with mdm2 and c-myc) to be reservoirs of amplified sequences. It will be interesting to know whether *BARF1* is amplified in the double minute chromosome and to underestand the significance of its specific localization in these chromosomes.

The introduction of another well known transforming gene, *LMP1*, into primary B cells or mouse primary fibroblasts led to the induction of cell proliferation and transient activation of DNA synthesis. However, on its own, *LMP1* was unable alone to immortalize these cells (Peng and Lundgren., 1992). Interestingly co-expression of *LMP1* and *CDK4* in primary mouse MEF cells induced a tumoral transformation directly by activating telomerase (Yang et al., 2000). However *LMP1* is unlikely to activate telomerase.

The activation of telomerase was found in *BARF1*-immortalized monkey primary epithelial cells (PATAS) (Cabras et al., submitted). BARF1 could activate telomerase in HeLa cells and monkey primary epithelial cells when transfected in these cell lines. As early as six passages after transfection by *BARF1*, the PATAS cells expressed telomerase activity, while non-transfected cells did not show any telomerase activity. Just as *BARF1* can activate c-myc, *BARF1* could acivate telomerase in the E-box of telomerase promoter region. However recent data has shown that *BARF1* is unlikely to activate telomerase using the E-box site, because telomerase remained activated even after this site was inactivated. We need further study to determine which site is utilised by *BARF1* to activate telomerase. In *BARF1*-immortalized PATAS cells, not only telomerase, but also Bcl2 and Ha-*ras* protein were expressed. We however do not know whether *LMP1* alone is capable of immortalizing the primary epithelial cells.

BARF1 and malignant transformation

Cord blood B cells immortalized *in vitro* by EBV do not generally give any tumor when injected in nude mice. However some *in vitro* immortalized B cells are occasionally tumorigenic in SCID mice (48). Introduction of *BARF1* into rodent fibroblastes such as Balb/c3T3 or NIH3T3 cell line (rodent fibroblasts already immortalized) could create a tumor transformation in immunosupressed new-born rats and in nude mice (Wei and Ooka., 1989 ; Sheng et al., 2001). *BARF1* was therefore capable of converting the immortalized cells into tumoral cells. Among the 90 genes encoded by EBV, only two *LMP1* and *BARF1* genes have so far been shown to have a tumorigenic activity in rodent cells (Sheng et al., ; 2001 ; Wang et al., 1985). The tumoral transforming activity of *BARF1* was demontrated in human B cells like EBV-negative Louckes and EBV-negative AKATA B cell lines (both cell lines are already immortalized, and non-tumorigenic in new-born rat or SCID mice). In comparison with the rodent fibroblast model, *BARF1*-transfected Louckes and AKATA B cells induce either a small tumor in new-born rats after 2 weeks with Louckes cells which disappeared after 4th weeks without any mortality of rats (Wei et al., ; Sheng et al., 2003) or no tumorigenic in nude mice, but a small tumor formation after 2-3 months of injection in SCID mice with AKATA cells (Sheng et al., 2003). Tumorigenic activity of *BARF1* in these B cells is therefore weak. *EBER*s, one of EBV latent genes, transfected EBV negative AKATA cells was able to induce a tumor in SCID mice, but the appeerence of tumor is also very late (Ruf et al., 2000 ; Komano et al., 1999). AKATA EBV negative cells transfected with *EBER*s did not, however, develop tumors in nude mice (Dr. K. Takada, personal communication).

Using a rodent fibroblast model and deletion mutants of *BARF1*, a transforming region of BARF1 protein was determined (Sheng et al., 2001) (Figure 2). A malignant transforming region was localized in N-terminal (codon 1-54th a.a.) of BARF1 protein. The mutants lacking this region were unable to transform the cells in malignant state. Furthermore, only the mutants containing this region rendered the cells resistant to apoptosis induced by serum deprivation. These mutants grow as well at 1% as at 10%, while the mutants lacking N-terminal region entered in senescence at 1% of serum. The *BARF1* gene was capable of activating anti-apoptotic *Bcl2* expression (Sheng et al., 2001). Interestingly, we found that the N-terminal transforming region alone (from 1 to 56th a.a.) was responsible for the activation of *Bcl2* expression.

In epithelial cell line (already immortalized), *LMP1* induced the inhibition of differentiation (Dawson et al., 1990). Introduction of a *LMP1* gene isolated from the CAO EBV strain (from NPC) into a non-tumorigenic human keratinocyte line, Rhek-1, immortalized with an adenovirus 12-SV40 hybrid virus induced a tumor when they are injected in SCID mice (Hu et al., 1993). It's however important to recall that SCID mice is more sensitive than nude mice for tumor development. In this regard, it will be interesting to examine *LMP1* transformation activity in nude mice. Also tumor induction observed in the above system should likely be re-examined in viral gene (adeno/SV40) free epithelial cells to avoid any eventual interaction during cell transformation.

BARF1 and BCL2 activation

Previous data showed that the *BARF1* gene was capable of activating anti-apoptotic *Bcl2* expression (Sheng et al., 2001). In the EBV system, at least three genes are associated with Bcl2 activation : *LMP1*, *EBER*s and *BARF1*. *LMP1* can up-regulate the expression of the cellular oncogene *bcl*-2 in a transfected cell line (Henderson et al., 1991). This upregulation seems to be part of the panoply of cellular changes induced by *LMP1*. But its effect is cell type specific (Rowe et al., 1994). *LMP1* activates the expression of *Bcl2* in B cells, but not in rodent fibroblasts (Martin et al., 1993). Interestingly Bcl2 gene activation was observed only in 2/41 in follicular lymphoma which expressed LMP1 protein (Seite et al., 1993). These suggest that

Bcl2 activation is not automatically associated with *LMP1* expression. On the other hand, the activation of *Bcl2* expression was reported in EBERs-transfected AKATA EBV negative B cells. However we do not know its activation by EBERs in other cell types.

Such acitvation was also observed not only in *BARF1*-transfected AKATA EBV negative cells (Sheng et al., 2003), but also in *BARF1*-transfeted primary monkey epithelial cells (Cabras et al., submitted). *BARF1* is therefore likely to be capable of activating Bcl2 in different type of cells. This suggests that a malignant transformation induced in B cells by these EBV genes is associated with the activation of *Bcl2*. It will be interesting to determine whether the activation of *Bcl2* is due to the direct action of *BARF1* (for instance, nuclear localized BARF1 as a transcriptional activation factor) or an indirect action by the other cellular proteins which activate *Bcl2* expression (mediated by BARF1).

The cooperative action of several oncogenes was essential for passage a cell from beign to malignant state in rodent cells. A malignant transformation observed here in Balb/c3T3 transfected by *BARF1* gene may come from a cooperation between Bcl2 with *BARF1*. Previous data showed that cooperation of *Bcl2* with c-Ha-Ras oncogenes was essential for a malignant transformation of Rat embryo fibroblasts (Reed et al., 1990). It was also shown in the hematopoietic system that co-oporation of BCR-ABL oncogenes with *Bcl2* expression resulted to prevent apoptotic death in murine hematopoietic cells (Sanchez-Garcia and Grütz., 1995). A malignant transformation induced by *BARF1* in rodent cells is likely due to the cooperation between a viral gene and a cellular anti-apoptotic protein. In NPC biopsies, high expression of *Bcl2* protein was identified (Lu et al., 1993). Therefore it will be interesting to define which genes (*BARF1* or other viral and/or cellular genes) are responsible for the activation of *Bcl2* expression in NPC biopsies.

BARF1 protein as a growth factor

Previous work showed the secretion of BARF1 protein in culture medium of *BARF1*-transfected B cells and EBV-positive B cell line like B95-8 (Strockbine et al., 1998). The molecular weight of the secreted protein is about 29 kDa. Its secretion was also found in 293 epithelial cells infected BARF1 recombinant adenovirus and in BARF1-transfected Balb/c3T3 rodent fibroblast (Sall et al., 2004). Highly purified BARF1 protein from 293 cell culture has the same molecular weight (M.W.) as those identified from B cell culture in denatured condition. Its native molecular weight obtained from a sucrose gradient (without SDS) showed about 8S (about 180 kDa) corresponding probably to its oligomeric form (M.W. more higher than that of Immunoglobulin G) (Sall et al., 2004). Purified p29 protein from 293 human epithelial cells infected with BARF1-positive adenovirus recombinant was capable of activating cell cycle of Balb/c3T3, human Louckes B cells and primary monkey epithelial cells (Sall et al., 2004) when this protein was added in serum-free culture medium (Figure 4). Interestingly, our recent study showed that Bcl2 expression was activated in Balb/c3T3 when purified BARF1 was added in serum free medium. This suggests that activation of Bcl2 expression in BARF1 transfected cells come likely from para-/autocrine mechanism (Sall et al., in preparation).

Previous studies by several laboratories demonstrated the difficulty in detecting the presence of the BARF1 protein in cells. Almost all BARF1 is secreted from the cell making it almost impossible to detect in cell extracts.

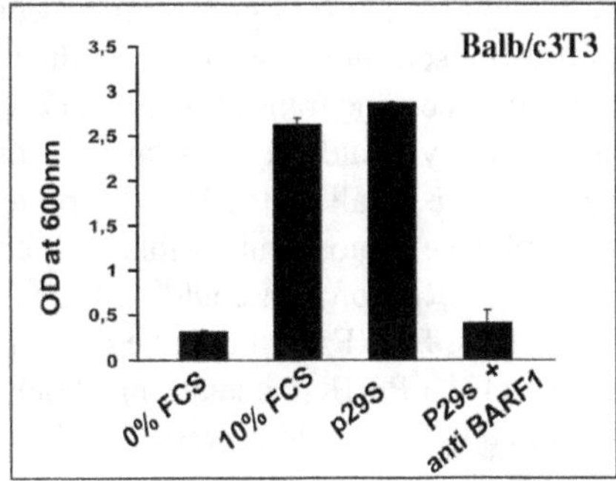

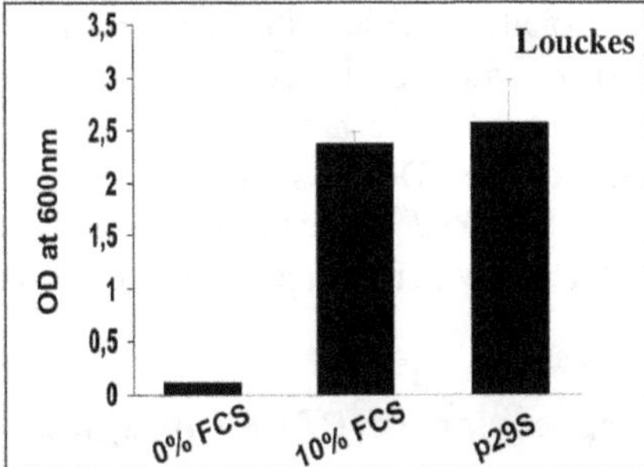

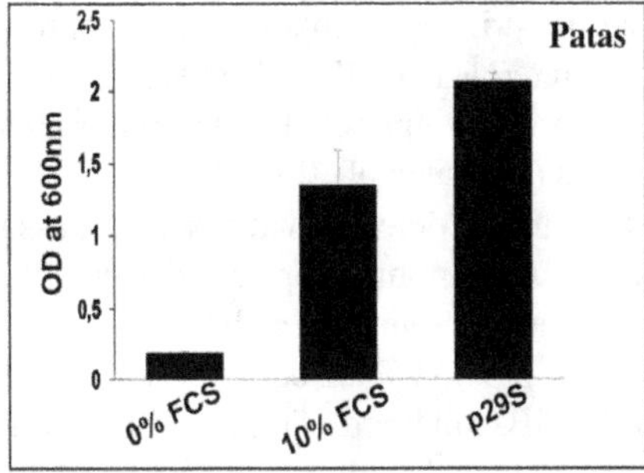

Figure 4. BARF1 protein secreted from human epithelial cells stimulates cell cycle. Secreted BARF1 protein (purified by sequencial chromotographies) was added in serum free culture medium and the activation of cell cycle was analysed by MTT test. Cell cycle of rodent fibroblasts, human B cell line and primary monkey epithelial cells was examined. The activation rate was compared with those of medium containing 10% FCS (Fetal Calf Serum) or without serum. BARF1 protein activated all type of cells more efficiently than 10% FCS.

Transcriptional and translational expression of BARF1 in BL biopsies and BL-derived B cell lines

A sensitive nucleic acid sequence-based amplification (NASBA) assays, was used to search for *BARF1* gene expression in the biopsies from BL and Hodgkin's disease (HD) (Zur Hausen et al., 2000). Its transcriptional activity was never detected in these biopsies (Zur Hausen et al., 2000). In contrast, *BARF1* transcription was detected in the lymphoma biopsies of tamarin (Zhang et al., 1992), induced by direct injection of concentrated EBV (Cleary et al., 1985). Interestingly, recent data showed that almost 95% of BL biopsies from Malawi in Africa contained the transcribed *BARF1* gene (Xue et al., 2002). Human BL cell lines that have been established for a long time like P3HR-1, Namalwa, AKATA, IB4 and other B cell lines transcribed its RNA which was detectable with RT-PCR. However BARF1 protein was undetectable on immunoblot assays in these cell lines. The *BARF1* transcription detected in P3HR-1 and AKATA cells was likely to have come from

0.04-2% of cell population expressing the lytic phase actively producing progeny (Decaussin et al., 2000). Interestingly *BARF1* transcription was also identified in latently infected Namalwa and IB4 cell lines, but the gene translation product was difficult to detect in these cell lines. This difficulty could probably may be due to production of the protein in its secreted form i.e. if all of the BARF1 protein was secreted into the culture medium, it would be almost impossible to detect intracellular BARF1 protein. In this regard, if BARF1 protein could be found in culture medium of Namalwa and/or IB4 cell lines, BARF1 can then be classified as a latent gene. In thymidine kinase negative (TK⁻) P3HR-1 containing about 11 EBV genome copy per cell in which 0.04% and 0.02% of cells expressed early and late antigens, respectively, *BARF1* gene was transcribed and translated (Decaussin et al., 2000). Its transcripional and translational expression increased in time of viral cycle induced by (TPA) in combination with SB (Decaussin et al., 2000). Its optimal translational expression was observed at 72 hours in asynchronized cells after the treatment of these chemical inducers (Decaussin et al., 2000). The transcriptional and translational expression of *BARF1* was detected in almost all BL-derived EBV-positive B cell lines in which few percentage of cells expressed viral lytic proteins .

Transcriptional and translational expression of *BARF1* in epithelial tumor

The *BARF1* gene was able to immortalize the primary monkey kidney epithelial cells (see next section), suggesting its oncogenic role in epithelial cells. Recent data showed that *BARF1* gene was expressed in primary epithelial cells immortalized with EBV virions isolated from NPC biopsies (Danve et al., 2001).

BARF1 was transcribed in both gastric cancer (detected with NASBA assay) (Zur Hausen et al., 2000) and nasopharyngeal carcinoma biopsies (detected with Northern blot, NASBA and RT-PCR methods) (Sbih-Lammali et al., 1996b; Decaussin et al., 2000; Hayes et al., 1999). Almost 90% of both epithelial tumor biopsies became positive to RT-PCR or NASBA method. The same proportion of NPC biopsies translated BARF1 protein detected with anti-BARF1 antibody (Decaussin et al., 2000). Its expression is therefore likely to be associated with epithelial tumor. In NPC biopsies, an immunohistochemical study revealed that a large proportion of tumor epithelial cells expressed BARF1 protein in the cytoplasm (Decaussin et al., 2000). As BARF1 protein is secreted into the culture medium (Sall et al., 2004), it will be interesting to examine whether this protein is also secreted outside of tumor cells and in serum.

BARF1 and immunomodulation

Secreted BARF1 protein could form a complex with purified CSF1 molecule, so that the activation of macrophage *in vitro* was inhibited by the formation of this complex (Strockbine et al., 1998). In combination with poly(I.C), CSF1 is known to induce alpha interferon in monocytes. BARF1 negative virus was signicantly capable of inhibiting secretion of interferon (Cohen and Lekstrom., 1999). These observations suggest that BARF1 protein could be intervened in immunomodulation. Interestingly, the BARF1 protein expressed in a human Raji B cell line (lacking the *BARF1* gene) by transfection was recognized by Natural Killer cells (NK) in the presence of either anti-BARF1 antibody or NPC-derived sera in ADCC (Antibody Dependent Cell Cytotoxity) test (Tanner et al., 1997). As BARF1 protein became a target for NK cells, this viral protein could likely be used for immunotherapy.

GENERAL CONCLUSION

In general, the EBV latent genes are thought to play an important role in the mechanism of B cell immortalization. In the EBV-epithelial cell model, as in the case of B95-8 cells (simian B cells immortalized by EBV), EBV virion isolated from NPC biopsy was able to immortalize *in vitro* primary simian epithelial cells. Among two oncogenes (*LMP1* and *BARF1*) capable of transforming rodent fibroblasts, no expression of *LMP1* was observed in those immortalized cells, while *BARF1*, one of viral early genes was expressed, suggesting its role in epithelial oncogenesis. This hypothesis would be reinforced by the observations that -1) *BARF1* gene alone was able to immortalize *in vitro* primary epithelial cells and -2) this gene was expressed in epithelial tumors (NPC and gastric cancer). *BARF1*-immortalized epithelial cells are however not tumorigenic and only immortalized. Therefore cellular and/or other viral genes will likely need for passage of these cells from benign state to malignantly transformed state. Finally *BARF1* acts all together as a tumoral transforming gene in already immortalized (non tumorigenic) cells and also as an immortalizing gene in primary epithelial cells (see Figure 3). Its epithelial immortalizing activity should however be examined in primary human epithelial cells. As BARF1 is secreted in culture medium and could activate cell cycle by para-and/or autocrine mechanism, its receptor would then be expressed (Figure 5). Among cellular proteins activated by BARF1, at least Bcl2 could be

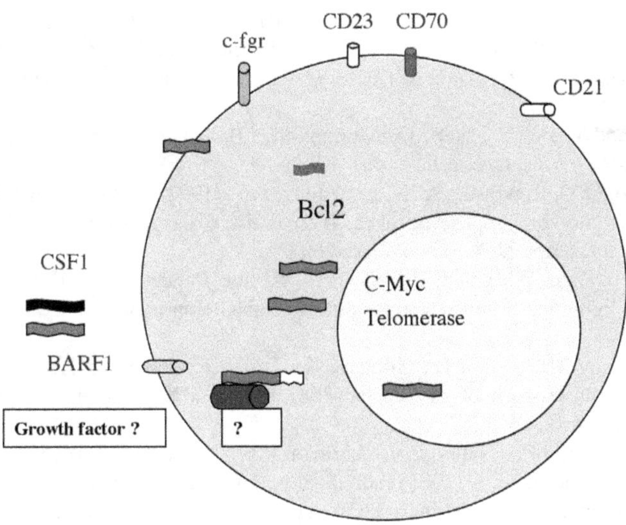

Figure 5. Biological properties of BARF1 protein: activation of cellular proteins, neutralization of CSF1 activity and as a growth factor.

activated by para-/autocrine mechanism. The mechanism of activation should be further analysed in detail to understand oncogenic activity of BARF1. It will be also an exciting question whether secreted BARF1 protein can be found in the serum of the majority of NPC patients. Further studies would be necessary to define the precise role of *BARF1* expressed in the tumorigenic epithelial cells of NPC.

PERSPECTIVE

An interesting question can be raised whether the *BARF1* gene belongs to the early or latent gene family. *BARF1* has been so far considered as an early gene, but the transcription of *BARF1* was detected in latently infected IB4 and Namalwa. It will be important to examine whether these latently infected cells secrete the p29 BARF1 protein. Secretion of BARF1 protein into the culture medium raises two questions: 1) Is its receptor involved in cell cycle activation? and 2) what are the cellular genes activated by this paracrine mechanism? To understand its oncogenic activity, the viral and/or cellular genes activated and/or suppressed by the BARF1 protein need to be identified. Inhibition of BARF1 (and/or Bcl2) expression can be attempted by siRNA in BARF1-transfected cells or an EBV positive NPC cell line to determine if the inhibition of its expression will result in suppression of the tumor. Interestingly, the BARF1 protein could be found in NPC serum, therefore, establishment of viable diagnostic and/or pronostic tests based on BARF1 protein will be of great benefit for treatment of EBV-associated epithelial cancers.

Acknowledgements

I thank Dr. Erle Robertson for his critical Reading and correction. The works were supported by ARC (Association pour la Recherche sur Cancer) and Ligue Nationale contre le cancer (Comité du Rhône),

Referencs

Baer, R., Bankier, A.T., Biggin, M.D., Feininger, P.L., Farrell, P.J., Gibson, T.J., Hatfull, G., Hudson, G.S., Satchnell, S.G., Seguin, C., Tuffnell, P.S., and Barrell, B.G. (1984). DNA sequence and expression of B95-8 Epstein-Barr virus genome. Nature, London, *310*, 207-211 .

Baichwal VR. and Sugden B. (1988) Transformation of Balb 3T3 cells by the BNLF-1 gene of Epstein-Barr virus. Oncogene *2*: 461-467.

Brooks, L., Yao, Q.Y., Rickinson, A.B., and Young, L.S. (1992). Epstein-Barr Virus latent gene transcription in nasopharyngeal carcinoma cells: Co-expression of EBNA1, *LMP1* and LMP2 transcripts. J. Virol. *66*, 2689-2697.

Bryan, T.M., Englezou, A., Gupta, J., Bacchetti, S., and Reddel, R.R. (1995). Telomere elongation in immortal human cells wiout detectable telomerase activity. EMBO J. *14*, 4240-4248.

Busson, P., McCoy, R., Sadler, R., Gilligan, K., Tursz, T., and Raab-Traub, N. (1992). Consistent transcription of the Epstein-Barr virus LMP2 gene in nasopharyngeal carcinoma. J. Virol. *66*: 3257-3262.

Cleary, M.L., Epstein, M.A., Finerty, S., Dorfman, R.F., Bornkamm, G.W., Kirwood, J.K., Morgan, A.J., and Sklar, J. (1985) Individual tumours of multifocal EB virus-induced malignant lymphoma in tamarins arise from different B-Cell clones. Science *228*: 722-724.

Cochet, C., Martel-Renoir, D., Grunewald, V., Bosq, J., Cochet, G., Schwaab, G., Bernaudin, J.F., and Joab, I. (1993). Expression of the Epstein-Barr virus immediate early gene, BZLF1, in nasopharyngeal carcinoma tumor cells. Virology *197*, 358-365.

Cohen, J. and Lekstrom K. (1999). Epstein-Barr virus BARF1 protein is dispensable for B-cell transformation and inhibits alpha interferon secretion from mononuclear cells. J. Virol. *73*, 7627-7632.

Cong, Y.S. and Bacchetti, S. (2000). Histone deacetylation is involved in the transcriptional repression of hTERT in normal human cells. J. Biol. Chem. *275*, 35665-35668.

Counter, C.M., Hahn, W.C., Wei, W., Caddle, S.D., Beijersbergen, R.L., Lansdorp, P.M., Sedivy, J.M., and Weinberg, R.A. (1998). Dissociation among *in vitro* telomerase activity, telomere maintenance, and cellular immortalization. Proc Natl Acad Sci U S A.. *95*, 14723-14728.

Danve, C., Decaussin, G., Busson, P., and Ooka, T. (2001). Immortalization of a monkey kidney epithelial cell by Epstein-Barr Virus from Nasopharyngeal carcinoma's biopsy. Virology *288*, 223-235.

Dawson CW., Rickinson AB. and Young LS. (1990) Epstein-Barr virus latent membrane protein inhibits human epithelial cell differentiation. Nature *344*: 777-780.

De Turenne-Tessier, M., Jolycoeur, P., and Ooka, T. (1997). Expression of the protein encoded by Epstein-Barr virus (EBV) *BARF1* open reading frame from a recombinant adenovirus system.Virus Res. *52*, 73-85.

de Turenne-Tessier, M., Jolicoeur, P., Middeldorp, J.M., and Ooka, T. (2005). Expression and analysis of the Epstein-Barr virus BARF1-encoded protein from a tetracycline-regulatable adenovirus system.Virus Res. *109*, 9-18.

Decaussin, G., Leclerc, V., and Ooka, T. (1995). The lytic cycle of Epstein-Barr virus in non-producer Raji line can be rescued by the expression of e 135 kDa protein encoded by BALF2 ORF deleted in the cells. J.Virol *69*, 7309-7314.

Decaussin, G., Sbih-Lammali, F., De Turenne-Tessier, M., Bouguermouh, A.M., and Ooka, T. (2000). Expression of *BARF1* RNA and protein in Nasopharyngeal Carcinoma. Cancer Res. *60*, 5584-5588.

Fahraeus, R., Fu, H.L., Ernberg, I., Finke, J., Rowe, M., Klein, G., Falk, K., Nilsson, E., Yadav, M., Busson, P., Turz, T., and Kallin, B. (1988). Expression of Epstein-Barr Virus-encoded proteins in nasopharyngeal carcinoma. Int. J. Cancer. *42*, 329-338.

Gao, Y., Lu, Y.J., Xue, S.A., Chen, H., Wedderburn, N., and Griffin, B.E. (2002). Hypothesis: a novel route for immortalization of epithelial cell from Epstein-Barr virus. Oncogene *21*, 825-835.

Gewin, L. and Galloway, D.A. (2001). E box-dependent activation of telomerase by human papillomavirus type 16 E6 does not require induction of c-myc. J Virol. *75*, 7198-7201.

Gilligan, K.J., Rajadurai, P., Lin, J.C., Busson, P., Abdelhamid, M., Prasad, U., Tursz, T., and Raab-Traub, N. (1991). Expression of the Epstein-Barr virus BamHI A fragment in nasopharyngeal carcinoma: evidence for a viral protein expressed *in vivo*. J. Virol. *65*, 6252-6259.

Griffin, B.E. and Karran, L. (1984). Immortalization of monkey epithelial cells by specific fragment of Epstein-Barr virus DNA. *Nature 309*, 78-82.

Hahn, W.C., Counter, C.M., Lundberg, A.S., Beijersbergen, R.L., Brooks, M.W., and Weinberg, R.A. (1999). Creation of human tumour cells with defined genetic elements. Nature *400*, 464-468

Hatfull, G., Bankier, A.T., Barrell, B.G., and Farrell, P.J. (1988). Sequence analysis of Raji Epstein-Barr virus DNA. Virology *164*, 334-340.

Hayes, D.P., Brink, A.A., Vervoort, M.B., Middeldorp, J.M., Meijer, C.J., and van den Brule, A.J. (1999). Expression of Epstein-Barr virus (EBV) transcripts encoding homologues to important human proteins in diverse EBV associated diseases. Mol pathol.. *52*, 97-103.

Henderson S, Rowe M, Gregory C, Croom-Carter D, Wang F, Longnecker R, Kieff E, Rickinson A. (1991) Induction of bcl-2 expression by Epstein-Barr virus latent membrane protein 1 protects infected B cells from programmed cell death. Cell. *65*:1107-15.

Hitt, M.M., Allday, M.J., Hara, T., Karran, L., Jones, M.D., Busson, P., Tursz, T., Ernberg, I., and Griffin, B. E. (1989). EBV gene expression in an NPC-related tumor. EMBO J. *8*, 2639-2651.

Hu LF, Chen F, Zheng X, Ernberg I, Cao SL, Christensson B, Klein G, Winberg G. (1993) Clonability and tumorigenicity of human epithelial cells expressing the EBV encoded membrane protein LMP1. Oncogene. *8*:1575-83.

Karran, L., Teo, C.G., King, D., Hitt, M.M., Gao, Y., Wedderburn, N., and Griffin, B.E., (1990). Establishment of immotalized primate epithelial cells with sub-genomic EBV DNA. Int. J. Cancer *45*, 763-772.

Kieff, E. (1996). Epstein-Barr Virus and its replication. Chapter 74: 2343-2394.In Fields Virology, Third Edition edited by B.N. Fields, D.M. Knipe, P.M. Howley et al. Lippincott-Raven Publishers, Philadelphia.

Klingelhutz, A.J., Foster, S.A., and McDougall, J.K. (1996). Telomerase activation by the E6 gene product of the human papillomavirus type 16. Nature *380*, 79-82.

Komano J, Maruo S, Kurozumi K, Oda T and Takada K. (1999) Oncogenic role of Epstein-Barr virus-encoded RNAs in Burkitt's lymphoma cell line Akata. J Virol. *73:* 9827-9831.

Ku, W.C., Cheng, A.J., and Wang, T.C.V. (1997). Inhibition of télomérase activity by PKC inhibitors in human nasopharyngeal cancer cells in culture. Biochem. Biophys. Res. Commun. *241*, 730-736.

Lu QL, Elia G, Lucas S and Thomas JA. (1993) *Bcl2* proto-oncogene expression in Epstein Barr virus-associated nasopharyngeal carcinoma. Int. J. Cancer *53*: 29-35.

Luka, J., Deeb, Z.E., Hartman, D.P., Jenson, B., and Pearson, G.R. (1988). Detection of antigens associated with Epstein-Barr Virus replication in extracts from biopsy specimens of nasopharyngeal carcinomas. J. Natl. Cancer Inst. *80*, 1164-1167.

Maida, Y., Kyo, S., Kanaya, T., Wang, Z., Yatabe, N., Tanaka, M., Nakamura, M., Ohmichi, M., Gotoh, N., Murakami, S., and Inoue, M. (2002). Direct activation of telomerase by EGF through Ets-mediated transactivation of TERT via MAP kinase signalling pathway. Oncogene *21*, 4071-4079.

Mansur, C.P., and Androphy, E.J. (1993). Cellular transformation by papillomavirus oncoproteins. Biochim. Biophys. Acta. *1155*, 323-345.

Martin JM, Veis D, Korsmeyer SJ and Sugden B. (1993) Latent membrane protein of Epstein-Barr virus induces cellular phenotypes independently of expression of *Bcl2*. J. Virol. *67:*5269-5278

Meyerson, M., Counter, C.M., Eaton, E.N., Ellisen, L.W., Steiner, P., Caddle, S.D., Ziagura, L., Bejersbergen, R.L., Davidoff, M.J., Liu, Q., Bacchetti, S., Haber, D.A., and Weinberg, R.A. (1997). hEST2, the putative human télomérase catalytic subunit gene, is up-regulated in tumor cells and during immortalization. *Cell 90*, 785-795.

Miller, G., Shop, T., Lisco, H., Stitt, D., and Lipman, M. (1972). Epstein-Barr Virus: transformation, cytopathic changes and viral antigens in squirrel monkey and marmoset leukocytes. Proc. Natl. Acad. Sci. USA. *69*, 383-387.

Nishikawa, J., Imai, S., Oda T., Kojima, T., Okita, K., and Takada, K. (1999). Epstein-Barr virus promotes epithelial cell growth in the absence of EBNA2 and LMP1. J. Virol., *73*, 1286-1292.

Oh, S.T., Kyo, S., and Laimins, L.A. (2001). Telomerase activation by human papillomavirus type 16 E6 protein: induction of human telomerase reverse transcriptase expression through Myc and GC-rich Sp1 binding sites. J Virol. *75*, 5559-5566.

Peng, M. and Lundgren E. (1992). Transient expression of the Epstein-Barr virus *LMP1* gene in human primary B cells induces cellular activation and DNA synthesis. Oncogene *9*, 1775-1782 .

Polack, A., Delius, H., Zimber, U., and Bornkamm, G.W. (1994). Two deletions in the Epstein-Barr virus genome of the Burkitt lymphoma nonproducer line Raji. Virology *133*, 146-157.

Raab-Traub, N., Hood, R., Yang, C.S., Henry, B., and Pagano, J.S. (1983). Epstein-Barr virus transcription in nasopharyngeal carcinoma. J. virol. *48*, 580-590.

Reed, J.C., Haldar, S., Croce, C.M. and Cuddy, M.P. (1990) Complementation by *BCL2* and C-HA-RAS oncogenes in malignant transformation of rat embryo fibroblasts Mol. Cell. Biol. *10*: 4370-4374.

Rickinson, A.B., and Kieff, E. (1996). Epstein-Barr virus. In: Fields BN, Knipe DM, Howley PM, editors. Virology. 3rd ed. Philadelphia: Lippincott-Raven, 2397-2446.

Robertson, E., Ooka, T., and Kieff, E. (1996). Epstein-Barr virus vectors for gene delivery to B lymphocytes Proc. Natl. Acad. Sci. USA *93*, 11334-11340.

Rowe M, Peng-Pilon M, Huen DS, Hardy R, Croom-Carter D, Lundgren E and Rickinson AB. (1994) Upregulation of *Bcl2* by the Epstein-Barr virus latent membrane protein LMP1: a B-cell-specific response that is delayed relative to NF-kappa B activation and to induction of cell surface markers.J Virol. *68*:5602-5612.

Ruf IK, Rhyne PW, Yang C, Cleveland JL and JT Sample. (2000) Epstein-barr virus small RNAs potentiate tumorigenicity of burkitt lymphoma cells independently of an effect on apoptosis. J Virol. *74*: 10223-10228.

Sall, A., Caserta, S., Jolicoeur, P., Franqueville, L., de Turenne-Tessier, M., and Ooka, T. (2004). Mitogenic activity of Epstein-Barr virus-encoded BARF1 protein. Oncogene. *23*, 4938-4944.

Sanchez-Garcia, I. and Grütz, G. (1995) Tumorigenic activity of the BCR-ABL oncogenes is mediated by *BCL2*. Proc. Natl. Acad. Sci. USA *92*: 5287-5291.

Sbih-Lammali, F., Berger, F., Busson, P., and Ooka, T. (1996a). Expression of EBV DNAase in the tumour cells of nasopharyngeal Carcinoma .Virology *222*, 64-74.

Sbih-Lammali, F., Bouguermouh, A.M., Decaussin, G., and Ooka, T. (1996b). Transcriptional expression of Epstein-Barr virus genes and proto-oncogenes in Algerian nasopharyngeal carcinoma. J. Med.Virol. *49*, 7-14.

Seite P, Hillion J, d'Agay MF, Gaulard P, Cazals D, Badoux F, Berger R, Larsen CJ. (1993) BCL2 gene activation and protein expression in follicular lymphoma: a report on 64 cases. Leukemia.7:410-417.

Shay, J.W. and Bacchetti, S. (1997). A survey of télomérase activity in human cancer. *Eur. J. Cancer 33*, 1787-1791.

Sheng W. Decaussin G. Ligou A. and Ooka T. (2003) Malignant transformation of EBV negative AKATA cells by introduction of *BARF1* gene encoded by Epstein-Barr virus J. Virol. *77* : 3859-65.

Sheng, W., Decaussin, G., Sumner, S. and Ooka, T. (2001). N-terminal domain of *BARF1* gene encoded by Epstein-Barr virus is essential for malignant transformation of rodent fibroblasts and activation of *Bcl2*. Oncogene *20*, 1176-1185.

Strockbine, L.D., Cohen, J.I., Farrah, T., Lyman, S.D., Wagener, F., DuBose, R.F., Armitage, R.J., and Spriggs, M.K. (1998). The Epstein-Barr virus *BARF1* gene encodes a novel, soluble colony-stimulating factor-1 receptor. J Virol. *72*, 4015-4021.

Tanner, JE., Wei MX., Ahamad A., Alfieri C., Tailor P., Ooka T. and Menezes J (1997) Epstein-Barr virus protein BARF1 expressed in lymphoid cell lines serves as a target for antibody-dependant cellular cytotoxicity. J. Infect. Dis. *175*: 38-46.

Wang, D., Liebowitz, D., and Kieff, E. (1985). An EBV membrane protein expressed in immortalized lymphocytes transforms established rodent cells. Cell *43*, 831-840.

Wei MX., Moulin JC., Decaussin G., Berger F and Ooka T. (1994) Expression of the *BARF1* gene encoded by Epstein-Barr Virus in human lymphoid cells and its tumorigenicity. Cancer Res *54*: 1843-1848.

Wei, M.X., de Turenne-Tessier, M., Decaussin, G., Benet, G. and Ooka, T. (1997). Establishment of a monkey kidney epithelial cell line with the Epstein-Barr virus *BARF1* gene. Oncogene *14*, 3073-3082.

Wei. M.X. and Ooka T. (1989). A transforming function of the BARF 1 gene encoded by Epstein-Barr Virus. EMBO J. *8*, 2897-2903.

Xue SA, Labrecque LG, Lu QL, Ong SK, Lampert IA, Kazembe P, Molyneux E, Broadhead RL, Borgstein E and Griffin BE. (2002) Promiscuous expression of Epstein-Barr virus genes in Burkitt's lymphoma from the central African country Malawi. Int J Cancer. *99*:635-643.

Yang, X., Sham, J.S., Ng, M.H., Tsao, S.W., Zhang, D., Lowe, S.W., and Cao, L. (2000). Epstein-Barr virus induces proliferation of primary mouse embryonic fibroblasts and cooperatively transforms the cells with a p16-insensitive CDK4 oncogene. J. Virol *74*, 883-891.

Young, L.S., Dawson, C.W., Clar, D., Rupani, H., Busson, P., Turz, T., Johnson, A., and Rickinson, A.B. (1988). Epstein-Barr Virus gene expression in nasopharyngeal carcinoma. J. Gen. Virol. *69*, 1051-1065.

Zhang CX., Decaussin G., Finerty S., Morgan A and Ooka T. (1992) Transcriptional expression of the viral genome in the Epstein-Barr Virus induced tamarin lymphoma and the corresponding lymphoblastoid tumour lines.Virus Res. *26*: 153-166.

Zhang, CX., Decaussin, G., Daillie, J., and Ooka, T. (1988). Altered expression of two Epstein-Barr virus early genes localized in BamHI-A in nonproducer Raji cells. J Virol. *62*, 1862-1869.

Zur Hausen, A., Brink, A.A., Craanen, M.E., Middeldorp, J., Meijer, C.J., and van Den Brule, A.J. (2000). Unique transcription pattern of Epstein-Barr virus (EBV) in EBV-carrying gastric adenocarcinomas: expression of the transforming *BARF1* gene. Cancer Res. *60*, 2745-2748.

zur Hausen, H. (2000). Papillomaviruses causing cancer: evasion from host-cell control in early events in carcinogenesis. J. Natl. Cancer Inst. *92*, 690-698.

From: Epstein-Barr Virus. Edited by: Erle S. Robertson

Chapter 29

Post-transcriptional Gene Regulation by EBV SM Protein

*Sankar Swaminathan**

ABSTRACT

The EBV SM protein is a member of a highly conserved family of proteins present in most herpesviruses known to infect mammals. There is a significant amount of functional and sequence divergence among the different homologs encoded by the human herpesviruses, including the stage of lytic replication during which they are expressed, differences in mechanism of action, and varying effects on splicing and transcription. SM is an early antigen having multiple post-transcriptional gene-regulatory functions shown to be essential for lytic EBV replication. SM enhances expression of EBV lytic genes and has both positive and negative effects on cellular gene expression. In addition to enhancing accumulation of EBV gene mRNAs, SM also has important effects on cellular mRNAs, significantly altering the host cell transcriptional profile.

INTRODUCTION

The SM protein is an early gene product with multiple post-transcriptional gene-regulatory functions that are essential for lytic EBV replication. In addition to enhancing expression of other EBV lytic genes, SM has both positive and negative effects on cellular gene expression. SM, also known as EB2, Mta and BMLF1, is a member of a highly conserved family of proteins represented in most human and primate herpesviruses as well as several other mammalian herpesviruses, including bovine, ovine, murine, equine and alcelaphine species. The homologs of SM in human herpesviruses include herpes simplex virus (HSV) ICP27, human cytomegalovirus (hCMV) UL69, varicella-zoster virus (VZV) ORF4 and Kaposi's sarcoma-associated herpesvirus (KSHV/HHV8) ORF57. There is a significant amount of functional and sequence divergence among these proteins, including the stage of lytic replication during which they are expressed, differences in mechanism of action, and varying effects on splicing and transcription (Albrecht et al., 1992; Bello et al., 1999; Chapman et al., 1992; Chee and Barrell, 1990; Gupta et al., 2000; Winkler et al., 1994). However, in both HSV and EBV, where recombinant

*For correspondence email sswamina@ufl.edu

viruses defective for the genes in question have been constructed, ICP27 and SM are essential for lytic virus replication.

GENE AND PROTEIN STRUCTURE

SM mRNA is spliced from an 1.8 kb transcript that spans the *Bam*HI restriction endonuclease site separating the Bam S and Bam M DNA fragments of the EBV genome (Buisson et al., 1989; Cho et al., 1985; Cook et al., 1994; Sample et al., 1986; Wong and Levine, 1986). A spliced 1.7 kb mRNA is generated from the BSLF2 exon in Bam S joined to the BMLF1 exon. The spliced gene product, SM, encodes a protein with a predicted size of 479 amino acids which is expressed as a heterogeneous species ranging in mw from 47-70 kDa, with a predominant size of approximately 55-60 kDa in EBV-infected cells (Marschall et al., 1989). Due to the presence of a methionine at aa 42, a shorter protein, referred to as M, can potentially be produced from the unspliced mRNA. Examination of the protein synthesized early in lytic replication of B95-8 and Akata strains of EBV indicated that SM is the predominant species, although small amounts of M are detectable (Cook et al., 1994). These additional amino acids present in SM but not in M are not critical for activity since several early publications using expression vectors capable of producing only M protein nevertheless demonstrated enhancement of reporter gene expression (Kenney et al., 1989a; Lieberman et al., 1986).

SM has several potential PKC and casein kinase II (CK II) phosphorylation sites located primarily in the amino terminal half of the molecule. SM is phosphorylated *in vivo* as demonstrated by ^{32}P incorporation and can also be phosphorylated *in vitro* by purified CKII (Cook et al., 1994; Wong and Levine, 1986). Mutation of serines 55-57 essentially abolished phosphorylation by CK II, but the mutant gene retained activity in transactivation assays, indicating that phosphorylation

Figure 1. Structural and functional domains of SM protein. The leucine rich region which binds REF/Aly and overlaps with the RNA binding domain is shown containing two potential nuclear export signals (NES1 and NES2). Another potential nuclear export signal may be present further upstream. The nuclear localization signal (NLS) lies upstream of the arginine-rich region containing the repeating RXP amino acid motifs. The conserved histidine and two cysteines (HCC) and the hydrophobic region in the carboxy-terminus that is important for proper folding and activity are shown. Amino acid numbers are shown below the diagram.

by CKII is not critical for activation or function (Cook et al., 1994). Whether phosphorylation regulates SM function *in vivo* remains to be further studied. There is one predicted potential asparagine glycosylation and two potential myristylation sites in SM but these have not been experimentally demonstrated to be modified *in vivo*. A consensus bipartite nuclear localization signal is located at aa 127-145, and deletion mutations including this region localize primarily to the cytoplasm (Figure 1). It has also been demonstrated that the first basic region can function independently as an NLS (Hiriart et al., 2003).

Comparison of the predicted protein sequence of SM homologs in human and primate herpesviruses reveals two strongly conserved motifs, both in the carboxy-terminal region. The first is an HX_3CX_4C motif, which has been referred to as a zinc finger-like motif. Each of these amino acids is essential for function of the protein as demonstrated by the inability of specific mutants to complement and rescue lytic replication of SM-deleted recombinant EBV (Ruvolo et al., 2004). This region may also be important in splicing regulation (See below). The second highly conserved motif is a GLFF motif found at the extreme carboxy terminus of the protein. These residues are at the center of a hydrophobic region which appears structurally critical for proper folding of the entire protein since mutations which alter this region lead to inactive and improperly localized and aggregated nuclear protein whereas mutations which are predicted to perturb the secondary structure less drastically maintain activity (Ruvolo et al., 2004).

REGULATION OF SM EXPRESSION

SM is expressed early in the lytic cycle, first becoming apparent at two hours after induction of lytic replication in EBV-infected Akata Burkitt lymphoma cells (Takada and Ono, 1989). SM expression rapidly follows expression of the immediate early transactivator genes Z and R and is cycloheximide sensitive, indicating that it is an early gene. Response elements for both Z and R are present in the SM promoter (Kenney et al., 1989a). Whereas Z and R expression declined by six hours post-induction, SM expression increased after 7 hours post-induction and was maintained for at least twenty-four hours post-induction (Takada and Ono, 1989).

ACTIVATION OF GENE EXPRESSION

SM increases mRNA accumulation of target genes by multiple mechanisms (See Figure 2). Early studies of SM function using reporter plasmids revealed that co-transfection of cloned genomic EBV DNA fragments encoding both the BSLF2 and BMLF1 exons of SM or BMLF1 alone led to increased reporter gene expression (Buisson et al., 1989; Kenney et al., 1989a; 1989b; 1988; Lieberman et al., 1986). Although several reports indicated that gene activation by SM was promoter-independent, it was initially assumed that the effect was transcriptional. One report suggested that SM did not significantly increase mRNA levels of target genes (Kenney et al., 1989a). However, it is now well established that in the case of chloramphenicol acetyl transferase (CAT) and several early EBV mRNAs, SM does lead to increased accumulation of its target mRNAs (Cook et al., 1994; Gao et al., 1998; Ruvolo et al., 1998). In the case of some EBV mRNAs, co-transfection

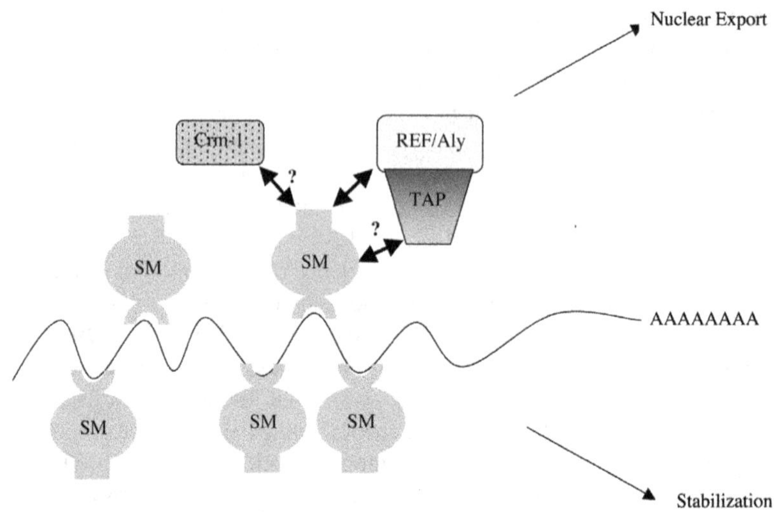

Figure 2. Enhancement of gene expression by SM. SM (red molecule) is shown binding at multiple sites to an intronless mRNA. SM bound to intronless EBV mRNAs is thought to increase stability of the mRNAs and lead to increased accumulation until the mRNAs are exported to the cytoplasm. Export is shown being mediated by SM interacting with cellular export factors. SM may also exert similar effects on specific cellular mRNAs that undergo splicing.

of SM leads to more than fifty-fold greater expression than in the absence of SM. Where the effect of SM on transcription has been directly measured using nuclear run-on assays, SM does not increase the rate of transcript initiation (Nicewonger et al., 2004; Ruvolo et al., 1998). These data clearly established that the major effect of SM is post-transcriptional.

Due to the difficulty in achieving high efficiency lytic gene expression in most EBV infected cells *in vitro*, the majority of evidence regarding SM mechanism derives from transfection experiments. Several targets, including non-EBV genes such as CAT and Renilla luciferase, as well as BMRF1 and other early EBV genes, have been tested by co-transfection and found to be responsive to SM (Nicewonger et al., 2004; Ruvolo et al., 2001). SM leads to increased accumulation of the mRNA transcripts of these genes, not only in the cytoplasm, but also in the nucleus. Increased nuclear stability in the presence of SM appears to be due at least in part, to enhanced 3' mRNA processing. Exogenous expression of SM led to increased processing of the EBV DNA polymerase mRNA, which contains a noncanonical poly A signal, and is cleaved and polyadenylated inefficiently (Furnari et al., 1993; Key et al., 1998). When the 3' UTR immediately following canonical polyadenylation sequences was replaced with different UTRs from EBV and HSV genes, the effect of SM varied, depending on the specific UTR employed. These data indicate that EBV, like HSV ICP27, may enhance 3' processing of viral mRNAs (Ruvolo et al., 1998). Whether SM protects mRNAs from nuclear decay pathways involving deadenylation and nucleolytic degradation remains to be determined.

SM, similar to HSV ICP27 and KS ORF57 proteins, shuttles between the nucleus and the cytoplasm. SM also interacts with components of cellular export pathways and binds RNA. It is therefore likely that SM also acts as an export factor facilitating nuclear export of intronless EBV mRNAs that may otherwise be poorly exported to the cytoplasm.

RNA BINDING

Initial reports that SM was capable of binding mRNA were based on *in vitro* experiments using bacterially derived fusions of SM with glutathione-S-transferase (GST). GST-SM fusion proteins immobilized on membranes bind to radiolabeled mRNA and GST-SM can also be cross-linked to mRNAs in solution (Gao et al., 1998; Ruvolo et al., 1998). Although one report suggested that SM preferentially interacted with specific EBV mRNAs, *in vitro* binding appears to be non-specific, and even antisense transcripts are capable of binding to SM *in vitro* (Gao et al., 1998; Ruvolo et al., 1998). Further detailed analysis of the requirements for *in vitro* RNA binding by SM revealed that such binding was dependent on the arginine-rich domain containing multiple amino acid RXP triplets consisting of arginine, a second amino acid (primarily alanine) and proline (See Figure 1) (Buisson et al., 1999; Ruvolo et al., 2001). This RXP motif is exposed at the carboxy terminus of bacterially derived SM proteins due to either proteolysis or early termination of translation in bacteria. In fact, little full-length protein can be made in bacteria unless the RXP domain is deleted. Such RXP-deleted bacterially synthesized proteins do not bind mRNA *in vitro* (Ruvolo et al., 2001). However, RXP-deleted SM functions normally when transfected and expressed in mammalian cells with respect to enhancement of reporter gene activity and cytoplasmic target mRNA accumulation (Ruvolo et al., 2001). The non-essential nature of the RXP motifs for SM-RNA interactions was conclusively demonstrated by experiments that examined the association of SM with mRNAs in cells (Ruvolo et al., 2001). When SM was immunoprecipitated from cellular lysates, specific co-immunoprecipitation of several mRNAs could be demonstrated. SM associated primarily with newly synthesized mRNAs. Whether such association was direct or as part of a multi-protein complex was not determined. However, deletion of the RXP domain did not affect the *in vivo* association of SM with mRNA (Ruvolo et al., 2001). These experiments taken together demonstrate that the RXP motifs are dispensible for *in vivo* mRNA binding and effects on mRNA export *in vivo*.

However, the RXP domain does appear to bind dsRNA in the context of intact SM protein produced in a baculovirus system (Poppers et al., 2003). The RXP domain is similar to a 21 triplet RXP domain in the HSV US11 protein (Roller et al., 1996). Both SM and US11 are capable of preventing activation of PKR *in vitro* and thereby inhibiting shut off of cellular protein translation (Poppers et al., 2000; Poppers et al., 2003). The ability of SM to inhibit PKR activation was demonstrable with full-length baculovirus-synthesized SM but not RXP-deleted SM. SM interacted directly with PKR; interaction was resistant to treatment with RNAse and was dependent on the presence of the RXP domain. Although immunoblotting has failed to reveal the presence of truncated SM peptides in SM-expressing cells, the PKR and dsRNA binding data suggest that the RXP domain

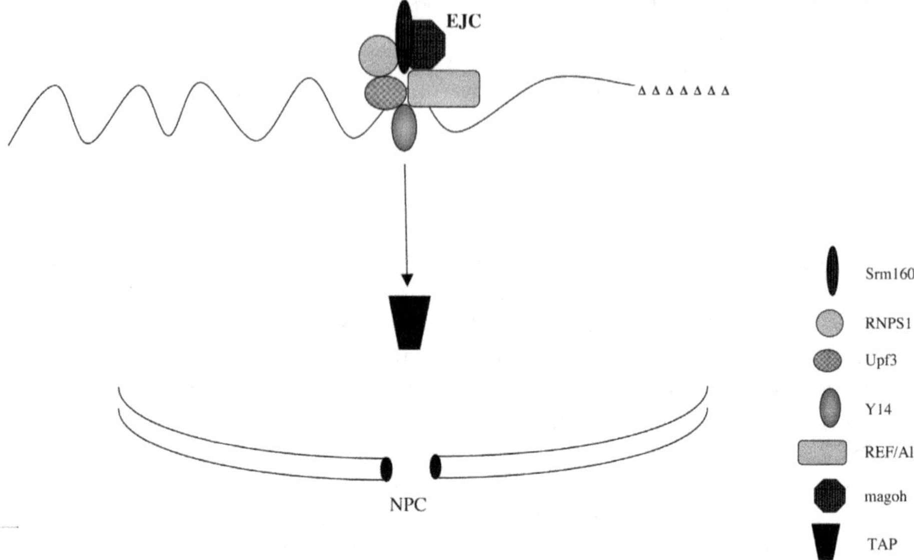

Figure 3. The exon junction complex and TAP in mRNA export. Six proteins known to associate with mRNA at sites proximal to the exon junction are shown with TAP acting as the essential export factor. One or more of these proteins are likely to be important in recruiting TAP to the spliced mRNA. SM associates with REF and TAP in RNA containing complexes. NPC (nuclear pore complex).

may be capable of binding RNAs *in vivo* and preventing activation of cellular defenses against viral replication.

In an elegant series of experiments, an additional arginine-rich region (ARM) immediately downstream of the RXP domain has been shown to mediate direct binding of SM to RNA (Hiriart et al., 2003). A 33 aa peptide encompassing this region bound to RNA targets in electrophoretic mobility shift assays and deletion of this region impaired *in vitro* and *in vivo* binding of SM to mRNA. Binding of the ARM peptide to mRNA was apparently non-specific since approximately 10 peptides bound at multiple site on human β–globin mRNA. Importantly, SM mutants deleted for the ARM, although capable of nuclear shuttling, were incapable of rescuing replication of recombinant SM-deleted virus in 293 cells.

SM, mRNA EXPORT AND NUCLEO-CYTOPLASMIC SHUTTLING

SM enhances expression of several intronless EBV lytic genes and intronless heterologous reporter genes. It has long been known that the presence of introns facilitates gene expression and that cDNAs are inherently poorly expressed from vectors in the absence of artificially added introns, although the mechanism for this effect was unclear (Huang and Gorman, 1990). Therefore, a common hurdle faced by replicating herpes viruses is the need to express unspliced (intronless) genes efficiently. It has recently been shown that during the process of pre-mRNA splicing, the mRNA becomes marked at exon-exon junctions with certain cellular proteins which earmark the processed mRNA for export (Luo and Reed, 1999)(for review, see Dreyfuss et al., 2002). There are at least six such proteins that comprise a complex, referred to as the exon-junction complex (EJC) that binds ~20 nt upstream of the exon junction: SRm160, RNPS1, REF/Aly, Y14, magoh, and Upf3 (See Figure 3). The role of these proteins in nuclear export was first suggested by experiments demonstrating that the protein REF/Aly enhances export of pre-mRNAs when injected into Xenopus oocyte nuclei (Zhou et al., 2000). REF/Aly binds to TAP, a central molecule in the nuclear export of vertebrate mRNA (Stutz et al., 2000). TAP binds to components of the nuclear pore, shuttles from nucleus to cytoplasm and is required for exporting the majority of mRNAs. REF/Aly, magoh, Y14 and Upf3 all can bind TAP and thus may serve as mediators of TAP recruitment to the mRNP, which enables its export. There are probably other unidentified RNA-binding proteins which can bind to TAP because depletion of the known EJC proteins does not completely block bulk mRNA export whereas TAP is essential (Gatfield and Izaurralde, 2002).

Another cellular pathway important for 5S RNA and U snRNA export utilizes Crm-1 (exportin-1), the first described nuclear export factor (Fornerod et al., 1997; Ossareh-Nazari et al., 1997; Stade et al., 1997). Crm-1 forms a complex with proteins containing a leucine-rich nuclear export signal (NES) and the small GTPase Ran bound to GTP (RAN-GTP) in the nucleus (Figure 4). Crm-1 interacts sequentially with nucleoporins, resulting in translocation through the nuclear pore. In the cytoplasm, RanGTP undergoes hydrolysis to yield Ran-GDP, leading to dissociation of the complex, allowing recycling of its components. The prototype of viral hijacking of this pathway is the HIV rev RNA export protein which contains a leucine rich NES, that interacts with Crm-1 and allows export of rev-bound HIV RNAs (for review, see Cullen, 2003).

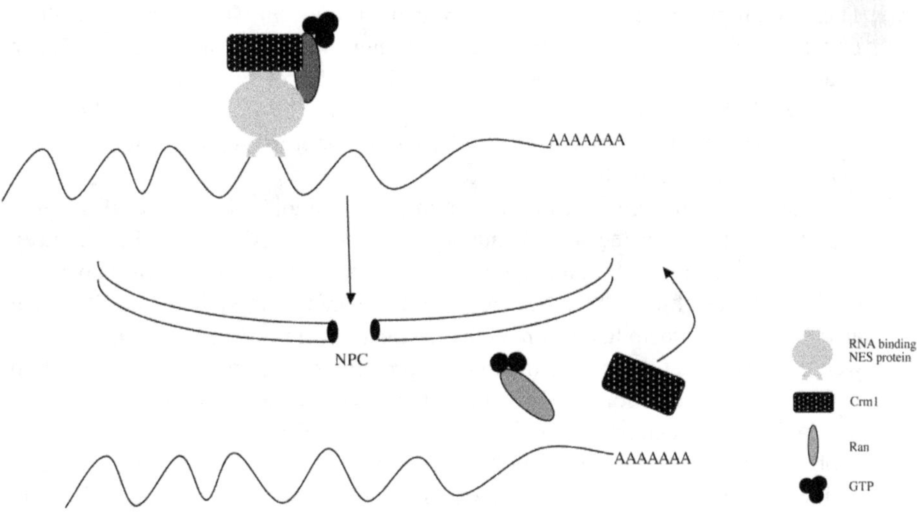

Figure 4. Crm1 mediated RNA export. Crm1 is shown in association with an RNA binding protein via its nuclear export signal (NES) and Ran-GTP. In the cytoplasm, GTP hydrolysis leads to dissociation of the complex and release of the protein and its bound RNA. NPC (nuclear pore complex).

Evidence exists that SM interacts with both of these RNA export pathways although many aspects of the relationship between SM and cellular export pathways remain to be characterized. The interaction of SM with Crm-1 is controversial, with two reports suggesting that CRM-1 is involved in SM shuttling and RNA export whereas another group failed to confirm such a connection (Boyle et al., 1999b; Chen et al., 2001; Farjot et al., 2000). A sequence of 10 amino acids in SM from aa 227-236, (see Figure 1) meets the general requirements for a Crm-1-interacting leucine-rich NES motif ($LX_{2-3}LX_{2-3}LXL$). In our initial studies, overexpression of Crm-1 in SM-transfected cells led to a marked translocation of SM to the cytoplasm and SM was co-immunoprecipitated with Crm-1 (Boyle et al., 1999b). Treatment with leptomycin B (LMB) a specific inhibitor of Crm-1, led to a moderate decrease in SM activity in reporter assays. Mutation of the core LXL motif or deletion of the entire sequence led to increased SM accumulation in detergent-insoluble structures, an impaired ability to be translocated to the cytoplasm, and reduced activity in reporter activation assays. Because of the relatively drastic effects of the mutations however, a conclusive role for this domain as a Crm-1 interacting region could not be established. This region was therefore referred to simply as the leucine rich region (LRR). A subsequent study identified an additional region immediately upstream (aa 218-227) similar to the first potential NES, which may also be involved in nuclear shuttling and mRNA export (Chen et al., 2001). Mutation of either this upstream sequence (NES2) or deletion of NES1 individually led to dramatic reduction of SM shuttling in a heterokaryon assay and mutation of either sequence also markedly reduced SM-mediated export of EBV BALF2 mRNA. Both sequences were capable of causing cytoplasmic redistribution of a nuclear GFP-tagged reporter protein and the redistribution was LMB-sensitive. These findings are in direct contrast to another study in which almost identical mutations of NES1 did not affect SM shuttling in a heterokaryon assay (Farjot et al., 2000). Furthermore, NES1 alone was incapable of substituting for the HIV Rev NES in Rev-mediated RNA export. However, mutations of either of these two putative NESs did lead to a loss of activity in reporter assays, consistent with the findings of previous studies.

Further complicating the interpretation of these studies, the double NES region has also been implicated in SM interactions with REF/Aly. GST-REF has been shown to interact with SM synthesized *in vitro* and this interaction was dependent on the presence of the double NES region (Hiriart et al., 2003). REF and TAP were also co-immunoprecipitated from cells transfected with SM but these interactions were RNAse-sensitive, suggesting that the proteins are part of a ribonucleoprotein complex. The interaction of SM with TAP was not dependent on the presence of the double NES region, and the interaction with REF, although reduced, was not abolished by deletion of this region, suggesting that association of these proteins with SM is at least partly RNA-dependent. Importantly, another region from aa 60-140, and therefore clearly distinct from the double NES, was also identified as containing a potential NES (See Figure 1). This portion of SM, when expressed as a fusion protein with GST, was capable of mediating nucleo-cytoplasmic shuttling when injected into the nucleus of a polykaryon (Hiriart et al., 2003).

While it is difficult to reconcile some of these contradictory findings, the following picture of SM shuttling and mRNA export emerges. SM is capable of nuclear shuttling and facilitating export of unspliced mRNAs and the region from aa 218-237 is clearly critical for these functions. Whether Crm-1 is involved in these functions remains somewhat open to debate although it is clear that this region does not operate as a classic Rev-type NES. It has been argued that the interpretation of the experiments with LMB is questionable due to the generalized toxic effects of the drug. On the other hand, certain observations strongly suggest an involvement of this pathway in SM function. Overexpression of Crm-1 clearly has effects on intracellular SM localization. Crm-1, Ran-GTP and nup214, a nucleoporin which interacts directly with Crm-1, can all be co-immunoprecipitated with SM, although these experiments did not address whether these interactions were RNA-dependent (Boyle et al., 1999a). In addition, dominant-negative fragments of the nucleoporin nup214, which inhibit Crm-1 export, also inhibit SM function although other export pathways may also interact with nup214 (Guzik et al., 2001).

The data that REF and TAP are complexed with SM in ribonucleoprotein complexes are likewise convincing and consistent with mechanisms of homologous proteins in other human herpesviruses. HSV ICP27 and KSHV ORF57 are also thought to interact with REF, although here again, redundant or cooperative mechanisms may operate (Chen et al., 2002; Malik et al., 2004). Mutants of ICP27 which do not bind REF are nevertheless viable, and ICP27 also interacts independently with TAP (Chen et al., 2002). Similarly, KSHV ORF57 protein appears to function independently of Crm1, but nevertheless competitively inhibits HIV Rev-mediated export, suggesting that components of both pathways may be used by SM homologs (Malik et al., 2004). There are other examples of specific mRNAs whose export may occur by more than one export pathway. Cellular cytokine mRNAs may be exported via HuR binding or via the Crm-1 pathway depending on environmental conditions (Gallouzi et al., 2001). Finally, the data that an NES in the amino-terminal portion of SM is capable of independently facilitating nuclear export of a fused heterologous protein is extremely intriguing and suggests that SM shuttling per se may be independent of any interaction with REF or Crm-1, although the region from aa 218-237 is ultimately critical for mRNA export and for EBV replication.

GENE-SPECIFIC ACTIVATION BY SM

One of the earliest observations and still unexplained aspect of SM function is the specificity of its effect on target mRNAs. SM enhances the expression of certain mRNAs whereas it has little or no effect on other mRNAs (Buisson et al., 1989; Kenney et al., 1989b; Markovitz et al., 1989). This phenomenon is most clearly obvious with various reporter genes that have been tested in co-transfection experiments with SM. For example, CAT is highly responsive to SM in B lymphocytes, with the levels of CAT protein activity and cytoplasmic mRNA increasing by ten to twenty fold in the presence of SM. On the other hand, β-galactosidase and firefly luciferase are not responsive to SM in the same lymphocyte cells (Ruvolo et al., 2001). Renilla luciferase is also activated by SM, but not as strongly as CAT in both lymphocytes and HeLa cells (Nicewonger et al., 2004). Similarly, although several of the early EBV genes comprising the DNA polymerase complex, when transfected into EBV-negative cells, are highly dependent on SM for expression, the the early BBLF2/3 is not (Semmes et al.,

1998). It does not appear that this phenomenon is due to preferential RNA binding since SM associates with both firefly luciferase and CAT mRNAs *in vivo* despite enhancing expression of CAT but not firefly luciferase (Ruvolo et al., 2001). As discussed previously, none of the data available to date have identified a specific SM RNA response element and most experiments have suggested a relatively non-specific association with mRNA. In addition, based on transcriptional profiling of SM expressing cells, SM does not appear to have a significant enhancing effect on the few intronless cellular genes that are known (Ruvolo et al., 2003). The simplest model consistent with these findings is that SM has the greatest effects on those unspliced transcripts that are otherwise inefficiently exported or are unstable. Clearly pathways must exist that allow the export and expression of cellular intronless genes, several of which are important in regulating cell proliferation. SM might have relatively little effect on such genes which have access to efficient cellular stabilization and export pathways although SM does provide a stimulus to expression of intronless EBV genes which may not have access to such pathways. It is also possible that depending on the cellular proteins that decorate a specific mRNA, SM interactions with individual cellular RNA-binding proteins could modulate the expression of cellular genes in a highly specific manner. The overall dependence of various intronless lytic EBV genes on SM for expression remains to be determined but should be simplified with the availability of both SM-deleted EBV recombinants and EBV gene microarrays.

INHIBITION OF SPLICED GENE EXPRESSION

In contrast to its enhancing effect on many unspliced genes, EBV SM inhibits the expression of target genes containing constitutive splicing signals (Ruvolo et al., 1998). It has been suggested that SM preferentially inhibits splicing directed by weak splicing signals since transport of some unspliced mRNAs expressed from plasmids is enhanced by SM expression (Buisson et al., 1999). Introduction of an intron either upstream or downstream of an intronless ORF such as CAT leads to inhibition rather than enhancement of expression by SM (Ruvolo et al., 1998). Thus, SM has diametrically opposite effects on the same open reading frame depending on the presence of introns. The only cellular gene with its native introns that has been tested in co-transfection assays with SM is the human growth hormone (hGH) gene, which contains four introns and five exons within 800 bp. The effect of SM on this gene is dramatic, virtually abrogating hGH expression at the mRNA level (Ruvolo et al., 1998). In the presence of SM, spliced hGH mRNA does not accumulate in either nucleus or cytoplasm and unspliced species accumulate in the nucleus (Ruvolo et al., 2004). SM does not inhibit transcription of the hGH gene in these experiments, indicating that the effect is post-transcriptional. These effects of SM are consistent with reports that SM affects the morphology of splicing factor SC35-containing speckles in transfected cells (Chen et al., 2001). When measured by microarray analysis, SM also decreases the amount of the majority of cellular mRNAs when expressed in EBV-negative B lymphocytes (Ruvolo et al., 2003). While the mechanism of this latter effect has not been proven to be post-transcriptional, it is consistent with SM having a globally repressive effect on cellular splicing or post-splicing processing, similar to HSV ICP27 (Sandri-Goldin, 1994).

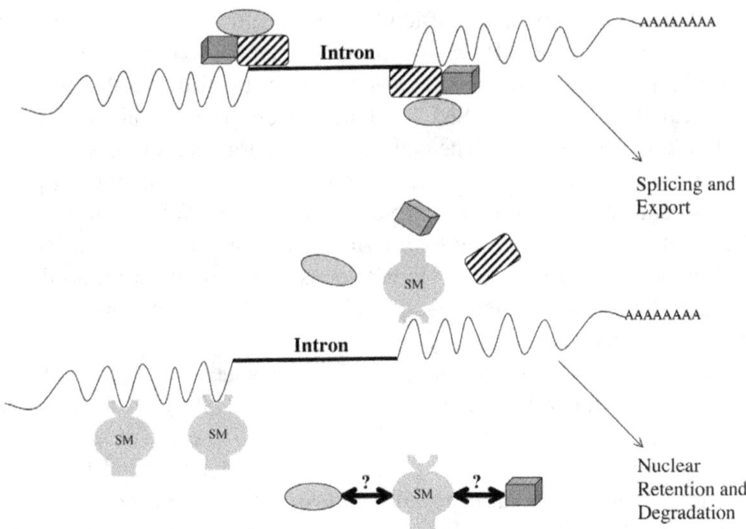

Figure 5. Splicing inhibition by SM. Normal mRNA splicing leading to nuclear export is diagrammed at top with splicing factors bound to intron-exon junctions. SM (red) is shown binding to pre-mRNA and disrupting formation of splicing complexes, leading to nuclear retention and degradation. SM is also shown possibly binding and sequestering one or more proteins important in splicing.

Several mechanisms, which are not mutually exclusive can be envisioned to explain the effect of SM on spliced mRNAs (See Figure 5). The first is that SM may bind and inhibit one or more proteins involved in splicing itself. The appearance of unspliced forms in the hGH transfection experiments described above indicate that SM must act prior to completion of splicing. Such a possibility is supported by the finding that specific mutation of C454 to alanine yields an SM mutant relatively unimpaired in the ability to activate reporter genes but does not inhibit splicing of hGH (Ruvolo et al., 2004). C454 may therefore be important for binding to, and inhibiting the function of a cellular splicing protein. SM could also inhibit splicing by interacting with unspliced pre-mRNA and preventing access of splicing factors. SM may bind relatively nonspecifically to mRNAs, including intron-containing and intronless species. Whereas binding may protect and enhance export of some, if not all, intronless mRNAs, such binding could have a deleterious effect on the normal pathways of post-splicing export mediated by the EJC and the TAP pathway. Alternatively, interaction of SM with REF/Aly and other proteins of the cellular export pathways could perturb the appropriate interaction of these proteins with the exon junctions required for export of spliced mRNAs whereas the same interactions generate functionally improved export substrates in the case of intronless mRNAs.

SM EFFECTS ON CELLULAR GENE EXPRESSION

Based on these various aspects of SM activity; enhancement of intronless gene expression; a gene-specific mode of action; and a general but perhaps not exclusive effect on intron-containing genes, it is clear that *a priori* prediction of effect of

SM on cellular genes would be difficult. The construction of an EBV-negative B lymphoma cell line, BJAB, that expresses a tamoxifen-inducible SM gene has allowed an analysis of the effects SM on the host cell in the absence of EBV infection (Ruvolo et al., 2003). SM has a significant growth inhibitory effect on cells within 24h of expression. The cells do not undergo cell cycle arrest, and do not exhibit any decrease in short-term viability. However, attempts to generate stable transfectants of SM in B cells have not been successful, possibly reflecting growth inhibition by SM over the long term. A comparison of the transcriptional profile of induced cells expressing SM with cells that were mock-induced and not expressing SM protein revealed that expression of the vast majority of cellular genes was either reduced or unchanged.

Surprisingly, of the approximately one dozen genes that were induced by SM, several were known interferon-stimulated genes (ISGs). In addition, STAT 1 mRNA levels were also significantly increased in SM-expressing cells. Since STAT 1 plays a central role in type I IFN signal transduction, and IFN-alfa and IFN-beta are intronless cellular genes, it was possible that SM exerted its effect by increasing type I IFN synthesis. However, IFN-alfa and IFN-beta transcript levels were not detectably increased in the microarray analyses of SM expressing cells, and sensitive ELISAs also did not reveal increased levels of IFN alfa or beta in the supernatants of cells expressing SM. Finally, when SM-expressing cells were co-cultivated with SM-negative cells separated by a semi-permeable membrane, ISGs were not induced in the SM-negative cells, but only in the SM-expressing cells, indicating that a diffusible factor such as IFN does not mediate ISG induction by SM. In fact, the level of ISG induction in the SM-expressing cells exceeded that induced by 1000 units/ml of type I IFN. While it is possible that small amounts of cell-associated IFNs induced by SM mediate these effects, the most likely explanation is that SM enhances expression of STAT1 mRNA, and thereby leads to induction of ISGs. Further, the pattern of STAT1 mRNA induced by SM differs from that induced by type I IFN treatment. Two isoforms of STAT1 mRNA, STAT1α and STAT1β, are known to be synthesized and are produced by alternative splice-site selection of the final exon (Schindler et al., 1992). The two isoforms are both capable of mediating type I IFN signal transduction when forming an active trimeric complex with STAT2 and IRF9 (p48), but only the STAT1α isoform is thught to be able to form active homodimers which can bind and activate GAS sequences which mediate IFN-γ signal transduction (Muller et al., 1993; Shuai et al., 1994; Shuai et al., 1992; Shuai et al., 1993). In SM-expressing BJAB cells, STAT1β becomes the predominant form of STAT1 expressed, in contrast to the results of IFN treatment. This change in the ratio of STAT1β to STAT1α is likely due to SM alteration of STAT1 splicing. One potential effect of SM therefore might be to inhibit the ability of the cell to respond to IFN-γ.

While it appears counter-intuitive that induction of an IFN signal transduction pathway by SM might be beneficial for EBV replication, several points are worth noting. First, based on the global effects of SM on cellular gene expression, it is likely that not every IFN-stimulated gene is upregulated by SM. While several ISGs have been demonstrated to be induced by SM, a comparison of transcriptional profiles of IFN-treated versus SM-expressing cells will be necessary to understand the differences between type I IFN and SM effects on cellular gene expression, particularly on the hundreds of ISGs. Second, the function of many ISGs remain unknown, and it is likely that one or more may be commandeered by incoming viruses to enhance virus replication. Precedent for this hypothesis exists with hCMV, where gB induces ISGs, including viperin (cig 5), a cellular protein which inhibits CMV replication (Boyle et al., 1999a; Chin and Cresswell, 2001; Navarro et al., 1998; Zhu et al., 1997). However, when induced by hCMV replication, viperin is localized differently than when induced by IFN, and it has been proposed that viperin is utilized to facilitate hCMV replication (Chin and Cresswell, 2001). As described in the following section, SM induces and physically associates with one cellular ISG which synergizes with SM in enhancing SM target gene expression.

FUNCTIONAL AND PHYSICAL INTERACTIONS OF SM WITH SP110B

SM has recently been found to physically interact with and induce expression of an ISG product that is also a component of PML bodies (Nicewonger et al., 2004). Several ISGs are known to be components of PML bodies, multi-protein nuclear structures usually present at 10-20 per cell (For review, see (Maul et al., 2000). PML bodies increase in size and number in response to interferon treatment. Infection by several viruses, including HSV, hCMV and EBV, leads to disruption of PML nuclear bodies, suggesting that the both viral replication and host defense are linked to the function of the PML nuclear body, which remains poorly understood (Adamson and Kenney, 1999; Bell et al., 2000; Everett et al., 1998; Korioth et al., 1996). One model of the relationship between viral infection and PML bodies is that the PML bodies exert an antiviral function and that their disruption is critical to allow efficient virus replication. A second hypothesis supported by findings that some protein components of the PML body are retained at sites of virus replication, is that the PML bodies serve as depots of essential factors that the virus must access in order to transcribe or replicate viral DNA. Sp110b, a cellular protein which is a component of the PML body, is induced by interferon and similar to several other ISGs, is induced by SM expression (Bloch et al., 2000; Nicewonger et al., 2004). Sp110b is also induced during EBV replication in P3HR-1 and Akata BL cells. Importantly, treatment of EBV-negative Akata cells, which have spontaneously lost EBV, with anti-IgG does not lead to Sp110b induction, as it does in EBV-positive cells, indicating that Sp110b is induced by EBV replication itself and not by the induction regimen. Sp110b was initially identified as an SM-binding protein in yeast two-hybrid screens and binds SM *in vitro* and *in vivo*. Sp110b binds to SM in an RNA-independent manner via two independent SM-interacting regions.

Sp110b, when co-transfected with SM in reporter assays, synergizes with SM to further enhance expression of CAT, Renilla luciferase or BMRF1 mRNAs although when transfected by itself, it has no effect. This effect is post-transcriptional, and is not dependent on an enhancement of nuclear export, since nuclear as well as cytoplasmic accumulation of the target mRNAs is increased by expression of Sp110b. The synergistic enhancement of SM function by Sp110b has been demonstrated to act at the level of mRNA stability, increasing the half-life of BMRF1 mRNAs more than two-fold. Finally, knockdown of Sp110b levels by the use of siRNA, in cells induced to lytically replicate EBV, leads to a decreased expression of BMRF1 mRNAs, indicating that Sp110b is a functional component of the cellular machinery that SM uses to facilitate EBV lytic gene expression

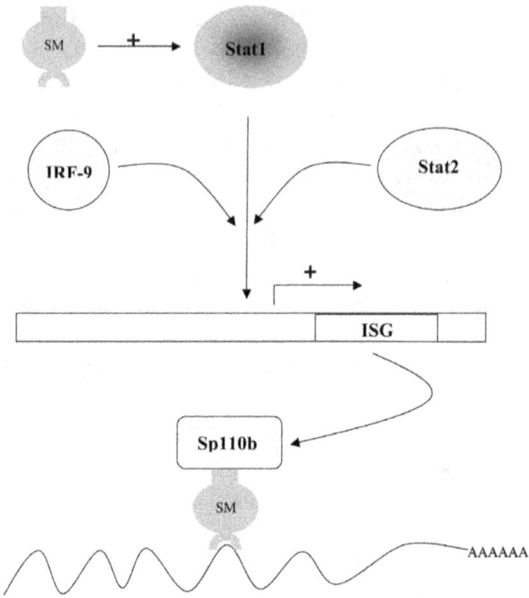

Figure 6. SM affects cellular gene expression and interacts with cellular gene products. SM is shown increasing cellular STAT1 levels which leads to transcriptional activation of many interferon-stimulated genes (ISGs) by the trimeric complex composed of STAT1, STAT2 and IRF-9. One such ISG, Sp110b, binds to SM and enhances stability of RNAs bound to SM in the nucleus.

required for EBV replication. These observations, taken together, indicate that Sp110b induction by SM during the course of EBV replication represents one example of an interferon-inducible protein that has been productively co-opted by the incoming virus (See Figure 6).

ROLE OF SM IN LYTIC REPLICATION

SM has been shown to be essential for lytic replication of EBV by the generation of an SM-deleted recombinant virus that is unable to replicate and produce infectious virions in 293 cells (Gruffat et al., 2002). Lytic gene expression is relatively, but not absolutely limited in the absence of SM and can be rescued by transfection of wt SM. EBV DNA replication is also restricted in the absence of SM, consistent with the dependence of the EBV DNA replication genes on SM for expression. Interestingly, a mutant deleted for the double NES region, which is predicted to be deficient in both REF binding and nuclear export, was still capable of rescuing SM-deleted virus replication, albeit at a reduced efficiency. Both ICP27 and UL69, the HSV and hCMV homologs of SM, although strong activators of gene expression in transfection assays were markedly inferior in rescuing virus production of SM-deleted EBV. Consistent with these latter observations, when the EBV SM gene was inserted into an ICP27-mutant HSV recombinant which is otherwise defective for replication, replication was only poorly rescued (Boyer et al., 2002). SM mutants which were as active as wt SM in enhancing reporter gene assays were nevertheless extremely inefficient at rescuing SM-deleted recombinant virus replication (Ruvolo et al., 2004). These findings underscore the multifunctional and specific actions of each of the SM homologs in the life cycle of the corresponding herpesvirus. Further, they suggest that SM and ICP27 are required for more than merely enhancing lytic virus gene expression but that they also play an important role in regulating host cell genes in a virus-specific fashion.

CONCLUSION

Although much has been learned about the function and mechanism of post-transcriptional gene regulation by SM, several important questions still remain to be explored. First, how does SM exert gene-specific effects on target genes? Are there structural elements or cis regulatory sequences in certain mRNAs that make them particularly responsive to SM? Alternatively, the responsiveness of particular genes may reflect the specific complement of cellular proteins that decorate their mRNAs. Thus specific cellular proteins that interact with SM may confer SM responsiveness to a subset of cellular or viral mRNAs. SM may also exert a stabilizing role that preferentially helps expression of genes with intrinsically vulnerable mRNAs such as intronless EBV lytic genes. Second, does SM act to repress spliced EBV genes such as Z and R, and how essential is this repressive function for the orderly progression of EBV replication? Finally and perhaps most interestingly, how does SM regulate cellular gene expression, and are there new insights into mechanisms of RNA processing that can be gleaned from investigating these interactions?

References

Adamson, A. L., and Kenney, S. (1999). The Epstein-Barr virus BZLF1 protein interacts physically and functionally with the histone acetylase CREB-binding protein. J. Virol. *73*, 6551-6558.

Albrecht, J. C., Nicholas, J., Billler, D., Cameron, K. R., Beisinger, B. C. N., Wittman, S., Craxton, M. A., and Coleman, H. (1992). Primary structure of the herpesvirus saimiri genome. J. Virol. *66*, 5047-5058.

Bell, P., Lieberman, P. M., and Maul, G. G. (2000). Lytic but not latent replication of epstein-barr virus is associated with PML and induces sequential release of nuclear domain 10 proteins. J. Virol. *74*, 11800-11810.

Bello, L. J., Davison, A. J., Glenn, M. A., Whitehouse, A., Rethmeier, N., Schulz, T. F., and Barklie Clements, J. (1999). The human herpesvirus-8 ORF 57 gene and its properties. J. Gen. Virol. *80*, 3207-3215.

Bloch, D. B., Nakajima, A., Gulick, T., Chiche, J. D., Orth, D., de La Monte, S. M., and Bloch, K. D. (2000). Sp110 localizes to the PML-Sp100 nuclear body and may function as a nuclear hormone receptor transcriptional coactivator. Mol. Cell Biol. *20*, 6138-6146.

Boyer, J. L., Swaminathan, S., and Silverstein, S. J. (2002). The Epstein-Barr virus SM protein is functionally similar to ICP27 from herpes simplex virus in viral infections. J. Virol. *76*, 9420-9433.

Boyle, K. A., Pietropaolo, R. L., and Compton, T. (1999a). Engagement of the cellular receptor for glycoprotein B of human cytomegalovirus activates the interferon-responsive pathway. Mol. Cell Biol. *19*, 3607-3613.

Boyle, S. M., Ruvolo, V., Gupta, A. K., and Swaminathan, S. (1999b). Association with the cellular export receptor CRM 1 mediates function and intracellular localization of Epstein-Barr virus SM protein, a regulator of gene expression. J. Virol. *73*, 6872-6881.

Buisson, M., Hans, F., Kusters, I., Duran, N., and Sergeant, A. (1999). The C-terminal region but not the Arg-X-Pro repeat of Epstein-Barr virus protein EB2 is required for its effect on RNA splicing and transport. J. Virol. *73*, 4090-4100.

Buisson, M., Manet, E., Trescol-Biemont, M. C., Gruffat, H., Durand, B., and Sergeant, A. (1989). The Epstein-Barr Virus (EBV) early protein EB2 is a posttranscriptional activator expressed under the control of EBV transcription factors EB1 and R. J. Virol. *63 (12)*, 5276-5284.

Chapman, C. J., Harris, J. D., Hardwicke, M. A., Sandri-Goldin, R. M., Collins, M. K., and Latchman, D. S. (1992). Promoter-independent activation of heterologous virus gene expression by the herpes simplex virus immediate-early protein ICP27. Virology *186*, 573-578.

Chee, M., and Barrell, B. (1990). Herpesviruses: a study of parts. Trends Genet. *6*, 86-91.

Chen, I. H., Sciabica, K. S., and Sandri-Goldin, R. M. (2002). ICP27 interacts with the RNA export factor Aly/REF to direct herpes simplex virus type 1 intronless mRNAs to the TAP export pathway. J. Virol. *76*, 12877-12889.

Chen, L., Liao, G., Fujimuro, M., Semmes, O. J., and Hayward, S. D. (2001). Properties of two EBV Mta nuclear export signal sequences. Virology *288*, 119-128.

Chin, K. C., and Cresswell, P. (2001). Viperin (cig5), an IFN-inducible antiviral protein directly induced by human cytomegalovirus. Proc. Natl. Acad. Sci. USA. *98*, 15125-15130.

Cho, M. S., Jeang, K. T., and Hayward, S. D. (1985). Localization of the coding region for an Epstein-Barr virus early antigen and inducible expression of this 60-kilodalton nuclear protein in transfected fibroblast cell lines. J. Virol. *56*, 852-859.

Cook, I. D., Shanahan, F., and Farrell, P. J. (1994). Epstein-Barr Virus SM protein. Virology *205*, 217-227.

Cullen, B. R. (2003). Nuclear mRNA export: insights from virology. Trends Biochem. Sci. *28*, 419-424.

Dreyfuss, G., Kim, V. N., and Kataoka, N. (2002). Messenger-RNA-binding proteins and the messages they carry. Nat. Rev. Mol. Cell Biol. *3*, 195-205.

Everett, R. D., Freemont, P., Saitoh, H., Dasso, M., Orr, A., Kathoria, M., and Parkinson, J. (1998). The disruption of ND10 during herpes simplex virus infection correlates with the Vmw110- and proteasome-dependent loss of several PML isoforms. J. Virol. *72*, 6581-6591.

Farjot, G., Buisson, M., Duc Dodon, M., Gazzolo, L., Sergeant, A., and Mikaelian, I. (2000). Epstein-Barr virus EB2 protein exports unspliced RNA via a Crm-1- independent pathway. J. Virol. *74*, 6068-6076.

Fornerod, M., Ohno, M., and Yoshida, M. (1997). CRM1 is an export receptor for leucine-rich nuclear export signals. Cell *90*, 1051-1060.

Furnari, F. B., Adams, M. D., and Pagano, J. S. (1993). Unconventional processing of the 3' termini of the Epstein-Barr virus DNA polymerase mRNA. Proc. Natl. Acad. Sci. USA. *90*, 378-382.

Gallouzi, I. E., Brennan, C. M., and Steitz, J. A. (2001). Protein ligands mediate the CRM1-dependent export of HuR in response to heat shock. Rna *7*, 1348-1361.

Gao, Z., Krithivas, A., Finan, J. E., Semmes, O. J., Zhou, S., Wang, Y., and Hayward, S. D. (1998). The Epstein-Barr virus lytic transactivator Zta interacts with the helicase-primase replication proteins. J. Virol. *72*, 8559-8567.

Gatfield, D., and Izaurralde, E. (2002). REF1/Aly and the additional exon junction complex proteins are dispensable for nuclear mRNA export. J. Cell Biol. *159*, 579-588.

Gruffat, H., Batisse, J., Pich, D., Neuhierl, B., Manet, E., Hammerschmidt, W., and Sergeant, A. (2002). Epstein-Barr virus mRNA export factor EB2 is essential for production of infectious virus. J. Virol. *76*, 9635-9644.

Gupta, A. K., Ruvolo, V., Patterson, C., and Swaminathan, S. (2000). The human herpesvirus 8 homolog of Epstein-Barr virus SM protein (KS- SM) is a posttranscriptional activator of gene expression. J. Virol. *74*, 1038-1044.

Guzik, B. W., Levesque, L., Prasad, S., Bor, Y. C., Black, B. E., Paschal, B. M., Rekosh, D., and Hammarskjold, M. L. (2001). NXT1 (p15) is a crucial cellular cofactor in TAP-dependent export of intron-containing RNA in mammalian cells. Mol. Cell Biol. *21*, 2545-2554.

Hiriart, E., Bardouillet, L., Manet, E., Gruffat, H., Penin, F., Montserret, R., Farjot, G., and Sergeant, A. (2003). A region of the Epstein-Barr virus (EBV) mRNA export factor EB2 containing an arginine-rich motif mediates direct binding to RNA. J. Biol. Chem. *278*, 37790-37798.

Huang, M. T. F., and Gorman, C. M. (1990). Intervening sequences increase efficiency of RNA 3' processing and accumulation of cytoplasmic RNA. Nucleic Acids Research *18*, 937-947.

Kenney, S., Holley-Guthrie, E., Mar, E. C., and Smith, M. (1989a). The Epstein-Barr Virus BMLF1 promoter contains an enhancer element that is responsive to the BZLF1 and BRLF1 transactivators. J. Virol. *63*, 3878-3883.

Kenney, S., Kamine, J., Holley-Guthrie, E., Mar, E. C., Lin, J. C., Markovitz, D., and Pagano, J. (1989b). The Epstein-Barr Virus immediate-early gene product, BMLF1, acts in trans by a posttranscriptional mechanism which Is reporter gene dependent. J. Virol. *63*, 3870-3877.

Kenney, S., Kamine, J., Markovitz, D., Fenrick, R., and Pagano, J. (1988). An Epstein-Barr virus immediate-early gene product trans-activates gene expression from the human immunodeficiency virus long terminal repeat. Proc. Natl. Acad. Sci. USA. *85*, 1652-1656.

Key, S. C., Yoshizaki, T., and Pagano, J. S. (1998). The Epstein-Barr virus (EBV) SM protein enhances pre-mRNA processing of the EBV DNA polymerase transcript. J. Virol. *72*, 8485-8492.

Korioth, F., Maul, G. G., Plachter, B., Stamminger, T., and Frey, J. (1996). The nuclear domain 10 (ND10) is disrupted by the human cytomegalovirus gene product IE1. Exp. Cell Res. *229*, 155-158.

Lieberman, P. M., O'Hare, P., Hayward, G. S., and Hayward, S. D. (1986). Promiscuous trans activation of gene expression by an Epstein-Barr Virus-encoded early nuclear protein. J. Virol. *60*, 140-148.

Luo, M. J., and Reed, R. (1999). Splicing is required for rapid and efficient mRNA export in metazoans. Proc. Natl. Acad. Sci. USA. *96*, 14937-14942.

Malik, P., Blackbourn, D. J., and Clements, J. B. (2004). The evolutionarily conserved Kaposi's sarcoma-associated herpesvirus ORF57 protein interacts with REF protein and acts as an RNA export factor. J. Biol. Chem.

Markovitz, D. M., Kenney, S., Kamine, J., Smith, M. S., Davis, M., and Huang, E.-S. (1989). Disparate effects of two herpesviruses immediate-early gene trans-activators on the HIV LTR. Virology *173*, 750-754.

Marschall, M., Leser, U., Seibl, R., and Wolf, H. (1989). Identification of proteins encoded by Epstein-Barr virus trans-activator genes. J. Virol. *63*, 938-942.

Maul, G. G., Negorev, D., Bell, P., and Ishov, A. M. (2000). Review: properties and assembly mechanisms of ND10, PML bodies, or PODs. J. Struct Biol. *129*, 278-287.

Muller, M., Laxton, C., Briscoe, J., Schindler, C., Improta, T., Darnell, J. E., Jr., Stark, G. R., and Kerr, I. M. (1993). Complementation of a mutant cell line: central role of the 91 kDa polypeptide of ISGF3 in the interferon-alpha and -gamma signal transduction pathways. Embo J. *12*, 4221-4228.

Navarro, L., Mowen, K., Rodems, S., Weaver, B., Reich, N., Spector, D., and David, M. (1998). Cytomegalovirus activates interferon immediate-early response gene expression and an interferon regulatory factor 3-containing interferon- stimulated response element-binding complex. Mol. Cell Biol. *18*, 3796-3802.

Nicewonger, J., Suck, G., Bloch, D., and Swaminathan, S. (2004). Epstein-Barr Virus (EBV) SM Protein induces and recruits cellular Sp110b to stabilize mRNAs and enhance EBV lytic gene expression. J. Virol. *78,* 340-352.

Ossareh-Nazari, B., Bachelerie, F., and Dargemont, C. (1997). Evidence for a role of CRM1 in signal-mediated nuclear protein export. Science *278*, 141-144.

Poppers, J., Mulvey, M., Khoo, D., and Mohr, I. (2000). Inhibition of PKR activation by the proline-rich RNA binding domain of the herpes simplex virus type 1 Us11 protein. J. Virol. *74*, 11215-11221.

Poppers, J., Mulvey, M., Perez, C., Khoo, D., and Mohr, I. (2003). Identification of a lytic-cycle Epstein-Barr virus gene product that can regulate PKR activation. J. Virol. *77*, 228-236.

Roller, R. J., Monk, L. L., Stuart, D., and Roizman, B. (1996). Structure and function in the herpes simplex virus 1 RNA-binding protein U(s)11: mapping of the domain required for ribosomal and nucleolar association and RNA binding *in vitro*. J. Virol. *70*, 2842-2851.

Ruvolo, V., Gupta, A. K., and Swaminathan, S. (2001). Epstein-Barr virus SM protein interacts with mRNA *in vivo* and mediates a gene-specific increase in cytoplasmic mRNA. J. Virol. *75*, 6033-6041.

Ruvolo, V., Navarro, L., Sample, C. E., David, M., Sung, S., and Swaminathan, S. (2003). The Epstein-Barr virus SM protein induces STAT1 and interferon-stimulated gene expression. J. Virol. *77*, 3690-3701.

Ruvolo, V., Sun, L., Howard, K., Sung, S., Delecluse, H. J., Hammerschmidt, W., and Swaminathan, S. (2004). Functional analysis of Epstein-Barr virus SM protein: identification of amino acids essential for structure, transactivation, splicing inhibition, and virion production. J. Virol. *78*, 340-352.

Ruvolo, V., Wang, E., Boyle, S., and Swaminathan, S. (1998). The Epstein-Barr virus nuclear protein SM is both a post-transcriptional inhibitor and activator of gene expression. Proceedings of the National Academy of Sciences of the United States of America *95*, 8852-8857.

Sample, J., Lancz, G., and Nonoyama, M. (1986). Mapping of Genes in BamHI Fragment M of Epstein-Barr Virus DNA That May Determine the Fate of Viral Infection. J. Virol. *57*, 145-154.

Sandri-Goldin, R. M. (1994). Properties of an HSV-1 regulatory protein that appears to impair host cell splicing. Infect. Agents Dis. *3*, 59-67.

Schindler, C., Fu, X. Y., Improta, T., Aebersold, R., and Darnell, J. E., Jr. (1992). Proteins of transcription factor ISGF-3: one gene encodes the 91-and 84- kDa ISGF-3 proteins that are activated by interferon alpha. Proc. Natl. Acad. Sci. USA. *89*, 7836-7839.

Semmes, O. J., Chen, L., Sarisky, R. T., Gao, Z., Zhong, L., and Hayward, S. D. (1998). Mta has properties of an RNA export protein and increases cytoplasmic accumulation of Epstein-Barr virus replication gene mRNA. J. Virol. *72*, 9526-9534.

Shuai, K., Horvath, C. M., Huang, L. H., Qureshi, S. A., Cowburn, D., and Darnell, J. E., Jr. (1994). Interferon activation of the transcription factor Stat91 involves dimerization through SH2-phosphotyrosyl peptide interactions. Cell *76*, 821-828.

Shuai, K., Schindler, C., Prezioso, V. R., and Darnell, J. E., Jr. (1992). Activation of transcription by IFN-gamma: tyrosine phosphorylation of a 91-kD DNA binding protein. Science *258*, 1808-1812.

Shuai, K., Stark, G. R., Kerr, I. M., and Darnell, J. E., Jr. (1993). A single phosphotyrosine residue of Stat91 required for gene activation by interferon-gamma. Science *261*, 1744-1746.

Stade, K., Ford, C. S., Guthrie, C., and Weis, K. (1997). Exportin 1 (Crm1p) is an essential nuclear export factor. Cell *90*, 1041-1050.

Stutz, F., Bachi, A., Doerks, T., Braun, I. C., Seraphin, B., Wilm, M., Bork, P., and Izaurralde, E. (2000). REF, an evolutionary conserved family of hnRNP-like proteins, interacts with TAP/Mex67p and participates in mRNA nuclear export. RNA *6*, 638-650.

Takada, K., and Ono, Y. (1989). Synchronous and sequential activation of latently infected Epstein-Barr virus genomes. J. Virol. *63*, 445-449.

Winkler, M., Rice, S. A., and Stamminger, T. (1994). UL69 of Human Cytomegalovirus, an open reading frame with homology to ICP27 of Herpes simplex virus, encodes a transactivator of gene expression. J. Virol. *68*, 3943-3954.

Wong, K.-M., and Levine, A. J. (1986). Identification and mapping of Epstein-Barr virus early antigens and demonstration of a viral gene activator that functions in trans. J. Virol. *60*, 149-156.

Zhou, Z., Luo, M. J., Straesser, K., Katahira, J., Hurt, E., and Reed, R. (2000). The protein Aly links pre-messenger-RNA splicing to nuclear export in metazoans. Nature *407*, 401-405.

Zhu, H., Cong, J. P., and Shenk, T. (1997). Use of differential display analysis to assess the effect of human cytomegalovirus infection on the accumulation of cellular RNAs: induction of interferon-responsive RNAs. Proc. Natl. Acad. Sci. USA. *94*, 13985-13990.

From: Epstein-Barr Virus. Edited by: Erle S. Robertson

Chapter 30

Developing Vaccines Against EBV-Associated Diseases

Denis J. Moss, Scott R. Burrows and Rajiv Khanna*

ABSTRACT

The past ten years have seen a dramatic accumulation of insights into the biology, immunology and virology of EBV so that for the first time rational vaccine development can begin. Thus the targets molecules present on latency I, II and III diseases have been well defined and the relative importance of both the cellular and the humoral responses are understood at least in the broad sense. Several vaccine trials towards infectious mononucleosis (IM) and post-transplant lymphoproliferative disease (PTLD) have been conducted and others are planned. While formulating vaccines for nasopharyngeal carcinoma (NPC) and Hodgkin's lymphoma (HL) are more speculative, there is good reason to believe that formulations encompassing cytotoxic T cell (CTL) epitopes encoded by LMP1 and LMP2 should have therapeutic benefit. On the other hand, there is little prospect for a vaccine to reverse the rapid growth of Burkitt's lymphoma.

INTRODUCTION

There is a strong scientific and commercial focus on developing vaccines and immunotherapeutic strategies for the treatment of EBV-associated diseases. However, it is unlikely that a single vaccine that is applicable to all EBV-associated diseases will be developed. Given the variety of potential EBV targets in latency III diseases and the problems of immune recognition of latency II and latency I diseases, vaccines against infectious mononucleosis (IM) and post transplant lymphoproliferative disease (PTLD) would seem to offer the best opportunity for early development. However, in spite of this caveat, early human trials aimed at nasopharyngeal carcinoma (NPC) and Hodgkin's lymphoma (HL) deserve serious consideration while there remain significant obstacles towards the reality of a therapeutic vaccine against Burkitt's lymphoma (BL).

M.A. Epstein first proposed the concept of the development of EBV vaccines as long ago as 1976. These original proposals were based on the notion that vaccination might prevent EBV infection and break the link in the complex chains of events that lead to EBV-associated disease. Since that time, a better

*For correspondence email denisM@qimr.edu.au

understanding of EBV biology has led to the postulation of more sophisticated but, as yet, untested vaccination strategies. Presently, it seems most unlikely that vaccination of any kind will achieve sterilizing immunity against herpesviruses. The murine gamma-herpesvirus, MHV68, establishes the same steady-state levels of lytic and latent infection whatever the route of infection or dose. It may be that a single virus particle successfully infecting a single target cell will be enough to establish persistent infection in a susceptible subject (Tibbetts et al., 2003). There is a general consensus that the goal of EBV vaccination is the prevention of disease and not of infection. Vaccination that could modify infection, or at least the subsequent immunological status of the infected person with respect to EBV, may minimize disease or reverse the expansion of EBV-associated tumors.

It is unlikely that EBV has evolved to avoid immune recognition. On the contrary, it is likely that there may well be selection towards immune recognition since this is likely to favor host and viral survival. An important precedent for herpesvirus vaccination is the attenuated varicella zoster virus (VZV) Oka strain vaccine that does not prevent infection but is able to prevent disease. More recently the concept of therapeutic vaccination to treat EBV-associated tumors themselves has begun to emerge (Khanna et al., 2001, Ong et al., 2003).

The lack of a complete understanding of the biology of EBV *in vivo* remains an unfortunate reality and the various approaches to EBV vaccine design discussed below are, of necessity, based on a number of unproven assumptions. The virus is an orally transmitted infection of B-cells both peripherally and in lymphoid tissue in the oropharyngeal region. Although it has not been possible to convincingly demonstrate the presence of EBV in oropharyngeal epithelial, it has been shown recently, using polarized tongue and oropharyngeal epithelial cells *in vitro*, that EBV-infected donor cells in saliva are very efficient at infecting recipient epithelial cells at their apical surface by cell-cell contact. However, these same epithelial cells are refractory to infection with free virus at their apical surface. It has also been shown that neighboring epithelial cells are infected by cell-cell transmission and free virus is produced at both the apical and basolateral epithelial surfaces (Tugizov et al., 2003). Presumably, it is the latter cell-free virus that subsequently infects B-cells circulating within the oropharyngeal epithelium and oropharyngeal lymphoid tissues.

LESSONS FROM ANIMAL MODELS

It is unfortunate that there is no model of EBV infection in animals which entirely reproduces the cell-virus relationships seen in either symptomatic primary infection or in EBV-associated malignancies in humans upon which vaccine formulation decisions could be based. However, there are now available a range of model systems in mice and primates which provide an experimental basis for testing vaccine formulations and providing pre-clinical data to support justification for human trials.

Murine models

Striking similarities between MHV-68 and human γ-herpesviruses in both genetic sequence and immune control means MHV-68 provides a powerful mouse model to analyze the *in vivo* function of herpesvirus genes and host control; aspects that are greatly limited by the high degree of species specificity of human viruses. MHV-68 is a natural γ_2-herpesvirus of small rodents, which like EBV, manifests symptoms of IM with splenomegaly, polyclonal B cell activation with associated autoantibody production, and a $CD8^+$ T cell dominated peripheral lymphocytosis (Nash et al., 2001).

Recently, the Rag2-/- γ_c-/-mutant strain of mice that lacks B, T, and NK cells, were engineered to develop a functional human immune system following intrahepatic transfer of $CD34^+$ human cord blood cells into newborns (Traggiai et al., 2004). EBV successfully infected the human immune cells in these Hu-Rag2-γ_c mice. EBV infected mice developed T cells that proliferated in response to EBV antigen *in-vitro*. While it remains to be clarified how human T cell selection on a mouse thymic background occurs, the T cells generated discriminate self from allogeneic MHC and these xenotransplanted mice raised human IgG to tetanus toxoid antigen. Hu-Rag2-γ_c mice thus offer an opportunity to test EBV vaccine formulations in a human cell background that supports efficient high level viral gene expression. With the Hu-Rag2-γ_c mice xenotransplant model it may be feasible to evaluate (a) the immune response to EBV in the context of the human immune response over a relatively short period of time and (b) the safety and efficacy of novel vaccine formulations in humans.

SCID mice, which lack both functional B and T cells due to a deficiency in double-stranded break repair activity in the proper joining of V(D)J coding sequences offers an alternative model for the testing of vaccine formulations. Such mice are able to sustain human PBMC for some time and when injected with lymphocytes from EBV immune individuals, EBV positive lymphomas arise. This model has recently been modified (Islas-Ohlmayer et al., 2004) using nonobese diabetic SCID (NOD/SCID) mice populated with human haemopoietic stem cells. Such mice are susceptible to EBV infection *in vivo* resulting in high viral loads and the development of malignant lymphomas.

Human HLA transgenic mice have been used to test the ability of various EBV vaccine formulations to induce CTL responses. Particular use has been made of HLA A2 transgenic mice which express a chimeric class I molecule composed of the alpha 1 and 2 domains of the human A*0201 allele and the alpha 3 domains of the mouse H-2Kb class I molecules (Duraiswamy et al., 2003). These mice are valuable in several regards. Firstly, they can be used to assess the ability of potential formulations to activate EBV-specific CTL responses and secondly these activated T cells are subsequently available to determine the efficacy of these CTL to resist the expansion of EBV-driven malignancies expanding in SCID mice. However, it should not be assumed that the relative dominance between individual CTL epitopes seen in response to immunization of human HLA transgenic mice will reflect immunodominance seen in response to vaccination of humans.

Primate models

Three species of new world monkeys that can be experimentally infected have been studied. Cottontop tamarins (Saguinus oedipus) inoculated with high titred EBV develop multiple B-cell lymphomas. Protection from the formation of these lymphomas has been used as a basis for testing various EBV formulations. The cottontop tamarin has a number of shortcomings as a model of EBV infection and disease. This species is not infected by the oral route and does not sustain a persistent infection. A further complication is that the tamarin has an unusually restricted histocompatibility complex polymorphism and only expresses the alleles G, F and E, associated primarily with NK cell function (Cadavid et al., 1999). Moreover, contrary to earlier beliefs, tamarins and marmosets have been found to carry their own resident γ-herpesviruses (de Thoisy et al., 2003) which might complicate the results of immunogenicity studies in these species. Other primate models include the common marmoset (Callithrix jaccus) which presents a rather ill-defined IM-like syndrome and the owl monkey (Aotus trivirgatus) which is susceptible to EBV-induced lymphomas (Epstein et al., 1973; Shope et al., 1973). However none of these models are ideal as there are no demonstrable asymptomatic virus persistence and/or replication in these primate models. Furthermore the route of experimental challenge in these models is generally intra-peritoneal and the virus-cell relationships established in human mucosal surfaces might be quite different.

Perhaps the best primate model is that based on infection of rhesus monkeys with the rhesus lymphocryptovirus (LCV). This virus, which shares significant sequence homology with EBV, reproduces many of the key events associated with primary EBV infection when these primates are infected orally. Moreover, this model appears to be relevant to the testing of vaccine formulations since animals show resistance to a second challenge following primary infection (Moghaddam et al., 1997). Although, this model is likely to prove expensive when used to screen potential formulations, it may provide an excellent system for demonstrating final confirmation of the efficacy of a formulation before human trials begin.

STRATEGIES FOR EBV VACCINE DEVELOPMENT

There are two schools of thought in regard to the development of an EBV vaccine. On the one hand there are those who would seek to develop a vaccine capable of preventing primary infection by inducing sterilizing immunity. The alternative view is towards a CTL-based vaccine formulation capable of modifying clinical symptoms (in the case of IM) or by specifically targeting EBV-infected tumor cells. The latter strategy might not only involve the development of formulations capable of boosting CTL numbers but might also include immune response modification and targeting to the site of tumor formation. The fact that a single virus particle successfully infecting a single target cell will be enough to establish persistent infection raises serious doubts concerning the feasibility of sterilizing immunity (Tibbetts et al., 2003). This is particularly highlighted in the case of individuals residing in developing countries where primary infection is acquired soon after birth.

Given the fact that the individuals at greatest risk of developing PTLD are EBV seronegative graft recipients, there is reason to believe that a vaccine that protects from the clinical consequences of primary EBV infection would largely protect individuals from the development of PTLD. Thus it might be anticipated that such a vaccine might be given to transplant patients at the same time they are receiving a cocktail of other vaccines.

One of the earliest and most successful methods of creating vaccines has been the use of live, attenuated forms of the pathogen. This approach evokes both humoral and cellular immunity because the full repertoire of viral proteins is presented in the same quantity and context as in the virulent form. This method has proven successful in developing the chickenpox vaccine, currently the only herpesvirus vaccine licensed by the FDA. However, in the case of EBV, the potential oncogenicity of the virus precludes its use particularly in the case of a vaccine to IM. Another obstacle towards developing such an attenuated vaccine is the absence of an efficient *in vitro* culture system capable of yielding the quantities of virus required for a commercially viable vaccine.

DEVELOPMENT OF A VACCINE TO PREVENT INFECTIOUS MONONUCLEOSIS

Commercial considerations have meant that the focus of attention in developing a vaccine to EBV has been dominated by IM. IM affects an estimated 250,000 young adults in the USA and Europe annually. Approximately 90 percent of adults in the USA have been infected with EBV and between 35 and 50 percent of young adults will develop IM when exposed to the virus for the first time. Prophylactic EBV vaccine development has focused on the gp350 and EBNA3 proteins. The major EBV envelope glycoprotein, gp350, binds to the CR2 complement receptor on B-cells and is consistent with it being a target for neutralizing antibodies. Following attachment, the virus infects the target cell through envelope fusion events involving other EBV envelope glycoproteins gp85, gp42 and gp25 (Borza and Hutt-Fletcher, 2002). However, more recent work has shown that gp350 is not an absolute requirement for infection to occur (Janz et al., 2000).

The other strategy involves targeting CTL epitope determinants either on recently infected B cells in the case of IM or on tumor cells in the case of EBV-associated malignancies.

GP350-based vaccine

Oligoclonal B-cell lymphomas closely resembling those seen in PTLD can be routinely induced in cottontop tamarins following infection with EBV. Recombinant gp350 in combination with adjuvants or when expressed in vaccinia or adenovirus vectors induced protective immunity in these primates. Protective immunity was not dependent on the induction of gp350-specific neutralizing antibodies but was achieved through cell-mediated immune responses (Morgan, 1992). Gp350 vaccine formulations have since been shown to induce CTL responses as well as neutralizing antibody (Khanna et al.,1999A; Wilson et al.,1996). The mechanism of protection in this animal model is unknown. Since the tumor cells have a Latency III phenotype and do not express gp350, the induced immune responses were probably able to reduce the effective virus dose on challenge since the induction of tumors by EBV is dose-dependent. Derivatives of the gp350 vaccines described above (Jackman et al.,1999) have entered small-scale human trials (Gilbert, 1999). The fact that little attention has been placed on developing DNA vaccines coding for gp350 may be related to complications surrounding the commercial exploitation of this technology. Immunization of mice with such a vector gave rise to antibodies to gp350, antibody-dependent cellular cytotoxicity (ADCC) and gp350-specific CTLs (Jung et al., 2001).

Some development work has been carried out on the generation of a recombinant VZV vaccine vector for the delivery of EBV genes, and VZV recombinants were produced which are able to express EBV gp350 (Lowe et al., 1987). It may be timely to explore this option further given the success of the Oka VZV vaccine strain and its incorporation into national vaccine programs in the USA, Japan and elsewhere.

There have been two EBV phase-I vaccine trials to-date involving gp350. The first (Gu et al., 1995) was conducted in China using a recombinant vaccinia vector gp220/350 in three distinct human populations: EBV-positive adults pre-exposed to vaccinia virus, EBV-positive children not pre-exposed to vaccinia virus, and EBV-negative infants not exposed to the vaccinia virus. The vaccination resulted in (a) no significant impact on the level of EBV antibody in the adults; (b) an increase in the level of EBV-neutralizing antibody titers in the vaccinated children, while antibodies to the viral capsid antigen remained the same; and (c) development of gp350 antibodies with neutralizing properties *in vitro* in all vaccinated infants. The second phase I clinical trial conducted by Aviron (Gilbert, 1999) demonstrated safety and immunogenicity for a subunit vaccine containing the gp220/350 surface glycoprotein. The trial was a randomized, double-blind study, containing 67 healthy adult participants.

CTL-based vaccine for IM

A large number of CTL epitopes encoded within EBV proteins have now been defined. In particular, there is known to be a focus of epitopes within EBNA3 proteins. The efficacy of using a CTL-based approach came from the demonstration that autologous $CD8^+$ T cells against EBNA3, propagated *ex vivo* and introduced into immunosuppressed patients at high risk of PTLD, were able to prevent PTLD and in some cases cause PTLD to regress (Gottschalk et al., 2002). This observation raises the likelihood that CTLs specific for some EBV latent proteins can control the propagation of EBV-infected B-cells *in vivo*. This background has justified a clinical trial based on a peptide restricted by HLA B8. This approach utilized an EBNA3 peptide epitope (FLRGRAYGL), incorporated into a water-in-oil emulsion adjuvant, and demonstrated that such a formulation was safe and induced an epitope-specific CTL response (Moss et al.,1998). Ultimately, a collection of peptide epitopes may be used to span the majority of HLA types and encompass strain variation in target populations. Until it is known which T-cell specificities are important in protecting against IM, there may be a case for constructing an epitope vaccine using epitopes from both lytic and latent proteins. Vaccination with a latency sequence inserted into a DNA plasmid expression vector has been tested in the MHV 68 murine γ-herpesvirus model using the M2 latency-associated gene. M2 DNA vaccination had no effect on virus replication in the lung but did reduce the latently infected cell burden in the early, but not the later, stages of infection (Usherwood et al., 2001).

An EBV vaccine based on CTL epitopes will need to incorporate protective determinants restricted by a range of class I MHC to provide cover for all human populations. In considering this selection it should also be borne in mind that the response seen during natural primary infection may be different from the immunogenic potential of CTL epitopes delivered in an effective formulation. For example, a hierarchy of HLA-A2-restricted CTL responses was established in acute IM and in healthy seropositive individuals and the relative immunogenicity of these epitopes assessed in HLA-A2 transgenic mice (Bharadwaj et al., 2001). When analyzed using the ELISPOT assay of interferon-γ release, there was no obvious correlation between the strength of the response seen to individual epitopes in transgenic mice and that seen in healthy sero-positive individuals or during acute IM. This is an important observation and serves to highlight the difference in responses seen as a result of natural infection and that deliverable by a formulation in which all of the epitopes are presented to the immune system at the same concentration and at the same site. It might thus be possible to boost protective responses to individual epitopes in a vaccine preparation to levels not seen in naturally infected individuals.

As with any subunit vaccine, critical decisions need to be reached not only on the choice of immunogen but on the means of delivering a comparatively large number of CTL epitopes that might be needed to span the MHC diversity seen in man. Although, formally this requirement might be satisfied by mixing peptide epitopes, technical and regulatory considerations are likely to restrict the use of this approach. An alternative approach under active consideration is to make use of a technical advance (Thomson et al., 1995; 1998) in which minimal EBV CTL epitopes are encoded in a recombinant or synthetic polyepitope protein. The constituent epitopes are processed and presented on the cell surface and can efficiently activate CTL responses. Thus, for example, it might be possible to link an array of dominant CTL epitopes from EBV latent proteins as a polyepitope formulated in an adjuvant capable of inducing a CTL response. It is unlikely that regulatory authorities will permit the use of either purified DNA or a live vector to deliver such a formulation given that the recipients are likely to be healthy adolescents.

EBV-ASSOCIATED TUMORS

EBV is associated with a broad range of human malignancies of diverse cellular origin that arise in both immunosuppressed and immunocompetent individuals. These include BL, NPC, PTLD and HL (reviewed in Rickinson and Kieff, 1996; Khanna et al., 1995). The unlikely prospect of a prophylactic EBV vaccine that protects individuals against primary infection would obviously reduce the burden of EBV-associated cancers worldwide. A more feasible possibility is that a prophylactic vaccine, given before or after primary EBV-infection, could modify the virus life-cycle such that the incidence of certain EBV-associated malignancies might be reduced.

Although no tumors to date have consistently been controlled in humans by vaccine-induced immune responses, success with adoptive transfer of tumor-antigen-specific CTLs (Savoldo et al., 2000; Khanna et al., 1999B) have encouraged vaccine development efforts by demonstrating that CD8$^+$ EBV-specific T cells alone can reverse the growth of EBV-associated malignancies. Reconstitution or enhancement of T cell immunity using low cost prophylactic/therapeutic vaccine formulations incorporating immunogenic determinants has obvious practical advantages over cellular immunotherapy which requires specialized facilities that are likely to preclude their widespread use, particularly in developing countries.

A realistic aim for many current researchers in the field is to develop strategies to boost immunity in patients already harboring EBV-associated tumors. In most cases, these alternative vaccine approaches are aimed at generating a therapeutic CTL response by enhancing the existing immune response towards the particular viral antigens expressed by tumor cells. Preclinical studies in animal models based on recombinant viral vectors or plasmid DNA expressing either full length EBV antigens or CTL epitopes linked together as a polyepitope have shown encouraging results (Taylor et al., 2004; Duraiswamy et al., 2003; 2004). Different vaccine strategies are likely to be required to accommodate the unique characteristics of each tumor type. Although all the malignancies express some EBV latent antigens, most express a restricted array thereby limiting the potential viral antigens that can be targeted in vaccine design. Each EBV-associated malignancy is also characterized by a unique cellular phenotype which, to some extent, is influenced by the pattern of EBV gene expression, and this too presents problems for vaccine design. Furthermore, vaccination strategies will need to counter tumor-induced immune evasion strategies such as the expression of immunosuppressive cytokines, the presence of regulatory T cells and/or loss of antigen processing function. Co-delivery of immune potentiating cytokines and specific targeting of recombinant antigens through the MHC class I/II pathways is likely to enhance the efficacy of these vaccines (Thomson et al., 1998).

Burkitt's lymphoma

Perhaps the most challenging EBV-associated malignancy to control with a vaccine-based approach is BL due to a variety of features that reduce tumor cell susceptibility to EBV-specific CTL surveillance. Firstly viral gene expression is very limited, with many of most important target antigens of the CTL response not expressed. Until recently, EBNA1 was thought to be the only viral antigen expressed in endemic BL cells; however, the recent finding that three out of fifteen BL samples also expressed a truncated EBNA LP, EBNA3A, 3B and 3C has raised questions about this generalization (Kelly et al., 2002). Secondly, BL cells frequently show selective down-regulated expression of certain HLA class I alleles which also reduces the potential for CD8[+] T cells to control tumor outgrowth (Masucci et al., 1987; Torsteindottir et al., 1988). Thirdly, the absence of LMP1 expression in BL cells reduces susceptibility to CTL lysis because this antigen has the capacity to upregulates the antigen processing machinery in EBV-infected normal B cells (Pai et al., 2002). For example, BL cells express very low levels of the transporters associated with antigen processing (TAP1 and TAP2) which are responsible for transporting antigenic peptides from the cytoplasm into the ER, and this was shown to prevent recognition by virus-specific CTLs *in vitro* (Khanna et al., 1994; Rowe et al., 1995).

Is the EBNA1 antigen a potential target antigen for a vaccine designed to control BL and other EBV-associated malignancies? For many years, EBNA1 was thought to escape CTL recognition through a number of mechanisms. It was proposed that the glycine-alanine repeat domain within the antigen forms β sheets that are resistant to unfolding, thereby affecting its capacity to associate with various components of the ubiquitin/proteasome pathway, leading to inefficient processing for MHC class I-restricted CTL recognition (Levitskaya et al., 1997).

More recently, the glycine-alanine repeat domain was also shown to inhibit mRNA translation of EBNA1, leading to low antigen expression in virus-infected cells (Yin et al., 2003). Recent work, however, has shown that some EBNA1 epitopes are presented on EBV-infected cells at sufficient level for CTL recognition (Jones et al., 2003; Tellam et al., 2004; Lee et al., 2004; Voo et al., 2004). Furthermore, some EBNA1 CTL epitopes are highly immunogenic, stimulating large numbers of CTLs in healthy donors (Blake et al., 2000). These apparently contradictory results were explained by evidence that defective ribosomal products of the EBNA1 gene (missing the glycine-alanine repeat) are the primary source of $CD8^+$ T cell epitopes from this antigen (Tellam et al., 2004). This data therefore raises the exciting possibility that EBNA1 could be utilized in the development of novel therapeutic strategies against EBV-associated malignancies, all of which express EBNA1.

If class 1-binding CTL epitopes from EBNA1 are to be exploited in a therapeutic vaccine for BL, the low expression of proteins involved in antigen processing observed in these tumor cells will need to be addressed. Interestingly, it has been shown that treatment of the BL cells with soluble CD40 ligand (CD40L) is highly effective in reversing the down-regulated expression of TAP1 and TAP2 (Khanna et al., 1997A). Moreover, CD40L-treated BL cells regain susceptibility to EBV-specific CTL-mediated lysis. These data suggest that direct infusion of soluble CD40L at tumor sites, or cytokine-mediated induction of CD40 ligand on bystander lymphocytes should be considered as an alternative approach to restoring the immunogenicity of these malignant cells.

Another novel strategy that could be utilized in conjunction with CD40L treatment of BL cells is to modulate the stability of EBNA1 in normal and malignant cells. This could be used to boost $CD8^+$ T cell responses to EBNA1 or to enhance presentation of EBNA1 epitopes on tumor cells. These schemes may involve treatment of virus-infected cells with synthetic or biological mediators capable of enhancing stable ubiquitination and rapid intracellular degradation of EBNA1 *in vivo*. Since the substrate specificity of the ubiquitin-proteasome pathway is conferred by the E3 ubiquitin-protein ligases, one approach may involve manipulation of the ubiquitin-dependent proteolytic machinery by targeting specific E3 ubiquitin-protein ligases to direct the degradation of EBNA1 (Tellam et al., 2003).

The MHV-68 model of γ-herpesvirus infection has indicted a role for $CD4^+$ T cells in both the control of infection (Hogan et al., 2001) and in the control of tumor cells expressing MHV-68 antigens. When MHV-68-infected S11 cells were injected subcutaneously into nude mice, adoptively transferred lymphocytes caused the regression of S11 tumors, and $CD4^+$ T cells were most effective in preventing tumor formation. $CD4^+$ T cells were also present in the regressing tumors (Robertson et al., 2001). These studies raise the possibility that manipulation of the EBV-specific $CD4^+$ T cell response could also be exploited to treat EBV-associated malignancies, and this could be particularly useful for BL due to the down-regulated expression of proteins involved in class I antigen processing. Supporting this notion, recent studies have shown that BL cell lines displaying antigen processing defects through the MHC class I pathway are efficiently recognized by EBV-specific $CD4^+$ CTLs. Furthermore, these tumor cells express normal levels of all the essential components involved in the processing of T cell epitopes through the MHC class II pathway

(Khanna et al., 1997B). The importance of these studies has been strengthened by the observation that CD4[+] EBNA1-specific CTLs from healthy virus carriers can efficiently recognize virus-infected normal B cells as well as BL cells expressing only EBNA1 (Munz et al., 2000; Paludan et al., 2002). These observations raise the possibility that a vaccine based on EBNA1 that induces a strong CD4[+] T cell response may provide therapeutic benefit to BL patients.

One attractive way to enhance the delivery of EBNA1 through the MHC class II pathway which could be exploited to boost EBNA-1 specific CD4[+] T cell responses in BL patients, is to direct the antigen to the endosomal or lysosomal compartments. MHC class II-restricted presentation of endogenously synthesized proteins mainly involves membrane antigens that are thought to enter the endosomal or lysosomal pathway by internalization from the cell surface. The lysosome-associated membrane protein (LAMP-1) and the invariant chain are transmembrane proteins, which are predominantly localized in the lysosomes and endosomes, respectively. The cytoplasmic domains of these proteins contain specific targeting or address signals that mediate their translocation to the specific compartments. Previous studies have shown that these targeting signal sequences can be utilized to direct multiple MHC class II-restricted CTL epitopes into the endosomal and lysosomal compartments (Thomson et al., 1998). This approach not only preferentially translocates the polyepitope protein to these compartments but also enhances endogenous presentation of CTL epitopes. Furthermore, this strategy was successfully used to activate a virus-specific memory CTL response from peripheral blood lymphocytes. Endosome/lysosome-targeted EBNA1 is currently being tested in our laboratories as a therapeutic vaccine in a non-immunogenic murine B cell tumor model in which the tumor cells express EBNA1. If this vaccine strategy is successful in preventing the outgrowth of these tumor cells, it is possible that an EBNA1 vaccine may not only be applicable to BL but also against other EBV-associated malignancies such as HD and NPC.

Hodgkin's lymphoma and nasopharyngeal carcinoma

EBV is detected in up to half of all HL tumors from people in developing countries and most undifferentiated NPC cases. Current evidence suggests that the overall EBV-specific immune response in patients with these malignancies is relatively normal, although there is evidence of a depressed response in late-stage NPC patients (Moss et al., 1983). Furthermore, an immunohistochemical analysis of fresh biopsies and laboratory-established tumor lines indicates that malignant cells in both NPC and HL express normal levels of HLA class I and TAP-1/2 (Khanna et al., 1998A; Lee et al., 1998). Despite arising in different cell types and at different locations, HL and NPC tumor cells express the same restricted subset of EBV antigens, EBNA1, LMP1 and LMP2. Since all these antigens have been shown to include T cell epitopes, it seems likely that immune evasion mechanisms are operating. It is, however, notable that the number of CTLs in the circulation of healthy individuals that recognize LMP1 and LMP2 is relatively low in comparison to responses to EBNA3 and BZLF1 (Khanna et al., 1998B). Both LMP1 and LMP2 may interfere with their own MHC class I presentation and may be poor targets and/or weak inducers of CTLs (Dukers et al., 2000; Ong et al., 2003). Moreover, independent studies by different groups have also shown that Reed-Sternberg cells in HL secrete anti-inflammatory cytokines (such as IL10, IL13, TARC and TGF-β), which could block tumor-infiltrating virus-specific T cell responses (Herbst et al., 1996; Hsu et al., 1993). These effects could be mediated through chronic or aberrant STAT activation in EBV-associated tumors that express LMP1 (Chen et al., 2001; Wang et al., 2004).

Another important observation in regard to CTL recognition of LMP1 and LMP2 is that the majority of the T cell epitopes from these proteins are processed through a TAP-independent pathway (Khanna et al., 1996; Lee et al., 1996). Interestingly, TAP-deficient LCLs expressing LMP proteins are more efficiently recognized by LMP-specific CTLs than TAP$^+$ LCLs. These observations suggest that in the absence of TAP, ER-generated LMP epitopes are presented more efficiently. Thus presentation of these epitopes may be significantly reduced if TAP is expressed, as the peptides originating from the cytoplasmic compartment compete with ER generated LMP epitopes. It is therefore tempting to speculate that LMP$^+$ NPC and HD cells may have evolved to maintain TAP expression which limits the presentation of CTL epitopes from LMP1 and LMP2 and thus allows these to escape CTL recognition *in vivo*.

Recent promising results using adoptively transferred CTLs to treat a limited number of HL and NPC patients (Straathof et al., 2004; Roskrow et al., 1998) have indicated that it might be possible to control these malignancies by boosting CTL responses to EBNA1 and/or the LMP proteins using an appropriately designed vaccine formulation. Since LMP1 has the capacity to initiate an independent neoplastic process in normal cells, it will not be possible to immunize using full-length protein expression. A strategy to overcome this problem involves the delivery of immunogenic determinants from LMP1, LMP2 and EBNA1 as a "polyepitope" vaccine (Thomson et al., 1995). Indeed, recent studies have shown that multiple HLA A2-restricted LMP1 CTL epitopes, when used as a polyepitope vaccine in a poxvirus vector, efficiently induced a strong CTL response, and this response could reverse the outgrowth of LMP1-expressing tumors in HLA-A2/Kb mice (Duraiswamy et al., 2003A). The potent responses to polyepitope proteins may reflect their lack of stability within the cytoplasm as a result of limited secondary and tertiary structure. Polyepitope vaccines can therefore be used to induce long-term protective CTL responses against a large number of CTL epitopes using a relatively small construct without any obvious need for a cognate help. This could be exploited to overcome any potential problem with EBV genetic variants in different ethnic groups of the world (Duraiswamy et al., 2003B).

Although the poxvirus polyepitope vaccine vector provides long-lived expression of encoded epitopes, there are concerns in terms of its safety profile, with adverse side-effects including post-vaccine encephalitis when used in humans. Moreover, the poxvirus-based LMP1-polyepitope vaccine tested in these studies contained only HLA A2-restricted epitopes, and the HLA A2 allele is carried by only about 55% of the individuals in most populations. If a CTL-based therapy for NPC and HL is to be applicable to a significant number of patients, the target epitopes must bind to HLA alleles present at high frequency in the patient population. In this context, in addition to LMP1, previously defined LMP2-specific responses restricted through A11, A24 and B40 are of particular interest because these alleles are very common in the Southern Chinese population where NPC is endemic (A11, 56%; A24, 27%; B40, 28%). To address these problems, a novel approach has been devised in our laboratories to activating LMP-specific CTL responses with a replication-incompetent adenovirus encoding multiple epitopes from LMP1 and LMP2 (Duraiswamy et al., 2004). This vaccine includes both LMP1 and LMP2 epitopes restricted through HLA alleles common in different ethnic groups including NPC endemic regions (HLA A2, A11, A23, A24, B27, B40 and B57). It has been estimated that these optimally selected MHC class I-restricted epitopes would include more than 90% of the Asian, African and Caucasian populations. Attractive features of adenovirus-based vaccines are their well-characterized genetic arrangement and function, as well as their extensive and safe usage in North American army recruits without inducing adverse side effects (Imler, 1995). Adenovirus-based vectors are being increasingly recognized for their high efficiency and low toxicity and have been used in multiple human gene therapy clinical trials and preclinical vaccine applications (Kusumoto et al., 2001). Our studies with the adenovirus-based LMP polyepitope vaccine have shown that each of the epitopes in this vaccine is not only efficiently processed endogenously by the human cells but also recalls memory CTL responses specific for LMP antigens in healthy virus carriers and HD patients. Furthermore, the adenoviral polyepitope vaccine is capable of inducing a primary T cell response in mice, which was shown to be therapeutic in a tumor challenge system. Other vaccine strategies showing promising results include the use of recombinant viral vectors expressing full-length LMP2 fused to the CD4 epitope-rich C-terminal domain of EBNA1 (Taylor et al., 2004).

CONCLUDING REMARKS

Current knowledge of the immune control of EBV infection has reached a level where vaccine trials aimed at controlling IM, PTLD, NPC and HL are justified, particularly in light of the recent success with adoptive immunotherapy to a number of these conditions. Recent advances in characterizing the T cell response to the EBNA1 gene suggest that it may also be possible to treat BL with a vaccine derived from this antigen; however it would be advisable to attempt adoptive immunotherapy before vaccine trials are considered. One of the key challenges facing researchers attempting pioneering vaccine trials are the increasingly stringent regulatory issues involved. Too often, regulatory authorities apply restrictive criteria based on a model of product development which is better suited to the drug industry than to the academic research institute. For the future benefit of patients with EBV-associated malignancies, it is vital that regulatory authorities do not apply stringent controls on small-scale trials applied to patients with late stage malignancies, so that academic-led interventional research can advance rapidly.

References

Bharadwaj, M., Sherritt, M., Khanna, R., and Moss, D.J. (2001). Contrasting Epstein-Barr virus specific cytotoxic T cell responses to HLA A2 restricted epitopes in humans and HLA transgenic mice: implications for vaccine design. Vaccine *19*, 3769-3777.

Blake, N., Haigh, T., Shaka'a, G., Croom-Carter, D., and Rickinson, A. (2000). The importance of exogenous antigen in priming the human CD8[+] T cell response: lessons from the EBV nuclear antigen EBNA1. J Immunol. *165*, 7078-7087.

Borza, C.M., and Hutt-Fletcher, L.M. (2002). Alternate replication in B cells and epithelial cells switches tropism of Epstein-Barr virus. Nat. Med. *8*, 594-599.

Cadavid, L.F., Mejia, B.E., and Watkins, D.I. (1999). MHC class I genes in a New World primate, the cotton-top tamarin (Saguinus oedipus), have evolved by an active process of loci turnover. Immunogenetics *49*, 196-205.

Chen, H., Lee, J.M., Zong, Y., Borowitz, M., Ng, M.H, Ambinder, R.F., and Hayward, S.D. (2001). Linkage between STAT regulation and Epstein-Barr virus gene expression in tumors. J. Virol. *75*, 2929-2937.

de Thoisy, B., Pouliquen, J.F., Lacoste, V., Gessain, A., and Kazanji, M. (2003). Novel gamma-1 herpesviruses identified in free-ranging new world monkeys (golden-handed tamarin [Saguinus midas], squirrel monkey [Saimiri sciureus], and white-faced saki [Pithecia pithecia]) in French Guiana. J. Virol. *77*, 9099-9105.

Dukers, D.F., Meij, P., Vervoort, M.B., Vos, W., Scheper, R.J., Meijer, C.J., Bloemena, E., and Middeldorp, J.M. (2000). Direct immunosuppressive effects of EBV-encoded latent membrane protein 1. J. Immunol. *165*, 663-670.

Duraiswamy, J., Sherritt, M., Thomson, S., Tellam, J., Cooper, L., Connolly, G., Bharadwaj, M., and Khanna, R. (2003). Therapeutic LMP1 polyepitope vaccine for EBV-associated Hodgkin disease and nasopharyngeal carcinoma. Blood *101*, 3150-3156.

Duraiswamy, J., Burrows, J.M., Bharadwaj, M., Burrows, S.R., Cooper, L., Pimtanothai, N., and Khanna, R. (2003). Ex vivo analysis of T-cell responses to Epstein-Barr virus-encoded oncogene latent membrane protein 1 reveals highly conserved epitope sequences in virus isolates from diverse geographic regions. J. Virol. *77*, 7401-7410.

Duraiswamy, J., Bharadwaj, M., Tellam, J., Connolly, G., Cooper, L., Moss, D., Thomson, S., Yotnda, P., and Khanna, R. (2004). Induction of therapeutic T-cell responses to subdominant tumor-associated viral oncogene after immunization with replication-incompetent polyepitope adenovirus vaccine. Cancer Res. *64*, 1483-1489.

Epstein, M.A., Rabin, H., Ball, G., Rickinson, A.B., Jarvis, J., and Melendez, L.V. (1973) Pilot experiments with EB virus in owl monkeys (Aotus trivirgatus). II. EB virus in a cell line from an animal with reticuloproliferative disease. Int. J. Cancer. *12*, 319-332.

Gu, S.Y., Huang, T.M., Ruan, L., Miao, Y.H., Lu, H., Chu, C.M., Motz, M., and Wolf, H. (1995). First EBV vaccine trial in humans using recombinant vaccinia virus expressing the major membrane antigen. Dev. Biol. Stand. *84*, 171-177.

Gilbert, K. (1999). Results of a Phase I clinical trial of an Epstein-Barr virus vaccine. In Aviron Press Release. Mountain View, CA, USA.

Gottschalk, S., Heslop, H.E., and Roon, C.M. (2002). Treatment of Epstein-Barr virus-associated malignancies with specific T cells. Adv. Cancer Res. *84*, 175-201.

Herbst, H., Foss, H.D., Samol, J., Araujo, I., Klotzbach, H., Krause, H., Agathanggelou, A., Niedobitek, G., and Stein, H. (1996). Frequent expression of interleukin-10 by Epstein-Barr virus-harboring tumor cells of Hodgkin's disease. Blood. *87*, 2918-2929.

Hogan, R.J., Zhong, W., Usherwood, E.J., Cookenham, T., Roberts, A.D., and Woodland, D.L. (2001). Protection from respiratory virus infections can be mediated by antigen-specific CD4[+] T cells that persist in the lungs. J. Exp. Med. *193*, 981-986.

Hsu, S.M., Lin, J., Xie, S.S., Hsu, P.L., and Rich, S. (1993). Abundant expression of transforming growth factor-beta 1 and -beta 2 by Hodgkin's Reed-Sternberg cells and by reactive T lymphocytes in Hodgkin's disease. Hum. Pathol. *24*, 249-255.

Imler, J.L. (1995). Adenovirus vectors as recombinant viral vaccines. Vaccine. *13*, 1143-1151.

Islas-Ohlmayer, M., Padgett-Thomas, A., Domiati-Saad, R., Melkus, M.W., Cravens, P.D., Martin, M.P., Netto, G., and Garcia, J.V. (2004). Experimental infection of NOD/SCID mice reconstituted with human CD34+ cells with Epstein-Barr virus. J. Virol. *78*, 13891-13900.

Janz, A., Oezel, M., Kurzeder, C., Mautner, J., Pich, D., Kost, M., Hammerschmidt, W., and Delecluse, H.J. (2000). Infectious Epstein-Barr virus lacking major glycoprotein BLLF1 (gp350/220) demonstrates the existence of additional viral ligands. J. Virol. *74*, 10142-10152.

Jackman, W.T., Mann, K.A., Hoffmann, H.J., and Spaete, R.R. (1999). Expression of Epstein-Barr virus gp350 as a single chain glycoprotein for an EBV subunit vaccine. Vaccine *17*, 660-668.

Jones, R.J., Smith, L.J., Dawson, C. W., Haigh, T., Blake, N.W., and Young, L.S. (2003). Epstein-Barr virus nuclear antigen 1 (EBNA1) induced cytotoxicity in epithelial cells is associated with EBNA1 degradation and processing. Virology *313*, 663-676.

Jung, S., Chung, Y.K., Chang, S.H., Kim, J., Kim, H.R., Jang, H.S., Lee, J.C., Chung, G.H., and Jang, Y.S. (2001). DNA-mediated immunization of glycoprotein 350 of Epstein-Barr virus induces the effective humoral and cellular immune responses against the antigen. Mol. Cells *12*, 41-49.

Kelly, G., Bell, A. and Rickinson, A. (2002). Epstein-Barr virus-associated Burkitt lymphomagenesis selects for downregulation of the nuclear antigen EBNA2. Nat. Med. *8*, 1098-1104.

Khanna, R., Burrows, S.R., Argaet, V., and Moss, D.J. (1994). Endoplasmic reticulum signal sequence facilitated transport of peptide epitopes restores immunogenicity of an antigen processing defective tumor cell line. Int. Immunol. *6*, 639-645.

Khanna, R., Burrows, S.R., Moss D.J. (1995). Immune regulation in Epstein–Barr virus associated diseases. Microbiol. Rev. *59*, 387–405.

Khanna, R., Burrows, S.R., Moss, D.J., and Silins, S.L. (1996). Peptide transporter (TAP-1 and TAP-2)-independent endogenous processing of Epstein-Barr virus (EBV) latent membrane protein 2A: implications for cytotoxic T-lymphocyte control of EBV-associated malignancies. J. Virol. *70*, 5357-5362.

Khanna, R., Cooper, L., Kienzle, N., Moss, D.J., Burrows, S.R., and Khanna, K.K. (1997A). Engagement of CD40 antigen with soluble CD40 ligand up-regulates peptide transporter expression and restores endogenous processing function in Burkitt's lymphoma cells. J. Immunol. *159*, 5782-5785.

Khanna, R., Burrows, S.R., Thomson, S.A., Moss, D.J., Cresswell, P., Poulsen, L.M., and Cooper, L. (1997B). Class I processing-defective Burkitt's lymphoma cells are recognized efficiently by CD4+ EBV-specific CTLs. J. Immunol. *158*, 3619-3625.

Khanna, R., Busson, P., Burrows, S.R., Raffoux, C., Moss, D.J., Nicholls, J.M., and Cooper, L. (1998A). Molecular characterization of antigen-processing function in nasopharyngeal carcinoma (NPC): evidence for efficient presentation of Epstein-Barr virus cytotoxic T-cell epitopes by NPC cells. Cancer Res. *58*, 310-314.

Khanna, R., Burrows, S.R., Nicholls, J., and Poulsen, L.M. (1998B). Identification of cytotoxic T cell epitopes within Epstein-Barr virus (EBV) oncogene latent membrane protein 1 (LMP1): evidence for HLA A2 supertype-restricted immune recognition of EBV-infected cells by LMP1-specific cytotoxic T lymphocytes. Eur. J. Immunol. *28*, 451-458.

Khanna, R., Moss, D.J., and Burrows, S.R. (1999A). Vaccine strategies against Epstein-Barr virus-associated diseases: lessons from studies on cytotoxic T-cell-mediated immune regulation. Immunol. Rev. *170*, 49-64.

Khanna, R., Bell, S., Sherritt, M., Galbraith, A., Burrows, S.R., Rafter, L., Clarke, B., Slaughter, R., Falk, M.C., Douglass, J., Williams, T., Elliott, S.L., and Moss, D.J. (1999B). Activation and adoptive transfer of Epstein-Barr virus-specific cytotoxic T cells in solid organ transplant patients with posttransplant lymphoproliferative disease. Proc. Natl. Acad. Sci. USA. *96*, 10391-10396.

Khanna, R., Tellam, J., Duraiswamy, J., and Cooper, L. (2001). Immunotherapeutic strategies for EBV-associated malignancies. Trends Mol. Med. *7*, 270-276.

Kusumoto, M., Umeda, S., Ikubo, A., Aoki, Y., Tawfik, O., Oben, R., Williamson, S., Jewell, W., and Suzuki, T. (2001). Phase 1 clinical trial of irradiated autologous melanoma cells adenovirally transduced with human GM-CSF gene. Cancer Immunol. Immunother. *50*, 373-381.

Lee, S.P., Thomas, W.A., Blake, N.W., and Rickinson, A.B. (1996). Transporter (TAP)-independent processing of a multiple membrane-spanning protein, the Epstein-Barr virus latent membrane protein 2. Eur. J. Immunol. *26*, 1875-1883.

Lee, S.P., Constandinou, C.M., Thomas, W.A., Croom-Carter, D., Blake, N.W., Murray, P.G., Crocker, J., and Rickinson, A.B. (1998). Antigen presenting phenotype of Hodgkin Reed-Sternberg cells: analysis of the HLA class I processing pathway and the effects of interleukin-10 on Epstein-Barr virus-specific cytotoxic T-cell recognition. Blood. 92, 1020-1030.

Lee, S.P., Brooks, J.M., Al-Jarrah, H., Thomas, W.A., Haigh, T.A., Taylor, G.S., Humme, S., Schepers, A., Hammerschmidt, W., Yates, J.L., Rickinson, A.B., and Blake, N.W. (2004). CD8 T cell recognition of endogenously expressed Epstein-Barr virus nuclear antigen 1. J. Exp. Med. *199*, 1409-1420.

Levitskaya, J., Sharipo, A., Leonchiks, A., Ciechanover, A., and Masucci, M.G. (1997). Inhibition of ubiquitin/proteasome-dependent protein degradation by the Gly-Ala repeat domain of the Epstein-Barr virus nuclear antigen 1. Proc. Natl. Acad. Sci. USA. *94*, 12616-12621.

Lowe, R.S., Keller, P.M., Keech, B.J., Davison, A.J., Whang, Y., Morgan, A.J., Kieff, E., and Ellis, R.W. (1987). Varicella-zoster virus as a live vector for the expression of foreign genes. Proc. Natl. Acad. Sci. U.S.A. *84*, 3896-3900.

Masucci, M.G., Torsteinsdottir, S., Colombani, J., Brautbar, C., Klein, E., and Klein, G. (1987). Down-regulation of class I HLA antigens and of the Epstein-Barr virus-encoded latent membrane protein in Burkitt lymphoma line. Proc. Natl. Acad. Sci. USA. *84*, 4567-4571.

Moghaddam, A., Rosenzweig, M., Lee-Parritz, D., Annis, B., Johnson, R.P., and Wang, F. (1997). An animal model for acute and persistent Epstein-Barr virus infection. Science. *276*, 2030-2033.

Morgan, A.J. (1992). Epstein-Barr virus vaccines. Vaccine *10*, 563-571.

Moss, D.J., Chan, S.H., Burrows, S.R., Chew, T.S., Kane, R.G., Staples, J.A., and Kunaratnam, N. (1983). Epstein-Barr virus specific T-cell response in nasopharyngeal carcinoma patients. Int. J. Cancer. *32*, 301-305.

Moss, D.J., Suhrbier, A., and Elliott, S.L. (1998). Candidate vaccines for Epstein-Barr virus. Brit. Med. Journal *317*, 423-424.

Munz, C., Bickham, K.L., Subklewe, M., Tsang, M.L., Chahroudi, A., Kurilla, M.G., Zhang, D., O'Donnell, M. and Steinman, R.M. (2000). Human CD4+ T lymphocytes consistently respond to the latent Epstein-Barr virus nuclear antigen EBNA1. J. Exp. Med. *191*, 1649-1660.

Nash, A.A., Dutia, B.M., Stewart, J.P., and Davison, A.J. (2001). Natural history of murine gamma-herpesvirus infection. Philos. Trans. R. Soc. Lond. B. Biol. Sci. *356*, 569-579.

Ong, K.W., Wilson, A.D., Hirst, T.R., and Morgan, A.J. (2003). The B Subunit of Escherichia coli Heat-Labile Enterotoxin Enhances CD8+ Cytotoxic-T-Lymphocyte Killing of Epstein-Barr Virus-Infected Cell Lines. J. Virol. *77*, 4298-4305.

Pai, S., O'Sullivan, B.J., Cooper, L., Thomas, R., and Khanna, R. (2002). RelB nuclear translocation mediated by C-terminal activator regions of Epstein-Barr virus-encoded latent membrane protein 1 and its effect on antigen-presenting function in B cells. J. Virol. *76*, 1914-1921.

Paludan, C., Bickham, K., Nikiforow, S., Tsang, M.L., Goodman, K., Hanekom, W.A., Fonteneau, J.F., Stevanovic, S., and Munz, C. (2002). Epstein-Barr nuclear antigen 1-specific CD4+ Th1 cells kill Burkitt's lymphoma cells. J. Immunol. *169*, 1593-1603.

Rickinson, A.B. and Kieff, E. (1996). Epstein-Barr virus. In: Fields Virology, B.N. Fields et al.,., ed., (Lippincott–Raven Press), pp. 2397–2446.

Robertson, K.A., Usherwood, E.J., and Nash, A.A. (2001). Regression of a murine gammaherpesvirus 68-positive B-cell lymphoma mediated by CD4 T lymphocytes. J. Virol. *75*, 3480-3482.

Roskrow, M.A., Suzuki, N., Gan, Y., Sixbey, J.W., Ng, C.Y., Kimbrough, S., Hudson, M., Brenner, M.K., Heslop, H.E., and Rooney, C.M. (1998). Epstein-Barr virus (EBV)-specific cytotoxic T lymphocytes for the treatment of patients with EBV-positive relapsed Hodgkin's disease. Blood. *91*, 2925-2934.

Rowe, M., Khanna, R., Jacob, C.A., Argaet, V., Kelly, A., Powis, S., Belich, M., Croom-Carter, D., Lee, S., Burrows, S.R., Moss, D.J., and Rickinson, A.B. (1995). Restoration of endogenous antigen processing in Burkitt's lymphoma cells by Epstein-Barr virus latent membrane protein-1: coordinate up-regulation of peptide transporters and HLA-class I antigen expression. Eur. J. Immunol. *25*, 1374-1384.

Savoldo, B., Heslop, H.E., and Rooney, C.M. (2000). The use of cytotoxic T cells for the prevention and treatment of Epstein-Barr virus induced lymphoma in transplant recipients. Leuk. Lymphoma. *39*, 455-464.

Straathof, K.C., Bollard, C.M., Popat, U., Huls, M.H., Lopez, T., Morriss, M.C., Gresik, M.V., Gee, A.P., Russell, H.V., Brenner, M.K., Rooney, C.M., and Heslop, H.E. (2004). Treatment of Nasopharyngeal Carcinoma with Epstein-Barr virus-specific T lymphocytes. Blood. Nov 12

Shope, T., Dechairo, D., and Miller, G. (1973). Malignant lymphoma in cottontop marmosets after inoculation with Epstein-Barr virus. Proc. Natl. Acad. Sci. U.S.A. *70*, 2487-2491.

Taylor, G.S., Haigh, T.A., Gudgeon, N.H., Phelps, R.J., Lee, S.P., Steven, N.M., and Rickinson, A.B. (2004). Dual stimulation of Epstein-Barr Virus (EBV)-specific CD4+- and CD8+-T-cell responses by a chimeric antigen construct: potential therapeutic vaccine for EBV-positive nasopharyngeal carcinoma. J. Virol. *78*, 768-778.

Tellam, J., Connolly, G., Webb, N., Duraiswamy, J., and Khanna, R. (2003). Proteasomal targeting of a viral oncogene abrogates oncogenic phenotype and enhances immunogenicity. Blood. *102*, 4535-4540.

Tellam, J., Connolly, G., Green, K.J., Miles, J.J., Moss, D.J., Burrows, S.R., and Khanna, R. (2004) Endogenous presentation of CD8+ T cell epitopes from Epstein-Barr virus-encoded nuclear antigen 1. J. Exp. Med. *199*, 1421-1431.

Thomson, S.A., Khanna, R., Gardner, J., Burrows, S.R., Coupar, B., Moss, D.J., and Suhrbier, A. (1995). Minimal epitopes expressed in a recombinant polyepitope protein are processed and presented to CD8+ cytotoxic T cells: implications for vaccine design. Proc. Natl. Acad. Sci. U.S.A. *92*, 5845-5849.

Thomson, S.A., Burrows, S.R., Misko, I.S., Moss, D.J., Coupar, B.E., and Khanna, R. (1998). Targeting a polyepitope protein incorporating multiple class II-restricted viral epitopes to the secretory/endocytic pathway facilitates immune recognition by CD4+ cytotoxic T lymphocytes: a novel approach to vaccine design. J. Virol. *72*, 2246-2252.

Thomson, S.A., Elliott, S.L., Sherritt, M.A., Sproat, K.W., Coupar, B.E., Scalzo, A.A., Forbes, C.A., Ladhams, A.M., Mo, X.Y., Tripp, R.A., Doherty, P.C., Moss, D.J., and Suhrbier, A. (1996). Recombinant polyepitope vaccines for the delivery of multiple CD8 cytotoxic T cell epitopes. J. Immunol. *157*, 822-826.

Tibbetts S.A., Loh, J., van Berkel V., McClellan J.S., Jacoby M.A., Kapadia S.B., Speck S.H., and Virgin H.W. 4th. (2003). Establishment and maintenance of gamma herpes virus latency are independent of infective dose and route of infection. J. Virol. *77*, 7696-7701.

Torsteindottir, S., Brautbar, G., Klein, G., Klein, E., and Masucci, M.G. (1988). Differential expression of HLA antigens on human B-cell lines of normal and malignant origin: a consequence of immune surveillance or a phenotype vestige of the progenitor cells? Int. J. Cancer *41*, 913-919.

Traggiai, E., Chicha, L., Mazzucchelli, L., Bronz, L., Piffaretti, J.C., Lanzavecchia, A., and Manz, M.G. (2004). Development of a human adaptive immune system in cord blood cell-transplanted mice. Science. *304*, 104-107

Tugizov, S.M., Berline, J.W., and Palefsky, J.M. (2003). Epstein-Barr virus infection of polarized tongue and nasopharyngeal epithelial cells. Nat. Med. *9*, 307-314.

Usherwood, E.J., Ward, K.A., Blackman, M.A., Stewart, J.P., and Woodland, D.L. (2001). Latent antigen vaccination in a model gammaherpesvirus infection. J. Virol. *75*, 8283-8288.

Voo, K.S., Fu, T., Wang, H.Y., Tellam, J., Heslop, H.E., Brenner, M.K., Rooney C.M., and Wang, R.F. (2004). Evidence for the presentation of major histocompatibility complex class I-restricted Epstein-Barr virus nuclear antigen 1 peptides to CD8+ T lymphocytes. J. Exp. Med. *199*, 459-470.

Wang, T., Niu, G., Kortylewski, M., Burdelya, L., Shain, K., Zhang, S., Bhattacharya, R., Gabrilovich, D., Heller, R., Coppola, D., Dalton, W., Jove, R., Pardoll, D., and Yu, H. (2004). Regulation of the innate and adaptive immune responses by Stat-3 signaling in tumor cells. Nat. Med. *10,* 48-54.

Wilson, A.D., Shooshstari, M., Finerty, S., Watkins, P., and Morgan, A.J. (1996). Virus-specific cytotoxic T cell responses are associated with immunity of the cottontop tamarin to Epstein-Barr virus (EBV). Clin. Exp. Immunol. *103*, 199-205.

Yin, Y., Manoury, B., and Fahraeus, R. (2003). Self-inhibition of synthesis and antigen presentation by Epstein-Barr virus-encoded EBNA1. Science. *301*, 1371-1374.

From: Epstein-Barr Virus. Edited by: Erle S. Robertson

Chapter 31

Epstein Barr Virus Cell Based Immunotherapeutics

Gianpietro Dotti, Helen Heslop and Cliona Rooney*

ABSTRACT

The idea of using the immune system to control cancer in human derives from the concept of "immune surveillance" proposed by Burnet in early 1950. The adoptive transfer of antigen specific cytotoxic T-cells represent one the most advanced efforts to translate this concept in a clinical application. New tumor-associated antigens are continuously being identified and proposed as potential targets for immunotherapy of human malignancies. Other tumors are clearly associated with viral infections, and Epstein Barr Virus (EBV) related tumors express immunogenic viral proteins that make them particularly attractive for immunotherapy approaches. Over the last ten years major efforts to optimize clinical grade protocols for *ex vivo* expansion of EBV-specific cytotoxic T-lymphocytes (CTL) have been made. In this chapter we will summarize the clinical experiences of CTL adoptive transfer in different EBV-related diseases including Post Transplant Lymphoproliferative Disorders (PTLD), Hodgkin's lymphoma (HD), Nasopharyngeal carcinoma (NPC) and Severe Chronic EBV infection (SCAEBV). We also will discuss the limitations and the future strategies to improve the efficiency of this approach as well as potential applications in non-EBV related tumors.

INTRODUCTION: POST-TRANSPLANT LYMPHOPROLIFERATIVE DISORDERS (PTLD)

Organ transplantation is one of the more advanced frontiers of modern medicine. Dr Barnard reported the first heart transplantation in 1967 (Barnard, 1967). In the early 60's the first allogeneic stem cell transplantations (allo-SCT) were also performed (Buckner et al., 1970; Good et al., 1969). More recently, the application of orthotopic transplant has been expanded to many organs and to different pathologies (Gridelli and Remuzzi, 2000). However, new diseases, secondary to the immunosuppression required to maintain the graft, have emerged, including the peculiar lymphoid proliferations called Post-Transplant Lymphoproliferative Disorders (PTLD). The first case of "reticulum sarcoma" occurring in a patient with kidney transplant was described by Doak et al. (1968). These disorders were

*For correspondence email gdotti@bcm.tmc.edu

systematically characterized in 1981 by Frizzera et al. (1981). Soon after the first reported cases, the association between PTLD and EBV in patients chronically treated with immunosuppressive drugs became clear (Nagington and Gray, 1980), and currently PTLD are considered an EBV-related disease.

Incidence and risk factors

The overall incidence of PTLD varies according to different series, but is generally higher in solid organ transplant (SOT) recipients (1-20%) compared to allo-SCT recipients (1%). For SOT recipients, the incidence of PTLD depends on the type of transplant (kidney <1%, liver 2-8%, heart or heart-lung 2-10%, lung and small bowel 10-20%) and the type of immunosuppression (higher incidence in patients receiving combinations of immunosuppressive drugs or monoclonal antilymphocytic agents) (Penn, 1991). The highest incidence is reported in EBV-seronegative transplant recipients, which likely explains the overall higher incidence of PTLD in pediatric recipients, who are frequently EBV seronegative at time of transplant (Ho et al., 1988). For allo-SCT recipients, the risk for PTLD is significantly increased for recipients of T-cell depleted stem cell grafts (8-10%) or recipients of unrelated or human leukocyte antigen (HLA) mismatched transplant (10-20%) (Rooney et al., 1998; Curtis et al., 1999). However, the improved management of immunosuppression in the last years has significantly reduced the incidence of PTLD in both allo-SCT and SOT recipients.

EBV and PTLD

Most PTLD are EBV associated, even though tumors occurring late after transplant may have a different pathogenesis (Paya et al., 1999; Dotti et al., 2000). A commonly accepted pathogenetic model for EBV-related PTLD assumes that the impairment of the cytotoxic immune response secondary to the immunosuppressive treatment allows the uncontrolled expansion of EBV-transformed B lymphocytes expressing a full spectrum of viral latency associated proteins (EBV nuclear antigens EBNA1, -2, -3A, -3B, -3C and Latent membrane proteins LMP1 and LMP2) known as latency III, which is also observed in EBV transformed B lymphoblastoid cell line (LCL) (Frizzera et al., 1981; Klein and Purtilo, 1981). The subsequent, prolonged and uncontrolled expansion of polyclonal B cells predisposes to additional genetic mutations in some clones and to final emergence of a truly neoplastic proliferation (Frizzera et al., 1981; Klein and Purtilo, 1981).

PTLD have recently been included in the World Health Organization (WHO) classification of neoplastic diseases of the hematopoietic and lymphoid tissues (Harris et al., 1999). They include early lesion/mononucleosis-like, polymorphic and monomorphic PTLD. The majority of PTLD express B-cell markers and the EBV genome is almost invariably detected in early and polymorphic lesions, while monomorphic PTLD can be EBV negative especially when occurring late (more then 2 years) after transplantation (Dotti et al., 2000). Early lesions occur in the first years after transplant, typically in children and young patients, and are generally polyclonal proliferations. They may have a very aggressive and rapidly fatal course, but can regress after the reduction of immunosuppression (Harris et al., 1997). Polymorphic- and monomorphic-PTLD may also occur close to transplant, although monomorphic-PTLD is more frequent after the first year. They are monoclonal or oligoclonal tumors, frequently associated with additional genetic abnormalities and rarely regress after modulation of immunosuppression (Harris et al., 1997; Dotti et al., 2002).

Cell therapy for EBV-related PTLD

Currently, despite the general effort to produce guidelines for the treatment of PTLD (Paya et al., 1999), the management of this complex disorder remains controversial. However, the crucial role of the immune system became immediately evident from the pioneering experience of the Pittsburgh group that reported in 1984 the regression of lymphoid proliferations in solid organ transplant recipients after reduction or suspension of the immunosuppressive treatment (Starzl et al., 1984). This simple strategy is still recommended as first line approach in PTLD patients since remission can be obtained in 25-50% of patients (Paya et al., 1999). Analogously, in the allo-SCT setting, regressions of PTLD were achieved after the infusion of lymphocytes obtained from the bone marrow donor (donor lymphocytes infusion or DLI) (Papadopoulos et al., 1994), suggesting that a competent immune system can successfully control virus-driven lymphoproliferation. Both these approaches are simple and inexpensive, thus they can be adopted for the daily clinical practice even in non-specialized departments. However, in SOT recipients reactivation of the immune system induced by withdrawal of immunosuppression can be complicated by graft rejection in up to 50% of patients (Starzl et al., 1984; Paya et al., 1999). By comparison, in SCT-recipients DLI can promote the occurrence of acute or chronic graft-versus-host-disease (GVHD) in up to 40-50% of patients (Luznik and Fuchs, 2002). Despite careful monitoring and early intervention, these complications can be fatal in a significant number of patients. The adoptive transfer of EBV-specific T cells expanded *ex vivo* represents a rational approach to rebalance the impairment of the EBV immunity in these patients, minimizing the risk of GVHD and the risk of damaging the grafted organ.

Selection and expansion of clinical grade EBV-specific CTL

In healthy EBV seropositive individuals the frequency of circulating T-lymphocyte precursors, specific for EBV antigens ranges from 1/400 to 1/14,000. In 1993, our group established a procedure for the *ex vivo* expansion of clinical grade EBV-specific CTL in compliance with recommendations for good manufacture practices (GMP) (Figure 1) (Rooney et al., 1998; 1995). The first step in this procedure consists of the generation of an EBV transformed B lymphoblastoid cell line (LCL) from each donor/patient to use as EBV specific antigen presenting cells. For this step, peripheral blood-derived mononuclear cells (PBMC) (5×10^6) are incubated for 1 hour with concentrated supernatant from a master cell bank of the EBV producer cell line B95-8. After the infection, the cells are plated at 10^5 cells per well in presence of 1 µg/mL Cyclosporin A and fed weekly with media until LCL are established and expanded in a sufficient number for the stimulations of T-lymphocytes. Acyclovir is added to the culture immediately after the LCL establishment to inhibit the production of infectious virus. Overall this procedure takes 4-5 weeks from normal donors, but can be longer for patients with EBV-

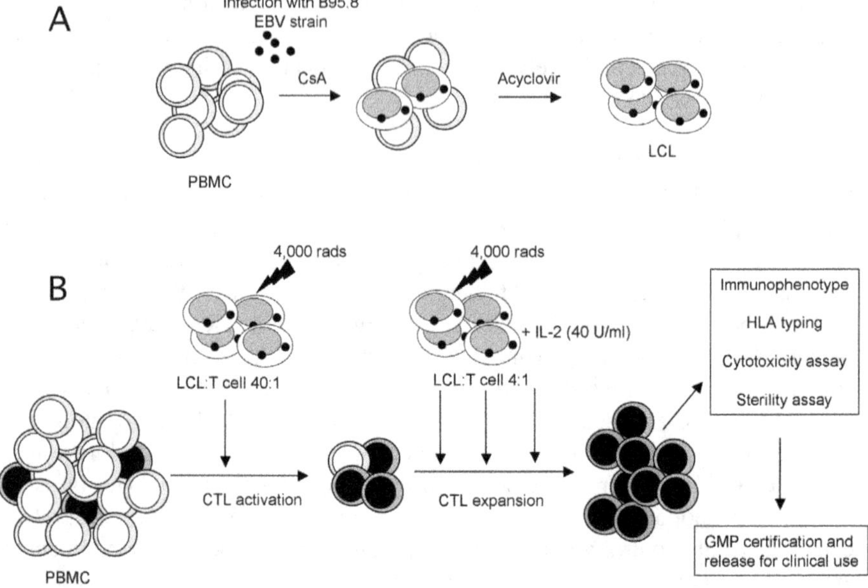

Figure 1. Selection and expansion of EBV-specific CTL for clinical use. The upper panel (A) illustrates the generation of EBV-LCL from the PBMC using the B95-8 EBV strain. PBMC infected with the virus are selected and expanded *in vitro* in presence of Cyclosporin-A (1 μg/mL) to inhibit the growth of T-cells, and Acyclovir after the establishment to inhibit the virus replication. The lower panel (B) illustrates the activation and expansion of EBV-specific CTL using PBMC as effector cells and autologous LCL as antigen presenting cells.

related diseases. Typically, LCL carry multiple copies of the viral episome and constitutively express six nuclear antigens (EBNA1, -2, -3A, -3B, -3C and -LP), three latent membrane proteins (LMP1, -2a and -2b) (latency III) and a small percentage of cells within each LCL express lytic antigens.

To generate EBV-specific T-cell lines, peripheral blood (20-40 mL) is collected from the patient for generation of autologous CTL or from the donor in case of patients receiving allo-SCT. PBMC are plated in growth medium at 2×10^6 cells per well and stimulated with 5×10^4 irradiated (4,000 rads) autologous LCL. Live cells isolated 10 days later are subcultured at 10^6 per well and stimulated with irradiated LCL at 2.5×10^5 per well. Four days later, recombinant human interleukin-2 (rIL2) is added to the culture at 20 - 40 U/mL. Then CTL are stimulated weekly with irradiated autologous LCL at a T lymphocyte:LCL cell ratio of 4:1 and fed with rIL2 twice a week, each time at 40-50 U/mL. Depending on the starting number of PBMC, a sufficient number of CTL for adoptive transfer (up to $3 \times 10^8/m^2$) is usually reached after 4-6 weeks. For patients with EBV related diseases, additional time may be required to reach such CTL numbers (Roskrow et al., 1998). To increase the rate of expansion when the cell line is growing poorly "superexpansion" can be used after 3 to 4 stimulations with autologous LCL. This superexpansion consists of the addition of a submitogenic (0.05 µg/mL) dose of anti-CD3 monoclonal antibody and allogeneic irradiated PBMC from cytomegalovirus-negative, blood-bank-approved, healthy donors (T cell:PBMC ratio, 1:1) as feeder cells, in addition to the autologous LCL and rIL2.

Before cryopreservation, CTL lines are evaluated for immunophenotype, EBV specificity, identity by HLA (Human Leucocytes Antigens) typing and sterility. Most CTL lines are predominantly CD3+/CD8+ T-cells (mean 71%, range 3-99) expressing the T-cell receptor (TcR) $\alpha\beta$. However, CTL lines also contain variable amounts of CD3+/CD4+ T-cells (mean 20%, range 2-98) and lymphokine-activated cells CD3+/CD56+ (mean 13%, range 1-49) (Savoldo et al., 2000). The EBV specificity is determined using a standard cytotoxicity assay (chromium release assay) using autologous LCL, HLA class I and class II mismatched LCL, autologous PHA stimulated T-blasts and the HSB-2 cell line as target cells. Significant lysis of autologous LCL and less than 10% lysis of autologous PHA blasts (at an effector target ratio of 20:1) are required as release criteria for the CTL line. The specific lysis of autologous LCL is characteristically HLA class I restricted since it is significantly inhibited by the coincubation of the target cells with an anti-HLA class I antibody, however, the presence of class II-restricted CD4+ CTLs is not uncommon. HLA typing of the CTL lines is performed to confirm identity of autologous CTL and sterility tests are performed to exclude contamination by bacteria, mycoplasma and fungi. These analyses require 2 weeks, after which the product can be released with a certificate of analysis and CTL can be infused in the patient. Samples with satisfactory test results are thawed rapidly at 37°C and administered intravenously without further manipulation.

Adoptive immunotherapy of EBV-specific CTLs in allo-SCT recipients

The great majority of PTLD occurring after allo-SCT is of donor origin and adoptive transfer of selectively expanded donor-derived EBV-specific CTL has been explored in this clinical setting (Table 1) (Rooney et al., 1995; 1998; Gustafsson et al., 2000).

Our group several years ago used donor-derived EBV-specific T-cell lines as prophylactic treatment in children receiving T-cell depleted SCT or in children showing increased EBV-DNA viral load in the peripheral blood after SCT, both situations known to be associated with a high risk to develop EBV-related PTLD (Rooney et al., 1998; 1995). EBV-specific CTL were transfused from day 45 after transplant without significant side effects. In particular, no patients experienced de novo occurrence of GVHD. When CTL were transfused in patients with high viral load before the CTL injection, they clearly showed an antiviral effect because a significant reduction of EBV-DNA was observed within 2-3 weeks after the CTL infusion. Moreover, the long-term follow up of all patients receiving CTL demonstrated that this prophylactic approach significantly prevented the occurrence of PTLD, because none of the patients transfused with CTL developed PTLD compared to 7 out of 61 (11.5%) in a matched historical control group (Rooney et al., 1995; 1998). Similar results were subsequently reported by another group in a small series of patients (Gustafsson et al., 2000). To evaluate the antitumor activity of CTL, we also treated 6 patients with EBV-driven lymphoma, and in 5 out of 6 patients, CTL infusion induced tumor regression (Rooney et al., 1995) [and unpublished results]. Interestingly, from this small experience, we observed that the inflammatory response determined by CTL infiltration within the tumor

Table 1. Treatment of PTLD using adoptive immunotherapy

	Type of transplant	No of patients treated	Age (years)	Type of treatment	Source of CTL	Doses	Results
Rooney, 1998	Allo-BMT T-cell depleted	39 + 17	0.9-20	Prophylaxis or high EBV-DNA load	Donor	$10\text{-}20\text{x}10^6$ 2-4 doses	No PTLD compared to 11.5% control group
Rooney, 1998	Allo-BMT T-cell depleted	3		EBV+ lymphoma	Donor	$10\text{-}20\text{x}10^6$ 2-4 doses	CR 2/3, PD 1/3
Gustafsson, 2000	Allo-BMT T-cell depleted	5	1-39	High EBV DNA load	Donor	$10\text{x}10^6$ 2-4 doses	4/5 decrease of EBV-DNA loa
Haque, 1998	SOT	3	29-54	Prophylaxis	Autologous	$50\text{x}10^6$ to $200\text{x}10^6$ escalating dose	3/3 decrease EBV-DNA load
Khanna, 1999	SOT	1	39	EBV+ lymphoma	Autologous	$35\text{x}10^6$ 2+2 doses	Tumor regression
Comoli, 2001	SOT	7	4-60	High EBV-DNA load	Autologous	$20\text{x}10^6/\text{m}^2$ 1-5 doses	5/7 decrease EBV-DNA load. occurrence of PTLD
Savoldo, ongoing trial	SOT	7	1-4	High EBV-DNA load	Autologous	$20\text{-}50\text{x}10^6/\text{m}^2$ escalating dose	5/7 decrease EBV-DNA load. occurrence of PTLD
Haque, 2002	SOT	7	1.5-60	Refractory PTLD	Partly HLA-matched CTL	10^6 1-6 doses	3/7 complete and persistent regression

could induce serious adverse events, like airways obstruction in a case with bulky nasopharyngeal PTLD. In a patient who progressed despite CTL transfer, we demonstrated a deletion within the virus genome that significantly reduced the sensitivity of tumor cells to CTL cytolysis, indicating that escape mutants may be a limitation of virus specific CTL adoptive transfer (Gottschalk et al., 2001). Both these observations suggest that prophylactic treatment or early intervention is preferable, reducing the risk of side effects and potentially reducing the chance of developing tumor mutations.

The overall clinical success of EBV-specific CTL adoptive transfer in allo-SCT recipients was enhanced by important biological indications that can be useful for future applications of adoptive immunotherapy. Antigen specific-CTLs expanded *ex vivo* are typically effector-memory CTLs, and the *in vivo* behaviour of these cells after infusion represents an important biological aspect of this procedure. Thus, while transient biological activity of adoptively transferred CTL might be sufficient to protect allo-SCT patients from a fatal viral infection, and the subsequent reconstitution of the immune system from the engrafted stem cells will provide a long term immune competence, a more durable immune response would be desirable when adoptive transfer is proposed for the treatment of patients with cancer. In our studies, EBV-specific CTL were gene-marked with the G1Na retroviral vector containing the Escherichia Coli-derived neomycin resistant gene (neoR) to allow tracking of these cells *in vivo* by polymerase chain reaction (PCR). Using this strategy, we were able to demonstrate that infused CTL expanded *in vivo* up to 4 logs and that they persisted for up 85 months after infusion (in the peripheral blood or in regenerated EBV-specific CD8+ and CD4+ CTL subpopulations) (Heslop et al., 1996), encouraging the development of this approach for patients with other cancers. In addition, the detection of neo positive CTL within a lymphoma biopsy in one patient, suggests that the CTL have the ability to migrate to the tumor and that the tumor regression was likely induced by the cytotoxic effect of these cells.

Currently, the incidence of PTLD in patients receiving T-cell depleted SCT can be significantly reduced if elimination of B-lymphocytes from the graft is included in the depletion procedure. Indeed, it has been reported that techniques of T-cell depletion like elutriation and incubation with CAMPATH-1 monoclonal antibody, which reduce the load of B cell in the graft, are associated with a lower incidence of PTLD (Curtis et al., 1999). Most importantly, in recent years, the robust correlation between persistent high EBV viral load in the peripheral blood and occurrence of PTLD in allo-SCT patients (Wagner et al., 2004b; van Esser et al., 2002), and the introduction of the FDA approved humanized chimeric anti-CD20 monoclonal antibody MabThera (Rituximab), have significantly changed the prophylactic approach of PTLD in allo-SCT patients. The administration of the antibody is generally well tolerated and rapidly induces the depletion of mature B-lymphocytes in the peripheral blood reducing the compartment of EBV-infected cells and normalizing the viral load (Kuehnle et al., 2000; van Esser et al., 2002). The *in vivo* depletion of B cells can persist for several months likely allowing enough time for the reconstitution of T-cell immunity after the stem cell engraftment. This primary intervention has been adopted in our institution in patients with high viral load or clinical symptoms related to PTLD after allo-SCT (Kuehnle et al., 2000), reserving the generation and infusion of EBV-specific CTL as salvage treatment.

Adoptive immunotherapy with EBV-specific CTL in SOT recipients

PTLD occurring in SOT recipients, unlike allo-SCT recipients, usually originate from recipient B-cells. For this reason, most of the studies of EBV-specific CTL transfer in SOT recipients used autologous CTL in the attempt to guarantee the appropriate recognition of EBV-infected cells in the contest of HLA class I molecules and restore the immunocompetence for EBV while minimizing the risk of graft damage. Autologous EBV-specific CTL can be expanded from peripheral blood of EBV-seropositve SOT recipients collected before transplant (Haque et al., 1998). However, in our and other experiences the immunosuppression does not compromise the expansion of EBV-specific CTL from patients developing EBV reactivation after transplant, eliminating the requirement to collect and store material from patients before transplant (Savoldo et al., 2001; Comoli et al., 2002). CTL lines reactivated from these patients have the same biological characteristics as those generated from healthy EBV-seropositve donors. In addition, they retain the ability to recognize and kill the autologous LCL in presence of immunosuppressive drugs, even though their proliferative activity is significantly suppressed (Savoldo et al., 2001).

EBV-seronegative patients have a very high risk to develop primary EBV infection after SOT, because they generally receive their graft from EBV seropositive donors, and the course of this infection can be rapid and aggressive. For these patients a "prophylactic" generation and infusion of autologous EBV-specific CTL is highly desirable. However, in EBV-seronegative individual, the protocol used to reactivate memory T-cells in EBV-seropositve individuals is unsuccessful. A more sophisticated procedure, based on the selection of activated T-cells, is required to generate CTL from EBV-seronegative patients and this procedure is still not approved for a GMP application (Savoldo et al., 2002a). As an alternative to generation of autologous EBV-specific CTL, Crawford et al. have generated a bank of allogeneic partially matched EBV-specific CTL obtained from healthy donors (Haque et al., 2002).

Several groups, including ours, recently reported preliminary experiences of adoptive EBV-CTL transfer in SOT recipients (Table 1) (Haque et al., 1998; 2001; 2002; Comoli et al., 2002; Khanna et al., 1999; Savoldo manuscript in preparation). Most groups chose a prophylactic approach and infused patients with a persistent high level of EBV-DNA in the peripheral blood, but without clinical evidence of PTLD. Overall these studies indicate that CTL infusion is generally well tolerated and no graft toxicity was reported. Most studies evaluated the reduction of the viral load in the peripheral blood as marker of the activity of the CTL infused and some patients experienced a significant decrease of the viral load. Remarkably, the antiviral effect was obtained even though most of the patients were still receiving immunosuppressive treatment to prevent the graft rejection. Although, EBV-DNA did not decreased uniformly in all patients, none of the patients treated for high viral load developed PTLD. The most impressive illustration of the efficacy of the adoptive transfer of CTL in SOT recipients comes from the experiences reported by Khanna et al. (1999) who described the achievement of complete remission after multiple CTL infusion in patients treated for refractory PTLD. A massive necrosis in a lung lesion caused a fatal haemorrhage in one patient, suggesting that

CTL should be preferably used in patients with minimal disease. Promising results were also reported by Crawford et al. using partly HLA-matched EBV-specific CTL (Haque et al., 2002).

Compared to allo-SCT patients, SOT recipients require the maintenance of an adequate immunosuppression treatment to prevent the graft rejection. For this reason the adoptive transfer of EBV-specific CTL should provide not only a transient clearance of the viral infected cells, but should also rebalance the immune system to allow a more efficient and persistent control of the virus reactivation. This effect cannot be achieved using partly HLA-matched EBV-specific CTL because the *in vivo* survival of these cells in very short (Haque et al., 2002). The studies so far available in the autologous setting showed that CTL infusion increases the frequencies of EBV-specific CTL precursors at least transiently (Haque et al., 1998; Comoli et al., 2002).

In SOT recipients PTLD can be very aggressive and rapidly fatal. However, the expansion of autologous EBV-specific CTL requires several weeks, limiting the feasibility of this approach in patients with a clinical diagnosis of PTLD. On the other hand, the "prophylactic" generation of CTL for all patients undergoing a high-risk transplant like heart, liver, lung or small bowel is not feasible. The use of partly HLA-matched EBV-specific CTL (Haque et al., 2002) or alternatively the identification of prognostic factors highly predictive of the occurrence of PTLD in SOT recipients may significantly reduce the number of patients for which a prophylactic generation of CTL and an early intervention is indicated, and thus balance the cost/benefit of this procedure. However, the identification of patients

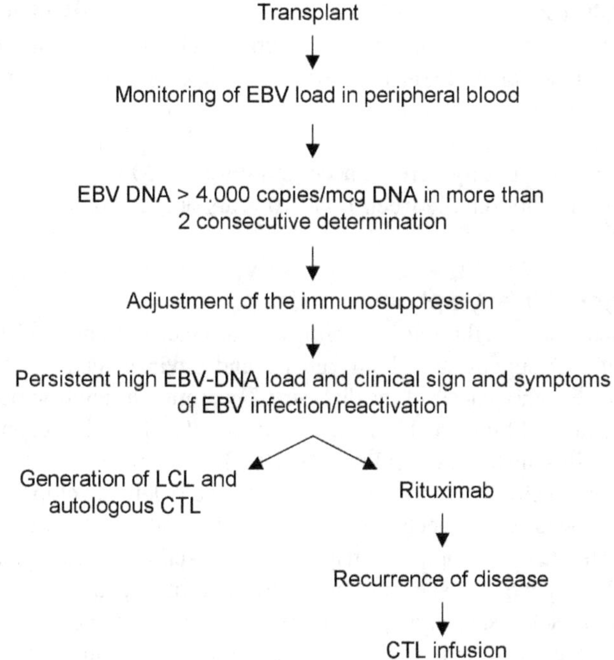

Figure 2. Algorithm for the treatment of PTLD after SOT.

at high risk for PTLD is problematic in SOT recipients because the high EBV DNA viral load in the peripheral blood is not highly specific as more frequently reported in allo-SCT recipients (Wagner et al., 2004b; Gustafsson et al., 2000; van Esser et al., 2002), thus the power of prediction of this test is still matter of discussion in SOT recipients (Green and Webber, 2002). Other predictive markers such as production of cytokines, genetic polymorphisms and immune parameters are currently investigated to better identify the patients at high risk.

Rituximab has significantly improved the immediate treatment of SOT patients with clinical diagnosis of EBV-related PTLD (Milpied et al., 2000; Cook et al., 1999; Dotti et al., 2001). The administration of the antibody is well tolerated and the clinical remission is variable between 30 to 70% (Milpied et al., 2000). In contrast to allo-SCT patients, duration of remission can be relatively short, since the recovery of B cells in presence of a impaired T cell immunity can be accompanied by a new increase in EBV load and eventually relapse of the PTLD (Savoldo et al., 2005). However, this treatment may help to identify patients who can benefit from adoptive transfer of autologous EBV-specific CTL to restore the balance of the immune compartment and at the same time allows enough time to generate autologous CTL. Based on these considerations in Figure 2 we illustrate our current approach for cellular therapy in SOT transplant recipients.

HODGKIN'S LYMPHOMA

Incidence

The incidence of Hodgkin's lymphoma (HD) in Western countries is about 2.4 per 100,000 per year and presents a typical bimodal age distribution with a first peak in young adults (between 15 and 35 years old) and a second peak in older subjects (>60 years old) (Yung and Linch, 2003; Hjalgrim et al., 2001). The age distribution of HD has been attributed to different pathogenetic events and reflects a different distribution of the histological subtype of HD. Indeed, the nodular sclerosis variant, accounting for more than 75% of all cases of HD, is typical of the young adults, while the mixed cellularity variant, accounting for 15-20% of all cases of HD is more common in children (between 1 and 10 years old) and older subjects (Glaser et al., 1997).

EBV and Hodgkin's lymphoma

The increased risk of HD among siblings of affected patients and the familiar clustering of HD suggest that both genetic and environmental factors may be important in the pathogenesis of the disease. Concerning the environmental factors, epidemiological and biological evidences indicate that EBV infection could play a role at least in a histological subtype of HD. Indeed, patients with a history of infectious mononucleosis have a three fold increased risk to develop HD compared to EBV sero-negative subjects (Hjalgrim et al., 2003). Moreover, molecular evaluations showed that approximately 75% of mixed cellularity subtype of HD contain the EBV genome, especially in children and older patients, while less than 25% of nodular sclerosing subtype (the most common variant in adults) contain the EBV genome, even though the majority of the patients are EBV seropositive

(Glaser et al., 1997). Reed-Sternberg cells, unlike PTLD and LCL, express a limited number of viral proteins namely EBNA1, LMP1 and LMP2, known as latency pattern II.

Adoptive transfer of EBV-specific CTL

Modern radiotherapy and/or chemotherapy regimens have dramatically improved the cure rate of patients with HD in the past four decades, reducing the chance of recurrence or progressive disease to less than 30% (Canellos et al., 1992; Bonadonna et al., 2004). Moreover, second line treatments based on the use of high dose chemotherapy and autologous stem cell transplantation rescue 25-60% of patients with refractory/relapsing disease (Gribben et al., 1989). Despite the identification of clinical prognostic factors (Hasenclever and Diehl, 1998), and the optimal use of primary and secondary treatments, HD remains fatal for more than 15% of patients (Yung and Linch, 2003). Therefore new therapeutic agents are required for primary refractory patients and also biological strategies are desirable to maintain the remission in high-risk patients after conventional treatment. In addition, biological treatments may reduce the serious long term side effects correlated with radiation and chemotherapy treatment (Ng and Mauch, 2004)

The association between EBV and HD, the expression of viral latency proteins by tumor cells and the encouraging results obtained in the setting of allo-SCT, constituted the rational to assess the feasibility of EBV-specific CTL adoptive transfer for patients with EBV-related HD. In a phase I clinical trial involving 13 patients, we showed that using autologous LCL as antigen presenting cells, EBV-specific CTL can successfully be generated in a majority of patients even when in clinical relapse. Despite a slow growth rate and a more frequent requirement of superexpansion, CTL generated from these patients showed the same phenotype and antigen specificity observed in the CTL generated from healthy donors. Thirteen patients, nine with measurable disease and four in remission after autologous SCT, received between $2 \times 10^7/m^2$ and 1.2×10^8 CTL per m^2. CTL lines transduced with the G1Na neo retroviral vector were infused in 7 patients. The infusions were well tolerated and the reduction of the EBV-DNA load in most patients suggested that CTL had *in vivo* antiviral activity as previously observed in allo-SCT patients. CTL produced mixed tumor responses in five patients and two complete remissions. Those who received CTL as adjuvant therapy after autologous SCT remain in remission 11 to 45 months post infusion as of July 2004. Gene marked CTL were detectable in the peripheral blood for up to 12 months after infusion. Moreover, they homed to the tumor sites as demonstrated in a patient with a malignant pleural effusion where the neo signal was over 100 fold greater than in peripheral blood and by in situ hybridization in a patient with mediastinal disease (Roskrow et al., 1998; Bollard et al. 2004).

NASOPHARYNGEAL CARCINOMA

Incidence

Nasopharyngeal carcinoma (NPC) is a rare malignancy in most populations, with an annual incidence of 1.8/100,000. However, the incidence is approximately 30-80/100,000 per year in endemic areas of Southern China, Taiwan, Vietnam and The

Philippines. Intermediate incidence (4/100,000 per year) is observed in population in North African, Saudi Arabia and Greenland. The age distribution also varies in different areas. In Asia the disease occurs most frequently in adult individuals (50-60 years old), while in North Africa a peak is observed in younger patients (10-25 years old) (Spano et al., 2003).

EBV and nasopharyngeal carcinoma

WHO has categorized NPC into three histological types: keratinizing squamous cell carcinoma (type I), non-keratinizing carcinoma (type II) and the undifferentiated carcinoma (UCTN or type III) (Spano et al., 2003). The association between EBV and NPC derives from epidemiological observations, and by the formal detection of the EBV genome in epithelial cells of type II and type III NPC (zur et al., 1970). The association between EBV and type I NPC remains unclear. EBV genome is detected in NPC regardless of the geographical area (endemic or non endemic) suggesting a real causative role of the pathogen. However, epidemiological data indicate that genetic and environmental factors, like the consumption of salted fish and other preserved foods containing an elevated amount of nitrosamines, may play a significant role in the Chinese population (Yu et al., 1990).

NPC cells express a limited number of viral proteins namely EBNA1, LMP1 and LMP2 as previously described for the Reed-Sternberg cells of HD (Spano et al., 2003). However, the pathogenetic role of viral proteins in epithelial cells is not yet formally demonstrated. Gene deletions as well as DNA hypermethylation have been reported in NPC suggesting a multi step process for the establishment of the tumor.

Adoptive transfer of EBV-specific CTL

Radiotherapy remains the gold standard treatment for NPC with 80-90% of the localized stages being cured (Perez et al., 1992). However, for more advanced stages radiotherapy alone is significantly less efficient since up to 40% of patients can relapse. Although concomitant use of radiotherapy and chemotherapy enhances the rate of disease free survival even in patients with more advanced stages, relapse at distant site from the original lesions remains the major cause of treatment failure (Teo et al., 1999). Thus, novel therapies that could reduce the disease dissemination are desirable.

Considering that the tumor cells express latent proteins LMP1 and LMP2, and that CTL specific for these antigens have been detected in patients with NPC (Chua et al., 2001), albeit at low level compared to normal controls, the ex-vivo reactivation and adoptive transfer of EBV-specific CTL has been proposed as alternative treatment for NPC patients.

Chua et al. (2001) reported their experience in four patients with advanced NPC who received $5 \times 10^7 - 3 \times 10^8$ autologous EBV-specific CTL. The treatment was well tolerated without immediate side effects. The authors observed an increase in the EBV-specific CTL precursors 2-3 weeks after the CTL infusion. An antiviral effect by the infused CTL was suggested by the decrease of the EBV-DNA viral load in the peripheral blood. However, the anti-tumor effect of the CTL infusion was not clearly evaluable possibly because patients enrolled in the study presented with very advanced stages of disease. Comoli et al. (2004) recently reported a single patient with relapsed NPC who was treated with multiple doses of EBV-specific CTL generated ex-vivo from a HLA-identical sibling. Despite the short survival of the CTL *in vivo*, a temporary stabilization of disease was obtained. Our group has recently concluded a phase I/II study of adoptive transfer of autologous EBV-specific CTL in patients with relapsed or refractory NPC. Cell lines generated from 12 patients presented biological characteristic similar to the CTL lines generated from normal EBV seropositive donors and roughly 5% of the CTL recognized one or two LMP2 epitopes. Eleven patients received $2 \times 10^7 - 3 \times 10^8/m^2$ CTL according to a dose escalation protocol. The treatment was generally well tolerated even at the higher dose level. Of 6 patients with active disease, one patient rapidly progressed, but remarkably 2 patients obtained a complete remission and 3 patients achieved a partial remission or stabilization of the disease up to 7 months after CTL infusion (Straathof et al., 2005).

FUTURE STRATEGY FOR ADOPTIVE IMMUNOTHERAPY IN EBV POSITIVE HD AND NPC

The experience with EBV specific CTL in HD and NPC patients has been less successful than that in allo-SCT patients. A number of obstacles may explain these results. Unlike, LCL and the majority of PTLD, the Reed-Sternberg cells and NPC tumor cells express a limited number of viral proteins namely EBNA1, LMP1 and LMP2 (latency pattern II). Both LMP1 and LMP2 antigens are weakly immunogenic in the context of most HLA types, while EBNA1 is typically poorly presented in the contest of HLA class I molecules. Obviously, the type II latency of EBV positive HD and NPC significantly limits the spectrum of targets for the immune system.

Our clinical studies have demonstrated that CTL infused in patients migrate to the tumor. However, tumor microenvironment, at least for the HD, may be particularly "hostile" for anti tumor immunity. Indeed, the Reed-Stermberg cells that comprise the malignant cells of HD use several strategies to create a protected immunological environment. They secretes TGFß and IL-10 that can induce anergy of professional antigen presenting cells (APC) and effector T cells (Bollard et al., 2002); they can produce the chemokine TARC that attracts effector T-cells with Th2-like phenotype (van den et al., 1999), and they can express Fas Ligand that can induce apoptosis of activated T cells (Verbeke et al., 2001). More recently, it has been reported that "regulatory T-cells" surrounding the tumor may play a significant protective role against the immune mediated elimination of tumor cells (Marshall et al., 2004).

Our group and others have recently proposed several strategies aimed to overcome the tumor escape mechanisms through the genetic manipulation of professional APC or EBV-specific CTL (Table 2). To force the presentation of LMP2 or LMP1 epitopes we have transfected dendritic cells with adenoviral vectors expressing LMP-2 or a deleted mutant of LMP-1 proteins. Using these APC we have increased the frequency of LMP2 and LMP1 specific CTL compared to CTL generated with the LCL (Gahn et al., 2001; Gottschalk et al., 2003). At our institution we have recently initiated a phase I study in which we generate and infuse CTL enriched in LMP2 precursors in patients with relapsed HD to test the

Table 2. Mechanisms that limit the immune response and potential overcoming strategies

Escape mechanism	Overcoming strategy
Insufficient number of antigen specific CTL precursors	
- Viral protein epitopes expression limited to LMP1 and LMP2	Force the expression of the viral proteins in professional antigen presenting cells
Tumor mechanisms limiting the effector phase of CTL	
- Production of inhibitory factors by the tumor cells (TGFβ)	Genetic modification of CTL to express a dominant negative receptor
- Th2-like tumor microenvironment	Genetic modification of CTL to autonomously produce IL12
- Fas/FasL induced apoptosis of CTL	Down modulation of Fas in CTL
Limited *in vivo* expansion of CTL	Infusion of CTL in a lympho-depleted host
	Genetic modification of CTL to autonomously produce growth factors

hypothesis whether the higher frequency of LMP2 specific CTL provide a more efficient recognition of tumor cells *in vivo*.

We also showed that EBV-specific CTL could be genetically modified to overcome some of the inhibitory effects of the tumor microenvironment. CTL retrovirally transduced to express a mutant dominant-negative TGFß type II receptor (DNR) became more resistant to the inhibitory effects of TGFß (Bollard et al., 2002). CTL have also been retrovirally transduced to produce IL12 in order both to be more functional in a Th2-lyke microenvironment and to induce a breakdown in this microenvironment that both protects and supports the tumor cells (Wagner et al., 2004a). Finally, CTL can also be induced to become more resistant to the Fas/FasL induced apoptosis, by a retrovirally mediated production of small interfering RNA (siRNA) down modulating the Fas receptor (Dotti et al., 2005). All these modifications significantly improved the functionality of CTL maintaining their antigen specificity and dependence upon physiological growth signals. The possibility of combining two or more molecules enhancing the survival of the CTL may significantly improve the anti-tumor effect *in vivo*.

Methods to increase the capacity of CTL to expand *in vitro* and to maintain proliferative potential *in vivo* may also enhance the anti-tumor effect of CTL. Evidence from animal models suggests that the depletion of the T-cell compartment before the CTL transfer may enhance their *in vivo* expansion. Rosenberg et al. recently reported promising results in melanoma patients infused with CD8+ clones immediately after a non-myeloablative treatment based on the administration of cyclophosphamide and fludarabine. They observed an *in vivo* expansion of anti-melanoma CTL and tumor regression in several patients (Dudley et al., 2002). Alternatively, the *in vivo* expansion of the CTL can be promoted by the genetic manipulation of CTL to produce their own growth factors (Liu and Rosenberg, 2001), even though a more controlled cytokine production would be desirable.

CHRONIC ACTIVE EPSTEIN-BARR VIRUS (CAEBV)

A minority of patients can develop a chronic fatigue syndrome following primary infection with EBV. The exact incidence and the pathogenesis of this disorder are not well established (Okano, 2002). Authors have described patients with mild/moderate CAEBV characterized by fever, malaise, arthralgia, myalgia and lymphadenopathy, persisting for at least 6 months, associated with abnormally high titers of antibodies to EBV-capsid antigen (VCA-IgG) and early antigen (EA-IgG), with little or no antibody to EBV nuclear antigens (EBNA) (Okano, 2002). More aggressive forms are reported in Japanese patients in whom also T-cell and natural killer compartments are involved with subsequent evolution into lethal lymphomas (Jones et al., 1988).

EBV specific CTL for CAEBV

Even when classified as mild/moderate, CAEBV can be a severely debilitating disorder since patients can develop depression and side effects determined by prolonged treatment with anti-inflammatory drugs and steroids (Jones and Straus, 1987).

Considering that chronic fatigue might be determined by an unbalanced immunity to EBV, we have evaluated whether adoptive transfer of EBV-specific CTL could restore anti-viral immunity and improve signs and symptoms in these patients. Five patients were enrolled in a phase I study. All patients had a diagnosis of CAEBV. Even though only one out of seven presented with severe CAEBV, all patients had been treated symptomatically without long-term beneficial effects. EBV-specific CTL were successfully generated from all patients and the expansion rate, immunophenotype and cytotoxic activity were comparable to those observed in normal healthy donors. Five patients were infused with CTL in a dose escalation protocol from $20 \times 10^6/m^2$ to $50 \times 10^6/m^2$ with no immediate or delayed adverse events. Although the clinical benefit of CTL infusion is difficult to assess objectively in these patients, we observed improvement of subjective symptoms in some patients. Moreover, objective reduction of lymphadenopathy, or EBV-related oral lesions was observed in patients with these abnormalities. Normalization of the immune serology to the virus and increase in the frequency of circulating EBV-specific T-cells were also observed in all cases (Savoldo et al., 2002b). We are currently evaluating if this approach can be beneficial in patients with more aggressive forms of CAEBV.

REDIRECT THE FUNCTION OF EBV SPECIFIC CTL WITH CHIMERIC T CELL RECEPTORS

The clinical experience with adoptive transfer of EBV-specific CTL has demonstrated that: (1) the procedure to expand EBV-specific CTL from EBV-seropositve donors is highly reproducible; (2) CTL line are polyclonal allowing a broad antigen recognition; (3) CTL persist long-term after infusion; (4) CTL lines contain CD4+ cells that can help to enhance the survival of CD8+ CTLs *in vivo*; (5) CTL obtained from MHC matched stem cell donors can be infused after allo-SCT avoiding the risk of GVHD. Based on this evidence we are currently evaluating whether EBV-specific can be used as platform for the immunotherapy of non-EBV related tumors.

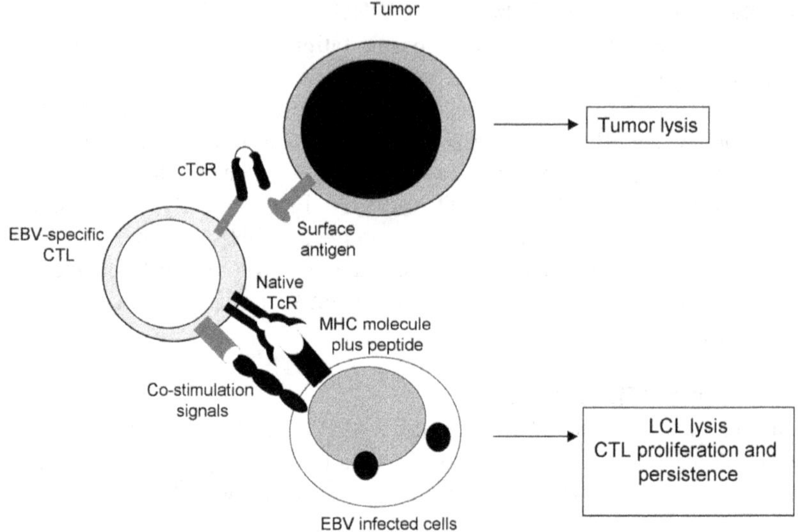

Figure 3. Redirection of EBV-specific CTL through chimeric T cell receptor transfer. CTL are genetically modified to express a chimeric T-cell receptor. This strategy allows the killing of tumor cells expressing the antigen recognized by the chimeric receptor. The expression of the native T-cell receptor and CD28 by the CTL should allow an appropriate co-stimulation in presence of EBV infected cells and thus the survival of the CTL.

An attractive strategy to target tumor associated antigens or self-restricted antigens is represented by the genetic modification of T-lymphocytes to recognize a cell surface antigen through the expression of a chimeric T-cell receptor (cTcR) (Eshhar et al., 1993). These artificial molecules are generated by fusing the antigen-binding domains of the variable region (V_H-V_L) of an antibody [single-chain-Fv (scFV)] to a signaling region from either the ζ-chain of the TcR/CD3 complex, or the γ-chain from FcϵRI receptor (Pule et al., 2003). When expressed by T cells, chimeric receptor binds the antigen expressed on the cell surface of the target cells, avoiding MHC restriction, and activates the lytic pathway of the T-cells that in turn determine the lysis of the target cells. In the last ten years many cTcR have been constructed targeting different molecules (Pule et al., 2003). Although adoptive transfer of primary T-cells expressing a cTcR has induced some antitumor activity in mice (Hwu et al., 1995; Altenschmidt et al., 1997), the results of the phase I clinical trials have been disappointing (Mitsuyasu et al., 2000; Walker et al., 2000) at least in part because of the short persistence *in vivo* of T-cell expressing the transgene. This defect is predominantly determined by the inability of cTcR to provide co-stimulation signals to T-cells that are crucial to sustain an active immune response and that are usually provided by professional APC through the expression of molecules belong to the B7 family. T-cells expressing the cTcR can efficiently bind the antigen expressed by the tumor cells, but the general lack of expression of co-stimulatory molecules by tumor cells impairs the maintenance of an efficient immune response (Krause et al., 1998). We demonstrated that the use of antigen specific CTL as carrier for the cTcR, instead of primary T-lymphocytes may overcome this limitation (Rossig et al., 2002). EBV-specific CTL can be efficiently transduced to express a cTcR. As a consequence they became bi-specific since they maintain the antigen specificity for autologous LCL through the native TcR, and can kill the target tumor cells through the cTcR (Figure 3). Based on these results, the latent EBV infection in EBV seropositve individuals may guarantee the *in vivo* persistence and potentially the expansion of EBV-specific CTL expressing the transgene (Rossig et al., 2002). In a current ongoing phase I clinical trial, we are testing whether, co-stimulation by latently infected B-lymphocytes in EBV-seropositive patients will be sufficient to guarantee the *in vivo* survival of EBV-specific CTL transduced with a cTcR targeting the G(D2a) molecule expressed on neuroblastoma cells. In the future, the potential use of an "EBV vaccine" to boost the CTL *in vivo* may further enhance the CTL persistence.

SUMMARY

Immunotherapy of human cancer remains an exiting challenge for basic immunologists and clinicians. Adoptive transfer of EBV-specific CTL to prevent or treat the occurrence of PTLD after allo-SCT is one the most successful examples in which the intense laboratory activity has been translated in a clinical protocol. Our preliminary experience with HD and NPC patients is also important to underline that a continuous effort to better understand the biology of tumor cells and their interaction with the immune system is required to counterattack tumor cells. Moreover, the integration of cell therapy and gene transfer technologies might offer a new opportunity to enhance the efficiency of the immune system against the tumor cells.

References

Altenschmidt, U., Klundt, E., and Groner, B. (1997). Adoptive transfer of *in vitro*-targeted, activated T lymphocytes results in total tumor regression. J. Immunol. *159*, 5509-5515.

Barnard, C.N. (1967). The operation. A human cardiac transplant: an interim report of a successful operation performed at Groote Schuur Hospital, Cape Town. S. Afr. Med. J. *41*, 1271-1274.

Bollard, C.M., Rossig, C., Calonge, M.J., Huls, M.H., Wagner, H.J., Massague, J., Brenner, M.K., Heslop, H.E., and Rooney, C.M. (2002). Adapting a transforming growth factor beta-related tumor protection strategy to enhance antitumor immunity. Blood *99*, 3179-3187.

Bollard, C.M., Aguilar, L., Straathof, K.C., Gahn, B., Huls, M.H., Rousseau, A., Sixbey, J., Gresik, M.V., Carrum, G., Hudson, M., Dilloo, D., Gee, A., Brenner, M.K., Rooney, C.M., and Heslop, H.E. (2004). Cytotoxic T lymphocyte therapy for Epstein-Barr virus+ Hodgkin's disease.J. Exp. Med. *200*, 1623-1633.

Bonadonna, G., Bonfante, V., Viviani, S., Di Russo, A., Villani, F., and Valagussa, P. (2004). ABVD Plus Subtotal Nodal Versus Involved-Field Radiotherapy in Early-Stage Hodgkin's Disease: Long-Term Results. J. Clin. Oncol.

Buckner, C.D., Epstein, R.B., Rudolph, R.H., Clift, R.A., Storb, R., and Thomas, E.D. (1970). Allogeneic marrow engraftment following whole body irradiation in a patient with leukemia. Blood *35*, 741-750.

Canellos, G.P., Anderson, J.R., Propert, K.J., Nissen, N., Cooper, M.R., Henderson, E.S., Green, M.R., Gottlieb, A., and Peterson, B.A. (1992). Chemotherapy of advanced Hodgkin's disease with MOPP, ABVD, or MOPP alternating with ABVD. N. Engl. J. Med. *327*, 1478-1484.

Chua, D., Huang, J., Zheng, B., Lau, S.Y., Luk, W., Kwong, D.L., Sham, J.S., Moss, D., Yuen, K.Y., Im, S.W., and Ng, M.H. (2001). Adoptive transfer of autologous Epstein-Barr virus-specific cytotoxic T cells for nasopharyngeal carcinoma. Int. J. Cancer *94*, 73-80.

Comoli, P., De Palma, R., Siena, S., Nocera, A., Basso, S., Del Galdo, F., Schiavo, R., Carminati, O., Tagliamacco, A., Abbate, G.F., Locatelli, F., Maccario, R., and Pedrazzoli, P. (2004). Adoptive transfer of allogeneic Epstein-Barr virus (EBV)-specific cytotoxic T cells with *in vitro* antitumor activity boosts LMP2-specific immune response in a patient with EBV-related nasopharyngeal carcinoma. Ann. Oncol. *15*, 113-117.

Comoli, P., Labirio, M., Basso, S., Baldanti, F., Grossi, P., Furione, M., Vigano, M., Fiocchi, R., Rossi, G., Ginevri, F., Gridelli, B., Moretta, A., Montagna, D., Locatelli, F., Gerna, G., and Maccario, R. (2002). Infusion of autologous Epstein-Barr virus (EBV)-specific cytotoxic T cells for prevention of EBV-related lymphoproliferative disorder in solid organ transplant recipients with evidence of active virus replication. Blood *99*, 2592-2598.

Cook, R.C., Connors, J.M., Gascoyne, R.D., Fradet, G., and Levy, R.D. (1999). Treatment of post-transplant lymphoproliferative disease with rituximab monoclonal antibody after lung transplantation. Lancet *354*, 1698-1699.

Curtis, R.E., Travis, L.B., Rowlings, P.A., Socie, G., Kingma, D.W., Banks, P.M., Jaffe, E.S., Sale, G.E., Horowitz, M.M., Witherspoon, R.P., Shriner, D.A., Weisdorf, D.J., Kolb, H.J., Sullivan, K.M., Sobocinski, K.A., Gale, R.P., Hoover, R.N., Fraumeni, J.F., Jr., and Deeg, H.J. (1999). Risk of lymphoproliferative disorders after bone marrow transplantation: a multi-institutional study. Blood *94*, 2208-2216.

Doak, P.B., Montgomerie, J.Z., North, J.D., and Smith, F. (1968). Reticulum cell sarcoma after renal homotransplantation and azathioprine and prednisone therapy. Br. Med. J. *4*, 746-748.

Dotti, G., Fiocchi, R., Motta, T., Gamba, A., Gotti, E., Gridelli, B., Borleri, G., Manzoni, C., Viero, P., Remuzzi, G., Barbui, T., and Rambaldi, A. (2000). Epstein-Barr virus-negative lymphoproliferate disorders in long-term survivors after heart, kidney, and liver transplant. Transplantation *69*, 827-833.

Dotti, G., Fiocchi, R., Motta, T., Mammana, C., Gotti, E., Riva, S., Cornelli, P., Gridelli, B., Viero, P., Oldani, E., Ferrazzi, P., Remuzzi, G., Barbui, T., and Rambaldi, A. (2002). Lymphomas occurring late after solid-organ transplantation: influence of treatment on the clinical outcome. Transplantation *74*, 1095-1102.

Dotti, G., Rambaldi, A., Fiocchi, R., Motta, T., Torre, G., Viero, P., Gridelli, B., and Barbui, T. (2001). Anti-CD20 antibody (rituximab) administration in patients with late occurring lymphomas after solid organ transplant. Haematologica *86*, 618-623.

Dotti, G., Savoldo, B., Pule, M., Straathof, K.C., Biagi, E., Yvon, E., Vigouroux, S., Brenner, M.K., and Rooney, C.M. (2005). Human cytotoxic T lymphocytes with reduced sensitivity to Fas-induced apoptosis. Blood *105*, 4677-4684. Epub 2005 Feb 15.

Dudley, M.E., Wunderlich, J.R., Robbins, P.F., Yang, J.C., Hwu, P., Schwartzentruber, D.J., Topalian, S.L., Sherry, R., Restifo, N.P., Hubicki, A.M., Robinson, M.R., Raffeld, M., Duray, P., Seipp, C.A., Rogers-Freezer, L., Morton, K.E., Mavroukakis, S.A., White, D.E., and Rosenberg, S.A. (2002). Cancer regression and autoimmunity in patients after clonal repopulation with antitumor lymphocytes. Science *298*, 850-854.

Eshhar, Z., Waks, T., Gross, G., and Schindler, D.G. (1993). Specific activation and targeting of cytotoxic lymphocytes through chimeric single chains consisting of antibody-binding domains and the gamma or zeta subunits of the immunoglobulin and T-cell receptors. Proc. Natl. Acad. Sci. U. S. A *90*, 720-724.

Frizzera, G., Hanto, D.W., Gajl-Peczalska, K.J., Rosai, J., McKenna, R.W., Sibley, R.K., Holahan, K.P., and Lindquist, L.L. (1981). Polymorphic diffuse B-cell hyperplasias and lymphomas in renal transplant recipients. Cancer Res. *41*, 4262-4279.

Gahn, B., Siller-Lopez, F., Pirooz, A.D., Yvon, E., Gottschalk, S., Longnecker, R., Brenner, M.K., Heslop, H.E., Aguilar-Cordova, E., and Rooney, C.M. (2001). Adenoviral gene transfer into dendritic cells efficiently amplifies the immune response to LMP2A antigen: a potential treatment strategy for Epstein-Barr virus--positive Hodgkin's lymphoma. Int. J. Cancer *93*, 706-713.

Glaser, S.L., Lin, R.J., Stewart, S.L., Ambinder, R.F., Jarrett, R.F., Brousset, P., Pallesen, G., Gulley, M.L., Khan, G., O'Grady, J., Hummel, M., Preciado, M.V., Knecht, H., Chan, J.K., and Claviez, A. (1997). Epstein-Barr virus-associated Hodgkin's disease: epidemiologic characteristics in international data. Int. J. Cancer *70*, 375-382.

Good, R.A., Meuwissen, H.J., Hong, R., and Gatti, R.A. (1969). Bone marrow transplantation: correction of immune deficit in lymphopenic immunologic deficiency and correction of an immunologically induced pancytopenia. Trans. Assoc. Am. Physicians *82*, 278-285.

Gottschalk, S., Edwards, O.L., Sili, U., Huls, M.H., Goltsova, T., Davis, A.R., Heslop, H.E., and Rooney, C.M. (2003). Generating CTLs against the subdominant Epstein-Barr virus LMP1 antigen for the adoptive immunotherapy of EBV-associated malignancies. Blood *101*, 1905-1912.

Gottschalk, S., Ng, C.Y., Perez, M., Smith, C.A., Sample, C., Brenner, M.K., Heslop, H.E., and Rooney, C.M. (2001). An Epstein-Barr virus deletion mutant associated with fatal lymphoproliferative disease unresponsive to therapy with virus-specific CTLs. Blood *97*, 835-843.

Green, M. and Webber, S.A. (2002). EBV viral load monitoring: unanswered questions. Am. J. Transplant. *2*, 894-895.

Gribben, J.G., Linch, D.C., Singer, C.R., McMillan, A.K., Jarrett, M., and Goldstone, A.H. (1989). Successful treatment of refractory Hodgkin's disease by high-dose combination chemotherapy and autologous bone marrow transplantation. Blood *73*, 340-344.

Gridelli, B. and Remuzzi, G. (2000). Strategies for making more organs available for transplantation. N. Engl. J. Med. *343*, 404-410.

Gustafsson, A., Levitsky, V., Zou, J.Z., Frisan, T., Dalianis, T., Ljungman, P., Ringden, O., Winiarski, J., Ernberg, I., and Masucci, M.G. (2000). Epstein-Barr virus (EBV) load in bone marrow transplant recipients at risk to develop posttransplant lymphoproliferative disease: prophylactic infusion of EBV-specific cytotoxic T cells. Blood *95*, 807-814.

Haque, T., Amlot, P.L., Helling, N., Thomas, J.A., Sweny, P., Rolles, K., Burroughs, A.K., Prentice, H.G., and Crawford, D.H. (1998). Reconstitution of EBV-specific T cell immunity in solid organ transplant recipients. J. Immunol. *160*, 6204-6209.

Haque, T., Taylor, C., Wilkie, G.M., Murad, P., Amlot, P.L., Beath, S., McKiernan, P.J., and Crawford, D.H. (2001). Complete regression of posttransplant lymphoproliferative disease using partially HLA-matched Epstein Barr virus-specific cytotoxic T cells. Transplantation *72*, 1399-1402.

Haque, T., Wilkie, G.M., Taylor, C., Amlot, P.L., Murad, P., Iley, A., Dombagoda, D., Britton, K.M., Swerdlow, A.J., and Crawford, D.H. (2002). Treatment of Epstein-Barr-virus-positive post-transplantation lymphoproliferative disease with partly HLA-matched allogeneic cytotoxic T cells. Lancet *360*, 436-442.

Harris, N.L., Ferry, J.A., and Swerdlow, S.H. (1997). Posttransplant lymphoproliferative disorders: summary of Society for Hematopathology Workshop. Semin. Diagn. Pathol. *14*, 8-14.

Harris, N.L., Jaffe, E.S., Diebold, J., Flandrin, G., Muller-Hermelink, H.K., Vardiman, J., Lister, T.A., and Bloomfield, C.D. (1999). The World Health Organization classification of neoplastic diseases of the hematopoietic and lymphoid tissues. Report of the Clinical Advisory Committee meeting, Airlie House, Virginia, November, 1997. Ann. Oncol. *10*, 1419-1432.

Hasenclever, D. and Diehl, V. (1998). A prognostic score for advanced Hodgkin's disease. International Prognostic Factors Project on Advanced Hodgkin's Disease. N. Engl. J. Med. *339*, 1506-1514.

Heslop, H.E., Ng, C.Y., Li, C., Smith, C.A., Loftin, S.K., Krance, R.A., Brenner, M.K., and Rooney, C.M. (1996). Long-term restoration of immunity against Epstein-Barr virus infection by adoptive transfer of gene-modified virus-specific T lymphocytes. Nat. Med. 2, 551-555.

Hjalgrim, H., Askling, J., Pukkala, E., Hansen, S., Munksgaard, L., and Frisch, M. (2001). Incidence of Hodgkin's disease in Nordic countries. Lancet 358, 297-298.

Hjalgrim, H., Askling, J., Rostgaard, K., Hamilton-Dutoit, S., Frisch, M., Zhang, J.S., Madsen, M., Rosdahl, N., Konradsen, H.B., Storm, H.H., and Melbye, M. (2003). Characteristics of Hodgkin's lymphoma after infectious mononucleosis. N. Engl. J. Med. 349, 1324-1332.

Ho, M., Jaffe, R., Miller, G., Breinig, M.K., Dummer, J.S., Makowka, L., Atchison, R.W., Karrer, F., Nalesnik, M.A., and Starzl, T.E. (1988). The frequency of Epstein-Barr virus infection and associated lymphoproliferative syndrome after transplantation and its manifestations in children. Transplantation 45, 719-727.

Hwu, P., Yang, J.C., Cowherd, R., Treisman, J., Shafer, G.E., Eshhar, Z., and Rosenberg, S.A. (1995). *In vivo* antitumor activity of T cells redirected with chimeric antibody/T-cell receptor genes. Cancer Res. 55, 3369-3373.

Jones, J.F., Shurin, S., Abramowsky, C., Tubbs, R.R., Sciotto, C.G., Wahl, R., Sands, J., Gottman, D., Katz, B.Z., and Sklar, J. (1988). T-cell lymphomas containing Epstein-Barr viral DNA in patients with chronic Epstein-Barr virus infections. N. Engl. J. Med. 318, 733-741.

Jones, J.F. and Straus, S.E. (1987). Chronic Epstein-Barr virus infection. Annu. Rev. Med. 38, 195-209.

Khanna, R., Bell, S., Sherritt, M., Galbraith, A., Burrows, S.R., Rafter, L., Clarke, B., Slaughter, R., Falk, M.C., Douglass, J., Williams, T., Elliott, S.L., and Moss, D.J. (1999). Activation and adoptive transfer of Epstein-Barr virus-specific cytotoxic T cells in solid organ transplant patients with posttransplant lymphoproliferative disease. Proc. Natl. Acad. Sci. U. S. A 96, 10391-10396.

Klein, G. and Purtilo, D. (1981). Summary: symposium on Epstein-Barr virus-induced lymphoproliferative diseases in immunodeficient patients. Cancer Res. 41, 4302-4304.

Krause, A., Guo, H.F., Latouche, J.B., Tan, C., Cheung, N.K., and Sadelain, M. (1998). Antigen-dependent CD28 signaling selectively enhances survival and proliferation in genetically modified activated human primary T lymphocytes. J. Exp. Med. 188, 619-626.

Kuehnle, I., Huls, M.H., Liu, Z., Semmelmann, M., Krance, R.A., Brenner, M.K., Rooney, C.M., and Heslop, H.E. (2000). CD20 monoclonal antibody (rituximab) for therapy of Epstein-Barr virus lymphoma after hemopoietic stem-cell transplantation. Blood 95, 1502-1505.

Liu, K. and Rosenberg, S.A. (2001). Transduction of an IL-2 gene into human melanoma-reactive lymphocytes results in their continued growth in the absence of exogenous IL-2 and maintenance of specific antitumor activity. J. Immunol. 167, 6356-6365.

Luznik, L. and Fuchs, E.J. (2002). Donor lymphocyte infusions to treat hematologic malignancies in relapse after allogeneic blood or marrow transplantation. Cancer Control 9, 123-137.

Marshall, N.A., Christie, L.E., Munro, L.R., Culligan, D.J., Johnston, P.W., Barker, R.N., and Vickers, M.A. (2004). Immunosuppressive regulatory T cells are abundant in the reactive lymphocytes of Hodgkin lymphoma. Blood 103, 1755-1762.

Milpied, N., Vasseur, B., Parquet, N., Garnier, J.L., Antoine, C., Quartier, P., Carret, A.S., Bouscary, D., Faye, A., Bourbigot, B., Reguerre, Y., Stoppa, A.M., Bourquard, P., Hurault, d.L., Dubief, F., Mathieu-Boue, A., and Leblond, V. (2000). Humanized anti-CD20 monoclonal antibody (Rituximab) in post transplant B-lymphoproliferative disorder: a retrospective analysis on 32 patients. Ann. Oncol. 11 Suppl 1, 113-116.

Mitsuyasu, R.T., Anton, P.A., Deeks, S.G., Scadden, D.T., Connick, E., Downs, M.T., Bakker, A., Roberts, M.R., June, C.H., Jalali, S., Lin, A.A., Pennathur-Das, R., and Hege, K.M. (2000). Prolonged survival and tissue trafficking following adoptive transfer of CD4zeta gene-modified autologous CD4(+) and CD8(+) T cells in human immunodeficiency virus-infected subjects. Blood *96*, 785-793.

Nagington, J. and Gray, J. (1980). Cyclosporin A immunosuppression, Epstein-Barr antibody, and lymphoma. Lancet *1*, 536-537.

Ng, A.K. and Mauch, P.M. (2004). Late complications of therapy of Hodgkin's disease: prevention and management. Curr. Hematol. Rep. *3*, 27-33.

Okano, M. (2002). Overview and problematic standpoints of severe chronic active Epstein-Barr virus infection syndrome. Crit Rev. Oncol. Hematol. *44*, 273-282.

Papadopoulos, E.B., Ladanyi, M., Emanuel, D., Mackinnon, S., Boulad, F., Carabasi, M.H., Castro-Malaspina, H., Childs, B.H., Gillio, A.P., Small, T.N., et al. (1994). Infusions of donor leukocytes to treat Epstein-Barr virus-associated lymphoproliferative disorders after allogeneic bone marrow transplantation. N. Engl. J. Med. *330*, 1185-1191.

Paya, C.V., Fung, J.J., Nalesnik, M.A., Kieff, E., Green, M., Gores, G., Habermann, T.M., Wiesner, P.H., Swinnen, J.L., Woodle, E.S., and Bromberg, J.S. (1999). Epstein-Barr virus-induced posttransplant lymphoproliferative disorders. ASTS/ASTP EBV-PTLD Task Force and The Mayo Clinic Organized International Consensus Development Meeting. Transplantation *68*, 1517-1525.

Penn, I. (1991). The changing pattern of posttransplant malignancies. Transplant. Proc. *23*, 1101-1103.

Perez, C.A., Devineni, V.R., Marcial-Vega, V., Marks, J.E., Simpson, J.R., and Kucik, N. (1992). Carcinoma of the nasopharynx: factors affecting prognosis. Int. J. Radiat. Oncol. Biol. Phys. *23*, 271-280.

Pule, M., Finney, H., and Lawson, A. (2003). Artificial T-cell receptors. Cytotherapy. *5*, 211-226.

Rooney, C.M., Smith, C.A., Ng, C.Y., Loftin, S., Li, C., Krance, R.A., Brenner, M.K., and Heslop, H.E. (1995). Use of gene-modified virus-specific T lymphocytes to control Epstein-Barr-virus-related lymphoproliferation. Lancet *345*, 9-13.

Rooney, C.M., Smith, C.A., Ng, C.Y., Loftin, S.K., Sixbey, J.W., Gan, Y., Srivastava, D.K., Bowman, L.C., Krance, R.A., Brenner, M.K., and Heslop, H.E. (1998). Infusion of cytotoxic T cells for the prevention and treatment of Epstein-Barr virus-induced lymphoma in allogeneic transplant recipients. Blood *92*, 1549-1555.

Roskrow, M.A., Suzuki, N., Gan, Y., Sixbey, J.W., Ng, C.Y., Kimbrough, S., Hudson, M., Brenner, M.K., Heslop, H.E., and Rooney, C.M. (1998). Epstein-Barr virus (EBV)-specific cytotoxic T lymphocytes for the treatment of patients with EBV-positive relapsed Hodgkin's disease. Blood *91*, 2925-2934.

Rossig, C., Bollard, C.M., Nuchtern, J.G., Rooney, C.M., and Brenner, M.K. (2002). Epstein-Barr virus-specific human T lymphocytes expressing antitumor chimeric T-cell receptors: potential for improved immunotherapy. Blood *99*, 2009-2016.

Savoldo, B., Cubbage, M.L., Durett, A.G., Goss, J., Huls, M.H., Liu, Z., Teresita, L., Gee, A.P., Ling, P.D., Brenner, M.K., Heslop, H.E., and Rooney, C.M. (2002a). Generation of EBV-specific CD4+ cytotoxic T cells from virus naive individuals. J. Immunol. *168*, 909-918.

Savoldo, B., Goss, J., Liu, Z., Huls, M.H., Doster, S., Gee, A.P., Brenner, M.K., Heslop, H.E., and Rooney, C.M. (2001). Generation of autologous Epstein-Barr virus-specific cytotoxic T cells for adoptive immunotherapy in solid organ transplant recipients. Transplantation *72*, 1078-1086.

Savoldo, B., Heslop, H.E., and Rooney, C.M. (2000). The use of cytotoxic t cells for the prevention and treatment of epstein-barr virus induced lymphoma in transplant recipients. Leuk. Lymphoma *39*, 455-464.

Savoldo, B., Huls, M.H., Liu, Z., Okamura, T., Volk, H.D., Reinke, P., Sabat, R., Babel, N., Jones, J.F., Webster-Cyriaque, J., Gee, A.P., Brenner, M.K., Heslop, H.E., and Rooney, C.M. (2002b). Autologous Epstein-Barr virus (EBV)-specific cytotoxic T cells for the treatment of persistent active EBV infection. Blood *100*, 4059-4066.

Savoldo, B., Rooney, C.M., Quiros-Tejeira, R.E., Caldwell, Y., Wagner, H.J., Lee, T., Finegold, M.J., Dotti, G., Heslop, H.E., and Goss, J.A. (2005). Cellular immunity to Epstein-Barr virus in liver transplant recipients treated with rituximab for post-transplant lymphoproliferative disease. Am. J. Transplant. *5*, 566-572.

Spano, J.P., Busson, P., Atlan, D., Bourhis, J., Pignon, J.P., Esteban, C., and Armand, J.P. (2003). Nasopharyngeal carcinomas: an update. Eur. J. Cancer *39*, 2121-2135.

Starzl, T.E., Nalesnik, M.A., Porter, K.A., Ho, M., Iwatsuki, S., Griffith, B.P., Rosenthal, J.T., Hakala, T.R., Shaw, B.W., Jr., Hardesty, R.L., et al. (1984). Reversibility of lymphomas and lymphoproliferative lesions developing under cyclosporin-steroid therapy. Lancet *1*, 583-587.

Straathof, K.C., Bollard, C.M., Popat, U., Huls, M.H., Lopez, T., Morriss, M.C., Gresik, M.V., Gee, A.P., Russell, H.V., Brenner, M.K., Rooney, C.M., and Heslop, H.E. (2005). Treatment of nasopharyngeal carcinoma with Epstein-Barr virus--specific T lymphocytes. Blood. *105*, 1898-1904. Epub 2004 Nov 12.

Teo, P.M., Chan, A.T., Lee, W.Y., Leung, T.W., and Johnson, P.J. (1999). Enhancement of local control in locally advanced node-positive nasopharyngeal carcinoma by adjunctive chemotherapy. Int. J. Radiat. Oncol. Biol. Phys. *43*, 261-271.

van den, B.A., Visser, L., and Poppema, S. (1999). High expression of the CC chemokine TARC in Reed-Sternberg cells. A possible explanation for the characteristic T-cell infiltratein Hodgkin's lymphoma. Am. J. Pathol. *154*, 1685-1691.

van Esser, J.W., Niesters, H.G., van der, H.B., Meijer, E., Osterhaus, A.D., Gratama, J.W., Verdonck, L.F., Lowenberg, B., and Cornelissen, J.J. (2002). Prevention of Epstein-Barr virus-lymphoproliferative disease by molecular monitoring and preemptive rituximab in high-risk patients after allogeneic stem cell transplantation. Blood *99*, 4364-4369.

Verbeke, C.S., Wenthe, U., Grobholz, R., and Zentgraf, H. (2001). Fas ligand expression in Hodgkin lymphoma. Am. J. Surg. Pathol. *25*, 388-394.

Wagner, H.J., Bollard, C.M., Vigouroux, S., Huls, M.H., Anderson, R., Prentice, H.G., Brenner, M.K., Heslop, H.E., and Rooney, C.M. (2004a). A strategy for treatment of Epstein-Barr virus-positive Hodgkin's disease by targeting interleukin 12 to the tumor environment using tumor antigen-specific T cells. Cancer Gene Ther. *11*, 81-91.

Wagner, H.J., Cheng, Y.C., Huls, M.H., Gee, A.P., Kuehnle, I., Krance, R.A., Brenner, M.K., Rooney, C.M., and Heslop, H.E. (2004b). Prompt versus preemptive intervention for EBV lymphoproliferative disease. Blood *103*, 3979-3981.

Walker, R.E., Bechtel, C.M., Natarajan, V., Baseler, M., Hege, K.M., Metcalf, J.A., Stevens, R., Hazen, A., Blaese, R.M., Chen, C.C., Leitman, S.F., Palensky, J., Wittes, J., Davey, R.T., Jr., Falloon, J., Polis, M.A., Kovacs, J.A., Broad, D.F., Levine, B.L., Roberts, M.R., Masur, H., and Lane, H.C. (2000). Long-term *in vivo* survival of receptor-modified syngeneic T cells in patients with human immunodeficiency virus infection. Blood *96*, 467-474.

Yu, M.C., Garabrant, D.H., Huang, T.B., and Henderson, B.E. (1990). Occupational and other non-dietary risk factors for nasopharyngeal carcinoma in Guangzhou, China. Int. J. Cancer *45*, 1033-1039.

Yung, L. and Linch, D. (2003). Hodgkin's lymphoma. Lancet *361*, 943-951.

zur, H.H., Schulte-Holthausen, H., Klein, G., Henle, W., Henle, G., Clifford, P., and Santesson, L. (1970). EBV DNA in biopsies of Burkitt tumours and anaplastic carcinomas of the nasopharynx. Nature *228*, 1056-1058.

From: Epstein-Barr Virus. Edited by: Erle S. Robertson

Chapter 32

Epstein-Barr Virus Related Lymphocryptoviruses of Old and New World Nonhuman Primates

*Fred Wang**

ABSTRACT

There is a renewed interest in the EBV-related herpesviruses belonging to the lymphocryptovirus (LCV) genus that naturally infect nonhuman primates. It has been long recognized that virtually all species of Old World nonhuman primates are naturally infected with their own LCV, but interest has been restimulated by the ability to experimentally infect naive rhesus macaques with rhesus LCV and the development of an animal model that accurately reproduces many aspects of EBV infection in humans. The recent derivation of the complete rhesus LCV genome sequence shows identity of the viral gene repertoire between EBV and rhesus LCV and provides genetic validation for rhesus LCV infection as an experimental system to model and study EBV pathogenesis. The recent discovery of LCV infecting New World primates indicates that EBV-related herpesviruses evolved earlier than previously believed, and the full length marmoset LCV genome sequence provides unique insight into the evolution of this tumor-associated herpesvirus genus. These recent advances have provided a much better understanding of EBV's closest relatives, as well as new laboratory and animal model systems for studying the molecular biology and pathogenesis of EBV infection.

INTRODUCTION

Closely related herpesviruses in the same gamma-1 herpesvirus, or lymphocryptovirus (LCV), genus as Epstein-Barr virus were known to infect Old World nonhuman primates as early as the 1970's. Serologic assays using EBV infected human cell lines could detect cross-reactive antibodies to viral capsid antigens, early antigens and nuclear antigens in sera from Old World monkeys and great apes (Dillner et al., 1987; Dunkel et al., 1972; Kalter et al., 1972; 1973; Landon and Malan, 1971). Evidence for LCV infection has now been established in species from all families and subfamilies of Old World primates (Ablashi et al., 1979; Frank et al., 1976).

LCV were not believed to naturally infect New World nonhuman primates because they lacked cross-reactive antibodies to EBV infected B cells. This

*For correspondence email fwang@rics.bwh.harvard.edu

suggested that EBV-related herpesviruses arose quite late in primate evolution, perhaps only 25 million years ago between the appearance of New and Old World nonhuman primates. However, an EBV-related herpesvirus was recently discovered infecting common marmosets (Cho et al., 2001), and along with other studies it is now evident that LCV are probably present in most, if not all, New World primate species (Ehlers et al., 2003; Jenson et al., 2002). Thus, the EBV-related LCV genus is older than previously believed appearing approximately 40 million years ago with the evolution of New World primates.

Full length genomes for a prototypic Old World LCV and New World LCV have now been sequenced. In this chapter, we will review the molecular biology of these viruses, their relationship to EBV, the evolution of the LCV genus, and how studies of simian LCV might benefit EBV research. Historically, these viruses have been given a variety of different names as they have been discovered, such as rhesus EBV (rhesus macaque), cyno EBV (cynomolgous macaque), Herpesvirus papio (baboon), and Herpevirus pan (chimpanzee). Unfortunately, these names are often neither correct nor very specific. Epstein-Barr virus is the name applied to the LCV naturally infecting humans, and the simian LCVs described subsequently, although similar to EBV, are genetically distinct. Many Old World nonhuman primates are infected with alpha- and betaherpesviruses, as well as gammaherpesviruses, so that names such as herpesvirus papio are not very specific. Official nomenclature recommended by the Study Group for the International Committee on the Taxonomy of Viruses (http://www.ncbi.nlm.nih.gov/ICTV/) uses a sequential number in order of discovery combined with the family or subfamily of the natural host, *e.g.* cercopithicine herpesvirus 15 for "rhesus EBV" as the 15th herpesvirus found to naturally infect species in the family Cercopithicidae. This system recognizes the unique identity of these various virus species, but fails to provide a very useful and easily recognizable name. Thus, we support the formal adoption of the ICTV nomenclature while using the combination of the host species and LCV for vernacular usage, *e.g.* rhesus LCV, baboon LCV, chimpanzee LCV.

LYMPHOCRYPTOVIRUSES NATURALLY INFECTING OLD WORLD NONHUMAN PRIMATES

Many studies suggest that Old World LCV infection is very similar to EBV infection. LCV-infected cell lines were established from healthy and diseased Old World nonhuman primates including baboons (Djatchenko et al., 1976; Falk et al., 1976; Rabin et al., 1977), chimpanzees (Gerber et al., 1976; Landon et al., 1968), gorillas (Neubauer et al., 1979), orangutans (Rasheed et al., 1977), and various macaque species (Bocker et al., 1980; Fujimoto et al., 1990; Heberling et al., 1981; Lapin et al., 1985; Rangan et al., 1986). Simian LCV infected cell lines produced virus that could immortalize B cells from autologous and closely related species *in vitro* (Falk et al., 1977; Fujimoto et al., 1990; Gerber et al., 1977; Ishida and Yamamoto, 1987; Neubauer et al., 1979; Rabin et al., 1977; 1978; Rangan et al., 1986). These cell lines also expressed a latent infection nuclear antigen similar to the EBNAs that could be detected by immune simian sera, but the simian latent infection nuclear antigens did not cross-react well with EBV immune human sera (Dillner et al., 1987; Gerber et al., 1976; Li et al., 1993; Ohno et al., 1977) providing the earliest clues that the latent infection genes may not be as well conserved as the lytic infection genes.

In vivo, LCV infection in nonhuman primates mirrors EBV infection in humans. Newborn animals are born seropositive for antibodies to the viral capsid antigen due to transplacental transfer of maternal antibodies. Within four to six months after birth maternal antibodies disappear, but most animals seroconvert again within the year indicating a high prevalence of naturally acquired infection (Frank et al., 1976; Fujimoto and Honjo, 1991; Jenson et al., 2000; Landon and Malan, 1971; Rao et al., 2000). Once infected, animals have lifelong antibody responses and asymptomatic persistent infection in peripheral blood lymphocytes and the oropharynx similar to humans (Moghaddam et al., 1997).

Simian LCV infection is also associated with the development of B cell malignancies. In the late 1960's an epidemic of lymphoma in over 30 baboons was reported at the Institute of Experimental Pathology, USSR Academy of Medical Sciences in Sukhumi (Lapin, 1974). This was most likely associated with the baboon LCV because a herpesvirus was detected by electron microscopy and nucleic acid homology to human EBV was demonstrated in cell lines established from lymphomatous baboons. Interestingly, a primate retrovirus was later isolated from a lymphomatous baboon in the Sukhumi colony suggesting that this malignant epidemic may have been associated with an outbreak of a simian immunodeficiency virus (SIV) infection (Goldberg et al., 1974).

More recent work has confirmed the association of LCV associated disease in immunosuppressed Old World primates. Feichtinger, et al. reported a high frequency of LCV positive B cell lymphomas (62%; 8 of 13 LCV seropositive animals) appearing 5-15 months after SIV infection of cynomolgus macaques (Feichtinger et al., 1990). LCV positive B cell lymphomas in association with SIV infection (Cho et al., 1999; Feichtinger et al., 1990; Habis et al., 1999; Pingel et al., 1997) or drug-induced immunosuppression (Schmidtko et al., 2002) have now been reported from a number of different primate centers. Recent work has also demonstrated that immunosuppression is associated with simian LCV infection of epithelial cells with LCV positive, proliferative epithelial cell lesions arising in SIV-infected rhesus macaques similar to Oral Hairy Leukoplakia in AIDS patients (Baskin et al., 1995; Kutok et al., 2004).

MOLECULAR BIOLOGY OF OLD WORLD LYMPHOCRYPTOVIRUSES

In the 1980's, cross-hybridization studies with EBV DNA showed that the baboon and chimpanzee LCV genomes were collinear and shared a similar structural format with tandem repeats at both ends and tandem direct repeats separating long and short unique regions (Heller et al., 1981; Heller and Kieff, 1981). Almost all EBV DNA fragments were found to cross react with viral DNA from baboon and chimpanzee LCVs suggesting approximately 40% nucleotide homology. It is interesting to note that two regions of the EBV genome did not cross-hybridize well to the simian LCV genomes, specifically the BamHI E and Nhet DNA fragments that were subsequently identified to encode the latent infection nuclear antigen 3 family and the latent infection membrane proteins 1 and 2 respectively. In retrospect, this provided further evidence that the latent infection genes were not as well conserved as the lytic infection genes.

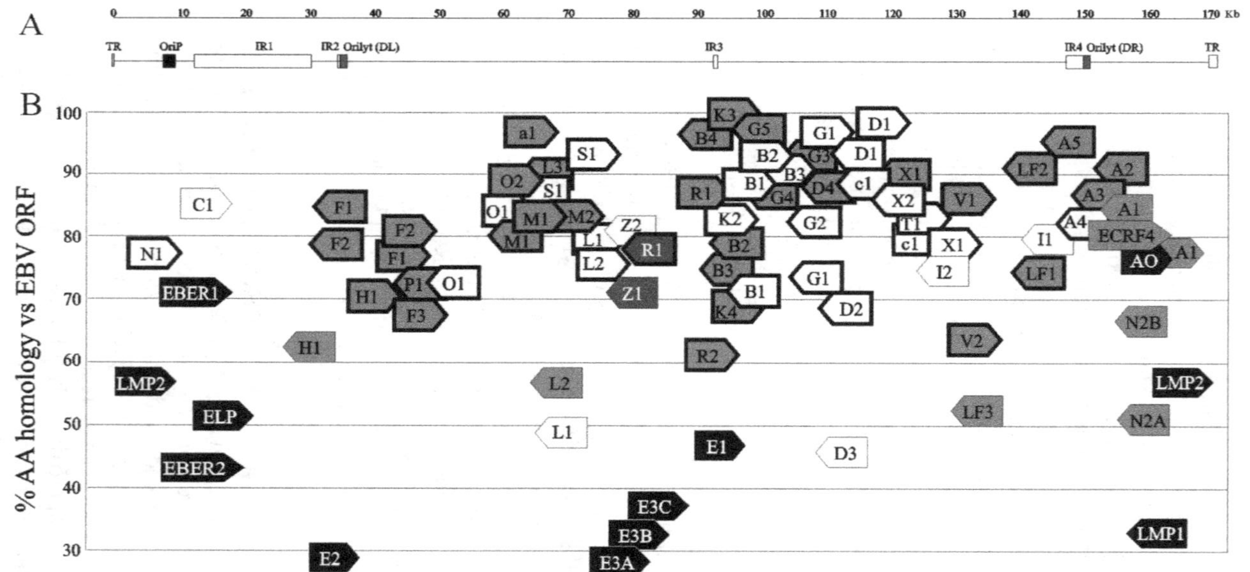

Figure 1. The rhesus LCV genome, open reading frames, and homology with EBV ORFs. (A) Organization of the rhesus LCV genome. Homologues for the EBV lytic and latent origins of replication (ori-p; 7,511-9,357), ori-lyt DL (34,141-35,138), ori-lyt DR (138,080-139,080), major repeat regions (IR1; 12,240-29,750, IR2; 33674-34,047, IR3; 89,780-90,460, IR4; 13,5263-137,761), and terminal repeats (TR;167,326-171,106) are identified in the rhesus LCV genome as shown. (B) Rhesus LCV ORFs and amino acid homology with EBV ORFs. The per cent amino acid similarity is shown on the Y axis. Latent, immediate early, early and late lytic ORFs are in black, dark grey, light grey and white respectively. Latent infection genes are identified by name (LMPs, EBERs, EBNAs (E), BARF0 (B0)). Each lytic infection ORF is identified using the EBV nomenclature for BamHI ORFs. The orientations of the ORF is shown by the direction of the arrow (*i.e.* right or left). The EBV BamHI fragment is indicated by the letter within the arrow, and the number of the ORF in the EBV BamHI fragment is given last, eg the rhesus LCV BCRF1 homologue is indicated by the rightward C1 arrow with approximately 85% amino acid similarity. (The ECRF4 Orf is the only exception to these abbreviations). ORFs common to other herpesviruses are boxed. The initiator codon for each ORF is positioned accurately, but the ORF size is not drawn to scale. Reproduced with permission. Copyright 2002, The American Society for Microbiology. All rights reserved.

The complete rhesus LCV genome: a prototype for an Old World LCV

The recently completed rhesus LCV genome sequence provides more detailed insight into a prototype for the Old World LCV (Rivailler et al., 2002). Like EBV there are four internal repeat regions and the major internal repeat of the sequenced virus contains 5.7 copies of a 3,072 bp motif that is 61.5% homologous to the EBV repeat sequence (Figure 1). The rhesus LCV terminal repeat sequence contains 4 repeats of a 933 bp motif compared to the 538 bp EBV TR, and there is no significant sequence homology besides a similarly high GC content in both (75%). The overall size of the rhesus LCV genome is 171,096 bp (5.7 IR1 copies and 4 TP copies) compared to 172,231 for EBV (11.3 IR1 copies and 4 TR copies), and overall there is 65% nucleotide homology with EBV.

There is identity in the coding capacity of the rhesus LCV and EBV. Eighty open reading frames were identified. Every ORF in rhesus LCV has an amino acid sequence homologue at the same relative position in EBV and vice versa. Thus, to date, no genes have been identified in rhesus LCV that do not exist in EBV, and similarly there is no EBV gene that is not present in rhesus LCV. An overall average of each rhesus LCV gene compared its EBV counterpart shows a 75.6% amino acid homology, but there is considerable variability when one examines different classes of viral genes. In general, those genes that have homologues in all herpesviruses, ie alpha, beta and gamma herpesviruses, are more well conserved, and those genes that are found only among viruses in the LCV genus tend to be much less well conserved.

The rhesus LCV ORFs have been named by the same convention as EBV ORFs, eg rhBHRF1 for rhesus BamHI H rightward reading frame 1, because of the high likelihood that the corresponding rhesus LCV gene is a functional homologue of the EBV gene, ie the identity in viral gene repertoire and the high degree of amino acid sequence homology. The advantage of this nomenclature is that it provides easy identification of the rhesus LCV gene based on its EBV counterpart. The disadvantage is that it is technically a misnomer since the gene is most likely not encoded in the same BamHI DNA fragment from rhesus LCV as in EBV.

Rhesus LCV lytic infection genes

56 of the ORFs that have homologues in other alpha, beta, and gamma herpesviruses are lytic infection genes, and these have an overall homology of 82.8% amino acid homology with the EBV homologues. Fifteen ORFs are lytic infection genes that are found only in gamma-1 herpesviruses. These genes have appeared later in herpesvirus evolution and have an average homology of 60.3% with the EBV homologues. Of the 10 glycoproteins encoded by EBV, 5 are conserved in all herpesviruses (gB, gH, gI, gM and gN), and these rhesus LCV glycoproteins have an overall homology of 83.4% with the EBV homologues. Five glycoproteins are generally restricted to gamma-1 herpesviruses. Three of these, BILF1, BILF2, and BZLF2 (gp64, gp78, and gp42) are reasonably well conserved in rhesus LCV (77.7% overall homology versus EBV). The major membrane glycoprotein, gp350 (BLLF1) and BDLF3 (gp150) are much less conserved compared to EBV (49.3% and 46.6%).

Rhesus LCV latent infection genes

The latent infection proteins are much less well conserved. These viral genes associated with immortalizing B cell infection are restricted to gamma-1 herpesviruses and fit the paradigm of less sequence homology among those LCV genes that have evolved more recently. For example, the 6 EBNA proteins have an overall amino acid homology of 39% to each of the EBV nuclear proteins with EBNA-1 being the most well conserved (range 29.4-60.6%, Figure 1). The transmembrane domains of LMP1 and LMP2 are reasonably well conserved probably due to the biologic selection for hydrophobic residues, but the cytoplasmic domains have only around 30% homology.

However, in all instances studied so far, the rhesus LCV latent infection genes can substitute for the EBV gene for *in vitro* functional assays. For example, the rhesus LCV EBNA-2, 3A, -3B, and -3C homologues can bind to human RBP-Jk/CBF1 (Jiang et al., 2000), the rhesus LCV EBNA-2 can transactivate the Cp and LMP1 promoters (Fuentes-Panana et al., 1999; Peng et al., 2000), the rhesus LCV EBNA-LP can co-activate with EBV EBNA-2 (McCann et al., 2001; Peng et al., 2000), and the rhesus LCV LMP1 activates NF-κB (Franken et al., 1996). One would predict that viruses in the same LCV genus would utilize the same molecular pathways for B cell immortalization and that these cell signaling pathways would be very well conserved among human and nonhuman primates. This functional conservation in the setting of low sequence homology can prove to be useful. For example, the rhesus LCV LMP1 activates NF-κB and interacts with TNF receptor associated factors like the EBV LMP1. However, there is little significant sequence homology except for the isolated domains that identified the PXQXT/S binding motif for TRAFs (Franken et al., 1996). Conserved sequence domains have also been used to identify important interactions and functional domains among the rhesus LCV, baboon LCV, and EBV EBNA-2 and EBNA-LP (Ling et al., 1993; McCann et al., 2001; Peng et al., 2000).

A species restricted block for LCV-induced B cell immortalization

Despite the experimental evidence for common molecular mechanisms and interchangeability for *in vitro* assays among LCV latent infection genes from different species, there appears to be a species restriction for LCV-induced B cell immortalization. Specifically, EBV is unable to immortalize rhesus macaque B cells, and vice versa, rhesus LCV is unable to immortalize human B cells (Moghaddam et al., 1998). This restriction is not absolute but appears to be associated with greater evolutionary divergence in Old World primates. For example, baboon LCV can immortalize rhesus macaque, but not human B cells (Moghaddam et al., 1998), and EBV has been used to immortalize chimpanzee B cells (Shimizu et al., 1985). Laboratory studies suggest that the block is beyond the point of virus entry since baboon LCV can infect, persist, and replicate in human B cells (Moghaddam et al., 1998). One potential hypothesis is that one or more of the latent infection genes has a critical interaction with a cell protein, but is capable of doing so only in the same or closely related species. Thus, understanding exactly why rhesus LCV is unable to immortalize human B cells may identify cell signaling pathways that are intrinsically important for B cell immortalization.

Genetic evidence for two major types of rhesus LCV

There is also evidence for two major types of rhesus LCV similar to type 1 and 2 EBV. EBNA-2 immunoblotting of a rhesus LCV infected cell line, 208-95, derived from a B cell tumor in a SIV infected macaque at the New England Primate Research Center (NEPRC) demonstrated an EBNA-2 protein migrating significantly more rapidly than the EBNA-2 from the prototypic rhesus LCV infected cell line, LCL8664 (Cho et al., 1999). The 208-95 EBNA-2 gene was cloned and sequenced showing a predicted EBNA-2 protein (464 amino acids) with only 41% amino acid homology to the LCL8664 derived EBNA-2 (605 amino acids) (Cho et al., 1999). For comparison, the EBV type 1 and type 2 EBNA-2s have 55% homology. The most significant region of divergence resides in the central portion of the 208-95 and LCL8664 derived EBNA-2s similar to the most divergent region between the EBV EBNA-2s that defines the different types. Other previously recognized domains such as the C-terminal acidic domain, the RBP-Jk/CBF-1 binding region, and N-terminus were more well conserved between the 208-95 and LCL8664 EBNA-2s as they are in the type 1 and 2 EBV EBNA-2s. Thus, the structural similarities of the 208-95 and LCL8664 EBNA-2s suggest that there are two types of rhesus LCV defined by allelic variation in the EBNA-2 genes, just as there are two types of EBV. Whether this allelic variation extends to the EBNA-3 genes in rhesus LCV as in EBV has not been determined, and what biologic advantage the different EBNA-2 alleles provide is uncertain. PCR studies using rhesus LCV EBNA-2 type specific PCR primers to detect viral shedding in oral secretions indicate that infections with either type of rhesus LCV are present in naturally infected macaques at the NEPRC at similar frequencies (Cho et al., 1999). This suggests that the same biologic pressures selecting for type 1 and type 2 EBV infection in humans have also selected for two similar types of rhesus LCV infection in macaques.

RHESUS LCV ANIMAL MODEL FOR EBV PATHOGENESIS

The biologic similarities of natural rhesus LCV infection in macaques to EBV infection in humans and the identical repertoire of viral genes in rhesus LCV and EBV suggest that rhesus macaques could provide a valuable animal model for studying EBV infection. The evidence for a species restricted block to EBV-induced immortalization of rhesus macaque B cells suggests that using EBV to infect rhesus macaques would not be effective, and this may explain the lack of success with this experimental approach in the late 1970's and 1980's. Using rhesus LCV to experimentally infect rhesus macaques provides a model of LCV infection in a natural host that is closely related to humans. The characterization of the rhesus LCV genome also provides genetic validation that rhesus LCV infection is a good model for EBV infection.

Rhesus LCV naive macaques for experimental infection

The most powerful aspect of an experimental animal model in rhesus macaques is the ability to control the type and timing of LCV infection, ie experimental inoculation and infection of naive animals versus studying macaques with naturally acquired LCV infection. Since natural rhesus LCV infection spreads rapidly within a domestic colony, obtaining rhesus LCV naive animals requires some special handling. One approach is to identify seronegative animals at 6 months of age as maternal antibody is lost and to house these animals individually to prevent natural rhesus LCV infection. Alternatively, the NEPRC has developed a specific pathogen free colony, originally designed to raise macaques free of herpesvirus B infection that can be dangerous for humans (Desrosiers, 1997). The NEPRC hand raised newborn macaques, screened them to ensure that they were free of specific pathogens, and housed them separately from other macaques. Although rhesus LCV was not specifically screened for, animals raised under these conditions were found to be LCV naive (Rao et al., 2000), as well as naive for beta and gamma-2 herpesvirus infections. Efforts are also underway to establish a self-sustaining breeding colony of specific pathogen free animals using similar handling and screening procedures. This type of domestic primate colony would be extremely valuable for research on EBV, as well as other herpesvirus, infections.

Experimental infection of immunocompetent rhesus macaques

Naive rhesus macaques can be successfully infected by experimental oral inoculation with rhesus LCV (Moghaddam et al., 1997). This reproduces the natural route of transmission and many of aspects of acute and persistent EBV infection in humans. DNA viremia in peripheral blood lymphocytes can be detected by PCR around 21 days after oral experimental infection (Rivailler et al., 2004). Some animals develop an atypical lymphocytosis, lymphadenopathy, and splenomegaly (Moghaddam et al., 1997), but like in humans, the magnitude of the response can be variable. It remains to be determined whether the variability in responses is age related like the infectious mononucleosis syndrome in humans. Viremia is cleared to undetectable levels around 10-12 weeks after inoculation (Rivailler et al., 2004), but persistent infection can be detected in the peripheral blood by RT-PCR for EBERs expression (Rivailler et al., 2004) and in the oral cavity by PCR from oral secretions (Moghaddam et al., 1997). Persistent infection has been documented for at least 5 years in 2 animals suggesting that persistent infection is as durable after experimental inoculation as from natural infection. This experimental infection model provides the opportunity to examine early time points in infection before the development of mono-like symptoms that are required to identify potential human subjects with primary EBV infection. It also provides a model system where mutant rhesus LCV can be used for infection in order to study the potential contributions of specific viral genes for successful acute or persistent LCV infection *in vivo*.

Experimental infection of immunosuppressed rhesus macaques

The rhesus LCV animal model can also be used to study the pathogenesis of LCV infection in the setting of immunosuppression (Rivailler et al., 2004) as occurs in AIDS and transplant patients. In these experiments, LCV naive macaques were first infected with a simian-human chimeric immunodeficiency virus (SHIV) that routinely depletes CD4+ T lymphocytes 2 weeks after infection resulting in a retrovirus-induced immunodeficiency (Reimann et al., 1996) similar to HIV infection in humans. Oral LCV infection in this setting resulted in an earlier and approximately one log higher peripheral blood DNA viremia, and the CD4+ T cell immunodeficiency was sufficiently severe to prevent seroconversion after primary LCV infection. Most animals succumbed to SIV infection or associated complications, with no animals experiencing uncontrolled lymphoproliferation during primary LCV infection in the setting of an immunosuppressed host. In one animal, DNA viremia was controlled to undetectable levels for over one year despite a CD4+ T cell deficit that prevented seroconversion. Thus, an intact CD4+ T cell response does not appear to be essential for control of primary LCV infection.

In order to determine whether a more potent viral challenge might result in a LCV-induced lymphoproliferation, SHIV-infected macaques were challenged with an intravenous infusion of autologous B cells that had previously been immortalized with rhesus LCV *in vitro* (Rivailler et al., 2004). This route of inoculation was associated with an extremely high transient DNA viremia. Two animals with moderate CD4+ T cell depletion but >50 CD4+ T cells/μl seroconverted to rhesus LCV infection and were able to control the acute infection. Thus, autologous, LCV-immortalized B cells derived *in vitro* may not be inherently tumorigenic *in vivo* after intravenous inoculation, even in immunosuppressed hosts. In two animals with severe CD4+ T cell depletion, <50 CD4+ T cells/μl, there was evidence for tissue invasion with LCV infected B cells and development of a malignant lymphoma in one animal. The growth of this lymphoma was unifocal and aggressive. Histopathology showed an infiltrate of large immunoblastic cells with prominent nucleoli and frequent mitotic figures. Immunohistochemistry showed that the vast majority of cells were B cells and various assays for LCV showed that the malignant cells were virus infected. There was also a modest amount of T cell infiltration as frequently seen in EBV-associated malignancies. These studies demonstrate that the experimental virus isolate used has the potential to cause virus induced malignancies in the setting of immunosuppression. They also demonstrate the depth and strength of the host immune response required before there is loss of control over LCV infection and may provide a system for identifying the specific components of the host immune response critical for control of LCV infection, the potential changes to LCV-infected cells that may contribute to malignant conversion, and strategies to reduce the risk of LCV-induced malignancies.

LYMPHOCRYPTOVIRUSES NATURALLY INFECTING NEW WORLD PRIMATES

EBV-related herpesviruses were thought to have developed relatively late in primate evolution since they were restricted to humans and Old World non-human primates. However, in 2000, Ramer and colleagues reported an unusually high frequency of abdominal B cell lymphomas in the common marmoset colony at the Wisconsin Primate Research Center (WPRC) (Ramer et al., 2000). The lymphomas arose in a subset of animals that suffered from a diarrheal wasting disease of unknown etiology. Using a set of degenerative PCR primers capable of amplifying segments of the DNA polymerase and terminase genes from all known herpesviruses, the WPRC investigators isolated novel herpesvirus DNA fragments from the common marmoset lymphomas that were most closely related to EBV suggesting the possible association of the lymphomas with LCV infection (Ramer et al., 2000).

A tumor cell line was derived from one of the WPRC common marmoset tumor tissues (Cho et al., 2001). This cell line was PCR positive for the same herpesvirus DNA polymerase fragment identified by Ramer et al. A cosmid clone identified by

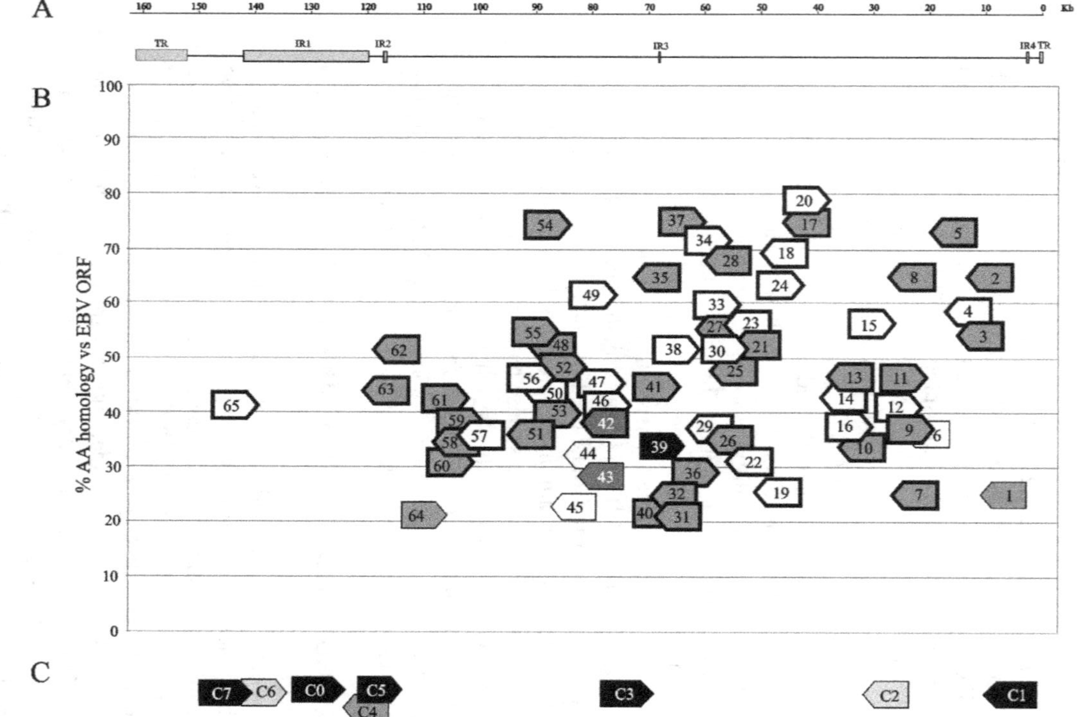

Figure 2. The marmoset LCV genome, open reading frames, and homology with EBV ORFs. (A) Organization of the marmoset LCV genome. Major repeat regions IR1 (142,148 to 120,197), IR2 (117,265 to 116,619), IR3 (69,982 to 69,841), IR4 (2,792 to 2,469) and terminal repeats TR (161,345 to 152,171) are identified in the marmoset LCV genome as shown. (B) Marmoset LCV ORFs and amino acid homology with EBV ORFs. The percent amino acid similarity is shown on the y axis. Putative latent, immediate-early, early and late lytic ORFs are represented by black, dark grey, light grey and white arrows, respectively. The ORFs are numbered from the right to the left and the orientation of the ORFs is shown by the direction of the arrow (*i.e.*, right or left). ORFs common to other herpesviruses are shown with a bold outline. The initiator codon for each ORF is positioned accurately, but the ORF size is not drawn to scale. (C) Marmoset LCV unique genes. ORFs are represented as defined in (B). Reproduced with permission. Copyright 2002, The American Society for Microbiology. All rights reserved.

hybridization with the PCR fragment and two adjacent viral cosmid clones were derived to yield 105 kb of viral genome sequence (Cho et al., 2001). The partial genome sequence showed that this marmoset herpesvirus was unequivocally a LCV based on collinear organization with the last 100 kb at the right end of the EBV genome and 57 ORFs that were more similar to EBV homologues than KSHV, ie gamma-2 herpesvirus, homologues.

The same herpesvirus DNA polymerase PCR fragment could be amplified from the peripheral blood of healthy common marmosets at the WPRC and NEPRC indicating that LCV infection was prevalent in two different marmoset colonies (Cho et al., 2001). Furthermore, a closely related herpesvirus DNA polymerase PCR fragment could be amplified from the peripheral blood of squirrel monkeys indicating that LCV infection is present in species from both major families of New World primates (Cho et al., 2001). Subsequently, other investigators have confirmed the presence of LCV infection in marmosets and other New World primates species (Ehlers et al., 2003; Jenson et al., 2002). The presence of LCV infection in New World primates indicates that this family of tumor viruses evolved earlier than previously believed, perhaps some 40 million years ago concomitant with the appearance of New World primates versus around 25 million years ago with the evolution of the Old World primates.

MOLECULAR BIOLOGY OF NEW WORLD LCV
The complete marmoset LCV genome: A prototype for a New World LCV

The full genome sequence from the marmoset LCV has been derived as a prototype for a New World LCV (Cho et al., 2001; Rivailler et al., 2002). The genome is significantly different from EBV and rhesus LCV reflecting the much earlier lineage of this virus and providing an interesting perspective on the evolution of the EBV-related herpesviruses.

The marmoset LCV genome is 161,345 bp with 7.5 copies of a 2,912 bp major internal repeat and 35 copies of a GC rich 328 bp terminal repeat (Cho et al., 2001; Rivailler et al., 2002). The size of the major internal repeat motif is similar to EBV and rhesus LCV, but the large number of terminal repeats is more reminiscent of the KSHV-related gamma-2 herpesviruses, or Rhadinovirus, than EBV and rhesus LCV. Since the marmoset LCV genome was cloned and sequenced starting from the opposite end relative to EBV and rhesus LCV, the actual nucleotide numbering of the marmoset LCV genome in Genbank (AF319782) is in the opposite orientation relative to EBV and rhesus LCV (Figure 2).

There are 73 ORFs predicted in the marmoset LCV (Figure 2). 59 of these ORFs are homologous to genes present in all herpesviruses, and six ORFs have recognizable homology to genes present in other LCV only (BALF1, BILF1, EBNA-1, BZLF-1 and BZLF-2). These 65 marmoset LCV genes can be compared to the EBV homologues and overall they have an amino acid homology of 47% to the EBV ORFs. In comparison, the rhesus LCV ORFs had an overall amino acid homology of 75% to EBV consistent with the closer evolutionary relationship of macaques versus marmosets to humans. Since the marmoset LCV ORFs are more divergent from EBV and the viral gene repertoire is not identical, marmoset LCV ORFs were named by the same convention typically used for gamma-2 herpesviruses, ie genes present in other herpesviruses are numbered consecutively from one end of the genome and unique genes are given the prefix from the virus name and numbered sequentially (Figure 2).

Unique marmoset LCV genes

There are eight marmoset LCV ORFs that have no sequence homology to any known cell or viral genes (Cho et al., 2001; Rivailler et al., 2002). These are named C0 through C7, with the prefix derived from the formal name for the marmoset LCV, Callitrichine Herpesvirus 3 . Five of these genes (C0, C1, C3, C5, and C7) are located in the same position in the marmoset LCV genome as the latent infection genes EBNA-LP, LMP1, EBNA-3, EBNA-2, and LMP2 in the EBV genome, and these are presumed to be functional, if not sequence, homologues.

C1: A LMP1 homologue

This has formally been shown for the marmoset LCV C1 gene. Even though it has no sequence homology with the EBV LMP1, C1 is predicted to have six hydrophobic transmembrane domains and a large carboxy terminal cytoplasmic domain (Cho et al., 2001). Like EBV LMP1, C1 can bind TRAFs and can induce NF-kB activity when transfected and expressed in human cells (Cho et al., 2001). The TRAF interaction domain in the C1 carboxy terminus appears to be a P/AxEE consensus sequence rather than the PxQxT/S motif conserved in the EBV and rhesus LCV sequence. However, C1 function is not identical to EBV LMP1 since C1 transfection is not able to induce AP-1 activity like the EBV and rhesus LMP1 (Cho et al., 2001).

C3: An EBNA-3 homologue

The marmoset LCV C3 gene is located between the gp350 and gp42 homologue similar to the location of EBNA-3A, -3B, and -3C in EBV (Cho et al., 2001). In EBV, there is almost 10 kb of DNA between the two glycoprotein genes, but in the marmoset LCV there is only 2.7 kb. This region of the EBV genome was thought to be created by gene duplication since the EBNA-3A, -3B, and -3C genes are distantly related, are all relatively large (900+ amino acids), and have a similar short first exon, long second exon structure. RT-PCR analysis indicates that the C3 gene encodes an 825 amino acid protein derived from a spliced transcript with a similar short first exon, long second exon structure (Cho et al., 2001). These similarities suggest that the marmoset LCV C3 gene may be the predecessor of the subsequently duplicated EBNA-3A, -3B, and -3C genes, but whether C3 shows functional similarities to the EBNA-3s remains to be determined.

C0: An EBNA-LP homologue and conserved EBNA transcription

The marmoset LCV C0 gene is presumed to be an EBNA-LP homologue and was identified as a common transcriptional leader sequence by 5′ RACE analysis from ORF39, C3, and C5 mRNAs (the EBNA-1, EBNA-2, and EBNA-3 homologues) (Rivailler et al., 2002). Like EBNA-LP, C0 is composed of two repeating exons derived from alternate repeats of the major internal repeat region and two unique exons from immediately 3′ of the major internal repeat. A translational initiation site was not present in the longest 5′ RACE products, so it is still a presumption that a C0 protein is translated from a transcript that has successfully spliced a translational start site. The presence of the C0 leader sequence on the transcripts for the EBNA-1, EBNA-2, and EBNA-3 homologues indicates that the marmoset LCV probably uses a common promoter for these putative latent infection genes. Thus, this transcriptional pattern using a common promoter for bicistronic EBNA transcripts is probably a *sine quo non* for the LCV genus.

C7: A LMP2 homologue and a potential link to gamma-2 herpesviruses

The marmoset LCV C7 gene was first identified as a multiply spliced transcript using a computer program to detect potential exons (Rivailler et al., 2002). 10 exons were subsequently identified by 5′ and 3′ RACE with the 5′ most exon encoded on the opposite side of the TR from the other 9 exons, similar to the EBV LMP2 (Rivailler et al., 2002). However, unlike the EBV LMP2A or 2B, the most 5′ exon is located inbetween the second and third exons of C1 (LMP1 homologue). It is likely that the 5′ end of the transcript has not yet been identified since there is no strong consensus translational start codon, the longest ORF remains in frame up to the 5′ end, and there is no obvious promoter like sequence upstream of the longest 5′ RACE product. The longest open reading frame is predicted to encode a 413 amino acid protein with no homology to any known cell or viral gene, but this sequence is predicted to encode 12 transmembrane domains like the EBV LMP2. In addition, there are three tyrosine residues that are predicted to phosphorylated as src homology (SH2) domains reminiscent of the tyrosine phosphorylation sites in the amino terminus of LMP2A. Curiously, the tyrosine phosphorylation sites are in the C7 carboxy terminus, not the amino terminus as in LMP2A. The multiple transmembrane domains and carboxy terminal tyrosine phosphorylation sites are more analogous to the KSHV K15 membrane protein that is also encoded by a highly spliced gene near the termini of the human rhadinovirus. Thus, the marmoset LCV C7 gene, as well as the high number of terminal repeats in the marmoset LCV, may be vestiges from the evolution of a LCV from a Rhadinovirus precursor.

C2 and C4

The marmoset LCV C2 and C4 genes have no sequence homology but are in the same position as the EBV BILF2 and BHLF1 (Rivailler et al., 2002). Like BILF2, C2 is predicted to encode a transmembrane protein, but since little is known about EBV BILF2 or BHLF1 it is difficult to know whether the marmoset LCV genes represent functional homologues. The marmoset LCV C6 is a putative ORF that is encoded across the repeat elements in the lytic origin of replication. There is no positional homologue for this ORF in EBV or rhesus LCV, and it is not known whether C6 is expressed.

EBV genes not present in marmoset LCV: Acquired LCV genes

Eleven genes that are present in EBV and rhesus LCV do not have apparent positional or sequence homologues in the marmoset LCV (Table 1; Acquired genes). Thus, these genes appear to have been acquired during the evolution of EBV and rhesus LCV, sometime after the appearance of marmoset LCV. Two of these genes are immunomodulators that may have been acquired from the cell genome; BCRF1, a homologue of cellular IL-10, and BARF1, a homologue of

Table 1. Classification of LCV specific genes based on evolutionary analysis

ANCESTRAL GENES (present in all LCVs; marmoset LCV, rhesus LCV, and EBV)

Latent Infection genes

EBNA-LP	Transcriptional co-activator
EBNA-1	Episomal maintenance
EBNA-2	Transcriptional activator
EBNA-3s	Transcriptional regulator
LMP1	NF-kB activation
LMP2	Alter tyrosine kinase activity, regulate latency

Lytic Infection genes

BLLF1-gp350	Major membrane glycoprotein
BZLF2-gp42	Binds to HLA-II coreceptor
BILF1-gp64	Membrane glycoprotein, unknown function
BILF2-gp78	Membrane glycoprotein, unknown function
BZLF1	Zebra Transactivator
BHRF1	Bcl-2 homologues
BALF1	Bcl-2 homologues
LF1	Unknown function
LF2	Unknown function
BHLF1	Unknown function

ACQUIRED GENES (present only in Old World LCVs; rhesus LCV and EBV)

Latent Infection genes

EBERs	Alter IFN responses
BARF0	unknown function

Lytic Infection genes

BCRF1 vIL-10	Immunomodulator
BARF1 CSF-1R	Immunomodulator
BDLF3-gp150	Membrane glycoprotein, unknown function
BLLF2	Unknown function
LF3	Unknown function
ECRF4	Unknown function
BNLF2s	Unknown function

cellular CSF-1R. EBERs are latent infection RNAs that have been implicated in modulating interferon induced responses. BDLF3 encodes a lytic infection glycoprotein of unknown function, but its deletion can enhance EBV infection of epithelial cells. BARF0, BLLF2, LF3, ECRF4, and BNLF2 all encode proteins of unknown function. It is interesting to note that all of these acquired genes are unique to LCV (with the exception that IL-10 homologues may be encoded by human and primate CMVs), and none of these genes appears to be essential for EBV replication or EBV-induced B cell immortalization *in vitro*. Thus, phylogenetically, these genes may have been acquired by EBV and Old World LCV because they make LCV infection more successful *in vivo* in higher order primates. For example, the development of more mature immune responses in humans and Old World primates may require additional viral immunomodulators and immune evasion strategies in order to successfully persist in the host.

Ancestral LCV genes: Present in EBV, rhesus LCV and marmoset LCV

It is also interesting phylogenetically to examine the genes found only in LCV and present from the earliest member of the genus (Table 1; Ancestral genes). These are the genes that contribute most significantly to making LCV different from other types of herpesviruses, and their presence in the earliest LCV would suggest that they confer core functions essential and intrinsic to all LCV. Thus, it is interesting to note that each latent infection gene essential, or contributing significantly, to the efficiency of EBV-induced B cell immortalization is present in this class of ancestral LCV genes, *i.e.* EBNA-1, -2, -3, -LP, and LMP1. LMP2 may be included in this group because the ability to prevent reactivation in virus activated B cells could be essential for LCV survival *in vivo*. Similarly, the major membrane glycoprotein, gp350, that confers B cell tropism and the viral glycoprotein, gp42 that binds to the HLA class II co-receptor are present in this group. Reactivation of lytic infection is another key aspect of the LCV life cycle, so that the presence of BZLF1, the lytic transactivator important for activating lytic EBV infection, in all three LCV genomes fits the paradigm. Functions for the other ancestral genes are less well defined, and it will be interesting to apply a phylogenetic perspective as more is learned about how these LCV specific genes contribute to successful LCV infection.

SUMMARY

The renewed interest in simian LCV has provided a better appreciation of EBV's closest relatives. The discovery of LCV infecting New World primates and the full genome sequences from EBV, rhesus LCV, and marmoset LCV provide a snapshot of how a Rhadinovirus precursor may have evolved into a genus of novel herpesviruses uniquely successful for ubiquitous and persistent B cell infection in humans. LCV genes from different host species may show significant sequence divergence, but they use the same molecular mechanisms to successfully infect their hosts. Thus, in a sense, the simian LCV provide a set of "natural mutants" for EBV genes, ie gene products with altered sequences that use the same molecular strategy to perform similar biologic functions. In addition, the genetic and biologic similarities between rhesus LCV and EBV make rhesus LCV infection of naive rhesus macaques a unique animal model system for studying EBV pathogenesis. It appears that much can be learned about a family by studying its relatives.

References

Ablashi, D. V., Gerber, P., and Easton, J. (1979). Oncogenic herpesviruses of nonhuman primates. Comp Immunol. Microbiol. Infect. Dis. *2*, 229-241.

Baskin, G. B., Roberts, E. D., Kuebler, D., Martin, L. N., Blauw, B., Heeney, J., and Zurcher, C. (1995). Squamous epithelial proliferative lesions associated with rhesus Epstein-Barr virus in simian immunodeficiency virus-infected rhesus monkeys. J. Infect. Dis. *172*, 535-539.

Bocker, J. F., Tiedemann, K. H., Bornkamm, G. W., and zur, H. H. (1980). Characterization of an EBV-like virus from African green monkey lymphoblasts. Virology *101*, 291-295.

Cho, Y., Ramer, J., Rivailler, P., Quink, C., Garber, R. L., Beier, D. R., and Wang, F. (2001). An Epstein-Barr-related herpesvirus from marmoset lymphomas. Proc. Natl. Acad. Sci. USA. *98*, 1224-1229.

Cho, Y. G., Gordadze, A. V., Ling, P. D., and Wang, F. (1999). Evolution of two types of rhesus lymphocryptovirus similar to type 1 and type 2 Epstein-Barr virus. J. Virol. *73*, 9206-9212.

Desrosiers, R. C. (1997). The value of specific pathogen-free rhesus monkey breeding colonies for AIDS research. AIDS Res. Hum. Retroviruses *13*, 5-6.

Dillner, J., Rabin, H., Letvin, N., Henle, W., Henle, G., and Klein, G. (1987). Nuclear DNA-binding proteins determined by the Epstein-Barr virus- related simian lymphotropic herpesviruses H. gorilla, H. pan, H. pongo and H. papio. J. Gen. Virol. *68*, 1587-1596.

Djatchenko, A. G., Kakubava, V. V., Lapin, B. A., Agrba, V. Z., Yakovleva, L. A., and Samilchuk, E. I. (1976). Continuous lymphoblastoid suspension cultures from cells of haematopoietic organs of baboons with malignant lymphoma--biological characterization and biological properties of the herpes virus associated with culture cells. Exp. Pathol. *12*, 163-168.

Dunkel, V. C., Pry, T. W., Henle, G., and Henle, W. (1972). Immunofluorescence tests for antibodies to Epstein-Barr virus with sera of lower primates. J. Natl. Cancer Inst *49*, 435-440.

Ehlers, B., Ochs, A., Leendertz, F., Goltz, M., Boesch, C., and Matz-Rensing, K. (2003). Novel simian homologues of Epstein-Barr virus. J. Virol. *77*, 10695-10699.

Falk, L., Deinhardt, F., Nonoyama, M., Wolfe, L. G., and Bergholz, C. (1976). Properties of a baboon lymphotropic herpesvirus related to Epstein-Barr virus. Int. J. Cancer *18*, 798-807.

Falk, L. A., Henle, G., Henle, W., Deinhardt, F., and Schudel, A. (1977). Transformation of lymphocytes by Herpesvirus papio. Int. J. Cancer *20*, 219-226.

Feichtinger, H., Putkonen, P., Parravicini, C., Li, S. T., Kaya, E. E., Bottiger, D., and Biberfeld, P. (1990). Malignant lymphomas in cynomolgus monkeys infected with simian immunodeficiency virus. Am. J. Pathol. *137*, 1311-1315.

Frank, A., Andiman, W. A., and Miller, G. (1976). Epstein-Barr virus and nonhuman primates: natural and experimental infection. Adv. Cancer Res. *23*, 171-201.

Franken, M., Devergne, O., Rosenzweig, M., Annis, B., Kieff, E., and Wang, F. (1996). Comparative analysis identifies conserved tumor necrosis factor receptor-associated factor 3 binding sites in the human and simian Epstein-Barr virus oncogene LMP1. J. Virol. *70*, 7819-7826.

Fuentes-Panana, E. M., Swaminathan, S., and Ling, P. D. (1999). Transcriptional activation signals found in the Epstein-Barr virus (EBV) latency C promoter are conserved in the latency C promoter sequences from baboon and Rhesus monkey EBV-like lymphocryptoviruses (cercopithicine herpesviruses 12 and 15). J. Virol. *73*, 826-833.

Fujimoto, K., and Honjo, S. (1991). Presence of antibody to Cyno-EBV in domestically bred cynomolgus monkeys (*Macaca fascicularis*). J. Med. Primatol *20*, 42-45.

Fujimoto, K., Terato, K., Miyamoto, J., Ishiko, H., Fujisaki, M., Cho, F., and Honjo, S. (1990). Establishment of a B-lymphoblastoid cell line infected with Epstein-Barr-related virus from a cynomolgus monkey (Macaca fascicularis). J. Med. Primatol *19*, 21-30.

Gerber, P., Kalter, S. S., Schidlovsky, G., Peterson, W. J., and Daniel, M. D. (1977). Biologic and antigenic characteristics of Epstein-Barr virus-related herpesviruses of chimpanzees and baboons. Int. J. Cancer *20*, 448-459.

Gerber, P., Pritchett, R. F., and Kieff, E. D. (1976). Antigens and DNA of a chimpanzee agent related to Epstein-Barr virus. J. Virol. *19*, 1090-1099.

Goldberg, R. J., Scolnick, E. M., Parks, W. P., Yakovleva, L. A., and Lapin, B. A. (1974). Isolation of a primate type-C virus from a lymphomatous baboon. Int. J. Cancer *14*, 722-730.

Habis, A., Baskin, G. B., Murphey-Corb, M., and Levy, L. S. (1999). Simian AIDS-associated lymphoma in rhesus and cynomolgus monkeys recapitulates the primary pathobiological features of AIDS-associated non-Hodgkin's lymphoma. AIDS Res. Hum. Retroviruses *15*, 1389-1398.

Heberling, R. L., Bieber, C. P., and Kalter, S. S. (1981). Establishment of a lymphoblastoid cell line from a lymphomous cynomolgus monkey. In Advances in comparative leukemia research, D. S. Yohn and J. R. Blakeslee, eds. (Amsterdam: Elsiever), pp. 385-386.

Heller, M., Gerber, P., and Kieff, E. (1981). Herpesvirus papio DNA is similar in organization to Epstein-Barr virus DNA. J. Virol. *37*, 698-709.

Heller, M., and Kieff, E. (1981). Colinearity between the DNAs of Epstein-Barr virus and herpesvirus papio. J. Virol. *37*, 821-826.

Ishida, T., and Yamamoto, K. (1987). Survey of nonhuman primates for antibodies reactive with Epstein-Barr virus (EBV) antigens and susceptibility of their lymphocytes for immortalization with EBV. J. Med. Primatol *16*, 359-71.

Jenson, H. B., Ench, Y., Gao, S. J., Rice, K., Carey, D., Kennedy, R. C., Arrand, J. R., and Mackett, M. (2000). Epidemiology of herpesvirus papio infection in a large captive baboon colony: similarities to Epstein-Barr virus infection in humans. J. Infect. Dis. *181*, 1462-1466.

Jenson, H. B., Ench, Y., Zhang, Y., Gao, S. J., Arrand, J. R., and Mackett, M. (2002). Characterization of an Epstein-Barr virus-related gammaherpesvirus from common marmoset (Callithrix jacchus). J. Gen. Virol. *83*, 1621-3163.

Jiang, H., Cho, Y. G., and Wang, F. (2000). Structural, functional, and genetic comparisons of Epstein-Barr virus nuclear antigen 3A, 3B, and 3C homologues encoded by the rhesus lymphocryptovirus. J. Virol. *74*, 5921-5932.

Kalter, S. S., Heberling, R. L., and Ratner, J. J. (1972). EBV antibody in sera of non-human primates. Nature *238*, 353-354.

Kalter, S. S., Herberling, R. L., and Ratner, J. J. (1973). EBV antibody in monkeys and apes. Bibl Haematol. *39*, 871-875.

Kutok, J. L., Klumpp, S., Simon, M., MacKey, J. J., Nguyen, V., Middeldorp, J. M., Aster, J. C., and Wang, F. (2004). Molecular evidence for rhesus lymphocryptovirus infection of epithelial cells in immunosuppressed rhesus macaques. J. Virol. *78*, 3455-3461.

Landon, J. C., Ellis, L. B., Zeve, V. H., and Fabrizio, D. P. (1968). Herpes-type virus in cultured leukocytes from chimpanzees. J. Natl. Cancer Inst *40*, 181-192.

Landon, J. C., and Malan, L. B. (1971). Seroepidemiologic studies of Epstein-Barr virus antibody in monkeys. J. Natl. Cancer Inst *46*, 881-884.

Lapin, B. A. (1974). The epidemiologic and genetic aspect of an outbreak of leukemia among baboons of the Sukhumi monkey colony in Dutcher and Chieco-Bianchi. Unifying concepts of leukemia. Biblphyl. Haematol. *39*, 263-268.

Lapin, B. A., Timanovskaya, V. V., and Yakovleva, L. A. (1985). Herpesvirus HVMA: a new representative in the group of the EBV-like B- lymphotropic herpesviruses of primates. Hamatol Bluttransfus *29*, 312-313.

Li, S. L., Feichtinger, H., Kaaya, E., Migliorini, P., Putkonen, P., Biberfeld, G., Middeldorp, J. M., Biberfeld, P., and Ernberg, I. (1993). Expression of Epstein-Barr-virus-related nuclear antigens and B-cell markers in lymphomas of SIV-immunosuppressed monkeys. Int. J. Cancer *55*, 609-615.

Ling, P. D., Ryon, J. J., and Hayward, S. D. (1993). EBNA-2 of herpesvirus papio diverges significantly from the type A and type B EBNA-2 proteins of Epstein-Barr virus but retains an efficient transactivation domain with a conserved hydrophobic motif. J. Virol. *67*, 2990-3003.

McCann, E. M., Kelly, G. L., Rickinson, A. B., and Bell, A. I. (2001). Genetic analysis of the Epstein-Barr virus-coded leader protein EBNA-LP as a co-activator of EBNA2 function. J. Gen. Virol. *82*, 3067-3079.

Moghaddam, A., Koch, J., Annis, B., and Wang, F. (1998). Infection of human B lymphocytes with lymphocryptoviruses related to Epstein-Barr virus. J. Virol. *72*, 3205-3212.

Moghaddam, A., Rosenzweig, M., Lee-Parritz, D., Annis, B., Johnson, R. P., and Wang, F. (1997). An animal model for acute and persistent Epstein-Barr virus infection. Science *276*, 2030-2033.

Neubauer, R. H., Rabin, H., Strnad, B. C., Nonoyama, M., and Nelson, R. W. (1979). Establishment of a lymphoblastoid cell line and isolation of an Epstein-Barr-related virus of gorilla origin. J. Virol. *31*, 845-848.

Ohno, S., Luka, J., Falk, L., and Klein, G. (1977). Detection of a nuclear, EBNA-type antigen in apparently EBNA-negative Herpesvirus papio (HVP)-transformed lymphoid lines by the acid-fixed nuclear binding technique. Int. J. Cancer *20*, 941-946.

Peng, R., Gordadze, A. V., Fuentes Panana, E. M., Wang, F., Zong, J., Hayward, G. S., Tan, J., and Ling, P. D. (2000). Sequence and functional analysis of EBNA-LP and EBNA2 proteins from nonhuman primate lymphocryptoviruses. J. Virol. *74*, 379-389.

Peng, R., Tan, J., and Ling, P. D. (2000). Conserved regions in the Epstein-Barr virus leader protein define distinct domains required for nuclear localization and transcriptional cooperation with EBNA2. J. Virol. *74*, 9953-9963.

Pingel, S., Hannig, H., Matz-Rensing, K., Kaup, F. J., Hunsmann, G., and Bodemer, W. (1997). Detection of Epstein-Barr virus small RNAs EBER1 and EBER2 in lymphomas of SIV-infected rhesus monkeys by in situ hybridization. Int. J. Cancer *72*, 160-165.

Rabin, H., Neubauer, R. H., Hopkins, R. F., Dzhikidze, E. K., Shevtsova, Z. V., and Lapin, B. A. (1977). Transforming activity and antigenicity of an Epstein-Barr-like virus from lymphoblastoid cell lines of baboons with lymphoid disease. Intervirology *8*, 240-249.

Rabin, H., Neubauer, R. H., Hopkins, R. F. d., and Nonoyama, M. (1978). Further characterization of a herpesvirus-positive orang-utan cell line and comparative aspects of *in vitro* transformation with lymphotropic old world primate herpesviruses. Int. J. Cancer *21*, 762-767.

Ramer, J. C., Garber, R. L., Steele, K. E., Boyson, J. F., O'Rourke, C., and Thomson, J. (2000). Fatal lymphoproliferative disease associated with a novel gammaherpesvirus in captive population of common marmosets. Comparative Medicine *50*, 59-68.

Rangan, S. R., Martin, L. N., Bozelka, B. E., Wang, N., and Gormus, B. J. (1986). Epstein-Barr virus-related herpesvirus from a rhesus monkey (Macaca mulatta) with malignant lymphoma. Int. J. Cancer *38*, 425-432.

Rao, P., Jiang, H., and Wang, F. (2000). Cloning of the rhesus lymphocryptovirus viral capsid antigen and Epstein-Barr virus-encoded small RNA homologues and use in diagnosis of acute and persistent infections. J. Clin. Microbiol. *38*, 3219-3225.

Rasheed, S., Rongey, R. W., Bruszweski, J., Nelson-Rees, W. A., Rabin, H., Neubauer, R. H., Esra, G., and Gardner, M. B. (1977). Establishment of a cell line with associated Epstein-Barr-like virus from a leukemic orangutan. Science *198*, 407-409.

Reimann, K. A., Li, J. T., Veazey, R., Halloran, M., Park, I. W., Karlsson, G. B., Sodroski, J., and Letvin, N. L. (1996). A chimeric simian/human immunodeficiency virus expressing a primary patient human immunodeficiency virus type 1 isolate env causes an AIDS-like disease after *in vivo* passage in rhesus monkeys. J. Virol. *70*, 6922-6928.

Rivailler, P., Carville, A., Kaur, A., Rao, P., Quink, C., Kutok, J. L., Westmoreland, S., Klumpp, S., Simon, M., Aster, J. C., and Wang, F. (2004). Experimental Rhesus Lymphocryptovirus Infection in Immunosuppressed Macaques: An Animal Model for Epstein-Barr Virus Pathogenesis in the Immunosuppressed Host. Blood *18*, 18.

Rivailler, P., Cho, Y. G., and Wang, F. (2002). Complete genomic sequence of an Epstein-Barr virus-related herpesvirus naturally infecting a new world primate: a defining point in the evolution of oncogenic lymphocryptoviruses. J. Virol. *76*, 12055-12068.

Rivailler, P., Jiang, H., Cho, Y. G., Quink, C., and Wang, F. (2002). Complete nucleotide sequence of the rhesus lymphocryptovirus: genetic validation for an Epstein-Barr virus animal model. J. Virol. *76*, 421-426.

Schmidtko, J., Wang, R., Wu, C. L., Mauiyyedi, S., Harris, N. L., Della Pelle, P., Brousaides, N., Zagachin, L., Ferry, J. A., Wang, F., Kawai, T., Sachs, D. H., Cosimi, B. A., and Colvin, R. B. (2002). Posttransplant lymphoproliferative disorder associated with an Epstein-Barr-related virus in cynomolgus monkeys. Transplantation *73*, 1431-9.

Shimizu, Y. K., Oomura, M., Abe, K., Uno, M., Yamada, E., Ono, Y., and Shikata, T. (1985). Production of antibody associated with non-A, non-B hepatitis in a chimpanzee lymphoblastoid cell line established by *in vitro* transformation with Epstein-Barr virus. Proc. Natl. Acad. Sci. USA. *82*, 2138-42.

Chapter 33

Epstein-Barr Virus: Summary, Conclusions and a Forward Look

*Alan Rickinson**

ABSTRACT

By drawing together some of the key findings and issues of debate described in the earlier chapters, this summary aims to show how our current understanding of EBV biology and disease pathogenesis was gained, to identify the strengths and weaknesses of our current position, and to consider what are the most important challenges for the future.

THE HISTORICAL PERSPECTIVE

Epstein-Barr virus (EBV) was discovered some 40 years ago in electron microscopic images of cells cultured from an obscure malignancy of African children (Epstein et al., 1964). With the benefit of hindsight, the full significance of this observation is now apparent and the tumour, now known as Burkitt's lymphoma (BL), takes its place as the first documented example of a virus-associated human malignancy and EBV as the first example of a human tumour virus. In Chapter 1, Tony Epstein recounts in fascinating and hitherto unpublished detail the sequence of events that led to that discovery. In so doing, he pays tribute to Denis Burkitt's extraordinary combination of clinical insight in first recognizing this childhood tumour as a distinct clinical entity, and tenacity in then bringing it to the attention of the wider research community. The discovery of EBV in BL cell cultures was serendipitous only in the sense that Epstein had the electron microscope as a tool and was not relying on the then available cell culture assays for virus detection, assays in which EBV proved biologically inert. In other respects the search for viruses in association with BL was a logical progression from Burkitt's original work indicating that the geographical distribution of the tumour resembled that of certain arthropod-borne infectious diseases. The irony is that the virus first visualized in BL proved to be neither arthropod-borne nor restricted to Africa, but an orally-transmitted herpesvirus that was widespread in all human populations and in most people carried for life as a seemingly innocuous infection of lymphoid tissues.

Reviews by Harald zur Hausen, George Klein and Guy de The (Chapters 2, 3 and 4 respectively) chart some of the key findings that strengthened EBV's

*For correspondence email A.B.Rickinson@bham.ac.uk

credentials as a *bona fide* tumour virus and not just a casual passenger in BL cells. Centrally important were two laboratory findings; firstly that EBV was capable of transforming human B cells into permanent lymphoblastoid cell lines (LCLs) in culture (Henle et al., 1967; Pope et al., 1968), and secondly that all LCLs as well as African BL lines maintained the viral genome in episomal form and expressed at least one viral protein, a nuclear antigen (EBNA), detectable in a complement-enhanced immunofluorescence test using sera from EBV carriers (Reedman and Klein, 1973). The transformation system at once provided both a dramatic illustration of EBV's potential for cell growth deregulation and an *in vitro* model through which to identify the viral effectors of cellular change and their mechanisms of action. While it has become clear that EBV's B cell transforming capacity *in vitro* may not recapitulate all aspects of EBV's aetiologic role in relation to B cell malignancies, the discovery of EBV-induced transformation in the laboratories of Werner Henle and John Pope in 1967/68 laid the foundation for a large part of what we now know about virus latency and latent gene function in relation to cell growth. Equally important to EBV's candidature as a tumour virus in those early years were two sero-epidemiological findings. The first showed that BL patients had unusually elevated anti-EBV serum antibody titres (Henle et al., 1969), markers of an atypical virus-host balance that large scale prospective studies later found to pre-date tumour development (de The et al., 1978). The second identified another set of patients with an atypical antibody profile (Old et al., 1966); quite unexpectedly, these were patients with an epithelial tumour, nasopharyngeal carcinoma (NPC). Not only was this the first hint that the virus' association with human cancer would extend beyond BL but also, with the subsequent detection of the virus genome in NPC cells, the first demonstration that EBV infection *in vivo* was not exclusively B-lymphotropic. As is clear from many of the contributions to the present volume, vigorous debates about the definitive range of EBV-associated human tumours and about the physiologic importance of epithelial cells as targets for EBV infection *in vivo* continue to reverberate through to the present day (see Chapters 6, 7, 10, 11, 17 and 18).

THE EVOLUTIONARY PERSPECTIVE

While our familiarity with EBV as an object of scientific obsession goes back only 40 years, the virus' familiarity with man goes back to the very origins of our species. This is clear from the study of lymphocryptovirus (or gamma-1 herpesvirus) evolution, an area of considerable interest which is being re-shaped by recent findings, as described by Fred Wang (Chapter 32).

Within a decade of the discovery of EBV, closely related members of the lymphocryptovirus genus were being identified in several other species of Old World primates, largely through access to the virus-transformed B cell lines that, as with EBV, were found to grow out spontaneously from peripheral blood lymphocyte cultures of the persistently infected host (Frank et al., 1976). These Old World agents show the same genomic organisation and gene content as EBV, are similarly B-lymphotropic, have the same complex array of latent genes and the same growth transforming ability. Although the latent genes show quite significant sequence divergence from their EBV homologues, certain conserved sequences reflect the retention of key functional domains and the essential similarity of the homologues' transforming functions. Importantly these Old World primate viruses can be pathogenic in their natural hosts, as is clear from their association with the B lymphoproliferative disease/lymphoma arising in animals immunocompromised as a result of simian immunodeficiency virus infection (Feichtinger et al., 1992). Such tumours are directly analogous to the EBV-driven lymphoproliferative lesions, discussed below, that are seen both in immunosuppressed transplant patients and in end-stage AIDS patients. It is entirely possible that, like EBV, the Old World primate viruses are also associated with other types on malignancy in their natural hosts, but the chances of detecting such rare accidents of virus carriage either in wild or in laboratory animals must be extremely low.

Natural infection with their own lymphocryptovirus, leading to the induction of cross-reactive antibodies, probably explains the resistance of Old World primates to experimental EBV infection noted in the early literature. However, important studies in the laboratories of Epstein (Epstein et al., 1973) and George Miller (Shope et al., 1973) showed that EBV could infect certain New World primates such as owl monkeys and cotton top tamarins, and indeed could induce fatal B lymphoproliferative lesions that again resembled the EBV-positive lesions associated with immune impairment in man. These animal studies formally confirmed EBV's oncogenic potential *in vivo* and at the same time provided a disease model which was subsequently used to test the effectiveness of a prophylactic EBV vaccine using different formulations of the major glycoprotein gp340 as the vaccine antigen (see Chapter 1). The susceptibility of these New World primates to EBV infection strengthened the prevailing view that lymphocryptoviruses were restricted to Old World primate species. However in recent years this view has been completely overturned by the identification of an EBV-related agent in a spontaneously arising B cell lymphoma of a common marmoset (Cho et al., 2001). This virus proved to be widespread in marmosets and subsequent work went on to show that unique lymphocryptovirus sequences could be PCR amplified from the blood of several other New World primate species. As fully described by Wang (Chapter 32), complete sequencing of the marmoset virus genome clearly identifies this as an agent with a genomic organization essentially similar to that of Old World viruses but with what appears to be a minimalised set of latent growth-transforming genes identifiable through positional (rather than obvious sequence) homology with their Old World counterparts. On-going work on this expanding group of New World lymphocryptoviruses will no doubt continue to shed valuable light on the evolution of this genus. Furthermore, comparison of the marmoset virus sequence with that of EBV and its close relative from the rhesus macaque has identified a set of genes that are unique to the Old World lymphocryptoviruses i.e genes that have been acquired since the separation of the Old and New World primate lineages some 40 million years ago (Rivailler et al., 2002). Two of these "recently acquired" genes, BCRF1 (an IL10 homologue) and BARF1 (a distant homologue of colony stimulating factor 1 receptor), are predicted to have immunomodulatory function. Hence it seems likely that some of the other genes in this group, currently of unknown function, may also reflect the incremental acquisition of properties favouring virus persistence in the immune host. In that regard, one recalls that many of the immune evasion genes of alpha, beta and gamma-2 herpesviruses that act to impair T cell control are unique to each particular virus sub-family or genus (Vossen et al., 2002), and it is likely that lymphocryptoviruses will also have acquired their own unique strategies to that end. We know next to nothing about the *in vivo* significance of many lymphocryptovirus genes, and it will be interesting to compare the fine detail of Old World and New World lymphocryptovirus infections in their natural hosts, given the simpler complement of latent cycle and also immunomodulatory genes that New World viruses possess.

POLYMORPHISM AMONG EBV STRAINS

As with all herpeviruses, there are many different strains of EBV circulating in the wild, distinguishable from one another by sequence polymorphisms at particular genomic loci. By far the most marked example, and the only one known to be reflected in an *in vitro* assay of viral activity, is the broad division of EBV isolates into Type 1 and Type 2 strains (also called Types A and B in some publications). These are characterised by a set of linked polymorphisms at the EBNA2, 3A, 3B and 3C latent gene loci; the Type 1 and Type 2 proteins encoded by these alleles share 55, 84, 80 and 72 % amino acid identity respectively and are to varying degrees antigenically distinct. There is still much to be learnt about the relative prevalence of the two virus types in different human populations. Most studies have used relatively small study cohorts and much of the early work was based on the study of viruses rescued as LCL isolates in culture (Gratama and Ernberg, 1995), a potential problem because Type 1 virus-infected B cells are known to grow out to LCLs more readily than Type 2-infectants *in vitro*. With the switch to direct PCR amplification of *ex vivo* samples as the method of choice for epidemiological work, Type 1 strains still appear to be dominant in both Western and S.E.Asian populations, although a significant minority of healthy people (10-20%) in both countries carry a Type 2 strain and the prevalence may be higher in certain social groups (Rickinson and Kieff, 2001). The two virus types appear to be more evenly represented in equatorial Africa, but here the evidence is particularly sparse. It would be interesting to know much more about virus type distribution in different human populations worldwide since one possibility, albeit speculative, is that Type 2 viruses (with apparently weaker growth transforming capacity) may compete for transmission to naïve hosts more effectively where a pre-existing infection, for instance malaria, is already acting as a source of chronic B cell stimulation. The evolutionary origins of the two EBV types remains shrouded in mystery, but it may stem from a very ancient split since two types of rhesus lymphocryptovirus have also been distinguished by sequence divergence at the rhesus EBNA2 locus (Cho et al., 1999).

Superimposed on this broad division of viruses into Types I and II, sequence variation at multiple polymorphic sites in the genome allows one to distinguish between individual virus strains. As pointed out by Paul Farrell (Chapter 15), no virus strain directly isolated from a normal healthy carrier has ever been sequenced in its entirety. Thus the Type 1 prototype B95.8 (originally derived from a mononucleosis patient) represents just one of many contemporary Caucasian strains in circulation and indeed is now recognized to have a significant deletion around the second origin of lytic replication, *ori-lyt*. It is interesting that most of the best known "strain-specific" markers are found in latent gene sequences. This has fuelled continuing debate as to whether particular markers, for example in the latent membrane protein LMP1 gene, might be associated with heightened oncogenic risk (see Chapter 7 by Nancy Raab-Traub). There is as yet no consensus on this point and further work is needed in several different disease settings to compare disease-associated isolates with appropriate geographically matched controls. The need for such controls is particularly important because many "strain-specific"markers, or combinations of such markers, identify strains that are either unusually prevalent or indeed unique to particular human populations. Thus Type 1 strains from indigenous European, South East Asian and African populations can usually be distinguished from one another by the different range of sequence variations apparent at specific sites in EBNA1, EBNA3B and LMP1 (Rickinson and Kieff, 2001). This clearly indicates that, since the radiation of human migration out of Africa began some 100, 000 years ago, EBV has continued to evolve within its separate host populations; such evolution has produced the geographic markers of virus polymorphism apparent today. Whether this diversification is entirely occurring through random genetic drift, or is to some extent driven by immunological pressure from the particular host population in question, remains an interesting but unresolved question (Midgley et al., 2003; Burrows et al., 2004). Another fascinating but still unexplained feature of lymphocryptovirus evolution, which is hinted at among EBV isolates themselves but becomes much clearer when comparing EBV with other primate virus sequences, is the greater drive to sequence divergence among latent cycle genes compared to those of the lytic cycle (see Chapter 32). Is this because, compared to the virus-coded enzymes and virion structural proteins of the lytic cycle, latent cycle proteins are functionally much more dependent upon specific physical interactions with proteins encoded by the host?

VIRUS GENETICS AND THE REGULATION OF LATENT CYCLE GENES

The publication of the EBV genome sequence in 1984, coming exactly 20 years after the discovery of the virus, was a landmark event (Baer et al., 1984). At that time, EBV was not only the first herpesvirus but also by far the largest organism of any type to have been sequenced. For a virus whose lack of a fully permissive *in vitro* system rendered it impervious to conventional genetic approaches, access to the genomic sequence opened the door to molecular analysis in a quite spectacular way. Paul Farrell (Chapter 15) introduces us to an updated map of the EBV genome and provides a valuable overview of gene transcription which sets the scene for later chapters on the regulation of latent and lytic gene expression. With the genome map in hand, Ken Izumi (Chapter 16) describes in detail the subsequent development of EBV genetics. It is interesting to reflect how much of the early progress in this area depended upon a fortuitous accident of history, the chance isolation of a BL cell subclone which carried a replication-competent but transformation-defective (EBNA2 gene-deleted) EBV genome, the P3HR1 strain (Hinuma et al., 1967; Miller et al., 1974). Complementation with EBNA2 and selection for recombinants with B cell transforming ability provided the first genetic analysis of EBNA2 function

(Hammerschmidt and Sugden, 1989; Cohen et al., 1989) and, when later combined with second site recombination, allowed the identification of EBNA3A, EBNA3C and LMP1 also as essential transforming genes. The subsequent development of methods to manipulate the EBV genome as a bacterial artificial chromosome and to rescue the recombinant genome into mature virions (Delecluse et al., 1998) has given new impetus to the genetic analysis of both latent and lytic gene function. In particular, the ability to produce genetically pure virus preparations, titrated for virion number by PCR, should allow the contribution of other genes such as EBNA-LP and the non-coding EBERs to transformation to be accurately quantified. There are no doubt many surprises still in store, as exemplified by the recent finding that deletion of EBNA1, the genome maintenance protein, impairs but does not completely abolish transformation; rare EBNA1-negative LCLs do arise in which chance integration of the viral genome into cellular DNA has rendered EBNA1's maintenance function redundant (Humme et al., 2003).

Access to the EBV genome sequence was key to the efforts of many laboratories characterizing the highly sliced viral transcripts expressed in LCL cells and to the subsequent mapping of the viral promoters active in this form of latency. These studies are authoritatively reviewed by Sam Speck (Chapter 20). There he draws attention to the complexity of events occurring early in the transformation process, where the pan-EBNA BamH1W promoter (Wp) is activated first and preferentially drives expression of EBNA2 and EBNA-LP. Thereafter its function is largely subsumed by a second such promoter upstream in BamH1C (Cp); this is activated by EBNA2 and drives expression of all six EBNAs (Woisetschlaeger et al., 1990; Woisetschlaeger et al., 1991). The bi-directional LMP1/LMP2B promoter and the separate LMP2A promoter are also activated, again under the influence of EBNA2 and its co-activator EBNA-LP, leading to full latent cycle antigen expression (Wang et al., 1990; Zimber-Strobl et al., 1991). Although the regulation of all four of these latent promoters by immediate upstream sequences has been dissected in reporter gene and factor binding assays, many of the definitive experiments examining promoter function in the context of the complete viral genome remain to be carried out. This should help to determine how important these proximal control elements are vis-à-vis longer range effects, particularly those that might be mediated by EBNA1 binding at the origin of plasmid maintenance (*ori-p*) region that lies within 12kb of the Wp, Cp and LMP promoters. One interesting conundrum is why EBV should have acquired two alternative pan-EBNA promoters when one of these, Cp, is known to be dispensable for transformation *in vitro*? Such apparent redundancy almost certainly illustrates the fact that our *in vitro* models are limited and do not reflect the virus' need for greater flexibility of transcriptional control *in vivo*. In that context Wp, a promoter highly responsive to transcription factors in resting B cells, could provide the most efficient means of initiating the infection since there are multiple copies of Wp, one from each of the tandem BamH1W repeats, within the EBV genome and all are potentially active at the onset of infection. By contrast, Cp regulation is more complex with influences from both EBNA1 (binding to *ori-p*) and EBNA2 (binding to a Cp regulatory site through its interaction with cellular RBP-Jk) as well as from additional cellular factors. The possibility that Cp could be subject to negative control by the EBNA3 proteins through competition for RBPJk binding would provide a mechanism for feedback inhibition of the transformation process *in vivo,* thereby allowing the infected B cell to escape from the proliferative LCL-like state and persist as a resting cell.

Central to all discussions of viral gene regulation is the realization that there are forms of latent infection other than the so-called Latency III programme seen in LCLs. Some insight into these alternative forms came originally from two EBV-associated B cell tumours, BL, where classically EBNA1 is the only viral antigen detectably expressed (so-called Latency I) (Rowe et al., 1987), and Hodgkin's Lymphoma (HL) where the malignant cells express EBNA1, LMP1 and LMP2 (so-called Latency II) (Pallesen et al., 1991; Herbst et al., 1991). Selective expression of EBNA1 in the absence of the other EBNAs is achieved through a promoter in Bam HIQ (Qp) that appears to be spontaneously activated when Wp and Cp are silent (Sample et al., 1991; Schaefer et al., 1991). Identifying the cellular and viral factors that determine the choice between Wp/Cp usage in Latency III and Qp usage in Latency I and II remains an important objective. As discussed in David Thorley-Lawson's detailed analysis of EBV infections *in vivo* (Chapter 17), arguably Qp has evolved as a failsafe promoter that can be used in specific situations during EBV's persistence in the B cell system, for example in germinal centres where antigen-driven (rather than virus-driven) cell proliferation requires that the virus amplify and segregate its genome with each cell cycle. In the same way, the expression of LMP1 and LMP2 in the absence of the EBNA2 transactivator (Latency II) demands explanation. As discussed by Rymo and Zetterburg (Chapter 22) and Kempkes et al. (Chapter 23), the LMP promoters are complex and require cellular factors as well as EBNA2 for their activation; the possibility of Notch proteins replacing EBNA2 in such circumstances should be borne in mind, as should the detection (for LMP1 only) of an alternative upstream promoter that could itself account EBNA2-independent expression in some circumstances (Sadler and Raab-Traub, 1995). Once again such flexibility of LMP gene transcription is very likely to have been acquired by EBV and its close relatives as part of their strategy for persistence in the B cell system. As discussed by Thorley-Lawson (Chapter 17) and Richard Longnecker (Chapter 25), the capacity to express LMP1 and LMP2 independently of the full B cell growth transforming programme could be of significant advantage to infected B cells at the culmination of the germinal centre reaction; thus LMP1 (a CD40 surrogate) and LMP2 (a B cell receptor surrogate) could mimic the physiologic signals that normally determine selection of germinal centre progeny cells into the long-lived memory pool.

LATENT GENE FUNCTIONS

A great deal of effort has been directed towards understanding the function of the latent cycle proteins at a molecular level, since here almost certainly lies the key to understanding much of EBV's aetiologic role in human malignancies. The first protein to be assigned a specific function was EBNA1 (see Chapter19 by Mack and Sugden); its capacity to bind to specific DNA sequences within *ori-p* (Yates et al., 1984) underpins its role both in the replication of EBV episome with each cell cycle and in the segregation of episomes with cellular chromosomes at mitosis. Whether EBNA1 makes additional contributions to B cell growth transformation through the direct or indirect transactivation of cellular genes is still unclear (Kang et al., 2001; Kennedy et al., 2003). By contrast, other latent cycle proteins such as EBNA2 and LMP1 are more obvious effectors of cellular change and achieve their effects by engaging B cell growth pathways at strategically important points, leading in each case to constitutive activation of a number of downstream signaling pathways. Thus EBNA2 engages the Notch signaling pathway at the point of RBPJk interaction (Grossman et al., 1994; Henkel et al., 1994; Waltzer et al., 1994; Zimber-Strobl et al., 1994), with EBNA-LP and the EBNA3 proteins respectively acting as positive and negative regulators of this engagement (see Chapter 23 by Kempkes et al). Likewise LMP1 mimics many aspects of signaling from the TNF receptor family member CD40 through recruitment of similar sets of TRAF molecules ((Mosialos et al., 1995); see Chapter 26 by Cahir-McFarland and Kieff). The lymphocryptovirus strategy of targeting particular cellular pathways at key gatekeeper points therefore has parallels with the transforming activities of the small DNA tumour viruses; however, as George Klein (Chapter 3) argues, the points of engagement and even the particular pathways themselves appear to be different between the two sets of transforming agent. It is not clear to what extent this reflects the different types of cell which these different agents infect and/or the fact that, compared to small DNA tumour viruses, the lymphocryptoviruses have a more complex life cycle *in vivo*. Thus EBV, the model lymphocryptovirus, uses growth transforming infection to establish a sufficiently large reservoir of latently-infected cells that then persist in the host; small DNA tumour viruses drive infected cells into cycle only as a means of rendering them immediately able to support lytic virus replication. In Chapter 24, Knight and Robertson draw together some of the detail from earlier chapters and review our current understanding of EBV's interaction with cell cycle checkpoints and cell growth control (see also (O'Nions and Allday, 2004)). Many questions still remain. For example, what are the most important cellular genes regulated by EBNA2 in the transformation process? What contributions to sustained cell growth come from the EBNA3 proteins, beyond their role as EBNA2 regulators? How important is LMP1 as a driver of cell growth as opposed to a mediator of enhanced cell survival? How can the data on EBV protein function be integrated with the larger body of literature now available on eukaryotic cell cycle progression?

Arguably one of the most unexpected findings to come from the study of EBV recombinants was that LMP2 appeared dispensable for transformation. In one sense this makes the protein even more interesting since one needs to go to other models to examine its potential *in vivo* function. As Ikeda et al. (Chapter 25) describe, biochemical studies showed that the full length LMP2A molecule engages the B cell receptor (BCR) pathway at the point of tyrosine kinase recruitment and activation. The functional consequences of such engagement were first observed as an ability of the protein to inhibit the activation of lytic cycle that is induced in the Akata –BL cell line by antibody-mediated BCR cross-linking (Miller et al., 1994). This effect, ascribed to a possible sequestering of signal transducing molecules away from the BCR, implied a physiologic role for LMP2A in regulating the switch from latency into lytic cycle. Subsequently, studies in a transgenic mouse models have shown that the protein can in fact mimic rather than block BCR signaling. Thus reconstitution of RAG-knockout or immunoglobulin heavy chain (IgH)-knockout mice with an IgH enhancer-driven LMP2A transgene allows the unscheduled maturation of progenitor B cells from bone marrow into the periphery without the usual requirement for surface BCR expression (Caldwell et al., 1998; Casola et al., 2004). Recently this model has been used to show how the strength of antigen-independent signaling from the BCR determines the differentiation of peripheral B cells into different functional compartments in the periphery (Casola et al., 2004). It is important to note that in this study LMP2A itself did not itself induce germinal centre formation (*i.e.* it could not mimic the antigen-driven germinal centre reaction in follicular B cells) but it did allow germinal centre formation in gut-associated lymphoid tissues, a unique effect which is driven by innate signals from gut flora and occurs independent of antigen/BCR engagement.

To date all aspects of the phenotype of these LMP2 transgenic animals can be explained through LMP2A acting as a BCR surrogate. Though LMP2 did alter mouse B cell transcription patterns in ways that superficially resembled aspects of the HL cell phenotype (Portis et al., 2003), even high level expression of the protein did not carry an obvious oncogenic risk. The contrast with LMP1 (a CD40 signal surrogate) is quite interesting in this context. Thus IgH enhancer-driven expression of LMP1 cannot mimic physiologic CD40 function in a CD40-deficient animal and thus cannot restore the ability to form germinal centres in response to antigenic challenge; indeed, in a wild type host, LMP1 actively disturbs mature B cell function by preventing antigen-driven germinal centre formation (Uchida et al., 1999) and many LMP1 transgenic animals eventually succumb to B cell lymphoma (Kulwichit et al., 1998). One important lesson to learn here is that the effects of EBV latent protein expression in B cells are not restricted to those apparent in the conventional B cell transformation system. Physiologically important effects on B cell survival/B cell differentiation may not be apparent in any of the available *in vitro* models, but may be revealed in transgenic animal systems if expression of the protein can be achieved at relevant levels in cells at the right stage of differentiation.

Two other sets of latent cycle genes that appear to be dispensable for B cell transformation but whose *in vivo* functions have yet to be revealed are those expressing the small non-coding nuclear RNAs, EBERs 1 and 2, and those in the BamH1 A region of the genome encoding the multiply-spliced BamH1 A rightward transcripts (BARTs), RNAs that are processed as conventional mRNAs but whose postulated protein products have still not been identified in infected cells. Both sets of transcripts appear to be ubiquitous markers of latent EBV infections and, at least for EBERs, are downregulated in lytic cycle. The EBERs, described by Kenzo Takada (Chapter 21), have the potential to interfere with the interferon alpha/beta response of a cell to viral infection by inhibiting the double-stranded RNA-activated protein kinase PKR. Furthermore, through an unknown mechanism but independent of their PKR interaction, EBER expression in different cell types *in vitro* leads to the activation of a number of cellular genes that may be relevant to EBV's contribution to tumourigenesis in BL, in T cell lymphomas and in gastric carcinomas. The BARTs, described by Nancy Raab-Traub (Chapter 7), are predicted to encode a number of small proteins which, when used as bait in yeast two hybrid assays, reveal a number of potentially interesting interactions with cellular proteins, particularly of the Notch signaling pathway. However the physiologic relevance of these predicted proteins and their proposed engagements with cellular partners remain in doubt.

VIRUS LYTIC CYCLE

Discussion so far has focused on latent gene regulation but, as the contributions from Shannon Kenney (Chapter 27) and Shankar Swaminathan (Chapter 29) make clear, considerable efforts have also been made to understand the control of lytic gene transcription. Particular attention has been given to the initiation of the lytic cycle, mainly using as a model the Akata-BL cell line that can be induced from Latency I into lytic cycle by BCR cross-linking (Takada and Ono, 1989). This leads first to the activation of the two immediate early genes BZLF1 and BRLF1 that lie close to one another in the genome and form an interdependent transcription unit. The BZLF1 promoter in particular has been extensively studied in reporter gene assays and, more significantly, after stable introduction on an episomal plasmid into Akata cells (Binne et al., 2002). In such cell culture models, and arguably in physiologic activation for latency, lytic cycle induction involves an initial reversal of the epigenetic modifications of the viral genome such as DNA methylation and histone deactylation that characterize the BZLF1/BRLF1 locus in latently infected cells. Signalling through BCR ligation leads to the activation of several cell transcription factors that drive BZLF1/BRLF1 expression. The BZLF1 and BRLF1 proteins amplify their own expression in a positive feedback loop and activate a number of early gene promoters, after which the lytic cycle proceeds eventually through to virus DNA replication, late gene expression and the assembly of mature virions. BZLF1, a member of the bZIP family of leucine zipper transactivators (Farrell et al., 1989), also has a number of other interesting effects on cells themselves, activating the expression of genes such as TGFbeta and IL10 and downregulating others such as the interferon gamma and TNFalpha receptors, changes that may have immunomodulatory consequences. These implications have yet to be tested experimentally. It will also be interesting to see whether BZLF1 or another gene is principally responsible for the down-regulation of HLA class I and class II expression seen on the surface of infected cells early in lytic cycle (Keating et al., 2002). Still unknown are the signals that induce activation of the latent to lytic cycle switch *in vivo* but there is some evidence to suggest that within B cells these are linked to the cell's commitment to antigen-driven plasmacytoid differentiation (see Chapter 17). It may be that the induction of virus replication by BCR ligation of Akata-BL, an unusual IgG-positive BL cell line, reflects aspects of that process.

One should also remember that the distinction between latent and lytic cycle genes in the EBV system may not be as strict as conventionally thought. Thus the human gamma-2 herpesvirus HHV8 has a subset of genes which appear to be intermediate between the two categories, being expressed in some latent infections *in vitro* yet also further upregulated in lytic cycle (Boshoff and Weiss, 2002). The same may prove true of EBV in certain circumstances. For example the BARF1 gene, which encodes a surrogate colony stimulating factor 1 receptor, is classified as a lytic gene in the LCL culture system yet appears to be a feature of the latent infection seen in EBV-positive epithelial malignancies (see Chapter 28). Likewise during EBV-induced B cell transformation *in vitro*, mRNA transcripts of BZLF1 and certain other lytic genes such as BCRF1 are reportedly detectable within the first few days post-infection; it is also unlikely to be an accident that another early lytic gene, the bcl2 homologue BHRF1, can be transcribed by differential splicing of the long primary transcripts from the latent cycle-associated promoter Cp. While viruses deleted for any one of these individual lytic genes are still transforming, there has been no detailed study of their relative efficiency of transformation vis-a-vis wild type virus; subtle effects may not have been detected in conventional *in vitro* assays.

VIRUS TROPISM: B CELLS, EPITHELIAL CELLS AND THE BIOLOGY OF VIRUS INFECTION *IN VIVO*

The ease with which EBV infects B cells *in vitro,* and the difficulties encountered in trying to infect other cell types experimentally, reinforced the view that this is an essentially B-lymphotropic agent. Yet epithelial cell infections clearly do occur under some circumstances *in vivo.* The evidence for this does not solely come from the existence of EBV-positive epithelial tumours, which it might be argued represent aberrant cul-de-sacs of virus infection; importantly a fully productive infection of epithelial cells is seen in oral hairy leukoplakia (Greenspan et al., 1985), a wart-like lesion of tongue epithelium first recognized in AIDS patients (as described by Scott and colleagues, Chapter 6). Many efforts have since been made to establish equivalent *in vitro* models of epithelial cell infection. None have achieved an efficient fully productive system, probably because (as with human papilloma viruses) virus replication appears to be dependent on ordered epithelial cell differentiation which is difficult to reproduce in culture. However work with marker gene-tagged recombinant viruses has shown that epithelial cell infection can be achieved with reasonable efficiency by co-culturing the target cells with semi-permissive B cell lines rather than by exposure to cell-free virus (Imai et al., 1998). Co-cultures may well recapitulate the intimate cell contacts that exist between lymphoid and epithelial cells in the oropharyngeal mucosa. Despite these observations, there is still no consensus view as to whether virus replication in oral/pharyngeal epithelium is an integral part of the EBV-host interaction and necessary for efficient transfer of this orally-transmitted agent into the B cell system or just an epiphenomenon of the very high virus loads seen in immunocompromised AIDS patients. Lindsey Hutt-Fletcher (Chapter 18) charts the history of opinion on this issue and cites the arguments for and against each position. In my own view, the recent evidence indicating that virus entry into epithelial cells involves specific receptor-ligand interactions that are distinct from those operating in B cells and that virions produced in one cell type might preferentially infect the other (Borza and Hutt-Fletcher, 2002) implies an important physiologic role for epithelial cell infection *in vivo,* albeit at an oral/pharyngeal site that is still to be identified. Furthermore the fact that shedding of infectious virions into the throat remains high for some months after primary infection, despite the rapid fall in latent virus load that occurs in the circulating B cell pool (Fafi-Kremer et al., 2005), suggests that the virus is replicating in a niche which is both distinct from that B cell pool and also in some way protected from immune T cell surveillance. Hutt-Fletcher's survey of the viral glycoproteins (Chapter 18) emphasizes how much there is still to learn about the role of individual glycoproteins in virus binding and penetration into its target cells, and about the route whereby viral capsids entering the cytosol then track to nucleopores and deliver the virus genome into the cell nucleus.

Recent work in the B cell system suggests that the principal rate-limiting step in EBV infection is the loss that occurs between virus binding at the cell surface and successful virus genome delivery to the cell nucleus (Shannon-Lowe et al., 2005). This highlights the importance of studying these post-binding events in greater detail.

While the physiological importance of virus replication at epithelial sites remains unclear, there is general acceptance of the view that B cells are the essential reservoir of virus latency and persistence *in vivo*. New impetus to the analysis of *in vivo* latency came with the finding that latently-infected cells in the blood of virus carriers were confined to the IgD-negative, CD27-positive memory B cell subset and not the numerically larger subset of naïve cells (Babcock et al., 1998). Furthermore this selective colonization of the memory B cell pool was established very early following primary infection, since it was even apparent in the blood of infectious mononucleosis (IM) patients (Hochberg et al., 2004). Sometimes virus-infected cells were detectable in the IgD-positive, CD27-negative naïve B cell fraction of tonsillar lymphocyte suspensions from long-term virus carriers. However this only occurred in those tonsils where there was also evidence of on-going virus replication, suggesting that such naïve B cells had been the targets of recent infection by newly produced progeny virions. Indeed, by RT/PCR amplification only the naïve B cell subset expressed EBV transcripts suggestive of recent growth transforming (Latency III) infection; by contrast virus-infected cells were always present within the other tonsil B cell subsets (germinal centre and memory cells) but they appeared to show more restricted forms of latency classically associated with long-term persistence in the immune host (Babcock et al., 2000).

This led to the idea, expounded in some detail by David Thorley-Lawson (Chapter 17), that EBV colonises the human B cell system by preferentially infecting naïve B cells in the oropharyngeal lymphoid tissues and driving their differentiation into memory via the physiologic route of memory cell selection, *i.e.* through a germinal centre reaction. During physiologic responses to antigen challenge, naïve B cells of the appropriate specificity will bind antigen through their BCR and, after receipt of antigen-specific T cell help delivered via a CD40 pathway-dependent B cell/T cell interaction, will proliferate to form germinal centres within the follicles (B cell areas) of lymphoid tissues. The clonal descendents of these germinal centre founder cells diversify their BCR by somatic hypermutation of immunoglobulin genes and the few cells then re-expressing a BCR of higher antigen affinity can, again with T cell help, be positively selected into memory, while the non-selected progeny with lower affinity BCRs are lost by apoptosis. One possibility therefore is that the initial growth-transforming (Latency III) infection of a naive B cell mimics antigen stimulation and commits the infected cell to the germinal centre reaction. Here a switch to selective expression of EBNA1 only (Latency I) guarantees maintenance of the virus genome during clonal cell expansion while subsequent broadening of gene expression to include the LMPs (Latency II) guarantees successful selection into memory through mimicry of the required BCR- and CD40-mediated rescue signals. Subsequently virus antigen expression is largely if not completely extinguished (Latency 0) as the rescued cell enters the long-lived recirculating memory B cell pool. Therafter viral gene expression will be reactivated to an appropriate extent only in response to a change in the activity of the host cell, for example during the homeostatically controlled turnover of cells within the memory B cell pool or following a chance BCR encounter with specific antigen that might lead either to re-entry into a germinal centre reaction or to plasma cell differentiation and activation of the lytic cycle.

These ideas represent an important first attempt to align what is known about EBV colonization of the B cell pool *in vivo* with the physiologic controls normally determining B cell fate. But they are just a first attempt and, in the absence as yet of any detailed analysis in primate models of lymphocryptovirus infection, are necessarily based on the limited amount of information available from *ex vivo* studies on human material. Even in that context there are findings that are difficult to reconcile with the above scenario. In particular, although there are large numbers of infected B cells in the tonsils of IM patients, they are almost all located in extra-follicular areas and not in germinal centres (Anagnostopoulos et al., 1995). This contrasts markedly with the situation seen during primary infection with the mouse gamma-2 herpesvirus MHV68, a B-lymphotropic but non-transforming agent, which specifically does use the T cell-dependent germinal centre reaction as a way of amplifying the latently-infected B cell pool; in this model, MHV68 genome-positive germinal centre B cells are the dominant population at the height of the acute infection (Marques et al., 2003). There is also evidence from the genotypic analysis of cells micromanipulated from IM tonsils that the EBV-infected population is mainly composed of memory cell clones with no on-going hypermutation, as if the virus had directly driven the expansion of pre-existing memory cells; furthermore even the rare EBV-positive cells present in germinal centre areas were found not to be part of the resident germinal centre population (Kurth et al., 2000; 2003). As discussed by Thorley-Lawson (Chapter 17), there may be ways to reconcile the apparently discordant sets of results obtained by these two different approaches, studying on the one hand separated B cell subsets and on the other individual micromanipulated cells. To stand back from these differences for a moment, there is an important lesson to be learnt from the single cell analysis of IM tonsil sections; namely that, even within a genotypically-defined clone of EBV-infected B cells, there is heterogeneity both in cellular phenotype and viral antigen status (Kurth et al., 2000). This dramatically illustrates one of the central tenets of all models of EBV infection *in vivo*, namely the capacity of the virus to adopt different forms of latency that are both responsive to and also influence the state of differentiation of the infected B cell.

While the emphasis above has been solely on the EBV-B cell interaction, Sutkowski and Huber (Chapter 14) challenge the tacit assumption that EBV can achieve entry into the memory B cell pool *in vivo* without T cell help. Thus their *in vitro* studies have shown that EBV infection of B cells activates expression of an endogenous human retrovirus (HERV-K18) sequence encoding an envelope protein which can act as a superantigen engaging all T cells with Vbeta13-bearing T cell receptors (TCRs) (Sutkowski et al., 1996; 2001). They draw an interesting analogy between EBV and the mouse mammary tumour virus MMTV, a B-lymphotropic retrovirus whose colonization of the mouse B cell system is dependent on T cell help induced by an MMTV-coded superantigen (Acha-Orbea et al., 1993). Pursuing this analogy, they argue that *in vivo* induction of the K18 superantigen may be important in recruiting non-antigen-specific T cell help both to a freshly-infected B cell (since one K18 activating signal is delivered by virus binding to CR2) and to cells at the end of the germinal centre reaction (since additional K18 activating signals come from the LMPs). The physiological significance of the K18 superantigen effect, which in the *in vitro* model system requires additional phorbol ester treatment of infected cells to induce T cell activation, remains to be determined. However it does open up the possibility that EBV has evolved to exploit both B cell and T cell biology in its strategy for persistence.

Understanding EBV's interaction with its host remains one of the most interesting challenges for future work, and one that will almost certainly continue to spring surprises as the analytic tools become more precise. This is best illustrated with reference to recent work, described by Nancy Raab-Traub (Chapter 7), on the number of viral strains carried by the individual host. Earlier studies based on isolating resident EBV strains through their B cell transforming capacity *in vitro* had suggested that most immunocompetent individuals, as opposed to transplant recipients and HIV-positive patients, carried a single dominant virus. This has now been re-investigated using PCR to amplify viral DNA from the blood or throat washings of EBV-infected individuals and the heteroduplex tracking assay to screen the products for the presence of allelic LMP1 sequences. Such heterodupex analysis clearly shows that most healthy virus carriers and, perhaps more surprisingly, most IM patients harbour two, three or more distinct virus strains (Sitki-Green et al., 2003; 2004). These findings, now confirmed for IM patients by amplification across a second polymorphic site in the EBNA2 gene (Tierney et al., 2005), suggest that primary infection often involves the coincident transmission of two or more independent EBV strains. A number of important questions now follow. Does the preferential *in vitro* isolation of just one of the co-resident strains simply reflect the inherent insensitivity of the transformation assay or are many carriers also harbouring transformation-defective strains that are being transmitted in the wild? Do all the strains present at the time of primary infection persist in the longer term and, if so, how does their relative representation vary with time? Most importantly from the perspective of vaccine design, does primary infection absolutely protect the immune host from re-infection or can an existing virus carrier acquire additional strains serially over time?

IMMUNE CONTROL OF VIRUS INFECTION

EBV provides a fascinating model through which to probe the workings of the human immune system. Early work concentrated on the antibody response to the virus and provided a number of serologic markers that have proved to be useful indicators of the general virus-host balance and, in some circumstances, important diagnostic tools. For example, the serologic profile can unequivocally diagnose classical IM and can help to identify those rare cases where primary exposure leads to chronic active EBV infection (see Chapter 5 by Cohen); furthermore in S.E.Asian populations where NPC occurs at high incidence, an EBV-specific IgA antibody profile is predictive of elevated disease risk (see Chapter 4 by de The). However, since congenitally B cell-deficient patients lack not only serum antibodies but also the main target cell for EBV infection, clinical experience tells us nothing about role of humoral responses in EBV control. Arguably neutralizing antibodies, principally directed against the major viral envelope glycoprotein gp340, will be protective against the spread of cell-free virions *in vivo* but, as such antibodies are slow to develop during primary infection, their main role is likely to be as a protection against new externally acquired virus or against reactivation of the existing infection to a full-blown viraemic phase. By contrast, the fact that T cell-compromised patients develop much higher viral loads both in the blood and in the throat, and are at particular risk of EBV-positive lymphoproliferative disease, indicates the critical importance of this arm of the immune system in the long-term control of the virus-host balance.

Our understanding of cellular responses to primary EBV infection comes almost entirely from the study of IM patients. In that regard we do not know whether IM simply exaggerates the cellular events occurring during asymptomatic infection or actually distorts them. Furthermore, although a highly expanded T cell response is apparent at the time of acute IM itself, this tells us nothing about events occurring in the 3-5 week period between virus transmission and the onset of disease symptoms. It may be that other controls operate at this time; indeed one extreme view would be that the full-blown IM syndrome only develops when those earlier protective responses fail. Christian Munz (Chapter 13) draws attention to this important gap in our knowledge and discusses how innate immune responses, mediated by interferon alpha produced in direct response to virus infection and/or by interferon gamma produced locally by activated natural killer (NK) cells, could serve to delay the initial virus-induced growth transformation of B cells in the oropharyngeal region. As seen generally for virus infections in animal models, such activity would effectively buy time during which the adaptive immune response can develop. Innate immune responses may also help to control the primary focus of virus replication in the oropharynx since cells progressing through lytic cycle lose surface HLA class I expression and are therefore likely to become sensitive to NK killing. Circumstantial evidence supporting a role for NK cells in early EBV control comes from the study of the X-linked lymphoproliferative (XLP) syndrome. Boys with this single X-linked gene defect are hyper-sensitive to EBV such that primary infection is frequently fatal with severe IM-like symptoms progressing to haemophagocytosis and bone marrow failure (Bar et al., 1974; Purtilo et al., 1975). The gene involved, SH2D1A, encodes a protein, SAP, which serves as an adaptor involved in the control of signaling through a family of lymphocyte surface molecules that includes SLAM, 2B4 and NTB-A (Coffey et al., 1998; Nichols et al., 1998; Sayos et al., 1998). Such molecules are involved in a number of lymphocyte-lymphocyte interactions, of which the engagement of 2B4 on NK cells with its ligand CD48 on the B cell surface is of particular interest. This interaction appears to be essential to license NK killing of B cell targets, and thus NK cell clones derived from XLP patients fail to kill HLA class I-deficient B cell lines *in vitro* (Parolini et al., 2000). This strongly suggests that the NK deficiency renders XLP patients hyper-sensitive to EBV infection. It is important to note, however, that 2B4 is also expressed on some CD8+ T cells, including many of the activated T cells in IM blood, and so the SAP defect and its clinical consequences for EBV sensitivity may reflect the combined effect of a failure of early NK function and also some derangement, as yet undefined, of the EBV-induced CD8+ T cell response.

By the time classical IM patients develop symptoms, the leukocyte count is dominated by a highly expanded population of activated CD8+ T cells accompanied by some CD56+ NK cells. So large is the T cell expansion that it led to speculation about the involvement of a virus-coded or virus-induced superantigen, albeit one activating the CD8+ as opposed to more usual CD4+ T cell subset. In fact the CD8+ T cell population in IM blood has none of the properties of a superantigen-induced response and instead contains large numbers of EBV epitope-specific effectors displaying classical restriction through HLA I alleles and identifiable by staining with HLA class I/ peptide tetramers (Callan et al., 1998). Many of these cells are directed against immunodominant epitopes drawn from lytic cycle antigens (Pudney et al., 2005), especially immediate early and early proteins such as BZLF1, BRLF1, BMLF1 and BMRF1, with smaller numbers being directed against latent cycle epitopes, drawn especially from the EBNA3 proteins. Interestingly the kinetics of the lytic and latent cycle responses are different, with the former rising to an early peak before being culled to a lower level as the acute disease subsides and the latter rising more slowly to a steady state (Hislop et al., 2002). These memory populations are then maintained for life in the re-circulating CD8+ T cell pool and, in total, typically make up 1-5% of the entire CD8 population in the blood of healthy virus carriers. Clones derived from these lytic and latent epitope-specific memory cells respectively are able to recognize and kill lytic- and latently-infected target cells in *in vitro* assays. The implication is that *in vivo* such cells act throughout life to control sites of chronic EBV replication in the throat and also foci of growth-transformed B cells arising from new infections with virions produced at these sites. Interestingly recent work has shown that EBV-specific CD8+ memory cells are 4 to 10-fold more abundant in tonsillar CD8 populations than in the blood, consistent with their active recruitment to the main site of EBV persistence in oral/pharyngeal lymphoid tissues. Surprisingly this was not the case during the acute primary infection seen in IM, where the activated EBV-specific population (dominated by lytic antigen reactivities) was poorly represented in the tonsil and, in contrast to the developing latent antigen-specific response, took several months to accumulate at that site after disease resolution (Hislop et al., 2005). These differences in migratory behaviour appeared to be determined by the phenotype of the relevant populations in the blood since the CCR7 and CD62L homing markers that normally mediate the re-circulation of "central memory" cells through lymphoid tissues are not expressed on the effector cells of acute IM and are particularly slow to develop on lytic antigen-specific memory cells. The apparently inappropriate homing of the highly expanded primary T cell response in acute IM adds to the circumstantial evidence (see below) that this response contributes to rather than alleviates the disease symptoms.

The sheer size of the CD8 response to EBV infection attracted such attention that for many years the accompanying CD4 response was not seriously investigated. As Munz (Chapter 13) describes, this has been corrected in recent years and much more is now known about CD4 reactivities, at least in terms of the response to latent cycle antigens. In many viral systems T-helper 1 (Th1) CD4 responses are thought to be important in maintaining effective CD8+ T cell immunity and, in line with this, EBV carriers possess latent epitope-specific CD4+ T cells with a Th1 (interferon-gamma secreting) cytokine profile in peripheral blood, albeit in numbers 10-fold lower than their CD8 counterparts. The primary CD4 response is less well analysed but again seems to be numerically small, consistent with its presumed helper role. Several EBNA proteins, in particular EBNA1, are proven sources of immunodominant epitopes presented in the context of defined HLA II alleles. Somewhat unexpectedly, many EBNA-specific CD4+ T cell clones reactivated *in vitro* from the memory pool display cytotoxic as well as interferon-gamma secretor function. Indeed clones to some (but not all) epitopes are, like their CD8 counterparts, even able to recognize and kill latently-infected B cell line targets (Munz et al., 2000; Omiya et al., 2002; Long et al., 2005). Precisely how the endogenously expressed EBNA proteins are being processed and presented via the HLA class II pathway to CD4+ T cells is an important current issue. Driving much of this work is the goal of exploiting latent antigen-specific CD4+, in addition to CD8+, T cells as effectors that could be used therapeutically to target EBV-positive malignancies. As Denis Moss and colleagues (Chapter 30) describe, this could be of particular relevance to BL, a tumour which is defective in HLA class I-restricted antigen presentation but in which the HLA class II pathway is still intact (Paludan et al., 2002; Fu et al., 2004).

EBV INFECTION, T CELLS AND THE PATHOGENESIS OF MONONUCLEOSIS

In his review of clinical aspects of EBV infection, Jeffrey Cohen (Chapter 5) draws attention to the clinical parallels between three disease states all initiated by primary EBV infection. These are the classical acute IM seen in Western counties in individuals infected for the first time in adolescence, the familial XLP syndrome (also called "fatal IM") specifically affecting boys with SAP gene mutations, and a range of life-threatening conditions variously described as virus-associated haemophagocytic syndrome (VAHS) or chronic active EBV infection that are particularly evident among young children in S.E.Asian countries. Besides the characteristic triad of fever, sore throat and lymphadenopathy, some acute IM patients can transiently suffer symptoms such as haemophagocytosis or hepatitis that are also characteristic features of end-stage XLP and VAHS. A picture is now emerging in which inappropriate T cell activation plays a key pathogenetic role in all three situations.

In acute IM itself, it is clear that the resolution of disease symptoms is coincident with loss of the CD8+ T cell expansion and return to a normal CD4/CD8 ratio in the circulating T cell pool. While viral loads in circulating B cells are also falling at this time, the shedding of infectious virus in the throat can remain high long after the illness has resolved (Fafi-Kremer et al., 2005); therefore clinical symptoms do not necessarily titrate the overall position of the virus-host balance. Indeed a few cases of sub-clinical primary infection have been identified where viral loads in the blood were high yet there were no disease symptoms; importantly, these individuals showed no evidence of CD8+ T cell expansion in the blood (Silins et al., 2001). It therefore seems likely that many of the clinical features of acute IM are induced by the highly amplified CD8+ T cell response, either by cytokines such as interferon gamma and TNFalpha derived from the reactive cells and leading to macrophage activation, or by the direct infiltration of large numbers of T cells into vital organs.

Information on the acute phase of XLP syndrome is very sparse. However, that which does exist describes the presence of large numbers of activated CD8+ T cells and NK cells in the blood before the final phase of haemophagocytosis and bone marrow destruction (Sullivan and Woda, 1989). This implies that the fatal course of primary EBV infection in XLP patients cannot be ascribed to any inability to mount a T cell response; indeed XLP patients who have survived a similarly severe episode of primary infection do possess EBV-specific CD8+ T cell memory. The argument, therefore, is that XLP patients mount a primary T cell response to EBV but the response never resolves, either because of a deficiency of effector function in SAP-negative cells or, more likely, because SAP deficiency (as seen in a mouse knockout model (Wu et al., 2001; Engel et al., 2003)) amplifies the Th1-mediated arm of the T cell response to viral infection at the expense of the Th2-mediated humoral response. The pathogenetic effects of a hyper-expanded T cell response are therefore exaggerated beyond those seen in classical IM; at the same time, antibody responses to EBV are either absent or atypical. A number of important questions now need to be answered. Thus, in acute XLP can one detect the presence of an EBV epitope-specific CD8+ T cell response using appropriate HLA class I/ peptide tetramers? Are such cells, cloned *in vitro* either from the primary infection or from memory, functionally impaired in any way; if so, is this analogous to the impairment that is specifically seen when XLP NK cells are tested against B cell targets? What is the cytokine profile of T cell clones derived from XLP patients versus clones derived from classical IM patients or healthy virus carriers?

The third set of clinical syndromes, incorporating EBV-associated VAHS and chronic active EBV infection, are perhaps the most interesting in that the patients present with symptoms ranging from acute VAHS to chronic IM-like disease apparently in the absence of an on-going virus-specific CD8+ T cell response. Such conditions are distinct from classical IM, and apparently also from XLP, in one important respect. In this case, EBV infection is detectable within either the T cell or NK cell compartment in the blood. Furthermore the infected T or NK cell population, often oligo- or mono-clonal , is activated and apparently expanding under the influence of the resident infection, therefore explaining why viral loads in the blood are so unusually high (Kikuta et al., 1988; Su and Chen, 1997). Infected cells are variously reported to express EBNA1 accompanied by LMP1 and/or LMP2, and probably are the source of the battery of both Th1 and Th2 cytokines reportedly present in patients' serum. In that regard, although direct EBV infection of resting human T cells is very difficult to achieve *in vitro*, transfection of T cell lines with an LMP1 expression vector can activate cytokine production. Among these are cytokines such as interferon gamma and TNFalpha which *in vivo* are thought to mediate macrophage activation leading to haemophagocytosis. (Lay et al., 1997). It is not known how often EBV accidentally infects T and NK cells, or by what route this occurs, but only very rarely has evidence of such infection ever been observed in classical IM. Arguably, such an event is very rare *in vivo* but, when it occurs, carries a high pathogenic risk. Moreover the risk is not just confined to cytokine-mediated pathology because many survivors of the acute VAHS syndrome and many longer term chronic active EBV patients eventually develop EBV-positive T or NK cell lymphomas.

EBV-ASSOCIATED TUMOURS

EBV is associated with a large range of tumour types and many of the Chapters in this volume deal with these associations individually. In validating the strength of such associations, *in situ* hybridization for the EBER RNAs provides a highly sensitive marker of latent EBV infection that is detectable at the single cell level on archival biopsy material. Hence both the numbers of cases of a given tumour that are EBER-positive, and the proportion of cells within any one tumour that are positive, become important criteria in deciding biological significance. These issues are fully discussed by Scott and colleagues (Chapter 6). If a particular tumour type shows other evidence of an EBV connection yet does not satisfy the EBER criteria, then one needs to retain a healthy scepticism until more information comes to light. The present controversy regarding EBV and breast carcinoma, fully reviewed by Irene Joab (Chapter 10), is one such case and is reminiscent of an earlier, as yet unsubstantiated, suggestion of an EBER-negative form of EBV latency in certain cases of hepatocellular carcinoma. I will not consider these contentious examples further but instead focus on the already substantial range of tumours whose link to EBV is properly documented.

To review EBV's credentials as a human tumor virus and to make sense of the sheer breadth of its tumour associations, let us first reiterate some of the questions that immediately arise. Thus, how do we reconcile EBV's suspected aetiologic role with the fact that, in several tumour contexts, only a subset of cases is EBV-positive? Why does the epidemiology of EBV-positive disease vary in different human populations? What is the significance of the different forms of latent gene expression seen in different tumour types, and do they imply a different role for the virus in each case? If EBV has an aetiologic role, is it confined to tumour initiation or does the virus still contribute to the growth/survival of the malignant clone itself? I will address some of these issues at relevant points in the following overview, where I have attempted to classify EBV-positive malignancies into what I consider to be pathogenetically coherent groups.

TUMOURS SPECIFIC TO THE IMMUNOCOMPROMISED HOST

EBV-positive tumours that arise only in the setting of the T cell-immunocompromised host represent a special group where, arguably, malignant cells are growing out opportunistically in the absence of virus-specific T cell surveillance. Certainly in one case, B lymphoproliferative disease (LPD), and possibly in the other, leiomyosarcoma, EBV appears to be both a necessary and sufficient cause of tumour growth.

B lymphoproliferative disease

Congenitally T cell-deficient patients, transplant recipients and other patients on immunosuppressive therapy, and end-stage AIDS patients with profound T cell impairment are all at risk of developing EBV-positive LPD lesions (see Cohen, Chapter 5). Because such lesions are not always clearly distinguished from other lymphomas arising in AIDS patients (see below), LPDs are better characterised from studies of the disease as seen in transplant recipients. These tumours (often described as post-transplant lymphoproliferative disease or post-transplant lymphoma) may present as a single focus or independently at several sites, individual lesions may be oligo- or mono-clonal in origin, and may be polymorphic or monomorphic in histologic appearance. Importantly most arise within the first 6-12 months post-transplant, when T cell suppressive therapy is most intense in solid organ transplant patients and when the peripheral T cell system has not yet reconstituted in bone marrow transplant patients. This early onset disease is classically EBV-positive and the tumour cells show the same full expression of virus latent cycle antigens as do *in vitro*-transformed LCLs (Latency III). Tumours arising in later years post-transplant are more heterogeneous as a group. They include some classical LPDs of the above type, other monoclonal lesions which are also EBV-positive but where viral antigen expression appears to be more limited, usually to EBNA1 only, and a minority of EBV genome-negative lymphomas of varied type (Nalesnik, 1998; Rickinson and Kieff, 2001). The range of lymphomas seen in end-stage AIDS patients is similar to that seen later post-transplant but with a greater representation of EBV-negative tumours, mainly diffuse large B cell lymphomas; interestingly the classical EBV-positive Latency III LPD lesions of AIDS (often described as AIDS immunoblastic lymphomas) are most prevalent among tumours presenting in the central nervous system, as if immune surveillance is particularly compromised at this site (Boshoff and Weiss, 2002; Carbone, 2003).

I would argue that the Latency III LPDs should be considered a distinct clinical entity, representing B cells directly transformed by EBV *in vivo* and growing out opportunistically in the absence of T cell surveillance. Their existence both proves the oncogenic potential of the virus and also highlights the parallels that exist between EBV and the other primate lymphocryptoviruses, where immunosuppression predisposes to LPD-like lesions in their natural hosts (see Chapter 32). The other EBV-positive B lymphomas that arise in immunocompromised patients but show more restricted antigen expression are aetiologically more complex. For the moment I believe they need to be considered as a separate entity but, interestingly, one that could have evolved from a progenitor population of classical LPD-like cells; in that regard, there is a need for prospective banking of tumour tissue from post-transplant patients so that, in rare cases where both early and late onset tumours occur, their cellular origins can be compared. Recent Ig genotyping studies have shown that EBV-positive post-transplant lymphomas can arise from both naïve and conventional memory B cells. However an unexpectedly large number, of early LPDs as well as late onset tumours, arise from atypical post-germinal centre B cells. These show patterns of Ig gene mutation which are not only unlike those found in conventional antigen-selected memory cells but also in some cases have completely inactivated the reading frame such that the cells are surface Ig-negative (Timms et al., 2003; Brauninger et al., 2003; Capello et al., 2003). This suggests that non-antigen-selected B cells, that would otherwise have been lost as failed products of germinal centre reactions, have been rescued by EBV infection. An alternative explanation, that EBV infection/growth transformation *in vivo* has itself induced unscheduled mutation of the host cell's Ig gene sequences, seems less likely, if only because there is no evidence of ongoing mutation in most of these tumours. Other work has indeed detected the presence of an unusual subset of virus-infected B cells in the blood of immunosuppressed transplant recipients which is CD19-positive but surface Ig-negative (Schauer et al., 2004), suggesting that T cell impairment increases EBV's chances of rescuing B cells that normally would have been eliminated. An interesting question is whether such cells are for some reason more susceptible than conventional B cells to subsequent oncogenic change, since surface Ig-negative tumours are well represented among the later onset post-transplant lymphomas with restricted EBV antigen expression. (Timms et al., 2003)

Smooth muscle cell tumours

The discovery of an association between EBV and smooth muscle cell tumours, both in post-transplant and in AIDS patients, came as a complete surprise (Lee et al., 1995; McClain et al., 1995; Timmons et al., 1995). Leiomyosarcomas are extremely rare in the general population and, when they do occur, are never associated with EBV; by contrast, such tumours occur more frequently in immunocompromised settings where they are consistently EBV genome-positive. Very little is known about the pathogenesis of this tumour. One interpretation is that EBV has direct growth transforming ability in smooth muscle cells but only ever accesses such cells under conditions of extremely high viral loads *in vivo*; the same logic might also apply to another rare tumour of the immunosuppressed, oral T cell lymphoma, which again is frequently EBV-positive. There is still no rigorous study of viral gene expression in EBV-positive leimyosarcomas. These tumours therefore provide a very interesting test case; is the smooth muscle cell a non-lymphoid cell in which, uniquely, the full EBV growth transforming programme is activated, or is tumour growth driven by a more limited set of latent cycle proteins, possibly acting in concert with cellular change?

TUMOURS OF THE IMMUNOCOMPETENT HOST

Most EBV-associated tumours arise in apparently immunocompetent individuals and their incidence is not obviously increased in T cell-compromised settings. All such tumours display restricted forms of latent gene expression and have complex multi-step pathogeneses involving EBV-mediated effects and complementing cellular genetic changes. In each case the nature and timing of EBV's contribution is not fully understood. Furthermore, though we still know rather little about their interactions with the host immune system *in vivo*, it appears that these tumours are either immunologically silent or employ immune evasion strategies to avoid detection

(i) B cell malignancies of the immunocompetent host

I contend that Hodgkin's Lymphoma (HL) and BL form a coherent group because they are both B cell malignancies that are seen worldwide but whose incidence is increased in settings of chronic immune stimulation. Put another way, both arise from rare accidents of EBV's normal lifestyle within the B cell system and are more likely to develop if the B cell system itself is disturbed by co-resident chronic infections. To justify this argument, we first have to deal with the objection, which others might raise, that some loss of virus-specific T cell function is important to the development of both tumours.

There is evidence from certain non-specific assays of T cell function that all HL patients (irrespective of the EBV status of their tumour) suffer a degree of global immune impairment when compared to healthy controls, and so the tumour never arises in a truly immunocompetent host (Poppema et al., 1999). It is not clear whether these global effects precede tumour development or are somehow a product of the malignant condition itself. Either way, as yet I see no clear indication that the EBV-host balance (a rather sensitive index of overall T cell competence) is significantly altered in the many patients who present with EBV-negative HL but who happen to be EBV carriers. Perhaps this has not yet been looked at sufficiently closely but, as things currently stand, the subtle immune impairment that is described as a general feature of HL therefore does not appear to affect the overall set point of EBV infection. Now turning to BL, as Griffin and Rochford discuss (Chapter 9), one view is that one holoendemic malaria, a well-known co-factor for BL, increases tumour risk by suppressing immune control of EBV infection. There is indeed *in vitro* evidence both for a transient deregulation of the virus-host balance during bouts of acute malaria (though possibly this is a non-specific feature of any severe febrile illness) and for a mild shift in the stable set point of EBV infection in populations where malaria is endemic. However these issues have not yet been re-investigated using contemporary assays of EBV-specific T cell number and function. Furthermore limited studies using *in vitro* reactivation assays (C.M Rooney and A.B.Rickinson, unpublished) did not reveal any gross impairment of EBV-specific T cell responses in the blood of endemic BL patients themselves. More work is required but, provisionally, endemic BL patients do not appear to be obviously immunocompromised.

Perhaps the most powerful argument against T cell impairment as a contributor to HL and BL pathogenesis is the epidemiologic evidence; thus the incidence of these tumours is not significantly raised in chronically immunosuppressed transplant patients, in stark contrast to the same patients' elevated risk of LPD (Penn, 1986; Nalesnik, 1998). On the other hand HIV-infected patients are at increased risk of both HL and BL (Levine, 1998; Carbone, 2003) but, as the evidence (see below) suggests, this occurs early in the course of HIV carriage and not as a result of immune T cell impairment. Indeed the introduction of effective anti-retroviral therapy for HIV infection has emphasized the fundamental distinction between HL and BL versus LPD in their relation to host immune status; thus restoring the immune competence of HIV-infected patients with anti-retroviral drugs has dramatically reduced the incidence of LPD lesions but has left HL and BL incidence largely unaffected (Thirlwell et al., 2003).

Hodgkin's lymphoma

HL is an unusual tumour where the malignant population of Hodgkin's and Reed Sternberg cells is greatly outnumbered by non-malignant infiltrating cells, the nature of the infiltrate determining the different histologic subtypes of the tumour. Importantly Ig genotyping shows that the malignant population arises from an atypical post-germinal centre B cell, where Ig genes have been inactivated by somatic mutation or silenced by other means (Kuppers and Rajewsky, 1998) and also where many other B lineage markers have been down-regulated; it was this atypical phenotype that for many years obscured the true B cell origin of HL. The link with EBV is discussed in detail by Murray and Young (Chapter 8). Thus in the general population in the West, some 30-40% HLs are EBV genome-positive.

Most of these are tumours of the mixed cellularity and (rarer) lymphocyte-depleted subtypes that are the commonest forms of the disease in children and older adults; by contrast the marked peak of HL incidence seen among young adults is mainly EBV-negative and preferentially involves tumours of the nodular sclerosing subtype. The situation is different in the developing world where there is no young adult peak of nodular sclerosing disease; instead the two peak incidences are seen in childhood (mainly in boys) and again in older adults. In these countries, mixed cellularity and lymphocyte depleted subtypes are most common and at least 80% of all tumours are EBV genome-positive.

EBV's association with HL highlights a number of the general issues raised at the start of our discussion of EBV as a tumour virus. The first (one that also arises in the context of BL) is how to interpret the significance of such an association when the majority of HLs, at least in the Western world, do not carry the viral genome in the tumour cells and some even arise in EBV-uninfected individuals. The question is especially relevant when, in HL and also BL, there are very few if any discernible differences in tumour cell phenotype between virus-positive and –negative tumours. So, when present in HL, why should the virus not simply be an innocent passenger? An important finding in that regard is that, in EBV-associated cases of HL, all the tumour cells are EBV genome-positive and indeed the tumour population is monoclonal both by cellular (Ig gene) and viral (EBV terminal repeat) genomic markers. Thus the virus is present in the progenitor cell at the initiation of malignant growth, consistent with its postulated aetiologic role, and is not casually acquired by a sub-population of cells during subsequent growth of the tumour. If such casual acquisition were common, we would indeed expect to see HLs where the malignant population was mosaic for EBV genome status, yet this has rarely if ever been reported. Another sceptical position is to argue that the tumour arises from a particular B cell subset *in vivo* and that the incidence of EBV-positive HLs simply reflects the frequency of naturally-infected cells in that subset; however, levels of EBV carriage in blood and lymphoid tissue B cell populations rarely if ever exceed 0.1% cells and infected cells are equally distributed between memory and germinal centre populations (see Chapter 17). The possibility of HL arising from an EBV-positive target cell purely by chance is therefore commensurately low. Finally, if the virus were a casual passenger in HL cells, it is a remarkable coincidence that its resident pattern of antigen expression (Latency II: EBNA1, LMP1 and LMP2-positive) includes proteins whose known effects upon the B cell phenotype (for example, activation of NFkB, JNK, p38 and JAK/STAT signalling pathways by LMP1, tolerance of loss of surface Ig by LMP2) are entirely consistent with their having an active role in growth and survival of the tumour clone. Unfortunately, in the absence of cell lines derived from EBV-positive tumours, one cannot formally test *in vitro* whether HL cell growth is dependent upon continued EBV gene expression. However, compelling evidence for a direct effector role for the virus in EBV-positive HL comes from parallel studies on EBV-negative disease where, in the absence of the viral effector LMP1, constitutive activation of the NFkB and JAK/STAT pathway appears to be achieved by cellular genetic change ((Kuppers, 2003); see also Murray and Young, Chapter 8). These findings strongly suggest that the pathogenesis of HL requires the serial acquisition of several key cellular changes that lead to a common endpoint or at least to one of a set of closely related endpoints (assuming that subtle differences in tumour cell phenotype influence the precise nature of the infiltrate). A cell may arrive at that endpoint by different routes; in one case some of the key changes are induced by EBV, in another these changes accumulate by other means.

A second issue raised by EBV's association with HL (and reflected in the context of several other malignancies) is how we interpret the different epidemiology of the tumour in different host populations. The existence of an early adult peak of nodular sclerosing disease in the general Western population, but not in the developing world, was interpreted as indicative of a disease peak caused by delayed exposure to a common infectious agent. The irony is that EBV, an agent that would fit that description, is indeed linked to the tumour, but specifically not to the bulk of early adult disease seen in the West. The peak of early adult disease therefore remains unexplained. Yet at the same time epidemiological evidence was showing a significant increase in HL incidence among patients with a history of IM (Hjalgrim et al., 2000). In that regard an important recent study compared subsequent HL incidence in serologically confirmed (Paul-Bunnell-positive) IM patients versus patients where IM had been a possible diagnosis but who proved Paul-Bunnell-negative (most of whom would therefore not have been primary EBV infections). This confirmed a 4-fold increase in HL incidence specifically in the IM group and found that this increase specifically involved EBV-positive tumours, occurring within a median interval of just 4 years (Hjalgrim et al., 2003). Primary EBV infection, at least when manifest as IM, therefore does predispose to the subsequent development of EBV-positive HL. However, still only 1 in 1000 IM patients are affected in this way and so, even though IM is a disease of early adulthood, its overall impact on the early adult peak of HL incidence will be marginal. Nevertheless the result is important since it demonstrates a clear link between primary EBV infection and an increased risk of EBV-positive HL arising within the next few years. This could help to explain why, in developing countries where primary EBV infection invariably occurs early in life, there is a peak of EBV-positive HL incidence in childhood.

How and when EBV infects the progenitor cell of the malignant HL clone is simply not known. It may be interesting that within the EBV-infected B cell clones in IM tonsil sections there are some cells that morphologically resemble the malignant Reed-Sternberg cells of HL and like them stain as EBNA2-negative but LMP1-positive, equivalent to Latency II (Kurth et al., 2000). Such cells are also visible within some EBV-driven LPD lesions, albeit retaining typical B cell markers rather than mimicking the atypical HL cell phenotype (Ranganathan et al., 2004). We do not know what happens to these cells over time. In principle, Latency II infections should be recognized by CD8+ T cells in individuals with responses to one of the three relevant antigens, most likely LMP2 or EBNA1 given the general paucity of epitopes in LMP1 (Rickinson and Moss, 1997). However, in practice we have no true *in vitro* correlate of the HL-like cell for laboratory assays and no information as to whether the very high level of LMP1 expression in such cells could either directly, or through inducing regulatory cytokine expression, provide an immune evasion mechanism that would render the cells less susceptible to T cell control *in vivo* (Dukers et al., 2000; Marshall et al., 2003). It is therefore possible that such cells may persist and constitute a precursor population at risk for subsequent cellular genetic changes en route to HL, perhaps changes later acquired as by-products of antigen-driven somatic hypermutation in germinal centres. The postulated effector role of the EBV latent proteins, LMPs 1 and 2, in mediating unscheduled survival of germinal centre cells that have failed the affinity maturation process are discussed fully by Murray and Young (Chapter 8) among others.

The identification of HL cells as atypical products of germinal centre reactions is particularly interesting when set aside the epidemiology of the tumour in HIV-positive patient cohorts. Since HL incidence rates are not elevated in transplant patients, it was initially a surprise to find that tumour incidence is around 8-fold higher in HIV-positive patients than in the general population in the West. Moreover all of this increment represents EBV-positive disease, and most of it tumours of the mixed cellularity and lymphocyte-depleted subtypes (Levine, 1998). Most importantly, this so-called AIDS-associated HL presented early in the course of HIV infection, when CD4+ T cell counts were still reasonably high, and in most cases arose in patients with persistent generalized lymphadenopathy (Vaccher et al., 2001). Such lymphadenopathy is known to be associated with highly exaggerated germinal centre activity, thereby expanding the very population from which the HL progenitor cell arises. These findings strengthen the view that HIV infection predisposes to HL through its (still poorly understood) ability to as a chronic B cell stimulus.

Burkitt's lymphoma

BL is a histologically and cytogenetically well defined tumour that occurs in three distinct epidemiologic contexts. The high incidence "endemic" form of the disease, seen in holoendemic malarial areas of equatorial Africa and New Guinea, is consistently EBV-positive. Elsewhere "sporadic" BL incidence is 10-100-fold lower and EBV association varies from 85% in intermediate incidence areas to only 15% in low incidence Western countries. Both endemic and sporadic BL are childhood malignancies with broadly overlapping clinical presentations, and their age peaks suggest that EBV-positive tumours mainly arise within 3-5 years of primary EBV infection. A third form of BL is seen in HIV-positive patients in the West, with a remarkably high incidence well above that of endemic BL in Africa; some 30-40% of these so-called AIDS-BLs are EBV-positive (Rickinson and Kieff, 2001).

The potential parallels between the Burkitt and Hodgkin tumours become obvious when one looks at the epidemiology of AIDS-BL. Again this typically arises well before the onset of gross immune impairment in HIV carriers, and often in individuals with persistent generalized lymphadenopathy (Knowles and Pirog, 2001). Since cell phenotype and Ig genotype data identify BL as a tumour of germinal centre origin (Kuppers, 2003), a more likely explanation of HIV's cofactor role in BL is again through chronic germinal centre activation. Holoendemic malaria, also known to cause chronic B cell stimulation, might act in a similar way (see Chapter 9). Hence it becomes important now to look specifically for effects of malarial infection on germinal centre activity. The parallels between these two co-factors are not absolute, however. Firstly, HIV is a much stronger than malaria in the magnitude of its co-factor effect. Secondly all of the malaria-linked increment in BL incidence is EBV-positive disease, whereas this is not the case for the HIV-linked increment in the West. It is not clear whether this difference relates to the identity of the co-factor or to the different host populations involved; a very important unknown here is the degree to which HIV infection has altered the incidence and epidemiology of BL in Africa, and in particular whether AIDS-BL in that population is a consistently EBV-positive disease.

Discussions of BL pathogenesis revolve around the one feature common to all forms of the tumour, deregulation of the c-myc oncogene through its translocation into one of the three Ig gene loci. The balance of evidence now favours such translocation occurring within germinal centre cells as an accident of somatic hypermutation rather than in pro-B cells during Ig gene rearrangement (Kuppers, 2003). It is not known whether the cell destined to become malignant acquires EBV infection before (the usually favoured scenario) or after the c-myc translocation. While the virus seems very unlikely to be merely a passenger in BL, for the reasons already outlined above for HL, we do not know how EBV might contribute to tumour development. It may act early as an initiator, Latency III infection driving B cell expansion to create a larger pool of cells at subsequent risk of translocation, or later by contributing to the growth/survival of the malignant clone. The fact that the episomal virus genome is faithfully maintained in the entire tumour population both *in vivo* and in the great majority of BL-derived cell lines *in vitro* indeed suggests that EBV continues to contribute to the tumour phenotype. In this context, the late passage Akata-BL line (derived from an EBV-positive sporadic tumour) is unusual in generating EBV genome-negative subclones; these continue to grow well under optimal conditions in suspension culture but do not form colonies in soft agar or grow as tumors in nude mice (Shimizu et al., 1994). Arguably these differences might reflect a role for viral gene products in reducing the susceptibility of BL cells to the strong apoptosis-sensitising effects of c-myc-driven proliferation. The original Akata-BL growth phenotype can be partially restored in negative subclones by vectored expression of the EBER RNAs, an effect coincident with the upregulation of endogenous levels of bcl2 and the IL10 cytokine, whereas expression of the EBNA1 protein had no effect (see Chapter 21). To judge the generality of these potentially important findings it will be important to look for other examples, preferably among early passage BL lines, where EBV genome-loss subclones can be can established and their growth phenotype determined.

BL represents the classical example of a tumour where viral infection and cellular genetic changes make complementary contributions to pathogenesis. One important general lesson we see illustrated here is the existence of constraints that govern such complementation. The key experiments (discussed by Kempkes et al., Chapter 23) involved B cells transformed by a recombinant EBV with regulatable EBNA2 function and then transfected with a c-myc expression construct. These showed that the EBV growth programmme, driven by Latency III infection and imposing an LCL-like cell phenotype, and the c-myc programme, driven by a c-myc/Ig fusion gene construct and imposing a BL-like phenotype, were incompatible with one another (Polack et al., 1996; Pajic et al., 2001). In this model the c-myc programme, which is essential for development of the BL tumour *in vivo*, could only be sustained in EBV-infected cells if EBNA2 and the EBNA2-activated LMPs were downregulated. This incompatibility may well be the pressure forcing the selection of Latency I in the tumour since this form of latency provides a means of retaining the EBV genome while suppressing EBNA2. The strength of selection against EBNA2 is further illustrated by the existence of a subset of endemic tumours where Wp rather than Qp is transcriptionally active and antigen expression includes EBNAs 1, 3A, 3B, 3C and –LP but not EBNA2 and the LMPs; interestingly these tumours carry both an EBNA2 gene-deleted and a wild-type virus genome but only the deleted genome is active. The broadening of latent gene expression to include the EBNA 3 proteins renders these cells less sensitive than conventional Latency I BL cells to apoptosis. (Kelly et al., 2002; Kelly et al., 2005). This, combined with the earlier evidence from the Akata-BL system, suggests that (in the absence of EBNA2 and the LMPs) a number of other EBV genes are able provide small increments in cell survival that can assist the development of a malignant cell *in vivo*.

The BL tumour is also frequently cited in the literature as an example of a virus-associated tumour that has evolved specifically to avoid host cell-mediated immune recognition. This would fit well with the concept of BL arising in the immunocompetent host. Antigen presentation to CD8+ T cells is indeed deficient in BL cells, such that even when the immunogenic EBNA3 proteins are endogenously expressed in tumour cells, they remain impervious to T cell detection (Kelly et al., 2002). However this antigen processing defect appears to be integral to the c-myc-associated growth phenotype and not determined by a need to avoid EBV detection. Thus the defect is a feature of both EBV-positive and EBV-negative BLs (Khanna et al., 1994) and indeed can be imposed on normal EBV-transformed B cells in the absence of any selective pressure by c-myc transfection *in vitro* (Staege et al., 2002). The ability of EBV-positive BL cells to evade the CD8+ T cell recognition (for example by EBNA1-specific effectors) is therefore a by-product of the fact that these cells are subject to the c-myc-driven growth programme. Interestingly, with serial passage *in vitro* some EBV-positive BL lines will spontaneously activate full EBV latent gene expression (Rowe et al., 1987); when this occurs, the cells assume an LCL-like phenotype with antigen processing function restored and the translocated c-myc allele is silenced. Like many aspects of the EBV/BL story, this process of BL cell progression to an LCL phenotype with suppression of the endogenous c-myc-driven programme is very poorly understood.

(ii) Non- B cell malignancies of the immunocompetent host

The non-B cell tumours arising in immunocompetent individuals, that is T and NK cell lymphomas and the range of EBV-positive carcinomas, form a third group of EBV-associated malignancies. I would argue that these all arise from rare accidents where the virus infects cell types that are normally either never accessed *in vivo* or, if they are, cannot maintain infection; the implication is that unscheduled entry and retention of the virus in these atypical target tissues carries a high oncogenic risk.

T and NK cell lymphomas

EBV is consistently linked to two specific types of T and NK cell lymphoma, described below. These authentic associations are sometimes obscured in the literature by reports of other putatively virus-associated T cell lymphomas where only a fraction of the tumour cells appear to be EBER-positive. It is possible that some of these cases do reflect actual EBV infection of a subset of tumour cells. However many of these initial reports involved angioimmunoblastic lymphadenopathy (AILD)-like tumours of T cell origin where, interestingly, EBER expression has now been localized to foci of EBV-infected B cells that are able to survive and expand within the niche of a co-resident T cell lymphoma (Brauninger et al., 2001).

Turning to the authentic associations with T and NK cell tumours, EBV is consistently linked to an angiocentric nasal lymphoma of adults (sometimes called lethal mid-line granuloma) that is most common in S.E. Asian populations and is mainly of NK cell (CD56+, CD3-) but occasionally of T cell (CD3+, CD56-) origin (Jaffe et al., 2003). EBV-positive tumours of an essentially similar phenotype can sometimes present in the skin or even as NK leukaemia. The natural history of this characteristic nasal tumour is not as well studied as that of a second EBV-associated lymphoma, again commonest in S.E.Asia but in this case usually of CD4+ or CD8+ T cell type (Su and Chen, 1997). These T lymphomas arise in patients either with a recent history of VAHS or with on-going chronic active EBV infection (described earlier in the section on T cells and the pathogenesis of mononucleosis); the high incidence among S.E Asian populations is likely to have a genetic basis, just as certain other forms of haemophagocytic syndrome and at least one unusual case of chronic active infection have been mapped to specific gene defects (Katano et al., 2004). In some VAHS patients, one can define the time of primary infection from the onset of VAHS symptoms, demonstrate the existence of oligoclonal T cell expansions in the blood at that time and find, within a period of 1 to 2 years, the emergence of a monoclonal EBV-positive T cell lymphoma with gross choromosomal changes. Both the early oligoclonal lesions and the monoclonal lymphoma display an intermediate Latency I/II form of infection with EBNA1 and LMP2 expression, sometimes accompanied by LMP1 in a small subpopulation of cells. These accidents of nature provide a classic illustration of how EBV accessing the wrong cell type *in vivo* can initiate proliferative lesions that rapidly progress to malignancy. The timing of EBV infection in the sequence of events leading to tumour development is therefore well known in this case but, in the absence of a tractable *in vitro* model of resting T cell infection, the biology of the EBV-T cell interaction is simply not understood.

Nasopharyngeal, salivary gland and gastric carcinomas

EBV is linked to carcinomas arising from nasopharyngeal, salivary gland and gastric epithelium, and detailed descriptions of these tumours are given by Raab-Traub (Chapter 7), Scott et al. (Chapter 6) and Iwakiri and Takada (Chapter 11) respectively. It is interesting to consider them collectively here because, although the target tissues are anatomically distinct, many of these tumours display similar histological features and there is growing evidence that the parallels might extend to EBV's role in the pathogenetic process.

NPC of undifferentiated or poorly differentiated cell type is a rare tumour in many countries but is unusually common in people of Southern Chinese origin and in certain other populations, notably the Inuit people of North America and Greenland. Irrespective of the local incidence rate, however, all cases of the tumour worldwide are EBV genome-positive and by EBER in situ hybridization the virus is present in every malignant cell. NPC is characteristically infiltrated by non-malignant cells, mainly T lymphocytes, and so is frequently referred to as a lymphoepithelioma. The same Southern Chinese and Inuit populations also show high incidence of a histologically similar salivary gland carcinoma, often described as lymphoepithelioma of the parotid, which is again consistently EBV-positive; of the rare cases of this tumour seen in the West, some but not all are virus-associated. Epidemiological evidence suggests that these high tumour incidences in particular populations reflect the combination of two non-viral factors: one is cultural and could involve chronic exposure to an inhaled dietary or other environmental carcinogen (nitrosamines in salted fish being one candidate), the other is an as yet undefined genetic predisposition peculiar to these racial groups. Rare tumours with an identical type of histology arise at other tissue sites but show no obvious geographic variation in incidence, the best example being lymphoepitheliomas of gastric epithelium; these tumours are also EBV-positive in most if not all cases. Somewhat surprisingly, a number of regular gastric adenocarcinomas are also linked to the virus and overall, depending upon the study population, EBER-positive cases are reported as constituting between 4 and 18% of all gastric tumours. Collectively, the worldwide burden of disease from these three EBV-associated carcinomas is even greater than that from EBV-associated lymphoid tumours.

Viral gene expression has been best studied in NPC and the gastric carcinomas. All tumours express the EBNA1 protein (in the absence of other EBNAs) as well as the EBER and BART RNAs; indeed the BARTs were first identified through cDNA cloning from NPC cells and this reflects their higher level of expression in carcinomas compared to B cell tumours. There is a more heterogeneous picture with respect to expression of the LMPs, such that most tumours seem to adopt an intermediate form of latency between Latency I (BL) and Latency II (HL) paradigms. Thus, depending on the series, only around 20-40% NPC tumours express LMP1 at levels detectable by sensitive monoclonal antibody staining. LMP2A and B transcripts can be amplified easily from most if not all cases of NPC but immunohistochemical staining for protein is relatively insensitive; so, even with improvements in staining protocols, the recent value of 45% tumours with detectable LMP2A protein (Heussinger et al., 2004) may still represent an underestimate. Gene expression is even more limited in EBV-positive gastric carcinomas, of both histological types. These tumours are LMP1-negative and only express detectable levels of LMP2A (not B) transcripts in around 40% cases; how this is reflected at the protein level remains to be seen.

Much effort has focused on the LMP1 and LMP2A proteins as potential effectors of malignant change in an epithelial cell context. Interestingly neither of these proteins induces tumours in mice when expressed from transgenes in skin epithelium (Curran et al., 2001; Longan and Longnecker, 2000). However there is a wealth of evidence from *in vitro* systems documenting LMP-induced effects on the epithelial cell phenotype Engaging many of the same signaling pathways as in B cells, LMP1 can induce major changes in epithelial cell morphology/growth phenotype, can enhance cell survival through induction of anti-apoptotoic proteins, and can block the cellular response to differentiation signals, possibly through upregulating inhibitor of differentiation (Id) family proteins (see Raab-Traub, Chapter 7). Furthermore, as described by Yoshizaki et al. (Chapter 12), LMP1 activates the expression of various angiogenic factors as well as certain integrins and matrix metalloproteinases that promote cell motility and tissue penetration, all changes which would be expected to facilitate tumour cell invasion and metastasis *in vivo*. More recently LMP2A has been of particular interest since, in contrast to its rather subtle effects on B cells *in vitro*, in epithelial cell systems it can promote cell growth both in monolayer and soft agar culture and can protect from some apoptotic signals. These effects appear to be mediated through LMP2A-induced activation of the PI3 kinase/Akt/beta-catenin signaling pathway and possibly also of MAP kinase cascades in epithelial cells (see Ikeda et al., Chapter 25). As to other potential contributions of the virus to epithelial cell growth, no specific effects have yet been observed for EBNA1 but vectored expression of the EBERs in a virus-negative gastric carcinoma line is reported to enhance growth through induction of the insulin-like growth factor IGF-1 (see Takada, Chapter 21). Finally recent work has found that, in contrast to B cell malignancies, latent EBV infection in epithelial tumours is accompanied by transcription of the BARF1 gene (zur Hausen et al., 2000). It is not known whether this is related in any way to the higher levels of BART transcripts seen in these tumours, since both sets of RNAs come from the BamH1A region of the genome and are in the same transcriptional orientation. If BARF1 expression can be confirmed at the protein level, then as Tadamasa Ooka describes (Chapter 28), there is already evidence from *in vitro* systems that expressing this distant viral homologue of the colony stimulating factor CSF1 receptor is able to promote epithelial cell growth.

There are therefore many potential candidates as EBV-coded effectors of carcinogenesis. How important a role any of these play in real life is difficult to determine experimentally, not least because we have no animal model that recapitulates the oncogenic process and also because attempts to establish EBV-positive carcinoma cell lines *in vitro* have met with little success. Some epithelial cell lines have reportedly been derived directly from NPC tumour biopsies but these are all EBV genome-negative and their provenance remains in doubt. However a small number of EBV-positive transplantable NPC lines have been established in nude mice (Busson et al., 1988) and one such line has since adapted to *in vitro* passage (Cheung et al., 1999); likewise the first EBV-positive gastric carcinoma lines have now been established in SCID mice (Iwasaki et al., 1998) and directly *in vitro* (Oh et al., 2004). These various tumour lines retain Latency I/II patterns of viral gene expression and are potentially useful targets in experiments attempting to block the function of particular viral proteins and then looking for effects on cell growth or tumourigenicity. Experimental infection of primary nasopharyngeal epithelial cells *in vitro* has led to transient infection at best, but similar studies using primary gastric epithelium and a virus with a drug-selectable marker have reported a significant increase in the cells' doubling capacity and, where the primary cells had a p53 mutation, a permanent EBV-positive cell line grew out (Nishikawa et al., 1999). A reproducible *in vitro* system of virus-induced epithelial cell immortalization remains an important goal.

It is clear that the pathogenesis of all three EBV-associated carcinomas involves numerous cellular genetic changes in addition to the viral contribution. Most work in this context has focused on NPC, where obvious candidate genes such as c-myc, ras, Rb and p53 appear to be unaffected (despite high level p53 protein levels). However, expression of the p16 tumour suppressor protein is consistently abrogated, by methylation and/or chromosome 9p21 loss, and p16 downregulation is also a specific feature of EBV-positive gastric carcinomas. In addition, NPCs show frequent loss of heterozygosity at chromosomes 3p24 (affecting the RASSF1a gene), 11q, 14q12-13 and 14q32, suggesting that several other tumour suppressor gene losses might play a role in tumourigenesis (Lo et al., 2004). It is interesting that, despite intense screening, pre-malignant lesions of the nasopharynx are very infrequent in high NPC-risk populations. However, apparently normal or early dysplastic lesions have recently been detected in nasopharyngeal epithelium that already show losses of heterozygosity at some of the above chromosomal loci in the complete absence of EBV infection (Chan et al., 2000; Chan et al., 2002), whereas later carcinoma in situ lesions were uniformly EBV-positive (Pathmanthan et al., 1995). Similar studies of pre-malignant foci in EBV-positive gastric carcinoma further support the view that EBV infection is a late event in the development of that tumour (zur Hausen et al., 2004). A picture is therefore emerging where the initial accumulation of cellular genetic changes in nasopharyngeal or gastric epithelium creates a focus of pre-malignant cells and that EBV infection of such cells provides the additional drive to malignancy. It is even possible that these pre-malignant changes create an atypical cellular environment (for example, through upregulation of the STAT 3 transcription factor to which the EBNA1 and terminal repeat LMP1 promoters respond) that now favours a latent viral infection being sustained. The monoclonal nature of the resultant tumour suggests either that EBV infection of the pre-malignant cells is itself a rare event or that there is further selection of a malignant clone from the EBV-positive carcinoma in situ lesion.

IMMUNE INTERVENTION AGAINST EBV-ASSOCIATED DISEASES

The very large burden of EBV-associated disease worldwide, reviewed by Cohen (Chapter 5), challenges us to find ways to alleviate that burden. The contributions from Denis Moss and colleagues (Chapter 30) and from Dotti et al. (Chapter 31) describe the different immunotherapeutic approaches that are now either in clinical trials or currently being developed with a view to trials. This is an area of work, particularly with respect to the immunotherapy of human malignancies, in which lessons from the EBV field are of widespread general interest.

Tony Epstein (see Chapter 1) was the person who first championed the idea of an EBV vaccine, in that case a vaccine based on the gp340 envelope glycoprotein which represents the principal target of the EBV neutralizing antibody response. The original aim was a prophylactic vaccine that could provide antibody-based sterile immunity to infection. Conventional wisdom in the field currently views sterile immunity as an unattainable goal since neutralizing antibodies cannot completely prevent direct cell-to-cell transfer of EBV and salivary transmission may well be mediated by productively-infected cells as well as cell-free virus. However that may be too negative a view in the long-term as prophylactic vaccine strategies improve to elicit both humoral and appropriate cell-mediated immune responses. First, as discussed earlier (p724), we need to know whether natural EBV infection renders the immunocompetent host resistant to subsequent infection with new orally-transmitted virus strains; if that is the case, then there is a chance that sterile immunity might eventually be achieved by vaccination. What we know so far from human trials is that a gp350-based sub-unit vaccine, of the kind that protected cotton-top tamarins from lymphomas induced by intraperitoneal injection of a large dose of transforming EBV, will not protect naïve human subjects from natural EBV infection. Very interestingly, however, this recent clinical trial was conducted in young adult volunteers not previously infected by the virus and, although levels of EBV acquisition did not significantly differ in the vaccinated group and placebo controls, there was a marked reduction in the incidence of IM symptoms among primary infections in the vaccinees (Denis et al., 2004). This suggests that pre-existing neutralising antibodies are able to limit the initial phase of oropharyngeal virus replication, thereby reducing lytic cycle antigen load and the chances of the lytic antigen-induced CD8+ T cell response expanding to a point where IM symptoms ensue. Arguably, this could be life-saving for boys with the XLP trait who are at special risk of fatal IM. Furthermore, by reducing the load of infectious virus available to initiate B cell infections, such a vaccine could also reduce the extent of EBV-induced B cell proliferation that normally occurs in lymphoid tissues during primary infection. Vaccination could therefore be very important in seronegative transplant recipients where primary infections occurring in an immunocompromised setting have a high risk of progressing to LPD. Such a trial is now on-going in one especially high risk group, paediatric liver transplant recipients, many of whom are young infants and seronegative at the time of transplantation. There it will be interesting to see how well neutralizing antibodies contain primary infection acquired naturally by the oral route versus primary infection acquired iatrogenically from latently-infected B cells in the graft.

As Moss et al. discuss (Chapter 30), there are reasonable grounds for hope that prophylactic vaccination might alleviate disease associated with acute infection, especially if vaccine constructs and delivery systems can be modified to prime for protective T cell as well as antibody responses. However, the effects of such vaccination on long-term virus carriage are more difficult to predict. In that context, the mouse gamma-2 herpesvirus MHV68 model is potentially instructive. Intranasal infection with MHV68 leads to a burst of acute virus replication in lung epithelium which is soon cleared, followed by a short proliferative burst of latently-infected B cells which also soon resolves; thereafter infected B cell numbers fall to a low steady state value which is then maintained for life (Woodland et al., 2001). Vaccines inducing T cell responses to lytic cycle antigens reduce both virus replication in the lung and the early B cell proliferation but do not affect the load of virus subsequently established in lymphoid tissues. Likewise a vaccine inducing T cell responses against a latent antigen expressed in the early B cell proliferative phase reduces that proliferation but not the long-term virus load. However, these approaches used heterologous vectors to prime the T cell system selectively against a subset of viral antigens, in the absence of other MHV68-specific responses (Woodland et al., 2001). Very interestingly, recent work has shown that vaccination with MHV68 recombinants defective either in latency establishment (and therefore non-persistent *in vivo*; Fowler and Efstathiou, 2004) or in reactivation from latency (Tibbetts et al., 2003) enable the host to completely clear a subsequent wild type virus challenge; this likely reflects the ability of such a vaccine to elicit a wide range of both humoral and cell-mediated immune responses. Whether a replication-competent, transformation-defective EBV strain could in principle achieve the same effect is an interesting debating point.

Given the scientific and logistic difficulties associated with prophylactic vaccination, a great deal of attention has focused on the possibility of direct immune intervention against existing EBV-positive disease. As described by Dotti and colleagues (Chapter 31), the particular circumstances of LPD arising in the context of haemopoietic stem cell transplantation proved ideal for treatment with EBV-specific T cell preparations. Such preparations, dominated by CD8+ cytotoxic T cells against the EBNA3 latent cycle proteins expressed in LPD lesions, could be produced *in vitro* by LCL stimulation of memory T cells in the blood of the stem cell donor. Following infusion, the cells expanded dramatically *in vivo* in the empty haematological space that exists post-stem cell transplant and proved effective both therapeutically against existing LPDs and also as a prophylactic measure against LPD arising in high risk patients (Rooney et al., 1995). Variants of this protocol, using either autologous or partially-matched effector T cell preparations, have since been adopted in order to treat LPDs arising in solid organ transplant patients (Khanna et al., 1999; Haque et al., 2002). In future, it is likely that these types of approaches will be targeted to specific clinical situations where high EBV DNA loads in blood identify transplant patients at particular risk of LPD and/or where alternative anti-CD20 monoclonal antibody treatment for the disease has failed.

The success of adoptive T cell transfer against post-transplant LPD lesions has inspired a number of laboratories to consider ways of targeting other EBV-positive tumours by immunological means. As Dotti et al. describe (Chapter 31), even though the malignant cells of both EBV-positive HL and NPC express just a subset of EBV latent cycle antigens as available targets, nevertheless there are some encouraging early signs to suggest that adoptive transfer with T cells of appropriate antigen specificity may be capable of homing to tumour sites and attacking tumour cells *in vivo* (Bollard et al., 2004; Straathof et al., 2004). In that regard, a number of challenges remain: to optimize *in vitro* protocols that specifically reactivate T cells against viral antigens expressed in the tumour, to develop pre-conditioning protocols giving the best expansion of infused cells *in vivo,* and to explore methods of genetically modifying T cells in ways that could overcome inhibitory mechanisms operative at the tumour site. While adoptive T cell transfer is enormously important for the lessons it reveals, if immunotherapy is to have any significant impact on disease worldwide then arguably we need to develop therapeutic vaccines that can be delivered at all treatment centres independent of specialist laboratory support. Such vaccines must be able both to activate the patient's own immune response against viral antigens expressed in the tumour cells and to direct those responses to the tumour. Both objectives are important since preliminary work to date suggests that patients with EBV-positive HL or NPC have some, albeit low level, CD8+ T cell memory to relevant target antigens but that these cells are not being recruited into the tumour-infiltrating lymphocyte population. Current efforts, as Moss and colleagues (Chapter 30) describe, have begun with epitope peptide-based vaccines but are now progressing to a variety chimaeric antigen and polyepitope constructs built into vector systems that are already approved for human use. These are the first steps on a road that will require the strongest possible collaborative links between research laboratories and centres of clinical excellence worldwide.

CONCLUSION

The 40 years since EBV's discovery in cells cultured from a little known African tumour have seen the agent move from relative obscurity to its current position as the most intensely studied and best characterised human tumour virus. As this volume illustrates, the virus has attracted an active research community that straddles many disciplines. These range from molecular virology and the biochemistry of cell transformation through the study of virus-host interactions and disease pathogenesis to the design and testing of novel tumour therapies. EBV continues to occupy that special place on the map where virology, immunology and oncology meet and where there are still swathes of open country to explore; go to it!

References

Acha-Orbea, H., Held, W., Waanders, G.A., Shakhov, A.N., Scarpellino, L., Lees, R.K., and MacDonald, H.R. (1993). Exogenous and endogenous mouse mammary tumour virus superantigens. Immunol. Rev. *131*, 5-25.

Anagnostopoulos, I., Hummel, M., Kreschel, C., and Stein, H. (1995). Morphology, immunophenotype and distribution of latently and/or productively Epstein-Barr virus-infected cells in acute infectious mononucleosis: Implications for the interindividual infection route of Epstein-Barr virus. Blood *85*, 744-750.

Babcock, G.J., Decker, L.L., Volk, M., and Thorley-Lawson, D.A. (1998). EBV persistence in memory B cells *in vivo*. Immunity *9*, 395-404.

Babcock, G.J., Hochberg, D., and Thorley-Lawson, D.A. (2000). The expression pattern of Epstein-Barr virus latent genes *in vivo* is dependent upon the differentiation stage of the infected B cell. Immunity *13*, 497-506.

Baer, R., Bankier, A.T., Biggin, M.D., Deininger, P.L., Farrell, P.J., Gibson, T.G., Hatfull, G., Hudson, G.S., Satchwell, S.C., Seguin, C., Tuffnel, P.S., and Barrell, B.G. (1984). DNA sequence and expression of the B95-8 Epstein-Barr virus genome. Nature *310*, 207-211.

Bar, R.S., Delor, C.J., Clausen, K.P., Hurtubise, P., Henle, W., and Hewetson, J.F. (1974). Fatal infectious mononucleosis in a family. N. Eng. J. Med. *290*, 363-367.

Binne, U.K., Amon, W., and Farrell, P.J. (2002). Promoter sequences required for reactivation of Epstein-Barr virus from latency. J. Virol. *76*, 10282-10289.

Bollard, C.M., Aguilar, L., Straathof, K.C., Gahn, B., Huls, M.H., Rousseau, A., Sixbey, J., Gresik, M.V., Carrum, G., Hudson, M., Dilloo, D., Gee, A., Brenner, M.K., and Rooney, C.M. (2004). Cytotoxic T Lymphocyte Therapy for Epstein-Barr Virus+ Hodgkin's Disease. J. Exp. Med. *200*, 1623-1633.

Borza, C.M. and Hutt-Fletcher, L.M. (2002). Alternate replication in B cells and epithelial cells switches tropism of Epstein-Barr virus. Nat. Med. *8*, 594-599.

Boshoff, C. and Weiss, R. (2002). AIDS-related malignancies. Nat. Rev. Can. *2*, 373-382.

Brauninger, A., Spieker, T., Mottok, A., Baur, A.S., Kuppers, R., and Hansmann, M.L. (2003). Epstein-Barr virus (EBV)-positive lymphoproliferations in post-transplant patients show immunoglobulin V gene mutation patterns suggesting interference of EBV with normal B cell differentiation processes. Eur. J. Immunol. *33*, 1593-1602.

Brauninger, A., Spieker, T., Willenbrock, K., Gaulard, P., Wacker, H.H., Rajewsky, K., Hansmann, M.L., and Kuppers, R. (2001). Survival and clonal expansion of mutating "forbidden" (immunoglobulin receptor-deficient) epstein-barr virus-infected b cells in angioimmunoblastic t cell lymphoma. J. Exp. Med. *194*, 927-940.

Burrows, J.M., Bromham, L., Woolfit, M., Piganeau, G., Tellam, J., Connolly, G., Webb, N., Poulsen, L., Cooper, L., Burrows, S.R., Moss, D.J., Haryana, S.M., Ng, M., Nicholls, J.M., and Khanna, R. (2004). Selection Pressure-Driven Evolution of the Epstein-Barr Virus-Encoded Oncogene LMP1 in Virus Isolates from Southeast Asia. J. Virol. *78*, 7131-7137.

Busson, P., Ganem, G., Flores, P., Mugneret, F., Clausse, B., Caillou, B., Braham, K., Wakasugi, H., Lipinski, M., and Tursz, T. (1988). Establishment and characterization of three transplantable EBV- containing nasopharyngeal carcinomas. Int. J. Can. *42*, 599-606.

Caldwell, R.G., Wilson, J.B., Anderson, S.J., and Longnecker, R. (1998). Epstein-Barr virus LMP2A drives B cell development and survival in the absence of normal B cell receptor signals. Immunity *9*, 405-411.

Callan, M.F.C., Tan, L., Annels, N., Ogg, G.S., Wilson, J.D.K., O'Callaghan, C.A., Steven, N., McMichael, A.J., and Rickinson, A.B. (1998). Direct visualization of antigen-specific CD8(+) T cells during the primary immune response to Epstein-Barr virus *in vivo*. J. Exp. Med. *187*, 1395-1402.

Capello, D., Cerri, M., Muti, G., Berra, E., Oreste, P., Deambrogi, C., Rossi, D., Dotti, G., Conconi, A., Vigano, M., Magrini, U., Ippoliti, G., Morra, E., Gloghini, A., Rambaldi, A., Paulli, M., Carbone, A., and Gaidano, G. (2003). Molecular histogenesis of posttransplantation lymphoproliferative disorders. Blood *102*, 3775-3785.

Carbone, A. (2003). Emerging pathways in the development of AIDS-related lyphomas. Lancet Oncology *4*, 22-29.

Casola, S., Otipoby, K.L., Alimzhanov, M., Humme, S., Uyttersprot, N., Kutok, J.L., Carroll, M.C., and Rajewsky, K. (2004). B cell receptor signal strength determines B cell fate. Nat. Immunol. *5*, 317-327.

Chan, A.S., To, K.F., Lo, K.W., Ding, M., Li, X., Johnson, P., and Huang, D.P. (2002). Frequent chromosome 9p losses in histologically normal nasopharyngeal epithelia from southern Chinese. Int. J. Can. *102*, 300-303.

Chan, A.S., To, K.F., Lo, K.W., Mak, K.F., Pak, W., Chiu, B., Tse, G.M., Ding, M., Li, X., Lee, J.C., and Huang, D.P. (2000). High Frequency of Chromosome 3p Deletion in Histologically Normal Nasopharyngeal Epithelia from Southern Chinese. Cancer Research *60*, 5365-5370.

Cheung, S.-T., Huang, D.P., Hui, A.B.Y., Lo, K.-W., Ko, C.W., Tsang, Y.S., Wong, N., Whitney, B.M., and Lee, J.C.K. (1999). Nasopharyngeal carcinoma cell line (C666-1) consistently harbouring Epstein-Barr virus. Int. J. Can. *83*, 121-126.

Cho, Y.G., Gordadze, A.V., Ling, P.D., and Wang, F. (1999). Evolution of two types of rhesus lymphocryptovirus similar to type 1 and type 2 Epstein-Barr virus. J. Virol. *73*, 9206-9212.

Cho, Y.G., Ramer, J., Rivailler, P., Quink, C., Garber, R.L., Beier, D., and Wang, F. (2001). An Epstein-Barr related herpesvirus from marmoset lymphomas. Proc. Natl. Acad. Sci. U. S. A *98*, 1224-1229.

Coffey, A.J., Brooksbank, R.A., Brandau, O., Oohashi, T., Howell, G.R., Bye, J.M., Cahn, A.P., Durhan, J., Heath, P., Wray, P., Pavitt, R., Wilkinson, J., Leversha, M., Huckle, E., Shaw-Smith, C.J., Dunham, A., Rhodes, S., Schuster, V., Porta, G., Yin, L., Serafini, P., Sylla, B., Zollo, M., Franco, B., Bolino, A., Seri, M., Lanyi, A., Davis, J.R., Webster, D., Harris, A., Lenoir, G., St Basile, G.D., Jones, A., Behloradsky, B.H., Achatz, H., Murken, J., Fassler, R., Sumegi, J., Romeo, G., Vaudin, M., Ross, M.T., Meindl, A., and Bentley, D.R. (1998). Host response to EBV infection in X-linked lymphoproliferative disease results from mutations in an SH2-domain encoding gene. Nat. Gen. *20*, 129-135.

Cohen, J.I., Wang, F., Mannick, J., and Kieff, E. (1989). Epstein-Barr virus nuclear protein 2 is a key determinant of lymphocyte transformation. Proc. Natl. Acad. Sci. U. S. A *86*, 9558-9562.

Curran, J.A., Laverty, F.S., Campbell, D., Macdiarmid, J., and Wilson, J.B. (2001). Epstein-Barr virus encoded latent membrane protein-1 induces epithelial cell proliferation and sensitizes transgenic mice to chemical carcinogenesis. Cancer Research *61*, 6730-6738.

de The, G., Geser, A., Day, N.E., Tukei, P.M., Williams, E.H., Beri, D.P., Smith, D.P., Dean, A.G., Bornkamm, G.W., Feorino, P., and Henle, W. (1978). Epidemiological evidence for causal relationship between Epstein- Barr virus and Burkitt's lymphoma and from Ugandan prospective study. Nature *274*, 756-761.

Delecluse, H.-J., Hilsendegen, T., Pich, D., Zeidler, R., and Hammerschmidt, W. (1998). Propagation and recovery of intact, infectious Epstein-Barr virus from prokaryotic to human cells. Proc. Natl. Acad. Sci. U. S. A *95*, 8245-8250.

Denis, Haumont, and Bollen. Vaccination against Epstein-Barr virus (EBV): report of phase II studies using recombinant viral glycoprotein gp350 in healthy adults. 2004. Regensburg, 11th Biennial Conference of the International Associating for Research on Epstein-Barr virus and associated diseases. 2004.
Ref Type: Conference Proceeding

Dukers, D.F., Meij, P., Vervoort, M.B., Vos, W., Scheper, R.J., Meijer, C.J., Bloemena, E., and Middeldorp, J.M. (2000). Direct immunosuppressive effects of EBV-encoded latent membrane protein 1. Journal of Immunology *165*, 663-670.

Engel, P., Eck, M.J., and Terhorst, C. (2003). The SAP and SLAM families in immune responses and x-linked lymphoproliferative disease. Nat. Rev. Immunol. *3*, 813-821.

Epstein, M.A., Barr, Y.M., and Achong, B.G. (1964). Virus particles in cultured lymphoblasts from Burkitt's lymphoma. Lancet *i*, 702-703.

Epstein, M.A., Hunt, R.D., and Rabin, H. (1973). Pilot experiments with EB virus in owl monkey (Aotus trivirgatus). I. Reticuloproliferative disease in an inoculated animal. Int. J. Can. *12*, 309-318.

Fafi-Kremer, S., Morand, P., Brion, J.P., Pavese, P., Baccard, M., Germi, R., Genoulaz, O., Nicod, S., Jolivet, M., Ruigrok, R.W.H., Stahl, J.P., and Seigneurin, J.M. (2005). Long-term shedding of infectious epstein-barr virus after infectious mononucleosis. J. Infect. Dis. *191*, 985-989.

Farrell, P.J., Rowe, D.T., Rooney, C.M., and Kouzarides, T. (1989). Epstein-Barr virus BZLF1 transactivator specifically binds to a consensus AP1 site and is related to c-fos. EMBO Journal *8*, 127-132.

Feichtinger, H., Li, S.-L., Kaaya, E., Putkonen, P., Grunewald, K., Weyrer, K., Boltiger, D., Ernberg, I., Linde, A., Biberfeld, G., and Biberfeld, P. (1992). A monkey model for Epstein-Barr virus-associated lymphomagenesis in human acquired immunodeficiency syndrome. J. Exp. Med. *176*, 281-286.

Fowler, P. and Efstathiou, S. (2004). Vaccine potential of a murine gammaherpesvirus-68 mutant deficient for ORF73. J. Gen. Virol. *85*, 609-613.

Frank, A., Andiman, W.A., and Miller, G. (1976). Epstein-Barr virus and nonhuman primates: natural and experimental infection. Adv. Can. Res. *23*, 171-201.

Fu, T., Voo, K.S., and Wang, R.F. (2004). Critical role of EBNA1-specific CD4+ T cells in the control of mouse Burkitt lymphoma *in vivo*. J. Clin. Inv. *114*, 542-550.

Gratama, J.W. and Ernberg, I. (1995). Molecular epidemiology of Epstein-Barr virus infection. Adv. Can. Res. *67*, 197-255.

Greenspan, J.S., Greenspan, D., Lennette, E.T., Abrams, D.I., Conant, M.A., Petersen, V., and Freese, U.K. (1985). Replication of Epstein-Barr virus within the epithelial cells of oral "hairy" leukoplakia, an AIDS-associated lesion. N. Eng. J. Med. *313*, 1564-1571.

Grossman, S.R., Johannsen, E., Tong, X., Yalamanchili, R., and Kieff, E. (1994). The Epstein-Barr virus nuclear antigen 2 transactivator is directed to response elements by the JK recombination signal binding protein. Proc. Natl. Acad. Sci. U. S. A *91*, 7568-7572.

Hammerschmidt, W. and Sugden, B. (1989). Genetic analysis of immortalising functions of Epstein-Barr virus in human B lymphocytes. Nature *340*, 393-397.

Haque, T., Wilkie, G.M., Taylor, C., Amlot, P.L., Murad, P., Iley, A., Dombagoda, D., Britton, K.M., Swerdlow, A.J., and Crawford, D.H. (2002). Treatment of Epstein-Barr-virus-positive post-transplantation lymphoproliferative disease with partly HLA-matched allogeneic cytotoxic T cells. Lancet *360*, 436-442.

Henkel, T., Ling, P.D., Hayward, S.D., and Petersen, M.G. (1994). Mediation of Epstein-Barr virus EBNA2 transactivation by recombination signal-binding protein JK. Science *265*, 92-95.

Henle, G., Henle, W., Clifford, P., Diehl, V., Kafuko, G., Kirya, B.G., Klein, G., Morrow, R.H., Manube, G.M.R., Pike, M.C., Tukei, P.M., and Ziegler, J.L. (1969). Antibodies to EB virus in Burkitt's lymphoma and control groups. J. Nat. Can. Inst. *43*, 1147-1157.

Henle, W., Diehl, V., Kohn, G., zur Hausen, H., and Henle, G. (1967). Herpes-type virus and chromosome marker in normal leukocytes after growth with irradiated Burkitt cells. Science *157*, 1064-1065.

Herbst, H., Dallenbach, F., Hummel, M., Niedobitek, G., Pileri, S., Muller-Lantzsch, N., and Stein, H. (1991). Epstein-Barr virus latent membrane protein expression in Hodgkin and Reed-Sternberg cells. Proc. Natl. Acad. Sci. U. S. A *88*, 4766-4770.

Heussinger, N., Buttner, M., Ott, G., Brachtel, E., Pilch, B.Z., Kremmer, E., and Niedobitek, G. (2004). Expression of the Epstein-Barr virus (EBV)-encoded latent membrane protein 2A (LMP2A) in EBV-associated nasopharyngeal carcinoma. Journal of Pathology *203*, 696-699.

Hinuma, Y., Konn, M., Yamaguchi, J., Wudarski, D.J., Blakeslee, J.R., Jr., and Grace, J.T., Jr. (1967). Immunofluorescence and herpes-type virus particles in the P3HR-1 Burkitt lymphoma cell line. J. Virol. *1*, 1045-1051.

Hislop, A.D., Annels, N.E., Gudgeon, N.H., Leese, A.M., and Rickinson, A.B. (2002). Epitope-specific evolution of human CD8(+) T cell responses from Primary to persistent phases of Epstein-Barr virus infection. J. Exp. Med. *195*, 893-905.

Hislop, A.D., Kuo, M., Drake-Lee, A., Bergler, W., Hammerschmitt, N., Akbar, A.N., Khan, N., Leese, A.M., Timms, J.M., Bell, A.I., and Rickinson, A.B. (2005). Tonsillar accumulation of Epstein-Barr virus-specific CD8+ T cells in persistent but not acute primary infection. J. Clin. Invest. In Press.

Hjalgrim, H., Askling, J., Rostgaard, K., Hamilton-Dutoit, S., Frisch, M., Zhang, J.S., Madsen, M., Rosdahl, N., Konradsen, H.B., Storm, H.H., and Melbye, M. (2003). Characteristics of Hodgkin's lymphoma after infectious mononucleosis. N. Eng. J. Med. *349*, 1324-1332.

Hjalgrim, H., Askling, J., Sorensen, P., Madsen, M., Rosdahl, N., Storm, H.H., Hamilton-Dutoit, S., Eriksen, L.S., Frisch, M., Ekbom, A., and Melbye, M. (2000). Risk of Hodgkin's disease and other cancers after infectious mononucleosis. J. Nat. Can. Inst. *92*, 1522-1528.

Hochberg, D., Souza, T., Catalina, M., Sullivan, J.L., Luzuriaga, K., and Thorley-Lawson, D.A. (2004). Acute Infection with Epstein-Barr Virus Targets and Overwhelms the Peripheral Memory B-Cell Compartment with Resting, Latently Infected Cells. J. Virol. *78*, 5194-5204.

Humme, S., Reisbach, G., Feederle, R., Delecluse, H.-J., Bousset, K., Hammerschmidt, W., and Schepers, A. (2003). The EBV nuclear antigen 1 (EBNA1) enhances B cell immortalization several thousandfold. Proc. Natl. Acad. Sci. U. S. A *100*, 10989-10994.

Imai, S., Nishikawa, J., and Takada, K. (1998). Cell-to-cell contact as an efficient mode of Epstein-Barr virus infection of diverse human epithelial cells. J. Virol. *72*, 4371-4378.

Iwasaki, Y., Chong, J.M., Hayashi, Y., Ikeno, R., Arai, K., Kitamura, M., Koike, M., Hirai, K., and Fukayama, M. (1998). Establishment and Characterization of a Human Epstein-Barr Virus-Associated Gastric Carcinoma in SCID Mice. J. Virol. *72*, 8321-8326.

Jaffe, E.S., Krenacs, L., and Raffeld, M. (2003). Classification of cytotoxic T-cell and natural killer cell lymphomas. Sem. Haematol. *40*, 175-184.

Kang, M.S., Hung, S.C., and Kieff, E. (2001). Epstein-Barr virus nuclear antigen 1 activates transcription from episomal but not integrated DNA and does not alter lymphocyte growth. Proc. Natl. Acad. Sci. U. S. A *98*, 15233-15238.

Katano, H., Ali, M.A., Patera, A.C., Catalfamo, M., Jaffe, E.S., Kimura, H., Dale, J.K., Straus, S.E., and Cohen, J.I. (2004). Chronic active Epstein-Barr virus infection associated with mutations in perforin that impair its maturation. Blood *103*, 1244-1252.

Keating, S., Prince, S., Jones, M., and Rowe, M. (2002). The lytic cycle of Epstein-Barr virus is associated with decreased expression of cell surface major histocompatibility complex class I and class II molecules. J. Virol. *76*, 8179-8188.

Kelly, G., Bell, A., and Rickinson, A. (2002). Epstein-Barr virus-associated Burkitt lymphomagenesis selects for downregulation of the nuclear antigen EBNA2. Nat. Med. *8*, 1098-1104.

Kelly, G.L., Milner, A.E., Tierney, R.J., Croom-Carter, D.S.G., Altmann, M., Hammerschmidt, W., Bell, A.I., and Rickinson, A.B. (2005). Epstein-Barr virus nuclear antigen (EBNA) 2 gene deletion is consistently linked with EBNA3A, 3B, 3C expression in Burkitt's Lymphoma (BL) cells and with increased resistance to apoptosis. J. Virol. In press.

Kennedy, G., Komano, J., and Sugden, B. (2003). Epstein-Barr virus provides a survival factor to Burkitt's lymphomas. Proc. Natl. Acad. Sci. U. S. A *100*, 14269-14274.

Khanna, R., Bell, S., Sherritt, M., Galbraith, A., Burrows, S.R., Rafter, L., Clarke, B., Slaughter, R., Falk, M.C., Douglass, J., Williams, T., Elliott, S.L., and Moss, D.J. (1999). Activation and adoptive transfer of Epstein-Barr virus-specific cytotoxic T cells in solid organ transplant patients with post-transplant lymphoproliferative disease. Proc. Natl. Acad. Sci. U. S. A *96*, 10391-10396.

Khanna, R., Burrows, S.R., Argaet, V., and Moss, D.J. (1994). Endoplasmic reticulum signal sequence facilitated transport of peptide epitopes restores immunogenicity of an antigen processing defective tumour cell line. International Immunology *6*, 639-645.

Kikuta, H., Taguchi, Y., Tomizawa, K., Kojima, K., Kawamura, N., Ishizaka, A., Sakiyama, Y., Matsumoto, S., Imai, S., Kinoshita, T., Koizumi, S., Osato, T., Kabayashi, I., Hamada, I., and Hirai, K. (1988). Epstein-Barr virus genome-positive T lymphocytes in a boy with chronic active EBV infection associated with Kawaski-like disease. Nature *333*, 455-457.

Knowles, D.M. and Pirog, E.C. (2001). Pathology of AIDS-related lymphomas and other AIDS-defining neoplasms. European Journal of Cancer *37*, 1236-1250.

Kulwichit, W., Edwards, R.H., Davenport, E.M., Baskar, J.F., Godfrey, V., and Raab-Traub, N. (1998). Expression of the Epstein-Barr virus latent membrane protein 1 induces B cell lymphoma in transgenic mice. Proc. Natl. Acad. Sci. U. S. A *95*, 11963-11968.

Kuppers, R. (2003). B cells under influence: transformation of B cells by Epstein-Barr virus. Nat. Rev. Immunol. *3*, 801-812.

Kuppers, R. and Rajewsky, K. (1998). The origin of Hodgkin and Reed/Sternberg cells in Hodgkin's disease. Annual Review of Immunology *16*, 471-493.

Kurth, J., Hansmann, M.L., Rajewsky, K., and Kuppers, R. (2003). Epstein-Barr virus-infected B cells expanding in germinal centers of infectious mononucleosis patients do not participate in the germinal center reaction. Proc. Natl. Acad. Sci. U. S. A *100*, 4730-4735.

Kurth, J., Spieker, T., Wustrow, J., Strickler, J.G., Hansmann, M.L., Rajewsky, K., and Kuppers, R. (2000). EBV-infected B cells in infectious mononucleosis: Viral strategies for spreading in the B cell compartment and establishing latency. Immunity *13*, 485-495.

Lay, J.-D., Tsao, C.J., Chen, J.Y., Kadin, M.E., and Su, I.-J. (1997). Upregulation of tumour necrosis factor-alpha gene by Epstein-Barr virus and activation of macrophages in Epstein-Barr virus-infected T cells in the pathogenesis of hemophagocytic syndrome. J. Clin. Inv. *100*, 1969-1979.

Lee, E.S., Locker, J., Nalesnik, M., Reyes, J., Jaffe, R., Alashari, M., Nour, B., Tzakis, A., and Dickman, P.S. (1995). The association of Epstein-Barr virus with smooth muscle tumours occurring after organ transplantation. N. Eng. J. Med. *332*, 19-25.

Levine, A.M. (1998). Hodgkin's Disease in the Setting of Human Immunodeficiency Virus Infection. J. Nat. Can. Inst. *23*, 37-42.

Lo, K.W., To, K.F., and Huang, D.P. (2004). Focus on nasopharyngeal carcinoma. Cancer Cell *5*, 423-428.

Long, H.M., Haigh, T.A., Gudgeon, N.H., Leen, A.M., Tsang, C.W., Brooks, E., Landais, E., Houssaint, E., Lee, S.P., Rickinson, A.B., and Taylor, G.S. (2005). CD4+ T cell responses to Epstein-Barr virus (EBV) latent cycle antigens and the recognition of EBV-transformed lymphoblastoid cell lines. J. Virol. *79*, 4896-4907.

Longan, L. and Longnecker, R. (2000). Epstein-Barr virus latent membrane protein 2A has no growth-altering effects when expressed in differentiating epithelia. J. Gen. Virol. *81*, 2245-2252.

Marques, S., Efstathiou, S., Smith, K.G., Haury, M., and Simas, J.P. (2003). Selective Gene Expression of Latent Murine Gammaherpesvirus 68 in B Lymphocytes. J. Virol. *77*, 7308-7318.

Marshall, N.A., Vickers, M.A., and Barker, R.N. (2003). Regulatory T cells secreting IL-10 dominate the immune response to EBV latent membrane protein 1. Journal of Immunology *170*, 6183-6189.

McClain, K.L., Leach, C.T., Jenson, H.B., Joshi, V.V., Pollock, B.H., Parmley, R.T., DiCarlo, F.J., Chadwick, E.G., and Murphy, S.B. (1995). Association of Epstein-Barr virus with leiomyosarcomas in young people with AIDS. N. Eng. J. Med. *332*, 12-18.

Midgley, R.S., Bell, A.I., McGeoch, D.J., and Rickinson, A.B. (2003). Latent gene sequencing reveals familial relationships among Chinese Epstein-Barr virus strains and evidence for positive selection of A11 epitope change. J. Virol. *77*, 11517-11530.

Miller, C.L., Lee, J.H., Kieff, E., and Longnecker, R. (1994). An integral membrane protein (LMP2) blocks reactivation of Epstein-Barr virus from latency following surface immunoglobulin cross-linking. Proc. Natl. Acad. Sci. U. S. A *91*, 772-776.

Miller, G., Robinson, J., Heston, L., and Lipman, M. (1974). Differences between laboratory strains of Epstein-Barr virus based on immortalization, abortive infection and interference. Proc. Natl. Acad. Sci. U. S. A *71*, 4006-4010.

Mosialos, G., Birkenbach, M., Yalamanchili, R., Van Arsdale, T., Ware, C., and Kieff, E. (1995). The Epstein-Barr virus transforming protein LMP1 engages signalling proteins from the tumor necrosis factor receptor family. Cell *80*, 389-399.

Munz, C., Bickham, K.L., Subklewe, M., Tsang, M.L., Chahroudi, A., Kurilla, M.G., Zhang, D., O'Donnell, M., and Steinman, R.M. (2000). Human CD4(+) T lymphocytes consistently respond to the latent Epstein-Barr virus nuclear antigen EBNA1. J. Exp. Med. *191*, 1649-1660.

Nalesnik, M.A. (1998). Clinical and pathological features of post-transplant lymphoproliferative disorders (PTLD). Springer Seminars in Immunopathology *20*, 325-342.

Nichols, K.E., Harkin, D.P., Levitz, S., Krainer, M., Kolquist, K.A., Genovese, C., Bernard, A., Ferguson, M., Zuo, L., Snyder, E., Buckler, A.J., Wise, C., Ashley, J., Lovett, M.,

Valentine, M.B., Look, A.T., Gerald, W., Housman, D.E., and Haber, D.A. (1998). Inactivating mutations in an SH2 domain-encoding gene in X-linked lymphoproliferative syndrome. Proc. Natl. Acad. Sci. U. S. A *95*, 13765-13770.

Nishikawa, J., Imai, S., Oda, T., Kojima, T., Okita, K., and Takada, K. (1999). Epstein-Barr virus promotes epithelial cell growth in the absence of EBNA2 and LMP1 expression. J. Virol. *73*, 1286-1292.

O'Nions, J. and Allday, M.J. (2004). Deregulation of the cell cycle by the Epstein-Barr virus. Adv. Can. Res. *92*, 119-186.

Oh, S.T., Seo, J.S., Moon, U.Y., Kang, K.H., Shin, D.-J., Yoon, S.K., Kim, W.H., Park, J.-G., and Lee, S.K. (2004). A naturally derived gastric cancer cell line shows latency I Epstein-Barr virus infection closely resembling EBV-associated gastric cancer. Virology *320*, 330-336.

Old, L.J., Clifford, P., Boyse, E.A., De Harven, E., Geering, G., Oettgen, H.F., and Williamson, B. (1966). Precipitating antibody in human serum to an antigen present in cultured Burkitt's lymphoma cells. Proc. Natl. Acad. Sci. U. S. A *56*, 1699-1704.

Omiya, R., Buteau, C., Kobayashi, H., Paya, C.V., and Celis, E. (2002). Inhibition of EBV-induced lymphoproliferation by CD4+ T cells specific for an MHC class II promiscuous epitope. J. Immunol. *169*, 2172-2179.

Pajic, A., Staege, M.S., Dudziak, D., Schuhmacher, M., Spitkovsky, D., Eissner, G., Brielmeier, M., Polack, A., and Bornkamm, G.W. (2001). Antagonistic effects of c-myc and Epstein-Barr virus latent genes on the phenotype of human B cells. Int. J. Can. *93*, 810-816.

Pallesen, G., Hamilton-Dutoit, S.J., Rowe, M., and Young, L.S. (1991). Expression of Epstein-Barr virus latent gene products in tumour cells of Hodgkin's disease. Lancet *337*, 320-322.

Paludan, C., Bickham, K., Nikiforow, S., Tsang, M.L., Goodman, K., Hanekom, W.A., Fonteneau, J.F., Stevanovic, S., and Munz, C. (2002). EBNA1 specific CD4+ Th1 cells kill Burkitt's lymphoma cells. J. Immunol. *169*, 1593-1603.

Parolini, S., Bottino, C., Falco, M., Augugliaro, R., Giliani, S., Franceschini, R., Ochs, H.D., Wolf, H., Bonnefoy, J.Y., Biassoni, R., Moretta, L., Notarangelo, L.D., and Moretta, A. (2000). X-linked lymphoproliferative disease: 2B4 molecules displaying inhibitory rather than activating function are responsible for the inability of natural killer cells to kill Epstein-Barr virus-infected cells. J. Exp. Med. *192*, 337-346.

Pathmanthan, R., Prasad, U., Sadler, R., Flynn, K., and Raab-Traub, N. (1995). Clonal proliferations of cells infected with Epstein-Barr virus in pre-invasive lesions related to nasopharyngeal carcinoma. N. Eng. J. Med. *333*, 693-698.

Penn, I. (1986). The occurrence of malignant tumours in immunosuppressed states. Prog. Allergy *37*, 259-300.

Polack, A., Hortnagal, K., Pajic, A., Christoph, B., Baier, B., Falk, M., Mautner, J., Geltinger, C., Bornkamm, G.W., and Kempkes, B. (1996). c-myc activation renders proliferation of Epstein-Barr virus (EBV)-transformed cells independent of EBV nuclear antigen 2 and latent membrane protein 1. Proc. Natl. Acad. Sci. U. S. A *93*, 10411-10416.

Pope, J.H., Horne, M.K., and Scott, W. (1968). Transformation of foetal human leukocytes *in vitro* by filtrates of a human leukaemic cell line containing herpes-like virus. Int. J. Can. *3*, 857-866.

Poppema, S., Potters, M., Emmens, R., Visser, L., and van den Berg, A. (1999). Immune reactions in classical Hodgkin's lymphoma. Sem. Haematol. *36*, 253-259.

Portis, T., Dyck, P., and Longnecker, R. (2003). Epstein-Barr Virus (EBV) LMP2A induces alterations in gene transcription similar to those observed in Reed-Sternberg cells of Hodgkin lymphoma. Blood *102*, 4166-4178.

Pudney, V.A., Leese, A.M., Rickinson, A.B., and Hislop, A.D. (2005). CD8+ immunodominance among Epstein-Barr virus lytic cycle antigens directly reflects the efficiency of antigen presentation in lytically-infected cells. J. Exp. Med. *201*, 349-360.

Purtilo, D.T., Cassel, C., Yang, J.P.S., Stephenson, S.R., Harper, R., Landing, G.H., and Vawter, G.F. (1975). X-linked recessive progressive combined variable immunodeficiency (Duncan's Disease). Lancet *i*, 935-941.

Ranganathan, S., Webber, S., Ahuja, S., and Jaffe, R. (2004). Hodgkin-like posttransplant lymphoproliferative disorder in children: does it differ from posttransplant Hodgkin lymphoma? Pediatr. Dev. Pathol. *7*, 348-360.

Reedman, B.M. and Klein, G. (1973). Cellular localisation of an Epstein-Barr virus (EBV_-associated complement-fixing antigen in producer and non-producer lymphoblastoid cell lines. Int. J. Can. *11*, 499-520.

Rickinson, A.B. and Kieff, E. (2001). Epstein-Barr virus. In Fields Virology, D.M.Knipe and P.M.Howley, eds. (Philadelphia: Lippincott Williams and Wilkins), pp. 2575-2627.

Rickinson, A.B. and Moss, D.J. (1997). Human cytotoxic T lymphocyte responses to Epstein-Barr virus infection. Annual Review of Immunology *15*, 405-431.

Rivailler, P., Cho, Y.G., and Wang, F. (2002). Complete genomic sequence of an Epstein-Barr virus-related herpesvirus naturally infecting a new world primate: a defining point in the evolution of oncogenic lymphocryptoviruses. J. Virol. *76*, 12055-12068.

Rooney, C.M., Smith, C.A., Ng, C.Y.C., Loftin, S., Li, C., Krance, R.A., Brenner, M.K., and Heslop, H.E. (1995). Use of gene-modified virus-specific T lymphocytes to control Epstein-Barr virus-related lymphoproliferation. Lancet *345*, 9-13.

Rowe, M., Rowe, D.T., Gregory, C.D., Young, L.S., Farrell, P.J., Rupani, H., and Rickinson, A.B. (1987). Differences in B cell growth phenotype reflect novel patterns of Epstein-Barr virus latent gene expression in Burkitt's lymphoma cells. EMBO Journal *6*, 2743-2751.

Sadler, R.H. and Raab-Traub, N. (1995). The Epstein-Barr virus 3.5-kilobase latent membrane protein 1 mRNA initiates from a TATA-Less promoter within the first terminal repeat. J. Virol. *69*, 4577-4581.

Sample, J., Brooks, L., Sample, C., Young, L., Rowe, M., Gregory, C., Rickinson, A., and Kieff, E. (1991). Restricted Epstein-Barr virus protein expression in Burkitt lymphoma is due to a different Epstein-Barr nuclear antigen 1 transcriptional initiation site. Proc. Natl. Acad. Sci. U. S. A *88*, 6343-6347.

Sayos, J., Wu, C., Morra, M., Wang, N., Zhang, X., Allen, D., van Schaik, S., Notarangelo, L.D., Geha, R., Roncarolo, M.G., Oettgen, H.F., De Vries, J.E., Aversa, G., and Terhorst, C. (1998). The X-linked lymphoproliferative-disease gene product SAP regulates signals induced through the co-receptor SLAM. Nature *395*, 462-469.

Schaefer, B.C., Woisetschlaeger, M., Strominger, J.L., and Speck, S.H. (1991). Exclusive expression of Epstein-Barr virus nuclear antigen 1 in Burkitt lymphoma arises from a third promoter, distinct from the promoters used in latently infected lymphocytes. Proc. Natl. Acad. Sci. U. S. A *88*, 6550-6554.

Schauer, E., Webber, S., Green, M., and Rowe, D. (2004). Surface immunoglobulin-deficient Epstein-Barr virus-infected B cells in the peripheral blood of pediatric solid-organ transplant recipients. J. Clin. Microbiol. *42*, 5802-5810.

Shannon-Lowe, C., Baldwin, G., Feederle, R., Bell, A., Rickinson, A.B., and Delecluse, H.-J. (2005).Epstein Barr Virus-induced B cell transformation: Quantitating events from binding to cell outgrowth. Submitted for publication.

Shimizu, N., Tanabe-Tochikuru, A., Kuroiwa, Y., and Takada, K. (1994). Isolation of Epstein-Barr virus (EBV)-negative cell clones from the EBV-positive Burkitt's lymphoma (BL) line Akata: Malignant phenotypes of BL cells are dependent on EBV. J. Virol. *68*, 6069-6073.

Shope, T.C., Dechairo, D., and Miller, G. (1973). Malignant lymphoma in cotton-top marmosets after inoculation with Epstein-Barr virus. Proc. Natl. Acad. Sci. U. S. A *70*, 2487-2491.

Silins, S.L., Sherritt, M.A., Silleri, J.M., Cross, S.M., Elliott, S.L., Bharadwaj, M., Le, T.T., Morrison, L.E., Khanna, R., Moss, D.J., Suhrbier, A., and Misko, I.S. (2001). Asymptomatic primary Epstein-Barr virus infection occurs in the absence of blood T-cell repertoire perturbations despite high levels of systemic viral load. Blood *98*, 3739-3744.

Sitki-Green, D., Covington, M., and Raab-Traub, N. (2003). Compartmentalization and transmission of multiple epstein-barr virus strains in asymptomatic carriers. J. Virol. *77*, 1840-1847.

Sitki-Green, D., Covington, M.M., and Raab-Traub, N. (2004). Biology of Epstein-Barr virus during infectious mononucleosis. J. Inf. Dis. *189*, 483-492.

Staege, M.S., Lee, S.P., Frisan, T., Mautner, J., Scholz, S., Pajic, A., Rickinson, A.B., Masucci, M.G., Polack, A., and Bornkamm, G.W. (2002). MYC overexpression imposes a nonimmunogenic phenotype on Epstein-Barr virus-infected B cells. Proc. Natl. Acad. Sci. U. S. A *99*, 4550-4555.

Straathof, K.C., Bollard, C.M., Popat, U., Huls, M.H., Lopez, T., Morriss, M.C., Gresik, M.V., Gee, A.P., Russell, H.V., Brenner, M.K., Rooney, C.M., and Heslop, H.E. (2004). Treatment of Nasopharyngeal Carcinoma with Epstein-Barr Virus-specific T Lymphocytes. Blood *105*, 1898-1904.

Su, I.-J. and Chen, J.Y. (1997). The role of Epstein-Barr virus in lymphoid malignancies. Critical Reviews in Oncology/Haematology *26*, 25-41.

Sullivan, J.L. and Woda, B.A. (1989). X-linked lymphoproliferative syndrome. Immunodeficiency Review *1*, 325-347.

Sutkowski, N., Conrad, B., Thorley-Lawson, D., and Huber, B.T. (2001). Epstein-Barr virus transactivates the human endogenous retrovirus HERV-K18 that encodes a superantigen. Immunity *15*, 579-589.

Sutkowski, N., Palkama, T., Ciurli, C., Sekaly, R.-P., Thorley-Lawson, D.A., and Huber, B.T. (1996). An Epstein-Barr virus-associated superantigen. J. Exp. Med. *184*, 971-980.

Takada, K. and Ono, Y. (1989). Synchronous and sequential activation of latently infected Epstein-Barr virus genomes. J. Virol. *63*, 445-449.

Thirlwell, C., Sarker, D., Stebbing, J., and Bower, M. (2003). Acquired immunodeficiency syndrome-related lymphoma in the era of highly active antiretroviral therapy. Clinical Lymphoma *4*, 86-92.

Tibbetts, S.A., McClellan, J.S., Gangappa, S., Speck, S.H., and Virgin, H.W., IV (2003). Effective Vaccination against Long-Term Gammaherpesvirus Latency. J. Virol. *77*, 2522-2529.

Tierney, R.J., Edwards, R.H., Sitki-Green, D., Croom-Carter, D., Yao, Q.-Y., Raab-Traub, N., and Rickinson, A.B. (2005). Multiple Epstein-Barr Virus strains in infectious mononucleosis patients: comparing ex vivo samples with *in vitro* isolates using heteroduplex tracking assays. Submitted for publication.

Timmons, C.F., Dawson, D.B., Richards, C.S., Andrews, W.S., and Katz, J.A. (1995). Epstein-Barr virus-associated leiomyosarcomas in liver transplantation recipients. Origin from either donor or recipient tissue. Cancer *76*, 1481-1489.

Timms, J.M., Bell, A., Flavell, J.R., Murray, P.G., Rickinson, A.B., Traverse-Glehen, A., Berger, F., and Delecluse, H.J. (2003). Target cells of Epstein-Barr-virus (EBV)-positive post-transplant lymphoproliferative disease: similarities to EBV-positive Hodgkin's lymphoma. Lancet *361*, 217-222.

Uchida, J., Yasui, T., Takaoka-Schichijo, Y., Muraoka, M., Kulwichit, W., Raab-Traub, N., and Kikutani, H. (1999). Mimicry of CD40 signals by Epstein-Barr virus LMP1 in B lymphocyte responses. Science *286*, 300-303.

Vaccher, E., Spina, M., and Tirelli, U. (2001). Clinical aspects and management of Hodgkin's disease and other tumours in HIV-infected individuals. European Journal of Cancer 1306-1315.

Vossen, M.T.M., Westerhout, E.M., Soderberg-Naucler, C., and Wiertz, E.JH.J. (2002). Viral immune evasion: a masterpiece of evolution. Immunogenetics *54*, 527-542.

Waltzer, L., Logeat, F., Brou, C., Israel, A., Sargeant, A., and Manet, E. (1994). The human Jκ recombination signal sequence binding protein (RBP-Jκ) targets the Epstein-Barr virus EBNA2 protein to its DNA responsive elements. EMBO Journal *13*, 5633-5638.

Wang, F., Tsang, S.-F., Kurilla, M.G., Cohen, J.J., and Kieff, E. (1990). Epstein-Barr virus nuclear antigen 2 transactivates latent membrane protein LMP1. J. Virol. *64*, 3407-3416.

Woisetschlaeger, M., Jin, X.W., Yandava, C.N., Furmanski, L.A., Strominger, J.L., and Speck, S.H. (1991). Role for the Epstein-Barr virus nuclear antigen 2 in viral promoter switching during initial stages of infection. Proc. Natl. Acad. Sci. U. S. A *88*, 3942-3946.

Woisetschlaeger, M., Yandava, C.N., Furmanski, L.A., Strominger, J.L., and Speck, S.H. (1990). Promoter switching in Epstein-Barr virus during the initial stages of infection of B lymphocytes. Proc. Natl. Acad. Sci. U. S. A *87*, 1725-1729.

Woodland, D.L., Usherwood, E.J., Liu, L., Flano, E., Kim, I.J., and Blackman, M.A. (2001). Vaccination against murine gamma-herpesvirus infection. Viral Immunology *14*, 217-226.

Wu, C., Nguyen, K.B., Pien, G.C., Wang, N., Gullo, C., Howie, D., Sosa, M.R., Edwards, M.J., Borrow, P., Satoskar, A.R., Sharpe, A.H., Biron, C.A., and Terhorst, C. (2001). SAP controls T cell responses to virus and terminal differentiation of TH2 cells. Nat. Immunol. *2*, 410-414.

Yates, J., Warren, N., Reisman, D., and Sugden, B. (1984). A cis-acting element from the Epstein-Barr viral genome that permits stable replication of recombinant plasmids in latently- infected cells. Proc. Natl. Acad. Sci. U. S. A *81*, 3806-3810.

Zimber-Strobl, U., Strobl, L.J., Meitinger, C., Hinrichs, R., Sakai, T., Furukawa, T., Hong, T., and Bornkamm, G.W. (1994). Epstein-Barr virus nuclear antigen 2 exerts its transactivating function through interaction with recombination signal binding protein RBP-Jk, the homologue of Drosophila suppressor of hairless. EMBO Journal *13*, 4973-4982.

Zimber-Strobl, U., Suentzenich, K.-O., Laux, G., Eick, E., Cordier, M., Calender, A., Billaud, M., Lenoir, G.M., and Bornkamm, G.W. (1991). Epstein-Barr virus nuclear antigen 2 activates transcription of the terminal protein gene. J. Virol. *65*, 415-423.

zur Hausen, A., Brink, A.A., Craanen, M.E., Middeldorp, J.M., Meijer, C.J., and van den Brule, A.J. (2000). Unique transcription pattern of Epstein-Barr virus (EBV) in EBV-carrying gastric adenocarcinomas: expression of the transforming BARF1 gene. Cancer Research *60*, 2745-2748.

zur Hausen, A., van Rees, B.P., van Beek, J., Craanen, M.E., Bloemena, E., Offerhaus, G.J.A., Meijer, C.J.L.M., and van den Brule, A.J.C. (2004). Epstein-Barr virus in gastric carcinomas and gastric stump carcinomas: a late event in gastric carcinogenesis. Journal of Clinical Pathology *57*, 487-491.

Z

Colour Plates

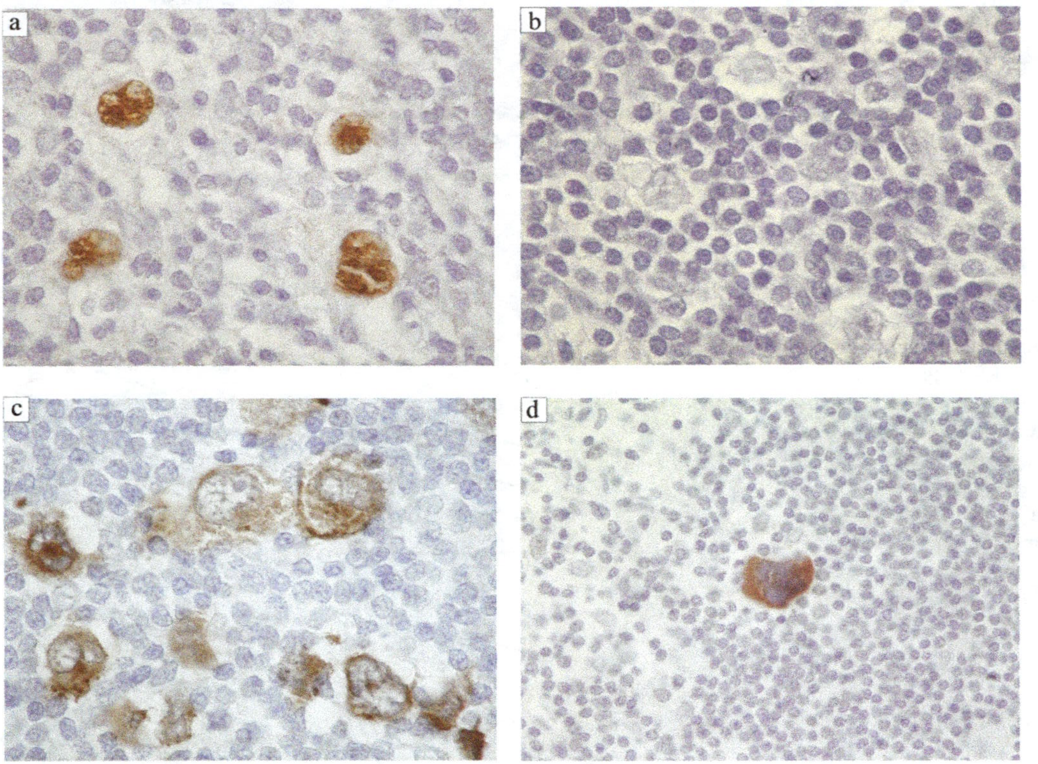

Figure 1, Chapter 8. Gene expression in EBV-associated Hodgkin's lymphoma. a) Demonstration of EBER expression (brown staining) in the nuclei of Hodgkin's/Reed-Sternberg (HRS) cells from an EBV-associated Hodgkin's lymphoma biopsy. The use of *in situ* hybridisation to detect EBER expression is the most reliable and sensitive method to detect the presence of latent EBV infection in clinical samples. b) EBV-positive HRS cells lack expression of the transforming EBNA2 protein, whereas c) the latent membrane protein-1 (LMP1) and d) LMP2 are both highly expressed. Not shown is the consistent expression of the EBV maintenance protein, EBNA1.

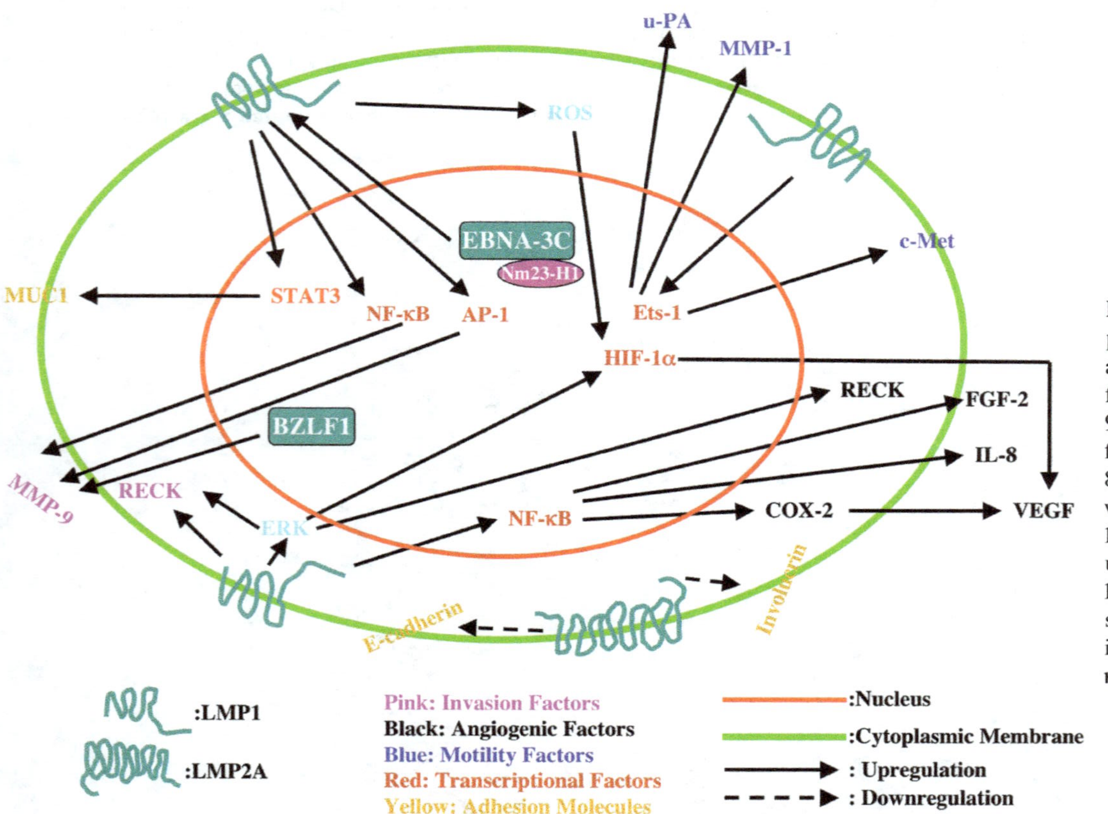

Figure 2, Chapter 12. EBV Latent membrane protein 1 (LMP-1) induces invasiveness, angiogenic, cell motility, and transcriptional factors. MMP-9, matrix metalloproteinase-9; COX-2, cyclooxygenase-2; FGF-2, fibroblast growth factor-2; IL-8, interleukin-8; ROS, reactive oxygen species; VEGF, vascular endothelial growth factor; MMP-1, matrix metalloproteinase-1; u-PA, urokinase-type plasminogen activator; AP-1, activator protein-1; ERK, extracellular signal-regulated kinase; HIF-1α, hypoxia-inducible factor-1α; NF-κB, nuclear factor-κB; MUC-1, mucin 1.

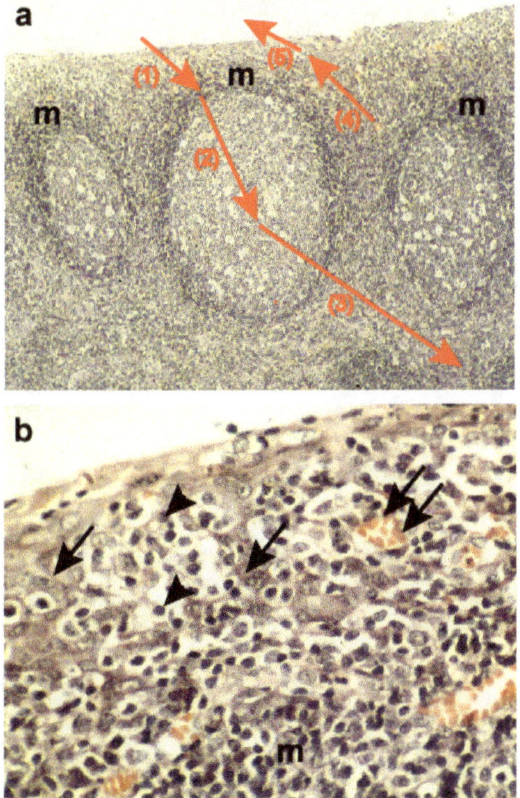

Figure 1, Chapter 17. The lymphoepithelium of the palatine tonsils from Waldeyer's ring. The tonsil consists of a highly involuted surface creating a large surface area with deep invaginations called crypts. The surface of the crypt lumen is at the top of both micrographs. a. Numerous follicles containing B cells are arranged parallel to the crypt surface. The mantle zone (m), containing naive B cells, is always facing the crypt and is continuous with the reticulated epithelium. b. A higher magnification reveals the sponge like structure of the epithelial cells in the lymphopepithelium (black arrows) that create spaces extending all the way to the mantle zone and which are filled with infiltrating lymphocytes (arrow heads) such that there is frequently only a single epithelial cell between the outer surface and the lymphocytes. The infiltrating lymphocytes include memory B cells, plasma cells and T cells. (The micrograpahs were provided by Dr. Marta Perry). The colored arrows indicate the pathway of the antigen response that is reiterated by the virus with the difference that the virus provides most if not all of the signals required for the process through the activity of its latent proteins. The parallel signaling steps are detailed diagrammatically in Figure 2. Antigen and EBV both enter through saliva and cross the epithelial barrier (1). Antigen binds to and EBV infects naïve B cells in the mantle zone causing them to become B cell blasts that migrate into the germinal center (2). Here the B cells receive rescue and survival signals that allow them to exit as memory cells into the peripheral circulation through the efferent lymphatics. Memory cells reenter the lymphoepithelium by passing through the walls of small venules termed HEV (4) and occasionally differentiate into plasma cells that migrate into the epithelium where they release antibody onto the mucosal surface (5). If the plasma cell contains EBV the virus will be triggered to replicate and infectious virus will be shed into the saliva (5). For more details see Figure 2.